Canada
Year
Book
1978~79

Canada
Year
Book
1978~79

An annual review
of economic, social and political
developments in Canada

Published by authority of the
Minister of Industry, Trade and Commerce

© Minister of Supply and Services Canada 1978

Available by mail from

Publications Distribution
Statistics Canada
Ottawa, Canada K1A 0T6

Printing and Publishing
Supply and Services Canada
Hull, Quebec, Canada K1A 0S9

or through your bookseller

Catalogue No. CS11-202/1978

ISBN 0-660-10014-2

Price, Canada: $15.00 .

Other countries: $18.00

Price subject to change without notice

Typesetting: Southam Business Publications, Contract OKP 8-0773
Printing: The Bryant Press Limited, Contract OKX8-0805

Preface

The *Canada Year Book 1978-79* follows a publication tradition dating back to 1905 which has furnished the public with an abundant source of information on social, economic and political conditions in Canada. This edition carries on the conversion to the metric system initiated in the *Canada Year Book 1976-77*. With few exceptions, units of measurement are given exclusively in metric for all sectors of the economy.

Every chapter has been revised to include the newest data available at the time of preparation, with sources listed at the end of text and tables for further reference. Chapters on health, education, housing and energy have been completely rewritten to reflect changing perspectives and new developments. A list of special articles from previous editions has been added, by popular request, to the appendices.

Material for the *Canada Year Book* is solicited from federal departments and agencies, provincial and territorial offices and numerous non-governmental organizations, as well as from our colleagues responsible for Statistics Canada surveys, census data and analyses. The book could never have been compiled in its present form were it not for the expertise of countless individuals and their willingness to provide concise information.

The budgetary pressures arising from continually rising prices, compounded with the recent federal government initiatives to effect general expenditure reductions, have forced Statistics Canada reluctantly to conclude that, henceforth, it can only publish the *Canada Year Book* every second year rather than annually. With the co-operation of our contributors, it will continue to record the Canadian scene, with the next edition due for release in 1980.

Peter Kirkham
Chief Statistician of Canada
Ottawa, December 1978

The *Canada Year Book* is planned and produced in the
Information Division, Statistics Canada.
Director: David Davidson

Assistant director, publishing: Tom Mitchell

Editor: Borgny Eileraas Pearson
French editor and technical adviser: Frances L. O'Malley
Assistant editors: Denis Schuthe, Diane Basso
Production supervisor and editor of tables: Ella Blair
Production liaison: Alice Guay
Production assistants: Patricia Harris, Audrey Miles,
                      Louise Cojbasic, Helen Wieland,
                      Victoria Perrior
Design co-ordinator: John MacCraken
Chart preparation: Charles Gravel, Gérard Montplaisir,
                  Jacques Fontaine

Contents

Progress in metric conversion

Since the *White paper on metric conversion in Canada* was tabled in the House of Commons on January 16, 1970, implementation of the SI metric system has been gradual but steady. Since 1972, under the guidance of Metric Commission Canada, such widely varied aspects of Canadian life as temperature, precipitation, grocery store products, road signs, much of the construction and automotive industries, grain sales, wines, seeds and fertilizers have been adapted to metric, and the process is continuing. In 1979 the petroleum industry, from oil well to service station, will convert to metric. It appears that the metric commission's original goal of Canada being substantially converted to the metric system by the end of 1980 will be realized.

Although conversion to metric is primarily voluntary in Canada, legislation has had to be introduced in Parliament to enable metric units to be used. In some cases legislation requiring only metric has been necessary to avoid the confusion of two systems. Conversion of the relevant acts is being generally accomplished by a series of four omnibus bills. The first of these, covering nine acts, received royal assent on August 5, 1977. The second omnibus bill, covering seven acts, received first reading in the House of Commons on December 20, 1977.

In view of the widespread implementation of metric conversion in Canada to date, most quantities in the *Canada Year Book 1978-79* appear only in SI metric or in neutral units such as dollars or dozens.

Relative weights and measures: SI Metric, Canadian Imperial and United States units

The following are conversion factors for units used in the present edition of the *Canada Year Book*, and some others in common use. Conversions are from SI metric to traditional units. For a full listing of the mathematical relationships between traditional units and SI metric, readers are referred to *Canadian Metric Practice Guide*, published by the Canadian Standards Association, 178 Rexdale Blvd., Rexdale, Ont. M9W 1R3. The same number of significant digits is used in the conversion factors which follow as is used in the *Canadian Metric Practice Guide*. If users do not need this level of accuracy, they can round off figures at any number of digits, either in the calculations or in the results. It is a requirement in SI metric to use spaces instead of commas to separate groups of three digits; a space is optional with a four-digit number. Although this practice is not imperative with neutral units, it is taking place in many cases now and will undoubtedly come about generally through standardization. In all Statistics Canada publications, a period is used as a decimal marker.

Area
1 km² (square kilometre) = 0.3861022 square miles
1 ha (hectare) = 2.471054 acres

Length
1 m (metre) = 39.37 inches
= 3.281 feet
= 1.094 yards
1 km (kilometre) = 0.6213712 statute miles = 3,280.840 feet
= 0.5399568 nautical miles = 3,282.937 feet

Volume and capacity

1 dm³ (cubic decimetre)	= 0.0353147 cubic feet
	= 0.4237760 board feet (for lumber)
	= 0.0274962 bushels (for grain)
	= 1 L (litre) (for liquids or, in some cases, for fine solids which pour)
	= 0.2199693 Canadian gallons
	= 35.1951 fluid ounces
	= 0.8798774 quarts
	= 1.75975 pints
	= 0.264172 US gallons
	= 1.05669 US quarts
	= 2.11338 US pints
1 imperial proof gallon	= 1.36 US proof gallons
1 m³ (cubic metre)	= 6.289811 barrels (petroleum or other liquid)
	= 0.3531466 registered tons (in shipping)*
	= 35.31466 cubic feet

Mass (weight)

1 g (gram)	= 0.03527396 ounces (avoirdupois)
	= 0.03215075 ounces (troy or apothecary)
1 kg (kilogram)	= 2.20462262 pounds (avoirdupois)
1 t (metric tonne)	= 1.10231131 tons (short)
	= 0.98420653 tons (long)

(For registered ton, see Volume and capacity above and footnote*)

Length and mass

1 t.km (tonne kilometre) = 0.6849446 short ton miles

Temperature

Fahrenheit temperature = 1.8 (Celsius temperature) +32
Celsius temperature = 5/9 (Fahrenheit temperature −32)

The following weights and measures are used in connection with the principal field crops and fruits:

Crops	Pounds per bushel	Kilograms per bushel	Bushels per 1 000 kg (1 t)
Wheat, potatoes and peas	60	27.215 5	36.7437
Wheat flour	43.48	19.721 4	50.7063
Oats	34	15.422 1	64.8418
Barley and buckwheat	48	21.772 4	45.9296
Rye, flaxseed and corn	56	25.401 2	39.3682
Mixed grains	45	20.411 7	48.9916
Rapeseed, mustard seed, pears, plums, cherries, peaches and apricots	50	22.679 6	44.0925
Sunflower seed	24	10.886 2	91.8593
Apples	42	19.050 9	52.4910

Strawberries and raspberries 1 kg = 1.47 quarts in BC
  = 1.76 quarts in all other provinces
To produce 100 kg of flour it takes 138 kg of wheat.

*Gross registered tonnage of a ship, as used by Lloyd's Register of Shipping, is a measurement of the total capacity of the ship and is not a measure of weight. Net registered tonnage equals gross registered tonnage minus space used for accommodation, machinery, engine area and fuel storage, and so states the cargo carrying ability of the ship.

Physical setting

Chapter 1

Tables

Physical setting   Chapter 1

Dimensions   1.1

Canada is the largest country in the Western Hemisphere and second largest in the world. Its territory of 9 922 330 square kilometres varies from the almost semi-tropical Great Lakes peninsula and southwest Pacific Coast to wide fertile prairies and great areas of mountains, rocks and lakes to northern wilderness and Arctic tundra. The farthest point south is Middle Island in Lake Erie, at 41°41′N; 4 627 km away in the Arctic is Cape Columbia on Ellesmere Island, the farthest point north, at 83°07′N. From east to west the greatest distance is 5 187 km — from Cape Spear, Nfld., at 52°37′W, to Mount St. Elias, YT, at 141°W.

Politically, Canada is divided into 10 provinces and two territories. Each province administers its own natural resources. The resources (except for game) of the Yukon Territory and Northwest Territories are administered by the federal government, because of the extent and remoteness of the territories and their sparse population. Land and freshwater areas of the provinces and territories are given in Table 1.1.

There is no permanent settlement in approximately 89% of Canada. Only the smallest province, Prince Edward Island, is completely occupied. Large parts of the interior of Nova Scotia, New Brunswick and the Gaspé Peninsula are vacant. Around the Newfoundland coast and on the shores of the St. Lawrence River below Quebec City there are only narrow bands of settlement.

About 58% of Canada's population lives between the American border and a 1 046 km east-west line from Quebec City to Sault Ste Marie. In this area, Montreal, Toronto, Hamilton, Ottawa, London, Windsor, Quebec City and Kitchener account for more than one-third of the Canadian population.

The largest tract of continuous settlement is in the Prairie provinces, running 1 448 km along the US border, north 161 km in Manitoba and west to the 55th parallel in Alberta, about 644 km north of the international border. This block occupies about 6.2% of Canada's area and contains four cities, Edmonton, Calgary, Winnipeg and Regina. North of this block, astride the Alberta–British Columbia border, the Peace River district is an agricultural area which reaches the 57th parallel.

Settlement is continuous through the southern half of British Columbia in interconnecting strips following mountain valleys and coastal plains. BC's population is most dense, however, in the Lower Fraser Valley, principally in the Vancouver area.

North of the areas already described are a number of unconnected settlements, the largest located in Ontario and Quebec between the 47th and 50th parallels. From east to west these are: the Lac St-Jean Lowland, some 161 km north of Quebec City, the Clay Belts astride the Ontario–Quebec border, the Lakehead, and the Dryden and Fort Frances areas near the Manitoba boundary. Outside these urban-rural blocks are numerous settlements related to mining, forest industries, transportation, administration, defence, hunting and fishing but with little or no agriculture.

Mountains and other heights   1.1.1

The great Cordilleran mountain system is Canada's most impressive physical feature. Many peaks in the various ranges of the Canadian Cordillera are over 4 500 metres high, and a total of approximately 1 502 square kilometres of territory lies above the 3 048 m mark. Mount Logan, 5 951 m above sea level, in the St. Elias Mountains of the Yukon Territory is the highest point in Canada.

The highest points in each province are: Newfoundland, 1 652 m; Prince Edward Island, 142 m; Nova Scotia, 532 m; New Brunswick, 820 m; Quebec, 1 652 m; Ontario, 693 m; Manitoba, 832 m; Saskatchewan, 1 392 m; Alberta, 3 747 m; British Columbia, 4 663 m; Yukon Territory, 5 951 m; and the Northwest Territories, 2 762 m.

Rossland, BC, is the highest city in Canada (1 056 m), and Banff, Alta., is the highest hamlet (1 396 m). Chilco Lake in British Columbia, with an area of 194 km², is the highest major lake (1 171 m). Heights of the more important Canadian mountains and other elevations are given in Table 1.2.

1.1.2   Inland waters

Abundant water supplies have been essential to the development of Canada's fisheries and wildlife resources, hydroelectric power, agriculture, recreational activities, navigation, domestic water supply and industrial production.

Each year 7 254 478 million tonnes of water fall on Canada as rain and snow. Much of it evaporates, some is stored temporarily in lakes, groundwater reservoirs and glaciers, and a large amount drains as surface runoff following streams and rivers to the oceans. Rapid melting of snow in spring causes floods, erosion and other problems. Most of Canada has ample precipitation averaging about 76 to 91 centimetres annually in many regions. In areas of little precipitation, greatest demand for water occurs in the hot summer weather; prolonged dry spells may mean water shortages. Drought conditions prevailed during some of 1977 over large parts of the Prairies.

Much of Canada's water is in undeveloped areas. Some other areas, such as the Prairies, have insufficient water for present needs.

About 755 165 square kilometres or 7.6% of Canada's total area is covered by lakes (Table 1.1). Lake storage provides water in time of drought that is later replenished. Lakes are natural regulators of river flow; they smooth out peak flows during flooding and sustain stream flow during dry seasons. Among the largest freshwater bodies in the world are the Great Lakes with an area of almost 258 999 km²; 37% is in Canada and 63% in the United States (Table 1.3). These lakes are sufficiently large to have measurable, although slight, tides. Other large lakes in Canada are Great Bear Lake, Great Slave Lake and Lake Winnipeg, with areas from 24 390 to 31 328 km². Countless smaller lakes are scattered throughout the country, particularly in the Canadian Shield. For example, southeast of Lake Winnipeg there are some 3,000 lakes in an area of 15 773 km²; and southeast of Reindeer Lake in Saskatchewan there are some 7,500 lakes in an area of 13 727 km². The size and elevation of Canada's lakes more than 388 km² in area are listed in Table 1.4.

Groundwater is another important source of freshwater supply for communities, industries and irrigators, contributing about 10% of the water supplied by municipal water systems. Although the quantities are much smaller than from rivers and lakes, many communities and some industries are completely dependent on groundwater supplies. In some areas, particularly the Prairies, groundwater is the principal source of water streams during extended dry weather.

The volume of water stored as snow and ice in North America's glaciers is many times greater than all the lakes, rivers and reservoirs. Most of this is permanently frozen in the polar ice caps and is inaccessible, but polar ice masses have a strong indirect influence on the hydrologic cycle through their effect on weather patterns. In the temperate regions, however, the alpine glaciers exert a direct influence on the hydrologic cycle as water from melting glaciers frequently sustains stream flow during dry seasons. In hot summer months, glaciers may contribute up to 25% of the flow in part of the Saskatchewan and Athabasca rivers. About 150 000 km² or 75% of the glaciated areas of Canada are in the Arctic islands and 50 000 km² or 25% on the mainland. Of the latter figure 38 000 km² are in the Pacific drainage basin and 10 500 km² in the Yukon drainage basin. The remaining 3 885 km² are shared among the Arctic, Great Slave, Saskatchewan–Nelson and Labrador drainage basins. The number of glaciers in Canada is estimated at 75,000.

In Canada 90% of water used comes from streams and other surface sources such as lakes and man-made reservoirs. The combined mean annual flow of all streams has been estimated at 99.1 million cubic decimetres per second, equivalent to about 60% of Canada's mean annual precipitation.

Canada's history and industrial development has been influenced by its great rivers. Earliest settlements centred around water supplies and water was essential for transportation. Canada's fur trade flourished because of the ready access to the interior provided by the St. Lawrence River, the Great Lakes and many other waterways. Plentiful water supplies in the fertile plains of Southern Ontario and Quebec attracted an industrious farming people. The river-borne transportation of lumber and later the power of water-driven turbines were vital factors in building an industrial base. Water remains a key to Canada's development, supplying renewable energy for industrial growth, providing easy and cheap transport for raw materials and playing a vital part in their processing.

Water problems in Canada are associated with storage, distribution and pollution. Current demands for greater and more diversified water use are complicated by a need to reverse the trend toward deterioration in water quality resulting from urbanization, industrialization and agricultural developments. Pollution and water quality are of major concern since they have a direct bearing on Canada's national well-being and economic growth.

The international boundary line between Canada and the United States, including Alaska, is 8 892 km long, of which 5 063 km lie along or across water bodies. Boundary water basins are of economic importance to both countries. Natural resources of the boundary basins and transportation and hydroelectric power resources of the waterways in these basins have helped foster population concentration and industrial development in Canada along a broad band bordering the 49th parallel.

Approximate population in some selected boundary basins is summarized in the following table. (Canadian statistics are compiled from census divisions that approximate the basin boundary; US statistics were published in 1974, prepared jointly by the US departments of commerce and agriculture for the US Water Resources Council. Both give 1971 figures.)

	Canada	United States
Saint John–St. Croix	450,000	125,000
Chaudière	215,000	395,000
St. François	295,000	20,000
Richelieu–Lake Champlain	325,000	335,000
Lake Ontario–Upper St. Lawrence	4,430,000	4,115,000
Lake Erie–Lake St. Clair	1,580,000	9,780,000
Lake Huron–Lake Michigan	690,000	14,900,000
Lake Superior	265,000	535,000
Lake of the Woods–Rainy River	80,000	20,000
Red River	715,000	545,000
Souris River	100,000	110,000
Missouri–Milk	200,000	225,000
Pend d'Oreille–Kootenay	86,000	225,000
Columbia River	190,000	195,000
Lower Mainland	1,490,000	1,190,000
Alaska Panhandle and Yukon	80,000	315,000

In 1909 Canada and the US signed the Boundary Waters Treaty which set out clear limitations on either country's freedom to act if such action might affect the other country. Under the treaty, the International Joint Commission was created to deal with problems that could arise along the boundary. Since then the commission has handled problems in international basins from the Pacific to the Atlantic Ocean, from small streams to the St. Lawrence River. More recently, the commission was given responsibility for overseeing implementation of the Canada–US agreement on Great Lakes water quality, with goals of improving water quality in polluted areas and ensuring future protection of water quality. Table 1.5 lists the principal rivers of Canada and their tributaries.

The accompanying map shows major drainage basins of Canada. The Atlantic drainage basin is dominated by the Great Lakes–St. Lawrence system which drains an area of approximately 1 756 012 km² and forms a navigable inland waterway through a region rich in natural and industrial resources. From the head of Lake Superior to Belle Isle at the entrance of the Gulf of St. Lawrence is 3 669 km. The entire drainage area north of the St. Lawrence and the Great Lakes is occupied by the southern fringe of the Canadian Shield, a rugged, rocky plateau with many tributaries. These rivers and the St. Lawrence provide much of the electric power for the area's industries. South of the St.

Lawrence, smaller rivers are important locally. The Saint John, for instance, drains a fertile area and provides most of New Brunswick's hydro power.

The Hudson Bay drainage basin is the largest and its main river is the Nelson. The Winnipeg River, a tributary of the Nelson via Lake Winnipeg, is completely developed for hydroelectric power but development of the Nelson itself is just beginning. The Saskatchewan River, tributary to the Nelson via Lake Winnipeg, drains the great agricultural region of the mid-west and is an important source for irrigation and hydroelectric power.

The Arctic drainage basin is dominated by the Mackenzie, one of the world's longest rivers. It flows 4 241 km from the head of the Finlay River to the Arctic Ocean and drains an area of approximately 1 812 992 km² in the three westernmost provinces and the two territories. Except for a 26 km portage in Alberta, barge navigation is possible from Waterways on the Athabasca River to the mouth of the Mackenzie, a distance of 2 736 km.

Rivers of the Pacific basin rise in the mountains of the Cordilleran region and flow to the Pacific Ocean through steep canyons and over innumerable falls and rapids. They provide power for large hydroelectric developments and in season swarm with salmon returning inland to their spawning grounds. The Fraser River rises in the Rocky Mountains and, toward its mouth, flows through a rich agricultural area. The Columbia is an international river which falls 808 metres during its course and thus has tremendous power potential. Although a considerable part of the United States potential has been developed, the Canadian portion of the basin remained relatively untouched until recent years when three large reservoirs were built under the terms of the Columbia River Treaty. These reservoirs make it possible for British Columbia to develop up to 4 000 megawatts of hydroelectric generating capacity in the Columbia basin. The Yukon River, also an international river and the largest on the Pacific slope, has not yet been developed in Canada.

Utilization of inland water. Over 43% of all water withdrawn in Canada (excluding withdrawals associated with hydro projects) is for condenser cooling in steam-electric plants. About 99% is returned. Municipal use, including small industrial processors served by municipal systems, accounts for 10.5% of current water withdrawals. On average, approximately 75% of the water pumped into the system is discharged as storm and sanitary sewage containing waste materials.

Other industrial users, manufacturing and mining firms, account for 38% of total withdrawals of water and about 10% of that intake is consumed or lost. Discharged water is frequently returned to source in a highly polluted condition and may be unfit for most uses downstream. Canadian agriculture depends largely upon supplies of water from melting snow and rainfall. In many regions, however, such natural sources are inadequate. Agriculture requires 7.7% of the nation's total withdrawals annually for irrigation, stock watering and rural domestic use.

Hydroelectric power generation uses the kinetic energy of falling water to produce electricity. Except for evaporation losses from the surface of reservoirs, the water is not consumed or changed in any way. However, flooding of land for storage and interference with natural flow may have adverse effects.

Water transport is no longer the principal mode of transportation, but competes with railways, pipelines, aircraft and motor carriers. Water is still the most economical means of transporting bulky raw materials such as wheat, pulp and paper, lumber and minerals for export, especially in the Great Lakes–St. Lawrence and Mackenzie River regions.

The popularity of water-oriented recreational activities, including swimming, boating, sightseeing, fishing, hunting, and water skiing, is growing as more leisure time becomes available. Although provincial and federal governments produce recreation data, co-ordinated national information on the role of water in outdoor recreation is not yet available.

Fish and wildlife from river and lake systems make a vital contribution to Canada's economy. In addition to sport-fishing and hunting, the inland waters support important commercial fisheries. Fish and wildlife require water of high quality. When water systems are put to multiple-purpose use, pollution can destroy these resources. Within government agencies there has been increased work on water pollution problems. Universities are also developing programs in environment-related water research.

Coastal waters   1.1.3

Canada's coastline, of over 241 402 kilometres, comprises the following measurements — Mainland: Atlantic 15 841 km; Pacific 7 022 km; Hudson Strait 4 253 km; Hudson Bay 12 268 km; Arctic 19 125 km; total 58 509 km. Islands: Atlantic 29 251 km; Pacific 18 704 km; Hudson Strait 8 594 km; Hudson Bay 14 775 km; Northwest Territories south of Arctic Circle 22 209 km; Arctic 91 755 km; total 185 289 km.

Atlantic. Along this coastal area, the sea has inundated valleys and lower parts of the Appalachian Mountains and the Canadian Shield. The submerged continental shelf is distinguished by great width and diversity of relief. From the coast of Nova Scotia its width varies from 60 to 100 nautical miles, from Newfoundland 100 to 280 nautical miles at the entrance of Hudson Strait, and northward it merges with the submerged shelf of the Arctic Ocean. The outer edge varies in depth from 183 to 366 metres. The overall gradient of the Atlantic continental shelf is slight but the whole area is studded with shoals, plateaus, banks, ridges and islands. The 73 m line is an average of 12 nautical miles from the Nova Scotia coast and is the danger line for shipping. The whole floor of the marginal sea is traversed by channels and gullies cutting deep into the shelf.

The topography of much of the Atlantic marginal sea floor was shaped by processes of glacial erosion and deposition. Large areas, however, undergo constant change because of continuous marine deposition of materials eroded by rivers, wave action, wind and ice.

Hudson Bay and Hudson Strait bite deeply into the continent. Hudson Bay is an inland sea 822 324 km² in area having an average depth of about 128 m; the greatest depth in the centre of the bay is 258 m. Hudson Strait separates Baffin Island from the continental coast and connects Hudson Bay with the Atlantic Ocean. It is 796 km long and from 69 to 222 km wide; its greatest depth of 880 m is close inside the Atlantic entrance. There are great irregularities in the sea floor but, except in inshore waters, few navigational hazards have been located.

Pacific. The marginal sea of the Pacific differs strikingly from the other marine zones of Canada. The hydrography of British Columbia is characterized by bold, abrupt relief — a repetition of the mountainous landscape. Numerous inlets penetrate the mountainous coasts for distances of 93 to 139 km. They are usually a nautical mile or two wide and very deep, with steep canyon-like sides. From the islet-strewn coast, the continental shelf extends from 50 to 100 nautical miles to its limit at depths of about 366 metres. The sea floor drops rapidly, parts of the western slopes of Vancouver Island and the Queen Charlotte Islands lying only four nautical miles and one nautical mile, respectively, from the declivity. These detached land masses are the dominant features of the Pacific marginal sea. The region's numerous shoals and pinnacle rocks necessitate cautious navigation.

Arctic. The submerged plateau extending from the northern coast of North America is a major part of the great continental shelf surrounding the Arctic Ocean, on which lie all the Arctic islands of Canada, Greenland, and most of the Arctic islands of Europe and Asia. This shelf is most uniformly developed north of Siberia where it is about 500 nautical miles wide; north of North America it surrounds the western islands of the archipelago and extends 50 to 300 nautical miles seaward from the outermost islands.

The floor of the submerged part of this continental margin is nearly flat to gently undulating, with isolated rises and hollows. Most of it has an average slope seaward of about one-half degree, with an abrupt break at the outer edge to the continental slope whose declivity is commonly six degrees or more. From the Alaskan border eastward to the mouth of the Mackenzie River the shelf is shallow and continuous with the coastal plain on the mainland; its outer edge is at a depth of about 64 m and about 40 nautical miles offshore. This shelf is continuous with that north of Alaska and Siberia. Near the western edge of the Mackenzie River delta, it is indented by the deep Herschel Sea Canyon, whose head comes within 15 nautical miles of the coast. Between Herschel Sea Canyon and Amundsen Gulf, the typical features of the continental shelf are replaced by the submerged portion of the Mackenzie River delta, which forms a great pock-marked undersea plain, most of it less than 55 m deep, up to 75 nautical miles wide and 250 miles long.

North and east of the submerged portion of the Mackenzie River delta, the continental shelf is more deeply submerged than that off the mainland and Alaska. Its gently undulating surface is generally 366 m or more below sea level, and most of the well-defined continental shoulder is over 549 m deep, giving way to the smooth continental slope which extends without significant interruption to the abyssal Canada Basin at about 3 658 m. The deeply submerged continental shelf runs along the entire west coast of the Canadian Arctic Archipelago from Banks Island to Greenland. All major channels between the islands — Amundsen Gulf, M'Clure Strait, Prince Gustav Adolf Sea, Peary Channel, Sverdrup Channel and Nansen Sound — have flat floors at about the same depth as the shelf and appear to enter it at grade, but a few local irregularities may be the result of glacial action. The only deep indentation known to cut the continental slope or continental shelf off the archipelago is one sinuous canyon that heads off Robeson Channel at the northeastern end, close to Greenland. Submerged sides of the channels of the archipelago, and slopes from the islands' western shores to the inner edge of the deeply submerged shelf, are marked in many places by a series of steps or terraces.

1.1.4 Islands

Canada's largest islands are in the North in an arctic climate. The northern group extends from the islands in James Bay to Ellesmere Island which reaches 83°07'N.

Those in the District of Franklin, north of the mainland of Canada, are generally referred to as the Canadian Arctic Archipelago; those in the extreme north — lying north of 73°30'N — are known as the Queen Elizabeth Islands.

The largest and most important islands on the West Coast are Vancouver Island and the Queen Charlotte Islands, but the coastal waters are studded with many small rocky islands. The largest off the East Coast are the island of Newfoundland, the province of Prince Edward Island, Cape Breton Island of Nova Scotia, Grand Manan and Campobello islands of New Brunswick and Anticosti Island and the Madeleine group of Quebec.

Notable islands of the inland waters include Manitoulin Island, 2 766 km² in area, in Lake Huron, the so-called Thirty Thousand Islands of Georgian Bay and the Thousand Islands in the outlet from Lake Ontario into the St. Lawrence River.

The areas of principal islands by region are given in Table 1.6.

Surveying and mapping   1.1.5

The surveys and mapping branch of the energy, mines and resources department is Canada's major mapping agency. The branch compiles topographic maps, aeronautical charts, thematic maps and base maps of various scales for specialized uses by other agencies to provide geological, aeromagnetic, electoral and land use information. The geodetic survey division establishes and maintains the national system of control surveys to serve the needs of mapping, charting and boundary surveys and geoscience research. Topographical surveys has completed the mapping of Canada at the scale of 2.0 cm to 5.0 km and is now mapping the country at the scale of 2.0 cm to 1.0 km. All of the settled areas and many regions of northern development, amounting to slightly more than half of the country, are mapped at this larger scale. There are 690 maps available at the scale of 4.0 cm to 1.0 km covering all major cities and their suburbs. Photomaps derived from air photographs using photogrammetric technology are also available covering some of the areas mapped at the two largest scales, 2.0 cm to 1.0 km and 4.0 cm to 1.0 km. The legal surveys division is responsible for the technical management of legal surveys of land under federal jurisdiction, such as the northern territories, national parks and Indian reserves. It also executes such surveys on behalf of administering departments, collaborates in the demarcation of provincial boundaries, prepares descriptions of electoral districts and generally provides land-surveying services to other departments.

A permanent committee on geographical names deals with all questions of geographical nomenclature affecting Canada and advises on research and investigation into the origin and usage of names. Its membership includes representatives of federal mapping agencies and other federal offices concerned with nomenclature and representatives appointed by each province. The committee's functions were redefined in 1969 (order-in-council PC 1969-1458). The order-in-council recognizes that the provinces have exclusive jurisdiction to make decisions on names in lands under their jurisdiction. The committee is administered by the energy, mines and resources department.

Geology   1.2

Canada is composed of 17 geological provinces that may be grouped under four main categories — continental shelf, platform, orogen and shield. The geologically youngest provinces, the Atlantic, Pacific and Arctic continental shelves, are made up of little-deformed sediments and volcanics, mainly of Mesozoic and Cenozoic age, which are still accumulating along the margins of the present continental mass. The St. Lawrence, Interior, Arctic and Hudson platforms are formed of thick flat-lying Phanerozoic strata covering large parts of the crystalline basement rocks of the continental interior, the extension of the Canadian Shield. The Appalachian, Cordilleran and Innuitian orogens are mountain belts of deformed and metamorphosed sedimentary and volcanic rocks mainly Phanerozoic and Proterozoic in age, intruded by granitic plutons. They were produced during the various Phanerozoic orogenies 50 to 500 million years ago. Of the seven provinces comprising the Precambrian Canadian Shield, the Grenville, Churchill,

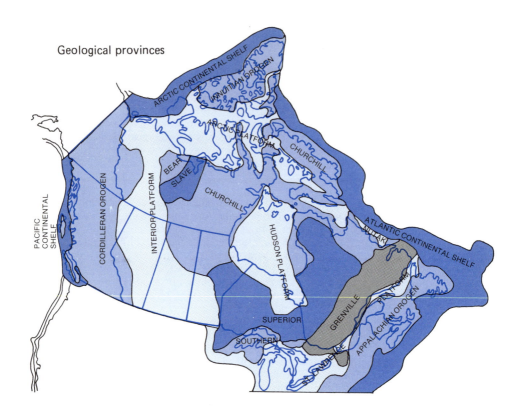

Geological provinces

Southern and Bear embrace the orogenic belts produced during the Proterozoic orogenies, 900 to 1,800 million years ago. The remaining three, the Superior, Slave and Nutak provinces, were deformed during the Archean Eon, and include the oldest continental crust known in Canada, 2,500 to 3,000 million years old. The Precambrian orogenic belts have many features in common with those of Phanerozoic age but are so deeply eroded that the mountainous parts have been reduced to plains or lowlands and in many places the basement crystalline rocks upon which the sediments and volcanics initially accumulated are now exposed.

The land and freshwater area of Canada is 9 922 330 square kilometres, but Canada also includes within the confines of this area some 2 222 210 km² of marine waters. The rocks beneath have geological features akin to the adjacent regions on shore. In addition, the submarine area of the bordering continental shelves is about 1 354 564 km² and of the continental slopes, 1 458 163 km². Altogether, this embraces 14 312 274 km², about 3% of the surface of the globe.

For an account of Canada's geology see the *Canada Year Book 1973* pp 8-14.

1.3 Climate

Climate depends primarily on radiative exchanges between the sun, the atmosphere and the surface of the earth. In addition, regional climates of Canada are controlled by the geography of North America and by the general movement of air from west to east. The Pacific Coast is cool and fairly dry in summer but mild, cloudy and wet in winter. Interior British Columbia has climates varying more with altitude than latitude: wet windward mountain slopes with heavy snows in winter, dry rainshadow valleys, hot in summer, and high plateaus with marked day to night temperature contrasts. Interior Canada, from the Rocky Mountains to the Great Lakes, has a continental-type climate with long cold winters, short but warm summers and scanty precipitation. Southern portions of Ontario and Quebec have a humid climate with cold winters, hot summers and generally ample precipitation all year. The four Atlantic provinces have a humid continental-type climate although in the immediate coastal areas there is a marked maritime effect. On the northern islands, along the Arctic Coast and around Hudson Bay, arctic conditions persist, with long frigid winters and only a few months with temperatures averaging above freezing. Precipitation is light in the tundra area north of the treeline. Between the arctic and southern climates boreal Canada has a transitional type climate with bitter long winters but appreciable summer periods. Precipitation is light in the west, but heavier in the Ungava Peninsula.

Climatic data. Temperature and precipitation data for various districts are shown in Table 1.7. Additional data from hundreds of stations and reports concerning the climates of Canada and the regions are available from the atmospheric environment service of the fisheries and environment department. Definitions, methods of observation, the instrumentation used and other information are included in the department's publications.

Time zones   1.4

Based on atomic clocks, Canada's time is established by the National Research Council with a precision of one ten-millionth of a second per day, and co-ordination with other countries is maintained to the same precision through the Bureau international de l'Heure in Paris. Irregularities in the rotation of the earth give rise to a difference between mean solar time and atomic time, and a leap second is introduced to ensure that this difference, called DUT1, does not exceed 0.8 seconds. At present DUT1 is decreasing by about one-twelfth of a second per month, and positive leap seconds were necessary on June 30, 1972 and on December 31 of each year from 1972 to 1977.

A continuous broadcast of Canadian time is made on station CHU, Ottawa (3 330 kHz, 7 335 kHz, 14 670 kHz), with a bilingual voice announcement each minute, and with a split pulse code to give the value of DUT1. Once a day the time signals are broadcast across Canada on the CBC networks.

Standard Time, adopted at a world conference at Washington, DC, in 1884, sets the number of time zones in the world at 24, each zone ideally extending over one twenty-fourth of the surface of the earth and including all the territory between two meridians 15° of longitude apart. In practice, the zone boundaries are quite irregular for geographic and political reasons. Universal Time (UT) is the time of the zone centred on the zero meridian through Greenwich, England. Each of the other time zones is a definite number of hours ahead of or behind UT to a total of 12 hours, at which limit the international date-line runs roughly north-south through the mid-Pacific.

Canada has six time zones. The most easterly, Newfoundland standard time, is three hours and 30 minutes behind UT, and the most westerly, Pacific standard time, is eight hours behind UT. From east to west, the remaining zones are called Atlantic, Eastern, Central and Mountain. In October 1973 the nine hour Western Yukon time zone was eliminated by order of the Yukon Territorial Council, placing the entire Yukon eight hours behind UT.

Legal authority for the time zones. Time in Canada has been of provincial rather than federal jurisdiction. Each of the provinces and territories has enacted laws governing standard time and these laws determine the time zone boundaries. Lines of communication, however, have sometimes caused communities near the boundary of a time zone to adopt the time of the adjacent zone, and in most cases these changes are acknowledged by amendments to provincial legislation. Official time for dominion official purposes is the responsibility of the National Research Council of Canada.

Daylight saving time. Most provinces have legislation controlling the provincial or municipal adoption (or rejection) of daylight saving time; in the other provinces

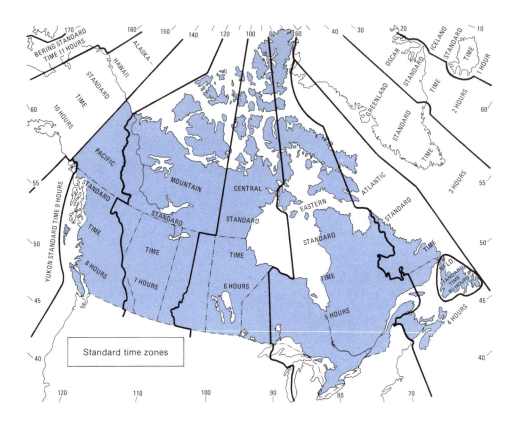

Standard time zones

authority is left to the municipalities. By general agreement, daylight saving time, where it is observed, is in force from the last Sunday in April until the last Sunday in October.

1.5 Public land

The total area of Canada and areas of individual provinces and territories are classified by tenure in Table 1.8. All lands, except those privately owned or in process of alienation, are Crown lands under the jurisdiction of either federal or provincial governments.

Federal public land. Public lands under federal government administration comprise lands in the Northwest Territories including the Arctic Archipelago and the islands in Hudson Strait, Hudson Bay, James Bay and Ungava Bay, lands in the Yukon Territory, ordnance and admiralty lands, national parks and national historic parks and sites, forest experiment stations, experimental farms, Indian reserves and, in general, all public lands held by the several departments of the federal government for various purposes connected with federal administration. These lands are administered under the Territorial Lands Act (RSC 1970, c.T-6) and the Public Lands Grants Act (RSC 1970, c.P-29).

The largest areas under federal jurisdiction are in the Northwest Territories and the Yukon Territory where only 241 square kilometres of a total area of 3 916 007 km² are privately owned and 3 877.5 km² are administered by the territorial governments.

Provincial and territorial public land. Public lands of Nova Scotia, New Brunswick, Quebec, Ontario and British Columbia (except the "railway belt" and Peace River block) have been administered since Confederation by the provincial governments. In

1930, the federal government transferred the unalienated portions of the natural resources of Manitoba, Saskatchewan and Alberta and of sections of British Columbia to the respective governments, and all unalienated lands in Newfoundland, except those administered by the federal government, became provincial public lands under the terms-of-union on March 31, 1949. All land in Prince Edward Island has been alienated except 344 km² under federal or provincial administration.

Transfer by the federal government of land within and immediately surrounding established communities in the Northwest Territories and the Yukon Territory to the respective territorial governments began in September 1970 when four such transfers were completed, three in the Northwest Territories and one in the Yukon Territory, for a total of 1 722 km². Since then transfers were completed in the following areas: Yukon Territory: Faro 236 km², Beaver Creek 5 km², Mayo 10 km², Teslin 2.6 km², Carmacks 31 km², Destruction Bay 5 km², Carcross 24.9 km², Watson Lake 5 km², Northwest Territories: Frobisher Bay 132 km², Aklavik 21 km², Fort Simpson 363 km², Fort Smith 57 km², Fort Providence 210 km², Hay River–Enterprise 368 km², Norman Wells 453 km², Fort McPherson 80 km², Fort Franklin 65 km², Fort Good Hope 57 km², Pine Point 29.8 km².

Federal parks 1.5.1

Parks Canada, a program of the Indian affairs and northern development department, includes national parks, national historic parks and sites, and agreements for recreation and conservation. Parks Canada has its headquarters in Ottawa but operational responsibility in five regions: the Atlantic regional office in Halifax, the Quebec regional office in Quebec City, the Ontario regional office in Cornwall, the Prairie regional office in Winnipeg and the Western regional office in Calgary.

National historic parks and sites. National historic parks and sites commemorate persons, places and events of major significance in Canada's historical development.

The National Parks Act of 1930 provided that the Governor-in-Council may set apart any land as a national historic park to commemorate a historic event, or preserve any historic landmark or any object of historic, prehistoric or scientific interest of national importance. The Historic Sites and Monuments Act of 1953 provided the statutory base for the operation of the historic sites and monuments board and defined the role of the board as adviser to the minister. Further legislation in 1955 and 1959 amended and broadened the scope of the original act. The Canadian historic sites division, now the national historic parks and sites branch, was created in the northern affairs and national resources department in 1955 to develop, interpret, operate and maintain historic parks and sites and to act as secretariat for the board.

A policy statement in 1967 specified that for commemoration, a site or structure must be closely associated with a person, place or event of national historical importance, or it must illustrate the cultural, social, political, economic or military patterns of history or of a prehistoric people or archaeological discovery, or be valuable as an example of architecture. The statement included guidelines for provision of visitor services, interpretative programs and information to the public. Standards were established for preservation, restoration and reconstruction of structures which stressed authenticity in the materials used and in the furnishings and artifacts. The policy recognized the need for a comprehensive program to give full thematic and geographical representation and to establish a long-range planning program.

The Historic Sites and Monuments Act provides for a board of 17 members: two representatives each from Ontario and Quebec and one from each of the 10 other provinces and territories appointed by the Governor-in-Council, the dominion archivist, one representative from the National Museums of Canada and one from the Indian affairs and northern development department as ex officio members. Members are generally historians of distinction. The board may recommend that sites, buildings and other structures of national importance be developed as national historic parks or historic sites or that commemoration be carried out by the erection of plaques or distinctive monuments. Suggestions for establishment of historic sites and parks come from many sources — the general public, members of Parliament, historical societies and other groups, department staff and board members themselves. Before a suggestion is considered, a background paper is prepared by the national historic parks and sites branch research staff. The board then determines the significance of the site and makes its recommendation to the minister. If approved, a development plan is prepared.

The national historic parks and sites branch has been instrumental in the creation of 80 national historic parks and major sites, over 53 operational, and in the commemoration with plaques of more than 700 persons and events of national (as opposed to local or regional) significance. Negotiations are conducted with provinces for acquiring other sites. The department has entered into 40 cost-sharing agreements with provincial and municipal governments and with incorporated non-profit societies for acquisition and restoration of architecturally or historically significant buildings and structures on the understanding that the other party will pay the balance of the acquisition and restoration costs and will maintain the buildings in perpetuity. A number of monuments are maintained by the national historic parks and sites branch.

From April 1, 1976 to March 31, 1977 there were 4.5 million visits to Canada's national historic parks and sites. Details on location and characteristics of national historic parks and sites may be obtained from Parks Canada.

The Canadian inventory of historic building begun in 1970 is a computerized program to survey, analyze and categorize old buildings. Exteriors of more than 165,000 buildings have been surveyed and almost all have been indexed; interiors of approximately 1,800 of these structures have been surveyed.

Heritage Canada is an independent corporation concerned with conservation of buildings, sites and natural and scenic areas. It received an initial federal capital endowment of $12 million in 1972 and the interest on this fund is used to further its work. Heritage Canada enlists the support of the general public and of foundations and corporations.

National parks. Canada's national parks system, encompassing more than 129 499 square kilometres, is the largest and most rapidly expanding in the world. It has grown from the federal government's efforts, with co-operation of provincial and territorial governments, to preserve natural areas of scenic and biological interest for the public.

In 1885 the Canadian government reserved from private ownership the mineral hot springs of Sulphur Mountain in what is now Banff National Park. Two years later this 26 km² reserve was extended to 673 km² and named Rocky Mountain Park, the first federal park in Canada. Two land reserves in southern British Columbia — Yoho and Glacier — were made by the federal government in 1886, a reserve of 140 km² in the Waterton Lakes area of southern Alberta in 1895, and an area of 12 950 km² around Jasper, Alta., in 1907. These four western mountain reserves, together with Rocky Mountain Park, formed the nucleus of the national park system after the Dominion Forest Reserves and Parks Act was passed in May 1911. A national parks branch was created to protect, administer and develop the parks.

By 1930 there were nine more national parks. Three in Ontario consisted of federal Crown land or land held in trust for Indians: St. Lawrence Islands National Park, Point Pelee National Park and Georgian Bay Islands National Park. Prince Albert National Park in Saskatchewan and Riding Mountain National Park in Manitoba were former federal forest reserves. Elk Island National Park near Edmonton was established as a preserve for buffalo and Wood Buffalo National Park, a 44 807 km² area straddling the Alberta–Northwest Territories border, as a refuge for the largest surviving herd of buffalo in North America. In British Columbia two scenic areas were preserved — Mount Revelstoke National Park and Kootenay National Park.

Between 1930 and 1973 the following new parks were added: Northwest Territories: Nahanni and Auyuittuq; Yukon Territory: Kluane; British Columbia: Pacific Rim; Ontario: Pukaskwa; Quebec: La Mauricie and Forillon; New Brunswick: Kouchibouguac and Fundy; Nova Scotia: Cape Breton Highlands and Kejimkujik; Prince Edward Island: Prince Edward Island National Park; Newfoundland: Terra Nova and Gros Morne.

For parks in the Yukon Territory and Northwest Territories, lands have been reserved from all alternative disposition by orders-in-council and proclamation. Within provinces, land is acquired by the province acting within a federal-provincial agreement to establish a national park. These lands are transferred to Canada and the establishment of the park is made formal by Parliament.

In 1971, *A national parks system planning manual* was published, in recognition that new and comprehensive measures are needed to preserve Canada's natural heritage. With a view to protecting not only unique and outstanding areas of the Canadian land and seascapes but also those representative of its physical, biological, and oceanographic characteristics, 48 distinctive natural regions were identified for which natural history themes were defined.

From April 1, 1976 to March 31, 1977, there were 16.5 million visits to the national parks. Visitors to the parks can participate in activities ranging from guided walks and canoeing in summer to cross-country skiing and snowshoeing in winter.

A detailed list of national parks was included in the 1972 and 1973 editions of the *Canada Year Book,* and a location map and details of these parks are available in *Canada's national parks* published by Parks Canada.

National marine parks. Canada is bounded by three oceans and has the largest volume of fresh water in the world. The national parks system will be extended to represent the Pacific, Arctic and Atlantic coasts and inland waters, with identification of the marine natural regions and marine natural history themes.

National landmarks. Preservation of specific natural wonders, such as the Chub crater in Northern Quebec, the frozen pingoes of the Arctic, semi-desert and eroded hills of the Prairies and mountain caves and seascapes, allow on-site interpretation of Canada's natural evolution.

Agreements for Recreation and Conservation (ARC). Public agencies, organizations and individuals are actively protecting and preserving heritage resources. To provide Canadians new opportunities to appreciate and understand their natural, cultural and historical heritage, Parks Canada created a co-operative program, agreements for recreation and conservation, which focuses principally on heritage canals and co-operative heritage areas. Initiatives would be established, developed and managed according to terms and conditions of co-operative agreements between Parks Canada and other agencies.

Exemplified by the Rideau-Trent-Severn waterway, the contemporary importance of heritage canals as recreational waterways emphasizes not only navigation but also visitor participation in a diversity of recreational activities. Heritage canals have acquired new significance by illustrating historical development and early engineering technology.

Co-operative heritage areas contain natural and cultural heritage resources which are nationally significant. These resources must exist in a condition and setting permitting continued co-operative protection. Such initiatives may have a concentration of distinctive natural and cultural resources or display examples of one particular type of heritage resource such as historic land and water routes, urban conservation areas, rural cultural landscapes or wild rivers. Each agreement for a co-operative heritage area concentrates on a combination of resources and is designed to achieve the joint objectives of the participants.

Gatineau Park. In addition to the national parks a 356 km² recreation area, Gatineau Park north of Ottawa and Hull, is being developed by the federal government as part of the national capital region under the National Capital Commission. It is a wilderness area of great potential, extending northward from Hull for 56 km. With 40 km of parkway, magnificent lookouts, lakes, fishing streams, beaches, picnic areas, camping sites, skiing and walking trails it is one of the finest recreation areas in Canada, enjoyed by 1.8 million visitors each year. A master plan for further development is under way.

Provincial parks   1.5.2

All provincial governments have established parks within their boundaries. Some are wilderness areas set aside so that portions of the country might be retained in their natural state. Most of them, however, are smaller areas of scenic interest, easily accessible and equipped or slated for future development as recreational parks with camping and picnic facilities.

Newfoundland. In the early 1950s Newfoundland's first organized recreational facilities were roadside picnic sites. The department of tourism currently administers 72 units including camping parks, day-use parks, 17 inland and coastal public beaches, a wilderness area, and internationally significant bird sanctuaries in the North Atlantic. Sir Richard Squires Memorial Park and River of Ponds Park are located on famous Atlantic salmon rivers while others such as Bellevue Beach Park and Chance Cove Park are on Newfoundland's dramatic coastline.

Prince Edward Island. In the provincial park system, 40 areas comprise five classes of parks: nature preserves, natural environment parks, recreation parks, wayside/beach access, and historic parks. The parks, maintained by the tourism, parks and conservation department, enhance the scenic drives which loop coastal areas.

Two recreation parks are resort complexes: Brudenell in Kings County and Mill River (under development) in Prince County. These sites include specialty facilities and accommodations, including 18-hole golf courses.

Nova Scotia. The provincial parks system, administered by the lands and forests department, parks and recreation division, started in the late 1950s with roadside sites. This has expanded to 19 overnight campgrounds, 61 day-use picnic and roadside parks, and 20 day-use beach parks. Most of the parks are easily accessible from main highways. Campgrounds contain from 16 to 165 sites in parks of from 12 to 675 hectares. The picnic and beach parks range from one to more than 117 hectares. Picnic and campground facilities are primitive with only tables, water and pit privies.

New Brunswick. The provincial park system, administered by the tourism department, includes 24 recreational parks ranging from 10 to 567 hectares, 21 rest areas, seven campground parks, seven beach parks, a marine park, a wildlife park and a resource park. Most are in rural areas adjacent to or easily accessible from main roads.

Several parks have organized activity, lifeguards and interpretation programs. Mactaquac, near Fredericton, one of two year-round parks, boasts a championship 18-hole, 6 428-metre golf course and two marinas. During the winter there are facilities for snowmobiling, cross-country skiing, snowshoeing, tobogganing, skating, sleigh rides and camping. Sugarloaf, near Campbellton, the other year-round park, features an alpine ski hill with three lifts, cross-country skiing, skating, snowmobiling, tobogganing and tennis. Campobello provincial park on Campobello Island has a nine-hole, 3 008-metre golf course and lodge.

Quebec. The provincial parks branch administers Quebec's parks and reserves. During 1976-77 the operations division of the parks branch was responsible for 53 parks and 47 reserves, including 24 salmon streams and 62 campgrounds with more than 7,506 campsites.

Legislation governing creation of the parks dates back to 1894 but was updated in 1941. More recently, the Provincial Parks Act was passed in 1964 and amended in 1971.

The first park, Mont-Tremblant, was established in 1894. It covers 2 564 square kilometres and is located north of Montreal. Five years later Laurentides provincial park was established north of Quebec City covering 6 967 km². Three new parks were in operation in 1976-77: Mount St. Bruno in the Montreal area; Baie de Tadoussac, a marine park in the Quebec district; and Fauvel in the lower St. Lawrence–Gaspé area.

The parks and reserves are important to tourism. In 1976-77 an unprecedented total of more than 5.8 million visitor-days was recorded compared with 5.5 million in 1975-76. Increases were most noticeable in the following activities: hiking, 315,740 persons; canoeing, 61,840. Snowshoeing, cross-country and downhill skiing drew more than 1,305,610 visitors compared with 1,050,000 in 1975-76.

Ontario. There are 127 provincial parks for public use in Ontario and 138 special recreation areas or areas held in reserve for development. The provincial park system, begun in 1893 with Algonquin Park, now comprises 53 105 square kilometres.

The goal of the provincial park system is to provide a variety of outdoor recreation opportunities and to preserve provincially significant natural, cultural and recreational environments. Parks are classified into five categories: primitive, natural environment, recreation, nature reserve and wild river. The following are examples of each class: Polar Bear provincial park, occupying 24 087 km² in the Hudson Bay lowland bordering Hudson Bay and James Bay, is a primitive park containing boreal forest, tundra and arctic flora and fauna. Algonquin provincial park, a natural environment park, has 17 picnic and camping areas accessible by car and 7 537 km² offering canoeing and hiking opportunities. In Southern Ontario, the recreational park Bronte Creek provincial park provides tennis courts, man-made swimming lake, outdoor artificial skating rink, toboggan hill and bike paths, hiking and cross-country ski trails. Ouimet Canyon, a nature reserve park, preserves a 150-metre wide, 100-metre deep canyon in which arctic plants flourish, kilometres from their usual habitat. The Mattawa wild river park follows 40 km of an old Indian and voyageur route.

In 1977 there were 11 million visitors to provincial parks including 1.5 million campers using 20,000 campsites.

Manitoba. The parks branch of the tourism, recreation and cultural affairs department was established in 1966. It is responsible for the administration and management of 12 provincial natural parks, 45 provincial recreation parks, two provincial heritage parks, 104 provincial wayside parks and numerous special use parks, all of which comprise over 10 230 square kilometres. The basic objective of the parks branch is to constitute, establish and maintain a system of resource-based parks and related land-use areas for the use, benefit, health, enjoyment, recreation and education of Manitoba citizens and visitors to the province.

Saskatchewan. The Saskatchewan park system was established in 1931 when Duck Mountain, Cypress Hills and Moose Mountain became the first provincial parks. Today 17 provincial parks represent all regions in the province and are classified for development as wilderness, natural environment or recreation. The social importance of outdoor recreation and culture is reflected in regional and historic parks. Regional parks were designed to fulfil the outdoor recreational aspirations of communities; there are 96 of these throughout southern Saskatchewan. Nine historic parks are monuments to early trade, conflict and settlement of the territory.

Alberta. Provincial parks first came into existence in Alberta in 1932 under the jurisdiction of a provincial parks board. In 1964, Alberta Parks emerged as a separate division with its own director. Provincial park resources are now administered by a provincial parks branch and consist of 54 provincial parks, six natural areas, and the three wilderness areas of Whitegoat, Siffleur, and Ghost River, along with Willmore wilderness park which is administered by the energy and natural resources department. Major provincial parks include Kananaskis, Cypress Hills, Dinosaur, Lesser Slave Lake and Writing-on-Stone.

British Columbia. At the end of 1977 British Columbia had 320 parks, 22 recreation areas, and one wilderness conservancy, totalling 4 531 562 hectares in area. BC's park system began in 1911 with the establishment of Strathcona provincial park, 231 384 hectares, in central Vancouver Island. Since then the park system has steadily expanded to include vast wilderness areas, camping and picnicking sites, downhill and cross-country ski areas, a comprehensive marine park system, historic parks and sites, a famous canoe circuit, wildlife sanctuaries, and outstanding examples of the province's physical features. In 1977 there were 9.5 million visits to the parks.

The national capital region   1.5.3
Canada's capital lies on the Ottawa River below the Chaudière Falls and just above the confluence of the Rideau and Gatineau rivers. Ottawa comes from Outaouac or Outaouais, an Indian tribe from Lake Huron which traded with the French in the 17th century.

The United Province of Canada, following its formation in 1841, shuttled its capital among Kingston, Toronto, Montreal and Quebec while trying to agree on a permanent site. Queen Victoria settled the dispute by choosing Ottawa in 1858. In 1866 the government of the Province of Canada moved to Ottawa. The next year the Parliament of the new Dominion of Canada met for the first time.

Little effort was made to preserve the capital's natural beauty until the Ottawa Improvement Commission was formed in 1899. The present National Capital Commission was formed in 1959 to carry out the master plan conceived for the national capital region by town planner Jacques Gréber.

Ottawa and Hull comprise the core of the national capital region, an area of about 4 662 km² in Ontario and Quebec with a population of about 660,000. Industrial development in the region is limited. A large proportion of the work force is employed by the federal government.

Although the terms of reference of the National Capital Commission are "to prepare plans for and assist in the development, conservation and improvement of the national capital region in order that the nature and character of the seat of the Government of Canada may be in accordance with its national significance," the commission does not have jurisdictional authority over any municipal or regional authorities or the two provincial governments concerned. Most matters affecting the municipalities — planning, zoning, land use, building density, public transit, parking and construction of streets, arterial roads and highways — are within their sole jurisdiction, subject only to provincial government approval. The commission in its development efforts depends essentially upon the co-operation of each municipality and provincial government.

In recent years, the efforts of the commission have focused on development of a unified and lively core for the capital. At a constitutional conference in Ottawa in 1969, the federal and provincial first ministers declared the cities of Ottawa and Hull and their surrounding areas to be the Canadian capital region. Almost immediately, work began to remove the longstanding economic disparity between Hull and Ottawa; land was acquired in Hull for a federal building program to house various government departments.

1.6 The environment

The environment department was created in June 1971 to ensure the management and development of Canada's renewable natural resources and to head the attack on pollution. It has the responsibility to initiate government-wide programs and co-ordinate efforts related to environmental protection. It also provides specialist advisory services to other departments in setting up programs and in development of regulations under federal acts assigned to other ministries.

The department's environmental protection service is responsible for developing and enforcing environmental protection regulations and other instruments to implement federal laws relating to the environment. It is also an information source for other federal departments.

The water pollution control program's main objectives are: reduction of existing pollution and prevention of new problems; achievement of regional water quality objectives; and development of technologies to solve water pollution control problems more economically. Regulations and guidelines have been developed for the pulp and paper industry, for mercury from the chlor-alkali industry and the petroleum refinery industry, and for the meat and poultry, potato processing, fish processing and base metal, uranium and iron ore mining and metal finishing industries. These regulations under the Fisheries Act restrict the amount of effluents an industry can discharge into waters inhabited by aquatic life. Regulations are being developed for other industries. Other programs include water pollution surveys in shellfish growing areas; phosphorus concentration control regulations; analytical and advisory services to other federal agencies; inventories of water pollution problems in Canada and annual assessments of pollution control costs and studies on the treatment of municipal and industrial waste waters.

Broad objectives of the air pollution control program are to preserve, restore or enhance the quality of air in Canada. Programs include: collection and evaluation of information regarding air pollution sources; development of abatement and compliance programs for stationary and mobile air pollution sources; preparation of regulations controlling the amount of lead in leaded gases and defining the limits of lead and phosphorus in unleaded grades of gasoline, as well as regulations defining national emission standards for secondary lead smelters, emissions of mercury from mercury cell chlor-alkali plants and emissions of asbestos from asbestos mines and mills; maintenance of a mobile motor vehicle emission testing facility and promulgation of national air quality objectives.

The environmental conservation directorate develops and implements programs to protect and conserve the environment. The environmental emergency program co-ordinates activities associated with environmental threats, such as oil spills, and examines the national state of preparedness to cope with such accidents.

The environmental contaminants program involves protection from adverse effects of substances produced by industry. This program is responsible for management of hazardous materials and development of codes of good practice and guidelines for identifying, transporting, storing and disposing of hazardous materials. Under the Environmental Contaminants Act the federal government may provide for control of chemicals that may be disseminated, are persistent and are harmful to human health or the environment. Under the Ocean Dumping Control Act deliberate dumping of certain substances from ships, aircraft and platforms at sea is subject to a permit. The solid waste management program's objectives are to reduce impact on the environment and increase resource recovery and energy conservation from solid wastes.

The federal activities environmental protection program deals with activities of all federal government agencies and Crown corporations, concerning both land installations and vessels. These include treatment and disposal of waste water, solid waste management, air pollution, noise pollution and other threats to environmental quality. This program is responsible for a national approach to noise management and for developing and implementing ecological protection regulations, guidelines and codes; analysis and appraisal of ecological impact studies; implementation of control measures; and enforcement and surveillance activities.

The federal government is committed to cleaning up pollution at federal facilities within a reasonable time. Clean-up projects have dealt with water, air, noise, dust and solid waste pollution problems at airports, government offices, laboratories, grain elevators, defence bases, parks, ships and harbours.

An interdepartmental committee on the environment was established in 1973. This is the primary forum for interdepartmental consultation on environmental and related resource issues and assists the fisheries and environment department in co-ordinating development and implementation of environmental policies and programs.

Federal-provincial programs   1.6.1

Responsibility for renewable resources and environmental matters is shared by federal and provincial governments. In some areas, such as fisheries, legislative jurisdiction rests with the federal government, although management and administration have been delegated to certain provinces; in other areas, such as forest resources, the provinces have legislative jurisdiction.

To develop further co-operative action, the fisheries and environment department is developing, with the provinces, federal-provincial accords for protection and enhancement of environmental quality. These are viewed as umbrella agreements under which specific agreements on environmental action may be signed. Examples of some current federal-provincial programs follow:

Assessments of potential environmental impact of major projects in which the federal government has an interest are carried out by federal departments in consultation and in co-operation with provincial and territorial governments. Procedures ensure consideration of environmental matters.

A national air pollution surveillance network established under the Clean Air Act consists, with certain exceptions, of monitoring stations operated by provincial governments using equipment loaned by the federal government.

Management of Canada's water resources requires continuing arrangements within all jurisdictions. Federal-provincial agreements under the Canada Water Act provide for water basin management programs and include agreements for joint study of water management in specific areas. Arrangements may be made to co-ordinate federal and provincial water quality monitoring programs and to exchange data.

Many other formal and informal federal-provincial programs are related to specific aspects of renewable resources and the environment such as fisheries, forest and wildlife management, hydrometric data gathering, flood damage assistance and flood control, forest pest control and weather forecasting.

Specific non-recurring joint programs are developed from time to time. An example is the Canada/Ontario study to determine Great Lakes shoreline damage resulting from high water levels and to provide the basis for recommendations on long-term remedial and protective measures.

1.6.2 International programs

Canada takes part in two distinct types of multilateral programs — those primarily environmental and those which focus primarily on resource conservation and management, mainly of the aquatic environment. For example, Canada belongs to a 17-member international council for exploration of the sea which encourages and co-ordinates studies of marine environment with particular reference to living resources of the sea, primarily in the North Sea and North Atlantic. Canada is also a member of 10 international fisheries commissions established under formal conventions. These commissions investigate specific living marine resources in defined areas.

Canada participates in the international hydrological program set up under UNESCO to facilitate a better scientific understanding of hydrological phenomena, and provided the chairman of the intergovernmental council for the program during the biennium 1975-77. Canada also participates in the operational hydrological program set up under World Meteorological Organization auspices. These two international programs are closely co-ordinated.

In 1974 the World Health Organization designated the inland waters directorate's principal research laboratory, the Canada Centre for Inland Waters (CCIW), as its international collaboration centre for surface and groundwater quality; and CCIW continued to act as lead institute in the lake eutrophication measurement program for the Organization for Economic Co-operation and Development.

Canada played a leading role in the United Nations water conference in Argentina in March 1977, a policy conference on co-ordinated water management at the national level. Canada also took part in the UN conference on growth of desert lands in Nairobi in August 1977; this examined the extent of desert encroachment on other land and sought ways to restore affected areas.

Canada has also been active in the Intergovernmental Maritime Consultative Organization (IMCO), a specialized agency of the UN, particularly on the marine environment protection committee. In October 1973, IMCO sponsored an international conference on marine pollution which drafted an agreement regulating intentional and negligent discharges of oil and other harmful substances by ships and other equipment. The international convention on dumping of wastes at sea, formulated in accordance with a recommendation of the Stockholm UN conference and opened for signature in December 1972, has been signed and ratified by Canada. Canada has been an active member in meetings concerning that convention.

Canada was deeply involved in preparations for the third UN conference on the law of the sea. The procedural session opened in New York in December 1973 and substantive sessions took place in Caracas in 1974, Geneva in 1975 and New York in the summer of 1976. Among issues dealt with were sovereign rights over resources of the continental shelf, rights concerning management and conservation of living resources in coastal waters, rights of coastal states to protect their marine environment, rights to control scientific research in zones of maritime jurisdiction, and rights of all states to the disposition of seabed riches beyond national jurisdiction for the benefit of all mankind.

Among broad multilateral environmental organizations, the United Nations Environment Program has a governing council composed of representatives from developed and developing countries from both centrally-planned and free enterprise economics; it is the only multilateral intergovernmental body established for the sole purpose of dealing with global and regional environmental issues and problems; and it is essentially a co-ordinating body rather than an operational one. It resulted from the Stockholm UN conference on human environment in 1972 and it has its headquarters in Nairobi, Kenya. Canada is a member of the governing council, has participated in the development of its program, and has contributed to the environment fund. At the fifth session of the governing council in Nairobi, May 1977, a work program in the following areas was approved: human settlements and human health, ecosystems, environment and development, oceans, energy and natural disasters.

Canada also participates in activities of the senior advisers on environmental problems, one of the principal subsidiary bodies of the UN Economic Commission for Europe (ECE). The senior advisers assemble information on the environment in the ECE region, analyze international implications of national policies, encourage studies of selected problems of environmental policy, and promote relevant intergovernmental agreements. The senior advisers in Geneva in February 1977, in a policy discussion on environmental impact assessment practices in different countries, agreed to undertake a co-operative program on the long-range transport of air pollutants and to set up a task force on recycling and recovery of municipal and industrial solid wastes. The secretariat was asked to prepare a proposal for a future work program on low- and non-waste technology.

Canada has continued to participate in the UNESCO program on man and the biosphere, a research program to improve knowledge of terrestrial biological resources and of inter-relationships between human activities and ecosystems. As a member of the international co-ordinating council, Canada has been influential in injecting social science concerns into the program. The main focus for Canadian input is related to four themes: urbanization and industrialization, agricultural and forestry management practices, coastal ecosystems, and Arctic and isolated area development.

The Organization for Economic Co-operation and Development established an environment committee to examine common problems related to protection and enhancement of the natural and urban environment. This committee recommended adoption of the polluter pays principle as well as limitations in use of polychlorinated biphenyls (PCBs). It approved work in recycling and waste prevention, and economic incentives in waste management, air pollution, water pollution, urban environment and energy. Canada has participated actively in this committee.

Canada is also a member of the North Atlantic Treaty Organization's committee on challenges of modern society. Work of this committee includes pilot projects in which one country accepts a leadership role for a particular problem and only those countries with sufficient interest participate. Canada was the lead country in a project on inland water pollution, completed in 1974, and in 1978 was the lead country in a nutrition and health project. Canada also participates in other projects on waste water treatment, disposal of hazardous wastes and air pollution assessment.

Canada is a member of the World Weather Watch, and has set up nine stations of a planned network of 11 for monitoring air pollution in non-urban areas. In co-operation with some provinces, Canada is providing air quality data for Canadian cities as part of a World Health Organization program. Canada also contributed to a program of the integrated global ocean station system, a global investigation of the pollution of the marine environment of the Intergovernmental Oceanographic Commission, participated in a group of experts on the scientific aspects of marine pollution and co-operated in the development of a global environmental monitoring system, an integral part of the Earthwatch program.

Sources

1.1 National Geographical Mapping Division, Surveys and Mapping Branch, Department of Energy, Mines and Resources.

1.1.1 Topographical Survey Division, Surveys and Mapping Branch, Department of Energy, Mines and Resources.

1.1.2 - 1.1.3 Information Services Directorate, Department of Fisheries and the Environment.

1.1.4 - 1.1.5 Surveys and Mapping Branch, Department of Energy, Mines and Resources.

1.2 Geological Survey of Canada, Department of Energy, Mines and Resources.

1.3 Information Services Directorate, Department of Fisheries and the Environment.

1.4 Division of Physics, National Research Council.

1.5 - 1.5.1 Canada Year Book Section, Information Division, Statistics Canada; Information Services, Department of Indian Affairs and Northern Development; except Gatineau Park supplied by the National Capital Commission.

1.5.2 Supplied by the respective provincial government departments.

1.5.3 National Capital Commission.

1.6 Information Services Directorate, Department of Fisheries and the Environment.

Tables

1.1   Approximate land and freshwater areas, by province

Province or territory	Land km²	Freshwater km²	Total km²	Percentage of total area
Newfoundland	370 485	34 032	404 517	4.1
Island of Newfoundland	*106 614*	*5 685*	*112 299*	*1.1*
Labrador	*263 871*	*28 347*	*292 218*	*3.0*
Prince Edward Island	5 657	—	5 657	0.1
Nova Scotia	52 841	2 650	55 491	0.6
New Brunswick	72 092	1 344	73 436	0.7
Quebec	1 356 791	183 889	1 540 680	15.5
Ontario	891 194	177 388	1 068 582	10.8
Manitoba	548 495	101 592	650 087	6.5
Saskatchewan	570 269	82 631	651 900	6.6
Alberta	644 389	16 796	661 185	6.6
British Columbia	930 528	18 068	948 596	9.5
Yukon Territory	478 034[1]	4 481	482 515[1]	4.9
Northwest Territories	3 246 390	133 294	3 379 684	34.1
Franklin	*1 403 134*	*19 425*	*1 422 559*	*14.3*
Keewatin	*565 809*	*25 123*	*590 932*	*6.0*
Mackenzie	*1 277 447*	*88 746*	*1 366 193*	*13.8*
Canada	9 167 165[1]	755 165	9 922 330[1]	100.0

[1] Recalculated figures 1977.

1.2   Principal heights in each province

Province and height	Elevation m	Province and height	Elevation m
NEWFOUNDLAND		**QUEBEC** (concluded)	
		Albert Nord Summit	1 083
Long Range Mountains		Matawees Mountain	1 074
Lewis Hills	814	Rond Summit (Sutton Mountains)	968
Gros Morne	806	Mount Bayfield	892
Table Mountain (St. Barbe District)	724	Mount Orford	876
Mount St. Gregory	686	Hereford Mountain	846
Gros Paté	656	Barn Mountain	846
Blue Mountain	649	Le Pinacle Mountain	709
Blue Hills of Couteau		The Laurentians	
Peter Snout	495	Mont-Tremblant	968
Central Highlands		Mont Sainte-Anne	800
Main Topsail	555	Mont Sir Wilfrid	783
Mizzen Topsail	537	Monteregian Hills	
Torngat Mountains		Brome Mountain	533
Unnamed peak (58°57' 63°47')	*1 652*	Shefford Mountain	518
Cirque Mountain	1 568	Mont Saint-Hilaire	411
Mount Cladonia	1 453	Yamaska Mountain	411
Mount Eliot	1 388	Rougemont	396
Mount Tetragona	1 356		
Quartzite Mountain	1 186	**ONTARIO**	
Blow-Me-Down Mountain	1 183		
Kaumajet Mountains		Highest point, Timiskaming District	
Bishops Mitre	1 113	*(47°20' 80°44')*	*693*
Finger Hill	1 033	Ogidaki Mountain	665
		Batchawana Mountain	653
PRINCE EDWARD ISLAND		Tip Top Mountain	640
		Niagara Escarpment	
Highest point on the Island, Queens		Blue Mountains	541
County (46°20' 63°27')	*142*	Osler Bluff	526
		Caledon Mountain	427
NOVA SCOTIA		High Hill	354
		Mount Nemo	305
Highest point, Cape Breton (46°42' 60°36')	*532*		
Franey Mountain	428	**MANITOBA**	
Nuttby Mountain (Cobequid)	367		
Dalhousie Mountain (Cobequid)	340	Baldy Mountain	*832*
		Porcupine Hills	823
NEW BRUNSWICK		Riding Mountain	610
Mount Carleton	*820*	**SASKATCHEWAN**	
Moose Mountain	404		
		Cypress Hills	*1 392*
QUEBEC		Wood Mountain	1 013
		Vermilion Hills	785
Mont D'Iberville (Torngat Mountains)	*1 652*		
Appalachian Mountains		**ALBERTA**	
Mont Jacques-Cartier (Shickshock			
Mountains)	1 268	Mount Columbia	*3 747*
Mount Richardson	1 185	The Twins	3 734
Mount Albert District		Mount Alberta	3 619
Albert Sud Summit	1 151	Mount Assiniboine	3 618
Mount Logan	1 135	Mount Forbes	3 612
Mont Mégantic	1 105		

1.2   Principal heights in each province (concluded)

Province and height	Elevation m	Province and height	Elevation m
ALBERTA (concluded)		**BRITISH COLUMBIA** (concluded)	
Mount Temple	3 544	Mount Ball	3 312
Mount Lyell	3 520	Bush Mountain	3 307
Mount Hungabee	3 520	Mount Geikie	3 305
Snow Dome	3 520	Mount Sir Alexander	3 274
Mount Kitchener	3 505	Fresnoy Mountain	3 271
Mount Athabasca	3 491	Mount Gordon	3 216
Mount King Edward	3 475	Mount Stephen	3 199
Mount Brazeau	3 470	Cathedral Mountain	3 189
Mount Victoria	3 464	Odaray Mountain	3 155
Stutfield Peak	3 450	The President	3 139
Mount Joffre	3 449	Mount Laussedat	3 059
Deltaform Mountain	3 424		
Mount Lefroy	3 423	**YUKON TERRITORY**	
Mount Alexandra	3 418	St. Elias Mountains	
Mount Sir Douglas	3 406	Mount Logan	*5 951*
Mount Woolley	3 405	Mount St. Elias	5 489
Lunette Peak	3 399	Mount Lucania	5 226
Mount Hector	3 398	King Peak	5 173
Diadem Peak	3 371	Mount Steele	5 073
Mount Edith Cavell	3 363	Mount Wood	4 842
Mount Fryatt	3 361	Mount Vancouver	4 785
Mount Chown	3 331	Mount Hubbard	4 577
Mount Wilson	3 261	Mount Walsh	4 505
Clearwater Mountain	3 176	Mount Alverstone	4 439
Mount Coleman	3 135	McArthur Peak	4 344
Eiffel Peak	3 079	Mount Augusta	4 289
Pinnacle Mountain	3 067	Mount Kennedy	4 238
		Mount Strickland	4 212
BRITISH COLUMBIA		Mount Newton	4 210
Vancouver Island Ranges		Mount Cook	4 194
Golden Hinde	2 200	Mount Craig	4 039
Mount Albert Edward	2 081	Mount Malaspina	3 886
Mount Arrowsmith	1 817	Mount Badham	3 848
Coast Mountains		Mount Seattle	3 073
Mount Waddington	3 994		
St. Elias Mountains		**NORTHWEST TERRITORIES**	
Fairweather Mountain	*4 663*	Arctic Islands	
Mount Root	3 920	Baffin	
Monashee Mountains		Penny Ice Cap	2 057
Mount Begbie	2 732	Mount Thule	1 711
Storm Hill	1 615	Cockscomb Mountain	1 625
Selkirk Mountains		Barnes Ice Cap	1 123
Mount Sir Sandford	3 522	Knife Edge Mountain	760
Mount Dawson	3 390	Banks	
Adamant Mountain	3 356	Durham Heights	732
Grand Mountain	3 305	Devon	
Iconoclast Mountain	3 251	Ice Cap	1 920
Rogers Peak	3 214	Ellesmere	
Purcell Mountains		Barbeau Peak, highest point in	
Mount Farnham	3 457	Arctic Islands	2 616
Mount Karnak	3 383	Commonwealth Mountain	2 210
Columbia (Cariboo) Mountains		Mount Jeffers	1 905
Sir Wilfrid Laurier	3 444	Mount Wood	1 448
Rocky Mountains		Mount Cheops	1 448
Mount Robson	3 954	Victoria	
Mount Clemenceau	3 658	Shaler Mountains	655
Mount Goodsir	3 581	Mount Bumpus	503
Mount Bryce	3 507	Mainland	
The Helmet	3 429	Mount Sir James MacBrien	*2 762*
Resplendent Mountain	3 426	Franklin Mountains	
Mount King George	3 422	Cap Mountain	1 577
Whitehorn Mountain	3 395	Mount Clark	1 462
Mount Huber	3 368	Pointed Mountain	1 405
Mount Freshfield	3 336	Nahanni Butte	1 396
Mount Mummery	3 328	Richardson Mountains	
Mount Vaux	3 320	Mount Goodenough	981

1.3   Elevations, areas and depths of the Great Lakes

Lake	Elevation[1] m	Length km	Breadth km	Maximum depth m	Total area km²	Area on Canadian side of boundary km²
Superior	183	563	257	405	82 103	28 749
Michigan	176	494	190	281	57 757	—
Huron	176	332	295	229	59 570	36 001
St. Clair	175	42	39	6	1 114	694
Erie	174	388	92	64	25 667	12 769
Ontario	75	311	85	244	19 011	10 049

[1]Long-term mean, 1860-1972; International Great Lakes Datum, 1955.

1.4 Elevations and areas of principal lakes[1] (exceeding 388 km²), by province

Province and lake	Elevation m	Area km²	Province and lake	Elevation m	Area km²
NEWFOUNDLAND AND LABRADOR			**SASKATCHEWAN** (concluded)		
Ashuanipi	529	598	Montreal	490	456
Atikoniak	518	433	Peter Pond	421	777
Grand	87	539	Pinehouse	385	404
Joseph	518	451	Primrose	599	448
Melville	tidal	3 069	Reindeer[2]	337	6 651
Michikamau	460	2 031	Scott	444	394
Lobstick	457	510	Tazin	344	391
Ossokmanuan Reservoir	479	834	Wollaston	398	2 681
Smallwood Reservoir	471	6 475	**ALBERTA**		
NOVA SCOTIA			Bistcho	552	427
Bras d'Or	tidal	1 098	Claire	213	1 437
			Lesser Slave	577	1 168
QUEBEC			**BRITISH COLUMBIA**		
Albanel	389	445	Atlin[2]	668	774
Bienville	427	1 248	Babine	711	495
Cabonga Reservoir	361	679	Kootenay	532	407
Dozois Reservoir	346	404	Ootsa	853	404
Eau Claire	241	1 383	Williston	664	1 761
Evans	241	546	**YUKON TERRITORY**		
Gouin Reservoir	404	1 570	Kluane	781	409
Kaniapiskau	564	471			
Leaf	tidal	453	**NORTHWEST TERRITORIES**		
Lower Seal	262	578	Aberdeen	80	1 101
Manouane	494	585	Amadjuak	113	3 116
Minto	168	761	Angikuni	257	510
Mistassini	372	2 336	Artillery	365	552
Payne	130	534	Aubry	258	391
Pipmuacan	396	979	Aylmer	375	847
Saint-Jean	98	1 002	Baker	2	1 888
Sakami	195	593	Bluenose	557	401
			Buffalo	265	614
ONTARIO			Clinton Colden	375	736
Abitibi[2]	265	932	Colville	245	456
Big Trout	213	660	Contwoyto	445	958
Lake of the Woods[2] (total			De Gras	416	632
4 349) Canadian part 3 149	323	3 149	Des Bois	297	469
Nipigon	261	4 848	Dubawnt	236	3 833
Nipissing	196	831	Ennadai	311	681
Rainy (total 932)			Eskimo North	0.3	839
Canadian part 741	338	741	Eskimo South	2	629
St. Joseph	371	492	Faber	213	440
Sandy	276	526	Ferguson	11	588
Seul	357	1 658	Garry	148	976
Simcoe	219	743	Great Bear	156	31 328
Trout (English River)	394	414	Great Slave	156	28 570
			Hall	6	492
MANITOBA			Hazen	158	541
Cedar	253	1 352	Hottah	180	917
Cross	207	756	Kamilukuak	266	635
Dauphin	260	521	Kaminak	53	601
Gods	178	1 150	Kaminuriak	92	549
Granville	258	490	Kasba	336	1 342
Island	227	1 222	Keller	247	394
Manitoba	248	4 659	La Martre	265	1 777
Molson	221	399	Mac Alpine	176	448
Moose	255	1 368	Mackay	431	1 062
Oxford	187	401	Mallery	158	479
Playgreen	217	658	Netilling	29	5 543
Sipiwesk	183	456	Netsilik	8	391
Southern Indian	255	2 248	Nonacho	319	785
Winnipeg	217	24 390	Nueltin[2]	278	2 279
Winnipegosis	253	5 374	Point	375	702
			Princess Mary	116	523
SASKATCHEWAN			Selwyn[2]	398	717
Amisk	294	430	Snowbird	359	505
Athabasca[2]	213	7 936	South Henik	184	513
Black	281	464	Takiyuak	381	1 080
Churchill	421	559	Tathlina	280	572
Cree	487	1 435	Tebesjuak	146	575
Deschambault	324	541	Tehek	133	482
Doré	459	642	Trout	503	505
Frobisher	421	515	Tulemalu	279	668
Ile à la Crosse	421	391	Wholdaia	364	679
La Ronge	364	1 414	Yathkyed	141	1 448

Areas given are for mean water levels. All elevations are in metres above mean sea level.
[1] Excludes Great Lakes, see Table 1.3.
[2] Spans provincial or territorial boundary. Listed under province or territory containing larger portion. Area given is total area.

1.5   Lengths of principal rivers and their tributaries

Drainage basin and river	Length km	Drainage basin and river	Length km
FLOWING INTO THE PACIFIC OCEAN		**FLOWING INTO HUDSON BAY**	
Yukon (mouth to head of Nisutlin)	3 185	**AND HUDSON STRAIT** (concluded)	
(International Boundary to head of		Koksoak (to head of Caniapiscau)	874
Nisutlin)	1 149	Nottaway (via Bell to head of Mégiscane)	776
Porcupine	721	Rupert (to head of Témiscamie)	763
Stewart	644	Eastmain	756
Pelly	608	Attawapiskat (to head of Bow Lake)	748
Teslin	393	Kazan (to head of Ennadai Lake)	732
Columbia (mouth to head of Columbia Lake)	2 000	Grande rivière de la Baleine (Great Whale)	724
(International Boundary to head of		George	563
Columbia Lake)	801	Moose (to head of Mattagami)	547
Kootenay	781	Abitibi (to head of Louis Lake)	547
Elk (to head of Elk Lake)	220	Mattagami (to head of Minisinakwa Lake)	443
St. Mary	117	Missinaibi	426
Slocan (to head of Slocan Lake)	97	Harricana	533
Kettle (to head of Holmes Lake)	336	Hayes	483
Okanagan (to head of Okanagan Lake)	314	Aux Feuilles (Leaf)	480
Similkameen	251	Winisk	475
Canoe	169	Broadback	451
Spillimacheen	84	A la Baleine (Whale)	428
Kicking Horse (to head of Wapta Lake)	84	de Povungnituk	389
Illecillewaet	77	Innuksuac	385
Fraser	1 368	Petite rivière de la Baleine (Little Whale River)	380
Thompson (to head of North Thompson)	489	Arnaud	377
North Thompson	338	Nastapoca	360
South Thompson (to head of Shuswap)	332	Kogaluc	304
Shuswap	185		
Nechako (to head of Eutsuk Lake)	462	**FLOWING INTO THE ATLANTIC OCEAN**	
Stuart (to head of Driftwood)	415	St. Lawrence River	3 058
Chilcotin	235	Lake Superior	
West Road	227	Nipigon (to head of Ombabika)	209
Quesnel (to head of Mitchell Lake)	203	Magpie (to head of Merekeme Lake)	114
Lilooet	177	Lake Huron	
Bridge	142	Spanish	338
Skeena	579	French (to head of Sturgeon)	290
Bulkley (to head of Maxam Creek)	257	Mississagi	266
Stikine	539	Saugeen	161
Nass	380	Lake St. Clair	
Homathko	137	Thames	262
		Lake Erie	
FLOWING INTO THE ARCTIC OCEAN		Grand	266
Mackenzie (to head of Finlay)	4 241	Lake Ontario	
Peace (to head of Finlay)	1 923	Trent (to head of Irondale)	402
Smoky	492	Moira	124
Finlay	402	Ottawa River	1 271
Parsnip	233	Gatineau	386
Athabasca	1 231	du Lièvre	330
Pembina	547	Madawaska (to head of Madawaska Lake)	230
Liard	1 115	Coulonge	217
South Nahanni	563	Petawawa (to head of Butt Lake)	187
Fort Nelson (to head of Sikanni Chief)	517	Rouge	185
Petitot	404	Mississippi (to head of Mazinaw Lake)	169
Hay	702	South Nation	161
Peel (mouth of west channel to head of Ogilvie)	684	Rideau	146
Arctic Red	499	Dumoine	129
Slave (from Peace River to Great Slave Lake)	415	du Nord	113
Fond du Lac (to outlet of Wollaston Lake)	277	de la Petite Nation	97
Back (to outlet of Muskox Lake)	974	Saguenay (to head of Péribonca)	698
Coppermine	845	Péribonca	451
Anderson	692	Mistassini	298
Horton	618	Ashuapmuchuan	266
		Saint-Maurice	563
		Matawin	161
FLOWING INTO HUDSON BAY		Manicouagan (to head of Mouchalagane)	560
AND HUDSON STRAIT		aux Outardes	499
Nelson (to head of Bow)	2 575	Romaine	496
(to outlet of Lake Winnipeg)	644	Betsiamites (to head of Kanouanis)	444
Saskatchewan (to head of Bow)	1 939	Moisie	410
South Saskatchewan (to head of Bow)	1 392	Bersimis	386
Red Deer	724	St-François	280
Bow	587	St-Augustin	233
Oldman	362	Chaudière	193
North Saskatchewan	1 287	Richelieu (to mouth of Lake Champlain)	171
Battle (to head of Pigeon Lake)	570	Churchill (to head of Ashuanipi)	856
Red (to head of Sheyenne)	877	Saint John	673
Assiniboine	1 070	Tobique (to outlet of Nictau Lake)	148
Winnipeg (to head of Firesteel)	813	du Petit-Mécatina	547
English	615	Natashquan	410
Fairford (to head of Manitoba Red Deer)	684	Exploits	246
Churchill (to head of Churchill Lake)	1 609	Eagle	233
Beaver (to outlet of Beaver Lake)	491	Miramichi	217
Severn (to head of Black Birch)	982	Gander (to head of Northwest Gander River)	175
Albany (to head of Cat)	982	Nepisiguit (to outlet of Nepisiguit Lake)	121
Thelon	904	St. Mary's (to head of North Nelson)	95
Dubawnt	842	Mersey (to outlet of 11 Mile Lake)	93
La Grande-Rivière (Fort George River)	893	Bay du Nord	66
		Pipers Hole	37

1.6   Areas of major islands, by region

Region and island	Area km²	Region and island	Area km²
Baffin Island	507 451	HUDSON BAY AND HUDSON STRAIT	
		Southampton	41 214
QUEEN ELIZABETH ISLANDS		Coats	5 499
Ellesmere	196 236	Mansel	3 181
Devon	55 247	Akimiski	3 002
Axel Heiberg	43 178	Flaherty	1 585
Melville	42 149	Nottingham	1 373
Bathurst	16 042	Resolution	1 015
Prince Patrick	15 848	Vansittart	997
Ellef Ringnes	11 295	Akpatok	904
Cornwallis	6 996	Salisbury	805
Amund Ringnes	5 255	Big	803
Mackenzie King	5 048	White	790
Borden	2 795	Loks Land	420
Cornwall	2 258		
Eglinton	1 541	PACIFIC COAST	
Graham	1 378		
Lougheed	1 308	Vancouver	31 284
Byam Martin	1 150	Graham	6 361
Ile Vanier	1 127	Moresby	2 608
Cameron	1 059	Princess Royal	2 251
Meighen	956	Pitt	1 375
Brock	764	Banks	989
King Christian	645	King	808
North Kent	591	Porcher	521
Emerald	549	Nootka	510
Alexander	484	Aristazabal	420
Massey	433	Gilford	383
Little Cornwallis	412	Hawkesbury	365
		Hunter	363
ARCTIC ISLANDS SOUTH OF		Calvert	329
QUEEN ELIZABETH ISLANDS		Texada	300
Victoria	217 290	Swindle	285
Banks	70 028	McCauley	275
Prince of Wales	33 338	Louise	275
Somerset	24 786	Quadra	269
King William	13 111		
Bylot	11 067	ATLANTIC COAST	
Prince Charles	9 521	Newfoundland and Labrador	
Stefansson	4 463	Newfoundland (main island)	108 860
Richards	2 165	South Aulatsivik	456
Air Force	1 720	Killinek	269
Wales	1 137	Fogo	254
Rowley	1 090	Random	249
Russell	940	New World	189
Jens Munk	919	Tunungayualok	186
Langley and Ellice	780	West Okak	179
Bray	689	Paul	179
Foley	637	Gulf of St. Lawrence	
Royal Geographical Society Islands	609	Cape Breton	10 311
Sillem	482	Anticosti	7 941
Matty	477	Prince Edward	5 657
Spicer Islands	458	Madeleine Islands	202
Koch	458	Boularderie	192
Jenny Lind	420	Shippegan	150
Prescott	412	Bay of Fundy	
Crown Prince Frederick	401	Grand Manan	137

1.7   Temperature and precipitation data for typical stations in various districts

District and station	Temperatures (Celsius)						Precipitation		
	Mean Jan.	Mean July	Highest on record	Lowest on record	Av. dates of freezing temperatures (0°C or lower)		Total (all forms) mm	Snowfall cm	Av. number of days (all forms)
					Last in spring	First in autumn			
NEWFOUNDLAND									
Island									
Belle Isle	-9.6	9.4	22.8	-35.0	June 21	Sept. 26	893.1	240.0	149
Gander A	-6.1	16.5	35.6	-31.1	June 4	Oct. 5	1 078.2	354.8	204
St. Andrew's	-3.6	15.0	27.2	-23.9	June 3	Sept. 24	1 112.3	196.3	176
St. John's A	-3.8	15.3	30.6	-23.3	June 3	Oct. 12	1 511.5	363.7	210
Labrador									
Cartwright	-13.1	12.9	36.1	-37.8	June 20	Sept. 9	946.4	433.8	179
Goose A	-16.3	15.8	37.8	-39.4	June 6	Sept. 17	876.8	409.2	176

1.7 Temperature and precipitation data for typical stations in various districts (continued)

District and station	Temperatures (Celsius)								Precipitation		
	Mean Jan.	Mean July	Highest on record	Lowest on record	Av. dates of freezing temperatures (0°C or lower)				Total (all forms) mm	Snowfall cm	Av. number of days (all forms)
					Last in spring		First in autumn				
MARITIME PROVINCES											
Prince Edward Island											
Charlottetown A	-6.7	18.4	34.4	-27.8	May	17	Oct.	15	1 127.8	305.1	169
Nova Scotia											
Annapolis Royal	-3.9	18.3	32.8	-27.2	May	19	Oct.	2	1 204.5	218.2	149
Halifax	-3.2	18.3	34.4	-25.0	May	1	Nov.	1	1 318.8	210.8	152
Sydney A	-4.4	17.9	35.0	-25.6	May	23	Oct.	16	1 340.9	288.0	179
Yarmouth A	-2.7	16.4	30.0	-21.0	May	2	Oct.	24	1 283.2	204.5	157
New Brunswick											
Chatham A	-9.3	19.2	37.8	-35.0	May	22	Sept.	21	1 051.2	309.4	152
Grand Falls	-11.9	18.3	36.7	-43.3	May	24	Sept.	21	1 021.6	265.2	105
Moncton A	-7.9	18.6	37.2	-32.2	May	23	Sept.	23	1 099.3	313.7	156
Saint John A	-7.1	17.1	34.4	-36.7	May	18	Oct.	2	1 400.3	204.7	149
QUEBEC											
Northern											
Fort Chimo A	-23.4	11.4	32.2	-46.7	June	27	Aug.	30	483.8	236.7	155
Inoucdjouac (Port Harrison)	-24.7	8.9	30.0	-46.1	July	1	Sept.	4	355.6	122.9	133
Nitchequon	-22.9	13.6	32.2	-49.4	June	13	Sept.	13	764.5	284.7	192
Schefferville A	-22.7	12.6	31.7	-50.6	June	18	Aug.	31	722.5	335.5	188
Southern											
Bagotville A	-15.7	17.8	36.1	-43.3	May	26	Sept.	18	936.6	341.6	177
Montreal McGill	-8.9	21.6	36.1	-33.9	Apr.	22	Oct.	23	999.0	243.1	164
Pointe au Père	-10.9	15.4	32.2	-36.1	May	19	Sept.	28	848.6	285.8	135
Quebec A	-11.6	19.2	35.6	-36.1	May	18	Sept.	28	1 088.6	326.6	164
Sept-Îles A	-13.9	15.1	32.2	-43.3	May	30	Sept.	17	1 090.3	423.2	146
Sherbrooke	-9.6	20.1	36.7	-41.1	May	12	Sept.	27	972.6	244.6	170
ONTARIO											
Northern											
Kapuskasing A	-18.2	17.0	36.7	-44.4	June	13	Sept.	5	871.5	321.8	186
Sioux Lookout A	-18.7	18.4	36.1	-46.1	May	29	Sept.	20	741.5	236.7	165
Thunder Bay A	-14.8	17.5	37.2	-41.1	May	31	Sept.	10	738.5	222.0	141
Trout Lake	-24.1	15.9	35.6	-47.8	June	11	Sept.	16	597.3	212.3	158
Southern											
London A	-6.0	20.5	36.7	-31.7	May	9	Oct.	6	924.5	201.2	165
Ottawa A	-10.9	20.7	37.8	-36.1	May	11	Oct.	1	850.9	215.6	152
Parry Sound	-9.5	19.3	37.8	-41.1	May	14	Oct.	2	1 020.1	296.7	158
Toronto	-4.4	21.8	40.6	-32.8	Apr.	20	Oct.	30	789.9	141.0	134
Windsor A	-4.3	22.3	38.3	-26.1	Apr.	29	Oct.	20	836.1	103.6	137
PRAIRIE PROVINCES											
Manitoba											
Churchill A	-27.6	12.0	33.9	-45.0	June	22	Sept.	12	396.6	183.9	141
The Pas A	-22.4	17.9	36.7	-49.4	May	28	Sept.	20	449.7	157.2	128
Winnipeg A	-18.3	19.7	40.6	-45.0	May	25	Sept.	21	535.2	131.3	121
Saskatchewan											
Regina A	-17.3	18.9	43.3	-50.0	May	27	Sept.	12	397.9	114.8	114
Saskatoon A	-18.7	18.8	40.0	-47.8	May	27	Sept.	15	352.6	112.5	103
Swift Current A	-13.9	18.7	38.9	-42.8	May	28	Sept.	19	389.9	123.7	112
Alberta											
Beaverlodge CDA	-14.9	15.6	36.7	-47.8	May	22	Sept.	7	454.7	183.6	129
Calgary A	-10.9	16.5	36.1	-45.0	May	28	Sept.	12	437.1	153.9	113
Edmonton Ind. A	-14.7	17.5	34.4	-48.3	May	14	Sept.	19	446.5	132.1	121
Medicine Hat A	-12.1	20.2	42.2	-46.1	May	17	Sept.	20	347.8	121.7	89
BRITISH COLUMBIA											
Pacific Coast and Coastal Valleys											
Estevan Point	4.5	13.8	28.9	-13.9	Apr.	5	Nov.	18	3 027.9	34.3	203
Langara	2.5	12.4	25.6	-14.4	Apr.	3	Nov.	26	1 675.6	61.2	248
Prince Rupert	1.8	13.6	32.2	-21.1	Apr.	19	Nov.	5	2 414.5	113.0	227
Vancouver A	2.4	17.4	33.3	-17.8	Mar.	31	Oct.	30	1 068.1	52.3	161
Victoria											
Gonzale Hts	4.1	15.7	35.0	-15.6	Feb.	28	Dec.	9	657.1	32.8	142
Southern Interior											
Glacier	-11.3	14.4	36.7	-35.6	June	12	Sept.	6	1 492.8	969.5	192
Kamloops A	-6.0	20.9	40.6	-37.2	May	5	Sept.	28	260.6	77.0	90
Penticton A	-2.9	20.1	40.6	-27.2	May	10	Oct.	1	296.2	69.1	100
Princeton A	-8.1	17.6	41.7	-42.8	June	3	Sept.	12	359.1	157.0	115
Central Interior											
Barkerville	-9.8	12.3	35.6	-46.7	June	29	Aug.	18	1 148.8	581.4	185
McBride	-9.1	15.9	37.8	-46.7	June	9	Sept.	1	524.5	197.4	128
Prince George A	-11.8	14.9	34.4	-50.0	June	10	Aug.	28	620.7	233.4	162
Smithers A	-10.6	14.6	34.4	-43.9	June	10	Sept.	1	512.2	197.4	158
Northern Interior											
Atlin	-16.6	12.6	30.6	-50.0	June	5	Aug.	28	283.2	121.4	86
Dease Lake	-19.3	12.6	33.9	-51.1	June	29	Aug.	13	394.5	186.7	143
Fort Nelson A	-23.2	16.7	36.7	-51.7	May	24	Sept.	5	446.4	191.5	130
Fort St. John A	-17.2	15.9	33.3	-47.2	May	20	Sept.	9	449.8	206.2	128
Smith River A	-24.5	14.1	33.3	-58.9	June	21	Aug.	11	465.3	211.6	148

1.7   Temperature and precipitation data for typical stations in various districts (concluded)

District and station	Temperatures (Celsius)						Precipitation		
	Mean Jan.	Mean July	Highest on record	Lowest on record	Av. dates of freezing temperatures (0°C or lower)		Total (all forms) mm	Snowfall cm	Av. number of days (all forms)
					Last in spring	First in autumn			
YUKON TERRITORY									
Dawson	-28.6	15.5	35.0	-58.3	May 26	Aug. 27	325.5	136.4	120
Snag A	-28.2	13.9	31.7	-62.8	June 18	Aug. 9	359.7	140.5	118
Watson Lake A	-25.3	14.9	33.9	-58.9	May 30	Sept. 3	432.3	227.3	153
Whitehorse A	-18.9	14.1	34.4	-52.2	June 5	Sept. 1	260.3	127.8	118
NORTHWEST TERRITORIES									
Mackenzie Basin									
Fort Good Hope	-31.0	15.9	34.4	-55.6	June 3	Aug. 19	283.7	124.0	101
Fort Simpson A	-27.6	16.1	35.0	-53.3	May 31	Aug. 29	343.2	137.9	126
Hay River A	-25.5	15.6	35.6	-48.3	June 6	Sept. 11	339.8	165.1	109
Barrens									
Baker Lake	-33.6	10.7	30.6	-50.6	June 25	Aug. 31	213.0	88.9	96
Chesterfield	-31.8	8.7	30.6	-51.1	June 29	Sept. 6	263.5	112.8	98
Coppermine	-29.4	9.3	32.2	-50.0	June 27	Aug. 21	216.3	101.9	110
Arctic Archipelago									
Clyde	-26.9	4.6	22.2	-45.6	July 13	July 18	206.3	152.9	94
Eureka	-36.6	5.5	19.4	-53.9	June 27	Aug. 5	58.4	38.4	52
Frobisher Bay A	-26.2	7.9	24.4	-45.6	June 30	Aug. 29	415.2	246.9	135
Mould Bay	-33.8	3.7	16.1	-53.9	July 12	July 19	86.4	59.9	73
Resolute A	-32.6	4.3	18.3	-52.2	July 10	July 20	136.4	78.7	94

A = Airport, Ind. A = Industrial Airport.
CDA = Canada Department of Agriculture.

1.8   Total area classified by tenure, 1976 (km²)

Item	Province or territory						
	Nfld.	PEI	NS	NB	Que.	Ont.	Man.
Federal Crown lands other than national parks, Indian reserves and forest experiment stations	440	16	181	1 489	1 295[1]	1 158	259
National parks	2 339	21	1 331	433	790	1 922	2 978
Indian reserves	—	8	114	168	4 077[2]	6 703	2 383
Federal forest experiment stations	—	—	—	91	28	103	
Privately owned land or land in process of alienation from the Crown	17 788	4 944	37 438	39 754	112 664	119 023	138 008
Provincial or territorial area other than provincial parks and provincial forests[3]	382 842	435	2 652	28 495	1 210 799	891 261	482 204[4]
Provincial parks	805	31	109	215	194 249	48 412	10 230
Provincial forests	303	202	13 665	2 792	16 778	—	14 025
Total area	404 517	5 657	55 490	73 437	1 540 680	1 068 582	650 087

Item	Sask.	Alta.	BC	YT	NWT	Canada
Federal Crown lands other than national parks, Indian reserves and forest experiment stations	5 452	2 896[5]	904	513 193	3 340 849	3 868 132
National parks	3 875	54 084	4 690	22 015	35 690	130 168
Indian reserves	5 688	6 566	3 390	5	135	29 237
Federal forest experiment stations	—	155	—	—	—	377[6]
Privately owned land or land in process of alienation from the Crown	247 662	181 925	55 040	168	72	795 800[6]
Provincial or territorial area other than provincial parks and provincial forests[3]	34 758	63 525[7]	539 280	943	2 937	3 621 560
Provincial parks	4 944	7 700	41 629	—	—	308 187
Provincial forests	349 521	344 334[8]	303 663	—	—	1 084 669
Total area	651 900	661 185	948 596	536 324	3 379 683	9 976 138

[1]Includes Gatineau Park (356.1 km²) and Quebec Battlefields Park (0.93 km²) which are under federal jurisdiction but are not technically national parks.
[2]Includes increase awarded by the James Bay Agreement.
[3]Includes freshwater area.
[4]Includes only those provincial lands held under Crown Lands Act, of which 7 280 km² are under lease.
[5]Excludes Department of National Defence agreement areas.
[6]Excludes area for Manitoba (federal forest experiment stations are combined with privately owned land or land in process of alienation from the Crown, for that province).
[7]Includes lands held by the federal government under agreement with Alberta (one national defence area and one agriculture experiment station.
[8]Includes Department of National Defence agreement area.

1.9 Provincial parks, by province, 1977

Province	Total area km²	Developed area km²	Parks No.	Type of park	Accommodation and facilities	Activities	Visitors	Rates
Newfoundland and Labrador	8 026.3	1064.0	72	Wilderness area Public beaches Seabird sanctuaries Camping Day-use Fishery access roads	Overnight camping — picnic tables — fireplaces — firewood — potable water — pit privies Day-use — picnic tables — fireplaces — beach — boat launch — change houses	Fishing Swimming Hiking Canoeing and boating Photography Interpretative programs Snowshoeing Skiing	3.0 - 4.0 million	$3.00 for seasonal park entry permit. $1.00 daily fee. $2.50 a night for camping permit.
Prince Edward Island	31.3	16.3	40	Nature preserves Natural environment Recreation Beach access Historic	Picnic sites Sandy beaches Campgrounds — fresh spring water Serviced tent and trailer sites	Museum Swimming Golf Tennis Interpretative programs Camping Skiing	1.0 million	$5.00 for serviced campsites, $3.75 for unserviced. No charge for day use.
Nova Scotia	72.7	50.8	100	Campgrounds Picnic Beach Roadside rest sites	Day-use picnic parks Roadside rest sites Day-use beach parks Overnight campgrounds — tables — water — pit privies	Swimming Picnicking Camping	145,869 (campers)	$4.00 a night for camping permit. No charge for day use.
New Brunswick	223.7	31.3	62	Recreation Rest areas Campgrounds Beach Marine Resource Wildlife	Lodge Marinas Campgrounds — tables — some form of toilet facility — potable water	Swimming Boating Camping Golfing Interpretative programs with naturalists Tennis Snowshoeing Skiing — cross-country — downhill Skating Tobogganing Sleigh rides	3.1 million	$3.50 to $4.50 daily camping fee.
Quebec	..	..	100	Parks and reserves	Campgrounds Other accommodation	Hunting Fishing Hiking Swimming Canoeing Snowshoeing Skiing	5.8 million	

1.9 Provincial parks, by province, 1977 (continued)

Province	Total area km²	Developed area km²	Parks No.	Type of park	Accommodation and facilities	Activities	Visitors	Rates
Ontario	53 105.0	43 762.0	127	Primitive Natural environment Wild river Nature reserves Recreation	Picnic and camping areas — beaches — picnic tables — fireplaces — firewood — electricity — tested drinking water — washrooms/comfort stations — trailer sanitation stations	Museums Outdoor exhibits Nature trails Swimming Canoeing Boating Fishing Hiking Snowmobiling Skiing — cross-country	11.0 million	Effective April 1, 1978, $1.00 a day for vehicles plus $0.50 passenger charge (age 16-64) or $10.00 a year for vehicle plus passenger charge each visit. For camping including vehicle, $5.00 a night, $5.50 with comfort station, $7.00 with electricity.
Manitoba	10 230.5	..	163	Wilderness area Recreation Wayside Heritage Marine Natural Historic sites	Motels Hotels Cabins Fishing lodges Campgrounds Space available for building summer homes	Swimming Camping Hunting Fishing Hiking Canoeing Boating Picnicking Snowmobiling	5.0 million	$1.25 daily camping fee. $6.00 for the season.
Saskatchewan	5 055.6	..	122	Provincial — Wilderness area — Natural environment — Recreation Regional — Historic	Campgrounds Modern cabins Chalet	Skiing Camping Picnicking Swimming Historic interest sites Snowmobiling Hikes on nature trails Arts and crafts Social functions Cafeteria Team sports Hunting, fishing, boating and sailing Snorkelling Auto touring Horseback riding Tennis Golf Cycling	3.5 million	$5.00 for a seasonal park entry permit. $2.00 daily camping fee. $4.00 a day for serviced campsites, $2.00 for unserviced. Free entry for senior citizens.

1.9 Provincial parks, by province, 1977 (concluded)

Province	Total area km²	Developed area km²	Parks No.	Type of park	Accommodation and facilities	Activities	Visitors	Rates
Alberta	6 733.5	1069.8	54	Wilderness area Recreation Preservation Natural environment	Campgrounds Playgrounds Picnic areas Beaches Trails	Camping Picnicking Fishing Hiking Swimming Boating Interpretative programs Skiing – cross-country	5.0 million	$1.50 per night for camping plus $0.25 each additional service.
British Columbia	45 315.6	..	343	Wilderness area Recreation Natural Marine Historic sites Restored gold town	Campgrounds Picnic areas Mooring facilities Hiking trails Nature trails Boat ramps Recreation vehicle sani-stations	Boating Camping Picnicking Nature houses[1] Interpretative programs with naturalists Winter sports Skiing Canoeing Mountain climbing Swimming Hiking	9.5 million	$2.00 to $4.00 daily camping fee. No charge for day use. Some free camping.

[1]A building in the park for discussions, lectures and the showing of films on the natural history of the park.

Sources

1.1 - 1.2, 1.6 Topographical Survey, Surveys and Mapping Branch, Department of Energy, Mines and Resources.
1.3 - 1.4 *Inventory of Canadian freshwater lakes*, Inland Waters Directorate, Water Resources Branch, Department of Fisheries and the Environment.
1.5, 1.7 Information Services Directorate, Department of Fisheries and the Environment.
1.8 - 1.9 Respective provincial government departments.

The constitution and the legal system

Chapter 2

Tables

The constitution and the legal system

The constitution   2.1

The Canadian federal state of 10 provinces and two territories had its foundation in an act of the British Parliament, the British North America Act, 1867, fashioned for the most part from Seventy-two Resolutions drafted by the Fathers of Confederation at Quebec in 1864. The BNA Act provided for the federal union of three British North American provinces — Canada (Ontario and Quebec), Nova Scotia and New Brunswick — into one dominion under the name Canada. The act made provision for possible future entry into Confederation of the colonies or provinces of Newfoundland, Prince Edward Island and British Columbia, and of Rupert's Land and the North-Western Territory, a vast expanse then held by the Hudson's Bay Company. In 1870, the company surrendered its territories to the British Crown which transferred them to Canada. In exchange it received a cash payment from the Canadian government of £300,000, one-twentieth of the lands in the southern part, "the fertile belt", of the territory, and designated blocks of land around its trading posts. From this new territory was carved Manitoba in 1870, much smaller at its inception than now, and later, in 1905, Saskatchewan and Alberta. British Columbia entered Confederation in 1871 on condition that a railway linking it with Eastern Canada be commenced within two years. It was not until 1873 that Prince Edward Island entered the union, and much later, 1949, that Newfoundland joined (see Table 2.1).

Although the BNA Act of 1867 and its amendments contain a substantial portion of Canada's constitution, it is not a comprehensive constitutional document. There are unwritten and equally important parts such as common law, convention and usage transplanted from Britain over 200 years ago and basic to the Canadian style of democratic government. Among these are the principles of the Cabinet system of responsible government with close relationship between executive and legislative branches.

The constitution, in its broadest sense, also includes other Imperial statutes (Statute of Westminster, 1931) and Imperial orders-in-council admitting various provinces and territories to the federation; statutes of the Parliament of Canada pertaining to such matters as succession to the throne, the royal style and title, the Governor General, the Senate, the House of Commons, the creation of courts, the franchise and elections, as well as judicial decisions that interpret the BNA Act and other statutes of a constitutional nature. The constitutions of the provinces of Canada form part of the overall Canadian constitution, and provincial acts which are of a fundamental constitutional nature similar to those listed above are also part of the constitution. The same can be said of both federal and provincial orders-in-council that are of a similar fundamental nature.

Although the essential principles of Cabinet government are based on custom or usage, the federal structure of Canadian government rests on written provisions of the BNA Act. A dominant feature of the act was the distribution of powers between the central or federal government and the component provincial governments, granting to the Parliament of Canada legislative jurisdiction over all subjects of general or common interest while giving provincial legislatures jurisdiction over all matters of local or particular interest.

Unlike the written constitutions of many nations, the BNA Act lacks comprehensive "bill of rights" clauses, although it does accord specific although limited constitutional protection to the use of the English and French languages and special safeguards for sectarian or denominational schools. Freedom of speech, freedom of assembly, freedom of religion, freedom of the press, trial by jury and similar liberties enjoyed by the individual citizen are not recorded in the BNA Act but rather depended on the statute law and the common law inheritance until these rights were confirmed, as far as federal law is concerned, by the passage of a Canadian bill of rights — An Act for the Recognition and Protection of Human Rights and Fundamental Freedoms (SC 1960, c.44) assented to August 10, 1960.

The right to use either the English or the French language in the House of Commons, the Senate, the legislature of Quebec and the federal and Quebec courts is constitutionally guaranteed by Section 133 of the BNA Act. The use of English and French in the administration of the federal government and its Crown corporations is dealt with in the Official Languages Act (RSC 1970, c.O-2). That act provides that government notices to the public, certain orders and regulations, and final decisions of federal courts are to be made or issued in both languages and that, in the national capital region and in federal bilingual districts, government services are to be available in both languages. The commissioner of official languages is responsible for ensuring compliance with this act.

2.1.1 Amendment of the constitution

No provision was made in the BNA Act for its amendment by any legislative body in Canada but both the Parliament of Canada and the provincial legislatures were given legislative jurisdiction with respect to certain matters relating to government. Thus, for example, the Parliament of Canada was given jurisdiction with respect to the establishment of electoral districts and election laws and the privileges and immunities of members of the Senate and House of Commons. Each provincial legislature was empowered to amend the constitution of its province except as regards the office of Lieutenant-Governor. Amendments to the BNA Act have been made by the British Parliament on 14 occasions since 1867. By an amendment to the BNA Act in 1949, the authority of the Parliament of Canada to legislate with respect to constitutional matters was considerably enlarged and it may now amend the constitution of Canada except as regards the legislative authority of the provinces, the rights and privileges of provincial legislatures or governments, schools, the use of English or French and the provision that no House of Commons shall continue for more than five years other than in time of real or apprehended war, invasion or insurrection.

The search for a satisfactory procedure for amending the constitution in Canada which satisfies the need to safeguard basic provincial and minority rights and yet possesses sufficient flexibility to ensure that the constitution can be altered to meet changing circumstances has been the subject of repeated consideration in Parliament as well as in a series of federal-provincial conferences and meetings held in 1927, 1935-36, 1950 and 1960-61. In October 1964 the text of a draft bill "to provide for the amendment in Canada of the constitution of Canada," which embodied an amending procedure or formula and was recommended by a conference of attorneys general, was unanimously accepted by a conference of the prime minister and the premiers. However, Quebec subsequently withdrew its approval of the formula and it was never adopted.

Between February 1968 and June 1971, eight federal-provincial conferences were held to study the drafting of a new constitution. A committee was established to help study constitutional questions. The provincial governments, with one exception, and the federal government submitted proposals for a new constitution. The discussions culminated in the drafting of a constitutional charter which set out specific constitutional reforms, including a revised amendment procedure. The charter was considered at a constitutional conference in Victoria in June 1971 but was not accepted.

2.1.2 Treaty-making powers

The federal government has primary responsibility for the conduct of external affairs. The policy in discharging this responsibility is to promote the interest of the entire country and of all Canadians.

In matters of specific concern to the provinces, it is Canadian government policy to assist them in achieving their particular aspirations and goals, as illustrated by the "entente" signed by Quebec and France in the field of education in February 1965. Provincial and federal authorities co-operated in a procedure that enabled Quebec, within the framework of the constitution and national policy, to participate in international arrangements. Once it is determined that what a province wishes to achieve in the field of provincial jurisdiction falls within the framework of Canadian foreign policy, the provinces may discuss arrangements with the authorities of the country concerned. For a formal international agreement the federal signature of treaties and conduct of overall foreign policy must come into operation.

Distribution of federal and provincial powers   2.2

Since the purpose of the BNA Act was to create a federal system of government, important provisions of that document deal with the division of powers between the federal and provincial governments. Each level of government is virtually sovereign with respect to the powers it exercises. While the federal government under the BNA Act has the power to disallow provincial legislation, this power has not been exercised in recent years.

Section 91 of the BNA Act gives the Parliament of Canada a general power to "make laws for the peace, order and good government of Canada" and gives a list of classes of subjects over which Parliament has exclusive authority which illustrate but do not restrict the general power. The list contains 31 classes of federal powers such as regulation of trade and commerce, defence, currency, raising money by any mode or system of taxation, postal services, navigation and shipping, weights and measures and criminal law. Section 92 assigns to the provinces the power to legislate regarding direct taxation within the province, the management and sale of public lands and timber belonging to the province, municipal institutions, laws relating to property and civil rights and all matters of a merely local or private nature. (For details see *Canada Year Book 1973* pp 71-73.) Section 95 of the BNA Act gives the federal government and the provinces concurrent powers over agriculture and immigration but federal law prevails in cases where the laws of both levels of government are in conflict. Similar concurrent powers exist in respect of old age pensions and supplementary benefits, including survivors and disability benefits, but no federal legislation affects the operation of provincial laws in this field if a conflict occurs with provincial legislation.

The drafters of the BNA Act in 1867 probably thought that such a division of powers was so definite and precise that no future difficulties would arise in deciding what subjects were under federal or provincial legislative control. However, the powers enumerated in Sections 91 and 92 are not mutually exclusive and sometimes overlap. Interpretation on the division of powers has given rise to many legal disputes, parliamentary discussions, royal commission inquiries and federal-provincial conferences.

Difficulty in interpreting the division of powers has also resulted from new social, technological and political conditions, unforeseen at the time of Confederation. Social welfare legislation, such as unemployment insurance, and legislation concerning modern communication facilities were not contemplated by the drafters of the BNA Act. But power to legislate on these subjects had to be assigned either to the federal or provincial governments by reference to the BNA Act. Canada's emergence into the international community as an independent nation, also not foreseen in 1867, required an allocation of responsibility for aviation, broadcasting and citizenship between the two levels of government or in some cases to one or the other government.

One significant outcome of the allocation of powers under the BNA Act has been that expenditures of the provincial governments have often outstripped their tax resources. In 1867, the provinces were assigned responsibility for social services such as hospitals and schools as well as for municipal institutions. At that time this did not involve major expenditure of public funds. However, changing demands of society and entry of government into the field of social welfare led to expenditure of large sums. The provinces have the power to levy direct taxation within the province for provincial purposes while the federal government has a broader authority to levy taxes by "any means of taxation." The federal government therefore has substantial tax resources. While the provinces have responsibility for many costly public institutions they often lack the necessary financial resources. In order to redress this, numerous federal-provincial tax-sharing agreements and shared-cost programs have been reached by the federal and provincial governments. Such agreements were not, of course, anticipated by the original drafters of the BNA Act. Nevertheless these agreements have resulted in new constitutional arrangements and techniques for dealing with federal-provincial economic relations and have come to be known collectively as co-operative federalism.

2.3   The legal system

2.3.1   Common law and Quebec civil law

The legal system in the provinces and territories is derived from the common law system of England with the exception of Quebec, where the system has been influenced by the legal developments of France. Quebec has its own civil code and code of civil procedure. However, in the field of public law the principles of common law apply. Over the years, both Canadian common law and Quebec civil law have developed unique characteristics. The body of law changes as society changes. In many provinces there are now law reform commissions which have been charged with the function of inquiring into matters relating to the reform of the law having regard to both the statute law and the common law. A general revision of the civil code is taking place in Quebec under the auspices of the Civil Code Revision Office. At the federal level there is the Law Reform Commission of Canada whose purpose is "to study and keep under review on a continuing basis the statutes and other laws comprising the law of Canada with a view to making recommendations for their improvement, modernization and reform."

2.3.2   Criminal law

Criminal law is that branch or division of law which treats crimes and their punishment. A crime may be described as an act against society, as distinct from a dispute between individuals. It has been defined as any act done in violation of those duties which an individual owes to the community and for the breach of which the law has provided that the offender shall make restitution to the public.

Canada's criminal law has as its foundation the criminal law of England built up through the ages and consisting first of customs and usages and later expanded by principles enunciated by generations of judges. There is no statutory declaration of the introduction of English criminal law into those parts of Canada that are now New Brunswick, Nova Scotia and Prince Edward Island. Its introduction there depends upon a principle of the common law itself by which English law was declared to be in force in uninhabited territory discovered and planted by British subjects, except insofar as local conditions made it inapplicable. The same may be said of Newfoundland although the colony dealt with the subject in a statute of 1837. In Quebec, its reception depends upon the Royal Proclamation of 1763 and the Quebec Act of 1774. In each of the other provinces and in the Yukon Territory and Northwest Territories, the matter has been dealt with by statute.

The criminal law systems of the provinces as they exist today are based on the British North America Act of 1867. Section 91 of the act provides that "exclusive legislative authority of the Parliament of Canada extends to . . . the criminal law, except the constitution of courts of criminal jurisdiction but including the procedure in criminal matters." By Section 92, the legislature of the province exclusively may make laws in relation to "the administration of justice in the province, including the constitution, maintenance, and organization of provincial courts, both of civil and criminal jurisdiction and including procedure in civil matters in those courts." The Parliament of Canada may, however (Section 101), establish any additional courts for the better administration of the laws of Canada. It should be noted that the Statute of Westminster, 1931 effected important changes, particularly by abrogating in part the Colonial Laws Validity Act, 1865 (British) and confirming the right of a dominion to make laws having extraterritorial operation.

At the time of Confederation each of the colonies affected had its own body of statutes relating to criminal law. In 1869, in an endeavour to assimilate them into a uniform system applicable throughout Canada, Parliament passed a series of acts, some of which dealt with specific offences and others with procedure. Most notable of the latter was the Criminal Procedure Act, but other acts provided for the speedy trial or summary trial of indictable offences, the powers and jurisdiction of justices of the peace in summary conviction matters and otherwise, and the procedure in respect of juvenile offenders.

Codification of the criminal law through a criminal code bill founded on the English draft code of 1878, Stephen's *Digest of criminal law,* Burbidge's *Digest of the Canadian criminal law,* and the relevant Canadian statutes was brought about by the justice minister, Sir John Thompson, in 1892. This bill became the Criminal Code of Canada and came into force on July 1, 1893. It must be remembered, however, that the criminal code was not exhaustive of the criminal law. It was still necessary to refer to English law in certain matters of procedure and it was still possible to prosecute for offences at common law. Moreover, Parliament has declared offences under certain other acts such as the Narcotic Control Act, to be criminal offences.

An examination and study of the criminal code was authorized by order-in-council dated February 3, 1949, and the commission which had been assigned the task of revising the code presented its report with a draft bill in February 1952. After coming before successive sessions of Parliament it was finally enacted on June 15, 1954 and the new criminal code (RSC 1970, c.C-34) came into effect on April 1, 1955. Since then a number of important amendments have been made. These include inter alia, provision for motions for leave to appeal to the Supreme Court of Canada in criminal cases to be heard by a quorum of at least five judges of that court instead of by a single judge; a statutory extension of the definition of obscenity and authorization of the seizure and condemnation of offending material without a charge necessarily being laid against any person; crimes of genocide and public incitement of hatred; offences committed in aircraft in flight over the high seas; procedures relating to the invasion of privacy and interception of communications; the forbidding of publication in a newspaper or broadcast of any evidence tendered at a preliminary inquiry unless and until the accused has been discharged or, if the accused has been committed for trial, the trial has ended; the elimination of the death penalty for all offences except certain ones under the National Defence Act; the modifying of offences relating to gaming and lotteries, drinking and driving, homosexual acts and therapeutic abortion; the reforming of the jail system; offences relating to hijacking and endangering the safety of aircraft; the abolishing of offences of vagrancy and attempted suicide; and conditional discharges for convicted persons.

Human rights   2.3.3

In 1960 (SC 1960, c.44) Parliament enacted what is known as the Canadian Bill of Rights. Although the act sets out further details, its general scope appears in Section 1, as follows: "It is hereby recognized and declared that in Canada there have existed and shall continue to exist without discrimination by reason of race, national origin, colour, religion or sex, the following human rights and fundamental freedoms, namely, (a) the right of the individual to life, liberty, security of the person and enjoyment of property, and the right not to be deprived thereof except by due process of law; (b) the right of the individual to equality before the law and the protection of the law; (c) freedom of religion; (d) freedom of speech; (e) freedom of assembly and association; and (f) freedom of the press."

In 1977, the Canadian Human Rights Act was passed which, within the federal area of legislative competence, outlawed discrimination on grounds of race, national or ethnic origin, colour, religion, age, sex, marital status, conviction for which a pardon has been granted and, with respect to employment, physical handicap in such areas as provision of goods, services, facilities or accommodation, employment, trade union membership, wages, publication of notices and hate messages. Privacy provisions in the act give an individual a right of access to personal information held by government on that individual. The act also established the Canadian Human Rights Commission and a privacy commissioner to administer the rights and obligations in this legislation.

2.4 Courts and the judiciary

2.4.1 The federal judiciary

The Parliament of Canada is empowered by Section 101 of the British North America Act from time to time to provide for the constitution, maintenance and organization of a general court of appeal for Canada and for the establishment of any additional courts for the better administration of the laws of Canada. Under this provision, Parliament has established the Supreme Court of Canada, the Federal Court of Canada and certain specialized courts.

Supreme Court of Canada. This court, first established in 1875 and now governed by the Supreme Court Act (RSC 1970, c.S-19), consists of a chief justice, who is called the chief justice of Canada, and eight puisne judges. The chief justice and the puisne judges are appointed by the Governor-in-Council and hold office during good behaviour but are removable by the Governor General on address of the Senate and the House of Commons. They cease to hold office on attaining the age of 75 years. The court sits at Ottawa and exercises general appellate jurisdiction throughout Canada in civil and criminal cases. The court is also required to consider and advise on questions referred to it by the Governor-in-Council and it may also advise the Senate or the House of Commons on private bills referred to the court under any rules or orders of the Senate or of the House of Commons.

Appeals may be brought from any final judgment of the highest court of final resort in a province by obtaining leave to do so from that court or from the Supreme Court itself. The Supreme Court may grant leave to appeal from any judgment whether final or not, and as well there is provision for appeals whereby the highest court of final resort in a province may grant leave on a question of law alone from a final judgment of some other court in that province. Appeals in respect of indictable offences are regulated by the criminal code. Appeals from federal courts are regulated by the statute establishing such courts. The judgment of the Supreme Court of Canada in all cases is final and conclusive.

Chief Justice and Judges of the Supreme Court of Canada as at January 1, 1978

Chief Justice of Canada, Rt. Hon. Bora Laskin *(appointed December 27, 1973, first appointed a Judge of the Supreme Court, March 23, 1970)*
Hon. Mr. Justice Ronald Martland *(appointed January 15, 1958)*
Hon. Mr. Justice Roland Almon Ritchie *(appointed May 5, 1959)*
Hon. Mr. Justice Wishart Flett Spence *(appointed May 30, 1963)*
Hon. Mr. Justice Louis-Philippe Pigeon *(appointed September 21, 1967)*
Hon. Mr. Justice Robert George Brian Dickson *(appointed March 26, 1973)*
Hon. Mr. Justice Joseph Philemon Jean Marie Beetz *(appointed January 1, 1974)*
Hon. Mr. Justice Willard Zebedee Estey *(appointed September 29, 1977)*
Hon. Mr. Justice Yves Pratte *(appointed October 1, 1977).*

Federal Court of Canada. The Federal Court of Canada was constituted by an act of the Parliament of Canada under Section 101 of the British North America Act, 1867, which, after authorizing the creation of the Supreme Court of Canada, confers on Parliament authority to constitute other courts for the better administration of the laws of Canada. The Federal Court of Canada is a court of law, equity and admiralty and it is a superior court of record having civil and criminal jurisdiction (Section 3 of the act). The Exchequer Court of Canada, (established in 1875), was replaced in December 1970 by the Federal Court of Canada (SC 1970-71, c.1).

The court has two divisions called the Federal Court — Appeal Division, and the Federal Court — Trial Division. The appeal division may be called the Court of Appeal or Federal Court of Appeal (Section 4 of the act). The Court of Appeal consists of the chief justice of the Federal Court of Canada and five other judges. The trial division consists of the associate chief justice of the Federal Court of Canada and nine other judges. Every judge is an ex officio member of the division of which he is not a regular member (Section 5). In addition to the establishment of full-time judges, an added capacity to cope with the purely judicial work of the court is provided by the authority to invite retired federally appointed judges to act as deputy judges of the court (Section 10). This authority extends also to federally appointed judges who are still in office, but only with the consent of the appropriate chief justice or attorney general. Former district judges in admiralty are also deputy judges of the court and their services can be used on a limited basis (Section 60).

Provision is also made in the act for quasi-judicial officers called prothonotaries (Section 12). Their duties are defined by the rules and may be of a judicial nature (Section 46). In addition to being taxing-masters, they can, subject to supervision by the court, deal with interlocutory work, and even take trials in minor matters as the associate chief justice may find expedient in order to ensure the expeditious dispatch of the court's business.

While all the full-time judges must live in or near the national capital region (Section 7), each division of the court can sit any place in Canada and the place and time of the sittings must be arranged to suit the convenience of the litigants (Sections 15 and 16). In addition, there is authority in the statute (Section 7) for a rotation of judges to provide for a continuity of judicial availability in any place where the volume of work, or other circumstances, makes such an arrangement expedient.

Judges of the Federal Court of Canada as at January 1, 1978

Chief Justice, Hon. Wilbur Roy Jackett *(appointed June 1, 1971)*
Associate Chief Justice, Hon. Arthur Louis Thurlow *(appointed to Court of Appeal, June 1, 1971; appointed Associate Chief Justice, December 4, 1975)*
Court of Appeal Judges: Hon. Louis Pratte *(appointed to Trial Division, June 10, 1971; appointed to Court of Appeal, March 5, 1973),* Hon. Darrel Verner Heald *(appointed to Trial Division, July 9, 1971; appointed to Court of Appeal, December 4, 1975),* Hon. John J. Urie *(appointed June 8, 1973),* Hon. William F. Ryan *(appointed April 11, 1974),* Hon. Gerald Eric Le Dain *(appointed September 1, 1975)*
Trial Division Judges: Hon. Angus Alexander Cattanach *(appointed June 1, 1971),* Hon. Hugh Francis Gibson *(appointed June 1, 1971),* Hon. Allison Arthur Mariotti Walsh *(appointed June 1, 1971),* Hon. Frank U. Collier *(appointed September 16, 1971),* Hon. George A. Addy *(appointed September 17, 1973),* Hon. Patrick M. Mahoney PC *(appointed September 17, 1973),* Hon. Raymond G. Decary *(appointed September 17, 1973),* Hon. Jean-Eudes Dubé PC *(appointed April 24, 1975),* Hon. Louis Marceau *(appointed December 23, 1975)*
Deputy Judges of the Federal Court (Section 60(3), Federal Court Act): Hon. Robert S. Furlong, Hon. Dalton C. Wells.

The provincial judiciary   2.4.2

Certain provisions of the British North America Act govern to some extent the provincial judiciary. Under Section 92(14) the legislature of each province exclusively may make laws in relation to the administration of justice in the province including the constitution, maintenance and organization of provincial courts of both civil and criminal jurisdiction. Section 96 provides that the Governor General shall appoint the judges of the superior, district and county courts in each province, except those of the courts of probate in Nova Scotia and New Brunswick.

The territorial judiciary   2.4.3

In 1971 amendments [now cited as RSC 1970, c.48 (1st Supplement)] to the Yukon Act and the Northwest Territories Act were proclaimed in force, simultaneously with certain ordinances of the Yukon Territory and the Northwest Territories, allowing the territorial governments to assume responsibility for the administration of justice other than the conduct of criminal prosecutions.

In the Yukon Territory, provision was made for a territorial (now supreme) court, a magistrate's court, justices of the peace and a court of appeal. The supreme court consists of a single judge of superior court rank and the magistrate's court. Both are located in Whitehorse, although from time to time magistrate's court sittings are held in other communities. There are 32 justices of the peace, appointed by the commissioner, located at 15 points in the Yukon Territory. The judge of the Supreme Court of the

Northwest Territories is ex officio judge in the Yukon Territory and vice versa. The court of appeal consists of the chief justices of British Columbia, the justices of appeal of British Columbia and the judge of the Supreme Court of the Northwest Territories.

The court system in the Northwest Territories consists of a superior court called the Supreme Court of the Northwest Territories, presided over by one judge located in Yellowknife. The Court of Appeal of the Territories consists of the justices of appeal of Alberta and the judges of the Yukon Territory and Northwest Territories supreme courts. There are also two full-time magistrates appointed by the commissioner who have jurisdiction similar to provincial judges; a number of justices of the peace, also appointed by the commissioner, serve in widely scattered settlements.

2.4.4   Salaries, allowances and pensions of judges

Section 100 of the British North America Act provides that the salaries, allowances, and pensions of the judges of the superior, district, and county courts (except the courts of probate in Nova Scotia and New Brunswick) and of the admiralty courts in cases where the judges thereof are for the time being paid by salary, shall be fixed and provided by the Parliament of Canada. These are provided under the Judges Act (RSC 1970, c.J-1 as amended by SC 1970-71, c.55, SC 1973-74, c.17, SC 1974-75, c.48).

The salary of the chief justice of Canada is $65,000 a year and those of the puisne judges of the Supreme Court of Canada $60,000. The salaries of the chief justice and the associate chief justice of the Federal Court of Canada are $55,000 a year and of the other judges of the court $50,000.

All chief justices of provincial superior courts, the senior associate chief justice and the associate chief justice of the Superior Court of Quebec receive annual salaries of $55,000; the puisne judges of these courts and the judges of the two territorial courts receive $50,000. Where judicial offices are created for supernumerary judges, the incumbents receive the salary of a puisne judge. Supernumerary judges are those judges of a superior court of a province who have given up their regular judicial duties to make themselves available to perform such special judicial duties as may be assigned to them from time to time by the chief justice or associate chief justice of the court of which they are a member. The chief judges of county and district courts receive salaries of $48,000 a year and the remaining judges and junior judges of all county and district courts $43,000.

Every judge who receives a salary under the Judges Act is paid an additional salary of $3,000 a year as compensation for any extra-judicial services that he may be called upon to perform by the federal government or the government of a province, and for incidental expenditures that proper execution of his office may require. In the case of each judge of the Federal Court of Canada and of the territorial courts of the Yukon Territory and the Northwest Territories an additional allowance of $3,000 a year is paid as compensation for special incidental expenditures.

The Judges Act provides that a judge of a superior or county court, required to perform duties outside the immediate vicinity where he is by law obliged to reside, is entitled to be paid moving or transportation expenses and reasonable travelling and other expenses. There is also provision for the payment of reasonable expenses incurred in the discharge of special extra-judicial obligations and responsibilities that devolve on a chief justice, puisne judge of the Supreme Court of Canada or chief judge.

One of the cornerstones of Canadian parliamentary democracy lies in the independence of the judiciary. Because the person responsible for litigating matters on behalf of the Canadian government (the attorney general of Canada) is the same as the one responsible for administering the provisions of the Judges Act (the minister of justice), there has been some concern expressed that the judges before whom the attorney general appears may not seem to be as independent as they ought to be. Therefore, in 1977 the Judges Act was amended to provide for an independent commissioner for federal judicial affairs who is to act independently of the justice department in carrying out ministerial responsibilities with respect to matters in the Judges Act, and personnel, financial and accommodation arrangements on behalf of the Federal Court and the Canadian Judicial Council. The registrar of the Supreme Court of Canada is entrusted with the same responsibilities on behalf of that court.

Legal services   2.5

The legal profession   2.5.1
The adjective "fused" is sometimes used to describe the legal profession in common law Canada since practising lawyers are both called as barristers and admitted as solicitors. Admission to practise is a provincial matter. Statutes setting out the powers and responsibilities of the provincial organizations are: (Alberta) The Legal Profession Act RSA 1970, c.203; (British Columbia) The Legal Professions Act RSBC 1960, c.214; (Manitoba) The Law Society Act RSM 1970, c.L-100; (New Brunswick) The Barristers' Society Act, 1973, SNB 1973, c.80; (Newfoundland) The Law Society Act RSN 1970, c.201; (Nova Scotia) Barristers and Solicitors Act RSNS 1967, c.18; (Ontario) The Law Society Act RSO 1970, c.238; (Prince Edward Island) The Law Society and Legal Profession Act RSPEI 1974, c.L-9; (Saskatchewan) The Legal Profession Act RSS 1965, c.301; (Northwest Territories) The Legal Profession Ordinance RONWT 1956, c.57; (Yukon) The Legal Profession Ordinance ROY 1971, c.L-4. In Quebec the legal profession is divided into the separate branches of advocate and notary and their statutes are the Bar Act, SQ 1966/67, c.77 and the Notarial Act, SQ 1968, c.70.

Legal aid   2.5.2
For many years the provision of legal services to persons unable to afford the fees normally charged by a lawyer was viewed as a responsibility to be assumed by individual lawyers on a voluntary basis as a form of charity. In more recent times all provincial governments have moved to establish publicly funded legal aid programs under which persons of limited means may obtain the services of a lawyer in a number of criminal and civil matters at either no cost or modest cost to themselves depending upon the client's financial circumstances. The lawyers who act for clients in matters covered by a provincial legal aid program are then paid by the government, usually at a reduced rate, on a fee-for-services basis or by salary depending upon the type of legal aid program operated in the province. The provincial legal aid programs vary considerably in terms of formalities, scope of coverage and methods of providing services. Some are established by legislative enactment while others exist and operate by way of informal agreements between the provincial government and the law society. Some programs provide for fairly comprehensive coverage in both criminal and civil matters while others at present encompass only criminal offences. In some provinces a mixed system is in operation.

In 1971 the federal government entered the field and concluded an agreement with the government of the Northwest Territories for sharing the costs of providing legal aid in both criminal and civil matters for persons in the territories financially unable to retain the services of a lawyer. This program was implemented on August 17, 1971. In the Yukon Territory the legal aid program is a service operated by the territorial bar with the government paying fees to lawyers who act for legal aid clients charged with criminal offences.

In August 1972, the federal government announced that it was prepared to enter into agreements with the provincial governments under which federal funds would be paid to the provinces to assist them in developing or expanding their legal aid programs in matters related to criminal law. Agreements have since been concluded with all provincial governments. Amendments to these agreements provided that the federal government would contribute the lesser of 75 cents per capita of the provincial population or 90% of the program expenditures toward the costs of providing lawyers' services to eligible persons subject to criminal charges or proceedings under federal laws. These federal-provincial agreements enable the provincial governments to determine the method or methods by which legal services will be provided to persons who qualify for assistance, but in cases where an individual is charged with a criminal offence carrying a penalty of mandatory life imprisonment that person is entitled to retain a lawyer of his or her own choice. The agreements also ensure that a person otherwise eligible to receive legal aid will not be disqualified as a recipient only because he or she is not a resident of the province in which the criminal proceedings take place.

2.6 The federal Department of Justice

The department is divided, for administrative and functional purposes, into a number of service areas. Lawyers in the department may be assigned as legal advisers to other government departments or agencies as part of departmental legal services, or to offices in Vancouver, Edmonton, Saskatoon, Winnipeg, Toronto, Montreal and Halifax as part of regional legal services. The sections in headquarters legal services are described below.

Advisory and research services. This section prepares legal opinions requested by the federal government and its departments and agencies.

Civil law. This section conducts litigation and gives legal advice to the government on all matters of a non-criminal nature arising in Quebec.

Civil litigation. The lawyers in this section are responsible for the conduct of the non-criminal litigation involving the federal government originating in those provinces where the common law prevails. This litigation includes customs and excise tax matters, expropriation cases, disputes over contracts, accident claims, suits for defamation and claims for breach of copyright.

Constitutional, administrative and international law. This section co-ordinates and provides legal advice in the general fields of constitutional and administrative law within the federal government and its various departments and agencies. It is concerned with long-term policy in constitutional affairs and problems of federal-provincial relations. The section also deals with the areas of public and private international law. Canada became a member of The Hague Conference on Private International Law in 1968 and the department is responsible for Canadian participation. This section co-ordinates Canadian activities in the conference, which meets every other year, and has a similar role with regard to Unidroit, the International Institute for the Unification of Private Law. In both public and private international law this section has a particular interest in matters concerning countries of the British Commonwealth.

Criminal law. Lawyers in this section participate in criminal litigation in every jurisdiction. They co-operate with members of the department's six regional offices in prosecution of violations of federal statutes and regulations and are involved in extradition of persons to and from Canada. An additional and important function is work on criminal law amendment which involves considering and assessing suggestions for amendment of the criminal code and certain other statutes received from many sources. The section advises the justice minister on these recommendations.

Legislation. This section is concerned with preparation of legislation from the time a topic is given approval in principle by Cabinet until the resulting enactment receives royal assent. Periodic revisions of the *Statutes of Canada* are also compiled here.

Policy planning. This section develops legal initiatives and responses to emerging social problems. In co-operation with other departments and levels of government, it assesses recommendations for changes in the law proposed by the Law Reform Commission of Canada and other groups.

Privy council. This section examines what is sometimes called subordinate legislation. Parliament often delegates certain legislative functions to other bodies and officials and it is the responsibility of this section to consult with the clerk of the privy council in order to maintain general supervision over the legislative product resulting from this delegation and to consider whether it is within the authority conferred by Parliament (see Statutory Instruments Act, 1970-71, c.38). The section is asked to assume responsibility for the actual drafting of certain subordinate legislation. Lawyers in this section also act as legal advisers to the clerk of the privy council and his staff.

Programs and law information development. This section develops and administers service, research or information programs with respect to such matters as legal aid, compensation for crime victims, native court workers and law for the layman.

Property and commercial law. This section handles all work involved when land is required for public purposes and deals with contracts and commercial agreements and relations to which the federal government or its departments or agencies are party.

Tax litigation. The lawyers in this section represent the Crown in all aspects of most federal tax litigation. The section has an advisory function on tax matters with the national revenue department.

Police forces   2.7

Organization of police forces   2.7.1

The police forces of Canada are organized in three groups: (1) the federal force, the Royal Canadian Mounted Police (RCMP); (2) provincial police forces — Ontario and Quebec have their own police forces; the RCMP performs parallel functions in all other provinces; and (3) municipal police forces — most urban centres have their own police forces or provincial police, under contract, to attend to police matters. In addition, the Canadian National Railways, the Canadian Pacific Railway Company and the National Harbours Board have their own police forces.

The Royal Canadian Mounted Police. This is a civil force maintained by the federal government. It was established in 1873 as the North-West Mounted Police for service in what was then the North-West Territories and, in recognition of its services, was granted the prefix "Royal" by King Edward VII in 1904. Its sphere of operations was expanded in 1918 to include all of Canada west of Port Arthur and Fort William (now Thunder Bay). In 1920 it absorbed the Dominion Police, its headquarters was transferred from Regina to Ottawa and its title changed to Royal Canadian Mounted Police.

The force operates under authority of the Royal Canadian Mounted Police Act (RSC 1970, c.R-9). It is responsible to the solicitor general and is controlled and managed by a commissioner who holds the rank and status of a deputy minister and is empowered to appoint members to be peace officers in all provinces and territories.

Administration of justice within the provinces, including enforcement of the Criminal Code of Canada, is part of the power and duty delegated to the provincial governments. All provinces except Ontario and Quebec have entered into contracts with the RCMP to enforce criminal and provincial laws, under direction of the respective attorneys general. In these eight provinces, the force is under agreement to provide police services to 192 municipalities, assuming enforcement responsibility of municipal as well as criminal and provincial laws within these communities. The Yukon Territory and Northwest Territories are policed exclusively by the RCMP and therefore criminal offences, federal statutes and all ordinances of the territories fall within their responsibility. The force maintains liaison officers in London, Paris, Bonn, Rome, Hong Kong and Washington, and represents Canada in the International Criminal Police Organization, which has headquarters in Paris.

Thirteen operational divisions make up the strength of the force across Canada; they comprise two districts and 41 subdivisions which include 702 detachments. Headquarters division, as well as the office of the commissioner, is in Ottawa. Divisional headquarters, for the most part, are located in the provincial or territorial capitals.

A police information centre at RCMP headquarters is staffed and operated by the force. Law enforcement agencies throughout Canada have access via remote terminals to information on stolen vehicles, licences, wanted persons and stolen property.

The RCMP operates the Canadian Police College at which force members and selected representatives of other Canadian and foreign forces may study crime prevention and detection.

As of October 30, 1977 the force had a total authorized strength of 19,004 including regular members, special constables, civilian members and public service employees.

Ontario Provincial Police. The Ontario Provincial Police, a Crown force, is the third largest deployed force in North America with an authorized strength of more than 5,000 (1977) uniformed and civilian personnel.

The OPP is administered from general headquarters at Toronto by the commissioner, under the solicitor general's ministry. Other senior executive officers include two deputy commissioners and six assistant commissioners. The force has two principal sides — operations and services — each administered by a deputy commissioner. In turn, six divisions at the next level — field, traffic, management, staff services, special services, and staff development — are administered by assistant commissioners.

Under provisions of the Ontario Police Act, the force is responsible for: enforcing federal and provincial statutes in those areas that are not required to maintain their own police departments; maintaining a traffic patrol on the more than 21 000 kilometres of highways and 104 607 km of secondary county and township roads; enforcing the Liquor Licence Act and the Liquor Control Act for Ontario; and maintaining a criminal investigation branch and other specialized branches to assist all other forces in investigation of major crimes.

A central records and communications branch offers continuous service to all police departments in Ontario on such matters as criminal and fingerprint records.

The OPP operates one of the largest frequency-modulation radio networks in the world, with 107 fixed radio stations and more than 1,532 radio-equipped mobile units including motorcycles, boats and aircraft. It also operates a telecommunications network connecting all 16 districts as well as other police departments on a local, national and international basis.

Quebec Police Force. Under the authority of the attorney general, the Quebec Police Force is responsible for maintaining peace, order and public safety throughout the province, and for prevention and investigation of criminal offences and violations of provincial law. The force is under the command of a director general assisted by five assistant directors general and a director of personnel and communications.

For police purposes, the province is divided into nine districts each under the command of a chief inspector or an inspector and named as follows: Bas St-Laurent, Saguenay–Lac St-Jean, Quebec, Mauricie, Estrie, Montreal, Outaouais, Nord-Ouest and Côte-Nord. Strength of the force at the end of November 1977 was 4,340 officers, non-commissioned officers and constables and 1,004 civilian employees.

Municipal police forces. Provincial legislation makes it mandatory for cities and towns to furnish adequate municipal policing for the maintenance of law and order in their communities. Also, all villages and townships or parts of townships having a population density and a real property assessment sufficient to warrant maintenance of a police force, and having been so designated by order-in-council, are responsible for policing their municipalities.

2.7.2 Uniform crime reporting

The present method of reporting police statistics, known as the Uniform Crime Reporting Program, was started on January 1, 1962.

As shown in Table 2.2, police personnel in Canada numbered 63,675 at the end of 1976, including 51,629 sworn-in police officers, 11,503 other full-time employees serving as clerks, technicians, artisans, commissionaires, guards, special constables and 543 cadets. The ratio of police personnel per 1,000 population was 2.8 and the ratio of police was 2.3. Comparable statistics for 1975 are also given in Table 2.2. In 1975 provincial and territorial ratios for police personnel ranged from 1.5 to 5.9 per 1,000 persons and for police only from 1.4 to 5.0. Total municipal police personnel numbered 34,911 made up of 32,182 members of municipal forces, 2,667 Royal Canadian Mounted Police and 62 Ontario Provincial Police under municipal contracts.

Two policemen were killed by criminal action during 1975. Police facilities at the end of the year included 11,900 automobiles, 845 motorcycles, 1,089 other motor vehicles, 437 boats, 31 aircraft, 237 horses and 145 service dogs.

Table 2.3 shows the number of crimes dealt with by police in 1975 and 1976 including offences under the criminal code, federal statutes, provincial statutes and municipal bylaws other than traffic. In 1975 offences reported or known to police which investigation proved unfounded are not shown but numbered 118,329 including 94,149 under criminal code classifications; 14,786 under federal statutes; 7,642 under provincial statutes; and 1,752 under municipal bylaws.

During 1975, police reported 114,125 offences against persons including 633 murders, 642 attempted murders, 63 manslaughters, one infanticide, 10,900 rape and other sexual offences, and 101,886 offences of wounding and other assaults (not indecent). All offences against the person resulted in the charging of 39,249 persons, 2,277 of them juveniles. During the year there were 1,062,335 cases of robbery, breaking and entering, theft, fraud and other offences against property resulting in 203,447 persons charged, 52,124 of them juvenile males and 6,968 juvenile females. There were 3,409 cases of prostitution, 3,619 gaming and betting, 12,578 offensive weapons and 389,739 other criminal code offences. In addition to 44,972 offences under various federal statutes, there were 50,081 under the Narcotic Control Act and 5,461 under the controlled and restricted drug parts of the Food and Drugs Act. These two classifications resulted in the charging of 46,515 persons including 2,205 juvenile males and 368 juvenile females.

There were 87,193 motor vehicles stolen (an estimated 762.0 per 100,000 registered vehicles); 73,578 or 84.4% of these vehicles were recovered.

During 1975, police departments reported 252,734 (239,737 in 1974) criminal code traffic offences resulting in 182,545 (174,559) persons charged, 6,493 (5,867) of them females. Total traffic charges under other federal statutes numbered 13,280 (10,395); 2,371,492 under provincial statutes (other than the three selected offences almost identical to those under the criminal code that are shown separately in Table 2.4) (2,269,590 in 1974) and 323,404 (318,690) under municipal bylaws excluding parking.

Crime and delinquency   2.8

Adult offenders and convictions   2.8.1

Offences may be classified under two headings, indictable offences and offences punishable on summary conviction. Indictable offences are grouped in two main categories: offences that violate the criminal code and offences against federal statutes. These include the more serious crimes. Offences punishable on summary conviction — those not expressly made indictable — include offences against the criminal code, federal statutes, provincial statutes and municipal bylaws. Increases in the total number of summary conviction offences do not measure adequately the increase in the seriousness of crime. Many summary conviction offences amount to mere disturbances of the peace, minor upsets to public safety, health and comfort such as parking violations, intoxication and practising trades without licence. Nevertheless, summary conviction offences may include more serious charges such as assault and contribution to juvenile delinquency.

Adults convicted of indictable offences. Statistics are available for persons convicted of indictable offences. Although individuals may be charged with more than one offence, only one is tabulated for each person and is selected according to the following criteria: if the person was tried on several charges, the offence is that for which proceedings were carried to the farthest stage — conviction and sentence; if there were several convictions, the offence is that for which the heaviest punishment was awarded; if the final result of proceedings on two or more charges was the same, the offence is the more serious one, as measured by the maximum penalty allowed by the law; and if a person was prosecuted for one offence and convicted of another, such as a person charged with murder and convicted of manslaughter, the offence is the one for which the person was convicted.

In 1973 there were 53,964 adults charged with 95,045 indictable offences and 40,761 of them were found guilty of 72,430 offences (see Table 2.5). All data for 1972 and 1973 exclude returns for Quebec and Alberta. Figures given in Tables 2.5 - 2.8 and 2.11 are based on information received through the provincial judicial systems and consequently cannot be compared with figures reported by police under the Uniform Crime Reporting Program (Tables 2.2 - 2.4) which include these two provinces.

Table 2.6 classifies indictable offences by type of offence for 1972 and 1973. Class I covers offences against the person and in 1973, 3,348 males and 237 females were convicted in this category, mostly for assaults of various kinds. Classes II to IV deal with offences against property. Thefts predominate among the offences in these classes, and breaking and entering, extortion and robbery — serious crimes which involve acts of violence — are the next most numerous. Class V deals with offences relating to currency and Class VI with miscellaneous offences; among the latter, the most numerous convictions are for offences connected with gaming, betting and lotteries. In 1973 there were 2,316 men and 224 women convicted under federal statutes of whom 1,962 men and 195 women were offenders under the Narcotic Control Act.

The number of female offenders convicted of indictable offences decreased from 7,283 in 1972 to 6,706 in 1973. Table 2.7 summarizes the most serious court sentences given for indictable offences in 1972 and 1973, and Table 2.8 shows the method of trial and disposition of cases in 1973.

Two kinds of sentences — probation and commitment to an institution — link the person dealt with by the court and the legal institutions of a community. There are several types of institutions — penitentiaries, reformatories, jails and industrial farms. Theoretically, each has a specific purpose which is supposed to be taken into account when arriving at a legal decision. In practice, however, the availability of an institution in a given community is a factor in determining the court decision.

Convictions for summary conviction offences. Offences punishable on summary conviction under the criminal code or under the provincial summary conviction acts can be tried by magistrates and justices of the peace. Data relating to these offences are based on convictions; no information is available on either the number of persons involved in these offences or the number of charges (see Table 2.9).

Appeals. The conviction or the sentence pronounced by a judge of a first instance court may be appealed on the grounds that the verdict was unreasonable, that there was a wrong decision on some question of law or that there was a miscarriage of justice. In 1973 (excluding Alberta) there were 3,260 appeals in indictable cases disposed of by the courts, of which 325 were Crown appeals and 2,935 appeals of the accused. Of the Crown appeals, 96 were from acquittal and 229 from sentence. Appeals in summary conviction cases disposed of by the courts numbered 2,734 in 1973. Of these, 324 were appeals of the informant and 2,410 appeals of the accused. The informant appeals comprised 263 from acquittal and 61 from sentence, and appeals of the accused comprised 1,757 from conviction and 653 from sentence.

2.8.2   Juvenile delinquents

Juvenile delinquent, as defined in the Juvenile Delinquents Act, means any child who violates any provision of the criminal code, any federal or provincial statute, any bylaw or ordinance of any municipality, who is guilty of sexual immorality or any similar form of vice, or who is liable by reason of any other act to be committed to an industrial school or juvenile reformatory under the provision of any federal or provincial statute. The commission by a child of any of these acts constitutes an offence known as a delinquency. The upper age limit of children brought before the juvenile courts in the provinces varies. The Juvenile Delinquents Act defines a child as meaning any boy or girl apparently or actually under the age of 16 or such other age as may be directed in any province. In Prince Edward Island, Nova Scotia, New Brunswick, Ontario and Saskatchewan under 16 is the official age; in Alberta under 16 for boys and under 18 for girls; in Newfoundland and British Columbia under 17; in Quebec and Manitoba under 18 years. Up to 1967, it was the practice of Statistics Canada to publish information about juvenile delinquents 16 years of age and over separate from that of juveniles under 16 years of age. From 1968 on, the figures include all those considered as juveniles by the respective provinces, regardless of the differing upper age limits.

The figures in Tables 2.10 - 2.12 represent the number of juveniles brought before the courts. If a juvenile was charged with committing more than one offence during the year, only one delinquency — the most serious — was selected for tabulation. With the exception of Manitoba, juveniles involved only in informal hearings are not included in these statistics. Also excluded are children presenting conduct problems which were not brought to court or which were dealt with by the police, social agencies, schools or youth-serving agencies. Thus, community facilities for dealing with children's problems may influence the number of cases referred to court and, therefore, the statistics.

Correctional institutions   2.9

Correctional institutions may be classified under three headings: training schools — operated by the provinces or private organizations under provincial charter for juvenile offenders serving indefinite terms up to the legal age for children in the particular province; provincial adult institutions; and penitentiaries — operated for adult offenders by the federal government in which sentences of over two years are served.

Canadian Penitentiary Service   2.9.1

The Canadian Penitentiary Service operates under the Penitentiary Act (RSC 1970, c.P-6) and is under the jurisdiction of the solicitor general. It is responsible for all federal penitentiaries and for care and training of persons committed to those institutions. The commissioner of penitentiaries, under direction of the solicitor general, is responsible for control and management of the service.

Headquarters of the penitentiary service is in Ottawa. Regional directorates are located in Vancouver, BC; Kingston, Ont.; Ville de Laval, Que.; Saskatoon, Sask.; and Moncton, NB. There are five correctional staff colleges, at Kingston, Ville de Laval, New Westminster, Edmonton and Moncton, where personnel are trained and given refresher courses.

In the year ended December 31, 1977, the penitentiary service controlled 56 institutions at three security levels: 14 maximum, 13 medium, and 29 minimum security. Maximum security institutions include psychiatric centres where specialized medical service is given to inmates. Total inmate population was 9,376, of whom 171 were female offenders; 3,703 males and 170 females were in maximum security; 4,339 males and one female were in medium security, and 1,163 males were in minimum security institutions. New, smaller institutions have been built and others designed to house inmates, providing vocational and academic training and indoor and outdoor recreation. The present construction program will provide nine new institutions. Maximum security penitentiaries will be located in Renous, NB; Mirabel and Ste-Anne-des-Plaines, Que.; and Agassiz, BC. Medium security facilities will be built in Dorchester, NB; Donnacona and Drummondville, Que.; and Kamloops, BC. A new regional psychiatric centre will be constructed at Collins Bay, Ont.

After sentence by the court, prisoners are received at a reception centre, a maximum security institution, where security and training classification is carried out. Based on the results of diagnostic tests at this centre, inmates are placed in an institution which provides the best training program and degree of security required. Minimum stay at the centre is usually six weeks.

Some inmates sentenced to federal penitentiary terms in Newfoundland are held in the provincial centre at St. John's under provisions of the Penitentiary Act. It allows contracts governing the exchange of services between the federal government and some provinces.

Minimum security institutions include community correctional centres, forestry camps and farms. Community correctional centres are located in urban communities across Canada and offer parolees contact with potential employers and access to communities as a rehabilitative measure.

In 1977 close to one-third of the inmate population was enrolled full time in educational programs and technical training. The occupational development program provided academic courses at all levels up to university graduation for 2,100 inmates and technical training for 1,100 inmates. Vocational education offered more than 100 courses in 15 occupations. Almost all academic and technical courses are recognized for accreditation or trade certification by provincial authorities.

In the temporary absence program 48,246 permits were granted in 1977; 48,043 inmates returned, making the success rate better than 99%. Temporary absence is granted for periods up to three days by heads of institutions and up to 15 days by the commissioner of penitentiaries for humanitarian, rehabilitative or medical reasons. Evening and weekend activities involving the outside community were continued. Twenty-one citizen participation committees comprising 210 citizens operated in the institutions. More than 4,000 citizen volunteers were involved in inmate programs in the institutions and outside; these included ex-inmates. Community-based programs, such as Alcoholics Anonymous, drama, music instruction, public speaking, lectures, films, recreation, discussion groups led by private agencies, professionals, citizen volunteers, and community groups all have a part in the inmate's life. Most institutional chapels have multi-purpose programs where religious instruction is provided and other activities are available. Community participation in the programs is encouraged.

2.9.2 The parole system

Significant changes to the National Parole Board's operations were made with the passing of two bills by Parliament in 1976 and 1977. The Criminal Law Amendment Act (No. 2), 1976 (SC 1974-75-76, c.105), proclaimed July 26, 1976, changed the parole eligibility dates for anyone sentenced to life for murder on or after that date and required the board to approve temporary absences for certain inmates. The Criminal Law Amendment Act, 1977 (SC 1977 c.53) was proclaimed, in part, October 15, 1977; the remaining sections would come into effect in 1978. Amendments to the Parole Act expanded the board's membership, brought community participation in the parole review for murderers or those serving indeterminate sentences, and introduced procedural safeguards for inmates being considered for parole. The responsibility for the national parole service was transferred to the commissioner of corrections who is also responsible for the Canadian Penitentiary Service. It was expected that during 1978 the sections allowing the transfer of the responsibility for unescorted temporary absences from the penitentiary service to the board would be proclaimed.

The board's headquarters is in Ottawa with regional offices in Moncton, Montreal, Kingston, Saskatoon and Vancouver. There are 26 full-time board members, including a chairman and vice-chairman, all appointed by the Governor-in-Council for a period of up to 10 years. All may be reappointed. The government may also appoint temporary members for a maximum period of one year and a temporary substitute member for a member who is absent or unable to act. Representatives of police forces, of provincial, municipal, or other local governments, of local professional, trade, or community associations in any region may be designated to act as regular members in the review of cases of inmates serving life sentences for murder or indeterminate sentences as dangerous offenders.

The National Parole Board has exclusive jurisdiction and absolute discretion to grant, refuse, or revoke full parole or day parole for any person serving a sentence of imprisonment imposed under an act of Parliament or for criminal contempt of court. The board has no jurisdiction over a child under the Juvenile Delinquents Act or a person serving an intermittent sentence under Section 663 of the criminal code.

Parole is a conditional release of a prison inmate who has served a specific portion of the sentence as laid down by law, meets certain criteria and, following a review of the case, is considered ready to finish the sentence in the community. The inmate is released with definite conditions and is under supervision.

The board is also involved in granting temporary absences, a short-term release of penitentiary inmates, given before the eligibility date for parole. It must approve any escorted temporary absence for anyone serving a sentence of murder. With the expected proclamation of additional sections of the Criminal Law Amendment Act, 1977 in 1978, it would be given the responsibility to authorize all unescorted temporary absences for medical, humanitarian or rehabilitative reasons.

During the period before the inmate is eligible for full parole consideration, the board may grant longer part-time releases, known as day parole, for education or training not available in the institution or for counselling. Inmates return to the institution or to a special centre at specific times during the period of release, which may last four months. Most inmates may start a day parole program two years before the full parole eligibility date. Inmates serving life sentences for murder become eligible three years before their full parole eligibility date. Day parole normally leads to release on full parole.

Full parole is a full-time release that lasts to the end of the sentence, including remission periods. Inmates who are not serving a life or indeterminate term become eligible for consideration after serving one-third of the sentence or after seven years, whichever comes first. The eligibility date is set by the Parole Act regulations and the criminal code.

The parole system provides a means of reintegrating an offender into the community. The board is as concerned with the protection of society as with the regeneration of the offender and, therefore, supervision is as much a part of the parole system as are assistance and guidance. All parolees and those not paroled but released under mandatory supervision because of remission of sentence are subject to conditions that, if violated, may result in a return to prison. Remission, time off the sentence for good behaviour, may amount to one-third of the sentence.

Since October 15, 1977 the provinces may appoint parole boards to deal with inmates in accordance with the Parole Act and its regulations, except those inmates serving life imprisonment for murder or detention for an indeterminate period. If this is not done, the National Parole Board continues to assume the responsibility for inmates in provincial prisons under federal law. British Columbia and Ontario have had their own parole boards for some years. In these provinces an inmate may be serving a definite or fixed term plus an indeterminate term. The National Parole Board may grant parole during the definite term and the provincial board during the indeterminate term.

Anyone serving a sentence of preventive detention as an habitual criminal or dangerous sexual offender has his case reviewed at least once a year under the criminal code to see if he should be granted parole. This type of offender is classified as a dangerous offender and is sentenced to a period of detention for an indeterminate period. He becomes eligible for parole three years after being taken into custody and thereafter the case is reviewed every two years. The board has found that few such inmates are ready for release before eight to 10 years have been served. An offender sentenced to life for a crime other than murder becomes eligible for parole after serving seven years.

Inmates sentenced to life for murder before July 26, 1976 may become eligible after a minimum of 10 years. For those sentenced to life terms after January 1974, the eligibility date may follow the jury's recommendation but the judge's pronouncement of the earliest possible date for eligibility is ultimately binding. This may be set at any time between 10 and 20 years. Inmates sentenced to life imprisonment on or after July 26, 1976 for first degree murder are not eligible for parole consideration before they have served 25 years. First degree murder covers all planned and deliberate murders; contracted murders; murder of police officers, prison employees, or others authorized to work in a prison; and murder while committing or attempting to commit rape, indecent assault on a male or female, kidnapping and forcible confinement, or hijacking. Anyone who commits a second murder, no matter of what nature, is considered to have committed a first degree murder.

Any other murder is second degree murder and the mandatory period to be served before parole eligibility is between 10 and 25 years, as indicated by the sentencing judge after the view of the convicting jury has been sought. A person convicted of second degree murder and sentenced to serve more than the minimum 10 years, before becoming eligible for parole, may appeal this additional period of ineligibility to a court of appeal.

Anyone convicted of first degree murder who has served 15 years of the 25-year mandatory period before parole eligibility or anyone convicted of second degree murder, whose mandatory term exceeds 15 years and who has served 15 years of the sentence, may apply for a judicial review by a superior court judge and a jury to either reduce the remaining period of ineligibility or to be declared immediately eligible for parole.

The board is also involved in another type of release, mandatory supervision. Anyone who is not paroled and is released from a federal institution more than 60 days before the end of his sentence, because of remission of the sentence, is subject to supervision for the full period of that remission. The release is made by law, not by a decision of the board. Release conditions are the same as those for parole. An inmate may choose to remain in the institution to complete his sentence; if he later changes his mind he will be released on mandatory supervision.

The decision of the board about any inmate is based on reports it receives from the police, the judge and various people at the institution who deal with him. Reports may also be obtained from a psychologist or a psychiatrist. A community investigation is made to gather information about his family and background, his work record and his relationship with the community. These reports help the board assess whether the offender can lead a law abiding life.

In 1976 the board granted 2,136 full paroles. This number together with those already on parole meant that there were 5,694 inmates at liberty in Canada for part or all of the year. Similarly, with the granting of 2,027 day paroles there were 2,697 inmates on day parole during the year, and with 2,531 inmates released on mandatory supervision there were 4,245 offenders at liberty for part or all of the year. There were 336 revocations of full parole and 579 of mandatory supervision in 1976. A total of 1,976 day paroles were either terminated or completed. There were 444 forfeitures of full paroles, 56 of day paroles and 699 of mandatory supervision releases.

Sources
2.1 - 2.6 Advisory and Research Services Section, Public Law Branch, Department of Justice.
2.7 Justice Statistics Division, Institutional and Public Finance Statistics Branch, Statistics Canada; Royal Canadian Mounted Police; Ontario Provincial Police; Quebec Police Force.
2.8 - 2.9 Justice Statistics Division, Institutional and Public Finance Statistics Branch, Statistics Canada.
2.9.1 Canadian Penitentiary Service.
2.9.2 National Parole Board.

Tables

..	not available	e	estimate
...	not appropriate or not applicable	p	preliminary
—	nil or zero	r	revised
- -	too small to be expressed		certain tables may not add due to rounding

2.1 Provinces and territories of Canada, dates of admission to Confederation, legislative processes by which admission was effected, present area and seat of government

Province, territory or district	Date of admission or creation	Legislative process	Present area km²	Seat of provincial or territorial government
Ontario[1]	July 1, 1867	Act of Imperial Parliament — The British North America Act, 1867 (Br. Stat. 1867, c. 3) and Imperial Order in Council, May 22, 1867	1 068 582	Toronto
Quebec[2]	July 1, 1867		1 540 680	Quebec
Nova Scotia	July 1, 1867		55 491	Halifax
New Brunswick	July 1, 1867		73 437	Fredericton
Manitoba[3]	July 15, 1870	Manitoba Act, 1870 (SC 1870, c. 3) and Imperial Order in Council, June 23, 1870	650 087	Winnipeg
British Columbia	July 20, 1871	Imperial Order in Council, May 16, 1871	948 596	Victoria
Prince Edward Island	July 1, 1873	Imperial Order in Council, June 26, 1873	5 657	Charlottetown
Saskatchewan[4]	Sept. 1, 1905	Saskatchewan Act, 1905 (SC 1905, c. 42)	651 900	Regina
Alberta[4]	Sept. 1, 1905	Alberta Act, 1905 (SC 1905, c. 3)	661 185	Edmonton
Newfoundland	Mar. 31, 1949	The British North America Act, 1949 (Br. Stat. 1949, c. 22)	404 517	St. John's
Northwest Territories[5]	July 15, 1870	Act of Imperial Parliament — Rupert's Land Act, 1868 (Br. Stat. 1868, c. 105) and Imperial Order in Council, June 23, 1870	3 379 683	
Mackenzie[6]	Jan. 1, 1920	Order in Council, Mar. 16, 1918	1 366 193	Yellowknife
Keewatin[6]	Jan. 1, 1920		590 931	
Franklin[6]	Jan. 1, 1920		1 422 559	
Yukon Territory[7]	June 13, 1898	Yukon Territory Act, 1898 (SC 1898, c. 6)	536 324	Whitehorse
Canada			9 976 139	

[1]The area of Ontario was extended by the Ontario Boundaries Extension Act, 1912 (SC 1912, c. 40).

[2]Extended by Quebec Boundaries Extension Act, 1912 (SC 1912, c. 45).

[3]Extended by the Extension of Boundaries Act of Manitoba, 1881 and the Manitoba Boundaries Extension Act, 1912 (SC 1912, c. 32).

[4]Saskatchewan and Alberta created as provinces in 1905 from the area formerly comprised in the provisional districts of Assiniboia, Athabaska, Alberta and Athabaska established May 17, 1882 by minute of Canadian Privy Council concurred in by Dominion Parliament and Order in Council, Oct. 2, 1895.

[5]By an Imperial Order in Council passed on June 23, 1870 pursuant to the Rupert's Land Act, 1868 (Br. Stat. 1868, c. 105), the former territories of the Hudson's Bay Company known as Rupert's Land and the North-Western Territory were transferred to Canada effective July 15, 1870. These territories were designated as the North-West Territories by the Act of SC 1869, c. 3, and as the Northwest Territories by RSC 1906, c. 62. By Imperial Order in Council of July 31, 1880 (effective Sept. 1, 1880), all British territories and possessions in North America not already included within Canada and all islands adjacent thereto (with the exception of the Colony of Newfoundland and its dependencies) were annexed to Canada and these additional territories were formally included in the North-West Territories by SC 1905, c. 27. The province of Manitoba was formed out of a portion of the territories by the Manitoba Act, 1870 (SC 1870, c. 3) and a further portion was added to Manitoba in 1881 by SC 1881, c. 14. The provinces of Alberta and Saskatchewan were formed out of portions of the territories in 1905 and in 1912 other portions were added to Manitoba, Ontario and Quebec.

[6]By SC 1876, c. 21, a separate district to be known as the District of Keewatin was established and provision was made for the local government thereof. The Act was expressed to come into force by proclamation. It provided that portions of the District might be re-annexed to the North-West Territories by proclamation; in 1886 a portion of the District of Keewatin was re-annexed and in 1905 the entire Keewatin District was re-annexed. The Act of 1876 was never proclaimed. By Order in Council of May 8, 1882 the provisional districts of Assiniboia, Saskatchewan, Alberta and Athabaska were created for the convenience of settlers and for postal purposes. By Order in Council of Oct. 2, 1895 the further provisional districts of Ungava, Franklin, Mackenzie and Yukon were created. The boundaries of these provisional districts were re-defined by Order in Council of Dec. 18, 1897. Subsequently the Yukon Territory was formed, the provinces of Alberta and Saskatchewan were created and other portions of the territories were annexed to Quebec, Ontario and Manitoba. By Order in Council dated Mar. 16, 1918 (effective Jan. 1, 1920) the remaining portions of the Northwest Territories were divided into three provisional districts known as Mackenzie, Keewatin and Franklin.

[7]The provisional district of Yukon established in 1895 was created a judicial district of the North-West Territories by proclamation issued pursuant to Sect. 51 of the North-West Territories Act (RSC 1886, c. 50) on Aug. 16, 1897 and, by the Yukon Territory Act (SC 1898, c. 6), was declared to be a separate territory.

2.2 Police personnel, actual strength, 1975 and 1976

Force	1975				1976			
	Police	Cadets	Other full-time employees	Total	Police	Cadets	Other full-time employees	Total
Royal Canadian Mounted Police	14,072	—	4,365	18,437	14,012	—	4,650	18,662
Ontario Provincial Police	4,046	—	1,166	5,212	4,064	—	1,169	5,233
Quebec Police Force	4,107	1	964	5,072	4,194	1	975	5,170
Municipal Police (excl. RCMP and OPP contracts)	27,430	575	4,177	32,182	28,372	542	4,498	33,412
Canadian National Railways Police	458	—	28	486	432	—	25	457
Canadian Pacific Railway Company Police	318	—	92	410	318	—	89	407
National Harbours Board Police	236	—	90	326	237	—	97	334
Total	50,667	576	10,882	62,125	51,629	543	11,503	63,675

2.3 Crime statistics, by type of offence, 1975 and 1976 (based on Uniform Crime Reporting Program)

Year and offence	Actual offences[1]	Offences cleared		Persons charged			
		By charge	Other-wise	Adults		Juveniles	
				Male	Female	Male	Female
1975							
Criminal code	1,585,804	329,259	209,379	235,461	38,425	62,832	8,380
Murder, capital and non-capital	633	433	47	405	61	21	7
Attempted murder	642	468	38	428	53	14	—
Manslaughter	63	54	2	45	10	4	2
Rape	1,848	805	319	935	3	65	3
Other sexual offences	9,052	3,044	1,671	2,532	41	321	23
Wounding	2,128	999	494	792	187	61	9
Assaults (not indecent)	99,758	31,240	46,372	28,778	2,702	1,397	350
Robbery	21,299	5,726	701	5,549	398	1,384	91
Breaking and entering	260,652	42,221	17,698	30,381	1,098	21,695	1,061
Theft, motor vehicle	90,791	16,025	6,208	11,412	458	7,867	340
Theft over $200	94,957	8,248	4,651	7,107	947	2,008	177
Theft $200 or under	492,372	67,898	48,511	38,741	17,426	16,593	4,905
Having stolen goods	16,240	14,299	1,344	9,040	1,056	1,747	220
Fraud	86,024	37,843	12,057	16,788	3,954	830	174
Prostitution	3,409	3,271	56	696	2,372	12	38
Gaming and betting	3,619	3,265	166	3,467	187	10	—
Offensive weapons	12,578	8,110	2,347	6,733	352	603	34
Other criminal code[1]	389,739	85,310	66,697	71,632	7,120	8,200	946
Federal statutes[2]	44,972	29,541	9,762	18,417	1,550	1,123	449
Narcotic Control Act	50,081	39,346	4,667	36,512	3,829	1,959	320
Controlled drugs under the Food and Drugs Act	5,461	3,788	479	3,138	463	246	48
Provincial statutes[1]	381,388	271,346	94,897	249,803	17,537	7,467	4,350
Municipal bylaws[1]	64,800	27,864	19,628	23,440	2,960	1,420	197
1976P							
Criminal code	1,647,171	356,502	221,694	255,351	43,441	62,177	9,294
Murder, capital and non-capital	597	423	57	396	57	18	4
Attempted murder	694	533	36	481	55	34	4
Manslaughter	46	41	2	43	8	3	—
Rape	1,834	825	323	877	1	66	—
Other sexual offences	8,813	3,127	1,786	2,600	36	360	8
Wounding	1,993	998	432	804	161	62	7
Assaults (not indecent)	103,280	33,079	47,318	30,101	3,008	1,565	361
Robbery	20,095	5,646	775	5,410	410	1,243	96
Breaking and entering	269,599	45,335	18,585	32,917	1,313	22,026	1,186
Theft, motor vehicle	88,129	16,163	6,484	11,357	468	7,351	329
Theft over $200	106,217	9,315	5,558	7,808	1,095	2,081	182
Theft $200 or under	501,104	75,129	52,509	42,043	19,878	15,936	5,551
Having stolen goods	17,759	15,667	1,545	9,865	1,216	1,820	296
Fraud	86,548	40,709	13,198	18,627	4,745	533	237
Prostitution	2,842	2,702	20	901	2,038	16	50
Gaming and betting	3,755	3,457	111	3,249	155	8	1
Offensive weapons	13,544	8,796	2,418	7,130	395	656	44
Other criminal code[1]	420,322	94,557	70,537	80,742	8,402	8,399	938
Federal statutes[2]	50,627	34,408	9,590	20,825	2,116	1,529	358
Narcotic Control Act	59,974	49,402	5,892	46,033	4,968	2,184	555
Controlled drugs under the Food and Drugs Act	3,192	2,282	373	1,948	309	62	22
Provincial statutes[1]	369,106	251,853	103,523	232,456	16,002	7,731	4,331
Municipal bylaws[1]	64,448	27,279	20,579	22,822	3,532	949	74

[1]Except traffic.
[2]Except traffic, Narcotic Control Act and Food and Drugs Act.

2.4 Traffic enforcement statistics, by type of offence, 1975 and 1976 (based on Uniform Crime Reporting Program)

Offence	Actual offences	Offences cleared		Persons charged	
		By charge	Otherwise	Male	Female
1975					
Criminal code	252,734	186,796	10,886	176,052	6,493
Criminal negligence					
Causing death	245	235	4	227	7
Causing bodily harm	149	133	6	108	5
Operating motor vehicle	770	744	13	640	10
Failing to stop or remain at scene of accident	74,792	12,177	8,304	10,484	887
Dangerous driving	6,658	6,158	206	5,723	127
Failure or refusal to provide breath sample	12,378	12,446	88	11,817	364
Driving while impaired	134,936	132,275	2,037	126,072	4,784
Driving while disqualified	22,806	22,628	228	20,981	309
Federal statutes (except parking)				13,280	
Provincial statutes (except parking)				2,371,492	
Municipal bylaws (except parking)				323,404	
Provincial statutes[1]	115,341	65,094	13,268	59,044	5,634
Failing to stop or remain at scene of accident	38,809	10,666	3,490	9,303	938
Dangerous driving	68,978	52,089	9,717	47,456	4,653
Driving while disqualified	7,554	2,339	61	2,285	43

2.4   Traffic enforcement statistics, by type of offence, 1975 and 1976 (based on Uniform Crime Reporting Program) (concluded)

Offence	Actual offences	Offences cleared		Persons charged	
		By charge	Otherwise	Male	Female
1976P					
Criminal code	257,619	189,841	10,894	177,931	7,286
Criminal negligence					
Causing death	228	209	2	193	12
Causing bodily harm	140	136	3	110	4
Operating motor vehicle	784	758	11	678	18
Failing to stop or remain at scene of accident	75,115	11,524	8,400	9,918	886
Dangerous driving	6,613	5,987	315	5,529	137
Failure or refusal to provide breath sample	12,767	12,676	67	11,808	447
Driving while impaired	136,003	132,844	1,897	126,036	5,344
Driving while disqualified	25,969	25,707	199	23,659	438
Federal statutes (except parking)				27,864	
Provincial statutes (except parking)				2,666,636	
Municipal bylaws (except parking)				358,381	
Provincial statutes[1]	125,353	69,239	12,745	62,838	6,177
Failing to stop or remain at scene of accident	44,460	12,248	4,095	10,811	1,165
Dangerous driving	69,482	53,740	8,578	49,003	4,949
Driving while disqualified	11,411	3,251	72	3,024	63

[1]Provincial traffic offences almost identical to those under the criminal code.

2.5   Persons charged and persons convicted of indictable offences, with ratio per 100,000 population 16 years of age and over, by province and total, 1972 and 1973

Province or territory	Persons charged		Persons convicted				Persons convicted per 100,000 population 16 years of age and over	
	1972 No.	1973 No.	1972 No.	%	1973 No.	%	1972 No.	1973 No.
Newfoundland	946	615	877	92.7	527	85.7	269	158
Prince Edward Island	14	42	13	92.9	33	78.6	17	42
Nova Scotia	2,541	2,213	2,260	88.9	1,891	85.4	417	342
New Brunswick	1,985	1,093	1,803	90.3	993	90.9	422	226
Ontario	29,634	30,470	23,985	80.9	23,408	76.8	437	417
Manitoba	4,588	4,745	3,416	74.5	2,789	58.8	495	395
Saskatchewan	3,933	3,258	3,541	90.0	2,225	68.3	565	354
British Columbia	11,426	11,010	9,309	81.5	8,491	77.1	585	514
Yukon Territory and Northwest Territories	474	518	410	86.5	404	78.0	1,277	1,213
Total[1]	55,541	53,964	45,614	82.1	40,761	75.5	466	406

[1]Excludes Quebec and Alberta.

2.6   Persons charged and convicted of indictable offences, by class of offence, 1972 and 1973

Class of offence	1972			1973		
	Persons charged	Persons convicted		Persons charged	Persons convicted	
		Male	Female		Male	Female
Criminal code						
Class I. Offences against the person	6,526	4,693	316	5,097	3,348	237
Class II. Offences against property with violence	7,740	6,665	162	7,556	6,296	202
Class III. Offences against property without violence	29,483	18,719	5,900	30,884	17,261	5,495
Class IV. Malicious offences against property	1,711	1,310	102	1,236	863	52
Class V. Forgery and other offences relating to currency	1,273	903	236	1,142	798	193
Class VI. Other offences	4,356	3,288	286	4,446	3,173	303
Total, criminal code	51,089	35,578	7,002	50,361	31,739	6,482
Federal statutes	4,452	2,753	281	3,603	2,316	224
Total[1]	55,541	38,331	7,283	53,964	34,055	6,706

[1]Excludes Quebec and Alberta.

2.7   Court sentences given for indictable offences, by province, 1972 and 1973

Year and sentence	Nfld.	PEI	NS	NB	Ont.	Man.	Sask.	BC	YT and NWT	Canada[1]
1972										
Option of fine	324	—	877	573	8,573	837	817	3,067	94	15,162
Jail										
Under one year	176	5	501	451	5,147	694	981	2,794	138	10,887
One year and over	35	—	34	101	949	228	193	546	43	2,129
Reformatory and training school	—	—	2	1	1,038	—	—	—	—	1,041
Penitentiary										
Under two years	2	—	2	1	8	2	—	4	—	19
Two years and under five	21	1	149	91	526	159	80	253	8	1,288
Five years and under ten	2	—	12	13	150	27	16	93	2	315
Ten years and under fourteen	—	—	7	2	30	7	2	35	—	83
Fourteen years and over	—	—	—	3	2	6	—	10	—	21
Life	2	—	7	—	29	3	6	21	—	68
Preventive	—	—	—	—	1	—	—	3	—	4
Death	—	—	—	—	—	—	—	1	—	1
Suspended sentence without probation	19	—	40	249	1,428	793	519	387	14	3,449
Suspended sentence with probation	296	7	629	318	6,104	660	927	2,095	111	11,147
Total	877	13	2,260	1,803	23,985	3,416	3,541	9,309	410	45,614
1973										
Option of fine	137	3	730	324	8,559	810	343	2,957	120	13,983
Jail										
Under one year	209	10	453	264	5,049	714	749	2,414	146	10,008
One year and over	17	2	62	56	1,038	182	181	524	24	2,086
Reformatory and training school	—	—	4	—	1,007	—	—	—	—	1,011
Penitentiary										
Under two years	—	—	—	—	5	—	1	1	—	7
Two years and under five	10	6	141	61	582	125	39	294	9	1,267
Five years and under ten	—	—	10	3	154	19	19	99	1	305
Ten years and under fourteen	—	—	1	2	30	3	1	29	—	66
Fourteen years and over	—	—	—	1	13	1	—	15	—	30
Life	—	—	2	1	19	1	4	13	—	40
Preventive	—	—	—	—	2	—	—	1	—	3
Death	—	—	—	—	1	—	—	—	—	1
Suspended sentence without probation	81	3	40	93	1,094	320	311	396	17	2,355
Suspended sentence with probation	73	9	448	188	5,855	614	577	1,748	87	9,599
Total	527	33	1,891	993	23,408	2,789	2,225	8,491	404	40,761

[1]Excludes Quebec and Alberta.

2.8   Method of trial of persons charged with indictable offences, showing disposition of cases, by province and total, 1973

Method of trial	Nfld.	PEI	NS	NB	Ont.	Man.	Sask.	BC	YT and NWT	Canada[1]
By judge and jury										
Convicted	1	—	19	8	492	28	56	159	2	765
Acquitted	—	—	6	2	202	8	14	61	2	295
Detained because of insanity	—	—	—	—	14	—	—	5	—	19
Disagreement of jury	—	—	—	—	5	—	2	4	—	11
Stay of proceedings	—	—	1	1	12	3	12	18	—	47
No bill	—	—	—	—	59	—	—	—	—	59
Conditional discharge	—	—	—	—	6	2	—	—	—	8
Absolute discharge	—	—	1	—	—	—	4	2	—	7
By a judge without jury										
Convicted	3	3	124	10	1,448	49	207	249	8	2,101
Acquitted	—	1	30	2	424	6	47	64	3	577
Detained because of insanity	—	—	—	1	3	—	—	—	—	4
Disagreement of jury	—	—	—	—	—	—	—	—	—	—
Stay of proceedings	—	—	—	—	18	3	21	36	—	78
No bill	—	—	—	—	—	—	—	—	—	—
Conditional discharge	1	—	1	—	60	1	29	5	—	97
Absolute discharge	—	—	—	1	16	—	10	5	—	32

2.8   Method of trial of persons charged with indictable offences, showing disposition of cases, by province and total, 1973 (concluded)

Method of trial	Nfld.	PEI	NS	NB	Ont.	Man.	Sask.	BC	YT and NWT	Canada[1]
By a magistrate with consent										
Convicted	266	21	881	508	11,336	1,597	1,239	4,154	249	20,251
Acquitted	3	—	47	24	1,151	75	50	505	10	1,865
Detained because of insanity	1	—	1	—	6	—	1	9	—	18
Disagreement of jury	—	—	—	—	1	—	—	—	—	1
Stay of proceedings	—	—	1	5	41	463	20	617	36	1,183
No bill	—	—	—	—	—	—	—	—	—	—
Conditional discharge	21	3	39	11	749	307	291	140	8	1,569
Absolute discharge	4	2	24	2	279	99	142	29	4	585
By a magistrate, absolute jurisdiction										
Convicted	257	9	867	467	10,132	1,115	723	3,929	145	17,644
Acquitted	7	—	52	14	1,306	59	72	413	11	1,934
Detained because of insanity	1	—	1	1	1	1	—	3	—	8
Disagreement of jury	—	—	—	—	1	—	—	—	—	1
Stay of proceedings	1	—	—	2	10	266	—	226	22	527
No bill	—	—	—	—	—	—	—	—	—	—
Conditional discharge	40	2	59	11	1,751	391	249	269	12	2,784
Absolute discharge	9	1	59	23	947	272	69	108	6	1,494
Total, persons charged	615	42	2,213	1,093	30,470	4,745	3,258	11,010	518	53,964
Total, persons convicted	527	33	1,891	993	23,408	2,789	2,225	8,491	404	40,761

[1]Excludes Quebec and Alberta.

2.9   Convictions for summary conviction offences[1], by type, 1972 and 1973

Type of offence	1972	1973
CRIMINAL CODE	*104,825*	*107,688*
Assault with intent		534
Attempt to commit suicide	21	2
Attempts, conspiracies, accessories, counselling	134	92
Bawdy house	126	137
Being at large and failing to appear	..	3,225
Causing disturbance by being drunk	1,724	1,982
Common assault	7,965	7,126
Communicating venereal disease	28	2
Contempt of court	32	69
Corrupting morals	253	313
Cruelty to animals	48	46
Damage not exceeding $50 and other interference with property	3,113	2,422
Disclosure of jury proceedings	..	5
Disorderly conduct	8,705	8,021
Duty of persons to provide necessaries	74	115
Duty to safeguard dangerous places	2	—
Failing to comply with order	..	218
Fraudulently obtaining food or lodging	715	727
Fraudulently obtaining transportation	154	134
Gaming, betting, lotteries	556	407
Intimidation	151	189
Killing or injuring bird or animal other than cattle	46	49
Mischief in relation to private property	..	878
Mischief in relation to property endangering life	..	158
Mischief in relation to public property	..	442
Motor vehicle		
Criminal negligence in operation	252	325
Dangerous driving	2,035	1,967
Dangerous operation of vessel	135	144
Driving while impaired	25,392	26,026
Driving while disqualified	5,640	5,992
Driving with more than 80 mg in blood	27,502	26,893
Failing to stop at scene of accident	3,346	3,286
Failure or refusal to provide breath sample	5,801	5,781
Motor vehicle equipped with smoke screen	70	63
Taking motor vehicle without consent	1,506	1,218
Offences relating to public or police officer	..	782
Offensive weapons	1,202	1,020
Personating peace officer	101	36
Public mischief	..	282
Recognizance, breach of	2,938	2,551
Sample of breath (vessel)	..	6
Soliciting	..	744
Vagrancy	883	5
Other criminal code	4,175	3,276

2.9   Convictions for summary conviction offences[1], by type, 1972 and 1973 (concluded)

Type of offence	1972	1973
FEDERAL STATUTES	*28,849*	*38,228*
Customs	63	618
Excise	421	434
Fisheries	952	786
Food and drugs	2,932	1,871
Harbour board and merchant seamen's	193	506
Immigration	352	370
Income tax	8,957	7,698
Indian		
Intoxication	201	143
Other	236	194
Juvenile delinquents		
Adults who contribute to delinquency	474	738
Inducing child to leave home	20	174
Sexual immorality	142	57
Other	..	1,087
Lord's day	343	188
Narcotic Control Act	..	8,823
National defence	255	29
Railway	562	198
Unemployment insurance	1,805	1,938
Weights and measures	248	146
Other federal statutes	10,693	12,230
PROVINCIAL STATUTES	*1,281,582*	*1,151,535*
Children of unmarried parents	1,104	1,157
Child welfare (Protection Act)		
Maintenance	..	553
Cruelty	..	88
Wardship	..	4,591
Deserted wives and children's maintenance	10,762	9,557
Game and fisheries	7,026	5,660
Highway traffic		
Driving without due care and attention	49,027	36,347
Other traffic (excludes parking)	1,046,581	937,910
Liquor control	125,862	125,910
Master and servant	435	380
Medical, dentistry and pharmacy	62	115
Mental diseases	120	247
Prairie and forest fire prevention	48	269
Protection of children	4,989	1,023
Public health	523	222
School laws	209	475
Other provincial statutes	34,834	27,031
MUNICIPAL BYLAWS	*102,206*	*86,566*
Intoxication	522	971
Traffic	75,088	62,598
Other	26,596	22,997
Total convictions	1,517,462	1,384,017

[1]Excludes Quebec, Alberta and Yukon Territory, in 1972, and Quebec and Alberta in 1973.
[2]Attempted suicide not a criminal offence after July 1972.

2.10   Juvenile delinquents, by group of offence, and ratio per 100,000 juvenile population, 1971-73[1]

Group of offence		1971	1972	1973
Delinquencies against the person	*No.*	1,723	1,726	1,963
	Ratio	38	38	43
Delinquencies against property with violence	*No.*	8,790	9,064	9,335
	Ratio	192	199	205
Delinquencies against property without violence	*No.*	15,093	15,116	15,298
	Ratio	330	331	335
Wilful and forbidden acts in respect of certain property	*No.*	2,112	2,640	2,668
	Ratio	46	58	58
Forgery and delinquencies relating to currency	*No.*	136	174	156
	Ratio	3	4	3
Other delinquencies	*No.*	10,944	13,463	14,731
	Ratio	239	295	323
Total	*No.*	38,798	42,183	44,151
	Ratio	848	924	968

[1]Canada total juvenile population figure in 1971 was 4,576,700, 4,564,000 in 1972, and 4,562,000 in 1973.

2.11   Juvenile delinquents classified by type of delinquency and percentage distribution, 1971-73

Delinquency		1971	1972	1973
Manslaughter and murder and causing death by criminal negligence	No.	24	21	25
	%	0.06	0.05	0.06
Murder, attempt	No.	11	8	5
	%	0.03	0.02	0.01
Common assault	No.	771	747	862
	%	1.99	1.77	1.95
Other delinquencies against the person	No.	315	313	404
	%	0.81	0.74	0.92
Breaking and entering a place	No.	8,407	8,694	8,905
	%	21.67	20.61	20.17
Robbery and extortion	No.	383	370	430
	%	0.99	0.88	0.97
Theft and having in possession[1]	No.	13,159	13,033	12,984
	%	33.92	30.90	29.41
False pretences and fraud and corruption	No.	219	213	302
	%	0.57	0.51	0.68
Arson and other fires[2]	No.	194	222	244
	%	0.50	0.53	0.55
Other interference with property	No.	2,295	4,053	4,182
	%	5.92	9.61	9.47
Incorrigibility and vagrancy	No.	1,431	657	444
	%	3.69	1.56	1.01
Immorality	No.	409	390	320
	%	1.06	0.92	0.72
Theft, auto	No.	1,126	1,246	1,437
	%	2.90	2.95	3.26
Various other delinquencies	No.	10,054	12,216	13,607
	%	25.89	28.95	30.82
Total	No.	38,798	42,183	44,151
	%	100.00	100.00	100.00

[1]Includes having in possession, theft, theft from mail, theft of bicycle.
[2]Includes false alarm of fire.

2.12   Disposition of delinquents, by type of sentence and percentage distribution, 1971-73

Type of sentence		1971	1972	1973
Reprimanded	No.	524	791	1,076
	%	1.35	1.88	2.44
Probation[1]	No.	12,422	12,053	11,990
	%	32.02	28.57	27.16
Fined or made restitution	No.	4,318	5,220	6,237
	%	11.13	12.37	14.13
Detained indefinitely	No.	207	129	77
	%	0.53	0.31	0.17
Sent to training school	No.	1,641	1,190	1,108
	%	4.23	2.82	2.51
Final disposition suspended	No.	5,897	7,179	7,763
	%	15.20	17.02	17.58
Mental hospital	No.	14	35	20
	%	0.04	0.08	0.04
Other	No.	13,775	15,586	15,880
	%	35.50	36.95	35.97
Total dispositions	No.	38,798	42,183	44,151
	%	100.00	100.00	100.00

[1]Includes probation of court and probation of parents.

Sources

2.1 Advisory and Research Services Section, Public Law Branch, Department of Justice.
2.2 - 2.12 Justice Statistics Division, Institutional and Public Finance Statistics Branch, Statistics Canada.

Government

Chapter 3

Tables

Government

Organization of the federal government

3.1

In Canada the legal framework within which political processes take place is provided through a written constitution embodied in the British North America acts. The first of these acts, passed by the British Parliament in 1867, not only established the institutions through which legislative, executive and judicial powers are exercised in Canada but also established a federal form of government. A central government — the federal government — has legislative jurisdiction primarily over matters of national concern and over those matters not assigned to the provinces. The 10 provincial governments are assigned specific areas of legislative jurisdiction, including municipal institutions.

In Canada there is a fusion of executive and legislative powers. Formal executive power is vested in the Queen, whose authority is delegated to the Governor General, her representative. Legislative power is vested in the Parliament of Canada which consists of the Queen, an appointed upper house called the Senate and a lower house called the House of Commons, elected by universal adult suffrage. The independence of the judiciary is safeguarded through the constitutional provision that superior court judges are appointed by the Governor-in-Council, that is, by the Governor General on advice of the Cabinet, and that they hold office during good behaviour and cannot be removed unless both houses of Parliament, the Cabinet and the Governor General agree.

In the Canadian system, where the executive is part of Parliament, democratic principles could not be adhered to without the constitutional convention that the government is responsible to the House of Commons. When the government loses the confidence of the House of Commons, it must resign or the prime minister must ask the Governor General to dissolve Parliament and call a general election. Although there are conventions that help in deciding when the government has lost the confidence of the House, all doubt is removed when the government is defeated on a motion on which it had explicitly staked its life or when a motion of non-confidence in the government is passed. If the government resigns, the Governor General can call on the leader of the opposition (who is usually the leader of the political party that has the second largest number of seats in the House of Commons) to form a new government. If a government that has lost the confidence of the House of Commons and has been granted a dissolution is defeated in the ensuing general election and if no clear majority is elected, the government has two choices — it can remain in office and seek the confidence of the Commons when it meets or it can resign at once. If it resigns, the Governor General will normally ask the leader of another party, usually the one that has won the most seats, to form a new government. The primary responsibility of the Governor General in either circumstance is to provide the nation with a government capable of carrying on with the support of the House of Commons.

The prime minister and his Cabinet, who with one or two exceptions are members of the House of Commons, are, formally speaking, the Queen's advisers. In fact virtually no significant actions can be taken by the Queen or her representative in Canada, the Governor General, without Cabinet advice. The prime minister and his Cabinet determine executive policies and are responsible for them to the House of Commons. The Queen and the Governor General have the traditional rights to be consulted, to encourage and to warn the government.

The demands of citizens are conveyed primarily to members of Parliament, directly to Cabinet ministers or indirectly to Cabinet ministers through the public service. Demands may originate from individuals, political parties or pressure groups; members of Parliament, Cabinet ministers and public servants may take the initiative in suggesting the adoption of policies and programs in the public interest. Although the roles of Parliament, the public service and the Cabinet cannot be defined precisely, the following description deals with the obvious and primary roles of each.

Determination of public policy rests with the Cabinet but begins generally with the formulation of policy by individual ministers. Working in co-operation with public servants, a minister formulates policy proposals for consideration by the Cabinet, which chooses those policies it wishes to implement. The Cabinet may itself formulate policies, but may also select a policy from among the alternatives submitted. A Cabinet committee system and, more especially, a Cabinet committee on priorities and planning, enhance the capacity of Cabinet in policy determination and priority setting.

In conformity with the principle of the rule of law, all executive acts must be authorized by law, and laws are enacted by Parliament. Executive acts may be carried out under a statute which specifies how a policy is to be implemented, or under a statute which authorizes the Governor-in-Council to undertake specific acts. Much of the activity of the public service is authorized through the yearly enactment of appropriation acts approving the expenditure of public funds for specific purposes. Apart from its concern with the appropriation of funds, Parliament is concerned with the discussion and authorization of policy submitted for its approval by the government. Approval of policies is mainly through the enactment of legislation. To enable the Commons to perform this role more efficiently, numerous changes in the rules of procedure were adopted in 1969 and are included in the standing orders of the House of Commons.

The most significant feature of these processes is that Cabinet ministers have seats in Parliament and thus share in the exercise of legislative power. In fact, the majority of legislation enacted by Parliament is submitted by the government; the British North America Act provides that all financial measures must originate in the Commons.

The role of the judiciary is to apply the laws enacted by Parliament. Because Parliament is supreme in the Canadian government, the judiciary must apply the law as Parliament has enacted it and cannot declare laws to be unconstitutional if they are within the legislative jurisdiction of Parliament or of the legislature that enacted them.

The administration of legislation and of the government's policies is carried out through a public service comprising employees organized as of 1977 in 25 departments of government and a large number of special boards, commissions and Crown corporations or other agencies. Legislation and tradition have combined to develop a non-partisan public service, whose employees' tenure is unaffected by changes in government. The only direct contact between public servants and Parliament occurs when they are called as witnesses before parliamentary committees; public servants do not, by convention, express opinions on public policy but usually appear as experts and to explain existing policy. Public servants who head agencies such as the Public Service Commission, the office of the auditor general, the office of the commissioner of official languages, the Library of Parliament or the office of the chief electoral officer are responsible directly to Parliament. They are not subject to direction by the government on matters of policy and may appear before parliamentary committees to explain the policies of their agencies.

Growth in number, variety and complexity of the demands placed on the government requires it not only to adjust its policies but to make significant changes in the organization of the public service so that required policies can be implemented. Major reorganization of the public service was authorized by the passage of a series of government organization acts in 1966, 1969 and 1970.

3.1.1 The executive

The Crown. The British North America (BNA) Act of 1867 (Sect. 9) provides that the executive government and authority of and over Canada is vested in the Queen. The functions of the Crown (that is, the formal executive represented by the Queen), which are substantially the same as those of the Crown in relation to the British government, are discharged in Canada by the Governor General.

The Sovereign. Since Confederation Canada has had six sovereigns: Victoria, Edward VII, George V, Edward VIII, George VI and Elizabeth II. The present sovereign is not only Queen of Canada but of other countries in the Commonwealth. Her title for Canada was approved by Parliament and established by a royal proclamation on May 28, 1953: Elizabeth the Second, by the grace of God of the United Kingdom, Canada and her other realms and territories, Queen, Head of the Commonwealth, Defender of the Faith.

From time to time the Queen personally discharges the Crown's functions in Canada, such as the appointment of the Governor General, which Her Majesty does on the recommendation of the prime minister of Canada. During a royal visit, the Queen may participate in ceremonies normally carried out in her name by the Governor General, such as the opening and dissolution of Parliament, the assent to bills passed by the House of Commons and the Senate, and the granting of a general amnesty.

The Governor General is the representative of the Crown in Canada. The Right Honourable Jules Léger, the 21st Governor General since Confederation, was appointed by Queen Elizabeth on October 5, 1973 and took office on January 14, 1974. Constitutionally, the Queen of Canada is the Canadian head of state but the Governor General fulfils her role on her behalf. The letters patent revised and issued under the Great Seal of Canada on October 1, 1947 authorized the Governor General "to exercise on the advice of his Canadian ministers, all Her Majesty's powers and authorities in respect of Canada."

Following are the Governors General of Canada since Confederation, with dates of appointment:

The Viscount Monck of Ballytrammon, June 1, 1867
The Baron Lisgar of Lisgar and Bailieborough, December 29, 1868
The Earl of Dufferin, May 22, 1872
The Marquis of Lorne, October 5, 1878
The Marquis of Lansdowne, August 18, 1883
The Baron Stanley of Preston, May 1, 1888
The Earl of Aberdeen, May 22, 1893
The Earl of Minto, July 30, 1898
The Earl Grey, September 26, 1904
Field Marshal HRH The Duke of Connaught, March 21, 1911
The Duke of Devonshire, August 19, 1916
General The Baron Byng of Vimy, August 2, 1921
The Viscount Willingdon of Ratton, August 5, 1926
The Earl of Bessborough, February 9, 1931
The Baron Tweedsmuir of Elsfield, August 10, 1935
Major General The Earl of Athlone, April 3, 1940
Field Marshal The Viscount Alexander of Tunis, March 21, 1946
The Right Honourable Vincent Massey, January 24, 1952
General The Right Honourable Georges P. Vanier, August 1, 1959
The Right Honourable Roland Michener, March 29, 1967
The Right Honourable Jules Léger, October 5, 1973.

One of the most important responsibilities of the Governor General is to ensure that the country always has a government. If the office of the prime minister becomes vacant because of death, resignation or defeat of the government in the Commons, the Governor General must see that the office of the prime minister is filled and that a new government is formed.

As the representative of the Queen, the Governor General summons, prorogues and dissolves Parliament on the advice of the prime minister. He signs orders-in-council, commissions and other state documents, and gives his assent to bills that have been passed in both Houses of Parliament and which thereby become acts of Parliament with the force of law (unless Parliament prescribes specifically otherwise). Like the Queen, in virtually all cases he is bound by constitutional convention to carry out these duties in accordance with the advice of his responsible ministers. Should he not wish to accept their advice, and should they maintain that advice, his only alternative is to replace the existing government with a new government. This alternative could be exercised only if, at the same time, the principle of responsible government could be upheld. This means that the Governor General's discretion in choosing another government is strictly limited to a situation in which a person other than the existing prime minister could command the confidence of the House of Commons.

Canadian honours system. An exclusively Canadian honours system was introduced in 1967 with the establishment of the Order of Canada. The honours system was enlarged in 1972 with the addition of the Order of Military Merit and three decorations to be awarded in recognition of acts of bravery. A complete description of these awards and a list of the recipients during 1977 are given in Appendix 4.

The Privy Council. The BNA Act of 1867 (Sect. 11) provides for "a council to aid and advise in the Government of Canada," called the Queen's Privy Council for Canada. The council that in fact advises the Queen's representative, the Governor General, is the committee of the Privy Council whose membership comprises the Cabinet.

Membership in the Privy Council is for life and includes Cabinet ministers of the government of the day, former Cabinet ministers, various members of the royal family, past and present Commonwealth prime ministers, premiers of provinces, former speakers of the Senate and the House of Commons of Canada and a few other distinguished persons. As a condition of office, all ministers must first be sworn into the Privy Council. A member is styled "Honourable" and may use the initials PC after his name. A member of the Privy Council of Britain is styled "Right Honourable"; the Governor General, the chief justice of Canada and the prime minister of Canada assume the title "Right Honourable" when they take office.

The Privy Council as a whole has met on only a few ceremonial occasions; its constitutional responsibilities to advise the Crown on government matters are discharged exclusively by the committee of the Privy Council. The legal instruments through which executive authority is exercised are called orders-in-council. The committee makes a submission to the Governor General for his approval which he is obliged to give in almost all circumstances; with this approval, the submission becomes an order-in-council. Meetings of the committee or a subcommittee of it are held without formal ceremony.

The office of president of the Privy Council was formerly occupied, more often than not, by the prime minister but from time to time, especially in recent years, it has been occupied by another minister. On July 5, 1968, the prime minister explained that the incumbent of the office of president of the Privy Council would also be the government's leader in the House of Commons, with the broad responsibility of directing House business, including supervision of the government's replies to questions in the House and of parliamentary returns in general, and a special responsibility of ensuring that Parliament, through its operations and organization of business, can effectively function under the increasing pressure of modern government.

The following, with the dates when they were sworn in, were members of the Queen's Privy Council for Canada in November 1977:

Hon. William Earl Rowe, August 30, 1935
Hon. Joseph Thorarinn Thorson, June 11, 1941
Hon. Lionel Chevrier, April 18, 1945
Hon. Paul Joseph James Martin, April 18, 1945
Hon. Douglas Charles Abbott, April 18, 1945
Hon. Hugues Lapointe, August 25, 1949
Hon. Gabriel-Édouard Rinfret, August 25, 1949
Hon. Walter Edward Harris, January 18, 1950
Hon. James Sinclair, October 15, 1952
Hon. John Whitney Pickersgill, June 12, 1953
Hon. Jean Lesage, September 17, 1953
Hon. George Carlyle Marler, July 1, 1954
Hon. Paul Theodore Hellyer, April 26, 1957
Rt. Hon. John George Diefenbaker, June 21, 1957
Hon. Howard Charles Green, June 21, 1957
Hon. Donald Methuen Fleming, June 21, 1957
Hon. George Hees, June 21, 1957
Hon. Léon Balcer, June 21, 1957
Hon. George Randolph Pearkes, June 21, 1957
Hon. Gordon Churchill, June 21, 1957

Hon. Edmund Davie Fulton, June 21, 1957
Hon. Douglas Scott Harkness, June 21, 1957
Hon. Ellen Louks Fairclough, June 21, 1957
Hon. John Angus MacLean, June 21, 1957
Hon. Michael Starr, June 21, 1957
Hon. William McLean Hamilton, June 21, 1957
Hon. William Joseph Browne, June 21, 1957
Hon. Jay Waldo Monteith, August 22, 1957
Hon. Francis Alvin George Hamilton, August 22, 1957
HRH The Prince Philip, Duke of Edinburgh, October 14, 1957
Hon. Henri Courtemanche, May 12, 1958
Hon. David James Walker, August 20, 1959
Hon. Joseph-Pierre-Albert Sévigny, August 20, 1959
Hon. Hugh John Flemming, October 11, 1960
Hon. Noël Dorion, October 11, 1960
Hon. Walter Dinsdale, October 11, 1960
Hon. Jacques Flynn, December 28, 1961
Hon. Paul Martineau, August 9, 1962

Hon. Richard Albert Bell, August 9, 1962
Rt. Hon. Roland Michener, October 15, 1962
Hon. Marcel-Joseph-Aimé Lambert, February 12, 1963
Hon. Théogène Ricard, March 18, 1963
Hon. Frank Charles McGee, March 18, 1963
Hon. Martial Asselin, March 18, 1963
Hon. Walter Lockhart Gordon, April 22, 1963
Hon. Mitchell William Sharp, April 22, 1963
Hon. Azellus Denis, April 22, 1963
Hon. George James McIlraith, April 22, 1963
Hon. William Moore Benidickson, April 22, 1963
Hon. Maurice Lamontagne, April 22, 1963
Hon. Lucien Cardin, April 22, 1963
Hon. Allan Joseph MacEachen, April 22, 1963
Hon. Jean-Paul Deschatelets, April 22, 1963
Hon. Hédard Robichaud, April 22, 1963
Hon. John Watson MacNaught, April 22, 1963
Hon. Roger Teillet, April 22, 1963
Hon. Judy V. LaMarsh, April 22, 1963
Hon. Charles Mills Drury, April 22, 1963
Hon. John Robert Nicholson, April 22, 1963
Hon. Harry Hays, April 22, 1963
Hon. John Joseph Connolly, February 3, 1964
Hon. Maurice Sauvé, February 3, 1964
Hon. Yvon Dupuis, February 3, 1964
Hon. George Stanley White, June 25, 1964
Hon. Edgar John Benson, June 29, 1964
Hon. Léo Alphonse Joseph Cadieux, February 15, 1965
Hon. Lawrence T. Pennell, July 7, 1965
Hon. Jean-Luc Pepin, July 7, 1965
Hon. Alan Aylesworth Macnaughton, October 25, 1965
Hon. Jean Marchand, December 18, 1965
Hon. John James Greene, December 18, 1965
Hon. Joseph Julien Jean-Pierre Côté, December 18, 1965
Hon. John Napier Turner, December 18, 1965
Hon. Maurice Bourget, February 22, 1966
Rt. Hon. Pierre Elliott Trudeau, April 4, 1967
Hon. Joseph-Jacques-Jean Chrétien, April 4, 1967
Hon. Pauline Vanier, April 11, 1967
Hon. John Parmenter Robarts, July 5, 1967
Hon. Louis-J. Robichaud, July 5, 1967
Hon. Dufferin Roblin, July 5, 1967
Hon. William Andrew Cecil Bennett, July 5, 1967
Hon. Alexander B. Campbell, July 5, 1967
Hon. Ernest Charles Manning, July 5, 1967
Hon. Joseph Robert Smallwood, July 5, 1967
Hon. Robert L. Stanfield, July 7, 1967
Rt. Hon. John Robert Cartwright, September 4, 1967
Hon. Charles Ronald McKay Granger, September 25, 1967

Hon. Bryce Stuart Mackasey, February 9, 1968
Hon. Donald Stovel Macdonald, April 20, 1968
Hon. John Carr Munro, April 20, 1968
Hon. Gérard Pelletier, April 20, 1968
Hon. Jack Davis, April 26, 1968
Hon. Horace Andrew Olson, July 6, 1968
Hon. Jean-Eudes Dubé, July 6, 1968
Hon. Stanley Ronald Basford, July 6, 1968
Hon. Donald Campbell Jamieson, July 6, 1968
Hon. Eric William Kierans, July 6, 1968
Hon. Robert Knight Andras, July 6, 1968
Hon. James Armstrong Richardson, July 6, 1968
Hon. Otto Emil Lang, July 6, 1968
Hon. Herbert Eser Gray, October 20, 1969
Hon. Robert Douglas George Stanbury, October 20, 1969
Rt. Hon. Joseph Honoré Gérald Fauteux, March 23, 1970
Hon. Jean-Pierre Goyer, December 22, 1970
Hon. Alastair William Gillespie, August 11, 1971
Hon. Martin Patrick O'Connell, August 11, 1971
Hon. Patrick Morgan Mahoney, January 21, 1972
Hon. Stanley Haidasz, November 27, 1972
Hon. Eugene F. Whelan, November 27, 1972
Hon. Warren Allmand, November 27, 1972
Hon. J. Hugh Faulkner, November 27, 1972
Hon. André Ouellet, November 27, 1972
Hon. Daniel J. MacDonald, November 27, 1972
Hon. Marc Lalonde, November 27, 1972
Hon. Jeanne Sauvé, November 27, 1972
Rt. Hon. Bora Laskin, January 7, 1974
Hon. Lucien Lamoureux, August 8, 1974
Hon. Raymond Joseph Perrault, August 8, 1974
Hon. Barnett Jerome Danson, August 8, 1974
Hon. J. Judd Buchanan, August 8, 1974
Hon. Roméo LeBlanc, August 8, 1974
Hon. Muriel McQueen Fergusson, November 7, 1974
Hon. Pierre Juneau, August 29, 1975
Hon. Marcel Lessard, September 26, 1975
Hon. Jack Sydney George Cullen, September 26, 1975
Hon. Leonard Stephen Marchand, September 15, 1976
Hon. John Roberts, September 15, 1976
Hon. Monique Bégin, September 15, 1976
Hon. Jean-Jacques Blais, September 15, 1976
Hon. Francis Fox, September 15, 1976
Hon. Anthony Chisholm Abbott, September 15, 1976
Hon. Iona Campagnolo, September 15, 1976
Hon. Joseph-Philippe Guay, November 3, 1976
Hon. John Henry Horner, April 21, 1977
Hon. Norman A. Cafik, September 16, 1977.

The prime minister is the leader of the political party requested by the Governor General to form the government, which almost always means that he is the leader of the party with the strongest representation in the Commons. His position is one of exceptional authority stemming in part from the success of the party at an election. The prime minister chooses his Cabinet. When a member of Cabinet resigns, the remainder of the Cabinet is undisturbed; when the prime minister vacates his office, this act normally carries with it the resignation of all those in the Cabinet.

Part of the prime minister's authority lies in his prerogative to recommend dissolution of Parliament. This prerogative, which in most circumstances permits him to precipitate an election, is a source of considerable power both in his dealings with colleagues and with the opposition parties in the House. The prime minister is also responsible for organization of the Cabinet and its committees; for the organization and functions of his own office, as well as the Privy Council and federal-provincial relations offices; and for the allocation of responsibilities between ministers.

Another source of the prime minister's authority derives from the appointments which he recommends including privy councillors, Cabinet ministers, lieutenant-governors of the provinces, provincial administrators, speakers of the Senate, chief justices of all courts, senators and certain senior executives of the public service. The prime minister also recommends the appointment of a new Governor General to the Sovereign, although this normally follows consultation with his Cabinet.

Following are the prime ministers since Confederation, with dates of administrations:

Rt. Hon. Sir John Alexander Macdonald, July 1, 1867 — November 5, 1873
Hon. Alexander Mackenzie, November 7, 1873 — October 9, 1878
Rt. Hon. Sir John Alexander Macdonald, October 17, 1878 — June 6, 1891
Hon. Sir John Joseph Caldwell Abbott, June 16, 1891 — November 24, 1892
Rt. Hon. Sir John Sparrow David Thompson, December 5, 1892 — December 12, 1894
Hon. Sir Mackenzie Bowell, December 21, 1894 — April 27, 1896
Rt. Hon. Sir Charles Tupper, May 1, 1896 — July 8, 1896
Rt. Hon. Sir Wilfrid Laurier, July 11, 1896 — October 6, 1911
Rt. Hon. Sir Robert Laird Borden, October 10, 1911 — October 12, 1917 (Conservative Administration)
Rt. Hon. Sir Robert Laird Borden, October 12, 1917 — July 10, 1920 (Unionist Administration)
Rt. Hon. Arthur Meighen, July 10, 1920 — December 29, 1921 (Unionist — National Liberal and Conservative Party)
Rt. Hon. William Lyon Mackenzie King, December 29, 1921 — June 28, 1926
Rt. Hon. Arthur Meighen, June 29, 1926 — September 25, 1926
Rt. Hon. William Lyon Mackenzie King, September 25, 1926 — August 6, 1930
Rt. Hon. Richard Bedford Bennett, August 7, 1930 — October 23, 1935
Rt. Hon. William Lyon Mackenzie King, October 23, 1935 — November 15, 1948
Rt. Hon. Louis Stephen St-Laurent, November 15, 1948 — June 21, 1957
Rt. Hon. John George Diefenbaker, June 21, 1957 — April 22, 1963
Rt. Hon. Lester Bowles Pearson, April 22, 1963 — April 20, 1968
Rt. Hon. Pierre Elliott Trudeau, April 20, 1968 — ... .

The Cabinet. The Cabinet's primary responsibility in the Canadian political system is to determine priorities among the demands expressed by the people and to define policies to meet those demands. The Cabinet is a committee of ministers chosen by the prime minister, generally from among members of the House of Commons, although one or two Cabinet ministers are usually chosen from the Senate including the leader of the government in the Senate. It is unusual for a senator to head a government department because the constitution provides that measures for appropriating public funds or imposing taxes must originate in the Commons. If a senator headed a department, another minister in the Commons would have to speak on his behalf on its affairs.

In May 1978, the following were members of the 20th ministry according to precedence:

Prime Minister, Rt. Hon. Pierre Elliott Trudeau
President of the Queen's Privy Council for Canada, Hon. Allan Joseph MacEachen
Minister of Finance, Hon. Jean Chrétien
Minister of Labour, Hon. John Carr Munro
Minister of Justice and Attorney General of Canada, Hon. Stanley Ronald Basford

Secretary of State for External Affairs, Hon. Donald Campbell Jamieson
President of the Treasury Board, Hon. Robert Knight Andras
Minister of Transport, Hon. Otto Emil Lang
Minister of Supply and Services, Hon. Jean-Pierre Goyer
Minister of Energy, Mines and Resources, Hon. Alastair William Gillespie
Minister of Agriculture, Hon. Eugene Francis Whelan
Minister of Consumer and Corporate Affairs, Hon. W. Warren Allmand
Minister of Indian Affairs and Northern Development, Hon. James Hugh Faulkner
Minister of State for Urban Affairs, Hon. André Ouellet
Minister of Veterans Affairs, Hon. Daniel Joseph MacDonald
Minister of State for Federal-Provincial Relations, Hon. Marc Lalonde
Minister of Communications, Hon. Jeanne Sauvé
Leader of the Government in the Senate, Hon. Raymond Joseph Perrault
Minister of National Defence, Hon. Barnett Jerome Danson
Minister of Public Works and Minister of State for Science and Technology, Hon. J. Judd Buchanan
Minister of Fisheries and the Environment, Hon. Roméo LeBlanc
Minister of Regional Economic Expansion, Hon. Marcel Lessard
Minister of Employment and Immigration, Hon. Jack Sydney George Cullen
Minister of State (Environment), Hon. Leonard Stephen Marchand
Secretary of State of Canada, Hon. John Roberts
Minister of National Health and Welfare, Hon. Monique Bégin
Solicitor General, Hon. Jean-Jacques Blais
Minister of State (Small Business), Hon. Anthony Chisholm Abbott
Minister of State (Fitness and Amateur Sport), Hon. Iona Campagnolo
Minister of National Revenue, Hon. Joseph-Philippe Guay
Minister of Industry, Trade and Commerce, Hon. Jack Henry Horner
Minister of State (Multiculturalism), Hon. Norman A. Cafik
Postmaster General, Hon. J. Gilles Lamontagne.

Each Cabinet minister usually assumes responsibility for one of the departments of government, although a minister may hold more than one portfolio at the same time or he may hold one or more portfolios and one or more acting portfolios. A minister without portfolio may be invited to join the Cabinet because the prime minister wishes to have him or her in the Cabinet without the heavy duties of running a department, or to provide a suitable balance of regional representation, or for any other reason that the prime minister sees fit. Because of Canada's cultural and geographical diversity, the prime minister must see that his Cabinet is representative.

With the enactment of the Ministries and Ministers of State Act (Government Organization Act, 1970), five categories of ministers of the Crown may be identified: departmental ministers, ministers with special parliamentary responsibilities, ministers without portfolio, and two types of ministers of state. Ministers of state for designated purposes may head a ministry of state created by proclamation. They are charged with developing new and comprehensive policies in areas of particular urgency and importance and have a mandate determined by the Governor-in-Council. They may have powers, duties and functions and exercise supervision and control of elements of the public service, and may seek parliamentary appropriations independently of any minister to cover the cost of their staff and operations. Other ministers of state, usually undesignated, may be appointed to assist a departmental minister with his responsibilities. They may have statutory powers, duties and functions and are limited in number by the appropriations that Parliament is willing to pass. They receive the same salary as a minister without portfolio, as provided for in the estimates of the minister with whom they are associated. All ministers are appointed on the advice of the prime minister by commissions of office issued by the Governor General under the Great Seal of Canada, to serve and to be accountable to Parliament as members of the government and for any responsibility that might be assigned to them by law or otherwise.

In Canada, almost all executive acts of the government are carried out in the name of the Governor-in-Council. The Privy Council committee makes submissions to the Governor General for his approval, and he is bound by the constitution in nearly all circumstances to accept them. About 3,326 such orders-in-council were enacted in 1976 compared with 3,417 in 1975. Although some were routine and required little discussion in Cabinet, others were of major significance and required extensive deliberation, sometimes covering months of meetings of officials, Cabinet committees, and the full Cabinet.

The Cabinet must consider and approve the policy underlying each piece of proposed legislation. After proposed legislation is drafted it must be examined in detail. Recently, between 40 and 60 bills have been considered by Cabinet during a parliamentary session. Proposals for reform of government organization or administration, and policy to be adopted in fundamental constitutional changes or at a major international conference are among the issues which, on occasion, demand this extensive and detailed consideration.

The Cabinet committee system. The nature and large volume of policy issues to be decided on by Cabinet do not lend themselves to discussion by 25 or 30 ministers. The first Cabinet committee system was established after the outbreak of World War II. Since then, growing demands on the executive have further stimulated delegation of some Cabinet functions to its committees.

Cabinet committees usually have less than 10 ministers, providing a forum for thorough study of policy proposals. Membership of Cabinet committees is confidential and the same rules of secrecy that apply to Cabinet apply to Cabinet committees. Otherwise, these committees might develop an importance and authority inconsistent with the principle of collective responsibility of ministers. The prime minister determines the establishment of Cabinet committees, their membership, and terms of reference. Ministers may invite one or two officials as advisers during Cabinet committee meetings. The secretariats of the committees are provided by the Privy Council office and the secretary of a Cabinet committee is usually also an assistant secretary to the Cabinet. Treasury Board, which is a Cabinet committee — or more precisely a subcommittee of the Privy Council committee — is the only exception; it has its own secretariat headed by a secretary who has the status of a deputy minister.

Under the direction of the prime minister, the secretary to the Cabinet prepares agenda and refers memoranda to Cabinet to the appropriate committee for study and report to the full Cabinet. Except where the prime minister instructs otherwise, all memoranda to Cabinet are submitted over the signature of the minister concerned.

The terms of reference of Cabinet committees cover virtually all government responsibility. All memoranda to Cabinet are first considered by a Cabinet committee, except when they are of exceptional urgency or when the prime minister directs otherwise, in which case an item may be considered immediately by the full Cabinet.

In 1977 there were four co-ordinating committees: priorities and planning; legislation and house planning; federal-provincial relations; and the Treasury Board and five subject-matter committees: economic policy; external policy and defence; social policy; culture and native affairs; and government operations. These committees meet regularly.

In addition there were four special and ad hoc committees of the Cabinet that met as required: the Cabinet committees on the public service, security and intelligence, labour relations and the special committee of council which considers all submissions to the Governor-in-Council on behalf of the Privy Council committee. The accompanying chart indicates the relationship of these committees to the Cabinet process.

Growing reliance on the Cabinet committee system since World War II is evidence of its usefulness. The following is a brief outline of the involvement of Cabinet and Cabinet committees with a piece of legislation that the government ultimately introduces in the Commons or the Senate.

On the initiative of a minister a policy proposal is prepared, the implementation of which will require new legislation or the amendment of existing legislation. The proposal is addressed formally to Cabinet, but is considered first by a subject-matter committee. If approved, the proposal goes forward for consideration by Cabinet. Proposals with financial implications are considered by the Treasury Board before going forward to Cabinet. If Cabinet confirms the committee's decision or makes a revision, the justice department is instructed by the minister who made the proposal to prepare a draft bill expressing in legal terms the intent of the policy proposal. If the draft bill has the

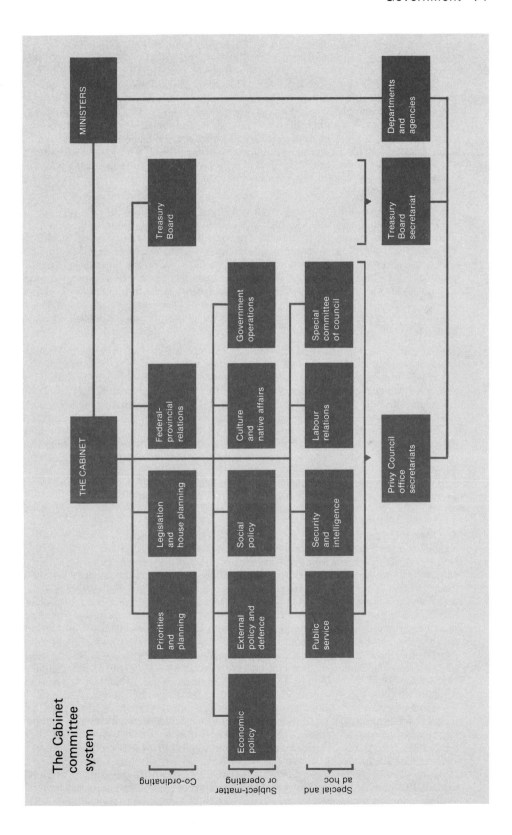

The Cabinet committee system

MINISTERS

Treasury Board

Departments and agencies

Treasury Board secretariat

THE CABINET

Federal-provincial relations

Government operations

Special committee of council

Legislation and house planning

Culture and native affairs

Labour relations

Priorities and planning

Social policy

Security and intelligence

Privy Council office secretariats

External policy and defence

Public service

Economic policy

Co-ordinating

Subject-matter or operating

Special and ad hoc

minister's approval, he submits it to the Cabinet committee on legislation and house planning and it is examined from a legal rather than a policy point of view. If this committee agrees that the bill is acceptable and could be introduced in Parliament, it reports this to Cabinet, which decides on the committee's decision. If confirmation is given, the prime minister initials the bill and it is then introduced either in the Senate or the House of Commons, depending on constitutional and political considerations.

The order and manner in which a bill is considered in Parliament is the responsibility of the president of the Privy Council and house leader who negotiates these matters with his counterparts in the opposition parties. If a bill is to be introduced in the Senate, the house leader will discuss questions such as timing and tactics with the leader of the government in the Senate, who in turn will negotiate consideration of the bill with his counterpart in the Senate.

The Privy Council office is a secretariat providing staff support to the Privy Council committee and to the Cabinet. For the purposes of the Financial Administration Act it is considered a government department. The office provides secretariats to serve the Cabinet, the Privy Council committee and their various subcommittees except the Cabinet committee on federal-provincial relations, which is served by a federal-provincial relations office. Since the prime minister is, in effect, chairman of the Cabinet, he is the minister responsible for the Privy Council office. The work of the Privy Council office is directed by a public servant known as the clerk of the Privy Council and secretary to the Cabinet. He is the senior member of the public service of Canada.

Parliamentary secretaries. The Parliamentary Secretaries Act of June 1959 provided for the appointment of 16 parliamentary secretaries from among the members of the Commons to assist ministers. That act was amended by the Government Organization Act, 1970, which allows the number of parliamentary secretaries to equal the number of ministers who hold offices listed in Section 4 of the Salaries Act, that is, ministers with departmental responsibilities, the prime minister, the leader of the government in the Senate and the president of the Privy Council. A parliamentary secretary works under direction of his minister, but has no legal authority in his association with the department, nor is he given acting responsibility or any of the powers, duties and functions of a minister in the event of his minister's absence or incapacity. Parliamentary secretaries are appointed by the prime minister and hold office for 12 months.

3.1.2 The Legislature

The federal legislative authority is vested in the Parliament of Canada — the Queen, the Senate, and the House of Commons. Bills may originate in either the Senate or the House, subject to the provisions of Section 53 of the British North America (BNA) Act, 1867, which provides that bills for the appropriation of any part of the public revenue or the imposition of any tax or impost shall originate in the House of Commons. Bills must pass both houses and receive royal assent before becoming law. In practice, most public bills originate in the House of Commons although, at the request of the government, more have recently been introduced and dealt with in the Senate while the Commons is engaged in other matters such as the debate on the speech from the throne. Private bills usually originate in the Senate. The Senate may delay, amend or even refuse to pass bills sent to it from the Commons, but differences are usually settled without serious conflict.

Section 91 of the BNA acts, 1867 to 1964, assigns to the Parliament of Canada legislative authority in very clearly specified areas. These are discussed in Chapter 2.

Under Section 95, the Parliament of Canada may make laws in relation to agriculture and immigration concurrently with provincial legislatures although federal legislation is paramount in any conflict. An amendment to the BNA Act in 1951 (Br. Stat. 1950-51, c.32) authorized the Parliament of Canada to make laws in relation to old age pensions subject to the proviso that no such law should affect the operation of any provincial laws in relation to such pensions. By the BNA Act, 1964 this amendment was extended to permit the payment of supplementary benefits, including survivors and disability benefits irrespective of age, under a contributory pension plan.

The passage of legislation. If a bill is introduced and approved in the House of Commons, it is then introduced in the Senate and follows a similar procedure. If a bill is first introduced in the Senate, the reverse procedure is followed. There are three types of bills: public bills introduced by the government; public bills introduced by private members of Parliament; and private bills introduced by private members of Parliament. Each type is treated in a slightly different manner, and there are even differences in procedure when the house deals with government bills introduced pursuant to supply and ways and means motions on the one hand, and other government bills on the other. The following outline describes the procedure for a government bill introduced in the House of Commons.

The sponsoring minister gives notice that he intends to introduce a bill on a given subject. Not less than 48 hours later he moves for leave to introduce the bill and that the bill be given first reading. This is granted automatically because this first step does not imply approval of any sort. It is only after first reading that the bill is ordered printed for distribution to the members.

At a later sitting the minister moves that the bill be given second reading and that it be referred to an appropriate committee of the House of Commons. A favourable vote on the motion for second reading represents approval of the bill in principle so there is often an extensive debate, which, according to the procedures of the Commons, must be confined to the principle of the bill. The debate culminates in a vote which, if favourable, results in the bill being referred to the appropriate committee of the House, where it is given clause-by-clause consideration.

At the committee stage, expert witnesses and interested parties may be invited to give testimony pertaining to the bill, and the proceedings may cover many weeks.

The house committee prepares and submits a report to the House of Commons which must decide whether to accept the report, including any amendments the committee has made to the bill. At the report stage any member may, on giving 24 hours notice, move an amendment to the bill. All such amendments are debated and are usually put to a vote. Following that, a motion "that the bill be concurred in" or "that the bill, as amended, be concurred in", is put to the vote.

After this report stage, the minister moves that the bill be given third reading and passage. Debate of this motion is limited to whether the bill should be given third reading. Amendments are permitted at this stage but they must be of a general nature, similar to those allowed on second reading. If the vote is favourable, the bill is introduced in the Senate where it goes through a somewhat similar though not identical process, since each chamber has its own rules of procedure. The bill is then presented to the Governor General for royal assent and signature. Depending on the provisions in the bill it may come into force when it is signed by the Governor General, on an appointed day, or when it is officially proclaimed.

Duration and sessions of Parliaments. The length and sessions of the first to the 12th Parliaments, covering the period from Confederation to 1917, are given in the *Canada Year Book 1940* p 46; of the 13th to the 17th Parliaments in the 1945 edition, p 53; of the 18th and 19th Parliaments in the 1957-58 edition, p 46; of the 20th to the 23rd Parliaments in the 1965 edition, p 65; of the 24th to the 26th Parliaments in the 1975 edition, p 132; and of the 27th to the 30th Parliaments in this edition, Table 3.1.

The Senate has grown from an original membership of 72 at Confederation, through the addition of members to represent new provinces and the general increase in population, to a total of 104 members; the latest change in representation was made on June 19, 1975 when an act of Parliament (SC 1974-75-76, c.53) amended the Canadian Constitution to entitle the Yukon Territory and the Northwest Territories to be represented by one senator each. The growth of representation in the Senate is summarized in Table 3.2.

Senators are appointed by the Governor General by instrument under the Great Seal of Canada. By constitutional usage the actual power of nominating senators resides in the prime minister whose advice the Governor General accepts. Until the passage of an act providing for the retirement of senators (SC 1965, c.4), assented to on June 2,

Passage of legislation

Policy proposal requiring legislation (submitted to Cabinet by a minister)

Consideration of policy proposal in a subject-matter committee and decision or recommendation

Cabinet confirmation of committee decision

Responsible minister issues drafting instructions for legislation to Department of Justice

Cabinet confirmation of committee decision and prime minister's signature

Consideration of draft bill by Cabinet committee on legislation and house planning

Draft bill prepared by Department of Justice and approved by responsible minister

The Cabinet process

Parliament first reading in either Senate or House of Commons[1] (reading of title and brief explanation of bill)

Parliament second reading in same House of Parliament (debate and vote on principle of bill)

Consideration by appropriate parliamentary committee (clause by clause examination of bill)

Introduction of bill into other House of Parliament and repetition of the process

Parliament third reading and vote

Parliament report stage and vote on any amendments prepared by committee

The legislative process

The Governor General in presence of Senate and House of Commons assents to bill and signs it into law

1. All money bills must be introduced in the House of Commons.

1965, senators were appointed for life; that act set 75 years as the age at which any person appointed to the Senate after the coming into force of the act would cease to hold his place in the Senate.

In each of the four main areas of Canada (Ontario, Quebec, Atlantic provinces and western provinces) except Quebec, senators represent the whole of the province for which they are appointed; in Quebec, one senator is appointed for each of the 24 electoral divisions of what was formerly Lower Canada. The deliberations of the Senate are presided over by a speaker appointed by the Governor-in-Council (in effect by the government) and government business in the Senate is sponsored by the government leader in the Senate.

The Senate's traditional role in legislation originating in the House of Commons is to take a sober second look at such legislation and amend it if necessary; such amendments are often concurred in by the Commons. If representatives of the two houses cannot resolve disagreements arising from Senate amendments, the legislation cannot be further considered.

The Senate provides a national forum for discussion of public issues and airing of grievances from any part of Canada. Through its own committees and its participation in joint committees of both houses, the Senate is particularly active in making studies in depth on matters of public concern.

Since 1971, Senate committees have been performing a new function, that of studying the subject matter of government bills, including money bills, in advance of their formal introduction in the Senate. Under this procedure, amendments to a bill suggested by a Senate committee are often accepted by the government and by the House of Commons before the bill itself actually reaches the Senate.

In May 1978 the representation in the Senate was as follows:

Newfoundland
Eric Cook
William John Petten
Frederick William Rowe
Philip Derek Lewis
Jack Marshall
1 vacancy

Prince Edward Island
Florence Elsie Inman
Orville Howard Phillips
Mark Lorne Bonnell
1 vacancy

Nova Scotia
Donald Smith
Harold Connolly
John Michael Macdonald
Margaret Norrie
Henry D. Hicks
Bernard Alasdair Graham
Augustus Irvine Barrow
Ernest George Cottreau
George Isaac Smith
1 vacancy

New Brunswick
Fred A. McGrand
Edgar Fournier
Charles Robert McElman
Hervé J. Michaud
Michel Fournier
Louis-J. Robichaud, PC
Daniel Riley
Margaret Jean Anderson
2 vacancies

Quebec
Sarto Fournier
Hartland de Montarville Molson
Josie Alice Dinan Quart
Louis Philippe Beaubien
Jacques Flynn, PC
Maurice Bourget, PC
Azellus Denis, PC
Jean-Paul Deschatelets, PC
Alan Aylesworth Macnaughton, PC
J.G. Léopold Langlois
Paul Desruisseaux
Maurice Lamontagne, PC
Raymond Eudes
Louis de Gonzague Giguère
Paul C. Lafond
H. Carl Goldenberg
Renaude Lapointe (Speaker)
Martial Asselin, PC
Maurice Riel
Jean Marchand, PC
Pietro Rizzuto
Joseph Napoléon Claude Wagner
2 vacancies

Ontario
Salter Adrian Hayden
Norman McLeod Paterson
John J. Connolly, PC
David A. Croll
Joseph A. Sullivan
Lionel Choquette
Allister Grosart
David James Walker, PC
Rhéal Bélisle

Daniel Aiken Lang
William Moore Benidickson, PC
Douglas Keith Davey
Andrew Ernest Thompson
Keith Laird
Richard James Stanbury
Eugene A. Forsey
George James McIlraith, PC
John James Greene, PC
Joan Neiman
John Morrow Godfrey
Royce Frith
Peter Bosa
Stanley Haidasz, PC
Florence Bayard Bird

Manitoba
Paul Yuzyk
Douglas Donald Everett
Gildas L. Molgat
William C. McNamara
Duff Roblin, PC
Joseph-Philippe Guay, PC

Saskatchewan
Alexander Hamilton McDonald
Hazen Robert Argue

Herbert O. Sparrow
Sidney L. Buckwold
David Gordon Steuart
1 vacancy

Alberta
Donald Cameron
Earl Adam Hastings
Harry William Hays, PC
Ernest C. Manning, PC
Horace Andrew (Bud) Olson, PC
1 vacancy

British Columbia
Ann Elizabeth Bell
Edward M. Lawson
George Clifford van Roggen
Guy Williams
Raymond J. Perrault, PC
Jack Austin

Yukon Territory
Paul Lucier

Northwest Territories
Willie Adams.

The principal positions in the Senate are occupied as follows: Speaker of the Senate, Hon. Renaude Lapointe; Leader of the Government in the Senate, Hon. Raymond J. Perrault, PC; Leader of the Opposition in the Senate, Hon. Jacques Flynn, PC. The principal officer of the Senate is Mr. Robert Fortier, QC, Clerk of the Senate and Clerk of the Parliaments.

The House of Commons. Following the 1971 Census the number of members in the House of Commons was determined by the representation commissioner in accordance with Section 51 of the British North America Act, with the total representation at 264.

The readjustment of the federal electoral districts was carried out during 1972 and 1973 in accordance with the Electoral Boundaries Readjustment Act. The last of 10 reports was submitted to the Commons in July 1973. After debate it was agreed to suspend the readjustment until January 1, 1975. To this effect, Bill C-208 Electoral Boundaries Readjustment Suspension Act was given royal assent on July 27, 1973.

On December 20, 1974 royal assent was given to the Representation Act, 1974 which removed the temporary suspension of the Electoral Boundaries Readjustment Act and provided for representation in the Commons under a revised formula awarding each province the following number of members: Ontario 95, Quebec 75, British Columbia 28, Alberta 21, Manitoba 14, Saskatchewan 14, Nova Scotia 11, New Brunswick 10, Newfoundland seven, Prince Edward Island four, the Northwest Territories two and the Yukon Territory one. The revised number of members in the House of Commons, following the next general election, would be 282.

The number of representatives of each province elected at each of the 30 general elections since Confederation is given in Table 3.3. Historical data concerning representation may be found in the *Canada Year Book 1973*.

Salaries, allowances and pensions. Members of the Senate and House of Commons receive sessional allowances of $26,900 per annum. This rate is subject to an annual adjustment based on the industrial composite index or 7%, whichever is less. For each session of Parliament, they may also be paid such travelling expenses between their home or constituency and Ottawa as may be required to perform their duties. A senator receives an annual expense allowance of $5,900 and a member of the House of Commons receives an expense allowance of $12,000 to $15,875 dependent upon the electoral district represented; neither is subject to income tax and is payable monthly. Members of the House of Commons may receive up to $9,116 annually for the payment of staff in their constituency, and up to $4,800 annually for rental of premises in their constituency. Since the present member of the Senate occupying the position of leader of the government in the Senate is a member of the Cabinet and receives a salary as a minister under the Salaries Act, he does not receive a special allowance as leader of the government. The member of the Senate occupying the position of opposition leader in the Senate is paid an annual allowance of $9,000. The deputy leaders of the government and of the opposition in the Senate receive additional annual allowances of $4,000 and $3,200 respectively. The remuneration of the prime minister is $33,300 a year and of a Cabinet minister and the leader of the opposition in the House of Commons $20,000 a year in addition to the sessional and expense allowances each receives as a member of Parliament. The chief government whip, the chief opposition whip, the opposition house leader and the leader of a party having a recognized membership of 12 or more in the House of Commons, other than the prime minister and the leader of the opposition, each receives an annual allowance of $5,300 in addition to the sessional allowance and expense allowance. In addition to sessional and expense allowances, the speaker of the Senate receives a salary of $13,300 per annum, the speaker of the House of Commons, $20,000 per annum and the deputy speaker of the House of Commons, $8,000 per annum. The speakers of the Senate and of the House of Commons are also each entitled to $3,000 and the deputy speaker of the House of Commons to $1,500, in lieu of residence; these allowances are not taxable. The deputy chairman of committees receives an annual allowance of $5,300. Parliamentary secretaries to ministers of the Crown receive an annual allowance of $5,300, in addition to their sessional and expense allowances. Motor vehicle allowances of $2,000 are paid to ministers of the Crown and to the leader of the opposition in the House of Commons, and motor vehicle allowances of $1,000 are paid to the speakers of the Senate and of the House of Commons; these allowances are not taxable. The sessional and expense allowances of a senator are subject to a total deduction of $120 per day ($60 from each allowance) for every day beyond 21 on which the senator does not attend a sitting of the Senate, unless he is unable to attend because of illness or public or official business.

A member of Parliament contributes 7.5% of his sessional indemnity toward his retirement allowance, which is based on the average of the sessional indemnity received over the best consecutive six years of his pensionable service accumulated as follows: 3.5% of this six-year average for each of the first 10 years of pensionable service; 3% of this average for each of the next 10 years; 2% of this average for each of the next five years; and 2% of this average for each of the years of pensionable service earned by his contributions from salary for extra duties performed, for example, as a minister; subject to an overall maximum of 75% of that best six-year average. The member holding the office of prime minister contributes from the salary payable to him under the Salaries Act an amount equal to 6% of that salary to the consolidated revenue fund. Survivors' benefits are as follows: 60% of the member's pension entitlement to the widow or widower; if there is a surviving parent, 10% of the member's pension entitlement for each child up to three; and if there is no surviving parent, 20% of the member's pension entitlement for each child up to four. A member who was a member on March 31, 1970 had a year in which to elect to come under the plan described here or to remain under a previous plan, described in the *Canada Year Book 1969* p 75.

An act to make provision for the retirement of members of the Senate (SC 1965, c.4) entitles a senator appointed after June 2, 1965 to become a contributor under the provisions of the Members of Parliament Retiring Allowances Act. Senators appointed prior to that date and who have not reached age 75, who wish to come under the provisions of this act, are also entitled to become contributors. Under the provisions of the Retirement Act, as amended, a senator contributes 6% of $24,000. A senator appointed before June 2, 1965 who within one year of reaching age 75 resigns, or who resigns because of some permanent infirmity disabling him from performing his duties, may be granted an annuity equal to $16,000. The widow of a senator granted such an annuity may receive an annuity equal to three-fifths of it.

Every former prime minister who held office for four years will receive from the consolidated revenue fund two-thirds the annual salary provided for prime ministers under the Salaries Act beginning when he ceases to hold any office in Parliament, or reaches age 65, whichever is the later. The allowance continues during his lifetime. The widow of a prime minister will receive annually one-half the allowance that was being paid or that would have been paid in the event that he died before receiving it. This allowance would commence immediately after her husband's death and continue during her natural life or until her remarriage.

None of these allowances is payable while the recipient remains a member of Parliament.

The federal franchise. The present federal franchise laws are contained in the Canada Elections Act (RSC 1970, c.14, 1st Supp.) as amended by the Election Expenses Act (SC 1973-74, c.51). Generally, the franchise is conferred upon all Canadian citizens who have reached age 18 and ordinarily live in the electoral district on the date fixed for the beginning of the enumeration at the election. Persons denied the right to vote are: the chief electoral officer and the assistant chief electoral officer; judges appointed by the Governor-in-Council; the returning officer for each electoral district; inmates of any penal institution; persons whose liberty of movement is restricted or who are deprived of the management of their property because of mental disease; and persons disqualified by law for corrupt or illegal practices.

The special voting rules set out in Schedule II to the Canada Elections Act prescribe voting procedures for members of the Canadian forces, for members of the federal public service posted abroad, and also for veterans receiving treatment or domiciliary care in certain institutions.

Electoral districts, voters on list, votes polled and names and addresses of members of the House of Commons elected at the 30th general election, July 8, 1974 are given in Table 3.4. Table 3.5 indicates voters on the lists and votes polled at federal general elections in 1965, 1968, 1972 and 1974.

3.1.3 The Judiciary

Parliament is empowered by Section 101 of the British North America Act to provide for the constitution, maintenance and organization of a general court of appeal for Canada and for the establishment of any additional courts for the better administration of Canada's laws. Under this provision Parliament has established the Supreme Court of Canada, the Federal Court of Canada and certain miscellaneous courts. A detailed discussion of the judiciary and legal system of Canada is presented in Chapter 2.

3.2 Federal government administration

3.2.1 Financial administration and control

The financial affairs of the Government of Canada are administered under the basic principle set out in the British North America Act, that no tax shall be imposed and no money spent without the authority of Parliament and that expenditures shall be made only for the purposes authorized by Parliament. The government introduces all money bills and exercises financial control through a budgetary system based on the principle that all the financial needs of the government for each fiscal year should be considered at one time so that both the current and prospective conditions of the public treasury may be clearly evident.

Estimates and appropriations. The Treasury Board, whose secretariat is a separate department of government under the president of the Treasury Board, co-ordinates the estimates process. Under the Financial Administration Act, the board may act for the Privy Council in all matters of financial management (including estimates, expenditures, financial commitments, establishments, revenues and accounts), personnel management and general administrative policy in the public service.

Departments submit forecasts of their requirements about 12 months before a new fiscal year. These are divided into "A Budgets" for the next three years, to maintain the current levels of service, and "B Budgets" for new activities or expansion of existing activities. These proposals are reviewed by Treasury Board in the light of Cabinet expenditure guidelines that express government priorities. The Treasury Board secretariat prepares recommendations for the budgetary and non-budgetary allocations to each program for Treasury Board and Cabinet review. In August of the year preceding the fiscal year, departments are advised of the allocations approved by Cabinet. Departments then develop detailed estimates of their resource requirements for the new year and submit them at the end of October. Following review by Treasury Board and approval by Cabinet, they are tabled in Parliament in February.

Main estimates and supplementary estimates are referred to committees of the House of Commons. The standing orders of the house (March 1975) call for the referral of the new year main estimates to standing committees of the house by March 1 of the expiring fiscal year. The committees must report back to the house not later than May 31. Supplementary estimates are referred to standing committees immediately after they are tabled and reporting dates are stipulated.

There are three supply periods that end December 10, March 26 and June 30. The first supplementary estimates for a year are usually dealt with in the December period and the final supplementary estimates in the March period. In addition, interim supply (consisting of 3/12ths for all voted items in main estimates and extra 12ths for some voted items) is dealt with in the March period. In the June period the house is asked to provide full supply on main estimates. In each supply period a number of days are allotted to the business of supply. Opposition motions have precedence over all government supply motions on allotted days and opportunities to put forward motions of non-confidence in the government are provided. On the last allotted day in each period, the appropriation acts then before the house must be voted on. These acts authorize payments out of the consolidated revenue fund of the amounts included in the estimates, whether main or supplementary, subject to the conditions stated in them.

The budget. The finance minister usually presents his annual budget speech in the House of Commons some time after the main estimates have been introduced. The budget speech reviews the state of the national economy and the financial operations of the government in the previous fiscal year and gives a forecast of the probable financial requirements for the year ahead, taking into account the main estimates and allowing for supplementary estimates. At the close of his address, the minister tables the formal notices of ways and means motions for any changes in the existing tax rates or rules and customs tariff which, in accordance with parliamentary procedure, must precede the introduction of any money bills. These resolutions give notice of the amendments which the government intends to ask Parliament to make in the taxation statutes. However if a change is proposed in a commodity tax, such as a sales tax or excise duty on a particular item, it is usually effective immediately; the legislation, when passed, is retroactive to the date of the speech.

The budget speech supports a motion that the house approve in general a budgetary policy of the government; debate on this motion may take up six sitting days, but once it is passed the way is clear for consideration of the budget resolutions. When these have been approved by the committee, a report to this effect is made to the house, and the tax bills are introduced and dealt with in the same manner as all other government financial legislation.

Revenues and expenditures. Administrative procedures for revenues and expenditures are, for the most part, contained in the Financial Administration Act.

The basic requirement for revenues is that all public money shall be paid into the consolidated revenue fund, which is the aggregate of all public money on deposit to the credit of the Receiver General for Canada, who is the supply and services minister. Treasury Board has prescribed detailed regulations for the receipt and deposit of this money. The Bank of Canada and the chartered banks are the custodians of public money. Balances are apportioned among the various chartered banks according to a percentage allocation established by agreement among all the banks and communicated to the finance department by the Canadian Bankers' Association. The daily operating account is maintained with the Bank of Canada and the division of funds between it and the chartered banks takes into account the immediate cash requirements of the government and consideration of monetary policy. The finance minister may purchase and hold securities of, or guaranteed by, Canada and pay for them out of the consolidated revenue fund or may sell such securities and pay the proceeds into the fund. Thus, if cash balances in the fund exceed immediate requirements, they may be invested in interest-earning assets. In addition, the finance minister has established a purchase fund to assist in the orderly retirement of the public debt.

Treasury Board has central control over the budgets of departments and over financial administrative matters generally, principally during the annual consideration of departmental long-range plans and of the estimates. The board also has the right to maintain continuous control over certain types of expenditure to ensure that activities and commitments for the future are held within approved policies, and that the government is informed of and approves any major development of policy or significant transaction that might give rise to public or parliamentary criticism.

To ensure enforcement of the expenditure decisions of Parliament, the government and ministers, the Financial Administration Act provides that no payment shall be made out of the consolidated revenue fund without the authority of Parliament and no charge shall be made against an appropriation except on the requisition of the appropriate minister or a person authorized by him in writing. These requisitions, which must meet certain standards prescribed by Treasury Board regulation, are presented to the receiver general, who makes the payment.

At the beginning of each fiscal year, or whenever Treasury Board may direct, each department submits a division of each vote included in its estimates into allotments. Once approved, they cannot be varied or amended without the consent of the board. To avoid overexpenditures, commitments due to be paid within a fiscal year are recorded and controlled by the departments concerned. Commitments made under contract that will fall due in succeeding years are recorded since the government must be prepared in the future to ask Parliament for appropriations to cover them. Any unspent amounts in the annual appropriations lapse at the end of the fiscal year, but for 30 days subsequent to March 31 payments may be made and charged to the previous year's appropriations for work performed, goods received or services rendered prior to the end of that fiscal year.

Public debt. In addition to collecting and disbursing public money, the government receives and pays out substantial sums in connection with its public debt operations. The finance minister is authorized to borrow money by the issue and sale of securities at whatever rate of interest and under whatever terms and conditions the Governor-in-Council approves. Although new borrowings require specific authority of Parliament, the Financial Administration Act authorizes the Governor-in-Council to approve borrowings, as required, to redeem maturing or called securities. To ensure that the consolidated revenue fund will be sufficient to meet lawfully authorized disbursements, he may also approve the temporary borrowing of necessary sums for periods not exceeding six months. The Bank of Canada acts as the fiscal agent of the government in the management of the public debt.

Accounts and financial statements. Under the Financial Administration Act, Treasury Board may prescribe the manner and form in which the accounts of Canada and the accounts of individual departments shall be kept. Annually, on or before December 31 or, if Parliament is not then sitting, within any of the first 15 days after Parliament resumes, the *Public accounts,* prepared by the receiver general, are laid before the Commons by the minister of finance. The *Public accounts* contain a survey of the financial transactions of the fiscal year ended the previous March 31 and statements of revenues and expenditures, assets and direct and contingent liabilities, together with other accounts and information required to show the financial position of Canada. The statement of assets and liabilities is designed to disclose the net debt, which is determined by offsetting against the gross liabilities only those assets regarded as readily realizable or interest- or revenue-producing. Fixed capital assets, such as government buildings and public works, are charged to budgetary expenditures at the time of acquisition or construction and are shown on the statement of assets and liabilities at a nominal value of $1.00. Monthly financial statements are also published in the *Canada Gazette.*

The auditor general. The government's accounts are subject to an independent examination by the auditor general who is an officer of Parliament. With respect to expenditures, this examination is a post-audit to report whether the accounts have been properly kept, the money spent for the purposes for which it was appropriated by Parliament and as authorized; any audit before payment is the responsibility of the requisitioning department or agency. With respect to revenues, the auditor general must ascertain that all public money is fully accounted for and that the rules and procedures applied ensure an effective check on the assessment, collection and proper allocation of the revenue. With respect to public property, he must satisfy himself that essential records are maintained and that the rules and procedures applied are sufficient to safeguard and control it. The auditor general reports the results of his examination to the Commons, calling attention to any case which he considers should be brought to the notice of the house. He also reports to ministers, the Treasury Board or the government any matter which in his opinion calls for attention so that remedial action may be taken promptly. It is the usual practice to refer the *Public accounts* and the *Auditor General's report* to the House of Commons Standing Committee on Public Accounts, which may review them and report the findings and recommendations to the Commons.

Government employment 3.2.2

Treasury Board (a statutory committee of Cabinet) has overall responsibility for personnel management in the federal public service. It is responsible for development and application of personnel policies, systems and methods to ensure that the people needed to carry out programs effectively are obtained at competitive wages and put to efficient use with consideration for the individual and collective rights of employees.

Under provisions of the amended Financial Administration Act and the Public Service Staff Relations Act, both proclaimed in March 1967, Treasury Board is responsible for the development of policy guidelines, regulations, standards and programs in the areas of classification and pay, conditions of employment, collective bargaining and staff relations, official languages, human resources training, development and utilization, pensions, insurance and other employee benefits and allowances, and other personnel management matters affecting the public service. Treasury Board is also responsible for making recommendations on organization development, human resources planning, the determination and evaluation of training needs and education programs, and standards governing health and safety. It advises departments and agencies on the design and implementation of systems to improve personnel management.

Responsibility for classification and the administration of salaries has, with a few exceptions, been delegated to departments, subject to a monitoring process. Benefit programs and allowance policies approved by the board are designed to give departments maximum responsibility for administration.

Under the system of collective bargaining established by the Public Service Staff Relations Act, Treasury Board is the employer for employees in the public service, except for certain separate employers such as the National Research Council and the National Film Board. The board negotiates collective agreements with unions representing 81 bargaining units and advises departments on their administration. Consultations are held with representatives of bargaining agents, directly or through the National Joint Council, on matters which are not subject to bargaining or which have wide application in the public service. The board determines terms and conditions of employment of employees excluded from collective bargaining, and develops policy guidelines and standards to govern physical working conditions and occupational health and safety. It determines the employer's position on grievances referred to adjudication, and advises or assists departmental management regarding discipline and grievance cases. The board presents the position of the employer in applications for certification by employee organizations and in hearings before the Public Service Staff Relations Board on applications for the exclusion of employees from bargaining units.

The board develops policy guidelines for public service pension, insurance and related programs, co-ordinates their administration and recommends periodic revisions. It negotiates reciprocal pension transfer agreements with other public and private employers. It also studies and proposes means of ensuring compatibility between public service employee benefits and social security programs such as medicare and the Canada and Quebec pension plans.

Public Service Commission. The Public Service Employment Act, which became effective in March 1967, continues the status of the Public Service Commission as an independent agency responsible to Parliament. The commission has the exclusive right and authority to make appointments to and from within the public service. The commission is also empowered to operate staff development and training programs, to assist deputy heads in carrying out training and development and in 1972 was charged to investigate cases of alleged discrimination on grounds of sex, race, national origin, colour or religion in the application and operation of the Public Service Employment Act. Age and marital status were added to these grounds by amendment to the Public Service Employment Act in 1975.

It may establish boards to decide on appeals against appointments made from within the public service and against release or demotion for incompetence or incapacity; to make recommendations on the revocation of appointments improperly made under delegated authority; and to decide on allegations of political partisanship.

The commission grants or withholds approval of applications for leave of absence from public servants who wish to be candidates in federal, provincial or territorial elections and investigates allegations of improper political activities by public servants.

The act authorizes the commission to delegate to deputy heads any of its powers, except those relating to appeals and inquiries. The commission has delegated powers to make appointments in operational and administrative support categories; employing departments are required to use Canada Manpower Centres as their recruitment agency for appointments from outside the public service. Appointing authority has been delegated in the administrative and foreign service, technical, and scientific and professional categories under conditions which preserve the commission's authority as central recruiting agency for the public service of Canada with a few exceptions, that is, those cases where a department is virtually the sole employer of a particular occupational specialty. The commission ensures that appointments made under delegated authority comply with the law and commission policies.

The Public Service Commission is guardian of the merit principle, ensuring that high standards are maintained in the service, consistent with adequate representation of the two official language groups, a bilingual capability to the extent prescribed by the government, equal employment and career development opportunities irrespective of sex, race, national origin, colour or religion, and encouragement of opportunities for the disadvantaged.

Every citizen may apply for positions. Competitive examinations are announced through the news media and posters displayed on public notice boards of major post offices, Canada Manpower Centres, Public Service Commission offices and elsewhere.

The commission's major task — staffing the public service according to merit — is done on an occupational basis. The classification system divides the service into six broad occupational categories which are further divided into groups of occupationally similar jobs. For each major category or group there is a program of recruitment, selection and placement. Appointments are made from within the service except where the commission believes it is in the best interests of the service to do otherwise. Appointments from within are made either through a formal competition or from an employee inventory. The commission's computerized manpower inventory is the primary employee inventory for the executive, scientific and professional, technical and administrative and foreign service categories. Under the Public Service Employment Act, public servants who are candidates in a competition open to all or part of the service may appeal the selections to the Public Service Commission.

When a promotion is made without competition, those who would have been eligible to apply if a competition had been held may appeal. Public servants may also appeal the decision of a deputy head to recommend release or demotion because of incompetence or incapacity.

The Public Service Commission offers interdepartmental courses in government administration, occupational training and management improvement. The commission acts as the consultant and adviser to deputy heads, and training and development facilities are made available to train employees for specific occupations or for promotion in administrative and managerial ranks.

In order that departments may serve the public in accordance with the Official Languages Act, the commission ensures that employees appointed are qualified to meet the linguistic requirements of positions and, in situations where they do not qualify, that incumbents or winners of competitions for bilingual positions receive training in their second official language. Part-time language training is also available to other public servants.

The commission has specific responsibilities in language training, research and the development of selection standards for the linguistic requirements of positions and groups of positions within the federal public service. It must establish both the method of assessing language knowledge and the degree of language knowledge or proficiency of candidates.

Appropriate selection standards are formulated from the decisions of deputy heads based on the linguistic requirements of positions and groups of positions.

Native peoples   3.2.3

Indians   3.2.3.1
The federal Indian affairs and northern development department is responsible for meeting statutory obligations to Indians registered under the Indian Act and for programs approved specifically for them.

The department's local government branch assists with the physical development of Indian communities which involves planning, housing, water, sanitation, electricity and the construction and maintenance of roads on reserves. Indian participation in these activities and in services such as school maintenance, fire and police protection and local government is increasing as band management is extended. For more than 10 years the department has assisted Indians to develop the expertise to manage their communities. Under departmental programs, capital and operating funds are provided to bands. In 1976-77 Indian and Inuit councils administered the expenditure of approximately $147 million in public funds and more than $55 million in band funds on a variety of local government projects. Along with capital, operation and maintenance funds, the department provides core funds to band councils. Core funds are based on total band membership and are used to finance items such as band offices and associated operating costs, support staff, salaries, professional advisory services and honoraria, and travel costs for chiefs and councillors. Additional funds to administer various programs are provided at levels negotiated for each activity.

The role of the federal government in programs for Indians has changed from direct program management at the local level to an advisory and consultative capacity as Indians assume responsibility for managing their own affairs. Emphasis is placed on definition of needs and priorities with the department and Indian bands working jointly, and on development of close consultation in both policy and administrative matters.

Under agreements with the federal government, provincial Indian associations receive funds to administer community development programs planned jointly with government officials, but administered by the associations themselves. These programs are intended to help Indians to improve social, economic and cultural conditions in their communities.

Since the first such agreement with the Manitoba Indian Brotherhood in 1969, others have been made with Indian associations in Nova Scotia, New Brunswick, Ontario, Saskatchewan, BC, the Yukon Territory and the Northwest Territories.

Establishment of the Indian Economic Development Fund (IEDF) in 1970 was a landmark in Indian economic development. The fund formed a financial base for the department's mandate to assist Indians to develop income opportunities and create employment. Capital was provided to Indian businessmen and businesses and basic management skills and technical expertise were made available. From the outset, an important provision of the fund was that Indians be involved in the design and delivery of economic programs.

During the fiscal year 1976-77, the Indian and Inuit program approved 1,021 loans for $17.1 million and guaranteed 51 loans from private sources totalling $2.5 million.

Main estimates during the fiscal year authorized about $15.0 million in grants and contributions. The economic development branch has provided to Indian enterprises 588 grants and contributions for start-up costs totalling $7.9 million. To help Indians establish their own enterprises, the branch, through the IEDF, assisted them with business planning and helped provide other professional and technical services. A total of 2,786 jobs were created in 1976-77. In the first six years of the fund's operation about 8,400 employment man-years were created. Financing was advanced to enterprises in agriculture, forestry, fishing and trapping, construction, real estate, manufacturing, transportation, communications and wholesale-retail operations.

The lands and membership branch is responsible for ensuring that treaty obligations covering lands and memberships are met and that statutory responsibilities under the Indian Act for membership and the administration and management of Indian lands are fulfilled. The branch also helps bands obtain maximum benefits from mineral resources on their own reserves.

Since 1969 the government has provided financial assistance to Indians and Inuit for research to support their claims to traditional interests in lands, and their rights under treaty or the Indian Act. Recognizing its lawful obligations, the government has undertaken negotiations with Indians and Inuit. Claims may be based on traditional use and occupancy of land in areas where the Indian interest has not been extinguished by treaty or superseded by law (comprehensive claims), or they may be based on interpretation of treaties and legislation, or the administration of assets (specific claims).

In 1974 an office of native claims was established within the department to represent the government in negotiating claims settlements; to advise the minister on the further development of claims policy and the processes for settling both comprehensive and specific claims; and to co-ordinate the government's response to claim proposals.

Comprehensive claims are based on the traditional use and occupancy of land. With such claims the government tries to redefine in contemporary terms the relationship between native people and the government. The settlement process for these claims has been under way since the James Bay and Northern Quebec Agreement in 1975. An agreement was also signed on January 31, 1978 with the Naskapis of Schefferville, providing them with rights and benefits similar to those gained by the James Bay Crees and Inuit of Quebec under the James Bay agreement, but adapted to the special circumstances of the Naskapis. Other major activity areas of comprehensive claims include the Yukon Territory, the Northwest Territories, British Columbia and Labrador.

Specific claims cover every aspect of the government's past administration of band lands and other assets, and the fulfilment of the terms of treaties. A major preoccupation in 1977 was treaty land entitlements in the Prairie provinces, the result of some bands not having received their full entitlement under treaties signed between 1871 and 1906. In Saskatchewan, the Federation of Saskatchewan Indians and the federal and provincial governments reached agreement in August 1977 on basic principles for settling outstanding treaty land entitlements. Efforts to reach similar agreements with Alberta and Manitoba continue.

Another specific claim was settled in December 1977 with the Manitoba Northern Flood Agreement which provides rights and benefits to five bands as compensation for the adverse effects of Manitoba's Lake Winnipeg and Churchill River diversion project. Other ongoing specific claims issues include lands cut off from reserves in BC since 1916, and alleged improper alienation of reserve land.

Inuit

Canada's 18,000 Inuit, most of whom live in the Northwest Territories, Quebec and Labrador, are the concern of the federal Indian affairs and northern development department, the government of the Northwest Territories and provincial governments.

From 1966 to 1975 a northern rental housing program provided 1,505 three-bedroom houses for the Inuit. In April 1975 the Northwest Territories Housing Corporation became responsible for Inuit housing, and new accommodation since then has been supplied under National Housing Act building programs.

The national defence department offers employment at its station at Alert to civilian Inuit. A student centre for Inuit was established in Ottawa in 1974 and an Inuit orientation centre was planned for 1976. Inuit are involved in a departmental on-the-job training program to place them in middle management positions related to resource development and the environment. A special northern unit has been established by the Public Service Commission to improve employment and career possibilities for northern native people.

The Inuit Tapirisat of Canada (Eskimo Brotherhood) was founded in 1971 with financial assistance from the secretary of state department. Affiliated with Inuit Tapirisat are the Committee of Original Peoples Entitlement (COPE), serving native people in the Mackenzie Delta and the Western Arctic, the Labrador Inuit Association (LIA), and the Northern Quebec Inuit Association (NQIA). Other regional associations in the Northwest Territories have been established by Inuit Tapirisat in the Central Arctic, Baffin and Keewatin regions to facilitate local participation in domestic affairs.

In early 1976 the Tapirisat presented the Government of Canada *Nunavut*, a proposal for land claims settlement. In preparing this document, the Inuit Tapirisat did considerable land claim research, and produced the Inuit land-use and occupancy study, an environmental, geographical and historical work which the government has published.

Similar studies were undertaken by the Indian Brotherhood and the Métis Association in the Mackenzie region of the Northwest Territories, and by the Labrador Inuit Association. Northern native associations were provided financial assistance to participate in matters of northern development, such as the Mackenzie Valley pipeline inquiry.

Additional programs included an Inuit language commission, established to make recommendations on the revision and standardization of Inuktitut (the Inuit language) orthographies; the publication of a layman's guide to Canadian law entitled *Inuit and the law;* legal services centres in Frobisher Bay, NWT and Happy Valley, Labrador, to provide counsel and guidance for the Inuit; the support of an Inuit film-making society in Frobisher Bay formed to produce native language programs for broadcast on the CBC northern service television; and the development of a syllabic character typewriter element to meet the increasing need for written material in Inuktitut.

Inuit art and crafts are promoted by preparing interpretive exhibits for circulation to museums, universities and other institutions in Canada and abroad. Artists are protected against copyright infringement and competitive mass reproductions through a program of information to artists and the public, promotion of the "Canada Eskimo Art" trademark and support of legal action where infringements occur. Information on art and culture is conveyed to the public through booklets, articles and lectures.

The Inuit Cultural Institute based at Eskimo Point, NWT is a focal point for cultural concerns and programs related to traditional and present-day Inuit life. The institute also administers and oversees the work of the Inuit language commission.

As a result of the search for oil, gas and minerals in the Arctic, many Inuit are finding employment in petroleum and related industries. The petroleum industry reports for the 1974/1975 seasons that 761 northern residents accepted employment. Studies have been undertaken with a view to increasing native involvement in the mining industry, which is showing an overall decline. However, many Inuit still live by their traditional skills of hunting, trapping and fishing. One of the most successful enterprises is the production and sale of Inuit artwork — stone, bone and ivory sculpture and graphics. The industry is expanding and co-operatives are run by the Inuit.

3.2.4 Departments, boards, commissions and corporations

In Canada the work of government is conducted by federal departments, special boards, commissions and corporations owned or controlled by the Government of Canada, as well as several corporations in which the government holds a minority interest. Of the corporations owned by the Government of Canada, the Crown corporation mode of organization is the most common. The government has resorted to Crown corporations with increasing frequency to administer and manage public services, many of which require the combination of business enterprise and public accountability. The historical evolution of Crown corporations is described in the government's proposals on the control, direction and accountability of Crown corporations published in August 1977. Chapter I of that paper describes the historical and constitutional background of the Crown corporation form. Copies are available free of charge from the Privy Council office in Ottawa. Part VIII of the Financial Administration Act (RSC 1970, c.F-10) provides a uniform system of financial and budgetary control and of accounting, auditing and reporting for the majority of Crown corporations. In addition, that legislation defines a Crown corporation as a corporation that is ultimately accountable, through a minister, to Parliament for the conduct of its affairs and establishes three classes of Crown corporation — departmental, agency and proprietary.

Departmental corporations. A departmental corporation is defined as a corporation that is a servant or agent of Her Majesty in right of Canada and is responsible for administrative, supervisory or regulatory services of a governmental nature. The following corporations are classified as departmental corporations in Schedule B to the Financial Administration Act:

Agricultural Stabilization Board
Atomic Energy Control Board (The government has proposed through
   the Nuclear Control and Administration Bill tabled November 24, 1977,
   that the AECB be removed from the schedules.)
Director of Soldier Settlement
The Director, The Veterans' Land Act
Economic Council of Canada
Fisheries Prices Support Board
Medical Research Council
National Museums of Canada
National Research Council
Science Council of Canada
Canada Employment and Immigration Commission.

Agency corporations. An agency corporation is defined as a corporation that is an agent of Her Majesty in right of Canada and is responsible for the management of trading or service operations on a quasi-commercial basis or for the management of procurement, construction or disposal activities on behalf of Her Majesty in right of Canada. The following corporations are classified as agency corporations in Schedule C to the Financial Administration Act:

Atomic Energy of Canada Limited
Canadian Arsenals Limited
Canadian Commercial Corporation
Canadian Dairy Commission
Canadian Film Development Corporation
Canadian Livestock Feed Board
Canadian National (West Indies) Steamships Limited (inactive)
Canadian Patents and Development Limited
Canadian Saltfish Corporation
Crown Assets Disposal Corporation
Defence Construction (1951) Limited
Loto Canada
National Battlefields Commission
National Capital Commission
National Harbours Board

Northern Canada Power Commission
Royal Canadian Mint
Uranium Canada Limited.

Proprietary corporations. A proprietary corporation is defined as a corporation that is responsible for the management of lending or financial operations, or for the management of commercial or industrial operations involving the production of or dealing in goods and the supplying of services to the public, and is ordinarily required to conduct its operations without parliamentary appropriations. The following corporations are classified as proprietary corporations in Schedule D to the act:

Air Canada
Canada Deposit Insurance Corporation
Canadian Broadcasting Corporation
Cape Breton Development Corporation
Central Mortgage and Housing Corporation
Eldorado Aviation Limited
Eldorado Nuclear Limited
Export Development Corporation
Farm Credit Corporation
Federal Business Development Bank
Federal Mortgage Exchange Corporation (inactive)
Freshwater Fish Marketing Corporation
National Railways, as defined in the Canadian National–Canadian Pacific Act
Northern Transportation Company Limited
Petro-Canada
Pilotage Authorities
    Atlantic Pilotage Authority
    Great Lakes Pilotage Authority
    Laurentian Pilotage Authority
    Pacific Pilotage Authority
St. Lawrence Seaway Authority
Seaway International Bridge Corporation Limited (formerly Cornwall International Bridge
    Company Limited)
Teleglobe Canada.

Departmental corporations are governed by the provisions of the Financial Administration Act that apply to departments generally. Agency and proprietary corporations are subject only to the provisions of Part VIII of the act; if there is any inconsistency between its provisions and those of any other act applicable to a corporation, the latter prevail. The same part provides for the regulation and process of approval of corporation budgets and the control of bank accounts, turning over surplus money to the receiver general, providing loans for limited working-capital purposes, awarding contracts and establishing reserves, keeping and auditing accounts, and preparing financial statements and reports for submission to Parliament through the appropriate minister.

A further form of control is exercised by Parliament through the power to vote financial assistance to a corporation, which may secure financing through parliamentary grants, loans or advances, by the issue of capital stock to the government, or by borrowings from the capital markets, often with a government guarantee. Most corporations in Schedule D endeavour to finance themselves through government-held equity, repayable loans, and charges for goods and services.

Unclassified corporations. The following government-owned corporations are not listed in schedules to the Financial Administration Act but are governed by their own special act, letters patent or articles of incorporation: the Bank of Canada, the Canada Council, the Canadian National Railways Securities Trust, the Canadian Wheat Board, and the National Arts Centre Corporation. The only provision of the Financial Administration Act to which they are subject is that governing the appointment of auditors, although the Governor-in-Council has the power in some instances to add an unclassified corporation to one of the schedules to the Financial Administration Act.

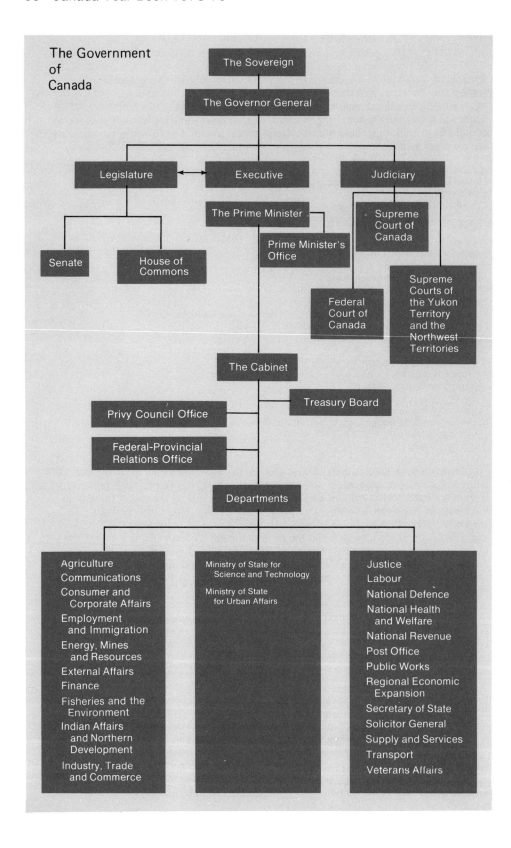

The Government of Canada

Other corporations. The federal government has established or assisted in the establishment of a number of corporations in which it holds a portion of the equity. In most cases, private sector investors hold the remaining shares; in several cases shares are held by provincial or other governments. These corporations, known as mixed enterprises, have been established either by a special act of Parliament, for example, the Canada Development Corporation or Telesat Canada, or by letters patent or articles of incorporation, for example, Panarctic Oils Ltd. Such corporations are not listed in the schedules of the Financial Administration Act and are not subject to its general provisions.

In May 1977, Treasury Board's secretariat tabled before the public accounts committee a list of all corporations owned or controlled by the Government of Canada. The list included 53 corporations that are given in Schedule B, C or D of the Financial Administration Act (12 Bs, 19 Cs, 22 Ds), 27 corporations wholly owned by the Government of Canada, yet not listed in any schedule, 24 mixed enterprises, 152 subsidiaries and their subsidiaries, 93 associated corporations and 17 other corporate entities for a total of 366 corporations owned or controlled by the federal government.

An alphabetical list of federal ministers and the departments and other agencies for which they report to Parliament follows (as in effect in May 1978). Brief descriptions of the functions of many of these government organizations and related agencies will be found in Appendix 1. The accompanying organization chart illustrates the federal structure to the departmental level.

Minister of Agriculture
Department of Agriculture
Agricultural Products Board
Agricultural Stabilization Board
Canadian Dairy Commission
Canadian Grain Commission
Canadian Livestock Feed Board
Crop Insurance Administration
Farm Credit Corporation
National Farm Products Marketing Council

Minister of Communications
Department of Communications
Canadian Radio-television and
   Telecommunications Commission
Teleglobe Canada

Minister of Consumer and Corporate Affairs
Department of Consumer and Corporate
   Affairs
Canadian Consumer Council
Copyright Appeal Board
Patent Appeal Board
Restrictive Trade Practices Commission

Minister of Employment and Immigration
Department of Employment and Immigration
Canada Employment and Immigration
   Advisory Council
Canada Employment and Immigration
   Commission
Immigration Appeal Board

Minister of Energy, Mines and Resources
Department of Energy, Mines and Resources
Atomic Energy Control Board
Atomic Energy of Canada Limited
Board of Examiners for Dominion Land
   Surveyors

Canadian Permanent Committee on
   Geographical Names
Columbia River Treaty Permanent Engineering
   Board
Eldorado Aviation Limited
Eldorado Nuclear Limited
Energy Supplies Allocation Board
Interprovincial and Territorial Boundary
   Commissions
National Energy Board
Oil and Gas Committee
Petro-Canada
Uranium Canada Limited

Secretary of State for External Affairs
Department of External Affairs
Canadian International Development Agency
Foreign Claims Commission
International Boundary Commission
International Development Research Centre
International Joint Commission
Roosevelt Campobello International Park
   Commission

Minister of State for Federal-Provincial Relations
Federal-Provincial Relations Office

Minister of Finance
Department of Finance
Anti-dumping Tribunal
Bank of Canada
Canada Deposit Insurance Corporation
Department of Insurance
Tariff Board
Anti-Inflation Board

Minister of Fisheries and the Environment
Department of the Environment
Canadian Saltfish Corporation

Environmental Assessment Panel
Fisheries Prices Support Board
Fisheries Research Board of Canada
Freshwater Fish Marketing Corporation
International Fisheries Commissions
Minister of State (Environment)

*Minister of Indian Affairs and Northern
Development*
Department of Indian Affairs and Northern
Development
Government of the Northwest Territories
Government of the Yukon Territory
Historic Sites and Monuments Board of Canada
National Battlefields Commission
Northern Canada Power Commission
Northwest Territories Water Board
Oil and Gas Committee
Yukon Territory Water Board

Minister of Industry, Trade and Commerce
Department of Industry, Trade and Commerce
Export Development Corporation
Federal Business Development Bank
Foreign Investment Review Agency
General Adjustment Assistance Board
Loto Canada
Machinery and Equipment Advisory Board
Metric Commission Canada
National Design Council
Standards Council of Canada
Statistics Canada
Textile and Clothing Board
Minister of State (Small Business)

Minister of Justice and Attorney General of Canada
Department of Justice
Anti-Inflation Appeal Tribunal
Law Reform Commission
Tax Review Board

Minister of Labour
Department of Labour
Canada Labour Relations Board
Merchant Seamen Compensation Board

Minister of National Defence
Department of National Defence
Defence Construction (1951) Limited
Defence Research Board
National Emergency Planning Establishment

Minister of National Health and Welfare
Department of National Health and Welfare
Advisory Council on the Status of Women
Medical Research Council
National Advisory Council on Fitness and
Amateur Sport
National Council of Welfare
Office of the Co-ordinator, Status of Women
Pension Appeals Board
Minister of State (Fitness and Amateur Sport)

Minister of National Revenue
Department of National Revenue (Customs
and Excise)
Department of National Revenue (Taxation)
Office of the Administrator, Anti-Inflation Act

Postmaster General
Post Office Department

Minister of Public Works
Department of Public Works

Minister of Regional Economic Expansion
Department of Regional Economic Expansion
Atlantic Development Council
Canadian Council on Rural Development
Cape Breton Development Corporation
Prairie Farm Rehabilitation Administration
Regional Development Incentives Board

Minister of State for Science and Technology
Ministry of State for Science and Technology
Canadian Patents and Development Limited
National Research Council of Canada
Science Council of Canada

Secretary of State of Canada
Department of the Secretary of State
Canada Council
Canadian Broadcasting Corporation
Canadian Film Development Corporation
National Arts Centre Corporation
National Film Board
National Library of Canada
National Museums of Canada
Public Archives of Canada
Minister of State (Multiculturalism)

Solicitor General
Department of the Solicitor General
Canadian Penitentiary Service
National Parole Board
Royal Canadian Mounted Police
Correctional Investigator

Minister of Supply and Services
Department of Supply and Services
Canadian Arsenals Limited
Canadian Commercial Corporation
Canadian Government Specifications Board
Crown Assets Disposal Corporation
Royal Canadian Mint
Office of the Custodian of Enemy Property

Minister of Transport
Department of Transport
Air Canada
Canadian National Railways
Canadian Transport Commission
Canadian Wheat Board
Maritime Pollution Claims Fund
National Harbours Board
Northern Transportation Company Limited

Pilotage Authorities
  Atlantic Pilotage Authority
  Great Lakes Pilotage Authority, Limited
  Laurentian Pilotage Authority
  Pacific Pilotage Authority
St. Lawrence Seaway Authority
Seaway International Bridge Corporation
  Limited

President of Treasury Board
Treasury Board Secretariat

Minister of State for Urban Affairs
Ministry of State for Urban Affairs
Central Mortgage and Housing Corporation
National Capital Commission

Minister of Veterans Affairs
Department of Veterans Affairs
Army Benevolent Fund Board
Bureau of Pensions Advocates
Canadian Pension Commission
Pension Review Board
War Veterans Allowance Board

Federal Identity Program   3.2.5

The use of identifying titles as alternatives to the statute names of departments, for example, Labour Canada, reflects the policy of the Federal Identity Program (FIP) stemming from a 1969 study conclusion that visual communications of the federal government were in urgent need of improvement. The heart of the FIP is the consistent application of specific identifying symbols by all organizations of the government. These symbols, organized into systematic formats with distinctive typography and colours form the visual identity of the government.

Policy direction for the program emanates from a Cabinet committee on science, culture and information. Program co-ordination is the responsibility of the Treasury Board secretariat's administrative standards division. An interdepartmental advisory committee provides advice on implementation details and the management of the program.

The departments, agencies and other government organizations to which the program applies are required not only to implement it, but to assist with its further development through active participation. These bodies are now in the process of adopting identifying titles where appropriate. These titles, such as Revenue Canada for Department of National Revenue and Health and Welfare Canada for Department of National Health and Welfare, will not replace the formal names which may be required for contracts, federal-provincial agreements and other legal applications. However, on such documents, the title shall appear as the principal identifying device.

Provincial and territorial governments   3.3

Provincial governments   3.3.1

In each of the provinces, the Queen is represented by a lieutenant-governor appointed by the Governor General-in-Council. The lieutenant-governor acts on the advice and with the assistance of his ministry or an executive council which is responsible to the legislature and resigns office under circumstances similar to those described concerning the federal government.

The legislature of each province is unicameral, consisting of the lieutenant-governor and a legislative assembly. The assembly is elected by the people for a statutory term of five years but may be dissolved within that period by the lieutenant-governor on the advice of the premier of the province.

Sections 92, 93 and 95 of the British North America Act, 1867 (Br. Stat. 1867, c.3 and amendments) assign legislative authority in certain areas to the provincial governments (see Chapter 2).

Details regarding qualifications and disqualifications of the franchise are contained in the elections act of each province. In general, every person at a specified age who is a Canadian citizen or (in certain provinces) other British subject, who complies with certain residence requirements in the province and the electoral district of polling and who falls under no statutory disqualifications, is entitled to vote. Persons can vote in Prince Edward Island, New Brunswick, Quebec, Ontario, Manitoba, Saskatchewan and Alberta at age 18 and in Newfoundland, Nova Scotia and British Columbia at 19.

3.3.1.1   Newfoundland

The government of Newfoundland has a lieutenant-governor, an executive council and a House of Assembly made up of 51 members elected for a term not to exceed five years. On July 4, 1974 the Honourable Gordon A. Winter became the lieutenant-governor. The 37th Legislature in the history of Newfoundland and the ninth since Confederation, elected September 16, 1975, comprised 30 Progressive Conservatives, 16 Liberals, four members of the Liberal Reform party and one Independent Liberal; as at November 1, 1977 party standings had changed to 30 Progressive Conservatives, 19 Liberals, one independent member with one seat vacant.

The premier receives a salary of $22,555 and cabinet ministers $14,245 per annum, plus a car allowance of $3,180, sessional indemnity of $11,970 and a travelling expense allowance of $5,985. Each member of the House of Assembly receives a sessional indemnity of $11,970 plus a travelling and expense allowance of $5,985. The leader of the opposition receives an additional allowance of $14,245.

The Executive Council of Newfoundland in November 1977

Premier, Hon. F.D. Moores
President of the Council, Hon. Dr. T.C. Farrell
Minister of Justice and Minister responsible for Intergovernmental Affairs, Hon. T. Alex Hickman
Minister of Finance and President of the Treasury Board, Hon. C.W. Doody
Minister of Education, Hon. Wallace House
Minister of Transportation and Communications, Hon. James Morgan
Minister of Social Services, Hon. R.C. Brett
Minister of Health and Minister of Rehabilitation and Recreation, Hon. Harold Collins

Minister of Industrial Development and Rural Development, Hon. John Lundrigan
Minister of Municipal Affairs and Housing, Hon. J.W. Dinn
Minister of Public Works and Services and Minister of Labour and Manpower, Hon. Joseph G. Rousseau
Minister of Consumer Affairs and Environment, Hon. A.J. Murphy
Minister of Fisheries, Hon. Walter Carter
Minister of Tourism, Hon. T.V. Hickey
Minister of Mines and Energy, Hon. A.B. Peckford
Minister of Forestry and Agriculture, Hon. Edward Maynard.

3.3.1.2   Prince Edward Island

The government of Prince Edward Island consists of a lieutenant-governor, an executive council and a Legislative Assembly. The Honourable Gordon L. Bennett was appointed lieutenant-governor effective October 24, 1974. The Legislative Assembly has 32 members from 16 electoral districts who may serve for a statutory term not exceeding five years. Each district elects two representatives. The 53rd Assembly elected April 29, 1974 consisted of 26 Liberals and six Progressive Conservatives; as at November 1, 1977 party standings had changed to 24 Liberals and eight Progressive Conservatives.

A member of the assembly receives $7,000 per annum and an additional $3,500 tax-free for travelling and other expenses incurred in attending sessions and representing his district. In addition the premier receives a salary of $24,000, a cabinet minister $14,500 and a minister without portfolio $14,500. The speaker of the assembly is paid an additional indemnity of $2,000 and his tax-free allowance is $1,000; higher sessional indemnities and allowances are also available to the deputy speaker in the amounts of $1,000 and $500, respectively, and to the leader of the opposition in the amount of $10,000. All indemnities and allowances accrue from the date of election to the legislature and are paid monthly. No sessional indemnity or expenses are paid for any special session of the legislature.

The Executive Council of Prince Edward Island in November 1977

Premier, President of the Executive Council, Minister of Justice, Attorney and Advocate General, and Minister responsible for Cultural Affairs, Hon. Alexander B. Campbell

Minister of Development and Minister of Industry and Commerce, Hon. John H. Maloney
Minister of Public Works and Minister of Highways, Hon. Bruce L. Stewart

Provincial Secretary, Hon. Arthur J. MacDonald

Minister of Education and Minister of Finance, Hon. W. Bennett Campbell

Minister of Municipal Affairs, Minister of Tourism, Parks and Conservation and Minister of the Environment, Hon. Gilbert Clements

Minister of Health and Minister of Social Services, Hon. Catherine Callbeck

Minister of Fisheries and Minister of Labour, Hon. George Henderson

Minister of Agriculture and Forestry, Hon. A.E. Ings

Minister without portfolio (and responsible for the PEI Housing Corporation), Hon. George Proud.

Nova Scotia 3.3.1.3

The government of Nova Scotia consists of a lieutenant-governor, an executive council and a House of Assembly. The Honourable Clarence L. Gosse became lieutenant-governor on October 1, 1973. The legislature has 46 members elected for a maximum term of five years. On April 2, 1974, 31 Liberals, 12 Progressive Conservatives and three New Democrats were elected to the province's 51st Legislature and 28th since Confederation.

Each member of the House of Assembly is paid an annual indemnity of $9,600 and an annual expense allowance of $4,800. In addition to the amounts to which they are entitled under the House of Assembly Act, the premier of the province receives an annual salary of $25,000, all other cabinet members $21,000 or less as the Governor-in-Council may determine, the leader of the opposition $21,000, the speaker $11,000, the deputy speaker $3,500, and any member occupying the recognized position of leader of a recognized party other than the premier and leader of the opposition $6,000.

The Executive Council of Nova Scotia in November 1977

Premier and President of the Executive Council, Hon. Gerald A. Regan

Deputy Premier and Minister of Finance, Hon. Peter Nicholson

Minister of Recreation, Hon. A. Garnet Brown

Attorney General and Minister in charge of administration of the Human Rights Act, Hon. Leonard L. Pace

Minister of Public Works and Minister in charge of administration of the Liquor Control Act, Hon. Benoit Comeau

Minister of Mines and Minister in charge of the Nova Scotia Energy Council, Hon. J. William Gillis

Minister of Highways, Hon. J. Fraser Mooney

Minister of Municipal Affairs, Hon. Glen M. Bagnell

Minister of Tourism, Hon. Maurice E. DeLory

Minister of Agriculture and Marketing and acting Chairman of the Treasury Board, Hon. John Hawkins

Minister of Education, Hon. George M. Mitchell

Provincial Secretary and Minister in charge of administration of the Communications and Information Act, Hon. Harold M. Huskilson

Minister of Development and Minister in charge of administration of the Civil Service Act, the Civil Service Joint Council Act and the Research Foundation Corporation Act, Hon. Alexander M. Cameron

Minister of Labour and Minister in charge of administration of the Housing Development Act, Hon. Walter R. Fitzgerald

Minister of Social Services and Minister responsible for Status of Women, Hon. William M. MacEachern

Minister of Health, Minister in charge of administration of the Drug Dependency Act and Registrar General, Hon. Maynard C. MacAskill

Minister of Lands and Forests, Minister of the Environment and Minister in charge of administration of the EMO (NS) Act and Regulations, Hon. Vincent J. MacLean

Minister of Consumer Affairs and Minister in charge of administration of the Residential Tenancies Act, Hon. Guy A.C. Brown

Minister of Fisheries, Hon. Daniel S. Reid.

New Brunswick 3.3.1.4

The government of New Brunswick has a lieutenant-governor, an executive council and a Legislative Assembly. The Honourable H.J. Robichaud was sworn in October 8, 1971, as lieutenant-governor. The legislature elected November 18, 1974, the 48th in New Brunswick's history and 27th since Confederation, had 58 members, including 33 Progressive Conservatives and 25 Liberals, elected for a statutory term not to exceed five years.

The premier receives $25,000 per annum in addition to the salary for any other portfolio he may hold. Each cabinet minister is paid $16,000; each member of the Legislative Assembly receives $8,000 and a $2,500 allowance for expenses. The leader of the opposition receives an additional $16,000. The speaker and deputy speaker are paid $5,000 and $2,500, respectively, in addition to the regular indemnity.

The Executive Council of New Brunswick in November 1977

Premier, Hon. Richard Hatfield

Acting Provincial Secretary and Minister of Justice, Hon. Rodman E. Logan, QC

Chairman of the New Brunswick Electric Power Commission, Hon. G.W.N. Cockburn

Minister of Finance, Hon. Lawrence Garvie

Chairman of Treasury Board, Hon. Jean Maurice Simard

Minister of Fisheries, Hon. Omer Léger

Minister of Supply and Services, Hon. Harold Fanjoy

Minister of Transportation, Hon. Wilfred Bishop

Minister of Natural Resources, Hon. Roland Boudreau

Minister of Agriculture and Rural Development, Hon. Malcolm MacLeod

Minister of Health, Hon. Brenda Robertson

Minister of Social Services, Hon. Leslie Hull

Acting Minister of Labour and Manpower, Hon. Rodman E. Logan, QC

Minister of Education, Hon. Charles Gallagher

Minister of Municipal Affairs, Hon. Horace B. Smith

Minister of Commerce and Development, Hon. Gerald S. Merrithew

Minister of Youth, Recreation and Cultural Resources, Hon. Jean-Pierre Ouellet

Minister of Tourism and Environment, Hon. Fernand Dubé.

3.3.1.5   Quebec

In Quebec, legislative and executive powers are vested in the National Assembly and an executive council. As the representative of the Crown, the lieutenant-governor plays a role in the functioning of both branches. (The Honourable Jean-Pierre Côté assumed the office on April 27, 1978 replacing The Honourable Hugues Lapointe.)

The National Assembly consists of 110 members elected for a maximum term of five years. Party standings as at November 1, 1977, following the general election of November 15, 1976, were as follows: Parti Québécois 71, Liberals 26, Union nationale 11, Ralliement des créditistes, one, and Independent, one.

All members receive an annual indemnity of $27,800 and a tax-free representation allowance of $7,000. In addition, the Executive Council Act and the Legislative Assembly Act provide for additional taxable allowances for the prime minister ($41,700), ministers ($30,580), ministers without portfolio ($30,580), the speaker of the National Assembly ($30,580), deputy speakers ($13,900), parliamentary assistants ($8,340), the leader of the official opposition ($30,580), leaders of other recognized parties ($12,510), the house leader of the official opposition ($12,510), house leaders of other recognized parties ($11,120), chief government whip ($12,510), chief whip of the official opposition ($8,340), whips of other recognized parties and deputy whips ($6,950) and the chairmen of elected commissions ($4,170). Internal regulations also provide for allowances for specified travelling by a member, for maintaining an office in his constituency and for a second residence in Quebec in cases where the member represents a riding outside the capital area.

Members of the Executive Council of Quebec in November 1977

Prime Minister, Hon. René Lévesque

Vice-Prime Minister and Minister of Education, Hon. Jacques-Yvan Morin

House Leader and Minister of State for Electoral and Parliamentary Reform, Hon. Robert Burns

Minister of Intergovernmental Affairs, Hon. Claude Morin

Minister of Finance and Revenue Minister, Hon. Jacques Parizeau

Minister of State for Cultural Development, Hon. Camille Laurin

Minister of State for Social Development, Hon. Pierre Marois

Minister of State for Economic Development, Hon. Bernard Landry

Minister of State for Planning, Hon. Jacques Léonard

Minister of Justice, Hon. Marc-André Bédard

Minister of Transport, Hon. Lucien Lessard

Minister responsible for the Environment, Hon. Marcel Léger

Minister responsible for Youth, Sports and Leisure, Hon. Claude Charron

Minister responsible for Energy, Hon. Guy Joron

Minister of Consumer Affairs, Co-operatives and Financial Institutions, Hon. Lise Payette

Minister of Agriculture, Hon. Jean Garon

Minister of Social Affairs, Hon. Denis Lazure

Minister of Municipal Affairs, Hon. Guy Tardif

Minister of Labour and Manpower, Hon. Pierre-Marc Johnson

Minister of Immigration, Hon. Jacques Couture

Minister of Cultural Affairs and Minister of Communications, Hon. Louis O'Neill

Minister of Natural Resources and Minister of Lands and Forests, Hon. Yves Bérubé

Minister of Industry and Commerce, Hon. Rodrigue Tremblay

Minister of Tourism, Fish and Game, Hon. Yves Duhaime

Public Service Minister and Vice-President of the Treasury Board, Hon. Denis De Belleval

Minister of Public Works and Supply, Hon. Jocelyne Ouellette.

Ontario   3.3.1.6

The government of Ontario consists of a lieutenant-governor, an executive council and a Legislative Assembly. In April 1974 the Honourable Pauline McGibbon took office as lieutenant-governor. The Legislative Assembly is composed of 125 members elected for a statutory term not to exceed five years. At the provincial election June 9, 1977, 58 Progressive Conservatives, 34 Liberals and 33 New Democrats were elected to the province's 31st Legislature.

In addition to the regular ministries are the following provincial agencies: the Niagara Parks Commission, the Ontario Municipal Board, Ontario Hydro, the St. Lawrence Development Commission, the Ontario Northland Transportation Commission, the Liquor Control Board and the Liquor Licence Board.

Under the provisions of the Legislative Assembly Act (RSO 1970, c.240 as amended) each member of the assembly is paid an annual indemnity of $15,000 and an expense allowance of $7,500. In addition, the speaker receives a special annual indemnity of $9,000, the chairman of the committee of the whole house $5,000 and the leader of the opposition a salary of $18,000. Each member of the cabinet having charge of a ministry receives the ordinary indemnity as a member of the legislature in addition to his salary as a minister of the Crown. The salary provided in the Executive Council Act for the premier is $25,000 and for a cabinet minister having charge of a ministry $18,000. The leader of the opposition receives a representation allowance of $3,000 per annum. Each minister without portfolio receives an annual salary of $7,500.

The Executive Council of Ontario in November 1977

Premier and President of the Council, Hon. William G. Davis

Minister of Culture and Recreation and Deputy Premier, Hon. Robert Welch

Chairman of Management Board of Cabinet, Hon. James A.C. Auld

Provincial Secretary for Resources Development, Hon. René Brunelle

Minister of Education, Hon. Thomas L. Wells

Minister of the Environment, Hon. George A. Kerr

Minister of Northern Affairs, Hon. Leo Bernier

Minister of Transportation and Communications, Hon. James W. Snow

Provincial Secretary for Social Development, Hon. Margaret Birch

Minister of Industry and Tourism, Hon. Claude Bennett

Treasurer of Ontario, Minister of Economics and Intergovernmental Affairs, Hon. W. Darcy McKeough

Minister of Agriculture and Food, Hon. William Newman

Minister of Natural Resources, Hon. Frank S. Miller

Minister of Housing, Hon. John R. Rhodes

Minister of Health, Hon. Dennis R. Timbrell

Provincial Secretary for Justice and Solicitor General, Hon. John P. MacBeth

Minister of Revenue, Hon. Margaret Scrivener

Minister of Colleges and Universities, Hon. Harry C. Parrott

Minister of Energy, Hon. James A. Taylor

Minister of Labour, Hon. Bette M. Stephenson

Attorney General, Hon. Roy McMurtry

Minister without portfolio and Chairman of Cabinet, Hon. Lorne C. Henderson

Minister of Community and Social Services, Hon. Keith C. Norton

Minister of Correctional Services, Hon. Frank Drea

Minister of Consumer and Commercial Relations, Hon. Larry Grossman

Minister of Government Services, Hon. George McCague.

3.3.1.7 Manitoba

In addition to a lieutenant-governor, Manitoba has an executive council composed of 15 members and a Legislative Assembly of 57 members elected for a maximum term of five years. The Honourable Francis L. Jobin became lieutenant-governor on March 15, 1976. In the general election October 11, 1977, 33 Progressive Conservatives, 23 New Democrats and one Liberal were elected to the 31st Legislature.

The premier of the province is paid a salary of $16,600 a year and each of the other members of the cabinet $15,600. Members of the legislature were each paid a sessional indemnity of $12,199 and a tax-free expense allowance of $6,099 for the fiscal year ended March 31, 1978. Each member attending the session receives an additional allowance of $900 for expenses incidental to the discharge of his duties as member. The leader of the opposition is paid $15,600. The speaker of the Legislative Assembly receives an additional indemnity of $5,000 and expenses not exceeding $1,500 in aggregate. The deputy speaker receives an additional indemnity of $2,500 and expenses not exceeding $500 in aggregate. Members required to live away from home receive a per diem allowance of $25 from the opening of the session to prorogation excepting days during an adjournment for a period of four or more continuous days.

The Executive Council of Manitoba in November 1977

Premier, President of the Executive Council, Minister of Dominion-Provincial Relations and Chairman of Management Committee of Cabinet (Treasury Board), Hon. Sterling Lyon

Minister of Finance, Minister charged with the administration of the Manitoba Hydro Act and Chairman of the Manitoba Energy Council, Hon. Donald Craik

Minister of Consumer, Corporate and Internal Services, Minister of Co-operative Development, Minister responsible for Manitoba Telephone System and Communications and Minister responsible for administration of Manitoba Lotteries Act, Hon. Ed McGill

Minister without portfolio and Government House Leader, Hon. Warner Jorgenson

Minister of Health and Social Development and Minister responsible for Corrections and Rehabilitation, Hon. L.R. (Bud) Sherman

Minister without portfolio and Co-Chairman of Task Force on Government Organization and Economy, Hon. Sidney Spivak

Minister of Public Works, Minister of Highways and Minister responsible for Manitoba Public Insurance Corporation, Hon. Harry Enns

Minister without portfolio and Minister responsible for Manitoba Housing and Renewal Corporation, Hon. J. Frank Johnston

Minister of Agriculture, Hon. James E. Downey

Minister of Education and Minister of Continuing Education and Manpower, Hon. Keith A. Cosens

Attorney-General, Keeper of the Great Seal, Minister of Municipal Affairs, Minister for Urban Affairs and Minister responsible for administration of the Liquor Control Act, Hon. Gerald W.J. Mercier

Minister of Industry and Commerce, Minister of Tourism, Recreation and Cultural Affairs and Minister responsible for the administration of the Manitoba Development Corporation Act, Hon. Robert Banman

Minister of Labour, Minister responsible for the Civil Service Act, the Civil Service Superannuation Act, the Public Servants Insurance Act and the Pensions Benefits Act, Hon. Norma L. Price

Minister of Northern Affairs and Minister of Renewable Resources and Transportation Services, Hon. Ken MacMaster

Minister of Mines, Resources and Environmental Management, Hon. Brian Ransom.

3.3.1.8 Saskatchewan

The government of Saskatchewan consists of a lieutenant-governor, an executive council and a Legislative Assembly. On March 3, 1976 the Honourable George Porteous became lieutenant-governor. The statutory number of members of the Legislative Assembly is 61, elected for a maximum term of five years. As a result of the general election June 11, 1975, 39 New Democrats, 15 Liberals and seven Progressive Conservatives were elected to form Saskatchewan's 18th Legislature.

The premier receives $24,580 and each cabinet minister $18,205 annually in addition to a sessional indemnity and allowance. The leader of the opposition receives $18,205 plus an office allowance of $35,000 per annum. The leader of a third party receives $9,102 plus an office allowance of $17,500 per annum, the speaker $6,535 and the deputy speaker $3,925.

Each member of the legislature is paid an annual indemnity of $7,840, an expense allowance of $7,000 and a sessional allowance of $3,270. Each of the members for the two northernmost constituencies of Athabasca and Cumberland receives an $8,840 annual indemnity and a $7,250 expense allowance. Government and opposition whips are paid an annual allowance of $1,310 each and legislative secretaries an annual allowance of $3,000 each. The third party whip is paid an annual allowance of $708.

The Executive Council of Saskatchewan in November 1977

Premier and President of the Executive Council, Hon. A.E. Blakeney
Attorney General, Hon. R.J. Romanow
Minister of Mineral Resources, Hon. J.R. Messer
Minister of Finance, Hon. W.E. Smishek
Minister of Labour, Hon. G.T. Snyder
Minister of Northern Saskatchewan, Hon. G.R. Bowerman
Minister of the Environment and Minister of Telephones, Hon. N.E. Byers
Minister of Municipal Affairs, Hon. G. MacMurchy
Minister of Highways and Transportation, Hon. E. Kramer
Provincial Secretary, Hon. E. Cowley

Minister of Health, Hon. E. Tchorzewski
Minister of Revenue, and Minister of Co-operation and Co-operative Development, Hon. W.A. Robbins
Minister of Consumer Affairs, Hon. E.C. Whelan
Minister of Agriculture, Hon. E.E. Kaeding
Minister of Tourism and Renewable Resources, Hon. A. Matsalla
Minister of Social Services, Hon. H.H. Rolfes
Minister of Culture and Youth and Minister of Government Services, Hon. E.B. Shillington
Minister of Education and Minister of Continuing Education, Hon. D.L. Faris
Minister of Industry and Commerce, Hon. N. Vickar.

Alberta   3.3.1.9

In addition to the lieutenant-governor (since April 1974 the Honourable Ralph Steinhauer) the government of Alberta is composed of an executive council and a Legislative Assembly of 75 members elected for a maximum of five years. On March 26, 1975, 69 Progressive Conservatives, four members of the Social Credit party, one of the New Democratic Party and one Independent-Social Credit member were elected to form the 18th Legislature.

Each member of the Legislative Assembly receives a sessional indemnity of $12,469, a $6,234 expense allowance and an amount not to exceed $40 for each day during the session when the member is necessarily absent from his ordinary place of residence. In addition to the indemnity and expense allowance, the speaker receives a salary of $18,500 and the deputy speaker $6,735. The salary of the leader of the opposition, in addition to the indemnity and expense allowance, is $30,483. The speaker, deputy speaker and leader of the opposition also receive an amount not to exceed $40 for each day during the session when they are necessarily absent from their ordinary place of residence. In addition to the sessional indemnity and allowance the premier receives $37,410, other ministers $30,483 and ministers without portfolio $22,169. These figures have been in effect since April 1, 1978.

The Executive Council of Alberta in November 1977

Premier and President of the Executive Council, Hon. Peter Lougheed
Deputy Premier and Minister of Transportation, Hon. Hugh M. Horner
Minister of Energy and Natural Resources, Hon. Donald R. Getty
Minister of Federal and Intergovernmental Affairs and Government House Leader, Hon. Louis D. Hyndman
Minister of Hospitals and Medical Care, Hon. Gordon T.W. Miniely
Provincial Treasurer, Hon. C. Mervin Leitch
Minister of Labour, Hon. Neil S. Crawford

Minister of Advanced Education and Manpower, Hon. Bert Hohol
Minister of Housing and Public Works, Hon. William J. Yurko
Minister of Environment, Hon. David J. Russell
Attorney General, Hon. James L. Foster
Solicitor General, Hon. Roy A. Farran
Minister of Utilities and Telephones, Hon. Allan A. Warrack
Minister of Government Services, also responsible for Culture, Hon. Horst A. Schmid

Minister of Social Services and Community Health, Hon. Helen Hunley
Minister of Business Development and Tourism, Hon. Robert W. Dowling
Minister of Recreation, Parks and Wildlife, Hon. J.A. Adair
Minister of Municipal Affairs, Hon. Dick Johnston
Minister of Consumer and Corporate Affairs, Hon. Graham Harle

Minister of Agriculture, Hon. Marvin Moore
Minister of Education, Hon. Julian Koziac
Minister without portfolio (Native Affairs), Hon. Robert Bogle
Minister without portfolio, Hon. Stu McCrae
Associate Minister for Energy and Natural Resources in charge of Public Lands and Minister without portfolio, Hon. Dallas Schmidt.

3.3.1.10   British Columbia

The government of British Columbia consists of a lieutenant-governor, an executive council and a Legislative Assembly. On March 19, 1973 the Honourable Walter Stewart Owen took office as lieutenant-governor. The Legislative Assembly has 55 members who are elected for a term not to exceed five years. As at November 1, 1977 the assembly consisted of 35 Social Credit members, 18 New Democrats, one Liberal and one Progressive Conservative.

Each member of the executive council and the Legislative Assembly receives an annual allowance of $16,000 and an annual expense allowance of $8,000. In addition, the premier is paid an annual salary of $28,000, each cabinet minister (with portfolio) $24,000 and each member of the executive council without portfolio $21,000. The leader of the opposition and the speaker receive special expense allowances of $19,000; the deputy speaker and the leader of a recognized political party, $8,500.

The Executive Council of British Columbia in November 1977

Premier, President of the Council, Hon. William R. Bennett
Provincial Secretary, Deputy Premier and Minister of Travel Industry, Hon. Grace M. McCarthy
Minister of Finance, Hon. Evan Wolfe
Attorney General, Hon. Garde B. Gardom
Minister of Health, Hon. Robert H. McClelland
Minister of Human Resources, Hon. William N. Vander Zalm
Minister of Agriculture, Hon. James J. Hewitt
Minister of Economic Development, Hon. Donald M. Phillips
Minister of Labour, Hon. L. Allan Williams
Minister of Mines and Petroleum Resources, Hon. James R. Chabot

Minister of Forests, Hon. Thomas M. Waterland
Minister of Municipal Affairs and Housing, Hon. Hugh A. Curtis
Minister of Education, Hon. Pat L. McGeer
Minister of Environment, Hon. James A. Nielsen
Minister of Consumer and Corporate Affairs, Hon. Kenneth Rafe Mair
Minister of Energy, Transport and Communication, Hon. John (Jack) Davis
Minister of Highways and Public Works, Hon. Alex V. Fraser
Minister of Recreation and Conservation, Hon. Robert Samuel Bawlf.

3.3.2   Territorial governments

3.3.2.1   Yukon Territory

The constitution for the government of the Yukon Territory is based on two federal statutes: the Yukon Act (RSC 1970, c.Y-2) and the Government Organization Act (SC 1966, c.25). The Yukon Act provides for a commissioner as head of government and for a legislative body called the Yukon Legislative Council. Under the Government Organization Act, the minister of Indian affairs and northern development is responsible (with the Governor-in-Council) for directing the commissioner in the administration of the territory.

The executive level of the Yukon government consists of the commissioner and an executive committee. The Office of the Commissioner incorporates several functions: head of the territorial government and senior representative of the Indian affairs and northern development department in the Yukon Territory. The commissioner performs duties similar to those of a lieutenant-governor in relation to the legislature.

In administering the territorial government the commissioner is assisted by the executive committee, which is modelled on a cabinet structure. The committee is composed of the commissioner, as chairman, with the deputy commissioner and three councillors as members. Each of the members is assigned portfolios by the chairman.

The territorial government forgoes its taxing authority on private and corporate incomes and collection of corporate taxes and succession duties in deference to annual federal-territorial financial agreements. Under these agreements the federal government contributes the funds necessary to cover the deficit arising from the forecast of revenues available to the territory and the forecast cost of services to be provided.

Administration. The territorial public service, comprising approximately 1,200 employees, has 12 conventional administrative departments and a number of special service departments. Whitehorse is the administrative centre of the government. A few departments have necessary regional postings and territorial agents represent the government in outlying communities.

Health services and land are administered jointly by the territorial and federal governments. Health services are administered and operated by a Yukon hospital and health care insurance services department in conjunction with the national health and welfare department. Health responsibility in the territory was to be transferred from federal to territorial control April 1, 1978.

Certain areas have been designated to the commissioner for administration under the territory's Lands Ordinance. The remaining land is under the jurisdiction of the Indian affairs and northern development department.

In addition to these shared responsibilities, the federal government, through the Indian affairs and northern development department, retains control of the natural resources of the Yukon Territory, except game. Local administration is carried out by federal public servants.

Legislature. The Yukon Act delineates the jurisdiction of the legislative council. It is like those of provincial assemblies with two exceptions: matters touching on natural resources, except legislation concerning the preservation of game, are reserved to the federal government, and budgetary matters are reserved to the commissioner. Council is called into session and prorogued by the commissioner.

Legislative authority for the Yukon Territory is vested in the Commissioner-in-Council. All bills must be approved by council and assented to by the commissioner before becoming law. As in other jurisdictions, the Governor-in-Council may disallow any ordinance within one year. Ordinances are printed on a sessional basis and consolidated annually.

Amendments to the Yukon Act passed by Parliament in 1974 provided for an immediate expansion of council membership from seven to 12 and for future expansion to 20. Redistribution and expansion for the 1978 territory-wide election was expected to result in 16 members. Members are elected for four years. The council nominates three members to an executive committee which administers the following portfolios: education, Yukon Housing Corporation, local government, highways and public works, health and human resources. A fourth elected member is expected to be appointed to the executive committee following the 1978 election.

The council meets at least twice a year usually in the territorial capital, Whitehorse.

Commissioner, council and council staff of the Yukon Territory in November 1977

Commissioner, A.M. Pearson
Deputy Commissioner, D. Bell
Clerk of Council, L.J. Adams
Legal Adviser to the Commissioner and
Council, P. O'Donoghue

Executive Committee: A.M. Pearson, chairman; D. Bell, F. Whyard, D. Lang, J.K. McKinnon, members; L.J. Adams, secretary
Members of Council: A. Berger, B. Fleming, J. Hibberd, D. Lang, E. Millard, S. McCall, W. Lengerke, D. Taylor, F. Whyard, G. McIntyre, J.K. McKinnon, H. Watson.

Northwest Territories   3.3.2.2

The Northwest Territories Act (RSC 1970, c.N-22) provides for an executive, legislative and judicial structure. The commissioner is the chief executive officer, appointed by the federal government and responsible for the administration of the Northwest Territories under the direction of the minister of Indian affairs and northern development. The commissioner spends funds voted by council and all new revenue measures are subject to council approval. Normally the commissioner obtains federal approval of proposed legislation and budgetary measures before submitting them to council.

The Council of the Northwest Territories consists of 15 members elected for four years. It meets at least twice a year, usually for four weeks at a winter session and two weeks at a spring session, but more often if required. A clerk of council and a legal adviser provide the main administrative assistance. Debates are recorded verbatim.

The Northwest Territories Act gives the territorial council authority to legislate in most areas of government activity except for natural resources other than game; these are reserved to the federal government. Legislation must receive three readings and have the assent of the commissioner. The federal government may disallow any ordinance within one year. The commissioner proposes most legislation but private members' bills are allowed, except for money matters, which are the prerogative of the commissioner. Besides draft legislation, the council gives considerable time to policy papers in which the commissioner seeks advice or authority to take a particular course of action.

Parliament approved legislation in 1974 for the political development of the Northwest Territories. Amendments to the Northwest Territories Act increased the number of elected members of the territorial council from 10 to 15 and eliminated appointed members. Elections were held in March 1975 for the first entirely elected council. The new council selects its speaker from among its members; previously the commissioner was the presiding officer. Council also nominates three other members to the executive committee along with the commissioner, who is chairman, the deputy commissioner and an assistant commissioner. This committee advises the commissioner on broad policy matters and acts as a consultative body for him.

The justice minister is the attorney general of the Northwest Territories under the Criminal Code of Canada, with responsibility for criminal but not for civil matters or the constitution or organization of the courts (see Chapter 2). Law enforcement is provided by the Royal Canadian Mounted Police.

Administration. In 1963 a full-time commissioner was appointed and charged with building up a territorial administration located initially in Ottawa. In September 1967 the commissioner and about 50 staff members moved to Yellowknife and assumed responsibility for the game management service, municipal affairs, the issuing of all licences, tax collection and the operation of the liquor system (already staffed by territorial contract employees). Operational responsibility for other government services was transferred from federal to territorial control in the Mackenzie District in April 1969, and in the Eastern Arctic in April 1970. The territorial government carries out its administration through five program and six service departments, each under the direction of a senior public servant reporting to one member of the executive. The field staff is organized into four regions with regional directors at Fort Smith, Inuvik, Frobisher Bay and Rankin Inlet.

Continuing federal responsibility. The Government Organization Act charges the minister of Indian affairs and northern development with responsibility for the development of the North and for the general co-ordination of federal activities in the area. Other federal government agencies, such as the northern health service of the national health and welfare department and the Royal Canadian Mounted Police, are responsible for health and police services with the territorial government sharing their costs. The ministry of transport operates main line airports throughout the North; the Canadian Broadcasting Corporation provides live radio and television service via Anik and special shortwave northern broadcasts, and maintains local stations in the territories. Federal cost-shared national assistance programs, appropriate to territorial needs, are available to it on the same conditions as they are to the provinces.

Extensive financial assistance is given to the territorial government under special federal-territorial agreements. These agreements allocate the financial responsibility of each government for the provision of services in the territories.

Commissioner, council and council staff of the Northwest Territories in November 1977

Commissioner, S.M. Hodgson
Deputy Commissioner, J.H. Parker
Assistant Commissioner, Gary Mullins
Members of the Council: Bill Lyall, Ipeelee
Kilabuk, Mark Evaluarjuk, Don M. Stewart,
Ludy Pudluk, Tom Butters, Peter Ernerk,

William Lafferty, Arnold McCallum, Bryan
Pearson, John Steen, Dave Nickerson, Peter
Fraser, Richard Whitford
Clerk of the Council, W.H. Remnant
Legal Adviser, Ms. P. Flieger.

Royal commissions and commissions of inquiry   3.4

Federal commissions   3.4.1

Royal commissions, now generally called commissions of inquiry, established up to
April 30, 1975 under Part I of the Inquiries Act are described in previous editions of the
Canada Year Book beginning with the 1940 edition. The following list presents the
federal commissions established between that date and October 3, 1977, and the name
of the chief commissioner or chairman.

Financial Claims Commission, Hon. Thane A. Campbell
Bilingual Districts Advisory Board, Paul Fox
Inquiry into rail needs of Manitoba, Saskatchewan and Alberta, Emmett Hall
Inquiry into costs and revenue of grain (rail) traffic, Carl M. Snavely
Commission of inquiry into financial controls of Air Canada, Hon. Mr. Justice W.Z. Estey
Inquiry into crash of Lockheed aircraft near Rea Point, NWT, Judge W.A. Stevenson
Commission of inquiry into bilingual air traffic services in Quebec, Co-Commissioners: Hon. Mr.
Justice W.R. Sinclair, Hon. Mr. Justice Julien Chouinard, Hon. Mr. Justice D.V. Heald
Correctional investigator, penitentiary problems, Miss Inger Hansen
Royal Commission on financial management and accountability, Allen T. Lambert
Commission of inquiry into establishment of a marine (oil) terminal at Kitimat, BC, Dr. Andrew R.
Thompson
Special inquirer for elder Indians' testimony, Dr. Lloyd I. Barber
Canadian Indian Rights Commission, Hon. Mr. Justice E. Patrick Hartt
Newfoundland Transportation Commission, Dr. Arthur Sullivan
Task force on Canadian unity, Hon. Jean-Luc Pepin and Hon. John P. Robarts
Commission of inquiry concerning certain activities of the Royal Canadian Mounted Police, Hon.
Mr. Justice David C. McDonald.

Provincial commissions   3.4.2

The following list presents provincial commissions established between July 1976 and
November 1977, the name of the chief commissioner or chairman, and the date each
was established:

Nova Scotia
To inquire into and make recommendations to the Governor-in-Council respecting the appropriate
form of local government organization for the area of the County of Halifax lying in the vicinity of
Sackville, H.M. Nason, September 8, 1977.

Quebec
To inquire into and advise on the systems of collective bargaining in the public and para-public
sectors, Yves Martin, July 1977
To inquire into the cost of the 21st Olympic games and facilities, Albert H. Malouf, July 1977
To inquire into the search during the night of October 6/7, 1972 at 3459 St-Hubert Street in
Montreal, Jean Keable, September 1977.

Ontario
To inquire into the North Pickering Project, Hon. J.F. Donnelly, October 26, 1976
To inquire into the state of freedom of information and individual privacy as it pertains to the
government of Ontario, D. Carlton Williams, March 30, 1977
To investigate the safety and reliability of aluminum-wired electrical circuits for residential use,
Tuzo Wilson, April 6, 1977
To inquire into the status of pensions in Ontario, Miss Donna J. Haley, QC, April 20, 1977
To inquire into any improper influence being brought to bear on members of the Ontario
government or its public service by officials of Waste Management Inc., Mr. Justice Sam Hughes,
May 12, 1977

To determine the effects on the environment of major enterprises north of the 50th parallel in Ontario, Mr. Justice Patrick Hartt, July 13, 1977.

Manitoba
To review and study the organization of the executive government of the province and the various departments thereof, the Crown agencies and the boards and commissions that perform duties and functions under various acts of the legislature to ascertain whether any improvement in the administration of government can be achieved, Hon. Sidney J. Spivak, QC and Conrad S. Riley, November 16, 1977.

British Columbia
To inquire into methods of improving the scope and effectiveness of technical, vocational and trade training in British Columbia, Dean Goard, August 12, 1976
To inquire into the conduct of the public business relating to the proposed development and construction of the Grizzly Valley natural gas pipeline, Hon. Mr. Justice Walter Kirke Smith, January 11, 1977
To inquire into all aspects of the management and development of the British Columbia Railway, Hon. Mr. Justice Lloyd George McKenzie, February 7, 1977.

3.5 Local government

Local government in Canada comprises all government entities created by the provinces and territories to provide services that can be more effectively discharged through local control. Broadly speaking, local government services are identified in terms of seven main functions: protection, transportation, environmental health, environmental development, recreation, community services and education. In addition local government may operate such facilities as public transit and the supply of electricity and gas. Education is normally administered separately from the other local functions.

Under the British North America Act local government was made a responsibility of the provincial legislatures, a responsibility extended to the territories when their governments were constituted in their present forms. The unit of local government, apart from the school board, is usually the municipality which is incorporated as a city, town, village, township or other designation. The powers and responsibilities of municipalities are delegated to them by statutes passed by their respective provincial or territorial legislatures.

An increasing number of special agencies or joint boards and commissions have been created to provide certain services for groups of municipalities. Local government revenue has been supplemented by provincial grants, either unconditional or for specific purposes. Certain functions traditionally assigned to local government have been assumed in whole or in part by the provinces. Besides encouraging the amalgamation of small units, the provinces have established new levels of local government to provide services which can be better discharged at a regional level. Second-tier local governments now cover the whole of British Columbia and are planned for all of Ontario, where several now exist, and for Quebec, where three have been established. In Manitoba the Metropolitan Corporation of Greater Winnipeg and its constituent municipalities were amalgamated into a single city in January 1972.

The major revenue source available to local government is the taxation of real property, supplemented by taxation of personal property, businesses and amusements. Revenue is also derived from licences, permits, rents, concessions, franchises, fines and surplus funds from municipal enterprises.

The structure of local government in Canada varies widely. Table 3.6 gives the types of municipal organization in each province and territory.

3.6 External relations

3.6.1 Canada's international status

The growth of Canada's international status is reflected in the development of the external affairs department since its establishment in 1909. Until the 20th century

Canadian negotiations with foreign countries were conducted through the British foreign office and dealings with other parts of the Empire through a Colonial office. The gradual recognition of Canadian autonomy in international affairs and increased Canadian responsibilities abroad made expansion of services and representation after World War I not only inevitable but imperative. British diplomatic and consular authorities could no longer conveniently look after all Canadian interests. An important step in the evolution of the external affairs department as the foreign service arm of the Canadian government resulted from an agreement reached at the 1926 Imperial Conference allowing for Canadian sovereignty in international negotiations and affairs.

In the 1920s and 1930s Canada established its own diplomatic relations with other Commonwealth, European, African and Latin American countries and the United States. This independent expansion has continued to the point where Canada now has formal diplomatic relations with countries around the world, maintaining diplomatic, consular or trade representation in 140 countries.

Membership in international organizations has entailed establishment of a permanent Canadian delegation to the United Nations in New York and a Canadian office at the organization's European headquarters in Geneva in 1949. These permanent missions have since been expanded to include UN agencies in Paris and Vienna. Canada was one of the founding members of the North Atlantic Treaty Organization (NATO) in 1949 and when the NATO permanent council was established in 1952 a Canadian permanent delegation was set up in Paris (transferred to Brussels in 1967). Canada maintains a permanent delegation to the Organization for Economic Co-operation and Development in Paris, and a mission of Canada to the European communities in Brussels is responsible for Canada's relations with the European Economic Community, the European Atomic Energy Community and the European Coal and Steel Community. Canada also maintains a permanent observer mission to the Organization of American States in Washington, DC. In addition, officials of the external affairs department represent Canada at many international conferences.

Today Canada's status is reflected in its role in international negotiations over such vital issues as law of the sea, energy reserves, nuclear non-proliferation, north-south economic dialogue and human rights.

Diplomatic and/or consular representation   3.6.1.1
The addresses of Canadian representatives abroad and representatives of other countries in Canada may be found in Appendix 5.

Federal-provincial aspects of Canada's international relations   3.6.1.2
As a result of the growing international dimension of provincial interests, a federal-provincial co-ordination division was established in 1967 in the external affairs department. Its purpose was to maintain liaison with the provinces to facilitate their legitimate international activities in a manner that would meet provincial objectives consistent with a unified Canadian foreign policy.

The federal government's position on provincial international relations was outlined in the 1968 white paper *Federalism and international relations,* which emphasizes that Canada's foreign relations must serve and reflect the interests of all provinces as well as those of its two major linguistic communities. The federal government's international policies include recognition of legitimate provincial interests beyond national borders and continued promotion of national unity through adequate projection internationally of Canada's bilingual character.

Provincial participation at international conferences and in the work of international organizations is assured by including provincial officials on Canadian delegations and by canvassing provincial governments for their views on positions and attitudes which Canada might adopt on the subjects treated by these organizations. These include areas of particular interest to the provinces such as human and civil rights, education, health, agriculture, labour and environment.

Other aspects of Canada's international relations of particular interest to the provinces include promotion of trade, investment, industrial development, immigration, tourism, cultural exchanges, environmental questions, science and technology, bilateral and multilateral agreements, and assistance to developing countries. In matters of aid, the federal government encourages a detailed federal-provincial consultation to ensure that specific projects are co-ordinated. The promotional activities of the provinces coupled with their increased interests in international activities have led to a greatly increased number of provincial visits abroad. The federal government assists provincial officials by making arrangements and appropriate appointments for their visits abroad and in co-ordinating visits of foreign personalities to provincial capitals.

Treaty-making powers. The federal government has exclusive responsibility for external affairs. It promotes the interest of the entire country within the overall framework of a national policy.

In matters of specific concern to the provinces, it is the policy of the Canadian government to do its utmost to assist the provinces in achieving their particular aspirations and goals. The attitude of the federal government is reflected in the frequent consultations between federal and provincial levels of government regarding treaties bearing on areas of provincial interest and responsibility. A variety of methods have been developed which can allow for full expression of provincial interests in treaty-making.

Once it has been determined that what a province seeks through agreements, in fields of provincial jurisdiction, falls within the framework of Canadian foreign policy, provision is made for direct provincial participation in negotiating with the authorities of the foreign country. When these arrangements are to be incorporated in an international agreement having legal effect, however, this can be achieved only through the federal power to conclude treaties.

3.6.2 International activities

3.6.2.1 Canada and the Commonwealth

The present-day Commonwealth has evolved into an international association of 36 sovereign states embracing approximately one-quarter of the earth's surface and one billion of its people, who are diverse in race, colour, creed and language. Comprising both developed and developing countries, the Commonwealth represents a unique association whose members share many of the same traditions, political and social values, attitudes and institutions. All members collectively subcribe to certain common ideals known as the Declaration of Commonwealth Principles. Commonwealth membership, however, is not an alternative, but a complement to other forms of international co-operation — its members believe in and work for the success of the United Nations and together belong to a wide range of international organizations.

Commonwealth members (with the year when membership was proclaimed in parentheses if post-1931) are as follows: Australia, Britain, Canada, New Zealand, the Bahamas (1973), Bangladesh (1972), Barbados (1966), Botswana (1966), Cyprus (1961), Fiji (1970), Gambia (1965), Ghana (1957), Grenada (1974), Guyana (1966), India (1947), Jamaica (1962), Kenya (1963), Lesotho (1966), Malawi (1964), Malaysia (1963), Malta (1964), Mauritius (1968), Nauru (Special Member), Nigeria (1960), Papua New Guinea (1975), Seychelles (1976), Sierra Leone (1961), Singapore (1965), Sri Lanka (Ceylon) (1948), Swaziland (1968), Tanzania (1964), Tonga (1970), Trinidad and Tobago (1962), Uganda (1962), Western Samoa (1970) and Zambia (1964). Nauru has special membership in the Commonwealth with all the advantages of membership except attendance at heads of government meetings. Through their association with Britain, which has retained responsibility for foreign affairs and defence, the five West Indies Associated States of Antigua, Dominica, St. Kitts–Nevis–Anguilla, St. Lucia and St. Vincent are also associated with the Commonwealth, as are the British dependencies and the external territories of Australia and New Zealand in the Caribbean, the Atlantic and the Pacific.

Membership in the Commonwealth is an important aspect of Canadian foreign policy. Canada has consistently supported its expansion and development as a vigorous and effective association working for international peace and progress. Canadian objectives have remained constant: to strengthen the association, to encourage more active participation in it by members, and to assist its development as a vehicle for practical co-operation. The organization has no binding rules and decisions are by consensus rather than formal vote.

A Commonwealth secretariat in London organizes and services official Commonwealth conferences, facilitates exchanges of information between member countries and collates their views. Canada's assessment to the 1976-77 budget of the secretariat was 20.6% of the total, or approximately $912,000. In addition, in 1977 Canada contributed over $11.0 million to many other Commonwealth institutions and programs, with particular emphasis on a Commonwealth fund for technical co-operation ($6.5 million), a Commonwealth youth program ($521,000), a Commonwealth foundation ($401,000) and a Commonwealth scholarship and fellowship plan ($2.2 million).

An important duty of the secretariat is organization of Commonwealth heads of government meetings such as the one in London in June 1977 and the next to be held in Lusaka, Zambia in mid-1979. Of approximately 50 Commonwealth conferences in 1977, almost half were in the non-governmental sector, such as the Commonwealth parliamentary conference in Ottawa. Major governmental meetings in 1978 include the Commonwealth senior officials meeting in Kuala Lumpur, Commonwealth finance ministers meeting in Montreal, and a Commonwealth youth affairs council meeting in Ottawa. In addition, Canada hosted the Eleventh Commonwealth Games in Edmonton, August 3-12, 1978.

Canada and "la Francophonie"   3.6.2.2

The term "la Francophonie" generally describes countries whose language is wholly or partly French, that is, the French-language community. This term has also been used to designate a movement aimed at providing the French-language world with an organized framework and functional structures.

To demonstrate abroad the French aspect of Canadian society, the federal government fosters the strengthening of ties with francophone countries. In the last few years relations with French-language countries of Europe have been considerably expanded and diversified, complemented by the establishment of ties with the French-language countries of the Third World. Development aid remains an important activity.

Canada also participates in multilateral organizations such as the Agency for Cultural and Technical Co-operation, of which it is a founding member. At the agency's fifth general conference, in Abidjan, Ivory Coast, in December 1977, it was agreed to develop scientific and technical co-operation within the agency, following the resolutions prepared at a conference of ministers responsible for science and technology held in Luxembourg in September 1977. Furthermore, the conference agreed to launch a special development program based on voluntary contributions from approximately half of the participating countries.

The conference also reaffirmed the principle of regrouping the agency's programs around three main cores — development, education and scientific and technical co-operation, as well as promotion of national cultures and languages.

Canada is a member of the conference of ministers of education of French-language countries; at the annual session which was held in Ouagadougou, Upper Volta in April 1978, Quebec, Ontario and New Brunswick participated as part of a Canadian delegation led by the Canadian ambassador to the Ivory Coast and Upper Volta. Canada also participated in the conference of ministers of youth and sports of French-language countries in Lomey, Togo in September 1977; the minister responsible for these matters in New Brunswick led the Canadian delegation.

The federal government is not alone in its efforts to draw francophone countries closer. On the bilateral level, the provinces take part in joint commissions and in the implementation of Canadian government aid programs. On the multilateral level, New Brunswick, Ontario, Manitoba and Quebec participate in some of the agency's activities. Since 1971, Quebec has had the status of a participating government within the agency's institutions, activities and programs. On a proposal of the Canadian government, the general conference in Abidjan approved New Brunswick's request for similar recognition.

Various private French-language associations also work to develop relations between their members around the world. The agency has stimulated their activities and led to the creation of a number of new organizations. The Canadian government supports several that are either Canadian or have significant Canadian participation. The most recent of these institutions is the International Council of French speaking Radio and Television which was to be formally constituted in June 1978.

3.6.2.3   Canada and the United Nations

Since the inception of the United Nations, support for the UN system has been an integral part of Canadian foreign policy. Canada has played a significant role in the General Assembly and is a member of a number of its subsidiary bodies including the special committee on peacekeeping operations, the conference of the committee on disarmament, the committee on the peaceful uses of outer space, the United Nations scientific committee on the effects of atomic radiation, the committee on contributions and the board of auditors. At the beginning of 1977, the General Assembly had 147 members and was close to achieving universal membership.

In 1977 Canada served on the Security Council for the fourth time. Canada was previously on the Security Council in 1948-49, 1958-59 and 1967-68. Each term is two years long; Canada's present term ends December 31, 1978.

On the 12 occasions that UN troops have been dispatched to deal with threats to peace and security, Canada has actively participated. In 1976 over 1,500 Canadians were involved in UN peacekeeping, the largest commitment being to the United Nations force in the Sinai, where over 850 specialists of the Canadian forces were employed in logistics support. A similar role was being performed by more than 150 Canadians in the United Nations force in the Golan Heights area. In Cyprus, Canada provided infantry to patrol and monitor existing arrangements between the disputants.

Canada contributed over $6 million to peacekeeping in 1977. At the same time, Canada actively sought equitable reimbursement arrangements for countries which were participants in UN peacekeeping forces. Standard scales of reimbursement for each of the troop-contributing countries have been adopted for the UN forces in the Sinai and the Golan Heights. This is a significant advance over the uneven reimbursement scales of previous peacekeeping operations.

Canada has also served at regular intervals on the third principal organ of the UN, the Economic and Social Council. Canada's most recent term on ECOSOC was 1975-77. Generally, two sessions of the council are held annually, one in New York to discuss social and humanitarian questions, and one in Geneva to examine economic questions including, for example, food problems and international co-operation. The council is also charged with co-ordinating the work of some 167 subsidiary bodies of the UN system. Examples of those on which Canada is represented are: the governing council of the UN environment program, the commission on narcotic drugs and the committee on science and technology for development.

In recent years the UN has devoted more time to human rights, and new declarations, conventions and covenants have been promulgated. In 1976 four international human rights instruments came into force: the international covenant on economic, social and cultural rights; the international covenant on civil and political rights; the latter's related optional protocol; and the international convention on the suppression and punishment of the crime of apartheid. Canada has encouraged the preparation of such instruments and has stressed building better mechanisms for effective enforcement of standards. To emphasize Canada's commitment, special importance has been placed on securing membership on UN human rights bodies. During 1977 Canadians served on the commission on human rights and the UN human rights committee.

Canada is the ninth largest contributor to the UN, and in 1977 was assessed 2.96% of the regular budget or in dollar terms nearly $10 million (Table 3.7). Canada also makes voluntary contributions to the United Nations development program, the United

Nations commissioner for refugees, the United Nations children's fund, the United Nations relief and works agency for Palestine refugees, the world food program, the United Nations institute for training and research, the United Nations education and training plan for southern Africa, the United Nations fund for population activities, the committee on racial discrimination, the trust fund for South Africa and the fund for drug abuse control. The United Nations development program is the largest of these, and has a team leadership function in co-ordinating development activities in the UN system. Canada's voluntary donations in both cash and commodities to various UN programs totalled approximately $165 million in the 1976-77 fiscal year.

Specialized agencies. Canada is a member of all 14 specialized agencies of the UN, and is the host country of one, the International Civil Aviation Organization. Canada maintains permanent missions to the UN headquarters in both New York and Geneva, and has accredited representatives to agencies located in Paris (UNESCO), Rome (FAO), Nairobi (UNEP) and Vienna (IAEA and UNIDO). The contributions of these agencies have been one of the greatest strengths of the UN system.

The World Bank Group, consisting of the International Bank for Reconstruction and Development (IBRD) or World Bank, the International Finance Corporation and the International Development Association, is by far the largest of the multilateral aid-giving institutions. A brief summary of the agencies follows:

The International Labour Organization (ILO), originally established with the League of Nations in 1919, became a specialized agency of the UN in 1946. It brings together representatives of governments, employers and workers from 133(1977) member states in an attempt to promote social justice by improving living and working conditions in all parts of the world. Canada has been a member of the ILO from its inception and as a leading industrial state has been assigned one of the 10 non-elective seats on the governing body.

The Food and Agriculture Organization (FAO), established in 1945, is one of the largest of the specialized agencies, with 136 members. Raising the nutrition levels and living standards of its member countries and improving production and distribution techniques for food, agriculture, fishery and forest products are two of its objectives. The FAO secretariat provides advisory services, collects and publishes agricultural and fisheries statistics, and organizes international conferences and meetings of experts.

FAO has headquarters in Rome and regional offices in Washington, Bangkok, Rio de Janeiro, Santiago and Cairo. Canada participates in FAO functions and is a member of the FAO council, the committee on commodity problems, the committee on fisheries, the consultative subcommittee on surplus disposals, the FAO group on grains, the North American forestry commission and other FAO bodies. The joint FAO-WHO food standards program is administered by an executive committee of which Canada is a member.

The world food program was established under the joint auspices of the FAO and the UN to provide food aid on a multilateral basis for emergency relief, including the feeding of children, and to promote economic and social development. Its approved target for pledges for 1978-79 was $950 million. Canada pledged $190 million to the two-year program and is the second largest contributor. A Canadian, Mr. G. Vogel, former chief commissioner of the wheat board, is executive director of the program.

The World Health Organization (WHO), with 150 members and two associate members, is a directing and co-ordinating authority on international health matters. The objective is the attainment by all peoples of the highest possible level of health; WHO provides advisory and technical services from its Geneva headquarters to help countries develop and improve their national health services. At the 28th world health assembly in Geneva in May 1975, Canada was elected to the WHO executive board for a three-year term.

The United Nations Educational, Scientific and Cultural Organization (UNESCO) was established in 1946 to contribute to peace and security by promoting collaboration among nations through education, science and culture to further universal respect for justice, the rule of law, human rights and fundamental freedoms. Its headquarters is in Paris and membership is 141 states.

UNESCO has three main components — a general conference which is the policy-making body, an executive board and a secretariat. Representatives from member states make up the general conference which meets every two years. The 19th session of the general conference in Paris in 1976 approved a budget of approximately US$224 million for 1977-78, giving priority to the educational needs of the developing countries and to science activities, particularly the application of science to development.

The International Civil Aviation Organization (ICAO), with headquarters in Montreal, was established in 1947 to promote the safe, orderly and economic development of international civil aviation. It has a membership of 138 (1978). Canada has been a member of the 30-nation council, the governing body of ICAO , since its inception, as a state of chief importance in air transport.

The International Telecommunication Union (ITU), founded to oversee application of the international telegraph convention of 1865 and the international radio telegraph convention of 1906, is concerned with international co-operation for improvement and use of telecommunications for the benefit of the general public; it has 153 member countries. Canada is represented on the 36-member administrative council, the executive organ of the ITU.

The World Intellectual Property Organization (WIPO) came into being in January 1974 to protect intellectual property, such as patents and copyright, and to ensure administrative co-operation among the 11 organizations or unions established for these purposes previously.

The World Meteorological Organization (WMO), a specialized agency of the UN since 1951, has evolved from an international meteorological organization founded in 1878; in 1976 WMO had 145 members. One of its major programs is the world weather watch for developing an improved worldwide meteorological system and environment. Canada is represented on the organization's executive committee.

The Inter-Governmental Maritime Consultative Organization (IMCO) was established in 1959 to promote international co-operation on technical shipping problems and the adoption of the highest standards of safety and navigation; its membership in 1976 was 101. IMCO exercises bureau functions for international conventions of safety of life at sea, prevention of pollution of the sea by oil, and facilitation of international maritime traffic. Canada is a member of both the IMCO council and the maritime safety committee.

The Universal Postal Union (UPU), one of the oldest and largest of the specialized agencies, was founded in Berne in 1874 with the principal aim of improving postal services throughout the world and promoting international collaboration. It has 157 members. A universal postal congress meets every five years to review the universal postal convention and its subsidiary instruments. In the interim, UPU activities are carried on by an executive council, a consultative committee on postal studies and an international bureau. Canada was elected to the executive council in 1974 during the 17th congress in Lausanne.

The International Monetary Fund (IMF), created at the Bretton Woods conference in 1944 and established in 1945, was designed to facilitate expansion of world trade and payments as a means of raising world standards of living and of fostering economic development. It promotes stability and order of exchange rates, and provides financial mechanisms for balance of payments assistance to enable member countries to correct temporary imbalances with a minimum of disturbance to the international monetary system. The original membership of 45 countries has grown to 130, of which over 100 are classified as developing.

Canada's participation in the IMF is authorized under the Bretton Woods Agreement Act of 1945. The SDR (special drawing rights) has been defined as being equal in value to a fixed basket of 16 currencies, one of which is the Canadian dollar.

The Canadian quota and subscription is SDR1.1 billion in 1978. Fund holdings of Canadian dollars as of December 31, 1977 amounted to the equivalent of SDR604 million or approximately 54.9% of the Canadian quota. The reserve position of Canada in the IMF at the end of 1977 amounted to SDR701.5 million of which SDR169.4 million represented loans by Canada to the oil facility. The oil facility assists member countries in financing deficits arising from cost increases of imports of petroleum products.

The International Bank for Reconstruction and Development (World Bank), also originated at the Bretton Woods conference of 1944. Its early loans were made to assist in post-war reconstruction of Europe but it has played an increasingly important role in providing financial assistance and economic advice to less-developed countries. It has become the world's largest multilateral source of development finance. As of June 30, 1976 effective loans made by the World Bank totalled US$23 billion.

Most World Bank loans are made to finance roads, rails, ports and electricity generation and transmission which provide the framework basic to a country's economy but which generally do not attract private investors. Increasingly, however, more emphasis has been given to other sectors such as agriculture, rural development, telecommunications, education, water supply and sewage.

In 1977 Canada's subscription to the World Bank was the equivalent of $1,136.1 million in current US dollars out of a total for all countries of US$30,861.0 million. Only 10% of each subscription is paid in, however, with the balance remaining as a guarantee against which the bank is able to sell its own bonds in world capital markets.

The International Development Association (IDA) was established as an affiliate of the IBRD in 1960. Its resources come mainly from governments in the form of interest-free advances, enabling it to make loans on very soft terms (no interest and 50 years to repay). IDA lends to member countries with per capita income less than $375 a year.

Since IDA cannot borrow from world capital markets, its loanable resources have been derived largely from budgetary allocations from its member governments, principally the developed-country members. Total resources made available or committed to IDA from the beginning of its operations to the end of June 30, 1977 were approximately US$11.7 billion, and as a developed country, Canada had paid in US$631.7 million.

The International Finance Corporation (IFC), established in 1956 as an affiliate of the IBRD assists less-developed member countries to promote growth of the private sector of their economies. IFC provides risk capital for productive private enterprises in association with private investors and management, encourages development of local capital markets, and stimulates international flow of private capital. IFC makes investments in the form of share subscriptions and long-term loans, carries out standby and underwriting arrangements and provides financial and technical assistance to privately controlled development finance companies. Of IFC's total subscribed capital of US$1.33 billion, Canada provided US$3.6 million. In addition to its subscribed capital, IFC is able to finance its activities through loans from its parent institution, the World Bank. Total investment and underwriting commitments by IFC to December 31, 1975 amounted to US$1.33 billion in 57 countries. Commitments made during 1976 were US$196 million.

The International Atomic Energy Agency (IAEA) was created in 1957 as an autonomous international organization under the aegis of the UN which has empowered it to try to accelerate and enlarge the contribution of atomic energy to peace, health and prosperity throughout the world. In 1976, membership consisted of 106 states. Because Canada has been designated as one of the members most advanced in nuclear technology, including the production of source materials, a Canadian representative has served on the board of governors since the agency's inception.

Conferences and symposia, dissemination of information and provision of technical assistance are among the methods adopted to carry out the IAEA's functions. With rapid expansion in the use of nuclear power, much activity is devoted to this field as well as to the use of isotopes in agriculture and medicine. Another significant role is development and application of safeguards to ensure that nuclear materials supplied for peaceful purposes are not diverted to military uses. Under terms of a treaty for the non-proliferation of nuclear weapons, each non-nuclear weapons state adhering to the treaty was to conclude an agreement with the IAEA providing for safeguards on its entire nuclear program. The IAEA also imposes safeguards pursuant to agreements relating to individual nuclear facilities. Agency inspectors have carried out safeguard inspections in Canada and more than 60 other countries.

3.6.2.4   Canada and disarmament

Canada is an active member of the Conference of the Committee on Disarmament (CCD), a 31-nation negotiating body. This committee, with the United States and the Soviet Union as co-chairmen, represents the worldwide concern with the arms race. The CCD is seeking a comprehensive prohibition of nuclear weapons testing, including underground tests, and a ban on the development, production and stockpiling of chemical weapons.

3.6.2.5   Canada and force reductions in Central Europe

Canada continues to participate in the conference on the mutual reduction of forces and armaments and associated measures in Central Europe, which opened officially in Vienna in October 1973.

3.6.2.6   Canada, NATO and NORAD

NATO. Canada was one of the 12 original signatories of the North Atlantic Treaty Organization (NATO) in 1949. Successive Canadian governments reaffirmed the view that Canada's security remains linked to that of Europe and the United States. Canada is committed to the principle of collective defence and remains convinced of the importance of NATO's role in reducing, and eventually removing, the underlying causes of potential East–West conflict through negotiation, reconciliation and settlement.

A number of major equipment procurement decisions arising from the defence structure review of 1975 were made during 1977, the most important of which was to purchase 130-150 new fighter aircraft to replace the outdated CF-101s, CF-104s and CF-5s currently in service. In December 1977, Cabinet approved funds for the project definition phase leading to the acquisition in the early 1980s of six new patrol frigates.

Canada participates in the mutual and balanced force reductions negotiations in Vienna. These negotiations are generally recognized as difficult because they touch on vital security interests of both NATO and the Warsaw Pact nations. Now in their fifth year, the negotiations have not yet resulted in agreement, largely because of differing perceptions of the actual size of the military forces on each side. Similarly, little progress was recorded in the strategic arms limitation talks between the US and the USSR, although events in 1977 led to optimism that progress might be made in 1978.

Members of the alliance continue to experience, in varying degrees, the impact of severe inflation and other economic problems. Under these circumstances, particular attention has been paid to the problem of maintaining an adequate defence capability in the face of serious strains on the economies of some of the allies. Alliance members, including Canada, continue to seek economies by increasing specialization in the development, production and acquisition of military equipment to avoid costly duplication of efforts.

Canada's membership in NATO continues to be a factor in the development of its political, economic and scientific-technological relations with Europe, by which Canada seeks to balance its relations with the United States. The alliance obliges both Canada and the United States to maintain a deep interest in European affairs and exemplifies the interdependence of Europe and North America. It also provides Canada with an opportunity to consult with 14 other countries (including eight of the nine members of the European community) continuously and regularly on a variety of political and military questions.

North American defence co-operation. Canada's support of collective security is not limited to its role in NATO. Through its continuing co-operative defence arrangements with the United States it participates in aerospace surveillance and warning systems, active air defence, anti-submarine defence and measures designed to protect the deterrent capacity of the United States.

Canada and the United States   3.6.2.7

There is no more important external relationship for Canada than that with the United States. As a result of geography and economic and social patterns, the two countries frequently meet to discuss various aspects of governmental policies and programs. In addition to informal consultations, there are official and technical committees in which Canadian and US officials discuss such bilateral questions as defence and transboundary environmental matters. For example, the International Joint Commission, an independent agency, was established by the US and Canada to deal with regulation of flows of boundary waters and the abatement of transboundary air and water pollution. Canada and the US have a long history of defence co-operation through a permanent joint board on defence and through NATO.

Canada and the US also work together on international questions in multilateral organizations such as the UN, the OECD, GATT, the IMF and others in which both countries are active members.

In trade, each is the other's best customer, and in 1977 two-way trade was approximately $60 billion. Canada sells to the United States about 70% of all exports and buys from the US about 20% of all US exports.

Canada and the Commonwealth Caribbean   3.6.2.8

Canada has long enjoyed close relations with the countries of the Commonwealth Caribbean. The current phase began with the Commonwealth Caribbean–Canada conference of 1966, followed by a special Canadian mission to the area in 1970. In April 1975 the prime minister visited Trinidad and Tobago, Barbados and Guyana, then went to the Commonwealth heads of government meeting in Kingston, Jamaica. State visits were made to Canada by the prime minister of Jamaica in 1976 and the prime minister of Guyana in 1977.

In 1977 Canadian investment in the region was estimated at approximately $350.0 million; Canadian imports from the region totalled $139.6 million while exports were valued at $148.6 million. Canadian bilateral development assistance to the Caribbean, begun in 1958, has averaged approximately $22.0 million a year in loans and grants in recent years. It has been concentrated in the sectors of education, air transport, water supply and agriculture (including forestry). Funds have also been made available on a multilateral basis through various organizations including the United Nations and the Caribbean Development Bank.

More than 3,000 Canadians are permanent residents in the region and over 250,000 visit the islands annually. There are Canadian high commissions in Jamaica, Trinidad and Tobago, Guyana and Barbados, and these four countries and Grenada maintain high commissions in Ottawa. There is also a commissioner for the Eastern Caribbean in Montreal who represents the five West Indies Associated States (Antigua, Dominica, St. Kitts–Nevis–Anguilla, St. Lucia and St. Vincent) and Montserrat.

Canada and Latin America   3.6.2.9

Canada maintains diplomatic relations with all Latin American countries through 13 resident missions and dual or multiple accreditation from those missions. In addition, Canada is associated with the inter-American system through membership of observer status in many inter-American institutions including a permanent observer mission to the Organization of American States.

The growth of Canadian relations with Latin America has led to contacts and exchanges in many fields. The secretary of state for external affairs visited Brazil, Peru and Colombia in January 1977, and Mexico in April, to further relations in the region following the prime minister's 1976 visit to Mexico, Cuba and Venezuela. The deputy prime minister, the minister of trade and commerce, and the premier of New Brunswick travelled to Venezuela in 1977 and the minister of fisheries and the environment visited

Cuba. The Brazilian agriculture minister visited Canada in 1977 and the Mexican foreign minister led the Mexican delegation to the third meeting of the Canada–Mexico ministerial committee held in Ottawa. The Canada–Cuba joint economic committee met in Havana in June 1977 and it was expected the Canada–Brazil joint economic committee would meet in Brasilia in the first half of 1978.

During 1977 Canadian trade with Latin America increased slightly in comparison with 1976. Canadian imports from Latin America in 1977 amounted to $2,200 million ($1,938 million in 1976) while Canadian exports to the area amounted to $1,587 million ($1,480 million in 1976). Canada's trade deficit increased from $457 million in 1976 to $613 million in 1977, largely because of higher costs of imported oil.

In 1977 Canada, through the Canadian International Development Agency (CIDA), continued to provide substantial assistance to various countries of the region to help them, particularly the poorer countries, achieve their social and economic development objectives. The program expanded during 1977 and CIDA devoted a larger proportion of its global resources to Latin America. CIDA also participated in multilateral regional projects with a variety of inter-American institutions.

At the multilateral level, Canada is an active member of many inter-American organizations, namely: the Pan American Institute of Geography and History, the Pan American Health Organization, the Inter-American Institute for Agricultural Sciences, the Inter-American Statistical Institute, the Inter-American Centre for Tax Administrators, the Centre for Latin American Monetary Studies and the Postal Union of the Americas and Spain. Canada supports various technical and professional inter-American organizations. In April 1978, Canada hosted the annual meeting of the Inter-American Bank.

3.6.2.10 Canada and Europe, the Middle East, Africa, and the Far East

Canada and Europe. Canadian activities in 1977 in Western Europe were directed toward the development and strengthening of political, economic and commercial relations. In political matters, co-operation was actively promoted through regional and international conferences, official visits and ministerial meetings. Exchanges centred chiefly on peacekeeping, the use of atomic energy for peaceful purposes, increased security in Europe and improvement of East–West relations. In the commercial and economic sphere, Canada's participation in the London economic summit and active part in the North–South conference, the meetings of joint economic commissions and exchanges of industrial missions, all resulted in a strengthening of ties with Western Europe. The framework agreement for commercial and economic co-operation with the European communities spearheads Canada's third-option policy in Europe.

Canada and the Eastern European states have in recent years increased trade, scientific and technological co-operation as well as cultural exchanges. Canada participated with the 35 signatory states of the Helsinki Final Act in the follow-up meeting of the conference on co-operation and security in Europe which was held in Belgrade, Yugoslavia from October 4, 1977 to March 9, 1978.

In January 1978 a Canada–USSR mixed commission met in Ottawa to draw up a new program of scientific, academic and cultural exchanges and co-operation for 1978-79. This is the fourth program of exchanges under the terms of a Canada–USSR general exchanges agreement since it was signed in Ottawa in 1971. With other Eastern European countries Canada has worked for mutually beneficial bilateral relations through resident diplomatic missions in Prague, Warsaw, Belgrade, Budapest and Bucharest and through non-resident ambassadors accredited to Bulgaria and the German Democratic Republic.

Canada and the Middle East. Canada has consistently attempted to follow a policy of balance and objectivity between the parties to the Arab–Israeli dispute. Over the years, Canada has supported the efforts of the UN Relief and Works Agency to alleviate the plight of Palestine refugees and has contributed to the maintenance of the ceasefire that followed the war of October 1973 by providing the largest national contingent to the United Nations peacekeeping forces.

Many of the major oil-exporting countries of the Middle East have put their increased revenues to use by expanding their developmental projects. In addition, some have sought to employ a part of their surpluses in assisting other countries that lack such valuable resources. These countries are becoming more aware of Canada's potential as a reliable supplier not only of traditional but also of more sophisticated goods and services. In 1977 Canadian exports to the Middle East increased by some 16.0% for a total value of $635 million, while the value of Canada's imports from this region, mainly of oil, rose by about 16.5%, to reach $1,469 million.

Canada and Africa. Direct relations were established with former British colonies in Africa as they became independent members of the Commonwealth. Increasing contacts and diplomatic relations with the newly independent French-language African states soon followed. Canada now maintains diplomatic relations with almost all the independent African states and through resident Canadian missions in 15 countries. The development of diplomatic and commercial relations has been accompanied by a significant and growing program of Canadian development assistance to Africa. This program directed $210 million in assistance to the African continent in 1976-77 and approximately the same amount in 1977-78.

Canada and the Asian and Pacific Region. Relations with the countries of Asia and the Pacific are diverse, for the region includes some of the oldest and most varied civilizations in the world, some of the most highly industrialized nations, and some of the least developed economies. The countries of the region hold over half the world's population and consequently their governments are faced with daunting administrative and political problems. Some of Canada's earliest forays into external relations were with countries of the region. Commonwealth ties with many remain important and Canada's commercial links go far back. Canadians over several generations have lived and worked in the area. In turn, Canada has over the past two decades become the new home of many emigrants from the region. Development assistance programs with some of the countries are the oldest and the largest in which Canada has engaged. Some of Canada's best customers are in the region and in turn, imports from it have steadily increased. Canada is a Pacific nation and consciousness of this is growing among Canadians.

Japan is Canada's second largest national trading partner and bought $2.5 billion of Canadian exports in 1977. Both countries have, during the last three years, exerted special efforts to broaden and deepen the relationship. The conclusion in 1976 of a framework for economic co-operation and a cultural agreement are concrete signs of each country's willingness to understand the needs and aspirations of the other and to co-operate to achieve these goals.

Relations between Canada and China continue to develop in commercial and political spheres. (In 1977 Canada exported $369 million worth of goods to China.) A great deal of Canada's regular intercourse with China is in the form of exchange groups and delegations. In 1977 these exchanges ranged from sport teams to music groups to missions studying agriculture, pulp and paper technology, occupational health and geology.

Canada's bilateral relations with the individual countries of South East Asia remain important for both development assistance and commercial interest. A further dimension has been added in the growth of Canada's relationship with the Association of South East Asian Nations (ASEAN). Indonesia, Singapore, Malaysia, Thailand and the Philippines have, through their participation in ASEAN, indicated an increased willingness to co-operate for their mutual benefit. In two formal meetings with ASEAN representatives since 1976 Canada has expressed interest and support for this organization in its efforts to promote broad regional development and increase stability in the area. Two-way trade with the ASEAN countries totalled close to one-half billion dollars in 1977, a 20% increase over 1976.

Relations with Australia and New Zealand are deeply rooted in similar legislative and judicial experience as well as in shared problems and common action over several generations. More recently, new and rapidly developing mutual interests have arisen over a wide range of government activity including domestic issues, the export of uranium and nuclear safeguards, the exploration and marketing of raw materials and multilateral trade questions. Two-way trade with Australia totalled almost $750 million in 1977 and with New Zealand the figure reached $130 million. In both cases approximately 85% of Canada's exports were manufactured goods.

In 1977, as in previous years, Canada's three largest development assistance programs were in India, Pakistan and Bangladesh. The nature of these programs varied widely from immediate disaster relief for victims of the Bay of Bengal cyclone to large amounts of food aid to continuing development assistance. Canada is one of the largest non-regional contributors to the Asian Development Bank whose financial assistance has done much to promote development throughout Asian and Pacific regions.

3.6.2.11   Canada and the OECD

The Organization for Economic Co-operation and Development (OECD) was established in Paris in September 1961 as successor to the Organization for European Economic Co-operation (OEEC) founded in 1948 by the countries of Western Europe to facilitate reconstruction of their war-shattered economies and to administer the Marshall Plan. With the establishment of the OECD, Canada and the United States and later Japan (May 1964), Australia (June 1971) and New Zealand (May 1973) joined with the countries of Western Europe to form a major intergovernmental forum for consultation and co-operation among the advanced industrialized nations in virtually every major field of economic activity. At present 24 countries are full members while Yugoslavia has a special status entitling it to participate in certain activities.

The aim of the OECD is to facilitate the formulation of policy approaches which are conducive to stability, balanced economic growth and social progress of both member and non-member countries. The organization assembles and examines knowledge relevant to policy-making and is a forum, meeting the year round, for exchange and analysis of ideas and experiences from all member countries.

The organization plays a significant role in harmonizing international economic and financial policy and is the main area where industrialized nations hold consultations on questions of development assistance. The original focus on more traditional economic, trade and development matters has altered and new activities have been undertaken in agriculture, the environment, industry, science and technology, international investment and multinational enterprises, social affairs, manpower and education. The International Energy Agency (IEA) established within the framework of the OECD in November 1974, plays an important role in four main areas: emergency oil sharing, consultations on the oil market, promotion of the accelerated development of new sources of energy, and relations between oil consuming and oil producing countries. Another agency of the OECD, the Nuclear Energy Agency which celebrates its 20th anniversary in 1978, has been involved in the co-ordination and exchange of views of the technical aspects of nuclear power. This broader orientation places increasing emphasis on qualitative as well as quantitative aspects of world economic growth.

The OECD brings together government officials and representatives of private business, labour unions, universities and other non-governmental bodies at the international level. Within Canada, the Canadian Business and Industry International Advisory Committee, comprising representatives of the Canadian Chamber of Commerce, the Canadian Manufacturers' Association, the Canadian chapter of the International Chamber of Commerce, the Canadian Association for Latin America, and the Pacific Basin Economic Council, was established in 1962 to ensure input from the business community. Arrangements exist for consultation with Canadian labour organizations, universities and other non-governmental bodies. Representatives of provincial governments attend OECD meetings when subjects of particular interest to the provinces are being discussed.

3.6.2.12   Canadian development assistance programs

The Canadian International Development Agency (CIDA) is responsible for the operation and administration of Canada's international development assistance programs. In 1976-77 Canada spent $963.3 million on foreign aid, an increase of $60.0 million over the previous year. Of that amount $416.6 million went to multilateral assistance programs and $477.7 million to bilateral assistance programs. The remaining funds were divided among non-governmental organizations working in international development, international emergency relief programs, an international development research centre, incentives to Canadian private investment in developing countries and the CIDA scholarship fund for Canadians taking postgraduate degrees in international development and related fields. The authorized level for 1977-78 was $1.1 billion.

CIDA's multilateral assistance programs are directed toward the United Nations and its affiliated organizations, the World Bank Group, the regional development banks and several regional institutions.

CIDA's bilateral development program is divided into three types of aid — technical assistance, economic assistance and international food aid — and into five regional programs. During 1976, 1,472 students and trainees from developing countries studied in Canada under CIDA's technical assistance program and 1,605 Canadian advisers and educators worked overseas. Under a unique feature of Canada's technical training program more than 650 students and trainees studied in developing countries other than their own.

Canadian bilateral economic assistance is divided almost evenly between grants and loans. Most loans are extended for 50 years and are without interest, with no repayment required for the first 10 years. Spending for bilateral food programs totalled $149.4 million in 1976-77.

Canada's development aid program to Asia is the largest and oldest regional bilateral aid program administered by CIDA. It received $237.2 million in 1976-77.

Since 1951 Canada has provided more than $2.4 billion in bilateral aid, most of it directed to Bangladesh, India, Indonesia, Pakistan and Sri Lanka. In recent years Canada's program in this area has changed considerably. Capital assistance, in the form of loans and grants, is now provided for specific economic sectors given priority by the recipient countries in fields such as communications, transportation, electric power development, agriculture, fisheries, mining, lumbering, medicine and public health.

CIDA's programs in francophone Africa, which includes eight of the least developed nations of the world, were initially concentrated on technical assistance projects particularly in education and health. Since 1970, however, Canada has broadened the scope of its assistance and increased its support in the area from $29.7 million to $88.7 million in 1976-77. Canada has become increasingly involved in the economic development of the region through projects that combine capital and technical assistance, in accordance with the priorities set by the recipient countries.

Canadian assistance to Commonwealth Africa has grown from an initial provision for technical and educational assistance in 1960 to include a variety of capital projects and pre-investment surveys. Undertakings in energy, transportation, communications, agriculture and economic planning in eastern and southern Africa have balanced an original focus on West Africa. Between 1960 and March 1977, Canada contributed $550 million to bilateral development programs in the region. In 1976-77 expenditures for Commonwealth Africa totalled $93 million, for a variety of projects ranging from mining to beekeeping.

Canadian economic and technical assistance to the Commonwealth Caribbean began in 1958. Since then the region has received more Canadian aid per capita than any other area of the world. Canada's bilateral allocations, amounting to more than $180 million since 1964 including $23.4 million in 1976-77, have contributed to construction projects, transportation surveys, water systems, medical assistance, support for the University of the West Indies and other development projects.

In 1971 CIDA began a bilateral technical assistance program for Latin America concentrating on agriculture, forestry, fisheries, education and community development. In 1974-75 a bilateral loan program was introduced; in 1976-77 spending in Latin America was $26.5 million.

CIDA is also involved with non-government aid organizations and business and industry. In the 1968-69 fiscal year $4 million was spent to help voluntary agencies increase their contribution to international development. This figure had risen to $38.1 million by 1976-77.

CIDA has become involved in the private sector of developing countries' economies and in expanding suitable Canadian enterprises overseas. The organization works with Canadian business, the industry, trade and commerce department, international finance corporations, development banks, and overseas corporations to identify and help finance worthwhile investment opportunities in all types of secondary industry in the developing world.

The International Development Research Centre (IDRC) is an international organization supported financially by Canada. Established in 1970 to initiate and encourage research focused on the problems of the world's developing regions, it fosters co-operation between developing nations as well as between the developed and the developing world. In its role as co-ordinator of international development research, it helps developing regions to build up research capabilities, innovative skills and institutions to solve their own problems. The centre offers research awards to PhD candidates and mid-career professionals who are Canadian citizens or landed immigrants with three years residence.

IDRC's chairman, vice-chairman and nine of the other 19 governors are Canadian citizens. There is a strong international element. In 1977 six governors were from developing nations (Jamaica, Mexico, Ethiopia, Zaire, Iran and Indonesia) and one from each of Britain, France, the United States and Australia. Professional staff included citizens of 14 countries.

Operations are conducted under five programs: agriculture, food and nutrition sciences; information sciences; population and health sciences; social sciences and human resources; and publications. As at March 31, 1978 IDRC had approved 694 projects worth $123.8 million involving grantees in 75 countries. Most of the research activities and related seminars were conducted in developing countries by their research organizations. The Canadian government's contribution to IDRC was $27.0 million in 1975-76, $29.7 million in 1976-77 and $34.5 million in 1977-78.

3.7   Defence

3.7.1   The Department of National Defence

The national defence department was created by the National Defence Act, 1922, which established one civil department of government in place of the previous departments of Militia and Defence, Naval Service and the Air Board. The department now operates under authority of RSC 1970, c.N-4.

The defence minister controls and manages the Canadian forces, the Defence Research Board and all matters relating to national defence establishments. He is responsible for presenting to Cabinet matters of major defence policy for which Cabinet direction is required. He is also responsible for the National Emergency Planning Establishment. The minister continues to be responsible for certain civil emergency powers and duties as outlined in order-in-council PC 1965-1041, June 8, 1965.

The chief of the defence staff is the senior military adviser to the minister and is charged with the control and administration of the Canadian forces. He is responsible for the effective conduct of military operations and the readiness of the forces to meet the commitments assigned to the department.

The Defence Research Board is responsible for advising the minister on scientific matters relating to defence and for evaluating the contribution of science and technology to defence.

The minister of national defence is responsible for administering the following laws which relate to the national defence department: National Defence Act (RSC 1970, c.N-4), Defence Services Pension Continuation Act (RSC 1970, c.D-3), Canadian Forces Superannuation Act (RSC 1970, c.C-9) and Visiting Forces Act (RSC 1970, c.V-6).

Liaison in other countries. The chief of the defence staff, the Canadian military representative to the North Atlantic Treaty Organization, is responsible for advice on all

NATO military matters and acts as a military adviser to the government and to Canadian delegations to NATO. For purposes of liaison and international co-operation in defence, Canada also maintains: the Canadian defence liaison staffs in London and Washington, two logistic liaison units in the United States, a Canadian member of a NATO military committee in permanent session in Brussels, a military adviser to the Canadian permanent representative to the North Atlantic Council and also a Canadian national military representative to Supreme Headquarters Allied Powers Europe (SHAPE), and Canadian forces attachés in various countries throughout the world. In addition, a number of defence matters of concern to both Canada and the United States are considered by a permanent joint board on defence, which provides advice on such matters to the respective governments.

The command structure of the Canadian forces 3.7.2

The Canadian forces are organized on a functional basis to reflect the major commitments assigned by the government. All forces devoted to a primary mission are grouped under a single commander. Specifically, the Canadian forces are formed into National Defence Headquarters and five major commands reporting to the chief of the defence staff.

Mobile command 3.7.2.1

The role of mobile command is to provide military units suitably trained and equipped for the protection of Canadian territory, to maintain operational readiness of combat formations in Canada required for overseas commitments, and to support United Nations or other peacekeeping operations.

The forces assigned include: three airportable combat groups in Canada, the special service force, the Canadian contingent of the United Nations force in Cyprus, the Canadian contingent of the United Nations Middle East, and one combat training centre.

The militia is assigned its traditional role as a component in support of the regular force. Under the present organization, militia units have been placed under either the commander of mobile command or Canadian forces communication command.

Mobile command exercises command and control of 99 militia combat units plus administrative and service units through five militia area headquarters and 21 militia districts across the country. Mobile command militia is charged with providing trained individuals for augmentation and reinforcement of the regular force, providing trained units to support the field force for the defence of Canada and the maintenance of internal security, providing trained personnel for the augmentation of the civil emergency operations organization, and forming the base on which the regular force could be expanded for service in an emergency.

Maritime command 3.7.2.2

All Canadian maritime forces are under the commander, maritime command, whose headquarters is in Halifax. The deputy commander is the commander, maritime forces Pacific, with headquarters in Esquimalt. The role of maritime command is to defend Canadian interests from assault by sea, to support measures to protect Canadian sovereignty, to support Canadian military operations as required and to conduct search and rescue operations within the Atlantic and Pacific search and rescue areas (roughly the Atlantic provinces and British Columbia) and to contribute to NATO defence.

As at December 1977, the following vessels were in service in maritime command: 19 frigates (eight helicopter equipped) and four Iroquois class helicopter destroyers, three operational support ships, three Oberon class submarines, six Bay class coastal patrol vessels (employed as training vessels), and two escort repair vessels (retained in service as alongside workshops and temporary accommodation vessels). Three of the 19 frigates are held in reserve.

The naval reserve is an essential component of maritime command. Its primary function is to provide trained personnel to augment the fleet in emergencies. Another essential role is to provide and maintain naval control of shipping in time of emergency or war to meet national and NATO requirements. There are 16 naval reserve units in major Canadian cities.

3.7.2.3   Air command

With the formation of an air command on September 2, 1975 overall responsibility for Canada's military air forces was again vested in one senior commander to provide greater flexibility in the employment of air power as well as to increase operational effectiveness, safety and economy.

The command's principal function is to provide operationally-ready regular and reserve air forces to meet Canada's national, continental and international commitments, and to carry out regional commitments within the Prairie region — Saskatchewan, Alberta and Manitoba.

Air command, with headquarters at Winnipeg, consists of four operational groups: air defence group, air transport group, maritime air group and 10 tactical air group. Air command also exercises control over the air training schools and the reserve.

Air defence group, with headquarters at North Bay, Ont., is responsible for maintaining sovereignty of Canada's airspace. In addition, the group provides Canada's contribution to NORAD, the joint Canada–US North American Air Defence Command. It has command of three all-weather fighter squadrons, a training squadron, two transcontinental radar lines, a satellite tracking unit and an electronic warfare squadron.

Air transport group provides airlift resources to enable the Canadian forces to meet their commitments. It also undertakes national and international tasks as directed by the government. The group provides search and rescue service for downed aircraft and co-ordinates marine search and rescue operations. Heavy transport resources consist of 24 C-130 Hercules aircraft and five Boeing 707 aircraft. A squadron at Ottawa provides medium-range passenger transport with seven Cosmopolitan and seven Falcon aircraft.

Transport and rescue squadrons at Comox, BC, Edmonton, Alta., Trenton, Ont., and Summerside, PEI, are equipped with Buffalo and Twin Otter fixed-wing aircraft, and some with Labrador and Voyageur helicopters. Rescue co-ordination centres at Trenton and Edmonton co-ordinate search and rescue activities. They work closely with maritime command in Victoria and Halifax.

Air movements units at Ottawa, Trenton, Edmonton and Lahr, Federal Republic of Germany, with detachments at Comox and Vancouver, BC, Winnipeg, Man., and Greenwood and Shearwater, NS, provide passenger and cargo-processing services.

In 1977 strategic and tactical airlift by 10 tactical air group aircraft enabled other elements of the forces to participate in a wide range of activities embracing national sovereignty, North American defence, NATO, humanitarian missions and contributions to hemispheric security.

About half of the group's flying is devoted to joint exercises with mobile command and other Canadian forces commands, often in conjunction with NATO allies. The remainder is used to support Canadian forces in Europe, isolated bases in the North, Canadian military and civil missions abroad, and department and other government agencies in Canada.

Maritime air group (MAG) is responsible for management of all air resources engaged in maritime patrol, maritime surveillance and anti-submarine warfare.

The commander of maritime air group, while responsible to the commander of air command, is under the operational control of the commander of maritime command while carrying out surveillance roles. A close working relationship between maritime command and maritime air group enables them to use a common operations centre.

The group conducts surveillance flights over Canada's coastal waters and the Arctic Archipelago. It also provides anti-submarine air forces as part of Canada's contribution to NATO.

Air reserves. The air reserve is organized into four air reserve wing headquarters at Montreal, Toronto, Winnipeg and Edmonton and six flying squadrons of six DHC Otter aircraft each, at Montreal (two), Toronto (two), Winnipeg (one) and Edmonton (one). The air reserve is required to provide light tactical air transport support to the regular force and in particular to mobile command ground forces. Air reserve tasks include logistic airlift, air evacuation of patients, aerial surveillance and photography, and communications and liaison.

The Canadian forces training system 3.7.2.4

The system was created in September 1975 with formation of air command and the realignment of the Canadian forces command structure. With headquarters at Trenton, Ont., it plans and conducts all recruit, trades, specialist and officer classification training common to more than one command.

The commander of Canadian forces training system also assumes regional commitments in Ontario, including responsibility for planning and implementing aid to the civil power, assistance to civil authorities and other federal departments, liaison with the provincial government and its agencies, and provision of support services to selected units of other commands.

Information on recruit and trades training, training for officers, flying training, the three Canadian military colleges, the cadet movement and other related programs is included in Chapter 7, Education, training and cultural activities.

The Canadian forces communication command 3.7.2.5

This command maintains strategic communications for the forces and, in emergencies, for the federal and provincial governments. The command also provides points for interconnecting strategic and tactical networks. It also operates the major defence department automatic data processing centres.

The 12 Canadian forces communication command militia units are centred in: Vancouver, Edmonton and Calgary; Regina and Winnipeg; Toronto and Ottawa; Montreal and Quebec City; and Saint John, Halifax and Charlottetown. Their tasks collectively include the augmentation of Canadian forces communication command in an emergency, provision of communications support to mobile command militia in peacetime emergency operations, provision of instructors for the training of mobile command unit signalers, and provision of communications support for control of mobile command militia tactical exercises.

Canadian forces in Europe. Canadian forces allocated to support NATO in Europe consist of land and air elements. The land element is a mechanized brigade group. The air element consists of three CF-104 Starfighter squadrons. These elements are located in the Baden-Baden area of the Federal Republic of Germany.

Administration of military bases in Canada. Staffs and services required below command headquarters level to administer and support units based in a particular locality have been organized on Canadian forces bases. Each base has been allocated to a functional commander to whom the base commander reports.

Function/regional organization. Functional commanders have been assigned a regional as well as a functional responsibility for representation to provincial governments, aid to the civil power, emergency and survival operations, administration of cadets, and provision of regional support services for all units in the region.

Emergency planning 3.7.3

An emergency measures organization was created to co-ordinate the civil aspects of defence policy delegated to federal departments and agencies to meet the threat of nuclear war on Canada. In late 1973 certain changes were made to ensure an effective response to any emergency. The organization was renamed the National Emergency Planning Establishment in 1974. The new organization works under the direction of the Privy Council with its main function to mitigate the effects of disasters in Canada. It will continue to have regional offices in each provincial capital to ensure continuing support for provincial authorities in development of mutual emergency capabilities.

Sources

3.1 Government Organization Division, Privy Council Office; Clerk of the Senate; House of Commons Division, Department of Supply and Services; Office of the Chief Electoral Officer.

3.2.1 Secretary of the Treasury Board.

3.2.2 Secretary of the Treasury Board; Information Services Directorate, Public Service Commission.

3.2.3 Deputy Minister, Department of Indian Affairs and Northern Development.

3.2.4 Government Organization Division, Privy Council Office; *Canada Year Book* staff.

3.2.5 Communications Division, Treasury Board.

3.3 Supplied by the respective provincial and territorial governments.

3.4.1 Privy Council Office.

3.4.2 Supplied by the respective provincial governments.

3.5 Public Finance Division, Institutional and Public Finance Statistics Branch, Statistics Canada.

3.6 Information Services Division, Department of External Affairs; Information Division, Canadian International Development Agency; International Development Research Centre.

3.7 Directorate of Parliamentary Affairs, Department of National Defence.

Tables

3.1   Duration and sessions of Parliaments, 1965-78

Order of Parliament	Session	Date of opening	Date of prorogation	Days of session	Sitting days of House of Commons	Date of election, writs returnable, dissolution, and length of Parliament[1,2]
27th Parliament	1st	Jan. 18, 1966	May 8, 1967	476[6]	250	Nov. 8, 1965[3]
	2nd	May 8, 1967	Apr. 23, 1968	352[7]	155	Dec. 9, 1965[4] Apr. 23, 1968[5] 867 d
28th Parliament	1st	Sept. 12, 1968	Oct. 22, 1969	406[8]	197	June 25, 1968[3]
	2nd	Oct. 23, 1969	Oct. 7, 1970	350[9]	155	July 25, 1968[4]
	3rd	Oct. 8, 1970	Feb. 16, 1972	497[10]	244	Sept. 1, 1972[5]
	4th	Feb. 17, 1972	Sept. 1, 1972	198[11]	91	1,500 d
29th Parliament	1st	Jan. 4, 1973	Feb. 26, 1974	419[12]	206	Oct. 30, 1972[3] Nov. 20, 1972[4]
	2nd	Feb. 27, 1974	May 9, 1974	72	50	May 9, 1974[5] 536 d
30th Parliament	1st	Sept. 30, 1974	Oct. 12, 1976	744[13]	343	July 8, 1974[3]
	2nd	Oct. 12, 1976	Oct. 17, 1977	371[14]	175	July 31, 1974[4]
	3rd	Oct. 18, 1977	...	[15]	...	...

[1]The ordinary legal limit of duration for each Parliament is five years.

[2]Duration of Parliament in days. The life of a Parliament is counted from the date of return of election writs to the date of dissolution, both days inclusive (BNA Act, Sect. 50).

[3]Date of general election.

[4]Writs returnable.

[5]Dissolution of Parliament.

[6]Includes Easter adjournment from Apr. 6, 1966 to Apr. 19, 1966; two summer adjournments from July 14, 1966 to Aug. 29, 1966 and Sept. 9, 1966 to Oct. 5, 1966; Christmas adjournment from Dec. 21, 1966 to Jan. 9, 1967 and Easter adjournment from Mar. 22, 1967 to Apr. 3, 1967.

[7]Includes summer adjournment from July 7, 1967 to Sept. 25, 1967; Christmas adjournment from Dec. 21, 1967 to Jan. 22, 1968; and Easter (Liberal Convention) Mar. 28, 1968 to Apr. 23, 1968.

[8]Includes Christmas adjournment from Dec. 20, 1968 to Jan. 14, 1969; Easter adjournment from Apr. 2, 1969 to Apr. 14, 1969; and summer adjournment from July 25, 1969 to Oct. 22, 1969.

[9]Includes Christmas adjournment from Dec. 19, 1969 to Jan. 12, 1970; Easter adjournment from Mar. 25, 1970 to Apr. 6, 1970; and summer adjournment from June 26, 1970 to Oct. 5, 1970.

[10]Includes Christmas adjournment from Dec. 18, 1970 to Jan. 11, 1971; Easter adjournment from Apr. 7, 1971 to Apr. 19, 1971; summer adjournment from June 30, 1971 to Sept. 7, 1971; Christmas adjournments from Dec. 23, 1971 to Dec. 28, 1971; and Dec. 31, 1971 to Jan. 12, 1972.

[11]Includes Easter adjournment from Mar. 29, 1972 to Apr. 13, 1972; and summer adjournment from July 7, 1972 to Aug. 31, 1972.

[12]Includes Easter adjournment from Apr. 19, 1973 to May 6, 1973; summer adjournments from July 27, 1973 to Aug. 30, 1973 and Sept. 21, 1973 to Oct. 15, 1973; Christmas adjournments from Dec. 22, 1973 to Jan. 2, 1974 and Jan. 14, 1974 to Feb. 26, 1974.

[13]Includes Christmas adjournment from Dec. 20, 1974 to Jan. 22, 1975; Easter adjournment from Mar. 27, 1975 to Apr. 7, 1975; summer adjournment from July 31, 1975 to Oct. 13, 1975; Christmas adjournment from Dec. 20, 1975 to Jan. 26, 1976; Easter adjournment from Apr. 14, 1976 to Apr. 26, 1976; and summer adjournment from July 16, 1976 to Oct. 12, 1976.

[14]Includes Christmas adjournment from Dec. 22, 1976 to Jan. 24, 1977; Easter adjournment from Apr. 6, 1977 to Apr. 18, 1977; summer adjournments from July 25, 1977 to Aug. 4, 1977; Aug. 5, 1977 to Aug. 9, 1977; and Aug. 9, 1977 to Oct. 17, 1977.

[15]Parliament still in session Apr. 27, 1978. Includes Christmas adjournment from Dec. 20, 1977 to Jan. 23, 1978; and Easter adjournment from Mar. 22, 1978 to Apr. 3, 1978.

3.2   Representation in the Senate since Confederation, 1867

Province or territory	1867	1870	1871	1873	1882	1887	1892	1903	1905	1915-1948	1949-1974	1975-1978
Ontario	24	24	24	24	24	24	24	24	24	24	24	24
Quebec	24	24	24	24	24	24	24	24	24	24	24	24
Atlantic provinces	24	24	24	24	24	24	24	24	24	24	30	30
Nova Scotia	12	12	12	10	10	10	10	10	10	10	10	10
New Brunswick	12	12	12	10	10	10	10	10	10	10	10	10
Prince Edward Island	...	...	...	4	4	4	4	4	4	4	4	4
Newfoundland	...	...	...	...	...	...	...	...	...	...	6	6
Western provinces	...	2	5	5	6	8	9	11	15	24	24	24
Manitoba	...	2	2	2	3	3	4	4	4	6	6	6
British Columbia	...	...	3	3	3	3	3	3	3	6	6	6
Saskatchewan	...	...	...	...	...	2	2	4	4	6	6	6
Alberta	...	...	...	...	...				4	6	6	6
Territories	...	...	...	...	...	...	...	...	...	...	...	2
Yukon Territory	...	...	...	...	...	...	...	...	...	...	...	1
Northwest Territories	...	...	...	...	...	...	...	...	...	...	...	1
Total	72	74	77	77	78	80	81	83	87	96	102	104

3.3 Representation in the House of Commons, as at federal general elections 1867-1974

Province or territory	1867	1872	1874 1878	1882	1887 1891	1896 1900	1904	1908 1911	1917 1921	1925 1926 1930	1935 1940 1945	1949	1953 1957 1958 1962 1963 1965	1968 1972 1974
Ontario	82	88	88	92	92	92	86	86	82	82	82	83	85	88
Quebec	65	65	65	65	65	65	65	65	65	65	65	73	75	74
Nova Scotia	19	21	21	21	21	20	18	18	16	14	12	13	12	11
New Brunswick	15	16	16	16	16	14	13	13	11	11	10	10	10	10
Manitoba	...	4	4	5	5	7	10	10	15	17	17	16	14	13
British Columbia	...	6	6	6	6	6	7	7	13	14	16	18	22	23
Prince Edward Island	...	...	6	6	6	5	4	4	4	4	4	4	4	4
Saskatchewan	}				4	4	10	{ 10	16	21	21	20	17	13
Alberta	}	...	...	...				{ 7	12	16	17	17	17	19
Yukon Territory													1	1
Mackenzie River	}						1	1	1	1	1	1	{	
NWT[1]	}	...	...	...	...	...							{ 1	1
Newfoundland	...	...	...	...	...	...	...	...	...	...	...	7	7	7
Total	181	200	206	211	215	213	214	221	235	245	245	262	265	264

[1]Electoral district of Northwest Territories in 1963, 1965, 1968, 1972 and 1974.

3.4 Electoral districts, voters on the list, votes polled and names and addresses of members of the House of Commons as elected at the thirtieth general election, July 8, 1974

Province and electoral district	Population, Census 1971	Voters on list	Total votes polled (incl. rejections)	Votes polled by member	Name of member	Postal address	Party affiliation[1]
NEWFOUNDLAND (7 members)							
Bonavista—Trinity— Conception	69,543	44,012	26,839	13,258	D. Rooney	Lower Island Cove	Lib.
Burin—Burgeo	54,044	31,149	16,698	13,550	Hon. D.C. Jamieson	Swift Current	Lib.
Gander—Twillingate	71,480	40,875	23,035	12,721	G.S. Baker	Gander	Lib.
Grand Falls—White Bay—Labrador	75,106	41,250	23,290	12,689	B.Rompkey	Grand Falls	Lib.
Humber—St. George's —St. Barbe	82,263	44,731	28,104	16,500	J. Marshall	Corner Brook	PC
St. John's East	87,477	52,148	30,371	16,941	J.A. McGrath	Portugal Cove	PC
St. John's West	82,191	50,205	27,197	14,550	W. Carter	St. John's	PC
PRINCE EDWARD ISLAND (4 members)							
Cardigan	23,363	15,212	13,043	6,958	Hon. D.J. MacDonald	Bothwell	Lib.
Egmont	30,629	18,583	14,581	7,583	D. MacDonald	Alberton	PC
Hillsborough	35,639	25,322	19,801	9,917	H. Macquarrie	Victoria	PC
Malpeque	22,010	13,952	11,224	5,649	Hon. J.A. MacLean	Belle River	PC
NOVA SCOTIA (11 members)							
Annapolis Valley	74,123	48,463	36,702	19,174	P. Nowlan	Wolfville	PC
Cape Breton—East Richmond	64,371	40,428	31,923	14,192	A. Hogan	Glace Bay	NDP
Cape Breton Highlands— Canso	62,550	41,476	32,406	17,977	Hon. A.J. MacEachen	Inverness	Lib.
Cape Breton— The Sydneys	68,135	43,553	32,682	14,371	R. Muir	Sydney Mines	PC
Central Nova	62,726	42,956	32,532	17,459	E.M. MacKay	Lorne	PC
Cumberland— Colchester North	65,899	46,059	34,388	18,078	R.C. Coates	Amherst	PC
Dartmouth—Halifax East	98,399	61,917	43,030	22,090	M. Forrestall	Dartmouth	PC
Halifax	64,523	42,970	30,339	14,865	Hon. R.L. Stanfield[2]	Ottawa	PC
Halifax—East Hants	100,637	70,222	50,199	25,563	B. McCleave	Halifax	PC
South Shore	65,420	46,174	33,660	18,206	L.R. Crouse	Lunenburg	PC
South Western Nova	62,177	40,549	30,969	15,067	C. Campbell	Yarmouth	Lib.

3.4   Electoral districts, voters on the list, votes polled and names and addresses of members of the House of Commons as elected at the thirtieth general election, July 8, 1974 (continued)

Province and electoral district	Population, Census 1971	Voters on list	Total votes polled (incl. rejections)	Votes polled by member	Name of member	Postal address	Party affiliation[1]
NEW BRUNSWICK (10 members)							
Carleton—Charlotte	59,244	38,195	24,541	12,315	F.A. McCain	Florenceville	PC
Fundy—Royal	70,316	48,459	31,594	13,631	G. Fairweather	Rothesay	PC
Gloucester	63,556	39,011	28,598	16,031	H. Breau	Tracadie	Lib.
Madawaska—Victoria	54,772	33,187	22,395	14,310	E. Corbin	Edmundston	Lib.
Moncton	80,188	56,121	45,433	20,671	L. Jones	Moncton	Ind.
Northumberland—Miramichi	54,094	32,320	24,186	12,648	M. Dionne	Millerton	Lib.
Restigouche	52,485	31,208	22,709	12,492	Hon. J.-E. Dubé	Campbellton	Lib.
Saint John—Lancaster	68,460	42,945	28,024	12,860	M. Landers	Saint John	Lib.
Westmorland—Kent	51,856	34,805	26,686	16,340	R.-A. LeBlanc	Memramcook	Lib.
York—Sunbury	79,586	50,267	35,326	17,673	J.R. Howie	Fredericton	PC
QUEBEC (74 members)							
Abitibi	58,427	38,197	22,758	12,425	G. Laprise	LaSarre	SC
Argenteuil—Deux-Montagnes	80,574	54,218	39,071	20,414	F. Fox	St-Eustache	Lib.
Beauce	69,984	46,155	33,221	13,855	Y. Caron	St-Georges-Ouest	Lib.
Beauharnois—Salaberry	73,396	48,313	33,864	16,828	G. Laniel	Valleyfield	Lib.
Bellechasse	64,675	42,225	27,641	12,550	A. Lambert	Berthier-sur-Mer	SC
Berthier	62,521	42,574	29,050	15,266	A. Yanakis	St-Gabriel-de-Brandon	Lib.
Bonaventure-Îles-de-la-Madeleine	55,004	33,376	21,867	12,977	A. Béchard	Carleton	Lib.
Brome—Missisquoi	76,787	51,019	38,680	19,490	H. Grafftey	Knowlton	PC
Chambly	120,337	82,787	56,583	30,099	B. Loiselle	Beloeil	Lib.
Champlain	62,068	41,017	30,332	14,466	R. Matte	St-Marc-des-Carrières	SC
Charlevoix	59,686	39,016	25,595	10,372	C. Lapointe	Tadoussac	Lib.
Chicoutimi	82,658	51,405	35,405	17,096	P. Langlois	Chicoutimi	Lib.
Compton	62,197	39,068	27,198	11,474	C. Tessier	Lac-Mégantic	Lib.
Drummond	75,533	49,826	38,606	15,561	Y. Pinard	Drummondville	Lib.
Frontenac	67,991	42,338	29,817	14,236	L. Corriveau	Thetford Mines	Lib.
Gaspé	56,280	34,679	21,436	12,213	J.-A. Cyr	Chandler	Lib.
Gatineau	81,320	53,833	33,404	19,513	G. Clermont	Thurso	Lib.
Hull	93,804	61,475	38,837	26,872	G. Isabelle	Lucerne	Lib.
Joliette	83,417	57,126	42,603	21,935	R. LaSalle	Crabtree	PC
Kamouraska	63,228	39,885	23,805	11,664	C.-E. Dionne	St-Pascal	SC
Labelle	82,228	55,680	39,534	16,224	M. Dupras	St-Jérôme	Lib.
Lac-Saint-Jean	56,862	33,618	24,483	11,162	M. Lessard	Alma	Lib.
Langelier	58,559	36,584	22,463	13,616	Hon. J. Marchand	Quebec	Lib.
Lapointe	72,451	44,318	28,634	16,617	G. Marceau	Jonquière	Lib
Laprairie	131,675	89,276	60,419	35,276	I. Watson	Laprairie	Lib.
Lévis	80,037	56,925	38,538	20,348	R. Guay	Lauzon	Lib.
Longueuil	112,703	79,375	50,876	24,500	J. Olivier	Longueuil	Lib.
Lotbinière	70,964	46,948	36,505	21,448	A. Fortin	Victoriaville	SC
Louis-Hébert	106,928	74,430	51,825	32,441	Albanie Morin	Sillery	Lib.
Manicouagan	80,461	57,879	25,945	16,229	G. Blouin	Sept-Îles	Lib.
Matane	48,373	29,449	17,133	11,194	P. De Bané	Matane	Lib.
Montmorency	116,204	79,714	54,754	27,082	L. Duclos	Ste-Foy	Lib.
Pontiac	59,956	37,761	23,249	12,642	T. Lefebvre	Davidson	Lib.
Portneuf	116,079	80,495	55,065	25,630	P. Bussières	Charlesbourg	Lib.
Québec-Est	81,782	52,449	34,032	19,019	G. Duquet	Quebec	Lib.
Richelieu	77,197	53,200	38,334	20,801	F. Côté	Ste-Brigitte-des-Saults	Lib.
Richmond	62,741	38,576	28,305	11,825	L. Beaudoin	Bromptonville	SC
Rimouski	69,276	44,692	31,515	15,085	E. Allard	Rimouski	SC
Rivière-du-Loup—Témiscouata	59,816	37,302	25,714	11,071	R. Gendron	Rivière-du-Loup	Lib.
Roberval	53,671	33,606	23,464	12,877	C.-A. Gauthier	Mistassini	SC
Saint-Hyacinthe	82,540	56,801	43,602	21,453	C. Wagner	Saint-Hyacinthe	PC
Saint-Jean	83,274	54,334	37,571	18,798	W.B. Smith	Hemmingford	Lib.
Saint-Maurice	71,147	47,117	33,009	20,468	Hon. J. Chrétien	Shawinigan	Lib.
Shefford	79,083	51,856	38,397	15,572	G. Rondeau	Granby	SC
Sherbrooke	97,550	66,005	44,944	23,903	Irénée Pelletier	Sherbrooke	Lib.
Témiscamingue	54,545	32,226	22,911	14,026	R. Caouette[2]	Rouyn	SC
Terrebonne	122,332	83,020	55,124	28,651	J.-R. Comtois	Repentigny	Lib.
Trois-Rivières métropolitain	95,389	64,050	42,278	24,335	C.-G. Lajoie	Cap-de-la-Madeleine	Lib.
Villeneuve	58,859	38,259	24,695	10,452	A. Caouette	Val-d'Or	SC
Island of Montreal and Île-Jésus							
Ahuntsic	90,537	56,170	39,547	24,041	Hon. Jeanne Sauvé	Montreal	Lib.
Dollard	123,429	79,270	57,138	37,200	Hon. J.-P. Goyer	St-Laurent	Lib.
Duvernay	112,102	69,264	47,526	25,674	Y. Demers	Ville-de-Laval	Lib.
Gamelin	92,533	57,861	37,993	20,625	A. Portelance	Montreal	Lib.
Hochelaga	65,393	38,884	22,991	10,561	Hon. G. Pelletier	Westmount	Lib.
Lachine-Lakeshore	92,202	57,365	42,591	22,068	R. Blaker	Pointe-Claire	Lib.
Lafontaine	70,166	42,952	25,890	11,429	C.A. Lachance	Montreal	Lib.
LaSalle—Émard—Côte Saint-Paul	116,235	72,724	47,179	28,146	J. Campbell	LaSalle	Lib.
Laurier	67,023	31,634	18,573	10,085	F.-E. Leblanc	Montreal	Lib.
Laval	115,908	71,697	48,729	29,715	M. Roy	Laval-des-Rapides	Lib.

3.4   Electoral districts, voters on the list, votes polled and names and addresses of members of the House of Commons as elected at the thirtieth general election, July 8, 1974 (continued)

Province and electoral district	Population, Census 1971	Voters on list	Total votes polled (incl. rejections)	Votes polled by member	Name of member	Postal address	Party affiliation[1]
QUEBEC (concluded)							
Maisonneuve— Rosemont	74,499	44,444	27,576	13,817	S. Joyal	Montreal	Lib.
Mercier	118,807	74,272	44,505	22,545	P. Boulanger	Pointe-aux-Trembles	Lib.
Montréal—Bourassa	124,746	78,559	48,646	26,550	J.-L. Trudel	Montreal	Lib.
Mount Royal	90,844	59,202	43,525	32,166	Rt. Hon. P.-E. Trudeau[2]	Ottawa	Lib.
Notre-Dame-de-Grâce	77,052	45,939	33,177	20,151	Hon. W. Allmand	Montreal	Lib.
Outremont	75,621	44,931	30,725	20,400	Hon. M. Lalonde	Montreal	Lib.
Papineau	73,439	41,921	25,583	14,532	Hon. A. Ouellet	Montreal	Lib.
Saint-Denis	77,362	36,737	25,640	15,310	M. Prud'homme	Montreal	Lib.
Saint-Henri	57,162	29,790	17,664	8,813	G. Loiselle	Montreal	Lib.
Saint-Jacques	53,179	24,845	14,740	7,709	J. Guilbault	Montreal	Lib.
Sainte-Marie	58,381	34,418	20,595	8,300	R. Dupont	Longueuil	Lib.
Saint-Michel	138,109	81,077	49,012	29,822	Monique Bégin	Saint Léonard	Lib.
Vaudreuil	112,103	71,417	51,630	29,685	H.T. Herbert	Hudson	Lib.
Verdun	74,718	47,567	31,286	17,633	Hon. B. Mackasey	Verdun	Lib.
Westmount	83,645	51,592	37,127	20,816	Hon. C.M. Drury	Ste-Cécile-de-Masham	Lib.
ONTARIO (88 members)							
Algoma	52,746	30,937	21,927	11,360	M. Foster	Desbarats	Lib.
Brant	97,549	62,333	46,444	19,453	D. Blackburn	Brantford	NDP
Bruce—Gray	63,308	46,111	36,225	17,158	C. Douglas	Wingham	Lib.
Cochrane	54,786	31,793	21,073	11,379	R.W. Stewart	Moonbeam	Lib.
Elgin	66,608	43,088	33,799	15,851	J. Wise	St. Thomas	PC
Essex—Windsor	94,846	60,901	44,305	24,357	Hon. E.F. Whelan	Amherstburg	Lib.
Fort William	60,207	38,892	28,336	13,789	P. McRae	Thunder Bay "F"	Lib.
Frontenac—Lennox and Addington	61,668	40,801	29,329	14,102	D. Alkenbrack	Napanee	PC
Glengarry—Prescott— Russell	62,599	41,302	30,714	18,478	D. Éthier	Dalkeith	Lib.
Grenville—Carleton	119,408	79,873	62,620	33,946	W. Baker	Ottawa	PC
Grey—Simcoe	67,997	47,652	34,454	15,917	G. Mitges	Owen Sound	PC
Halton	105,801	68,063	52,145	23,520	F. Philbrook	Oakville	Lib.
Halton—Wentworth	124,390	82,370	61,964	26,798	B. Kempling	Dundas	PC
Hamilton East	74,709	39,992	28,356	15,298	Hon. J.C. Munro	Hamilton	Lib.
Hamilton Mountain	106,266	67,718	50,965	22,217	G. MacFarlane	Hamilton	Lib.
Hamilton—Wentworth	99,169	67,067	48,002	18,874	S. O'Sullivan	Hamilton	PC
Hamilton West	81,664	49,060	34,372	15,421	L.M. Alexander	Hamilton	PC
Hastings	64,328	42,657	31,076	14,893	J. Ellis	Belleville	PC
Huron—Middlesex	58,515	38,560	29,122	17,185	R.E. McKinley	Zurich	PC
Kenora—Rainy River	54,853	33,993	23,197	10,317	J.M. Reid	Kenora	Lib.
Kent—Essex	85,580	53,937	36,198	17,800	B. Daudlin	Leamington	Lib.
Kingston and The Islands	82,907	53,642	38,820	17,844	Flora MacDonald	Kingston	PC
Kitchener	106,127	66,445	48,427	21,091	P. Flynn	Kitchener	Lib.
Lambton—Kent	67,892	43,021	29,687	14,315	J.R. Holmes	Wallaceburg	PC
Lanark—Renfrew— Carleton	63,818	43,953	34,816	18,242	P. Dick	Kanata	PC
Leeds	66,263	43,597	33,375	17,724	T. Cossitt	Brockville	PC
Lincoln	84,935	52,148	39,087	17,499	B. Andres	Niagara-on-the-Lake	Lib.
London East	89,221	53,199	36,788	18,429	C.R. Turner	London	Lib.
London West	106,317	76,796	57,593	32,188	J. Buchanan	London	Lib.
Middlesex—London— Lambton	92,814	58,435	44,306	20,703	L. Condon	Strathroy	Lib.
Mississauga	172,532	118,909	87,492	38,517	T. Abbott	Oakville	Lib.
Niagara Falls	89,537	55,345	39,004	20,618	R. Young	Niagara Falls	Lib.
Nickel Belt	85,577	46,001	35,587	17,668	J. Rodriguez	Capreol	NDP
Nipissing	67,312	41,154	30,743	16,549	J.-J. Blais	North Bay	Lib.
Norfolk—Haldimand	74,568	47,842	37,046	17,867	W. Knowles	Langton	PC
Northumberland— Durham	73,705	50,584	38,965	16,824	A. Lawrence	Janetville	PC
Ontario	87,842	54,783	42,440	20,096	N.A. Cafik	Pickering	Lib.
Oshawa—Whitby	111,361	72,407	51,661	25,013	E. Broadbent[2]	Oshawa	NDP
Ottawa—Carleton	130,906	93,141	72,344	38,465	Hon. J. Turner	Ottawa	Lib.
Ottawa Centre	70,584	46,561	35,640	15,308	H. Poulin	Ottawa	Lib.
Ottawa—Vanier	71,277	45,372	32,365	21,773	J.-R. Gauthier	Ottawa	Lib.
Ottawa West	98,956	68,085	52,907	23,604	L. Francis	Ottawa	Lib.
Oxford	80,336	51,752	40,937	18,934	B. Halliday	Tavistock	PC
Parry Sound—Muskoka	62,162	43,279	31,807	14,030	S. Darling	Burks Falls	PC
Peel—Dufferin— Simcoe	119,885	79,104	58,927	27,298	R. Milne	Brampton	Lib.
Perth—Wilmot	72,996	48,075	35,226	17,636	B. Jarvis	Stratford	PC
Peterborough	85,064	58,250	45,235	23,865	Hon. J.H. Faulkner	Lakefield	Lib.
Port Arthur	57,456	36,647	26,756	14,523	Hon. R.K. Andras	Thunder Bay	Lib.
Prince Edward— Hastings	74,856	48,835	35,844	19,219	Hon. G. Hees	Cobourg	PC
Renfrew North— Nipissing East	61,707	35,060	26,701	14,613	L.D. Hopkins	Petawawa	Lib.
Sarnia—Lambton	83,461	51,728	36,917	20,661	J. Cullen	Sarnia	Lib.
Sault Ste Marie	81,002	47,643	37,898	19,050	C. Symes	Sault Ste Marie	NDP
St. Catharines	101,448	68,509	49,011	22,526	G. Parent	St. Catharines	Lib.
Simcoe North	93,655	64,616	47,326	18,857	P.B. Rynard	Orillia	PC
Stormont—Dundas	72,052	46,332	34,669	18,047	E. Lumley	Cornwall	Lib.
Sudbury	94,624	57,344	44,579	23,374	J. Jerome[3]	Sudbury	Lib.
Thunder Bay	53,214	32,842	21,029	11,435	K. Penner	Dryden	Lib.

3.4 Electoral districts, voters on the list, votes polled and names and addresses of members of the House of Commons as elected at the thirtieth general election, July 8, 1974 (continued)

Province and electoral district	Population, Census 1971	Voters on list	Total votes polled (incl. rejections)	Votes polled by member	Name of member	Postal address	Party affiliation[1]
ONTARIO (concluded)							
Timiskaming	49,870	30,205	22,061	10,263	A. Peters	New Liskeard	NDP
Timmins	53,616	33,297	25,379	12,904	J. Roy	Timmins	Lib.
Victoria—Haliburton	60,996	44,758	32,932	17,570	W.C. Scott	Kinmount	PC
Waterloo—Cambridge	120,719	81,248	61,466	25,479	M. Saltsman	Cambridge	NDP
Welland	82,860	53,088	37,845	21,228	V. Railton	Port Colborne	Lib.
Wellington	75,989	50,983	39,421	18,139	F. Maine	Guelph	Lib.
Wellington—Grey—Dufferin—Waterloo	73,846	48,948	34,125	17,253	P. Beatty	Fergus	PC
Windsor—Walkerville	87,514	52,447	40,099	18,977	M. MacGuigan	Windsor	Lib.
Windsor West	90,466	51,150	35,056	19,474	Hon. H. Gray	Windsor	Lib.
York North	125,296	87,554	68,819	34,179	B.J. Danson	Willowdale	Lib.
York—Simcoe	99,624	67,128	50,331	23,591	S. Stevens	King City	PC
Metropolitan Toronto							
Broadview	78,601	35,119	23,754	9,392	J. Gilbert	Toronto	NDP
Davenport	84,780	27,622	20,851	12,294	C.L. Caccia	Toronto	Lib.
Don Valley	104,606	71,345	57,702	29,180	J. Gillies	Ottawa	PC
Eglinton	78,314	52,008	41,133	19,951	Hon. M. Sharp	Ottawa	Lib.
Etobicoke	135,971	92,769	73,969	37,847	Hon. A. Gillespie	Toronto	Lib.
Greenwood	80,797	43,720	31,992	11,038	A.F. Brewin	Toronto	NDP
High Park—Humber Valley	86,050	53,016	40,997	17,389	O. Jelinek	Toronto	PC
Parkdale	82,207	36,612	25,307	13,134	Hon. S. Haidasz	Toronto	Lib.
Rosedale	81,265	47,645	35,723	17,227	Hon. D.S. Macdonald	Toronto	Lib.
St. Paul's	72,174	46,761	35,561	16,100	J. Roberts	Toronto	Lib.
Scarborough East	149,514	89,142	65,578	30,586	M. O'Connell	Scarborough	Lib.
Scarborough West	87,383	51,012	37,691	13,702	A. Martin	Scarborough	Lib.
Spadina	75,487	24,711	17,506	9,393	P. Stollery	Toronto	Lib.
Toronto—Lakeshore	77,227	46,537	35,713	14,241	K. Robinson	Toronto	Lib.
Trinity	81,073	27,364	20,395	10,683	A. Nicholson	Toronto	Lib.
York Centre	160,051	88,379	64,350	31,792	B. Kaplan	Toronto	Lib.
York East	102,910	65,188	48,709	20,682	D. Collenette	Toronto	Lib.
York—Scarborough	193,156	124,638	98,594	47,450	Hon. R. Stanbury	Don Mills	Lib.
York South	85,768	38,715	29,191	12,473	U. Appolloni	Toronto	Lib.
York West	139,650	72,177	53,391	28,075	J. Fleming	Weston	Lib.
MANITOBA (13 members)							
Brandon—Souris	62,547	41,314	28,856	16,624	Hon. W.G. Dinsdale	Brandon	PC
Churchill	77,507	45,107	27,476	11,415	C. Smith	Thompson	PC
Dauphin	54,110	34,440	24,611	11,439	G. Ritchie	Dauphin	PC
Lisgar	56,974	35,657	23,375	16,478	J.B. Murta	Graysville	PC
Marquette	54,070	33,253	23,885	16,033	C. Stewart	Minnedosa	PC
Portage	51,951	31,138	21,731	11,829	P.P. Masniuk	Inwood	PC
Provencher	62,089	37,738	24,590	13,405	J. Epp	Steinbach	PC
St. Boniface	103,943	69,572	51,522	21,812	J.-P. Guay	St. Boniface	Lib.
Selkirk	98,106	66,551	50,643	22,441	D. Whiteway	Winnipeg	PC
Winnipeg North	83,845	54,039	37,006	15,026	D. Orlikow	Winnipeg	NDP
Winnipeg North Centre	73,559	43,086	27,160	12,023	S.H. Knowles	Winnipeg	NDP
Winnipeg South	94,743	65,017	50,888	23,297	Hon. J. Richardson	Winnipeg	Lib.
Winnipeg South Centre	114,803	76,499	56,688	32,277	D. McKenzie	Winnipeg	PC
SASKATCHEWAN (13 members)							
Assiniboia	57,131	33,321	26,852	9,986	R. Goodale	Wilcox	Lib.
Battleford—Kindersley	66,855	38,453	29,201	10,751	C. McIsaac	North Battleford	Lib.
Mackenzie	47,919	29,049	20,600	8,292	S.J. Korchinski	Rama	PC
Meadow Lake	50,391	28,281	19,363	7,419	B. Cadieu	Spiritwood	PC
Moose Jaw	61,810	37,703	28,282	11,678	D. Neil	Moose Jaw	PC
Prince Albert	72,195	44,292	30,167	17,787	Rt. Hon. J.G. Diefenbaker	Prince Albert	PC
Qu'Appelle—Moose Mountain	64,000	37,174	27,813	13,124	Hon. A. Hamilton	Estevan	PC
Regina East	89,048	56,872	40,883	15,030	J. Balfour	Regina	PC
Regina—Lake Centre	97,537	64,657	47,449	16,874	L. Benjamin	Regina	NDP
Saskatoon—Biggar	87,303	56,160	37,781	14,296	R. Hnatyshyn	Saskatoon	PC
Saskatoon—Humboldt	102,185	63,682	47,362	23,242	Hon. O.E. Lang	Saskatoon	Lib.
Swift Current—Maple Creek	60,972	36,309	27,360	11,336	F. Hamilton	Swift Current	PC
Yorkton—Melville	68,896	43,363	32,155	14,586	L. Nystrom	Yorkton	NDP
ALBERTA (19 members)							
Athabasca	67,746	36,601	20,887	13,157	P. Yewchuk	Lac La Biche	PC
Battle River	59,545	35,138	24,770	16,819	A. Malone	Rosalind	PC
Calgary Centre	87,346	56,802	37,834	23,810	H. Andre	Calgary	PC
Calgary North	118,118	67,720	46,796	30,102	E. Woolliams	Calgary	PC
Calgary South	133,796	86,392	61,108	41,530	P. Bawden	Calgary	PC
Crowfoot	55,672	34,635	23,956	18,048	J. Horner	Pollockville	PC
Edmonton Centre	94,410	55,929	33,576	18,165	S.E. Paproski	Edmonton	PC
Edmonton East	105,904	57,434	35,444	18,293	W.M. Skoreyko	Sherwood Park	PC
Edmonton—Strathcona	109,725	69,820	47,278	25,808	D. Roche	Edmonton	PC
Edmonton West	126,765	86,100	57,430	29,990	Hon. M. Lambert	Edmonton	PC
Lethbridge	75,795	47,857	32,545	20,602	K. Hurlburt	Fort Macleod	PC

3.4 Electoral districts, voters on the list, votes polled and names and addresses of members of the House of Commons as elected at the thirtieth general election, July 8, 1974 (concluded)

Province and electoral district	Population, Census 1971	Voters on list	Total votes polled (incl. rejections)	Votes polled by member	Name of member	Postal address	Party affiliation[1]
ALBERTA (concluded)							
Medicine Hat	62,697	39,647	28,391	15,525	B. Hargrave	Walsh	PC
Palliser	100,115	72,591	49,750	34,184	S. Schumacher	Drumheller	PC
Peace River	62,413	36,588	23,430	14,153	G. Baldwin	West Peace River	PC
Pembina	94,678	61,731	44,161	19,172	P. Elzinga	Sherwood Park	PC
Red Deer	78,792	48,123	33,438	22,251	T.G. Towers	Red Deer	PC
Rocky Mountain	63,834	40,711	26,322	16,042	J. Clark	Edson	PC
Vegreville	58,986	35,509	25,537	18,328	D. Mazankowski	Vegreville	PC
Wetaskiwin	71,537	46,718	31,996	21,341	S. Schellenberger	Spruce Grove	PC
BRITISH COLUMBIA (23 members)							
Burnaby—Richmond— Delta	123,381	82,889	62,180	34,013	J. Reynolds	Delta	PC
Burnaby—Seymour	103,410	66,348	49,521	18,058	M. Raines	Burnaby	Lib.
Capilano	103,918	66,491	52,838	25,797	R. Huntington	West Vancouver	PC
Coast Chilcotin	67,858	43,695	29,423	10,336	J. Pearsall	Powell River	Lib.
Comox—Alberni	89,644	56,598	38,158	13,594	H. Anderson	Port Alberni	Lib.
Esquimalt—Saanich	105,411	74,919	55,299	27,571	D. Munro	Victoria	PC
Fraser Valley East	85,401	55,196	40,656	18,780	A.B. Patterson	Abbotsford	PC
Fraser Valley West	117,467	77,253	55,522	22,925	R. Wenman	Surrey	PC
Kamloops—Cariboo	104,739	68,925	49,196	20,474	L.S. Marchand	Kamloops	Lib.
Kootenay West	67,513	42,245	30,373	12,575	B. Brisco	Trail	PC
Nanaimo—Cowichan— The Islands	97,106	68,325	50,125	20,434	T.C. Douglas	Nanaimo	NDP
New Westminster	106,331	66,956	47,000	15,397	S.M. Leggatt	Port Coquitlam	NDP
Okanagan Boundary	101,304	74,876	53,665	23,044	G.H. Whittaker	Kelowna	PC
Okanagan—Kootenay	92,717	61,905	43,269	17,164	H. Johnston	Salmon Arm	PC
Prince George— Peace River	108,022	60,773	40,186	18,769	F. Oberle	Chetwynd	PC
Skeena	87,917	43,901	30,498	12,218	I. Campagnolo	Prince Rupert	Lib.
Surrey—White Rock	104,072	68,892	49,472	21,540	B. Friesen	White Rock	PC
Vancouver Centre	91,473	67,222	45,929	19,064	Hon. R. Basford	Vancouver	Lib.
Vancouver East	85,071	41,537	26,710	9,671	A. Lee	Vancouver	Lib.
Vancouver Kingsway	85,005	46,322	32,268	12,002	S. Holt	Vancouver	Lib.
Vancouver Quadra	79,949	49,692	39,277	18,892	B. Clarke	Vancouver	PC
Vancouver South	88,701	57,141	44,078	23,247	J.A. Fraser	Vancouver	PC
Victoria	88,211	64,965	48,576	26,771	A.B. McKinnon	Victoria	PC
YUKON TERRITORY (1 member)							
Yukon	18,388	12,312	8,354	3,913	E. Nielsen	Whitehorse	PC
NORTHWEST TERRITORIES (1 member)							
Northwest Territories	34,807	21,299	13,008	5,410	W. Firth	Yellowknife	NDP

[1]Party standings as a result of the general election, July 8, 1974: Liberal 141, Progressive Conservative 95, New Democratic 16, Social Credit 11, Independent 1.
[2]Leader of a political party.
[3]Speaker of the House of Commons.

3.5 Voters on the lists and votes polled at the federal general elections of 1965, 1968, 1972 and 1974

Province or territory	Voters and votes polled			
	Voters on the lists			
	1965	1968	1972	1974
Newfoundland	226,082	237,594	289,294	304,370
Prince Edward Island	56,484	58,216	68,992	73,069
Nova Scotia	401,521	412,791	492,001	524,767
New Brunswick	304,734	317,912	387,136	406,518
Quebec	2,933,031	3,083,260	3,693,918	3,849,009
Ontario	3,609,895	3,846,064	4,601,282	4,803,822
Manitoba	517,928	531,563	610,568	633,411
Saskatchewan	508,733	517,598	558,876	569,316
Alberta	725,447	774,565	955,531	1,016,046
British Columbia	972,063	1,059,959	1,312,832	1,407,066
Yukon Territory[3]	6,660	7,559	10,857	12,312
Northwest Territories[4]	12,326	13,807	19,491	21,299
Total	10,274,904	10,860,888	13,000,778	13,621,005

3.5 Voters on the lists and votes polled at the federal general elections of 1965, 1968, 1972 and 1974 (concluded)

Province or territory	Voters and votes polled			
	Votes polled			
	1965	1968	1972	1974
Newfoundland	148,392	161,570	182,482	175,534
Prince Edward Island	72,006[1]	51,225	59,078	58,649
Nova Scotia	420,146[2]	339,600	391,590	388,830
New Brunswick	244,184	254,716	298,164	289,492
Quebec	2,073,314	2,229,345	2,790,172	2,592,679
Ontario	2,770,222	2,973,745	3,650,542	3,582,489
Manitoba	382,362	403,272	453,642	448,431
Saskatchewan	404,631	416,793	442,246	415,268
Alberta	534,870	567,416	722,338	684,649
British Columbia	731,438	804,108	961,441	1,014,219
Yukon Territory[3]	5,760	6,563	8,638	8,354
Northwest Territories[4]	9,403	9,563	14,328	13,008
Total	7,796,728	8,217,916	9,974,661	9,671,602

[1]Each voter in the double-member constituency of Queens County, PEI had two votes; in 1963, 26,472 voters on the list cast 42,703 votes; in 1965, 26,250 voters on the list cast 44,895 votes.
[2]Each voter in the double-member constituency of Halifax, NS had two votes; in 1963, 122,846 voters on the list cast 183,402 votes; in 1965, 124,633 voters on the list cast 184,153 votes.
[3]Electoral district of Yukon.
[4]Electoral district of Northwest Territories.

3.6 Number of municipalities classified by type and size group, by province, as at Jan. 1, 1976 and 1977

Year, type and size group	Nfld.	PEI	NS	NB	Que.	Ont.	Man.	Sask.	Alta.	BC	YT	NWT	Canada
1976													
TYPE													
Regional municipalities	—	—	—	—	75	39	—	—	—	28	—	—	142
Metropolitan and regional municipalities[1]	—	—	—	—	3	12	—	—	—	—	—	—	15
Counties and regional districts	—	—	—	—	72	27	—	—	—	28	—	—	127
Unitary municipalities	127	35	65	111	1,500	782	183	784	327	140	3	7	4,064
Cities[2]	2	1	3	6	64	45	5	11	9	33	2	1	182
Towns	125[3]	8	38	20	195	142	33	132	103	10	1	4	811
Villages	—	26	—	85	242	121	40	348	167	60	—	2	1,091
Rural municipalities[4]	—	—	24	—	999	474	105	293	48	37	—	—	1,980
Quasi-municipalities[5]	174	—	—	—	—	13	19	9	22	—	4	10	251
Total	301	35	65	111	1,575	834	202	793	349	168	7	17	4,457
POPULATION SIZE GROUP (1971 Census)													
Unitary municipalities													
Over 100,000	—	—	1	—	3	14	1	2	2	2	—	—	25
50,000-99,999	1	—	2	1	10	13	—	—	—	6	—	—	33
10,000-49,999	1	1	16	6	70	66	3	5	12	29	1	—	210
Under 10,000	125	34	46	104	1,417	689	179	777	313	103	2	7	3,796
Total	127	35	65	111	1,500	782	183	784	327	140	3	7	4,064
1977[6]													
TYPE													
Regional municipalities	—	—	—	—	75	39	—	—	—	28	—	—	142
Metropolitan and regional municipalities[1]	—	—	—	—	3	12	—	—	—	—	—	—	15
Counties and regional districts	—	—	—	—	72	27	—	—	327	28	—	—	127
Unitary municipalities	129	36	65	112	1,500	784	185	783	327	140	3	7	4,071
Cities[2]	2	1	3	6	64	45	5	11	10	33	2	1	183
Towns	127[3]	8	38	21	195	144	35	135	102	10	1	4	820
Villages	—	27	—	85	242	120	40	344	167	59	—	2	1,086
Rural municipalities[4]	—	—	24	—	999	475	105	293	48	38	—	—	1,982
Quasi-municipalities[5]	171	—	—	—	—	13	17	7	22	—	4	10	244
Total	300	36	65	112	1,575	836	202	790	349	168	7	17	4,457

3.6   Number of municipalities classified by type and size group, by province, as at Jan. 1, 1976 and 1977 (concluded)

Year, type and size group	Nfld.	PEI	NS	NB	Que.	Ont.	Man.	Sask.	Alta.	BC	YT	NWT	Canada
1977[6] (concluded)													
POPULATION SIZE GROUP (1976 Census)													
Unitary municipalities													
Over 100,000	—	—	1	—	4	17	1	2	2	3	—	—	30
50,000-99,999	1	—	2	2	14	14	—	—	—	9	—	—	42
10,000-49,999	5	1	17	5	72	76	3	6	14	26	1	—	227
Under 10,000	123	35	45	105	1,410	677	181	775	311	102	2	7	3,772
Total	129	36	65	112	1,500	784	185	783	327	140	3	7	4,071

[1]Includes urban communities in Quebec; and Metropolitan Toronto, regional municipalities and the district municipality in Ontario.
[2]Includes the five boroughs of Metropolitan Toronto.
[3]Includes 11 rural districts.
[4]Includes municipalities in Nova Scotia; parishes, townships, united townships and municipalities in Quebec; townships in Ontario; rural municipalities in Manitoba and Saskatchewan; municipal districts and counties in Alberta; and districts in British Columbia.
[5]Includes local government communities, local improvement districts and the metropolitan area in Newfoundland; improvement districts in Ontario and Alberta; local government districts in Manitoba; local improvement districts in Saskatchewan and the Yukon Territory; and hamlets in the Northwest Territories.
[6]Information for Quebec municipalities not available for 1977. Data for 1976 has been repeated.

3.7   Canadian financial contributions[1] to the United Nations and specialized agencies, years ended Mar. 31, 1976 and 1977 with totals for 1945-77 (thousand dollars Canadian)

Agency	1976	1977	Total 1945-77
UN regular budget	9,856	9,593	101,450
Peacekeeping			
UNEF I	—	—	5,910
ONUC	—	—	9,187
UNFICYP	1,930	2,600	30,412
UNEF II	4,620	2,844	11,221
Social and economic programs			
UNDP	24,500	29,435	190,506
UNHCR	750[r,2]	750	40,082
UNICEF	3,500	5,000	38,875
UNRWA	1,350	1,650	34,403
UNITAR	60	70	730
UNETPSA	175	225	874
WFP	10,000	10,998[3]	152,250
UNFPA	3,500	5,000	18,052
Committee on Elimination of Racial Discrimination	3	4	17
Trust Fund for South Africa	10	10	70
UN Fund for Drug Abuse Control	200	200	1,150
UN Voluntary Fund for Environment	—	3,618	3,622
Specialized agencies, IAEA and GATT			
ILO	2,761	2,386	21,953
FAO	3,321	3,293	28,971
WHO	3,676[r]	2,650	35,799
UNESCO	2,491	2,523	25,140
ICAO	443	478	7,077
IMCO	42	50	349
ITU	690	928	5,719
WMO	243	149	1,538
UPU	130	292	1,298
IAEA regular and operational budgets	1,155	1,514	8,734
GATT	651	697	4,595
WIPO	145[r]	207	465
UN Association in Canada	35	55	525
Total	76,237[r]	87,219	780,974

[1]Canada ranks sixth to eighth largest contributor to the budget of the United Nations and its related agencies.
[2]Does not include special appeals.
[3]Does not include a contribution of approximately $78 million (Canadian).

3.8 Federal government, employees of general government by metropolitan area

Metropolitan area	September 1975[1]			September 1976[1]		
	Male	Female	Total	Male	Female	Total
St. John's, Nfld.	2,286	544	2,830	2,340	565	2,905
Halifax, NS	9,510	2,795	12,305	9,159	2,849	12,008
Saint John, NB	1,142	366	1,508	1,180	395	1,575
Quebec, Que.	5,344	1,610	6,954	5,894	1,940	7,834
Montreal, Que.	20,511	7,663	28,174	20,939	8,179	29,118
Chicoutimi-Jonquière, Que.	544	134	678	278	72	350
Ottawa, Ont.-Hull, Que.	53,223	35,933	89,156	52,539	36,230	88,769
Toronto, Ont.	17,143	7,666	24,809	18,622	8,938	27,560
London, Ont.	2,536	1,531	4,067	2,463	1,452	3,915
Hamilton, Ont.	1,976	901	2,877	2,094	963	3,057
Windsor, Ont.	1,173	483	1,656	1,151	499	1,650
Kitchener, Ont.	928	430	1,358	984	433	1,417
St. Catharines-Niagara, Ont.	1,150	461	1,611	895	439	1,334
Thunder Bay, Ont.	710	258	968	869	288	1,157
Sudbury, Ont.	389	297	686	422	321	743
Oshawa, Ont.	[2]	[2]	[2]	273	151	424
Winnipeg, Man.	5,674	2,739	8,413	6,044	3,228	9,272
Regina, Sask.	2,490	860	3,350	2,253	904	3,157
Saskatoon, Sask.	1,319	571	1,890	1,356	598	1,954
Edmonton, Alta.	4,632	2,704	7,336	4,537	2,840	7,377
Calgary, Alta.	2,964	1,598	4,562	2,879	1,683	4,562
Vancouver, BC	9,451	4,169	13,620	9,972	4,429	14,401
Victoria, BC	4,826	1,246	6,072	4,293	1,239	5,532
Total	149,921	74,959	224,880	151,436	78,635	230,071

[1]Excludes Canadian Armed Forces.
[2]Oshawa became a metropolitan area in 1976.

Sources

3.1 Committees Branch, House of Commons.
3.2 Clerk of the Senate.
3.3 Office of the Representation Commissioner.
3.4 - 3.5 Office of the Chief Electoral Officer.
3.6 Public Finance Division, Institutional and Public Finance Statistics Branch, Statistics Canada.
3.7 Information Services Division, Department of External Affairs.
3.8 Public Finance Division, Institutional and Public Finance Statistics Branch, Statistics Canada.

Demography   Chapter 4

Tables

Demography

Chapter 4

Population growth   4.1

The most fundamental fact about a population is its rate of growth which affects almost
every aspect of the national life. The opening up of a new continent and the gradual
evolution of an industrial and urban economy form the historical background for
population growth in Canada. Several demographic elements have combined to produce
this growth: births, deaths, immigration and emigration, which are the processes, or
components, of population change.

The early period   4.1.1

The growth of Canada's population today is the culmination of a trend which began
early in the 17th century with the arrival of the first French settlers. From this
beginning, the population of the area now known as Canada (excluding Newfoundland)
grew from a handful of colonists and an unknown number of native Indians and
Eskimos in 1611 to about 2.4 million in 1851 and 3.7 million at the first census of
Canada in 1871. Rough estimates suggest there were about 136,000 Indians in 1851.

Growth rates in the early settlement years were irregular. The immigrant
population grew rapidly while the native population remained almost stationary or
declined as a result of attrition from warfare and disease. Between 1681, when the
number of settlers passed the 10,000 mark, and 1851 the average annual growth rate of
the non-native population in each decade varied between 1.6% and 4.5%; the average
annual growth rate for the whole period was 3.2%. The small size of the initial
population and the continuous expansion into empty lands were contributing factors in
the rapid growth rates in the early periods.

The decade 1851-61 was one of surging growth, second only to the growth rate in
the first decade of the present century (Table 4.1). The average annual growth rate
during this period was 2.9%, with about 23.0% of the population increase due to net
migration; over 350,000 immigrants arrived and there was very little emigration. A long
period of slow growth followed and lasted until the beginning of the 20th century.
Between 1861 and 1901 the average growth rate was closer to 1.0%, matched only by the
rate during the depression period of the 1930s. This slow growth toward the end of the
last century was due to heavy emigration resulting in a net migration loss (Table 4.2).
Emigrants included elements of both the Canadian and foreign-born populations. While
many immigrants continued to come to Canada during this period a large number of
them re-emigrated to the United States where prospective settlers found more
favourable economic and climatic conditions. The westward movement in the United
States attracted not only settlers from many parts of that country, but from Canada as
well.

Recent trends   4.1.2

The beginning of the present century witnessed a flood of immigrants which helped to
raise the growth rate to 3% per annum during 1901-11, the highest rate since 1851. Over
1.5 million immigrants entered Canada in this decade, as many as had arrived during the
previous 40 years. Over 44% of the total population increase during this period was due
to migration gain.

Following this phenomenal increase, the intercensal rate of increase dropped
during each successive decade until it reached a low of 10.9% during 1931-41 when
reduced birth rates during the economic depression seriously affected population
growth; immigration was negligible, and there was a net migration loss of about 92,000
persons.

After 1941, population growth again accelerated, reaching a near-record expansion
rate of 30.2% in 1951-61, nearly three times that in 1931-41. Part of the increase after
1941 was due to the addition of Newfoundland in 1949, but the surge in birth rates

(commonly referred to as the baby boom) and the upswing in immigration during the immediate post-war years were the main factors.

After 1956 population growth declined from an average 2.8% per annum in 1951-56 to 1.5% in 1966-71. This gradual fall in the growth rate, the lowest except during the depression decade, has evoked special interest mainly because it occurred after the peak of 3.3% in 1956-57 and at a time when the economic outlook was favourable for high growth rates. Despite this trend, with a further drop to 1.3% per annum in 1971-76, Canada still ranks as one of the industrialized countries with the fastest population growth. During the 12-month period following the 1976 Census, the Canadian population increased by 298,500 persons, at a rate nearly equalling the average for the five-year period 1971-76, despite a decrease in the number of immigrants entering Canada.

4.1.3 Future prospects

The dominant component of population growth in Canada since 1851 has been natural increase (births minus deaths). This trend is likely to continue with a modest contribution from migration. Of the two components of natural increase, the birth rate will continue to be the dynamic and crucial factor of growth. Moreover, fluctuations in birth rates can create major economic and social problems as society adjusts to the effects of such fluctuations. For example, although the post-war baby boom is long past, society is now feeling the impact of this generation on the labour market and other aspects of the national economy. Similarly, problems associated with the sharp drop in the birth rate since 1957 are being felt by school systems as fewer children enter school.

Because of the importance of the fertility factor, the tempo of future growth depends mainly on whether the total fertility rate of 2.19 births (1971), which is close to the replacement level of 2.13 births under existing mortality conditions, will remain constant, fall or rise. A fertility rate close to the replacement level does not mean that Canada will soon reach zero population growth. Calculations show that even if immigration ceased and the average fertility rate were only 2.13, the population would continue to grow until about the year 2040, when birth and death rates would each stabilize at about 13 per 1,000 population. This long delay in achieving zero growth may be attributed to the current high percentage of young people who are moving into the child-bearing age groups.

Table 4.3 summarizes the results of population projections for Canada and the provinces prepared under different assumptions of fertility and migration. For a full account of the methodology and results of these projections, see *Population projections for Canada and the provinces, 1972-2001* (Statistics Canada Catalogue 91-514).

Projection A uses the highest fertility assumption of 2.60 children by 1985, and a net migration gain of 100,000 a year. Under these assumptions, the total population would increase from 21.6 million in 1971 to 27.8 million in 1986 and 34.6 million by 2001. On the other hand, projection C is a low projection based on an assumed fertility rate of 1.80 by 1985, and a net migration gain of 60,000 a year. This projection yields a total population of 25.4 million by 1986 and 28.4 million by 2001.

These projections indicate that after a short phase of increase in the population growth rate (between 1976 and 1986), the rate will gradually decline toward the end of the century to about 1.3% per annum according to projection A, and to 0.6% under projection C. The slowdown in population growth and fertility rates will cause some aging of Canada's population. With an upward shift in the age structure, there will be a steady decline in the child-dependency ratio and an increase in the old age dependency ratio.

4.2 Population distribution

Decennial and quinquennial censuses of Canada make possible periodic assessments of the nation's human resources. They provide data on the distribution of population for many types of geographical, political and statistical entities. Used as benchmarks, the census counts enable annual estimates to be made for some of the larger areas such as provinces, counties and metropolitan areas. A small selection of these data is presented in this section, embodying results of the 1971 Census and some published data from the 1976 Census.

The 1976 Census   4.2.1

Canada took its third quinquennial census on June 1, 1976. The aim of this census was to keep statistical information abreast of the demographic and socio-economic factors that form the foundation for decision making in both the private and public sectors. The census is a principal source of information for measuring social and economic progress, and for detecting those needs which necessitate the development and implementation of policies and programs such as regional development, health and welfare programs, education facilities, immigration, low income housing and transportation networks.

Questions of the 1976 Census covered age, sex, marital status, relationship to head of household, and mother tongue (the language first learned and still understood). These were asked of every person whose usual place of residence on June 1, 1976 was in Canada. In addition there were questions on type and tenure of housing of Canadian households. Sampling was also used in the 1976 Census. Persons 15 years of age and over of every third private household provided answers for school attendance, level of schooling, labour force activity and migration (place of residence five years ago).

The population of Canada at June 1, 1976 was 22,992,604, an increase of 6.6% or 1,424,693 from the previous census in 1971. Alberta and British Columbia recorded the largest growth rates at 12.9% each. Ontario was next at 7.3% (Table 4.4).

Provincial and sub-provincial areas   4.2.2

The basic legal reason for decennial censuses is to enable a redistribution of seats in the House of Commons. Under the terms of the Electoral Boundaries Readjustment Act, the census must provide population counts by electoral districts. Those from the 1971 Census are shown in Chapter 3, Table 3.4, according to the electoral district boundaries established by the 1966 Representation Order (the redistribution following the 1961 Census).

Provincial trends, 1951-76. Ontario, Alberta, British Columbia and the Northwest Territories had growth rates higher than national figures in all five-year periods between 1951 and 1976 (Table 4.4). However, a decline in the rate of growth occurred in all provinces as birth rates began to fall in the mid-1950s. The most spectacular change took place in Quebec where the rate of growth declined by about 76% between 1951-56 and 1971-76 (from 14.1% to 3.4%).

The lowest rates for 1971-76 were in the Atlantic provinces, Quebec, Manitoba and Saskatchewan. Saskatchewan registered a decline while the Northwest Territories had the highest growth rate (22.4%) followed by the Yukon Territory, Alberta and BC.

Provincial estimates. In addition to the five-year census, estimates are constructed for the total population of Canada and for each province on both an annual and quarterly basis. The estimates of population begin with the preceding census counts, the births of each year are added and the deaths subtracted; immigrants are added and an estimate of emigrants subtracted. Family allowance statistics showing the number of migrant families by province are used in estimating interprovincial shifts in population. The next census serves as a basis for revision of all annual estimates of each intercensal period. Table 4.6 shows population by province for selective years, with 1977 preliminary estimates.

Cities, towns and villages. As at June 1, 1976, some 67.4% of Canada's population lived in 2,079 centres classified as incorporated cities, towns and villages. These are grouped into 13 broad size categories in Table 4.7. Canadian cities and towns having a population of over 50,000 in 1976 are listed in Table 4.8 together with figures for 1966 and 1971. The date of incorporation to their present status is indicated also.

Census terms. A census agglomeration (CA) is an area comprised of at least two adjacent municipal entities, each at least partly urban. Its urbanized core is a continuous

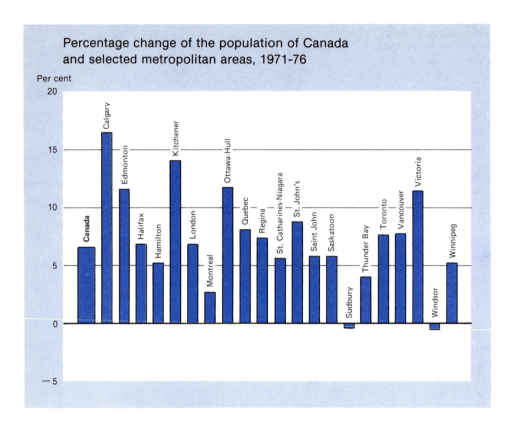

Percentage change of the population of Canada
and selected metropolitan areas, 1971-76

built-up area including the largest city and, where applicable, the urban part of surrounding municipalities, the urban fringe and rural fringe. A CA with an urbanized core of 100,000 or more, based on previous census figures, is called a census metropolitan area (CMA). Usually the CMA or CA takes the name of its largest component city.

Metropolitan areas. For census purposes a metropolitan area represents the main labour market of a continuous built-up area having a population of 100,000 or more. The growth of 22 census metropolitan areas over the period 1951-71 appears in Table 4.9. Populations of these areas in earlier censuses were adjusted to conform to the boundaries delineated for the 1971 Census. The 1976 Census population figures have been added, based on 1976 census metropolitan areas. The 1976 Census saw the addition of a 23rd census metropolitan area, Oshawa. Population figures from the 1971 Census have been adjusted to conform to the boundaries delineated for the 1976 Census for this metropolitan area.

The proportion of Canada's population in the major metropolitan centres increased steadily and over one-half (55.7%) resided in the 23 metropolitan areas as defined for the 1976 Census. Calgary CMA showed the highest rate of growth in the period 1971-76 at 16.5%, followed by Kitchener at 14.1% (based on 1976 areas). The greatest gains in numbers were registered by Toronto at 175,058 and Vancouver at 83,996 (based on 1976 areas). The Toronto CMA became the largest in Canada, with a population of 2,803,101, some 7.7% more than in 1971, while Montreal was close behind at 2,802,485, showing a growth of 2.7% since 1971 (Table 4.9).

Because of the growing interest in the expanding metropolitan areas a series of intercensal estimates was begun in 1957. As in the preparation of intercensal population estimates for provinces, the births in the metropolitan areas were added to the census population and deaths subtracted. Immigrants reporting these metropolitan areas as places of destination were added and allowances made for losses by emigration. Also, the net in-movement by internal migration was calculated from family allowances and other data.

Population density   4.2.3
At 2.49 persons a square kilometre in 1976, Canada's average population density still ranks among the lowest in the world. Table 4.10 shows that if the Yukon Territory and Northwest Territories were omitted from this calculation, there would be 4.21 persons/km² in 1976 compared to 3.67 persons/km² in 1966 and 2.95 persons/km² in 1956. However, such average density figures over all types of land terrain and open spaces in the country or in individual provinces obscure the high urban densities which can reach close to 7,722 persons/km² as in Montreal and Toronto. Moreover, the highest provincial densities are not necessarily found among the provinces with the largest populations. For example, the highest average density of any province is that of Prince Edward Island (20.90 persons/km²) which has the smallest population and represents an anomaly resulting from its limited land area rather than from heavy concentrations of population. In contrast, the far more populous British Columbia, with its vast mountainous regions and areas of sparse population, has an average density of only 2.65 persons/km².

Urban and rural   4.2.4
The urban population was defined in the 1976 Census as all persons living in an area having a population concentration of 1,000 or more and a population density of at least 386 a square kilometre. All the remaining population was classified as rural.

Over 75.5% of Canada's population lived in an urban environment, with the degree of urbanization ranging from 37.1% in Prince Edward Island to 81.2% in Ontario. In comparison with the national average, only Ontario, Quebec and British Columbia were more highly urbanized (Table 4.11).

The rural population, 24.5% of the Canadian total in 1976, is classified in Table 4.11 as non-farm or farm. The rural farm population was defined for census purposes as persons living in rural areas on an agricultural holding of at least 0.4 hectares (one or more acres) with sales of agricultural products amounting to $1,200 or more in the previous year. The rural non-farm category in 1976 accounted for 20.0% of the population, compared to 4.5% for the rural farm segment.

Demographic and social characteristics   4.3

Sex, age and marital status   4.3.1
The distribution of a population by age, sex and marital status represents the effect of the most fundamental variables of vital trends: births, deaths, marriages, and dissolutions of marriages. Social and economic factors, by their effects on vital events and migration, also influence this distribution. An unbroken series of census data is available as far back as the first census of Canada in 1871; only recent trends are summarized here.

Sex ratios. The demographic history of Canada has been characterized by an excess of males until recently. Thus, over the past century the sex ratio (number of males per 100 females) reached a peak of 113 in 1911 following a decade of heavy immigration in which males have traditionally predominated. By 1971 the sex ratio had almost evened out at 100.2 with only 22,425 more males than females in a total population of over 21.5 million. The 1976 Census was the first Canadian census to record more females than males, the sex ratio being 99.2 (Table 4.12). In the older settled provinces the sex ratio has varied between Nova Scotia's 104 in 1911, and Quebec's and Ontario's 1976 ratios of 98. In the West, which was being rapidly settled early in this century, the sex ratio has ranged between Alberta's 1911 high of 149 and Manitoba's 1976 figure of 99.

Age structure. Age composition is a reflection of past trends in vital rates and immigration. Lower birth rates of the 1961-76 period compared with those in the 1950s had a great impact on the population under 15 years of age in 1976. This group decreased by 485,000 or 7.6% between 1971 and 1976, compared with a loss of only 3.2% in 1966-71 and a gain of 6.4% in 1961-66 (the 1971-76 increase of the total population was 6.6%). The proportion of the age group 0-14 in the total population fell from 29.6% in 1971 to 25.6% in 1976 (Table 4.13).

The adult population (generally regarded as 15-64 years) increased substantially, with a gain of 1,651,000 or 12.3% in the 1971-76 period. This group comprised 65.6% of the total population in 1976 compared with 62.3% in 1971 and 59.4% in 1966.

The growth of the junior working ages (20-34) is of particular significance in the context of Canada's employment situation. In 1976 the count for this population group was 5,754,000, compared with 4,779,000 in 1971, a 20.4% increase. Furthermore, of the total 1971-76 population increase of 1,424,000 persons, the increase in the junior working ages accounts for as much as 68.5%. This group in 1976 corresponds mainly to the children born in the high-birth-rate years following World War II.

The aged population, persons 65 and over, increased 14.8% in the period 1971-76 from 1,744,000 to 2,002,000. In 1976 the proportion of aged persons in the total population was 8.7%, compared with 8.1% in 1971.

The population by age group and sex as at June 1, 1976 is shown in Table 4.14.

Marital status. The marital status composition of the 1976 population of Canada indicates increasing proportions of persons married (1976, 47.7%; 1971, 45.3%) and divorced (1976, 1.3%; 1971, 0.8%); a decreasing proportion of persons never married (1976, 46.4%; 1971, 49.5%); and a fairly stable proportion of persons widowed (1976, 4.5%; 1971, 4.4%). Generally, these trends also apply to each sex separately.

The most dramatic change concerns Canada's divorced population, the number of whom increased from 175,100 in 1971 to 302,500 in 1976 (a 73% increase); particularly, the number of divorced females increased from 100,800 in 1971 to 183,500 in 1976 (an 82% increase). This trend, the beginning of which was observed in 1971, reflects the more liberal divorce laws of 1968.

The overwhelming preponderance of females among widowed persons (widows, 853,900 or 82%; widowers, 189,700 or 18%) is a consequence mainly of higher age-specific rates of both mortality and remarriage among males than among females.

Analyses of marital status composition are most instructive when conducted in conjunction with sex and age, as in Table 4.15. Two major findings follow.

The years 1971-76 saw a decline in the proportion of married persons in the age range 25-34 and a corresponding increase in the proportion of single persons in that age range. (Data by quinquennial age group confirm this finding for virtually the entire young adult population.) The implication of this finding is that the 1971-76 increase in the proportion married in the population as a whole (as noted above) is a consequence of the changing age structure, rather than of a tendency to marry at a younger age. The same applies to the 1971-76 decrease in the proportion single in the population as a whole.

In connection with the recent decline in birth rates, the percentages of married women in the prime child-bearing ages 20-39 are: 1976, 11.6%; 1971, 10.6% and 1966, 10.2%. (Table 4.15 reports the figures for the age group 25-34, which show a similar trend.) That birth rates have been declining even though the population in the prime child-bearing group has increased further emphasizes the drop in birth rates over the last decade.

4.3.2   Language

In the 1976 Census, a question on language asked for mother tongue, the language first spoken in childhood and still understood, with spaces to mark English, French, German, Italian, Ukrainian, or other languages.

Mother tongue. Summary figures on mother tongue in Table 4.16 show the principal languages reported in the 1976 Census with comparative figures for 1971. The proportion of the Canadian population reporting English mother tongue increased from 60.2% in 1971 to 61.4% in 1976, while those reporting French declined from 26.9% to 25.6%. Chinese and Portuguese showed significant advances, while Ukrainian, German, Dutch, Polish and Yiddish were among those registering declines.

Table 4.17 shows the number and proportion of the population reporting English or French as their mother tongue, by province. The relative gains in English mother tongue over the 1971-76 period occurred mostly in the western provinces at the expense of others such as Ukrainian, German and Polish, as descendants of earlier immigrants reported English as their mother tongue to a greater extent than in previous decades.

It should be noted that the not stated category makes direct comparisons between 1971 and 1976 data problematic. In 1971, persons who did not report a mother tongue were assigned a language as mother tongue. Consequently, the 1976 counts are lower for any given language than they would have been if the 1971 procedure had been followed.

Official language. Table 4.18 shows 1971 Census figures on the population reporting the ability to speak one or both of Canada's two official languages, with comparative data for 1961. In 1971 a total of 67.1% were able to speak English only, 18.0% French only, and 13.4% were bilingual. These ratios represent a slight increase in the proportion able to speak both English and French over 1961, when the percentage was 12.2.

Language spoken in the home. This inquiry was introduced in the 1971 Census on the recommendation of the Royal Commission on Bilingualism and Biculturalism and other groups. It added insight into the languages of Canada since some persons, particularly immigrants, did not indicate either of the two official languages as the one they spoke most often in their homes. Conversely, many with a non-English mother tongue no longer used their mother tongue. Table 4.16 indicates that 67.0% of the population spoke English most often in their homes, whereas only 60.2% reported English as their mother tongue.

Ethnicity, religion, birthplace   4.3.3

Because of the varied nature of Canada's population, the measurements provided by decennial censuses on such subjects as ethnic and religious composition are of widespread interest and in continuous demand. Tables 4.19 to 4.21 show summary figures from the 1971 Census, with comparative data for earlier years.

Ethnic groups. The ethnic composition of Canada has changed considerably because of many factors, including differences in the flow and source of immigrants. Trends in recent years have been characterized by a decline in the proportions of British Isles groups and a corresponding increase in European ethnic groups other than French. For example, the former groups had dropped from 57.0% of the total population in 1901 to 44.6% by 1971, whereas other European groups rose from 8.5% to 23.0%. The French ethnic group remained relatively stable, varying from 30.7% in 1901 to 28.7% by 1971. Table 4.19 provides 1971 figures for the larger ethnic groups, together with data from 1951 and 1961.

Religious denominations. Census figures do not measure church membership or the degree of affiliation with a particular religious body. Respondents were asked to enter a specific religious denomination, sect or community, with the opportunity to report no religion if so desired. As shown in Table 4.20, three out of every four persons in Canada in 1971 reported one of the three numerically largest denominations — Roman Catholic, United Church or Anglican. Largest relative gains since 1961 occurred in such groups as Jehovah's Witnesses and Pentecostal. None of the major denominations registered numerical declines in the 1961-71 period, but the Anglican, Baptist, Lutheran, Presbyterian and United Church groups were among those showing percentage losses relative to the total population.

Country of birth. The proportion of the population born outside Canada ranged from a high of 22% throughout the period 1911-31 to a low of 15% in 1951 following a period of lower immigration and rising birth rates. Persons born in the United Kingdom comprised over 11% of the population in 1911 and 1921, but this declined gradually to 4.3% by 1971 because of the rising proportions of Canadian-born and immigration from other European countries. Persons born in the latter countries rose from 5.6% of Canada's population in 1911 to 7.8% in 1971 (Table 4.21).

4.3.4   The native peoples

Many centuries before the first European settlers arrived in what is now Canada, the country received immigrants in the prehistoric period. Present-day Inuit and Indians are the descendants of these early settlers but as a result of heavy immigration by other groups they now represent less than 2% of Canada's population. Administration relating to the affairs of the Indian and Inuit peoples is described in Chapter 3. Demographic data on their numbers and locations, from the 1971 Census summary figures, show a total of 295,215 native Indians and 17,550 Inuit. The former figure includes both registered or status Indians and non-status.

From a later source, there were 288,938 persons registered as status Indians by the Indian affairs and northern development department at December 31, 1976. These persons are entitled to registration in accordance with the terms of the Indian Act. They comprise 568 bands who occupy or have access to some 2,230 reserves having a combined area of about 2.6 million hectares. Membership of these bands is distributed among the provinces and territories as shown in Table 4.22. The 29 Indian bands in the Yukon Territory and Northwest Territories are located in seven reserves and in 46 settlements that have not been formally designated as reserves. There are at present no Indian bands in Newfoundland.

About two-thirds of Canada's Inuit reported in the 1971 Census live in communities in the Northwest Territories (11,400), and the remainder mainly in Arctic Quebec (3,800), Labrador (1,000), and Northern Ontario (800). As in the rest of Canada, the Inuit birth rate has been declining, but at a faster rate and from a much higher level. By 1971 the birth rate for the Inuit population in the Northwest Territories had decreased to about 38 per 1,000 as compared with the Canadian average of 17 per 1,000.

4.4   Households and families

4.4.1   Household size and type

A household, as defined in the census, consists of a person or a group of persons occupying one dwelling, usually a family with or without lodgers or employees. However, it may consist of a group of unrelated persons, of two or more families sharing a dwelling, or of one person living alone. The statistics presented in this section pertain to private households only. Collective households such as hotels, motels, institutions of various types (usually considered to contain 10 or more persons unrelated to the household head) have been excluded as well as households outside Canada for the 1971 and 1976 data.

The number of private households in Canada increased to 7.2 million in 1976 from 5.2 million a decade earlier, a gain of 38%. The population rate of increase was considerably lower at 14%. This difference reflects the marked rise in the number of households of only one or two persons. Table 4.23 shows that the rate of growth in the number of households was not uniform across the country. During the 1971-76 period, growth rates ranged from 8.7% in Saskatchewan to 23.9% in British Columbia and 30.2% in the Yukon Territory and Northwest Territories. New Brunswick and Alberta had growth rates higher than the national average of 18.6%.

Households by size. Table 4.23 also shows the average size of households by province for 1966, 1971 and 1976. In the 1976 Census, the average Canadian household had 3.1 persons as compared to 3.5 in 1971 and 3.7 in 1966. In all these censuses, the average number of persons per household was highest in Newfoundland. While the decline in average size of households during 1966-71 was seen mainly in the Maritime provinces and Quebec, a further decline during 1971-76 was realized in all provinces.

Households by type. All private-type households are divided for census purposes into two basic categories: family and non-family households. Table 4.24 shows the distribution on this basis for 1966, 1971 and 1976.

Family households increased from 4.4 million in 1966 to almost 4.9 million in 1971 and to 5.6 million in 1976, but dropped proportionately from 84.5% in 1966 to 81.7% in 1971 and to 78.6% in 1976. The proportion consisting of two or more families dropped from 2.5% in 1966 to 1.3% in 1976, indicating a decrease in overcrowding in households. Non-family households, on the other hand, increased in number and in proportion to the total number of households; this is mainly due to the increase in the proportion of one-person households from 11.4% in 1966 to 13.4% in 1971 and to 16.8% in 1976. Thus, new family formation alone was not responsible for the overall increase in the number of households; some families and family persons who previously shared accommodation now maintain their own households.

Households by age and marital status of head. The upward trend in households headed by persons under 25 years of age is indicated in Table 4.25. Although total households increased by 18.6% between 1971 and 1976, the number of households with heads under 25 years of age grew by 41.0%, reaching 584,270 in 1976 from 414,470 in 1971. By province, this group increased by as much as 60.3% in Alberta and 54.5% in Saskatchewan. The Yukon Territory and the Northwest Territories registered the largest growth in households with heads 65 years of age and over, increasing 36.6%. Quebec followed with an increase of 25.0% from 1971. Nationally, the proportion of households with heads 65 and over increased from 16.2% to 16.4%.

Growth in the number of households analyzed by marital status of head is seen in Table 4.26. The most significant increase during the 1966-76 period was recorded by households with divorced heads, at 452.9%; the proportion of households with divorced heads more than tripled from 0.8% in 1966 to 3.1% in 1976. At 99.3%, the increase in households with single never-married heads was the next highest. The rate of increase over the decade was 34.8% for households with widowed heads and 29.1% for households with married heads (including separated).

Family size and composition   4.4.2

A family, as defined in the census, consists of a husband and wife without children or with children who have never married, regardless of age, or a lone parent with one or more children who have never married, regardless of age, living together in the same dwelling. Adopted children and stepchildren have the same status as own children.

The 1976 family data pertain to families in private households only. The number of families in Canada increased to 5.7 million in 1976 from 4.5 million in 1966. Following the patterns of provincial population growth, and reflecting the factors of migration, the largest rate of increase occurred in the Yukon Territory and the Northwest Territories (69.3% in the 1966-76 period), followed by British Columbia (41.4%) and Alberta (35.5%).

Families by size. The number and average size of families are given in Table 4.27 by province for 1966, 1971 and 1976. The average size dropped to 3.7 persons between 1966 and 1971, and to 3.5 in 1976 reflecting declining birth rates. The largest reductions in average family size occurred in Quebec (from 4.2 persons in 1966 to 3.5 in 1976), New Brunswick (from 4.3 to 3.7), Newfoundland (from 4.6 to 4.0) and the Yukon Territory and the Northwest Territories (from 4.5 to 3.9).

Family structure refers to the classification of census families into husband-wife families and lone parent families. Husband-wife families consist of a husband and a wife (with or without children), or persons who live common law (with or without children). Lone parent families consist of a parent, regardless of marital status, with one or more children living in the same dwelling.

Table 4.28 shows that 90.2% of Canadian families in 1976 were husband-wife families, a decrease of 1.6% from 1966. Lone parent families increased both in number (371,885 to 559,335) and in proportion of total families (8.2% to 9.8%) over the decade.

The proportion of female lone parent families increased from 6.6% to 8.1% while the proportion of male lone parent families only increased from 1.6% to 1.7%.

Husband-wife families. For the analysis of family data, a subdivision into husband-wife families and lone parent families in Table 4.29 shows the distribution by age of husbands, wives and lone parent families are further classified into male lone parent and female lone parent families. In husband-wife families, for statistical tabulating purposes, the husband was considered the head of the family in 1971 and earlier censuses.

Most lone parent families were headed by a female parent; these families increased both in number and proportion from 6.6% in 1966 to 8.1% in 1976. This reflects an increase in broken families in Canada because the proportion of female lone parent families increased 171.6% in the age group 25-34 and 68.2% in the age group 35-44, the ages at which most divorces were granted. An increase of 163.8% was recorded for the under 25 age group. The greatest proportion of male lone parent families fell in the 45-54 age group followed by the 35-44 age group.

Families by mother tongue of husband, wife and lone parent. For census purposes, mother tongue is defined as the first language learned that is still understood. The proportion of husbands, wives and lone parents reporting English, French, or other mother tongue in the 1976 Census showed a pattern fairly similar to that for the population as a whole (Table 4.30). For example, 61.4% of the Canadian population reported English as the mother tongue, as compared with 58.7% of all husbands, wives and lone parents. The corresponding proportions for French mother tongue were 25.6% and 25.4%. However, mother tongues other than English or French were reported by only 11.0% of the total population, but by 14.1% of all husbands, wives and lone parents. For the Canadian population, 1.9% did not state mother tongue while the not stated category was 1.8% for husbands, wives and lone parents.

Children in families. In 1976 the definition of children was expanded from the 1971 definition to include all persons, regardless of age, who were living with their parent(s) or guardian(s) at the time of the census. Unrelated wards or foster children however, were designated as lodgers rather than children. The 1971 age restriction of under 25 years was deleted. Children are classified into selected age groups which roughly correspond to pre-school age (under six), elementary school age (6-14), secondary school age (15-17), college or working age (18-24) and 25 years and over (Table 4.31).

The 1976 data showed the following distribution by age group of children; under six, 23.0%; 6-14 years, 41.4%; 15-17 years, 14.9%; 18-25 years, 16.6%; and 25 years and over, 4.1%. Total children in families in private households within Canada for all age groups decreased from 9.2 million in 1971 to 8.9 million in 1976, and the average number of children per family declined from 1.8 to 1.6.

4.5 The vital components of population change

Vital statistics are indispensable to the measurement and interpretation of population change. They provide such information as the rate at which population increases by natural means, women marry and have children, or marriages are dissolved. The statistics are derived from the records of births, deaths, marriages and divorces registered in the provinces and territories.

4.5.1 History of vital statistics

A historical summary of vital statistics data for Canada and the provinces back to 1921 is contained in *Vital statistics, annual reports* (Statistics Canada 84-203 to 206). Some estimates of birth, natural increase, and death rates back to the mid-1800s by 10-year periods are given in Sections 4.6.1, 4.6.3 and 4.7.1, which follow.

4.5.2 Summary of principal data

Table 4.32 provides a summary of the principal vital statistics for each year from 1971 to 1975 for Canada, the provinces and territories, with comparative figures by five-year periods back to 1951-55. Table 4.33 shows similar data for urban centres of 50,000 population and over for 1975 with comparative averages for 1961-65 and 1966-70. More detailed information on vital statistics, including analyses of recent trends, is published annually in the Statistics Canada reports *Vital statistics, volume I, births* (Catalogue 84-204), *Vital statistics, volume II, marriages and divorces* (Catalogue 84-205), *Vital statistics, volume III, deaths* (Catalogue 84-206) and *Causes of death, provinces by sex and Canada by sex and age* (Catalogue 84-203). Certain unpublished data are available on request.

Fertility 4.6

Of all the demographic factors which introduce changes in population (fertility, mortality, nuptiality, immigration, emigration), none exerts greater influence than the rate of reproduction or fertility. By comparison, the nation's death rate, which has reached low levels, could be considered far more stable; it is the birth and fertility rates that may well continue to be the dominant factor in the near future in shaping the demographic profile of Canada (see Section 4.1).

Births 4.6.1

No accurate figures on Canadian crude birth rates are available prior to 1921 when the annual collection of official national figures was initiated. However the following rough estimates of the average annual crude rates of live births (per 1,000 total population) for each 10-year intercensal period between 1851 and 1921 may be inferred from studies of early Canadian census data: 1851-61, 45; 1861-71, 40; 1871-81, 37; 1881-91, 34; 1891-1901, 30; 1901-11, 31; 1911-21, 29.

The general trend in the national crude birth rate since 1951 is shown in Table 4.32. The annual rates declined steadily from 29.3 in 1921 to a record low of 20.1 in 1937, recovered sharply in the late 1930s and rose during World War II to 24.3 in 1945. Following the war the birth rate rose to a high of 28.9 in 1947. Between 1948 and 1959 it remained remarkably stable at between 27.1 and 28.5, but has since declined dramatically to a record low of 15.4 by 1974. The rate for 1975 shows a slight increase to 15.7. Provincial rates have followed this trend with some regional differences.

Since these crude birth rates are based on the total population they do not reflect the true fertility of the women of reproductive ages in the different provinces. A more accurate measure of the true birth rate is one based on the number of women by age between the ages of 15 and 49 (Table 4.38; Section 4.6.2).

Age of mothers. The distribution of infants born alive in 1975 by age of the mother is given in Table 4.34. It shows that 69.1% of the live births in 1975 to all mothers were among women 20-29, another 13.6% to women 30-34, and only about 12.8% of births were to mothers under 20.

Order of birth. Table 4.34 also shows the order of birth of all live-born infants in 1975 according to the age of the mother. In 1975 the first births for mothers of all ages constituted 44.1% of all live births; births of fourth or higher order constituted 8.4%.

Table 4.35 summarizes the pattern of family formation since 1951 and shows that the percentages of first and second children have been increasing in recent years. This has been accompanied generally by a reduction in the proportion of third and higher birth orders.

Stillbirths. The 2,627 stillbirths of at least 28 weeks gestation that were delivered in 1975 represented a ratio of about seven for every 1,000 live births. The stillbirth ratio, decreasing steadily, has been cut by more than half over the past quarter-century (Table 4.36). Ratios in some provinces have been reduced more than in others.

Table 4.37 illustrates that the risk of having a stillborn child increases with the age of the mother. Although stillbirth rates for mothers of all ages have been declining, they continue to be much higher for older than for younger mothers.

Fertility rates 4.6.2

The sex and age composition of a population is a fundamental factor affecting its birth rates. Since almost all children are born to women between the ages of 15 and 49,

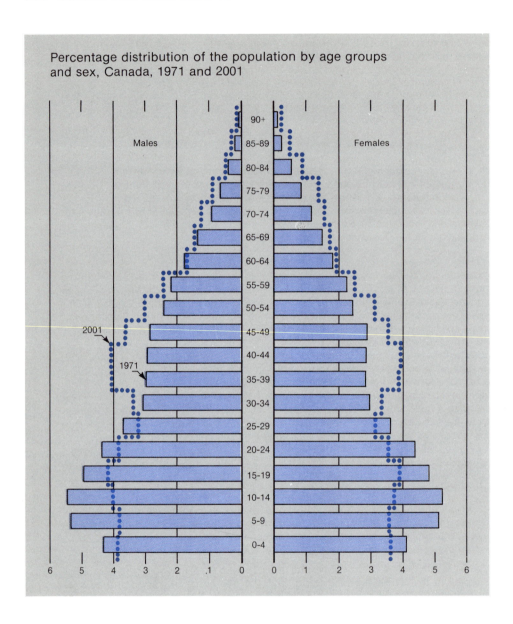

Percentage distribution of the population by age groups and sex, Canada, 1971 and 2001

variations in the proportion of women of these ages to the total population will cause variations in the crude birth rate of different countries, or of different regions, even though the actual rates of reproduction or fertility of the women are identical. It is therefore conventional practice to calculate age-specific fertility rates, the number of infants born annually to every 1,000 women in each of the reproductive age groups.

Table 4.38 indicates that women in their 20s are the most reproductive. On the average, for every 1,000 women between age 20 and 24, there were 112 infants born during 1975. Expressed another way, about one woman out of eight in that age group gave birth to a live-born infant. The highest rate is found in the 25-29 age group.

Another measure of fertility is the gross reproduction rate (Table 4.38) that indicates the average number of female children born to each woman in the child-bearing ages. In other words, the gross reproduction rate represents the average number of daughters that would be born to each woman throughout her child-bearing ages (15 to 49) if the fertility rate of the given year remained unchanged during the whole of her child-bearing period. A rate of 1.000 indicates that, on the basis of current fertility and without making any allowance for mortality among mothers during their child-bearing years, the present generation of child-bearing women would exactly maintain itself.

Canada has always had one of the highest gross reproduction rates among the industrialized countries of the world. Even at low birth rates in the 1930s the rate varied between 1.300 and 1.500 and since World War II has ranged from 1.640 in 1946 to a high of 1.915 in 1959. However, since 1963 the national gross reproduction rate has dropped sharply from 1.788 to 0.881 in 1975 — appreciably below the replacement level of 1.000. Among the provinces, Quebec, British Columbia and Ontario had the lowest gross reproduction rates in 1975, all below the replacement level.

Natural increase   4.6.3

The excess of births over deaths, or natural increase, has been the main factor in the growth of Canada's population. Some idea of the rate of natural increase back to the mid-1800s may be obtained from the estimates of births and deaths (see Sections 4.6.1 and 4.7.1) which produce the following natural increase rates (per 1,000 population): 1851-61, 23; 1861-71, 19; 1871-81, 18; 1881-91, 16; 1891-1901, 14; 1901-11, 18; 1911-21, 16.

During the 1920s and early 1930s the birth rate declined much more rapidly than the death rate and the natural increase rate dropped to a record low of 9.7 in 1937. Higher birth rates during and after World War II and a continued declining death rate caused the natural increase rate to rise steadily from 10.9 in 1939 to a record 20.3 in 1954. After that there was a steady drop due to declining birth rates and the natural increase rate fell below 10 for the first time in 1971 at 9.5 and dropped still further to 8.0 in 1974. It edged up slightly in 1975 to 8.4. Table 4.32 gives average rates of natural increase in the provinces for five-year periods 1951-70 and for individual years from 1971 to 1975.

Mortality   4.7

The Canadian crude death rate is one of the lowest in the world (7.3 per 1,000 population in 1975). After a gradual decline over the past century, the rate appears to have levelled off since 1967. In the opinion of demographers, further reductions in the crude death rate are likely to be small and to affect primarily persons in the older age groups.

General mortality   4.7.1

No official crude death rates (rates per 1,000 total population) are available prior to 1921. However studies of the early Canadian censuses resulted in the following estimated annual crude rates: 1851-61, 22; 1861-71, 21; 1871-81, 19; 1881-91, 18; 1891-1901, 16; 1901-11, 13; 1911-21, 13.

Typical of pioneer populations, Canada had high death rates in the mid-1800s with the crude death rate estimated between 22 and 25. It is assumed that while mortality was high at all ages, the rate among infants, children and young adults must have been particularly high. Even in 1921 the Canadian infant mortality rate was 102.1 per 1,000 live births. With increasing urbanization and improved sanitation and medical services, the crude death rate dropped by 50% from 22 to 11 between 1851 and 1930. It continued to decline to a low of 7.3 in 1970 and 1971, rising slightly to 7.4 in 1973 and 1974 and declining back to 7.3 in 1975. Table 4.32 also shows trends in crude death rates in the provinces and territories.

Table 4.33 shows numbers of deaths in urban centres of 50,000 population and over in 1975, and average deaths a year for the periods 1961-65 and 1966-70.

Age and sex distribution of deaths. Since 1921 the mortality trend at all ages has been downward. However, one of the contributing factors in lowering the general death rate has been the reduction in the mortality of infants and children. Between 1951 and 1975, death rates for infants and for children under five years of age dropped by more than 60% (Table 4.39). Rates for the 5-14 age group also declined. However, rates for boys and young men from 15 to 24 were actually higher than in 1951. Death rates for males over 25 were appreciably lower in 1975 than in 1951, except for men of 65-74 years, for whom there was little change. Rates for females of all ages declined substantially between 1951 and 1975.

Sharp reductions in male infant and child mortality and substantial declines in the female rates for all younger age groups have tended to raise the average age at death. Over the 1966-75 period the average for males rose 1.8 years from 62.0 to 63.8, while that for females advanced 3.5 years, from 65.9 to 69.4. The male median age at death remained unchanged at about 68.5, but the gain for females was 1.5 years, from 73.5 to 75.0. Thus half the females who died in 1975 were more than 75.0 years old.

Causes of death. Table 4.41 gives details of the 1975 Canadian deaths and death rates based on 50 causes as given in the international abbreviated list (international classification of diseases, 8th revision).

The proportion of older people in the population has been rising in recent years. Consequently, cancer and cardiovascular diseases account for a larger proportion of all deaths than formerly. On the other hand, deaths of infants, children and young adults from such diseases as pneumonia and tuberculosis have sharply declined.

Table 4.42 shows that the leading causes of infant mortality are radically different from the main causes of death at later periods. Accidents are the primary cause of death for males between one and 44 years of age. The majority of deaths among older males are due either to cardiovascular diseases or to cancer.

Accidents are also the primary cause of mortality among girls, with cancer being the leading cause of death for young and middle-aged women. Cardiovascular diseases and cancer are the leading causes of death for elderly women.

4.7.2 Infant mortality

Table 4.43 shows that mortality rates for both male and female infants (under one year of age) have been reduced by more than 60% since 1951. The improvement is due to many factors including better prenatal and postnatal care, improved sanitation, the use of antibiotics and higher living standards. In recent years, also, older women (a high-risk group) have been having fewer babies.

The 1975 provincial mortality rates for infants of both sexes vary substantially among the provinces and territories with the rates for the Northwest Territories being substantially higher. The national death rate for all infants was 14.3, the lowest on record (Table 4.43).

4.7.3 Life expectancy

Life tables are measures of life expectancy compiled from the death rates prevailing over a period. They assume that a given group of people (usually 100,000) are born simultaneously in a particular year and continue to be subject all their lives to the death rates prevailing in that year, or perhaps to the average death rates for a three-year period centred around that year. The expected deaths in the group are calculated (in the case of a complete life table) for the first year of life, second year of life, and the diminishing group is followed for 100 or more years until it has been virtually eliminated. Life expectancy at birth is calculated for the entire group and, subsequently, remaining life expectancy is calculated for the survivors at one year and subsequent years. It should be noted that the assumptions of such a life table are never fulfilled in practice and that the hypothetical groups in life tables do not represent any actual population. Usually, the persons in an actual group born in the life-table year will have a higher life expectancy than those in the life-table group because during their lifetimes public health conditions will presumably improve and standards of medical care will also presumably advance.

Seven official sets of life tables were published, based on deaths in the three year period around each of the censuses of 1931, 1941, 1951, 1956, 1961, 1966 and 1971. The Canadian life-table values for the 1971 period are given for selected ages in Table

4.45. This table shows that at 1970-72 mortality rates 2,002 of 100,000 males born would have died in their first year with 97,998 surviving to one year of age, that 126 more would have died in their second year with 97,872 reaching their second birthday and so on, with 191 survivors at age 100. The probability of dying column represents the ratio between the population at each age and the number of expected deaths in the coming year. The expectation of life column shows the number of remaining years of life expected at each age, given the 1970-72 mortality rates.

Male probabilities of dying were higher than the corresponding female probabilities at all ages. Mortality rates and the probabilities of dying were lowest at the age of about 10 for both sexes. Then male probabilities rose rapidly, reflecting accidents to teen-age boys; female probabilities rose more gradually. Male mortality was fairly constant from age 20 to the late 30s, and then increased steadily with advancing age. Female mortality rose slowly between 10 and 25 years, then more rapidly. About 11,200 of the male group would have died by age 50 as compared with roughly 6,600 of the female group, and 58,575 males would reach age 70 as compared with 75,995 females.

Life expectancy values over the 1951-71 period are shown in Table 4.46. By 1971 Canadian life expectancy at birth had reached an all-time high of 69.3 years for males and nearly 76.4 years for females. These figures are roughly comparable to the expectancies of other countries with highly developed programs of medical care. Because infant mortality is still quite substantial, life expectancies for male and female infants one year old were only slightly higher than expectancies at birth. Male expectancy at age 20 was 51.7 years, or 6.5 years below the corresponding female expectancy of 58.2. At age 40 the comparative expectancies were 33.2 for men and 39.0 for women. By age 65 the male expectancy had dropped to 13.7 years, with the female expectancy 3.8 years higher at 17.5 years.

Table 4.47 shows the life expectancies for five Canadian regions for 1951 and 1961, and expectancies by province for 1966 and 1971. The steady widening of the gap between male and female expectancies, evident at the national level, seemed to be continuing in every province to judge from the 1966 and 1971 figures. For the periods around the 1956 and 1961 censuses, the Prairie region showed the highest life expectancies, both male and female. Through 1951-61, Quebec life expectancies were the lowest, although they showed marked improvement over the decade.

In both the 1966 and 1971 periods, Saskatchewan life expectancies at birth were the highest for males and females alike, and Quebec expectancies, although increasing, were still the lowest. In 1971, the Saskatchewan male expectancy at birth (71.1 years) was 2.8 years above the corresponding Quebec expectancy of 68.3. For female expectancies at birth, Saskatchewan set a record of 77.6, about 2.3 years above the corresponding Quebec expectancy of 75.3. All the 1971 life expectancies in the four western provinces were above the national average. Ontario male rates at the younger ages were around the Canadian average, dropping slightly below for older men. Ontario female expectancies were fractionally above the national level.

In the Atlantic provinces, 1971 life expectancies for the very young did not differ greatly from the national average except in Nova Scotia for both males and females, which were somewhat below it, and Prince Edward Island females which were above it. This was also true for the expectancies at 20 years of age. The male expectancies at 40 were around the Canadian figure with Prince Edward Island a little higher and Nova Scotia slightly lower. Prince Edward Island showed the female expectancy at 40, a year above the national average, while Newfoundland and Nova Scotia were slightly below it. The Prince Edward Island expectancies at age 65 were above the Canadian level with the male expectancies at 65 for the other Atlantic provinces clustered around it. The Newfoundland female expectancy at 65 was somewhat below the Canadian level.

Nuptiality   4.8

This section includes statistics on marriages and marriage rates and also on dissolutions of marriages. In the *Canada Year Book 1973* pp 201-204, expanded coverage of the latter topic focused attention on dramatic increases in the number of divorces resulting from changes in 1968 to Canada's divorce laws.

4.8.1   Marriages

In 1975, there were 197,585 marriages solemnized in Canada. The rate of marriage remained unchanged at 8.9 per 1,000 population. In 1975, Alberta recorded 9.9 marriages per 1,000 population, the highest of any province (Table 4.32).

In 1975 the median age at marriage — the age above and below which half the marriages occurred — was 23.6 for bachelors and 21.5 for spinsters. In terms of averages, bachelors averaged 24.9 years of age, and spinsters 22.5.

Religious denomination. Some indication of the influence that religion has in selecting marriage partners is shown in Table 4.50. The majority of marriages in Canada were between persons of the same religious denomination. The proportions were higher for such denominations as Jewish and Roman Catholic and lower for others: Anglican, Baptist, Presbyterian and United Church.

4.8.2   Divorces

The number of decrees absolute granted in Canada has risen sharply as a result of the 1968 changes in divorce legislation. The number of divorces rose to over 50,600 in 1975 compared to an average of about 11,000 divorces per year over the period 1966-68. In 1975 Alberta's divorce rate was 309.7 per 100,000 population, and British Columbia 306.6, the highest rates among the provinces. By comparison, Newfoundland and Prince Edward Island had the lowest rates, 69.2 and 63.7 (Table 4.32).

Sex of petitioners. Table 4.51 shows that almost twice as many divorces were granted in 1975 to female petitioners (33,538), as to males (17,073). This represents a ratio of 51 divorces to male petitioners for every 100 to females.

Grounds for divorce. A cause for divorce can be mentioned singly in a decree or in combination with others. Table 4.52 indicates that in 1975 separation for not less than three years was the most frequent cause of divorce, accounting for 33.0% of all causes reported. This was followed by adultery (30.0%), mental cruelty (16.5%) and physical cruelty (13.9%).

Dependent children. Of the 50,611 divorces granted in 1975, 42.4% involved no dependent children. Table 4.53 shows increases in the proportion of divorces involving dependent children from 55.8% in 1972 to 57.6% in 1975. Almost two out of every five of the latter cases involved one child only, and almost one-third of them involved two children.

Duration of marriage. The duration of marriage in 15.6% of the divorces in Canada in 1975 was less than five years and in 45.0% of the cases it was less than 10 years. The short-term trend over five years indicates a relative shortening of the average marriage period before divorce. Table 4.54 shows that in 1972 only 15.3% of the divorces involved marriages of less than five years and 26.7% to those of less than 10 years. The median duration of marriage for 1972 divorces was 12.1 years compared to 11.4 in 1975.

Marital status. More than nine out of every 10 persons divorced in 1975 were involved in a first divorce. Slightly over 5% of the divorces were to persons who were divorced at the time of their last marriage and a little less than 2% to those who were widowed (Table 4.55).

4.9   Migration

Besides the vital components of population change (fertility, mortality, nuptiality), the flows of population across national borders (immigration and emigration) also affect the country's growth and demographic structure. This section provides data on the numbers and characteristics of immigrants entering Canada (Tables 4.56 to 4.62), as well as estimates of the numbers of emigrants leaving (Table 4.63). The relative influence of net migration (excess of immigrants over emigrants) compared to natural increase (excess of births over deaths) in past growth rates of Canada over the period 1851-1976 is shown in Table 4.2.

On the subject of internal migration within Canada, some estimates of total net migration by province in the 1966-76 period are given in Table 4.5. Brief summary data from the 1971 and 1976 censuses are provided in Tables 4.64 to 4.67.

Immigration   4.9.1

Canada's immigration policy is based on the principle of non-discrimination and emphasizes the selection of immigrants who are likely to adapt to the Canadian way of life, making a positive contribution to economic and cultural development in Canada.

Canadian immigration officers apply standard norms of assessment to applicants from all parts of the world and, apart from sponsored relatives and refugees, select those with skills in short supply in Canada.

The employment and immigration commission also regulates the entry of temporary workers and foreign students enrolled in public or private institutions and examines millions of others who come to Canada as tourists or for family, social, cultural or other visits. The commission facilitates the return of Canadian residents and implements enforcement and control measures that apply to visitors and immigrants whose presence in Canada may endanger the public health or welfare of Canadians or threaten national security.

In September 1973, the federal government began a review of immigration policy as the first step toward a new national policy. Briefs and letters submitted by national organizations, provincial authorities and members of the public were studied and a series of discussion documents, known collectively as the green paper on immigration, was prepared. It explained domestic and international challenges facing future immigration programs and became the focal point for a national debate on immigration objectives and policy.

Following the release of the green paper, a special joint committee of Parliament on immigration policy was set up to conduct a country-wide program of hearings, conferences and seminars and report its findings to Parliament. Over 90% of the committee's recommendations were later incorporated into an immigration bill tabled before Parliament in November 1976 and subsequently passed by the House of Commons and the Senate. On August 5, 1977, the bill received royal assent and was proclaimed early in 1978.

The resulting Immigration Act, 1976, brings Canada's immigration policy into sharper focus than before, introducing many new features and reinforcing, expanding or clarifying parts of the previous law. It states, for the first time in Canadian law, the basic principles underlying immigration policy — non-discrimination, family reunion, humanitarian concern for refugees and promotion of national goals. It contains provisions that link the immigration movement to Canada's population and labour market needs and provides for an annual forecast of the number of immigrants Canada can comfortably absorb, to be made in consultation with the provinces. The act establishes a new family class, allowing Canadian citizens and permanent residents to sponsor a wide range of relatives, confirms Canada's protective obligations to refugees under the United Nations Convention and establishes a new refugee class. It requires that immigrant and visitor visas and authorizations be obtained abroad and prohibits visitors from changing their status from within Canada.

In recent years, Canada's concern for the displaced and the persecuted has been manifested in two types of refugee resettlement programs: an ongoing program operating continuously around the world, and special programs in response to urgent refugee or humanitarian situations. Two special programs are a Chilean movement which began in November 1973 following the coup d'état in Chile, and a Vietnamese/ Cambodian program initiated in April 1975; as of October 31, 1977, approximately 5,600 South American refugees and 7,000 Vietnamese and Cambodians had been admitted to Canada.

There are employment and immigration offices in more than 60 cities throughout the world, and examination of immigrants and visitors is carried out at more than 500 ports of entry in Canada.

The extent of immigration to Canada in any period is affected by conditions at home and abroad. A review of these factors, with an analysis of trends, may be found in the *Canada Year Book 1972* pp 222-225. Immigrant arrivals for each year over the period 1951-77 are shown in Table 4.56.

Origin of immigrants. In 1976 Canada received 149,429 immigrants from various countries of origin, a decrease of 38,452 or 20.5% from the 1975 total of 187,881. Tables 4.57 and 4.58, showing the country of last permanent residence and of citizenship of immigrants, indicate that by world area Europe, Asia, North and Central America, Africa and Australasia contributed a lower proportion of the total immigration in 1976 than in the previous year. The British Isles was the largest source area for immigrants with 21,548 in 1976, followed by the United States with 17,315.

Destination of immigrants. On arrival in Canada, immigrants are asked to state their intended destinations. According to these records, Ontario absorbed by far the highest proportion of arrivals during 1976 — 48.2% of both males and females. Quebec was the second most-favoured province of destination, receiving 20.5% of males and 18.7% of females, followed by British Columbia with 13.1% of males and 14.3% of females. The proportions intending to settle in the Prairie provinces were 15.9% for males and 14.6% for females, and in the Atlantic provinces, 3.2% for males and 3.0% for females.

The provincial distribution as shown in Table 4.59 for 1975 and 1976 has changed little from year to year over the past two decades.

Sex, age and marital status. The sex distribution of immigrants for 1972-76 is shown in Table 4.60. In the period 1974-76 adult males constituted 34.5% of the immigrants, adult females 35.7%, and children under age 18 the remaining 29.8%. The number of female immigrants coming into Canada was higher than the number of male immigrants in every year from 1957 to 1964; since then, with the exception of 1969, 1971, 1972, 1975 and 1976 the trend has been in favour of males. There was an excess of 5,336 males over females in 1973 and 3,779 in 1974. In 1975, females exceeded males by 2,515 and in 1976 by 4,219. Single males as shown in Table 4.61 surpassed single females by 5,942 in 1976, but in all ever-married categories women outnumbered men: married females exceeded married males by 5,285; there were 5,384 widows compared to 932 widowers; divorcees outnumbered divorced men by 299; and there were 428 separated females, compared to 303 males. Of the total immigration for 1976, persons under age 15 totalled 25.5% and of those over 14 years, 31.3% were single, 60.9% married, and 7.8% were widowed, divorced or separated.

Intended occupations. Some 61,461 persons were added to the labour force in 1976 compared with 81,189 in 1975. The remainder, those not destined to the labour force, were mainly dependents of immigrants or close relatives sponsored by individuals in Canada. Persons employed in clerical occupations represented the largest occupations group with 9,345 workers. Other major groups were: product fabricating, and assembling and repairing with 8,380; managerial and administrative with 5,655; professional and technical with 5,648; services with 5,640; and construction with 4,008.

4.9.2   Emigration

Since the only statement a Canadian resident may be required to file on leaving the country is his income tax return, one cannot know the actual number of Canadian residents who emigrate from Canada, nor their previous province of origin. Three data sources are used to estimate the number of people leaving Canada each year. First, quarterly reports containing data on the number of immigrants to the United States from Canada are provided by the US immigration and naturalization service. Second, quarterly estimates of the number of persons entering the United Kingdom from Canada are taken from an international passenger survey based on a stratified sample of all passengers using the main sea and air routes between the United Kingdom and other countries. Third, an assumed level of the number of emigrants moving from Canada to countries other than the United States and the United Kingdom is added to the data described above. On the basis of Canadian census data, vital statistics, immigration figures and information provided by the United States and the United Kingdom, it has been estimated that 20,000 persons emigrate from Canada to countries other than the United States and the United Kingdom each year.

During each five-year period between June 1, 1961 and May 31, 1976, the total number of emigrants decreased: an estimated 377,700 people emigrated from Canada during the 1961-66 period; 327,400 during 1966-71; and 237,600 during 1971-76 (Table 4.63).

The annual number of emigrants for the 15-year period varied between 38,000 and 81,000. Although the decline in the estimated number of emigrants was hardly significant at first, it increased markedly during the 1971-76 period. The average annual estimated number of emigrants was 75,500 during the 1961-66 period; 63,400 during 1966-71; and 43,400 during 1971-76. The principal reason for this marked decline was the overall decline in the number of persons admitted to the United States from Canada: during the 1961-66 period, 62% of the emigrants from Canada went to the United States, 11% to the United Kingdom and the remaining 27% to other countries; during 1971-76, only 32% went to the United States, 22% to the United Kingdom and the remaining 46% migrated to other countries.

Although various socio-economic and political reasons contributed to the drop in emigration to the United States, it was also partly due to new US immigration laws. As of July 1968, the annual quota of immigrant visas allocated to the nationals of North and Latin American countries was set at 120,000. The number of applications therefore greatly exceeded the number of available visas with the result that the waiting periods in 1971 and 1976 were approximately 15 and 30 months respectively; a number of Canadians who might otherwise have emigrated to the United States were discouraged from doing so.

Of the 11,215 persons entering the United States from Canada in the year ended June 30, 1975 claiming Canada as country of last permanent residence — which includes native-born persons and those born in other countries who have resided in Canada — the US immigration and naturalization service lists 1,937 as professional, technical and kindred workers; 548 as craftsmen, foremen and kindred workers; and 630 as clerical and kindred workers. Housewives, children and others not in the labour force amounted to 6,255 or 55.8% of the total.

Of the 7,308 Canadian-born persons entering the United States in the year ended June 30, 1975 with the intention of remaining permanently, 3,141 were males and 4,167 females. Approximately one-fifth, or 1,570, of the native-born emigrants were males in the productive age group, 20-59 years. By occupation, the largest group of the total of 7,308 native-born persons was the professional or technical group which numbered 947; clerical and kindred workers numbered 384; and craftsmen and kindred workers numbered 229. On the other hand 4,819 persons, or 65.9% of the total, were classed as housewives, children and others with no reported occupation. Altogether, 39.7% of the total were persons under age 20.

Internal migration   4.9.3

As people move from one place of residence to another within the nation's borders, they set up varying patterns of migration which differ in intensity and directional flow. These internal movements have marked effects on regional economies and exert an influence on future population growth. Thus it is important that attempts be made to measure these various migration streams, such as from rural to urban centres, from cities to suburbs, from one province or economic region to another.

Migration by province of birth. Census figures on province of birth shed some light on internal migration flows by comparing the numbers of persons born in a given province with their province of present residence. However, such figures give no indication of the periodicity of the migrating process, and they apply only to the Canadian-born population presently living in a given province. Nevertheless they do reflect something of the accumulated results of the major patterns of interprovincial movement of native-born persons over the years.

Data from the 1971 Census (Table 4.64) show that Ontario, Alberta and British Columbia have been net gainers of Canadian-born migrants from other provinces, while the remaining provinces have been net losers (these data were not asked in the 1976 Census). Newfoundland and Quebec showed the fewest numbers of their 1971 population as having been born in another province (3.3% and 4.6%, respectively), while British Columbia showed the highest (37.4%). These percentages reflect low rates of Canadian-born migrants to the former two provinces and a high rate to the latter one. On the other hand, Saskatchewan-born persons were the most likely to be found living in a province other than that of their birth (40.0%), while Quebec-born and Ontario-born persons were the least likely (6.1% and 7.4%, respectively). This reflects a high rate of migration of native-born persons from the former province and low rates from the latter two.

Migration by residence five years ago. Perhaps the most useful estimates on internal migration are those resulting from the inclusion in national censuses of questions which seek to determine the exact place of residence of each person on the date of the preceding census five years earlier. From a comparison with the location of their present residence, it is possible to estimate the size, directional flows and characteristics of the migrating population over the period. Such questions were included in the 1961, 1971 and 1976 censuses of Canada. Migration trends in recent years as revealed by the 1961 and 1971 censuses were issued in monographs (Catalogues 99-548 and 99-557), profiles (Catalogue 99-705) and other studies (Catalogues 99-513 and 99-154). A few basic summary results of the 1976 Census, which included questions on place of residence five years earlier at June 1, 1971, are presented in Tables 4.65 and 4.66.

The migration status of the population in 1976 in terms of residence in 1971 is summarized in Table 4.65. It shows that almost half (48.5%) of Canada's population age five and over in 1976 were living in a different dwelling than five years earlier (movers). A total of 23.5% had moved within the same municipality (non-migrants), while 25.1% had moved from one municipality to another (migrants). The latter group consisted of 16.5% who were movers within the same province, 4.3% from one province to another, 0.9% whose province of residence in 1971 was not stated and 3.4% from outside Canada.

Table 4.66 summarizes the effect migration has had on provincial populations for 1971-76. For instance, Ontario was the most favoured province of destination for migrants from other provinces (22.3%) but it was also the largest single provincial source of interprovincial migrants (28.0%). This resulted in a net internal migration loss of 52,505 people for the province. Net internal migration losses were also experienced by Newfoundland, Quebec, Manitoba and Saskatchewan for 1971-76 while the Maritime provinces, traditional losers in net internal migration, recorded gains along with British Columbia, Alberta, the Yukon Territory and the Northwest Territories.

The overall effect of immigration on net migration for 1971-76 was a population gain for each province and territory except Newfoundland and Saskatchewan. In terms of individual provinces, Ontario, British Columbia and Quebec received the largest share of immigrants during this period — 50.5%, 15.9% and 15.0%, respectively.

Migration by type of locality. Table 4.67 compares the type of locality of residence on June 1, 1971 and June 1, 1976 for migrants who had moved from one municipality to another between these two dates. Metropolitan areas were the favourite destination of these migrants, as well as being the major locality of origin. A total of 55.2% of all migrants moved to a municipality within a metropolitan area, including those moving from one municipality to another within the same metropolitan area. Immigrants from outside Canada showed the greatest propensity to locate in a metropolitan area and this was the destination of 80.0% of their numbers. By place of residence, 1971, 52.7% of the migrants residing in census metropolitan areas (CMAs) in 1976 were also living in a metropolitan area in 1971 (in a different municipality of the same or a different metropolitan area), 5.6% were from a census agglomeration (CA), 17.4% from a non-CMA and 19.6% were from outside Canada. For migrants residing in CAs in 1976, the largest proportion were residing in non-CMA localities in 1971 (36.4%) while in the case of migrants residing in non-CMA localities in 1976 the majority (46.9%) lived in a similar locality in 1971.

Citizenship   4.9.4

Citizenship statistics. In 1976, citizenship certificates were granted to 117,276 new Canadian citizens and certificates of proof of citizenship were issued to 148,539 who were already Canadians.

The Citizenship Act (SC 1974-75-76, c. 108) came into effect on February 15, 1977. It replaced the Canadian Citizenship Act, passed in 1947, which was the first independent naturalization law to be enacted in the Commonwealth. By it the concept of a Canadian citizen distinct from a British subject came into being.

The Citizenship Act, administered by the secretary of state department, defines who is a citizen and covers loss and resumption of citizenship, requirements for acquiring citizenship, offences against the act and the taking of an oath of citizenship. It shortens the required length of residence in Canada prior to application from five to three years. The age of majority for citizenship purposes is reduced from 21 to 18. It removes various provisions of previous statutes and states that all persons are treated equally regardless of age, marital status, sex or country of origin.

Some of the requirements for an adult to become a Canadian citizen are: legal admission to Canada and three years residence in Canada; basic knowledge of the Canadian political system and of one of Canada's official languages; and compliance with the national security and criminal record provisions of the Citizenship Act.

Sources
4.1 Population Estimates and Projections Division, Content and Analysis Branch, Census and Household Surveys Field, Statistics Canada.
4.2 - 4.3.3 Census Characteristics Division, Content and Analysis Branch, Census and Household Surveys Field, Statistics Canada; Population Estimates and Projections Division, Content and Analysis Branch, Census and Household Surveys Field, Statistics Canada.
4.3.4 Statistics Section, Finance and Management Branch, Department of Indian Affairs and Northern Development.
4.4 Census Characteristics Division, Content and Analysis Branch, Census and Household Surveys Field, Statistics Canada.
4.5 - 4.8 Health Division, Institutional and Public Finance Statistics Branch, Institutions and Agriculture Statistics Field, Statistics Canada.
4.9 - 4.9.1 Canada Employment and Immigration Commission.
4.9.2 Population Estimates and Projections Division, Content and Analysis Branch, Census and Household Surveys Field, Statistics Canada.
4.9.3 Census Characteristics Division, Census and Household Surveys Field, Statistics Canada.
4.9.4 Communications Branch, Department of the Secretary of State of Canada.

Tables

All figures of the 1971 and 1976 censuses in Tables 4.11 to 4.13, 4.15 to 4.21, and 4.23 to 4.31, have been subjected to a confidentiality procedure to prevent the possibility of associating small figures with an identifiable individual. The particular technique used is known as "random rounding". Under this method, all last or "unit" digits in a table (including all totals) are randomly rounded (either up or down) to "0" or "5". This technique provides the strongest possible protection against direct, residual, or negative disclosures without adding any significant error to the census data. However, since totals are independently rounded they do not necessarily equal the sum of individual rounded figures in distributions. Also, minor differences can be expected for corresponding totals and cell values in various census tabulations.

4.1 Growth of the population of Canada, 1851-1976

Census year	Population No.	Increase during intercensal period No.	%	Average annual rate of population growth %
1851	2,436,297			
1861	3,229,633	793,336	32.6	2.9
1871	3,689,257	459,624	14.2	1.3
1881	4,324,810	635,553	17.2	1.6
1891	4,833,239	508,429	11.8	1.1
1901	5,371,315	538,076	11.1	1.1
1911	7,206,643	1,835,328	34.2	3.0
1921	8,787,949	1,581,306	21.9	2.0
1931	10,376,786	1,588,837	18.1	1.7
1941	11,506,655	1,129,869	10.9	1.0
1951[1]	14,009,429	2,502,774	21.8	1.7
1956	16,080,791	2,071,362	14.8	2.8
1961	18,238,247	2,157,456r	13.4r	2.5r
1966	20,014,880	1,776,633	9.7	1.9
1971	21,568,311	1,553,431r	7.8r	1.5r
1976	22,992,604	1,424,293	6.6	1.3

[1]Newfoundland included for the first time. Excluding Newfoundland the increase would have been 2,141,358 or 18.6%.

4.2 Growth components of Canada's population, 1851-1976[1]

Period	Total population growth '000	Births '000	Deaths '000	Natural increase '000	Ratio of natural increase to total growth %	Immigration '000	Emigration '000	Net migration '000	Ratio of net migration to total growth %	Population at the end of the census period '000
1851-1861	793	1,281	670	611	77.0	352	170	182	23.0	3,230
1861-1871	460	1,370	760	610	132.6	260	410	-150	-32.6	3,689
1871-1881	636	1,480	790	690	108.5	350	404	-54	-8.5	4,325
1881-1891	508	1,524	870	654	128.7	680	826	-146	-28.7	4,833
1891-1901	538	1,548	880	668	124.2	250	380	-130	-24.2	5,371
1901-1911	1,835	1,925	900	1,025	55.9	1,550	740	810	44.1	7,207
1911-1921	1,581	2,340	1,070	1,270	80.3	1,400	1,089	311	19.7	8,788
1921-1931	1,589	2,420	1,060	1,360	85.5	1,200‾	970	230	14.5	10,377
1931-1941	1,130	2,294	1,072	1,222	108.1	149	241	-92	-8.1	11,507
1941-1951[2]	2,503	3,212	1,220	1,992	92.3	548	382	166	7.7	14,009
1951-1956	2,071	2,106	633	1,473	71.1	783	185	598	28.9	16,081
1956-1961	2,157	2,362	687	1,675	77.7	760	278	482	22.3	18,238
1961-1966	1,777	2,249	731	1,518	85.4	539	280	259	14.6	20,015
1966-1971	1,553	1,856	766	1,090	70.2	890	427	463	29.8	21,568
1971-1976	1,424	1,756	822	934	65.6	841	351	490	34.4	22,993

[1]Includes Newfoundland since 1951.
[2]Data on growth components shown for 1941-51 were obtained by including data for Newfoundland for 1949-50 and 1950-51 only.

4.3 Projected population of Canada up to 2001

Year	Population as at June 1 '000	Annual rate of population growth %	Distribution by age 0-19 %	20-44 %	45-64 %	65+ %
Projection A[1]						
1971	21,568.3	...	39.4	33.9	18.6	8.1
1976	23,086.1	1.4	36.1	36.4	18.9	8.6
1981	25,311.5	1.8	34.3	38.6	18.1	9.0
1986	27,810.9	1.9	33.9	39.6	17.2	9.3
1991	30,177.6	1.6	34.9	38.1	17.2	9.8
1996	32,347.1	1.4	35.6	36.2	18.3	9.9
2001	34,611.4	1.3	34.6	35.4	20.2	9.8

4.3   Projected population of Canada up to 2001 (concluded)

Year	Population as at June 1 '000	Annual rate of population growth %	Distribution by age 0-19 %	20-44 %	45-64 %	65 + %
Projection B[2]						
1971	21,568.3	. . .	39.4	33.9	18.6	8.1
1976	22,846.3	1.2	35.9	36.4	19.1	8.6
1981	24,472.5	1.4	33.2	38.9	18.6	9.3
1986	26,258.6	1.4	31.6	40.5	18.1	9.8
1991	27,902.1	1.2	31.8	39.6	18.2	10.4
1996	29,317.0	1.0	31.8	37.7	19.7	10.8
2001	30,655.5	0.9	30.8	36.4	21.9	10.9
Projection C[3]						
1971	21,568.3	. . .	39.4	33.9	18.6	8.1
1976	22,772.4	1.1	35.7	36.5	19.1	8.7
1981	24,041.4	1.1	32.0	39.6	19.0	9.4
1986	25,382.9	1.1	29.3	41.9	18.7	10.1
1991	26,591.4	0.9	28.4	41.5	19.1	11.0
1996	27,569.7	0.7	27.8	39.8	20.9	11.5
2001	28,369.7	0.6	26.7	37.9	23.6	11.8

The above figures represent new series of projections replacing those in the *Canada Year Book 1974* (see text in Section 4.1.3).
[1]Projection A assumptions: total fertility will change from 2.19 children in 1971 to 2.60 by 1985 and then remain constant through 2001; net migration gain of 100,000 per year; and expectation of life at birth will increase gradually to 70.2 years for males and 78.4 for females by 1986 and then remain constant through 2001.
[2]Projection B assumptions: total fertility will change to 2.20 children by 1985 and then remain constant through 2001; net migration gain of 60,000 per year; and mortality same as Projection A.
[3]Projection C assumptions: total fertility will change to 1.80 children by 1985 and then remain constant through 2001; net migration is same as Projection B; and mortality is same as Projection A.

4.4   Population and percentage change of population, by province, 1951-76

Province or territory	Population and percentage change					
	Population					
	1951	1956	1961	1966	1971	1976
Newfoundland	361,416	415,074	457,853	493,396	522,104	557,725
Prince Edward Island	98,429	99,285	104,629	108,535	111,641	118,229
Nova Scotia	642,584	694,717	737,007	756,039	788,960	828,571
New Brunswick	515,697	554,616	597,936	616,788	634,557	677,250
Quebec	4,055,681	4,628,378	5,259,211	5,780,845	6,027,764	6,234,445
Ontario	4,597,542	5,404,933	6,236,092	6,960,870	7,703,106	8,264,465
Manitoba	776,541	850,040	921,686	963,066	988,247	1,021,506
Saskatchewan	831,728	880,665	925,181	955,344	926,242	921,323
Alberta	939,501	1,123,116	1,331,944	1,463,203	1,627,874	1,838,037
British Columbia	1,165,210	1,398,464	1,629,082	1,873,674	2,184,621	2,466,608
Yukon Territory	9,096	12,190	14,628	14,382	18,388	21,836
Northwest Territories	16,004	19,313	22,998	28,738	34,807	42,609
Canada	14,009,429	16,080,791	18,238,247	20,014,880	21,568,311	22,992,604
	Percentage change					
	1951-56	1956-61	1961-66	1966-71	1971-76	
Newfoundland	14.8	10.3	7.8	5.8	6.8	
Prince Edward Island	0.9	5.4	3.7	2.9	5.9	
Nova Scotia	8.1	6.1	2.6	4.4	5.0	
New Brunswick	7.5	7.8	3.2	2.9	6.7	
Quebec	14.1	13.6	9.9	4.3	3.4	
Ontario	17.6	15.4	11.6	10.7	7.3	
Manitoba	9.5	8.4	4.5	2.6	3.4	
Saskatchewan	5.9	5.1	3.3	-3.0	-0.5	
Alberta	19.5	18.6	9.9	11.2	12.9	
British Columbia	20.0	16.5	15.0	16.6	12.9	
Yukon Territory	34.0	20.0	-1.7	27.8	18.8	
Northwest Territories	20.7	19.1	25.0	21.1	22.4	
Canada	14.8	13.4	9.7	7.8	6.6	

4.5   Components of population change, by province, 1966-71 and 1971-76

Province or territory	Total population change 1966-71	1971-76	Natural increase 1966-71	1971-76	Net migration 1966-71	1971-76
Newfoundland	28,708	35,621	49,096	44,615	-20,388	-8,994
Prince Edward Island	3,106	6,588	5,207	4,498	-2,101	2,090
Nova Scotia	32,921	39,611	37,418	32,338	-4,497	7,273
New Brunswick	17,769	42,693	35,233	33,225	-17,464	9,468
Quebec	246,919	206,681	288,727	222,900	-41,808	-16,219
Ontario	742,236	561,359	373,072	327,861	369,164	233,498
Manitoba	25,181	33,259	49,260	45,160	-24,079	-11,901
Saskatchewan	-29,102	-4,919	50,867	38,123	-79,969	-43,042
Alberta	164,671	210,163	105,293	95,729	59,378	114,434
British Columbia	310,947	281,987	88,494	82,774	222,453	199,213
Yukon Territory and Northwest Territories	10,075	11,250	6,720	6,404	3,355	4,846
Canada	1,553,431	1,424,293	1,089,387	933,627	464,044	490,666

4.6 Population of Canada, by province, selected years, as at June 1, 1921-77 (thousands)

Year	Nfld.	PEI	NS	NB	Que.	Ont.	Man.	Sask.	Alta.	BC	YT	NWT	Canada
1921	—	88.6	523.8	387.9	2,360.5	2,933.7	610.1	757.5	588.5	524.6	4.1	8.1	8,787.8
1931	—	88.0	512.8	408.2	2,874.7	3,431.7	700.1	921.8	731.6	694.3	4.2	9.3	10,376.7
1941	—	95.0	578.0	457.4	3,331.9	3,787.7	729.7	896.0	796.2	817.8	5.0	12.0	11,506.7
1951	361.4	98.4	642.6	515.7	4,055.7	4,597.6	776.5	831.7	939.5	1,165.2	9.1	16.0	14,009.4
1956	415.1	99.3	694.7	554.6	4,628.4	5,404.9	850.0	880.7	1,123.1	1,398.5	12.2	19.3	16,080.8
1961	457.9	104.6	737.0	597.9	5,259.2	6,236.1	921.7	925.2	1,332.0	1,629.1	14.6	23.0	18,238.3
1966	493.4	108.5	756.0	616.8	5,780.8	6,960.9	963.1	955.4	1,463.2	1,873.7	14.4	28.7	20,014.9
1971	522.1	111.6	789.0	634.6	6,027.8	7,703.1	988.2	926.2	1,627.9	2,184.6	18.4	34.8	21,568.3
1972	530.0	112.6	794.6	640.1	6,053.6	7,809.9	991.2	914.0	1,657.2	2,241.4	19.5	37.3	21,801.5
1973	537.2	114.0	804.3	647.1	6,078.9	7,908.8	996.2	904.5	1,689.5	2,302.4	20.5	39.4	22,042.8
1974	541.5	115.2	811.5	653.6	6,122.7	8,054.1	1,007.5	899.7	1,722.4	2,375.7	20.5	39.6	22,364.0
1975	549.1	117.1	819.5	665.2	6,179.0	8,172.2	1,013.6	907.4	1,778.3	2,433.2	21.3	41.2	22,697.1
1976	557.7	118.2	828.6	677.3	6,234.5	8,264.5	1,021.5	921.3	1,838.0	2,466.6	21.8	42.6	22,992.6
1977P	562.5	120.3	835.4	686.4	6,283.1	8,373.5	1,031.3	936.5	1,899.7	2,497.6	21.5	43.3	23,291.1

4.7 Population of incorporated cities, towns and villages classified by size group, 1966, 1971 and 1976

Size group	1966			1971			1976		
	Incorporated centres	Population	% of total population	Incorporated centres	Population	% of total population	Incorporated centres	Population	% of total population
Over 500,000	2	1,886,839	9.4	2	1,927,138	8.9	3	2,274,738	9.9
Between									
400,000 and 500,000	1	410,375	2.1	3	1,267,727	5.9	3	1,341,466	5.9
300,000 and 400,000	2	707,500	3.5	2	611,514	2.8	2	616,465	2.7
200,000 and 300,000	3	845,867	4.2	4	900,778	4.2	3	736,652	3.2
100,000 and 200,000	6	997,051	5.0	8	1,060,048	4.9	12	1,578,755	6.9
50,000 and 100,000	26	1,740,446	8.7	26	1,870,435	8.7	34	2,286,408	10.0
25,000 and 50,000	43	1,438,388	7.2	49	1,633,969	7.6	52	1,795,675	7.8
15,000 and 25,000	52	1,019,205	5.1	59	1,150,768	5.3	50	978,090	4.3
10,000 and 15,000	65	781,611	3.9	55	675,748	3.1	69	849,488	3.7
5,000 and 10,000	125	898,136	4.5	144	1,028,412	4.8	149	1,051,844	4.6
3,000 and 5,000	165	637,117	3.2	173	670,537	3.1	179	694,881	3.0
1,000 and 3,000	471	818,003	4.1	502	866,086	4.0	490	827,807	3.6
Under 1,000	1,057	445,246	2.2	1,093	451,810	2.1	1,033	419,078	1.8
Total	2,018	12,625,784	63.1	2,120	14,114,970	65.4	2,079	15,451,347	67.4

4.8 Population of incorporated cities and towns of 50,000 and over, 1966, 1971 and 1976

Incorporated city or town	Year of incorporation	1966	1971	1976
Beauport, Que.	1976	11,742*	14,681*	55,339
Brampton, Ont.	1974	36,264	41,211*	103,459
Brantford, Ont.	1877	59,854*	64,421	66,950*
Burlington, Ont.	1915	65,941*	87,023	104,314*
Calgary, Alta.	1893	330,575*	403,319*	469,917*
Cambridge, Ont.	1973	—	64,794*	72,383
Charlesbourg, Que.	1976	24,926*	33,443*	63,147
Chicoutimi, Que.	1976	32,526*	33,893	57,737
Dartmouth, NS	1961	58,745	64,770	65,341
Edmonton, Alta.	1904	376,925*	438,152*	461,361*
Gatineau, Que.	1975	17,727*	22,321*	73,479
Guelph, Ont.	1879	51,377*	60,087*	67,538
Halifax, NS	1841	86,792	122,035*	117,882
Hamilton, Ont.	1846	298,121*	309,173	312,003
Hull, Que.	1875	60,176*	63,580*	61,039
Jonquière, Que.	1976	29,663	28,430*	60,691
Kamloops, BC	1973	10,759	26,168*	58,311
Kelowna, BC	1973	17,006*	19,412*	51,955
Kingston, Ont.	1846	59,004	59,047*	56,032
Kitchener, Ont.	1912	93,255*	111,804*	131,870*
LaSalle, Que.	1958	48,322	72,912	76,713
Laval, Que.	1965	196,088	228,010	246,243
London, Ont.	1855	194,416	223,222*	240,392
Longueuil, Que.	1920	25,593	97,590*	122,429
Markham, Ont.	1971	7,769	36,684*	56,206
Mississauga, Ont.	1968	93,492*	156,070*	250,017*
Moncton, NB	1973	45,847	47,891	55,934
Montreal, Que.	1832	1,222,255*	1,214,352*	1,080,546
Montreal N., Que.	1959	67,806	89,139*	97,250
Niagara Falls, Ont.	1903	56,891*	67,163*	69,423
North Bay, Ont.	1925	23,635	49,187*	51,639
Oakville, Ont.	1857	52,793*	61,483*	68,950*
Oshawa, Ont.	1924	78,082	91,587	107,023*
Ottawa, Ont.	1854	290,741	302,341	304,462
Peterborough, Ont.	1905	56,177*	58,111*	59,683
Prince George, BC	1915	24,471*	33,101*	59,929
Quebec, Que.	1832	166,984	186,088*	177,082*
Regina, Sask.	1903	131,127*	139,469*	149,593*

4.8   Population of incorporated cities and towns of 50,000 and over, 1966, 1971 and 1976 (concluded)

Incorporated city or town	Year of incor- poration	1966	1971	1976
Saint John, NB	1785	51,567	89,039*	85,956
St. Catharines, Ont.	1876	97,101	109,722*	123,351
Ste-Foy, Que.	1955	48,298*	68,385*	71,237
St. John's, Nfld.	1888	79,884*	88,102*	86,576*
St-Laurent, Que.	1955	59,479*	62,955*	64,404
St-Léonard, Que.	1963	25,328*	52,040*	78,452
Sarnia, Ont.	1914	54,552	57,644	55,576
Saskatoon, Sask.	1906	115,892*	126,449*	133,750
Sault Ste Marie, Ont.	1912	74,594*	80,332	81,048
Sherbrooke, Que.	1875	75,690	80,711	76,804*
Sudbury, Ont.	1930	84,888*	90,535	97,604*
Thunder Bay, Ont.	1970	104,539	108,411	111,476
Toronto, Ont.	1834	664,584	712,786*	633,318
Trois-Rivières, Que.	1857	57,540*	55,869	52,518
Vancouver, BC	1886	410,375	426,256	410,188
Verdun, Que.	1912	76,832	74,718	68,013
Victoria, BC	1862	57,453	61,761	62,551
Windsor, Ont.	1892	192,544*	203,300*	196,526
Winnipeg, Man.[1]	1972r	257,005*	246,246	560,874*

*Indicates a boundary change since the preceding census. Population totals in these cases are based on a different area, the boundaries at that particular census year.
[1]Includes St. James-Assiniboia, Man.

4.9   Population of census metropolitan areas, 1951-71 (based on 1971 boundaries) and 1976[1]

Census metropolitan area	1951	1956	1961	1966	1971	1976[1]
Calgary	142,315	201,022	279,062	330,575	403,319	469,917
Chicoutimi–Jonquière	91,161	110,317	127,616	132,954	133,703	128,643
Edmonton	193,622	275,182	359,821	425,370	495,702	554,228
Halifax	138,427	170,481	193,353	209,901	222,637	267,991
Hamilton	281,901	341,513	401,071	457,410	498,523	529,371
Kitchener	107,474	128,722	154,864	192,275	226,846	272,158
London	167,724	196,338	226,669	253,701	286,011	270,383
Montreal	1,539,308	1,830,232	2,215,627	2,570,982	2,743,208	2,802,485
Oshawa	...	...	...	...	120,318[1]	135,196
Ottawa–Hull	311,587	367,756	457,038	528,774	602,510	693,288
Quebec	289,294	328,405	379,067	436,918	480,502	542,158
Regina	72,731	91,215	113,749	132,432	140,734	151,191
Saint John, NB	80,689	88,375	98,083	104,195	106,744	112,974
St. Catharines–Niagara	189,046	233,034	257,796	285,453	303,429	301,921
St. John's, Nfld.	80,869	92,565	106,666	117,533	131,814	143,390
Saskatoon	55,679	72,930	95,564	115,900	126,449	133,750
Sudbury	80,543	107,889	127,446	136,739	155,424	157,030
Thunder Bay	73,713	87,624	102,085	108,035	112,093	119,253
Toronto	1,261,861	1,571,952	1,919,409	2,289,900	2,628,043	2,803,101
Vancouver	586,172	694,425	826,798	933,091	1,082,352	1,166,348
Victoria	114,859	136,127	155,763	175,262	195,800	18,250
Windsor	182,619	208,456	217,215	238,323	258,643	247,582
Winnipeg	357,229	412,741	476,543	508,759	540,262	578,217

[1]Based on 1976 census metropolitan area.

4.10   Land area and density of population, by province, 1956-76

Province or territory	Land area km²	Population per km²				
		1956	1961	1966	1971	1976
Newfoundland	370 485	1.12	1.24	1.33	1.41	1.51
Prince Edward Island	5 657	17.55	18.50	19.19	19.73	20.90
Nova Scotia	52 841	13.15	13.95	14.31	14.93	15.87
New Brunswick	72 092	7.70	8.29	8.56	8.80	9.39
Quebec	1 356 791	3.41	3.88	4.26	4.44	4.59
Ontario	891 194	6.07	7.00	7.81	8.64	9.27
Manitoba	548 495	1.55	1.68	1.76	1.80	1.86
Saskatchewan	570 269	1.54	1.62	1.68	1.63	1.62
Alberta	644 389	1.74	2.07	2.27	2.53	2.85
British Columbia	930 528	1.50	1.75	2.02	2.35	2.65
Canada (excl. the territories)	5 442 741	2.95	3.34	3.67	3.95	4.21
Yukon Territory	531 844	0.02	0.03	0.03	0.03	0.04
Northwest Territories	3 246 389	0.01	0.01	0.01	0.01	0.01
Canada	9 220 974	1.75	1.98	2.17	2.34	2.49

4.11 Number and percentage of the population classified as urban, and rural by non-farm and farm, by province, 1976

Province or territory	Urban		Rural								Total population
			Non-farm		Farm		Total				
	No.	%	No.	%	No.	%	No.	%			No.
Newfoundland	328,270	58.9	228,365	40.9	1,095	0.2	229,460	41.1			557,725
Prince Edward Island	43,880	37.1	62,155	52.6	12,190	10.3	74,345	62.9			118,230
Nova Scotia	462,590	55.8	353,815	42.7	12,170	1.5	365,985	44.2			828,570
New Brunswick	354,420	52.3	311,150	46.0	11,685	1.7	322,835	47.7			677,250
Quebec	4,932,755	79.1	1,110,580	17.8	191,110	3.1	1,301,690	20.9			6,234,445
Ontario	6,708,520	81.2	1,276,890	15.4	279,055	3.4	1,555,945	18.8			8,264,465
Manitoba	714,480	69.9	205,570	20.2	101,455	9.9	307,025	30.1			1,021,510
Saskatchewan	511,330	55.5	217,425	23.6	192,570	20.9	409,995	44.5			921,325
Alberta	1,379,165	75.0	269,225	14.7	189,650	10.3	458,875	25.0			1,838,035
British Columbia	1,897,085	76.9	525,950	21.3	43,575	1.8	569,525	23.1			2,466,605
Yukon Territory	13,310	61.0	8,525	39.0	—	—	8,525	39.0			21,835
Northwest Territories	21,165	49.7	21,435	50.3	10	—	21,445	50.3			42,610
Canada	17,366,970	75.5	4,591,070	20.0	1,034,560	4.5	5,625,630	24.5			22,992,605

4.12 Sex distribution of the population, by province, 1976 and sex ratios, 1966, 1971 and 1976

Province or territory	Population, 1976		Males to 100 females		
	Male	Female	1966	1971	1976
Newfoundland	283,385	274,340	104	104	103
Prince Edward Island	59,325	58,900	103	101	101
Nova Scotia	414,150	414,420	101	101	100
New Brunswick	339,335	337,915	101	101	100
Quebec	3,084,645	3,149,800	100	99	98
Ontario	4,096,865	4,167,600	100	99	98
Manitoba	508,010	513,495	101	100	99
Saskatchewan	464,770	456,550	105	103	102
Alberta	932,370	905,670	104	103	103
British Columbia	1,232,510	1,234,095	103	101	100
Yukon Territory	11,705	10,135	119	117	115
Northwest Territories	22,450	20,160	118	111	111
Canada	11,449,525	11,543,080	101	100	99

4.13 Age distribution of the population, 1966, 1971 and 1976

Age group	Number			Percentage		
	1966	1971	1976	1966	1971	1976
0- 4 years	2,197,387	1,816,155	1,731,995	11.0	8.4	7.5
5- 9 "	2,300,857	2,254,005	1,887,805	11.5	10.4	8.2
10-14 "	2,093,513	2,310,740	2,276,375	10.5	10.7	9.9
15-19 "	1,837,725	2,114,345	2,345,255	9.2	9.8	10.3
20-24 "	1,461,298	1,889,400	2,133,805	7.3	8.8	9.3
25-29 "	1,241,794	1,584,125	1,993,060	6.2	7.3	8.7
30-34 "	1,241,697	1,305,425	1,627,485	6.2	6.1	7.1
35-39 "	1,286,144	1,263,870	1,328,790	6.4	5.9	5.8
40-44 "	1,257,028	1,262,530	1,268,220	6.3	5.9	5.5
45-49 "	1,089,915	1,239,040	1,252,845	5.4	5.7	5.4
50-54 "	988,264	1,052,540	1,220,180	4.9	4.9	5.3
55-59 "	816,300	954,725	1,019,035	4.1	4.4	4.4
60-64 "	663,410	777,020	905,400	3.3	3.6	3.9
65-69 "	531,709	619,960	720,815	2.7	2.9	3.1
70-74 "	427,207	457,380	533,725	2.1	2.1	2.3
75-79 "	300,365	325,510	362,705	1.5	1.5	1.6
80-84 "	177,319	204,170	220,560	0.9	0.9	1.0
85-89 "	76,790	100,010	112,380	0.4	0.5	0.5
90 years and over	26,158	37,380	52,160	0.1	0.2	0.2
Total	20,014,880	21,568,310	22,992,605	100.0	100.0	100.0

4.14 Estimated population by age group and sex, by province, as at June 1, 1976[1] (thousands)

Province or territory	Age group and sex							
	0-4 years		5-9 years		10-14 years		15-19 years	
	Male	Female	Male	Female	Male	Female	Male	Female
Newfoundland	29.7	28.1	32.1	30.8	34.3	32.7	32.1	30.6
Prince Edward Island	5.0	4.6	5.4	5.1	6.8	6.3	6.5	6.4
Nova Scotia	33.7	31.9	36.9	34.9	44.1	42.2	44.6	42.1
New Brunswick	29.7	28.5	31.7	29.8	37.6	35.8	38.0	35.8
Quebec	227.3	215.4	248.6	237.0	318.7	303.4	338.4	327.9
Ontario	311.7	295.5	342.0	325.8	409.3	389.5	412.7	395.3
Manitoba	42.5	39.8	43.3	41.6	49.8	48.0	51.0	49.3
Saskatchewan	38.0	36.7	39.9	38.1	48.6	46.8	49.4	47.5
Alberta	78.4	74.6	83.5	79.5	95.6	91.6	99.0	94.3
British Columbia	88.6	84.6	99.2	94.8	116.2	111.7	121.0	116.9
Yukon Territory	1.1	1.0	1.1	1.0	1.2	1.1	1.1	1.0
Northwest Territories	2.8	2.7	2.9	2.7	2.7	2.6	2.2	2.1
Canada	886.6	843.4	966.7	921.1	1,164.6	1,111.7	1,196.0	1,149.3

4.14   Estimated population by age group and sex, by province, as at June 1, 1976[1] (thousands) (concluded)

Province or territory	20-24 years		25-34 years		35-44 years		45-54 years	
	Male	Female	Male	Female	Male	Female	Male	Female
Newfoundland	26.0	26.1	41.6	40.4	26.9	25.1	23.4	22.0
Prince Edward Island	5.0	5.0	8.3	8.0	6.0	5.7	5.3	5.3
Nova Scotia	38.2	37.2	62.3	60.4	42.9	42.0	39.1	40.8
New Brunswick	32.0	31.8	51.1	48.7	33.5	33.1	31.1	32.3
Quebec	299.2	299.3	515.6	514.2	363.9	365.0	330.3	346.3
Ontario	368.3	376.1	652.0	650.2	488.0	476.5	466.1	470.8
Manitoba	47.0	47.0	75.8	74.0	53.0	51.8	52.0	54.5
Saskatchewan	41.7	39.7	59.3	56.7	45.9	44.9	48.6	48.3
Alberta	94.6	91.4	150.0	144.0	106.0	99.8	93.2	90.2
British Columbia	110.3	111.2	200.8	194.9	144.8	135.1	134.4	134.3
Yukon Territory	1.2	1.2	2.5	2.2	1.5	1.1	1.1	0.8
Northwest Territories	2.2	2.1	4.0	3.5	2.4	2.0	1.6	1.3
Canada	1,065.8	1,068.0	1,823.2	1,797.3	1,314.9	1,282.1	1,226.2	1,246.8

	55-64 years		65-69 years		70+ years		All ages	
	Male	Female	Male	Female	Male	Female	Male	Female
Newfoundland	20.0	19.0	7.0	6.8	10.2	12.6	283.4	274.3
Prince Edward Island	5.0	5.3	2.1	2.1	3.9	5.1	59.3	59.0
Nova Scotia	36.2	38.3	14.0	14.5	22.0	30.3	414.2	414.4
New Brunswick	27.3	28.5	10.6	11.1	16.8	22.6	339.3	337.9
Quebec	237.9	264.7	84.5	101.6	120.3	175.0	3,084.6	3,149.8
Ontario	336.5	359.2	120.5	140.4	189.8	288.2	4,096.9	4,167.6
Manitoba	45.8	48.6	17.5	19.3	30.4	39.4	508.0	513.5
Saskatchewan	44.1	45.0	17.4	17.4	31.9	35.5	464.8	456.6
Alberta	66.4	68.2	23.9	24.6	41.8	47.6	932.4	905.7
British Columbia	107.3	118.4	40.7	44.2	69.2	88.0	1,232.5	1,234.1
Yukon Territory	0.6	0.4	0.2	0.1	0.2	0.2	11.7	10.1
Northwest Territories	0.9	0.7	0.3	0.2	0.3	0.3	22.4	20.2
Canada	928.1	996.4	338.5	382.3	536.9	744.6	11,449.5	11,543.1

[1]Totals do not necessarily add, due to rounding.

4.15   Marital status of the population 15 years and over, by age group and sex, 1976

Age group	Sex	Single[1]	Married	Widowed	Divorced	Total
15-24 years	M	1,893,215	364,765	900	2,860	2,261,745
	F	1,538,485	668,300	2,400	8,135	2,217,320
	T	3,431,700	1,033,065	3,300	10,995	4,479,060
25-34 years	M	377,610	1,415,415	1,925	28,260	1,823,210
	F	234,945	1,501,895	9,130	51,370	1,797,335
	T	612,555	2,917,310	11,050	79,630	3,620,545
35-44 years	M	113,900	1,165,525	5,065	30,425	1,314,915
	F	82,940	1,126,120	24,450	48,580	1,282,095
	T	196,845	2,291,645	29,510	79,005	2,597,005
45-54 years	M	101,895	1,079,460	16,035	28,800	1,226,190
	F	79,180	1,047,270	79,650	40,735	1,246,840
	T	181,075	2,126,730	95,685	69,535	2,473,030
55-64 years	M	76,200	801,400	32,400	18,055	928,050
	F	79,030	717,055	177,200	23,100	996,385
	T	155,230	1,518,455	209,600	41,155	1,924,440
65-69 years	M	30,640	278,595	24,010	5,275	338,520
	F	37,280	213,995	124,905	6,120	382,300
	T	67,920	492,590	148,915	11,395	720,815
70 years and over	M	53,115	369,075	109,330	5,360	536,885
	F	77,980	225,035	436,165	5,460	744,645
	T	131,100	594,110	545,500	10,820	1,281,525
Total	M	2,646,580	5,474,235	189,665	119,035	8,429,510
	F	2,129,840	5,499,670	853,900	183,505	8,666,915
	T	4,776,420	10,973,905	1,043,565	302,535	17,096,425

[1]The total number of single persons of all ages (including those under 15) amounted to 10,672,600, comprising 5,666,590 males and 5,006,005 females.

4.16 Population by mother tongue, 1971 and 1976, and language most often spoken in the home, 1971

Language	Mother tongue				Language most often spoken in the home	
	1971		1976		1971[1]	
	No.	%	No.	%	No.	%
English	12,973,810	60.2	14,122,765	61.4	14,446,235	67.0
French	5,793,650	26.9	5,887,205	25.6	5,546,025	25.7
Baltic[2]	43,385	0.2	34,190	0.1	29,345	0.1
Celtic	24,360	0.1	10,060	- -	1,545	- -
Chinese	94,855	0.4	132,560	0.6	77,890	0.4
Croatian, Serbian, etc.	74,190	0.3	77,570	0.3	29,310	0.1
Czech and Slovak	45,145	0.2	34,955	0.2	24,555	0.1
Finnish	36,725	0.2	28,470	0.1	18,280	0.1
German	561,085	2.6	476,715	2.1	213,350	1.0
Greek	104,455	0.5	91,530	0.4	86,830	0.4
Indo-Pakistani	32,555	0.2	58,420	0.3	23,110	0.1
Inuit (Eskimo)	15,295	0.1	15,900	0.1	15,080	0.1
Italian	538,360	2.5	484,045	2.1	425,235	2.0
Japanese	16,890	0.1	15,525	0.1	10,500	- -
Magyar (Hungarian)	86,835	0.4	69,305	0.3	50,670	0.2
Native Indian	164,525	0.8	117,110	0.6	122,205	0.6
Netherlandic and Flemish	159,165	0.7	122,555	0.5	39,360	0.2
Polish	134,780	0.6	99,845	0.4	70,960	0.3
Portuguese	86,925	0.4	126,535	0.5	74,765	0.3
Romanian	11,300	0.1	8,755	- -	4,455	- -
Russian	31,745	0.1	23,480	0.1	12,590	0.1
Scandinavian	84,335	0.4	59,410	0.3	10,055	- -
Semitic languages	28,550	0.1	37,100	0.2	15,260	0.1
Spanish	23,815	0.1	44,130	0.2	17,710	0.1
Ukrainian	309,855	1.4	282,060	1.2	144,760	0.7
Yiddish	49,890	0.2	23,440	0.1	26,330	0.1
Other	41,830	0.2	63,950	0.3	31,900	0.1
Not stated	...	...	445,020	1.9	...	...
Total	21,568,310	100.0	22,992,605	100.0	21,568,310	100.0

[1]Data not available for 1976.
[2]Includes Lithuanian, Estonian and Lettish.

4.17 Numerical and percentage distribution of the population by mother tongue, by province, 1976[1]

Province or territory		1976				
		English	French	Other	Not stated	Total
Nfld.	No.	545,340	2,760	3,965	5,665	557,725
	%	97.8	0.5	0.7	1.0	100.0
PEI	No.	109,745	6,545	935	1,005	118,230
	%	92.8	5.5	0.8	0.8	100.0
NS	No.	768,070	36,870	13,625	10,010	828,570
	%	92.7	4.5	1.6	1.2	100.0
NB	No.	435,975	223,780	6,925	10,565	677,250
	%	64.4	33.0	1.0	1.6	100.0
Que.	No.	800,680	4,989,245	334,055	110,470	6,234,445
	%	12.8	80.0	5.4	1.8	100.0
Ont.	No.	6,457,645	462,070	1,178,670	166,080	8,264,465
	%	78.1	5.6	14.3	2.0	100.0
Man.	No.	727,240	54,745	218,875	20,645	1,021,510
	%	71.2	5.4	21.4	2.0	100.0
Sask.	No.	715,685	26,710	163,935	14,995	921,325
	%	77.7	2.9	17.8	1.6	100.0
Alta.	No.	1,482,725	44,440	272,395	38,480	1,838,040
	%	80.7	2.4	14.8	2.1	100.0
BC	No.	2,037,645	38,430	325,610	64,930	2,466,610
	%	82.6	1.6	13.2	2.6	100.0
YT	No.	18,940	525	1,630	745	21,840
	%	86.7	2.4	7.5	3.4	100.0
NWT	No.	23,085	1,095	16,995	1,435	42,610
	%	54.2	2.6	39.9	3.4	100.0
Canada	No.	14,122,770	5,887,205	2,537,615	445,020	22,992,605
	%	61.4	25.6	11.0	1.9	100.0

[1]Based on 100% data.

4.18   Numerical and percentage distribution of the population speaking one or both of the official languages, by province, 1961 and 1971

Year and province or territory	English only		French only		English and French		Neither English nor French	
	No.	%	No.	%	No.	%	No.	%
1961								
Newfoundland	450,945	98.5	522	0.1	5,299	1.2	1,087	0.2
Prince Edward Island	95,296	91.1	1,219	1.2	7,938	7.6	176	0.2
Nova Scotia	684,805	92.9	5,938	0.8	44,987	6.1	1,277	0.2
New Brunswick	370,922	62.0	112,054	18.7	113,495	19.0	1,465	0.2
Quebec	608,635	11.6	3,254,850	61.9	1,338,878	25.5	56,848	1.1
Ontario	5,548,766	89.0	95,236	1.5	493,270	7.9	98,820	1.6
Manitoba	825,955	89.6	7,954	0.9	68,368	7.4	19,409	2.1
Saskatchewan	865,821	93.6	3,853	0.4	42,074	4.5	13,433	1.5
Alberta	1,253,824	94.1	5,534	0.4	56,920	4.3	15,666	1.2
British Columbia	1,552,560	95.3	2,559	0.2	57,504	3.5	16,459	1.0
Yukon Territory	13,679	93.5	38	0.3	825	5.6	86	0.6
Northwest Territories	13,554	58.9	109	0.5	1,614	7.0	7,721	33.6
Canada	12,284,762	67.4	3,489,866	19.1	2,231,172	12.2	232,447	1.3
1971								
Newfoundland	511,620	98.0	510	0.1	9,350	1.8	625	0.1
Prince Edward Island	101,820	91.2	680	0.6	9,110	8.2	30	- -
Nova Scotia	730,700	92.6	4,185	0.5	53,035	6.7	1,035	0.1
New Brunswick	396,855	62.5	100,985	15.9	136,115	21.5	600	0.1
Quebec	632,515	10.5	3,668,020	60.9	1,663,790	27.6	63,445	1.1
Ontario	6,724,100	87.3	92,840	1.2	716,065	9.3	170,090	2.2
Manitoba	881,715	89.2	5,020	0.5	80,935	8.2	20,585	2.1
Saskatchewan	867,315	93.6	1,825	0.2	45,985	5.0	11,110	1.2
Alberta	1,525,575	93.7	3,310	0.2	81,000	5.0	17,990	1.1
British Columbia	2,054,690	94.1	1,775	0.1	101,435	4.6	26,725	1.2
Yukon Territory	17,130	93.2	5	- -	1,210	6.6	35	0.2
Northwest Territories	25,500	73.3	100	0.3	2,120	6.1	7,085	20.4
Canada	14,469,540	67.1	3,879,255	18.0	2,900,155	13.4	319,360	1.5

4.19   Population by ethnic group, 1951, 1961 and 1971

Ethnic group	1951		1961		1971	
	No.	%	No.	%	No.	%
British Isles	6,709,685	47.9	7,996,669	43.8		
English	3,630,344	25.9	4,195,175	23.0		
Irish	1,439,635	10.3	1,753,351	9.6	9,624,115	44.6
Scottish	1,547,470	11.0	1,902,302	10.4		
Welsh and other	92,236	0.7	145,841	0.8		
French	4,319,167	30.8	5,540,346	30.4	6,180,120	28.7
Other European	2,553,722	18.2	4,116,849	22.6	4,959,680	23.0
Austrian	32,231	0.2	106,535	0.6	42,120	0.2
Belgian	35,148	0.2	61,382	0.3	51,135	0.2
Czech and Slovak	63,959	0.5	73,061	0.4	81,870	0.4
Danish	42,671	0.3	85,473	0.5	75,725	0.4
Finnish	43,745	0.3	59,436	0.3	59,215	0.3
German	619,995	4.4	1,049,599	5.8	1,317,200	6.1
Greek	13,966	0.1	56,475	0.3	124,475	0.6
Hungarian	60,460	0.4	126,220	0.7	131,890	0.6
Icelandic	23,307	0.2	30,623	0.2	27,905	0.1
Italian	152,245	1.1	450,351	2.5	730,820	3.4
Jewish	181,670	1.3	173,344	1.0	296,945	1.4
Lithuanian	16,224	0.1	27,629	0.2	24,535	0.1
Netherlands	264,267	1.9	429,679	2.4	425,945	2.0
Norwegian	119,266	0.8	148,681	0.8	179,290	0.8
Polish	219,845	1.6	323,517	1.8	316,425	1.5
Portuguese	. .		. .		96,875	0.4
Romanian	23,601	0.2	43,805	0.2	27,375	0.1
Russian	91,279	0.7	119,168	0.7	64,475	0.3
Spanish	. .		. .		27,515	0.1
Swedish	97,780	0.7	121,757	0.6	101,870	0.5
Ukrainian	395,043	2.8	473,337	2.6	580,660	2.7
Yugoslavic	21,404	0.2	68,587	0.4	104,950	0.5
Other	35,616	0.2	88,190	0.5	70,460	0.3
Asiatic	72,827	0.5	121,753	0.7	285,540	1.3
Chinese	32,528	0.2	58,197	0.3	118,815	0.6
Japanese	21,663	0.2	29,157	0.2	37,260	0.2
Other	18,636	0.1	34,399	0.2	129,460	0.6
Other	354,028	2.5	462,630	2.5	518,850	2.4
Eskimo	9,733	0.1	11,835	0.1	17,550	0.1
Native Indian	155,874	1.1	208,286	1.1	295,215	1.4
Negro	18,020	0.1	32,127	0.2	34,445	0.2
West Indian	. .		. .		28,025	0.1
Other and not stated	170,401	1.2	210,382	1.2	143,620	0.7
Total	14,009,429	100.0	18,238,247	100.0	21,568,310	100.0

4.20   Principal religious denominations of the population, 1951, 1961 and 1971

Religious denomination	1951		1961		1971	
	No.	%	No.	%	No.	%
Adventist	21,398	0.2	25,999	0.1	28,590	0.1
Anglican Church of Canada	2,060,720	14.7	2,409,068	13.2	2,543,180	11.8
Baptist	519,585	3.7	593,553	3.3	667,245	3.1
Christian Reformed	..	..	62,257	0.3	83,390	0.4
Greek Orthodox	172,271	1.2	239,766	1.3	316,605	1.5
Jehovah's Witnesses	34,596	0.2	68,018	0.4	174,810	0.8
Jewish	204,836	1.5	254,368	1.4	276,025	1.3
Lutheran	444,923	3.2	662,744	3.6	715,740	3.3
Mennonite [1]	125,938	0.9	152,452	0.8	181,800	0.8
Mormon	32,888	0.2	50,016	0.3	66,635	0.3
Pentecostal	95,131	0.7	143,877	0.8	220,390	1.0
Presbyterian	781,747	5.6	818,558	4.5	872,335	4.0
Roman Catholic	6,069,496	43.3	8,342,826	45.7	9,974,895	46.2
Salvation Army	70,275	0.5	92,054	0.5	119,665	0.6
Ukrainian (Greek) Catholic [2]	191,051	1.4	189,653	1.0	227,730	1.1
United Church of Canada	2,867,271	20.5	3,664,008	20.1	3,768,800	17.5
Other	317,303	2.2	469,030	2.6	1,330,480	6.2
Total	14,009,429	100.0	18,238,247	100.0	21,568,310	100.0

[1]Includes "Hutterites".
[2]Includes "Other Greek Catholic".

4.21   Country of birth of the population, 1951, 1961 and 1971

Country of birth	1951		1961		1971[1]	
	No.	%	No.	%	No.	%
Canada[1]	11,949,518	85.3	15,393,984	84.4	18,272,780	84.7
United Kingdom	912,482	6.5	969,715	5.3	933,040	4.3
Other Commonwealth countries	20,567	0.1	47,887	0.3	170,100	0.8
United States	282,010	2.0	283,908	1.6	309,640	1.4
European countries	*801,618*	*5.7*	*1,468,058*	*8.0*	*1,684,510*	*7.8*
Germany	42,693	0.3	189,131	1.0	211,060	1.0
Italy	57,789	0.4	258,071	1.4	385,755	1.8
Netherlands	41,457	0.3	135,033	0.7	133,525	0.6
Poland	164,474	1.2	171,467	0.9	160,040	0.7
USSR	188,292	1.3	186,653	1.0	160,120	0.7
Other	306,913	2.2	527,703	2.9	634,010	2.9
Asiatic countries	37,145	0.3	57,761	0.3	119,425	0.6
Other	6,089	- -	16,934	0.1	78,800	0.4
Total	14,009,429	100.0	18,238,247	100.0	21,568,310	100.0

[1]For figures on province of birth, see Table 4.64.

4.22   Indian bands and registered population, by province and type of residence, Dec. 31, 1975 and 1976

Year and province or territory	Number of bands[1]	Registered band membership			
		On reserves	Off reserves	Crown land	Total
1975					
Prince Edward Island	2	293	160	8	461
Nova Scotia	12	3,757	1,465	30	5,252
New Brunswick	15	3,738	1,226	28	4,992
Quebec	39	19,929	5,480	5,233	30,642
Ontario	112	36,317	20,226	5,078	61,621
Manitoba	55	28,270	9,836	3,081	41,187
Saskatchewan	68	28,513	12,442	1,465	42,420
Alberta	41	24,637	7,010	1,904	33,551
British Columbia	193	33,145	18,536	1,043	52,724
Yukon Territory	13	7	422	2,299	2,728
Northwest Territories	16	229	423	6,532	7,184
Canada, 1975	566	178,835	77,226	26,701	282,762
1976					
Prince Edward Island	2	287	168	12	467
Nova Scotia	12	3,899	1,427	38	5,364
New Brunswick	15	3,749	1,230	81	5,060
Quebec	39	20,153	5,446	5,480	31,079
Ontario	113	37,648	20,399	5,144	63,191
Manitoba	56	28,571	10,588	3,152	42,311
Saskatchewan	68	29,359	12,656	1,303	43,318
Alberta	41	24,891	7,307	2,079	34,277
British Columbia	193	33,253	19,393	1,130	53,776
Yukon Territory	16	14	251	7,084	7,349
Northwest Territories	13	45	436	2,265	2,746
Canada, 1976	568	181,869	79,301	27,768	288,938

[1]Bands whose members were known to reside in more than one province or territory were allocated to that province or territory in which the majority was known to reside.

4.23   Households and average persons per household, by province, 1966, 1971 and 1976

Province or territory	Households				Average persons per household		
	1966	1971	1976	% increase 1971-76	1966	1971	1976
Newfoundland	96,632	110,475	131,665	19.2	5.0	4.6	4.1
Prince Edward Island	25,360	27,895	32,930	18.0	4.2	3.9	3.5
Nova Scotia	185,245	208,425	243,100	16.6	4.0	3.7	3.3
New Brunswick	141,761	158,100	190,435	20.4	4.2	3.9	3.5
Quebec	1,389,115	1,605,750	1,894,110	18.0	4.0	3.7	3.2
Ontario	1,876,545	2,228,160	2,634,620	18.2	3.6	3.4	3.1
Manitoba	259,280	288,720	328,005	13.6	3.6	3.3	3.0
Saskatchewan	260,822	267,845	291,155	8.7	3.6	3.4	3.1
Alberta	393,707	464,945	575,280	23.7	3.6	3.4	3.1
British Columbia	543,075	668,305	828,285	23.9	3.3	3.2	2.9
Yukon Territory	8,931	12,685	{ 6,495 10,020 }	30.2	4.3	4.0	{ 3.1 4.1 }
Northwest Territories							
Canada	5,180,473	6,041,305	7,166,095	18.6	3.7	3.5	3.1

4.24   Households by type, 1966, 1971 and 1976

Type of household	Number			Percentage		
	1966	1971	1976	1966	1971	1976
Family households	4,376,409	4,933,450	5,633,945	84.5	81.7	78.6
One-family households	4,246,753	4,812,360	5,542,295	82.0	79.7	77.3
Family of household head	4,209,549	4,773,900	5,513,765	81.3	79.0	76.9
Without additional persons	3,754,530	4,285,960	5,025,815	72.5	70.9	70.1
With additional persons	455,019	487,935	487,950	8.8	8.1	6.8
Family other than that of household head	37,204	38,465	28,525	0.7	0.6	0.4
Multiple family households[1]	129,656	121,085	91,650	2.5	2.0	1.3
Including family of household head	128,325	120,000	91,185	2.5	2.0	1.3
With no family of household head	1,331	1,090	465	- -	- -	- -
Non-family households	804,064	1,107,855	1,532,150	15.5	18.3	21.4
One person only	589,571	811,835	1,205,340	11.4	13.4	16.8
Two or more persons	214,493	296,020	326,810	4.1	4.9	4.6
Total households	5,180,473	6,041,300	7,166,095	100.0	100.0	100.0

[1]Consists of two or more families in one dwelling.

4.25   Households by age and sex of head, 1966, 1971 and 1976

Age and sex of head	Number			Percentage		
	1966	1971	1976	1966	1971	1976
Under 25 years	269,065	414,470	584,270	5.2	6.9	8.1
Male	227,040	334,750	430,640	4.4	5.5	6.0
Female	42,025	79,720	153,630	0.8	1.3	2.1
25-34 years	1,014,676	1,265,290	1,678,965	19.6	20.9	23.4
Male	954,508	1,154,085	1,457,230	18.4	19.1	20.3
Female	60,168	111,205	221,735	1.2	1.8	3.1
35-44 years	1,190,133	1,252,500	1,339,420	23.0	20.7	18.7
Male	1,102,647	1,142,540	1,184,990	21.3	18.9	16.5
Female	87,486	109,960	154,430	1.7	1.8	2.2
45-54 years	1,052,705	1,173,055	1,305,650	20.3	19.4	18.2
Male	928,751	1,022,330	1,115,730	17.9	16.9	15.6
Female	123,954	150,725	189,925	2.4	2.5	2.6
55-64 years	803,338	955,995	1,079,005	15.5	15.8	15.1
Male	655,003	764,230	841,245	12.6	12.6	11.7
Female	148,335	191,765	237,755	2.9	3.2	3.3
65 years and over	850,556	979,995	1,178,775	16.4	16.2	16.4
Male	562,131	626,130	720,380	10.8	10.4	10.0
Female	288,425	353,870	458,390	5.6	5.9	6.4
Total household heads	5,180,473	6,041,300	7,166,095	100.0	100.0	100.0
Male	4,430,080	5,044,065	5,750,225	85.5	83.5	80.2
Female	750,393	997,240	1,415,875	14.5	16.5	19.8

4.26   Households by marital status of head, 1966, 1971 and 1976

Marital status of head	Number			Percentage			Percentage increase		
	1966	1971	1976	1966	1971	1976	1966-71	1971-76	1966-76
Married[1]	4,196,595	4,745,795	5,419,745	81.0	78.6	75.6	13.1	14.2	29.1
Widowed	553,119	629,670	745,645	10.7	10.4	10.4	13.8	18.4	34.8
Divorced	40,263	112,665	222,625	0.8	1.9	3.1	179.8	97.6	452.9
Single, never married	390,496	553,170	778,085	7.5	9.2	10.9	41.7	40.7	99.3
Total households	5,180,473	6,041,300	7,166,095	100.0	100.0	100.0	16.6	18.6	38.3

[1]Includes household heads who are married but separated.

4.27 Families and persons per family, by province, 1966, 1971 and 1976[1]

Province or territory	Families			Average persons per family		
	1966	1971[r]	1976	1966	1971	1976
Newfoundland	97,011	107,960	124,655	4.6	4.4	4.0
Prince Edward Island	22,728	24,170	27,560	4.2	4.0	3.7
Nova Scotia	166,237	179,595	200,480	4.0	3.8	3.5
New Brunswick	129,307	139,720	162,035	4.3	4.0	3.7
Quebec	1,229,301	1,353,655	1,540,400	4.2	3.9	3.5
Ontario	1,657,933	1,877,055	2,104,540	3.7	3.6	3.4
Manitoba	222,735	234,595	251,975	3.8	3.6	3.4
Saskatchewan	216,674	214,840	225,685	3.9	3.7	3.5
Alberta	331,158	380,220	448,765	3.9	3.7	3.5
British Columbia	445,297	530,830	628,445	3.6	3.5	3.3
Yukon Territory	} 7,885	} 10,530	{ 4,930	} 4.5	} 4.3	{ 3.5
Northwest Territories			8,425 }			4.3 }
Canada	4,526,266	5,053,170	5,727,875	3.9	3.7	3.5

[1]1966 figures include families in collective households and households outside Canada; 1971 and 1976 figures exclude families in collective households and households outside Canada.

4.28 Families by family structure, 1966, 1971 and 1976[1]

Family structure	Number			Percentage		
	1966	1971	1976	1966	1971	1976
Husband-wife families	4,154,381	4,575,640	5,168,565	91.8	90.6	90.2
Lone parent families	371,885	477,525	559,335	8.2	9.4	9.8
Male parent	71,502	100,355	94,990	1.6	2.0	1.7
Female parent	300,383	377,165	464,345	6.6	7.4	8.1
Total families	4,526,266	5,053,170	5,727,895	100.0	100.0	100.0

[1]See footnote 1, Table 4.27.

4.29 Husband-wife and lone parent families by age of husband, wife and lone parent, for Canada, 1966, 1971 and 1976[1]

Age	1966		1971		1976	
	No.	%	No.	%	No.	%
Husband-wife families	4,154,381		4,605,485		5,168,565	
Husbands	4,154,381	100.0	4,605,485	100.0	5,168,565	100.0
Under 25 years	214,742	5.2	294,450	6.4	338,625	6.6
25-34 years	945,374	22.8	1,092,435	23.7	1,335,830	25.8
35-44 years	1,069,471	25.7	1,075,205	23.3	1,109,470	21.5
45-54 years	874,492	21.0	941,910	20.5	1,025,585	19.8
55-64 years	593,864	14.3	686,270	14.9	760,230	14.7
65 years and over	456,438	11.0	515,215	11.2	598,820	11.6
Wives	4,154,381	100.0	4,605,485	100.0	5,168,560	100.0
Under 25 years	451,829	10.8	567,440	12.3	623,490	12.1
25-34 years	1,029,701	24.8	1,154,550	25.1	1,416,840	27.4
35-44 years	1,080,354	26.0	1,043,710	22.7	1,065,965	20.6
45-54 years	816,643	19.7	920,470	20.0	987,600	19.1
55-64 years	481,260	11.6	581,495	12.6	674,715	13.1
65 years and over	294,594	7.1	337,805	7.3	399,960	7.7
Lone parent families	371,885		478,745		559,335	
Male	71,502	100.0	100,680	100.0	94,990	100.0
Under 25 years	2,407	3.4	4,225	4.2	3,280	3.5
25-34 years	5,559	7.8	16,535	16.4	12,275	12.9
35-44 years	12,176	17.0	22,210	22.1	21,565	22.7
45-54 years	15,918	22.3	22,525	22.3	24,730	26.0
55-64 years	13,313	18.6	16,375	16.3	16,065	16.9
65 years and over	22,129	30.9	18,805	18.7	17,075	18.0
Female	300,383	100.0	378,065	100.0	464,345	100.0
Under 25 years	12,542	4.2	25,295	6.7	33,080	7.1
25-34 years	36,327	12.1	66,665	17.6	98,660	21.2
35-44 years	59,515	19.8	78,350	20.7	100,100	21.6
45-54 years	68,592	22.8	85,160	22.6	99,155	21.4
55-64 years	50,480	16.8	59,500	15.7	65,765	14.2
65 years and over	72,927	24.3	63,090	16.7	67,595	14.6

[1]1966 and 1971 figures include families in collective households and households outside Canada. 1976 figures include families in private households within Canada.

4.30   Husband-wife and lone parent families by mother tongue of husband, wife and lone parent, 1976[1]

Province or territory	Mother tongue											
	Mother tongue of husband						Total husbands	Mother tongue of wife						Total wives
	English		French		Other			English		French		Other		
	No.	%	No.	%	No.	%		No.	%	No.	%	No.	%	
Newfoundland	111,285	97.7	735	0.6	1,005	0.9	113,855	111,300	97.8	710	0.6	905	0.8	113,855
Prince Edward Island	22,710	92.0	1,470	6.0	325	1.3	24,685	22,665	91.8	1,560	6.3	290	1.2	24,685
Nova Scotia	163,565	91.4	9,480	5.3	4,015	2.2	179,010	163,925	91.6	9,475	5.3	3,515	2.0	179,010
New Brunswick	94,290	64.6	47,485	32.6	1,965	1.3	145,875	93,730	64.3	48,155	33.0	1,760	1.2	145,880
Quebec	174,560	12.6	1,090,545	78.9	92,170	6.7	1,381,505	172,080	12.5	1,099,535	79.6	84,675	6.1	1,381,505
Ontario	1,391,155	73.1	108,325	5.7	364,935	19.2	1,902,090	1,407,615	74.0	113,395	6.0	342,700	18.0	1,902,090
Manitoba	149,995	66.0	13,140	5.8	60,290	26.5	227,240	152,600	67.2	13,390	5.9	57,335	25.2	227,240
Saskatchewan	147,695	71.5	7,055	3.4	49,485	24.0	206,585	151,675	73.4	7,165	3.5	45,250	21.9	206,580
Alberta	304,645	74.7	11,690	2.9	83,870	20.6	407,570	311,620	76.5	11,630	2.9	76,755	18.8	407,570
British Columbia	441,990	77.8	11,065	1.9	102,245	18.0	568,245	450,320	79.2	10,925	1.9	93,580	16.5	568,250
Yukon Territory	3,695	83.3	145	3.3	475	10.7	4,435	3,740	84.3	125	2.8	425	9.6	4,435
Northwest Territories	4,095	54.9	275	3.7	2,880	38.6	7,465	4,050	54.2	250	3.3	2,920	39.1	7,470
Canada	3,009,675	58.2	1,301,405	25.2	763,635	14.8	5,168,565	3,045,310	58.9	1,316,330	25.5	710,115	13.7	5,168,560

	Mother tongue of male lone parent						Mother tongue of female lone parent						Total lone parents
	English		French		Other		English		French		Other		
	No.	%	No.	%	No.	%	No.	%	No.	%	No.	%	
Newfoundland	2,110	19.5	20	0.2	15	0.1	8,455	78.3	50	0.5	35	0.3	10,800
Prince Edward Island	450	15.7	40	1.4	5	0.2	2,195	76.4	145	5.0	20	0.7	2,875
Nova Scotia	3,560	16.6	185	0.9	95	0.4	16,370	76.2	815	3.8	315	1.5	21,470
New Brunswick	1,950	12.1	980	6.1	45	0.3	8,790	54.4	4,150	25.7	120	0.8	16,160
Quebec	3,385	2.1	21,105	13.3	1,445	0.9	18,125	11.4	108,215	68.1	5,350	3.4	158,900
Ontario	25,565	12.6	2,270	1.1	5,560	2.8	134,675	66.5	11,420	5.6	20,580	10.2	202,450
Manitoba	2,605	10.5	255	1.0	1,140	4.6	14,425	58.3	1,280	5.2	4,745	19.2	24,730
Saskatchewan	2,355	12.3	130	0.7	1,020	5.3	11,140	58.3	565	3.0	3,710	19.4	19,100
Alberta	4,825	11.7	205	0.5	1,485	3.6	27,280	66.2	980	2.4	5,945	14.4	41,200
British Columbia	8,110	13.5	255	0.4	1,830	8.0	41,265	68.6	975	1.6	6,950	11.6	60,200
Yukon Territory	100	20.0	—	—	20	4.0	310	62.0	5	1.0	50	10.0	500
Northwest Territories	95	10.0	5	0.5	155	16.2	285	29.8	15	1.6	365	3.8	955
Canada	55,115	9.9	25,460	4.6	12,820	2.3	283,310	50.7	128,615	23.0	48,200	8.6	559,330

[1]Includes persons whose mother tongue was not stated.

4.31   Children living at home in private households by age group and province, 1976[1]

Province or territory	Under 6 years	6-14 years	15-17 years	18-24 years	25 years and over	Total children living at home
Newfoundland	66,660	113,565	36,035	37,620	10,315	264,195
Prince Edward Island	11,275	20,940	7,515	7,965	2,765	50,455
Nova Scotia	76,725	137,930	48,210	52,535	15,475	330,875
New Brunswick	67,960	118,405	41,340	46,260	12,825	286,795
Quebec	523,695	985,635	379,880	483,895	139,775	2,512,885
Ontario	724,820	1,303,370	459,640	517,485	112,680	3,118,000
Manitoba	94,980	158,545	55,460	56,255	15,130	380,365
Saskatchewan	85,675	151,800	54,385	46,500	13,180	351,540
Alberta	178,570	306,250	104,910	92,330	17,610	699,670
British Columbia	205,890	371,045	133,350	127,835	25,510	863,630
Yukon Territory	2,390	3,655	1,105	770	135	8,055
Northwest Territories	6,255	9,210	2,285	1,915	615	20,280
Canada	2,044,890	3,680,345	1,324,110	1,471,380	366,020	8,886,745

[1]Excludes children in collective households and households outside Canada.

4.32   Summary of principal vital statistics, by province, 1951-75

Province or territory and year	Live births		Deaths		Natural increase[1]		Marriages		Divorces	
	No.	Rate[2]	No.	Rate[2]	No.	Rate[2]	No.	Rate[2]	No.	Rate[3]
NEWFOUNDLAND										
Av. 1951-55	13,101	34.1	2,926	7.6	10,175	26.5	2,836	7.4	5	1.3
" 1956-60	14,934	34.6	3,114	7.2	11,820	27.4	3,032	7.0	5	1.2
" 1961-65	15,104	31.8	3,142	6.6	11,962	25.2	3,331	7.0	5	1.0
" 1966-70	13,057	25.8	3,122	6.2	9,935	19.6	4,147	8.2	56	11.1
1971	12,767	24.5	3,199	6.1	9,568	18.4	4,685	9.0	150	28.7
1972	12,898	24.2	3,349	6.3	9,549	17.9	5,106	9.6	177	33.3
1973	11,906	22.0	3,405	6.3	8,501	15.7	5,048	9.3	224	41.4
1974	10,236	18.9	3,286	6.1	6,950	12.8	4,276	7.9	301	55.5
1975	11,213	20.4	3,219	5.9	7,994	14.5	4,313	7.5	380	69.2

4.32   Summary of principal vital statistics, by province, 1951-75 (continued)

Province or territory and year	Live births No.	Live births Rate[2]	Deaths No.	Deaths Rate[2]	Natural increase[1] No.	Natural increase[1] Rate[2]	Marriages No.	Marriages Rate[2]	Divorces No.	Divorces Rate[3]
PRINCE EDWARD ISLAND										
Av. 1951-55	2,720	27.2	923	9.2	1,797	18.0	623	6.2	10	9.8
" 1956-60	2,674	26.6	953	9.5	1,721	17.1	645	6.4	4	3.9
" 1961-65	2,767	25.7	1,006	9.3	1,761	16.4	672	6.2	8	7.8
" 1966-70	2,063	18.9	1,020	9.4	1,044	9.6	817	7.5	45	41.3
1971	2,103	18.8	1,007	9.0	1,096	9.8	961	8.6	61	54.7
1972	2,010	17.8	1,052	9.3	958	8.5	1,013	9.0	65	57.5
1973	1,886	16.4	1,020	8.9	866	7.5	1,014	8.8	54	47.0
1974	1,939	16.6	1,088	9.3	851	7.3	990	8.5	96	82.3
1975	1,928	16.2	1,057	8.9	871	7.3	936	7.9	75	63.7
NOVA SCOTIA										
Av. 1951-55	18,246	27.5	5,802	8.8	12,444	18.7	5,283	8.0	212	32.0
" 1956-60	19,097	26.9	6,062	8.5	13,035	18.4	5,289	7.4	227	32.0
" 1961-65	18,526	24.7	6,312	8.4	12,214	16.3	5,313	7.1	277	36.9
" 1966-70	14,217	18.7	6,622	8.7	7,594	10.0	6,335	8.3	573	75.4
1971	14,250	18.1	6,682	8.5	7,568	9.6	6,883	8.7	721	91.4
1972	13,536	17.0	6,904	8.7	6,632	8.3	7,291	9.2	927	116.7
1973	13,289	16.5	6,928	8.6	6,361	7.9	7,273	9.0	1,249	155.2
1974	12,941	15.9	6,899	8.5	6,042	7.4	7,112	8.7	1,591	195.6
1975	13,119	16.0	6,799	8.3	6,320	7.7	7,059	8.6	1,597	194.2
NEW BRUNSWICK										
Av. 1951-55	16,496	31.0	4,576	8.6	11,920	22.4	4,306	8.1	167	31.4
" 1956-60	16,567	29.0	4,640	8.1	11,927	20.9	4,357	7.6	194	34.0
" 1961-65	15,668	25.8	4,749	7.8	10,919	18.0	4,531	7.5	199	32.7
" 1966-70	11,984	19.3	4,873	7.8	7,112	11.4	5,481	8.8	267	42.9
1971	12,187	19.2	4,943	7.8	7,244	11.4	6,149	9.7	482	76.0
1972	11,806	18.4	4,982	7.8	6,824	10.6	6,455	10.0	466	72.5
1973	11,425	17.5	5,084	7.8	6,341	9.7	6,357	9.8	574	88.0
1974	11,444	17.3	5,205	7.9	6,239	9.4	6,108	9.2	755	114.1
1975	11,775	17.4	5,150	7.6	6,625	9.8	5,945	8.8	758	112.3
QUEBEC										
Av. 1951-55	128,523	30.0	34,269	8.0	94,254	22.0	35,584	8.3	327	7.6
" 1956-60	139,844	28.6	35,714	7.3	104,130	21.3	36,798	7.5	403	8.2
" 1961-65	131,453	24.0	37,698	6.9	93,755	17.1	38,126	7.0	380	6.9
" 1966-70	99,068	16.8	39,475	6.7	59,592	10.1	46,768	7.9	2,018	34.1
1971	89,210	14.8	40,738	6.8	48,472	8.0	49,695	8.2	5,201	86.3
1972	83,603	13.8	42,311	7.0	41,292	6.8	53,830	8.9	6,421	106.1
1973	84,057	13.8	42,666	7.0	41,391	6.8	51,943	8.5	8,091	133.1
1974	85,627	14.0	42,767	7.0	42,860	7.0	51,532	8.4	12,272	200.1
1975e	93,000	15.0	43,642	7.1	49,358	8.0	50,377	8.1	14,093	227.8
ONTARIO										
Av. 1951-55	128,861	26.1	44,715	9.0	84,146	17.1	45,213	9.1	2,430	49.2
" 1956-60	152,688	26.4	49,431	8.5	103,257	17.9	46,482	8.0	2,801	48.4
" 1961-65	152,629	23.5	52,664	8.1	99,965	15.4	46,794	7.2	3,342	51.3
" 1966-70	130,166	17.8	55,415	7.6	74,751	10.2	62,216	8.5	7,452	102.1
1971	130,395	16.9	56,623	7.4	73,772	9.5	69,590	9.0	12,205	158.4
1972	125,060	16.0	58,905	7.5	66,155	8.5	72,278	9.2	13,183	168.5
1973	123,776	15.6	59,876	7.5	63,900	8.0	72,371	9.1	13,781	173.6
1974	124,229	15.3	60,556	7.5	63,673	7.9	72,716	9.9	15,277	188.7
1975	125,708	15.0	60,604	7.4	65,104	7.9	72,209	8.8	17,485	212.6
MANITOBA										
Av. 1951-55	21,321	26.4	6,775	8.4	14,546	18.0	7,104	8.8	356	44.1
" 1956-60	22,408	25.6	7,293	8.3	15,115	17.3	6,600	7.5	315	35.9
" 1961-65	22,137	23.4	7,637	8.1	14,500	15.3	6,674	7.1	376	39.7
" 1966-70	17,734	18.3	7,868	8.1	9,865	10.2	8,283	8.5	805	82.9
1971	18,031	18.2	8,025	8.1	10,006	10.1	9,127	9.2	1,383	140.0
1972	17,398	17.6	8,225	8.3	9,173	9.3	9,181	9.3	1,413	142.5
1973	16,964	17.0	8,196	8.2	8,768	8.8	9,196	9.2	1,620	162.3
1974	17,311	17.1	8,430	8.3	8,881	8.8	9,231	9.1	1,796	177.6
1975	17,144	16.8	8,385	8.2	8,759	8.6	8,915	8.8	1,984	194.8
SASKATCHEWAN										
Av. 1951-55	23,554	27.5	6,547	7.6	17,007	19.9	6,876	8.0	231	26.9
" 1956-60	24,046	26.9	6,753	7.5	17,293	19.4	6,395	7.1	247	27.6
" 1961-65	22,811	24.4	7,268	7.8	15,543	16.6	6,316	6.7	298	31.8
" 1966-70	17,852	18.7	7,466	7.8	10,386	10.9	7,460	7.8	566	59.3
1971	16,054	17.3	7,413	8.0	8,641	9.3	7,813	8.4	815	88.0
1972	15,473	16.9	7,590	8.3	7,883	8.6	7,877	8.6	826	90.1
1973	14,806	16.3	7,646	8.4	7,160	7.9	7,847	8.6	887	97.7
1974	15,118	16.7	7,814	8.6	7,304	8.1	7,988	8.8	1,039	114.6
1975	15,260	16.6	7,672	8.4	7,588	8.3	8,066	8.8	1,131	123.2
ALBERTA										
Av. 1951-55	31,087	30.6	7,527	7.4	23,560	23.2	9,750	9.6	612	60.4
" 1956-60	36,920	30.6	8,329	6.9	28,591	23.7	10,230	8.5	788	65.1
" 1961-65	37,004	26.5	9,317	6.7	27,687	19.8	10,581	7.6	1,226	87.5
" 1966-70	30,851	20.2	9,839	6.4	21,012	13.8	13,711	9.0	2,481	162.4
1971	30,545	18.8	10,525	6.5	20,020	12.3	15,614	9.6	3,656	224.6
1972	29,282	17.7	10,699	6.5	18,583	11.2	16,345	9.9	3,767	227.8
1973	29,288	17.4	10,763	6.4	18,525	11.0	16,280	9.7	4,435	263.5
1974	29,813	17.4	11,252	6.6	18,561	10.8	16,691	9.7	4,947	288.6
1975	31,618	17.9	11,397	6.4	20,221	11.4	17,520	9.9	5,475	309.7

4.32 Summary of principal vital statistics, by province, 1951-75 (concluded)

Province or territory and year	Live births No.	Rate[2]	Deaths No.	Rate[2]	Natural increase[1] No.	Rate[2]	Marriages No.	Rate[2]	Divorces No.	Rate[3]
BRITISH COLUMBIA										
Av. 1951-55	31,347	25.1	12,233	9.8	19,114	15.3	11,131	8.9	1,461	116.9
" 1956-60	38,930	25.7	13,980	9.2	24,950	16.5	11,955	7.9	1,514	100.0
" 1961-65	36,753	21.5	15,236	8.9	21,517	12.6	11,927	7.0	1,592	93.1
" 1966-70	34,266	17.1	16,737	8.3	17,529	8.7	17,186	8.6	3,272	163.1
1971	34,852	16.0	17,783	8.1	17,069	7.9	20,389	9.3	4,926	225.5
1972	34,563	15.4	18,021	8.0	16,542	7.4	20,659	9.2	5,036	224.1
1973	34,352	14.8	18,095	7.8	16,257	7.0	21,303	9.2	5,687	245.7
1974	35,450	14.8	19,177	8.0	16,273	6.8	21,734	9.1	6,840	285.6
1975	36,277	14.8	19,151	7.8	17,126	7.0	21,824	8.9	7,534	306.6
YUKON TERRITORY										
Av. 1951-55	413	43.0	90	9.4	323	33.6	94	9.8	..	..
" 1956-60	505	39.4	91	7.1	414	32.3	109	8.5	..	..
" 1961-65	509	34.9	87	6.0	422	28.9	107	7.3	17	118.0
" 1966-70	407	27.1	89	5.9	319	21.3	153	10.2	31	206.7
1971	506	27.5	104	5.7	402	21.8	166	9.0	47	255.6
1972	451	23.9	103	5.5	348	18.4	181	9.5	47	248.7
1973	420	21.3	111	5.6	309	15.7	206	10.3	60	300.0
1974	495	25.5	114	5.9	381	19.6	190	9.8	46	237.1
1975	408	19.4	112	5.4	296	14.2	201	9.7	43	206.7
NORTHWEST TERRITORIES										
Av. 1951-55	666	40.1	284	17.1	382	23.0	115	6.9	..	..
" 1956-60	943	46.7	310	15.3	633	31.4	155	7.7	1	..
" 1961-65	1,174	45.9	250	9.8	924	36.1	154	6.0	3	11.5
" 1966-70	1,244	41.5	229	7.6	1,015	33.8	212	7.1	13	43.3
1971	1,287	37.0	230	6.6	1,057	30.4	252	7.2	25	71.8
1972	1,239	34.4	272	7.6	967	26.8	254	7.1	36	100.0
1973	1,204	31.9	249	6.6	955	25.3	226	5.9	42	110.5
1974	1,042	27.8	206	5.5	836	22.3	256	6.8	59	157.3
1975	1,171	30.8	216	5.7	955	25.3	220	5.8	56	148.1
CANADA										
Av. 1951-55	416,334	28.0	126,666	8.5	289,668	19.5	128,915	8.7	5,811	39.1
" 1956-60	469,555	27.6	136,669	8.0	332,886	19.6	132,047	7.8	6,498	38.2
" 1961-65	456,534	24.1	145,368	7.7	311,166	16.4	134,524	7.1	7,723	40.7
" 1966-70	372,909	18.0	152,755	7.4	220,154	10.6	172,769	8.3	17,579	84.8
1971	362,187	16.8	157,272	7.3	204,915	9.5	191,324	8.9	29,672	137.6
1972	347,319	15.9	162,413	7.4	184,906	8.5	200,470	9.2	32,364	148.3
1973	343,373	15.5	164,039	7.4	179,334	8.1	199,064	9.0	36,704	166.1
1974	345,645	15.4	166,794	7.4	178,851	8.0	198,824	8.9	45,019	200.6
1975e	358,621	15.7	167,404	7.3	191,217	8.4	197,585	8.9	50,611	222.0

[1]Excess births over deaths.
[2]Per 1,000 population.
[3]Per 100,000 population.

4.33 Summary of principal vital statistics for cities, towns and other municipal subdivisions[1] of 50,000 population and over[2], 1975 with averages for 1961-65 and 1966-70

Province and urban centre	Live births Av. 1961-65	Av. 1966-70	1975	Deaths Av. 1961-65	Av. 1966-70	1975	Marriages[3] Av. 1961-65	Av. 1966-70	1975
NEWFOUNDLAND									
*St. John's, c	1,966	1,812	1,386	542	608	648	736	932	961
PRINCE EDWARD ISLAND									
Charlottetown, c[4]	417	307	254	232	246	250	157	157	225
NOVA SCOTIA									
Dartmouth, c	1,700	1,409	1,170	230	261	296	287	379	430
*Halifax, c	2,109	1,791	1,743	736	855	1,002	1,047	1,281	1,611
NEW BRUNSWICK									
*Saint John, c	1,743	1,598	1,528	690	766	816	607	728	865
QUEBEC									
*Hull, c	1,640	1,289	1,025	419	423	442	430	520	675
LaSalle, c	1,062	1,371	1,218	210	295	377	128	222	314
Laval, c	3,939	3,372	3,195	669	902	1,197	599	968	1,175
*Longueuil, c	639	1,674	2,339	225	442	550	185	567	663
*Montreal, c	28,576	20,066	13,693	10,309	10,462	10,888	10,548	11,766	11,241
*Montreal North, c	1,453	1,490	1,506	343	382	483	215	386	408
*Quebec, c	3,601	2,672	2,346	1,612	1,587	1,642	1,536	1,683	1,812
*Ste-Foy, c	1,038	1,098	887	158	190	228	130	389	420
*St-Laurent, t	1,059	948	716	272	352	409	287	401	429
*St-Léonard, c	316	834	1,579	45	124	244	22	170	269
Sherbrooke, c	1,812	1,505	1,294	590	592	609	572	698	754
*Trois-Rivières, c	1,384	947	691	438	433	447	447	475	574
Verdun, c	1,547	1,119	697	606	665	769	528	636	470

4.33 Summary of principal vital statistics for cities, towns and other municipal subdivisions[1] of 50,000 population and over[2], 1975 with averages for 1961-65 and 1966-70 (concluded)

Province and urban centre	Live births			Deaths			Marriages[3]		
	Av. 1961-65	Av. 1966-70	1975	Av. 1961-65	Av. 1966-70	1975	Av. 1961-65	Av. 1966-70	1975
ONTARIO									
*Brantford, c	1,191	1,045	1,103	550	622	595	489	612	689
*Burlington, t	1,203	1,347	1,540	274	324	439	246	415	583
*Etobicoke, b	5,117	4,081	3,488	1,311	1,502	1,886	881	1,215	1,701
*Guelph, c	1,010	996	1,054	364	405	439	352	509	612
*Hamilton, c	6,467	5,475	4,531	2,447	2,495	2,604	2,351	2,946	2,936
*Kingston, c	1,363	1,126	787	481	523	602	527	718	824
*Kitchener, c	2,081	2,131	2,293	564	670	858	655	1,009	1,016
*London, c	4,129	3,752	3,617	1,482	1,576	1,651	1,387	1,914	2,159
*Mississauga, t	1,697	2,402	4,546	344	458	901	287	511	1,180
*Niagara Falls, c	1,151	1,030	915	441	456	559	416	549	605
*Oakville, c	905	913	957	186	227	314	223	450	455
Oshawa, c	1,769	1,686	1,750	459	509	646	545	691	871
Ottawa, c	6,034	4,745	3,678	2,271	2,443	2,499	2,209	3,051	3,476
*Peterborough, c	1,035	890	812	442	489	598	384	571	681
*St. Catharines, c	1,910	1,764	1,794	696	791	927	666	909	1,069
Sarnia, c	1,220	1,034	889	358	399	391	373	537	648
*Sault Ste Marie, c	1,439	1,512	1,323	385	487	528	488	645	780
Scarborough, b	6,419	5,150	5,627	1,237	1,546	2,055	962	1,765	2,214
*Sudbury, c	2,353	1,881	1,649	525	560	641	706	928	991
Thunder Bay, c	1,998	1,682	1,616	835	909	1,032	745	951	1,025
*Toronto, c	15,362	13,680	10,057	7,354	6,737	6,045	10,293	13,413	15,217
*Windsor, c	2,498	3,654	2,996	1,274	1,708	1,691	1,217	1,907	1,924
*York, b	3,497	3,101	2,652	1,022	1,009	1,048	488	523	520
*York, E., b	1,852	1,693	1,708	853	865	1,003	184	185	301
York, N., b	7,967	8,547	8,217	1,551	2,134	2,897	943	1,617	2,089
MANITOBA									
*Winnipeg, c	5,788	4,631	8,721	2,672	2,708	4,528	2,620	3,180	5,386
SASKATCHEWAN									
*Regina, c	3,265	2,840	2,674	820	928	1,045	1,004	1,263	1,472
*Saskatoon, c	2,770	2,676	2,240	769	911	1,023	923	1,281	1,399
ALBERTA									
*Calgary, c	8,083	7,503	7,821	2,002	2,238	2,648	2,410	3,483	4,770
*Edmonton, c	9,704	8,776	7,809	2,014	2,224	2,715	3,209	4,428	5,289
BRITISH COLUMBIA									
*Burnaby, dm	2,057	1,837	1,589	769	857	1,021	530	798	748
Coquitlam, dm	745	793	745	147	181	269	105	175	206
*North Vancouver, dm	864	797	818	228	233	283	140	251	403
Richmond, twp	1,093	898	1,225	231	265	394	171	291	527
Saanich, dm	1,042	827	777	416	485	519	199	334	388
Surrey, dm	1,761	1,577	2,030	550	637	705	288	451	637
Vancouver, c	6,743	6,317	4,654	4,758	4,928	4,733	3,881	5,044	4,776
Victoria, c	972	804	683	898	1,001	1,097	671	925	1,194

[1]Figures for certain subdivisions may not be comparable for the periods shown because of changes in area boundaries, particularly for those indicated by an asterisk: c=city, t=town, b=borough, dm=district municipality and twp=township.
[2]As at the date of the 1971 Census.
[3]By place of occurrence.
[4]Population fewer than 50,000 at date of 1971 Census but included as the largest urban centre in Prince Edward Island.

4.34 Number of live-born children in order of live births, by age of mother, 1975[1]

Order of birth of child	Age of mother										% of total
	Under 15	15-19	20-24	25-29	30-34	35-39	40-44	45 and over	Age not stated	All ages	
1st child	301	26,498	46,238	30,627	7,019	1,364	205	11	1	112,264	44.1
2nd "	4	5,177	30,918	35,163	11,744	2,139	292	12	5	85,454	33.6
3rd "	—	518	7,975	15,876	8,549	2,083	302	12	5	35,320	13.9
4th "	—	45	1,579	4,695	3,972	1,510	289	16	—	12,106	4.8
5th "	—	2	324	1,292	1,607	851	226	18	1	4,321	1.7
6th "	—	—	70	509	721	579	185	14	3	2,081	0.8
7th "	—	—	17	198	418	315	149	9	—	1,106	0.4
8th "	—	—	1	65	250	218	126	11	—	671	0.3
9th "	—	—	—	26	111	160	81	10	—	388	0.1
10th and over	—	—	—	8	121	268	219	29	—	645	0.3
Not stated	—	5	10	8	6	2	—	—	21	52	- -
Total	305	32,245	87,132	88,467	34,518	9,489	2,074	142	36	254,408	100.0
% of total	0.1	12.7	34.3	34.8	13.6	3.7	0.8	- -	- -	100.0	...

[1]Excludes Newfoundland and Quebec.

4.35   Percentage distribution of total live births, by order of birth, 1951-75[1]

Year	1st child	2nd child	3rd child	4th and later children	Total
1951	28.3	25.4	17.2	29.1	100.0
1956	26.7	24.0	17.8	31.5	100.0
1961	25.9	23.2	18.0	32.9	100.0
1966	33.1	24.8	16.2	25.9	100.0
1967	36.0	25.7	15.4	22.9	100.0
1968	37.8	26.7	15.1	20.3	100.0
1969	38.5	27.5	15.2	18.8	100.0
1970	39.9	28.0	15.2	16.9	100.0
1971	40.6	29.3	14.9	15.2	100.0
1972	41.9	30.9	14.3	12.9	100.0
1973	43.3	32.3	13.7	10.7	100.0
1974	44.0	33.4	13.5	9.1	100.0
1975	44.1	33.6	13.9	8.4	100.0

[1]Excludes Newfoundland, 1951-75 and Quebec for 1975.

4.36   Stillbirths and ratio per 1,000 live births, by province, 1951-75

Year	Nfld.	PEI	NS	NB	Que.	Ont.	Man.	Sask.	Alta.	BC	YT	NWT	Canada
	Number (28 weeks or more gestation)												
Av. 1951-55	222	52	337	291	2,705	2,017	336	313	425	374	6	11	7,088
" 1956-60	274	46	304	267	2,446	1,992	301	262	388	418	5	12	6,714
" 1961-65	261	47	256	220	1,727	1,818	278	242	358	370	5	19	5,600
" 1966-70	171	28	181	158	1,074	1,402	195	174	278	311	4	20	3,996
1971	158	23	139	138	807	1,221	164	159	254	310	6	17	3,396
1972	121	22	124	137	685	1,159	173	140	229	238	6	12	3,046
1973	151	15	119	132	658	1,034	151	144	206	243	3	10	2,866
1974	127	20	123	125	629	1,019	128	132	188	252	7	16	2,766
1975	116	25	98	130	650	906	128	100	173	284	2	15	2,627
	Ratio												
Av. 1951-55	17.0	19.0	18.4	17.7	21.0	15.6	15.7	13.3	13.7	11.9	14.1	16.5	17.0
" 1956-60	18.3	17.1	15.9	16.1	17.5	13.0	13.4	10.9	10.5	10.7	10.7	12.3	14.3
" 1961-65	17.3	17.1	13.8	14.0	13.1	11.9	12.5	10.6	9.7	10.1	9.0	16.0	12.3
" 1966-70	13.0	13.5	12.7	13.1	10.8	10.7	10.9	9.7	9.0	9.0	9.8	16.0	10.7
1971	12.4	10.9	9.8	11.3	9.0	9.4	0.1	9.9	8.3	8.9	11.9	13.2	9.4
1972	9.4	10.9	9.2	11.6	8.2	9.3	9.9	9.0	7.8	6.9	13.3	9.7	8.8
1973	12.7	8.0	9.0	11.6	7.8	8.4	8.9	9.7	7.0	7.1	7.1	8.3	8.3
1974	12.4	10.3	9.5	10.9	7.3	8.2	7.4	8.7	6.3	7.1	14.1	15.4	8.0
1975	10.3	13.3	7.5	11.0	7.0	7.2	7.5	6.6	5.5	7.8	4.9	12.8	7.3

4.37   Stillbirths and ratio per 1,000 live births, by age of mother, 1975[1]

Age group of mother	Live births[2]	Stillbirths	Stillbirth ratio per 1,000 live births[2]
Under 20 years	38,240	286	7.5
20-24 years	115,452	740	6.4
25-29 "	126,377	770	6.1
30-34 "	50,658	381	7.5
35-39 "	13,349	165	12.4
40-44 "	2,854	61	21.4
45 years and over	202	11	54.5
Age not stated	276	25	90.6
Total, all ages	347,408	2,439	7.0

[1]Excludes Newfoundland.
[2]Includes an estimate for Quebec.

4.38   Age-specific fertility rate and gross reproduction rate per 1,000 women, 1926-75[1]

Year and province or territory	15-19	20-24	25-29	30-34	35-39	40-44	45-49	Total fertility rate	Gross reproduction rate
Canada									
1926	29.0	139.9	177.4	153.8	114.6	50.7	6.0	3,357	1.628
1931	29.9	137.1	175.1	145.3	103.1	44.0	5.5	3,200	1.555
1936	25.7	112.1	144.3	126.5	90.0	36.3	4.4	2,696	1.310
1941	30.7	138.4	159.8	122.3	80.0	31.6	3.7	2,832	1.377
1946	36.5	169.6	191.4	146.0	93.1	34.5	3.8	3,374	1.640
1951	48.1	188.7	198.8	144.5	86.5	30.9	3.1	3,503	1.701
1956	55.9	222.2	220.1	150.3	89.6	30.8	2.9	3,858	1.874
1961	58.2	233.6	219.2	144.9	81.1	28.5	2.4	3,840	1.868
1966	48.2	169.1	163.5	103.3	57.5	19.1	1.7	2,812	1.369
1971	40.1	134.4	142.0	77.3	33.6	9.4	0.6	2,187	1.060
1972	38.5	119.8	137.1	72.1	28.9	7.8	0.6	2,024	0.982
1973	37.2	117.7	131.6	67.1	25.7	6.4	0.4	2,931	0.937
1974	35.3	113.1	131.1	66.6	23.0	5.5	0.4	1,875	0.911
1975	34.8	112.5	133.3	66.2	21.3	4.7	0.3	1,866	0.881
1975									
Prince Edward Island	49.8	113.2	138.3	76.1	27.2	5.0	0.4	2,050	0.988
Nova Scotia	51.7	122.7	126.5	59.8	22.4	6.8	0.4	1,952	0.930
New Brunswick	55.1	133.8	133.4	62.9	21.3	4.3	0.6	2,057	1.012
Quebec	17.5	93.9	139.1	71.0	21.4	4.4	0.3	1,738	0.842
Ontario	36.5	112.3	128.2	64.5	21.2	4.4	0.3	1,837	0.896
Manitoba	51.6	125.7	137.9	68.8	27.1	6.1	0.3	2,088	1.027
Saskatchewan	46.5	132.2	162.7	77.2	26.5	7.7	0.6	2,267	1.119
Alberta	50.1	142.6	138.9	64.6	20.2	4.9	0.4	2,109	1.030
British Columbia	38.8	117.0	121.9	58.4	17.1	3.8	0.2	1,786	0.876
Yukon Territory	67.8	165.6	114.5	60.0	25.0	6.0	—	2,195	1.103
Northwest Territories	134.1	255.3	213.8	99.3	56.0	21.3	1.7	3,908	2.012

[1]Excludes Newfoundland, 1926-75; Quebec estimated for 1975.

4.39   Percentage change in death rate for each age group, by sex, 1951-75

Age group	Male	Female	Age group	Male	Female
Under 1 year	-62.6	-62.9	50-54 years	-12.5	-30.8
1- 4 years	-57.1	-61.1	55-59 "	-11.1	-30.3
5- 9 "	-50.0	-57.1	60-64 "	-8.6	-32.9
10-14 "	-37.5	-40.0	65-69 "	-1.9	-30.5
15-19 "	+14.3	-44.4	70-74 "	-5.7	-33.4
20-24 "	—	-40.0	75-79 "	-11.9	-37.9
25-29 "	-16.7	-45.5	80-84 "	-11.4	-36.1
30-34 "	-23.8	-46.7	85 years and over	-5.7	-29.7
35-39 "	-12.0	-40.0			
40-44 "	-12.8	-33.3	All ages	-15.8	-20.5
45-49 "	-14.1	-28.9			

4.40   Numerical and percentage distribution of deaths by age group and sex, 1961, 1966 and 1971

Age group	1961 No.	1961 %	1961 Rate[1]	1966 No.	1966 %	1966 Rate[1]	1971 No.	1971 %	1971 Rate[1]	% change in death rate 1961-71
Male										
Under 1 year	7,447	9.0	30.5	5,138	5.8	25.8	3,712	4.1	19.9	-34.8
1- 4 years	1,154	1.4	1.3	988	1.1	1.1	679	0.7	0.9	-30.8
5- 9 "	672	0.8	0.6	669	0.8	0.6	641	0.7	0.6	—
10-14 "	527	0.6	0.6	620	0.7	0.6	589	0.6	0.5	-16.7
15-19 "	840	1.0	1.2	1,212	1.4	1.3	1,489	1.6	1.4	+16.7
20-24 "	969	1.2	1.7	1,324	1.5	1.8	1,697	1.9	1.8	+5.9
25-29 "	895	1.1	1.5	980	1.1	1.6	1,176	1.3	1.5	—
30-34 "	1,041	1.3	1.6	1,054	1.2	1.7	1,090	1.2	1.6	—
35-39 "	1,422	1.7	2.3	1,456	1.7	2.2	1,416	1.5	2.2	-4.3
40-44 "	1,916	2.3	3.4	2,146	2.4	3.4	2,310	2.5	3.6	+5.9
45-49 "	2,993	3.6	5.8	3,111	3.5	5.7	3,523	3.8	5.7	-1.7
50-54 "	4,242	5.1	9.6	4,855	5.5	9.7	4,839	5.3	9.3	-3.1
55-59 "	5,494	6.6	15.2	6,352	7.2	15.4	6,887	7.5	14.6	-3.9
60-64 "	7,028	8.5	24.0	7,911	9.0	24.0	8,755	9.5	22.9	-4.6
65-69 "	8,545	10.3	35.7	9,226	10.5	36.2	10,279	11.2	34.7	-2.8
70-74 "	10,582	12.8	54.0	10,549	12.0	53.1	10,663	11.6	51.9	-3.9
75-79 "	10,970	13.3	81.8	11,102	12.6	79.9	11,058	12.1	79.0	-3.4
80-84 "	8,635	10.4	125.1	10,006	11.4	124.0	10,182	11.1	118.8	-5.0
85 years and over	7,337	8.9	208.9	9,214	10.5	213.4	10,838	11.8	198.5	-5.0
Total, all ages	82,709	100.0	9.0	87,913	100.0	8.7	91,823	100.0	8.5	-5.6

4.40   Numerical and percentage distribution of deaths by age group and sex, 1961, 1966 and 1971 (concluded)

Age group	Distribution									% change in death rate 1961-71
	1961			1966			1971			
	No.	%	Rate[1]	No.	%	Rate[1]	No.	%	Rate[1]	
Female										
Under 1 year	5,493	9.4	23.7	3,822	6.2	20.2	2,644	4.0	15.1	-36.3
1- 4 years	844	1.4	1.0	775	1.3	0.9	551	0.8	0.8	-20.0
5- 9 "	405	0.7	0.4	480	0.8	0.4	424	0.7	0.4	—
10-14 "	278	0.5	0.3	318	0.5	0.3	365	0.6	0.3	—
15-19 "	322	0.6	0.5	467	0.8	0.5	579	0.9	0.6	+20.0
20-24 "	342	0.6	0.6	403	0.7	0.5	559	0.9	0.6	—
25-29 "	418	0.7	0.7	384	0.6	0.6	485	0.7	0.6	-14.3
30-34 "	562	1.0	0.9	564	0.9	0.9	565	0.9	0.9	—
35-39 "	880	1.5	1.4	845	1.4	1.3	815	1.2	1.3	-7.1
40-44 "	1,099	1.9	2.0	1,293	2.1	2.0	1,290	2.0	2.1	+5.0
45-49 "	1,617	2.8	3.2	1,823	2.9	3.3	1,901	2.9	3.0	-6.2
50-54 "	2,237	3.8	5.3	2,434	3.9	5.0	2,480	3.8	4.6	-13.2
55-59 "	2,749	4.7	8.0	3,115	5.0	7.7	3,477	5.3	7.2	-10.0
60-64 "	3,725	6.4	12.8	4,064	6.6	12.2	4,345	6.6	11.0	-14.1
65-69 "	5,304	9.1	21.4	5,393	8.7	19.5	5,614	8.6	17.3	-19.2
70-74 "	7,058	12.1	34.2	7,063	11.4	30.9	7,138	10.9	28.3	-17.3
75-79 "	8,290	14.2	59.2	8,695	14.0	53.9	8,930	13.6	48.1	-18.7
80-84 "	7,871	13.5	101.2	9,048	14.6	93.6	9,763	14.9	82.4	-18.6
85 years and over	8,782	15.1	192.2	10,964	17.7	183.4	13,524	20.7	163.3	-15.0
Total, all ages	58,276	100.0	6.5	61,950	100.0	6.2	65,449	100.0	6.1	-6.2

	1961		1966		1971	
	Male	Female	Male	Female	Male	Female
Average age at death	59.7	63.1	62.0	65.9	63.3	68.2
Median age at death[2]	67.9	72.2	68.4	73.5	68.5	74.7

[1]Per 1,000 population per age group.
[2]The age above and below which half of the total number of annual deaths occurred.

4.41   Deaths and rate per 100,000 population according to the International Abbreviated List of 50 Causes, 1975

Abbreviated "B" List No.	Detailed List No.	Cause (eighth revision)	Deaths	Rate per 100,000 population
1	000	Cholera	—	—
2	001	Typhoid fever	—	—
3	004,006	Bacillary dysentery and amoebiasis	6	—
4	008,009	Enteritis and other diarrhoeal diseases	248	1.1
5	010-012	Tuberculosis of respiratory system	168	0.7
6	013-019	Other tuberculosis, including late effects	110	0.5
7	020	Plague	—	—
8	032	Diphtheria	2	—
9	033	Whooping cough	9	—
10	034	Streptococcal sore throat and scarlet fever	1	—
11	036	Meningococcal infection	39	0.2
12	040-043	Acute poliomyelitis	—	—
13	050	Smallpox	—	—
14	055	Measles	7	—
15	080-083	Typhus and other rickettsioses	—	—
16	084	Malaria	—	—
17	090-097	Syphilis and its sequelae	21	0.1
18	Remainder of 000-136	All other infective and parasitic diseases	424	1.9
19	140-209	Malignant neoplasms, including neoplasms of lymphatic and haematopoietic tissue	34,019	149.2
20	210-239	Benign neoplasms and neoplasms of unspecified nature	362	1.6
21	250	Diabetes mellitus	3,163	13.9
22	260-269	Avitaminoses and other nutritional deficiency	177	0.8
23	280-285	Anaemias	336	1.5
24	320	Meningitis	110	0.5
25	390-392	Active rheumatic fever	24	0.1
26	393-398	Chronic rheumatic heart disease	1,137	5.0
27	400-404	Hypertensive disease	1,394	6.1
28	410-414	Ischaemic heart disease	50,638	222.1
29	420-429	Other forms of heart disease	4,280	18.8
30	430-438	Cerebrovascular disease	16,434	72.1
31	470-474	Influenza	597	2.6
32	480-486	Pneumonia	5,454	23.9
33	490-493	Bronchitis, emphysema and asthma	3,051	13.4
34	531-533	Peptic ulcer	735	3.2
35	540-543	Appendicitis	82	0.4
36	550-553, 560	Intestinal obstruction and hernia	621	2.7
37	571	Cirrhosis of liver	2,725	12.0
38	580-584	Nephritis and nephrosis	628	2.8
39	600	Hyperplasia of prostate	148	0.6[1]
40	640-645	Abortion	3	—
41	630-639, 650-678	Other complications of pregnancy, childbirth and the puerperium	24	0.1
42	740-759	Congenital anomalies	1,635	7.2

4.41 Deaths and rate per 100,000 population according to the International Abbreviated List of 50 Causes, 1975 (concluded)

Abbreviated "B" List No.	Detailed List No.	Cause (eighth revision)	Deaths	Rate per 100,000 population
43	764-768, 772, 776	Birth injury, difficult labour and other anoxic and hypoxic conditions	1,257	5.5
44	760-763, 769-771, 773-775, 777-779	Other causes of perinatal mortality	1,075	4.7
45	780-796	Symptoms and ill-defined conditions	2,632	11.5
46	Remainder of 240-738	All other diseases	16,901	74.1
47	E810-E823	Motor vehicle accidents	5,896	25.9
48	E800-E807, E825-E949	All other accidents	6,681	29.3
49	E950-E959	Suicide and self-inflicted injuries	2,808	12.3
50	E960-E999	All other external causes	1,114	4.9
		All causes	167,176	733.2

[1] Per 100,000 males.

4.42 Leading causes of death by sex and age group, 1975 and 1976

Year and cause	Male No.	Rate[1]	Year and cause	Female No.	Rate[1]
1975			**1975**		
Under 1 year			Under 1 year		
Congenital anomalies	1,076	580.0	Congenital anomalies	941	538.4
Accidents (other than motor vehicle)	86	46.4	Nervous and sense organs	78	44.6
Nervous and sense organs	78	42.0	Accidents (other than motor vehicle)	66	37.8
Motor vehicle accidents	8	4.3	Motor vehicle accidents	8	4.6
Malignant neoplasms of lymphatic and haematopoietic tissue	2	1.1	Malignant neoplasms of lymphatic and haematopoietic tissue	6	3.4
Other causes	1,548	834.4	Other causes	1,005	575.0
All causes	2,798	1,508.1	All causes	2,104	1,203.7
1-4 years			1-4 years		
Accidents (other than motor vehicle)	172	24.1	Congenital anomalies	165	24.3
Congenital anomalies	169	23.7	Accidents (other than motor vehicle)	115	17.0
Motor vehicle accidents	98	13.7	Motor vehicle accidents	64	9.4
Nervous and sense organs	90	12.6	Nervous and sense organs	39	5.8
Malignant neoplasms of lymphatic and haematopoietic tissue	21	2.9	Malignant neoplasms of lymphatic and haematopoietic tissue	28	4.1
Other causes	70	9.8	Other causes	49	7.2
All causes	620	86.8	All causes	460	67.8
5-9 years			5-9 years		
Accidents (other than motor vehicle)	145	14.8	Motor vehicle accidents	95	10.2
Motor vehicle accidents	128	13.1	Accidents (other than motor vehicle)	67	7.2
Malignant neoplasms of lymphatic and haematopoietic tissue	52	5.3	Malignant neoplasms of lymphatic and haematopoietic tissue	34	3.6
Congenital anomalies	40	4.1	Congenital anomalies	25	2.7
Nervous and sense organs	24	2.5	Nervous and sense organs	17	1.8
Other causes	110	11.3	Other causes	89	9.5
All causes	499	51.0	All causes	327	34.9
10-14 years			10-14 years		
Motor vehicle accidents	168	14.0	Motor vehicle accidents	93	8.1
Accidents (other than motor vehicle)	166	13.8	Accidents (other than motor vehicle)	72	6.3
Leukemia and lymphosarcoma neoplasms	30	2.5	Leukemia and lymphosarcoma neoplasms	28	2.4
Congenital anomalies	30	2.5	Nervous and sense organs	18	1.6
Nervous and sense organs	30	2.5	Cystic fibrosis	18	1.6
Other causes	126	10.5	Other causes	118	10.3
All causes	550	45.8	All causes	347	30.3
15-19 years			15-19 years		
Motor vehicle accidents	962	81.6	Motor vehicle accidents	269	23.8
Accidents (other than motor vehicle)	407	34.5	Accidents (other than motor vehicle)	78	6.9
Suicide	186	15.8	Suicide	48	4.2
Malignant neoplasms of lymphatic and haematopoietic tissue	48	4.1	Homicide	25	2.2
Homicide	32	2.7	Diseases of the respiratory system	18	1.6
Diseases of the respiratory system	26	2.2	Other causes	181	16.0
Other causes	269	22.8			
All causes	1,930	163.8	All causes	619	54.7

4.42   Leading causes of death by sex and age group, 1975 and 1976 (continued)

Year and cause	Male		Year and cause	Female	
	No.	Rate[1]		No.	Rate[1]
1975 (continued)			1975 (continued)		
20-24 years			**20-24 years**		
Motor vehicle accidents	825	77.2	Motor vehicle accidents	180	17.1
Accidents (other than motor vehicle)	444	41.6	Suicide	83	7.9
Suicide	316	29.6	Accidents (other than motor vehicle)	71	6.7
Homicide	65	6.1	Homicide	28	2.7
Malignant neoplasms of lymphatic			Diseases of the respiratory system	20	1.9
and haematopoietic tissue	38	3.6	Other causes	228	21.7
Diseases of the respiratory system	33	3.1			
Other causes	317	29.7			
All causes	2,038	190.8	All causes	610	57.9
25-29 years			**25-29 years**		
Motor vehicle accidents	419	43.2	Motor vehicle accidents	106	10.9
Accidents (other than motor vehicle)	336	34.6	Suicide	82	8.5
Suicide	244	25.1	Accidents (other than motor vehicle)	64	6.6
Homicide	57	5.9	Homicide	25	2.6
Diseases of the respiratory system	29	3.0	Malignant neoplasms of lymphatic		
Other causes	401	41.3	and haematopoietic tissue	24	2.5
			Cerebrovascular disease	24	2.5
			Diseases of the respiratory system	19	2.0
			Other causes	239	24.7
All causes	1,486	153.1	All causes	583	60.1
30-34 years			**30-34 years**		
Motor vehicle accidents	267	33.2	Motor vehicle accidents	83	10.6
Accidents (other than motor vehicle)	265	32.9	Suicide	58	7.4
Suicide	150	18.6	Accidents (other than motor vehicle)	58	7.4
Ischaemic heart disease	89	11.1	Malignant neoplasms of lymphatic		
Other causes	503	62.5	and haematopoietic tissue	36	4.6
			Cerebrovascular disease	35	4.5
			Breast cancer	32	4.1
			Other causes	312	40.0
All causes	1,274	158.4	All causes	614	78.7
35-39 years			**35-39 years**		
Accidents (other than motor vehicle)	263	39.6	Motor vehicle accidents	77	12.0
Ischaemic heart disease	226	34.0	Suicide	74	11.6
Motor vehicle accidents	210	31.6	Breast cancer	67	10.5
Suicide	154	23.2	Accidents (other than motor vehicle)	65	10.2
Cirrhosis of liver	74	11.1	Cancer of digestive system	48	7.5
Cancer of digestive system	49	7.4	Cerebrovascular disease	44	6.9
Other causes	509	76.6	Ischaemic heart disease	36	5.6
			Other causes	375	58.6
All causes	1,485	223.4	All causes	786	122.8
40-44 years			**40-44 years**		
Ischaemic heart disease	540	83.2	Breast cancer	159	25.6
Accidents (other than motor vehicle)	266	41.0	Ischaemic heart disease	99	15.9
Motor vehicle accidents	182	28.1	Cerebrovascular disease	84	13.5
Suicide	161	24.8	Suicide	81	13.0
Cirrhosis of liver	149	23.0	Accidents (other than motor vehicle)	77	12.4
Cancer of digestive system	112	17.3	Malignant neoplasms of genito-urinary		
Lung cancer	100	15.4	organs	76	12.2
Other causes	701	108.1	Malignant neoplasms of digestive system	74	11.9
			Motor vehicle accidents	68	10.9
			Other causes	504	81.1
All causes	2,211	340.8	All causes	1,222	196.6
45-49 years			**45-49 years**		
Ischaemic heart disease	1,101	174.8	Breast cancer	289	46.5
Accidents (other than motor vehicle)	284	45.1	Ischaemic heart disease	216	34.7
Cancer of digestive organs	218	34.6	Cancer of digestive organs	175	28.1
Cirrhosis of liver	216	34.3	Cancer of genito-urinary organs	153	24.6
Lung cancer	210	33.3	Cerebrovascular disease	124	19.9
Motor vehicle accidents	167	26.5	Suicide	84	13.5
Suicide	152	24.1	Accidents (other than motor vehicle)	73	11.7
Other causes	1,128	179.1	Diseases of the respiratory system	70	11.3
			Other causes	818	131.6
All causes	3,476	551.9	All causes	2,002	322.0
50-54 years			**50-54 years**		
Ischaemic heart disease	1,916	325.6	Breast cancer	407	66.5
Lung cancer	408	69.3	Ischaemic heart disease	385	62.9
Cancer of digestive organs	369	62.7	Cancer of digestive organs	231	37.7
Accidents (other than motor vehicle)	293	49.8	Cancer of genito-urinary organs	222	36.3
Cirrhosis of liver	291	49.5	Cerebrovascular disease	192	31.4
Diseases of the respiratory system	226	38.4	Lung cancer	139	22.7
Other causes	1,871	318.0	Cirrhosis of liver	119	19.4
			Other causes	1,063	173.7
All causes	5,374	913.3	All causes	2,758	450.7

4.42   Leading causes of death by sex and age group, 1975 and 1976 (continued)

Year and cause	Male No.	Rate[1]	Year and cause	Female No.	Rate[1]
1975 (concluded)			**1975 (concluded)**		
55-59 years			**55-59 years**		
Ischaemic heart disease	2,561	534.7	Ischaemic heart disease	702	138.9
Lung cancer	675	140.9	Breast cancer	384	76.0
Cancer of digestive organs	519	108.4	Cancer of digestive organs	379	75.0
Diseases of the respiratory system	373	77.9	Cerebrovascular disease	286	56.6
Cerebrovascular disease	348	72.7	Cancer of genito-urinary organs	270	53.4
Cirrhosis of liver	281	58.7	Lung cancer	164	32.5
Accidents (other than motor vehicle)	248	51.8	Other causes	1,410	279.0
Other causes	1,901	396.9			
All causes	6,906	1,441.8	All causes	3,595	711.5
60-64 years			**60-64 years**		
Ischaemic heart disease	3,715	864.0	Ischaemic heart disease	1,253	274.4
Lung cancer	933	217.0	Cancer of digestive organs	485	106.2
Cancer of digestive organs	773	179.8	Breast cancer	413	90.4
Diseases of the respiratory system	595	138.4	Cerebrovascular disease	381	83.4
Cerebrovascular disease	538	125.1	Cancer of genito-urinary organs	331	72.5
Cirrhosis of liver	276	64.2	Diseases of the respiratory system	222	48.6
Cancer of genito-urinary organs	259	60.2	Lung cancer	213	46.6
Other causes	2,522	586.5	Other causes	1,647	360.6
All causes	9,611	2,235.1	All causes	4,945	1,082.8
65-69 years			**65-69 years**		
Ischaemic heart disease	4,389	1,343.4	Ischaemic heart disease	1,946	536.2
Lung cancer	992	303.6	Cancer of digestive organs	612	168.6
Cancer of digestive organs	930	284.7	Cerebrovascular disease	565	155.7
Diseases of the respiratory system	819	250.7	Breast cancer	385	106.1
Cerebrovascular disease	767	234.8	Cancer of genito-urinary organs	323	89.0
Cancer of genito-urinary organs	369	112.9	Diseases of the respiratory system	309	85.1
Other forms of heart disease	240	73.5	Diabetes mellitus	232	63.9
Other causes	2,720	832.6	Other causes	1,910	526.3
All causes	11,226	3,436.2	All causes	6,282	1,731.1
70-79 years			**70-79 years**		
Ischaemic heart disease	8,468	2,217.3	Ischaemic heart disease	5,917	1,208.0
Cerebrovascular disease	2,351	615.6	Cerebrovascular disease	2,454	501.0
Diseases of the respiratory system	2,234	585.0	Cancer of digestive organs	1,539	314.2
Cancer of digestive organs	1,776	465.0	Diseases of the respiratory system	939	191.7
Lung cancer	1,581	414.0	Diabetes mellitus	628	128.2
Cancer of genito-urinary organs	1,035	271.0	Cancer of genito-urinary organs	617	126.0
Other forms of heart disease	571	149.5	Breast cancer	570	116.4
Other causes	5,388	1,410.8	Other causes	4,598	938.8
All causes	23,404	6,128.3	All causes	17,262	3,524.3
80 years and over			**80 years and over**		
Ischaemic heart disease	7,704	5,615.2	Ischaemic heart disease	9,260	3,923.7
Cerebrovascular disease	3,073	2,239.8	Cerebrovascular disease	4,619	1,957.2
Diseases of the respiratory system	2,546	1,855.7	Diseases of the respiratory system	1,876	794.9
Cancer of digestive organs	1,088	793.0	Disease of arteries	1,834	777.1
Cancer of genito-urinary organs	1,034	753.6	Cancer of digestive organs	1,330	563.6
Other forms of heart disease	769	560.5	Other forms of heart disease	1,040	440.7
Lung cancer	538	392.1	Accidents (other than motor vehicle)	676	286.4
Other causes	5,133	3,741.3	Other causes	5,017	2,125.9
All causes	21,885	15,951.2	All causes	25,652	10,869.5
1976			**1976**		
Under 1 year			**Under 1 year**		
Congenital anomalies	1,135	614.1	Congenital anomalies	950	542.4
Accidents (other than motor vehicle)	82	44.4	Nervous and sense organs	65	37.1
Nervous and sense organs	66	35.7	Accidents (other than motor vehicle)	48	27.4
Motor vehicle accidents	9	4.9	Motor vehicle accidents	11	6.3
Malignant neoplasms of lymphatic and haematopoietic tissue	3	1.6	Malignant neoplasms of lymphatic and haematopoietic tissue	3	1.7
Other causes	1,378	745.5	Other causes	931	531.5
All causes	2,673	1,446.2	All causes	2,008	1,146.4
1-4 years			**1-4 years**		
Accidents (other than motor vehicle)	172	24.2	Congenital anomalies	163	24.2
Congenital anomalies	156	21.9	Accidents (other than motor vehicle)	93	13.8
Motor vehicle accidents	92	13.8	Nervous and sense organs	58	8.6
Nervous and sense organs	70	9.8	Motor vehicle accidents	47	7.0
Malignant neoplasms of lymphatic and haematopoietic tissue	31	4.4	Malignant neoplasms of lymphatic and haematopoietic tissue	12	1.8
Other causes	71	10.0	Other causes	44	6.5
All causes	592	83.3	All causes	417	61.8
5-9 years			**5-9 years**		
Accidents (other than motor vehicle)	128	13.2	Motor vehicle accidents	78	8.5
Motor vehicle accidents	108	11.2	Accidents (other than motor vehicle)	58	6.3
Malignant neoplasms of lymphatic and haematopoietic tissue	42	4.3	Malignant neoplasms of lymphatic and haematopoietic tissue	27	2.9
Nervous and sense organs	25	2.6	Congenital anomalies	24	2.6
Congenital anomalies	23	2.4	Nervous and sense organs	16	1.7
Other causes	89	9.2	Other causes	80	8.7
All causes	415	42.9	All causes	283	30.7

4.42   Leading causes of death by sex and age group, 1975 and 1976 (continued)

Year and cause	Male		Year and cause	Female	
	No.	Rate[1]		No.	Rate[1]
1976 (continued)			**1976 (continued)**		
10-14 years			10-14 years		
Accidents (other than motor vehicle)	154	13.2	Motor vehicle accidents	67	6.0
Motor vehicle accidents	130	11.2	Accidents (other than motor vehicle)	38	3.4
Leukemia and lymphosarcoma neoplasms	38	3.3	Leukemia and lymphosarcoma neoplasms	24	2.2
Congenital anomalies	20	1.7	Nervous and sense organs	15	1.3
Nervous and sense organs	17	1.5	Cystic fibrosis	6	0.5
Other causes	122	10.5	Other causes	110	9.9
All causes	481	41.3	All causes	260	23.4
15-19 years			15-19 years		
Motor vehicle accidents	776	64.9	Motor vehicle accidents	248	21.6
Accidents (other than motor vehicle)	354	29.6	Accidents (other than motor vehicle)	57	5.0
Suicide	191	16.0	Suicide	47	4.1
Malignant neoplasms of lymphatic and haematopoietic tissue	52	4.3	Homicide	21	1.8
Homicide	30	2.5	Diseases of the respiratory system	12	1.0
Diseases of the respiratory system	30	2.5	Other causes	185	16.1
Other causes	224	18.7			
All causes	1,657	138.5	All causes	570	49.6
20-24 years			20-24 years		
Motor vehicle accidents	704	66.1	Motor vehicle accidents	151	14.1
Accidents (other than motor vehicle)	385	36.1	Suicide	83	7.8
Suicide	292	27.4	Accidents (other than motor vehicle)	52	4.9
Malignant neoplasms of lymphatic and haematopoietic tissue	47	4.4	Homicide	19	1.8
Homicide	40	3.8	Diseases of the respiratory system	12	1.1
Diseases of the respiratory system	26	2.4	Other causes	216	20.2
Other causes	294	27.6			
All causes	1,788	167.8	All causes	533	49.9
25-29 years			25-29 years		
Motor vehicle accidents	396	39.6	Motor vehicle accidents	101	10.2
Accidents (other than motor vehicle)	293	29.3	Suicide	67	6.8
Suicide	257	25.7	Accidents (other than motor vehicle)	38	3.8
Homicide	37	3.7	Malignant neoplasms of lymphatic and haematopoietic tissue	27	2.7
Diseases of the respiratory system	29	2.9	Homicide	20	2.0
Other causes	362	36.2	Cerebrovascular disease	19	1.9
			Diseases of the respiratory system	15	1.5
			Other causes	200	20.2
All causes	1,374	137.3	All causes	487	49.1
30-34 years			30-34 years		
Accidents (other than motor vehicle)	242	29.4	Suicide	86	10.7
Motor vehicle accidents	218	26.5	Motor vehicle accidents	59	7.3
Suicide	176	21.4	Accidents (other than motor vehicle)	51	6.3
Ischaemic heart disease	94	11.4	Breast cancer	41	5.1
Other causes	444	54.0	Cerebrovascular disease	27	3.4
			Malignant neoplasms of lymphatic and haematopoietic tissue	20	2.5
			Other causes	300	37.3
All causes	1,174	142.7	All causes	584	72.6
35-39 years			35-39 years		
Ischaemic heart disease	221	32.9	Breast cancer	84	12.8
Accidents (other than motor vehicle)	206	30.7	Suicide	67	10.2
Suicide	137	20.4	Motor vehicle accidents	57	8.7
Motor vehicle accidents	136	20.3	Accidents (other than motor vehicle)	52	7.9
Cirrhosis of liver	80	11.9	Cancer of digestive system	52	7.9
Cancer of digestive system	45	6.7	Cerebrovascular disease	40	6.1
Other causes	451	67.2	Ischaemic heart disease	33	5.0
			Other causes	365	55.0
All causes	1,276	190.1	All causes	750	114.1
40-44 years			40-44 years		
Ischaemic heart disease	471	73.2	Breast cancer	131	21.0
Accidents (other than motor vehicle)	224	34.8	Ischaemic heart disease	81	13.0
Motor vehicle accidents	170	26.4	Suicide	71	11.4
Suicide	165	25.6	Cerebrovascular disease	62	9.9
Cirrhosis of liver	127	19.7	Malignant neoplasms of genito-urinary organs	61	9.8
Cancer of digestive system	122	19.0	Malignant neoplasms of digestive system	60	9.6
Lung cancer	77	12.0	Accidents (other than motor vehicle)	56	9.0
Other causes	669	104.0	Motor vehicle accidents	53	8.5
			Other causes	457	73.2
All causes	2,025	314.6	All causes	1,032	165.2

4.42 Leading causes of death by sex and age group, 1975 and 1976 (concluded)

Year and cause	Male		Year and cause	Female	
	No.	Rate[1]		No.	Rate[1]
1976 (concluded)			**1976 (concluded)**		
45-49 years			**45-49 years**		
Ischaemic heart disease	1,090	172.9	Breast cancer	256	41.1
Accidents (other than motor vehicle)	232	36.8	Ischaemic heart disease	179	28.8
Cirrhosis of liver	225	35.7	Cancer of genito-urinary organs	149	23.9
Cancer of digestive organs	220	34.9	Cancer of digestive organs	131	21.0
Lung cancer	203	32.2	Cerebrovascular disease	131	21.0
Suicide	168	26.6	Suicide	80	12.9
Motor vehicle accidents	155	24.6	Diseases of the respiratory system	74	11.9
Other causes	1,095	173.7	Accidents (other than motor vehicle)	60	9.6
			Other causes	730	117.3
All causes	3,388	537.4	All causes	1,790	287.6
50-54 years			**50-54 years**		
Ischaemic heart disease	1,916	321.6	Breast cancer	375	60.1
Lung cancer	469	78.7	Ischaemic heart disease	354	56.7
Cancer of digestive organs	347	58.2	Cancer of digestive organs	242	38.8
Cirrhosis of liver	316	53.0	Cancer of genito-urinary organs	213	34.1
Accidents (other than motor vehicle)	259	43.5	Cerebrovascular disease	190	30.4
Diseases of the respiratory system	190	31.9	Lung cancer	145	23.2
Other causes	1,648	276.6	Cirrhosis of liver	129	20.7
			Other causes	988	158.2
All causes	5,145	863.7	All causes	2,636	422.1
55-59 years			**55-59 years**		
Ischaemic heart disease	2,628	533.9	Ischaemic heart disease	620	117.7
Lung cancer	675	137.1	Breast cancer	400	75.9
Cancer of digestive organs	512	104.0	Cancer of digestive organs	353	67.0
Diseases of the respiratory system	343	69.7	Cancer of genito-urinary organs	277	52.6
Cerebrovascular disease	291	59.1	Cerebrovascular disease	224	42.5
Cirrhosis of liver	256	52.0	Lung cancer	177	33.6
Accidents (other than motor vehicle)	250	50.8	Other causes	1,368	259.7
Other causes	1,859	377.6			
All causes	6,814	1,384.2	All causes	3,419	649.0
60-64 years			**60-64 years**		
Ischaemic heart disease	3,541	812.5	Ischaemic heart disease	1,168	248.7
Lung cancer	951	218.2	Cancer of digestive organs	505	107.5
Cancer of digestive organs	765	175.5	Breast cancer	413	87.9
Diseases of the respiratory system	611	140.2	Cerebrovascular disease	365	77.7
Cerebrovascular disease	502	115.2	Cancer of genito-urinary organs	338	72.0
Cirrhosis of liver	326	74.8	Lung cancer	209	44.5
Cancer of genito-urinary organs	275	63.1	Diseases of the respiratory system	194	41.3
Other causes	2,390	548.4	Other causes	1,598	340.3
All causes	9,361	2,148.1	All causes	4,790	1,020.0
65-69 years			**65-69 years**		
Ischaemic heart disease	4,193	1,238.6	Ischaemic heart disease	1,909	499.3
Lung cancer	1,062	313.7	Cancer of digestive organs	656	171.6
Cancer of digestive organs	937	276.8	Cerebrovascular disease	554	144.9
Diseases of the respiratory system	814	240.5	Breast cancer	358	93.6
Cerebrovascular disease	708	209.1	Cancer of genito-urinary organs	305	79.8
Cancer of genito-urinary organs	387	114.3	Diseases of the respiratory system	301	78.7
Other forms of heart disease	265	78.3	Diabetes mellitus	187	48.9
Other causes	2,617	773.1	Other causes	1,871	489.4
All causes	10,983	3,244.4	All causes	6,141	1,606.3
70-79 years			**70-79 years**		
Ischaemic heart disease	8,628	2,202.2	Ischaemic heart disease	5,834	1,156.1
Cerebrovascular disease	2,351	600.1	Cerebrovascular disease	2,291	454.0
Diseases of the respiratory system	2,195	560.2	Cancer of digestive organs	1,511	299.4
Cancer of digestive organs	1,855	473.5	Diseases of the respiratory system	926	183.5
Lung cancer	1,663	424.5	Breast cancer	613	121.5
Cancer of genito-urinary organs	985	251.4	Cancer of genito-urinary organs	594	117.7
Other forms of heart disease	616	157.2	Diabetes mellitus	553	109.6
Other causes	5,197	1,326.5	Other causes	4,455	882.8
All causes	23,490	5,995.5	All causes	16,777	3,324.6
80 years and over			**80 years and over**		
Ischaemic heart disease	7,569	5,216.6	Ischaemic heart disease	9,498	3,957.4
Cerebrovascular disease	2,863	1,973.2	Cerebrovascular disease	4,508	1,878.3
Diseases of the respiratory system	2,630	1,812.6	Diseases of the respiratory system	2,146	894.1
Cancer of digestive organs	1,098	756.7	Disease of arteries	1,852	771.7
Cancer of genito-urinary organs	933	643.0	Cancer of digestive organs	1,378	574.2
Other forms of heart disease	852	587.2	Other forms of heart disease	1,087	452.9
Lung cancer	539	371.5	Accidents (other than motor vehicle)	601	250.4
Other causes	4,964	3,421.2	Other causes	5,076	2,115.0
All causes	21,448	14,782.0	All causes	26,146	10,893.9

[1]Under one year rates are per 100,000 live births; all other age group rates are per 100,000 population.

4.43   Infant deaths and stillbirths, by province and sex, 1951-75

Province or territory and year	Infant deaths (<1 yr)			Neonatal deaths (<28 days)					Post-neo-natal deaths (28 days to 1 yr)	Still-births (28+ weeks gesta-tion)	Perinatal deaths (Stillbirths plus deaths <7 days)[1]
	Male	Female	Total	Male	Female	Total	<7 days	7-27 days			
NEWFOUNDLAND											
Number											
1951	361	276	637	176	112	288	209	79	349	189	398
1961	335	253	588	197	128	325	269	56	263	281	550
1971	173	120	293	123	84	207	175	32	86	158	333
1973	135	95	230	98	66	164	140	24	66	151	291
1974	108	73	181	70	48	118	101	17	63	127	228
1975	98	78	176	58	51	109	97	12	67	116	213
Rate											
1951	60.3	48.0	54.3	29.4	19.5	24.5	17.8	6.7	29.8	16.1	33.4
1961	41.7	33.5	37.7	24.5	16.9	20.8	17.3	3.6	16.9	18.0	34.7
1971	26.3	19.4	22.9	18.7	13.5	16.2	13.7	2.5	6.7	12.4	25.8
1973	22.1	16.4	19.3	16.0	11.4	13.8	11.8	2.0	5.5	12.7	24.1
1974	20.2	15.0	17.7	13.1	9.8	11.5	9.9	1.7	6.2	12.4	22.0
1975	16.8	14.5	15.7	9.9	9.5	9.7	8.7	1.1	6.0	10.3	18.8
PRINCE EDWARD ISLAND											
Number											
1951	60	30	90	29	17	46	33	13	44	56	89
1961	55	38	93	29	25	54	47	7	39	46	93
1971	29	17	46	24	13	37	34	3	9	23	57
1973	18	12	30	16	3	19	15	4	11	15	30
1974	21	13	34	18	12	30	29	1	4	20	49
1975	15	22	37	12	14	26	23	3	11	25	48
Rate											
1951	43.7	23.5	33.9	21.1	13.3	17.4	12.4	4.9	16.5	21.1	32.9
1961	37.4	27.8	32.8	19.7	18.3	19.0	16.6	2.5	13.8	16.2	32.2
1971	26.1	17.1	21.9	21.6	13.1	17.6	16.2	1.4	4.3	10.9	26.8
1973	17.9	13.6	15.9	15.9	3.4	10.1	8.0	2.1	5.8	8.0	15.8
1974	20.5	14.2	17.5	17.6	13.1	15.5	15.0	0.5	2.1	10.3	24.9
1975	15.0	23.7	19.2	12.0	15.1	13.5	11.9	1.6	5.7	13.0	24.6
NOVA SCOTIA											
Number											
1951	344	250	594	218	134	352	298	54	242	319	617
1961	309	229	538	187	140	327	280	47	211	300	580
1971	160	105	265	108	72	180	148	32	85	139	287
1973	113	93	206	78	57	135	107	28	71	119	226
1974	98	87	185	59	56	115	94	21	70	123	217
1975	131	82	213	73	51	124	99	25	89	98	197
Rate											
1951	38.9	30.2	34.7	24.7	16.2	20.6	17.4	3.2	14.1	18.6	35.4
1961	31.0	24.3	27.8	18.8	14.9	16.9	14.4	2.4	10.9	15.5	29.5
1971	21.8	15.2	18.6	14.7	10.4	12.6	10.4	2.2	6.0	9.8	19.9
1973	16.6	14.3	15.5	11.5	8.8	10.2	8.1	2.1	5.3	9.0	16.9
1974	14.7	13.9	14.3	8.8	8.9	8.9	7.3	1.6	5.4	9.5	16.6
1975	19.1	13.1	16.2	10.6	8.2	9.5	7.6	1.9	6.8	7.5	14.9
NEW BRUNSWICK											
Number											
1951	472	363	835	241	199	440	334	106	395	293	627
1961	248	186	434	145	105	250	217	33	184	222	439
1971	130	74	204	97	48	145	135	10	59	138	273
1973	105	68	173	75	47	122	113	9	51	132	245
1974	93	80	173	65	57	122	114	8	51	125	239
1975	114	69	183	78	47	125	105	20	58	130	235
Rate											
1951	57.6	46.0	51.9	29.4	25.2	27.4	20.8	6.6	24.5	18.2	38.3
1961	29.1	23.0	26.2	17.0	13.0	15.1	13.1	2.0	11.1	13.4	26.1
1971	20.7	12.5	16.7	15.5	8.1	11.9	11.1	0.8	4.8	11.3	22.2
1973	17.9	12.2	15.1	12.8	8.4	10.7	9.9	0.8	4.5	11.6	21.2
1974	16.0	14.2	15.1	11.2	10.1	10.7	10.0	0.7	4.5	10.9	20.6
1975	19.1	11.9	15.5	13.0	8.1	10.6	8.9	1.7	4.9	11.0	19.7
QUEBEC											
Number											
1951	3,335	2,486	5,821	1,864	1,311	3,175	2,398	777	2,646	2,768	5,166
1961	2,464	1,855	4,319	1,666	1,189	2,855	2,489	366	1,464	1,929	4,418
1971	948	692	1,640	690	500	1,190	1,059	131	450	807	1,866
1973	795	583	1,378	570	425	995	874	121	383	658	1,532
1974	733	558	1,291	531	393	924	811	113	367	629	1,440
1975e	774	556	1,330	562	396	958	..	..	372	650	..
Rate											
1951	53.7	42.3	48.1	30.0	22.3	26.3	19.8	6.4	21.8	22.9	41.8
1961	34.7	28.0	31.5	23.5	18.0	20.8	18.1	2.7	10.7	14.1	31.8
1971	20.6	16.0	18.4	15.0	11.6	13.3	11.9	1.5	5.0	9.0	20.7
1973	18.3	14.4	16.4	13.1	10.5	11.8	10.4	1.4	4.6	7.8	18.1
1974	16.7	13.4	15.1	12.1	9.4	10.8	9.5	1.3	4.3	7.3	16.7
1975e	16.1	12.3	14.3	11.7	8.8	10.3	..	..	4.0	7.0	..

4.43 Infant deaths and stillbirths, by province and sex, 1951-75 (continued)

Province or territory and year	Infant deaths (<1 yr) Male	Female	Total	Neonatal deaths (<28 days) Male	Female	Total	<7 days	7-27 days	Post-neonatal deaths (28 days to 1 yr)	Still-births (28+ weeks gestation)	Perinatal deaths (Stillbirths plus deaths <7 days)[1]
ONTARIO											
Number											
1951	2,010	1,535	3,545	1,389	1,040	2,429	2,033	396	1,116	1,975	4,008
1961	2,090	1,536	3,626	1,507	1,120	2,627	2,378	249	999	1,870	4,248
1971	1,146	844	1,990	821	603	1,424	1,255	169	566	1,221	2,476
1973	979	761	1,740	693	545	1,238	1,093	145	502	1,034	2,127
1974	915	751	1,666	621	542	1,163	1,020	143	503	1,019	2,039
1975	926	688	1,614	645	503	1,148	1,006	142	466	906	1,912
Rate[d]											
1951	33.9	27.6	30.9	23.5	18.7	21.2	17.7	3.4	9.7	17.2	34.3
1961	25.9	20.0	23.0	18.7	14.6	16.7	15.1	1.6	6.3	11.9	26.6
1971	17.1	13.3	15.3	12.2	9.5	10.9	9.6	1.3	4.3	9.4	18.8
1973	15.4	12.7	14.1	10.9	9.1	10.0	8.8	1.2	4.1	8.4	17.0
1974	14.3	12.4	13.4	9.7	9.0	9.4	8.2	1.2	4.0	8.2	16.2
1975	14.4	11.2	12.8	10.0	8.2	9.1	8.0	1.1	3.7	7.2	15.1
MANITOBA											
Number											
1951	369	289	658	209	169	378	301	77	280	340	641
1961	341	247	588	211	169	380	336	44	208	301	637
1971	184	132	316	117	87	204	177	27	112	164	341
1973	164	114	278	105	64	169	145	24	109	151	296
1974	152	120	272	97	79	176	146	30	96	128	274
1975	141	117	258	97	73	170	156	14	88	128	284
Rate[d]											
1951	35.6	30.2	33.0	20.1	17.7	19.0	15.1	3.9	14.0	17.0	31.6
1961	28.6	21.7	25.2	17.7	14.9	16.3	14.4	1.9	8.9	12.9	27.0
1971	20.0	15.0	17.5	12.7	9.9	11.3	9.8	1.5	6.2	9.1	18.7
1973	18.7	13.9	16.4	12.0	7.8	10.0	8.5	1.4	6.4	8.9	17.3
1974	16.9	14.4	15.7	10.8	9.5	10.2	8.4	1.7	5.5	7.4	15.7
1975	16.2	13.9	15.0	11.1	8.7	9.9	9.1	0.8	5.1	7.5	16.4
SASKATCHEWAN											
Number											
1951	353	323	676	226	175	401	338	63	275	303	641
1961	373	245	618	244	151	395	334	61	223	266	600
1971	189	136	325	132	90	222	199	23	103	159	358
1973	161	100	261	112	68	180	156	24	81	144	300
1974	181	132	313	112	79	191	172	19	122	132	304
1975	137	135	272	85	79	164	143	21	108	100	243
Rate[d]											
1951	31.8	30.4	31.1	20.3	16.5	18.5	15.6	2.9	12.6	13.9	29.1
1961	30.3	21.0	25.8	19.8	12.9	16.5	13.9	2.5	9.3	11.1	24.7
1971	23.2	17.2	20.2	16.2	11.4	13.8	12.4	1.4	6.4	9.9	22.1
1973	21.3	13.8	17.6	14.8	9.4	12.2	10.5	1.6	5.5	9.7	20.1
1974	23.5	17.8	20.7	14.5	10.7	12.6	11.4	1.3	8.1	8.7	19.9
1975	17.7	17.9	17.8	11.0	10.5	10.7	9.4	1.4	7.1	6.6	15.8
ALBERTA											
Number											
1951	531	358	889	345	212	557	462	95	332	402	864
1961	612	432	1,044	418	289	707	629	78	337	372	1,001
1971	325	223	548	226	162	388	334	54	160	254	588
1973	242	174	416	153	113	266	227	39	150	206	433
1974	264	185	449	157	121	278	234	44	171	188	422
1975	264	207	471	159	132	291	251	40	180	173	424
Rate[d]											
1951	38.6	27.0	32.9	25.1	16.0	20.6	17.1	3.5	12.3	14.9	31.5
1961	30.8	22.7	26.8	21.0	15.2	18.2	16.2	2.0	8.6	9.6	25.5
1971	20.5	15.2	17.9	14.3	11.0	12.7	10.9	1.8	5.2	8.3	19.1
1973	16.1	12.2	14.2	10.2	7.9	9.1	7.8	1.3	5.1	7.0	14.7
1974	17.4	12.7	15.1	10.3	8.3	9.3	7.8	1.5	5.7	6.3	14.0
1975	16.3	13.4	14.9	9.8	8.5	9.2	7.9	1.3	5.7	5.5	13.3
BRITISH COLUMBIA											
Number											
1951	487	352	839	299	214	513	435	78	326	365	800
1961	534	411	945	331	264	595	515	80	350	412	927
1971	381	272	653	271	188	459	417	42	194	310	727
1973	319	256	575	211	169	380	333	47	195	243	576
1974	349	223	572	226	138	364	325	39	208	252	577
1975	295	229	524	188	149	337	289	48	187	284	573
Rate[d]											
1951	33.8	25.8	29.9	20.7	15.7	18.3	15.5	2.8	11.6	13.0	28.1
1961	27.1	21.8	24.5	16.8	14.0	15.4	13.3	2.1	9.1	10.7	23.8
1971	21.2	16.1	18.7	15.1	11.2	13.2	12.0	1.2	5.6	8.9	20.7
1973	18.1	15.3	16.7	12.0	10.1	11.1	9.7	1.4	5.7	7.1	16.6
1974	19.1	12.9	16.1	12.4	8.0	10.3	9.2	1.1	5.9	7.1	16.1
1975	16.0	12.9	14.4	10.2	8.4	9.3	8.0	1.3	5.2	7.8	15.7

4.43 Infant deaths and stillbirths, by province and sex, 1951-75 (concluded)

Province or territory and year	Infant deaths (<1 yr)			Neonatal deaths (<28 days)			<7 days	7-27 days	Post-neonatal deaths (28 days to 1 yr)	Stillbirths (28+ weeks gestation)	Perinatal deaths (Stillbirths plus deaths <7 days)[1]
	Male	Female	Total	Male	Female	Total					
YUKON TERRITORY											
Number											
1951	10	9	19	4	2	6	..	..	13	2	8
1961	13	10	23	6	4	10	7	3	13	4	11
1971	8	5	13	3	3	6	4	2	7	6	10
1973	4	3	7	2	2	4	2	2	3	3	5
1974	7	5	12	3	2	5	4	1	7	7	11
1975	7	3	10	5	2	7	7	—	3	2	9
Rate[a]											
1951	57.8	53.3	55.6	23.1	11.8	17.5	..	..	38.1	5.8	23.3
1961	45.8	36.5	41.2	21.1	14.6	17.9	12.5	5.4	23.3	7.2	19.6
1971	29.0	21.7	25.7	10.9	13.0	11.9	7.9	4.0	13.8	11.9	19.5
1973	19.2	14.2	16.7	9.6	9.4	9.5	4.8	4.8	7.1	7.1	11.8
1974	26.9	21.3	24.2	11.5	8.5	10.1	8.1	2.0	14.1	14.1	21.8
1975	34.5	14.6	24.5	24.6	9.8	17.2	17.2	—	7.4	4.9	22.0
NORTHWEST TERRITORIES											
Number											
1951	43	27	70	20	14	34	..	..	36	11	26
1961	73	51	124	25	14	39	22	17	85	16	38
1971	39	24	63	11	12	23	19	4	40	17	36
1973	34	11	45	15	5	20	16	4	25	10	26
1974	25	19	44	14	6	20	19	1	24	16	35
1975	26	16	42	12	9	21	18	3	21	15	33
Rate[a]											
1951	135.6	81.3	107.9	63.1	42.2	52.4	..	..	55.5	16.9	39.4
1961	128.1	93.2	111.0	43.9	25.6	34.9	19.7	15.2	76.1	14.3	33.5
1971	58.8	38.4	49.0	16.6	19.2	17.9	14.8	3.1	31.1	13.2	27.6
1973	52.8	19.6	37.4	23.3	8.9	16.6	13.3	3.3	20.8	8.3	21.4
1974	45.3	38.8	42.2	25.4	12.2	19.2	18.2	1.0	23.0	15.4	33.0
1975	45.8	26.5	35.9	21.1	14.9	17.9	15.4	2.6	17.9	12.8	27.8
CANADA											
Number											
1951	8,375	6,298	14,673	5,020	3,599	8,619	6,862	1,757	6,054	7,023	13,885
1961	7,447	5,493	12,940	4,966	3,598	8,564	7,523	1,041	4,376	6,019	13,542
1971	3,712	2,644	6,356	2,623	1,862	4,485	3,956	529	1,871	3,396	7,352
1973	3,069	2,270	5,339	2,128	1,564	3,692	3,221	471	1,647	2,866	6,087
1974	2,946	2,246	5,192	1,973	1,533	3,506	3,069	437	1,686	2,766	5,835
1975[e]	2,928	2,202	5,130	1,974	1,506	3,480	..	..	1,650	2,627	..
Rate[a]											
1951	42.7	34.0	38.5	25.6	19.4	22.6	18.0	4.6	15.9	18.4	35.8
1961	30.5	23.7	27.2	20.3	15.6	18.0	15.8	2.2	9.2	12.7	28.1
1971	19.9	15.1	17.5	14.1	10.6	12.4	10.9	1.5	5.2	9.4	20.1
1973	17.4	13.6	15.5	12.0	9.4	10.8	9.4	1.4	4.8	8.3	17.6
1974	16.6	13.4	15.0	11.1	9.1	10.1	8.9	1.3	4.9	8.0	16.7
1975[e]	15.9	12.6	14.3	10.7	8.6	9.7	..	..	4.6	7.3	..

[1] Perinatal rates per 1,000 live-born and stillborn infants; all other rates per 1,000 live births.

4.44 Infant deaths, by age, 1974[1]

Time of death	Deaths		Cumulative deaths		Time of death	Deaths		Cumulative deaths	
	No.	%	No.	%		No.	%	No.	%
1st day	2,033	39.2	2,033	39.2	1st month	3,506	67.5	3,506	67.5
2nd "	376	7.2	2,409	46.4	2nd "	410	7.9	3,916	75.4
3rd "	302	5.8	2,711	52.2	3rd "	329	6.3	4,245	81.8
4th "	147	2.8	2,858	55.0	4th "	280	5.4	4,525	87.2
5th "	95	1.8	2,953	56.9	5th "	153	2.9	4,678	90.1
6th "	73	1.4	3,026	58.3	6th "	129	2.5	4,807	92.6
7th "	43	0.8	3,069	59.1	7th "	88	1.7	4,895	94.3
					8th "	83	1.6	4,978	95.9
1st week	3,069	59.1	3,069	59.1	9th "	60	1.2	5,038	97.0
2nd "	209	4.0	3,278	63.1	10th "	63	1.2	5,101	98.2
3rd "	131	2.5	3,409	65.7	11th "	50	1.0	5,151	99.2
4th "	97	1.9	3,506	67.5	12th "	41	0.8	5,192	100.0

[1] Later figures not available.

4.45   Canadian life table, 1971

Age	Male				Female			
	Number living at each age	Number dying between each age and the next	Probability of dying before reaching next birthday	Expectation of life yr	Number living at each age	Number dying between each age and the next	Probability of dying before reaching next birthday	Expectation of life yr
At birth	100,000	2,002	.02002	69.34	100,000	1,544	.01544	76.36
1 year	97,998	126	.00128	69.76	98,456	113	.00115	76.56
2 years	97,872	92	.00094	68.85	98,343	72	.00073	75.64
3 "	97,780	83	.00084	67.91	98,271	60	.00061	74.70
4 "	97,697	69	.00071	66.97	98,211	56	.00057	73.74
5 "	97,628	232	.00061	66.02	98,155	179	.00050	72.79
10 "	97,396	267	.00039	61.17	97,976	157	.00028	67.91
15 "	97,129	682	.00106	56.33	97,819	262	.00046	63.02
20 "	96,447	872	.00178	51.71	97,557	279	.00057	58.18
25 "	95,575	730	.00164	47.16	97,278	315	.00060	53.34
30 "	94,845	773	.00152	42.50	96,963	433	.00077	48.51
35 "	94,072	1,037	.00188	37.83	96,530	644	.00112	43.71
40 "	93,035	1,645	.00291	33.22	95,886	988	.00173	38.99
45 "	91,390	2,569	.00464	28.77	94,898	1,465	.00260	34.37
50 "	88,821	4,060	.00761	24.52	93,433	2,236	.00403	29.86
55 "	84,761	6,042	.01213	20.57	91,197	3,301	.00618	25.53
60 "	78,719	8,675	.01918	16.95	87,896	4,804	.00931	21.39
65 "	70,044	11,469	.02961	13.72	83,092	7,097	.01449	17.47
70 "	58,575	13,787	.04436	10.90	75,995	10,371	.02337	13.85
75 "	44,788	14,812	.06552	8.47	65,624	14,387	.03876	10.63
80 "	29,976	13,644	.09701	6.41	51,237	17,609	.06514	7.88
85 "	16,332	9,841	.14355	4.74	33,628	17,008	.10766	5.67
90 "	6,491	4,891	.20977	3.43	16,620	11,358	.17137	3.99
95 "	1,600	1,409	.30027	2.45	5,262	4,427	.26132	2.76
100 "	191		.41969	1.71	835		.38255	1.89

4.46   Expectation of life, 1951, 1961, 1966 and 1971 (years)

Age	1951		1961		1966		1971	
	Male	Female	Male	Female	Male	Female	Male	Female
At birth	66.33	70.83	68.35	74.17	68.75	75.18	69.34	76.36
1 year	68.33	72.33	69.50	74.98	69.53	75.71	69.76	76.56
2 years	67.56	71.55	68.63	74.11	68.64	74.81	68.85	75.64
3 "	66.68	70.66	67.71	73.18	67.71	73.88	67.91	74.70
4 "	65.79	69.74	66.78	72.23	66.77	72.93	66.97	73.74
5 "	64.86	68.80	65.83	71.27	65.82	71.97	66.02	72.79
10 "	60.15	64.02	61.02	66.41	61.00	67.12	61.17	67.91
15 "	55.39	59.19	56.20	61.51	56.16	62.22	56.33	63.02
20 "	50.76	54.41	51.51	56.65	51.50	57.37	51.71	58.18
25 "	46.20	49.67	46.91	51.80	46.94	52.52	47.16	53.34
30 "	41.60	44.94	42.24	46.98	42.29	47.68	42.50	48.51
35 "	37.00	40.24	37.56	42.18	37.62	42.88	37.83	43.71
40 "	32.45	35.63	32.96	37.45	33.01	38.15	33.22	38.99
45 "	28.05	31.14	28.49	32.82	28.55	33.51	28.77	34.37
50 "	23.88	26.80	24.25	28.33	24.31	29.02	24.52	29.86
55 "	20.02	22.61	20.30	24.01	20.38	24.70	20.57	25.53
60 "	16.49	18.64	16.73	19.90	16.81	20.58	16.95	21.39
65 "	13.31	14.97	13.53	16.07	13.63	16.71	13.72	17.47
70 "	10.41	11.62	10.67	12.58	10.83	13.14	10.90	13.85
75 "	7.89	8.73	8.21	9.48	8.37	9.94	8.47	10.63
80 "	5.84	6.38	6.14	6.90	6.36	7.26	6.41	7.88
85 "	4.27	4.57	4.46	4.89	4.79	5.16	4.74	5.67
90 "	3.10	3.24	3.16	3.39	3.60	3.60	3.43	3.99
95 "	2.24	2.27	2.20	2.32	2.71	2.48	2.45	2.76
100 "	1.60	1.59	1.49	1.56	2.04	1.69	1.71	1.89

4.47   Expectation of life at selected ages, by region or province, 1951, 1961, 1966 and 1971 (years)

Region or province and age	1951 Male	1951 Female	1961 Male	1961 Female	1966 Male	1966 Female	1971 Male	1971 Female
ATLANTIC PROVINCES								
At birth	66.57	70.50	68.58	73.92	..	..	..	..
1 year	69.08	72.41	70.06	75.10	..	..	..	..
20 years	51.59	54.52	52.17	56.82	..	..	..	..
40 ``	33.48	35.99	33.76	37.70	..	..	..	..
65 ``	13.90	15.42	14.16	16.35	..	..	..	..
NEWFOUNDLAND								
At birth	..	..	..	..	68.94	74.43	69.28	75.72
1 year	..	..	..	..	70.22	75.41	69.99	76.22
20 years	..	..	..	..	52.27	57.08	51.90	57.86
40 ``	..	..	..	..	33.78	37.83	33.22	38.58
65 ``	..	..	..	..	14.31	16.22	13.52	16.91
PRINCE EDWARD ISLAND								
At birth	..	..	..	..	68.32	75.51	69.30	77.35
1 year	..	..	..	..	69.43	76.22	70.10	77.64
20 years	..	..	..	..	51.56	57.88	52.04	59.36
40 ``	..	..	..	..	33.49	38.77	33.83	39.96
65 ``	..	..	..	..	14.43	17.57	14.40	18.41
NOVA SCOTIA								
At birth	..	..	..	..	68.34	74.80	68.66	75.97
1 year	..	..	..	..	69.16	75.43	69.05	76.13
20 years	..	..	..	..	51.32	57.16	51.02	57.77
40 ``	..	..	..	..	32.99	37.96	32.80	38.50
65 ``	..	..	..	..	13.80	16.75	13.58	17.14
NEW BRUNSWICK								
At birth	..	..	..	..	68.53	75.26	69.07	76.41
1 year	..	..	..	..	69.30	75.97	69.49	76.61
20 years	..	..	..	..	51.58	57.79	51.59	58.36
40 ``	..	..	..	..	33.35	38.53	33.23	39.15
65 ``	..	..	..	..	14.01	17.04	13.78	17.56
QUEBEC								
At birth	64.42	68.58	67.28	72.77	67.88	73.91	68.28	75.25
1 year	67.19	70.71	68.71	73.80	68.77	74.57	68.74	75.52
20 years	49.76	52.92	50.82	55.54	50.81	56.25	50.74	57.18
40 ``	31.54	34.36	32.29	36.38	32.33	37.05	32.30	38.02
65 ``	12.81	14.17	13.16	15.27	13.24	15.79	13.08	16.62
ONTARIO								
At birth	66.87	71.85	68.32	74.40	68.71	75.53	69.55	76.76
1 year	68.34	72.91	69.14	74.95	69.29	75.87	69.82	76.81
20 years	50.58	54.76	51.03	56.53	51.14	57.45	51.63	58.35
40 ``	32.03	35.75	32.35	37.27	32.44	38.17	32.91	39.08
65 ``	13.07	14.92	13.05	15.90	13.10	16.72	13.37	17.57
PRAIRIE PROVINCES								
At birth	68.36	72.28	69.79	75.66	..	..	..	..
1 year	69.90	73.43	70.96	76.40	..	..	..	..
20 years	52.24	55.53	52.90	58.08	..	..	..	..
40 ``	33.86	36.63	34.37	38.83	..	..	..	..
65 ``	13.88	15.51	14.22	17.00	..	..	..	..
MANITOBA								
At birth	..	..	..	..	69.80	76.11	70.16	76.93
1 year	..	..	..	..	70.54	76.57	70.60	77.21
20 years	..	..	..	..	52.48	58.25	52.67	58.88
40 ``	..	..	..	..	34.11	39.10	34.18	39 66
65 ``	..	..	..	..	14.18	17.42	14.32	18.02
SASKATCHEWAN								
At birth	..	..	..	..	70.45	76.45	71.05	77.59
1 year	..	..	..	..	71.49	77.06	71.76	77.98
20 years	..	..	..	..	53.50	58.80	53.82	59.62
40 ``	..	..	..	..	35.22	39.61	35.59	40.51
65 ``	..	..	..	..	15.00	17.59	15.44	18.54
ALBERTA								
At birth	..	..	..	..	70.10	76.24	70.42	77.30
1 year	..	..	..	..	70.82	76.72	70.90	77.52
20 years	..	..	..	..	52.70	58.30	52.94	59.17
40 ``	..	..	..	..	34.36	39.09	34.60	40.06
65 ``	..	..	..	..	14.46	17.34	14.64	18.24
BRITISH COLUMBIA								
At birth	66.73	72.37	68.94	75.42	69.21	75.84	69.85	76.69
1 year	67.97	73.32	69.83	76.00	69.94	76.33	70.26	76.85
20 years	50.41	55.51	51.85	57.61	51.91	58.01	52.29	58.53
40 ``	32.45	36.72	33.56	38.46	33.70	38.93	34.10	39.49
65 ``	13.50	15.86	13.98	16.94	14.20	17.41	14.50	18.00

4.48   Marriages and rate per 1,000 population, by province, with percentage distribution of bridegrooms and brides by birthplace, 1951, 1961, 1971, 1974 and 1975

Province or territory	Year	Total marriages	Rate per 1,000 popu- lation	Born in province where married		Born in other provinces		Born outside Canada	
				Grooms %	Brides %	Grooms %	Brides %	Grooms %	Brides %
Newfoundland	1951	2,517	7.0	85.2	96.7	2.4	1.9	12.4	1.4
	1961	3,306	7.2	88.0	97.2	3.8	1.6	8.2	1.2
	1971	4,685	9.0	91.6	95.3	4.2	3.3	4.2	1.4
	1974	4,276	7.9	..	..	..	..	..	..
	1975	4,313	7.8	..	..	..	..	..	..
Prince Edward Island	1951	583	5.9	82.3	91.1	12.9	6.0	4.8	2.9
	1961	624	6.0	81.7	89.6	15.4	7.2	2.9	3.2
	1971	961	8.6	78.9	88.1	17.4	9.5	3.7	2.4
	1974	990	8.5	..	..	..	..	..	..
	1975	936	7.9	..	..	..	..	..	..
Nova Scotia	1951	5,094	7.9	78.2	86.7	15.9	9.0	6.0	4.3
	1961	5,292	7.2	75.2	87.8	18.8	8.8	6.0	3.4
	1971	6,883	8.7	77.9	85.6	16.2	10.9	5.9	3.5
	1974	7,112	8.7	..	..	..	..	..	..
	1975	7,059	8.6	..	..	..	..	..	..
New Brunswick	1951	4,386	8.5	80.0	86.9	10.1	6.7	9.8	6.4
	1961	4,504	7.5	75.4	86.3	14.9	7.9	9.7	5.8
	1971	6,149	9.7	77.9	86.2	14.8	9.5	7.2	4.3
	1974	6,108	9.2	..	..	..	..	..	..
	1975	5,945	8.8	..	..	..	..	..	..
Quebec	1951	35,704	8.8	86.7	89.5	6.1	5.5	7.2	5.0
	1961	35,943	6.8	83.6	87.4	5.7	4.8	10.7	7.8
	1971	49,695	8.2	84.2	88.1	5.2	4.3	10.7	7.6
	1974	51,532	8.4	..	..	..	..	..	..
	1975	50,377	8.1	..	..	..	..	..	..
Ontario	1951	45,198	9.8	65.9	72.4	14.6	12.2	19.5	15.4
	1961	44,434	7.1	61.5	67.2	12.9	11.0	25.6	21.8
	1971	69,590	9.0	60.1	66.1	12.7	11.2	27.2	22.7
	1974	72,716	9.0	..	..	..	..	..	..
	1975	72,209	8.8	..	..	..	..	..	..
Manitoba	1951	7,366	9.5	67.9	75.1	15.4	13.3	16.8	11.6
	1961	6,512	7.1	66.6	74.5	18.5	14.5	14.8	11.0
	1971	9,127	9.2	67.1	75.2	17.7	13.8	15.1	11.0
	1974	9,231	9.1	..	..	..	..	..	..
	1975	8,915	8.8	..	..	..	..	..	..
Saskatchewan	1951	6,805	8.2	78.3	86.4	10.7	6.4	11.1	7.2
	1961	6,149	6.6	79.3	85.8	11.9	8.7	8.8	5.5
	1971	7,813	8.4	78.9	85.3	14.4	10.1	6.6	4.6
	1974	7,988	8.8	..	..	..	..	..	..
	1975	8,066	8.8	..	..	..	..	..	..
Alberta	1951	9,305	9.9	56.0	67.4	25.7	19.6	18.3	13.0
	1961	10,474	7.9	54.4	62.3	25.8	21.8	19.8	15.9
	1971	15,614	9.6	54.8	62.0	28.5	24.4	16.6	13.5
	1974	16,691	9.7	..	..	..	..	..	..
	1975	17,520	9.9	..	..	..	..	..	..
British Columbia	1951	11,272	9.7	35.5	41.6	43.1	43.0	21.3	15.5
	1961	10,964	6.7	36.4	45.9	35.9	32.4	27.7	21.8
	1971	20,389	9.3	43.1	50.5	32.1	29.3	24.9	20.3
	1974	21,734	9.1	..	..	..	..	..	..
	1975	21,824	8.9	..	..	..	..	..	..
Yukon Territory	1961	128	8.8	12.5	24.2	63.3	52.3	24.2	23.4
	1971	166	9.0	10.2	20.5	67.5	59.0	22.3	20.5
	1974	190	9.8	..	..	..	..	..	..
	1975	201	9.7	..	..	..	..	..	..
Northwest Territories	1961	145	6.3	54.5	61.4	35.9	31.7	9.7	6.9
	1971	252	7.2	41.3	55.6	43.3	33.3	15.5	11.1
	1974	256	6.8	..	..	..	..	..	..
	1975	220	5.8	..	..	..	..	..	..
Canada	1951[1]	128,230	9.2	70.5	76.5	15.1	12.8	14.5	10.6
	1961	128,475	7.0	67.9	74.2	14.3	11.7	17.9	14.1
	1971	191,324	8.9	67.2	73.2	14.5	12.3	18.3	14.5
	1974	198,824	8.9	..	..	..	..	..	..
	1975	197,585	8.7	..	..	..	..	..	..

[1]Excludes the Yukon Territory and Northwest Territories.

4.49 Brides and bridegrooms, by age and marital status, 1975

Age group	Marital status								
	Brides								
	Number				*Percentage*				
	Spinsters	Widows	Divorced	Total	Spinsters	Widows	Divorced	Total	
Under 15 years	—	—	—	—	—	—	—	—	
15-19 "	50,213	14	84	50,311	29.7	0.2	0.4	25.5	
20-24 "	87,433	186	2,422	90,041	51.8	2.5	11.4	45.6	
25-29 "	21,551	384	5,983	27,918	12.8	5.2	28.1	14.1	
30-34 "	5,030	417	4,385	9,832	3.0	5.6	20.6	5.0	
35-39 "	1,916	408	2,814	5,138	1.1	5.5	13.2	2.6	
40-44 "	868	574	2,023	3,465	0.5	7.7	9.5	1.8	
45-49 "	578	798	1,568	2,944	0.3	10.7	7.4	1.5	
50-54 "	390	1,010	1,072	2,472	0.2	13.5	5.0	1.2	
55-59 "	272	1,021	541	1,834	0.2	13.7	2.5	0.9	
60-64 "	175	1,099	287	1,561	0.1	14.7	1.3	0.8	
65 years and over	136	1,519	116	1,771	0.1	20.4	0.5	0.9	
Total, stated ages	168,562	7,430	21,295	197,287	99.8	99.7	99.9	99.9	
Age not stated	255	26	17	298	0.2	0.3	0.1	0.1	
Total, all ages	168,817	7,456	21,312	197,585	100.0	100.0	100.0	100.0	
Average age yr	22.5	53.1	34.9	25.0	...	...	...	...	
	Bridegrooms								
	Number				*Percentage*				
	Bachelors	Widowers	Divorced	Total	Bachelors	Widowers	Divorced	Total	
Under 15 years	—	—	—	—	—	—	—	—	
15-19 "	14,249	3	8	14,260	8.5	- -	- -	7.2	
20-24 "	93,304	28	891	94,223	55.9	0.4	3.7	47.7	
25-29 "	41,649	145	4,937	46,731	24.9	2.2	20.6	23.7	
30-34 "	9,878	224	5,450	15,552	5.9	3.4	22.8	7.9	
35-39 "	3,327	267	3,900	7,494	2.0	4.0	16.3	3.8	
40-44 "	1,753	374	2,886	5,013	1.1	5.7	12.0	2.5	
45-49 "	1,075	523	2,246	3,844	0.6	7.9	9.4	2.0	
50-54 "	635	759	1,621	3,015	0.4	11.5	6.8	1.5	
55-59 "	403	853	971	2,227	0.2	12.9	4.1	1.1	
60-64 "	252	1,028	572	1,852	0.2	15.6	2.4	0.9	
65 years and over	256	2,322	413	2,991	0.2	35.1	1.7	1.5	
Total, stated ages	166,781	6,526	23,895	197,202	99.9	98.7	99.8	99.8	
Age not stated	241	89	53	383	0.1	1.3	0.2	0.2	
Total, all ages	167,022	6,615	23,948	197,585	100.0	100.0	100.0	100.0	
Average age yr	24.9	58.8	38.3	27.6	...	...	...	...	

4.50 Marriages by religious denominations of brides and bridegrooms, 1975

Denomination of bridegroom	Denomination of bride										Total marriages	Percentage of grooms
	Angli-can	Bap-tist	Greek Orth-odox	Jewish	Lu-ther-an	Pres-by-terian	Roman Catho-lic	United Church	Other sects	Not stated		
Anglican	7,343	707	137	45	613	792	4,655	4,224	1,606	25	20,147	13.7
Baptist	687	2,125	21	5	141	229	923	1,013	541	8	5,693	3.9
Greek Orthodox	156	29	1,215	6	72	38	449	259	207	3	2,434	1.7
Jewish	61	10	2	1,119	9	11	104	67	113	—	1,496	1.0
Lutheran	657	139	50	11	1,358	182	1,132	1,188	522	11	5,250	3.6
Presbyterian	868	176	32	15	181	1,484	1,128	1,206	367	4	5,461	3.7
Roman Catholic	4,481	857	296	75	1,130	1,103	27,125	6,004	3,780	57	44,908	30.5
United Church	4,070	913	213	37	1,048	1,096	5,970	14,644	2,006	24	30,021	20.4
Other sects	2,109	649	209	94	760	477	4,736	2,884	19,441	25	31,384	21.3
Not stated	32	10	1	2	14	3	73	35	36	208	414	0.2
Total	20,464	5,615	2,176	1,409	5,326	5,415	46,295	31,524	28,619	365	147,208[1]	100.0
Percentage of brides	13.9	3.8	1.5	1.0	3.6	3.7	31.5	21.4	19.4	0.2	100.0	51.7[2]

[1]Excludes 50,377 marriages in the province of Quebec.
[2]Percentage of marriages between persons of the same religious denomination.

4.51   Divorces[1] granted to male and female petitioners, by province, 1972-75

Province or territory	1972		1973		1974		1975	
	Male	Female	Male	Female	Male	Female	Male	Female
Newfoundland	61	116	84	140	90	211	103	277
Prince Edward Island	23	42	20	34	36	60	18	57
Nova Scotia	325	602	429	820	517	1,074	519	1,078
New Brunswick	182	284	189	385	295	460	274	484
Quebec	2,504	3,917	3,055	5,036	3,723	8,549	4,391	9,702
Ontario	4,823	8,360	5,087	8,694	5,747	9,530	6,493	10,992
Manitoba	489	924	535	1,085	594	1,202	638	1,346
Saskatchewan	296	530	298	589	369	670	369	762
Alberta	1,057	2,710	1,178	3,357	1,368	3,579	1,457	4,018
British Columbia	1,879	3,157	2,177	3,510	2,554	4,286	2,772	4,762
Yukon Territory	18	29	24	36	14	32	18	25
Northwest Territories	13	23	13	29	21	38	21	35
Canada	11,670	20,694	13,089	23,615	15,328	29,691	17,073	33,538

[1]Only those filed under the new divorce laws of July 2, 1968.

4.52   Alleged grounds for divorce[1] by type of offence, 1972-75

Alleged grounds	1972		1973		1974		1975	
	No.	%	No.	%	No.	%	No.	%
Marital offence								
Adultery	12,645	29.6	14,853	30.3	17,806	29.4	20,356	30.0
Physical cruelty	5,579	13.0	6,598	13.5	8,340	13.8	9,387	13.9
Mental cruelty	6,308	14.7	7,734	15.8	9,774	16.2	11,204	16.5
Other	152	0.4	142	0.3	174	0.3	185	0.3
Total	24,684	57.7	29,327	59.9	36,094	59.7	41,132	60.7
Marriage breakdown by reason of:								
Addiction to alcohol	859	2.0	1,032	2.1	1,607	2.7	1,658	2.4
Separation for not less than 3 years	14,803	34.6	16,197	33.1	20,415	33.8	22,376	33.0
Desertion by petitioner for not less than 5 years	1,919	4.5	1,806	3.7	1,778	2.9	2,001	3.0
Other	504	1.2	583	1.2	535	0.9	602	0.9
Total	18,085	42.3	19,618	40.1	24,335	40.3	26,637	39.3
Total, alleged grounds[2]	42,769	100.0	48,945	100.0	60,429	100.0	67,769	100.0

[1]See footnote to Table 4.51.
[2]Totals are higher than the numbers of divorces because some divorce decrees involve more than one alleged ground.

4.53   Divorces[1] by number of dependent children, 1972-75

Number of children	1972		1973		1974		1975	
	No.	%	No.	%	No.	%	No.	%
0	14,305	44.2	15,890	43.3	18,588	41.3	21,458	42.4
1	7,078	21.9	8,209	22.4	10,277	22.8	11,523	22.7
2	5,956	18.4	6,842	18.6	8,817	19.6	9,985	19.7
3	2,963	9.2	3,384	9.2	4,349	9.7	4,643	9.2
4	1,294	3.9	1,520	4.2	1,850	4.1	1,904	3.8
5 and more	768	2.4	859	2.3	1,138	2.5	1,098	2.2
Total, divorces	32,364	100.0	36,704	100.0	45,019	100.0	50,611	100.0
Average number of children	1.15	...	1.16	...	1.22	...	1.17	...

[1]See footnote to Table 4.51.

4.54   Divorces[1] by duration of marriage, 1972-75

Duration of marriage	1972		1973		1974		1975	
	No.	%	No.	%	No.	%	No.	%
Less than 1 year	84	0.3	99	0.3	105	0.2	127	0.2
1 year	524	1.6	645	1.8	716	1.6	872	1.7
2 years	1,022	3.2	1,165	3.2	1,457	3.2	1,662	3.3
3 "	1,465	4.5	1,712	4.7	2,019	4.5	2,285	4.5
4 "	1,948	6.0	2,152	5.9	2,794	6.2	3,063	6.1
Total, 1-4 years	4,959	15.3	5,674	15.6	6,986	15.5	7,882	15.6
5 years	2,020	6.2	2,403	6.5	2,797	6.3	3,277	6.5
6 "	1,924	5.9	2,237	6.1	2,730	6.1	3,216	6.4
7 "	1,717	5.3	2,146	5.8	2,674	5.9	3,096	6.1
8 "	1,523	4.7	1,900	5.2	2,356	5.2	2,839	5.6
9 "	1,465	4.5	1,664	4.5	2,129	4.7	2,435	4.8
Total, 5-9 years	8,649	26.7	10,350	28.1	12,686	28.2	14,863	29.4
10-14 years	5,905	18.2	6,490	17.7	8,164	18.1	8,987	17.7
15-19 "	4,442	13.7	4,930	13.4	5,923	13.2	6,757	13.4
20-24 "	3,518	10.9	3,896	10.6	4,755	10.6	4,996	9.9
25-29 "	2,445	7.6	2,734	7.4	3,263	7.3	3,583	7.1
30 years and over	2,323	7.2	2,502	6.8	3,034	6.7	3,356	6.6
Not stated	39	0.1	29	0.1	103	0.2	60	0.1
Total, divorces	32,364	100.0	36,704	100.0	45,019	100.0	50,611	100.0
Median duration of marriage	12.1	...	11.8	...	11.7	...	11.4	...

[1]See footnote to Table 4.51.

4.55   Divorces[1] by marital status of husband and wife at time of marriage, 1972-75

Marital status	1972		1973		1974		1975	
	No.	%	No.	%	No.	%	No.	%
Husband								
Single	30,151	93.2	34,205	93.2	42,120	93.6	47,149	93.2
Widowed	470	1.4	522	1.4	624	1.4	693	1.4
Divorced	1,739	5.4	1,973	5.4	2,268	5.0	2,762	5.4
Not stated	4	- -	4	- -	7	- -	7	- -
Total	32,364	100.0	36,704	100.0	45,019	100.0	50,611	100.0
Wife								
Single	29,914	92.4	33,972	92.6	41,799	92.9	46,843	92.5
Widowed	727	2.3	734	2.0	866	1.9	942	1.9
Divorced	1,722	5.3	1,997	5.4	2,349	5.2	2,823	5.6
Not stated	1	- -	1	- -	5	- -	3	- -

[1]See footnote to Table 4.51.

4.56   Immigrant arrivals, 1951-77

Year	Arrivals	Year	Arrivals	Year	Arrivals
1951	194,391	1960	104,111	1969	161,531
1952	164,498	1961	71,689	1970	147,713
1953	168,868	1962	74,586	1971	121,900
1954	154,227	1963	93,151	1972	122,006
1955	109,946	1964	112,606	1973	184,200
1956	164,857	1965	146,758	1974	218,465
1957	282,164	1966	194,743	1975	187,881
1958	124,851	1967	222,876	1976	149,429
1959	106,928	1968	183,974	1977	114,914

4.57 Immigrant arrivals, by country of last permanent residence, 1976 and 1977

Country of last permanent residence	1976	1977	Country of last permanent residence	1976	1977
Europe	49,908	40,748	Indonesia	122	116
Austria	753	564	Iran	500	440
Belgium	532	436	Iraq	25	155
British Isles	21,548	17,997	Israel	1,201	957
England	16,759	13,648	Japan	498	412
Northern Ireland	1,536	1,391	Jordan	130	104
Scotland	2,343	2,284	Korea, South	2,249	1,243
Wales	890	659	Lebanon	7,161	3,847
Channel Islands	20	15	Malaysia	511	590
Czechoslovakia	85	118	Pakistan	2,173	1,575
Denmark	353	250	Philippines	5,939	6,232
Finland	237	177	Singapore	424	317
France	3,251	2,757	Sri Lanka	235	168
Germany, Federal Republic of	2,672	2,254	Syria	131	104
Greece	2,487	1,960	Taiwan	1,178	899
Hungary	195	287	Vietnam, South	2,269	243
Ireland	639	571	Other Asia	809	840
Italy	4,530	3,411	North and Central America	33,513	26,129
Malta	164	161	Antigua	149	139
Netherlands	1,359	1,247	Bahamas	122	129
Norway	144	100	Barbados	554	634
Poland	903	902	Bermuda	119	111
Portugal	5,344	3,579	Grenada	314	197
Spain	633	356	Haiti	3,061	2,026
Sweden	269	260	Jamaica	7,282	6,291
Switzerland	1,192	944	Mexico	757	794
Turkey	246	311	St. Kitts, Nevis and Anguilla	115	112
USSR	315	299	St. Vincent	322	266
Yugoslavia	1,741	1,408	Trinidad and Tobago	2,359	1,552
Other Europe	316	399	United States	17,315	12,888
Africa	7,752	6,372	Other North and Central America	1,044	990
Angola	912	602	South America	10,628	7,840
Egypt, Republic of	728	598	Argentina	1,463	1,499
Ghana	220	145	Bolivia	49	32
Kenya	1,202	379	Brazil	267	308
Morocco	325	220	Chile	2,082	1,546
Mozambique	252	202	Colombia	797	541
Nigeria	194	146	Ecuador	1,128	548
South Africa, Republic of	1,611	2,458	French Guiana	7	6
Tanzania	1,299	605	Guyana	3,430	2,472
Uganda	29	34	Paraguay	199	128
Zambia	193	130	Peru	552	388
Other Africa	785	853	Suriname	10	11
Australasia	1,886	1,545	Uruguay	486	266
Australia	1,387	1,063	Venezuela	158	95
New Zealand	487	475	Oceania	1,414	912
Other Australasia	12	7	Fiji	1,081	710
Asia	44,328	31,368	Mauritius	286	173
Bangladesh	79	114	Other Oceania	47	29
China	833	798	Total, all countries	149,429	114,914
Cyprus	403	288			
Hong Kong	10,725	6,371			
India	6,733	5,555			

4.58 Immigrant arrivals, by country of citizenship, 1975 and 1976

Country of citizenship	1975	1976	Country of citizenship	1975	1976
Australia	1,358	1,170	New Zealand	457	435
Austria	291	277	Norway	188	158
Belgium	332	375	Pakistan	2,520	2,202
Britain and colonies	40,495	28,308	Philippines	7,576	5,997
Central America	556	549	Poland	935	1,087
China	1,166	1,039	Portugal	9,530	6,904
Czechoslovakia	161	145	South Africa	1,272	1,283
Denmark	494	377	South America	12,790	10,259
Egypt	954	734	Spain	646	548
Finland	301	251	Sri Lanka	536	330
France	3,196	2,795	Sweden	395	281
Germany, Federal Republic of	2,387	2,159	Switzerland	902	921
Greece	4,032	2,457	Trinidad and Tobago	3,936	2,407
Haiti	3,533	3,172	Turkey	383	310
Hungary	432	323	Union of Soviet Socialist Republics	230	319
India	11,600	7,410	United States	18,912	16,236
Ireland	1,495	896	Yugoslavia	3,676	2,138
Israel	1,668	1,284	Other African	5,861	3,423
Italy	4,940	4,059	Other Asian	10,786	8,663
Jamaica	8,519	7,256	Other European	757	487
Japan	587	474	Stateless	8,643	7,188
Lebanon	1,227	6,795	Other	4,357	3,109
Mexico	776	707			
Morocco	625	379			
Netherlands	1,468	1,353	Total	187,881	149,429

4.59 Intended province of destination of male and female immigrants, 1975 and 1976

Province or territory	1975			1976		
	Male	Female	Total	Male	Female	Total
Newfoundland	606	500	1,106	353	372	725
Prince Edward Island	118	117	235	127	108	235
Nova Scotia	1,071	1,053	2,124	992	950	1,942
New Brunswick	1,044	1,049	2,093	885	867	1,752
Quebec	14,441	13,601	28,042	14,906	14,376	29,282
Ontario	48,021	50,450	98,471	34,209	37,822	72,031
Manitoba	3,560	3,574	7,134	2,736	2,773	5,509
Saskatchewan	1,360	1,477	2,837	1,167	1,156	2,323
Alberta	8,349	7,928	16,277	7,620	7,276	14,896
British Columbia	13,970	15,302	29,272	9,482	11,002	20,484
Yukon Territory and Northwest Territories	143	147	290	128	122	250
Canada	92,683	95,198	187,881	72,605	76,824	149,429

4.60 Sex distribution of immigrants, 1972-76

Year	Male	Female	Total
1972	60,070	61,936	122,006
1973	94,768	89,432	184,200
1974	111,122	107,343	218,465
1975	92,683	95,198	187,881
1976	72,605	76,824	149,429

4.61 Marital status of immigrant arrivals, by sex and age group, 1975 and 1976

Sex and age group	Single		Married		Widowed		Divorced		Separated		Total	
	1975	1976	1975	1976	1975	1976	1975	1976	1975	1976	1975	1976
Male												
0- 4 years	8,523	6,168	—	—	—	—	—	—	—	—	8,523	6,168
5- 9 "	9,907	7,429	—	—	—	—	—	—	—	—	9,907	7,429
10-14 "	7,748	6,081	—	—	—	—	—	—	—	—	7,748	6,081
15-19 "	6,570	5,602	88	96	—	—	—	—	1	—	6,659	5,698
20-24 "	8,691	6,613	3,210	2,955	—	2	14	15	11	6	11,926	9,591
25-29 "	7,564	4,833	9,280	7,154	7	8	118	99	92	46	17,061	12,140
30-34 "	2,287	1,716	8,099	5,908	9	15	209	153	76	78	10,680	7,870
35-39 "	740	537	5,896	4,249	12	13	142	127	69	53	6,859	4,979
40-44 "	267	208	3,293	2,601	18	12	114	84	36	27	3,728	2,932
45-49 "	117	99	2,008	1,695	24	18	64	49	34	25	2,247	1,886
50-54 "	66	56	1,393	1,243	42	58	52	43	31	20	1,584	1,420
55-59 "	31	35	898	985	70	74	21	22	6	11	1,026	1,127
60-64 "	32	28	1,762	2,054	175	149	15	22	13	16	1,997	2,269
65-69 "	23	13	1,175	1,367	187	181	17	18	13	8	1,415	1,587
70 years +	35	20	883	979	381	402	14	14	10	13	1,323	1,428
Total, male	52,601	39,438	37,985	31,286	925	932	780	646	392	303	92,683	72,605
Female												
0- 4 years	8,102	5,713	—	—	—	—	—	—	—	—	8,102	5,713
5- 9 "	9,362	6,890	—	—	—	—	—	—	—	—	9,362	6,890
10-14 "	7,475	5,833	5	5	—	—	—	—	—	—	7,480	5,838
15-19 "	6,137	4,914	1,721	1,433	—	3	1	1	1	—	7,860	6,351
20-24 "	6,523	5,119	9,030	7,884	8	8	51	31	30	24	15,642	13,066
25-29 "	4,197	3,023	11,303	8,727	24	16	187	148	68	43	15,779	11,957
30-34 "	1,260	933	7,416	5,690	35	27	203	162	63	43	8,977	6,855
35-39 "	437	339	4,449	3,331	49	49	146	124	38	32	5,119	3,875
40-44 "	216	130	2,600	2,091	131	83	111	76	49	26	3,107	2,406
45-49 "	138	81	1,818	1,622	327	259	88	81	36	25	2,407	2,068
50-54 "	115	87	1,564	1,550	597	603	86	87	52	40	2,414	2,367
55-59 "	82	64	1,368	1,546	842	795	74	62	34	38	2,400	2,505
60-64 "	107	131	1,317	1,480	1,054	1,010	87	80	61	86	2,626	2,787
65-69 "	102	114	663	786	912	971	53	55	47	35	1,777	1,961
70 years +	123	125	395	426	1,561	1,560	45	38	22	36	2,146	2,185
Total, female	44,376	33,496	43,649	36,571	5,540	5,384	1,132	945	501	428	95,198	76,824

4.62 Intended occupations of immigrants, 1974-76

Intended occupation	1974	1975	1976
WORKERS			
Managerial and administrative	*6,445*	*5,763*	*5,655*
Government administrators	70	65	66
Managerial (owners, managers, officials)	4,752	4,102	4,276
Other	1,623	1,596	1,313
Professional and technical			
Natural sciences, engineering and mathematics	*8,705*	*8,928*	*5,648*
Physical sciences			
Chemists	329	245	136
Geologists and related	145	102	81
Physicists	119	95	60
Other	491	384	258
Life sciences			
Agriculturists and related scientists	115	119	77
Biologists and related scientists	307	262	159
Other	86	86	67
Architects and engineers			
Architects	175	218	180
Engineers			
Chemical	233	207	129
Civil	509	530	425
Electrical	522	517	338
Industrial	157	129	79
Mechanical	427	379	305
Metallurgical	26	18	14
Mining	46	42	30
Petroleum	23	25	26
Aerospace	30	28	35
Nuclear	10	9	16
Other	75	81	78
Architecture and engineering			
Surveyors	118	112	93
Draughtsmen	1,551	2,005	1,030
Other	2,171	2,444	1,487
Mathematics, statistics, and systems analysis			
Mathematicians, statisticians and actuaries	141	110	48
Systems analysts	897	781	495
Other	2	—	2
Social sciences	*1,290*	*1,075*	*887*
Economists	241	244	188
Sociologists and anthropologists	51	32	43
Psychologists	139	90	101
Other	146	122	85
Social work			
Social workers	294	222	163
Other	130	112	91
Law and jurisprudence	140	124	112
Library, museum and archival sciences			
Librarians and archivists	125	113	79
Other	24	16	25
Religion	*504*	*409*	*476*
Teaching	*3,289*	*2,613*	*2,400*
University	1,100	766	614
Elementary and secondary	1,543	1,290	1,205
Other	646	557	581
Medicine and health	*6,349*	*5,604*	*3,750*
Health diagnosing and treating			
Physicians and surgeons	1,081	806	401
Dentists	83	83	102
Veterinarians	64	69	29
Osteopaths and chiropractors	8	8	6
Other	11	16	7
Nursing and therapy			
Supervisors of nurses	11	9	2
Graduate nurses (except supervisors)	1,702	1,839	1,130
Physiotherapists	331	334	275
Other	1,745	1,392	967
Other medicine and health			
Pharmacists	65	75	101
Optometrists	7	7	5
Other	1,241	966	725
Artistic, literary and performing arts	*1,462*	*1,192*	*1,217*
Advertising and illustrating artists	124	100	116
Fine and commercial art and photography	591	492	429
Performing and audiovisual arts	451	344	340
Writers and editors, publication	187	133	151
Writers and editors, radio, television, theatre and motion pictures	31	17	18
Other	78	106	163
Sport and recreation	*135*	*116*	*134*

4.62 Intended occupations of immigrants, 1974-76 (continued)

Intended occupation	1974	1975	1976
WORKERS (continued)			
Clerical and related	*15,660*	*11,803*	*9,345*
Stenographic and typing	6,251	4,340	3,571
Bookkeeping and accounting	3,356	2,618	1,906
Office machine and electronic data-processing equipment operators	731	600	399
Material recording, scheduling and distributing	895	532	326
Library, file and correspondence clerks	224	187	147
Reception, information, mail and message distribution	524	430	276
Other	3,679	3,096	2,720
Sales	*4,119*	*3,294*	*2,632*
Commodities	3,450	2,779	2,259
Services	566	435	314
Other	103	80	59
Services	*10,604*	*7,082*	*5,640*
Protective	386	388	336
Food and beverage preparation	4,053	3,034	2,317
Lodging and other accommodation	408	187	131
Personal	3,761	2,322	2,096
Apparel and furnishings	289	130	83
Other	1,707	1,021	677
Farming, horticultural and animal-husbandry	*2,637*	*1,511*	*1,162*
Farmers	98	53	66
Farm management	74	44	50
Other	2,465	1,414	1,046
Fishing, hunting and trapping	*50*	*27*	*24*
Forestry and logging	*243*	*144*	*36*
Mining and quarrying, including oil and gas field	*233*	*178*	*103*
Processing	*3,075*	*1,689*	*1,171*
Mineral ore treating	92	58	32
Metal processing	444	257	154
Clay, glass and stone processing	132	60	37
Chemicals, petroleum, rubber and plastic processing	94	73	93
Food and beverage processing	997	629	508
Wood processing (except paper pulp)	507	121	56
Pulp and paper making	41	23	11
Textile processing	716	443	259
Other	52	25	21
Machining and related	*7,603*	*5,178*	*3.019*
Metal machining	3,403	2,354	1,279
Metal shaping (except machining)	3,844	2,598	1,593
Wood machining	135	99	49
Clay, glass and stone machining	65	39	28
Other	156	88	70
Product fabricating, assembling and repairing	*15,466*	*11,936*	*8,380*
Metal products	410	285	186
Electrical and electronic equipment	1,539	1,566	1,059
Wood products	760	497	393
Textile, fur and leather products	5,658	3,721	2,763
Rubber, plastic and related products	209	171	107
Other, including mechanics and repairmen (n.e.c.)	6,890	5,696	3,872
Construction	*8,476*	*5,893*	*4,008*
Excavating, grading and paving	406	241	181
Electrical power linemen	35	52	45
Electricians and repairmen	820	648	498
Wire communications and electric power lighting equipment, installing and repairing	246	317	179
Carpenters	1,872	1,167	806
Brick and stone masons and tile setters	914	658	440
Concrete finishing	162	98	40
Plasterers	144	84	58
Painters and paperhangers	618	412	284
Roofing and waterproofing	57	36	25
Pipefitting and plumbing	494	383	264
Structural-metal erectors	65	64	32
Glaziers	25	35	23
Other	2,618	1,698	1,133
Transport equipment operating	*1,315*	*1,055*	*784*
Air	117	131	101
Railway	14	11	11
Water	167	168	145
Motor	1,012	742	525
Other	5	3	2
Material-handling and related (n.e.c.)	*1,119*	*694*	*490*

4.62   Intended occupation of immigrants, 1974-76 (concluded)

Intended occupation	1974	1975	1976
WORKERS (concluded)			
Other crafts and equipment operating	*668*	*489*	*425*
Printing and related	514	373	321
Stationary engine and utilities equipment operating	80	46	36
Electronic and related communications equipment operating	45	31	34
Other	29	39	34
Not stated and unknown	*6,636*	*4,516*	*4,075*
Total, workers	106,083	81,189	61,461
NON-WORKERS			
Spouses	32,470	30,175	25,330
Children	61,008	56,722	42,197
Other	18,904	19,795	20,441
Total non-workers	112,382	106,692	87,968
Total immigrants	218,465	187,881	149,429

4.63   Emigrants from Canada by country of destination, years ended May 31, 1962-76[e]

Year	United States		United Kingdom		Other countries		Total	
	No.	*%*	*No.*	*%*	*No.*	*%*	*No.*	*%*
1962	44,400	60.08	9,500	12.86	20,000	27.06	73,900	100
1963	50,000	65.10	6,800	8.86	20,000	26.04	76,800	100
1964	51,200	66.41	5,900	7.65	20,000	25.94	77,100	100
1965	50,000	61.65	11,100	13.69	20,000	24.66	81,100	100
1966	39,300	57.12	9,500	13.81	20,000	29.07	68,800	100
1967	33,800	53.82	9,000	14.33	20,000	31.85	62,800	100
1968	41,300	56.57	11,700	16.03	20,000	27.40	73,000	100
1969	30,200	48.16	12,400	19.84	20,000	32.00	62,600	100
1970	27,200	45.18	13,000	21.60	20,000	33.22	60,200	100
1971	22,900	39.08	15,700	26.79	20,000	34.13	58,600	100
1972	19,000	36.19	13,500	25.71	20,000	38.10	52,500	100
1973	15,000	33.26	10,100	22.39	20,000	44.35	45,100	100
1974	12,500	29.21	10,300	24.06	20,000	46.73	42,800	100
1975	11,200	29.02	7,400	19.17	20,000	51.81	38,600	100
1976	11,300	29.65	6,800	17.85	20,000	52.50	38,100	100

Statistics Canada publication, Catalogue 91-208; *International and interprovincial migration in Canada*, July 1977, Ottawa.

4.64   Internal migration of Canadian-born population, by province of birth and by province of residence, 1971 (thousands)

Province or territory of birth	Province or territory of residence, 1971											Total (place of birth)
	Nfld.	PEI	NS	NB	Que.	Ont.	Man.	Sask.	Alta.	BC	YT and NWT	
Newfoundland	496	1	17	4	7	60	2	1	2	5	- -	594
Prince Edward Island	- -	95	7	5	3	20	1	1	3	3	- -	139
Nova Scotia	4	4	661	26	21	116	5	3	12	21	- -	874
New Brunswick	2	3	20	539	49	81	3	2	7	12	- -	718
Quebec	3	1	10	16	5,303	252	9	6	18	31	1	5,650
Ontario	5	3	24	15	137	5,210	40	27	60	107	2	5,630
Manitoba	- -	- -	3	2	14	91	702	36	53	107	2	1,011
Saskatchewan	- -	- -	3	1	8	81	53	710	145	178	3	1,183
Alberta	- -	- -	3	2	7	43	13	20	1,003	162	6	1,260
British Columbia	- -	- -	3	1	7	39	8	9	41	1,056	5	1,170
Yukon Territory and Northwest Territories	- -	- -	- -	- -	2	4	1	1	3	4	28	43
Total (place of residence)	513	108	752	611	5,559	5,996	837	815	1,346	1,688	48	18,273

4.65   Population 5 years and over, by mobility status for the period 1971-76, by province, 1976

Mobility status (based on residence as at June 1, 1971)	Province of residence as at June 1, 1976											
	Nfld.		PEI		NS		NB		Que.		Ont.	
	'000	%	'000	%	'000	%	'000	%	'000	%	'000	%
Non-movers[1]	324	64.9	65	60.1	443	58.2	353	57.2	3,165	54.7	3,880	50.7
Movers[2]	175	35.1	43	39.9	318	41.8	264	42.8	2,624	45.3	3,769	49.3
Non-migrants[3]	90	18.1	16	15.2	152	20.0	128	20.8	1,367	23.6	1,840	24.1
Migrants[4]	85	17.0	27	24.7	166	21.8	136	22.0	1,257	21.7	1,929	25.2
Within same province	58	11.6	13	11.7	91	12.0	70	11.4	1,019	17.6	1,299	17.0
From different province	20	4.0	11	10.7	56	7.3	47	7.6	80	1.4	204	2.7
Province of residence in 1971 not stated	3	0.6	3	0.8	6	0.8	5	0.9	49	0.9	62	0.8
From outside Canada	4	0.8	2	1.6	13	1.7	13	2.1	108	1.9	364	4.8
Total[5]	499	100.0	108	100.0	761	100.0	618	100.0	5,788	100.0	7,649	100.0

	Man.		Sask.		Alta.		BC		YT		NWT		Canada	
	'000	%	'000	%	'000	%	'000	%	'000	%	'000	%	'000	%
Non-movers[1]	488	52.1	474	56.0	749	44.5	969	42.3	6	29.9	14	39.1	10,930	51.5
Movers[2]	450	47.9	372	44.0	935	55.5	1,323	57.7	14	70.1	23	60.9	10,309	48.5
Non-migrants[3]	259	27.6	172	20.4	426	25.3	520	22.7	5	26.5	9	24.9	4,985	23.5
Migrants[4]	191	20.3	200	23.6	509	30.2	803	35.1	9	43.6	13	36.0	5,324	25.1
Within same province	96	10.3	130	15.4	253	15.0	460	25.1	1	6.5	3	8.0	3,494	16.5
From different province	57	6.1	53	6.2	175	10.4	197	8.6	6	32.0	9	24.1	915	4.3
Province of residence in 1971 not stated	9	1.0	7	0.9	19	1.1	32	1.3	6	1.6	1	1.6	195	0.9
From outside Canada	28	3.0	10	1.2	62	3.7	115	5.0	1	3.5	1	2.2	720	3.4
Total[5]	938	100.0	846	100.0	1,684	100.0	2,291	100.0	20	100.0	37	100.0	21,239	100.0

[1]Persons living in same dwelling as on June 1, 1971 and June 1, 1976.
[2]Persons whose dwelling as of June 1, 1976 was in a different dwelling than that of June 1, 1971.
[3]Persons whose residence as of June 1, 1976 was in a different dwelling but in the same municipality as that of June 1, 1971.
[4]Persons whose residence as of June 1, 1976 was in a different municipality than that of June 1, 1971.
[5]Excludes persons in the armed forces or in other government service, stationed outside Canada.
[6]325 persons.

4.66   Population 5 years and over, net migration by province, 1971-76[1]

Province or territory	1971-76				
	In-migration[2]	Out-migration	Total net internal migration	Immigration[3]	Net migration[4]
Newfoundland	19,965	26,845	-6,880	4,180	-2,700
Prince Edward Island	11,560	9,175	2,385	1,695	4,080
Nova Scotia	55,775	50,380	5,395	12,535	17,930
New Brunswick	46,880	37,570	9,310	13,220	22,530
Quebec	79,680	139,480	-59,800	108,200	48,400
Ontario	203,895	256,400	-52,505	363,615	311,110
Manitoba	57,165	83,765	-26,600	28,265	1,665
Saskatchewan	52,555	82,700	-30,145	9,885	-20,260
Alberta	175,045	113,185	61,860	61,895	123,755
British Columbia	197,365	101,480	95,885	114,670	210,555
Yukon Territory	6,310	5,905	405	685	1,090
Northwest Territories	8,925	8,215	710	835	1,545

[1]Excludes persons in the armed forces or in other government service, stationed outside Canada.
[2]Excludes not-stated category for province of residence, 1971.
[3]Includes return migrants.
[4]Excludes emigrants.

4.67   Migrant population 5 years and over, by locality of residence in 1971 and 1976

Locality of residence in 1971	Locality of residence in 1976			Total migrants (by residence in 1971)
	Census metropolitan areas	Census agglomeration	Non-census metropolitan areas	
Census metropolitan areas (CMA)[1]	1,546,935	155,255	620,665	2,322,855
Census agglomeration (CA)[1]	163,915	116,730	183,260	463,910
Non-CMA or non-CA	511,235	191,295	872,690	1,575,220
Outside Canada	575,530	37,575	106,570	719,675
Residence of CMA or CA unknown	139,830	25,225	77,305	242,360
Total migrants[2] (by residence in 1976)	2,937,450	526,085	1,860,490	5,324,025

[1]As defined for the 1976 Census.
[2]Excludes persons in the armed forces or in other government service, stationed outside Canada.

Sources
4.1 - 4.3, 4.63 Population Estimates and Projections Division, Content and Analysis Branch, Census and Household Surveys Field, Statistics Canada.
4.4 - 4.21, 4.23 - 4.31, 4.64 - 4.67 Census Characteristics Division, Content and Analysis Branch, Census and Household Surveys Field, Statistics Canada.
4.22 Statistics Division, Finance and Management Branch, Indian and Inuit Affairs Program, Department of Indian Affairs and Northern Development.
4.32 - 4.55 Health Division, Institutional and Public Finance Statistics Branch, Institutions and Agriculture Statistics Field, Statistics Canada.
4.56 - 4.62 Canada Employment and Immigration Commission.

Health

Chapter 5

Tables

Health

Health status   5.1

To determine the state of health of a nation, one must first look at patterns of illness and what is done to prevent and deal with illness. Much of this chapter describes how the resources of Canadian society, on national, provincial and local levels, are used to combat illness.

There are many dimensions of health and illness in society. In April 1974, the national health and welfare minister tabled in Parliament and published *A new perspective on the health of Canadians*. This document suggests that what are usually called health services are really sickness or treatment services. It contends that further improvement in the health of Canadians will depend on a better knowledge of the human body, on the quality of the environment, and on individual lifestyles rather than on any improvements in health care services.

Despite the complexities of defining good health, there are various ways in which some fairly dependable measures of the state of health may be obtained. One attempt to gain information on this subject from a nationwide study of the population was a Canadian sickness survey of 1950-51. The Nutrition Canada study of 1970-72 was a national review of the impact of nutrition on health. A Canada health survey, which began in May 1978, will provide annual figures on many aspects of health, lifestyle, illness and use of health services. Preliminary data from the survey are expected to be available in 1979.

The most widely used measures of health status, based on available information, are life expectancy, infant mortality, causes of death, hospital and other morbidity data.

Life expectancy   5.1.1

Trends in life expectancy in Canada are depicted in Tables 4.46 and 4.47. Over the last 40 years expectation of life at birth has improved steadily for both males and females. In 1971 it reached 69.3 years for males and nearly 76.4 years for females. A major reason for the overall increase is the drop in infant mortality.

The difference between male and female life expectancy increased from 2.1 to 7.1 years between 1931 and 1971. This difference is reflected in lower death rates for women at all ages and a substantially larger decrease for female death rates compared to male death rates.

Infant mortality   5.1.2

Trends in infant mortality are presented in Table 4.43. Death rates for both male and female infants under one year of age declined about 63% between 1951 and 1975. The improvement is due to better health care before and after birth, improved nutrition and living standards and a decline in the number of children born to older mothers. However the death rate in Canada in recent years was still 20% to 25% higher for male infants than for females.

Causes of death   5.1.3

Table 4.42 presents causes of death by sex and age group for 1975 and 1976. The leading causes of infant mortality are considerably different from those at later ages. Most infant deaths occur during or shortly after delivery and most are caused by birth defects and conditions specific to the period immediately before and after birth. After the first week of life most infant deaths are due to congenital anomalies, acute infections of the respiratory tract or different kinds of accidents.

Between the ages of one and 14, death rates appear to be gradually stabilizing at low levels. However death rates for males in these age categories, as in all others up to age 80, are higher than those for females. More than half the deaths of children between age one and 14 are due to motor vehicle and other accidents.

In the age groups 15-19 and 20-24, a serious trend in the last 10 years has been an increase in death rates in both sexes. Furthermore, the rate for males is three times as great as for females. At these ages the most important causes of death are motor vehicle and other accidents and suicide.

Between ages 25 and 44 the death rate for men is twice that for women. For both sexes, but particularly for men, motor vehicle accidents, other accidents and suicides remain important causes of death. In this stage of life ischemic heart disease (in which the heart muscle has its own blood supply restricted) becomes a significant cause of death for men. For women, cancer of the breast, uterus, ovary and gastro-intestinal tract begin to contribute noticeably to the total number of deaths, as do cerebrovascular diseases (strokes).

With increased age the proportion of deaths due to cerebrovascular disease, respiratory diseases and various types of cancer increases. Until the most advanced age categories male deaths continue to exceed those of females. One of the most notable differences between males and females is the higher proportion of male deaths due to ischemic heart disease, respiratory diseases and lung cancer and cirrhosis of the liver, all of which are related to lifestyle factors including smoking, drinking, exercise and stress.

A review of causes of death, by sex and age, raises questions about what proportion of deaths at early ages might be prevented for males and females, but particularly for males. Many deaths might be prevented through attention to lifestyle factors and the potential for accidents. For females a number of deaths are the result of illnesses which may be treated if detected at early stages.

5.1.4 Hospital morbidity data

Hospital statistics compiled by the health division of Statistics Canada offer much detail about specific illnesses and disabilities and patterns of treatment. Some provinces also keep detailed records of diagnoses of patients' conditions derived from physicians' medical care insurance claim forms. It is not possible, however, to aggregate this diagnostic information on a national level.

Although hospital morbidity data remains the most comprehensive source of information on patterns of illness and disability in Canada, it has limitations. There are no data on illnesses which are self-treated or improved before admission to hospital. In addition there is little information about the chronically disabled and the number of days that Canadians remain at home in bed because of illness.

Some of these gaps in information will be filled by data from the Canada health survey, based on both self-reported conditions and some physical measurements performed by nurses. But, despite their limitations, in-patient hospital data will continue to be a useful source of information on illness in Canada.

Tables 5.1 and 5.2 include information on in-patient separations (deaths and discharges) from hospitals by diagnostic category. Table 5.1 shows relative importance of various diagnoses in terms of the number of separations and rate per 100,000 population, the number of days' stay per 100,000 population and the average length of stay. Patterns of hospital use differ by diagnostic categories. For example, complications of pregnancy resulted in the highest number of separations with 4,424 per 100,000 persons in 1975. However average length of stay was only 5.2 days. On the other hand, diseases of the circulatory system, with only 1,667 separations per 100,000 persons, resulted in the highest number of days per 100,000, an average of 23.3 days.

Table 5.2 shows hospital separations per 100,000 population by diagnostic category, sex and age group. These figures show the increase in hospital use with advancing age and a higher number of hospital separations for females than for males.

Tables 5.3 and 5.4 summarize cases undergoing surgery in relation to all separated cases and by type of primary operation, age group and sex. Overall, almost one-half of hospital cases result in surgery. Of the 1,898,008 operations in 1975, 15% were obstetrical procedures, 14% gynecological surgery, 13% abdominal surgery and 11% orthopedic surgery.

The 10 provinces and two territories reported to Statistics Canada that 54,478 therapeutic abortions were performed during the 12-month period January to

Trends in life expectancy and infant mortality[1], 1951-74

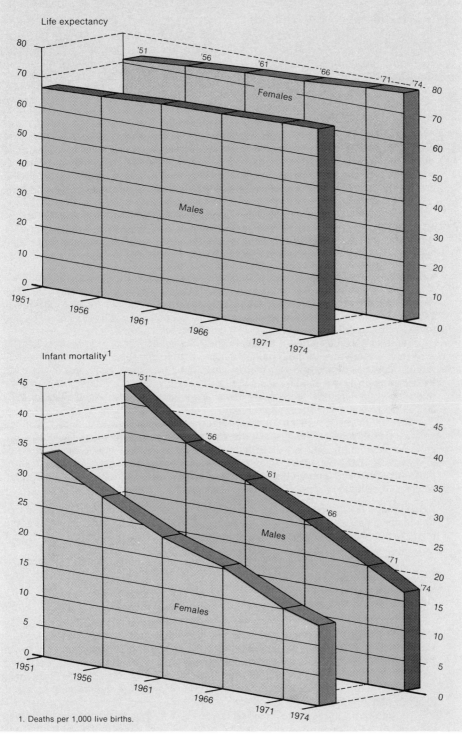

Life expectancy

Infant mortality[1]

1. Deaths per 1,000 live births.

December 1976. This represents 5,167 more than for 1975. There were 14.9 therapeutic abortions per 100 live births in 1976 compared to 13.8 in 1975. Data are in Table 5.7.

5.1.5 Specific diseases or disabilities

Besides general mortality and hospital morbidity data, Statistics Canada maintains registries and does special analyses that relate to particular disease conditions, their treatment and mortality resulting from them. Some of these information systems are developed in co-operation with voluntary agencies. Other data are derived from notifications which physicians are required by law to make to public health authorities. Although not all serious conditions are covered, these records are a valuable source of health status information.

Heart disease. The death toll from heart disease in Canada in 1975 was 56,970, or 250 deaths for each 100,000 persons. The male rate was higher than the female, 298 against 202. Among men aged 45 to 64, heart disease accounted for nearly 40% of all deaths, and the single diagnostic class ischemic heart disease killed 9,293 of the 25,367 men in this group. In 1975 heart disease required 3,840,000 days of care in general and allied special hospitals.

The Canadian Heart Foundation, inaugurated in 1955, had by mid-1977 devoted $57.1 million to cardiovascular research in Canada's universities and hospitals; its 1977-78 budget alone provided $8.7 million. The Medical Research Council spent $5.8 million on cardiovascular research in 1977-78.

Cancer. As the second leading cause of death in Canada, cancer accounts for about one of every five deaths, most of them occurring in the middle and later years of life. The death rate from cancer dropped slightly, from 150.4 per 100,000 population in 1974 to 149.2 in 1975. The rate for females decreased from 134.4 in 1974 to 131.1 in 1975, and for males increased from 166.3 in 1974 to 167.4 in 1975.

Statistics Canada started a national cancer incidence reporting system in January 1969 in co-operation with the National Cancer Institute and the nine existing provincial tumor registries; a centralized registry has not yet been organized in Ontario. Participating provinces send a simple notification card with basic patient and diagnostic information for each new primary site of malignant neoplasm discovered. Data for 1974 and 1975 are given in Tables 5.5 and 5.6.

Special provincial agencies for cancer control, usually in the health department or a separate cancer institute, carry out cancer detection and treatment, public education, professional training and research in co-operation with local public health services physicians and the voluntary Canadian Cancer Society branches. Provincial cancer programs operate both under the terms of provincial health insurance plans and through special supplementary services for cancer patients.

Renal failure. A Canadian renal failure register operated by Statistics Canada was started by and operates in co-operation with the Kidney Foundation of Canada. Its purpose is to register and follow all patients depending on artificial kidney treatment (chronic peritoneal or hemodialysis) or receiving kidney transplants since January 1973 in Canada. Table 5.8 reports the status of renal failure patients in Canada for 1974 and 1975. During 1974, 55 dialysis units reported on 1,776 patients; three other units have not submitted 1974 follow-up data; new chronic dialysis patients registered totalled 534. During 1975, 59 hospitals reported on 2,310 patients; new dialysis/transplant patients registered totalled 690.

Notifiable diseases. The number and rates per 100,000 population of notifiable diseases by province in 1975 and 1976 are shown in Table 5.24. Most predominant of these in 1976 were venereal diseases (56,344) and streptococcal sore throat and scarlet fever (18,512). There were 9,158 reported cases of measles and 4,167 of rubella or German measles. Reported rates for other diseases, although lower, are significant in terms of public health.

Of particular interest are venereal diseases, because public health authorities estimate that their real incidence may be three to four times the number of cases reported. The 1976 figure of 3,952 cases of syphilis, or 17.4 per 100,000 population, shows only a slight decrease from the 1975 figure of 3,967. Total gonorrhea cases in 1976 were 52,262, or 229.5 per 100,000. The 1975 figure was 50,752 or 222.6 per 100,000. This rise in incidence of gonorrhea is attributed to a supposed increase in sexual permissiveness, promiscuity and homosexuality, availability of the contraceptive pill, increased population mobility, changes in social values, lack of case reporting, and ignorance about venereal disease.

Provincial health departments have expanded public venereal disease clinics, which provide free diagnostic and treatment services. In some areas these departments pay private physicians to give free treatment to the poor. In addition, the provinces supply free drugs to physicians for treating private cases. Local departments or district health units carry out case finding, follow-up of contacts and health education programs, assisted by provincial directors of venereal disease control.

At one time tuberculosis was an extremely serious health problem. However the reported incidence of this disease has decreased steadily in recent years. In 1976 there were 2,601 reported cases, or 11.4 per 100,000 population. This may be compared with the 3,089 new active cases or 13.5 per 100,000 population in 1975. Most new cases of tuberculosis are discovered by practising physicians, but provincial health departments assisted by voluntary agencies continue to conduct anti-tuberculosis case finding programs through community tuberculin testing and X-ray surveys, with special attention to high risk groups, routine hospital admission X-rays and follow-up of arrested cases. Provincial tuberculosis programs include vaccination for children or high risk groups and free treatment, including hospital care, drugs and rehabilitation services.

Constitutional responsibilities in the health field   5.2

Government involvement in health care services in 1867, at Confederation, was minimal. For the most part the individual was compelled to rely on his own resources and those of his family group, and hospitals were administered and financed by private charities and religious organizations.

The only specific references to health in the distribution of legislative powers under the British North America Act allocate to Parliament jurisdiction over quarantine and the establishment and maintenance of marine hospitals, and to provincial legislatures jurisdiction over the establishment, maintenance and management of hospitals, asylums, charities and charitable institutions in and for the province, other than marine hospitals. In 1867 this latter reference probably was meant to cover most health care services. Since the provinces were assigned jurisdiction over generally all matters of a merely local or private nature in the province, it is probable that this power was deemed to cover health care, while the provincial power over municipal institutions provided a convenient means for dealing with such matters. Thus provision of health care services has been traditionally acknowledged as primarily a provincial responsibility. But a measure of responsibility in health matters has been expressed over the years in many federal programs and policies.

Federal-provincial co-operation   5.3

Since the federal and provincial governments share responsibility for dealing with health matters, a formal structure has been established for federal-provincial co-operation. It comprises the following: conference of ministers of health; conference of deputy ministers of health; federal-provincial advisory committees on institutional care services, community care services, health promotion and lifestyle, environmental and occupational health and health manpower. The conferences of ministers and deputy ministers of health involve matters of promotion, protection, maintenance and restoration of the health of the Canadian people. Normally, the conference of ministers meets annually and the conference of deputy ministers twice a year. The five advisory committees facilitate the work of the ministers and deputy ministers, and assist them in achieving objectives, identifying major issues and solving problems. They may set up groups to deal with particular subjects requiring more detailed study.

5.4   Federal health services

The national health and welfare department is the principal federal agency in health matters. It is responsible for the overall promotion, preservation, and restoration of the health of Canadians, and for their social security and social welfare. The department acts in conjunction with other federal agencies and with provincial and local services. The provincial governments actually administer health services. Although the patterns of health services are similar, their organization and administration vary from province to province.

Other federal agencies which carry out specialized health functions include, for example, the health division, Statistics Canada, which gathers health and vital statistics, the veterans affairs department, which administers hospitals and health services for war veterans, and the agriculture department, which has certain responsibilities for health aspects of food production.

Branches of the national health and welfare department are responsible for health protection, medical services, health programs, long-range health planning and fitness and amateur sport. The Medical Research Council supports research in health sciences in Canadian universities and affiliated institutions.

In the health and welfare department, an integrated program protects the public against unsafe foods, drugs, cosmetics, medical and radiation-emitting devices, harmful microbial agents and technological and social environments, environmental pollutants and contaminants of all kinds, and fraudulent drugs and devices.

Medical services include health care and public health services for registered Indians, Inuit and all residents of the Yukon Territory and Northwest Territories, as well as quarantine and regulatory services, immigration medical services, public service health, a national prosthetics service, civil aviation medicine, disability assessment and emergency health and welfare services. Long-range health planning assesses the orientation of health services and the organization of resources.

The fitness and amateur sport branch encourages excellence in Canada's athletes and participation of all Canadians in activities oriented toward fitness and recreation. The health programs branch administers federal aspects of Canada's two major health programs, hospital and medical insurance; supports health care delivery system and resource development; undertakes health promotion; and both supports and conducts research.

5.4.1   Health care

Medical care. Before the establishment of government-administered medical insurance, voluntary prepayment arrangements to cover the cost of physicians' services had developed in public and private sectors. By the end of 1968, basic medical or surgical coverage, or both, were being provided to about 17.2 million Canadians, 82% of the population. Voluntary plans in the private sector covered about 10.9 million, or 52%, and public plans covered 6.3 million, or 30%. By 1972 all 10 provinces and the two territories had met the criteria stipulated under the Medical Care Act as conditions for federal cost-sharing, and virtually the entire eligible population was insured for all required medical services plus a limited range of oral surgery. Members of the Canadian Armed Forces, the Royal Canadian Mounted Police, and inmates of federal penitentiaries whose medical care requirements are met under alternative provisions are excluded. Services by physicians that are not medically required, such as examinations for life insurance, services covered under other legislation, such as immunization where available through organized public health services, and services to treat work-related conditions already covered by worker compensation legislation are not covered.

Comprehensive coverage must be provided for all medically required services rendered by a physician or surgeon. There can be no dollar limit or exclusion except on the ground that the service was not medically required. The federal program includes not only those services that have been traditionally covered as benefits by the health insurance industry, but also those preventive and curative services that have been traditionally covered through the public sector in each province, such as medical care of patients in mental and tuberculosis hospitals and services of a preventive nature provided to individuals by physicians in public health agencies.

The plan must be universally available to all eligible residents and cover at least 95% of the total eligible provincial population (in fact the plans cover over 99%). A uniform terms and conditions clause is intended to ensure that all residents have access to coverage and to prevent discrimination in premiums because of previous health, age, non-membership in a group, or other considerations. If a premium system of financing is selected, subsidization in whole or in part for low-income groups is permitted. It has been left to the individual province to determine whether its residents should be insured on a voluntary or compulsory basis. Utilization charges at the time of service are not precluded by the federal legislation if they do not impede, either by their amount or by the manner of their applications, reasonable access to necessary medical care, particularly for low-income groups. The plan must provide portability of benefit coverage when the insured resident is temporarily absent from the province and when moving residence to another participating province. The provincial medical care insurance plan must be administered on a non-profit basis by a public authority that is accountable to the provincial government for its financial transactions. It is permissible for provinces to assign certain administrative functions to private agencies.

These criteria leave flexibility with each province to determine its own administrative arrangements for the operation of its medical care insurance plan and to choose the way in which it will be financed, that is, through premiums, sales tax, other provincial revenues, or by combination of methods.

Federal financial contributions to the provinces prior to April 1977 were based on half of the national per capita cost of the insured services of the national program, excluding administration, multiplied by the number of insured persons in each province. A 1976 amendment to the act established a ceiling of 113% on the per capita increase of the federal contribution for the fiscal year 1976-77.

Hospital insurance. The Hospital Insurance and Diagnostic Services Act which took effect on July 1, 1958, was designed to make available to all eligible residents a wide range of hospital and diagnostic services, subject to medical necessity, at little or no direct cost to the patient, thereby removing financial barriers to adequate care which existed for many residents prior to the introduction of the program.

Under the act, contributions by the federal government are authorized for programs administered by the provinces providing hospital insurance and laboratory and other services in aid of diagnosis.

The program incorporates five general principles: comprehensiveness of services; universal availability of coverage to all eligible residents; no barriers to reasonable accessibility of care; portability of benefits; and public administration of the provincial programs.

Facilities covered under the program include general, rehabilitation (convalescent), and extended care (chronic) hospitals together with specialized hospitals such as those providing maternity or pediatric care. The program may also cover diagnostic services in non-hospital facilities. Specifically excluded under the program are tuberculosis hospitals and sanatoria, hospitals or institutions for the mentally ill, and nursing homes, homes for the aged, infirmaries or other institutions whose purpose is to provide custodial care.

In development of hospital insurance legislation, existing traditions were maintained as far as possible. The pattern of hospital ownership and operation that existed before the act came into force was retained and provincial autonomy was not infringed. Consequently, even 20 years later, almost 90% of the beds covered by hospital insurance are located in facilities owned and operated by voluntary bodies and municipalities. The policy of provincial autonomy allows each province to decide on methods of administration and of financing its share of program costs while still ensuring a basic uniformity of coverage throughout the country. All provinces and territories have participated since 1961. Details of services provided are in Section 5.5.1, Provincial health insurance plans.

Insured in-patient services must include accommodation, meals, necessary nursing service, diagnostic procedures, most pharmaceuticals, the use of operating rooms, case rooms, anesthesia facilities, and radiotherapy and physiotherapy if available. Similar out-patient services may be included in provincial plans and authorized for contribution under the act. All provinces include a fairly comprehensive range of out-patient services.

The individual may select the hospital in which he will be treated provided his physician has admitting privileges, and the only limit to the duration of insured services is the extent of medical necessity. Moreover, during a temporary absence, coverage is portable anywhere in the world for in-patient services, and in the case of most provinces for out-patient services also, although such benefits are subject to provincially regulated maxima for rates of payment and length of hospital stay as set out in the summary of provincial programs.

Provinces may include additional benefits in their plans without affecting the federal-provincial agreements. Some provincial hospital plans provide additional services such as nursing home care and these are also mentioned in the provincial program summaries. These additional services are not cost-shared under hospital insurance.

The principles of universal availability of benefits to all eligible residents and portability of benefits are reflected in provisions of each provincial program. For many years, about 99% of all eligible residents have been insured persons. Although provincial plans in general stipulate a waiting period of three months, coverage may continue from the province of previous residence. First-day coverage is generally provided for the newborn, immigrants, and certain other categories of persons without prior coverage in other provinces. A health insurance supplementary fund has been established for residents who have been unable to obtain coverage or who have lost coverage through no fault of their own.

Until March 31, 1977 the federal government contributed approximately half the cost of insured in-patient and out-patient services for Canada as a whole. This included payments to Quebec under the Established Programs (Interim Arrangements) Act effective January 1965. The formula provided proportionately larger contributions in those provinces where per capita costs were below the national average and vice-versa.

Provinces may raise their portion of insurable costs as they wish, provided that access to services is not impaired. All provinces finance their share in whole or part from general revenue.

Established programs financing. Late in 1976, following several years of negotiations, the provinces and the federal government agreed to new financial arrangements for medical care and hospital insurance, among other fiscal matters. This led to the Federal-Provincial Fiscal Arrangements and Established Programs Financing Act, 1977, assented to on March 31, 1977, containing consequential amendments to the Medical Care Act and the Hospital Insurance and Diagnostic Services Act. Commencing April 1, 1977, federal contributions to the established programs of hospital insurance, medical care and post-secondary education are no longer directly related to provincial costs, but take the form of the transfer of a predetermined number of tax points, and related equalization and cash payments. Total federal contributions, in general terms, are now based on the current escalated value of the 1975-76 federal contributions for the programs in question. The tax room vacated by the federal government permitted the provinces to increase their tax rates so as to collect additional revenue without necessarily increasing the total tax burden on Canadians. The yield from the new provincial taxes will normally increase faster than the rate of growth of the Gross National Product (GNP). The cash payments are conditional upon the provincial health insurance plans meeting the criteria of the federal health insurance legislation. At the outset, the cash payments will approximate the value of the tax room transferred, and be in the form of per capita payments calculated in accordance with the Federal-Provincial Fiscal Arrangements and Established Programs Financing Act, 1977. These per capita payments will be escalated yearly in accordance with changes in the GNP, and adjusted gradually over time so that all provinces at the end of five years will be receiving equal per capita cash contributions.

Also under the act as of April 1, 1977, the federal government is making additional equal per capita cash contributions yearly to the provinces to contribute toward the costs of certain extended health care services.

Health resources fund. The Health Resources Fund Act of 1966 provided $500 million over 15 years (1966-80) for financial assistance in planning, acquisition, construction, renovation, and equipping health training and research facilities. Up to 50% of eligible costs of approved projects are supported by federal contributions. Of this total, $400 million is allocated to provinces on a per capita basis, $25 million is further allocated to the Atlantic provinces for joint projects, and $75 million for health training and research projects of national significance.

Professional training program. This program provides about $2.3 million a year to the provinces for training health and hospital personnel. Two types of training are funded by the federal government: bursaries for one academic year or longer, and short courses for up to three months. Assistance may also be given to the holding of, and attendance at, provincial and national conferences with emphasis on health manpower planning and development.

Health services for specific groups. Through medical services branch, the national health and welfare department provides or arranges health services for persons whose care is by custom or legislation a federal responsibility.

Indians and Inuit, as residents of a province or territory, are entitled to benefits of medical care and hospital insurance. These insured benefits are supplemented by the branch, which helps in arranging transportation and obtaining drugs and prostheses. A comprehensive public health program provides dental care for children, immunization, school health services, health education, and prenatal, postnatal and well-baby clinics. A native alcohol abuse program funds locally-run programs. Since Indians and Inuit comprise only 1.0% of the population and are distributed widely throughout Canada, a network of specially designed health facilities operates in almost 200 communities. Increasing numbers of Indians and Inuit are being trained and employed in public health and medical care programs to facilitate understanding and health activities in the communities.

With the exception of insured hospital and medical care programs, administered by the governments of the Yukon Territory and Northwest Territories, the national health and welfare department has for many years managed health services for all residents of the two northern territories. These comprise a comprehensive public health program, special arrangements to facilitate interstation communication, and the transportation of patients from isolated communities to referral medical centres. Several university groups provide, on a rotation basis for specified zones, medical personnel and students. Their activities are financed through government contracts and medical care insurance.

As of January 1978, departmental facilities included six hospitals, three health stations and nine health centres in the Yukon Territory and four hospitals, 39 nursing stations, six health stations and eight health centres in the Northwest Territories.

Under the Quarantine Act, all vessels, aircraft, and other conveyances and their crews and passengers arriving in Canada from foreign countries are subject to inspection to detect and correct conditions that could introduce such diseases as smallpox, cholera, plague and yellow fever. Quarantine stations are located at major seaports and airports. The branch enforces standards of hygiene on federal property including ports and terminals, interprovincial means of transport, and Canadian ships and aircraft.

Medical services branch determines the health status of all persons referred by the employment and immigration commission for Canadian immigration purposes. It also provides or arranges health care services for certain persons after arrival in Canada, including immigrants who become ill en route or while seeking employment. The branch is responsible for a comprehensive occupational health program for federal employees in Canada and abroad. This includes health counselling, surveillance of the occupational and working environment, pre-employment, periodic and special examinations, first aid and emergency treatment, advisory services and special health programs. Increased attention is given to pre-retirement and stress.

The department advises the ministry of transport on health and safety in Canadian civil aviation. Regional and headquarters aviation medical officers review medical examinations, participate in aviation safety programs, and assist in air accident investigations. There is close liaison with authorities in foreign aviation medicine, with standards usually based on international agreements.

Prosthetic services assists in prosthetic and corrective rehabilitation under agreements with most provinces and with the veterans affairs department, and provides a national focal point for related expertise. Discussions have been held on a plan to transfer this activity to provincial control.

Medical services physicians provide an assessment and advisory service to the employment and immigration commission on claims for benefits under the sickness and maternity benefit plan. The Canada Pension Plan maintains its own disability assessment service.

Emergency welfare services is responsible for a national capability, embracing government and welfare related non-government agencies of essential welfare services in any type of emergency in Canada.

In an effort to improve communication through new technology, the branch has participated in telemedicine experiments, with Moose Factory and Kashechewan, Ont. receiving direct consultation on medical and surgical matters through television.

The magnitude of health problems posed by environmental pollution has resulted in a number of activities. The environmental contaminants program is studying effects of mercury pollution from coast to coast. Other environmental contaminants such as cadmium, arsenic and mirex are of growing concern.

5.4.2   Health promotion and protection

5.4.2.1   Lifestyle and health promotion
Promotion of lifestyles that will improve personal health and development of comprehensive community health services readily accessible to all Canadians are major emphases of health programs branch.

The community health division of the health consultants directorate is concerned with consulting, planning, developing, and evaluating community health services and centres. The main thrust is to promote community health services; to facilitate co-ordination of community health services planning; and to encourage shifts in emphasis from institutional to ambulatory care, and from curative to health-promotional and preventive services.

The lifestyle and health promotion directorate has been established to help develop greater collaboration among government and other agencies in lifestyle and health promotion activities and to bring about better co-ordination of activities within the national health and welfare department.

5.4.2.2   Health protection
The health protection branch contains six operational directorates — food, drugs, environmental health, laboratory centre for disease control, non-medical use of drugs, and field operations.

Food. Standards of safety and purity are developed through laboratory research and maintained by means of a regular and widespread inspection program. The inspection of food-manufacturing establishments plays a major role in the production of clean, wholesome food containing ingredients that meet recognized standards. Changing food technology requires the development of methods of laboratory analysis to ensure the safety of new types of ingredients and packaging materials. The food and drug regulations list chemical additives that may be used in foods, the amounts that may be added to each food, and the underlying reason. Information on new additives must be submitted for review before they are included in the permitted list. Emphasis is placed on studies to ensure that the levels of pesticide residues in foods are not a health hazard. The effect of new packaging and processing techniques on the bacteria associated with food spoilage is also of special concern. A national reporting system for food-borne outbreaks of disease has been established.

A report by the committee on diet and cardiovascular disease was submitted in 1977 and many of its recommendations were adopted by the department. They will form the basis of nutrition education programs and have a marked influence on food regulations, labelling and advertising.

Drugs. The major activities are focused on the principle that Canadians should have access to drugs that are both safe and effective. A major part of the activity is devoted to clearing new drugs for marketing in Canada and maintaining a post-marketing surveillance over these products. For example, manufacturers of new drugs with unknown properties are required by law to submit extensive evidence of the safety and effectiveness of their products prior to marketing. This includes information about therapeutic properties and side effects. Continued surveillance of the new product is maintained. Additionally, the branch monitors manufacturers' compliance with official specifications and regulations setting standards for manufacturing facilities and quality control of drugs. Products such as serums and vaccines are subject to special licensing requirements for safety and effectiveness.

A quality assessment of drugs program integrates the above activities and provides information to provincial governments relating to manufacturers' facilities and their compliance with standards for their products.

Non-medical use of drugs. The non-medical use of drugs directorate administers three programs concerned with alcohol, tobacco and drugs, with its main focus on prevention. In addition, the directorate supports treatment and rehabilitation programs, and engages in information programs, community projects and research. Information programs include media campaigns, provision of informational materials and education programs to increase awareness, understanding and public concern about alcohol, tobacco and drug-related problems and responses, as well as increase knowledge, attitudes and skills which will improve public decision-making concerning substance abuse. Community projects are being funded through an alcohol-tobacco-drug resources fund. Projects are funded for up to three years in an effort to demonstrate new ways of preventing or treating problems or securing changes in existing policies and programs. Intramural and extramural research develops a knowledge base on alcohol, tobacco and drug use and of public health and socio-economic problems associated with them.

Environmental health. The environmental health directorate studies the adverse effects on human health of the chemical and physical environment, and ensures the safety, effectiveness, and non-fraudulent nature of medical devices. The directorate develops health hazard assessments for work and home environments, household products, and air and water criteria. Research on radiation hazards is conducted and environmental and occupational exposures are monitored. The directorate enforces the Radiation Emitting Devices Act and that portion of the Food and Drugs Act dealing with medical devices and radioactive pharmaceuticals, and it also administers the Hazardous Products Act jointly with the consumer and corporate affairs department. Under the Environmental Contaminants Act the directorate collects data and investigates substances that may be a danger to human health or the environment and, when necessary, recommends restricted or prohibited use of substances.

Disease control. The laboratory centre for disease control directorate is involved in the development and implementation of improved laboratory diagnostic procedures and other measures to combat communicable disease agents. Activities entail developing methods for detecting and preventing disease, and producing and distributing standardized diagnostic reagents to federal, provincial and other health organizations. A national reference service is provided for identification of disease-producing bacteria, viruses and parasites, and development of a co-operative federal-provincial program for laboratory quality assurance and proficiency testing. The directorate also maintains surveillance of birth defects, poisonings, and adverse drug reactions. Epidemiological research is carried out on communicable and non-communicable diseases.

5.4.3   Fitness and amateur sport

The fitness and amateur sport branch is comprised of four directorates. The two program directorates are: Fitness and Recreation Canada — concerned with physical recreation and fitness, and Sport Canada — concerned with the pursuit of excellence in amateur sport.

The two support directorates, program operations and administration and planning, research and evaluation, provide all backup services necessary for the branch's programs. The branch achieves its twin goals of fitness and recreation participation and sport excellence chiefly through financial contributions and a wide range of consultative services to national sport and recreation associations.

Sport Canada strives to upgrade the quality and improve the quantity of Canadian participation in sport by strengthening national sports governing bodies and other related national agencies. As well, the program initiates or supports specific activities to help Canadian athletes pursue national and international championships.

Game Plan, a project started in 1973, has developed programs of talent identification, athlete support, coaching and officiating development, and competition and training opportunities. This plan has set goals for Canadian athletes in national and international competition. A new athlete support program was implemented in 1978 which provides living expenses, training allowances, lost time payments, tuition fees and facility rental for top-calibre athletes.

The fitness and amateur sport branch administers a second program to support Canadian athletes. The grants-in-aid program assists athletes to continue both their educational and competitive careers by providing tuition and living expenses.

Sport Canada also funds an agency for coaching development. The Coaching Association of Canada encourages development of coaching skills through programs and services to all levels of coaches. A national coaching certification program trains thousands of volunteers to become qualified coaches; it affects almost every sport agency and government level and is Canada's first attempt to structure coaching development to standardize levels across the country. In 1977 the first step was taken in a program to train coaches to work at a national and international level by awarding grants enabling apprentice coaches to study under a master coach.

A national sport and recreation centre, funded primarily by the branch, provides accommodation and comprehensive administrative services for national sports and recreation associations. The branch also contributes toward the salaries of executive directors, technical co-ordinators and certain national coaches for these associations. As well, the associations receive funds on a project-by-project basis which allow them to conduct events such as national championships or conferences.

Sport Canada supports international sport exchanges with other countries to allow Canadian athletes to gain experience in international competition.

The Canada Games, held every second year, are the major multi-sport competition in this country. Over 2,500 athletes gathered in St. John's, Nfld. in 1977 and plans are under way for the upcoming 1979 winter games in Brandon, Man. and 1981 summer games in Thunder Bay, Ont. Sport Canada provides funds for the operating and capital costs of these games.

In 1978, the Commonwealth Games were held in Edmonton, Alta. The federal government, through the fitness and amateur sport branch, contributed to capital, operating and other indirect costs. National championships and single-sport world championships also receive support, as do the Arctic winter games and the northern games.

Fitness and Recreation Canada is mainly concerned with fitness and mass participation. It funds a variety of national agencies, which provide recreational opportunities to all Canadians. Activities focus on increasing the quantity and improving the quality of both human and physical resources, to provide recreational opportunities and increase the awareness of physical fitness. Financial contributions and consulting services are provided to groups such as the Girl Guides of Canada, the Canadian Red Cross, cycling, orienteering and a number of associations for native people and the handicapped.

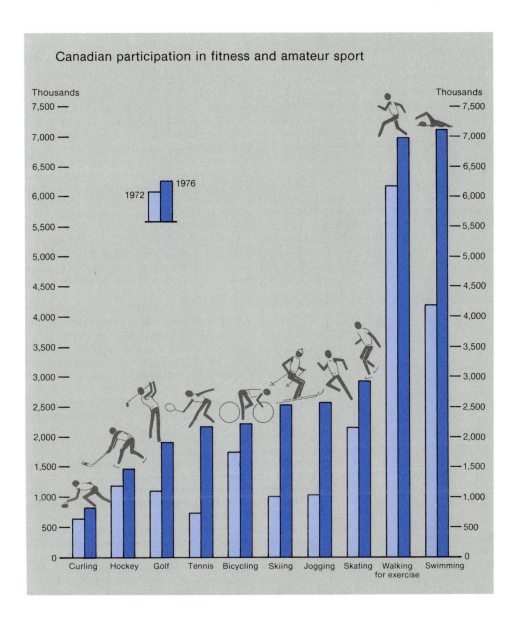

Canadian participation in fitness and amateur sport

Sport Participation Canada, known across Canada by the slogan ParticipAction, is funded by this directorate to promote physical activity among Canadians. It uses a variety of marketing techniques, such as television ads and brochures, to motivate involvement in physical activity.

The fitness section administers a number of programs and projects designed to promote an awareness of fitness and provide Canadians with information on fitness programs. Its major areas of interest have been fitness and health, employee fitness, fitness trails and the development of the Canadian home fitness test. An exercise break program has been developed to help fitness leaders and health professionals introduce the benefits of physical activity in business, industry and educational institutions.

The Canada fitness award program was developed to encourage boys and girls, aged 7 to 17, to strive for fitness and excellence. The program is administered by the branch and awards bronze, silver and gold crests and awards of excellence on the basis of test results. This popular test is widely used by schools and youth organizations. In 1977, Canada fitness award number 5,000,000 was presented to a Calgary student.

One of the more visible programs is a sports demonstration project. This project tours Canada each summer and gives residents the opportunity to try a variety of physical skills, such as skiing on artificial snow, skating on a specially-designed plastic surface or batting balls pitched from an automatic pitching machine.

The result of the 1976 fitness and sport survey conducted by the research section in collaboration with Statistics Canada will be used by all levels of government and by private agencies and organizations for the evaluation of existing sport and fitness programs and the development of new initiatives.

In October 1977, a discussion paper on the federal role in amateur sport was released for consideration by concerned organizations, individuals and levels of government. Meetings were held across Canada and numerous briefs submitted. This procedure should result in a policy paper which will ensure the continued development of sport in Canada. Similar discussions will be held on the future of fitness and recreation to provide a comprehensive approach to the participation of Canadians in fitness, recreational and sporting activities.

5.4.4 Research, planning, standards and consultation

Medical Research Council. Most federal grants supporting health science research in universities and hospitals are channelled through a medical research council which reports to Parliament through the minister of national health and welfare. The council provides grants in aid of operating and equipment requirements for research projects and direct support for a limited number of investigators and research trainees. It offers incentives for the development of research in highly productive fields where major contributions may be expected and in fields or regions where research is not adequately developed. Support is given for meetings, international scientific activities and exchange of scientists.

National health research and development program. The national health research and development program enables the national health and welfare department to obtain information and to evaluate and develop innovative options for the achievement of broad departmental objectives which embrace the promotion, protection, maintenance and restoration of the health of Canadians. These objectives include meeting similar provincial requirements having national interest or implications.

The program is designed to encourage and support the formulation, testing, evaluation and development of ideas and proposals generated by appropriately qualified individuals and agencies (other than federal government departments and employees); and the creation, development and maintenance of an adequate body of highly competent Canadian research investigators in the field of health care, including health hazards of the environment, the biology of human populations, lifestyle and the organization of health services.

Health statistics. The health division of Statistics Canada has established collection systems for data on vital statistics, special diseases, health manpower and hospital and institutional care. Units in the health protection branch of the national health and welfare department are concerned with data on health products, health hazards and certain disease areas. In the health programs branch, the health economics and statistics division operates a medical care data bank, and integrates health statistical data from various sources. The division undertakes socio-economic research in a variety of fields, including medical and hospital care, community health, health expenditures and resources and other matters relating to health costs and utilization. These studies support departmental health planning as well as the production of publications to increase public understanding of Canada's health services and resources.

Health planning. The long range health planning branch continuously assesses the overall orientation of health services and the organization of resources and factors influencing the health of Canadians. The branch collaborates with other branches of the department in developing proposals. It has specialists in the fields of medicine, pharmacy, epidemiology, economics, sociology, demography, political science, statistics and administration.

Standards and consultation. The national health and welfare department extends technical advisory services to provincial agencies, universities and other organizations for the development of health programs, health manpower and health research. Consultative services are available through the various administrative units of the department.

The health standards directorate co-operates with provinces, professional associations, universities and other organizations to establish and promote standards and directives for health services, clinical practice and health personnel. The health consultants directorate studies the health-related needs of Canadians, the means available to meet those needs, and the use of health resources. It provides information and consulting services particularly about health systems, plans and tenders for facilities, hospital administration, health personnel, community health and health promotion. Other technical advice is available through programs directly operated by the department for health protection including the safety of foods, drugs, and health appliances, environmental health, disease control and other specialized areas.

Canada health survey. The Canada health survey is a joint responsibility of the national health and welfare department and Statistics Canada. It was developed to obtain better data on the health status and risk exposure of the Canadian population and to complement existing information which comes primarily from vital statistics and medical care records. It is a continuing monitoring, not just of disease and disability, but also of relevant facts on lifestyle, environment and socio-economic factors. Information comes from an interview and questionnaire. Observed information comes from physical measurements and blood tests. The survey began in the spring of 1978. Approximately 12,000 homes (38,000 individuals) are visited each year. One-third of the households participate in the physical measures portion of the survey.

International health services   5.4.5

Through the national health and welfare department, Canada participates in the activities of the Commonwealth ministers of health, the Pan-American Health Organization, the World Health Organization, other United Nations specialized agencies and other intergovernmental organizations whose programs have a substantial health component. Similarly the department takes part in bilateral exchanges with countries that have scientific, technological or cultural arrangements with Canada.

In addition, each year Canadian experts in public health and in the health sciences undertake specific assignments abroad as special advisers or consultants at the request of the World Health Organization, the Pan-American Health Organization or one of the other agencies.

The department enforces regulations governing the handling and shipping of shellfish under an international shellfish agreement between Canada and the United States. Other responsibilities include the custody and distribution of biological, vitamin and hormone standards for the World Health Organization and certain duties in connection with an international convention on narcotic drugs.

Provincial and local health services   5.5

Regulation of health care, operation of health insurance programs and direct provision of some specialized services rest with the provincial governments; some health responsibilities are delegated to local authorities. Although provinces generally assign primary responsibility for health to one department, the distribution of function varies from one province to another. Some provinces have combined health and social services within the same department. Others maintain liaison between departments responsible for these related services.

In a number of provinces, health insurance programs are administered by semi-autonomous boards or commissions, or by a separate department. Some report directly to a minister of health; others are under the jurisdiction of a deputy minister. Several provincial health insurance programs are operated directly by health departments.

In each province both institutional and ambulatory care for tuberculosis and mental illness are provided by an agency of the department responsible for health, with increasing attention to preventive services. Programs related to other particular health problems such as cancer, alcoholism and drug addiction, venereal diseases and dental conditions have been developed by government agencies, often in co-operation with voluntary associations. A number of provincial programs serve specific population groups such as mothers and children, the aged, the needy and those requiring rehabilitation.

Environmental health, involving education, inspection and enforcement of standards, is frequently shared by health departments and other agencies.

Public health or community health units are among the most decentralized. Some are responsible for local health education, school health and organized home care. Although local and regional involvement in health services has been concentrated in hospital planning and some public health aspects, several provinces have inaugurated district and regional boards.

5.5.1   Provincial health insurance plans

Following is a summary of provincial health insurance plans. These cover benefits provided in accordance with the program criteria of the federal Hospital Insurance and Diagnostic Services Act and the Medical Care Act. Additional benefits are provided generally on a limited basis. Some such features of certain plans are: dental care for children, prescribed drugs for the elderly and persons with some particular illnesses, some services of health professionals other than physicians, some sight and hearing aids and rehabilitation services. The federal government is not contributing under federal health insurance legislation toward the costs of these additional benefits. However, it contributes toward the costs of certain health services under the extended health care services program such as nursing home and adult residential care, home care (health aspects) and ambulatory health care services.

This summary gives only the highlights of provincial plans and refers to the programs which were in effect on January 1, 1977. Standard medical and hospital benefits are listed, together with additional benefits. Information on details of the plans and on recent changes in coverage, premiums and authorized charges, if any, may be obtained from the provincial agencies responsible.

Except as otherwise indicated, there were no premiums or authorized charges. The provisions for assistance vary from province to province.

The summary does not include many services which are provided by provincial health departments on a universal basis (such as health unit services, institutional care for tuberculosis and mental patients, venereal disease control, some home care programs), nor does it include details of programs for social service recipients.

Newfoundland. *Medical care benefits:* all medically required services of medical practitioners and certain surgical-dental procedures undertaken by dental surgeons in hospitals. *Additional benefits:* children's dental health program available to children up to age 11. This program is administered by the health department.

Hospital in-patient benefits: standard ward and all approved available services. *Out-patient:* laboratory, radiological, and other diagnostic procedures, including the necessary interpretations; radiotherapy and physiotherapy where available, occupational therapy, where available, out-patient visits, emergency visits, operating room facilities including supplies, plaster casts, drugs and medical and surgical supplies administered in hospital.

Out-of-province benefits: same benefits as provided in the province.

Prince Edward Island. *Medical care benefits:* all medically required services of medical practitioners and certain surgical-dental procedures undertaken by dental surgeons in hospitals.

Hospital in-patient benefits: standard ward and all approved available services. *Out-patient:* laboratory procedures as specified, radiological procedures as specified, including use of radioactive isotopes; drugs, biologicals and related preparations for emergency diagnosis and treatment; all other services specified as in-patient services.

Out-of-province benefits: (In Canada) standard ward rate or rate authorized for out-patient services of host province in case of emergency, or referral with prior approval of commission for conditions that cannot be treated adequately in PEI; otherwise up to $65 per day for in-patient care. (Outside Canada) in-patient services only to maximum of: emergency, up to $100 per day toward costs of hospital room charges and 75% of balance of cost of insured services; referral with prior approval of commission for conditions that cannot be adequately treated in Canada, standard ward rate for hospital and all necessary essential services; otherwise, up to $65 per day for in-patient care.

Nova Scotia. *Medical care benefits:* all medically required services of medical practitioners and certain surgical-dental procedures undertaken by dental surgeons in hospitals. *Additional benefits:* optometric visual analysis; children's dental plan for children born after January 1, 1967; pharmacare plan for residents 65 years and over.

Hospital in-patient benefits: standard ward and all approved available services. *Out-patient:* broad range of essential services as approved by regulation including: medically necessary laboratory, electroencephalographic and radiological examinations, radiotherapy for malignant and non-malignant conditions, electrocardiograms, physiotherapy facilities where available, various drugs, hospital services including meals for day patient care for diabetes, hemodialysis, ultrasonic diagnostic procedures and interpretations, and electrocardiograms and interpretations, as well as specified hospital services when required for emergency diagnoses and treatment within 48 hours of an accident and specified hospital services in connection with various minor medical and surgical procedures.

Out-of-province benefits: (In Canada) in-patient only for emergencies, and with prior approval of commission of a medically necessary referral outside Nova Scotia from a Nova Scotia physician. (Outside Canada) ward rate up to $100 per day plus 75% of the remainder except for infants less than 15 days old, then up to $11 per day.

New Brunswick. *Medical care benefits:* all medically required services of medical practitioners and certain surgical-dental procedures undertaken by dental surgeons in hospitals. *Additional benefits:* prescription drug program for beneficiaries 65 years and over and for those with cystic fibrosis.

Hospital in-patient benefits: standard ward and all approved available services. *Out-patient:* all approved available services.

Out-of-province benefits: (In Canada) complete in-patient coverage at standard ward rate approved by hospital's provincial plan. Out-patient: total amount charged for entitled services at rates approved by hospital's provincial plan. (Outside Canada) in-patient coverage: all-inclusive rate not in excess of the average standard ward rate, rounded to the nearest dollar, of the three largest New Brunswick hospitals. Out-patient coverage: entitled out-patient services at New Brunswick rates.

The out-of-province rates apply only in the case of: emergency; temporary absence from province for education; referral by a New Brunswick physician with prior approval of the health department; special services not being available in New Brunswick; care and treatment received in a Canadian hospital where the medical component has been approved for payment under the Medical Services Payment Act; the treatment required up to the first day of the third month after the month of arrival at a new residence following a permanent move.

Quebec. *Medical care benefits:* all medically required services of medical practitioners and certain surgical-dental procedures undertaken by dental surgeons in hospitals. *Additional benefits:* optometry, oral surgery performed in a university establishment. Drugs and related professional services for recipients of social aid and recipients of certain governmental social aid measures: for persons aged 65 and over in receipt of a monthly guaranteed income supplement in addition to the Old Age Security pension; for persons between 60 and 64 who are eligible for an allowance under the Old Age Security Act and would otherwise, without that allowance, be eligible for social aid or certain governmental social aid measures. Dental services for children under age 10 (and under age 12 from May 1, 1977). Prostheses, orthopedic appliances or other appliances specified by regulation.

The legislation providing for health program financing has set the contribution of an individual at 1.5% of his net income for the year, to a maximum of $235 for salaried employees and $375 for self-employed persons. This individual contribution must not reduce net income to a figure below either $5,600 or $3,700 depending on whether the individual is married or single. The employer's contribution is set at 1.5% of an employee's salary; $\frac{8}{15}$ of the contributions thus collected are remitted to the Quebec Health Insurance Board and the remaining $\frac{7}{15}$ are turned over to the hospital services fund, which is used exclusively for the financing of hospital services in Quebec.

Hospital in-patient benefits: standard ward including all available services. *Out-patient:* clinical services of day or night psychiatric care, electroconvulsive therapy, insulin shock therapy, behavioural therapy, emergency care, minor surgery, radiotherapy, diagnostic services, phys-iotherapy, occupational therapy and inhalation therapy services, orthoptic services, services or examinations necessary for a resident to obtain employment, or those required in the course of employment or on the demand of the employer, provided that such examination or service is required by a law of Quebec other than the Collective Agreement Decrees Act.

Out-of-province benefits: (In Canada) in-patient: approved standard ward rate. Out-patient: insured services at the prevailing rate of the hospital where these services are received. (Outside Canada) in-patient: elective cases up to $25 per day. Emergency or sudden illness and referral cases receiving prior approval, room and board at the ward rate plus cost of other insured services. Excluded: spas, psychiatric and tuberculosis hospitals. Out-patient: at the prevailing rate of the hospital where the insured services are received provided that: the services were received during the 24 hours following an accident; the services became necessary due to a sudden illness or emergency.

Authorized charges: The payments authorized were $6 (and $7 from April 1, 1977) per day in extended care hospitals and in extended care units in short-term care hospitals. Children under age 18 are exempt. Low income individuals may benefit from total or partial exemption depending on their family and financial situation.

Ontario. *Medical care benefits:* all medically required services of medical practitioners and certa.n surgical-dental procedures undertaken by dental surgeons in hospitals. *Additional benefits:* optometry, chiropractic, podiatry, osteopathy. Also out-of-hospital benefit toward cost of physiotherapy and for ambulance services. Home care program services; home renal dialysis and home hyperalimentation equipment, supplies and medication. The provincial ministry of health administers a free drug benefit plan for persons 65 and over who are Canadian citizens or landed immigrants and who have lived in the province for the past 12 months, for disabled persons and persons with limited incomes.

Premium per month: single, $16; family of two or more, $32. The premiums are those for persons who do not qualify for premium assistance on account of limited income. Rates are for combined medical care and hospital insurance coverage. Premium exemption if member of premium unit is 65 or over and resided for at least the previous 12 months in province.

Hospital in-patient benefits: standard ward and all approved available services. *Out-patient:* broad range of essential services, physio-, occupational, speech, radio- and inhalation therapies, diet counselling services when prescribed by a physician, and other hospital services when medically necessary. The plan also provides an extensive nursing home benefit which is not eligible for a contribution to the province under the federal Hospital Insurance and Diagnostic Services Act. However, the province is being reimbursed under the Canada Assistance Plan for revenues lost due to implementation of universal nursing home coverage.

Out-of-province benefits: full rate in other Canadian provinces less any co-insurance or capital charges made by province concerned; 75% of standard ward for non-emergency admissions in the United States including room, board and all extras; 100% of emergencies except mental illness anywhere in the world; 100% of standard ward care in all other cases except mental illness occurring outside Canada or the US.

Manitoba. *Medical care benefits:* all medically required services of medical practitioners and certain surgical-dental procedures undertaken by dental surgeons in hospitals. *Additional benefits:* certain optometric and chiropractic services, prosthetic devices and certain limb and spinal orthotic devices and services when prescribed by a physician; contact lens following congenital cataract surgery; artificial eyes; a prescription drug program; ante-natal Rh immune globulin; a personal care program.

Hospital in-patient benefits: standard ward and all approved available services. *Out-patient:* all services except drugs and dressing in certain cases. The plan also provides an extensive nursing home benefit which is not eligible for a contribution to the province under the federal Hospital Insurance and Diagnostic Services Act. However, the province is being reimbursed under the Canada Assistance Plan for revenues lost due to implementation of universal nursing home coverage.

Out-of-province benefits: (In Canada) rate approved by hospital's provincial plan. (Outside Canada) the greater of 75% of hospital's charges or a daily allowance if: emergency; adequate care not available in Manitoba; during three months following permanent move; temporary employment or education. The lesser of 75% of hospital's charges or a daily allowance for elective cases.

Saskatchewan. *Medical care benefits:* all medically required services of medical practitioners and certain surgical-dental procedures undertaken by dental surgeons in hospitals. *Additional benefits:* optometry, chiropractic, referred services by dentist for care of cleft palate and for orthodontic oral surgery. With certain exceptions, Saskatchewan residents holding valid health services cards are eligible for the benefits of other plans administered by the provincial health department. These include a subsidized hearing aid plan; the provision of prosthetic and orthotic devices; provision of wheelchairs, walkers, commodes and other aids to daily living; a dental plan for children; a prescription drug plan.

Hospital in-patient benefits: standard ward and all approved available services. *Out-patient:* to the extent that a hospital is able to provide them.

Out-of-province benefits: (In-patient) in Canada: standard ward rate less co-insurance charge where applicable. Outside Canada: maximums apply as to rate and number of days of care. (Out-patient) in Canada: total amount charged. Outside Canada: total amount charged or a rate considered to be fair and reasonable.

Alberta. *Medical care benefits:* all medically required services of medical practitioners and certain surgical-dental procedures undertaken by dental surgeons in hospitals. *Additional benefits:* dental services rendered by dental surgeons as specified in regulations, optometric and chiropractic services and podiatric services and appliances. An optional health services contract is available through the commission providing Alberta Blue Cross Plan membership at reduced rates to residents who are not members of a group. For residents 65 and over and their dependents, the government provides a substantial portion of the cost of eyeglasses and a major portion of the cost of dentures and dental care; and assumes the cost of hearing aids and medical and surgical equipment, supplies and appliances.

Premium per month: single, $6.40; family of two or more, $12.80. The premiums are those for persons who do not qualify for premium assistance on account of limited income; premium exemption for basic, and for optional, coverage if member of a premium unit is 65 or over. Eligibility for hospital insurance depends on medical care insurance status.

Hospital in-patient benefits: standard ward and all approved available services. *Out-patient:* 100% of all out-patient procedures rendered by the hospital; 100% of all diagnostic and physiotherapy services rendered in approved facilities outside the hospital; 100% of all out-patient services provided by provincial cancer clinics; dietetic counselling services. The plan also provides an extensive nursing home benefit which is not eligible for a contribution to the province under the federal Hospital Insurance and Diagnostic Services Act. However, the province is being reimbursed under the Canada Assistance Plan for revenues lost due to implementation of universal nursing home coverage.

Out-of-province benefits: (In-patient) 100% of all approved in-patient charges in Canada. Outside Canada, at $50 per day or the actual cost whichever is the lesser, less the authorized charges. For the newborn — $9 per day or the actual cost, whichever is the lesser. (Out-patient) 100% of all services rendered by hospitals in Canada, at their respective approved rates. Outside Canada, for charges less than $25, the actual amount; for charges exceeding $25, the lesser of the amount charged by the hospital or the amount payable in Alberta, but not less than $25.

Authorized charges: adults (excluding residents 65 and over and their dependents) and children (excluding newborn): $5 for the first day in active treatment hospitals. Also excluded are inter-hospital transfers, admissions approved by cancer clinics, polio patients, recipients of social assistance from social services and community health department. Auxiliary hospitals: $5 per day after 120 days.

British Columbia. *Medical care benefits:* all medically required services of medical practitioners and certain surgical-dental procedures undertaken by dental surgeons in hospitals. *Additional benefits:* optometry, chiropractic, naturopathy, physiotherapy, podiatry, orthoptic treatment and services of Red Cross nurses, special nurses and the Victorian Order of Nurses, orthodontic services for harelip and cleft palate. Free prescription drug program for residents 65 and over, and a universal pharmacare plan effective June 1, 1977 which protects individuals from financial hardship as a result of high prescription drug expenses.

Premium per month: single, $7.50; two persons, $15.00; family of three or more, $18.75. The premiums are those for persons who do not qualify for premium assistance on account of limited income.

Hospital in-patient benefits: standard ward and all approved available services. *Out-patient:* emergency services, minor surgical procedures, day care surgical services, out-patient cancer therapy, psychiatric day care and night care services, day care rehabilitation services, narcotic addiction services, physiotherapy services, diabetic day care, and specified out-patient psychiatric services in designated hospitals, dietetic counselling services; cytology services operated by BC Cancer Institute and renal dialysis treatments in designated hospitals.

Out-of-province benefits: (in-patient) during a temporary period of absence that ends at midnight on the last day of the 12th month following the month of departure from province; maximum stay of 12 months unless otherwise approved; referral, if approved by deputy minister.

(Outside Canada) in-patient maximum $75 per day for adults and children; $12 per day for the newborn (Canadian funds).

Authorized charges: $4 per day in general hospitals, excluding newborn; $4 per day for adults and $1 per day for children under age 19 in extended care hospitals; $2 for each emergency or minor surgical out-patient treatment; $2 for day care surgical services; $1 for out-patient cancer therapy, psychiatric day care or night care and psychiatric out-patient services, out-patient physiotherapy services, diabetic day care services, day care rehabilitation services, each dietetic counselling session, renal dialysis treatment.

Northwest Territories. *Medical care benefits:* all medically required services of medical practitioners and certain surgical-dental procedures undertaken by dental surgeons in hospitals.

Hospital in-patient benefits: standard ward and all approved available services. *Out-patient:* emergency and follow-up treatment of injuries, medically necessary diagnostic radiological examinations with necessary interpretations; laboratory examinations; minor surgical procedures; physiotherapy and radiotherapy where available; and certain day care surgical procedures.

Out-of-territory benefits: (in-patient) rate approved for hospital by its own provincial plan. (Out-patient) same benefits as in Northwest Territories. (Outside Canada) up to a maximum specified rate.

Yukon Territory. *Medical care benefits:* all medically required services of medical practitioners and certain surgical-dental procedures undertaken by dental surgeons in hospitals.

Premium per month: single, $4.75; couple, $9.25; family, $11.00. Coverage depends on residency status rather than on payment of premiums. Persons 65 or over are premium-exempt. The premiums are those for persons who do not qualify for premium assistance on account of limited income.

Hospital in-patient benefits: standard ward rate and all approved available services. *Out-patient:* laboratory, radiological and other diagnostic procedures together with the necessary interpretations for the diagnosis and treatment of an injury, illness or disability excluding simple procedures which ordinarily form part of a physician's routine office examinations; day care surgical services.

Out-of-territory benefits: (in-patient) rate approved for hospital by its own provincial plan. (Out-patient) same benefits as in Yukon Territory. (Outside Canada) maximum applied as to rate.

5.5.2 Hospital statistics

Canadian hospitals can be categorized according to type of ownership: public, proprietary or federal; and type of service: general, allied special [extended care (chronic), rehabilitation (convalescent), maternity, communicable diseases, pediatric, orthopedic, neurological, cancer, nursing stations, outpost hospitals], mental or tuberculosis. General hospitals, which account for the largest proportion of beds, are divided into teaching (full and partial teaching) and non-teaching, which are further subdivided into varying bed-size groups based on rated bed capacity.

Tables 5.9 to 5.17 relate to hospital operation and patient movement. Tables 5.1 to 5.4, discussed in section 5.1, present hospital separations by diagnostic categories and primary operations in hospital.

As indicated in Table 5.9, the number and bed capacity of hospitals in Canada have remained relatively stable in recent years. In 1977 there were 1,389 hospitals with a total of 201,413 beds. Tables 5.10 and 5.11 show the distribution of hospitals and patient movement. The greatest concentration of beds is in public general and allied special hospitals. On a national level there were 6.7 beds in public general and allied special hospitals per 1,000 population. This ratio has increased slightly from 6.4 per 1,000 in 1975 and 6.5 per 1,000 in 1976. Although there was a fairly wide range of bed-population ratios from one province to another, there is considerable variation in type and level of care given by hospitals in the same category; in the Yukon Territory and Northwest Territories, federal hospitals provide most of the care that is comparable to that given in public hospitals in the provinces.

Tables 5.10 and 5.11 both reflect the decline in provision and use of beds in mental and tuberculosis hospitals, in contrast to the relative stability in statistics for general and allied special facilities. This decline in emphasis on the large, specialized facilities is matched by increased emphasis on care for mental patients and those with tuberculosis in general hospitals and in community programs. Despite this change in emphasis, the long-term nature of care in mental hospitals results in the accumulation of a large number of patient-days in those facilities. For example, in 1975 patient-days per 1,000 population were 647.4 for mental hospitals as opposed to 1,515.3 for public general hospitals.

There has been much discussion in recent years of the possibility of increasing efficiency of hospital care and limiting costs through reducing the length of stay in hospital. Suggestions for this include expanding home care programs to permit earlier release from hospital, particularly after surgery, and transferring some patients who require long-term care to less expensive rehabilitation and extended care facilities. Variations in length of stay between types of hospital and provinces are in Table 5.12.

The range in cost per patient-day is presented in Table 5.13. In 1975, for reporting public hospitals in Canada, expenditures per patient-day were $44.07 for chronic-extended care hospitals and $85.49 for convalescent-rehabilitation hospitals as opposed to $119.55 for general hospitals. Cost per patient-day was higher for some allied special hospitals, ranging up to $226.37 for pediatric facilities.

Revenue and expenditure for reporting public hospitals are shown per patient-day in Table 5.13 and in total dollar figures in Table 5.14. The labour-intensive nature of hospital care is reflected in the fact that, in 1975, of the $4.13 billion spent by public hospitals in Canada, 70.2% was for gross salaries and wages. Other expenditures were for medical and surgical supplies (3.3%), drugs (2.3%) and supplies and other expenses (24.2%). The increase in the proportion of hospital costs represented by salaries in recent years is due to increases in both hospital personnel and salaries in general and allied special and mental hospitals. Table 5.15 depicts the distribution of the 253,988 people employed full time in hospitals by province and by type of hospital.

Medical care statistics   5.5.3
Total cost of insured services, as shown in Table 5.16, represents all expenditures by provinces for services provided under terms of the Medical Care Act. Although the total cost per fiscal year increased from about $1.28 billion in 1972-73 to about $1.68 billion in 1975-76, the insured population has also grown. Annual percentage increases in the per capita cost of insured services were 7.2, 5.8, 4.8 and 12.9 during the four-year period. These increases may be compared with respective annual percentage increases in gross national product of 11.2, 17.1, 18.0 and 11.4 during the same period.

Table 5.17 depicts the variation in per capita cost, annual increase, and percentage contribution by the federal government for medical care programs of each province from 1972-73 to 1975-76. These figures illustrate the variability in per capita cost as well as patterns of annual change in that cost.

Residential special care facilities   5.5.4
The term special care facilities as used in this section refers to those residential facilities in Canada with four or more residents in which counselling, custodial, supervisory, personal, basic nursing or full nursing care is provided to at least one resident. Excluded are those facilities providing active medical treatment, that is, general and allied special hospitals. These facilities are commonly referred to by a variety of names, such as nursing home, convalescent home, home for the aged, rest home, home for incurables, home for crippled children and receiving home. Rather than using this popular nomenclature, Table 5.18 classifies the facilities by principal characteristics of the predominant group of residents: aged, physically handicapped or disabled, emotionally disturbed children, alcohol and drug addicts, delinquents, unmarried mothers, transients and others. Data are reported separately for Quebec, which uses a somewhat different classification of categories.

The facilities vary considerably in patterns of financing, ownership and management. Some services provided are supported in whole or in part by federal funds; others are funded entirely from provincial sources; others are largely financed privately or by voluntary associations.

The summary of provincial health insurance plans gives some information on coverage of care in certain special care facilities. Much of the care is the responsibility of provincial social service agencies. Further information for each province may be obtained directly from provincial health and social service departments. Federal funds to cover some of the services are provided under terms of the Canada Assistance Plan, vocational rehabilitation for disabled persons, established programs financing, and other federal programs.

5.5.5 Mental health and illness

Among provincially operated health services, mental health activities are one of the largest administrative areas in expenditure and employees. In 1974, mental institutions reported operating expenditures of $595.7 million, while their personnel numbered 52,814; corresponding figures for 1975 were $681.9 million and 51,582.

No adequate measure of mental disorders exists, but in 1977 there were 129,397 admissions to psychiatric in-patient facilities. Separations numbered 131,650, and the year-end census of patients on books totalled 48,238. There has been a gradual decline in all of these indicators in recent years. Table 5.19 contains information on patient movement in the various types of psychiatric facilities. Beyond these hospitals and clinics, however, are many other cases.

In 1977, 234 separate in-patient facilities and 148 psychiatric units in hospitals were caring for the mentally ill; most separate facilities are operated by the provinces. The majority of patients reside in the 42 public mental hospitals. Most mental hospitals have undergone successive additions to their original structures and many have pioneered new treatments for mental illness. Several provinces are arranging for boarding-home care with the federal government sharing the cost of maintaining needy patients in such homes under the Canada Assistance Plan. However, in each province most of the revenue of reporting mental institutions was provided by the provincial government or the provincial insurance plan.

Community mental health facilities are being extended beyond mental institutions to provide greater continuity of care, deal with incipient breakdown, and rehabilitate patients in the community. Psychiatric units in general hospitals contribute by integrating psychiatry with other medical care and making it available to patients in their own community. In 1977 the 148 psychiatric units, which had 4,349 patients as the year closed, admitted 54% of the total admissions to all kinds of mental institutions. In-patient services in psychiatric units are covered under all provincial hospital insurance plans. Some provinces have small regional psychiatric hospitals to facilitate patient access to treatment and the complete integration of medical services. Day-care centres, allowing patients to be in hospital during the day and at home at night, have been organized across the country. Community mental health clinics, some provincially operated, others municipally, and psychiatric out-patient services are open in all provinces.

Specialized rehabilitation services assist former patients to function more adequately and are operated by mental hospitals and community agencies. They include sheltered workshops that pay for work and provide training, and halfway houses in which patients can live and continue to receive treatment while becoming settled in a job.

Facilities for mentally retarded persons include day training schools or classes, summer camps and sheltered workshops as well as residential care in institutions. These facilities provide for social, academic, and vocational training. Manual skills are taught in the training-school workshops and some people are placed in jobs in the community.

Emotionally disturbed children presenting personality or behaviour disorders are treated at hospital units, community clinics, child guidance clinics and other out-patient facilities.

The mental health problems related to heavy alcohol use stem from brain damage due to toxic effects of alcohol and from associated nutritional deficiencies, as well as from related emotional difficulties. Of equal concern is the wide range of physical health problems often leading to death, and social problems resulting from excessive use of alcohol.

In 1975, alcoholic psychosis and alcoholism accounted for 11,626 (19%) of the first admissions to in-patient psychiatric facilities in Canada. Although it is difficult to define alcoholism and to estimate its prevalence, epidemiologists have suggested a strong relationship between overall patterns of alcohol consumption and alcohol-related problems. In Canada, for the population 15 years of age and older, per capita consumption in litres of absolute alcohol increased from 8.55 in 1966-67 to 11.55 in 1975-76, a rise of 35.1%.

Problems related to alcohol use are treated in hospitals, out-patient clinics, hostels, long-term residences or farms, and special facilities for the alcoholic offender. In each province, official and voluntary agencies carry out public education, treatment, rehabilitation and research.

Public health, rehabilitation and home care 5.5.6

Provincial and local structure. Provincial health departments, in co-operation with the regional and local health authorities, administer such services as environmental sanitation, communicable disease control, maternal and child health, school health, nutrition, dental health, occupational health, public health laboratories and vital statistics. Most provinces have delegated certain health responsibilities to health units in rural regions and to municipal health departments in urban centres. Several provinces also provide services directly to their thinly populated northern areas. Certain regulatory and preventive services, including case-findings, screening, diagnosis and referral, health education, personal health care, and supervision in certain areas of treatment services conducted through clinics and home visits, have continued to be the responsibility of local health authorities.

As metropolitan areas and population densities have increased, effective administration has required a broader geographical base. Some smaller local health services are provided or supervised by a regional health unit. A regional structure intermediate between provincial departments and local health units may provide technical advice. Some urban boards of health in metropolitan areas have been amalgamated to increase their effectiveness.

Maternal and child health. All provincial health departments have established maternal and child health consultant services that co-operate with the public health nursing services. The maternal and child health services also undertake studies in maternal and child care, including hospital care, and help train nursing personnel. At the local level, public health nurses provide preventive services to mothers, the newborn and children through clinics, home and hospital visits and school health services.

Nutrition and health education. Provincial health departments and some municipal or regional health departments employ nutrition consultants to extend technical guidance and education to health and welfare agencies, schools, nursing homes, various community service agencies and other institutions and hospitals. They also provide diet counselling to selected patient groups such as diabetics, and conduct nutritional surveys and other research. Most provincial health departments have a division or unit of health education under a full-time professional health educator to promote public knowledge of health needs and measures. These divisions provide educational materials to other divisions of the health department, local health authorities, schools, voluntary associations and the public. Many educational activities are directed to accident prevention and to changing habits harmful to health, such as smoking and the excessive use of alcohol and other drugs. All health workers carry out health education as part of their normal activities.

Dental health. Although public health programs at the provincial level have been largely preventive, increasing emphasis is now being given to dental treatment services. Dental clinics conducted by local health services are generally restricted to pre-school and younger school-age groups. A number of provinces send dental teams to remote areas lacking such services. All provinces have dental care schemes of varying coverage for welfare recipients. Other dental health programs are directed to the training of dentists, dental hygienists, dental nurses, dental therapists and dental assistants, the conducting of dental surveys and extension of water fluoridation.

Communicable disease control. The larger provincial health departments have separate divisions of communicable disease control headed by full-time epidemiologists; in others this function is combined with one or more community health services. Local health authorities organize public clinics for immunization against diphtheria, tetanus, poliomyelitis, whooping cough, smallpox and measles. They also engage in case-finding and diagnostic services in co-operation with public health laboratories and private physicians. Special services for tuberculosis and venereal disease have already been described.

Public health laboratories. All provinces maintain a central public health laboratory and most have branch laboratories to assist local health agencies and the medical profession in the protection of community health and the control of infectious diseases. Public health bacteriology (testing of milk, water and food), diagnostic bacteriology and pathology are the principal functions of the laboratory service, with medical testing for physicians and hospitals steadily increasing.

Rehabilitation and home care. Rehabilitation services are provided by a wide range of public and voluntary agencies. Physical medicine and rehabilitation services are based in several types of institution, including hospitals, separate in-patient facilities, worker compensation board centres, and out-patient centres for children. Financing is from various federal, provincial and voluntary agency sources. Every province includes some institution-based services under hospital and medical care insurance. Two provinces have recently extended this coverage to include the supply and fitting of certain prosthetic and corrective devices. Vocational rehabilitation for the disabled is also a joint federal-provincial activity.

Home care in Canada has developed in a variety of ways. Provincial home-care programs characterize the numerous approaches and organizational structures that exist in Canada today. Some programs are oriented to specific disease categories; some are attached to specific hospitals or community centres, while others are seen as integral parts of comprehensive health care delivery systems. The range of services delivered by the home-care programs varies from nursing services alone to a complete array of health and social services. Some programs concentrate on patients requiring short-term active treatment, while others treat convalescent or chronic patients. Some have as specific objectives the reduction of institutional costs and length of stay, and others aim for continuity of care and provision of co-ordinated health care services to patients for whom home care is the most appropriate level of care.

Most home care programs have two features: centralized control of the services within the program, and co-ordinated services to meet the changing needs of the patient. In some provinces the departments of health play an active role in the financing and administration of home-care programs, while in others local agencies, municipalities and hospitals assume major responsibility for home care.

Special schools or classes for various groups of handicapped children are usually operated by school boards, whereas most schools for the deaf and for the blind are residential schools operated by provincial governments.

5.5.7 Special programs for welfare recipients

All provinces pay all or part of the cost of additional services required by residents in financial need under their social assistance programs. These costs are shared equally with the federal government under the Canada Assistance Plan Act. The range of benefits varies from province to province, but may include such services as eyeglasses, prosthetic appliances, dental services, prescribed drugs, home care services, and nursing home care. Usually, if the benefit is universally available to insured residents under another program, this portion would not be administered under welfare auspices.

For several years federal and provincial governments have been discussing the possibility of replacement of the Canada Assistance Plan with a new social services act. The federal proposal for this new act broadens and re-defines services to be eligible for federal funding. It also recommends a block funding formula as a replacement for the present cost-sharing arrangements.

Health personnel   5.6

In terms of function, numbers and visibility, nurses and physicians may be seen as particularly significant categories of health personnel. However, because of increasing complexity of health care and a growing concern of efficiency in health services, other occupations have multiplied in number, size and importance in recent years. Tables 5.20, 5.21 and 5.22 present selected information on health personnel in Canada. Table 5.20 includes figures for interns and residents and those involved in administration, teaching, and research, as well as those in the clinical practice of medicine.

As of December 1976, there were 40,130 active civilian physicians in Canada. More important than the total number of physicians is the population/physician ratio. There is a greater concentration of the most highly qualified health personnel in urbanized areas. In 1976, the population/physician ratio ranged from 546 residents per physician in Ontario to 1,303 in the Northwest Territories. Nationally, this ratio has improved each year since 1966, reaching a level of 578 persons per active physician in 1976. This improvement has been particularly noteworthy in some of the less heavily populated provinces.

As of December 1976, there were 137,858 registered nurses working in Canada. In viewing the ratio of population to registered nurses, there is a range from 147 residents per RN working in Ontario, Manitoba and Nova Scotia to 234 in the Yukon Territory. These figures indicate that even the sparsely populated areas of Canada have good access to nursing personnel. Data are presented in Table 5.21.

The scope of health occupations in Canada is illustrated by Table 5.22. In addition to physicians and registered nurses, there were 157,948 persons listed in other health occupations in 1976. The importance of a wide range of professional, semi-professional and support occupations is reflected in the number of groups which have formed national associations. Provincial authorities have established registration and regulatory bodies for a number of health occupations.

Government expenditure   5.7

During the six-year period 1969-74, collective federal, provincial and local government expenditure on health more than doubled, expanding from $3,474.0 million to $7,357.5 million. When adjusted for population growth, per capita expenditure on health was almost twice as much in 1974 as in 1969, namely $328 compared with $165. The proportion of all levels of government expenditure on health, net of intergovernment transfers, in relation to total government expenditure, was about the same in 1974 as in 1969, at 12.4%. When only the year-to-year trend is considered, all governments' collective expenditure on health increased by $1,288.1 million between 1973 and 1974, compared with an increase of $591.4 million between 1972 and 1973.

Consolidated provincial-municipal expenditure on health, including outlays financed through federal government transfer payments, experienced a growth comparable to that described above. However, health expenditure is relatively more important in total provincial-municipal expenditure than is the case when all three levels of government are considered; for instance, in 1974 it was 20.3% at the provincial-municipal level, compared with 12.4% for all levels of government as an entity. Table 5.23 gives the relevant statistics.

Sources

5.1 - 5.6 Policy Development and Co-ordination Directorate, Health Programs Branch, Department of National Health and Welfare; Health Division, Institutional and Public Finance Statistics Branch, Statistics Canada.

5.7 Public Finance Division, Institutional and Public Finance Statistics Branch, Statistics Canada.

Tables

5.1 General and allied special hospital separations, days per 100,000 population, and average days of stay, by diagnostic category, 1974[1] and 1975[1]

Year and diagnostic category[2]	Separations	Separations per 100,000 population	Days per 100,000 population	Average days of stay
1974				
Infective and parasitic diseases	109,705	490	4,414	9.0
Neoplasms	217,703	972	15,844	16.3
Endocrine, nutritional, and metabolic diseases	72,620	324	5,534	17.1
Diseases of the blood and blood-forming organs	25,457	114	1,261	11.1
Mental disorders	152,070	679	12,256	18.1
Diseases of the nervous system and sense organs	155,536	694	12,614	18.2
Diseases of the circulatory system	377,599	1,685	39,884	23.7
Diseases of the respiratory system	465,330	2,077	14,599	7.0
Diseases of the digestive system	433,921	1,937	17,912	9.2
Diseases of the genito-urinary system	349,811	1,561	11,170	7.2
Complications of pregnancy, childbirth and the puerperium	496,633	4,430	23,286	5.3
Diseases of the skin and subcutaneous tissue	62,561	279	2,527	9.0
Diseases of the musculoskeletal system and connective tissue	162,111	724	10,278	14.2
Congenital anomalies	42,659	190	2,092	11.0
Symptoms and ill-defined conditions	148,675	664	4,452	6.7
Accidents, poisonings, and violence (nature of injury)	348,359	1,555	17,012	10.9
Supplementary classifications	94,182	420	3,264	7.8
All causes	3,714,932	16,582	186,766	11.3
1975				
Infective and parasitic diseases	103,619	456	3,939	8.6
Neoplasms	220,616	970	15,595	16.1
Endocrine, nutritional, and metabolic diseases	71,850	316	5,221	16.5
Diseases of the blood and blood-forming organs	24,833	109	1,229	11.3
Mental disorders	150,498	662	12,065	18.2
Diseases of the nervous system and sense organs	157,521	692	11,932	17.2
Diseases of the circulatory system	379,311	1,667	38,812	23.3
Diseases of the respiratory system	442,840	1,947	13,884	7.1
Diseases of the digestive system	421,261	1,852	16,715	9.0
Diseases of the genito-urinary system	342,831	1,507	10,620	7.0
Complications of pregnancy, childbirth and the puerperium	503,960	4,424	11,522	5.2
Diseases of the skin and subcutaneous tissue	60,776	267	2,406	9.0
Diseases of the musculoskeletal system and connective tissue	162,936	716	10,011	14.0
Congenital anomalies	42,807	188	2,002	10.6
Symptoms and ill-defined conditions	154,213	678	4,568	6.7
Accidents, poisonings, and violence (nature of injury)	345,100	1,517	16,183	10.7
Supplementary classifications	92,302	406	3,200	7.9
All causes	3,677,274	16,165	179,904	11.1

[1]Excludes newborn and data for the Yukon Territory and Northwest Territories.
[2]Major groupings of the International Classification of Diseases, Adapted — 8th Revision. More detailed information is available in Statistics Canada publication *Hospital morbidity* (Catalogue 82-206) and *Hospital morbidity* — *Canadian diagnostic list* (Catalogue 82-209).

5.2 Hospital separations per 100,000 population by diagnostic category, sex and age group, 1974[1] and 1975[1]

Year and diagnostic category[2]		Under 15	15-24	25-44	45-64	65+	Total
1974							
Infective and parasitic diseases	M	1,107	252	197	226	483	487
	F	987	342	247	273	520	492
	T	1,049	296	222	250	504	490
Neoplasms	M	116	157	269	1,423	4,779	752
	F	121	400	1,259	2,430	3,074	1,191
	T	118	277	757	1,934	3,822	972
Endocrine, nutritional, and metabolic diseases	M	189	93	150	428	954	261
	F	184	188	279	597	1,230	387
	T	187	140	214	514	1,109	324
Diseases of the blood and blood-forming organs	M	191	53	29	64	311	106
	F	153	73	64	94	344	121
	T	172	63	46	79	330	114
Mental disorders	M	108	459	869	1,182	832	633
	F	101	651	1,100	1,100	823	724
	T	105	554	983	1,141	827	679
Diseases of the nervous system and sense organs	M	813	296	410	820	1,824	681
	F	669	323	470	862	1,932	708
	T	743	309	440	841	1,884	694
Diseases of the circulatory system	M	53	148	773	4,091	10,973	1,822
	F	47	170	817	2,550	8,523	1,549
	T	50	159	795	3,309	9,598	1,685
Diseases of the respiratory system	M	4,711	915	743	1,461	4,330	2,280
	F	3,877	1,212	779	1,084	2,210	1,875
	T	4,304	1,061	761	1,270	3,140	2,077

5.2 Hospital separations per 100,000 population by diagnostic category, sex and age group, 1974[1] and 1975[1] (concluded)

Year and diagnostic category[2]		Under 15	15-24	25-44	45-64	65+	Total
1974 (concluded)							
Diseases of the digestive system	M	1,011	1,168	1,775	3,377	4,693	1,958
	F	685	1,649	2,060	2,852	3,606	1,915
	T	852	1,405	1,916	3,111	4,082	1,937
Diseases of the genito-urinary system	M	496	450	767	1,759	3,948	1,049
	F	458	1,720	3,445	3,020	1,605	2,072
	T	478	1,076	2,089	2,399	2,633	1,561
Complications of pregnancy, childbirth and the puerperium	F	34	11,021	8,964	38	—	4,430
Diseases of the skin and subcutaneous tissue	M	295	304	253	254	369	283
	F	265	331	225	266	348	275
	T	280	317	240	260	357	279
Diseases of the musculoskeletal system and connective tissue	M	179	491	863	1,280	1,275	708
	F	172	477	695	1,353	1,729	739
	T	176	484	780	1,317	1,530	724
Congenital anomalies	M	521	116	65	70	55	202
	F	366	166	119	80	49	179
	T	445	140	92	75	52	190
Symptoms and ill-defined conditions	M	591	323	512	915	1,406	638
	F	534	611	688	817	1,024	689
	T	563	465	599	869	1,191	664
Accidents, poisonings, and violence (nature of injury)	M	1,517	2,638	1,711	1,671	2,260	1,870
	F	925	1,152	982	1,203	3,092	1,240
	T	1,228	1,905	1,351	1,433	2,727	1,555
Supplementary classifications	M	58	104	145	198	412	142
	F	59	335	2,063	248	378	698
	T	58	218	1,092	223	393	420
All causes	M	11,955	7,966	9,530	19,217	38,904	13,874
	F	9,636	20,820	24,256	18,867	30,487	19,286
	T	10,823	14,308	16,801	19,039	34,179	16,582
1975							
Infective and parasitic diseases	M	1,048	234	187	223	462	456
	F	912	316	230	258	518	455
	T	982	272	208	241	494	456
Neoplasms	M	119	150	267	1,435	4,823	761
	F	119	393	1,219	2,387	3,043	1,178
	T	119	270	738	1,919	3,821	970
Endocrine, nutritional, and metabolic diseases	M	185	95	144	421	913	254
	F	179	177	270	580	1,199	377
	T	182	135	206	502	1,074	316
Diseases of the blood and blood-forming organs	M	178	52	28	64	303	100
	F	144	72	60	96	344	118
	T	161	62	44	80	326	109
Mental disorders	M	103	438	821	1,128	831	609
	F	94	608	1,086	1,080	824	714
	T	99	522	952	1,103	827	662
Diseases of the nervous system and sense organs	M	811	291	403	820	1,848	677
	F	672	322	462	856	1,960	708
	T	743	306	432	838	1,911	692
Diseases of the circulatory system	M	54	144	744	4,029	10,805	1,802
	F	43	162	767	2,544	8,363	1,533
	T	48	153	757	3,275	9,430	1,667
Diseases of the respiratory system	M	4,409	861	701	1,408	4,358	2,140
	F	3,612	1,140	739	1,039	2,231	1,754
	T	4,020	999	720	1,221	3,161	1,947
Diseases of the digestive system	M	946	1,105	1,676	3,268	4,573	1,879
	F	648	1,532	1,932	2,715	3,486	1,825
	T	801	1,316	1,802	2,987	3,961	1,852
Diseases of the genito-urinary system	M	501	419	690	1,667	3,924	1,006
	F	423	1,609	3,290	2,951	1,614	2,006
	T	458	1,005	1,974	2,320	2,623	1,507
Complications of pregnancy, childbirth and the puerperium	F	35	10,890	8,826 ·	36	—	4,424
Diseases of the skin and subcutaneous tissue	M	282	286	235	248	356	269
	F	250	324	217	256	341	265
	T	266	305	226	252	348	267
Diseases of the musculoskeletal system and connective tissue	M	186	489	836	1,257	1,262	702
	F	170	470	675	1,329	1,707	730
	T	178	479	756	1,293	1,512	716
Congenital anomalies	M	536	119	67	69	56	204
	F	362	158	112	77	53	173
	T	451	138	89	73	55	188
Symptoms and ill-defined conditions	M	585	323	504	971	1,498	651
	F	543	600	691	857	1,075	704
	T	565	459	597	913	1,260	678
Accidents, poisonings, and violence (nature of injury)	M	1,508	2,527	1,637	1,604	2,208	1,814
	F	930	1,117	967	1,166	3,026	1,221
	T	1,226	1,832	1,306	1,381	2,668	1,517
Supplementary classifications	M	60	101	134	205	408	141
	F	59	342	1,915	240	393	670
	T	60	220	1,013	223	399	406
All causes	M	11,501	7,635	9,078	18,816	38,628	13,467
	F	9,197	20,231	23,459	18,466	30,175	18,856
	T	10,376	13,844	16,180	18,638	33,869	16,165

[1]Excludes newborn and data for the Yukon Territory and Northwest Territories.
[2]See footnote 2 to Table 5.1.

5.3 Separated cases and operations in general and allied special hospitals, by age group, 1974[1] and 1975[1]

Year and item		Under 15	15-24	25-44	45-64	65+	Total
1974							
All separated cases							
Cases	*No.*	658,626	614,905	991,291	806,923	643,187	3,714,932
Days in hospital	''	4,126,724	3,778,150	7,451,151	10,283,858	16,201,606	41,841,489
Av. days per case	''	6.3	6.1	7.5	12.7	25.2	11.3
Separated cases undergoing surgery							
Cases (primary operations)	*No.*	258,409	363,292	631,194	417,635	223,587	1,894,117
Days in hospital	''	1,311,743	2,154,871	4,310,079	4,619,382	4,030,160	16,426,235
Av. days per case	''	5.1	5.9	6.8	11.1	18.0	8.7
Rate per 100,000 population							
All separated cases		10,823	14,308	16,801	19,039	34,179	16,582
All operated cases		4,246	8,453	10,698	9,854	11,882	8,455
Days of all separated cases		67,815	87,911	126,286	242,647	860,963	186,766
Days of all operated cases		21,556	50,140	73,050	108,994	214,165	73,321
Population[2]		6,085,300	4,297,700	5,900,200	4,238,200	1,881,800	22,403,200
1975							
All separated cases							
Cases	*No.*	622,653	610,801	984,853	804,377	654,590	3,677,274
Days in hospital	''	3,754,113	3,693,536	7,191,947	9,989,825	16,295,090	40,924,511
Av. days per case	''	6.0	6.0	4.3	12.4	24.9	11.1
Separated cases undergoing surgery							
Cases (primary operations)	*No.*	245,192	364,912	631,685	422,490	233,729	1,898,008
Days in hospital	''	1,230,529	2,115,948	4,182,776	4,561,529	4,253,176	16,343,958
Av. days per case	''	5.0	5.8	6.6	10.8	18.2	8.6
Rate per 100,000 population							
All separated cases		10,376	13,844	19,361	18,638	33,869	16,165
All operated cases		4,086	8,271	12,481	9,789	12,093	8,343
Days of all separated cases		62,561	83,714	82,410	231,471	843,125	179,903
Days of all operated cases		20,506	47,958	82,230	105,694	220,064	71,847
Population[3]		6,000,700	4,412,100	5,086,700	4,315,800	1,932,700	22,748,000

[1]Excludes newborn and data for the Yukon Territory and Northwest Territories.
[2]Estimate of Aug. 1, 1974.
[3]Estimate of Aug. 1, 1975.

5.4 Primary operations in general and allied special hospitals, by age group and by sex, 1974[1] and 1975[1]

Year and operation		Under 15	15-24	25-44	45-64	65+	Total
1974							
Neurosurgery	M	1,534	1,463	3,096	4,476	1,844	12,413
	F	1,042	830	2,526	3,856	1,403	9,657
Ophthalmology	M	6,782	2,334	3,081	6,432	8,685	27,314
	F	5,762	1,809	2,631	6,476	13,286	29,964
Otorhinolaryngology	M	59,141	16,706	16,627	8,549	2,388	103,411
	F	53,814	20,021	14,030	6,766	1,714	96,345
Thyroid, parathyroid and adrenals	M	289	186	395	418	112	1,400
	F	301	556	1,580	1,894	425	4,756
Vascular and cardiac surgery	M	1,765	1,132	4,669	13,458	6,306	27,330
	F	1,518	1,140	9,073	11,558	4,313	27,602
Thoracic surgery	M	451	863	1,191	2,477	1,733	6,715
	F	284	300	708	1,423	992	3,707
Abdominal surgery	M	21,262	13,838	27,407	41,192	22,281	125,980
	F	10,371	18,272	36,675	38,101	19,084	122,503
Proctological surgery	M	673	4,065	10,283	7,431	2,469	24,921
	F	464	4,015	6,956	5,445	2,158	19,038
Urological surgery	M	15,070	5,482	11,284	19,535	28,056	79,427
	F	2,983	1,699	5,552	6,989	4,401	21,624
Breast surgery	M	96	274	221	322	237	1,150
	F	111	2,244	7,833	8,625	3,259	22,072
Gynecological surgery	F	939	33,194	171,300	62,456	9,406	277,295
Obstetrical procedures	F	640	133,719	148,016	573	–	282,948
Orthopaedic surgery	M	16,769	27,992	34,138	26,209	10,154	115,262
	F	12,211	13,650	18,777	25,573	20,012	90,223
Plastic surgery	M	8,108	8,828	8,478	6,811	3,396	35,621
	F	6,387	6,814	9,200	6,828	3,745	32,974
Oral and maxillofacial surgery	M	1,281	2,276	2,245	1,324	519	7,645
	F	964	1,269	1,390	1,092	495	5,210
Dental surgery	M	4,071	6,588	6,484	4,010	1,089	22,242
	F	4,485	8,968	6,927	3,712	932	25,024
Biopsy	M	918	862	2,202	5,095	4,333	13,410
	F	737	1,926	5,795	6,309	3,446	18,213
Diagnostic endoscopy	M	3,262	1,899	6,224	12,452	11,008	34,845
	F	3,569	3,856	8,120	8,931	6,775	31,251
Diagnostic radiography	M	1,941	2,138	6,482	10,351	5,538	26,450
	F	2,181	2,822	6,141	7,619	4,514	23,277
Radiotherapy and related therapies	M	198	202	495	975	646	2,516
	F	153	125	777	2,364	1,240	4,659

5.4   Primary operations in general and allied special hospitals, by age group and by sex, 1974[1] and 1975[1] (concluded)

Year and operation		Under 15	15-24	25-44	45-64	65+	Total
1974 (concluded)							
Physical medicine and rehabilitation	M	1,503	943	1,771	2,845	2,751	9,813
	F	1,194	750	1,571	2,579	3,567	9,661
Other non-surgical procedures	M	1,628	3,539	10,316	14,523	2,844	32,850
	F	1,395	3,642	8,441	9,285	1,981	24,744
Other surgical and non-surgical	M	30	16	53	222	16	337
procedures	F	132	45	33	74	34	318
All operations	M	146,772	101,626	157,142	189,107	116,405	711,052
	F	111,637	261,666	474,052	228,528	107,182	1,183,065
	T	258,409	363,292	631,194	417,635	223,587	1,894,117
1975							
Neurosurgery	M	1,545	1,423	3,250	4,511	1,851	12,580
	F	1,034	863	2,597	4,041	1,534	10,069
Ophthalmology	M	6,528	2,256	3,221	6,662	9,286	27,953
	F	5,764	1,882	2,678	6,620	14,054	30,998
Otorhinolaryngology	M	54,202	16,587	16,812	8,782	2,491	98,874
	F	49,362	19,452	14,035	6,828	1,790	91,467
Thyroid, parathyroid and adrenals	M	305	239	454	473	135	1,606
	F	289	563	1,748	1,859	460	4,919
Vascular and cardiac surgery	M	1,772	1,072	4,800	14,459	6,895	28,998
	F	1,478	1,095	8,980	12,058	4,751	28,362
Thoracic surgery	M	506	836	1,163	2,590	1,805	6,900
	F	274	341	780	1,453	1,073	3,921
Abdominal surgery	M	20,086	13,471	26,541	41,051	22,909	124,058
	F	9,905	17,486	35,238	37,096	19,292	119,017
Proctological surgery	M	602	4,051	10,033	7,402	2,501	24,589
	F	443	4,047	7,003	5,659	2,238	19,390
Urological surgery	M	15,427	5,519	11,299	19,960	29,076	81,281
	F	2,972	1,721	5,720	7,136	4,707	22,256
Breast surgery	M	107	269	215	306	242	1,139
	F	127	2,252	8,413	9,282	3,457	23,531
Gynecological surgery	F	827	32,682	166,626	61,232	9,756	271,123
Obstetrical procedures	F	665	137,697	154,174	500	—	293,036
Orthopaedic surgery	M	16,753	28,558	34,480	26,626	10,512	116,929
	F	12,003	13,858	18,834	25,619	20,744	91,058
Plastic surgery	M	8,004	8,429	8,152	6,727	3,394	34,706
	F	6,271	6,732	9,266	6,629	3,808	32,706
Oral and maxillofacial surgery	M	1,315	2,518	2,365	1,359	516	8,073
	F	914	1,328	1,556	1,226	534	5,558
Dental surgery	M	3,311	6,391	6,290	3,975	1,072	21,039
	F	3,675	8,831	6,681	3,521	1,000	23,708
Biopsy	M	862	817	2,241	5,334	4,624	13,878
	F	663	1,821	5,956	6,916	3,794	19,150
Diagnostic endoscopy	M	3,308	2,045	6,460	12,945	11,171	35,929
	F	3,474	3,958	8,964	9,259	7,208	32,863
Diagnostic radiography	M	1,925	2,153	6,684	11,027	5,824	27,613
	F	2,078	2,906	6,543	8,166	4,816	24,509
Radiotherapy and related therapies	M	179	221	516	1,117	785	2,818
	F	168	157	849	2,450	1,222	4,846
Physical medicine and rehabilitation	M	1,609	937	1,726	2,847	2,921	10,040
	F	1,238	750	1,601	2,792	3,856	10,237
Other non-surgical procedures	M	1,676	3,435	9,266	14,281	3,613	32,271
	F	1,441	3,232	7,436	9,670	1,971	23,750
Other surgical and non-surgical	M	20	17	18	24	18	97
procedures	F	85	14	21	20	23	163
All operations	M	140,042	101,244	155,986	192,458	121,641	711,371
	F	105,150	263,668	475,699	230,032	112,088	1,186,637
	T	245,192	364,912	631,685	422,490	233,729	1,898,008

[1]Excludes newborn and data for the Yukon Territory and Northwest Territories.

5.5   Malignant neoplasms and rate per 100,000 population, 1974 and 1975

Year and province or territory of residence	Number of cases			Rate per 100,000 population		
	New primary sites	Deaths[1]	Hospital morbidity separations	New primary sites	Deaths	Hospital morbidity separations
1974						
Newfoundland	1,224	638	2,228	225.6	117.6	410.7
Prince Edward Island	403	209	635	345.0	179.1	552.7
Nova Scotia	2,204	1,385	6,105	271.0	170.3	750.7
New Brunswick	2,246	955	4,198	339.3	144.3	634.3
Quebec	15,976	9,348	29,461	260.4	152.4	480.3
Ontario	..	12,218	57,500	..	151.0	710.4
Manitoba	3,694	1,698	6,774	365.4	168.0	670.0
Saskatchewan	3,553	1,445	6,365	391.9	159.3	690.0
Alberta	4,885	2,094	10,627	285.0	122.2	620.0
British Columbia	9,875	3,716	16,676	412.3	155.1	696.2
Yukon Territory	29	19	..	149.5	97.9	..
Northwest Territories	80	26	..	213.3	69.3	..
Canada	44,169	33,751	140,569	307.8	150.4	627.5

5.5   Malignant neoplasms and rate per 100,000 population, 1974 and 1975 (concluded)

Year and province or territory of residence	Number of cases			Rate per 100,000 population		
	New primary sites	Deaths[1]	Hospital morbidity separations	New primary sites	Deaths	Hospital morbidity separations
1975						
Newfoundland	1,276	627	2,228	232.2	114.1	410.7
Prince Edward Island	433	185	751	364.2	155.6	631.1
Nova Scotia	2,175	1,388	6,494	264.5	168.8	789.9
New Brunswick	2,157	973	4,150	319.6	144.2	614.8
Quebec	13,617	9,189	30,287	220.1	148.5	489.5
Ontario	..	12,342	58,636	..	150.0	712.8
Manitoba	3,912	1,762	6,915	384.0	173.0	678.9
Saskatchewan	3,299	1,441	6,454	359.3	157.0	692.5
Alberta	5,367	2,211	10,677	303.6	125.1	603.9
British Columbia	10,205	3,838	17,674	415.3	156.2	719.2
Yukon Territory	47	14	..	226.0	67.3	..
Northwest Territories	60	28	..	158.7	74.1	..
Canada	42,548	33,998	144,266	291.9	149.1	634.2

[1]Includes only the deaths where underlying cause was stated to be due to malignant neoplasms.

5.6   Malignant neoplasms by A List Diagnosis and rate per 100,000 population, 1974 and 1975

Year and A List Diagnosis	Cases[1]	Rate per 100,000 population[1]	Deaths	Rate per 100,000 population	Hospital separations[2]	Rate per 100,000 population[2]
1974						
A45 Malignant neoplasm of buccal cavity and pharynx	1,461	10.2	642	2.9	3,344	14.9
A46 Malignant neoplasm of esophagus	354	2.5	610	2.7	1,532	6.8
A47 Malignant neoplasm of stomach	1,600	11.2	2,406	10.7	5,153	23.0
A48 Malignant neoplasm of intestine except rectum	3,314	23.2	3,914	17.4	10,512	46.9
A49 Malignant neoplasm of rectum and rectosigmoid junction	1,652	11.6	1,238	5.5	5,346	23.9
A50 Malignant neoplasm of larynx	487	3.4	297	1.3	1,870	8.3
A51 Malignant neoplasm of trachea, bronchus, and lung	3,977	27.8	6,602	29.4	16,023	71.5
A52 Malignant neoplasm of bone	135	0.9	170	0.8	1,310	5.8
A53 Malignant neoplasm of skin	7,881	55.1	396	1.8	4,142	18.5
A54 Malignant neoplasm of breast	5,185	36.3	3,131	13.9	16,975	75.8
A55 Malignant neoplasm of cervix uteri	901	12.6	552	4.9	4,920	22.0
A56 Other malignant neoplasm of uterus	1,345	18.8	469	4.2	4,380	19.6
A57 Malignant neoplasm of prostate	2,735	38.2	1,806	16.1	9,647	43.1
A58 Malignant neoplasm of other and unspecified sites	7,669	53.7	8,292	36.9	39,157	174.8
A59 Leukemia	1,014	7.1	1,364	6.1	6,820	30.4
A60 Other neoplasms of lymphatic and haematopoietic tissue	1,723	12.1	1,862	8.3	9,438	42.1
Total	41,433	289.8	33,751	150.4	140,569	627.5
1975						
A45 Malignant neoplasm of buccal cavity and pharynx	1,332	9.2	748	3.3	3,506	15.4
A46 Malignant neoplasm of esophagus	327	2.3	630	2.8	1,569	6.9
A47 Malignant neoplasm of stomach	1,674	11.5	2,410	10.6	5,161	22.7
A48 Malignant neoplasm of intestine except rectum	3,348	23.1	3,799	16.7	10,718	47.1
A49 Malignant neoplasm of rectum and rectosigmoid junction	1,837	12.7	1,268	5.6	5,588	24.6
A50 Malignant neoplasm of larynx	447	3.1	302	1.3	1,834	8.1
A51 Malignant neoplasm of trachea, bronchus, and lung	4,353	30.0	6,733	29.5	16,912	74.3
A52 Malignant neoplasm of bone	111	0.8	178	0.8	1,547	6.8
A53 Malignant neoplasm of skin	8,464	58.3	376	1.6	4,203	18.5
A54 Malignant neoplasm of breast	5,067	34.9	3,151	13.8	17,319	76.1
A55 Malignant neoplasm of cervix uteri	822	11.3	477	4.2	4,792	21.1
A56 Other malignant neoplasm of uterus	1,342	18.5	439	3.8	4,407	19.4
A57 Malignant neoplasm of prostate	2,846	39.2	1,896	16.7	9,866	43.4
A58 Malignant neoplasm of other and unspecified sites	7,941	54.7	8,443	37.0	40,382	177.5
A59 Leukemia	917	6.3	1,463	6.4	7,022	30.9
A60 Other neoplasms of lymphatic and haematopoietic tissue	1,613	11.1	1,706	7.5	9,440	41.5
Total	42,441	292.4	34,019	149.2	144,266	634.2

[1]Excludes Ontario, the Yukon Territory and Northwest Territories.
[2]Excludes the Yukon Territory and Northwest Territories.

5.7   Total therapeutic abortions[1] and abortion rate per 100 live births, by province, 1974-76

Province or territory	Number of therapeutic abortions			Rate per 100 live births		
	1974	1975	1976	1974	1975r	1976
Newfoundland	184	176	418	1.8	1.6	3.7
Prince Edward Island	50	77	57	2.6	4.0	3.0
Nova Scotia	1,062	1,017	1,247	8.2	7.8	9.4
New Brunswick	440	379	400	3.8	3.2	3.3
Quebec	4,453	5,579	7,249	5.2	6.0	7.6
Ontario	24,795	24,921	26,768	20.0	19.8	21.4
Manitoba	1,411	1,298	1,393	8.2	7.6	8.1
Saskatchewan	1,176	1,282	1,128	7.8	8.4	7.2
Alberta	4,391	4,333	4,943	14.7	13.7	15.0
British Columbia	10,024	10,076	10,704	28.3	27.8	27.7
Yukon Territory	63	77	79	12.7	18.9	17.6
Northwest Territories	75	95	90	7.2	8.1	8.1
Residence not reported	12	1	2	- -	- -	- -
Canada	48,136	49,311	54,478	13.9	13.8	14.9

[1]In addition 62 abortions were performed on non-residents in 1974, 79 in 1975 and 58 in 1976.

5.8   Summary of renal failure patients in Canada, years ended Dec. 31, 1974 and 1975

Status of patients	Number of cases	
	1974	1975
New patients entering program during year	534	690
On dialysis at December 31	1,234	1,488
Died during year	220	290
Received transplant during year	283	294
Alive with functioning transplant (performed after Dec. 31, 1972)	311	512
Lost to follow-up during year	11	20

5.9   Number and bed capacity of operating public, proprietary and federal hospitals as at Jan. 1, 1975-77

Type	1975		1976		1977	
	Hospitals	Beds	Hospitals	Beds	Hospitals	Beds
General	899	126,419	923	127,884	888	125,922
Allied special	356	30,115	347	32,480	371	39,289
Mental	122	41,821	127	36,652	126	35,969
Tuberculosis	7	542	7	418	4	233
Total	1,384	198,897	1,404	197,434	1,389	201,413

5.10   Number and bed capacity of operating public, proprietary and federal hospitals, by province and type, as at Jan. 1, 1977

Province or territory and category	Type of hospital								
	General			Allied special			Total, general and allied special		
	Hospitals	Beds	Beds per 1,000 population[1]	Hospitals	Beds	Beds per 1,000 population[1]	Hospitals	Beds	Beds per 1,000 population[1]
Newfoundland									
Public	34	2,896	5.1	13	507	0.9	47	3,403	6.0
Proprietary	—	—	—	—	—	—	—	—	—
Federal	—	—	—	—	—	—	—	—	—
Prince Edward Island									
Public	8	658	5.5	1	28	0.2	9	686	5.7
Proprietary	—	—	—	—	—	—	—	—	—
Federal	—	—	—	1	14	0.1	1	14	0.1
Nova Scotia									
Public	42	4,430	5.3	4	540	0.6	46	4,970	5.9
Proprietary	—	—	—	—	—	—	—	—	—
Federal	5	542	0.6	—	—	—	5	542	0.6
New Brunswick									
Public	34	4,319	6.3	1	20	- -	35	4,339	6.3
Proprietary	—	—	—	—	—	—	—	—	—
Federal	1	10	- -	—	—	—	1	10	- -

5.10　Number and bed capacity of operating public, proprietary and federal hospitals, by province and type, as at Jan. 1, 1977 (continued)

Province or territory and category	General			Allied special			Total, general and allied special		
	Hospitals	Beds	Beds per 1,000 population[1]	Hospitals	Beds	Beds per 1,000 population[1]	Hospitals	Beds	Beds per 1,000 population[1]
Quebec									
Public	125	27,083	4.3	66	18,025	2.9	191	45,108	7.2
Proprietary	1	183	--	39	2,438	0.4	40	2,621	0.4
Federal	2	401	0.1	10	1,198	0.2	12	1,599	0.3
Ontario									
Public	191	44,186	5.3	43	7,272	0.9	234	51,458	6.1
Proprietary	2	70	--	40	1,041	0.1	42	1,111	0.1
Federal	4	1,517	0.2	14	133	--	18	1,650	0.2
Manitoba									
Public	77	5,954	5.8	3	434	0.4	80	6,388	6.2
Proprietary	—	—	—	—	—	—	—	—	—
Federal	3	446	0.4	20	105	0.1	23	551	0.5
Saskatchewan									
Public	132	6,811	7.3	7	1,228	1.3	139	8,039	8.6
Proprietary	—	—	—	—	—	—	—	—	—
Federal	2	104	0.1	2	14	--	4	118	0.1
Alberta									
Public	117	10,796	5.7	29	3,364	1.8	146	14,160	7.5
Proprietary	—	—	—	—	—	—	—	—	—
Federal	4	843	0.4	6	41	--	10	884	0.5
British Columbia									
Public	92	14,188	5.7	24	2,643	1.1	116	16,831	6.7
Proprietary	2	16	--	—	—	—	2	16	--
Federal	1	60	--	4	43	--	5	103	--
Yukon Territory									
Proprietary	—	—	—	1	4	0.2	1	4	0.2
Federal	3	143	6.7	3	11	0.5	6	154	7.2
Northwest Territories									
Public	3	163	3.8	—	—	—	3	163	3.8
Proprietary	—	—	—	—	—	—	—	—	—
Federal	3	103	2.4	40	186	4.3	43	289	6.7
Canada									
Public	855	121,484	5.2	191	34,061	1.5	1,046	155,545	6.7
Proprietary	5	269	--	80	3,483	0.1	85	3,752	0.2
Federal	28	4,169	0.2	100	1,745	0.1	128	5,914	0.3

Province or territory and category	Mental			Tuberculosis			Total, all hospitals		
	Hospitals	Beds	Beds per 1,000 population[1]	Hospitals	Beds	Beds per 1,000 population[1]	Hospitals	Beds	Beds per 1,000 population[1]
Newfoundland									
Public	1	450	0.8	—	—	—	48	3,853	6.8
Proprietary	—	—	—	—	—	—	—	—	—
Federal	—	—	—	—	—	—	—	—	—
Prince Edward Island									
Public	2	296	2.5	1	30	0.2	12	1,012	8.4
Proprietary	—	—	—	—	—	—	—	—	—
Federal	—	—	—	—	—	—	1	14	0.1
Nova Scotia									
Public	7	1,656	2.0	—	—	—	53	6,626	7.9
Proprietary	—	—	—	—	—	—	—	—	—
Federal	—	—	—	—	—	—	5	542	0.6
New Brunswick									
Public	3	1,440	2.1	—	—	—	38	5,779	8.4
Proprietary	—	—	—	—	—	—	—	—	—
Federal	—	—	—	—	—	—	1	10	--
Quebec									
Public	19	5,468	0.9	—	—	—	210	50,576	8.0
Proprietary	—	—	—	—	—	—	40	2,621	0.4
Federal	—	—	—	—	—	—	12	1,599	0.3
Ontario									
Public	53	13,931	1.7	—	—	—	287	65,389	7.8
Proprietary	5	413	--	—	—	—	47	1,524	0.2
Federal	—	—	—	—	—	—	18	1,650	0.2
Manitoba									
Public	10	2,143	2.1	—	—	—	90	8,531	8.3
Proprietary	—	—	—	—	—	—	—	—	—
Federal	—	—	—	—	—	—	23	551	0.5
Saskatchewan									
Public	4	1,478	1.6	—	—	—	143	9,517	10.2
Proprietary	—	—	—	—	—	—	—	—	—
Federal	—	—	—	—	—	—	4	118	0.1

5.10   Number and bed capacity of operating public, proprietary and federal hospitals, by province and type, as at Jan. 1, 1977 (concluded)

Province or territory and category	Type of hospital								
	Mental			Tuberculosis			Total, all hospitals		
	Hos-pitals	Beds	Beds per 1,000 popu-lation[1]	Hos-pitals	Beds	Beds per 1,000 popu-lation[1]	Hos-pitals	Beds	Beds per 1,000 popu-lation[1]
Alberta									
Public	10	4,229	2.2	1	55	- -	157	18,444	9.7
Proprietary	—	—	—	—	—	—	—	—	—
Federal	—	—	—	—	—	—	10	884	0.5
British Columbia									
Public	12	4,465	1.8	2	148	0.1	130	21,444	8.6
Proprietary	—	—	—	—	—	—	2	16	- -
Federal	—	—	—	—	—	—	5	103	- -
Yukon Territory									
Proprietary	—	—	—	—	—	—	1	4	0.2
Federal	—	—	—	—	—	—	6	154	7.2
Northwest Territories									
Public	—	—	—	—	—	—	3	163	3.8
Proprietary	—	—	—	—	—	—	—	—	—
Federal	—	—	—	—	—	—	43	289	6.7
Canada									
Public	121	35,556	1.5	4	233	- -	1,171	191,334	8.2
Proprietary	5	413	- -	—	—	—	90	4,165	0.2
Federal	—	—	—	—	—	—	128	5,914	0.3

[1]Based on estimated population as at June 1, 1977.

5.11   Movement of patients[1] and patient-days in reporting public, proprietary and federal hospitals, 1974 and 1975

Type of service and item	1974r	1975
PUBLIC HOSPITALS		
General		
Beds set up at Dec. 31	120,436	121,600
Admissions	3,536,435	3,522,214
Per 1,000 population	157.6	154.5
Patient-days	34,263,829	34,549,666
Per 1,000 population	1,526.5	1,515.3
Av. daily no. of patients	93,874	94,657
Per 1,000 population	4.2	4.2
Percentage occupancy[2]	76.6	75.7
Allied special		
Beds set up at Dec. 31	23,506	25,981
Admissions	170,651	178,849
Per 1,000 population	7.6	7.8
Patient-days	7,494,675	8,295,233
Per 1,000 population	333.9	363.8
Av. daily no. of patients	20,533	22,727
Per 1,000 population	0.9	1.0
Percentage occupancy[2]	86.8	86.5
Mental		
Beds set up at Dec. 31	46,658	43,814
Admissions	55,778	55,702
Per 1,000 population	2.5	2.4
Patient-days	15,547,115	14,761,013
Per 1,000 population	692.6	647.4
Av. daily no. of patients	42,595	40,441
Per 1,000 population	1.9	1.8
Percentage occupancy[2]	90.1	88.9
Tuberculosis		
Beds set up at Dec. 31	415	340
Admissions	1,219	1,269
Per 1,000 population	0.1	0.1
Patient-days	113,410	108,757
Per 1,000 population	5.1	4.8
Av. daily no. of patients	311	298
Per 1,000 population	- -	- -
Percentage occupancy[2]	71.3	64.5
PROPRIETARY HOSPITALS		
General		
Beds set up at Dec. 31	357	346
Admissions	12,856	12,386
Per 1,000 population	0.6	0.5

Type of service and item	1974r	1975
Patient-days	104,706	97,514
Per 1,000 population	4.7	4.3
Av. daily no. of patients	287	267
Per 1,000 population	- -	- -
Percentage occupancy[2]	79.5	77.2
Allied special		
Beds set up at Dec. 31	2,918	3,028
Admissions	12,543	12,565
Per 1,000 population	0.6	0.6
Patient-days	1,159,684	1,185,552
Per 1,000 population	51.7	52.0
Av. daily no. of patients	3,177	3,248
Per 1,000 population	0.1	0.1
Percentage occupancy[2]	92.4	92.7
Mental		
Beds set up at Dec. 31	997	921
Admissions	4,124	2,254
Per 1,000 population	0.2	0.1
Patient-days	351,422	326,339
Per 1,000 population	15.7	14.3
Av. daily no. of patients	963	894
Per 1,000 population	- -	- -
Percentage occupancy[2]	96.1	95.9
FEDERAL HOSPITALS		
General		
Beds set up at Dec. 31	4,053	4,050
Admissions	46,140	47,018
Per 1,000 population	2.1	2.1
Patient-days	1,041,789	1,019,155
Per 1,000 population	46.4	44.7
Av. daily no. of patients	2,854	2,792
Per 1,000 population	0.1	0.1
Percentage occupancy[2]	63.2	62.4
Allied special		
Beds set up at Dec. 31	1,075	1,164
Admissions	3,664	4,600
Per 1,000 population	0.2	0.2
Patient-days	360,301	375,957
Per 1,000 population	16.1	16.5
Av. daily no. of patients	987	1,030
Per 1,000 population	- -	- -
Percentage occupancy[2]	72.1	71.8

[1]Patients refer to adults and children. All ratios are based on population estimates as at June 1 of the year concerned.
[2]Based on rated bed capacity.

5.12   Average length of stay of adults and children in public general and allied special hospitals, by province, 1974 and 1975 (days)

Year and type of hospital	Nfld.	PEI	NS	NB	Que.	Ont.	Man.	Sask.	Alta.	BC	Canada[1]
1974											
General											
Non-teaching with no long-term units											
1 - 24 beds	4.53	6.83	7.23	7.19	7.24	7.97	7.22	6.72	7.97	5.56	6.85
25 - 49 "	6.76	7.11	7.43	7.73	8.00	7.29	8.08	6.88	6.71	6.87	7.11
50 - 99 "	5.50	6.66	8.64	7.70	8.19	7.89	7.51	8.43	7.23	7.24	7.70
100 - 199 "	7.61	8.36	10.47	8.74	7.93	7.72	7.19	8.32	7.93	7.43	8.18
200 + "	8.94	9.19	9.59	9.18	9.57	8.31	9.00	13.17	8.03	7.47	8.74
Non-teaching with long-term units											
1 - 99 beds	14.55	—	—	—	11.31	10.00	8.86	7.20	6.76	9.11	9.89
100 - 199 "	—	7.06	—	—	9.16	9.44	7.65	10.99	—	9.96	9.37
200 + "	—	—	—	22.33	13.19	9.71	14.80	14.63	—	10.71	10.44
Total, non-teaching	7.55	7.83	9.34	8.84	9.47	8.93	8.52	7.85	7.31	9.08	8.80
Teaching, full											
1 - 499 beds	15.33	—	11.06	—	13.05	9.75	—	—	—	—	11.80
500 + "	—	—	13.73	—	12.42	10.23	10.80	12.35	9.77	9.96	10.81
Teaching, partial											
1 - 499 beds	10.57	—	—	—	9.46	15.02	—	10.38	—	8.87	9.72
500 + "	—	—	—	13.91	12.18	9.30	—	12.91	8.44	9.87	11.01
Total, general	9.88	7.83	10.17	9.84	10.47	9.29	9.22	9.29	8.41	9.27	9.50
Pediatric	9.76	—	8.15	—	8.58	8.49	—	—	10.42	5.15	8.52
Rehabilitation (convalescent)	38.52	31.09	24.27	62.40	44.70	33.51	—	—	42.30	59.17	41.75
Extended care (chronic)	272.25	—	—	—	186.65	181.65	178.03	581.26	376.50	533.83	218.19
Other	4.33	—	6.68	—	17.80	9.60	1.75	—	7.21	6.77	9.98
All public general and allied special hospitals	10.20	8.05	10.06	9.89	12.91	10.50	9.97	10.02	11.38	10.09	10.95
1975											
General											
Non-teaching with no long-term units											
1 - 24 beds	4.70	7.11	7.17	8.24	7.32	8.01	7.50	6.74	8.50	5.45	6.81
25 - 49 "	6.02	7.46	7.41	8.16	8.69	7.31	7.61	6.90	6.51	7.00	7.03
50 - 99 "	5.25	6.96	8.64	7.72	8.73	7.58	7.74	7.89	7.15	7.44	7.64
100 - 199 "	7.47	8.11	10.17	8.26	8.00	7.25	7.14	8.18	7.49	7.70	8.00
200 + "	8.44	8.91	9.59	9.08	9.52	8.28	9.63	12.12	7.62	7.51	8.70
Non-teaching with long-term units											
1 - 99 beds	15.07	—	—	—	11.02	9.43	8.75	8.17	11.69	8.49	9.45
100 - 199 "	—	6.59	—	—	11.29	9.14	8.83	12.90	—	10.07	9.64
200 + "	—	—	—	25.10	14.11	9.42	16.16	16.16	—	11.76	10.63
Total, non-teaching	6.80	7.65	9.19	8.87	10.02	8.79	8.85	7.88	7.19	9.82	8.88
Teaching, full											
1 - 499 beds	14.49	—	11.01	—	13.34	9.00	—	—	—	—	11.36
500 + "	—	—	13.81	—	12.07	10.18	11.59	12.23	9.33	10.53	10.75
Teaching, partial											
1 - 499 beds	10.10	—	—	—	9.66	16.20	—	10.41	—	9.37	9.96
500 + "	—	—	—	13.30	11.99	9.38	—	11.67	8.19	12.81	11.03
Total, general	8.83	7.65	10.07	9.79	10.70	9.17	9.69	9.11	8.05	10.07	9.52
Pediatric	9.56	—	8.30	—	8.23	7.35	—	—	9.22	4.91	7.87
Rehabilitation (convalescent)	36.36	29.27	23.48	43.90	46.19	34.87	—	—	46.54	66.64	41.74
Extended care (chronic)	293.98	—	—	—	252.26	182.16	158.59	435.64	460.40	182.84	241.79
Other	4.30	—	6.51	—	16.93	10.26	7.67	—	6.82	6.28	9.63
All public general and allied special hospitals	9.21	7.86	9.96	9.83	14.00	10.33	10.31	9.98	11.73	11.02	11.18

[1]Includes the Yukon Territory and Northwest Territories.

5.13 Patient-day revenue[1] and expenditure ratios of reporting public hospitals, by province and type of hospital, 1974 and 1975 (dollars)

Year, province and type of hospital	Revenue		Expenditure				
	Gross income from in-patient services	Total	Gross salaries and wages[2]	Medical and surgical supplies	Drugs	Supplies and other expenses	Total
1974[r]							
NEWFOUNDLAND							
General	79.58	93.52	62.29	3.25	2.88	29.29	97.71
Allied special							
Pediatric	90.61	104.55	68.91	4.13	2.54	32.90	108.48
Rehabilitation (convalescent)	76.96	89.51	63.89	0.99	0.44	26.48	91.80
Extended care (chronic)	19.03	23.57	19.37	0.14	0.15	5.98	25.65
PRINCE EDWARD ISLAND							
General	55.07	63.89	43.37	2.56	2.13	17.25	65.31
Allied special							
Rehabilitation (convalescent)	48.34	61.21	47.84	0.20	0.51	13.84	62.38
Mental	17.91	17.91	13.97	0.12	0.39	5.74	20.22
Tuberculosis	33.30	34.59	27.18	0.44	0.59	6.38	34.59
NOVA SCOTIA							
General	73.60	89.63	57.39	3.42	2.58	26.32	89.71
Allied special							
Pediatric	115.81	133.71	83.85	4.80	2.26	45.00	135.91
Rehabilitation (convalescent)	47.89	55.77	38.47	0.79	0.68	15.49	55.43
Other	104.19	112.41	73.48	3.38	1.97	32.94	111.78
Mental	35.35	38.12	28.40	0.11	0.49	9.31	38.31
Tuberculosis	57.15	65.80[r]	66.48	0.58	2.46	17.54	87.06
NEW BRUNSWICK							
General	68.38	81.69	54.05	3.23	1.96	23.69	82.93
Allied special							
Rehabilitation (convalescent)	52.38	59.29	43.36	0.68	0.48	16.01	60.53
Mental	27.17	27.44	21.15	0.10	0.34	5.85	27.44
QUEBEC							
General	112.55	120.84	86.75	3.89	2.86	28.91	122.41
Allied special							
Pediatric	164.99	187.02	136.21	5.23	3.63	43.31	188.37
Rehabilitation (convalescent)	55.83	68.14	47.29	1.10	0.52	18.59	67.50
Extended care (chronic)	40.02	41.42	30.11	0.41	0.57	9.85	40.94
Other	142.36	155.34	108.69	5.25	2.70	36.36	153.00
Mental	30.85	32.84	23.45	0.08	0.44	9.02	32.99
ONTARIO							
General	87.39	101.86	71.52	3.26	2.43	25.16	102.37
Allied special							
Pediatric	155.51	207.70	138.54	5.46	8.45	58.45	210.90
Rehabilitation (convalescent)	60.76	69.24	45.65	0.50	0.54	22.69	69.38
Extended care (chronic)	40.43	42.51	30.90	0.44	0.59	10.12	42.05
Other	138.15	197.01	133.00	2.76	6.24	64.15	206.16
Mental	48.76	51.89	40.88	0.10	0.38	10.42	51.78
MANITOBA							
General	80.42	92.13	61.35	3.10	2.53	25.14	92.13
Allied special							
Extended care (chronic)	46.11	49.09	38.61	0.44	1.22	8.82	49.09
Other	174.04	929.30	549.30	9.50	...	267.35	841.04
Mental	31.48	31.90	25.48	0.07	0.57	5.93	32.05
SASKATCHEWAN							
General	60.58	69.23	48.38	2.31	1.90	18.50	71.09
Mental	27.00	28.39	23.49	0.14	0.46	4.30	28.39
Tuberculosis	42.10	43.05	33.00	0.25	0.90	10.09	44.24
ALBERTA							
General	68.44	77.91	56.84	2.64	2.19	20.79	82.46
Allied special							
Pediatric	101.69	135.50	109.92	1.56	1.11	36.69	149.28
Rehabilitation (convalescent)	64.90	82.17	58.10	0.53	0.40	25.26	84.29
Extended care (chronic)	24.18	26.07	18.59	0.23	0.39	9.06	28.27
Other	76.79	160.69	115.55	2.44	7.50	35.87	161.36
Mental	16.00	27.59	20.66	0.09	0.25	6.37	27.37
Tuberculosis	...	...	...	...	...	...	...

5.13 Patient-day revenue[1] and expenditure ratios of reporting public hospitals, by province and type of hospital, 1974 and 1975 (dollars) (continued)

Year, province and type of hospital	Revenue		Expenditure				
	Gross income from in-patient services	Total	Gross salaries and wages[2]	Medical and surgical supplies	Drugs	Supplies and other expenses	Total
1974[r] (concluded)							
BRITISH COLUMBIA							
General	76.07	83.14	64.09	2.78	1.93	16.75	85.55
Allied special							
Pediatric	113.13	166.59	138.92	2.62	2.12	40.83	184.49
Rehabilitation (convalescent)	43.00	48.17	38.95	0.57	0.29	10.15	49.96
Extended care (chronic)	28.29	28.99	27.78	0.26	0.30	8.09	36.43
Other	93.22	157.48	123.23	2.45	4.26	32.08	162.02
Mental[e]	35.53	35.72	28.30	0.37	0.45	6.60	35.72
Tuberculosis	44.26	44.26	52.60	0.73	2.35	17.95	73.63
NORTHWEST TERRITORIES							
General	81.11	91.17	63.62	3.13	2.49	30.00	99.25[r]
CANADA							
General	86.94	98.21	69.68	3.24	2.42	24.34	99.68
Allied special							
Pediatric	145.92	177.72	123.96	4.96	4.82	47.12	180.86
Rehabilitation (convalescent)	54.41	64.49	45.91	0.75	0.46	17.92	65.03
Extended care (chronic)	36.73	38.46	28.35	0.38	0.54	9.65	38.92
Other	123.22	163.16	114.71	3.77	4.31	42.15	164.94
Mental[e]	35.66	38.58	29.52	0.12	0.41	8.56	38.61
Tuberculosis	47.11	50.36[r]	48.72	0.49	1.77	14.36	65.34
1975							
NEWFOUNDLAND							
General	101.89	119.29	80.20	4.20	3.20	36.12	123.72
Allied special							
Pediatric	121.18	136.80	88.54	6.71	3.17	42.51	140.93
Rehabilitation (convalescent)	90.52	105.40	77.59	1.30	0.28	28.69	107.86
Extended care (chronic)	24.33	25.68	25.10	0.25	0.27	4.50	30.11
PRINCE EDWARD ISLAND							
General	66.96	76.36	51.69	3.08	2.45	20.97	78.19
Allied special							
Rehabilitation (convalescent)	62.20	80.12	62.62	0.29	0.67	16.54	80.12
Mental	30.41	30.63	23.28	0.16	0.48	6.71	30.63
Tuberculosis	57.60	59.64	44.67	1.07	0.62	13.28	59.64
NOVA SCOTIA							
General	93.47	112.82	76.20	4.43	3.04	29.08	112.75
Allied special							
Pediatric	140.99	165.56	105.03	6.27	2.52	53.63	167.45
Rehabilitation (convalescent)	57.95	67.12	45.15	0.85	0.71	20.02	66.73
Other	130.86	143.54	96.59	5.06	2.35	37.86	141.86
Mental	43.98	48.16	35.12	0.19	0.59	11.66	47.56
Tuberculosis	90.27	99.24	74.33	1.47	3.10	20.34	99.24
NEW BRUNSWICK							
General	84.55	101.88	66.83	3.93	2.36	30.15	103.28
Allied special							
Rehabilitation (convalescent)	64.53	72.04	51.18	0.99	0.54	20.65	73.36
Mental	33.67	33.93	26.77	0.15	0.33	6.68	33.93
QUEBEC							
General	126.01	133.40	93.02	4.56	3.17	34.56	135.02
Allied special							
Pediatric	206.60	223.37	159.60	6.90	4.87	56.54	227.91
Rehabilitation (convalescent)	63.93	76.94	49.93	0.55	0.53	23.64	74.51
Extended care (chronic)	40.89	41.14	28.98	0.44	0.58	10.88	40.84
Other	160.20	175.74	118.17	5.93	3.23	47.29	174.62
Mental	32.35	34.39	23.92	0.12	0.44	9.71	34.19
ONTARIO							
General	107.35	124.83	88.33	4.12	2.89	29.87	125.21
Allied special							
Pediatric	194.54	249.46	173.89	6.90	8.23	68.88	257.91
Rehabilitation (convalescent)	74.06	84.11	54.26	0.79	0.62	29.31	84.98
Extended care (chronic)	49.94	52.36	38.07	0.56	0.64	12.70	51.96
Other	152.39	214.54	145.71	3.52	8.94	64.63	222.79
Mental	61.57	63.36	49.78	0.12	0.46	12.85	63.21

5.13   Patient-day revenue[1] and expenditure ratios of reporting public hospitals, by province and type of hospital, 1974 and 1975 (dollars) (concluded)

Year, province and type of hospital	Revenue		Expenditure				
	Gross income from in-patient services	Total	Gross salaries and wages[2]	Medical and surgical supplies	Drugs	Supplies and other expenses	Total
1975 (concluded)							
MANITOBA							
General	103.58	118.03	81.86	3.77	3.10	29.30	118.03
Allied special							
Extended care (chronic)	60.14	62.82	48.40	0.45	1.43	12.54	62.82
Other	99.13	332.13	211.01	5.37	3.24	112.51	332.13
Mental	41.29	41.83	33.25	0.09	0.59	7.66	41.59
SASKATCHEWAN							
General	70.77	81.46	57.95	3.03	2.23	22.88	86.09
Allied special							
Extended care (chronic)	35.19	38.76	27.01	0.55	1.03	9.28	37.87
Mental	26.08	27.22	24.99	0.13	0.48	6.03	31.63
Tuberculosis	51.81	52.96	37.00	0.37	0.80	12.74	50.91
ALBERTA							
General	96.94	108.24	80.81	3.50	2.72	25.85	112.88
Allied special							
Pediatric	185.80	242.57	195.50	2.77	1.10	57.37	256.74
Rehabilitation (convalescent)	96.55	116.43	80.85	0.74	0.53	32.17	114.29
Extended care (chronic)	33.29	35.64	26.80	0.31	0.43	11.25	38.79
Other	108.04	224.19	163.23	3.07	8.54	46.28	221.12
Mental	39.17	38.67	30.99	0.10	0.31	7.27	38.67
Tuberculosis	..	..	..	..	..	..	..
BRITISH COLUMBIA							
General	90.69	99.07	77.64	3.23	2.18	19.46	102.51
Allied special							
Pediatric	160.13	229.99	187.27	2.94	2.74	48.55	241.49
Rehabilitation (convalescent)	80.15	96.35	79.02	0.63	0.54	20.80	101.00
Extended care (chronic)	36.10	37.22	33.94	0.48	0.32	9.28	44.02
Other	128.34	210.83	164.47	3.22	6.38	41.19	215.25
Mental[e]	52.84	53.35	44.24	0.62	0.46	8.19	53.51
Tuberculosis	99.15	99.15	72.39	0.54	2.94	23.28	99.15
NORTHWEST TERRITORIES							
General	96.57	107.53	72.94	2.92	2.54	33.63	112.03
CANADA							
General	104.84	117.81	83.89	3.97	2.81	28.93	119.55
Allied special							
Pediatric	185.38	220.18	154.57	6.60	5.62	59.58	226.37
Rehabilitation (convalescent)	72.67	86.08	58.64	0.68	0.56	25.68	85.49
Extended care (chronic)	42.01	43.35	31.78	0.46	0.58	11.26	44.07
Other	144.93	194.57	135.67	4.50	5.76	50.46	196.39
Mental[e]	44.99	46.55	35.96	0.17	0.45	10.01	46.59
Tuberculosis	76.85	80.45	58.82	0.82	2.11	17.95	79.70

[1]Adults and children.
[2]Includes medical staff remuneration.

5.14   Revenue and expenditure of operating public general hospitals, by province, 1974 and 1975

Year and province or territory	Operating hospitals	Total revenue $'000	Expenditure				
			Gross salaries and wages[1] %	Medical and surgical supplies %	Drugs %	Supplies and other expenses %	Total $'000
1974							
Newfoundland	33	69,913	63.7ʳ	3.3	3.0	30.0	73,041
Prince Edward Island	8	12,089	66.4	3.9	3.3	26.4	12,358
Nova Scotia	43	107,916	64.0	3.8	2.9	29.3	108,012
New Brunswick	37	100,488	65.2	3.8ʳ	2.4	28.6	102,022
Quebec	127	954,722	70.9	3.2	2.3	23.6	967,190
Ontario	190	1,266,263	69.9ʳ	3.2	2.4	24.5ʳ	1,272,525
Manitoba	79	148,996	66.6ʳ	3.4	2.7ʳ	27.3	149,002
Saskatchewan	131	130,635	68.0	3.3ʳ	2.7	26.0	134,175
Alberta	117	225,231	68.9ʳ	3.2	2.7	25.2	238,366
British Columbia	92	339,999	75.0	3.2	2.2	19.6	349,643
Yukon Territory	—	—	—	—	—	—	—
Northwest Territories	3	2,834	64.1	3.2	2.5	30.2	3,085
Canada	860	3,359,086	69.9	3.2	2.4	24.4	3,409,419
1975							
Newfoundland	33	88,132	64.8	3.4	2.6	29.2	91,403
Prince Edward Island	8	14,258	66.1	3.9	3.1	26.8	14,598
Nova Scotia	43	133,741	67.6	3.9	2.7	25.8	133,658
New Brunswick	37	122,492	64.7	3.8	2.3	29.2	124,167
Quebec	127	1,060,147	68.9	3.2	2.3	25.6	1,072,748
Ontario	190	1,551,971	70.5	3.3	2.3	23.9	1,556,797
Manitoba	77	195,350	69.4	3.2	2.6	24.8	195,349
Saskatchewan	131	152,440	67.3	3.5	2.6	26.6	161,124
Alberta	116	308,632	71.6	3.1	2.4	22.9	321,873
British Columbia	92	443,166	75.7	3.2	2.1	19.0	458,558
Yukon Territory	—	—	—	—	—	—	—
Northwest Territories	3	3,589	65.1	2.6	2.3	30.0	3,739
Canada	857	4,073,919	70.2	3.3	2.3	24.2	4,134,018

[1]Includes medical staff remuneration.

5.15   Full-time personnel employed in reporting public, proprietary and federal hospitals, by province, 1974 and 1975

Year and province or territory	General[1]		General and allied special[1]		Mental		Tuberculosis	
	Number	Per 100 rated beds	Number	Per 100 rated beds	Number	Per 100 rated beds	Number	Per 100 rated beds
1974ʳ								
Newfoundland	6,383	242.0	7,189	228.7	645	143.3	—	—
Prince Edward Island	1,189	164.9	1,235	164.4	280	94.6	25	166.7
Nova Scotia	9,585	201.6	10,695	202.4	1,565	91.5	295	228.7
New Brunswick	8,272	191.1	8,307	191.1	1,264	92.3	—	—
Quebec	67,253	224.1	83,329	194.4	11,719	72.8	—	—
Ontario	87,041	191.0	98,265	182.7	19,938	125.7	—	—
Manitoba	13,380	198.7	14,103	195.0	2,456	105.6	—	—
Saskatchewan	9,900	140.0	10,494	133.8	1,176	72.5	134	93.1
Alberta	20,122	171.4	23,429	155.0	3,721	84.5	...	...
British Columbia	22,572	157.6	24,763	152.0	4,221	85.8	170	114.9
Yukon Territory	182	122.1	196	120.2	—	—	—	—
Northwest Territories	216	80.9	333	79.9	—	—	—	—
Canada	246,095	191.8	282,338	179.6	46,985	95.8	624	134.2
1975								
Newfoundland	6,863	257.4	7,655	241.3	615	151.9	—	—
Prince Edward Island	1,230	172.0	1,276	171.3	284	96.3	24	160.0
Nova Scotia	10,483	215.0	11,712	215.4	1,581	93.6	61	406.7
New Brunswick	8,264	187.4	8,301	187.4	1,205	98.2	—	—
Quebec	69,491	225.4	87,860	190.4	10,959	77.6	—	—
Ontario	88,062	191.3	99,581	183.8	19,136	122.6	—	—
Manitoba	14,157	211.4	14,980	207.6	2,306	95.0	—	—
Saskatchewan	10,312	143.3	10,900	137.4	1,328	79.0	114	79.2
Alberta	20,979	178.6	24,473	161.7	3,821	89.6	...	...
British Columbia	23,671	160.1	25,972	154.2	4,660	92.9	168	113.5
Yukon Territory	182	122.1	195	119.6	—	—	—	—
Northwest Territories	294	110.9	423	96.8	—	—	—	—
Canada	253,988	194.8	293,328	181.3	45,895	98.2	367	104.6

[1]Includes all medical interns and residents, other instructors, school staff and students of formally organized educational programs. Excludes all other medical staff.

5.16   Medical care, by cost of insured services, growth in GNP, population and per capita cost, 1972-73 to 1975-76

Item	1972-73		1973-74		1974-75		1975-76	
	Amount $	Increase %	Amount $	Increase %	Amount $	Increase %	Amount $	Increase %
Cost of insured services	1,280,363,649	9.3	1,372,941,490	7.2	1,463,045,729	6.6	1,677,766,606	14.7
Growth in GNP[1]	104,669,000,000	11.2	122,582,000,000	17.1	144,616,000,000	18.0	161,132,000,000	11.4
Insured population	21,869,511	1.9	22,172,500	1.4	22,536,395	1.6	22,886,480	1.6
Per capita cost of insured services	58.54	7.2	61.92	5.8	64.92	4.8	73.31	12.9

[1] Data for calendar years 1972-75.

5.17   Medical care, per capita cost of insured services, by province and percentage contribution, 1972-73 to 1975-76

Province or territory	1972-73			1973-74			1974-75			1975-76		
	Per capita cost $	In- crease %	Contri- bution %	Per capita cost $	In- crease %	Contri- bution %	Per capita cost $	In- crease %	Contri- bution %	Per capita cost $	In- crease %	Contri- bution %
Newfoundland	31.99	6.6	91.5	37.98	18.7	81.5	41.36	8.9	78.5	48.49	17.2	75.6
Prince Edward Island	40.67	5.1	72.0	41.68	2.5	74.3	49.44	18.6	65.6	54.18	9.6	67.7
Nova Scotia	46.21	8.9	63.4	48.99	6.0	63.2	56.27	14.9	57.7	68.55	21.8	53.5
New Brunswick	37.60	12.4	77.9	41.00	9.0	75.5	43.80	6.8	74.1	49.69	13.4	73.8
Quebec	55.98	11.2	52.3	61.17	9.3	50.6	66.50	8.7	48.8	75.75	13.9	48.4
Ontario	67.90	7.4	43.1	69.27	2.0	44.7	69.04	0.3[1]	47.0	74.57	8.0	49.2
Manitoba	46.48	14.6[1]	63.0	52.65	13.3	58.8	57.41	9.0	56.5	62.12	8.2	59.0
Saskatchewan	47.32	9.8	61.9	49.61	4.8	62.4	54.71	10.3	59.3	61.53	12.5	59.6
Alberta	55.19	4.7	53.0	54.96	0.8[1]	56.3	61.04	11.1	53.2	68.98	13.0	53.1
British Columbia	63.28	4.8	46.3	69.85	10.4	44.3	71.59	2.5	45.3	89.86	25.5	40.8
Yukon Territory	38.55	—	75.9	57.20	48.4	54.1	68.40	19.6	47.5	73.60	7.6	49.8
Northwest Territories	43.76	48.9	66.9	45.04	2.9	68.7	67.48	49.8	48.1	72.78	7.9	50.4
Canada	58.54	7.2	50.0	61.92	5.8	50.0	64.92	4.8	50.0	73.31	12.9	50.0

[1] Decrease.

5.18   Residential special care facilities and bed capacity, by province, 1975

Group of residents, facilities and bed capacity[1]	Nfld.	PEI	NS	NB	Ont.	Man.	Sask.	Alta.	BC	Yukon Territory	Northwest Territories	Total[2]
Aged												
Facilities	31	11	67	69	471	89	109	90	257	1	1	1,196
Beds	1,730	705	3,546	2,843	44,125	6,165	7,122	7,447	12,041	50	12	85,786
Physically handicapped and/or disabled												
Facilities	1	3	15	11	76	17	6	9	19	—	1	158
Beds	21	176	646	541	5,799	980	429	776	764	—	12	10,144
Blind												
Facilities	2	—	—	1	9	1	2	2	—	—	—	17
Beds	47	—	—	28	423	60	64	70	—	—	—	692
Mentally handicapped and/or disabled												
Facilities	23	1	39	43	350	73	13	42	120	—	1	705
Beds	498	8	1,931	799	8,264	939	246	690	2,209	—	20	15,604
Emotionally disturbed children												
Facilities	1	—	2	2	168	30	3	18	57	—	2	283
Beds	30	—	32	70	2,165	293	37	339	733	—	14	3,713
Alcohol/drug addicts												
Facilities	—	2	2	2	21	8	3	6	21	1	—	66
Beds	—	52	64	52	626	189	83	148	1,094	35	—	2,343
Delinquents												
Facilities	3	1	3	—	35	13	4	15	21	—	2	97
Beds	107	5	236	—	643	267	119	276	261	—	26	1,940
Unmarried mothers												
Facilities	—	—	3	—	13	2	3	1	2	—	—	24
Beds	—	—	61	—	343	40	56	38	43	—	—	581
Transients												
Facilities	—	—	—	1	8	—	—	1	4	—	—	14
Beds	—	—	—	71	347	—	—	50	342	—	—	810
Others												
Facilities	2	2	4	3	50	18	2	12	35	3	7	138
Beds	52	55	88	62	1,239	295	56	234	621	48	116	2,866
Total												
Facilities	63	20	135	132	1,201	251	145	196	536	5	14	2,698
Beds	2,485	1,001	6,604	4,466	63,974	9,228	8,212	10,068	18,108	133	200	124,479

[1] By principal characteristics of the predominant group of residents.
[2] Excludes Quebec, for which the following data is available: 31 transition facilities (1,658 beds); 122 rehabilitation facilities (12,203 beds) consisting of 52 facilities for the socially maladjusted (4,603 beds), 47 facilities for the mentally retarded (5,701 beds), 10 facilities for the physically handicapped (1,239 beds), 13 unclassified facilities (660 beds); and 476 facilities for domiciliary care (33,108 beds). Quebec has a total of 629 facilities (46,969 beds).

5.19   Psychiatric in-patient movement, by type of institution and sex, 1976 and 1977

Year and type of institution	Reporting institutions	Admissions[1]		Separations[2]		Patients on books, Dec. 31[3]	
		Male	Female	Male	Female	Male	Female
1976P							
Public mental hospital	43	19,633	13,002	20,624	13,675	12,488	9,351
Institution for the mentally retarded	91	2,435	1,597	2,880	2,134	10,928	8,022
Public psychiatric unit	137	25,799	38,596	25,679	38,617	1,723	2,229
Federal psychiatric unit	7	898	70	915	73	769	12
Psychiatric hospital	13	6,605	6,534	6,631	6,574	768	718
Aged and senile home	5	134	89	140	80	514	489
Hospital for addicts	19	7,114	1,330	7,185	1,330	305	59
Treatment centre for emotionally disturbed children	60	1,558	904	1,518	949	1,354	628
Epilepsy hospital	2	129	102	119	103	141	61
All institutions	377	64,305	62,224	65,691	63,535	28,990	21,569
1977P							
Public mental hospital	42	18,739	12,202	19,457	13,171	11,829	8,458
Institution for the mentally retarded	88	2,576	1,762	2,876	2,159	10,619	7,614
Public psychiatric unit	148	28,707	40,727	28,590	40,646	2,046	2,303
Federal psychiatric unit	6	721	70	711	72	512	3
Psychiatric hospital	13	6,347	5,846	6,317	5,807	772	714
Aged and senile home	5	128	83	147	106	494	464
Hospital for addicts	20	7,193	1,686	7,199	1,691	303	57
Treatment centre for emotionally disturbed children	58	1,462	915	1,553	924	1,242	600
Epilepsy hospital	2	120	113	115	109	136	72
All institutions	382	65,993	63,404	66,965	64,685	27,953	20,285

[1]Includes first admissions, readmissions and transfers-in.
[2]Includes discharges, deaths and transfers-out.
[3]Includes in addition to patients actually in residence those absent on probationary leave, boarding in approved homes, or otherwise absent from the institutions but not officially separated.

5.20   Physicians and population per physician, 1966-76, and by province, 1975 and 1976

Year and province or territory	Active civilian physicians			
	Including interns and residents[1]		Excluding interns and residents[1]	
	Number	Population per physician[2]	Number	Population per physician[2]
1966	26,528	763	21,615	936
1967	27,544	747	22,472	916
1968	28,209	740	22,969	909
1969	29,659	714	24,430	867
1970	31,166	689	25,657	837
1971	32,942	659	27,439	791
1972	34,508	636	28,606	767
1973	35,923	619	29,944	743
1974	37,297	605	31,108	726
1975	39,104	585	32,561	703
1976	40,130	578	33,754	687
1975				
Newfoundland	732	757	591	937
Prince Edward Island	120	1,000	118	1,017
Nova Scotia	1,388	598	1,086	764
New Brunswick	741	923	683	1,001
Quebec	10,846	574	8,752	711
Ontario	15,121	548	12,478	664
Manitoba	1,732	591	1,428	716
Saskatchewan	1,305	712	1,133	820
Alberta	2,737	659	2,310	781
British Columbia	4,328	573	3,928	632
Yukon Territory	23	913	23	913
Northwest Territories	30	1,267	30	1,267
Canada[3]	39,104	588	32,561	706

5.20   Physicians and population per physician, 1966-76, and by province, 1975 and 1976 (concluded)

Year and province or territory	Active civilian physicians			
	Including interns and residents[1]		Excluding interns and residents[1]	
	Number	Population per physician[2]	Number	Population per physician[2]
1976				
Newfoundland	779	720	634	885
Prince Edward Island	140	857	136	882
Nova Scotia	1,404	594	1,137	734
New Brunswick	773	884	713	958
Quebec	11,262	556	9,152	685
Ontario	15,251	546	12,801	651
Manitoba	1,769	581	1,477	695
Saskatchewan	1,315	708	1,138	818
Alberta	2,911	645	2,400	782
British Columbia	4,470	556	4,110	606
Yukon Territory	22	1,000	22	1,000
Northwest Territories	33	1,303	33	1,303
Canada[3]	40,130	578	33,754	687

[1]Based on data in Canada Health Manpower Inventory, 1976 and 1977; Health and Welfare Canada.
[2]Based on Statistics Canada estimates of the population as of Jan. 1 of the following year.
[3]Includes one physician, province unspecified.

5.21   Nurses registered and working in Canada, and population per nurse, 1966-76, and by province, 1975 and 1976[1]

Year and province or territory	Number	Population per nurse
1966	79,729	254
1967	81,904	251
1968	92,618	226
1969	100,003	212
1970	104,258	206
1971	108,630	200
1972	110,769	198
1973	115,929	192
1974	125,475	180
1975	140,388	163
1976	137,858	168
1975		
Newfoundland	2,648	209
Prince Edward Island	755	159
Nova Scotia	5,179	160
New Brunswick	3,676	186
Quebec	33,031	188
Ontario	58,087	143
Manitoba	5,906	173
Saskatchewan	5,857	159
Alberta	10,903	165
British Columbia	14,034	177
Yukon Territory and Northwest Territories	312	189
Canada	140,388	164
1976		
Newfoundland	2,815	199
Prince Edward Island	813	148
Nova Scotia	5,677	147
New Brunswick	4,272	160
Quebec	28,294	221
Ontario	56,717	147
Manitoba	6,978	147
Saskatchewan	6,303	148
Alberta	10,915	172
British Columbia	14,691	169
Yukon Territory	94	234
Northwest Territories	289	149
Canada	137,858	168

[1]Based on data in Canada Health Manpower Inventory, 1976 and 1977; Health and Welfare Canada.

5.22   Health professionals by selected categories of health manpower, for Canada, 1975[1] and 1976[1]

Category	Year	
	1975	1976
Active audiologists and speech therapists	973	1,048
Biomedical engineers	265	209
Licensed chiropractors	1,558	1,700
Dental assistants	1,519	2,928
Licensed dental hygienists	1,495	1,684
Active dentists	8,738	9,401
Registered dietitians	2,443	2,654
Electroencephalograph technologists	307	331
Health record administrators	2,274	2,055
Health service executives	1,118	1,198
Active laboratory technologists	13,303	13,976
Active nurses	140,388	137,858
Licensed nursing assistants	69,475	71,572
Active occupational therapists	1,178	1,384
Active opticians	1,910	2,303
Active optometrists	1,685	1,764
Active orderlies (1974)	12,266	12,996
Osteopaths	72	67
Licensed pharmacists	13,872	14,887
Active physicians	39,104	40,130
Active physiotherapists	2,802	3,065
Registered podiatrists	143	150
Public health inspectors	869	807
Registered radiological technicians	7,211	7,373
Registered respiratory technologists	1,054	1,421
Veterinarians	3,188	2,975
Total	329,210	335,936

[1]Excludes persons who are not active members of an occupational association nor included in a professional registry.

5.23   Total, per capita and percentage distribution of consolidated government expenditure on health, fiscal years ended nearest to Dec. 31, 1969-74

Year and item	All levels of government[1]	Provincial-local government[2]
Total expenditure *(million dollars)*		
1969r	3,474.0	3,366.1
1970r	4,262.4	4,144.6
1971r	4,886.2	4,752.6
1972	5,478.0	5,326.1
1973	6,069.4	5,902.7
1974[3]	7,357.5	7,177.6
Per capita expenditure *(dollars)*		
1969r	165	160
1970r	200	195
1971r	227	220
1972r	251	244
1973r	275	267
1974[3]	328	320
Percentage of health expenditure to total consolidated government expenditure		
1969r	12.4	19.7
1970r	13.5	21.0
1971r	13.5	20.9
1972	13.4	21.3
1973	12.9	20.8
1974[3]	12.4	20.3

[1]Excludes all intergovernment transactions.
[2]Excludes transactions occurring between provincial and local governments, but not between the latter and the federal government.
[3]More detailed information is available in Statistics Canada publication *Consolidated government finance,* (Catalogue 68-202).

5.24   Reported cases of selected notifiable diseases and rate per 100,000 population, by province, 1975 and 1976

1975 — *Number of cases*

International List No.	Disease	Nfld.	PEI	NS	NB	Que.	Ont.	Man.	Sask.	Alta.	BC	YT	NWT	Canada
009.1	Diarrhoea of the newborn, epidemic	—	1	30	—	78	—	1	—	—	—	—	—	110
032	Diphtheria	5	—	—	1	—	8	10	—	49	22	—	6	103
004	Dysentery, bacillary	39	6	1	16	182	365	162	486	311	184	—	455	2,207
062.1	Encephalitis, western equine	—	—	—	—	—	—	14	—	—	—	—	—	14
	Food poisoning, bacterial	—	3	2	70	95	—	32	—	9	—	1	3	215
005.0	Staphylococcal	—	3	2	70	88	—	32	—	9	—	1	—	205
005.1	Botulism	—	—	—	—	7	—	—	—	—	—	—	3	10
	Hepatitis, infectious (including serum hepatitis)	117	80	25	26	306	981	835	369	627	972	13	190	4,541
070	Hepatitis, infectious	113	80	19	21	216	723	787	353	598	962	10	182	4,064
999.2	Hepatitis, serum	4	—	6	5	90	258	48	16	29	10	3	8	477
055	Measles	37	—	646	46	1,394	3,641	693	989	4,380	1,149	33	135	13,143
	Meningitis, aseptic, due to enteroviruses	5	3	6	1	43	2	34	6	8	16	1	2	127
045.0	Coxsackie virus	—	2	2	—	12	2	3	—	—	—	—	—	21
045.1	ECHO virus	—	—	—	—	4	—	—	—	—	—	—	—	4
045.9	Not specified	5	1	4	1	27	—	31	6	8	16	1	2	102
036	Meningococcal infections	22	2	12	6	39	109	17	13	12	33	—	3	268
056	Rubella (German measles)	46	—	80	11	1,787	3,471	784	577	4,683	476	9	108	12,032
	Salmonella infections, other	64	1	46	20	1,224	1,064	61	166	331	456	3	20	3,456
003.0	With food as vehicle	—	—	22	2	504	—	4	—	82	51	—	—	666
003.9	Without mention of food as vehicle	64	1	24	18	720	1,064	57	166	249	405	3	20	2,790
034	Streptococcal sore throat and scarlet fever	51	4,784	1,964	27	758	1,944	1,874	698	7,682	739	60	1,062	21,643
010-018	Tuberculosis[2]	145	9	100	84	927	866	175	144	226	356	8	49	3,089
	Typhoid and paratyphoid fever	4	—	2	6	50	97	4	3	5	8	—	—	179
001	Typhoid	—	—	2	4	30	81	3	2	4	6	—	—	132
002	Paratyphoid	4	—	—	2	20	16	1	1	1	2	—	—	47
	Venereal diseases	647	49	1,223	474	4,649	18,868	4,364	3,960	7,537[3]	10,270[p]	449	2,268	54,758
098	Gonococcal infections	643	44	1,155	464	3,851	16,551	4,246	3,878	7,427	9,778	447	2,268	50,752
090-097	Syphilis	4	5	68	10	794	2,308	102	82	110	482	2	—	3,967
099.0,099.1,099.2	Other	—	—	—	—	4	9	16	—	—	10	—	—	39
033	Whooping cough	204	65	344	15	424	1,856	57	206	151	49	—	16	3,387

5.24 Reported cases of selected notifiable diseases and rate per 100,000 population, by province, 1975 and 1976 (continued)

1975 (concluded)

Rate per 100,000 population

Year and International List No.	Disease	Nfld.	PEI	NS	NB	Que.	Ont.	Man.	Sask.	Alta.	BC	YT	NWT	Canada
009.1	Diarrhoea of the newborn, epidemic	—	—	3.6	—	1.3	[1]	0.1	—	—	—	—	—	0.8
032	Diphtheria	0.9	—	—	0.1	[1]	0.1	1.0	—	2.8	0.9	—	15.9	0.5
004	Dysentery, bacillary	7.1	5.0	0.1	2.4	2.9	4.4	15.9	52.9	17.6	7.5	—	1,203.7	9.7
062.1	Encephalitis, western equine	—	[1]	—	—	—	—	1.4	—	—	—	—	—	0.1
	Food poisoning, bacterial													
005.0	Staphylococcal	—	2.5	0.2	10.4	1.5	[1]	3.1	—	0.5	—	4.8	7.9	0.9
005.1	Botulism	—	2.5	0.2	10.4	1.4	—	3.1	—	0.5	—	4.8	7.9	1.4
070	Hepatitis, infectious (including serum hepatitis)	21.3	67.3	3.0	3.9	4.9	11.9	82.0	40.2	35.5	39.6	62.5	502.6	19.9
	Hepatitis, infectious	20.6	67.3	2.3	3.1	3.5	8.8	77.3	38.4	33.8	39.2	48.1	481.5	17.8
999.2	Hepatitis, serum	0.7	[1]	0.7	0.7	1.5	3.1	4.7	1.7	1.6	0.4	14.4	21.2	2.1
055	Measles	6.7	—	78.6	6.8	22.5	44.3	68.0	107.7	247.7	46.8	158.7	357.1	57.9
	Meningitis, aseptic, due to enteroviruses	0.9	2.5	0.7	—	0.7	[1]	3.3	0.7	0.5	0.7	4.8	5.3	0.6
045.0	Coxsackie virus	—	1.7	0.2	—	0.2	—	0.3	—	—	—	—	—	0.1
045.1	ECHO virus	—	—	—	—	0.1	—	—	—	—	—	—	—	—
045.9	Not specified	0.9	0.8	0.5	0.1	0.4	—	3.0	0.7	0.5	0.7	4.8	5.3	0.4
036	Meningococcal infections	4.0	1.7	1.5	0.9	0.6	1.3	1.7	1.4	0.7	1.3	4.8	5.3	1.2
056	Rubella (German measles)	8.4	0.8	9.7	1.6	28.9	42.2	77.0	62.8	264.9	19.4	43.3	285.7	53.0
	Salmonella infections, other	11.6	0.8	5.6	3.0	19.8	12.9	6.0	18.1	18.7	18.6	14.4	52.9	15.2
003.0	With food as vehicle	—	0.8	2.7	0.3	8.1	—	0.4	—	4.6	2.1	—	—	4.6
003.9	Without mention of food as vehicle	11.6	—	2.9	2.7	11.6	12.9	5.6	18.1	14.1	16.5	14.4	52.9	12.2
034	Streptococcal sore throat and scarlet fever	9.3	4,023.5	238.5	4.0	12.2	23.6	184.0	76.0	434.5	30.1	288.5	2,809.5	94.9
010-018	Tuberculosis	26.4	7.6	12.2	12.4	15.0	10.5	17.2	15.7	12.8	14.5	38.5	129.6	13.5
	Typhoid and paratyphoid fever	0.7	—	0.2	0.9	0.8	1.2	0.4	0.3	0.3	0.3	—	—	0.8
001	Typhoid	—	—	0.2	0.6	0.5	1.0	0.3	0.2	0.2	0.2	—	—	0.6
002	Paratyphoid	0.7	—	—	0.3	0.3	0.2	0.1	0.1	0.1	0.1	—	—	0.2
098	Venereal diseases	117.7	41.2	148.7	70.2	75.1	229.4	428.4	431.3	426.3	418.0	2,158.7	6,000.0	240.2
	Gonococcal infections	117.0	37.0	140.5	68.8	62.2	201.2	416.8	422.4	420.1	397.9	2,149.0	6,000.0	222.6
090-097	Syphilis	0.7	4.2	8.3	1.5	12.8	28.1	10.0	8.9	6.2	19.6	9.6	—	17.4
099.0,099.1, 099.2	Other	—	[1]	—	—	0.1	0.1	1.6	—	—	0.4	—	—	0.2
033	Whooping cough	37.1	54.7	41.8	2.2	6.9	22.6	5.6	22.4	8.5	2.0	—	42.3	14.9

5.24   Reported cases of selected notifiable diseases and rate per 100,000 population, by province, 1975 and 1976 (continued)

1976

Number of cases

Year and International List No.	Disease	Nfld.	PEI	NS	NB	Que.	Ont.	Man.	Sask.	Alta.	BC	YT	NWT	Canada
009.1	Diarrhoea of the newborn, epidemic	–	4	–	–	3	–	–	–	–	–	–	1	8
032	Diphtheria	1	–	–	–	2	11	9	–	57	11	–	18	109
004	Dysentery, bacillary	62	1	–	4	85	267	193	148	499	120	50	225	1,654
062.1	Encephalitis, western equine	–	–	–	–	–	–	–	–	–	–	–	–	–
005.0	Food poisoning, bacterial	51	27	3	–	122	–	1	1	1	–	–	–	206
005.1	Staphylococcal	50	27	3	–	108	–	1	1	1	–	–	–	191
	Botulism	1	–	–	–	14	–	–	–	–	–	–	–	15
070	Hepatitis, infectious (including serum hepatitis)	238	85	22	10	307	952	735	278	585	757	7	189	4,165
	Hepatitis, infectious	212	85	20	10	189	622	683	262	538	745	6	178	3,550
999.2	Hepatitis, serum	26	–	2	–	118	330	52	16	47	12	1	11	615
055	Measles	479	–	1,041	14	4,095	1,890	381	190	833	181	12	42	9,158
045.0	Meningitis, aseptic, due to enteroviruses	6	2	8	3	156	10	29	9	7	11	1	157	399
045.1	Coxsackie virus	–	1	–	–	27	–	4	–	–	–	–	–	32
045.9	ECHO virus	–	–	–	–	37	–	3	1	1	–	–	–	42
	Not specified	19	1	8	3	92	96	22	8	7	69	–	–	325
036	Meningococcal infections	18	6	12	10	30	92	12	14	13	47	3	4	261
056	Rubella (German measles)	–	–	40	10	908	2,062	161	109	714	87	2	74	4,167
003.0	Salmonella infections, other	75	3	79	29	532	1,376	82	112	312	321	–	8	2,929
003.9	With food as vehicle	–	–	–	3	102	–	3	–	134	–	–	–	269
	Without mention of food as vehicle	75	1	53	26	430	1,376	79	112	178	321	–	8	2,660
034	Streptococcal sore throat and scarlet fever	251	4,296	2,248	3	443	2,015	2,117	336	4,956	698	89	1,060	18,512
010-018	Tuberculosis[2]	114	16	74	63	759	686	182	136	201	318	8	44	2,601
	Typhoid and paratyphoid fever	2	–	7	12	173	49	2	5	2	7	–	–	259
001	Typhoid	2	–	–	7	166	39	2	–	1	3	–	–	220
002	Paratyphoid	–	–	7	5	7	10	–	5	1	4	–	–	39
098	Venereal diseases	793	166	937	302	4,807	20,151	4,899	3,523	8,754*	10,085p	370	1,557	56,344
090-097	Gonococcal infections	781	162	898	291	4,115	17,480	4,728	3,461	8,666	9,757	370	1,553	52,262
	Syphilis	12	4	39	11	691	2,654	66	56	88	327	–	4	3,952
099.0, 099.1, 099.2	Other	–	–	–	–	1	17	105	6	–	1	–	–	130
033	Whooping cough	211	239	140	17	284	1,492	92	102	346	77	1	1	3,002

5.24   Reported cases of selected notifiable diseases and rate per 100,000 population, by province, 1975 and 1976 (concluded)

Year and International List No.	Disease	Nfld.	PEI	NS	NB	Que.	Ont.	Man.	Sask.	Alta.	BC	YT	NWT	Canada
1976 (concluded)		*Rate per 100,000 population*												
009.1	Diarrhoea of the newborn, epidemic	—	—	—	—	—	[1]	—	—	—	—	—	2.4	0.1
032	Diphtheria	0.2	—	—	—	—	0.1	0.9	—	3.2	0.4	—	43.6	0.5
004	Dysentery, bacillary	11.3	0.9	—	0.6	1.4	3.2	19.4	16.9	28.3	4.9	238.1	544.8	7.3
062.1	Encephalitis, western equine	—	[1]	—	—	—	—	—	—	—	—	—	—	—
	Food poisoning, bacterial	9.3	23.7	0.4	—	2.0	—	0.1	0.1	0.1	—	—	—	0.9
005.0	Staphylococcal	9.1	23.7	0.4	—	1.8	[1]	0.1	0.1	0.1	—	—	—	1.3
005.1	Botulism	0.2	—	—	—	0.2	—	—	—	—	—	—	—	0.1
070	Hepatitis, infectious (including serum hepatitis)	43.2	74.5	2.8	1.5	5.0	11.4	73.8	31.7	33.2	30.6	33.3	457.6	18.3
	Hepatitis, infectious	38.5	74.5	2.5	1.5	3.1	7.4	68.6	29.9	30.5	30.1	28.6	431.0	15.6
999.2	Hepatitis, serum	47	[1]	0.3	—	1.9	4.0	5.2	1.8	2.7	0.5	4.7	26.6	2.7
055	Measles	87.0	[1]	130.2	2.1	66.7	22.7	38.3	21.7	47.2	7.3	57.1	101.7	40.4
045.0	Meningitis, aseptic, due to enteroviruses	1.1	1.8	1.0	0.5	2.5	0.1	2.9	1.0	0.4	0.4	4.8	380.1	1.8
	Coxsackie virus	—	0.9	—	—	0.4	—	0.4	—	—	—	—	—	0.1
045.1	ECHO virus	—	—	—	—	0.6	—	0.3	0.1	0.1	—	—	—	0.2
045.9	Not specified	1.1	0.9	1.0	0.5	1.5	0.1	2.2	0.9	0.3	0.4	4.8	380.1	1.4
036	Meningococcal infections	3.5	5.3	1.5	0.8	0.5	1.2	1.2	1.6	0.7	1.9	14.3	9.7	1.1
056	Rubella (German measles)	3.3	[1]	5.0	1.5	14.8	24.7	16.2	12.4	40.5	2.8	9.5	179.2	18.4
	Salmonella infections, other	13.6	2.6	9.9	4.4	8.7	16.5	8.2	12.8	17.7	13.0	—	19.4	12.9
003.0	With food as vehicle	—	0.9	3.3	0.5	1.7	0.3	0.3	—	7.6	—	—	—	1.9
003.9	Without mention of food as vehicle	13.6	1.8	6.6	4.0	7.0	16.5	7.9	12.8	10.1	13.0	—	19.4	11.7
034	Streptococcal sore throat and scarlet fever	45.6	3,765.1	281.2	0.5	7.2	24.2	212.7	38.3	281.1	28.2	423.8	2,566.6	81.3
010-018	Tuberculosis	20.7	14.0	9.2	9.7	12.4	8.2	18.3	15.5	11.4	12.9	38.1	106.5	11.4
	Typhoid and paratyphoid fever	0.4	—	0.9	1.8	2.8	0.6	0.2	0.6	0.1	0.3	—	—	1.1
001	Typhoid	0.4	—	—	1.1	2.7	0.5	0.2	—	0.1	0.1	—	—	1.0
002	Paratyphoid	—	—	0.9	0.8	0.1	0.1	—	0.6	0.1	0.2	—	—	0.2
098	Venereal diseases	144.1	145.5	117.2	46.3	78.3	241.5	492.2	401.5	496.5	407.7	1,761.9	3,770.0	247.4
098	Gonococcal infections	141.9	142.0	112.3	44.6	67.0	209.5	475.0	394.4	491.5	394.5	1,761.9	3,760.3	229.5
090-097	Syphilis	2.2	3.5	4.9	1.7	11.3	31.8	6.6	6.4	5.0	13.2	—	9.7	17.4
099.0,099.1, 099.2	Other	—	[1]	—	—	—	0.2	10.5	0.7	—	—	—	—	0.6
033	Whooping cough	38.3	209.5	17.5	2.6	4.6	17.9	9.2	11.6	19.6	3.1	4.8	2.4	13.2

[1]Not reportable.
[2]Tuberculosis figures are provisional.
[3]Excludes 11 cases of syphilis, type undetermined.
[4]Excludes 22 cases of syphilis, type undetermined.

Sources
5.1 - 5.15 Health Division, Institutions and Agriculture Statistics Field, Statistics Canada.
5.16 - 5.17 Health Programs Branch, Department of National Health and Welfare.
5.18 - 5.19 Health Division, Institutions and Agriculture Statistics Field, Statistics Canada.
5.20 - 5.22 Health Programs Branch, Department of National Health and Welfare.
5.23 Public Finance Division, Institutions and Agriculture Statistics Field, Statistics Canada.
5.24 Health Division, Institutions and Agriculture Statistics Field, Statistics Canada.

Incomes and social security

Chapter 6

Tables

Incomes and social security   Chapter 6

Family incomes   6.1

Income distribution statistics for families and individuals in Canada have been available since the first survey of consumer finances was conducted in 1952. In early years of the survey, the sample was restricted to non-farm families with the sample size ranging between 5,000 and 10,000 families. Because of this limited sample, the amount of reliable data which could be tabulated was restricted. Regional distributions could not be further broken down to give provincial distributions, and different personal or labour force characteristics could not be simultaneously cross-tabulated.

In 1966, coverage was extended to the farm population. Today the only individuals excluded from the survey are residents of the Yukon Territory and Northwest Territories, persons living in institutions, on Indian reserves and in military camps. The survey was carried out every two years from 1966 until 1972 when it became annual. The sample gradually increased to 26,000 family units in 1976 and now fluctuates between a large sample every second year and a small sample (12,000 or so) in intervening years. Provincial distributions are still released only from the larger surveys. However, a much wider variety of tabulations is now published owing to the use of the computer and increased scope of the survey. In addition, special tabulations can usually be provided on request. For a more detailed description of the survey and a wider variety of tabulations than shown here, consult the annual report *Income distribution by size in Canada,* Statistics Canada Catalogue 13-207.

In addition to this main series of reports, an annual series entitled *Income after tax, distribution by size in Canada* (Statistics Canada Catalogue 13-210) became available in 1971 and other reports have also been published on special topics related to the survey of consumer finances (e.g. low income families, earnings and work experience, assets and debts of families, and health and education benefits).

Family and income concepts   6.1.1

Terms such as family, unattached individuals, and income are defined below as used in the annual survey of consumer finances.

Family. A family is defined as a group of individuals sharing a common dwelling unit and related by blood, marriage or adoption. This is often referred to as the economic family and is a broader definition than that employed by most demographic studies and the census, where a family is restricted to a married couple with or without unmarried children or a parent with unmarried children. Under the survey definition all relatives in a household, regardless of the degree of relationship, constitute a family.

Unattached individual. An unattached individual is a person living alone or in a household where he is not related to any other household member. The incomes of unattached individuals are different from those of families, particularly as a large portion of them are young entrants into the labour force or elderly persons living on pensions. Tabulations on unattached individuals are not included here but can be found in *Income distribution by size in Canada,* Statistics Canada Catalogue 13-207.

Income. Survey estimates relate to money income received from all sources before payment of taxes and such deductions as pension contributions and insurance premiums. This income may be composed of: wages and salaries; net income of the self-employed (e.g. partners in unincorporated businesses, professional practitioners and farmers); investment income (e.g. interest, dividends, and rents); transfer payments (e.g. old age pensions, family allowances); and other money income (e.g. retirement pensions, alimony). Thus the concept of income is similar to personal income in the national accounts except that, first, it covers only private households in the 10 provinces and not the non-commercial institutions such as churches and charitable organizations and, second, the survey estimates do not include imputed income such as the value of farm products produced and consumed on the farm. On the other hand, the survey income concept is broader than the income defined for the calculation of income tax since it includes such non-taxable money income as the Guaranteed Income Supplement and pensions to the blind.

6.1.2 Income trends, 1951-75

Tables 6.1 and 6.2 provide an indication of how family incomes changed over a period of years. The sample coverage changed in 1966 to include farm families, but this does not seriously affect the comparability of the data with earlier years. Although the first part of Table 6.1 indicates that the average income (in current dollars) from 1951 to 1975 increased four to five times in all regions, these changes do not reflect the decrease in the purchasing power of the dollar. The second part of Table 6.1 does take this into account and gives the average incomes in constant 1971 dollars. Averages in all regions have still at least doubled in constant dollar terms since 1951.

6.1.3 Major sources of income

The percentage distribution of families by major source of income within quintiles is shown in Table 6.7 for 1951-75. For this type of analysis families are arranged in an ascending order by size of income and divided into five equal groups or quintiles. The characteristics (e.g. major source of income) are then tabulated for each quintile.

Table 6.7 shows that while government transfer payments have replaced wages and salaries as the major source of income for the largest group of families in the lowest quintile, families in the other quintiles are still largely dependent on wages and salaries as their principal income source.

6.1.4 Regional income distributions

Although the average family income for all of Canada was $16,613 in 1975, as Table 6.3 shows the average for the different regions ranged from a low of $13,474 in the Atlantic provinces to a high of $18,047 in Ontario.

6.1.5 Income distributions by family characteristics

Income distributions are influenced by a variety of personal and labour force characteristics of the family and its head. While only three summary classifications of family income are presented here relating to age and sex of head, education of head and combination of income recipients, data on other variables may be found in the annual report *Income distribution by size in Canada,* Statistics Canada Catalogue 13-207.

Incomes by age and sex of family head. Table 6.4 shows that the average income of families headed by males ($17,293) was almost twice that of families headed by females ($9,291) in 1975. For the younger and middle age groups, in fact, the male-headed average was more than twice the female-headed one. However, the average income of female-headed families 65 years and over ($11,611) in 1975 exceeded the male-headed average for that age group ($10,391). While the average income for male-headed families increased with age to peak at 45-54, the average for female-headed families continued to increase into the older age groups (largely because of the more frequent presence of adult children or other working family members in female-headed families).

Incomes by education of family head. Education of the family head is another factor greatly affecting family income. As Table 6.5 shows, the average income of families whose head had a university degree was almost twice that of families whose head had only primary school education.

Incomes by combination of income recipients. The number and combination of family members receiving income obviously affects the family income. In Table 6.6 families are first divided into two groups: (a) husband-wife families and (b) all other families. This latter group includes single-parent families as well as groups of other relatives living together, e.g. brothers and sisters. As expected, the average income of

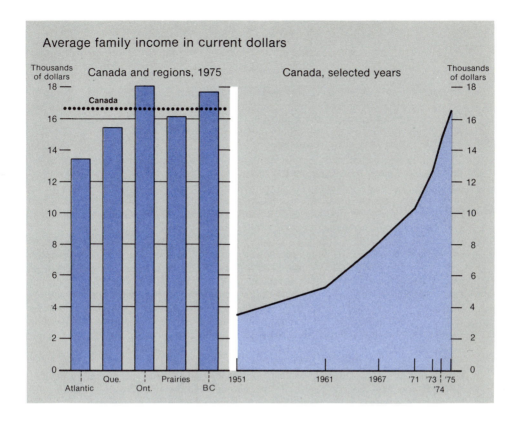

Average family income in current dollars

Thousands of dollars — Canada and regions, 1975 | Canada, selected years | Thousands of dollars

Canada

Atlantic · Que. · Ont. · Prairies · BC | 1951 · 1961 · 1967 · '71 '73 '75 '74

husband-wife families was greater than for other families and the average for both groups increased with the number of income recipients.

Family spending 6.2

Household surveys of family spending provide information on consumer spending that can be related to family characteristics such as geographic location, family size and income level. In general, the survey program has consisted of two phases: the collection, by means of monthly record-keeping surveys throughout the reference year, of detailed information on family food expenditures; and the collection of information by annual recall of all family expenditure, income and changes in assets and liabilities. The record-keeping phase was not featured in all the survey programs.

A primary use of such surveys is to provide information for constructing, reviewing and revising the weights of the Consumer Price Index (see Chapter 21). Initially these small-scale sample expenditure surveys carried out in selected Canadian urban centres since 1953 were designed to follow changes in the patterns of a well-defined group of middle-income urban families known as the "target group" of the Consumer Price Index. In recent years, demand for expenditure statistics to serve other needs of government, business, welfare organizations and academic research has resulted in a widening in the scope and size of the surveys. This culminated in the expansion of the biennial program for 1969 to provide a large-scale national survey for the first time since 1948-49 covering both urban and rural households in the 10 provinces.

The most recent survey program, the twelfth in the series, refers to 1976 and was carried out in two phases: the collection, by means of monthly record-keeping surveys throughout 1976, of detailed information on family purchases of food and non-alcoholic beverages; and the collection of information by annual recall of all money transactions of families and individuals — expenditures, income and changes in assets and liabilities — from a survey carried out in February and March 1977.

For the 1976 record-keeping survey, the usable sample of 3,536 non-boarding spending units was distributed in the cities of St. John's, Nfld., Halifax, Montreal, Ottawa, Toronto, Winnipeg, Edmonton and Vancouver.

The annual recall survey was conducted in the same eight cities. However, data from this survey are not yet available.

6.2.1 Family (spending unit) concept

The definition of a family or spending unit used in the family expenditure surveys is not the same as that of the census, or the economic family concept used in the surveys of consumer finances (see Section 6.1.1). The family or spending unit is defined as a group of persons dependent on a common or pooled income for the major items of expense and living in the same dwelling or one financially independent individual living alone. Never-married sons or daughters living with their parents are considered as part of their parents' spending unit. In the great majority of cases the members of spending units of two or more are related by blood, marriage or adoption, and are thus consistent with the economic family definition employed in surveys of family income, i.e. "a group of individuals sharing a common dwelling unit and related by blood, marriage or adoption". However, it should be noted that according to the economic family definition, unrelated persons living in the same household would be counted as unattached individuals. Under the definitions in the expenditure survey, it is possible for two or more unrelated persons to comprise one family or spending unit.

6.2.2 Family food expenditure patterns

It is useful to classify the expenditure patterns of families by a number of related variables (e.g. family income, size of family, family life cycle, and so on) to determine the influence and effects of these various factors on family spending habits. Such classifications will be available in the Statistics Canada report *Urban family food expenditure, 1976.* This section provides a classification of family food expenditures in 1976 by income quintiles, income being the most influential of all factors bearing on most items of family spending.

Food expenditure patterns by family income quintile, 1976. Table 6.8 shows the food expenditure patterns in 1976 of survey families of two or more persons arranged by income quintiles (families ranked in ascending order of income size and then divided into five equal groups). For example, the average net income before taxes of the 20% of all families comprising the lowest quintile was $7,077 as compared to an average of $34,159 for the 20% of families forming the highest quintile. Of the 2,931 families of two or more persons who provided at least one usable diary, 6.5% refused to give income information. Expenditures for these families are shown separately in Table 6.8. They are also included in the all classes column.

The percentages of total food expenditure on specific food items showed the most significant differences for the five family income quintile groups between food purchased for preparation at home and food in restaurants. The 20% of families in the lowest group spent on the average 13.7% of their total food expenditure in eating places such as restaurants, cafeterias, and drive-ins. The proportion ranged upwards to 33.8% for the 20% of families in the highest group. The offsetting difference was the amount for food purchased for preparation at home, which averaged 84.4% of total food expenditure for families in the lowest group compared with only 63.2% for those in the highest quintile.

The component items of food prepared at home as percentages of total food expenditure reflect a similar downward trend. Therefore, as proportions of total expenditure on food prepared at home, these components do not show any marked differences between income classes.

Successive income classes are not homogeneous with respect to family size; average family size rose from 2.84 persons in the lowest class to 3.55 persons in the highest. Total food expenditure per person, therefore, shows only a 40.4% increase between the lowest and the highest group compared with 83.9% for expenditure per family.

Social security programs   6.3

Federal, provincial and local governments provide a wide range of income security and social service programs to residents, and these publicly funded and administered programs are augmented by voluntary agencies.

The national health and welfare department administers the Canada Pension Plan, the Old Age Security pension program which includes the Guaranteed Income Supplement and Spouse's Allowance, Family Allowances and the Canada Assistance Plan. Other federal agencies in the social security field are the employment and immigration commission, which administers the unemployment insurance program and the employment, training and relocation programs; the veterans affairs department, which administers pensions and allowances, welfare, and treatment services for veterans and, where necessary, for their dependents; and the Indian affairs and northern development department which administers a number of welfare programs for Indian and Inuit people.

The provinces, in some instances assisted by municipalities, administer social assistance and social welfare service programs, the costs of which are sharable by the federal government under the Canada Assistance Plan. Quebec administers the Quebec Pension Plan which is comparable to the Canada Pension Plan, and Quebec and Prince Edward Island provide provincial family allowances which supplement federal family allowances. Several provinces provide income support to elderly persons which supplements that provided under the federal Old Age Security program.

The National Council of Welfare is an advisory body of citizens which provides the minister of national health and welfare with an independent source of advice reflecting the concerns and experience of low-income Canadians and those who work with them. The 21 members include past and present welfare recipients and other low-income citizens, as well as social workers and others involved in the social welfare field.

Types of programs. Income security programs provide cash payments directly to those persons eligible for such benefits. These programs include income insurance schemes such as the Canada and Quebec pension plans, unemployment insurance and worker compensation and income support measures such as the Old Age Security pension, the Guaranteed Income Supplement and the Spouse's Allowance, family allowances, and social assistance provided by the provinces and municipalities.

Social services programs provide some services, such as crisis intervention, information and referral, and family planning, to anyone who applies for them. Other specific services are provided to designated groups, and these include preventive, protective and supportive services to children, supportive services to the elderly, rehabilitative services to the disabled, residential services to those needing care in an institutional setting, and social integration services to persons who are or may become socially isolated from community life. Specific services also include development and preventive services to certain communities.

Income security programs of NHW   6.4

Family Allowances: the federal program   6.4.1

The Family Allowances Act, 1973, which came into effect on January 1, 1974, replaced the former Family Allowances Act of 1944 and the Youth Allowances Act, 1964. Section 6.7 describes Supplementary Family Allowances programs in Quebec and PEI.

Under the act of 1973, family allowances are payable monthly on behalf of a dependent child under 18 years of age who is resident in Canada and is maintained by a Canadian citizen or landed immigrant resident in Canada or a non-immigrant admitted to Canada under specific conditions.

The allowance may be paid when the child or parent is absent from Canada in prescribed circumstances. The allowance is normally paid to the mother of the child. Family allowances are taxable and must be included as income by the person who claims the child as a dependent.

Monthly, non-taxable special allowances are payable on behalf of a child under the age of 18 who is in the care of, and maintained by, a government, a government agency, or an approved private institution. Although normally paid to the institution which has care of the child, special allowances may be paid to the child's foster parent at the request of the institution.

The legislation provides for family allowances and special allowances to be increased at the beginning of each year if the Consumer Price Index for Canada increases. In 1974 the monthly allowances stood at $20, in 1975 at $22.08. In 1976, however, indexing was suspended as a result of the federal government's expenditure restraint program. Consequently, federal rates paid in 1976 were the same as those paid in 1975. In 1977 they were increased to $23.89 and in 1978 to $25.68.

The act permits a provincial legislature to specify family allowance rates to be paid to families resident in the province provided that: the rates are based only on the age of the child, the number of children in the family, or both; no monthly rate is less than 60% of the current federal monthly rate; and the total amount of family allowances paid in the province is, as far as practicable, the same as if the federal rates had been paid. Only Quebec and Alberta have specified their own rates. Provinces do not have the power to specify the special allowance rate.

Family allowance rates paid in Quebec monthly during 1977 and 1978 under the federal program were as follows:

Per family	Federal rate	
	1977	1978
1st child	$14.34	$15.42
2nd child	21.50	23.11
3rd child	35.53	47.45
Each additional child	50.75	59.13

An additional $6.42 is also paid on behalf of each child 12 years and over.

Monthly family allowances paid in Alberta, during 1977 and 1978 under the federal family allowance program were as follows:

Age of child	1977	1978
0-6 years	$18.00	$19.40
7-11 years	22.80	24.60
12-15 years	30.00	32.30
16-17 years	33.50	36.00

6.4.2   Canada Pension Plan

The Canada Pension Plan (CPP) is a compulsory, contributory, earnings-related pension plan that covers most employed members of the labour force between the ages of 18 and 70. The law was proclaimed in force May 5, 1965, collection of contributions started in January 1966, and first benefits, retirement pensions, were paid in January 1967. With the exception of Quebec where the Quebec Pension Plan is in effect, the plan covers all Canada. The Quebec and Canada pension plans have established administrative arrangements to deal with dual contributors so that pension credits under either plan are automatically taken into account for purposes of the other plan when a person moves from an area covered by one plan to that covered by the other (i.e. in or out of Quebec).

Contributions amounting to 1.8% of pensionable earnings falling within the range of $1,000 to $10,400 (1978) are made by employees and matched by the employer. Self-employed persons contribute 3.6% on the same range of earnings. The following types of benefits are payable.

Retirement pensions amount to 25% of a contributor's updated pensionable earnings averaged over the number of years contributions were required. Because the minimum period for averaging earnings is 120 months and applicants for pensions within the first

10 years could not contribute for 120 months, partial pensions were payable during this 10-year transitional period.

Although a retirement pension is payable on application as early as age 65, persons between the ages of 65 and 69 who are employed can postpone their applications and continue to contribute to the plan in order to increase future benefits. Once benefits are being paid, however, contributions cannot be continued.

Disability pensions are paid on application to contributors who, having contributed for at least five whole or part calendar years in the last 10-year period, have been determined to be suffering from a severe and prolonged mental or physical disability. This pension, commencing the fourth month after the month the person became disabled, consists of a fixed monthly amount ($48.19 in 1978) and 75% of the contributor's retirement pension calculated as though the contributor had reached the age of 65 when the disability pension commenced.

Children of persons receiving disability pensions receive benefits at the same monthly rate and under the same conditions of eligibility as those that apply to orphans (see below).

Survivors pensions are paid on application to the surviving spouse and orphans of a person who has contributed to the plan for at least one-third of the calendar years for which he or she could have contributed. The full survivor's benefit is payable to a disabled spouse, a spouse with dependent children, and a spouse 45 years of age or older. A partial survivor's pension is payable to a spouse between the ages of 35 and 45. The full survivor's pension for a spouse under the age of 65 includes a flat rate component ($48.19 in 1978) and 37.5% of the contributor's actual retirement pension or imputed pension if the contributor was not in receipt of a pension at the time of death. When such a spouse reaches the age of 65, and becomes eligible for the Old Age Security pension, the surviving spouse's pension changes to 60% of the deceased contributor's actual or imputed retirement pension.

Orphans benefits are paid on behalf of a deceased contributor's unmarried, dependent children up to the age of 18, or 25 if the orphan continues to attend school or university full time. The rate for each child equals the flat rate component of the survivor's pension, ($48.19 in 1978). An orphan may receive a benefit in respect of only one deceased contributor.

Death benefits. A lump-sum death benefit, equal to six times a contributor's monthly retirement pension, up to a maximum of 10% of that year's maximum pensionable earnings ($1,040 in 1978), is paid to the estate of a deceased contributor who has contributed to the plan for at least one-third of the calendar years for which he or she could have contributed.

Between 1966 and 1973, the annual cost of living increase paid to CPP beneficiaries was limited to 2% a year. Since the beginning of 1974, however, this ceiling has been removed and all benefits are adjusted annually to reflect full living cost increases.

Amendments to the Canada Pension Plan, effective January 1975, provided for: equal treatment for male and female contributors and beneficiaries; removal of the retirement and earnings test for persons aged 65 and over; fixing the rate of increase of the year's maximum pensionable earnings, i.e. it is to be increased each year by 12.5% until it is equal to the average annual wages and salaries of the industrial composite in Canada (for 1977 the maximum was $9,300); changing the basic exemption level of pensionable earnings from 12% to 10%; self-employed members of the labour force who are members of a prescribed religious sect to be exempted from contributions (and benefits) by filing their intentions with the national revenue department; and a series of technical changes designed to improve the administration of the plan, and further elaborating on the rights and procedures of appeal.

Effective January 1978, a number of additional amendments to the Canada Pension Plan were brought into effect. The most significant was a provision allowing CPP pension credits earned by a husband and wife during marriage to be divided equally between them if the marriage ends in divorce or annulment. This ensures that an asset accumulated through the efforts of both spouses while they were married can be shared equally by both when the marriage is dissolved. It also means that spouses who worked in the home and therefore did not contribute directly to the CPP will be protected by the plan for retirement, and may also be entitled to disability and survivorship coverage.

The split of pension credits would cover all years of a marriage after the inception of the CPP in 1966, up to the year the spouses ceased to cohabit. However, the spouses must have lived together for at least three consecutive years during the marriage, and application for the division of pension credits will have to be made within three years of marriage dissolution. The provision applies only to marriages dissolved on or after January 1, 1978.

A number of more technical CPP amendments also came into effect, including a provision which will allow up to 12 months retroactive payment of retirement benefits to late applicants. Previously, retroactive retirement benefits were only paid to applicants over 70 years of age. Another amendment eliminated reductions in payments which previously applied to the CPP benefits provided to orphans and children of disabled contributors in families with more than four children.

Excess funds collected by the plan are lent to a province under a formula based on the ratio of contributions from that province to total contributions. Any funds not borrowed by the provinces are invested in federal securities.

An advisory committee, representing employers, employees, self-employed persons and the public, regularly reviews the operation of the plan, the state of the investment fund and the adequacy of coverage and benefits, and reports to the health and welfare minister. The Canada Pension Plan authorizes reciprocal agreements to be made with other countries to achieve portability of pensions.

6.4.3 Old Age Security coverage

Persons aged 65 and over meeting either of the following residence conditions can receive the Old Age Security (OAS) pension, provided that they are Canadian citizens or legal residents of Canada immediately prior to approval of application. Applicants residing outside of Canada must have been Canadian citizens or legal residents in Canada at the time they ceased to reside in Canada.

The following requirements apply: residence in Canada for at least 10 consecutive years immediately prior to approval of application; gaps in this 10-year period can be offset by previous presence in Canada (since age 18) equal to three times the length of the gaps, as long as the person was resident in Canada for one full year immediately prior to approval of application; or residence in Canada for 40 years since age 18.

A person in receipt of the OAS pension may leave Canada and continue to receive the pension indefinitely if he has resided in Canada for a total of 20 years after age 18; if not, payment is made for only the month of departure plus an additional six months. In the case of a pensioner residing in Canada who is temporarily absent from the country, payment may be resumed when he returns to Canada. Where a pensioner has ceased to reside in Canada, payment may be resumed only when he again takes up residence in Canada.

The Old Age Security Act was amended effective July 1, 1977. Under these amendments, the Old Age Security pension is "earned" at the rate of 1/40th of the full pension for each year of residence in Canada after age 18. This means that residence in Canada for 40 years after age 18 qualifies a person for the full Old Age Security pension and residence for less than 40 years after 18 qualifies a person for a portion of the full pension. The partial pension is, as a minimum, 10/40ths of the full Old Age Security pension based upon a minimum requirement of 10 years residence in Canada after the age of 18. For indefinite payment outside of Canada the requirement of 20 years residence in Canada after age 18 remains.

The amendments to the legislation also protect those who now have the possibility of eventually qualifying for a full pension under the old rules. The new provisions are being implemented gradually over a 40-year period to ensure that everyone in Canada, and persons abroad with prior residence in Canada after age 18, will have the opportunity to earn a full Old Age Security pension. The new residence rules apply to everyone under age 25 on July 1, 1977 and to persons outside Canada on that date without prior residence here. However, persons over 25 on July 1, 1977 with prior residence in Canada after age 18 are able to obtain benefits under either the new or the old rules, whichever is more favourable.

Guaranteed Income Supplement 6.4.4

OAS pensioners with little or no income apart from OAS may, upon application, receive a full or partial supplement. Entitlement is normally based on the pensioner's income in the preceding year, calculated in accordance with the Income Tax Act. The maximum Guaranteed Income Supplement (GIS) is reduced by $1 a month for every $2 a month of other income. In the case of a married couple, each is considered to have one-half of their combined income. The GIS is added to the pensioner's OAS cheque. GIS is payable abroad for only six months following the month of departure from Canada. In the case of a pensioner residing in Canada who is temporarily absent from the country, payment may be resumed when he returns to Canada if the other conditions of eligibility are met. Where a pensioner has ceased to reside in Canada, payment may be resumed only when he again takes up residence in Canada.

Spouse's Allowance 6.4.5

The spouse of an OAS pensioner, aged 60 to 64, meeting the same residence requirements as those stipulated for OAS may be eligible for full or partial Spouse's Allowance (SA). As of January 1978, SA is payable, upon application, if the annual combined income of the couple is less than $7,104. This is subject to an income test which does not include the Old Age Security pension, the Guaranteed Income Supplement or the Spouse's Allowance.

The maximum SA is equal in amount to the OAS pension plus maximum GIS at the married rate. That portion of the Spouse's Allowance which is equivalent to OAS is reduced by $3 a month for every $4 a month of other income, until it reaches the maximum level of GIS. After that, the portion of the Spouse's Allowance which is equivalent to GIS and the GIS for the pensioner are each reduced by $1 for each $4 of other income.

Spouse's Allowance is payable outside of Canada for a period of six months following the month of departure of either the recipient or the pensioner spouse. In the case of a temporary absence from the country, payment may be resumed when the SA recipient or pensioner spouse returns to Canada if the other conditions of eligibility are met. Where the SA recipient or the pensioner spouse has ceased to reside in Canada, payment of the SA may be resumed only when the person again takes up residence in Canada.

Taxation and indexing. The OAS pension is taxable; GIS and SA are not taxable but must be included in computing the net income of a dependent for income tax purposes. OAS, GIS and SA are subject to an increase every January, April, July and October, to reflect increases in the Consumer Price Index. Administration is by Health and Welfare Canada, with funding from the Consolidated Revenue Fund.

Social services programs of NHW 6.5

National welfare grants 6.5.1

The National Welfare Grants Program was established in 1962 to help develop and strengthen welfare services in Canada. Project grants are made to provincial and municipal welfare departments, non-governmental welfare agencies, citizens organizations and universities. Fellowships are provided to individuals seeking advanced training in the social welfare field. The variety of provisions within the program, with its associated consultative services, allows it to operate as a flexible instrument in the development of welfare services and to give major emphasis to experimental activities. The allotment for the year ended March 31, 1978, was $4,468,000.

A wide range of demonstration, research and social development projects are eligible for grants, as are developmental projects related to welfare manpower. Fellowships are available for study at Canadian and foreign universities.

Expenditures under the National Welfare Grants Program for the year ended March 31, 1977 totalled $3,763,000. A sum of $2,017,000 was expended on demonstration projects; $656,000 on research projects; $409,000 on manpower utilization and development, including demonstration, curriculum review in schools of social work and fellowships; $429,000 on general national welfare agency projects; $252,000 on special welfare projects including provincially administered bursary and staff development programs.

6.5.2 New Horizons program

The New Horizons program for retired Canadians was announced by the health and welfare minister in July 1972. In January 1975 the Cabinet agreed to recommend that it be given continuing program status and that yearly contributions in grants be increased by $4 million, bringing the total annual amount for all costs to $14 million.

The program was designed to alleviate the loneliness and sense of isolation which characterize the lives of many older people by offering them opportunity to participate more actively in the life of the community. Grants are made available to groups of retired Canadians, consisting generally of no less than 10 members, to plan and operate projects in which their talent and skills are used for their own betterment, that of other older persons, or of the community. Projects must be non-profit. New Horizons is not an employment program in that participants receive no salary. Projects may be funded for up to 18 months. There is no fixed limit to the amount of a grant.

Projects funded include physical recreation; crafts and hobbies; historical, cultural and educational programs; social services; information services; and activity centres. As of December 7, 1977 a total of 8,443 projects had been awarded $51 million.

6.5.3 Family planning

A family planning division of national health and welfare was formed in January 1972 to provide a centre of responsibility for the federal family planning program. Its objective is to ensure, in co-operation with the provinces and territories, accessibility and availability of family planning services to all Canadians who want them. This is achieved by informing Canadians about the purpose and methods of family planning, promoting training of health and welfare professionals and other staff involved in family planning services, and aiding family planning programs operating under public or voluntary auspices through federal grants-in-aid and joint federal-provincial programs.

The division's major program activities include consultation, information, training and grants. Consultation is provided to a broad range of government and non-government organizations. Information on family planning, sex education and family life education is distributed free. Canadian material on these subjects is currently being developed. To the extent feasible, division consultants assist in training health, welfare and educational staff and others working in the area of family planning.

Since the inception of the family planning grants program in April 1972, a total of $7.6 million has been provided for support of innovative family planning services, demonstration, training and research projects and for university fellowships. Grant recipients have included provincial and municipal government departments, national and local voluntary family planning agencies, native community organizations and universities.

Government spending estimates for 1977-78 projected grants allocation of $2.1 million for a total of $9.7 million.

6.5.4 Emergency Welfare Services

The function of the emergency welfare services division of the health and welfare department is to develop and maintain community capability to provide basic survival and emergency social services in any emergency. The program includes emergency planning for special care facilities.

A program has been developed so that, in a national emergency, the division can co-ordinate the efforts of welfare departments at all levels of government, organizations, private social agencies, professional groups and volunteers to assist in recovery from a given situation and promote rehabilitation.

Adoption Desk   6.5.5

An adoption desk was established in the health department in 1975 to provide a co-ordinating and facilitating service to the provinces in the areas of international and interprovincial adoption. Approximately 120 children were placed interprovincially through the assistance of the desk by the end of 1977. The desk has also co-ordinated a number of international adoption placements.

Day Care Information Centre   6.5.6

This centre was established in the department in 1972 to act as a clearing house for materials on day care and to afford persons and groups involved in day care an opportunity to exchange information and to collaborate in that area. The centre has developed a series of pamphlets, information kits, bibliographies, and a quarterly newsletter and conducts an annual survey of day care services in Canada.

Rehabilitation services   6.5.7

The purpose of the resource centre of the social services programs branch is to compile and disseminate information related to rehabilitation of the physically, mentally or socially handicapped. The centre has published a bibliography on rehabilitation of the handicapped, a layman's guide to some of the literature and a bibliography on social service aspects of rehabilitation.

International welfare   6.5.8

Canada is actively involved in the social development activities of the United Nations, particularly with the executive board of the United Nations International Children's Emergency Fund (UNICEF). It participates in social programs of the Organization for Economic Co-operation and Development. Federal and provincial departments and agencies participate in the work of several international non-government organizations.

Program information is exchanged on social affairs with United Nations agencies, the Council of Europe, the Organization of Economic Co-operation and Development, the Overseas Development Institute and social affairs departments in other countries. Health and Welfare Canada arranges for the training in Canada of fellowship recipients, foreign students and government officials when recommended by their governments.

Canadian officials participate in the International Social Security Association and the social security program of the International Labour Organization. A convention on social security has been concluded with Italy and discussions have been held with other countries including the United Kingdom, the United States and France.

Federal-provincial cost-sharing programs   6.6

Canada Assistance Plan   6.6.1

The Canada Assistance Plan, 1966, was designed in consultation with the provinces as a comprehensive public assistance measure to support the integration and improvement of provincial and municipal assistance programs and encourage development and extension of welfare services which would lessen, remove or prevent the causes and effects of poverty, child neglect or dependence on public assistance. Under agreements with the provinces and territories, the federal government contributes 50% of sharable costs of provincial and municipal expenditures for public assistance and welfare services. Through the plan, Canada also shares in the cost of work activity projects designed to improve the employability of persons who have unusual difficulty in finding or retaining jobs or in undertaking job training.

The only eligibility requirement specified under the Canada Assistance Plan is that of need, *regardless of its cause,* determined through an assessment of budgetary requirements as well as of income and resources of the applicant. Previous residence in the province is not a condition of eligibility for assistance. Rates of assistance and eligibility requirements are set by the province so that they may be adjusted to local conditions and the needs of special groups. The provinces are required to establish procedures which enable applicants and recipients to appeal decisions relating to the provision of assistance.

Assistance includes any form of aid to, or on behalf of, persons in need for the purpose of providing basic requirements such as food, shelter, clothing, fuel, household and personal necessities; special items necessary for the safety, well-being, or rehabilitation of a person in need; non-insured health care services; and maintenance in a home for special care such as a home for the aged, a nursing home or a child care institution.

Welfare services, including counselling and assessment, casework, rehabilitation services, community development and day care, homemaker and adoption services, are provided to persons in need or to persons likely to become in need if they do not receive these services. The federal government shares in administration costs of assistance and welfare services programs, such as salaries and employee benefits, and in the costs of related staff training, research and consulting services. These may be provided by provincial or municipal governments or by provincially approved non-profit agencies.

Federal payments under the Canada Assistance Plan amounted to $1.6 billion in the fiscal year 1976-77. This figure includes payments made to Quebec through the finance department under terms of the Established Programs (Interim Arrangements) Act.

6.6.2   Allowances for the blind and disabled

Under the Blind Persons Act, 1951, and the Disabled Persons Act, 1954, the federal government shares in the cost to the provinces of providing assistance to blind and disabled persons, aged 18 and over, who meet certain income and residence requirements. However, most provinces have now ceased to accept applications under these categorical programs and provide assistance to all persons determined to be in need regardless of cause.

6.6.3   Vocational rehabilitation

Under the provisions of the Vocational Rehabilitation of Disabled Persons (VRDP) Act, the federal government contributes 50% of the costs incurred by a province in providing a program for vocational rehabilitation of physically and mentally disabled persons. A comprehensive program includes such services as medical, social and vocational assessment, counselling, restoration and placement services, the provision of prostheses, training, maintenance allowances and tools, books and other equipment. These services are provided directly by the provincial government or purchased from voluntary agencies. The disabled client participates in setting an employment objective for himself and in designing an appropriate program of services. His goal may be employment in the competitive labour market, a profession, homemaking, farm work, sheltered employment or homebound work of a remunerative nature. Sharable costs include the salary and necessary travelling costs of staff whose duties are directly related to this program and other administrative expenses necessary for the co-ordination and delivery of services to the disabled. Other rehabilitation services provided by agencies and voluntary organizations may be funded by a province and are eligible for 50% reimbursement from the federal government under the Canada Assistance Plan. All provinces and territories, except Quebec, participate in the program.

During fiscal 1976-77 the federal government contributed $23.6 million to the provinces under the act and 61,316 clients received services.

6.7   Provincial income security programs

6.7.1   Social assistance

All provinces make legislative provision for assistance to persons in need and their dependents. Eligibility for assistance toward basic needs is determined by the budget deficit method whereby the basic needs (food, clothing, personal and household needs) of the applicant and dependents are calculated according to a prescribed schedule. Assistance for shelter and utilities is paid according to actual costs, sometimes within stated ceilings. The amount of the allowance is the difference between calculated need and resources available to the applicant to meet that need. The maximum amount of monthly assistance paid toward items of basic need is subject to any ceilings which may be imposed by provincial legislation.

In addition to allowances to cover items of basic need, all provinces make provision for such special items as rehabilitation services, expenses due to education or to obtaining employment, counselling, homemaker services and institutional care. All provinces permit certain income or earnings exemptions and, under special circumstances, some provinces provide assistance to fully employed persons.

Provincial departments of welfare set rates of assistance and conditions of eligibility; they have regulatory and supervisory powers over municipal administration of assistance, and require certain standards as a condition of provincial aid. Municipal residence may determine the financially responsible authority within a province. The provincial authority takes responsibility for aid to persons residing outside municipal boundaries and for those who lack municipal residence.

The administration of assistance varies. In Nova Scotia, Ontario, Manitoba and Alberta, allowances to persons with long-term need, such as needy mothers with dependent children, disabled persons and the aged, are administered by the province, with other allowances administered by the municipalities. In Newfoundland, Prince Edward Island and New Brunswick, all assistance is administered by the provincial authority. In Quebec, the province administers assistance through regional and local offices except in Montreal where the municipality administers assistance on behalf of the province. In Saskatchewan, social assistance is administered by the province except in two municipalities. In British Columbia, social allowances are administered through regional and district offices of the provincial government and, in some municipalities, by municipal welfare departments.

In the seven provinces where municipalities have some administrative responsibility, the proportion of municipal costs borne by the province varies from 40% to 100% of assistance paid.

Quebec Pension Plan   6.7.2

The Quebec Pension Plan (QPP) was established in 1965 and is comparable to the Canada Pension Plan. Although the Canada and Quebec plans were introduced at the same time and are closely co-ordinated, a series of amendments to both plans have effected the following differences: the QPP's survivors' and disability benefits flat rate component is $123.59 compared to $48.19 under the CPP (in January 1978); since January 1974, the QPP's orphans' and children's benefits have been fixed at $29.00 a month whereas those of the CPP are increased annually in accordance with increases in the cost of living. As of January 1, 1977 QPP contributors who leave the labour force to raise young children under age seven can drop out these years of low or zero earnings from their contributory period in determining their eligibility for QPP benefits and in establishing the amount of any QPP entitlement. In October 1977, there were 293,648 QPP beneficiaries in receipt of slightly more than $30 million in monthly QPP benefits.

Supplementary family allowance programs   6.7.3

Quebec has a provincial program embodied in the Quebec Family Allowances Plan of 1973. Certain criteria of eligibility under this plan differ somewhat from those of the federal program. The monthly rates paid by the province, as of January 1, 1977 were $3.98 for the first child in a family, $5.32 for the second, $6.64 for the third, and $7.96 for each additional child. Effective May 1, 1977, these rates were increased to $5.05, $6.76, $8.43, and $10.11 respectively and, as of January 1, 1978, to $5.43, $7.27, $9.06 and $10.87 respectively.

Prince Edward Island passed a Family Allowances Act in 1973. Under this act, the province pays $10 a month on behalf of the fifth and each subsequent child in a family in addition to the federal payment of $23.89 in 1977 and $25.68 in 1978. The provincial supplement is included in the monthly federal cheque.

Payments[1] for family allowances as a percentage of family income[2], selected years, 1951-76

Atlantic provinces
1951
1961
1971
1976 P

Quebec

Ontario

Prairie provinces

British Columbia

Canada

1. Average monthly payment per family as of March.
2. Derived from annual income figures.

Provincial income supplementation programs   6.7.4

Several provinces administer programs designed to supplement existing programs for such groups as the elderly and the disabled.

In Nova Scotia, under a special social assistance program, residents who receive the federal Guaranteed Income Supplement (GIS) may be eligible to receive a non-taxable annual lump sum which is scaled to the amount of GIS received. Maximum payments (i.e. for a person receiving the full GIS payment) are $110 annually per person.

In July 1974, Ontario introduced the Guaranteed Annual Income System (GAINS) to ensure a basic income for qualifying residents 65 years of age or older. The income guaranteed is adjusted periodically and as of October 1977 was $294.82 a month.

In Manitoba, a supplement for the elderly, established in 1974 under the Social Services Administration Act, is designed to supplement the income of pensioners who receive the federal GIS and of spouses, 60 to 64 years old, of pensioners. Maximum payments are $23.46 every three months for a single, widowed or divorced pensioner or one whose spouse is not receiving OAS/GIS and $25.29 quarterly where both spouses are receiving OAS/GIS or where one spouse is a pensioner and the other is in receipt of Spouse's Allowance. The amount of benefit payable is determined according to the amount of GIS paid.

By regulations made under a Saskatchewan assistance act of 1975, the Saskatchewan income plan was established to provide senior citizens benefits to pensioners in receipt of the federal GIS. The program is income-tested and maximum benefits are $240 a year or $20 a month for a single person or for a person in receipt of GIS whose spouse is not in receipt of OAS/GIS, and $216 a year or $18 a month for each of a married couple, both in receipt of OAS/GIS. The amount of benefits payable is determined according to the amount of GIS payable.

In 1975, an amendment to the Senior Citizens Benefits Act in Alberta permitted the implementation of a guaranteed income program for senior citizens in receipt of the federal GIS. Benefits payable range between $10.00 and $45.01 a month for a person who is single, widowed, divorced or a married person whose spouse is not a pensioner. For a married couple (both receiving OAS/GIS), the payments range between $10.00 and $47.20 a month each; the actual amount payable is based on the amount of GIS payable to the pensioner(s).

Under the Guaranteed Available Income for Need Act (GAIN), British Columbia provides a guaranteed minimum income to individuals aged 60 and over and handicapped persons aged 18 to 59 inclusive. To be eligible for GAIN benefits, an applicant must meet prescribed age, residence, assets and income qualifications. Effective January 1978, the guaranteed minimum monthly income under the GAIN program for persons in receipt of OAS/GIS or Spouse's Allowance was $299.94 (single) and $597.66 (married) where both were eligible; the guaranteed minimum monthly income for persons aged 60 to 64 not in receipt of OAS/GIS/SA benefits and for handicapped persons was $265 (single) and $530 (married) where both were eligible.

Provincial social services programs   6.8

Services for children   6.8.1

All provinces and territories have legislation governing basic child welfare services which include the protection and care of children, adoption services, services to unmarried parents and, in most provinces, services designed to prevent child neglect or a need for protection. A number of provinces also offer help to families in emergency situations; this is provided for a limited time by agreement with the parents, and may take the form of special services to the child in his own home or a temporary foster home.

These services are administered by provincial departments of social services through a division of child welfare. Direct services are provided through regional or local offices or by approved agencies. In some provinces these include children's aid societies.

Protection services include supervision of a child in his own home when there is some element of identifiable neglect or need for protection. When it seems necessary for the protection of the child to remove him from home, the child welfare authority may take the child to a place of safety, but he must be brought before a court within a specified time. A child found to be neglected or in need of protection as defined in provincial law may be committed either temporarily or permanently to the care and custody of the provincial child welfare authority. Temporary commitment is for a limited time, after which the case is reviewed by the court. Permanent committal has the effect of transferring guardianship rights to the child welfare authority. Care is provided according to the needs of the child in a foster boarding home, group home or in a specialized institution.

The provincial child welfare authority arranges adoption placements where this appears appropriate. Children eligible for adoption placement are those legally free for adoption, that is, those in the permanent care and custody of the child welfare authority and those whose parents have formally relinquished them for the purpose of adoption.

Costs of maintenance of children in the care of the provincial authority or a provincially approved agency and of certain welfare services are sharable with the federal government under the Canada Assistance Plan.

As of March 31, 1977 there were approximately 81,651 day care spaces, including 5,534 family day care spaces. The number of day care spaces decreased by 1,869, a decrease of 2.24% from 1976. Family day care was the only service in 1977 to show a modest increase of 3.11%. Day care centres sponsored by public authority comprised 12.92%, those sponsored by community boards 40.44%, parent co-operatives 5.84% and commercially sponsored centres 40.80%. Subsidies for day care services for children in need, or likely to become in need if they do not receive the service, are provided by provincial or municipal authorities and are sharable under the Canada Assistance Plan.

6.8.2   Programs for the aged

Programs and services offered to the aged vary from province to province. Although by no means organized in all areas, such services as visiting nurse, homemaker, counselling, information and referral, meals-on-wheels, friendly visiting and housing registries have been established under public and voluntary auspices. Low-rental housing projects have been built in many communities; clubs and centres to provide recreation and social activities have been developed. Some provinces offer annual shelter assistance grants to senior citizens who are either tenants or home owners while others offer free prescription drugs.

In all provinces, homes for the aged and infirm are provided under provincial, municipal or voluntary auspices. These homes are required to meet standards set out in provincial legislation relating to homes for the aged, welfare institutions or public health. Homes for the aged, regardless of auspices, are usually inspected and in some provinces must be licensed.

Small proprietary boarding homes for the care of well elderly persons are found in some provinces. Those who suffer from long-term illnesses may be cared for in chronic or convalescent hospitals, private or public nursing homes or homes for the aged. Costs of care in the chronic or convalescent hospitals are paid through provincial hospital plans. In the case of needy persons, sharing is available toward the full costs of providing care in homes for special care which are not covered under the Federal-Provincial Fiscal Arrangements and Established Programs Financing Act (EPF). Under EPF effective April 1, 1977 the provinces receive a $20.00 annual per capita grant. In institutions covered by the per capita grant under EPF, federal sharing is still available under the Canada Assistance Plan up to the prevailing OAS/GIS maximum for a single person. The $20.00 grant is intended to cover costs over and above this maximum. The portion cost-shared under CAP is available toward the cost of room and board, clothing and comfort and non-insured health services on behalf of persons in need. Homes for special care under the Canada Assistance Plan include homes for the aged, nursing homes, hostels for transients, homes for unmarried mothers, child care institutions and others.

In varying degrees, all provinces make capital grants toward the construction or renovation of homes for the aged by municipalities or voluntary organizations and, generally speaking, such homes are exempt from municipal taxation.

Vocational rehabilitation of disabled persons 6.8.3

All provinces and territories except Quebec have specific programs for the provision of vocational rehabilitation services for physically or mentally disabled persons for which costs are shared with the federal government. These services are provided to enable the individual to become capable of pursuing a gainful occupation. The services are co-ordinated and administered by provincial departments but may be provided either directly by central or regional provincial offices or purchased from voluntary organizations. Vocational rehabilitation services include medical, social and vocational assessment to determine the individual's residual capacities. A suitable vocational plan is determined jointly by the individual and counselling staff.

Prosthetic and corrective appliances, wheelchairs and other mobility aids are provided so that the individual may participate in a vocational training program or undertake employment. Other remedial and restorative treatment is provided as necessary. Any vocational training is made available in regular municipal or provincial vocational schools, private trade schools or business colleges, special training centres such as rehabilitation workshops, universities, or through training on the job in business or industry. The provision of all equipment necessary for training is also covered, as well as any travel cost. Maintenance allowances are usually provided for the individuals and their dependents while participating in the program. Where employment placement outside the competitive labour market is indicated, such placement is arranged by the province. Provincial authorities also assist in regular employment placement when special problems arise and their help is required.

In Quebec, assistance of various types and rehabilitation services for disabled persons are provided through a variety of departments and agencies. Because Quebec does not participate in the particular cost-sharing program with the federal government for the provision of all these services, the province would receive some cost-shared benefits for persons in need under the Canada Assistance Plan while other costs are borne fully by the province.

Programs for Indians 6.9

As with other Canadians, Indians are entitled to the benefits of universal federal welfare schemes such as Family Allowances, Old Age Security pensions, and the Guaranteed Income Supplement. Subject to the standard qualifying conditions, Indians also receive Canada or Quebec Pension Plan payments, unemployment insurance, worker compensation and veterans benefits.

However, the extent to which provincial welfare benefits and services are available to Indians living on reserves and Crown land varies according to province. Similarly, the acceptance of financial responsibility for welfare assistance to Indians who do not live on reserves can vary. Most provinces seek recovery of the costs of assistance and services which are provided to such Indians if they have not acquired residence off a reserve in accordance with provincial requirements.

Federal-provincial arrangements. A number of individual arrangements have been worked out between the federal government and authorities at other levels. Under a 1965 agreement with Ontario, all provincial welfare programs are available to Indians living there, either on or off reserves. In Quebec, the federal government has contracts with eight private social agencies to furnish welfare service to Indians in their geographic jurisdictions. An agreement in 1973 between the federal and Alberta governments and the Blackfoot band permits the band to administer two programs offered by the province's health and social development department to members on the reserve. Similar agreements continue to be developed through federal-provincial negotiation and consultation with representatives of bands and associations. There are also the social assistance and other programs of the federal Indian affairs and northern development department aimed directly at assisting the Indian people.

Role of Indian and northern affairs. The department has four main objectives in the operation of welfare programs: to ensure that services available are comparable to those available to other Canadians in the province where they live; to increase Indian participation in the design and operation of social service programs; to strengthen family life and facilitate increased independence; and to facilitate the provision of social services by other government and private agencies to Indian people in their jurisdictions who request such service.

The department's social assistance program provides basic household essentials (food, clothing, shelter, fuel) to the needy. Scales of assistance and eligibility conditions are comparable to those of other residents of the provinces. Administration of this plan, as with other social services, is handled by departmental employees on some reserves, by employees of the band council on others.

Indian residents are subject to the child welfare legislation of the province in which they live. The aim of the federal department's child care program is to ensure the welfare of neglected, dependent, or delinquent Indian children living on reserves. In conformity with federal-provincial child welfare agreements, the department finances maintenance and protection services to Indian children in the Yukon Territory, Manitoba, Nova Scotia and British Columbia. In provinces where child care services are provided on a voluntary basis, the department pays administrative costs and per day rates for Indian children receiving care from foster homes or other agencies.

The department furnishes maintenance and care in homes for the aged and in other institutions for physically and socially handicapped adults. Indian recipients of benefits such as Old Age Security or the Guaranteed Income Supplement in amounts insufficient to meet their basic needs may get additional assistance from the department.

With departmental financial support, a growing number of bands now administer their own day care centres and senior citizens homes. The department also operates a rehabilitation program designed to prevent development of social problems. It attempts to reduce the effects of physical disabilities and emotional difficulties.

A work opportunity program was established in 1971 to give jobs to physically able social assistance recipients. Funds which would otherwise be spent on direct financial aid are used to provide native communities with facilities such as roads, and services such as day care, that they may lack. Each project is financed by a reallocation of social assistance funds amounting to the equivalent of what the participants would have been given had they remained in receipt of social assistance, plus funds from other sources (regional appropriations, provincial revenues, band revenues).

The program is an example of the transfer of social service administration from the government to the native people. Approval is granted only to projects that are planned, designed and operated by band councils or groups empowered by them. Bands are expected to contribute to the cost of projects in accordance with a schedule worked out on the basis of the band's annual revenue. Project approval is also contingent on employment of those without jobs who are receiving or are likely to need social assistance. Bands operating projects are expected to pay reasonable wages and to meet other employer requirements such as coverage for unemployment insurance and worker compensation.

6.10 Veterans programs

The legislation known collectively as the Veterans Charter is administered by the veterans affairs department and four affiliated independent agencies — Pension Review Board, Canadian Pension Commission, War Veterans Allowance Board and the Bureau of Pensions Advocates.

Changes in legislation through the years have endeavoured to keep pace with the changing economic and social circumstances of veterans, particularly in the areas of pensions and allowances. So too have the programs administered by the department which include medical treatment, housing, educational assistance, counselling and other services.

The work of the department is carried out through a network of regional and district offices across Canada.

Pensions and allowances 6.10.1

Disability and dependents pensions 6.10.1.1

Canadian Pension Commission. This commission administers the Pension Act, the Compensation for Former Prisoners of War Act and parts of the Civilian War Pensions and Allowances Act. It also has responsibility for adjudication of claims for pension made under the RCMP Superannuation Act, the RCMP Pension Continuation Act and the Flying Accidents Compensation Regulations. The commission reports to Parliament through the minister of veterans affairs.

The evolution of Canada's pension legislation can be traced through statistical presentations in earlier editions of the *Canada Year Book*. Major amendments were made to the Pension Act in 1971 and details of the principal changes are described in the *Canada Year Book 1972*. Effective August 1, 1975 the Pension Act was further amended to provide for equality of status and equal rights and obligations for male and female persons.

The act provides for payment of pensions in respect of disability or death resulting from injury incurred during or attributable to service with the Canadian forces in war or peace. Pensions may also be payable on behalf of the dependents of a disabled former member of the forces or to the surviving dependents of a deceased veteran.

The Compensation for Former Prisoners of War Act provides for the payment of compensation to former prisoners of war and their dependents. Under this legislation, all persons imprisoned by the Japanese for one year or more are eligible to receive compensation in an amount equal to a 50% disability pension. This compensation is paid in addition to any pension such persons may be awarded for assessable disabilities, up to a maximum of the equivalent of 100% pension. Persons imprisoned by the Japanese for not less than three months and not more than 12 receive compensation in an amount equal to a 20% disability pension, in addition to any disability pension they may be receiving. Former prisoners of war of other powers are compensated in the same manner and in amounts equal to a disability pension of 10% to 20% depending upon the length of the period of incarceration.

The amount of disability pension payable is set out in the Pension Act and is based on a basic rate of pension established in 1972. The determination of a basic rate is discussed in the *Canada Year Book 1976-77*. Pensions under the act are indexed in accordance with the Consumer Price Index.

In 1978 a single pensioner awarded a pension for a disability assessed at 100% would receive $596.70 monthly. The additional pension for a spouse was $149.18. The additional pension payable for children was calculated at $77.64 for the first child, $56.63 for the second child and $44.76 for each additional child. The pension payable to a surviving spouse was $447.53 with an additional amount payable for the surviving children and, under certain conditions, the parents and brothers and sisters of a pensioner or a member of the forces who died as a result of injuries or disease incurred during service.

Pension Review Board. The board was established under 1971 amendments to the Pension Act. It is exclusively an independent appeal tribunal reporting directly to Parliament through the minister of veterans affairs. It deals with appeals from decisions of the Canadian Pension Commission in matters of entitlement and the amount of awards under the Pension Act. The board is also the final authority on the interpretation of the act.

War veterans allowances and civilian war allowances 6.10.1.2

War Veterans Allowance Board. The allowance board is a quasi-judicial body consisting of eight members appointed by the Governor-in-Council. It is independent as far as its decisions are concerned and reports to Parliament through the minister of veterans affairs. It is administratively co-ordinated with the department which provides support services. The board is responsible, under the War Veterans Allowance Act and the Civilian War Pensions and Allowances Act, to advise the minister generally on the legislation and specifically on the regulations; to adjudicate pursuant to specific sections of the legislation where the board has sole jurisdiction; to act as a court of appeal for aggrieved applicants and recipients; and, on its own motion, to review decisions of district authorities to ensure that adjudication is consistent with the intent and purview of the legislation and that the legislation is applied uniformly throughout Canada. The board may at any time review and alter its own former decisions.

6.10.1.3   Bureau of Pensions Advocates

The Bureau of Pensions Advocates was established under the minister of veterans affairs by 1971 amendments to the Pension Act (SC 1970-71-72, c.31), effective March 30, 1971. The bureau provides an independent professional legal aid service to applicants for awards under the Pension Act. The chief pensions advocate is the chief executive officer and is assisted by pensions advocates, all of whom are lawyers, located at the bureau's head office in Ottawa and in district offices in major centres across Canada.

Pensions advocates prepare applications to the Canadian Pension Commission and represent applicants as counsel at entitlement board and pension review board hearings. No charge is made for these services.

During the fiscal year 1976-77 the bureau submitted 13,711 claims at all levels of adjudication. Of the 10,835 decisions rendered on bureau claims during the same period, 30.7% were wholly or partially granted.

6.10.2   Social and health services for veterans

6.10.2.1   Veterans services

In the closing months of 1977 a major organizational change occurred in the integration of health and social services in the veterans services branch of the department. The process of integration was expected to be completed in 1978.

Medical and dental services are provided for eligible veterans throughout Canada and to other persons for governments or federal departments at the request and expense of the authorities concerned. Prosthetic services provided to eligible veterans by the health and welfare department are paid for by the veterans affairs department.

The branch provides examination and treatment for pensionable disabilities to war veterans allowance recipients (but not to their dependents) and to veterans whose service and financial circumstances render them eligible. If a bed is available, any veteran may receive treatment in a departmental hospital on a guarantee of payment. Disability pensioners receive treatment for pensionable disabilities regardless of place of residence. Service to other veterans is available in Canada only. Home care may be provided to eligible veterans.

Treatment is provided in five active treatment hospitals at Halifax, NS, Montreal and Ste. Anne de Bellevue, Que., Winnipeg, Man., Calgary, Alta., and in three domiciliary care homes at Ottawa, Ont., Saskatoon, Sask., and Edmonton, Alta. The number of beds set up in these institutions at October 31, 1977, was 2,805. In Ottawa both acute and chronic cases requiring definitive treatment are admitted to the National Defence Medical Centre. A veterans' pavilion of 69 beds is located at St. John's General Hospital, St. John's, Nfld., 1,105 beds are available at Sunnybrook Hospital in Toronto, 150 beds in Quebec and 200 beds at West Saint John Community Hospital in Saint John, NB, for priority use of veterans, as well as some 2,590 beds in community hospitals in St. John's, Nfld., Fredericton, NB, Kingston, London and Thunder Bay, Ont., Regina and Saskatoon, Sask., Edmonton, Alta., and Vancouver and Victoria, BC.

The veterans services branch also administers a range of legislative measures providing social and financial benefits, provides support services to the Canadian Pension Commission, the War Veterans Allowance Board and services benevolent funds and extends counselling services to veterans and their dependents which include referral to other agencies.

Prominent among the financial benefits are those under the War Veterans Allowance Act and the Civilian War Pensions and Allowances Act consisting of monthly income maintenance payments based on service requirements, age or incapacity and an income test. Widows, widowers and orphans of qualified veterans are also eligible for benefits. Additional allowances are paid to eligible veterans and widows for dependent children. At the close of 1977 the maximum permissible income level for a single recipient was $287.53 monthly. For those receiving the allowance at married rates, the maximum income level was $492.24 monthly. Those with dependent children received an additional $64.28 (less family allowance) for each child. The maximum level of income for an orphan was $180.92 (less family allowance).

As of November 30, 1977, a total of 91,011 persons were receiving war veterans allowances: 51,848 veterans, 38,453 widows or widowers and 710 orphans. The monthly liability as of November 30, 1977, was estimated at $17.0 million.

Similar benefits are provided to civilians who served in close support of the armed forces during wartime and to their dependents. As of November 30, 1977, a total of 4,433 civilians, including 1,187 widows or widowers and 16 orphans, were receiving these allowances. Total monthly cost was estimated at $950,374.

Eligibility adjudications are made by 19 regional war veterans allowance district authorities consisting of branch employees appointed by the minister of veterans affairs with the approval of the Governor-in-Council.

Assistance fund. Recipients of benefits under the War Veterans Allowance Act and the Civilian War Pensions and Allowances Act living in Canada may be given help from the assistance fund if their total income is lower than the permitted maximum. The number of persons assisted in the year ended March 31, 1977 was 28,067, the number in receipt of monthly supplements at the end of 1977 was 23,283 and fund expenditures from April 1, 1976, to March 31, 1977, amounted to $12.3 million. Comparable statistics for one year earlier, in each case, were 26,233 persons assisted, 22,297 in receipt of monthly supplements and $10.95 million in fund expenditures.

Education assistance to children. The Children of War Dead (Education Assistance) Act provides help in the form of allowances and the payment of fees for the post-secondary education of children of persons whose deaths were due to military service. From its inception in July 1953 to March 31, 1977, expenditures totalled $16.1 million of which $9.2 million was spent in allowances and $6.9 million in fees. By the end of March 1977, training had been approved for 7,478 children; of these, 3,499 had successfully completed training; 794 students in university and non-university courses were receiving assistance.

Veterans insurance. Under the terms of the Returned Soldiers Insurance Act (SC 1920, c.54 as amended), any veteran of World War I became eligible to contract for life insurance with the federal government for a maximum of $5,000. During the eight years in which the act was open, 48,319 policies with a face value of $109.3 million were issued. On March 31, 1977, 2,034 policies with a value of $4.4 million were still in force.

The Veterans Insurance Act (RSC 1970, c.V-3) made life insurance up to a maximum of $10,000 available to veterans of World War II and Korean operations on their discharge as well as to widows of those who died. The period of eligibility to apply for this insurance ended October 31, 1968. By that date 56,148 policies amounting to $185.1 million had been issued and, of these, 17,687 policies with a value of $56.4 million were still in force on March 31, 1977.

Social and counselling services. Counsellors at district offices work closely with other branches of the department, with other public and private agencies and organizations in assisting veterans and their dependents to deal with problems of social adjustment. University, vocational, technical and home training with allowances is provided for disabled pensioned veterans. Sheltered workshops at Toronto and Montreal and home assembly work in other centres produce poppies and memorial wreaths associated with Remembrance Day observances.

Services benevolent funds. Veterans and their dependents receive considerable assistance through various services benevolent funds. These organizations work in co-operation with the department and veterans organizations in providing cash grants or loans to meet emergencies.

6.10.3   Land settlement and house construction

Because of the postwar rehabilitation nature and purpose of the legislation, March 31, 1975 was the final date for veterans of World War II or the Korean Special Force to apply for establishment under the various settlement plans of the Veterans' Land Act. Veterans with subsisting VLA contracts could apply for additional loans within the financial ceilings of the act to purchase land or effect improvements to their properties, up to March 31, 1977.

From enactment in 1942, loan and grant funds totalling more than $1.3 billion were issued to approximately 140,000 veterans. On March 31, 1977, more than 52,000 veterans still had subsisting VLA contracts representing a remaining principal indebtedness of $505 million.

The veterans' land administration also has operational responsibility for the veterans housing assistance program. Under this measure, veterans of moderate income who wish to build or purchase homes may receive assistance of up to $600 annually to reduce the portion of income required for principal, interest and taxes to a more affordable level. Since inception of the program in 1975, grants with an annual value of some $98,000 have been approved on behalf of nearly 200 veterans.

The program also authorizes the department to provide financial assistance to non-profit corporations which obtain loans under the National Housing Act for development of low-rental housing projects intended primarily but not necessarily exclusively for occupancy by veterans. In addition to the benefits available from Central Mortgage and Housing Corporation, the veterans affairs department may make a grant equal to 10% of the capital cost of such a project as determined by CMHC. To date grants totalling approximately $1,440,000 have been approved for 10 such projects involving 665 units.

6.10.4   Commonwealth War Graves Commission

The current charters of the Commonwealth War Graves Commission consist of two documents — the original charter of incorporation dated May 21, 1917, and a supplemental charter dated June 8, 1964. Under these charters the commission is entrusted with the marking and maintenance in perpetuity of the graves of those of the British Empire and Commonwealth armed forces who lost their lives between August 4, 1914, and August 31, 1921, and between September 3, 1939, and December 31, 1947, and with the erection of memorials to commemorate those with no known grave.

The Canadian high commissioner in London, England, is the official commission member for Canada and the minister of veterans affairs is the agent of the commission in Canada. The office of the secretary general of the Canadian agency is in Ottawa.

Sources

6.1 - 6.2 Consumer Income and Expenditure Division, Census and Household Surveys Field, Statistics Canada.

6.3 - 6.8 Welfare Information Systems Branch, Department of National Health and Welfare.

6.9 Deputy Minister, Department of Indian Affairs and Northern Development.

6.10 Public Relations, Department of Veterans Affairs.

Tables

6.1 Average income of families in current and constant dollars by region, selected years, 1951-75

Region	1951	1961	1967	1971	1973	1974	1975
Current dollars							
Atlantic provinces	2,515	4,156	5,767	7,936	9,965	11,647	13,474
Quebec	3,523	5,294	7,404	9,919	12,024	13,742	15,446
Ontario	3,903	5,773	8,438	11,483	13,912	16,144	18,047
Prairie provinces	3,261	4,836	6,908	9,309	11,760	14,755	16,177
British Columbia	3,669	5,491	7,829	11,212	13,942	15,620	17,746
Canada	3,535	5,317	7,602	10,368	12,716	14,833	16,613
Constant (1971) dollars							
Atlantic provinces	3,810	5,544	6,667	7,936	8,839	9,318	9,728
Quebec	5,337	7,062	8,559	9,919	10,665	10,994	11,152
Ontario	5,913	7,701	9,754	11,483	12,340	12,915	13,030
Prairie provinces	4,940	6,451	7,986	9,309	10,431	11,804	11,680
British Columbia	5,559	7,325	9,050	11,212	12,367	12,496	12,813
Canada	5,356	7,093	8,788	10,368	11,279	11,866	11,994

6.2 Percentage distribution of families in constant (1971) dollars, showing average and median incomes, selected years, 1965-75

Income group in constant (1971) dollars		1965	1967	1969	1971	1974	1975
Under $3,000		12.1	9.9	9.6	9.0	5.6	5.5
$ 3,000 - $ 4,999		14.9	13.5	13.0	11.5	9.7	10.1
5,000 - 6,999		19.5	17.6	14.6	12.2	10.2	9.5
7,000 - 9,999		26.4	26.9	25.2	22.0	18.5	18.8
10,000 - 11,999		10.9	12.0	13.0	14.0	14.3	13.2
12,000 - 14,999		8.6	10.2	11.5	14.2	17.0	16.3
15,000 - 19,999		4.6	6.4	8.1	10.9	14.9	15.8
20,000 and over		2.7	3.5	4.9	6.2	9.8	10.7
Total		100.0	100.0	100.0	100.0	100.0	100.0
Average income	$	8,127	8,788	9,490	10,368	11,866	11,994
Median income	$	7,320	7,906	8,465	9,347	10,827	10,881

Median income refers to the middle or central value when incomes are ranged in order of magnitude. Median income is lower than average income in these tables since it is not as affected by a few abnormally large values in the distribution.

6.3 Percentage distribution of families by income group, by region, 1975

Income group		Atlantic provinces	Quebec	Ontario	Prairie provinces	British Columbia	Canada
Under $2,000		1.4	1.2	1.2	3.1	1.6	1.6
$ 2,000 - $ 2,999		1.6	0.9	0.9	2.0	1.2	1.2
3,000 - 3,999		3.0	3.0	1.9	2.5	1.6	2.4
4,000 - 4,999		4.6	3.6	2.7	3.3	2.7	3.2
5,000 - 5,999		6.1	4.7	3.1	4.0	3.4	4.0
6,000 - 6,999		4.7	4.1	3.0	3.9	3.6	3.6
7,000 - 7,999		5.7	3.7	2.6	3.7	2.3	3.3
8,000 - 8,999		5.5	4.0	2.9	3.5	3.3	3.6
9,000 - 9,999		4.7	4.0	3.0	3.0	3.3	3.5
10,000 - 11,999		12.2	10.7	7.3	9.1	7.4	9.0
12,000 - 14,999		15.4	15.7	13.9	14.0	13.5	14.5
15,000 - 24,999		27.3	32.5	39.5	33.0	37.5	35.3
25,000 and over		7.8	12.0	18.0	14.8	18.4	15.0
Total		100.0	100.0	100.0	100.0	100.0	100.0
Average income	$	13,474	15,446	18,047	16,177	17,746	16,613
Median income	$	12,077	13,971	16,588	14,563	16,335	15,065
Sample size	No.	5,469	3,967	5,092	3,824	2,106	20,458

6.4 Percentage distribution of families by income group, age and sex of head, 1975

Income group		Age group							
		Under 25	25-34	35-44	45-54	55-64	65-69	70 and over	Total
Families with male heads									
Under $2,000		1.4	1.5	1.0	1.1	1.1	1.6	0.6	1.2
$ 2,000 - $ 2,999		1.4	0.5	0.4	0.4	1.1	3.0	2.9	0.9
3,000 - 3,999		1.5	0.8	0.5	1.0	2.1	5.6	8.9	1.8
4,000 - 4,999		3.5	1.3	0.9	1.2	2.4	7.1	11.7	2.5
5,000 - 5,999		2.3	1.8	1.4	1.1	2.5	9.5	21.0	3.4
6,000 - 6,999		5.2	2.0	1.7	1.6	3.6	7.6	12.3	3.3
7,000 - 7,999		4.9	2.4	1.6	1.7	3.7	7.7	7.0	3.0
8,000 - 8,999		5.7	3.2	2.7	2.0	4.1	5.7	5.1	3.4
9,000 - 9,999		4.6	2.8	2.7	2.5	4.1	6.6	3.4	3.3
10,000 - 11,999		14.9	9.6	7.7	6.6	9.0	11.0	7.2	8.8
12,000 - 14,999		22.6	17.4	14.3	13.4	15.1	10.8	6.5	14.9
15,000 - 24,999		29.6	46.0	44.2	39.1	32.9	15.9	9.5	37.4
25,000 and over		2.4	10.6	20.8	28.2	18.4	8.0	3.7	16.2
Total		100.0	100.0	100.0	100.0	100.0	100.0	100.0	100.0
Average income	$	12,826	16,674	19,571	21,159	17,602	12,174	9,043	17,293
Median income	$	12,601	16,044	18,009	19,170	15,300	9,344	6,402	15,726
Sample size	No.	1,213	4,874	3,937	3,472	2,724	1,067	1,443	18,730
Families with female heads									
Under $2,000		15.7	9.0	4.1	4.6	3.5	0.9		5.5
$ 2,000 - $ 2,999		11.9	4.9	2.3	3.1	4.9	3.8		4.3
3,000 - 3,999		14.3	13.1	5.3	7.1	6.1	7.5		8.5
4,000 - 4,999		18.0	15.8	9.7	7.7	6.6	8.5		10.6
5,000 - 5,999		8.0	10.4	10.8	7.4	8.0	10.9		9.5
6,000 - 6,999		5.9	8.7	9.3	5.7	6.8	6.9		7.5
7,000 - 7,999		1.1	7.4	8.7	4.8	8.5	5.8		6.6
8,000 - 8,999		3.8	7.2	5.1	4.4	4.2	4.8		5.2
9,000 - 9,999		5.3	7.0	8.3	5.1	2.1	3.7		5.5
10,000 - 11,999		4.6	7.0	13.1	17.0	8.3	10.4		10.8
12,000 - 14,999		2.4	5.0	10.7	13.5	13.5	14.3		10.5
15,000 - 24,999		8.9	4.2	10.9	16.9	22.2	17.1		13.1
25,000 and over		—	0.3	1.7	2.9	5.4	5.3		2.6
Total		100.0	100.0	100.0	100.0	100.0	100.0		100.0
Average income	$	5,756	6,555	9,113	10,468	11,231	11,611		9,291
Median income	$	4,452	5,685	7,968	10,012	9,709	9,221		7,633
Sample size	No.	128	332	351	344	244	329		1,728
All families									
Under $2,000		2.8	2.0	1.3	1.4	1.3	1.5	0.7	1.6
$ 2,000 - $ 2,999		2.4	0.8	0.6	0.7	1.4	3.2	2.9	1.2
3,000 - 3,999		2.8	1.6	0.9	1.5	2.4	5.7	8.8	2.4
4,000 - 4,999		4.9	2.3	1.7	1.8	2.7	6.9	11.5	3.2
5,000 - 5,999		2.9	2.4	2.2	1.6	3.0	9.2	19.8	4.0
6,000 - 6,999		5.3	2.5	2.3	2.0	3.9	7.7	11.4	3.6
7,000 - 7,999		4.6	2.7	2.2	2.0	4.1	7.7	6.8	3.3
8,000 - 8,999		5.5	3.5	2.9	2.2	4.1	5.4	5.1	3.6
9,000 - 9,999		4.7	3.1	3.2	2.7	3.9	6.5	3.4	3.5
10,000 - 11,999		13.9	9.5	8.1	7.6	8.8	11.3	7.4	9.0
12,000 - 14,999		20.6	16.5	14.0	13.4	14.9	11.4	7.4	14.5
15,000 - 24,999		27.5	43.1	41.4	37.1	32.0	15.8	10.6	35.3
25,000 and over		2.1	9.9	19.3	26.0	17.4	7.7	4.0	15.0
Total		100.0	100.0	100.0	100.0	100.0	100.0	100.0	100.0
Average income	$	12,132	15,972	18,712	20,218	17,084	12,149	9,384	16,613
Median income	$	12,038	15,503	17,210	18,276	14,884	9,421	6,550	15,065
Sample size	No.	1,341	5,206	4,288	3,816	2,968	1,162	1,677	20,458

6.5   Percentage distribution of families by income group and education of head[1], 1975

Income group		Elementary schooling 0-8 years	Some high school and no post-secondary[2]	Post-secondary (non-university)		University degree
				Some	Completed	
Under $2,000		1.8	1.5	1.4	0.9	1.7
$ 2,000 - $ 2,999		2.2	0.8	0.6	0.9	0.5
3,000 - 3,999		4.7	1.5	1.4	0.7	0.7
4,000 - 4,999		5.9	2.4	2.4	1.1	0.7
5,000 - 5,999		7.1	3.0	2.6	1.9	1.3
6,000 - 6,999		5.6	3.1	3.3	2.6	1.0
7,000 - 7,999		5.0	3.1	2.5	1.9	1.4
8,000 - 8,999		4.7	3.5	3.3	2.5	1.6
9,000 - 9,999		4.8	3.2	2.1	3.2	1.8
10,000 - 11,999		10.6	9.1	8.7	7.8	4.4
12,000 - 14,999		13.7	17.1	14.3	14.5	7.2
15,000 - 24,999		25.8	38.2	42.7	45.2	36.8
25,000 and over		8.1	13.5	14.5	16.8	40.9
Total		100.0	100.0	100.0	100.0	100.0
Average income	$	13,151	16,544	17,319	18,127	25,405
Median income	$	11,525	15,320	16,357	17,154	22,542
Sample size	No.	7,136	7,957	1,532	2,062	1,771

[1]Data by education are not comparable with previously published figures due to category revisions.
[2]Includes high school graduation.

6.6   Percentage distribution of families[1] by income group, family characteristics, and combination of income recipients, Canada, 1975

Income group		Income recipient[2]					Total
		Husband-wife families			All other families		
		Head only	Head and wife only	Head and other family members[3]	Head only	Head and other family members[3]	
Under $2,000		2.1	0.4	0.3	8.9	1.2	1.3
$ 2,000 - $ 2,999		1.4	0.9	0.1	6.9	1.7	1.2
3,000 - 3,999		2.4	2.1	0.3	13.6	2.9	2.4
4,000 - 4,999		3.2	2.9	0.6	14.5	6.0	3.2
5,000 - 5,999		4.1	4.4	0.7	9.3	7.6	4.0
6,000 - 6,999		3.7	3.9	1.2	8.3	5.8	3.7
7,000 - 7,999		4.2	3.1	1.5	7.2	5.0	3.3
8,000 - 8,999		5.2	3.0	1.9	5.5	5.0	3.6
9,000 - 9,999		5.1	2.8	1.8	7.1	5.0	3.5
10,000 - 11,999		12.8	8.4	5.0	8.9	11.7	9.0
12,000 - 14,999		19.7	14.6	10.3	4.5	15.9	14.5
15,000 - 24,999		29.7	40.3	41.7	4.6	24.3	35.4
25,000 and over		6.3	13.1	34.5	0.7	7.9	15.1
Total		100.0	100.0	100.0	100.0	100.0	100.0
Average income	$	13,983	16,616	23,044	6,782	13,101	16,656
Median income	$	12,822	15,655	21,369	5,654	11,618	15,093
Sample size	No.	5,076	8,584	4,436	855	1,361	20,406

[1]Excludes 15,000 families who received no cash income in 1975 (see footnote 2 to Table 6.7).
[2]Data not shown for income recipient group "Other than head only" due to the small number of cases in the sample.
[3]"Other family members" refers to any income of children or other relatives and may also include income of wife.

6.7   Percentage distribution of families by major source of income within income quintiles, selected years, 1951-75

Major source of income within quintiles[1]	1951	1961	1965	1967	1971	1973	1975
Lowest quintile							
No income[2]	2.0	2.4	0.8	1.6	1.6	1.1	1.3
Wages and salaries	48.2	46.3	39.9	36.3	33.6	35.5	31.5
Net income from self-employment	10.5	9.1	18.1	19.2	11.8	10.6	8.2
Transfer payments	26.6	34.6	34.6	35.9	46.2	45.5	51.5
Investment income	7.0	3.4	3.4	3.0	3.3	3.2	3.9
Miscellaneous income	5.7	4.2	3.2	4.1	3.4	4.0	3.6
Second-lowest quintile							
Wages and salaries	85.5	83.5	80.0	80.8	81.0	80.8	79.8
Net income from self-employment	10.7	11.6	13.0	10.0	7.6	7.9	6.3
Transfer payments	1.1	1.8	4.1	4.1	4.9	5.5	7.0
Investment income	1.7	1.0	1.3	2.1	2.6	2.8	3.0
Miscellaneous income	1.0	2.0	1.7	3.0	3.9	3.0	3.9

6.7 Percentage distribution of families by major source of income within income quintiles, selected years, 1951-75 (concluded)

Major source of income within quintiles[1]	1951	1961	1965	1967	1971	1973	1975
Middle quintile							
Wages and salaries	91.8	91.3	88.9	92.0	93.1	91.3	92.5
Net income from self-employment	7.2	7.0	8.9	6.0	3.9	5.2	3.8
Transfer payments	0.3	0.3	0.4	0.7	0.3	1.1	1.2
Investment income	0.5	1.0	0.9	0.7	1.3	1.1	1.3
Miscellaneous income	0.2	0.5	0.8	0.7	1.4	1.3	1.2
Second-highest quintile							
Wages and salaries	92.4	92.1	90.8	93.8	94.7	93.9	94.4
Net income from self-employment	6.9	6.2	7.5	4.5	2.8	3.9	3.8
Transfer payments	- -	0.1	0.3	0.3	0.2	0.3	0.4
Investment income	0.5	1.2	0.7	0.8	1.4	1.3	0.9
Miscellaneous income	0.3	0.5	0.7	0.6	0.9	0.7	0.5
Highest quintile							
Wages and salaries	85.9	86.8	86.7	89.1	91.2	90.0	91.0
Net income from self-employment	11.6	10.9	11.1	8.9	5.6	7.8	7.1
Transfer payments	—	—	—	—	0.1	—	—
Investment income	2.1	1.7	1.7	1.8	2.1	1.8	1.4
Miscellaneous income	0.4	0.6	0.6	0.2	0.9	0.3	0.4
All families							
No income[2]	0.4	0.5	1.1	0.3	0.3	0.2	0.3
Wages and salaries	80.8	80.0	73.6	78.4	78.7	78.3	77.9
Net income from self-employment	9.4	9.0	10.4	9.7	6.3	7.1	5.8
Transfer payments	5.6	7.3	10.7	8.2	10.3	10.5	12.0
Investment income	2.3	1.7	2.4	1.7	2.2	2.0	2.1
Miscellaneous income	1.5	1.6	1.8	1.7	2.1	1.9	2.0
Total	100.0	100.0	100.0	100.0	100.0	100.0	100.0

[1]Families are arranged in ascending order of size of total income and then divided into five equal groups, or quintiles.
[2]These are families who either immigrated in the survey year so that they had no income from Canadian sources in the previous year or are new families who had no income of their own in the previous year.

6.8 Patterns of food expenditure for families of two or more persons, by family income quintile group[1], based on survey of eight Canadian cities, 1976

Item		Lowest quintile	Second-lowest quintile	Middle quintile	Second-highest quintile	Highest quintile	Income not stated	All classes
Family characteristics								
Average								
Family size	No.	2.84	3.25	3.51	3.47	3.72	3.55	3.37
Persons at home	"	2.79	3.19	3.45	3.38	3.62	3.42	3.29
Children under 5	"	0.28	0.37	0.34	0.22	0.18	0.21	0.27
Children 5-15	"	0.51	0.64	0.78	0.72	0.67	0.69	0.67
Adults 16-64	"	1.55	2.08	2.29	2.47	2.76	2.49	2.25
Adults 65 and over	"	0.50	0.16	0.11	0.07	0.11	0.16	0.19
Age of head	yr	47.5	40.7	39.7	40.4	43.5	47.4	42.7
Net income before taxes	$	7,077	12,881	16,937	21,770	34,159	..	18,565[2]
Percentage								
Home-owners		39.9	45.7	58.0	66.4	78.2	76.8	58.8
Wife employed full time		4.6	10.7	20.4	32.9	41.2	19.7	21.8
Average weekly food expenditure per family	$	39.90	46.89	56.03	59.92	73.39	54.51	55.18
Percentage of total food expenditure								
Food expenditure at home								
Dairy products		12.8	12.6	11.4	10.7	9.8	11.9	11.3
Milk		6.3	6.2	5.4	5.0	4.4	6.0	5.3
Other dairy products		6.6	6.3	6.0	5.6	5.4	5.9	5.9
Eggs		2.0	1.8	1.6	1.5	1.4	1.7	1.6
Bakery products		7.1	6.7	6.5	5.8	5.3	6.3	6.2
Cereal products		3.2	2.5	2.1	2.4	1.8	2.5	2.3
Meat and poultry		24.7	24.0	22.8	22.1	19.2	23.3	22.2
Beef		10.6	10.5	9.8	9.7	8.2	9.9	9.6
Pork		6.3	6.1	5.8	5.5	5.0	5.8	5.6
Other meats		3.8	4.0	3.6	3.7	3.2	3.8	3.6
Poultry		4.0	3.5	3.6	3.3	2.8	3.9	3.4
Fish		2.8	2.0	1.8	1.9	2.0	2.6	2.1
Fats and oils		2.4	2.1	1.9	1.7	1.6	1.9	1.9
Beverages		5.7	5.4	5.1	4.7	4.1	4.8	4.9
Miscellaneous groceries		6.0	5.9	5.9	4.9	4.3	4.5	5.2
Canned and dried fruits		2.2	1.6	1.7	1.5	1.4	1.7	1.6
Canned and dried vegetables		1.7	1.8	1.7	1.6	1.4	1.5	1.6
Fresh fruits		4.9	4.4	3.9	3.7	3.9	4.4	4.1
Fresh vegetables		5.3	4.9	4.5	4.3	3.7	4.2	4.4
Frozen foods		1.7	1.7	1.9	1.7	1.8	2.1	1.8
Prepared and partially prepared foods		2.1	1.6	1.6	1.6	1.5	1.1	1.6
Total, food prepared at home		84.4	78.9	74.4	69.9	63.2	74.6	72.8

6.8 Patterns of food expenditure for families of two or more persons, by family income quintile group[1], based on survey of eight Canadian cities, 1976 (concluded)

Item	Lowest quintile	Second-lowest quintile	Middle quintile	Second-highest quintile	Highest quintile	Income not stated	All classes
Family characteristics (concluded)							
Board paid by family members	0.3	0.8	0.2	0.7	0.5	- -	0.5
Food and beverages in eating places	11.4	16.3	19.5	20.5	27.8	20.0	20.2
Total, food expenditure at home	96.2	96.0	94.1	91.2	91.4	94.6	93.5
Food expenditure away from home overnight or longer							
Board paid by family members	- -	0.4	- -	0.1	0.2	- -	0.1
Food and beverages in eating places	2.3	2.8	4.8	7.4	6.0	3.6	4.9
Meals prepared on a trip (n.e.s.)	1.5	0.8	1.1	1.3	2.4	1.8	1.5
Total, food expenditure away from home	3.8	4.0	5.9	8.8	8.6	5.4	6.5
Total, food expenditure	100.0	100.0	100.0	100.0	100.0	100.0	100.0

[1]Family income quintile groups were obtained by ranking records for each month in ascending order of total income. The records for each month were then partitioned into five groups so that the weighted number of records in each was the same. The lowest group from each month was averaged over the year to give the first quintile group; then the second lowest and so on.
[2]Average of the five quintiles.

6.9 Canada Pension Plan, number of recipients in March 1976 and 1977 and net benefits paid by province, fiscal years ended 1976 and 1977

Province or territory	Recipients in March 1976	1977	Net benefits paid during fiscal year 1976 $'000	1977 $'000
Newfoundland	16,929	19,693	11,517	16,456
Prince Edward Island	5,753	6,570	3,500	4,973
Nova Scotia	45,273	52,308	33,394	47,122
New Brunswick	30,063	34,816	20,627	29,935
Quebec[1]	3,157	3,514	2,548	3,417
Ontario	398,257	459,058	313,811	446,857
Manitoba	52,285	60,001	38,209	52,204
Saskatchewan	44,877	52,472	29,417	41,806
Alberta	68,385	78,605	49,275	68,066
British Columbia	112,351	129,752	84,891	121,572
Yukon Territory	491	581	390	540
Northwest Territories	354	361	257	305
Canada[2]	778,175	897,731	587,834	833,251

[1]Excludes recipients of benefits under the Quebec Pension Plan; benefits are paid to residents of Quebec where total or partial contributions were made to the Canada Pension Plan.
[2]Includes 3,285 recipients of death benefits in 1976, and 3,554 in 1977. Death benefits are lump sum payments made to the estate.

6.10 Old Age Security (OAS), Guaranteed Income Supplement (GIS), Spouse's Allowance (SA), number of recipients in pay[1] for March 1976 and 1977 and net benefits paid, by province, fiscal years ended Mar. 31, 1976 and 1977

Province or territory	OAS only	OAS and GIS	Spouse's Allowance[2]	Total recipients	OAS only $'000	GIS only $'000	Spouse's Allowance[2] $'000	Total $'000
Nfld.								
1976	6,778	29,713	2,379	38,870	55,470	28,048	1,719	85,237
1977	6,775	30,820	2,849	40,444	61,896	30,711	5,456	98,063
PEI								
1976	3,363	9,874	666	13,903	19,879	9,073	529	29,480
1977	3,455	9,966	696	14,117	22,148	9,487	1,158	32,793
NS								
1976	26,562	52,839	3,473	82,874	120,928	46,629	2,640	170,197
1977	26,915	54,461	4,104	85,480	134,248	51,461	6,264	191,973
NB								
1976	20,696	39,647	2,086	62,429	91,782	35,304	1,682	128,767
1977	20,535	41,379	3,027	64,941	103,542	39,434	4,397	147,373
Que.								
1976	181,036	296,977	15,905	493,918	727,954	261,998	10,480	1,000,433
1977	181,468	310,704	21,535	513,707	808,042	294,238	31,898	1,134,178
Ont.								
1976	378,677	338,404	12,434	729,515	1,087,498	275,027	6,768	1,369,293
1977	393,976	343,891	17,714	755,581	1,215,438	299,586	20,614	1,535,638
Man.								
1976	43,066	61,950	3,745	108,761	159,905	51,823	2,463	214,191
1977	45,660	62,231	4,215	112,106	177,809	55,999	6,096	239,904

6.10 Old Age Security (OAS), Guaranteed Income Supplement (GIS), Spouse's Allowance (SA), number of recipients in pay[1] for March 1976 and 1977 and net benefits paid, by province, fiscal years ended Mar. 31, 1976 and 1977 (concluded)

Province or territory	Recipients in March				Net benefits paid during fiscal year			
	OAS only	OAS and GIS	Spouse's Allowance[2]	Total recipients	OAS only $'000	GIS only $'000	Spouse's Allowance[2] $'000	Total $'000
Sask.								
1976	44,395	57,128	3,415	104,938	154,703	48,394	2,316	205,413
1977	46,636	57,002	3,998	107,636	172,011	51,598	5,916	229,525
Alta.								
1976	56,514	76,460	3,968	136,942	202,982	64,525	2,376	269,883
1977	58,896	78,510	5,371	142,777	226,901	71,789	7,443	306,133
BC								
1976	108,618	122,999	6,103	237,720	351,963	101,451	3,982	457,395
1977	114,828	124,505	8,215	247,548	393,961	111,487	11,272	516,720
YT								
1976	247	314	3	564	887	311	2	1,199
1977	246	346	12	604	1,015	361	22	1,398
NWT								
1976	223	808	17	1,048	1,613	831	14	2,457
1977	233	863	45	1,141	1,908	977	90	2,975
Canada								
1976	870,175	1,087,113	54,194	2,011,482	2,975,562	923,413	34,970	3,933,945
1977	899,623	1,114,678	71,781	2,086,082	3,318,919	1,017,128	100,626	4,436,673

[1]"In pay" is an administrative term which includes a few recipients who were not paid in March and excludes a few who were.
[2]Spouse's Allowance became payable in October 1975; the recipients are aged 60 to 64 (see text).

6.11 Family Allowances, number of children and accounts in pay[1] for March 1976 and 1977, and net benefits paid, by province, for the fiscal years ended Mar. 31, 1976 and 1977

Province or territory and year	Recipients as of March			Children	Net benefits paid during fiscal year $'000
	Accounts Regular	Agency[2]	Foster		
Newfoundland					
1976	90,302	939	273	225,904	60,222
1977	92,211	764	334	223,839	61,090
Prince Edward Island					
1976	17,200	125	118	41,107	10,967
1977	17,652	48	148	40,936	11,190
Nova Scotia					
1976	124,913	2,636	19	275,667	73,595
1977	126,363	2,484	21	272,078	74,374
New Brunswick					
1976	104,439	1,077	837	237,802	63,312
1977	106,577	1,152	715	235,937	64,391
Quebec					
1976	941,688	12,268	78	1,962,614	528,177
1977	953,293	12,201	32	1,927,051	526,763
Ontario					
1976	1,247,017	12,278	19	2,557,238	685,781
1977	1,262,485	11,975	15	2,534,919	694,318
Manitoba					
1976	148,991	3,250	193	325,853	87,064
1977	150,370	3,133	155	322,687	88,101
Saskatchewan					
1976	134,335	2,166	321	306,707	81,641
1977	137,012	1,900	326	306,186	83,404
Alberta					
1976	280,815	2,943	968	608,729	160,732
1977	291,749	4,566	—	617,441	167,537
British Columbia					
1976	361,927	5,936	491	743,018	198,706
1977	365,992	5,956	397	735,187	201,106
Yukon Territory					
1976	3,490	157	—	7,757	2,108
1977	3,523	136	—	7,570	2,108
Northwest Territories					
1976	7,279	231	27	19,488	5,208
1977	7,499	273	—	19,694	5,389
Canada					
1976	3,462,396	44,006	3,344	7,311,884	1,957,513
1977	3,514,726	44,588	2,143	7,243,525	1,979,770

[1]"In pay" is an administrative term which includes a few recipients who were not paid in March and excludes a few who were.
[2]Each account consists of one child.

6.12   Pensions in force under the Pension Act as at Mar. 31, 1976 and 1977

Service	Disability		Dependent		Total	
	Pensions No.	Liability $	Pensions No.	Liability $	Pensions No.	Liability $
1976						
World War I	12,404	32,173,808	10,695	48,023,015	23,099	80,196,823
World War II	96,776	203,857,818	15,317	63,129,460	112,093	266,987,278
Special Force	2,084	4,097,845	161	647,244	2,245	4,745,089
Regular Force	4,106	6,544,077	711	3,541,925	4,817	10,086,002
Total	115,370	246,673,548	26,884	115,341,644	142,254	362,015,192
1977						
World War I	10,829	30,040,123	10,247	50,011,266	21,076	80,051,389
World War II	95,235	222,069,611	15,307	69,477,619	110,542	291,547,230
Special Force	2,102	4,457,465	161	704,881	2,263	5,162,346
Regular Force	4,379	7,442,188	713	3,886,707	5,092	11,328,895
Total	112,545	264,009,387	26,428	124,080,473	138,973	388,089,860

Sources

6.1 - 6.8 Consumer Income and Expenditure Division, Census and Household Surveys Field, Statistics Canada.
6.9 - 6.11 Welfare Information Systems Branch, Welfare, Department of National Health and Welfare.
6.12 Public Relations, Department of Veterans Affairs.

Education, training and cultural activities

Chapter 7

Tables

Education, training and cultural activities

Chapter 7

Education in Canada

7.1

Statistical highlights

7.1.1

Education was the primary activity of 6,400,000 Canadians in 1976-77, about 28% of the total population. They included 325,000 full-time teachers and 6,100,000 full-time students in 15,500 educational institutions. Expenditures on education for 1976-77 amounted to an estimated $15 billion or 7.9% of the Gross National Product (GNP).

Enrolment. Total full-time enrolment at all levels rose steadily from 4,367,400 in 1960-61 to a record high of 6,363,900 in 1970-71, a 46% increase in 10 years. The subsequent 4% decrease to 6,099,900 in 1976-77 occurred at the elementary and secondary levels, reflecting the decline in the number of children aged 5-17. Post-secondary enrolment continued to grow.

Full-time post-secondary enrolment in 1976-77 was 603,500, a 2% increase over the 592,000 enrolled in 1975-76. University students made up 62% of the total, but their rate of increase over the past decade was lower than that of students in non-university institutions. Full-time enrolment in the latter almost tripled, rising from 80,200 in 1966-67 to 227,000 in 1976-77. Simultaneously, full-time university enrolment went from 230,300 to 376,500, an increase of 64%. The proportion of women enrolled full-time at the post-secondary level rose from 38% in 1966-67 to 45% in 1976-77. But while the percentage of female students in universities grew from 32% to 42%, their representation in non-university institutions fell from 55% to 49%.

Elementary-secondary enrolment was 5,496,400 in 1976-77, a loss of 2% from the previous year. Since the all-time 1970-71 high of 5,900,000, enrolment fell 7% because the young population had diminished. Likewise, the number of elementary-secondary schools was reduced, but this began much earlier. In 1960-61 there were 27,000 schools; by 1976-77 the total was 15,200. Whereas recent cuts in the number of schools were forced by the enrolment decline, the drop in the 1960s was a result of school consolidation. At that time small schools were being replaced by large buildings in response to the pressure of rising enrolment.

Enrolment in the future, as in the past, will be significantly affected by trends in the birth rate. At the elementary-secondary level where most attendance is compulsory the number of school-age children almost predetermines enrolment. And 18-24-year-olds make up about 80% of post-secondary students. Thus, the post-war baby boom, during which annual births soared from 300,600 in 1945 to 479,300 in 1959, meant that enrolment at all levels would increase as these children progressed through the education system. But starting in the 1960s annual births declined, falling to 343,400 in 1973. The low birth rates of these years will produce an enrolment slump at every educational level as the children mature.

Small annual reductions in elementary enrolment are expected for the rest of the decade. The drop from the 1968 high of 3,844,000 to the projected 1981 low of 3,011,000 is 22%. Enrolment will stabilize for several years and then rise until the mid-1990s. However, the next "peak" should be much below the baby boom maximum.

Secondary enrolment patterns resemble those at the elementary level but are delayed seven or eight years because of the age difference. From 1966-67 to 1974-75, enrolment rose 32% from 1,366,200 to 1,808,600. By 1976-77, it had fallen to 1,704,900. A continuing decline to about 1.4 million in the early 1990s is expected, followed by an increase peaking at about 1.7 million at the turn of the century.

The decrease in the post-secondary-age population will become evident in the mid-1980s. Since 1970-71 the full-time post-secondary enrolment rate rose from 18% to

20%. Assuming that the rate will hover around 20% for the rest of the century means that enrolment will follow the contours of the 18-24 age group. Thus, a high of about 673,000 should occur in the early 1980s and a low of 527,000 (a 22% drop) in the mid-1990s. By 2000, enrolment should again be rising.

Teachers. The number of full-time elementary-secondary teachers declined 0.8% from 276,900 in 1975-76 to 274,700 in 1976-77. Their ranks had been falling since 1972-73 when they totalled 278,300. However, because the number of teachers had not decreased as quickly as enrolment, the number of students in relation to teachers fell every year. Of the 1976-77 total, more than half (57%) were women. The majority were teaching grades one to eight, while 20% taught grade nine and higher. Only 14% of principals and 16% of vice-principals were women.

In contrast to the elementary-secondary situation, the number of full-time post-secondary teachers was still growing. The 1976-77 total of 50,100 represents a 4% increase over the previous year and a 117% jump since 1966-67. The rise occurred among both university and non-university teachers, although the latter increased more rapidly to keep pace with greater enrolment gains.

Graduates. More than a quarter of a million students (265,600) graduated from secondary school in 1975-76, a 4% increase over 1974-75. The size of these graduating classes reflects the high birth rates of the 1950s and a growing tendency of young people to remain in school at least until secondary graduation. About 60% of high school graduates enter a post-secondary institution.

For 1976 the number of earned degrees conferred by universities were: bachelor's and first professional 83,300; master's 11,560; and earned doctorates 1,690. Comparable figures for 1978 were expected to be 92,000, 12,200 and 1,820 — overall, more degrees than ever before. Annual increases are anticipated through the early 1980s. In 1974-75, non-university institutions awarded 53,200 diplomas, an increase of more than 10,000 over the 1970-71 total of 41,600.

Spending for education from kindergarten through graduate studies was estimated at $15.1 billion for 1976-77, and preliminary estimates place the 1977-78 figure at about $16.4 billion. Elementary-secondary education absorbed $10.0 billion of the 1976-77 total. Universities received $4.1 billion; non-university institutions $1.0 billion; and vocational training $883 million.

Between 1975-76 and 1976-77, education expenditures rose 15%. The greatest increase, 17%, was at the elementary-secondary level. Spending on universities rose 12% and on non-university institutions 15%, while the increase in expenditures for vocational training was 5.2%.

Total spending for education per capita of population and per capita of labour force has climbed steadily in the past decade. Spending per capita rose from $208 in 1966 to $656 in 1976. The increase per capita of labour force was from $560 to $1,462.

Nonetheless, other indicators point to a decline in education spending that coincides with the enrolment turnaround. In 1970 when total full-time enrolment reached a high, education expenditures were equivalent to 9% of GNP and absorbed 22% of government spending, more than any other major area. By 1976 expenditures on education represented only 7.9% of GNP and social welfare had assumed first place.

7.1.2 History of education

The earliest organized forms of education in the territory that was to become Canada were under church control. Quebec was founded as a colony of France in 1608 and the first school soon opened. But it was not until 1824 that Quebec passed an education act. Nova Scotia had done so in 1766, followed by New Brunswick in 1802 and Ontario in 1807. However, education at lower levels continued to be church-dominated until the mid-19th century.

During the 1840s and 1850s a public system of education was developed in Quebec (Canada East), supplemented by schools and colleges operated by Roman Catholic orders. At the same time, Ontario (Canada West) also established a public system, as did the Maritimes (New Brunswick, Nova Scotia and Prince Edward Island).

Higher education before Confederation was conducted in private institutions, most controlled by religious authorities. A Jesuit college that would evolve into the Université Laval began in 1635; the oldest English-language institution, King's College at Windsor, NS was founded by the Anglican Church in 1789. By 1867, Quebec had three universities and 712 classical colleges. There were three universities in New Brunswick, five in Nova Scotia and seven in Ontario. As well as in Nova Scotia, King's colleges had been established in New Brunswick and Ontario. Queen's and Victoria universities, supported by the Presbyterian and Methodist churches, had been chartered in Ontario. Only McGill in Montreal and Dalhousie in Halifax were non-sectarian.

Constitutional responsibility. The British North America Act, passed by the British Parliament in 1867, united four provinces, Ontario, Quebec, New Brunswick and Nova Scotia. Section 93 of the act placed education "exclusively" under the control of each province, confirming variations in the systems that already existed. As other provinces were admitted (Manitoba 1870, British Columbia 1871, Prince Edward Island 1873, Saskatchewan and Alberta 1905 and Newfoundland 1949) the provisions of the section were reaffirmed.

Officially the act recognized no federal presence in education. However, the federal government assumed direct responsibility for the education of persons beyond the bounds of provincial jurisdiction — Indians and Inuit, armed forces personnel and their families, and inmates of federal penal institutions. And as the education enterprise expanded, indirect federal participation in the form of financial assistance became extensive.

The education explosion. Until the late 1940s, Canada, according to a report by the Organization for Economic Co-operation and Development, was "one of the less educationally developed of the great democracies." Today it ranks among the world's educational leaders. This evolution was compelled by unprecedented population growth combined with the desire of students to continue to higher levels.

The population grew because of the post-war baby boom and sizable net immigration. Rising expectations and widespread belief in education as a means of upward mobility encouraged students to stay in school longer. Consequently, Canada's educational enrolment in the post-war period increased faster than that of any other industrialized country. Between 1951 and 1971 combined elementary-secondary enrolment more than doubled. The 1960s were the decade of fastest growth, with the number of elementary-secondary students increasing 40% and post-secondary enrolment 168%. Such growth necessitated construction of new schools, expansion of the post-secondary sector and a commensurate rise in numbers of teachers at all levels.

As well as increasing facilities and personnel, it was imperative to revise the curriculum to reflect new social and economic realities. A more industrialized and sophisticated economy imposed new standards on the labour force. The comprehensive secondary school, offering a wide range of options, was recognized as part of the answer to the need for versatility and choice.

Expansion of the education enterprise could not occur without a spending increase. In 1947 education expenditures totalled $350 million. By 1960 annual costs had risen to $1.7 billion. During the 1960s, expenditures grew at an average yearly rate of more than 10% (sometimes 20%) to $7.7 billion in 1970. The rate at which spending rose was faster than that of enrolment — costs tripled although total full-time enrolment at all levels increased by only 46%.

The decline in the birth rate and lower levels of immigration have produced an enrolment decline in elementary-secondary schools that is expected to persist into the 1980s. The 1970-71 peak is unlikely to be attained again this century. Post-secondary institutions will feel the effects of this decline.

Provincial administration   7.1.3

Each province and territory is responsible for its own education system. As a consequence, organization, policies and practices differ from one to another. A department of education in every province is headed by a minister who is an elected member of the provincial cabinet, or in the case of the territories, a councillor. Policy-making power rests with the department; the influence of the legislature is confined to formal matters such as passing budgets. Four provinces have established separate departments for post-secondary education. Where two departments exist there may be two ministers, or one may have dual jurisdiction.

While the education minister has general authority, day-to-day operation of the department is carried out by a deputy minister. The latter, a senior civil servant, is the permanent head of the department. The deputy minister advises the minister and supervises all functions of the department. These include: supervision and inspection of elementary and secondary schools; provision of curriculum and school organization guidelines; approval of new courses and textbooks; production of curriculum materials; finance; teacher training and certification; prescription of regulations for trustees and teachers; research; and support services such as libraries, health and transportation.

In most provinces, responsibility for teacher training has been transferred from teachers' colleges operated by the department to faculties or colleges of education in universities. Increasingly, this has meant that an elementary teacher must have a bachelor's degree.

The Nova Scotia Teachers' College is the only institution of its kind remaining in the country. Ontario teachers are trained in university faculties of education or the Ontario Teacher Education College, run directly by the province; to be admitted to it, students must have a degree, and graduates receive both a Bachelor of Education degree and a teaching certificate.

Other provincial departments have some responsibility for education. They operate apprenticeship programs, agricultural schools, reform schools and forest ranger schools.

Levels of education. Despite variations in matters such as the ages of compulsory attendance, course offerings and graduation prerequisites, the education systems that evolved in each province basically consist of three levels: elementary, secondary and post-secondary. The number of years required to complete each level and the dividing lines between them vary from province to province.

7.2 Elementary and secondary education

At the elementary and secondary level, most public schools are established and operated by local education authorities according to public school acts of the provinces. This category includes Protestant and Roman Catholic separate schools, and schools operated in Canada by the defence department within the framework of the public system. Private schools, church-affiliated or non-sectarian, are operated and administered by private individuals or groups. Private kindergartens and nursery schools for children of pre-elementary age offer education at that level only. These schools may be church-affiliated and are administered by private individuals or groups. Schools for the handicapped provide special facilities and training. Most are under direct provincial government administration. Federal schools are administered directly by the federal government including overseas schools operated by the defence department for dependents of servicemen, and Indian schools operated by the Indian and northern affairs department.

Local administration. Schools in all provinces are established under a public school act and operated by local authorities answering to the provincial government and resident ratepayers. Provincial authorities delineate school board areas. With the growth of cities and towns, and of educational facilities and requirements, small local boards have been consolidated into central, regional or county units with jurisdiction over both elementary and secondary schools in a wider area. The boards, composed of elected or appointed trustees or commissioners, are responsible for school management. Their powers are determined and delegated by the legislature or education departments and vary from province to province. Generally, they handle the business aspects of education — establishment and maintenance of schools, appointment of teachers, purchase of supplies and equipment, details of school construction and budget preparation. Boards are authorized to levy taxes and manage grants from the department.

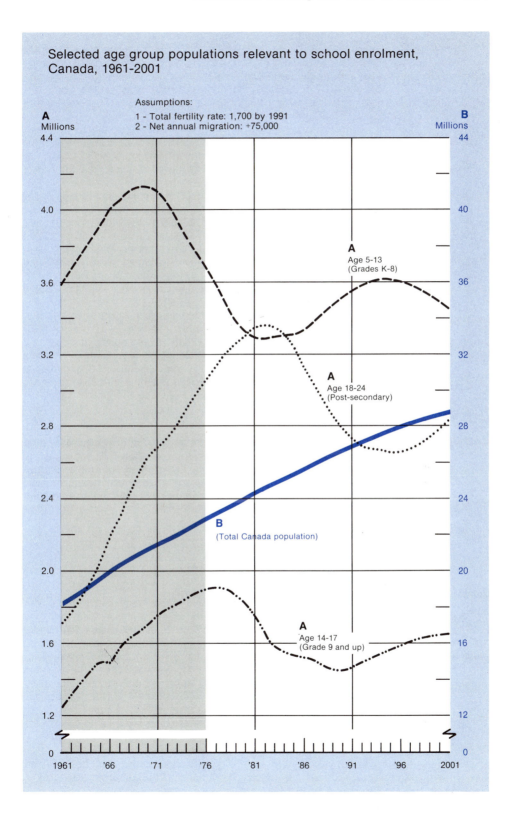

Selected age group populations relevant to school enrolment,
Canada, 1961-2001

Assumptions:
1 - Total fertility rate: 1,700 by 1991
2 - Net annual migration: +75,000

A
Millions

B
Millions

A
Age 5-13
(Grades K-8)

A
Age 18-24
(Post-secondary)

B
(Total Canada population)

A
Age 14-17
(Grade 9 and up)

Grade structure. School attendance is compulsory for about 10 years in every province — the starting age is 5, 6 or 7, and the minimum leaving age, 15 or 16. However, the elementary-secondary program usually extends over 12 years. Particularly in urban areas, local authorities may also provide an introductory year of education prior to grade one. More than 1,200 private kindergartens operate under varying degrees of provincial supervision. Some private kindergartens admit 3-year-olds. Total kindergarten enrolment in 1976-77 was 391,500.

In the past, eight elementary and four secondary grades constituted the basic structure of public education outside of Quebec. Many jurisdictions have modified this framework by adding a year of secondary school or introducing junior high. As a result, 6-3-3 and 6-3-4 grade plans are now common. Before the secondary level, education is general and basic.

High school students usually have a choice of two programs — academic or vocational. The latter range from one to four years. At one time secondary schools were predominantly academic and prepared students for university. Vocational schools were separate institutions, located only in large cities. Today, in addition to technical and commercial high schools, most secondary institutions are composite or comprehensive. Programs include both purely academic courses as a prelude to university, and vocational courses that prepare students either for an occupation or for further post-secondary non-university education. Vocational training covers such subjects as home economics, agriculture, shop-work and commercial skills.

The principle of continuous progress has been implemented to a greater extent in secondary schools than at the elementary level. Some jurisdictions have partially or entirely eliminated age-grouped classes. The length of schooling depends on accumulation of a requisite number of credits. Thus subject-promotion is replacing grade-promotion. Most provinces have abolished external graduating examinations administered by the education department; schools conduct their own. Diplomas are still issued by the province on the recommendation of individual schools.

7.2.1 Other types of schools

Separate schools. One of the most obvious differences among provincial education systems is in provision for separate schools. Some provinces allow religious groups to establish schools under the authority of the education department. They must conform to department regulations on curriculum, textbooks and teacher certification. As legal corporations, separate school boards can levy taxes and receive government grants but not always at the same level as the public system.

Private schools. Between 3% and 4% of all elementary-secondary students attend schools operated independently of the public systems. Provincial policies on private institutions vary from direct operating grants to minimum provincial control. Independent schools have been established as alternatives to the public system, based on religion, language, or academic or social status.

Special education. A number of strategies have been developed to educate children with special needs and abilities, an estimated 5%-10% of all students. They may be accommodated in separate institutions, public or private, or in special classes in regular schools. Interest in the education of exceptional children has resulted in enriched and accelerated elementary and secondary programs. Education for the handicapped varies from province to province, and is most common in city systems. Schools for the blind and deaf are sometimes administered directly by a province, sometimes by interprovincial agreement. Many local systems operate schools or classes for disabled children. Nonetheless, the trend, and the official goal in several provinces, is for handicapped students to stay in regular school as long as possible.

7.2.2 Federal schools

Although education is a provincial responsibility, the federal government has assumed direct control over the education of persons beyond the jurisdiction of the provinces — native peoples and armed forces personnel and their families.

Department of Indian and Northern Affairs 7.2.2.1

As of December 31, 1976 there were 288,938 registered Indians in Canada, and approximately 4,252 Inuit living in far northern areas. Registered Indians, organized in about 500 bands, are entitled to special protection and support from the federal government under the Indian Act. It does not cover non-status Indians and Métis.

Education of registered Indian and Inuit children is an obligation of the Indian and northern affairs department. The minister is authorized to maintain schools for Indian children directly or provide education services through a provincial government, the commissioners of the Yukon Territory and Northwest Territories, a public or separate school board, or a religious or charitable organization.

On Crown lands and reserves, the federal government owns and operates 264 schools. Attendance is compulsory from age 7 to 16 and children must attend the school designated by the minister. The minister makes regulations on matters such as buildings, inspection, teaching and discipline. Transportation and maintenance costs at residential schools are paid by the department. In recent years the policy has been to transfer control to native bands; 62 now manage their own schools.

About half the native children attend provincial public schools. The federal government reimburses the provinces, either by paying tuition or contributing to the school's capital costs. Most children of secondary age attend public schools. Indian representation is increasing on local provincial school boards — approximately 90 are now formal school board members in various provinces. In the Yukon Territory and Northwest Territories, Indian and northern affairs co-operates with territorial departments of education to educate native children. The last school there to be administered directly by the federal government closed in 1969.

In 1976-77 the department spent $196.6 million on education. Indian enrolment at the elementary level in federal schools was 33,187 and in non-federal schools 38,530, for a total of 71,717 pupils. Students in universities and post-secondary non-university institutions numbered 3,577; another 6,170 were taking vocational training.

Counselling units are maintained in Ottawa and Winnipeg to assist northern native students in southern Canada to attend high school, technical school, college and university. These units were established in the mid-1960s and have worked with an increasing number of students each year. In recent years the Ottawa unit has been involved with about 140 northern students and the Winnipeg unit with about 40 students. The dropout rate has been less than 5% and most students complete the program in which they enroll.

Department of National Defence 7.2.2.2

The defence department maintains schools for dependents of service personnel at military establishments in Canada and overseas. The policy is to avoid building schools wherever the children can attend existing institutions. Provinces are reimbursed on a per-pupil basis for armed service dependents in public schools. Federal and provincial governments share construction costs of new public schools according to the proportion of each category of student expected to enroll. The curriculum in such schools follows that of the province where they are located. There are 11 overseas schools in Belgium, the Netherlands and the Federal Republic of Germany. The elementary curriculum in these schools is a composite of various provincial programs; grades 7 to 13 follow the Ontario pattern.

Financing the elementary-secondary system 7.2.3

In 1976-77 total expenditures on the elementary-secondary level were an estimated $10 billion or about 66% of all education spending.

Financing elementary-secondary education was traditionally a municipal responsibility, local real estate taxes paying most of the cost of basic education. School boards determine their budgets and therefore the taxes required. In most cases municipalities levy and collect taxes for the boards. Where there is no municipal organization the boards have these powers. Taxes on real estate are still a vital element of elementary-secondary finance but the municipal share has declined in recent years. It represented 22.4% in 1970-71, but 18.7% in 1976-77.

Rapid post-war expansion of the need for educational services caused other levels of government to become more deeply involved in funding elementary and secondary schools. At the end of the 1940s provincial governments were contributing less than 20% of net general expenditures. During the next decade education spending tripled to more than $1 billion. This reflected rising enrolment, improvement of teacher salaries, large-scale building programs and the growth of special services. As budgets increased, municipal authorities requested more support from provincial governments.

By 1974 the provinces were paying about $2 for every dollar spent by local authorities on elementary-secondary education. The relative contributions of the two levels differ from province to province, each provincial authority deciding the magnitude of municipal responsibility. A system of formula financing determines how provincial funds are distributed. The intention is first to secure minimum standards, and second to moderate differences of wealth and income in different localities.

Part of this support actually comes from the federal government, but since the money is channelled through the provinces amounts are difficult to distinguish. Recognition of regional economic disparities eventuated in a system of grants to the provinces for education. In 1974-75 federal expenditures were $211.3 million or 2.9% of the elementary-secondary total, including what was spent on Indian and overseas schools. The federal government also contributes to elementary-secondary education under a federal-provincial program of co-operation for development of bilingualism in education. In 1974-75 contributions to the provinces under this program amounted to $75 million, raising the federal share of total elementary-secondary expenditures to nearly 5%.

7.3 Post-secondary education

By the 1960s the baby boom children were graduating from high school in record numbers. At the same time, the need for professional and technical manpower, as well as public expectations of education, were growing. Prospective post-secondary students could not be accommodated by existing institutions. In the five years from 1956-57 to 1961-62 enrolment almost doubled to 182,000; by 1966-67 it had risen to 310,500. That year a federal-provincial conference on higher education assigned high priority to post-secondary education. Co-operative action resulted and by 1976-77 a total of 603,500 students were attending post-secondary institutions full time.

7.3.1 Degree-granting institutions

Several types of degree-granting institutions exist in Canada. Universities have, as a minimum, degree programs in arts and sciences; liberal arts colleges are smaller institutions with degree programs, usually only in arts; theological colleges grant degrees exclusively in theology; other specialized colleges offer degree programs in a single field, such as engineering, art or education. In 1976-77 there were 47 universities in Canada, two liberal arts colleges, 12 theological colleges and five other specialized colleges. In this enumeration, affiliated and associated colleges are not counted separately from the parent institution. Fifteen such colleges grant degrees in theology.

History. The first institutions in Canada followed European models. The Séminaire de Québec, founded in 1663, was the base upon which Université Laval was established in 1852. The oldest English-language institution, King's College, at Windsor, NS opened in 1789. By 1867 there were 18 degree-granting institutions in the four provinces united by Confederation. Almost all were supported and controlled by religious groups. Their purpose was to train the clergy and a small, select group of laymen who wished to enter the professions. Teaching concentrated on theology, philosophy, the classics, medicine and law.

A reaction against these practices and attitudes began about the middle of the 19th century. McGill University introduced courses in natural sciences, opened a normal school for elementary teachers and pioneered instruction in applied science and engineering. Similar changes were taking place at other universities — Dalhousie in Halifax, Queen's in Kingston, and the University of Toronto.

While the trend in English-language institutions was toward practical and scientific studies and secular control, in the French-language sector emphasis continued on classical studies under clerical control. A system of classical colleges and seminaries, which became affiliated with Laval, was operated by various religious communities.

When the four western provinces were settled, other structures began to emerge. The American example of land-grant colleges led to a strong commitment to extension programs and community service. The University of Manitoba was granted a charter in 1877. In Saskatchewan and Alberta provincial universities were established in 1909 and 1908, respectively. The University of British Columbia, although chartered in 1908, did not open until 1915. By the outbreak of World War I, a score of universities in Canada had developed distinctive characteristics. To the traditional faculties of theology, law and medicine, schools of engineering, agriculture, forestry, education, dentistry and home economics had been added.

There was some institutional expansion after World War I. In 1939 Canada had 28 universities, varying in size from the University of Toronto with full-time enrolment of about 7,000 to institutions with fewer than 1,000 students. University expenditures, estimated at $11.87 million in 1926, had risen a decade later to nearly $14.15 million. There were about 40,000 students representing 5% of the population between the ages of 18 and 24.

Radical changes began after World War II. As a result of a veteran's rehabilitation program, 53,000 ex-soldiers entered the universities between 1944 and 1951. The immediate problem of space was solved by temporary buildings and creation of satellite colleges. By the mid-1950s places vacated by veterans had been filled with an increasing number of high school graduates. Demands for university expansion continued but the full force of this pressure came in the 1960s when enrolment rose from 128,600 in 1961-62 to 323,000 in 1971-72.

Governments in all provinces became increasingly involved in financing and planning university development. Federal concern was manifested by a system of grants inaugurated in 1951-52. Parliament allocated approximately $7 million (50 cents per capita) to be distributed to the provinces according to their population. They in turn distributed their portion among the universities according to full-time enrolment. The grant increased to $1 in 1957, $1.50 in 1958, $2 in 1962 and $5 in 1966. This amounted to $750 million in 1971-72. As a result, the universities, most of which had operated as private institutions before 1960, became heavily dependent on public funds. Religious sponsorship and control were modified to permit sectarian institutions public support.

In the early 1970s growth rates began to decline. Enrolment in most universities was below forecasts and larger numbers of students withdrew before completing their degrees. Part-time students began to increase in numbers more rapidly than those registered for full-time study. During the last years of the 1970s, full-time enrolment is expected to rise only slightly and to decline after 1980. At the same time, interest in part-time and extension study continues to grow.

Curriculum. Admission to university is usually after 11 to 13 years of schooling. Each institution controls its admission standards and policies. With discontinuation of provincial examinations in recent years, the school record has become the main basis for judging applicants. It is customary for students to enter directly from high school, except in Quebec where they qualify through the collèges d'enseignement général et professionnel (CEGEPs). Most universities provide for the admission of mature students. Those of age 21-24 may be accepted even though they do not meet normal entrance requirements.

The first or bachelor's degree is awarded after three or four years of full-time study. Admission to law, medicine, dentistry, business administration and theology is usually conditional upon completion of part or all the requirements for the first degree. A distinction may be made between general and honours degrees; the latter is more specialized and sometimes requires an additional year of study. A bachelor's degree at the honours level or the equivalent is necessary for acceptance into a master's program. Most entail one year of study, but some take two years. Entrants to doctoral studies must have a master's degree in the same field.

Some universities are bilingual, the University of Ottawa, Laurentian University of Sudbury and Université Sainte-Anne being notable examples. Instruction in these is offered in both English and French. Other universities conduct classes in one language only but permit students to submit term papers, examinations and theses in either French or English.

Higher education for women. Admission of women to undergraduate studies began in the 19th century but their numbers grew slowly. In the 1920s fewer than one-fifth of full-time students were women, and even after World War II the proportion had risen only to one-quarter. By 1970, however, it had increased to more than one-third and is currently 44%. To this must be added the growing number of part-time students. Women are now accepted in all faculties and with the integration of nursing, education and social work into universities they predominate in the social and health sciences. Their enrolment in graduate studies has risen less rapidly, and as a result the increase in women staff members has not been as noticeable.

Teaching staff. During the 1960s the demand for growth necessitated rapid and massive staff recruitment. From about 7,000 in 1960-61, the full-time teaching force increased to more than 30,700 in 1975-76. Most new appointees were Canadians but the number from other countries was significant. Canadian universties have never produced enough graduates to supply their own needs and about 30% of teachers are nationals of other countries. Recent changes in immigration and employment requirements are aimed at ensuring that foreign faculty are hired only after all efforts to recruit qualified Canadians have been exhausted.

Most Canadian universities have four teaching ranks: lecturer, assistant professor, associate professor and full professor. Although appointees are customarily expected to have a doctorate, in practice this applies only in the pure sciences. The percentage of full-time professors with doctorates rose from 40% in 1966-67 to 58% in 1975-76.

Students. There were 376,500 full-time students in Canadian universities in 1976-77. This represented 12% of the population age 18 to 24 and was about double the proportion in 1960. In addition, 188,890 part-time students were registered in degree programs. The number of graduate students (full- and part-time) increased 44% since 1970.

Tuition fees are charged, usually differing from one faculty to another. In Alberta and Ontario higher fees are required of foreign students. Quebec universities base tuition fees on the number of credits taken, irrespective of faculty. In the early 1960s one-quarter of university income was derived from student fees but with the increase in public funding this proportion has been reduced to approximately one-eighth. An estimated 40% of all students take advantage of the federal student loans plan.

Finance. The 1960s marked a turning point in higher education finance as governments began to assume a major share of support. From the beginning of that decade, expenditures rose from about $273 million to more than $1 billion in 1967-68, and to an estimated $3.1 billion in 1976-77. Together, federal and provincial governments contributed 82% of the total, so the importance of other sources declined.

Between 1958 and 1966 federal grants were distributed through the Canadian Universities Foundation, the executive agency of the National Conference of Canadian Universities, predecessor of the Association of Universities and Colleges of Canada. At a federal-provincial conference on university financing it was agreed that federal aid should be broadened to include all post-secondary institutions and that funds should be paid through the provincial governments. The provinces could choose between a per capita grant based on total provincial population or 50% of approved post-secondary operating expenditures. Newfoundland, Prince Edward Island and New Brunswick chose the former; the others, the latter. The original agreement, a section of the Federal-Provincial Fiscal Arrangements Act, extended over the five-year period 1967-72. It was renewed for two years in 1972 and for another three in 1974. A new condition stipulated that the total increase in the federal share for any given year would be limited to 15% of the preceding year.

This agreement expired on March 31, 1977. It has been replaced by the Established Programs Financing (EPF) plan covering education, hospital insurance and medicare. Half the federal payment consists of a transfer of tax points to the provinces (13.5 points of personal income tax and one point of corporation-tax). The other half is a per capita cash grant. The tax portion, based on 1975-76, will grow with the tax base, while per capita grants will increase in relation to the Gross National Product. EPF will be in effect for at least five years with a three-year notice of termination.

Colleges   7.3.2

Traditionally, higher education was the almost exclusive preserve of universities. Now, although universities still account for 62.2% of full-time students, post-secondary education is conducted in a variety of other institutions: regional colleges in British Columbia; public colleges in Alberta; institutes of applied arts and sciences in Saskatchewan; colleges of applied arts and technology in Ontario; collèges d'enseignement général et professionnel in Quebec; and institutes of technology, technical institutes and establishments providing training in the specialized fields of agriculture, fisheries, marine technologies and paramedical technologies.

In the past the term "college" applied to constituent parts of a university. However, it now generally refers to the community colleges which, with support from provincial and federal governments, have developed since 1960 as an alternative to university. A community college is any public or private non-degree-granting institution which provides post-secondary university transfer programs or semi-professional career programs, as well as other credit or non-credit educational programs oriented to community needs. In Quebec completion of a two-year college level program is required for university admission.

Hospital schools of nursing are not considered community colleges, but do comprise part of non-university enrolment. In any case, many provinces have transferred nursing training to community colleges.

History. Many of today's community colleges began as private church-related colleges, public technical schools or university affiliates. But not until the 1960s, often on the recommendation of special commissions, did the provinces attempt to organize post-secondary non-university education into a community college system, either by transforming older institutions or founding new ones. Colleges are based on the philosophy that educational opportunities should extend beyond existing schools and universities to include a broader segment of society. Criteria of admission are more flexible than those imposed by universities. Secondary school graduation is normally required but in some institutions mature student status allows otherwise ineligible applicants to enter. Qualifying programs are also offered to help them attain the appropriate academic level.

Organization. The recent development, structure and organization of post-secondary non-university education differ from province to province. Not all institutions were transformed into community colleges and amalgamated into a province-wide network. A number operate privately. However, the provinces are partially or totally responsible for co-ordinating, regulating and financing community colleges. Some provincial governments finance them completely, while others do so in part. Similarly, the colleges' local autonomy varies.

There are four main patterns of provincial government management: direct establishment and operation, largely confined to institutes of technology in the West and the Atlantic provinces; a triangular partnership between the government, colleges and school district boards, existing only in British Columbia; much delegation of provincial administrative responsibility to college boards, co-ordinated by a provincial commission or board, operating in Ontario and New Brunswick; a partnership between the department of education and college boards supplemented by non-governmental college associations, as in Quebec.

There were 189 institutions offering college-level programs in 1976-77: 30 in the Atlantic provinces; 76 in Quebec; 30 in Ontario; 31 in the Prairies; and 22 in BC.

Curriculum. Colleges offer two basic programs: university transfer and semi-professional career. The former enable students to proceed to university with degree credit of one or two years. The latter prepare them for direct entry into the labour force. Career programs take at least one academic year but more often two or three, sometimes four. Graduates of one-year programs receive certificates, those of longer programs, diplomas.

Quebec students who wish to attend university must first complete two preparatory years in a collège d'enseignement général et professionnel. By contrast, Ontario's colleges of applied arts and technology do not maintain a transfer program; however, universities in the province have agreed to admit with advanced standing college graduates on the basis of individual merit.

Staff. Unlike university faculty who are obliged to conduct scholarly research in addition to teaching, community college staff concentrate almost exclusively on instruction. From an estimated 4,900 in 1964-65, the number of full-time teachers in non-university institutions rose to 18,600 in 1976-77.

Students. Total full-time enrolment in post-secondary, non-university institutions in 1976-77 was 227,000, a 2% increase over 1975-76. About 63% of the students were in community college career programs. Another 33.8% were taking university transfer programs. The rest were in hospital schools of nursing (2.3%) and the Nova Scotia Teachers' College (0.2%). Almost half the students (49%) were female, but this reflects nursing enrolment. While women predominated in career programs (53%), they were outnumbered by men in transfer programs (42%). Quebec students accounted for more than half (53.5%) the total. Enrolment in Ontario represented more than one-fourth (26%), followed by British Columbia (7.5%) and Alberta (7%). In 1975-76 post-secondary non-university institutions awarded 55,400 diplomas: 31,200 to women (56%) and 24,200 to men (44%).

Nursing. Traditionally, registered nurse diploma courses have been conducted in hospital schools. In 1964 Toronto's Ryerson Institute of Technology became the first non-hospital institution to provide nurses' training. Since then the trend has been to transfer most diploma programs from hospital schools to community colleges. The former no longer exist in Quebec, Ontario or Saskatchewan. In the other western provinces most training still takes place in hospital schools but programs are also available in community colleges. Only in the Atlantic region is training carried out exclusively in hospital schools.

7.3.3 Technical and trades training

Technical and trades training varies between and within provinces. Beyond courses and programs available in high schools, students can continue this type of education at several levels in a variety of institutions.

History. Early in the 20th century, the accelerating pace of industrialization gave added importance to the acquisition of technical skills. Since public schools or universities rarely gave such instruction, this was one of the first areas of education in which the federal government became actively involved. In co-operation with several provinces, an agricultural training program was set up in 1913. Three years earlier the Royal Commission on Industrial Training and Vocational Education had made recommendations, a number of which were implemented in the Technical Education Act of 1919. Under the act, federal authorities offered to support provincial programs but few provinces were ready to participate. By World War II, however, enough programs had been instituted to warrant appointment of a national council of federal, provincial and public representatives to advise the labour minister on matters relating to vocational education. At that time most vocational institutions were administered by a variety of provincial government departments such as labour, agriculture, commerce and industry.

During the 1950s a critical shortage of technical manpower prompted federal officials to give the provinces greater assistance for vocational training. By 1960 about 30 technical institutes had been opened. The next year the Technical and Vocational Education Act was passed. It was designed to encourage the provinces to extend and improve facilities. Thereafter, new comprehensive schools frequently incorporated vocational programs. Federal participation increased after 1966 with adoption of the Adult Occupational Training Act and assistance through the purchase of courses given in various types of provincial institutions. Under this act, a training-in-industry program was inaugurated in 1967 and a training-on-the-job program in 1971. Both were superseded by an industrial training program which came into effect in 1974.

Institutions and programs. Technical and trades training is offered in public and private institutions such as community colleges, institutes of technology, trade schools and business colleges. It may also take place on the job, in apprenticeship programs or training in industry.

Technical career programs are conducted in community colleges and related post-secondary non-university institutions. High school graduation is usually required for admission. In programs lasting up to four years, students are trained to practice a career directly upon graduation. Some community colleges also give vocational instruction but graduates of career programs are generally qualified for semi-professional work.

Trades level courses emphasize manipulative skills and performance of established procedures and techniques. Less than one year is normally needed to complete them. Grade 9 or 10 is usually required for entrance but prerequisites vary.

Public trade schools and vocational centres concentrate on vocational skills and are administered by a provincial department. They may be separate establishments or divisions of a community college. Only persons who have left the regular school system and are older than the compulsory age may attend. High school graduation is not usually required although, depending on the province and the trade, admission standards can range from grade 8 to grade 12. Included in this group are adult vocational centres and schools related to specific occupations such as police work, forestry and nursing.

A number of institutions offer academic upgrading courses designed to raise a trainee's general level of education in one or a series of subjects. Courses may be taken to qualify for admission to higher academic studies or vocational training. The federal government sponsors basic training for skill development in public trade schools and adult vocational centres. However, completion of levels corresponding to the final grades of secondary school does not give high school graduation status.

Rather than attend an educational institution, individuals may acquire trades training as they work. Training on the job is organized instruction offered in a production environment. Skills and knowledge relating to a specific trade or occupation are imparted in a methodical, step-by-step approach.

Training in industry is provided by business and industrial establishments to train new employees, retrain experienced workers or upgrade their qualifications. It may be publicly supported, in full or in part, or entirely financed by the company. Training can be on the job, by classroom instruction, or a combination of the two. Under cost-sharing agreements the federal government reimburses companies that provide training. The provincial government monitors the company programs and approves them for federal support.

Apprenticeship programs combine on-the-job training with classroom instruction. Persons contract with an employer to learn a skilled trade and eventually reach journeyman status. Apprentices may be registered with a provincial or territorial labour department in order to train in a designated apprenticeship trade. The department sets standards for qualifying as a journeyman: minimum age, educational levels for admission, minimum wages, duration of apprenticeship and the ratio of apprentices to journeymen. Non-registered apprentices enter into a private agreement with an employer, perhaps in association with a labour union. They are not subject to regulations established by the provincial department for that trade.

The federal Vocational Rehabilitation for Disabled Persons Act was passed to facilitate trades-training for the handicapped. The federal government reimburses the provinces for 50% of the costs incurred for programs that enable disabled people to support themselves fully or partially. The provinces provide training directly in their community colleges and trade schools or purchase it from the private sector or voluntary agencies. Quebec does not participate in this agreement.

In co-operation with the provinces, the federal government has introduced standard interprovincial examinations to promote the mobility of journeymen. Those who pass examinations in certain apprenticeable trades have an interprovincial seal attached to their certificate, allowing them to work in any province.

Staff. In 1976-77 full-time educational staff administering and teaching trades level courses numbered 5,400. On the average, they had seven years of teaching experience and two years in industry. At the same time, 18,600 were administering and teaching in post-secondary technical programs.

Students. In 1975-76 an estimated 473,000 full-time students were enrolled in institutions providing technical and trades training. About three-quarters were in community colleges, and most of these were studying at the technical level. The rest were distributed between public trade schools (87,080) and hospital schools (6,600). The same year there were 116,100 registered apprentices, 57,000 participants in the federal manpower training program, 74,300 in Ontario's business and industry training program and 4,700 in vocational rehabilitation for disabled persons.

Business was the most popular field in both career programs and trades level courses (each with 28% of the students). Second place in career enrolment was taken by medical and dental fields (21%), while at the trades level, engineering and medical technologies and trades were chosen by the second highest percentage (16%).

7.3.4 Continuing education

Continuing or adult education is adapted to the needs of persons not in the regular system. Out-of-school adults (15 and older) are able to pursue accreditation at diverse levels or to advance their personal interests. Continuing education is given by school boards, provincial departments of education, community colleges and related institutions, and universities. Programs are also conducted or sponsored by non-profit organizations, professional associations, government departments, business and industry. However, it is not centred exclusively around institutions. As well as the time-honoured correspondence course, instruction is now available from travelling libraries, radio and television.

History. School boards and provincial departments of education have conducted evening classes for adults as far back as the turn of the century. However, rapid development occurred only after World War II. By the late 1950s more than 445,000 enrolments in academic and vocational courses were reported.

At the post-secondary level, extension programs have been part of some universities for many years. Probably most successful were those in the provincial universities of the West. Agricultural extension education was provided in Alberta and Saskatchewan and at St. Francis Xavier University in Nova Scotia fishermen's co-operatives were organized. Besides these practical and vocational programs, other cultural and recreational services were developed by several urban universities in Central Canada. Some courses were for academic credit, others were not. Many were offered only on campus, others were given in external centres as well.

Since the end of World War II, demand for continuing education has increased and new teaching media, such as television, have broadened the range of facilities. Extramural courses and degrees are now available from most universities. By 1971-72 about one million persons were taking courses offered by school boards, departments of education, provincial correspondence and vocational schools, colleges and universities.

Programs and courses. Individuals can participate in continuing education as part-time students in regular credit programs, or as students in the non-credit programs. Credit courses sponsored by school boards and departments of education may be applied toward a high school diploma. Credits in academic or vocational subjects can be acquired through evening classes or correspondence study. Post-secondary credit courses count toward a degree, diploma or certificate.

Non-credit programs consist of courses taken for personal enrichment or for leisure time use. Instruction is given in hobby skills (for example, arts and crafts), social education (health and family life), recreation (sports and games), and driver education. Professional development and refresher courses are also available to persons with prior training and experience.

Both programs include formal and non-formal courses. Formal courses are structured units of study presented systematically. Non-formal courses are activities for which registration is not required but where attendance or participation for a scheduled period is necessary.

Elementary-secondary institutions. Each province and territory has its own method of conducting continuing education in elementary-secondary schools. Administrative control in most provinces is assigned to individual school boards, but a variety of funding schemes has resulted in programs of different size. Continuing education is best developed in areas under the jurisdiction of large, urban-based boards.

In Saskatchewan all school board programs are now administered by a network of community colleges, although school board facilities are used. Likewise, Holland College in Prince Edward Island administers continuing education courses formerly provided by the education department. A similar administrative change from school boards to community colleges is occurring in British Columbia and New Brunswick. However, many boards will still offer some continuing education programs. The departments of education in Newfoundland and the Northwest Territories administer the programs from head office, while school board facilities are used for instruction. In the Yukon Territory, continuing education courses conducted by the education department are available through a vocational and technical training centre.

Community colleges and trade schools. Almost all community colleges and many public trade schools now provide part-time learning opportunities for adults. In 1975-76 more than 140 institutions operated extension courses ranging from academic upgrading and vocational programs to hobby courses. Part-time enrolment includes students in vocational training (trades level) and semi-professional career programs, academic upgrading, owner-manager supervisory courses sponsored by the employment and immigration department, and personal enrichment courses.

Universities. In 1975-76, 56 universities conducted non-credit programs. Moreover, part-time credit enrolment has always been substantial. In addition, the Banff Continuing Education Centre in Alberta has a program similar to that of the universities.

While the extent and type of involvement vary from one university to another, extension programs for students who cannot attend on-campus classes have become a recognized and accepted responsibility. Manitoba has established regional resource centres to which universities and other types of educational institutions contribute. Quebec has successfully developed Téléuniversité as a branch of the Université du Québec, and Memorial University has launched an experimental program on Fogo Island. In Alberta, Athabasca University is an open university sponsored by the provincial advanced education and manpower department to produce and deliver learning programs for adults who wish to study in their own communities or are unable to attend a traditional post-secondary institution.

Students. Overall, during 1975-76 more than 1.6 million students were taking continuing education courses at the various institutional levels. In relation to the out-of-school population 15 and over, 113 out of every 1,000 people were enrolled, up from 89 per 1,000 in 1972-73 when the total was 1.2 million.

There were 1,040,000 registrations, representing about 604,900 individuals, in school board and department of education courses. Credit registrations declined 55%, while non-credit rose 73%. Fine and applied arts were most popular (35.2% of all registrations) followed by the humanities (15.8%) and household sciences (11.7%).

Registrations in formal continuing education courses offered by community colleges and trade schools numbered 573,300 in 1975-76. This represented more than

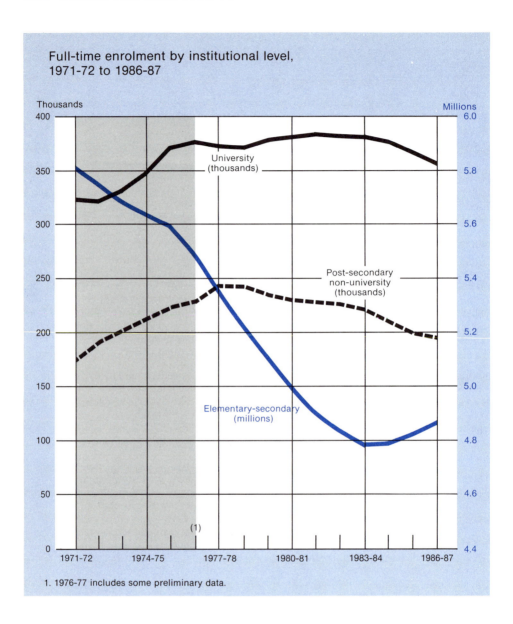

Full-time enrolment by institutional level,
1971-72 to 1986-87

1. 1976-77 includes some preliminary data.

431,100 students. In four years non-credit enrolment increased by more than 218%, compared with a 57% rise in credit courses. As in elementary-secondary institutions, fine and applied arts were predominant with 40% of all non-credit registrations. Business management (13%) and trade and technical courses (11%) ranked second and third.

In 1975-76 there were 748,100 registrations in university part-time credit and formal non-credit courses, a 42% increase since 1970-71. The 548,100 students outnumbered total full-time enrolment, which was 370,400. More than half (61%) the part-time students were enrolled in credit courses. But since 1970-71 the number taking such courses rose by only 20%, compared with an 82% jump in formal non-credit enrolment. Business and management were most popular non-credit courses, accounting for 24% of all registrations. Next were the social sciences (16%), health sciences (15%) and humanities (13%).

Education in the provinces and territories   7.4

Although a general structure of education prevails throughout the country, the system in each province is unique. Diverse historical developments, cultural traditions, geographic situations, and economic and social conditions have resulted in 12 education systems. Furthermore, even within a province school organization may vary from one region to another.

Newfoundland   7.4.1

Enrolment in Newfoundland's elementary and secondary schools was 158,100 in 1976-77; total full-time teachers numbered 7,740. The province's non-university institutions had 2,000 full-time students and 210 teachers, while full-time enrolment at Memorial University was 6,640 and teachers totalled 765. Education expenditures ($321 million) were 12.2% of the Gross Provincial Product, a larger percentage than was spent by any other province. Per capita this amounted to $576.

Established in 1874, the education system in Newfoundland was originally sectarian. As a result of a 1964 provincial royal commission on education and youth, the school systems of the major Protestant denominations were consolidated, although the Roman Catholic, Pentecostal and Seventh Day Adventist churches still manage their own schools. Reorganization in 1969 divided the province into six regions containing a total of 35 school districts. The Pentecostal Assemblies and the Seventh Day Adventists each operate a single district which theoretically covers the whole province. The Roman Catholic system is largest, but its school boards have been cut from more than 100 to 12.

Children age 7 to 15 must attend school. Enrolment before grade one is not compulsory, but with construction of larger, more centralized elementary schools, most 5-year-olds go to kindergarten. School organization follows two major patterns: elementary (kindergarten and grades 1-6) with central high schools (grades 7-11), and elementary (kindergarten and grades 1-8) with regional high schools (grades 9-11). There are a few junior high schools (grades 7-9) and about a dozen district vocational schools.

Technical training is provided by the Newfoundland College of Trades and Technology and the College of Fisheries, Navigation and Marine Engineering and Electronics. Registered Nurse diploma courses are conducted exclusively in hospital schools. Memorial University grants undergraduate and graduate degrees, diplomas and certificates. The campus is located in St. John's but extension programs operate in 26 centres. A junior college campus of Memorial has been established in Corner Brook.

Prince Edward Island   7.4.2

During the 1976-77 academic year, 28,000 elementary and secondary students were enrolled in Prince Edward Island's schools and were taught by 1,440 teachers. Full-time non-university enrolment was 750, and at the University of Prince Edward Island 1,480. There were 70 full-time non-university teachers and 110 at the university. The province devoted $67 million, 11.7% of its Gross Provincial Product, to education. Per capita this amounted to $569.

Throughout the 1960s small education units were consolidated. In 1972 the system was changed to one consisting of five administrative units with a school board in each. The age span of compulsory attendance is 7 to 15. Kindergarten is not part of the public system but private classes are available. The major pattern of organization is: elementary (grades 1-6), junior high (grades 7-9) and senior high (grades 10-12). In some cases, only two levels are distinguished: elementary (grades 1-8) and high school (grades 9-12). No provision is made for separate schools.

A network of regional high schools offers academic programs from grades 9-12 and a one- or two-year business education course. After grade 12, four additional years are required for a bachelor's degree. Two vocational high schools are operated by Holland College on behalf of the education department.

At the post-secondary level, Holland College offers non-degree programs, but because of the province's small population extensive technical education facilities have not been developed. Interprovincial arrangements allow students to attend the Nova

Scotia Agricultural College and institutions in Central Canada. The Prince Edward Island School of Nursing is the only establishment from which a nursing diploma may be obtained. In 1969 the University of Prince Edward Island was created as an amalgamation of St. Dunstan's University and Prince of Wales College.

7.4.3 Nova Scotia

In 1976-77 Nova Scotia's 203,900 elementary and secondary students were taught by 10,780 full-time teachers. At the post-secondary level full-time non-university enrolment totalled 3,000 and university enrolment 18,200. Full-time teaching staff in the two types of institution numbered 370 and 1,600, respectively. Education expenditures were $522 million or 10.4% of the Gross Provincial Product, representing an outlay of $630 per capita.

School board amalgamation occurred in 1968. For education purposes, the province is divided into municipalities, town sections, city sections and regional units, each with its own board.

Nova Scotia is the only province where pre-grade one (known in the province as primary) is obligatory, so the ages of compulsory attendance extend from 5 to 16. The predominant grade organization is: elementary (primary to grade 6), junior high (grades 7-9), and senior high (grades 10-12). Secondary graduation occurs after grade 11 (junior matriculation) or grade 12 (senior matriculation). A bachelor's degree requires three years of study beyond grade 12. The only vocational education given by regular high schools is in business and commercial subjects. Students in various secondary grades may choose one-, two- or three-year occupational training programs at regional vocational schools, most of which are operated by the province. Although there is no provision for separate schools, local boards can designate certain schools and teachers for instruction of religious or language groups.

Nova Scotia has two institutes of technology offering trades and post-secondary technical courses, an agricultural college providing terminal and university transfer programs, a land survey institute and a nautical institute. The Canadian Coast Guard College is located in Sydney. All diploma education for nurses is conducted by hospital schools. The Nova Scotia Teachers' College, with an enrolment of 550 in the fall of 1976, is the only independent normal school in Canada.

Graduate and undergraduate degree programs are available in the province's eight degree-granting institutions. The Nova Scotia government contributes to the support of Mount Allison University in New Brunswick as compensation for Nova Scotians who attend it. The government also supports the faculty of forestry at the University of New Brunswick. In turn, New Brunswick and Prince Edward Island students are accepted at the Nova Scotia Agricultural College, and by the faculties of medicine and dentistry at Dalhousie.

7.4.4 New Brunswick

New Brunswick's 1976-77 elementary-secondary enrolment totalled 164,500 and the teaching force numbered 7,760. The province's non-university institutions and four universities had 1,450 and 11,100 full-time students, respectively, taught by 180 and 1,100 teachers. Expenditures on all levels of education amounted to $371.5 million, representing 9.6% of the Gross Provincial Product and $549 per capita.

The province is divided into seven regions, each administered by a superintendent, and subdivided into 33 school districts. Every district is under the authority of a board of trustees. All board members serve a three-year term. Some are elected, others are appointed by the lieutenant-governor-in-council.

School attendance is compulsory from 7 to 16. Kindergarten is not offered in the public system. Students progress through 12 years of school leading to junior matriculation. The most common patterns of organization are: elementary (grades 1-6), junior high (grades 7-9) and senior high (grades 10-12) or elementary (grades 1-6) and high school (grades 7-12). Secondary students have a choice of three programs: college preparatory and technical; general educational and occupational; and practical. No provision is made for separate schools.

The New Brunswick Institute of Technology, the Saint John Institute of Technology and Northeastern Community College offer post-secondary vocational and technical programs. The Maritime Forest Ranger School is located in Fredericton. In addition to hospital schools, the Saint John School of Nursing provides RN training.

After grade 12, four years of study are required for a first degree. The province's four universities are the University of New Brunswick, St. Thomas University, Mount Allison University and Université de Moncton, the last providing higher education to the French-speaking population.

Quebec   7.4.5

Quebec spent $737 per capita on education in 1976, more than any other province. But the $4.6 billion represented only 10.2% of the Gross Provincial Product.

Enrolment in elementary and secondary schools was 1,398,600, about a quarter (25.5%) of the national total. Full-time elementary and secondary teachers numbered 72,600. At the post-secondary level, non-university enrolment was 121,400 in contrast to 77,600 in universities; Quebec is the only province where the former exceeds the latter, but this is because of the collèges d'enseignement général et professionnel (CEGEP) system. The relative sizes of the full-time staff in the two types of institution reflect this unique distribution: 8,990 non-university teachers; 6,810 in universities.

Much of the province's present education system resulted from a 1961 royal commission study of education. In 1964, acting on the commission's recommendations, the government passed legislation that created a ministry of education. The province was divided into nine administrative areas, each containing a regional education office headed by a director. A superior council of education was also created in 1964 as a public consultative body to supplement the department. Its 24 members are appointed by the government for a four-year term.

Each municipality has one or more public schools under the control of school commissioners or trustees. Elected five-member boards of school commissioners operate schools for an area's majority population, Roman Catholic or Protestant. However, a minority of ratepayers may constitute a separate school municipality under a board of school trustees. This three-member board, too, is elected and can own property, levy taxes, receive government grants, operate schools and hire teachers. The province, excluding Montreal, contains 189 school commissions grouped into 64 regional school boards; nine are Protestant. Montreal has an additional seven school commissions.

The ages of compulsory attendance are 6 to 15 although kindergartens admitting 5-year-olds are now part of the system. Elementary school consists of six years based on continuous progress. The comprehensive secondary program lasts five years. Promotion throughout is by subject and unlike most other provinces a final departmental exam is required for graduation.

Post-secondary education begins in the tuition-free CEGEPs. Inaugurated in 1967-68, CEGEPs were generally not created anew, but resulted from reorganization of existing institutions such as normal schools, classical colleges and technical institutes. They are administered by a public corporation composed of faculty, students, parents and community representatives but depend for revenue wholly on the education department. The department regulates budgets and issues guidelines for curriculum and administration.

Quebec is the only province where students must enrol in a community college before going to university. As well as the two-year preparatory academic program, CEGEPs provide three-year vocational programs that train students for direct labour market entry. Of the 37 CEGEPs, four are English-language institutions. Nursing diploma training takes place only in the CEGEPs.

Private or classical colleges offer the equivalent of the two-year CEGEP university transfer program. Students may, however, continue at the college and work toward a degree from the university with which it is affiliated.

The first degree requires three additional years of study after completion of two CEGEP years. The seven universities in the province (three of them English) have a variety of undergraduate and graduate degree, diploma and certificate programs. A semi-independent universities council plans their general development and makes recommendations on operating and capital budgets. It is chaired by a government official but includes representatives of the public and the universities.

7.4.6 Ontario

During 1976-77 more than one-third of Canada's elementary and secondary students were enrolled in Ontario schools — over 2 million students taught by nearly 100,000 full-time teachers. Full-time non-university enrolment was 58,900 — less than half the 164,000 students in the province's universities. Full-time teachers of the former numbered 5,010; of the latter 12,650. The $5.2 billion spent on education was 7.0% of the Gross Provincial Product. It amounted to $635 per capita.

Ontario was the first province to divide responsibility for education between two departments. A ministry of education is concerned with the elementary and secondary levels; a ministry of colleges and universities deals with post-secondary matters. Each department has its own minister.

Since 1966 the number of school boards in Ontario has been reduced from 1,600 to 193. Three types of boards exist: boards of education (76), non-sectarian bodies responsible for elementary and secondary education in large areas such as counties, districts, cities; boards which operate one type of school only (108), such as public elementary schools and Roman Catholic separate schools; and boards operating schools on Crown lands (9).

Roman Catholic schools provide tax-supported educational services for kindergarten through grade 10. In some schools of separate boards, grades 11, 12 and 13 are also offered but these grades constitute a private school and are not under the jurisdiction of the board.

Attendance is compulsory from 6 to 16 years. Most schools provide an optional year of kindergarten for 5-year-olds, and in some urban areas, junior kindergarten for 4-year-olds. Ontario has a 13-grade system: elementary school lasts eight years, secondary five. However, as continuous progress has become popular, the conventional grade pattern has been modified. The detailed standardized course of study has been replaced by curriculum guidelines issued by the ministry. The curriculum has been divided into four three-year segments: primary, junior, intermediate and senior.

Secondary education operates on a credit system; 27 credits are required for a graduation diploma (grade 12); six additional credits in honour level work for an honour graduation diploma (grade 13). The latter is necessary for university admission. High schools also offer trade, technical and business programs that prepare students for either immediate employment or entry to a college of applied arts and technology (CAAT) or other post-secondary non-university institution.

In the mid-1960s institutes of technology and provincial vocational centres were incorporated into CAATs. A network of 22 on more than 50 campuses provides technical and trades programs for students who do not intend to go to university. Although CAATs were not designed to accommodate prospective transfer students, universities do admit some graduates into the second or third year of degree courses. CAATs are completely under the jurisdiction of the ministry of colleges and universities. The Ontario Council of Regents, a 15-member body appointed by the government, advises on new programs and other matters. Each college is a separate corporation with a 12-member board of governors. In addition to CAATs, post-secondary non-university training is available in four colleges of agricultural technology, a school of horticulture, a chiropractic college and an institute of medical technology.

Ontario's 22 universities offer undergraduate and graduate programs leading to degrees, diplomas and certificates in a wide range of fields.

7.4.7 Manitoba

In 1976-77 enrolment in Manitoba's elementary and secondary schools was 241,000 and teachers numbered 12,180. Non-university institutions had 3,400 full-time students and 330 teachers, while the corresponding figures for universities were 18,300 and 1,600.

The $618 million spent on education was 7.8% of Manitoba's Gross Provincial Product and represented $605 per capita.

Two provincial departments have been established: a department of education and a department of college and university affairs. One minister may be responsible for both, but each structure has its own officials at the deputy minister level and below.

Local administration of elementary and secondary education is based on a variety of units: multi-district divisions, unitary school divisions, remote school districts, special revenue districts and special schools in sparse communities. School divisions and districts are under the jurisdiction of an elected board of trustees; special schools are administered by a trustee appointed by the provincial cabinet. No legal provision is made for separate schools.

The compulsory ages are 7 to 16. Elementary-secondary education lasts 12 years, and is organized into a six-year elementary segment, and three years each of junior and senior high school. However, where enrolment is low the pattern of eight elementary and four secondary years prevails. Final examinations are set and marked under the auspices of a high school examination board.

In high school, vocational students may take pre-employment commercial or industrial programs. There is also an occupational entrance program commencing at grade 7 and continuing until grade 10 or 11. A number of vocational secondary schools have been constructed in co-operation with federal authorities.

Support from the federal labour department helped establish an institute of technology in Winnipeg. This facility and vocational centres in Brandon and The Pas were designated community colleges in 1969. They offer post-secondary career and vocational courses. Although no provision is made for university transfer, in special circumstances graduates of career programs have been granted credits applicable to a degree. Registered nurses' training is provided at one college and at five hospital schools.

The province has three universities but only the largest, the University of Manitoba, has a faculty of graduate studies. In addition, four colleges (two associated with universities) grant degrees to students training for church ministry.

Saskatchewan   7.4.8

Saskatchewan's 1976-77 education expenditure of $543 million represented a per capita outlay of $589. It was equivalent to 7.5% of the Gross Provincial Product. Elementary and secondary enrolment totalled 226,700, and teachers numbered 11,130. In post-secondary non-university institutions, 350 teachers taught 2,390 full-time students, and in universities the corresponding figures were 1,370 and 15,000.

Two departments — education and continuing education — report through the same minister, but have separate structures. The latter was established in 1972 to handle all post-secondary matters.

The province is divided into eight education regions, and subdivided into 66 school units plus non-unit rural districts, villages, towns and cities. Local administration is based on districts, which may be set up for public and Protestant or Roman Catholic separate schools. School boards of five to eight members are elected in each district for three-year terms. Education in northern areas is administered by a department for Northern Saskatchewan.

Attendance is compulsory from 7 to 16 years, although kindergarten is available, particularly in larger centres. The traditional 12 elementary-secondary grades have been reorganized into four three-year divisions. Prospective grade 12 graduates must write standard departmental examinations.

High schools offer vocational subjects in general, industrial arts, commercial or special terminal programs, none of which qualify students for university entrance. The content of such courses is co-ordinated with the province's two community colleges and three technical institutes. Agricultural courses are given throughout the province in co-operation with the provincial agriculture department and apprenticeship training is provided in conjunction with the labour department. All nursing instruction is given by community colleges.

The University of Saskatchewan and the University of Regina operate undergraduate and graduate degree programs. In addition, four theological colleges (two associated with the University of Saskatchewan) have degree-granting power.

7.4.9 Alberta

Alberta's 1976-77 elementary and secondary enrolment totalled 451,400 and teachers numbered 22,270. Full-time university enrolment (33,500) was double that in post-secondary non-university institutions (16,600). The relative distribution of full-time teachers was 2,660 versus 1,730. Education expenditures of $1.2 billion amounted to 6.9% of the Gross Provincial Product. This represented an outlay of $655 per capita. Only in the North and Quebec was per capita spending greater.

In 1972 responsibility for education was divided between the education department and the department of advanced education and manpower, each with its own minister. The former deals with the elementary and secondary levels. The latter has jurisdiction over universities, post-secondary non-university institutions, vocational training centres, adult education provided by school boards and other public or private agencies, apprenticeship programs, federal manpower programs, and programs offered by licensed business or trade schools.

Local administration is based on the school district, although responsibility has largely been assumed by school divisions and counties. There are 60 school divisions and counties containing 4,191 districts and 149 districts not in divisions or counties. Districts and divisions have their own school boards. Religious minorities may establish separate school districts with the same rights and obligations as public ones.

Attendance is compulsory from ages 6 to 15. Kindergarten is not part of the provincial school system although some urban centres provide it. The predominant grade pattern is: elementary (grades 1-6), junior high (grades 7-9), and senior high (grades 10-12). Secondary schools operate on the comprehensive or composite principle. Thus academic and a wide range of vocational subjects are taught.

Post-secondary technical education is offered at two institutes of technology, and at agricultural and vocational colleges. Six community colleges and three other colleges have university transfer and technology programs, adult education, community service and academic upgrading. Nursing diploma programs are given in hospital schools and four community colleges.

University of Alberta programs lead to degrees, diplomas and certificates at the undergraduate and graduate levels. A constituent college, the Collège universitaire Saint-Jean, has a bilingual program toward the first degree. A number of affiliated colleges provide up to two years of university education. The university, located in Edmonton, operates extension programs in other centres. As well, there are two other universities, Calgary and Lethbridge, and the Newman Theological College.

7.4.10 British Columbia

British Columbia's 1976-77 expenditures on education ($1.4 billion) were third highest in the country after Ontario and Quebec but represented only 6.6% of the Gross Provincial Product, the lowest in any province. Spending per capita was $572.

More than half a million (562,140) students were enrolled in elementary and secondary schools with a full-time teaching staff of 27,780. The province's 17,090 full-time non-university students were taught by 1,360 teachers and the corresponding numbers of university students and faculty were 32,680 and 2,870.

The province is divided into about 90 school districts, each of which elects a board of trustees for a two-year term. As well as having jurisdiction over its district, a board may establish and operate regional colleges in conjunction with one or more other districts. No legal provision is made for separate schools.

Children age 7 to 15 must attend school but participation in an optional kindergarten year is almost universal. Elementary school extends over seven years, followed by three years of junior and two years of senior high. In the senior years employment-oriented courses are available. As well as high school courses, technical and vocational education is provided by programs run in co-operation with the federal government, and in vocational schools throughout the province.

Most of the 17 regional colleges are operated by groups of school boards, while the British Columbia Institute of Technology is maintained by the education department. The colleges conduct a variety of career and transfer programs, although some specialize in particular fields such as art or fashion design. BCIT provides career training only. As well as in hospital schools, students may earn nursing diplomas from BCIT and several community colleges.

The largest degree-granting institution, the University of British Columbia, has undergraduate and graduate programs in most major disciplines. There are three other universities and a number of colleges, most church-related. The universities are supervised by two regulatory bodies that advise the government on curriculum and finance. Another government-appointed board makes recommendations on development, co-ordination and financing of all types of post-secondary institutions.

Yukon Territory and Northwest Territories   7.4.11

Unlike the provincial governments that provide more than half the funds for education (virtually 100% in Prince Edward Island, New Brunswick and Newfoundland), in the Yukon Territory and Northwest Territories, federal sources predominate (64%). The territories' combined 1976-77 education expenditures of $178.5 million exceeded only those of Prince Edward Island, but on a per capita basis they were $858, higher than in any other part of the country.

Elementary and secondary enrolment in the Yukon Territory was 4,870 in 1976-77, and 12,920 in the Northwest Territories. There were 270 and 675 teachers, respectively. Neither region has post-secondary institutions.

The Yukon school system is administered by the territorial education department. The School Ordinance of 1962 recognized three types of school: public, separate and Indian. However, since the closing of the last Indian school in 1969, all native children have gone to public or separate schools. Attendance is compulsory from ages 7 to 16. Grades 1 to 12 follow the British Columbia organization and curriculum. Some secondary schools give commercial and technical courses, and advanced trades and technical training is available at the Yukon Vocational and Technical Training Centre. An aid program allows students to continue at the post-secondary level in one of the provinces.

In the Northwest Territories responsibility for education was transferred in 1969 and 1970 from the federal Indian and northern affairs department to the Territorial Council. The ages of compulsory attendance are 6 to 16. The region has developed its own curriculum, covering six elementary and four secondary grades. Because high schools are located only in the larger centres, residential facilities are available for children from outside the community. Some vocational training, too, is given in urban areas. As is the practice in the Yukon, the Territorial Council operates aid programs for students who wish to attend a post-secondary institution in another part of the country.

Council of Ministers of Education   7.4.12

An interprovincial council of ministers of education was established to facilitate co-operative action at the policy level by allowing the provinces to negotiate collectively with the federal government. The council grew out of discussions held in a standing committee of ministers of education established in 1960. An agreed memorandum was adopted in 1967 and amplified in 1974. The stated purpose of the council is to enable ministers of education to consult and act together on common interests. Provision is also made for consultation with other educational organizations. The council adheres to the principle that provincial ministries must remain autonomous; hence, no recommendation or decision is binding. Meetings are held at least twice a year. The council appoints an executive committee consisting of a chairman, vice-chairman and three other members representing all regions of the country.

Maritime Provinces Higher Education Commission   7.4.13

During the 1960s each of the Maritime provinces — Nova Scotia, New Brunswick and Prince Edward Island — appointed special committees or commissions to make recommendations on university development and finance. In 1975 a Maritime provinces higher education commission was created to advise the premiers, and through them the governments, on higher education in all three provinces. The commission dispenses operating and capital grants directly to universities and colleges in New Brunswick and Prince Edward Island. For Nova Scotia, grants are made by the education minister following the commission's recommendations.

7.5 Federal involvement in education

7.5.1 Educational services

Department of National Defence. As well as the schooling of children of service personnel in government quarters, the defence department is directly responsible for the instruction and training of those who join the armed forces.

The Canadian Forces Training System (CFTS) with headquarters in Trenton, Ont. plans, conducts and controls all recruit, trades, specialist and officer classification training. Five bases and 30 schools across Canada are under CFTS jurisdiction. CFTS trains an annual intake of about 11,000 recruits and provides continuing instruction to regular forces and reserve personnel. About 200 classification and trades qualification courses, more than 575 in-service specialty courses and 700 other courses are available. Canadian forces trades training is now accredited in most provinces. The average daily population at CFTS schools is 5,720 and in other schools 2,273. The average annual number of graduates is 37,158.

An agreement between the defence department and the University of Manitoba permits military personnel and their dependents to work toward a degree. Manitoba is the first university to award academic credit for training courses conducted at Canadian forces schools and for service experience.

A comprehensive system of educational courses and professional development programs prepares potential officers — the three-stage Officer Career Development Program. A combination of screening and self-selection, the program can cover participants throughout their years of service until retirement.

The department finances and controls three tuition-free colleges: the Royal Military College in Kingston, Ont., Royal Roads in Victoria, BC, and the Collège militaire royal de Saint-Jean in Saint-Jean, Que. Academic courses leading to degrees in arts, science or engineering are supplemented by military studies and practical training. Close to half of all graduates receive engineering degrees. Graduates are required to serve three to five years in the armed forces. The department also assists other educational institutions in carrying out certain specialized instruction and defence research. About 1,000 cadet corps are active in Canada. Most members are high school students who spend a limited amount of time learning military fundamentals.

Other federal instructional programs. The Public Service Commission provides federal public servants with refresher and upgrading courses, study grants, career development opportunities, and language training. The ministry of the solicitor general has an educational program for inmates of federal penal institutions. Full- and part-time instruction is offered in vocational and academic subjects, sometimes with credit given by provincial authorities. A day-parole system allows some prisoners to attend secondary schools, colleges and universities. The Canadian International Development Agency (CIDA) operates and administers a technical assistance program in developing countries.

7.5.2 Indirect participation

The growth of education, both in size and importance, made it almost inevitable that the federal government would play some role in its development even though the BNA Act restricts direct participation. Many departments have educational functions, but they tend to take a financial form. Grants for post-secondary and minority language education and sponsorship of manpower training programs have already been noted. A number of other federal bodies also make significant contributions.

Department of the Secretary of State. In 1963 the education support branch of the secretary of state department was established to advise the Cabinet on post-secondary education. In 1967 it became responsible for administering those parts of the Federal-Provincial Fiscal Arrangements Act that related to post-secondary finance. By 1973 the branch's authority had been enlarged to include the development, formulation, implementation and review of all federal policies and programs on education. This entailed communication with provincial governments, the academic community and national organizations, and co-operation with the external affairs department to co-ordinate Canada's international efforts.

In addition to administering post-secondary adjustment payments, the branch took over the student loans plan from the finance department in December 1977. Students had received direct aid since 1939. However, not until 1964 was a comprehensive scheme adopted to assist those whose financial circumstances would prevent them from carrying on full-time post-secondary studies.

Under the plan the government guarantees loans made by chartered banks and other designated lenders to students on the basis of certificates of eligibility issued by participating provinces. The federal government carries the cost of interest payments on these loans while students continue full-time studies and for six months after. There is no age limit for borrowing. All provinces participate except Quebec which has its own student assistance scheme. Effective July 1975 the maximum loan per academic year was $1,800 or $900 per semester to a total of $9,800. The repayment period may extend up to 10 years from the time a borrower leaves the educational institution. The act provides for basic allocations to each province and also supplementary allocations to compensate for differences in relative demand based on provincial populations between the ages of 18 and 24.

During 1975-76, 153,419 students received 190,893 certificates of eligibility with a total value of $157.4 million. They negotiated 179,735 certificates for loans amounting to $145.6 million. A small proportion of the students did not use their certificates or drew less than the authorized sum. From inception of the plan to June 30, 1976, lenders negotiated 1,369,884 certificates of eligibility for a total of $935.5 million.

The revenue department has given students further financial aid. Since 1961 they have been permitted deduction of tuition costs from taxable incomes and since 1972 education expenses up to $50 a month have also been deductible.

Health Resources Fund Act. In 1966 the federal government inaugurated a program of financial support to the provinces to provide facilities for training professional health services personnel. The Health Resources Fund Act (1970), administered by the health and welfare department, authorized establishment of a fund to be used for training or research facilities. Up to 50% of the cost of projects approved by an advisory committee is paid to the provinces. A total of $500 million was to be applied to costs incurred between 1966 and 1980. Of that amount, $400 million is available to the provinces on a per capita basis; $25 million is available to the Atlantic provinces for joint projects; and $75 million remains to be allocated by the Governor-in-Council.

Research support programs. The federal government operates a number of programs to promote research in the physical and natural sciences, social sciences, arts and humanities. The main channels for this support are the National Research Council, the Medical Research Council and the Canada Council. A tri-council committee co-ordinates their policies and programs. Other agencies and departments such as the Defence Research Board, the Atomic Energy Commission, Central Mortgage and Housing Corporation and the departments of agriculture and health and welfare also contribute to research. Support may consist of capital grants, operating grants, research grants, and contracts, scholarships and awards.

Other participants. The National Museums of Canada, the National Gallery, the National Film Board and the Canadian Broadcasting Corporation contribute directly or indirectly to various school programs.

7.6 Cultural activities

7.6.1 Education in the arts

In the past few years courses of artistic content have increased to some extent in universities but the main growth has taken place in the community colleges of Ontario and the collèges d'enseignement général et professionnel (CEGEPs) of Quebec.

Fine arts (architecture, painting and drawing, commercial and decorative arts, graphics, ceramics and sculpture) appears as an elective subject of the faculty of arts in a number of universities, where it may be taken as one of five, six or more subjects for a year or two. A number of universities offer a bachelor of arts degree with a major in fine arts. Others offer a bachelor of fine arts degree. It is also possible to complete a master's degree and doctorate in fine arts in some Canadian universities. There are many colleges and schools of art with varying academic requirements for admission. These offer diploma or certificate courses and are concerned largely with the technical development of the artist. Courses vary in length but may extend to four years. In some schools fine crafts as well as fine arts are taught. Summer schools of art are sponsored by some institutions, universities and independent groups.

Degree courses in music are offered at a number of Canadian universities. Opera may be studied at the Royal Conservatory Opera School of the University of Toronto where advanced students work in close collaboration with the Canadian Opera Company, and also at the Conservatoire de Musique et l'Art Dramatique in Montreal and Quebec City and at the Banff School of Fine Arts (summer). Degree courses in drama are given at several universities. The National Theatre School of Canada in Montreal offers a three-year acting course and two years for technical and production studies. The National Ballet School at Toronto is a residential ballet school. Professional instruction is also offered by Les Grands Ballets Canadiens, Montreal, and the Royal Winnipeg Ballet. Instruction in drama, ballet, opera, creative writing and fine arts is given in summer school courses at the Banff School of Fine Arts.

7.6.2 Performing arts

In recent years Statistics Canada has carried out an annual survey of performing arts groups in Canada. The organizations surveyed for 1976 include 45 theatre companies (43 in 1975), 14 orchestras and ensembles (14 in 1975), seven dance companies (six in 1975) and six opera companies (five in 1975). The following information is based on 1976 data with 1975 figures in parentheses.

During 1976 these 72 (68) companies gave a total of 14,612 (14,242) performances to a combined audience of 6.7 (6.8) million people. However, the revenues earned through these performances were only able to meet 46% (48%) of their expenses. The remaining gap was mainly filled by grants from the public and private sectors totalling $27.9 million ($21.6 million). These were supplemented by revenue from program sales, bar and concession sales, and other sidelines. More precisely grants accounted for 46% (42%) of all revenue for theatre companies, 54% (45%) for opera companies, 56% (53%) for orchestras and 58% (56%) for dance groups. In descending order of importance, the principal contributors were the federal government, provincial governments, private enterprise and municipal or regional governments.

On the expenditure side, personnel costs accounted for 60% to 64% (62% to 65%) of the total budget for theatre, dance or opera companies. This percentage rose to 75% (79%) for music groups. Publicity accounted for expenditures of 8% to 10% (6% to 11%) of total expenses, and administration from 4% to 7% (5% to 7%). Other production costs, such as expenditures on sets, costumes, props, technical equipment and printing of tickets accounted for 15% (13%) of expenditures for theatre, 8% (6%) for music, 17% (19%) for dance and 15% (19%) for opera.

Table 7.9 gives average grants, revenues and expenditures by type of company and by spectator for 1975 and 1976. Without financial help from both the public and private sectors, average deficits for 1976 (1975 in parentheses) would have reached $263,725 ($224,447) for theatre groups, $646,416 ($502,236) for musical organizations, $683,280 ($665,891) for dance groups and $443,726 ($572,715) for opera.

Attendance at large cultural institutions[1], by region

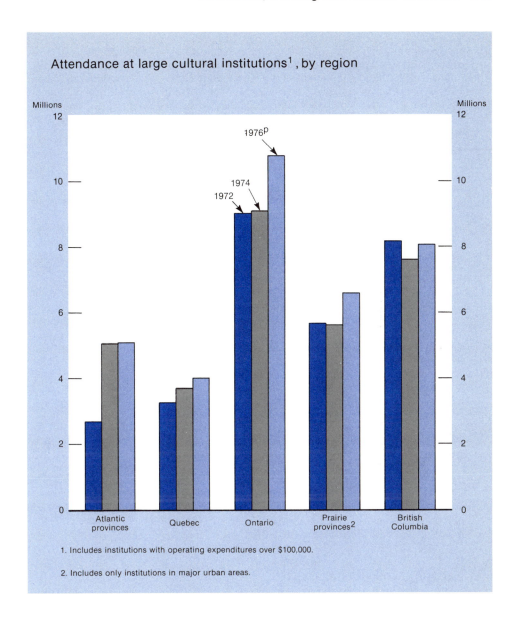

1. Includes institutions with operating expenditures over $100,000.

2. Includes only institutions in major urban areas.

Other data show that the federal government contributed, by spectator, an average of $1.51 ($1.02) for theatrical performances, $1.73 ($1.44) for music, $3.53 ($2.74) for dance and $2.78 ($3.11) for opera.

Art galleries   7.6.3

Public art galleries and art museums in the principal cities perform valuable educational services. Children's Saturday classes, conducted tours for school pupils and adults, radio talks, lectures and concerts are features of the programs of the various galleries. Many of these institutions supply travelling exhibitions for their surrounding areas or range even farther afield. Several organizations such as the Maritime Art Association, the Atlantic Provinces Art Circuit, the Western Canada Art Circuit, the Art Institute of

Ontario, the Art Gallery of Ontario and the Fédération des centres culturels du Québec have been founded to carry out this sort of travelling program on a regional basis. On a smaller scale, art circuits are organized to serve certain areas such as those around St. John's, Nfld., Charlottetown, PEI, Trois-Rivières and Hull, Que., and Winnipeg, Man. The National Gallery of Canada conducts a national program of this nature and is one of the largest art circulating agencies in North America. Several galleries maintain an art rental service. Table 7.13 gives the number of art galleries and museums and their location by region.

7.6.4 Museums

Museums of Canada range from small collections of locally-gathered historical artifacts and objects to large government-operated institutions which collect, classify and display such objects as may be useful to the study and teaching of natural history, human history, science and technology, with special but not exclusive reference to Canada. Many of these larger museums, especially the components of the National Museums of Canada and the Royal Ontario Museum, have a long, distinguished heritage in research and publication of scholarly works and are important educational and cultural centres. They offer many educational services to the public through exhibits, guided tours, lectures and scientific and popular publications.

Direct work with schools may involve holding classes in the museum or arranging visits of museum lecturers, with exhibits, to the schools. More informal are guided tours for visiting school classes, loans of specimens, slides, filmstrips or motion picture films to schools, and training student-teachers in the educational use of the museum. For children, a number of museums have special programs not directly associated with school work including Saturday lectures and film showings, activity groups, nature clubs and field excursions. At the higher educational level, museum field parties provide research training to university students in many disciplines and museum staffs act as professional consultants, answer inquiries on scientific and technical subjects, and serve as consultants or advisers to foreign scholars and institutions.

For adults, museums offer lectures, film shows and guided tours, the latter usually available throughout the year. Staff members may give lectures to service clubs, church groups, parent-teacher associations and hobby clubs. The latter, such as naturalists' groups, mineral clubs and astronomy societies, may be allowed to use the museum as their headquarters. Travelling exhibits are prepared for showing at local fairs, historical celebrations and conventions. Some Canadian museums have conducted regular radio or television programs and others have made occasional contributions. Some historical museums stage annual events during which the arts, crafts or industries represented by the exhibits are demonstrated to the public.

7.7 National Museums of Canada

The National Museums of Canada, a Crown corporation established in 1968 by the National Museums Act, reports to Parliament through the secretary of state but is administered by its own secretariat under the authority of a board of trustees. It incorporates in a single administration Canada's four major national museums, affiliated with a nationwide network of associate museums and exhibition centres. It administers a series of programs with main purposes to preserve and increase access to the treasures of the national heritage.

The four national museums in Ottawa are: the National Gallery; the National Museum of Man, which includes the Canadian War Museum; the National Museum of Natural Sciences; and the National Museum of Science and Technology, which includes the National Aeronautical Collection. In addition to these, the National Museums of Canada provides financial assistance to 21 associate museums, 28 national exhibition centres and almost 1,300 other museums and galleries across Canada.

The national museums policy, announced in a government declaration in 1972, calls for the "democratization and decentralization" of Canada's cultural heritage. In pursuit of this policy the National Museums of Canada plans the growth and development of its network of associate museums and national exhibition centres, and conducts programs which seek to bring museum services to all parts of the country.

Museum assistance programs provide financial and other assistance to museological institutions for projects to train and develop museum staff, improve facilities, support education and extension programs within museums, and for other projects such as travelling exhibitions, that increase public viewing of the collections. The conservation institute, with its regional centres, is engaged in national programs for the conservation of artifacts and works of art, technical consultation, conservation research and training of personnel. Museumobiles exhibit artifacts and related materials mainly in smaller communities across Canada which lack ready access to major museums. Each museumobile caravan, consisting of three 13.7-metre trailers, depicts the geographic, archeological, social and natural history of a region of Canada. A national inventory provides Canadian museums with an easily retrievable, bilingual record of the contents of collections in museums across Canada. The service, employing a computer with terminals in 20 locations from Victoria to St. John's, provides information on the fine and decorative arts, history (social), ethnology, ornithology and archeological sites and specimens. An international program serves as a focus for the exchange of exhibitions between Canadian museums and galleries and cultural institutions abroad.

The National Gallery of Canada   7.7.1

The beginnings of the National Gallery of Canada are associated with the founding of the Royal Canadian Academy of Arts in 1880. The Marquis of Lorne, then Governor General, had recommended and assisted in the founding of the academy and among the tasks he assigned to that institution was the establishment of a national gallery at the seat of government. Until 1907 the National Gallery was under the direct control of a minister but in that year, in response to public demand, an advisory arts council consisting of three persons outside government was appointed by the government to administer grants to the National Gallery. Three years later, the first professional curator was appointed.

In 1913, the National Gallery was incorporated by an act of Parliament and placed under the administration of a board of trustees appointed by the Governor-in-Council; its function was to encourage public interest in the arts and to promote the interests of art throughout the country. Under this management, the gallery increased its collections and developed into an internationally recognized art institution. Today, a board of trustees reporting to the secretary of state administers all the National Museums of Canada, including the National Gallery, under the National Museums Act (RSC 1970, c.N-12).

The gallery's collections have been built up along international lines and give the people of Canada an indication of the origins from which their own traditions are developing. The collection of Canadian art, the most extensive and important in existence, is continually being augmented. Over 60% of all acquisitions since 1966 have been Canadian. There are now more than 13,000 works of art in the collections, excluding photographs. Included are many old masters, 12 having been acquired from the famous Liechtenstein collection. The Massey collection was presented to the gallery during 1946-50 by the Massey Foundation. The Vincent Massey Bequest of 100 works was received in 1968. In 1974 an important gift of drawings was donated by Mrs. Samuel Bronfman of Montreal in memory of her husband. There is a growing collection of contemporary art, prints and drawings, and diploma works of the Royal Canadian Academy. The gallery's collection of photographs, built up since 1967, contains 6,000 works. The services of the gallery include the operation of a reference library open to the public containing more than 50,000 volumes and periodicals on the history of art and other related subjects.

A program of exhibitions, lectures, films and guided tours is maintained for visitors to the gallery in Ottawa. The interests of the country as a whole are served by circulating exhibitions, lecture tours, publications, reproductions and films prepared by the gallery staff. Promotion of and information on art films are distributed by the Canadian Film Institute. The gallery promotes interest in Canadian art abroad by participating in international exhibitions and by preparing major exhibitions of Canadian art for showing in other countries in collaboration with the external affairs department. It also brings important exhibitions from abroad for showing in Canada.

Major exhibitions in Ottawa in 1976-77 included Puvis de Chavannes, jointly organized by the Louvre and the National Gallery of Canada; works of the French photographer, Charles Nègre; British art in the collection; Guido Molinari, drawings from the collection of Mr. and Mrs. Eugene V. Thaw; and the architecture of the École des beaux-arts. Through the national program, the gallery circulated 13 exhibitions to 33 centres in Canada. Internationally, the gallery organized an exhibition of works by Greg Curnoe for the Venice Biennale which later was shown in London.

The major European acquisition of the year was *Salisbury Cathedral from the Bishop's Grounds* by John Constable, the important British landscape artist. In Canadian art, the gallery was fortunate to acquire *Mr. and Mrs. Croscup's Painted Room* which will have to be restored before being reassembled in Ottawa. It is an example of 19th-century naive art from the Maritimes.

Gifts received during the year included two paintings by Francesco Guardi, *The Dogana, Venice* and *Piazetta, Venice,* both given by Mrs. Dorothy Webb; *Déjeuner après le bain* donated by Dennis Molnar; a portrait of John G. McConnell by Kokoschka donated by Mrs. McConnell; and an 18th-century English coffee urn donated by the National Gallery Association.

7.7.2   The National Museum of Natural Sciences

This museum has seven divisions: botany, invertebrate zoology, vertebrate zoology, mineral sciences, paleobiology, interpretation and extension, and the Canadian aquatic identification centre. Thousands of specimens, originating from field trips, purchases, donations and exchanges, are added annually to the reference collections. The herbarium contains more than 375,000 sheets of vascular plants and 185,000 sheets of cryptogamic plants. National zoological collections include about 2,500,000 molluscs, 750,000 crustaceans, 662,000 other invertebrates, 215,000 fishes, 70,000 reptiles and amphibians, 65,000 birds and 43,000 mammals.

Only part of the collection of display minerals, containing approximately 17,000 catalogued specimens of gems and minerals, is displayed in the museum. Tens of thousands of specimens of minerals, rocks and ores from many regions of the world are contained in other collections of display and study materials. During 1976 almost 70 major research projects were undertaken by museum staff members or associated scientists from universities and other outside organizations. The museum provided financial assistance, facilities and field work for several National Research Council post-doctoral fellows, and collaborated with the National Parks Service in botanical and zoological surveys. The museum's Arctic research program was in turn supported by the energy, mines and resources department.

The vertebrate paleontology collection contains more than 12,600 fossils, which includes rare dinosaur specimens. Research was conducted on reptilian diversity and environments before extinction of the dinosaurs, the pleistocene fauna of Saskatchewan and the unglaciated region of the Yukon Territory, and fossilized pollen and spores.

The Canadian Aquatic Identification Centre (CAIC) sorts and identifies zoo-plankton, fish larvae and eggs, phytoplankton and other aquatic life. CAIC, with some 500 species in its reference collection, serves researchers, management, and survey and environmental agencies. Another unit identifies and interprets animal remains found in archeological investigations. This service studies both natural and human history, and can identify the species of an animal from a fragment of bone.

In the renovated Victoria Memorial Museum Building, shared by the National Museum of Natural Sciences and the National Museum of Man, two floors were opened to the public in October 1974. They featured audio-visual presentations, visitor-operated displays, drawings, models and specimens from the museum collections in four permanent exhibit halls, given to geology, paleontology and zoology.

A gallery entitled The Earth presents an explanation of the tectonic plate theory and shows how natural forces shaped and continue to shape the world. In Life Through the

Ages, drawings and specimens of early Canadian plants and animals show how some species adapted to change, whereas others died. The most impressive of these specimens are the dinosaurs of Western Canada's cretaceous period. The Birds in Canada exhibit features life-like dioramas showing the kinds of birds typical of nine of the major biological regions of Canada. Mammals in Canada presents dioramas of present-day Canadian mammals also in authentic settings.

The museum's newest exhibit hall, Animal Life, completed in October 1976, shows animal evolution through a 500 million-year period to the present. In addition, a visitor may be guided through the story of man's effort to understand the origin and diversity of animal life, to decipher its cryptic genetic codes, and to unravel the evolutionary process which relates all animals of the world to each other. A special exhibits hall displays temporary and travelling exhibits from the museum and other sources.

Public lectures, film presentations and special interpretative programs prepared by the interpretation and extension division are popular with school classes and the general public. In addition, the division prepares publications, provides educational materials for loans to schools and conducts a program of travelling exhibits.

The National Museum of Man   7.7.3

Through its seven divisions — archeological survey, ethnology service, centre for folk culture studies, the war museum, history, education and cultural affairs, and national programs — the National Museum of Man conducts research in Canadian studies, collects and preserves artifacts of material culture and extends the museum's programs across Canada through exhibits and educational loans.

The Canadian centre for folk cultural studies is both a research institute and a repository for collections of oral folklore and artifacts reflecting Canada's many folk cultures. The war museum houses an extensive collection of artifacts representative of Canada's military past ranging from military art and medals to tanks and bayonets from all wars involving Canadian participants. The archeological survey has undertaken more than 370 research and salvage projects in Canadian prehistory in the past 14 years, including 12 in 1976, which have dramatically altered the knowledge of Canada's past. The ethnology service collects and analyzes information pertinent to the traditional cultures of Canadian Indians, Inuit and Métis both through staff and contract ethnographic research and through artifact acquisition, research and conservation. This information is disseminated through scientific and popular publications, exhibits, lectures and loans. Research in Canadian history is conducted by the history division which has acquired for preservation over 40,000 items of period furnishings and Canadiana. The education and cultural affairs division designs and produces resources for communities across Canada in the form of educational publications, museum kits and films and provides a wide range of programs based on the museum's galleries. The national programs division co-ordinates an extension service of travelling exhibits and museum kits primarily related to Canadian archeology, ethnology, folk culture and history. Catalogues and monographs are part of the museum's diverse program of national and international travelling exhibitions, films and television programs.

The National Museum of Science and Technology   7.7.4

The newest of the four national museums, the National Museum of Science and Technology, opened in November 1967. This museum challenges visitors to climb, push, pull or just view its definitive collections. Thousands of others annually visit the aeronautical collection at Rockcliffe Airport.

The exhibit pavilions contain examples from the history of ground transportation such as sleighs, streetcars, steam locomotives and antique cars, to aviation and space, beginning with Canada's first powered heavier-than-air flight. Trains have figured prominently both as acquisitions and in programs. Steam train excursions, operated in collaboration with the National Capital Commission, are popular summer events. There are also experiments and skill-trying tests in the physics hall and exhibits on the history of agriculture, marine transport, meteorology, time pieces and astronomy.

In the aeronautical collection over 90 aircraft illustrate the progress of aviation from primitive to present times and the importance of the flying machine in the development of Canada. Included is one of the world's largest collections of aircraft engines.

Educational programs are developed and conducted by a staff of tour guides on general or special topics for all age groups. The museum's observatory houses Canada's largest refracting telescope which is used for evening educational programs. The museum's 16,500-volume library places special emphasis on a retrospective collection of Canadian aviation.

7.8 Aid to the arts

7.8.1 The Canada Council

The Canada Council was created in 1957 by an act of Parliament to foster and promote the study and enjoyment of, and the production of works in the arts, humanities and social sciences. It offers a broad range of grants and provides certain services in these and related fields. The council meets at least three times a year and is made up of a chairman, a vice-chairman and 19 other members, appointed by the Governor-in-Council.

Under the provisions of the Government Organization (Scientific Activities) Act passed by Parliament in June 1977, the council's work in the humanities and social sciences became the sole responsibility of a new body, the Social Sciences and Humanities Research Council, on April 1, 1978.

The Canada Council enjoys a large measure of autonomy, setting its own policies and developing and carrying out its own programs in consultation with the academic and artistic community. It reports to Parliament through the secretary of state and also appears before the standing committee on broadcasting, films and assistance to the arts.

The council's income is derived from three sources: an annual parliamentary grant which amounted to $59.7 million for the year ended March 31, 1977 ($54.7 million, 1976); interest from an endowment fund established by Parliament; and private funds willed or donated and used in accordance with the wishes of the donors.

Assistance to the humanities and social sciences accounted for $27.8 million in 1976-77, compared with $25.3 million in 1975-76 (further references to 1975-76 are in parentheses). In support of research training, the council awarded special MA scholarships, doctoral fellowships and postdoctoral research fellowships totalling $10.5 million ($9.5 million); leave fellowships totalling $3.8 million ($3.8 million); research grants to individuals totalling $5.2 million ($5.7 million); general research grants to universities totalling $1.0 million ($1.2 million); a total of $3.3 million ($3.3 million) in assistance to learned meetings, special grants and studies, attendance of Canadian scholars at international conferences and publication of learned journals and scholarly manuscripts; and $3.3 million ($1.2 million) in negotiated grants for large-scale research and editorial projects.

In the arts, the council spent $32.6 million (30.4 million), of which $4.5 million ($3.5 million) was used for grants to individuals and $28.1 million ($26.0 million) for grants to organizations. Total grants in the arts included $7.7 million ($7.0 million) for music and opera, $7.8 million ($7.2 million) for theatre, $3.5 million ($3.2 million) for dance, $3.6 million ($3.0 million) for visual arts and photography, $755,000 ($756,000) for an art bank, $5.8 million ($5.2 million) for writing, publication and translation; $1.5 million ($1.3 million) for film and video and $2.0 million ($1.2 million) for the touring office.

Through an explorations program, the council spent $1.3 million to assist projects on Canada's cultural and historical heritage and innovative projects which explore new forms of expression and creativity in the arts, humanities and social sciences.

The Canada Council administers, on behalf of the Canadian government, programs of cultural and academic exchanges with 37 countries. The council also conducts its own annual exchange programs for research scholars in the humanities and social sciences, under agreements with agencies in the USSR, France and Japan. In the arts, under the Canada–USSR general exchange agreement, two Canadian artists each year may study music, ballet or theatre in the Soviet Union.

The Killam Scholarships of the Canada Council were inaugurated in 1967 with funds from the I.W. Killam estate to help support scholars of exceptional ability engaged in research projects of far-reaching significance. In 1976-77, 34 awards were made under this program, totalling $1.1 million, and the previous year, 45 awards totalling $1.1 million.

The council annually awards three $20,000 prizes from a fund created by the Molson Foundation. The Governor General's Literary Awards, financed by the council, are awarded each year to six Canadian writers. The council also awards two translation prizes to the year's best English and French translations of Canadian works and two children's literature prizes to the year's best English and French books for young people. Co-sponsored by their respective governments and the Canadian government, the Canada–Australia Literary Prize and the Canada–Belgium Literary Prize are awarded by the council in alternate years to a writer from each country. The council also co-sponsors with the CBC an annual national competition for young composers.

The council provides the budget and secretariat for the Canadian Commission for UNESCO (United Nations Educational, Scientific and Cultural Organization). The commission serves as a non-political liaison agency between UNESCO and Canadian public and private bodies concerned with education, science, culture and communications. The commission also carries out a modest domestic program to further UNESCO objectives and co-operates with Canadian organizations in activities directed to the same ends. The commission operates the Canadian Communications Research Information Centre, a national clearing house of information on communications policy, research, resources and activities. It also administers the council's program of grants for international representation for travel of Canadians serving as senior officers or board members of international non-government organizations in the arts, humanities or social sciences.

National Arts Centre (NAC) 7.8.2

Parliament passed the National Arts Centre Act in 1966, creating a corporation to operate and maintain the centre, to develop the performing arts in the national capital area, and to assist the Canada Council in the development of the performing arts elsewhere in Canada. The building itself, designed by Montreal architect Fred Lebensold, was opened to the public on May 31, 1969. It stands on Confederation Square in the heart of Ottawa, and is in the form of a series of hexagonal halls built on landscaped terraces along the Rideau Canal.

The NAC has three main halls. The Opera, with 2,300 seats, was designed primarily for opera and ballet, with a full-size orchestra pit and the most advanced sound, lighting and other technical equipment available. Its stage is one of the largest in the world, measuring 58 by 34 metres, and the Opera's facilities can handle the most complicated changes required by the largest touring companies. The 950-seat Theatre is ideal for Greek, Elizabethan or contemporary plays, and its stage can easily be adjusted from the conventional rectangular style to the thrust stage style used for Shakespearean drama. Like the Opera, it is fully equipped for television, simultaneous translation and film projection, and its technical facilities are among the best available. The Studio is a hexagonal room which can seat up to 350 persons in a variety of seating plans. It is used for theatre productions, conferences and cabarets.

Other NAC facilities include: the Salon, a small hall seating up to 150 persons and used for chamber concerts, poetry readings, and receptions; a 900-car indoor garage; Le Restaurant, a restaurant and bar; Le Café, a smaller restaurant which, in the summer months, overflows on to the sidewalks along the Rideau Canal; and several large rehearsal halls. The building is richly ornamented with works of art. Its foyers are used for exhibits and public tours of the building are offered daily. On the terraces outside, the NAC plays host to art fairs, craft markets and summertime band concerts.

Under conductor and music director Mario Bernardi, the 46-member National Arts Centre Orchestra performs some 40 concerts a year in the centre and many more each year on tours in Canada and abroad. Music programming includes about 70 concerts a year, featuring distinguished soloists and guest orchestras from Canada and around the world.

The theatre department functions in both official languages. It offers several subscription series a year of English- and French-language plays, and also non-series productions. There are more than 400 performances of live theatre a year at the centre. Some of the plays are produced by the theatre department and others represent Canada's regional theatre or come from outside the country.

The theatre department also forms small companies which tour Canada, performing in high schools and elsewhere, and offering professional theatre to communities which would not otherwise have the opportunity to enjoy it. Workshops for students and teachers are among the other services offered.

The dance and variety department brings in some 100 different shows a year, including ballet, musical shows and comedy. The NAC is the only centre in Canada where every Canadian dance company of importance appears on a regular basis, and dance and variety programming has offered a showcase for performers from every part of the country. Each year during July, a festival of mainly musical entertainment is presented, centred around the NAC's own opera productions. Altogether, there are about 900 performances annually in the NAC, entertaining almost 800,000 people.

7.8.3 Provincial aid to the arts

All provinces except Prince Edward Island give some form of financial assistance to artists (writers, poets, painters and sculptors), cultural organizations or community councils.

Newfoundland. The tourism department contributes to the upkeep of cultural centres in the province and provides grants and awards to individuals or groups native to or touring the province for performances. Grants and awards enable local theatrical groups to produce and perform in centres around the province. For 1977 the provincial government budgeted $71,000 for these grants and a further $270,000 to enable locally, nationally and internationally known companies and groups to perform in the province at centres which normally would be unable to hire them. These funds also enable centres such as the Marine Museum in St. John's to purchase works to be exhibited.

Nova Scotia. In 1973, the Nova Scotia recreation department was given the legislative mandate to be responsible for cultural development, covering a broad range of activities consisting of music, theatre, crafts, multiculturalism, festivals, arts councils, dance, visual arts, writing, film, and photography, the Art Bank of Nova Scotia and the newly developing Art Gallery of Nova Scotia. The department supports eight cultural federations which act as service agencies for arts and cultural programs throughout the province.

In the year ended March 31, 1976, grants to major cultural institutions included a grant of $155,000 to the Neptune Theatre Foundation; $85,000 to the Atlantic Symphony Orchestra; and $150,000 to the Art Gallery of Nova Scotia. For the following year, the cultural component of the department had a budget of $859,500 with a projected budget of $1,035,800 for the year ended March 1978.

In addition, a number of program and leadership grants were made available to community-based programs in the performing, visual and literary arts. The tourism department granted $129,000 to cultural events and festivals.

A fine art and handcraft section of adult education was merged with the cultural section of the Nova Scotia recreation department to form the cultural affairs division of the department of recreation. The Nova Scotia museum section of the cultural services program of the education department, with a budget of $1,759,300, disbursed grants to local museums and historical societies. Other major cultural expenditures included $2,594,600 for a provincial library which serves communities with both permanent and mobile library facilities, and $315,000 for the operations of the archives, a contribution toward the preservation of the province's cultural heritage.

New Brunswick provides assistance to Le Centre de Promotion et de Diffusion de la Culture, a Moncton-based agency which co-ordinates Acadian cultural activities — choirs, theatre, and individual artist presentations. Assistance is also given to the provincial competitive music festival, folk song and band festivals, dance troupes, arts councils, a provincial youth orchestra, Atlantic Symphony, Theatre New Brunswick, various choral and drama groups, art associations and writers. In addition, the province helps sponsor tours by individual performing artists within and outside the province.

Quebec. The cultural affairs department was established in 1961 by an act of the provincial legislature. The department was charged with the administration of cultural organizations or institutions. Activities concentrate on five programs: books and other printed material, conservation and development of cultural properties, visual arts, performing arts and administration of the department.

The book program aims to promote literary creation and production, to stimulate distribution of Quebec writing and ensure that it is conserved and read. The program assists publishing and sale of books, develops a network of public libraries; and preserves and makes accessible the Quebec literary heritage through the National Library of Quebec.

Through the conservation and development of cultural properties, technical assistance and expertise are provided to preserve and develop the historical heritage. The program also aims to co-ordinate and supervise research in archeology and ethnology; to protect natural and historic sites; to preserve historic manuscripts, furniture and buildings; to make an inventory of private, semi-public and public archives; and to continue an inventory of art works.

Three separate components involve: the National Archives of Quebec; conservation of historical and archeological sites and properties; and recommendations on measures to be taken by the department for conservation of cultural properties. The last is the responsibility of an advisory commission, the Cultural Properties Commission.

The administrative program co-ordinates programs and provides advice on the use of human, physical and financial resources at departmental headquarters and eight regional offices.

The visual arts program has a dual objective: conservation and dissemination of visual arts and promotion of creativity. This program includes the operations of the Musée du Québec and the Musée d'art contemporain, and financial and technical assistance to private museums and promotion of visual arts.

The program on performing arts is designed to train professionals in conservatories of music and theatre in the fields of music, theatre and dance. This program also provides financial assistance to about 100 cultural organizations throughout Quebec.

Ontario. A new ministry of culture and recreation officially began operating in 1975. Through its arts division, the ministry has allocated the following amounts to major provincial agencies and institutions in 1977-78 (allocations for 1976-77 following): the Ontario Arts Council $11.5 million ($10.5 million), the Royal Ontario Museum $7.2 million ($6.9 million), the McMichael Canadian Collection of Art $566,000 ($556,500), the Art Gallery of Ontario $3.9 million ($3.6 million), the Royal Botanical Gardens $621,000 ($598,000), CJRT-FM Corporations $558,000 ($588,000), the Ontario Educational Communications Authority $12.8 million ($11.3 million), and the Ontario Science Centre $5.9 million ($5.9 million).

In addition, the ministry provided $855,000 in 1977-78 for OUTREACH programs ($770,000 in 1976-77) and Festival Ontario received $150,000 in 1977-78 ($210,000 in 1976-77). During 1977-78, the Wintario lottery made substantially more funding available to support cultural and recreational activities in the province.

The prime objectives of the arts division are the encouragement of the pursuit of excellence in the arts and the promotion of wider participation in and enjoyment of arts activities by the citizens of Ontario. Other divisions in the ministry are for heritage conservation, sports and fitness and citizenship and multiculturalism. The ministry has six regional offices with a network of field staff which provide consulting services to communities and local groups. Other ministries involved in cultural programming are agriculture, education, natural resources, industry and tourism, and treasury.

The Ontario Council for the Arts, an agency of the ministry, was established by legislation in 1963 to promote the study, enjoyment and production of works in the arts. It provides financial assistance to performing and creative arts groups and individuals; advises and consults with members of the arts community; and develops projects aimed at promoting and expanding the arts and the public interest in them throughout the province. The council also collaborates with other agencies and levels of government in encouraging support for the arts.

Manitoba. The Manitoba Arts Council was formed by legislation passed in 1965 providing for a chairman, vice-chairman and 10 appointed members with the objective of promoting the study, enjoyment, production and performance of works in the arts by assisting and co-operating with those organizations involved in cultural development, providing grants, scholarships or loans to Manitobans for study or research in the arts and making awards to citizens of Manitoba for outstanding accomplishment in the arts. Working on a budget of $636,000, the council in the year ended March 31, 1977, made grants to 31 organizations including $71,500 to the Manitoba Theatre Centre, $44,000 to Rainbow Stage, $103,000 to the Royal Winnipeg Ballet, and $79,200 to the Winnipeg Symphony Orchestra. The council's awards program provided assistance to individual Manitoba artists.

Saskatchewan. The Saskatchewan Arts Board was established in 1949 to make available to the people of Saskatchewan opportunities to engage in any of the following activities: drama, the visual arts, music, literature, handicrafts and other arts, to provide for the training of lecturers and instructors, to assist students ordinarily resident in Saskatchewan to pursue their studies in the arts, and to co-operate with organizations having similar objectives.

The board, funded by the provincial government but functioning independently, is composed of not less than seven nor more than 15 members appointed annually. These members are from all parts of the province but represent no specific areas or any specific discipline in the arts. The work of the board is carried out by a staff of consultants and office personnel under an executive director. Experts in various fields of the arts are engaged for specific projects.

The Saskatchewan School of the Arts, operated annually by the board at Echo Valley Centre offers one- to four-week courses during spring, summer and fall in music, creative writing, ballet, folk and modern dance, art and handicrafts.

Since 1950, the board has helped provincial artists through purchase of their works for the board's collection. This collection, which now contains over 700 items, is a valuable record of the development of Saskatchewan visual arts and crafts.

For several years an annual month-long Saskatchewan Festival of the Arts focused attention on the arts in communities throughout the province by presenting top-calibre artistic performances and exhibitions. This festival has been replaced by similar programs planned by individual communities to meet their specific needs, assisted by the organization of Saskatchewan arts councils, a provincial body funded in part by the Saskatchewan Arts Board.

The Saskatchewan Arts Board budget in 1976 was $794,000. Assistance to organizations included a direct grant to Globe Theatre of $64,000 plus $23,247 in deficit protection to 145 schools in 87 communities engaging Globe's School Company, $6,000 to the Mendel Art Gallery, $7,000 to the Norman Mackenzie Art Gallery, $3,500 to the Organization of Saskatchewan Arts Councils, $20,000 to Persephone Theatre, $3,300 to Photographers' Gallery, $12,000 to Regina Modern Dance Works, $47,500 to the Regina Symphony, $4,000 to the Saskatchewan Craft Council, $8,000 to the Saskatchewan Writers' Guild, $47,500 to the Saskatoon Symphony, $1,000 to the Shoestring Gallery, and $5,000 to Twenty-fifth Street House Theatre. Four film-makers received grants totalling $7,500 to continue work on individual film projects.

In 1976, the board's assistance to arts groups and organizations totalled $314,660. The board also helped administer funds provided by the Western Canada lottery to six performing arts companies. Aid to individual artists included $43,590 to 78 applicants of professional or near-professional status.

Alberta. After going through many changes since the creation of the Cultural Development Act of 1947, cultural development attained department status in its own right and had its first full year of operation in 1975-76. It was one of two departments created when the former culture, youth and recreation department was divided; the other is the recreation, parks and wildlife department. The two divisions of the culture department are historical resources and cultural development. Cultural development division is responsible for performing arts, visual arts, film and literary arts, library services, cultural facilities, cultural heritage, auditoriums and the censorship board.

The division's programs are intended to give the public opportunities to witness the best in performing arts tours and art exhibitions and to conduct training courses at regular intervals for a variety of leaders, particularly teachers, librarians, and music directors. Workshops and conferences are held to train people in all aspects of theatre. Awards to individuals wishing to further their training in some form of the arts amount to over $250,000 annually. The Alberta Art Foundation, established in 1972, has been allocated $50,000 annually for the purchase of Alberta arts and crafts. The budget for the year ended March 31, 1977 was in excess of $7 million.

A new $2.5 million program for library development provides for a system of central, regional, municipal and community libraries, bibliographic services and services for the handicapped. Grants are provided to advance creative writing through professional training and writers workshops. Competitions are sponsored for new novelists, best published non-fiction work and best regional history.

The Southern Alberta Jubilee Auditorium at Calgary and Northern Alberta Jubilee Auditorium at Edmonton are identical multi-purpose facilities each containing a 2,762-seat theatre, four meeting rooms and exhibit area. In April 1975 a major $200 million program was launched to develop cultural recreation facilities in Alberta communities over a 10-year period.

The cultural heritage branch provides grants to ethno-cultural groups; the division helped establish a Cultural Heritage Council. Under a language support program financial help is available to ethno-cultural schools operated outside school hours.

The Alberta Film Censorship Board, attached to the division for administrative purposes, issues permits for all films approved for exhibition in Alberta.

British Columbia. The British Columbia Cultural Fund was set up by statute in 1967. That act set aside $5 million in an endowment fund, the interest from which was to be spent to stimulate the cultural development of the people of the province. In September 1967 an advisory committee was established to receive applications for cultural grants and to report their recommendations to the finance department for the necessary funds. The amount of the endowment was raised to $10 million in 1969, to $15 million in 1972, and to $20 million in 1974. Proceeds from the Western Canada Lottery Foundation are used, in part, to support cultural activities.

An advisory body to the fund, the British Columbia Arts Board, was appointed in November 1974 to make recommendations to the provincial government on the allocation of cultural grants from the fund.

Up to December 31, 1976 grants totalling almost $9.5 million had been awarded by the fund to support cultural activities throughout the province. The fund also provides a small degree of financial support to the National Theatre School, the National Youth Orchestra and the Canadian Music Centre.

Grants totalling $2.4 million were made in the fiscal year 1976-77, of which about 70% went to major non-profit organizations such as symphony, drama and opera societies. About 15% of the grants went to community arts councils and the remainder to scholarships, seminars and miscellaneous grants.

Federal film agencies 7.9

National Film Board 7.9.1

The National Film Board (NFB), an agency of the federal government, was established by act of Parliament in 1939 and reconstituted by the National Film Act in 1950 to initiate and promote the production and distribution of films in the national interest. The board's films are produced in Canada's two official languages and have made a considerable contribution to the country's culture and to the national identity. The board's head office is in Ottawa, with operational headquarters in Montreal.

The growing sophistication of audiences and the increasing importance of film as a means of communication are reflected in the nature of the board's productions — features, documentaries, informational films, animation, films for the specific needs of government departments, and films designed for particular social purposes. In response to current high interest in fitness and amateur sport, the board in 1976 produced the official film of the Montreal Olympic Games, a critical and popular success around the world, and in 1978 the official film of the Commonwealth Games, held in Edmonton.

While production had been centred in Montreal, the rapid development of communication in Canada, coupled with the need to respond to cultural differences, prompted the NFB to accelerate its regionalization program. The objective is to provide each region of Canada with the opportunity to interpret a regional subject to a national audience or a national subject from a regional point of view. Regional production studios in Vancouver, Winnipeg, Toronto, Moncton and Halifax provide the NFB with access to the best creative and technical resources at the local level and in turn provide Canadian film-making talent access to the national agency.

NFB films are distributed in 16 mm or 35 mm. In addition, all films are now available in video cassettes. The board also produces and distributes other visual aids material such as silent and sound filmstrips, slide sets, overhead projectuals, multi-media kits and photo stories. In Canada, the board's productions are distributed through community outlets, schools and universities, television stations, theatres and commercial sales. A large part of the 16 mm community film audience is reached through film libraries, film councils and special interest groups. During 1976-77, community film distribution through the 27 NFB libraries in Canada rose to a new high of 456,344 bookings. Aside from the board's own film libraries, many public and school libraries across Canada distribute its films. New releases are shown regularly over English and French television networks in Canada and in theatres across the country.

NFB films are seen outside Canada on television, in theatres, in schools and in libraries, with distribution handled by the board's offices in New York, London, Paris, Tokyo and Sydney. As well, community distribution abroad is effected by 92 film libraries operated jointly with the external affairs department. For greater international distribution, many NFB films are versioned in foreign languages. The board, in co-operation with the federal government tourism office, distributes films supporting the travel industry to audiences throughout the world.

Each year, NFB films are presented at many international film festivals. In 1977, the board won 75 awards, including three Oscar nominations. Among the films which gained worldwide recognition were the features *J.A. Martin, Photographe* and *One Man,* and the animated short, *The Street.*

The film board now promotes and distributes selected CBC films across Canada, providing Canadians the opportunity to view these important programs. Started in 1975, there are now more than 125 English and 60 French titles, with many being added each year. Public acceptance at the community and school levels has been good, totalling more than 20,000 bookings during 1976-77.

7.9.2 Canadian Film Development Corporation

The Canadian Film Development Corporation (CFDC) was established in March 1967 to promote the development of a feature film industry in Canada, and in so doing it co-operates with federal and provincial departments and agencies with similar interests. It invests in Canadian productions in return for a share of the profits, makes loans to producers and assists financially in the promotion, marketing and distribution of feature films.

The corporation assisted in the production of 16 feature films during 1976-77, 12 in English and four in French. The cumulative cost of the 16 films was more than $12.0 million, as compared with nearly $6.2 million for 18 films in 1975-76. The corporation's share was $2.7 million or 22.6% as against $2.9 million or 46.8% the previous year. These were part of the total 98 English-language motion pictures and 76 French-language feature films for which the CFDC provided production assistance in its first nine years of operation.

The year's productions provided 130 assignments for Canadian writers, directors, producers and production assistants, 768 roles for performers and employment for 331 technicians in the Canadian film industry. Laboratories earned $876,500 and equipment rental accounted for $597,500.

More Canadian motion pictures were shown in more Canadian theatres in 1976-77 than in any other year. An agreement between the corporation and the two major theatre chains, Famous Players Ltd. with 234 outlets and Odeon Theatres (Canada) Ltd., with 123, provided for four weeks of screening time for Canadian films in all but their drive-in theatres. In Quebec, a new distribution network, Nouveau Réseau, extended exposure through new outlets.

Distribution. In 1976-77, 21 films were released in Canada with CFDC support, 13 in English and eight in French. *Shadow of the Hawk* realized $1.0 million in box-office receipts. Others with high receipts were: *Death Weekend* $850,000, *Je suis loin de toi mignonne* $470,000, *It Seemed Like A Good Idea At The Time* $400,000, *Parlez-nous d'amour* $382,000 and *Breaking Point* $315,000.

Internationally, Canadian pictures continued to make steady advances. Columbia Pictures had world rights to *Shadow of the Hawk,* 20th Century Fox played *Breaking Point* around the world, and such pictures as *Shivers, Death Weekend, Sudden Fury* and *Black Christmas* were shown internationally. *Death Weekend* and *Shivers* accounted for a good portion of the almost $2,000,000 in sales of Canadian films at Cannes. In addition to participation in the Cannes Film Festival and the Mifed Film Fair in Milan, Canadian cinema weeks were arranged for Japan, England and Australia.

At the first annual Festival of Festivals held in Toronto during October 1976, CFDC made available to film buyers 65 video-cassettes of Canadian films. CFDC representatives have also worked with the external affairs department to arrange meetings in New York, Washington, San Francisco and Los Angeles to discuss the United States market.

Television. The CBC French-language network (Radio-Canada) showed 35 French-language films and purchased a library of 20 films to be dubbed from English to French. The CBC English network presented four Canadian films, CTV showed five and Global television 14. The British Broadcasting Corporation bought five Canadian films and CBS in the United States telecast a second presentation of the Canadian feature *The Neptune Factor.*

Public archives and library services   7.10

The Public Archives of Canada   7.10.1

The archives, established in 1872, operates under the direction of the dominion archivist by authority of the Public Archives Act. As a research institution, it is responsible for acquiring all nationally significant documents relating to the development of Canada, and for providing research services and facilities to make this material available to the public. Administratively, it promotes efficiency and economy in the management of government records.

The archives branch is made up of eight divisions. The manuscript division contains manuscript collections; these include the private papers of statesmen and other distinguished citizens, records of cultural and commercial societies, and copies of records on Canada held in France, England and other countries. The holdings of the public records division consist of selected records of all the departments and agencies of the federal government. The picture division has charge of documentary paintings, water-colours, engravings, heraldry and medals. The photography collection is responsible for a national collection of historical photographs. The film archives holds a wide range of films and sound recordings. The map collection has custody of thousands of maps and plans pertaining to the discovery, exploration and settlement of Canada and its topography, as well as many current topographical maps of foreign countries. The archives library contains more than 80,000 volumes on Canadian history, including numerous pamphlets, periodicals and government publications. A machine-readable archives division holds selected automated public records and machine-readable archives of permanent value from the private sector.

Although documents in the archives may not be taken out on loan, they may be consulted in the building and a 24-hour-a-day service is provided for accredited research workers. Reproductions of material are available for a moderate fee and many documents on microfilm may be obtained on interlibrary loan. Archival material is also presented on microfilm, slides and microfiches, as well as in various publications and travelling exhibitions.

The records management branch helps departments and agencies to establish their own records management programs. Its service includes recommendations and advice on scheduling and disposal of records. At the Ottawa, Toronto, Montreal, Vancouver, Winnipeg, Edmonton and Halifax records centres, it provides storage, reference service and planned and economical disposal of dormant records. Other regional centres will be established in other major cities.

The administration and technical services branch, in addition to an extensive conservation and restoration program, provides a technical and advisory service on microfilming to government departments and agencies.

Branch offices of the archives are located in London, England and Paris, France. The archives also administers Laurier House in Ottawa as a historical museum.

7.10.2 The National Library of Canada

This library was formally established in 1953 by act of Parliament. On the same date it absorbed a Canadian bibliographic centre which had been engaged in preliminary work and planning since 1950. The library is governed by the National Library Act, 1969 which broadened the powers of the national librarian and established a national library advisory board consisting of 18 members. Under the act, the national librarian has responsibility for making the facilities of the library available to the government and people of Canada and for co-ordinating federal government library services. He also administers legal deposit regulations, which require two copies of current Canadian publications to be deposited with the library.

The library's collection consists of more than 800,000 volumes of monographs, supplemented by microcopies of about 800,000 additional titles and over 9 000 metres of periodicals. Newspaper files formerly in several locations have been brought together and now form the largest collection of Canadian newspapers in Canada. The library has important holdings of Canadian, foreign and international official publications, and an extensive collection of Canadian music scores, recordings and manuscripts.

The library compiles and publishes the national bibliography, *Canadiana,* available in tape, microfiche and printed editions. *Canadiana* lists new publications relating to Canada, and includes bibliographic descriptions of Canadian trade publications, official publications of the federal government and the 10 provinces, theses, films and phonograph records produced in Canada, works by Canadians and material on Canada published abroad. More than 27,000 titles were included in 1976. Retrospective bibliographies are planned or in progress.

The library maintains a Canadian union catalogue, which provides a key to the main library resources of the country. This catalogue lists about 4 million volumes in about 340 government, university, public and special libraries in all provinces. New accessions are reported regularly; these numbered over 1.4 million cards in 1976-77. The public service branch uses this catalogue to help it meet the requests sent in by Canadian libraries for location of materials. During the year ended March 31, 1977, the branch was asked to locate more than 150,000 titles; it found about 80% of them to be held in Canadian libraries. Automation of the union catalogue is in progress.

The library provides for Canadian subscribers a computerized literature search in the fields of the social and behavioural sciences and the humanities. This encompasses both a current awareness service and retrospective bibliographies prepared from various machine-readable data bases. In early 1974 the library began publishing a series of periodicals lists which would complement the computerized search service by providing library locations for journals indexed in specific data bases. These specialized lists will eventually be consolidated to form a full-scale union list of social science and humanities serials.

In addition, the library offers reference service on these subjects, and consultative services in such fields as library automation, Canadian library developments and rare books. It is developing a children's literature consultant service and a library service for the visually and physically handicapped. It provides to provincial library agencies loan collections of books in languages other than English and French, and assists Canadian libraries to develop their collections by redistributing library materials through a Canadian book exchange centre. It also plays an active co-ordinating role in attempts to develop national library and information networks, and is contributing to international efforts at universal bibliographic control.

A list of books about Canada, prepared by the national library, is published in Appendix 6.

Public libraries 7.10.3

Public libraries in Canada are organized under provincial legislation which specifies the method of establishment, the services to be provided and the means of support. Municipalities may organize and maintain public libraries or join together to form regional libraries according to provincial legislation. Provincial public library agencies advise local and regional libraries and distribute grants.

Table 7.10 gives summary statistics on nearly 753 public libraries providing over 2,660 service points. Book circulation was 112.2 million or 4.9 per capita in 1976. The operating payments of all public libraries amounted to $156.6 million or $6.81 per capita compared with $5.80 in 1975. The full-time professional librarians numbered 1,607 in 1976.

Book publishing 7.11

Books hold a prominent place in the realm of communication. They are an important tool in spreading knowledge, they play a vital role in our educational system, and they are both the keepers and messengers of culture.

Book exports and imports. In 1975 sales of books published in Canada reached $148 million, of which $29 million resulted from exports. However, the imports of books and pamphlets surpassed the sales of domestically produced books, reaching sales values estimated at $350 million. These imported books came mainly from the United States (76%), France (11%) and the United Kingdom (8%).

Publishing and sales. Referring to books produced domestically in 1975, there were 3,305 new titles published and 3,460 titles reprinted. Sales resulting from this production, along with backlist sales (sales of books previously published), were estimated at 107 million copies, which generated revenues of $148 million. The major category of book sales was trade or commercial books, which showed sales of 85.6 million copies, or 80% of the books sold. Textbooks were the next most popular category, showing sales of 19.3 million copies (18%), followed by scholarly books, general reference books, and professional and technical books which together formed only 2% of all copies sold.

Excluding imports, most of the books sold were in mass market paperback form (74%), while hardcover books were 9%, other forms of paperback formed 16% and unbound printed material made up less than 1% of total copies sold. However, due to the relatively higher prices of hardcover books, they brought in 39% of dollar sales ($57.7 million).

Of the total copies sold of books produced in Canada in 1975, 100 million (93%) were English, 6.4 million (6%) were French, and the rest in other languages; 342 titles published or reprinted were translations from other languages. Of these, 278 were translated to French and 54 to English. As well, 515 published or reprinted were adapted from books published outside Canada.

In an examination by category of copies sold of books produced in Canada, the novel/short story formed the largest group and accounted for 75.6% of the sales of books published or reprinted in 1975. Linguistics/philology formed 5.6% of copies sold, followed by general books (5.2%), mathematics/accounting (3.3%), entertainment/ hobbies (1.4%) and business management (1.1%). These six categories accounted for 92.5%.

Sources
7.1 - 7.6 Education, Science and Culture Division, Institutional and Public Finance Statistics Branch, Statistics Canada.
7.7 Information Services, National Museums of Canada.
7.8.1 Information Services, The Canada Council.
7.8.2 Communications Department, National Arts Centre.
7.8.3 Newfoundland Information Service, Nova Scotia Department of Recreation, New Brunswick Information Service, Quebec Cultural Affairs Office, Ontario Ministry of Culture and Recreation, The Manitoba Arts Council, Saskatchewan Arts Board, Alberta Culture, Deputy Provincial Secretary of British Columbia.
7.9.1 Information and Promotion Division, National Film Board.
7.9.2 Canadian Film Development Corporation.
7.10.1 Office of the Dominion Archivist, Public Archives of Canada.
7.10.2 Public Services Branch, National Library of Canada.
7.10.3 - 7.11 Education, Science and Culture Division, Institutional and Public Finance Statistics Branch, Statistics Canada.

Tables

7.1   Enrolment in elementary and secondary schools, by type of institution and by province, school years 1972-73 to 1976-77

Type of institution and year	Province or territory						
	Nfld.	PEI	NS	NB	Que.	Ont.	Man.
Public							
1972-73	161,723	29,340	211,262	173,851	1,514,512	2,028,114	238,861
1973-74	159,831	29,056	207,651	170,179	1,463,498	2,008,610	234,620
1974-75	158,014	28,149	204,280	166,550	1,419,997	1,994,489	229,552
1975-76	157,768	27,850	202,606	164,999	1,374,909	1,994,638	228,127
1976-77	157,686	27,903	201,279	163,317	1,318,350p	1,973,140	225,698
Private[1]							
1972-73	843	—	1,394	636	67,940r	44,826	7,224
1973-74	872	—	1,286	467	72,785r	47,500	6,912
1974-75	360	—	1,372	137	86,892	51,239	6,849
1975-76	280	—	1,418	421	87,987	54,598	7,122
1976-77	293	—	1,410	393	95,000p	58,226	7,642
Federal[2]							
1972-73	—	65	624	704	4,016	7,106	6,376
1973-74	—	53	645	728	5,059	7,149	6,830
1974-75	—	57	668	782	4,875	7,465	7,303
1975-76	—	57	665	838	4,751	7,391	7,254
1976-77	—	59	681	833	4,934	7,362	7,493
Schools for the blind and the deaf							
1972-73	134	10	488	—	1,059	1,175r	173
1973-74	138	8	472	—	981	1,175r	167
1974-75	131	16	439	—	1,141	1,165	161
1975-76	127	15	452	—	1,068	1,148	168
1976-77	117	16	480	—	1,040	1,126	156
Total							
1972-73	162,700	29,415	213,768	175,191	1,587,527r	2,081,221r	252,634
1973-74	160,841	29,117	210,054	171,374	1,542,323r	2,064,434r	248,529
1974-75	158,505	28,222	206,759	167,469	1,512,905	2,054,358	243,865
1975-76	158,175	27,922	205,141	166,258	1,468,715	2,057,775	242,671
1976-77	158,096	27,978	203,850	164,543	1,419,324p	2,039,854	240,989

	Sask.	Alta.	BC	YT	NWT	Canada
Public						
1972-73	234,152	425,251	537,067	4,749	11,369	5,570,251r
1973-74	223,798	419,737	549,019	4,957	12,627	5,483,583r
1974-75	224,176	432,177	541,575	4,903	12,504	5,416,366
1975-76	220,973	439,354	542,680	4,975	12,496	5,371,375
1976-77	219,191	441,070	536,237	4,866	12,916	5,281,653p
Private[1]						
1972-73	1,268	5,403	22,061	—	—	151,595r
1973-74	1,309	5,367	21,421	—	—	157,919r
1974-75	1,853	5,541	21,055	—	—	175,298
1975-76	1,453	5,651	23,071	—	—	182,001
1976-77	1,573	6,070	23,318	—	—	193,925p
Federal[2]						
1972-73	4,465	3,409	3,036	—	—	34,390[3]
1973-74	5,290	3,661	3,083	—	—	37,064[3]
1974-75	5,309	3,418	3,013	—	—	37,511[3]
1975-76	5,530	3,719	2,258	—	—	37,087[3]
1976-77	5,844	4,100	2,336	—	—	38,024[3]
Schools for the blind and the deaf						
1972-73	178	152	277	—	—	3,646r
1973-74	159	159	257	—	—	3,516r
1974-75	155	171	293	—	—	3,672
1975-76	150	176	278	—	—	3,582
1976-77	136	185	244	—	—	3,500
Total						
1972-73	240,063	434,215	562,441	4,749	11,369	5,759,882r,[3]
1973-74	230,556	428,924	573,780	4,957	12,627	5,682,082r,[3]
1974-75	231,493	441,307	565,936	4,903	12,504	5,632,847[3]
1975-76	228,106	448,900	568,287	4,975	12,496	5,594,045[3]
1976-77	226,744	451,425	562,135	4,866	12,916	5,517,102p,[3]

[1]Private kindergartens and nursery schools not included.
[2]Provincial figures are for federal schools for Indians and Inuit.
[3]Canada total also includes Department of National Defence schools overseas.

7.2 Full-time enrolment in community colleges and public trade schools, and in vocational programs in business and industry, by province, 1974-75 and 1975-76

Item	Province or territory and year							
	Newfoundland		Prince Edward Island		Nova Scotia		New Brunswick	
	1974-75	1975-76	1974-75	1975-76	1974-75	1975-76	1974-75	1975-76
Full-time enrolment								
Community colleges[1]								
Post-secondary level								
Career programs	1,861	1,964e	842	708e	2,526a	2,717a	1,187	1,309
University transfer programs	—	—	—	—	164	177	—	—
Trades-level vocational[a]								
Pre-employment courses	3,408	1,136	40	58	397	370	2,281	2,114
Skill upgrading courses	4	2,265	—	—	—	655	130	451
Secondary level pre-vocational								
Academic upgrading courses	440	315	—	—	—	—	1,097	1,187
Second language courses	—	—	—	—	—	—	—	—
Total, community colleges	5,709	5,680	882	766	3,087	3,919	4,695	5,061
Public trade schools								
Trades-level vocational								
Pre-employment courses	4,051	4,247	650e	764	6,912	7,183	906	2,025
Skill upgrading courses	76	30	50e	327	701	575	1,448	902
Secondary level pre-vocational								
Academic upgrading courses	1,939	1,014	700e	458	3,168	3,535	1,304	1,310
Second language courses	—	—	—	—	—	30	—	16
Total, public trade schools	6,066	5,291	1,400	1,549	10,811	11,326	3,658	4,253
Business and industry								
Registered apprentices	3,217	4,061	501	517	5,233	5,766	5,555	6,158
Canada Manpower Industrial Training Program	1,634	3,860	428	410	2,238	3,805	1,628	3,650
Total, business and industry	4,851	7,921	929	927	7,471	9,571	7,183	9,808
Total, all training	16,626	18,892	3,211	3,242	21,369	24,816	15,536	19,122

Item	Quebec		Ontario		Manitoba		Saskatchewan	
	1974-75	1975-76	1974-75	1975-76	1974-75	1975-76	1974-75	1975-76
Full-time enrolment								
Community colleges[1]								
Post-secondary level								
Career programs	50,341	53,371	56,642	59,640	2,792	3,177	2,293	2,334
University transfer programs	61,841	64,292	—	—	—	—	54	63
Trades-level vocational[a]								
Pre-employment courses	—	—	29,739	32,815	5,529	4,454	4,528	2,156
Skill upgrading courses	—	—	2,456	1,667	579	1,024	47	97
Secondary level pre-vocational								
Academic upgrading courses	—	—	24,192	26,305	2,326	2,884	1,019	1,040
Second language courses	—	—	9,320	9,532	280	312	29	119
Total, community colleges	112,182	117,663	122,349	129,959	11,506	11,851	7,970	5,809
Public trade schools								
Trades-level vocational								
Pre-employment courses	25,000e	20,595	4,110	205s	—	—	997	1,361
Skill upgrading courses	4	4	9	—	—	—	920	1,083
Secondary level pre-vocational								
Academic upgrading courses	25,000e	15,800e	12	88	—	—	1,246	1,406
Second language courses	—	—	—	—	—	—	—	—
Total, public trade schools	50,000	36,395	4,131	293	—	—	3,163	3,850
Business and industry								
Registered apprentices	19,000e	20,400e	30,967	33,398	3,579	3,597	3,822	4,259
Canada Manpower Industrial Training Program	13,421	14,130	14,800	16,560	1,534	1,410	2,160	1,940
Ontario Training in Business and Industry Program	...	...	73,551	74,336	...	...	...	...
Total, business and industry	32,421	34,530	119,318	124,294	5,113	5,007	5,982	6,199
Total, all training	194,603	188,588	245,798	254,546	16,619	16,858	17,115	15,858

Item	Alberta		British Columbia		Yukon		Northwest Territories		Canada	
	1974-75	1975-76	1974-75	1975-76	1974-75	1975-76	1974-75	1975-76	1974-75	1975-76
Full-time enrolment										
Community colleges[1]										
Post-secondary level										
Career programs	11,898	12,981	8,436	8,917	—	—	—	—	138,818	147,118
University transfer programs	2,193	2,282	8,135	7,662	—	—	—	—	72,387	74,476
Trades-level vocational[a]										
Pre-employment courses	1,956	2,559	11,158	13,114	—	—	—	—	59,036	58,776
Skill upgrading courses	397	397	2,514	2,357	—	—	—	—	6,123	8,913
Secondary level pre-vocational										
Academic upgrading courses	927	748	3,695	4,788	—	—	—	—	33,696	37,267
Second language courses	—	39	1,089	1,132	—	—	—	—	10,718	11,134
Total, community colleges	17,371	19,006	35,027	37,970	—	—	—	—	320,778	337,684

7.2 Full-time enrolment in community colleges and public trade schools, and in vocational programs in business and industry, by province, 1974-75 and 1975-76 (concluded)

Item	Province or territory and year									
	Alberta		British Columbia		Yukon		Northwest Territories		Canada	
	1974-75	1975-76	1974-75	1975-76	1974-75	1975-76	1974-75	1975-76	1974-75	1975-76
Full-time enrolment (concluded)										
Public trade schools										
Trades-level vocational										
Pre-employment courses	2,799	3,321	5,004	4,802	500e	215	221	150e	51,150	44,868
Skill upgrading courses	1,471	1,672	1,132	129	—	33	16	16	5,823	4,767
Secondary level pre-vocational										
Academic upgrading courses	3,952	4,253	1,299	876	100e	317	377	202	39,097	29,259
Second language courses	—	389	—	—	—	—	—	7	30	445
Total, public trade schools	*8,222*	*9,635*	*7,435*	*5,807*	*600*	*565*	*614*	*375*	*96,100*	*79,339*
Business and industry										
Registered apprentices	15,354	17,977	17,354	19,537	47	65	382	399	105,011	116,134
Canada Manpower Industrial Training Program	1,210	4,020	6,644	6,325	395	575	73	335	46,165	57,020
Ontario Training in Business and Industry Program	. . .	. . .	. . .	. . .	. . .	. . .	. . .	. . .	73,551	74,336
Total, business and industry	*16,564*	*21,997*	*23,998*	*25,862*	*442*	*640*	*455*	*734*	*224,727*	*247,490*
Total, all training	42,157	50,638	66,460	69,639	1,042	1,205	1,069	1,109	641,605	664,513

[1]Includes hospital schools of nursing.
[2]Includes Nova Scotia Teachers' College; students of Cape Breton College technical campus also shown in university enrolment statistics.
[3]Excludes registered apprentices attending formal classes for short periods, since they would be shown as well with registered apprentices under "Business and industry".
[4]Included in pre-employment courses.
[5]Decrease from previous year due to merger of some schools with community colleges.

7.3 Full- and part-time enrolment in universities, by level and province, 1974-75 to 1976-77[1]

Province and year		Undergraduate[2]		Graduate[3]		Non-university[4]	
		Full-time	Part-time	Full-time	Part-time	Full-time	Part-time
Newfoundland	1974-75	5,483	3,207	504	336	—	—
	1975-76	5,736	3,008	445	280	—	—
	1976-77	6,200	2,928	435	297	—	—
Prince Edward Island	1974-75	1,343	812	—	—	—	—
	1975-76	1,463	887	—	—	—	—
	1976-77	1,478	854	—	—	—	—
Nova Scotia	1974-75	15,428	4,731	1,010	656	370	—
	1975-76	16,422	5,088	1,125	687	417	—
	1976-77	16,635	5,254	1,253	854	436	—
New Brunswick	1974-75	9,972	4,928	504	232	25	7
	1975-76	10,476	4,925	508	411	183	106
	1976-77	10,393	3,947	513	395	156	111
Quebec	1974-75	59,896	43,318	8,487	7,555	3	3,443
	1975-76	65,578	51,898	9,213	8,421	102	100
	1976-77	66,561	56,391	9,414	8,500	99	88
Ontario	1974-75	131,498	57,485	15,286	10,865	1,037	436
	1975-76	140,252	62,042	16,153	11,973	892	73
	1976-77	144,481	60,721	16,122	12,468	900	45
Manitoba	1974-75	16,246	8,694	1,448	1,184	—	—
	1975-76	16,945	9,329	1,486	1,244	—	—
	1976-77	16,370	10,515	1,606	1,355	—	—
Saskatchewan	1974-75	12,724	5,100	604	653	251	784
	1975-76	13,389	6,100	676	682	285	144
	1976-77	13,576	5,452	785	662	603	1,336
Alberta	1974-75	26,978	5,994	2,698	1,703	—	384
	1975-76	28,893	6,579	2,876	1,784	—	410
	1976-77	29,032	6,607	3,058	1,699	307	345
British Columbia	1974-75	26,483	5,983	3,301	1,276	700	483
	1975-76	28,080	6,994	3,257	1,478	661	611
	1976-77	27,962	8,577	3,399	1,455	—	101
Total	1974-75	306,051	140,252	33,842	24,460	2,386	5,537
	1975-76	327,220	156,850	35,739	26,960	2,540	1,444
	1976-77	332,688	161,246	36,585	27,685	2,501	2,026

[1]Excludes 5,077 interns and residents in 1974-75, 5,334 in 1975-76 and 5,168 in 1976-77.
[2]Bachelors and first professional degrees, diplomas and certificates, auditors, special students.
[3]Masters, doctorates, diplomas and certificates, qualifying and special students.
[4]Diplomas, certificates.

7.4   Graduate-level degrees awarded by Canadian universities, by field of study, region and percentage distribution by sex, calendar years, 1975 and 1976

Degree and field of study	Atlantic provinces		Quebec		Ontario	
	1975	1976	1975	1976	1975	1976
Master						
Education	167	167	357	413	1,116	1,207
Fine and applied arts	4	3	29	38	54	71
Humanities	125	105	512	445	1,151	1,073
Social sciences	189	216	976	1,019	2,289	2,400
Agriculture and biological sciences	38	37	96	109	204	255
Engineering and applied sciences	42	46	216	192	537	559
Health professions	4	14	116	120	110	126
Mathematics and physical sciences	49	69	212	210	399	396
Total	618	657	2,514	2,546	5,860	6,087
Doctorate						
Education	—	—	17	15	74	74
Fine and applied arts	—	—	2	—	5	4
Humanities	9	15	66	37	166	156
Social sciences	6	12	69	68	203	247
Agriculture and biological sciences	16	13	47	29	80	74
Engineering and applied sciences	13	8	49	32	122	99
Health professions	2	2	32	43	50	39
Mathematics and physical sciences	30	27	72	75	206	186
Total	76	77	354	299	906	879

	Western provinces		Canada						
	1975	1976	1975			1976			
				M %	F %		M %	F %	
Master									
Education	521	567	2,161	69	31	2,354	63	37	
Fine and applied arts	43	46	130	50	50	158	42	58	
Humanities	293	294	2,081	54	46	1,917	53	47	
Social sciences	658	701	4,112	76	24	4,336	73	27	
Agriculture and biological sciences	152	193	490	72	28	594	73	27	
Engineering and applied sciences	168	218	963	96	4	1,015	97	3	
Health professions	73	61	303	49	51	321	52	48	
Mathematics and physical sciences	168	185	828	87	13	860	85	15	
Total	2,076	2,265	11,068	72	28	11,555	69	31	
Doctorate									
Education	81	68	172	71	29	157	71	29	
Fine and applied arts	—	1	7	71	29	5	60	40	
Humanities	36	40	277	78	22	248	65	35	
Social sciences	94	83	372	80	20	410	80	20	
Agriculture and biological sciences	98	83	241	82	18	199	80	20	
Engineering and applied sciences	43	50	227	96	4	189	94	6	
Health professions	38	21	122	80	20	105	80	20	
Mathematics and physical sciences	114	92	422	92	8	380	93	7	
Total	504	438	1,840	84	16	1,693	81	19	

7.5   Diplomas and certificates awarded by Canadian universities, by level and field of study, region and percentage distribution by sex, calendar years, 1975 and 1976

Level and field of study	Atlantic provinces		Quebec		Ontario	
	1975	1976	1975	1976	1975	1976
Undergraduate						
Education	73	69	2,212	2,994	107	287
Fine and applied arts	12	7	11	10	101	175
Humanities	1	2	223	313	120	104
Social sciences	141	165	676	1,087	586	557
Agriculture and biological sciences	—	1	24	24	12	148
Engineering and applied sciences	68	80	—	88	397	427
Health professions	125	115	185	606	235	441
Mathematics and physical sciences	5	5	40	16	25	36
Total	425	444	3,371	5,138	1,583	2,175
Graduate						
Education	79	90	117	138	—	—
Fine and applied arts	—	—	1	1	8	8
Humanities	—	—	2	6	13	26
Social sciences	10	10	166	282	164	181
Agriculture and biological sciences	—	—	2	2	12	22
Engineering and applied sciences	—	—	8	16	—	—
Health professions	—	2	170	162	91	107
Mathematics and physical sciences	—	—	—	2	—	—
Total	89	102	466	609	288	344

7.5   Diplomas and certificates awarded by Canadian universities, by level and field of study, region and percentage distribution by sex, calendar years, 1975 and 1976 (concluded)

Level and field of study	Region and year							
	Western provinces		Canada					
	1975	1976	1975	M %	F %	1976	M %	F %
Undergraduate								
Education	1,872	2,020	4,264	47	53	5,370	45	55
Fine and applied arts	9	21	133	28	72	213	38	62
Humanities	74	53	418	36	64	472	42	58
Social sciences	70	170	1,473	65	35	1,979	62	38
Agriculture and biological sciences	121	196	157	94	6	369	91	9
Engineering and applied sciences	–	–	465	98	2	595	98	2
Health professions	320	235	865	8	92	1,397	11	89
Mathematics and physical sciences	22	24	92	84	16	81	83	17
Total	2,488	2,719	7,867	50	50	10,476	49	51
Graduate								
Education	250	290	446	63	37	518	57	43
Fine and applied arts	–	–	9	44	56	9	33	67
Humanities	–	3	15	27	73	35	23	77
Social sciences	64	1	404	80	20	474	72	28
Agriculture and biological sciences	11	2	25	72	28	26	73	27
Engineering and applied sciences	8	5	16	94	6	21	100	–
Health professions	19	2	280	68	32	273	70	30
Mathematics and physical sciences	3	2	3	67	33	4	75	25
Total	355	305	1,198	70	30	1,360	65	35

7.6   Participation in continuing education courses per 1,000 population[1] by type of institution and by province, 1972-73 to 1975-76

Province or territory and year		School boards, departments of education		Department of education correspondence courses		Community colleges[2]		Universities		Total
		Part-time credit	Non-credit	Part-time credit	Non-credit	Part-time credit	Non-credit	Part-time credit	Non-credit	
Newfoundland	1972-73r	7.1	9.4	–	–	10.4	9.5	21.5	7.6	65.4
	1973-74r	7.6	7.5	–	–	9.5	16.9	21.0	8.0	70.5
	1974-75	10.6	17.4	–	–	10.6	17.5	19.9	8.0	83.9
	1975-76	10.6	16.8	–	–	6.9	15.4p	17.8	8.0	75.5p
Prince Edward Island	1972-73	9.9	43.3	–	–	15.9	3.9	33.9	2.8	109.7
	1973-74	9.5	51.0	–	–	11.8	5.2	35.1	2.0	114.6
	1974-75	8.0	45.4	–	–	11.6	4.6	27.8	1.9	99.4
	1975-76	[3]	[3]	–	–	10.5	19.0	28.9	1.1	59.5
Nova Scotia	1972-73	6.6	25.3	1.8	–	6.5	1.3	16.9	7.9	66.3
	1973-74	8.3	29.2	2.2	–	5.6	1.6	18.5	11.5	77.0
	1974-75	8.6	31.4	1.3	0.2	7.3	3.4	20.1	12.9	85.3
	1975-76	3.3	42.1	2.4	0.2	6.4	3.8	23.6	11.5	93.3
New Brunswick	1972-73	4.4	16.2	1.2	–	3.7	4.1r	22.5	7.3	59.3r
	1973-74r	3.0	19.5	1.0	–	4.2	3.5	22.6	7.3	61.3
	1974-75	2.4	21.5	1.0	–	6.0	6.7	21.7	8.7	68.1
	1975-76	1.9	16.2	0.8	–	4.6	4.5	23.1	10.5	61.6
Quebec	1972-73r	16.4	27.3	0.5	0.5	...	6.0	21.7	7.2	79.6
	1973-74r	13.0	29.7	0.4	0.6	...	5.4	18.7	10.2	78.0
	1974-75	11.2	28.8	0.8	0.8	...	5.2	21.1	9.3	77.3
	1975-76	12.2	29.0	1.0	0.6	...	6.7p	25.2	9.5	84.1p
Ontario	1972-73	4.2	25.3	9.4	–	12.0	8.4	23.5	8.3	91.0
	1973-74	4.4	24.9	8.8	–	12.5	11.3	24.0	12.2	98.3
	1974-75	4.0	27.8	9.6	–	16.5	16.8	23.8	14.3	112.8
	1975-76	5.0	36.1	10.1	–	15.4	18.3	25.1	13.2	123.2
Manitoba	1972-73	2.9	13.9	2.5	–	3.6	11.3	25.8	8.8	68.9
	1973-74	1.3	15.5	2.3	–	3.0	12.5	27.2	12.7	74.5
	1974-75	1.1	18.1	2.4	- -	4.6	15.2	28.3	14.9	84.6
	1975-76	1.3	25.2	2.3	0.2	5.5	10.0	27.7	14.0	86.2
Saskatchewan	1972-73	3.8	11.9	2.9	–	6.6	2.4	20.2	17.6	65.5
	1973-74	2.4	9.8	2.2	- -	5.1	1.9	20.4	23.2	65.1
	1974-75	1.6	11.8	2.2	–	4.6	37.6	22.0	19.4	98.7
	1975-76	[4]	[4]	3.0	0.1	8.2	84.6	22.7	23.8	142.4
Alberta	1972-73r	3.2	22.7	5.4	0.2	4.0	16.1	16.0	25.6	93.2
	1973-74r	5.4	23.2	4.9	0.4	4.8	12.4	14.5	29.1	94.7
	1974-75	4.3	30.9	6.5	0.8	6.7	24.1	15.0	28.3	116.6
	1975-76	7.9	49.7	6.9	1.3	8.0	36.3	18.1	31.6	159.8

7.6   Participation in continuing education courses per 1,000 population[1] by type of institution and by province, 1972-73 to 1975-76 (concluded)

Province or territory and year		School boards, departments of education		Department of education correspondence courses		Community colleges[2]		Universities		Total
		Part-time credit	Non-credit	Part-time credit	Non-credit	Part-time credit	Non-credit	Part-time credit	Non-credit	
British Columbia	1972-73	5.5	62.0	4.4	1.2	10.0ʳ	20.4	7.5	18.6	129.7ʳ
	1973-74	6.0	56.6	3.3	1.4	11.7ʳ	23.2	8.1	18.2	128.5ʳ
	1974-75	4.3	49.7	1.9	1.8	12.1	26.6	10.3	19.8	126.5
	1975-76	3.1	54.2	2.2	2.0	15.9	39.0	10.3	20.3	147.0
Yukon Territory and Northwest Territories	1972-73	7.2	27.7	—	—	17.6	—	—	—	52.5
	1973-74	7.9	57.8	—	—	22.6ʳ	10.4	—	—	98.7ʳ
	1974-75	1.3	22.2	—	—	17.6	7.8	—	—	48.9
	1975-76	4.0	73.6	—	—	9.9ᵖ	12.1	—	—	99.6ᵖ
Canada	1972-73ʳ	7.8	28.1	4.7	0.3	6.8	9.1	20.4	10.7	87.9
	1973-74ʳ	6.9	28.4	4.3	0.3	7.2	10.3	19.8	13.7	91.0
	1974-75	6.1	29.5	4.6	0.5	9.0	15.2	20.7	14.4	100.0
	1975-76	6.6	34.4	5.0	0.5	9.1	20.1	22.7	14.5	112.9

[1]Out-of-school population 15 years of age and over as of June 1, 1972-73 to 1975-76.
[2]Includes institutes of technology, community colleges, colleges of applied arts and technology, trade and vocational schools and collèges d'enseignement général et professionnel (CEGEPs) of Quebec.
[3]In 1975-76, Holland College assumed the responsibility for the continuing education programs previously administered by the Department of Education.
[4]In 1975-76, the community colleges assumed the responsibility for the continuing education programs previously offered by the local school boards.

7.7   Expenditures on education by level of study, and by province, 1974-75 and 1975-76 (million dollars)

Year and level of study	Province or region						
	Nfld.	PEI	NS	NB	Que.	Ont.	Man.
1974-75							
Elementary and secondary	160.6	34.9	209.1	174.0	2,135.4	2,599.3	304.3
Post-secondary							
Non-university	8.4	2.8	10.9	5.1	352.1	253.7	8.8
University	47.6	8.5	103.0	61.2	547.9	958.8	124.4
Vocational and occupational training	28.9	6.6	40.4	22.5	178.6	183.1	29.3
Total	245.5	52.8	363.4	262.8	3,214.0	3,994.9	466.8
1975-76							
Elementary and secondary	180.2	43.1	258.0	222.4	2,458.0	3,100.6	355.4
Post-secondary							
Non-university	9.9	3.1	14.1	5.9	425.6	290.0	16.9
University	61.6	9.5	115.5	75.2	651.9	1,075.3	136.7
Vocational and occupational training	38.3	9.2	52.9	34.0	202.2	211.5	33.1
Total	290.0	64.9	440.5	337.5	3,737.7	4,677.4	542.1

	Sask.	Alta.	BC	YT	NWT	Overseas and undistributed	Total
1974-75							
Elementary and secondary	273.5	553.6	688.2	10.3	27.4	20.2	7,190.8
Post-secondary							
Non-university	10.6	63.5	63.5	- -	- -	13.0	792.4
University	83.9	185.3	208.8	0.2	0.1	42.5	2,372.2
Vocational and occupational training	35.5	72.8	62.3	2.5	5.0	25.9	693.4
Total	403.5	875.2	1,022.8	13.0	32.5	101.6	11,048.8
1975-76							
Elementary and secondary	303.9	673.3	864.3	14.2	32.2	32.3	8,537.9
Post-secondary							
Non-university	12.7	77.9	76.2	0.1	- -	14.3	946.7
University	102.8	237.0	254.4	0.2	0.1	40.4	2,760.6
Vocational and occupational training	41.6	112.2	77.4	2.5	6.2	25.7	846.8
Total	461.0	1,100.4	1,272.3	17.0	38.5	112.7	13,092.0

7.8   Sources of funds for education at all levels, 1971-72 to 1975-76 (million dollars)

Year	Sources of funds			Fees	Other sources	Total
	Government					
	Federal	Provincial[1]	Municipal			
1971-72	924.0	4,966.7	1,713.6	386.8	358.6	8,349.7
1972-73	943.8	5,257.0	1,777.3	414.4	276.6	8,669.2
1973-74	984.8	5,847.2	1,940.0	436.6	426.6	9,635.2
1974-75	1,056.4	7,028.9	2,062.8	470.5	430.2	11,048.8
1975-76P	1,166.0	8,586.8	2,361.6	523.1	454.5	13,092.0

[1]Includes federal transfers to provinces for post-secondary education and for the minority language program, in the following amounts: $985,062 in 1971-72, $1,057,948 in 1972-73, $1,143,187 in 1973-74, $1,404,888 in 1974-75 and $1,658,670 in 1975-76.

7.9   Average grants, revenues and expenditures of the performing arts by company, spectator and by discipline, 1975 and 1976

Year and discipline	By company			By spectator		
	Grants $	Revenues[1] $	Expenditures $	Grants $	Revenues[1] $	Expenditures $
1975						
Theatre	202,920	491,936	513,463	2.31	5.61	5.85
Music	487,372	915,846	930,800	3.56	6.69	6.80
Dance	641,436	1,135,408	1,159,863	4.95	8.76	8.95
Opera	432,155	951,387	1,091,947	6.99	15.39	17.66
1976						
Theatre	259,413	565,442	569,754	3.25	7.09	7.15
Music	609,647	1,082,886	1,119,655	4.22	7.50	7.76
Dance	641,353	1,104,685	1,146,612	6.39	11.00	11.42
Opera	495,526	921,109	869,309	8.56	15.90	15.01

[1]Includes average grants.

7.10   Summary statistics of public libraries, 1975 and 1976

Year, province or territory	Population '000	Libraries reporting	Bookstock[1]	Circulation	Total operating expenditure $	Full-time professional librarians
1975						
Newfoundland	549	3	656,830	1,908,158	2,013,548	12
Prince Edward Island	119	1	161,181	459,766	599,500	9
Nova Scotia	822	12	859,537	3,065,293	2,843,018	49
New Brunswick	675	6	739,718	2,177,793	2,332,554	34
Quebec	6,188	102	5,441,131	11,981,858	13,532,919	116
Ontario	8,226	351	17,636,049	53,108,482	73,142,980	933
Manitoba	1,018	30	1,313,995	4,113,529	3,917,953	34
Saskatchewan	918	10	1,693,479	5,042,886	6,525,484	84
Alberta	1,768	156	2,828,749	8,502,414	8,721,559	79
British Columbia	2,457	66	3,980,174	17,459,366	17,746,644	222
Yukon Territory	21	1	104,221	126,242	519,638	3
Northwest Territories	38	1	75,473	98,712	276,600	2
Canada	22,799	739	35,490,537	108,044,499	132,172,397	1,577
1976						
Newfoundland	558	4	665,610	2,029,291	2,283,048	12
Prince Edward Island	118	1	162,755	514,036	675,000	9
Nova Scotia	829	12	907,968	3,290,939	3,265,415	50
New Brunswick	677	6	792,118	2,194,010	2,696,921	33
Quebec	6,234	107	5,674,470	13,108,524	17,129,284	121
Ontario	8,264	355	18,649,944	53,633,135	85,008,345	943
Manitoba	1,021	31	1,404,695	4,157,673	4,571,736	36
Saskatchewan	921	10	1,842,889	5,518,495	8,336,786	90
Alberta	1,838	160	3,030,641	9,006,157	10,442,151	76
British Columbia	2,467	65	4,203,472	18,556,479	21,297,159	232
Yukon Territory	43	1	116,824	124,416	659,914	2
Northwest Territories	22	1	82,462	107,191	269,221	3
Canada	22,992	753	37,533,848	112,240,346	156,634,980	1,607

[1]Books and other materials catalogued as books; does not include periodical and newspaper titles.

7.11   Titles published and reprinted[1], by language and commercial category, 1975

Item	Commercial category						
	Multi-volume reference	Textbook					Total
		Kinder-garten to grade 3	Grades 4-6	Grades 7-9	Grades 10-13	Post-second-ary	
Titles published							
English	78	43	94	106	99	74	416
French	47	40	28	23	30	78	199
Bilingual	5	—	1	—	3	2	6
Other	3	—	—	—	—	6	6
Total	133	83	123	129	132	160	627
Titles reprinted							
English	39	196	224	214	313	92	1,039
French	34	71	71	82	81	84	389
Bilingual	1	—	—	4	6	5	15
Other	—	1	—	—	1	4	6
Total	74	268	295	300	401	185	1,449

Item	Professional and technical	Schol-arly	Trade book				Total all categories
			Chil-dren's	Juvenile	Adult	Total	
Titles published							
English	117	208	23	23	1,092	1,138	1,957
French	43	48	50	14	214	278	615
Bilingual	4	13	2	—	8	10	38
Other	—	7	—	—	11	11	27
Total	164	276	75	37	1,325	1,437	2,637
Titles reprinted							
English	112	114	26	5	608	639	1,943
French	1	58	11	—	135	146	628
Bilingual	1	5	—	—	8	8	30
Other	—	2	—	—	3	3	11
Total	114	179	37	5	754	796	2,612

[1]Represents 75% to 80% of all titles actually published and reprinted. The text shows complete coverage.

7.12   Titles published and reprinted[1] and copies sold, by language, according to UNESCO classification, 1975

UNESCO classification	Titles							
	English		French		Bilingual		Other	
	Published	Copies sold	Published	Copies sold	Published	Copies sold	Published	Copies sold
General	351	3,292,976	30	120,453	5	2,411	5	4,486
Philosophy, psychology	14	13,107	21	37,604	—	—	—	—
Religion, theology	14	30,877	36	119,144	—	—	2	5,160
Sociology, statistics	24	49,797	13	8,538	4	334	—	—
Political science, economics	61	81,958	11	22,791	—	—	2	62
Law, public administration	113	165,589	30	56,132	5	1,695	—	—
Military science, national defence	4	2,392	—	—	—	—	—	—
Science of education	28	143,157	13	109,366	1	41	—	—
Political-economic aspects of trade	13	12,812	—	—	—	—	—	—
Ethnography, social anthropology	30	55,966	7	6,341	—	—	—	—
Linguistics, philology	45	206,740	33	25,553	6	36,004	6	678
Mathematics, accounting	89	589,418	86	416,474	—	—	—	—
Natural science	65	58,842	29	17,707	1	402	—	—
Medical science	19	62,804	10	10,838	1	168	—	—
Engineering, technology	11	17,155	28	8,543	2	165	—	—
Agriculture, stockbreeding	8	11,438	3	7,874	—	—	—	—
Domestic science, hotel management	8	42,366	5	35,422	—	—	—	—
Business management	46	184,925	46	66,668	—	—	—	—
Town planning, fine arts	63	230,306	8	21,179	9	14,476	2	1,235
Entertainment, hobbies	80	551,624	66	201,017	—	—	—	—
Literary history and criticism	28	18,744	19	7,011	—	—	1	77
Literature								
Novels, short stories	410	55,683,164	49	65,820	1	2,200	4	5,750
Poetry	65	129,413	28	16,023	—	—	4	992
Drama	10	6,984	6	1,488	—	—	—	—
Anthology	70	249,620	1	1,000	—	—	—	—
Geography, travel	43	135,376	5	8,517	2	3,593	—	—
History, biography	245	694,735	34	54,237	1	1,200	1	43
Non-specified	—	—	—	—	—	—	—	—
Total	1,957	62,722,285	617	1,445,740	38	62,689	27	18,483

7.12   Titles published and reprinted[1] and copies sold, by language, according to UNESCO classification, 1975 (concluded)

UNESCO classification	Titles							
	English		French		Bilingual		Other	
	Reprinted	Copies sold	Reprinted	Copies sold	Reprinted	Copies sold	Reprinted	Copies sold
General	157	314,872	49	423,261	5	6,938	—	—
Philosophy, psychology	15	17,566	16	41,341	—	—	—	—
Religion, theology	17	189,869	10	46,987	—	—	1	282
Sociology, statistics	22	117,221	5	3,315	1	4,100	—	—
Political science, economics	67	117,740	5	3,479	1	2,465	1	13
Law, public administration	59	160,663	11	10,027	—	—	—	—
Military science, national defence	1	762	—	—	—	—	—	—
Science of education	18	160,913	55	129,595	—	—	—	—
Political-economic aspects of trade	3	3,429	—	—	—	—	—	—
Ethnography, social anthropology	30	67,068	1	337	—	—	—	—
Linguistics, philology	322	3,201,194	118	1,137,269	12	68,776	6	7,045
Mathematics, accounting	133	1,351,302	95	320,897	—	—	—	—
Natural science	78	399,910	42	111,296	1	98	—	—
Medical science	23	77,721	47	206,451	—	—	—	—
Engineering, technology	31	118,155	32	40,532	—	—	—	—
Agriculture, stockbreeding	11	16,346	1	4,280	1	46	—	—
Domestic science, hotel management	14	70,646	17	51,607	—	—	—	—
Business management	99	517,638	25	65,988	7	46,755	—	—
Town planning, fine arts	27	55,009	6	5,134	—	—	—	—
Entertainment, hobbies	31	261,259	32	74,133	1	486	1	826
Literary history and criticism	15	17,854	6	1,223	—	—	—	—
Literature								
Novels, short stories	544	4,681,890	37	66,391	1	963	—	—
Poetry	11	5,664	2	1,895	—	—	—	—
Drama	8	17,474	1	587	—	—	—	—
Anthology	3	9,908	—	—	—	—	—	—
Geography, travel	81	494,593	8	26,113	—	—	—	—
History, biography	124	325,672	8	56,515	—	—	2	57
Non-specified	—	—	—	—	—	—	—	—
Total	1,944	12,772,338	629	2,828,653	30	130,627	11	8,223

[1] See footnote 1, Table 7.11.

7.13 Distribution of large cultural institutions[1] by region and type, 1976P

Region	Art museum or gallery	Restoration[2]	Science Technology[3]	Living[4]	History	General[5]	Community[6]	Archives	Other[7]	Total
Atlantic provinces	3	10	—	1	1	2	1	2	1	21
Quebec	4	2	1	7	1	1	—	—	—	16
Ontario	14	14	2	6	3	4	2	5	2	52
Prairie provinces	4	5	2	6	2	1	—	3	3	26
British Columbia	3	5	3	5	1	—	—	2	1	20
Yukon Territory and Northwest Territories	—	1	—	—	—	1	—	1	—	3
Canada	28	37	8	25	8	9	3	13	7	138

[1]Expenditures over $100,000.
[2]Includes individual building and historic community restorations.
[3]Includes science and technology museums, planetaria and observatories.
[4]Includes aquaria, botanical gardens, arboretums, conservatories and zoos.
[5]Includes more than one category of collection, e.g. archaeology, entomology, ethnology.
[6]Includes artifacts of relative recent history, from a specific geographic area.
[7]Includes nature park museums or nature centres.

7.14 Bachelors' and first professional degrees awarded by Canadian universities, by field of study, province, and percentage by sex, calendar years, 1975 and 1976

Field of study	Nfld. 1975	Nfld. 1976	PEI 1975	PEI 1976	NS 1975	NS 1976	NB 1975	NB 1976	Que. 1975	Que. 1976	Ont. 1975	Ont. 1976
Agriculture	—	—	—	—	—	—	—	—	92	114	214	268
Architecture	—	—	—	—	35	29	—	—	177	204	234	207
Arts	432	375	186	97	1,081	1,245	911	820	4,228	4,140	17,822	17,920
Commerce and business administration	81	59	45	48	477	514	188	238	1,914	2,166	1,399	1,612
Dentistry	—	—	—	—	22	29	—	—	147	146	164	178
Education	568	582	122	86	847	868	826	959	3,299	2,997	5,532	6,821
Engineering	77	56	—	—	123	137	146	127	1,041	1,002	1,851	1,845
Environmental studies	—	—	—	—	5	78	—	—	19	8	257	264
Fine and applied arts	—	—	—	—	70	74	12	25	267	341	160	132
Forestry	—	—	—	—	—	—	25	26	74	70	73	76
Household science	—	—	—	—	112	124	28	24	86	73	409	445
Journalism	—	—	—	—	—	—	—	—	—	—	—	—
Law	—	—	—	—	139	145	58	69	861	685	1,026	1,117
Library science	—	—	—	—	—	—	—	—	—	—	256	317
Medicine	109	102	—	—	92	92	—	—	686	729	556	573
Music	—	48	5	1	12	10	32	27	143	124	282	315
Nursing	28	—	—	—	126	87	95	78	192	176	407	487
Optometry	—	—	—	—	—	—	—	—	35	—	55	52
Pharmacy	—	—	—	—	51	46	—	—	165	245	145	144
Physical and health education	36	41	—	—	70	67	107	108	201	185	795	989
Rehabilitation medicine	—	—	—	—	—	—	—	—	248	258	178	204

7.14   Bachelors' and first professional degrees awarded by Canadian universities, by field of study, province, and percentage by sex, calendar years, 1975 and 1976 (concluded)

Field of study	Nfld. 1975	Nfld. 1976	PEI 1975	PEI 1976	NS 1975	NS 1976	NB 1975	NB 1976	Que. 1975	Que. 1976	Que. M %	Que. F %	Ont. 1975	Ont. 1976	Ont. M %	Ont. F %
Religion and theology	–	–	–	–	5	4	–	–	139	134	81	19	55	65	81	19
Science	218	188	61	50	754	796	260	222	2,862	2,801	89	11	4,314	4,501	85	15
Social work	18	32	–	–	–	–	22	17	209	182	49	51	223	250	47	53
Veterinary medicine	–	–	–	–	–	–	–	–	57	48			83	115		
Other	–	–	–	–	–	–	–	–	–	–			23	14		
Total	1,567	1,483	419	282	4,021	4,345	2,710	2,740	17,142	16,828	56	44	36,513	38,911	54	46

Field of study	Man. 1975	Man. 1976	Sask. 1975	Sask. 1976	Alta. 1975	Alta. 1976	BC 1975	BC 1976	Canada 1975	Canada 1976	M %	F %
Agriculture	82	98	61	68	102	91	62	78	613	717	81	19
Architecture	–	–	–	–	–	–	31	33	477	473	90	10
Arts	1,498	1,452	663	670	1,122	1,073	1,726	1,769	29,669	29,561	36	64
Commerce and business administration	207	263	221	258	469	543	245	282	5,246	5,983	83	17
Dentistry	28	29	10	15	45	43	40	38	456	478	90	10
Education	749	836	905	942	2,000	1,857	1,060	943	15,908	16,891	40	60
Engineering	157	147	168	154	308	204	186	180	4,057	3,852	98	2
Environmental studies	66	60	–	–	–	–	–	–	347	410	74	26
Fine and applied arts	101	107	9	27	69	75	56	67	744	848	42	58
Forestry	–	–	–	–	38	55	58	78	194	235	96	4
Household science	86	123	48	58	92	83	76	112	925	1,039	2	98
Journalism	–	–	–	–	–	–	–	–	342	390	61	39
Law	106	101	81	89	149	154	209	218	2,629	2,578	78	22
Library science	82	121	–	–	56	44	–	2	56	46	13	87
Medicine	27	46	84	80	253	261	74	84	1,936	2,042	73	27
Music	78	87	22	11	39	46	92	95	654	675	43	57
Nursing	–	–	76	80	144	121	94	85	1,240	1,249	3	97
Optometry	–	–	–	–	–	–	–	–	90	52	76	24
Pharmacy	38	40	65	63	82	80	93	92	639	710	49	51
Physical and health education	79	83	31	43	205	195	141	156	1,665	1,867	56	44
Rehabilitation medicine	29	24	3	2	51	59	31	34	540	581	8	92
Religion and theology	40	40	19	22	21	21	8	11	287	297	76	24
Science	529	520	246	286	924	992	852	878	11,020	11,234	69	31
Social work	78	92	77	51	116	117	–	–	743	773	32	68
Veterinary medicine	–	–	60	61	–	–	–	–	200	224	79	21
Other	9	29	–	–	–	–	28	28	60	71	62	38
Total	4,069	4,298	2,849	2,980	6,285	6,114	5,162	5,295	80,737	83,276	56	44

7.15   Students[1] in continuing education courses by type of institution and by province, 1972-73 to 1975-76

Province or territory and year	School boards, departments of education		Department of education correspondence courses		Community colleges[2]		Universities		Total
	Part-time credit	Non-credit	Part-time credit	Non-credit	Part-time credit	Non-credit	Part-time credit	Non-credit	
Newfoundland									
1972-73r	2,107	2,784	—	—	3,091	2,820	6,387	2,271	19,460
1973-74r	2,292	2,259	—	—	2,855	5,094	6,327	2,423	21,250
1974-75	3,269	5,370	—	—	3,278	5,418	6,142	2,472	25,949
1975-76	3,330	5,261	—	—	2,176	4,835p	5,573	2,473	23,648p
Prince Edward Island									
1972-73	677	2,966	—	—	1,086	269	2,319	189	7,506
1973-74	670	3,582	—	—	826	365	2,471	139	8,053
1974-75	581	3,290	—	—	843	334	2,017	138	7,203
1975-76	³		—	—	770	1,392	2,111	77	4,350
Nova Scotia									
1972-73	3,245	12,361	875	—	3,174	619	8,280	3,853	32,407
1973-74	4,164	14,580	1,104	—	2,821	818	9,252	5,741	38,480
1974-75	4,384	16,031	680	121	3,735	1,746	10,252	6,597	43,546
1975-76	1,724	21,760	1,238	127	3,303	1,949	12,200	5,929	48,230
New Brunswick									
1972-73r	1,683	6,245	452	—	1,418	1,579	8,682	2,823	22,882
1973-74r	1,208	7,721	402	—	1,670	1,410	8,963	2,904	24,278
1974-75	976	8,783	426	—	2,455	2,730	8,868	3,564	27,802
1975-76	787	6,766	322	—	1,937	1,871	9,610	4,354	25,647
Quebec									
1972-73r	62,904	104,525	1,777	1,961	...	23,018	82,818	27,428	304,431
1973-74r	50,748	116,243	1,624	2,236	...	21,048	73,256	40,033	305,188
1974-75	44,739	115,309	3,127	3,127	...	20,970	84,548	37,417	309,237
1975-76	49,463	116,907	4,011	2,259	...	27,103p	101,583	38,243	339,569p
Ontario									
1972-73	21,211	126,575	47,110	—	60,061	41,905	117,402	41,339	455,603
1973-74	22,480	128,428	45,433	—	64,666	58,480	123,815	63,068	506,370
1974-75	21,178	146,737	50,501	—	87,226	89,099	125,908	75,658	596,307
1975-76	26,859	193,458	54,107	—	82,249	97,932	134,421	70,853	659,879
Manitoba									
1972-73	1,802	8,744	1,561	—	2,249	7,112	16,179	5,523	43,170
1973-74	822	9,932	1,495	—	1,930	8,014	17,389	8,106	47,688
1974-75	721	11,767	1,531	5	2,963	9,846	18,384	9,704	54,921
1975-76	832	16,510	1,498	167	3,596	6,534	18,164	9,176	56,477
Saskatchewan									
1972-73	2,136	6,685	1,617	—	3,704	1,327	11,319	9,863	36,651
1973-74	1,380	5,531	1,259	6	2,867	1,103	11,550	13,146	36,842
1974-75	902	6,823	1,273	6	2,427	21,779	12,761	11,211	57,176
1975-76	⁴		1,771	62	4,873	49,904	13,415	14,025	84,050
Alberta									
1972-73r	3,237	22,968	5,453	229	4,004	16,321	16,179	25,979	94,370
1973-74r	5,586	24,227	5,160	421	5,029	12,895	15,159	30,401	98,878
1974-75	4,661	33,704	7,118	859	7,329	26,277	16,348	30,803	127,099
1975-76	8,919	55,938	7,747	1,471	9,005	40,861	20,317	35,498	179,756
British Columbia									
1972-73	8,256	92,725	6,575	1,822	14,983r	30,437	11,226	27,788	193,812r
1973-74	9,314	88,345	5,125	2,215	18,269r	36,159	12,634	28,487	200,548r
1974-75	7,023	80,362	3,011	2,872	19,607	42,967	16,639	32,033	204,514
1975-76	5,124	88,754	3,572	3,275	26,116	63,986	16,837	33,274	240,938

7.15 Students[1] in continuing education courses by type of institution and by province, 1972-73 to 1975-76 (concluded)

Province or territory and year	School boards, departments of education		Department of education correspondence courses		Community colleges[2]		Universities		Total
	Part-time credit	Non-credit	Part-time credit	Non-credit	Part-time credit	Non-credit	Part-time credit	Non-credit	
Yukon Territory and Northwest Territories									
1972-73	221	851	—	—	542		—	—	1,614
1973-74	242	1,771	—	—	692r	318	—	—	3,023r
1974-75	42	709	—	—	563	251	—	—	1,565
1975-76	130	2,369	—	—	320p	390	—	—	3,209p
Canada									
1972-73r	107,479	387,429	65,420	4,012	94,312	125,407	280,791	147,056	1,211,906
1973-74r	98,906	402,619	61,602	4,878	101,625	145,704	280,816	194,448	1,290,598
1974-75	88,476	428,885	67,667	6,984	130,426	221,417	301,867	209,597	1,455,319
1975-76	97,156	507,735	74,266	7,361	134,345	296,757	334,231	213,902	1,665,753

[1]Number of individuals enrolled in school board credit courses and in non-credit programs for all institutions are estimates based on course registrations.
[2]Includes institutes of technology, community colleges, colleges of applied arts and technology, trade and vocational schools, and collèges d'enseignement général et professionnel (CEGEPs) of Quebec.
[3]In 1975-76, Holland College assumed the responsibility for the continuing education programs previously administered by the Department of Education.
[4]In 1975-76, the community colleges assumed the responsibility for the continuing education programs previously offered by the local school boards.

7.16 Registrations in continuing education courses by type of institution and by province, 1972-73 to 1975-76

Province or territory and year	School boards, departments of education		Department of education correspondence courses		Community colleges[1]		Universities		Total
	Part-time credit	Non-credit	Part-time credit	Non-credit	Part-time credit	Non-credit	Part-time credit	Non-credit	
Newfoundland									
1972-73r	6,282	4,176	—	—	4,013	3,719	8,495	3,021	29,706
1973-74r	6,876	3,389	—	—	3,798	6,773	8,415	3,223	32,474
1974-75	9,807	8,055	—	—	4,360	7,207	8,169	3,288	40,886
1975-76	9,990	7,891	—	—	2,888	6,428p	8,777	3,289	39,263p
Prince Edward Island									
1972-73	2,708	4,449	—	—	1,444	358	3,084	251	12,294
1973-74	2,680	5,373	—	—	1,099	485	3,286	185	13,108
1974-75	2,322[2]	4,935[2]	—	—	1,121	445	2,683	183	11,689
1975-76			—	—	1,024	1,852	2,679	102	5,657
Nova Scotia									
1972-73	6,489	18,542	1,310	—	4,221	823	11,012	5,125	47,522
1973-74	8,328	21,870	1,684	—	3,752	1,088	12,305	7,636	56,663
1974-75	8,767	24,046	1,020	181	4,964	2,323	13,635	8,774	63,710
1975-76	3,447	32,640	1,844	190	4,390	2,591	15,069	7,886	68,057
New Brunswick									
1972-73r	3,366	9,366	1,220	—	1,886	2,100	11,547	3,754	33,239
1973-74r	2,416	11,581	1,003	—	2,220	1,875	11,920	3,862	34,877
1974-75	1,952	13,175	852	—	3,265	3,632	11,794	4,740	39,410
1975-76	1,575	10,149	447	—	2,581	2,489	16,530	5,791	39,562

7.16 Registrations in continuing education courses by type of institution and by province, 1972-73 to 1975-76 (concluded)

Province or territory and year	School boards, departments of education		Department of education correspondence courses		Community colleges[1]		Universities		Total
	Part-time credit	Non-credit	Part-time credit	Non-credit	Part-time credit	Non-credit	Part-time credit	Non-credit	
Quebec 1972-73r	101,211	237,731	2,665	2,942	...	30,614	110,147	36,479	521,789
1973-74r	96,203	229,838	3,436	3,433	...	27,994	97,429	53,244	511,577
1974-75	108,970	248,839	4,690	4,691	...	27,890	112,449	49,764	557,293
1975-76	109,074	245,840	6,017	3,389	...	36,047p	138,655	50,863	589,885p
Ontario 1972-73	42,422	189,862	71,428	—	79,881	55,733	156,144r	54,981	650,451r
1973-74	44,959	192,642	77,237	—	86,006	77,778	164,674	83,880	727,176
1974-75	42,355	220,105	80,801	—	116,011	118,501	167,458	100,625	845,856
1975-76	53,717	290,187	87,654	—	109,391	130,250	177,434	94,234	942,867
Manitoba 1972-73	3,603	13,116	2,654	—	2,991	9,459	21,518	7,345	60,686
1973-74	1,644	14,898	1,854	—	2,567	10,659	23,127	10,781	65,530
1974-75	1,441	17,651	2,296	7	3,941	13,095	24,451	12,906	75,788
1975-76	1,665	24,766	2,247	250	4,784	8,691	25,511	12,204	80,118
Saskatchewan 1972-73	4,272	10,027	2,506	—	4,926	1,765	15,054	13,118	51,668
1973-74	2,760	8,297	1,825	9	3,813	1,467	15,362	17,484	51,017
1974-75	1,803³		1,909	—	3,228	28,966	16,971	14,911	78,023
1975-76		10,235³	2,658	94	6,481	66,372	17,554	18,653	111,812
Alberta 1972-73r	6,472	34,453	7,907	332	5,325	21,707	21,518	34,552	132,266
1973-74r	11,172	36,341	7,688	612	6,689	17,150	20,164	40,433	140,249
1974-75	9,321	50,556	11,389	1,374	9,748	34,950	21,743	40,967	180,048
1975-76	17,838	83,907	12,861	2,442	11,977	34,345	33,246	47,212	263,828
British Columbia 1972-73r	16,512	139,088	8,123	2,250	19,927	40,481	14,930	36,958	278,269
1973-74r	18,627	132,518	6,355	2,747	24,298	48,091	16,803	37,888	287,327
1974-75	14,046	120,543	4,516	4,308	26,077	57,147	22,130	42,604	291,371
1975-76	10,412	133,132	5,358	4,912	34,734	85,101	28,149	44,255	346,053
Yukon Territory and Northwest Territories 1972-73	440	1,277	—	—	721	—	—	—	2,438
1973-74	484	2,656	—	—	920r	423	—	—	4,483r
1974-75	84	1,063	—	—	749	334	—	—	2,230
1975-76	234	3,572	—	—	425p	499	—	—	4,730p
Canada 1972-73r	193,777	662,087	97,813	5,524	125,335	166,759	373,449	195,584	1,820,328
1973-74r	196,149	659,403	101,082	6,801	135,162	193,783	373,485	258,616	1,924,481
1974-75	200,868	719,203	107,473	10,561	173,464	294,490	401,483	278,762	2,186,304
1975-76	207,952	832,084	119,086	11,277	178,675	394,665	463,604	284,489	2,491,832

¹See footnote 2, Table 7.15.
²See footnote 3, Table 7.15.
³See footnote 4, Table 7.15.

7.17   Registrations in formal non-credit courses by course, field of study and by province, 1974-75 and 1975-76

Year, province and type of course	Primary industries	Natural sciences	Business and management	Education	Engineering and applied sciences	Fine and applied arts	Health sciences	Household sciences	Humanities	Mathematics and computer science	Social sciences	Transportation and communication	Trade and technical	Special programs for the handicapped	Unclassified	Total
1974-75																
NEWFOUNDLAND																
Professional development																
Extension diploma	–	–	265	–	–	–	31	23	78	78	158	–	63	–	–	696
Association diploma	–	–	82	–	–	–	99	–	–	18	33	–	–	–	–	232
No diploma	649	–	1,137	14	49	74	115	59	55	52	74	484	1,760	–	111	4,633
General interest	45	–	309	14	46	9,034	135	1,798	755	193	88	180	392	–	–	12,989
Total	694	–	1,793	28	95	9,108	380	1,880	888	341	353	664	2,215	–	111	18,550
PRINCE EDWARD ISLAND																
Professional development																
Extension diploma	–	–	35	–	–	–	–	–	–	–	15	–	–	–	–	50
Association diploma	–	–	–	–	–	–	–	–	–	–	–	–	–	–	–	–
No diploma	–	3	430	28	21	128	10	–	11	30	–	–	92	–	–	753
General interest	55	–	–	107	–	3,019	136	822	245	261	83	–	32	–	–	4,760
Total	55	3	465	135	21	3,147	146	822	256	291	98	–	124	–	–	5,563
NOVA SCOTIA																
Professional development																
Extension diploma	–	–	249	6	6	–	–	–	51	–	15	–	–	–	–	327
Association diploma	–	–	308	–	–	–	113	–	26	–	68	–	–	–	–	515
No diploma	448	120	1,097	113	–	42	3,234	–	124	605	529	–	228	–	–	6,540
General interest	91	50	458	124	99	15,808	623	5,708	1,909	597	723	432	1,271	23	26	27,942
Total	539	170	2,112	243	105	15,850	3,970	5,708	2,110	1,202	1,335	432	1,499	23	26	35,324
NEW BRUNSWICK																
Professional development																
Extension diploma	–	–	–	–	–	–	–	–	74	–	38	–	–	–	–	112
Association diploma	–	–	19	–	–	–	–	–	–	–	97	–	–	–	–	116
No diploma	120	30	2,140	372	368	847	20	3,906	47	50	51	–	1,362	–	–	9,313
General interest	143	67	57	–	–	5,062	96	4,130	1,329	83	149	643	220	–	27	12,006
Total	263	97	2,216	372	368	5,909	116	8,036	1,450	133	335	643	1,582	–	27	21,547
QUEBEC																
Professional development																
Extension diploma	–	–	4,073	40	–	125	354	146	5,374	1,669	1,808	–	185	–	–	13,774
Association diploma	–	–	6,982	–	32	–	–	–	95	578	1,649	30	136	–	–	9,502
No diploma	1,008	672	17,528	1,350	822	1,905	4,599	490	1,764	1,658	6,802	61	36,889	–	285	75,833
General interest	634	683	10,293	88	25	83,572	32	83,205	16,604	63	2,634	–	5,414	–	28,713	231,960
Total	1,642	1,355	38,876	1,478	879	85,602	4,985	83,841	23,837	3,968	12,893	91	42,624	–	28,998	331,069

7.17   Registrations in formal non-credit courses by course, field of study and by province, 1974-75 and 1975-76 (continued)

Year, province and type of course	Field of study															Total
	Primary industries	Natural sciences	Business and management	Education	Engineering and applied sciences	Fine and applied arts	Health sciences	Household sciences	Humanities	Mathematics and computer science	Social sciences	Transportation and communication	Trade and technical	Special programs for the handicapped	Unclassified	
1974-75 (continued)																
ONTARIO																
Professional development																
Extension diploma	776	338	6,115	254	1,515	579	34	—	1,332	2,846	2,654	—	844	—	—	17,287
Association diploma	—	247	9,236	910	228	—	2,883	—	1,538	2,274	3,558	100	—	—	—	20,974
No diploma	1,881	969	26,895	4,267	675	1,866	12,475	1,359	3,257	1,380	4,589	121	3,940	244	120	64,038
General interest	6,431	1,184	12,549	3,101	544	162,006	3,067	42,386	38,676	3,915	13,693	16,703	24,465	123	8,089	336,932
Total	9,088	2,738	54,795	8,532	2,962	164,451	18,459	43,745	44,803	10,415	24,494	16,924	29,249	367	8,209	439,231
MANITOBA																
Professional development																
Extension diploma	—	—	774	—	—	—	—	—	212	41	269	—	—	—	6	1,302
Association diploma	—	—	1,011	100	30	—	—	—	187	35	324	—	—	—	—	1,687
No diploma	33	87	4,097	224	381	517	1,606	1,284	284	223	1,566	19	3,194	26	42	13,583
General interest	1,461	49	604	37	12	11,411	—	3,554	3,009	624	809	864	2,067	—	2,586	27,087
Total	1,494	136	6,486	361	423	11,928	1,606	4,838	3,692	923	2,968	883	5,261	26	2,634	43,659
SASKATCHEWAN																
Professional development																
Extension diploma	—	—	242	—	—	50	—	9	47	—	—	—	—	—	—	348
Association diploma	—	—	276	—	—	—	48	—	—	—	62	17	55	—	—	458
No diploma	3,353	—	2,368	280	278	291	4,467	98	204	87	1,050	—	3,281	—	23	15,780
General interest	1,315	36	2,247	326	—	14,823	328	5,059	1,823	4,061	1,835	1,164	4,273	—	236	37,526
Total	4,668	36	5,133	606	278	15,164	4,843	5,166	2,074	4,148	2,947	1,181	7,609	—	259	54,112
ALBERTA																
Professional development																
Extension diploma	—	—	1,913	—	79	—	118	—	611	150	734	—	—	—	—	3,605
Association diploma	—	—	2,111	—	34	—	—	—	236	781	795	—	30	—	—	3,987
No diploma	1,737	941	12,713	760	2,160	447	4,806	307	653	2,212	4,287	346	3,602	—	1,277	36,248
General interest	2,424	1,035	3,190	511	203	43,587	569	9,943	8,940	697	5,022	2,521	5,025	9	331	84,007
Total	4,161	1,976	19,927	1,271	2,476	44,034	5,493	10,250	10,440	3,840	10,838	2,867	8,657	9	1,608	127,847
BRITISH COLUMBIA																
Professional development																
Extension diploma	—	—	362	912	—	156	99	—	563	80	1,129	—	—	—	—	3,301
Association diploma	—	66	5,864	165	—	—	—	—	489	646	1,375	29	104	—	—	8,738
No diploma	1,557	466	10,816	2,111	821	2,444	5,817	427	1,182	928	3,925	305	11,669	42	172	42,682
General interest	5,415	2,384	2,292	691	185	78,986	2,187	21,110	20,348	3,149	11,960	6,829	6,905	85	7,355	169,881
Total	6,972	2,916	19,334	3,879	1,006	81,586	8,103	21,537	22,582	4,803	18,389	7,163	18,678	127	7,527	224,602

7.17   Registrations in formal non-credit courses by course, field of study and by province, 1974-75 and 1975-76 (continued)

Year, province and type of course	Field of study Primary industries	Natural sciences	Business and management	Education	Engineering and applied sciences	Fine and applied arts	Health sciences	Household sciences	Humanities	Mathematics and computer science	Social sciences	Transportation and communication	Trade and technical	Special programs for the handicapped	Unclassified	Total
1974-75 (concluded)																
YUKON TERRITORY AND NORTHWEST TERRITORIES																
Professional development																
Extension diploma	—	—	—	—	—	—	—	—	—	—	—	—	—	—	—	—
Association diploma	—	—	—	—	—	—	—	—	—	—	—	—	—	—	—	—
No diploma	—	12	68	—	10	—	—	—	33	11	—	—	73	—	25	232
General interest	—	—	170	—	10	403	—	260	156	10	17	48	91	—	—	1,165
Total	—	12	238	—	20	403	—	260	189	21	17	48	164	—	25	1,397
CANADA																
Professional development																
Extension diploma	776	338	13,993	1,212	1,600	910	636	178	8,342	4,864	6,805	—	1,092	—	6	40,752
Association diploma	—	313	25,924	1,175	324	—	3,143	—	2,571	4,332	7,976	176	325	—	—	46,259
No diploma	10,786	3,300	79,289	9,519	5,585	8,561	37,149	7,930	7,614	7,236	22,873	1,336	66,090	312	2,055	269,635
General interest	18,014	5,488	32,169	4,999	1,124	427,711	7,173	177,975	93,794	13,653	37,013	29,384	50,155	240	47,363	946,255
Total	29,576	9,439	151,375	16,905	8,633	437,182	48,101	186,083	112,321	30,085	74,667	30,896	117,662	552	49,424	1,302,901
1975-76																
NEWFOUNDLAND																
Professional development																
Extension diploma	—	—	357	80	—	—	87	—	—	40	428	—	—	—	—	992
Association diploma	—	—	212	—	14	—	—	—	—	8	—	—	—	—	—	234
No diploma	465	—	1,347	73	66	—	196	186	24	69	59	231	1,699	—	60	4,475
General interest	57	—	187	27	13	6,114	109	3,807	724	92	143	267	367	—	—	11,907
Total	522	—	2,103	180	93	6,114	392	3,993	748	209	630	498	2,066	—	60	17,608
PRINCE EDWARD ISLAND																
Professional development																
Extension diploma	—	—	8	—	—	—	—	—	—	—	—	—	—	—	—	8
Association diploma	—	—	20	—	255	—	—	—	—	—	—	—	243	—	—	518
No diploma	75	—	420	—	46	—	—	—	—	23	19	—	170	—	20	773
General interest	153	—	14	—	—	179	—	76	91	16	23	12	17	—	74	655
Total	228	—	462	—	301	179	—	76	91	39	42	12	430	—	94	1,954

7.17 Registrations in formal non-credit courses by course, field of study and by province, 1974-75 and 1975-76 (continued)

Year, province and type of course	Primary industries	Natural sciences	Business and management	Education	Engineering and applied sciences	Fine and applied arts	Health sciences	Household sciences	Humanities	Mathematics and computer science	Social sciences	Transportation and communication	Trade and technical	Special programs for the handicapped	Unclassified	Total
1975-76 (continued)																
NOVA SCOTIA																
Professional development																
Extension diploma	—	—	223	—	—	18	—	—	60	—	128	—	24	—	—	460
Association diploma	—	7	850	—	41	—	—	—	45	160	41	—	—	—	—	1,137
No diploma	15	215	2,498	75	80	89	1,864	43	114	442	694	15	520	20	—	6,684
General interest	148	54	2,966	32	115	19,141	878	6,000	2,517	576	290	482	1,775	—	52	35,026
Total	163	276	6,537	107	236	19,248	2,742	6,043	2,736	1,178	1,153	497	2,319	20	52	43,307
NEW BRUNSWICK																
Professional development																
Extension diploma	—	—	92	40	—	—	—	—	93	16	224	—	—	—	—	465
Association diploma	—	—	106	—	24	—	—	—	—	—	83	—	—	—	—	213
No diploma	—	—	989	174	161	1,162	53	1,403	64	312	61	47	438	—	—	4,864
General interest	26	47	406	20	55	5,969	—	3,337	1,114	82	350	644	827	—	10	12,887
Total	26	47	1,593	234	240	7,131	53	4,740	1,271	410	718	691	1,265	—	10	18,429
QUEBEC																
Professional development																
Extension diploma	—	—	5,671	—	1	434	323	20	5,530	1,983	2,513	27	131	—	—	16,633
Association diploma	—	33	6,079	—	39	—	28	—	49	481	893	17	117	—	—	7,736
No diploma	794	630	20,618	1,955	1,033	2,685	4,934	607	4,614	1,651	7,096	156	12,668	—	18,558	77,438
General interest	1,002	125	1,358	219	—	58,567	562	38,447	8,022	4,730	22,083	198	493	34	98,492	234,332
Total	1,796	788	33,726	2,174	1,073	61,125	5,847	39,074	18,215	8,845	32,585	398	13,409	34	117,050	336,139
ONTARIO																
Professional development																
Extension diploma	36	454	6,490	11	1,789	367	442	99	1,323	3,076	2,825	460	735	—	—	18,107
Association diploma	—	64	10,857	819	622	—	5,000	16	2,033	1,752	3,376	—	101	—	—	24,640
No diploma	3,637	695	21,428	2,522	662	2,685	10,053	448	4,302	2,925	2,637	66	5,065	12	151	57,288
General interest	4,565	1,855	11,483	3,741	592	217,913	3,218	50,657	44,759	2,052	10,814	19,894	28,559	16	14,518	414,636
Total	8,238	3,068	50,258	7,093	3,665	220,965	18,713	51,220	52,417	9,805	19,652	20,420	34,460	28	14,669	514,671
MANITOBA																
Professional development																
Extension diploma	—	—	1,428	—	—	—	8	—	—	31	—	—	—	—	—	1,467
Association diploma																
No diploma	12	52	3,317	1,146	395	508	2,163	636	305	800	1,069	29	2,647	—	51	13,130
General interest	751	123	587	98	16	16,478	135	3,958	3,954	312	1,104	1,243	2,410	—	145	31,314
Total	763	175	5,332	1,244	411	16,986	2,306	4,594	4,259	1,143	2,173	1,272	5,057	—	196	45,911

7.17   Registrations in formal non-credit courses by course, field of study and by province, 1974-75 and 1975-76 (concluded)

Year, province and type of course	Primary industries	Natural sciences	Business and management	Education	Engineering and applied sciences	Fine and applied arts	Health sciences	Household sciences	Humanities	Mathematics and computer science	Social sciences	Transportation and communication	Trade and technical	Special programs for the handicapped	Unclassified	Total
1975-76 (concluded)																
SASKATCHEWAN																
Professional development																
Extension diploma	—	—	147	—	—	—	154	—	—	—	—	—	—	—	—	301
Association diploma	—	—	140	—	—	—	—	—	—	—	—	—	—	—	—	140
No diploma	2,892	30	5,081	368	343	569	7,177	258	591	47	792	526	2,284	15	396	21,369
General interest	1,653	801	1,242	546	70	27,953	1,025	9,169	2,863	1,610	4,509	1,543	9,753	12	560	63,309
Total	4,545	831	6,610	914	413	28,522	8,356	9,427	3,454	1,657	5,301	2,069	12,037	27	956	85,119
ALBERTA																
Professional development																
Extension diploma	—	—	2,855	—	25	—	—	—	163	199	796	—	—	—	—	4,038
Association diploma	—	—	3,557	26	128	30	1,455	—	159	622	845	—	—	46	20	6,842
No diploma	854	1,399	10,598	3,112	2,959	2,725	3,739	1,706	1,732	2,419	4,612	591	6,372	46	647	43,511
General interest	6,087	1,183	3,604	387	133	63,834	1,003	12,237	20,629	970	10,305	2,673	7,019	8	3,443	133,515
Total	6,941	2,582	20,614	3,525	3,245	66,589	6,197	13,943	22,683	4,210	16,558	3,264	13,391	54	4,110	187,906
BRITISH COLUMBIA																
Professional development																
Extension diploma	194	—	—	1,649	—	—	—	—	354	—	691	—	—	—	—	2,888
Association diploma	—	261	4,712	—	—	—	53	—	171	418	625	—	—	67	—	5,979
No diploma	1,953	261	13,104	2,465	319	1,508	6,990	1,182	1,920	859	6,030	629	6,668	67	994	44,949
General interest	5,158	3,006	5,836	957	468	70,738	4,905	17,114	35,068	4,258	20,260	10,209	10,378	242	24,987	213,584
Total	7,305	3,267	23,652	5,071	787	72,246	11,948	18,296	37,513	5,535	27,606	10,838	17,046	309	25,981	267,400
YUKON TERRITORY AND NORTHWEST TERRITORIES																
Professional development																
Extension diploma	—	—	—	—	—	—	—	—	—	—	—	—	—	—	—	—
Association diploma	—	—	—	—	—	—	—	—	—	—	—	—	—	—	—	—
No diploma	—	—	179	—	—	—	6	—	16	41	30	157	141	—	32	602
General interest	—	22	412	—	—	1,130	144	816	401	24	10	100	165	—	245	3,469
Total	—	22	591	—	—	1,130	150	816	417	65	40	257	306	—	277	4,071
CANADA																
Professional development																
Extension diploma	230	461	17,271	1,780	1,815	819	1,014	119	7,523	5,345	7,605	487	890	—	—	45,359
Association diploma	—	97	26,533	845	1,123	30	6,536	16	2,457	3,441	5,863	17	461	—	20	47,439
No diploma	10,697	3,282	79,579	11,890	6,064	11,370	37,175	6,469	13,682	9,588	23,099	2,447	38,672	160	20,909	275,083
General interest	19,600	7,216	28,095	6,027	1,462	488,016	11,979	145,618	120,142	14,722	69,891	37,265	61,763	312	142,526	1,154,634
Total	30,527	11,056	151,478	20,542	10,464	500,235	56,704	152,222	143,804	33,096	106,458	40,216	101,786	472	163,455	1,522,515

Sources
7.1 - 7.17 Education, Science and Culture Division, Institutional and Public Finance Statistics Branch, Statistics Canada.

Labour

Chapter 8

Tables

Labour   Chapter 8

There were approximately 10,308,000 people in the civilian labour force in Canada in 1976 (Table 8.1). Of that number, an estimated 9,572,000 persons were employed; 736,000 or 7.1% were unemployed. The proportion of the female population in the labour force increased from 39.4% in 1971 to 45.0% in 1976 (Table 8.2). Of approximately 7,332,000 males over 15 years of age in the population in 1971, 77.3% were in the labour force; in 1976 the participation rate was 77.7% of 8,303,000.

The government in relation to labour   8.1

Labour Canada   8.1.1
The Canada Department of Labour (Labour Canada) was established by the Department of Labour Act (RSC 1970, c.L-2). The minister of labour's responsibilities include: collecting, digesting and publishing statistical and other information relating to labour conditions; conducting inquiries into industrial questions upon which adequate information may not be available; issuing monthly *The Labour Gazette* and *La Gazette du Travail* which give a perspective on the entire work-related scene.

The minister is responsible for the Canada Labour Code, in effect since July 1971, including sections on fair employment practices, labour standards, safety of employees, and industrial relations. The minister also administers acts covering fair wages and hours of work, and compensation for government employees and also for merchant seamen. The minister reports to Parliament on behalf of the Canada Labour Relations Board and the Merchant Seamen Compensation Board.

The industrial relations legislation applies to employers, employees and trade unions within federal jurisdiction. The department is responsible for conciliation procedures in industrial disputes; investigating complaints of unfair labour practices, refusals to bargain and violations of legislation, processing union applications for certification and decertification, and conducting representation votes. It determines wage rates and hours of work for federal government contracts for construction or supplies, and promotes improved industrial relations through union-management consultation and by preventive mediation through industrial relations consultants. The department administers assistance granted to workers in automotive manufacturing and a benefits program for displaced workers in textile and clothing, and footwear and tanning industries.

Reorganization of the department began in 1974-75. Its role is to promote and protect the rights of the parties involved in the world of work, a working environment conducive to physical and social well-being, and a fair return for efforts; and in all cases to ensure equitable access to employment opportunities. Reorganization included decentralization into five regions, with headquarters in Moncton, Montreal, Toronto, Winnipeg and Vancouver.

The department maintains records of labour legislation in the provinces and in other countries and provides liaison between the International Labour Organization and federal and provincial governments.

Canada Employment and Immigration Commission   8.1.2
The commission recruits and develops manpower resources in line with the needs of the economy. The prime goal of Canada's manpower policy is to contribute to the country's economic and social goals by making the best use of its work force. The commission's domestic field activities are carried out in 10 regions through more than 400 manpower and 97 immigration centres.

Broad objectives of the commission are: to provide an effective employment service for both workers and employers; to help workers attain their full potential through counselling or referral to training programs; to assist employers in recruiting skilled workers, and facilitate long-range manpower planning by providing occupational and labour market information; to help labour and management adapt to technological change through manpower adjustment programs; to provide reception, settlement and job placement services for immigrants; to process documents for international travellers and enforce the Immigration Act and its regulations.

In the fiscal year ended March 31, 1977, manpower centres assisted more than 863,000 persons, excluding casual workers, in finding continuing employment and referred 297,000 more to full- or part-time courses under the manpower training program. Under the Canada Manpower Mobility Program, 47,626 workers and trainees were granted moving and transportation assistance, and 77,703 received financial assistance.

A manpower division administers employment programs and services through the manpower centres. Dealing with the demand side of the labour market, it provides guidelines on employment services for employers and information on industrial needs. A consultative service helps industries undergoing manpower dislocation because of technological change. It also administers a mobility program to help workers move to areas of job opportunity. On the supply side of the labour market, client services provide employment counselling and aptitude and achievement tests. This branch also assists new members of the labour force and students seeking summer employment. Training programs help under-employed, unemployed or disadvantaged adult workers improve their qualifications through training courses bought from provincial or private schools or through contracts with employers. Participants receive wages or training allowances. A co-ordination branch is responsible for application of all manpower programs and services to the needs of disadvantaged unemployed persons in the labour force.

In 1976-77 the federal government continued its programs to alleviate seasonal unemployment through the Local Initiatives Program (LIP) and the Local Employment Assistance Program. A job creation branch directs these programs organized at regional, provincial and local levels. In the fiscal year ended March 31, 1977, LIP created more than 45,000 jobs through some 6,645 projects organized and run by communities. The employment assistance program, intended to assist severely disadvantaged groups, funds projects on a longer-term basis than LIP . The number of projects in operation at March 31, 1977 was 124 employing 1,479 persons.

A program of community employment strategy seeks new ways of opening up job opportunities for people who have difficulty in finding and keeping steady employment. The employment and immigration commission has spearheaded the program. Provincial and territorial governments work in conjunction with the commission and other federal bodies to develop employment opportunities for people who might otherwise have to depend on social assistance or unemployment insurance for most of their income. This concerted effort is being made in over 20 selected communities across Canada, using community initiative and resources and creating local involvement and responsibility.

The manpower delivery system provides three levels of service to people looking for work. The first level is an information centre where job vacancies are displayed. In addition, a library at the centre has information about commission programs and services of other departments and agencies. The second level of service is directed at people who are employable but who could benefit from counselling, from training courses or from assistance in finding and moving to jobs in other areas. The third level is designed for clients who require concentrated counselling. Counsellors may use outside agencies for special assistance in helping these people to become employable. They are then referred to a job or may make selections from a job bank.

The immigration sector is responsible for selection and reception of immigrants who will be able to establish themselves economically, culturally and socially. They include people whose skills are required by the Canadian economy, relatives of Canadian residents and refugees and non-immigrants entering on a short-term basis. The department is also responsible for enforcement and control measures to prevent immigration of undesirable persons.

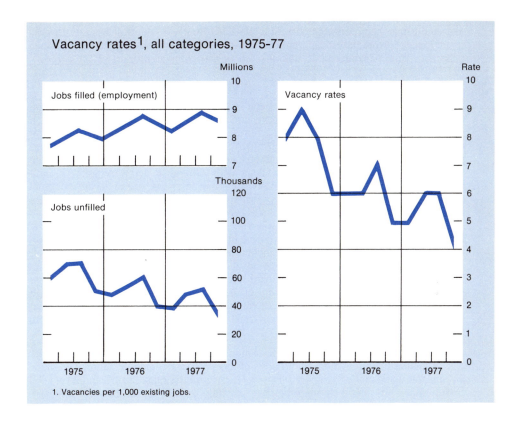

Vacancy rates[1], all categories, 1975-77

Jobs filled (employment)

Millions

Jobs unfilled

Thousands

Vacancy rates

Rate

1975   1976   1977

1975   1976   1977

1. Vacancies per 1,000 existing jobs.

Since January 1973, all non-immigrants entering Canada to take temporary work must have an employment visa. Visitors are not permitted to come to Canada to look for work. This regulation protects the Canadian labour force against unwarranted use of foreign labour.

To obtain an employment visa, the applicant must have pre-arranged employment and certification by a manpower centre that no Canadian citizen or landed immigrant is available for that job. Arrangements must be made at a Canadian immigration office in the person's own country.

A planning and research division collects and analyzes information on national, regional and local labour market conditions to give direction to the department's policies and programs. It carries out research programs in support of its own and other divisions' activities and develops career and occupational counselling and training materials.

Federal and provincial labour legislation   8.1.3

Jurisdictions   8.1.3.1
The Canada Labour Code (RSC 1970, c.L-1), which consolidates previous legislation regulating employment practices and labour standards, applies only to federal undertakings and any other operations that Parliament declares are for the general advantage of Canada or two or more of its provinces.

Because it imposes conditions on the rights of the employer and employee to enter into an employment contract, labour legislation is, generally speaking, law in relation to civil rights, and provincial legislatures are authorized to make laws in relation both to local works and to property and civil rights. Power to enact labour legislation has therefore become largely a provincial prerogative, under which a large body of legislation has been enacted affecting working hours, minimum wages, physical conditions of workplaces, apprenticeship and training, wage payment and wage collection, labour-management relations and worker compensation.

8.1.3.2 Federal labour legislation

Industrial relations. A mediation and conciliation branch of Labour Canada administers the industrial relations provisions of the Canada Labour Code relating to application of formal conciliation procedures, including appointment of conciliation officers and commissioners and establishment of conciliation boards. The branch also provides mediation services to parties in post-conciliation negotiations, including strike and lockout situations. If there is a dispute or difference between employer and employees in an industry, the labour minister may refer the matter to an industrial inquiry commission for investigation. On behalf of the minister, the branch administers the code's provisions relating to certain types of complaints which must receive ministerial consent before they can be referred to the Canada Labour Relations Board. It handles other violations of the code requiring ministerial consent for prosecution. When requested, the minister may appoint single arbitrators or arbitration board chairmen if parties or nominees are unable to agree on the selection.

The Canada Labour Relations Board administers provisions in the labour code governing acquisition and termination of bargaining rights, successor rights and obligations, disposition of applications relating to technological change and to illegal strikes and lockouts, complaints of unfair practices, and granting of access to employers premises.

Fair employment practices. The code section on fair employment practices prohibits discrimination in employment on grounds of race, colour, religion or national origin in any federal work, undertaking or business. It covers discrimination by employers, by trade unions in regard to membership or employment, by employers who use employment agencies that discriminate, and in the use of any form of application for employment, advertisement, written or oral inquiry that expresses directly or indirectly any limitation, specification or preference as to race, colour, religion or national origin.

Labour standards. The code sets minimum standards of employment for employers and employees in industries under the legislative authority of Parliament.

The code sets both standard and maximum hours of work. The overtime rate (one and a half times the regular rate) must be paid after eight hours in a day and 40 hours in a week, to a maximum of 48 hours in a week. Hours may be averaged when an employee's schedule of hours varies from day to day or week to week because of the nature of the work. If the labour minister is satisfied that exceptional circumstances justify it, he may issue a permit allowing an employee to exceed the maximum hours. The Governor-in-Council may make regulations varying standard and maximum hours for classes of employees in any industrial establishment where code standards would be unduly prejudicial to employees or seriously detrimental to operation of the establishment. An inquiry must be held before such regulations are made.

The minimum wage is $2.90 an hour for all persons 17 years of age and over and $2.65 an hour for persons under 17 as of April 1, 1976. The Governor-in-Council may issue orders from time to time changing the minimum rate.

Employees are entitled to two weeks vacation with pay each year and a holiday with pay on each of the nine general holidays or substitutes for them.

An employer must give advance notice to the labour minister and the union, with a copy to the employment and immigration commission, when dismissing 50 or more employees during a four-week period. The length of notice varies according to the number of employees being dismissed: 50-100 employees, eight weeks; 101-300 employees, 12 weeks; more than 300 employees, 16 weeks. In addition, the employer and the trade union must provide the employment and immigration commission with whatever information it requests to assist the employees. The requirement to give notice may be waived for an industrial establishment or a specified class of employees by an order of the labour minister, subject to any terms or conditions he may determine.

Under the code's provisions respecting individual dismissals, every employee with three months service (except a manager, superintendent or member of a profession) is entitled to two weeks notice of termination of his employment. In lieu of such notice, he is entitled to two weeks wages at his regular rate for his regular hours. In addition, an employee who has completed five consecutive years of continuous employment is entitled to severance pay based on two days wages at the regular rate for regular hours for each year of employment up to a maximum of 40 days wages. However, the employer is not required to give severance pay to an employee who is dismissed for just cause or to a person who, on termination of employment, is entitled to a retirement pension.

Maternity protection provisions grant 17 weeks of maternity leave — 11 weeks before and six weeks after childbirth — and ensure job security to women absent from work because of pregnancy. To be eligible a woman must have been continuously employed by her employer for 12 months. The code provides for voluntary prenatal leave up to 11 weeks before the anticipated date of delivery and this period extends to the actual date of confinement.

The code prohibits an employer from paying men and women employees at different rates if they work in the same establishment at equally demanding jobs under similar conditions. It also prohibits an employer from dismissing, laying off or suspending an employee solely because of garnishment.

Fair wages policy. Wages and hours on government construction contracts are regulated by the Fair Wages and Hours of Labour Act and its regulations. The rates are never less than the minimum hourly rate prescribed by labour standards in the labour code. Wages and hours of work on contracts for equipment and supplies are regulated by order-in-council.

Safety of employees. The code's safety section, incorporated in 1968, was the first general safety legislation passed by Parliament. To ensure safe working conditions for all employees in activities under federal jurisdiction, it provides for all elements of a complete industrial safety program; obliges employers and employees to perform their duties in a safe manner; authorizes the making of regulations to deal with safety problems; complements other federal laws and provincial legislation; authorizes advisory committees to assist in developing the program under consultation among federal and provincial government departments, industry and organized labour; and provides for research into causes and prevention of accidents and for an extended safety education program. Federal public service employees are given equivalent protection under Treasury Board standards complementary to the safety and health regulations of the code. Regional safety officers and federally authorized provincial inspectors enforce them.

As of January 31, 1975, regulations were in force governing coal mine safety, elevating devices, first aid, machine-guarding, noise control, hand tools, fire safety, temporary work structures, confined spaces, safe illumination, boilers and pressure vessels, building safety, dangerous substances, electrical safety, materials handling, protective clothing and equipment, sanitation, hours of service in the motor transport industry, and accident investigation and reporting.

Provincial labour legislation   8.1.3.3

Industrial relations. All provinces have legislation similar to the federal code designed to establish harmonious relations between employers and employees and facilitate settlement of industrial disputes. These laws guarantee freedom of association and the right to organize, establish labour relations boards or other administrative systems for certification of a trade union as exclusive bargaining agent of an appropriate unit of employees, and require an employer to bargain with the certified union representing his employees.

Alberta, Ontario, New Brunswick, Nova Scotia, Prince Edward Island and Newfoundland have special provisions in their general labour relations legislation dealing with accreditation of employers organizations in the construction industry. In

British Columbia accreditation provisions are not limited to the construction industry. The Quebec Construction Industry Labour Relations Act, 1968 provides for one employers association to represent all construction employers. Under every jurisdiction legislation requires that the parties comply with conciliation or mediation procedures before a strike or lockout may legally take place. Every collective agreement must provide for the final settlement, without stoppage of work, of disputes arising out of interpretation or application of the agreement. Strikes and lockouts are prohibited during the term of a collective agreement. Unfair labour practices are prohibited under every legislation. In some provinces labour relations for special groups such as teachers, municipal and provincial police personnel, municipal firemen, hospital workers, civil servants and employees of Crown corporations are regulated by special legislation.

Employment standards. Most provincial and territorial jurisdictions have legislated some or all of such recognized basic standards as: annual vacations with pay, statutory holidays, hours of work and overtime rates, maternity protection, minimum wage rates and termination of employment.

Hours of work. In Alberta and British Columbia hours are limited to eight a day and 44 a week, in Ontario to eight a day and 48 a week. One and a half times the regular rate is to be paid after eight and 44 hours in Alberta and after eight and 40 in British Columbia. The Ontario act requires, with some exceptions, that one and a half times the regular rate be paid for work done beyond 44 hours. The Saskatchewan act does not limit daily and weekly hours but requires the payment of one and a half times the regular rate if work is continued after eight and 40 hours; this provision applies to shop employees in Newfoundland, but for other employees it is effective after 44 hours. Manitoba does not require that an employee work overtime except in special circumstances and sets a rate of one and three-quarter times the regular rate after eight and 40 hours. One and a half times the regular rate must be paid in Nova Scotia after 48 hours a week, the Northwest Territories after eight hours a day and 44 a week, and Yukon Territory after eight a day and 40 a week. It must be one and a half times the minimum rate in Prince Edward Island after 48 hours, Quebec after 45 and New Brunswick after 44. Some exceptions occur in all acts. No general standard of hours of work is in effect in New Brunswick or Newfoundland.

Minimum wages. All jurisdictions have enacted minimum wage legislation to ensure adequate living standards for workers. These laws vest authority in a minimum-wage board or the Lieutenant-Governor-in-Council to set wages. Minimum wage orders are reviewed frequently. In most provinces such orders cover practically all employment. Domestic service in private homes is excluded in all provinces except Prince Edward Island and in Newfoundland where an employer may not pay less than $30 a week. Farm labour is also excluded except in Newfoundland but in several provinces people employed in farm-related occupations are covered. In Ontario and Nova Scotia this exclusion is limited to farming proper, although certain farm-related occupations are covered. Fruit, vegetable and tobacco harvesters are covered by Ontario's minimum wage. Minimum wage rates apply in Manitoba to those employed in selling horticultural or market garden products grown by another person, in Saskatchewan to those in egg hatcheries, greenhouses, nurseries and brush-clearing operations, and in Alberta and Prince Edward Island to farm workers employed in commercial undertakings. The wage rates set apply throughout the province and are the same for both sexes.

In the Northwest Territories and the Yukon Territory, labour standards regulations were issued under labour standards ordinances. Both require the payment of a minimum rate of wages to employees who are 17 and over.

Where employees are paid on a basis other than time, or on a combination of time and some other basis, they are required to receive the equivalent of the minimum wage. Provision is made in the legislation of almost all jurisdictions for employment of handicapped workers at rates below the established minimum, usually under a system of individual permits. Except in New Brunswick, Newfoundland, Ontario, Saskatchewan and the Yukon Territory, the orders set special minimum rates for young workers.

As of December 1, 1977, the minimum hourly wage rates for experienced adult workers were: Newfoundland $2.50, Prince Edward Island $2.70, Nova Scotia $2.75, New Brunswick $2.80, Quebec $3.15, Ontario $2.65, Manitoba $2.95, and Saskatchewan, Alberta, British Columbia, Northwest Territories and Yukon Territory all $3.00.

Regulation of wages and hours in certain industries. In five provinces, the general orders are supplemented by special orders, applying to a particular industry, occupation or class of workers and in some cases taking into account a special skill. British Columbia, which originally had a separate minimum wage order for each industry or occupation, has been consolidating its orders. One special order still remains and its minimum rate is the same as the rate set in the general order. Quebec has four industry orders, governing the retail food trade, public works, sawmills and forest operations. The rates set by all four are the same as the general rates.

The other three provinces set only a few special rates. Nova Scotia has established rates for employees in beauty parlours and province-wide rates for logging and forest operations and for road building and heavy construction. In New Brunswick special rates have been set for construction, mining, primary transportation and logging, forest and sawmill operations. In Alberta a weekly rate has been set for commercial agents and sales people. In Ontario special rates contained within the general regulation apply to the construction and ambulance service industries.

Under the Quebec Collective Agreement Decrees Act, certain terms of a collective agreement, including those dealing with hours and wages, may be made binding on all employers and employees in the industry concerned provided the parties to the agreement represent a sufficient proportion of the industry. The standards made binding under this procedure are contained in a decree which has the force of law. Approximately 54 decrees applying to the garment trades, barbering and hairdressing, commercial establishments, garages and service stations and other industries and services are in effect. A number apply throughout the province. In construction, working conditions are governed by a decree under the Construction Industry Labour Relations Act.

A construction wages act in Manitoba, applying to both private and public work, sets minimum wage rates and maximum hours of work at regular rates for employees in the industry on the recommendation of a board equally representing employers and employees, with a member of the public as chairman. Under this act annual schedules set the regular work week and hourly wage rates for various classifications of workers in the heavy construction industry, in the greater Winnipeg building construction industry and major building projects, and in rural areas.

Annual vacations and public holidays. All jurisdictions have annual vacations legislation applicable to most industries. The general standard is two weeks. In Manitoba and the Northwest Territories workers are entitled to three weeks after five years of service, and in Saskatchewan three weeks after one year and four weeks after 11 years (with a gradual reduction to four weeks after 10 years as of July 1, 1978). Several jurisdictions, including the federal, Alberta, British Columbia, Manitoba, New Brunswick, Nova Scotia, Ontario, Saskatchewan, the Yukon Territory and the Northwest Territories, have enacted general legislation dealing with public holidays. The number of holidays varies from five to nine and provisions for payment also vary.

Vacation pay equals: 4% of annual earnings in British Columbia, Newfoundland, Nova Scotia, Prince Edward Island, New Brunswick, the Yukon Territory, the Northwest Territories, Quebec and Ontario; regular pay in Manitoba and Alberta; and in Saskatchewan 3/52nds of annual earnings. The federal rate is 4%.

Termination of employment. As in the federal jurisdiction, eight provinces have legislation requiring an employer to give notice to the individual worker whose employment is terminated. In Saskatchewan and Prince Edward Island an employer must give an individual employee one week's written notice of termination. In Manitoba and Newfoundland it is one regular pay period. In Alberta, Nova Scotia and Ontario the length of notice varies with the period of employment. In Ontario and Nova Scotia:

three months to two years, one week; two to five years, two weeks; five to 10 years, four weeks; 10 years or more, eight weeks. In Alberta: three months but less than two years, seven days; two years or more, 14 days. Quebec requires the employer of a domestic, a servant, journeyman or labourer to give one week's notice of termination if the employee is hired by the week, two weeks notice if hired by the month and a month's notice if hired by the year. Alberta, Manitoba, Newfoundland, Nova Scotia, Prince Edward Island and Quebec require an employee to give similar notice before quitting a job.

As in the federal jurisdiction, five provinces require an employer to give advance notice of a planned termination of employment or layoff of a group of employees. Manitoba and Ontario notice requirements apply when an employer plans to terminate the employment of 50 or more persons within four weeks or less. Length of notice is related to the number of workers involved. Manitoba requirements are: 50-100 employees, eight weeks; 101-300, 12 weeks; over 300, 16 weeks. In Ontario and Newfoundland: 50-199, eight weeks; 200-499, 12 weeks; 500 or more, 16 weeks. Nova Scotia and Quebec group notice requirements apply when an employer contemplates dismissal of 10 or more employees within a period of two months. Again, length of notice required varies with the number of workers involved: 10-99, two months; 100-299, three months; 300 and over, four months.

Maternity protection. Several provinces have legislation to ensure job security of women workers before and after childbirth. Alberta and Saskatchewan provide for 12 weeks leave before childbirth and six weeks after. British Columbia and New Brunswick acts provide for six weeks leave before childbirth and six weeks after; in New Brunswick the leave may extend to 17 weeks; Manitoba allows 11 weeks before and six after. Ontario and Nova Scotia provide for a minimum of 17 weeks leave. Postnatal leave is compulsory, unless a medical doctor authorizes an earlier return to work. In all jurisdictions, the right to maternity leave is supplemented by a guarantee that an employee will not lose a job because of absence on maternity leave.

Human rights. Laws to ensure fair employment practices have been enacted throughout Canada. These include employment and employment-related subjects such as membership in trade unions. All provinces and the federal jurisdiction have augmented this legislation to form a human rights code. The Northwest Territories and Yukon Territory have enacted fair practices ordinances. Most of these codes cover general matters, employment and employment-related subjects, and occupancy and property matters.

Most jurisdictions prohibit discrimination on grounds of race, religion, national origin, colour, sex, age and marital status. In selected cases the prohibited grounds include political beliefs, ethnic origin, physical handicap, creed, source of income, ancestry, social condition, attachment or assignment of pay, and a conviction for which a pardon has been granted.

Equal pay provisions are in force across Canada. Criteria for determining the meaning of equal work vary from one act to another. Methods of enforcement also vary.

Apprenticeship. All provinces have apprenticeship laws providing for an organized procedure of on-the-job training and school instruction in designated skilled trades. Statutory provision exists in most for issuing qualification certificates, on application, to tradesmen in certain trades. In some provinces legislation is in effect making it mandatory for certain classes of tradesmen to hold a certificate of competency.

Accident prevention. In Canada both federal and provincial legislatures have the power to enact laws and regulations concerning the protection of workers against industrial accidents or diseases. However, the provinces have major jurisdiction in this field, with the federal authority limited to certain industries considered to be under federal regulation. Legal standards designed to ensure the safety, health and welfare of persons employed in resource, industrial and commercial establishments exist in all jurisdictions. Authorities responsible for administration of such standards are, in the main, the departments of labour, health, mines and worker compensation boards.

General safety laws and regulations cover most employment in the country. Safeguards for worker protection are established for fire safety, sanitation, heating, lighting, ventilation, protective equipment, materials handling, safety of tools, guarding of dangerous machinery, safe handling of explosives and protection against noise and radiation. In some jurisdictions, workers have the right to refuse work in certain circumstances where safety or health could be endangered.

Other safety laws and regulations are more specific. They concern hazardous equipment such as boilers and pressure vessels, electrical installations, elevating devices and equipment burning gas and oil. Others are directed toward hazardous industries such as mining, construction, demolition and logging.

Safety inspection is provided for in all provinces. An inspector can give directions on any matter regulated by legislation. Penalties exist where an employer contravenes any provision of an occupational safety act or regulation or fails or neglects to comply with a direction made by an inspector.

Worker compensation. In Canada, compensation laws are generally within the competence of provincial legislatures and apply to most employers in each province. In all provinces compensation is generally provided for personal injuries sustained at work unless the disablement is for less than a set number of days or where injury is due to the worker's serious and wilful misconduct and does not result in death or serious disablement. Compensation is also payable for industrial diseases arising from work.

Each act provides for an accident fund administered by a compensation board to which employers are required to contribute and through which compensation and medical benefits are paid. The acts thus provide for a system of compulsory collective liability, relieving employers of individual responsibility for accident costs. Assessment rates for each class of industry are fixed by the board according to hazards of the class.

Various types of benefits are provided for a worker protected by compensation legislation. Benefits for disability are based on a percentage of average weekly earnings subject to an annual ceiling. Persons with a permanent or temporary total disability are presumed not to be able to work at all and get 75% of average weekly earnings as long as the disability lasts. Partial disablement entitles a worker to proportionate compensation. Medical and hospital benefits are also provided.

A primary objective of compensation is rehabilitation of the injured worker. Boards may adopt any means considered expedient to help get workers back to work and to lessen any handicap.

When a worker dies from an industrial accident or disease, dependents are entitled to a monthly payment fixed by legislation. However, for recent cases in Alberta and Manitoba, a widow receives the permanent total disability pension the deceased worker would have been entitled to if he had lived. This is also true in British Columbia for a widow with two or more children. In all provinces payments are made in respect of children. In Ontario and Quebec such payments may continue for as long as the child is pursuing his studies.

The labour force 8.2

Labour force (monthly surveys) 8.2.1

Since 1946, statistics relating to employment and unemployment at the national level, and since 1966 at the provincial level, have been provided through a Statistics Canada labour force survey. From 1945 until 1952 it was conducted quarterly, and since November 1952 it has been carried out monthly. In 1976, after three years of developmental work, substantial revisions to the survey were made to enhance the quality and increase the range of data collected, particularly information relating to the dynamics of the labour market.

The survey sample was designed to represent all persons in the population 15 years of age and over residing in Canada with the exception of the following: residents of the Yukon Territory and Northwest Territories, persons living on Indian reserves, inmates of institutions and full-time members of the armed forces. Interviews are carried out in

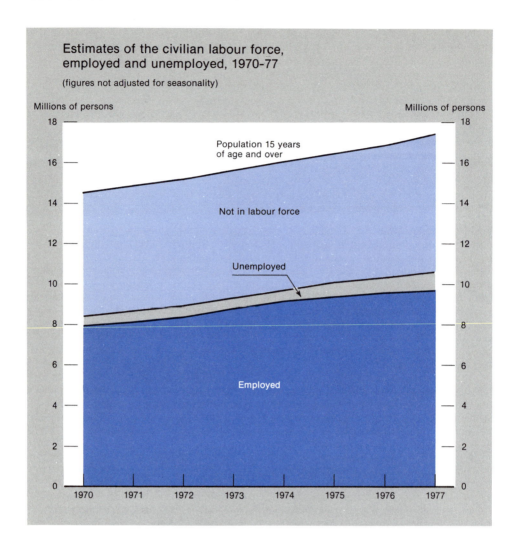

Estimates of the civilian labour force, employed and unemployed, 1970-77

(figures not adjusted for seasonality)

Millions of persons

Population 15 years of age and over

Not in labour force

Unemployed

Employed

approximately 55,000 households chosen by area sampling methods across the country. Until 1977 the sample size had been fixed at approximately 30,000 households. Estimates of employment, unemployment and non-labour force activity generated from the survey refer to a specific week each month, normally the week containing the 15th day. The labour force is composed of members of the civilian non-institutional population 15 years of age and over who, during reference week, were employed or unemployed.

The definition of employed includes all persons who, in reference week, did any work for pay or profit, either paid work in an employer-employee relationship or self-employment. Also included is unpaid family work contributing to the operation of a farm, business or professional practice owned or operated by a related member of the household. It also includes persons who had jobs but were not at work due to illness or disability, personal or family responsibilities, bad weather, labour disputes or other reason.

The unemployed are those who, in reference week, were without work, had actively looked for work in the past four weeks and were available for work; had not actively looked for work in the past four weeks but had been on layoff, with expectation of returning to work, for 26 weeks or less and were available for work; or had a new job to start in four weeks or less and were available for work. Persons not in the labour force are those defined as neither employed nor unemployed.

Because they are based on a sample of households, estimates derived from the survey are subject to sampling error. Extensive efforts are made to minimize the sampling error and in general the error, expressed as a percentage of the estimate, tends to decrease as the size of the estimate increases.

Revisions in the survey included introduction of an entirely new and expanded questionnaire, adjustment of some definitions, revision of the sample frame, change of population totals used to weight the sample and adoption of new methods of transmitting and processing the survey information. This revised survey was run in parallel with the former survey throughout 1975 and, as expected, some estimates from the two surveys differed significantly. By using relationships in estimates from the two surveys for 1975, estimates from the former survey have been revised for the period 1966 to 1974, allowing production of a consistent time series from 1966 to the present, except for the industry and paid worker series which were revised only to 1970.

In the period 1967-76, the Canadian labour force increased by 2.6 million persons or 33.1%. There was an increase of 55% in the number of women in the labour force and an increase in the number of men of only 23%. These increases resulted from an increase in the participation rate (the labour force as a percentage of the corresponding population aged 15 and over) for women from 36.5% in 1967 to 45.0% in 1976, and a small decrease for men from 79.3% to 77.7%.

The increase in the participation rate of young males, age 15-24, from 64.2% to 68.2% more than offset the slight decline for older males, 25 and over, from 84.5% to 81.2%. In the case of women, both age groups increased their participation although the rise was more pronounced among those aged 15-24.

The total number of persons employed in Canada rose by 2.1 million or 28.5% over the 1967-76 period. Although employment rose in all provinces the increases were not uniform, ranging from 17.3% in Saskatchewan to 44.2% in British Columbia. Other increases were 25.4% in Newfoundland, 25.7% in Prince Edward Island, 20.4% in Nova Scotia, 24.1% in New Brunswick, 18.9% in Quebec, 31.8% in Ontario, 20.9% in Manitoba and 44.0% in Alberta.

Unemployment as a percentage of the labour force varied from 3.8% in 1967 to 7.1% in 1976 with an average over the entire 1967-76 period of 5.7%. Throughout those years women had higher unemployment rates than men and persons aged 15-24 had considerably higher rates than persons 25 and over.

Labour force (1971 Census)   8.2.2

At each decennial census questions are asked of persons 15 and over relating to employment status and present work activities. These questions have the advantage of securing far more detailed information on the occupational and industrial structure and other characteristics of the labour force than the monthly surveys in terms of both geographical areas and classifications. Summary tabulations from the 1971 Census are presented in Tables 8.6-8.10. Further information is available in many census reports on these and other aspects of the labour force (see 1971 Census Publications, Statistics Canada Catalogue 94-701 to 94-789).

Because of differences in coverage, methodology and reference period, census information in some ways is not comparable with that collected by the monthly survey. Of particular importance among the differences are those of coverage and questions asked, even though the fundamental concepts are the same. As stated the smaller labour force survey sample included persons 15 and over but excluded the Yukon Territory and Northwest Territories, Indian reserves, members of the armed forces, overseas households and inmates of institutions. But the 1971 Census questions were asked of all persons 15 and over in a 33⅓% sample of households (about 2 million).

Foreign-born persons in the labour force. Results from the 1971 Census indicate that immigrants constituted 20% of the labour force. Table 8.6 presents data on persons in the total labour force by country of origin and region of Canadian residence in 1971. According to these figures, more than four-fifths of immigrants in the labour force came from Europe and one-quarter from Britain. Over half the immigrants are concentrated in Ontario where they form more than a quarter of the labour force, as they do in British Columbia, while they make up less than 5% in the Atlantic provinces.

Class of worker. In connection with questions on occupation and industry, the respondent was asked to report whether he was mainly working for wages and salary, was self-employed or was working without pay in a family business or farm. Table 8.8 provides 1961 and 1971 data.

Labour force by industry. In the 1971 Census respondents were asked for the name of their employer and the type of business, industry or service the firm was engaged in. For those self-employed, the name of the firm and type of activity were requested. Because a revised industrial classification was used (see *Standard industrial classification manual,* Statistics Canada Catalogue 12-501) special tabulations had to be made in order to compare 1971 data with 1961. Establishments owned and operated by government primarily engaged in activities assigned to other industries, such as transportation, communication (including the post office), liquor sales, health and educational services, were classified to those industries rather than to public administration. The division of public administration and defence covers establishments primarily engaged in activities such as enacting legislation, administering justice, collecting revenue and defence.

Table 8.9 shows the number of people 15 and over in the labour force by sex and major industrial group for 1961 and 1971. Two significant changes are illustrated: rapid growth of the service sector of the economy and increased participation of women in the labour force. From 1961 to 1971 the labour force in the industrial sector decreased by 174,335 or 20%, whereas the service sector rose by 1,398,091 persons or 39%; the other sector, manufacturing and construction, increased by 22% over the decade. The number of women in the labour force rose from 1,766,332 in 1961 to 2,961,210 in 1971 with increases in all sectors but most particularly in the service sector (71%).

Table 8.10 shows provincial distribution of main industrial sectors. The number of people in the labour force in the primary industries declined in all provinces except British Columbia (where it rose 24%), and most particularly in Quebec (−39%); even in Saskatchewan, where primary industries still account for 32% of the labour force, the decrease was 13%. In contrast to Saskatchewan, only 6% of Ontario's labour force is classified as being in the primary sector.

Manufacturing and construction accounted for 29% of Canada's labour force in 1961 and 28% in 1971, when most provinces had between 18% and 25% of their labour force employed in this sector. In Quebec and Ontario 32% and 33%, respectively, were employed in manufacturing and construction, but in Saskatchewan the proportion was only 11%.

Growth in the service sector has already been mentioned. Provincially, while both the number and proportion rose in all provinces after 1961, Prince Edward Island in 1971 had 61% of its labour force devoted to service industries compared to 50% in 1961. Quebec altered its proportion considerably from 54% to 62%. In their overall labour force breakdowns, Nova Scotia with 68% and British Columbia with 67% led the country with over two-thirds of their labour force in the service sector.

A new occupational classification was used for the 1971 Census (see *Occupational classification manual, Census of Canada 1971,* Volume I, Statistics Canada Catalogue 12-536). The new classification was designed to meet the need to standardize occupational data gathered from various sources. However, this standardization meant redesigning the existing classification, in terms of both the number of groupings of occupational titles and in assigning these titles to particular groups.

Therefore the use of the new classification in 1971 resulted in a complete break in the census time series on occupation. In an effort to reconcile the occupational data of 1971 with those of earlier censuses, occupations of a sample of individuals (nearly 110,000 experienced members of the labour force) were recoded based on the 1971 forms and the 1961 classification (see *Occupational classification manual, Census of Canada 1961,* Statistics Canada Catalogue 12-506).

Table 8.7 gives partial results of this recoding; it contains distributions by occupation divisions for the 1951, 1961 and 1971 censuses, all based on the 1961 classification. It can be seen that important changes occurred during these two intercensal periods.

The first notable change is the increasing importance of non-manual work. Professional and technical occupations have become much more important, increasing from 7.3% to 12.5% from 1951 to 1971, while office work moved up from 11.0% to 14.8% over the same period. There was also an increase in the proportion of service and recreation occupations to the overall labour force (11.6% in 1971 from 9.7% in 1951).

Other occupation divisions, however, lost ground. For instance, farmers and farm labourers, who made up 15.6% of the labour force in 1951, fell to 5.8% in 1971. Considerable decreases in manpower were recorded for primary occupations as a whole in the two decades. Despite their growth in size, manual occupations grouped under the heading of workers declined in relation to the overall labour force.

Employment statistics   8.3

Employment, earnings and hours   8.3.1

Monthly records of employment have been collected from larger business establishments since 1921. The surveys currently conducted by Statistics Canada collect employment, payroll and man-hours information; average weekly earnings, average hourly earnings, and average weekly hours are derived from this collected information. Employment indexes are based on 1961=100; the data are compiled on the 1960 Standard Industrial Classification (SIC).

Employment areas not covered are agriculture, fishing and trapping, education and related services, health and welfare services, religious organizations, public administration and defence and private households.

The monthly employment statistics relate to the number of employees drawing pay in the last pay period in the month. Data are requested for all classes of employees except homeworkers and casual employees working less than one day in the pay period. Working owners and partners of unincorporated business and professional practices are also excluded. The respondents report gross wages and salaries paid in the last pay period in the month, before deductions are made. The reported payrolls represent gross remuneration and paid absences in the period specified, including salaries, commissions, piece-work and time-work payments, and such items as shift premiums and regularly paid production, incentive and cost of living bonuses. Statistics on hours relate to the regular and overtime hours worked by those wage-earners for whom records of hours are maintained, and to hours credited to wage-earners absent on paid leave. If the report period exceeds one week, payroll and hours data are reduced to weekly equivalents.

Employment. Table 8.12 indicates that, over the 1972-76 period, the industrial composite index of employment (1961=100) for Canada rose by 10.9%. Among industry divisions showing gains over this period, services led with a 25.5% advance, followed by finance, insurance and real estate (23.7%), trade (17.8%), transportation, communication and other utilities (10.9%), mining (7.2%), construction (3.7%) and manufacturing (3.6%). A decline of 0.1% occurred in forestry during the same period. Compared with 1975, the industrial composite index for 1976 increased by 2.1%.

Annual average index numbers of employment for the years 1972-76 are shown by industrial division and group in Table 8.13, by province for 1972-76 and by month for 1975 and 1976 in Table 8.14 and by metropolitan area for 1972-76 and by month for 1975 and 1976 in Table 8.15.

Weekly earnings. Average weekly earnings at the national industrial composite level have increased substantially in the years for which current payroll statistics have been collected, rising from $23.44 in 1939 to $102.83 in 1967 and $228.03 in 1976. The upward movement gained momentum beginning in 1946 and average annual increases for the 1946-52 period were more than double those for 1939-45. After 1952 the rate of increase, in percentage terms, fell somewhat, particularly during 1959-62. In the recent period, gains have been 14.2% in 1975 and 12.1% in 1976. Annual index numbers of employment and average weekly earnings for 1974-76 are presented by industry, province and urban area in Table 8.16. Table 8.17 shows annual average weekly earnings by industrial division for the years 1972-76 and monthly averages for 1975 and 1976.

Hours and earnings of hourly-rated wage-earners. The monthly survey of employment, payrolls and man-hours covers statistics of hours of work and paid absence of those wage-earners for whom records of hours are maintained, plus corresponding totals of gross wages paid; these wage-earners are mainly hourly-rated production workers. Information on hours is frequently not kept by employers for ancillary workers nor, in many industries and establishments, for any wage-earners. Salaried employees are excluded by definition from the series. As a result of these exclusions, data are available for fewer industries and workers than are covered in the employment and average weekly earnings statistics.

During 1971-76 average weekly hours declined while average hourly earnings rose substantially. For the most part, upward wage-rate revisions in all industries were responsible. Technological changes, which in many cases involve the employment of more highly skilled workers at the expense of those in the lower-paid occupations, also contributed to the advance of average hourly earnings. As indicated in Table 8.18 from 1971 to 1976 average hourly earnings rose by 83.2% in mining, by 82.7% in construction and by 75.6% in manufacturing. During the same period, average weekly hours declined by 0.2% in mining, 0.8% in construction and 2.6% in manufacturing. Comparing 1976 to 1975 average hourly earnings increased by 15.3% in construction, by 13.8% in manufacturing and by 13.7% in mining; weekly hours decreased by 0.3% in construction and increased by 0.3% in manufacturing and by 0.8% in mining. Table 8.19 presents average weekly hours and hourly earnings in specified industries and selected urban areas for 1974-76.

8.3.2 Estimates of labour income

Labour income, comprising wages and salaries and supplementary labour income, is defined as all compensation paid to employees residing in Canada. By definition this includes Canadians who are employed abroad by the federal government. Not included are earnings received by self-employed persons such as independent professionals, proprietors of unincorporated businesses and farmers. Military pay and allowances which fit the definition of labour income are also excluded because they are shown as a separate item in the national income accounts.

Wages and salaries include directors fees, bonuses, commissions, gratuities, income in kind, taxable allowances and retroactive wage payments. Wages and salaries are estimated on a gross basis, before deductions for employees contributions to income tax, unemployment insurance and pension funds. Remuneration accumulating over time, for example, retroactive payments, are accounted for in the month and year in which they are paid.

Supplementary labour income, defined as payments made by employers for the future benefit of their employees, comprises employers contributions to employee welfare and pension funds, worker compensation funds and unemployment insurance.

Estimates of labour income based on the 1948 SIC have been released for the years 1926 to 1969. Data based on the 1960 SIC were first published in 1969 covering the period 1951 to 1968. These estimates were projected to the end of 1971. The entire series 1951 to 1971 has been revised and carried back to 1947 on the 1960 SIC.

8.3.3 Employer labour costs

The labour cost survey is designed to provide information on the composition of total employee compensation and to measure significance of fringe benefits. In Canada, as in most industrialized countries, there is an increasing awareness by employers, employees, labour unions and governments of the growing importance of fringe benefits. In a general way, the benefits comprise a wide variety of direct and indirect costs for non-wage items incurred by employers on behalf of their employees. At present the labour cost survey is limited to employer expenditures which comprise employee compensation — wages and salaries, additional cash payments, such as severance pay, and costs to employers of compulsory and voluntary welfare and benefit plans.

Since 1967 yearly labour cost surveys have covered one or more major industry divisions. Starting in 1976 the survey covered all industries. This all-industry survey was a sample of 7,600 reporting units representing all components of the economy with 20 or more employees except agriculture, fishing and trapping, but including government administration at federal, provincial and local levels.

Information from the all-industry survey for 1976 shows that total compensation amounted to $14,383 for each employee; $13,221 represented salaries, wages and other direct payments and the remaining $1,162 represented employer payments to employee welfare and benefit plans. These figures represent costs to the employer rather than benefits received by the employees.

Job vacancies   8.3.4

The job vacancy survey is a sample survey conducted by both mail and interviews among employers representing approximately 90% of employment. The survey covers all industrial sectors except agriculture, fishing and trapping, domestic service and the non-civilian component of public administration and defence. The industrial classification is based on the 1960 SIC. The basis of the occupational classification of published vacancy data is the *Canadian classification and dictionary of occupations (CCDO), 1971.*

The survey measures unfilled vacancies on six days in a quarter. From these a quarterly average is produced. The estimates should be interpreted as an approximation of the general level of vacancies at any day in the quarter.

For a job to be considered a vacancy, it must meet the following requirements: it must be available immediately; the employer must have undertaken, within four weeks prior to the reference date, some specific recruiting action to fill the vacancy; the job must be vacant for the entire reference day; and it must be available to persons outside the firm.

Data from the survey have been published since 1971. They are published by three categories of job vacancies: all categories, full-time, and longer-term; classified by industry division, by the four-digit CCDO level, by province and by quarter.

Occupational employment survey   8.3.5

The occupational employment survey is a sample survey of employers. It is designed to measure the occupational distribution of the paid worker portion of the labour force and to present such distributions by province and industry. During 1975 the survey covered approximately 80,000 reporting units in the public and private sectors. Job title information was collected for some 2,800,000 employees, about 36% of the total paid worker population.

The survey covered all industrial sectors except agriculture, fishing and trapping, domestic service and the non-civilian component of public administration and defence. The industrial classification is based on the 1960 SIC. The basis of the occupational classification of the data is the *Occupational classification manual, Census of Canada 1971,* Volume II, Statistics Canada Catalogue 12-537. The survey collected employment data reported to the monthly employment and payrolls surveys and job titles of paid workers as recorded in the end-month payroll.

Estimates of occupational employment in 1975 are available classified by industry division by the four-digit occupational classification level, for Canada and the provinces. These estimates have been published in Statistics Canada Catalogue 72-515.

Wage rates, salaries and working conditions   8.3.6

Statistics on occupational wage and salary rates by industry, locality and for all Canada, with standard weekly hours of work, are compiled by Labour Canada and published in an annual series of reports *Wage rates, salaries and hours of labour.* The statistics are based on an annual survey covering some 32,000 establishments in most industries and apply to the last normal pay period preceding October 1. Average wage and salary rates, number of employees, 1st and 9th deciles, 1st and 3rd quartiles and medians are shown for a number of office and service occupations, maintenance trades, labourers and specific industry occupations. Information on concepts and methods of producing these statistics is given in the reports.

Table 8.32 presents average wage and salary data on October 1, 1975 and 1976. Hourly and weekly rates of pay are listed for 19 occupations; salaries are shown separately for men and for women engaged in several office occupations.

Table 8.21 gives summary data on working conditions of office and non-office employees in manufacturing industries and in all industries for 1975 and 1976. The percentages denote proportions of office and non-office employees in establishments reporting specific items to the total number of such employees in all establishments replying to the survey; they are not necessarily the proportions of employees covered by the various items tabulated.

8.4   Pension plans

According to the Statistics Canada pensions data bank, membership in private pension plans in Canada increased to a record level of 3.9 million workers at the beginning of 1976, up considerably from 3.4 million in 1974. This increase is significant since it occurred over a period when the number of pension plans dropped slightly, from 15,853 in 1974 to 15,625 in 1976.

The 3.9 million plan members represented nearly 46% of the employed paid workers in the labour force. Excluded were unpaid family workers, the unemployed and self-employed who by definition are not eligible for employer-sponsored pension plans.

Pension plans were in operation in virtually all industries, but the degree of coverage varied widely. The most comprehensive coverage was in public administration and defence where nearly all employees were in a pension plan. In the private sector the most comprehensive coverage was in mining where 72% of the workers participated in a pension plan. Transportation and communication had fairly extensive coverage with nearly half the workers covered, followed closely by the manufacturing industry with over 48% coverage.

Of 15,625 plans in Canada at the beginning of 1976, 11,258 were funded with insurance companies, but these covered only 13% of the 3.9 million plan members. Small plans tended to be funded with insurance companies, but most large plans were funded on a trusteed basis. Although only one-quarter of all plans were trusteed, they covered two-thirds of all members, some 2.6 million of the 3.9 million members. Plans with the largest coverage were for federal and provincial public servants, with contributions paid into government consolidated revenue funds and not held in cash or securities. While only 21 in number, these plans covered 656,000 members.

Contributions from both employers and employees totalled nearly $4.6 billion for 1975, a record almost three times the amount contributed five years earlier. About two-thirds, $2.8 billion, was paid into trusteed pension funds which channel funds directly into the financial markets. With an annual cash flow of this magnitude, trusteed pension funds have become one of the largest single pools of investment capital in the country, reaching a total of $25.2 billion at book value by the end of 1976. Trusteed pension funds are surveyed annually and the results are published in *Trusteed pension plans, financial statistics,* Statistics Canada Catalogue 74-201. A summary tabulation of the key financial data related to these funds is presented in Table 8.22. The Canada and Quebec pension plans are discussed in Chapter 6.

Federal government annuities. The Government Annuities Act, which came into force September 1, 1908, authorized the sale of annuities. The object was to help Canadians provide for their later years. The act was one of the first significant pieces of social legislation in Canada. However, by the 1960s newer provisions such as the Canada Pension Plan and Old Age Security covered the public more effectively than government annuities. By Cabinet directive, at the close of 1967 the sales force was disbanded and active sales promotion of annuities ceased.

Inflationary pressures and rising interest rates had placed annuitants at a distinct disadvantage. The interest rate on government annuities ranged from 3% to 5¼% with the average rate about 4%.

The Government Annuities Improvement Act became law on December 20, 1975. It raised the interest rate on annuity contracts to 7% and a further percentage adjustment was provided for when the annuity came under payment. Greater flexibility was provided in annuities administration and the sale of new annuity contracts was terminated.

Unemployment insurance   8.5

Unemployment insurance has been part of Canada's social and economic life since the Unemployment Insurance Act was passed in 1940. Since that time various amendments brought new categories of workers into the plan and contributions and benefit rates were raised periodically to meet changing economic conditions. However, the basic structure of the plan remained unaltered until 1968, when Parliament approved upward revisions of both contributions and benefit rates and broadened the scope of coverage, and the Unemployment Insurance Commission was told to investigate the program and recommend appropriate changes. The Unemployment Insurance Act of 1971 was the result. Its basic objectives were to provide assistance to cope with interrupted earnings resulting from unemployment, including unemployment from illness.

The 1971 act extended coverage to all regular members of the labour force (effective January 2, 1972) for whom there existed an employer-employee relationship. The only non-insurable employees were those who earned less than 20% of the maximum weekly insurable earnings or less than 20 times the provincial hourly minimum wage, whichever was less.

Employers and employees would absorb the cost of initial benefits and administration with the employer rate at 1.4 times the employee rate. The government share was confined to the cost of extended benefits and the excess cost of initial benefits resulting from above-average rates of unemployment. In 1978 the rate of employee premiums was $1.50 per $100 of insurable earnings. The employer premium was $2.10 per $100. The revenue department collected the premiums.

Changes to the act have made the benefit structure more responsive to economic conditions. Under the 1971 program, the duration of benefit was not determined solely by how long a person had worked. A claimant could draw benefits for a maximum of 51 weeks depending on employment history and prevailing economic conditions, providing he contributed for at least eight weeks in the prior 52 and, now unemployed, was available, capable of and looking for work. Persons with 20 or more weeks of insured earnings (a major labour force attachment) were eligible for benefit payments when the interruption in earnings was caused by illness or pregnancy, and three weeks retirement benefit for older workers.

To be eligible in late 1977 the person must have worked from 10 to 14 weeks in insurable employment during his qualifying period, usually the 52 weeks prior to applying for benefits, or since the start of the last claim, whichever was more recent. The exact number of weeks required to be eligible depended on the unemployment rate in the economic region where the claimant normally lived. There were 16 economic regions in Canada. Benefits could be drawn for a maximum of 50 weeks.

Sickness benefit was available for a maximum of 15 weeks for persons with a major labour force attachment whose earnings were interrupted by illness, injury or quarantine. Previously, sickness benefit had been available only during the first 39 weeks of the claim. Now it was payable any time during the benefit period. If a person took ill while on regular claim, sickness benefit was available, limited by the weeks of regular benefit already received.

Maternity benefit was available for 15 weeks to women with a major labour force attachment. They must also have been part of the labour force for at least 10 weeks during the 20-week period between the 30th and 50th week before the expected date of birth. The claimant could take the 15 consecutive weeks anytime between the 10th week before the expected birth and the 17th week after.

Retirement benefit, available for three weeks, was paid in a lump sum to claimants with a major labour force attachment who were 65 years of age. They were eligible whether or not they continued working. The benefit was paid without a waiting period and without regard to earnings or availability.

The benefit rate for all claims was two-thirds of average insurable earnings in the qualifying period. This changes annually. In 1978, the weekly maximum insurable earnings were $240 and the maximum benefits two-thirds of that or $160.

In the case of regular benefit, employment income above 25% of the benefit rate was deducted. For sickness and maternity benefits, wage-loss income was deducted from the first benefit payments after the waiting period. All work-related income was deducted both during the waiting period and the benefit period.

The statistics in Table 8.23 summarize the insurance commission's activities in the years 1972-76. Figures prior to July 1971 are affected by the Unemployment Insurance Act of 1955, described in the *Canada Year Book 1973* p 352.

To assess the impact of changing economic conditions on the insurance program, current operational data, such as claims filed and processed and payments made, are collected and published monthly by Statistics Canada. Current claims and payment data are useful for administrative purposes and are also a source of information to the public regarding financial and other aspects of the program. In addition to monthly data on operation of the act, detailed figures on persons employed in insurable employment and benefit periods established and terminated are compiled annually and published in *Benefit periods established and terminated under the Unemployment Insurance Act,* Statistics Canada Catalogue 73-201.

8.6 Occupational injuries, illnesses and compensation

Fatal occupational injuries and illnesses. Data on fatal occupational injuries and illnesses compiled by the labour department are collected from provincial worker compensation boards. On the average annually in the period 1967-76, 1,162 industrial workers sustained fatal injuries and illnesses. Of 926 fatality reports received in 1976, collisions, derailments or wrecks caused 236 deaths; being struck by or against an object, 136; falls and slips, 93; drowning, 40; being caught in, on or between objects or vehicles, 68; occupational illnesses, 110; fire, explosion, temperature extremes, 37; and the remaining 206 resulted from miscellaneous accidents. Table 8.24 presents statistics on fatal occupational injuries and illnesses in 11 industries for 1974-76. Occupational injuries and illnesses, extent of disability and amount of compensation paid are reported by province for 1975 and 1976 in Table 8.25. In 1975, 985,317 injuries and illnesses resulted in $668 million in compensation compared with 1,047,033 injuries and illnesses and $526 million in compensation in 1974. Preliminary figures for 1976 show 1,042,940 injuries and illnesses resulted in $795 million in compensation.

8.7 Organized labour

8.7.1 Union membership

At January 1, 1977, labour unions reported a total of 3.1 million members in Canada, an increase of 3.5% over 1976 (Table 8.26). In 1977 union members consisted of 38.2% of non-agricultural paid workers and 31.0% of the total civilian labour force. Membership, by type of union and affiliation, is presented in Table 8.27. Canadian Labour Congress (CLC) affiliates, with 2.2 million members in 1977, accounted for 68.7% of total union membership compared with 71.3% in 1976. Of the total in CLC affiliates in 1977, 1.3 million members belonged to unions that were also affiliated with the American Federation of Labor and Congress of Industrial Organizations (AFL–CIO) in the United States; membership of unions affiliated with the CLC but not holding affiliation with the AFL–CIO totalled 884,705 or 28.1% of the total. Federations affiliated with the Quebec-based Confederation of National Trade Unions (CNTU) had 172,714 members or 5.5% of total union membership; the Confederation of Canadian Unions (CCU) represented 20,822 members or 0.7%; another 1.3% was reported by the Centrale des syndicats démocratiques with 39,663 members; and the remaining 23.7% belonged to various unaffiliated international and national unions and independent local organizations.

International unions with headquarters in the United States accounted for 49.0% of the 1977 membership compared with 49.6% in 1976; national and regional unions, which charter locals in Canada only, made up 47.4% (46.8% in 1976). Independent local organizations and local unions chartered by the CLC and the CNTU accounted for the remaining 3.6%.

In 1977, 18 unions reported 50,000 or more members, accounting for 52.1% of the total membership. The 10 largest, listed with their affiliation, ranked as follows in 1977 (1976 rank in parentheses):

1 (1) Canadian Union of Public Employees (CLC), 228,687
2 (2) United Steelworkers of America (AFL–CIO/CLC), 193,340
3 (3) Public Service Alliance of Canada (CLC), 159,499
4 (4) International Union, United Automobile, Aerospace and Agricultural Implement Workers of America (CLC), 130,000
5 (–) National Union of Provincial Government Employees (CLC), 101,131
6 (5) United Brotherhood of Carpenters and Joiners of America (AFL–CIO/CLC), 89,010
7 (7) International Brotherhood of Teamsters, Chauffeurs, Warehousemen and Helpers of America (Ind.), 86,603
8 (6) Quebec Teachers' Congress (Ind.), 85,000
9 (8) International Brotherhood of Electrical Workers (AFL–CIO/CLC), 63,914
10 (12) Ontario Public Service Employees' Union (Ind.) 63,340.

Wage trends in major collective agreements   8.7.2

Labour Canada publishes base rate settlement data for collective agreements on a quarterly basis. The agreements covered are limited to negotiating units of 500 or more employees in all industries except construction. The base rate for a negotiating unit is defined as the lowest rate of pay, expressed in hourly terms, for the lowest-paid classification used for qualified workers in the bargaining unit. The wage data are not necessarily representative of the average increases received by the workers in the whole negotiating unit. Nevertheless, the data are aggregated using the total number of employees in the negotiating unit.

In 1976, some 610 collective agreements covering 1,367,760 workers were settled. As shown in Table 8.28, the average annual percentage increase in base rates in these settlements was 10.5% compound over the term of the agreements. The comparable percentage for 1975 was 17.0% compound.

The 1976 settlements of one-year duration produced increases averaging 11.7%; those of two-year duration 12.7% and 8.1% for the first and second years, respectively; and those of three-year duration, 14.8%, 7.2% and 5.4% for the first, second and third years of the contracts. These increases compare with those of 1975 as follows: one-year agreements, average increase of 20.9%; two-year agreements, average increases of 22.1% and 11.2%; and three-year agreements, average increases of 19.4%, 8.6% and 4.0% for the first, second and third years, respectively.

A further breakdown reveals that of the 610 settlements in 1976, 157 covering 541,900 employees included a cost of living allowance (COLA). These 157 settlements produced an average increase, prior to the calculation of COLA, of 9.0% over the life of the agreements, whereas the remaining 453 agreements (825,860 employees) without a COLA clause produced an average increase of 11.5%.

Wage increase data given in Table 8.29 cover year-over-year percentage and cents-per-hour increases in base rates for all major collective agreements, excluding construction. During 1976, these rates rose from 12.4% (52.5 cents) in January to 15.0% (69.2 cents) in October, and then declined to 14.4% (67.5 cents) in December.

Strikes and lockouts   8.8

Statistical information on strikes and lockouts in Canada is compiled by Labour Canada on the basis of reports from manpower centres, provincial labour departments and other sources. Table 8.30 presents a breakdown by industry and jurisdiction of strikes and lockouts in 1976 involving three or more workers and amounting to 10 or more man-days. The 1,039 work stoppages reported involved 1,570,940 workers and 11.6 million man-days.

Time loss in 1976. Despite an October 14 national day of protest, Canada recorded less direct time loss from work stoppages due to strikes and lockouts in 1976 than it did in 1975. The 1976 figure was 11,609,890 — including 830,000 man-days lost by workers involved in the day of protest — compared with 10,908,810 the previous year.

There were 1,039 stoppages during 1976, involving 1,570,940 workers but more than half the workers — 830,000 — were accounted for by the day of protest. In 1975, there were 1,171 stoppages involving 506,443 workers.

In relation to total estimated working time of non-agricultural paid workers, the 1976 time loss was equivalent to 50 man-days per 10,000 man-days worked, down from 53 the previous year.

The number of workers involved includes all reported on strike or locked out, whether or not they belonged to the unions directly involved in the disputes leading to work stoppages. Workers indirectly affected, such as those laid off as a result of a work stoppage, are not included. Duration of strikes and lockouts in terms of man-days is calculated by multiplying the number of workers involved in each work stoppage by the number of working days the stoppage was in progress.

8.9   The Anti-Inflation Board

Canada's anti-inflation program went into effect October 14, 1975. The Anti-Inflation Board (AIB) was a part of this program, established to monitor compensation increases and profit margins. A gradual lifting of controls began April 14, 1978, with December 31, 1978, the designated expiry date.

The purpose of the compensation part of the anti-inflation program was to restrain the rate of increase in labour costs. The compensation regulations applied to employee groups rather than individuals. Employee groups generally were: bargaining units; groups established by the employer for purposes of determining salaries or wages; and executives of companies covered by the regulations.

More than 31,000 groups reported compensation increases during the first two years of the program to October 13, 1977. This represented approximately 3.5 million employees. During the second program year more than 67% of cases were either at or below the arithmetic guidelines compared with 59% during the first year.

In the first two years of the program, 16,950 compensation plans that were above the guidelines were submitted to the board. This represented 33% of all compensation plans received. The board ruled on 14,166 of these over-guideline compensation plans by the end of the second year of the program.

Sources
8.1.1 Public Relations Branch, Department of Labour.
8.1.2 Information Division, Canada Employment and Immigration Commission.
8.1.3 Public Relations Branch, Department of Labour.
8.2.1 Labour Force Survey Division, Census and Household Surveys Field, Statistics Canada.
8.2.2 Census Characteristics Division, Census and Household Surveys Field, Statistics Canada.
8.3.1 - 8.3.5 Labour Division, General Statistics Branch, Statistics Canada.
8.3.6 Labour Data Branch, Department of Labour.
8.4 Labour Division, General Statistics Branch, Statistics Canada; Information Division, Canada Employment and Immigration Commission.
8.5 Benefit Group, Public Affairs, Canada Employment and Immigration Commission.
8.6 Occupational Safety and Health Branch, Department of Labour.
8.7 - 8.8 Labour Data Branch, Department of Labour.
8.9 Editing Services, Communications, Anti-Inflation Board.

Tables

8.1 Estimates of the civilian labour force and its main components, annual averages 1971-76

Year	Civilian population (15 years of age and over) '000	Civilian labour force (15 years of age and over) Employed '000	Unemployed '000	Total labour force '000	Persons not in the labour force (15 years of age and over) '000	Participation rate %	Unemployment rate %
1971	14,878	8,107	536	8,643	6,235	58.1	6.2
1972	15,227	8,363	555	8,918	6,309	58.6	6.2
1973	15,608	8,802	519	9,321	6,287	59.7	5.6
1974	16,039	9,185	519	9,704	6,335	60.5	5.3
1975	16,470	9,363	697	10,060	6,410	61.1	6.9
1976	16,873	9,572	736	10,308	6,565	61.1	7.1

8.2 Distribution of population in the labour force and non-labour force categories, by age and sex, 1971-76

Sex, age and year	Population '000	Labour force Employed '000	Unemployed '000	Total '000	Not in labour force '000	Participation rate %	Unemployment rate %
Male							
1971	7,332	5,332	339	5,670	1,662	77.3	6.0
1972	7,501	5,476	338	5,814	1,687	77.5	5.8
1973	7,687	5,711	297	6,008	1,679	78.2	4.9
1974	7,901	5,919	297	6,216	1,685	78.7	4.8
1975	8,111	5,966	397	6,363	1,748	78.4	6.2
1976	8,303	6,038	411	6,449	1,854	77.7	6.4
15-24 years							
1971	1,967	1,084	149	1,233	734	62.7	12.1
1972	2,016	1,142	155	1,297	719	64.3	12.0
1973	2,071	1,243	139	1,382	689	66.7	10.1
1974	2,134	1,330	142	1,473	661	69.0	9.6
1975	2,194	1,325	190	1,515	680	69.0	12.6
1976	2,245	1,328	204	1,531	713	68.2	13.3
25 + years							
1971	5,365	4,247	190	4,437	928	82.7	4.3
1972	5,485	4,334	184	4,517	968	82.4	4.1
1973	5,617	4,467	158	4,626	991	82.4	3.4
1974	5,767	4,588	155	4,743	1,024	82.2	3.3
1975	5,916	4,641	206	4,848	1,069	81.9	4.3
1976	6,059	4,711	208	4,918	1,140	81.2	4.2
Female							
1971	7,546	2,775	197	2,972	4,574	39.4	6.6
1972	7,726	2,887	217	3,104	4,622	40.2	7.0
1973	7,921	3,091	221	3,313	4,608	41.8	6.7
1974	8,139	3,266	222	3,489	4,650	42.9	6.4
1975	8,359	3,397	301	3,697	4,662	44.2	8.1
1976	8,570	3,534	325	3,859	4,711	45.0	8.4
15-24 years							
1971	1,963	899	98	997	966	50.8	9.8
1972	1,999	936	99	1,035	964	51.8	9.6
1973	2,045	1,006	102	1,108	937	54.2	9.2
1974	2,102	1,071	106	1,177	925	56.0	9.0
1975	2,157	1,086	140	1,226	930	56.9	11.5
1976	2,203	1,102	152	1,253	950	56.9	12.1
25 + years							
1971	5,584	1,876	99	1,975	3,609	35.4	5.0
1972	5,727	1,951	117	2,068	3,659	36.1	5.7
1973	5,876	2,085	119	2,204	3,672	37.5	5.4
1974	6,037	2,195	117	2,312	3,725	38.3	5.1
1975	6,203	2,311	160	2,471	3,731	39.8	6.5
1976	6,367	2,432	173	2,605	3,761	40.9	6.7

8.3   Employment, unemployment and unemployment rates, by province, 1971-76

Year	Province				
	Nfld.	PEI	NS	NB	Que.
Employment ('000)					
1971	135	38	257	198	2,176
1972	141	37	261	206	2,208
1973	153	40	276	217	2,338
1974	149	41	294	226	2,415
1975	152	43	295	229	2,452
1976	158	44	295	232	2,479
Unemployment ('000)					
1971	13		19	13	171
1972	14		20	16	178
1973	17	¹	20	18	170
1974	23		22	19	171
1975	25		25	25	216
1976	25		31	29	236
Unemployment rates (%)					
1971	8.8		6.9	6.2	7.3
1972	9.0		7.1	7.2	7.5
1973	10.0	¹	6.8	7.7	6.8
1974	13.4		7.0	7.8	6.6
1975	14.2		7.8	9.9	8.1
1976	13.6		9.6	11.1	8.7

	Ont.	Man.	Sask.	Alta.	BC
Employment ('000)					
1971	3,114	379	334	643	835
1972	3,248	387	338	668	869
1973	3,400	403	347	702	928
1974	3,550	421	357	747	987
1975	3,613	420	373	778	1,009
1976	3,689	428	387	822	1,038
Unemployment ('000)					
1971	178	23	12	39	65
1972	172	22	16	40	74
1973	153	20	13	39	67
1974	165	16	10	27	65
1975	244	20	14	33	94
1976	242	21	16	33	98
Unemployment rates (%)					
1971	5.4	5.7	3.5	5.7	7.2
1972	5.0	5.4	4.5	5.6	7.9
1973	4.3	4.7	3.6	5.3	6.7
1974	4.4	3.7	2.7	3.5	6.2
1975	6.3	4.5	2.9	4.1	8.5
1976	6.2	4.7	4.0	3.9	8.6

¹Prior to April 1976, unemployment figures for Prince Edward Island were not published due to high sampling error.

8.4   Employees by industrial group, 1971-76

Industrial group (1970 Standard Industrial Classification)	Estimates in thousands					
	1971	1972	1973	1974	1975	1976
Agriculture	514	484	471	477	486	474
Other primary industries	221	215	226	231	222	237
Manufacturing	1,767	1,828	1,937	1,994	1,890	1,945
Construction	489	495	542	591	610	642
Trade	1,336	1,420	1,510	1,587	1,649	1,658
Transportation, communication and other utilities	707	736	779	797	820	834
Finance, insurance and real estate	399	399	426	465	478	501
Service	2,129	2,205	2,299	2,401	2,537	2,595
Public administration	545	580	612	644	670	685
All industries	8,107	8,363	8,802	9,185	9,363	9,572

8.5 Employees by occupation, 1975 and 1976

Occupation (1971 Census classification)	Estimates in thousands	
	1975	1976
Managerial, administrative	614	645
Natural sciences, engineering and mathematics	312	327
Social sciences	111	122
Religion	26	25
Teaching	420	431
Medicine and health	427	432
Art, literature and recreation	115	114
Clerical	1,640	1,680
Sales	1,039	1,035
Service	1,138	1,162
Farming, horticulture and animal husbandry	504	498
Fishing, hunting and trapping	20	20
Forestry and logging	49	53
Mining and quarrying	53	54
Processing	362	388
Machining	252	246
Product fabricating, assembling and repairing	872	894
Construction trades	652	679
Transport equipment operation	389	394
Materials handling	242	248
Other crafts and equipment operation	127	126
All occupations	9,363	9,572

8.6 Persons 15 years of age and over in the total labour force, by country of birth and region of residence, 1971

Country of birth	Canada[1]	Atlantic provinces	Quebec	Ontario	Prairie provinces	British Columbia
Total labour force	8,813,340	715,040	2,242,840	3,410,830	1,495,330	930,030
Born in Canada	7,049,575	682,545	1,989,720	2,430,865	1,241,715	688,765
Born outside Canada	1,763,770	32,495	253,115	979,960	253,610	241,265
US	124,405	7,575	18,020	45,390	30,460	22,590
Europe	1,429,985	21,130	194,180	830,880	199,030	182,160
Western Europe	*280,010*	*3,905*	*42,465*	*142,290*	*48,965*	*41,680*
Austria and Germany	144,730	1,450	13,355	75,920	29,170	24,400
Netherlands	85,774	1,875	2,795	52,020	15,515	13,425
France	26,405	315	18,895	4,415	1,510	1,200
Other Western Europe	23,100	265	7,425	9,935	2,770	2,655
Northern Europe	*513,355*	*14,055*	*33,595*	*298,545*	*72,415*	*93,450*
United Kingdom	450,850	12,835	30,185	268,380	60,270	78,110
Republic of Ireland	20,890	525	1,665	12,840	3,060	2,725
Scandinavia[2]	30,275	645	1,275	9,475	8,480	10,265
Finland	11,345	50	465	7,850	610	2,345
Southern Europe	*387,270*	*1,685*	*82,870*	*256,540*	*21,270*	*24,595*
Italy	234,360	710	53,150	156,485	10,755	13,120
Greece	50,995	530	17,315	29,145	2,015	1,965
Yugoslavia	48,805	120	3,210	35,190	5,130	5,035
Portugal	40,400	220	6,345	27,085	2,885	3,855
Other Southern Europe	12,710	100	2,850	8,640	485	620
Eastern Europe	*249,350*	*1,485*	*35,250*	*133,510*	*56,380*	*22,430*
Poland	87,835	555	12,510	46,420	22,035	6,230
USSR	78,605	355	8,305	41,605	20,850	7,440
Hungary	42,875	280	7,065	24,210	6,490	4,730
Czechoslovakia	25,345	230	3,270	14,370	4,515	2,900
Other Eastern Europe	14,690	65	4,095	6,900	2,495	1,135
Asia	97,700	2,450	14,080	43,935	13,815	23,285
Other	111,680	1,340	26,830	59,755	10,310	13,230

[1]Includes Yukon Territory and Northwest Territories.
[2]Includes Denmark, Iceland, Norway and Sweden.

8.7   Persons 15 years of age and over, in the labour force[1], by occupation[2], 1951-71 (based on the 1961 occupational classification)

Occupation	Labour force					
	1951		1961		1971[3]	
	No.	%	No.	%	No.	%
White collar workers	1,669,985	31.6	2,409,337	37.3	3,555,650	41.3
Managerial occupations	420,181	8.0	538,131	8.3	679,847	7.9
Professional and technical						
occupations	384,778	7.3	627,624	9.7	1,077,479	12.5
Clerical occupations	578,137	11.0	833,173	12.9	1,270,598	14.8
Sales occupations	286,889	5.4	410,409	6.4	527,726	6.1
Blue collar workers	1,654,767	31.4	1,871,562	29.0	2,200,156	25.6
Craftsmen, production process						
and related workers	1,303,559	24.7	1,527,129	23.6	1,792,607	20.8
Labourers	351,208	6.6	344,433	5.3	407,549	4.7
Transport and communication						
occupations	330,890	6.3	391,569	6.1	431,489	5.0
Service and recreation						
occupations	514,412	9.7	794,115	12.3	1,000,366	11.6
Primary occupations	1,042,639	19.8	826,072	12.8	642,573	7.5
Farmers and farm workers	826,093	15.6	648,910	10.0	500,356	5.8
Loggers and related workers	100,854	1.9	78,874	1.2	52,754	0.6
Fishermen, trappers and						
hunters	51,023	1.0	34,267	0.5	27,078	0.3
Miners, quarrymen and related						
workers	64,669	1.2	64,021	1.0	62,385	0.7
Occupations not stated	63,946	1.2	165,501	2.6	778,505	9.0
Total, occupations	5,276,639	100.0	6,458,156	100.0	8,608,739	100.0

[1]The labour force figures exclude a few persons seeking work who have never been employed.
[2]Excludes the Yukon Territory and Northwest Territories.
[3]The 1971 occupational data on a 1961 base was obtained from a recoded sample of the 1971 experienced labour force.

8.8   Persons 15 years of age and over in the labour force[1], by class of worker and sex, 1961 and 1971

Class of worker	1961			1971		
	Total	Male	Female	Total	Male	Female
Wage-earners	5,366,977	3,781,520	1,585,457	7,674,525	5,005,385	2,669,140
Paid workers	..	..	..	7,543,815	4,888,690	2,655,130
Self-employed in incorporated companies	..	..	..	130,705	116,695	14,010
Self-employed in unincorporated companies	940,488	846,467	94,021	668,850	586,130	82,720
Unpaid family workers	164,385	77,531	86,854	283,550	74,205	209,350
Total	6,471,850	4,705,518	1,766,332	8,626,925	5,665,715	2,961,210

[1]Includes the experienced labour force which is defined as the total labour force minus persons looking for work, who last worked prior to Jan. 1, 1970 or who never worked. For 1961, persons who never worked were excluded but persons looking for work who had not worked since Jan. 1, 1960 were included.

8.9   Labour force[1] 15 years of age and over, by major industrial group and sex, 1961 and 1971

Sector and industrial group[2]	1961		1971	
	Male	Female	Male	Female
Primary or extractive				
Agriculture	554,713	78,612	369,630	111,565
Forestry	106,387	2,193	71,025	3,355
Fishing and trapping	35,748	515	24,540	900
Mines, quarries and oil wells	112,254	3,963	129,675	9,360
Secondary or manufacturing				
Manufacturing	1,098,691	300,328	1,302,635	404,695
Construction	427,679	10,875	511,940	26,280
Tertiary or service				
Transportation, communication				
and other utilities	527,071	83,160	564,885	114,025
Trade	694,211	303,125	803,100	466,190
Finance, insurance and real				
estate	124,310	104,595	173,825	184,235
Community, business and				
personal service	518,515	750,332	865,345	1,176,045
Public administration and				
defence	389,360	86,620	468,420	163,320
Industry unspecified	116,579	42,014	380,700	301,240
Total	4,705,518	1,766,332	5,665,715	2,961,210

[1]Includes the experienced labour force which is defined as the total labour force minus persons looking for work, who last worked prior to Jan. 1, 1970 or who never worked. For 1961, persons who never worked were excluded but persons looking for work who had not worked since Jan. 1, 1960 were included.
[2]Data based on 1971 Standard Industrial Classification.

8.10   Labour force 15 years of age and over, by industrial sector and province, 1961 and 1971

Province	Primary		Secondary		Tertiary	
	1961	1971	1961	1971	1961	1971
Newfoundland	21,134	15,440	21,719	33,140	65,969	86,320
Prince Edward Island	11,386	8,130	5,242	7,020	16,790	24,010
Nova Scotia	33,689	21,980	49,693	62,790	149,301	181,165
New Brunswick	28,124	19,095	39,487	51,795	106,903	135,045
Quebec	199,943	122,110	593,646	621,435	923,263	1,218,450
Ontario	228,011	180,350	796,964	1,025,115	1,316,337	1,909,780
Manitoba	67,148	55,925	76,698	79,265	200,174	249,585
Saskatchewan	125,102	109,235	32,520	37,830	160,459	198,140
Alberta	120,187	115,575	79,817	114,850	278,396	400,625
British Columbia	55,891	69,285	149,818	210,830	355,172	558,305
Canada[1]	894,385	720,050	1,837,573	2,245,550	3,581,299	4,979,390

[1]Includes Yukon Territory and Northwest Territories.

8.11   Job vacancies, all categories[1], by province, annual averages 1971-76, and job vacancies (thousands), and vacancy rates[2] per 1,000 jobs, by quarter 1975 and 1976

Item and year	Nfld.	NS	NB	Que.	Ont.	Man.	Sask.	Alta.	BC	Canada[3]
Job vacancies										
Annual average										
1971	0.7	1.5	1.8	7.6	15.0	2.2	0.7	3.4	4.3	37.4
1972	1.0	2.0	1.7	17.3	26.4	2.8	1.6	7.0	5.8	66.2
1973	1.5	2.0	2.2	22.6	33.1	4.2	2.1	8.0	9.4	85.8
1974	1.4	2.4	2.2	21.4	41.0	5.7	4.1	13.1	9.8	101.7
1975	1.0	1.5	1.8	15.3	23.0	4.1	3.5	8.1	4.5	63.3
1976	0.6	1.1	1.2	11.9	17.6	2.9	2.5	9.2	4.1	51.4
By quarter										
1975										
1st quarter	0.8	1.5	2.0	13.6	23.7	4.6	2.6	7.9	4.3	61.3
2nd "	1.2	1.8	2.0	17.9	24.0	4.3	4.4	8.2	5.6	70.1
3rd "	1.4	1.7	1.7	18.4	24.1	5.2	3.8	9.1	4.7	71.0
4th "	0.5	1.0	1.4	11.2	20.1	2.4	3.2	7.3	3.4	50.8
1976										
1st quarter	0.5	1.3	1.4	13.7	15.4	2.7	2.2	7.3	3.5	48.1
2nd "	0.7	1.1	1.5	11.9	19.4	3.4	2.4	9.2	4.7	54.6
3rd "	0.7	1.3	1.2	13.2	22.1	3.6	3.2	11.6	5.0	62.3
4th "	0.5	0.6	0.8	8.9	13.7	1.7	2.1	8.5	3.3	40.3
Vacancy rates										
By quarter										
1975										
1st quarter	6	7	10	6	8	14	10	14	5	8
2nd "	9	8	10	8	7	13	17	14	6	9
3rd "	9	7	8	8	7	15	15	15	5	8
4th "	4	4	7	5	6	7	13	12	4	6
1976										
1st quarter	4	5	7	6	5	8	8	11	4	6
2nd "	5	4	7	5	6	9	9	14	5	6
3rd "	4	5	5	6	6	10	11	17	5	7
4th "	4	2	4	4	4	5	8	13	3	5

[1]Includes full-time, casual, part-time, seasonal and temporary jobs.
[2]A rate is obtained by expressing the number of vacancies per 1,000 existing jobs in all industries, except agriculture, fishing and trapping, domestic service and the non-civilian component of public administration and defence.
[3]Includes Prince Edward Island, the Yukon Territory and Northwest Territories.

8.12   Annual average index numbers[1] of employment, by industrial division, 1972-76, and monthly indexes 1975 and 1976

Year and month	Forestry	Mining (incl. milling)	Manufacturing	Construction	Transportation, communication and other utilities	Trade	Finance, insurance and real estate	Service[2]	Industrial composite
Averages									
1972	76.3	110.4	123.7	109.7	116.0	146.2	148.7	193.5	129.9
1973	86.4	111.4	129.9	109.9	118.0	155.3	157.1	206.1	135.9
1974	87.4	115.5	133.8	117.1	124.6	165.7	167.3	224.0	142.8
1975	76.0	114.1	126.3	117.1	125.8	168.5	175.0	231.9	141.1
1976P	76.2	118.3	128.1	113.8	128.7	172.2	183.9	242.8	144.1

8.12 Annual average index numbers[1] of employment, by industrial division, 1972-76, and monthly indexes 1975 and 1976 (concluded)

Year and month	Forestry	Mining (incl. milling)	Manufacturing	Construction	Transportation, communication and other utilities	Trade	Finance, insurance and real estate	Service[2]	Industrial composite
Averages									
1975									
January	77.2	114.9	125.2	101.2	124.3	166.7	170.3	218.8	138.0
February	73.0	115.1	124.7	101.0	123.7	163.4	170.5	220.0	137.0
March	66.8	113.1	125.5	101.9	124.8	163.4	172.2	219.2	137.5
April	58.9	111.1	125.6	107.6	125.1	166.8	171.5	224.5	138.7
May	77.4	113.7	128.1	119.1	129.3	168.5	174.9	234.0	142.7
June	96.0	115.8	130.3	122.7	131.8	168.6	176.1	241.3	145.4
July	88.4	117.3	126.3	126.8	131.8	165.2	177.9	240.9	143.4
August	87.6	112.6	127.7	131.6	129.6	165.3	176.5	243.7	143.8
September	79.8	110.1	126.5	132.6	128.6	168.5	177.6	239.7	143.3
October	79.5	113.3	127.0	130.7	118.2	173.3	177.1	236.8	142.2
November	68.6	116.3	125.8	122.9	116.3	175.4	177.5	233.3	140.9
December	58.6	116.1	123.3	107.5	125.1	177.0	177.5	230.9	140.3
1976P									
January	56.4	116.3	124.1	101.5	124.4	170.6	177.7	229.0	138.9
February	58.4	116.4	126.1	100.4	123.4	168.3	178.7	230.0	139.3
March	56.6	116.2	127.1	103.6	124.0	170.1	180.3	234.2	140.7
April	51.4	114.0	128.0	110.0	126.7	172.3	180.8	238.3	142.4
May	69.5	116.2	130.1	119.6	129.8	172.6	182.5	245.5	145.5
June	89.8	117.6	130.7	126.2	131.8	172.5	185.0	253.3	147.7
July	97.9	120.5	129.4	122.8	132.0	168.2	185.5	256.7	146.8
August	98.6	123.0	131.1	123.3	132.8	168.1	185 5	251.1	147.3
September	95.8	120.3	129.3	115.4	131.9	171.3	186.8	247.7	146.1
October	87.5	119.8	128.6	122.2	130.6	175.0	187.5	245.4	146.3
November	80.6	120.2	127.5	117.4	129.8	178.5	188.3	243.5	145.8
December	72.0	119.3	124.6	103.7	127.7	178.3	188.6	238.6	142.8

[1]Indexes are calculated as at the last pay period of each month (1961 = 100).
[2]Consists mainly of hotels, restaurants, laundries, dry-cleaning establishments and recreational and business services.

8.13 Annual average index numbers[1] of employment, by industrial division and group, 1972-76

Industry	1972	1973	1974	1975	1976
FORESTRY	76.3	86.4	87.4	76.0	76.2
MINING (incl. milling)	110.4	111.4	115.5	114.1	118.3
Metals	100.7	102.9	107.6	109.1	109.7
Gold	35.4	34.9	34.6	35.5	31.4
Copper-gold-silver	142.0	140.1	147.4	144.7	138.4
Iron	124.7	152.6	168.9	175.5	190.9
Mineral fuels	115.1	115.8	120.5	126.6	131.1
Coal	78.7	75.3	76.6	87.0	85.0
Crude petroleum and natural gas	151.1	155.7	164.0	165.7	176.7
Non-metals (except fuels)	131.5	130.2	140.2	117.6	139.1
Asbestos	117.2	114.0	117.3	83.9	117.9
MANUFACTURING	123.7	129.9	133.8	126.3	128.1
Durable goods	134.9	144.1	149.4	139.8	140.3
Non-durable goods	114.7	118.4	121.1	115.5	118.2
Foods and beverages	107.9	110.2	110.9	107.5	110.2
Slaughtering and meat processing	104.8	104.2	107.5	106.6	108.1
Dairy products	98.1	96.8	94.4	91.0	85.9
Fish products	137.9	149.8	147.2	129.8	152.4
Fish and vegetable processing	110.4	116.1	117.1	112.7	110.1
Grain mill products	102.8	103.0	109.7	110.6	109.5
Biscuits	101.7	102.5	100.0	98.4	96.0
Bakeries	88.7	88.1	85.3	84.7	87.3
Confectionery	96.4	96.2	94.8	87.3	86.7
Soft drinks	110.3	113.0	115.3	118.0	122.9
Distilleries	117.8	119.3	116.7	115.8	109.8
Breweries	104.6	112.3	115.9	117.1	118.5
Tobacco processing and products	92.8	92.1	91.2	85.3	86.3
Rubber products	118.4	125.5	120.8	123.0	125.9
Leather products	89.1	91.5	91.5	87.4	86.2
Shoes (except rubber)	81.7	82.2	81.8	78.4	79.6
Luggage, handbags and small leather goods	119.0	130.8	132.4	123.2	112.7
Textile products	122.3	127.8	126.2	112.3	110.7
Cotton yarn and cloth	75.1	73.7	72.3	60.2	63.4
Woollen yarn and cloth	87.3	94.1	88.9	76.6	72.2
Synthetic textiles	128.6	130.7	128.9	110.8	100.3
Knitting mills	116.4	118.7	116.8	114.0	108.5
Hosiery	92.5	87.2	83.6	68.8	66.8
Other knitting mills	131.0	137.7	136.8	140.6	132.9

8.13 Annual average index numbers[1] of employment, by industrial division and group, 1972-76 (continued)

Industry	1972	1973	1974	1975	1976
MANUFACTURING (concluded)					
Clothing	110.4	114.4	113.1	110.7	111.6
Men's clothing	124.1	128.1	128.2	123.9	125.5
Women's clothing	113.6	119.2	116.2	115.6	116.4
Wood products	121.3	132.7	132.4	117.2	130.2
Saw, shingle and planing mills	123.0	137.1	135.4	113.1	131.2
Furniture and fixtures	139.3	150.7	157.1	138.5	141.0
Household furniture	146.5	159.6	166.7	144.3	150.5
Paper and allied industries	119.6	121.7	133.0	117.2	128.3
Pulp and paper mills	114.9	115.5	127.1	108.1	123.7
Printing, publishing and allied industries	111.9	115.8	120.8	122.4	123.4
Commercial printing	114.4	121.4	128.8	128.9	131.9
Printing and publishing	106.3	104.6	108.0	109.4	109.4
Primary metal industries	127.5	132.4	138.3	135.0	128.1
Iron and steel mills	139.0	145.2	152.7	149.9	144.6
Iron foundries	136.0	144.6	155.1	142.4	136.7
Smelting and refining	109.0	112.8	115.9	116.0	104.4
Metal fabricating industries	133.3	141.8	148.4	139.1	139.7
Fabricated structural metals	98.3	103.5	113.1	118.6	115.5
Ornamental and architectural metals	123.2	142.5	149.1	131.1	138.2
Metal stamping, pressing and coating	147.7	151.8	155.2	140.7	146.8
Wire and wire products	136.7	147.9	152.0	137.6	136.7
Hardware, tools and cutlery	167.5	183.5	192.0	174.7	179.0
Heating equipment	98.6	101.5	105.3	99.0	101.9
Miscellaneous metal fabricating	134.1	145.2	152.2	136.4	134.7
Machinery (except electrical)	146.2	161.0	174.3	171.6	167.3
Agricultural implements	115.8	133.9	152.4	162.2	155.3
Miscellaneous machinery and equipment	148.5	167.6	185.4	178.5	176.1
Office and store machinery	190.7	158.1	141.6	150.3	141.1
Transportation equipment	152.5	161.3	161.2	151.1	156.0
Aircraft and parts	77.9	82.0	74.9	67.6	64.6
Motor vehicles	197.2	220.1	218.9	200.3	216.2
Assembling	180.0	194.2	202.2	186.7	200.2
Parts and accessories	194.2	218.1	205.4	184.2	206.5
Shipbuilding and repairing	113.0	106.6	100.9	111.4	106.4
Electrical products	135.8	142.6	154.2	140.0	137.2
Major appliances (incl. non-electrical)	113.3	120.6	120.8	100.1	102.8
Household radios and televisions	144.6	157.3	162.3	120.6	128.2
Communications equipment	143.3	155.1	169.5	147.7	137.8
Non-metallic mineral products	117.9	126.3	132.4	126.5	124.0
Concrete products	133.9	144.1	158.2	145.6	138.4
Clay products	93.1	99.9	103.0	97.1	97.2
Glass and glass products	125.9	137.2	138.3	129.3	131.3
Petroleum and coal products	105.9	105.7	113.5	115.0	114.9
Petroleum refineries	89.8	90.2	100.2	100.3	94.7
Chemicals and chemical products	114.3	117.8	123.7	124.6	127.0
Pharmaceuticals and medicines	154.2	164.2	170.3	164.5	163.0
Paints and varnishes	105.6	110.5	114.6	112.7	118.1
Soap and cleaning compounds	94.8	97.9	105.6	109.8	106.5
Industrial chemicals	105.8	104.5	109.9	114.9	122.7
Miscellaneous manufacturing industries	157.7	170.6	176.0	160.9	163.7
CONSTRUCTION	109.7	109.9	117.1	117.1	113.8
Building	115.4	117.1	124.7	121.5	117.6
General contractors	93.2	94.4	99.9	94.2	93.9
Special trade contractors	138.7	140.9	150.8	150.3	141.5
Engineering	99.0	97.7	104.1	109.7	107.3
Highways, bridges and streets	81.1	81.3	83.2	83.6	77.7
Other engineering	121.5	118.3	130.4	142.5	144.2
TRANSPORTATION, COMMUNICATION AND OTHER UTILITIES	116.0	118.0	124.6	125.8	128.7
Transportation	108.5	108.1	114.1	114.1	113.5
Air transport and services	166.8	180.7	206.0	213.1	211.8
Water transport and services	95.3	95.3	95.4	91.1	92.0
Railway transport	84.1	72.6	87.6	84.9	80.3
Truck transport	134.6	141.1	149.3	141.5	139.3
Bus transport, interurban and rural	141.8	147.0	156.8	161.7	171.7
Urban transit	116.4	124.2	132.8	143.8	154.9
Highway and bridge maintenance	121.1	116.0	114.4	119.7	121.6
Storage	111.6	112.4	111.4	112.2	112.8
Grain elevators	102.9	99.3	96.2	100.6	103.0
Other storage and warehousing	135.0	148.4	153.6	144.9	140.6
Communication	134.6	141.0	151.7	151.7	166.8
Radio and television broadcasting	138.8	143.6	153.0	162.9	177.4
Telephone	129.2	135.6	147.7	150.9	157.5
Telegraph and cable	77.8	72.6	76.6	78.4	74.0
Post office	160.7	171.5	182.3	171.2	209.0
ELECTRIC POWER, GAS AND WATER	122.5	128.9	133.2	141.3	142.4
Electric power	125.5	132.7	137.8	147.7	148.9
Gas distribution	109.3	113.5	113.8	124.1	125.4
TRADE	146.2	155.3	165.7	168.5	172.2
Wholesale	135.1	143.8	154.6	157.1	158.8
Retail	152.2	161.5	171.7	174.6	179.3
Food stores	151.7	157.1	164.1	170.8	175.6
Department stores	149.2	157.9	167.5	165.1	166.9
Variety stores	138.4	146.6	153.0	147.7	148.5
Automotive product stores	159.6	172.5	184.9	191.1	195.8

8.13   Annual average index numbers[1] of employment, by industrial division and group, 1972-76 (concluded)

Industry	1972	1973	1974	1975	1976
FINANCE, INSURANCE AND REAL ESTATE	148.7	157.1	167.3	175.0	183.9
Financial institutions	155.9	168.7	180.7	189.9	199.7
Insurance and real estate	139.0	141.2	149.1	154.8	162.5
Insurance carriers	121.3	116.6	119.0	120.6	124.5
SERVICE	193.5	206.1	224.0	231.9	242.8
Recreational services	174.8	186.2	197.8	204.6	230.5
Business services	220.5	234.9	271.2	286.1	294.7
Personal services	172.8	182.5	194.8	202.1	210.6
Miscellaneous services	239.2	256.5	273.9	277.0	285.0
Industrial composite	129.9	135.9	142.8	141.1	144.1

[1]Indexes refer to the last week of each month (1961 = 100).

8.14   Annual average index numbers[1] of employment for industrial composite, by province, 1972-76, and monthly indexes 1975 and 1976

Year and month	Nfld.	PEI	NS	NB	Que.	Ont.	Man.	Sask.	Alta.	BC	Canada
Averages
1972	126.5	140.6	116.7	122.7	120.1	134.2	117.6	116.8	143.6	148.4	129.9
1973	131.3	144.4	123.0	126.8	124.6	141.2	119.9	120.3	150.8	157.4	135.9
1974	139.5	152.0	130.2	134.4	129.9	147.5	128.4	130.0	163.1	167.2	142.8
1975	135.5	149.9	128.9	136.1	128.5	144.2	130.1	137.0	169.6	161.3	141.1
1976P	134.6	148.1	128.5	135.0	129.9	147.1	128.2	142.1	184.9	167.5	144.1
1975
January	126.4	132.1	124.6	127.1	125.5	142.6	127.3	129.2	160.5	158.9	138.0
February	124.8	128.8	123.2	125.1	124.8	141.5	127.0	128.1	160.6	158.0	137.0
March	126.1	128.7	124.4	125.3	124.5	141.8	126.9	129.2	160.9	161.8	137.5
April	128.3	134.5	126.4	127.5	124.5	143.2	127.8	131.6	162.2	164.3	138.7
May	131.6	159.0	133.1	137.8	129.3	145.8	130.6	137.7	168.3	168.7	142.7
June	140.9	170.8	133.0	145.4	132.3	147.8	133.2	141.4	173.1	169.7	145.4
July	145.5	169.6	133.9	147.5	130.6	145.7	133.1	144.4	174.9	158.9	143.4
August	145.7	171.5	132.6	149.0	131.1	146.5	133.3	143.4	176.3	157.7	143.8
September	149.6	163.1	129.0	148.3	132.6	145.8	133.8	141.6	176.0	151.9	143.3
October	147.4	156.9	131.5	138.1	130.4	143.7	131.1	138.0	174.5	161.9	142.2
November	135.0	143.9	129.1	133.7	128.4	143.3	129.3	139.3	173.4	161.4	140.9
December	124.7	140.3	126.4	128.4	127.8	143.2	127.6	140.0	174.5	161.9	140.3
1976P
January	119.6	130.1	124.1	126.7	126.3	142.2	125.6	136.0	174.1	160.6	138.9
February	123.2	126.0	123.2	124.9	126.3	143.0	124.8	136.7	174.3	161.2	139.3
March	123.8	124.4	124.1	126.3	127.5	144.0	125.8	137.4	176.8	164.6	140.7
April	126.0	135.8	126.3	127.7	129.2	145.9	126.9	139.0	177.7	166.9	142.4
May	132.4	162.7	129.7	137.2	132.5	148.2	129.4	142.5	183.9	168.4	145.5
June	143.7	162.4	133.1	144.1	134.5	150.0	130.9	145.4	187.8	168.3	147.7
July	147.1	165.3	133.6	144.4	133.4	148.3	130.8	143.2	190.2	168.2	146.8
August	147.5	169.6	132.0	146.5	131.9	149.7	130.9	145.7	192.2	170.2	147.3
September	146.3	163.2	130.4	143.2	129.3	149.1	130.5	145.4	191.7	170.5	146.1
October	139.7	153.9	130.5	137.3	130.1	149.4	130.0	146.3	191.3	171.9	146.3
November	136.4	146.2	129.6	133.9	130.0	149.3	128.2	145.4	190.2	170.9	145.8
December	129.4	137.9	124.9	128.2	127.2	146.5	124.9	142.3	188.0	167.9	142.8

[1]Indexes refer to the last week of each month (1961 = 100).

8.15   Annual average index numbers[1] of employment for industrial composite, by metropolitan area, 1972-76, and monthly indexes 1975 and 1976

Year and month	Montreal	Quebec	Toronto	Ottawa–Hull	Hamilton	Windsor	Winnipeg	Vancouver
Averages
1972	120.9	132.1	139.0	146.8	123.9	153.4	120.2	149.8
1973	126.0	134.4	147.2	156.9	129.8	160.4	123.0	157.6
1974	130.9	137.4	154.5	168.5	136.5	159.8	130.0	169.1
1975	130.8	137.2	152.4	167.1	134.1	148.1	132.3	166.3
1976P	132.3	138.4	154.9	170.8	134.3	163.0	132.4	168.8
1975
January	128.8	132.1	152.1	165.9	134.1	141.8	130.3	164.6
February	128.3	132.2	150.8	165.3	132.4	144.6	129.8	163.6
March	128.2	132.6	150.4	163.0	132.3	143.5	130.5	166.6
April	128.7	133.0	151.2	165.4	133.9	147.7	131.0	169.1
May	132.0	137.9	153.6	170.7	135.3	148.9	132.5	171.4
June	133.7	142.1	155.1	171.5	136.5	150.1	134.4	170.7
July	130.5	140.5	152.7	169.6	135.2	151.1	133.5	168.0
August	130.5	142.8	152.8	170.7	134.6	151.4	133.7	167.7
September	134.2	143.7	153.6	169.0	133.7	153.6	135.9	160.0
October	132.7	137.6	151.8	164.3	133.4	148.5	133.4	164.0
November	131.0	136.1	151.6	162.7	133.8	146.6	132.3	164.1
December	131.5	135.8	152.6	167.4	133.5	149.3	131.4	166.3

[1]Indexes refer to the last week of each month (1961 = 100).

8.15 Annual average index numbers[1] of employment for industrial composite, by metropolitan area, 1972-76, and monthly indexes 1975 and 1976 (concluded)

Year and month	Montreal	Quebec	Toronto	Ottawa–Hull	Hamilton	Windsor	Winnipeg	Vancouver
Averages								
1976P								
January	130.1	136.0	151.7	165.3	132.1	151.6	129.7	164.5
February	129.9	134.6	151.6	167.5	132.2	153.1	128.5	164.6
March	131.3	134.0	152.5	167.0	131.9	156.2	130.7	167.0
April	133.1	137.3	154.1	169.4	133.9	158.2	131.8	170.8
May	135.7	141.0	155.5	172.5	136.4	163.8	133.5	172.1
June	137.6	141.0	157.1	175.1	138.3	166.1	134.4	170.0
July	135.3	141.1	154.9	171.4	135.6	162.6	133.1	167.4
August	132.4	139.2	155.8	171.1	135.7	168.3	133.6	168.5
September	129.7	139.4	155.9	173.1	135.9	165.8	134.6	168.5
October	131.8	140.2	157.4	173.7	134.0	169.2	134.7	171.1
November	131.7	139.1	157.9	173.5	133.5	170.5	133.8	171.9
December	129.2	137.7	154.9	169.6	132.0	171.1	130.7	169.7

[1]Indexes refer to the last week of each month (1961 = 100).

8.16 Annual index numbers of employment and average weekly earnings, by industry, province and urban area, 1974-76

Industry, province and urban area	Employment (1961 = 100)			Average weekly wages and salaries (dollars)		
	1974	1975	1976P	1974	1975	1976P
INDUSTRY						
Forestry	87.4	76.0	76.2	219.64	249.58	287.36
Mining (incl. milling)	115.5	114.4	118.3	238.97	280.44	316.38
Manufacturing	133.8	126.3	128.1	185.62	213.43	241.19
Durable goods[1]	149.4	139.8	140.3	198.39	227.11	257.46
Non-durable goods[1]	121.1	115.5	118.2	172.86	199.98	225.60
Construction	117.1	117.1	113.8	250.30	290.95	331.02
Transportation, communication and other utilities	124.6	125.8	128.7	204.39	233.98	262.02
Trade	165.7	168.5	172.2	139.92	159.06	176.59
Finance, insurance and real estate	167.3	175.0	183.9	172.25	193.12	213.71
Service	224.0	232.0	242.8	126.08	143.69	160.49
Industrial composite	142.8	141.1	144.1	178.09	203.34	228.03
PROVINCE (industrial composite)						
Newfoundland	139.5	135.5	134.6	168.48	196.44	221.63
Prince Edward Island	152.0	149.9	148.1	126.92	149.84	170.88
Nova Scotia	130.2	128.9	128.5	149.98	172.40	193.21
New Brunswick	134.4	136.1	135.0	154.58	182.40	202.56
Quebec	129.9	128.5	129.9	172.89	199.22	222.41
Ontario	147.5	144.2	147.1	181.43	204.85	228.72
Manitoba	128.4	130.1	128.2	162.71	186.01	208.55
Saskatchewan	130.0	137.0	142.1	160.99	188.31	214.87
Alberta	163.1	169.6	184.9	178.72	207.38	236.89
British Columbia	167.2	161.3	167.5	200.55	230.01	259.52
URBAN AREA (industrial composite)						
Corner Brook, Nfld.	110.1	104.9	107.7	177.17	189.69	228.57
St. John's, Nfld.	171.7	165.0	161.6	150.10	175.17	198.25
Halifax, NS	146.3	146.7	145.7	148.56	171.23	190.41
Sydney, NS	97.9	92.1	87.4	167.27	196.37	217.58
Moncton, NB	152.7	153.1	150.2	143.72	166.04	185.20
Saint John, NB	129.6	135.3	133.2	162.92	192.38	214.61
Chicoutimi, Que.	119.3	116.2	99.0	199.87	226.83	251.18
Drummondville, Que.	139.1	130.1	122.1	138.67	159.34	182.03
Granby, Que.	110.5	113.9	117.2	139.26	162.09	183.66
Montreal, Que.	130.9	130.8	132.3	174.54	201.49	224.18
Ottawa, Ont.-Hull, Que.	168.5	167.1	170.8	165.13	186.03	208.62
Quebec, Que.	137.4	137.2	138.4	158.52	179.82	200.87
Rouyn-Noranda, Que.	107.8	101.0	94.6	182.99	221.52	253.05
Saint-Hyacinthe, Que.	123.6	118.4	128.0	133.92	157.16	177.69
Saint-Jean, Que.	151.3	138.0	137.3	148.35	170.62	192.50
Saint-Jérôme, Que.	149.3	136.1	150.3	150.10	176.70	194.58
Shawinigan, Que.	89.7	81.6	81.7	181.53	203.87	231.54
Sherbrooke, Que.	126.1	125.0	130.0	147.41	169.33	190.74
Sorel, Que.	184.0	186.7	190.8	197.97	228.30	259.34
Thetford Mines, Que.	126.7	79.2	117.1	174.40	196.18	251.65
Trois-Rivières, Que.	126.7	118.3	120.5	167.13	192.91	220.13
Valleyfield, Que.	142.7	139.1	144.7	179.99	203.80	233.90
Barrie, Ont.	222.7	219.2	228.7	155.96	175.46	192.82
Belleville, Ont.	142.7	137.5	141.6	151.03	171.09	192.75
Brampton, Ont.	360.1	353.2	362.2	186.95	211.71	228.79
Brantford, Ont.	152.4	153.2	152.5	170.50	195.25	218.12
Brockville, Ont.	141.2	133.7	140.7	168.30	192.46	221.39
Chatham, Ont.	184.5	179.1	183.1	185.32	208.19	228.10
Cornwall, Ont.	140.9	118.6	120.9	167.90	188.30	215.81
Guelph, Ont.	155.5	156.4	160.7	163.01	186.32	207.05

8.16   Annual index numbers of employment and average weekly earnings, by industry, province and urban area, 1974-76 (concluded)

Industry, province and urban area	Employment (1961 = 100)			Average weekly wages and salaries (dollars)		
	1974	1975	1976P	1974	1975	1976P
URBAN AREA (industrial composite) (concluded)						
Hamilton, Ont.	136.5	134.1	134.3	186.42	211.42	237.95
Kingston, Ont.	138.0	138.3	139.5	169.44	190.52	212.85
Kitchener, Ont.²	181.0	170.0	174.1	158.87	180.42	202.21
London, Ont.	138.5	132.9	134.7	171.57	193.92	213.28
Midland, Ont.	220.5	180.5	186.8	142.46	165.69	186.55
Niagara Falls, Ont.	125.7	125.7	122.4	161.15	179.39	203.04
North Bay, Ont.	134.1	132.6	131.9	178.71	198.35	218.11
Orillia, Ont.	161.5	154.9	154.5	137.65	163.66	174.46
Oshawa, Ont.	149.2	140.5	149.4	211.56	229.69	266.68
Owen Sound, Ont.	165.1	156.9	167.6	151.49	175.17	192.15
Pembroke, Ont.	119.7	111.0	114.6	134.09	158.41	171.00
Peterborough, Ont.	140.8	147.6	140.6	183.61	205.08	225.25
Port Hope, Ont.	189.3	180.5	183.5	172.54	199.22	222.99
St. Catharines, Ont.	145.6	136.1	142.2	196.10	216.46	254.17
St. Thomas, Ont.	225.5	197.3	207.3	194.12	204.09	220.88
Sarnia, Ont.	144.2	146.2	153.1	220.37	248.68	284.67
Sault Ste Marie, Ont.	136.0	137.2	136.3	204.14	223.10	254.11
Stratford, Ont.	187.7	173.3	180.7	149.66	172.07	189.78
Sudbury, Ont.	109.2	113.9	117.1	199.44	231.39	259.11
Thunder Bay, Ont.	145.6	138.9	151.1	178.31	203.29	240.11
Timmins, Ont.	82.2	87.1	99.7	182.42	214.54	241.83
Toronto, Ont.	154.5	152.4	154.9	182.36	206.08	228.10
Welland, Ont.	112.2	107.1	107.3	203.70	224.44	256.27
Windsor, Ont.	159.8	148.1	163.0	210.97	229.37	261.70
Woodstock, Ont.	152.5	133.4	140.2	163.65	182.64	211.64
Brandon, Man.	145.5	146.5	148.3	131.61	154.02	172.92
Winnipeg, Man.	130.0	132.4	132.4	150.81	172.86	196.08
Moose Jaw, Sask.	83.8	89.5	98.1	133.37	158.89	175.67
Prince Albert, Sask.	145.9	135.3	148.5	169.74	187.21	208.15
Regina, Sask.	141.2	152.9	158.7	156.57	184.39	213.80
Saskatoon, Sask.	152.8	158.2	163.2	147.32	171.61	199.14
Calgary, Alta.	179.7	189.6	201.5	175.66	203.18	228.96
Edmonton, Alta.	172.9	180.1	196.2	171.30	197.41	219.03
Lethbridge, Alta.	162.7	165.6	169.7	145.77	173.12	194.11
Medicine Hat, Alta.	122.8	143.4	159.7	153.98	209.29	233.61
Red Deer, Alta.	220.1	235.4	278.5	147.55	167.88	192.41
Kamloops, BC	314.4	301.0	313.8	180.67	203.86	234.67
Prince George, BC	302.3	274.4	290.5	205.66	233.32	271.12
Vancouver, BC	169.1	166.4	168.8	194.41	224.06	252.25
Victoria, BC	151.8	149.9	154.5	167.38	193.92	219.52

[1]Durable goods manufacturing includes wood products, furniture and fixtures, primary metal industries, metal fabricating industries, machinery (except electrical), transportation equipment, electrical products and non-metallic mineral products; non-durable goods manufacturing includes all other manufacturing industries.
[2]Kitchener, Cambridge and Waterloo.

8.17   Annual average weekly earnings, by industrial division, 1972-76, and monthly averages 1975 and 1976 (dollars)

Year and month	For-estry	Mining (incl. milling)	Manu-factur-ing	Con-struc-tion	Trans-portation, communi-cation and other utilities	Trade	Finance, insur-ance and real estate	Service[1]	Indus-trial com-posite
Averages									
1972	172.92	190.29	156.10	209.90	167.94	117.58	140.79	107.32	149.22
1973	197.04	211.42	167.48	225.45	181.89	126.49	154.54	114.53	160.46
1974	219.64	238.97	185.62	250.30	204.39	139.92	172.25	126.08	178.09
1975	249.58	280.44	213.43	290.95	233.98	159.06	193.12	143.69	203.34
1976P	287.36	316.38	241.19	331.02	262.02	176.59	213.17	160.49	228.03
1975									
January	245.16	263.29	202.66	267.48	222.10	148.18	185.53	136.09	192.08
February	251.78	265.63	204.91	271.53	224.22	150.73	185.67	136.42	194.18
March	264.37	273.47	205.79	265.78	224.25	153.17	188.88	138.98	195.60
April	266.79	272.25	209.68	280.08	227.33	153.91	189.71	139.48	197.44
May	255.62	273.13	210.77	284.18	228.42	157.36	191.65	140.79	200.44
June	255.46	273.61	212.47	288.72	230.80	161.53	193.39	143.25	203.13
July	229.37	278.24	212.75	295.79	234.59	164.11	195.25	146.41	205.25
August	242.44	281.64	213.98	303.55	235.67	163.32	193.85	147.27	206.38
September	241.01	289.31	218.59	314.13	237.72	162.02	195.17	146.19	208.83
October	263.05	296.32	222.80	321.57	248.02	163.37	197.80	148.08	212.90
November	265.77	299.15	224.48	318.51	249.75	162.90	198.56	149.79	213.37
December	214.18	299.20	222.30	280.03	244.89	168.08	201.96	151.42	210.43

8.17   Annual average weekly earnings, by industrial division, 1972-76, and monthly averages 1975 and 1976 (dollars) (concluded)

Year and month	For-estry	Mining (incl. milling)	Manu-factur-ing	Con-struc-tion	Trans-portation, communi-cation and other utilities	Trade	Finance, insur-ance and real estate	Service[1]	Indus-trial com-posite
Averages									
1976P									
January	279.03	306.10	229.34	317.46	250.39	167.12	203.93	155.35	217.03
February	291.30	310.03	232.57	328.55	252.51	169.13	207.40	155.03	220.02
March	291.81	310.78	235.52	328.68	251.44	171.13	212.70	156.14	221.76
April	297.65	311.84	238.07	322.92	255.32	174.59	214.62	158.63	224.40
May	290.00	312.67	238.43	326.41	255.61	175.13	212.94	159.49	225.37
June	283.30	315.15	242.23	335.61	259.24	179.34	215.54	162.04	229.50
July	276.99	311.81	239.98	334.39	265.92	181.13	216.37	163.52	230.11
August	285.64	310.86	241.87	332.78	266.55	178.98	214.12	161.83	230.49
September	297.34	320.77	245.76	337.73	268.51	178.10	215.47	160.85	232.12
October	301.31	327.29	248.16	346.70	271.26	180.33	216.64	163.84	235.03
November	313.40	334.16	252.00	350.86	273.31	179.81	215.89	163.52	236.59
December	240.51	334.14	250.36	310.11	274.23	184.29	218.93	165.61	233.99

[1]Mainly hotels, restaurants, laundries, dry-cleaning establishments and recreational and business services.

8.18   Annual average weekly hours and hourly earnings of hourly rated wage-earners in specified industries, 1971-76, and monthly averages 1975 and 1976

Year and month	All manufactures		Mining (incl. milling)		Construction	
	Average weekly hours	Average hourly earnings $	Average weekly hours	Average hourly earnings $	Average weekly hours	Average hourly earnings $
Averages						
1971	39.7	3.28	40.4	4.04	39.2	4.75
1972	40.0	3.54	40.3	4.34	40.1	5.15
1973	39.6	3.85	40.9	4.82	39.5	5.66
1974	38.9	4.37	40.4	5.50	39.1	6.43
1975	38.6	5.06	40.0	6.51	39.0	7.53
1976P	38.7	5.76	40.3	7.40	38.9	8.68
1975						
January	38.7	4.77	40.6	5.99	38.0	7.04
February	38.7	4.83	40.8	6.02	38.3	7.12
March	38.3	4.91	40.6	6.27	36.8	7.18
April	38.8	4.96	40.0	6.25	39.0	7.24
May	38.6	5.01	39.6	6.38	39.3	7.29
June	38.5	5.07	39.2	6.44	39.8	7.33
July	38.1	5.07	39.3	6.57	40.5	7.43
August	38.4	5.07	39.4	6.63	40.3	7.67
September	39.0	5.14	39.7	6.78	40.7	7.89
October	39.1	5.24	40.5	6.88	40.9	8.04
November	39.0	5.28	40.6	6.95	39.8	8.19
December	37.7	5.34	40.1	7.01	35.1	7.88
1976P						
January	38.7	5.42	40.7	7.10	38.6	8.39
February	38.9	5.51	40.7	7.16	39.8	8.50
March	39.0	5.58	40.1	7.26	39.6	8.53
April	38.9	5.65	39.9	7.29	38.7	8.48
May	38.5	5.73	40.2	7.27	38.6	8.58
June	38.7	5.80	40.4	7.32	39.8	8.61
July	38.2	5.79	39.8	7.34	39.9	8.55
August	38.5	5.80	39.4	7.37	39.6	8.57
September	38.9	5.87	40.2	7.51	39.2	8.80
October	38.9	5.92	40.7	7.63	39.6	8.97
November	39.1	5.98	41.2	7.72	39.3	9.17
December	38.2	6.02	40.5	7.85	34.1	8.96

8.19 Average weekly hours and hourly earnings of hourly rated wage-earners in specified industries and selected urban areas, 1974-76

Industry, province and urban area	Average weekly hours			Average hourly earnings ($)		
	1974	1975	1976P	1974	1975	1976P
INDUSTRY						
Mining (incl. milling)	40.4	40.0	40.3	5.50	6.51	7.40
Metal mining	39.4	39.4	39.6	5.65	6.60	7.48
Coal mining	39.8	38.2	39.5	5.25	6.38	6.94
Manufacturing	38.9	38.6	38.7	4.37	5.06	5.76
Durable goods[1]	39.5	39.1	39.5	4.69	5.41	6.13
Non-durable goods[1]	38.3	38.0	37.9	4.03	4.68	5.36
Construction	39.1	39.0	38.9	6.43	7.53	8.68
Building	37.5	37.4	37.4	6.63	7.68	8.73
Engineering	42.3	42.0	41.6	6.05	7.26	8.60
Other						
Urban transit	41.6	40.5	40.7	5.21	6.01	6.84
Highway and bridge maintenance	40.8	41.0	40.6	4.06	4.78	5.49
Hotels, restaurants and taverns	28.4	27.5	27.1	2.65	3.08	3.51
Laundries, cleaners and pressers	34.7	34.5	33.7	2.60	3.03	3.48
PROVINCE						
Manufacturing						
Newfoundland	38.4	37.8	36.9	4.17	4.81	5.55
Nova Scotia	38.2	38.5	37.9	3.87	4.57	5.07
New Brunswick	38.2	38.2	38.5	3.88	4.65	5.28
Quebec	39.6	39.1	38.9	3.87	4.56	5.16
Ontario	39.2	38.9	39.3	4.54	5.18	5.87
Manitoba	37.6	37.3	37.2	3.95	4.62	5.17
Saskatchewan	37.8	37.0	36.9	4.49	5.39	6.13
Alberta	37.4	37.4	37.5	4.66	5.53	6.25
British Columbia	36.4	36.1	36.5	5.66	6.55	7.55
SELECTED URBAN AREA						
Manufacturing						
Montreal	39.1	38.8	38.4	3.89	4.58	5.14
Toronto	39.3	38.9	39.4	4.31	4.94	5.52
Hamilton	38.8	38.7	38.9	4.95	5.68	6.53
Windsor	40.9	40.6	41.4	5.64	6.23	7.06
Winnipeg	37.5	37.2	37.2	3.87	4.54	5.09
Vancouver	36.0	35.7	36.0	5.43	6.37	7.25

[1]Durable goods manufacturing includes wood products, furniture and fixtures, primary metal industries, metal fabricating industries, machinery (except electrical), transportation equipment, electrical products and non-metallic mineral products; non-durable goods manufacturing includes all other manufacturing industries.

8.20 Wages and salaries, by industry and supplementary labour income, 1971-76, and by month 1975 and 1976 (million dollars)

Year and month	Industry						
	Agriculture	Forestry	Mining	Manufacturing	Construction	Transportation, communication and other utilities	Trade
Annual							
1971	389	552	1,240	12,293	4,215	5,191	6,562
1972	407	586	1,292	13,581	4,581	5,718	7,414
1973	494	744	1,515	15,528	5,611	6,504	8,524
1974	557	901	1,861	18,125	6,730	7,863	10,193
1975	685	905	2,195	19,933	8,121	8,968	11,986
1976P	850	1,080	2,540	22,717	8,492	10,432	13,581
1975							
January	32.5	73.6	171.6	1,561.9	526.6	694.2	913.3
February	33.0	71.2	172.7	1,560.1	535.8	695.7	910.9
March	37.5	71.5	176.7	1,592.2	552.6	713.5	931.4
April	44.8	63.5	171.5	1,605.7	604.2	718.5	957.9
May	54.8	77.8	182.5	1,661.5	676.6	747.1	989.9
June	66.6	96.2	184.2	1,704.7	696.4	772.3	1,018.8
July	79.3	87.5	180.8	1,669.4	746.6	793.6	1,012.9
August	90.3	82.7	184.2	1,683.2	785.8	786.2	1,006.7
September	81.0	78.0	181.6	1,710.2	808.4	769.7	1,020.2
October	66.1	79.4	189.9	1,724.6	812.4	763.4	1,047.3
November	53.7	73.3	198.9	1,731.5	750.5	734.9	1,066.2
December	45.5	50.6	200.5	1,727.8	625.3	779.1	1,110.1

8.20   Wages and salaries, by industry and supplementary labour income, 1971-76, and by month 1975 and 1976 (million dollars) (concluded)

Year and month	Industry						
	Agricul-ture	Forestry	Mining	Manufac-turing	Construc-tion	Trans-portation, commu-nication and other utilities	Trade
Annual							
1976ᴾ							
January	41.2	62.6	201.0	1,733.8	611.8	783.7	1,060.1
February	41.3	67.4	203.9	1,782.5	624.3	787.4	1,058.6
March	46.7	66.0	203.4	1,838.3	629.0	808.3	1,082.8
April	55.3	62.5	199.1	1,862.0	657.2	826.1	1,114.5
May	67.5	80.9	205.5	1,901.1	720.3	840.8	1,123.3
June	81.9	100.8	208.5	1,947.5	788.2	875.6	1,148.9
July	97.4	107.9	215.0	1,925.1	759.8	930.3	1,131.9
August	111.0	112.0	214.2	1,935.0	766.6	897.0	1,122.1
September	104.4	116.2	215.9	1,944.6	713.5	938.2	1,133.8
October	81.3	109.6	219.7	1,933.8	779.1	896.1	1,166.6
November	66.1	103.0	224.6	1,953.2	768.1	899.7	1,194.1
December	56.3	91.5	229.0	1,958.5	673.9	948.1	1,243.1

Year and month	Finance, insurance and real estate	Service	Public administra-tion and defence[1]	Total wages and salaries[2]	Supple-mentary labour income	Total labour income
Annual						
1971	2,579	11,576	3,944	48,591	2,938	51,528
1972	3,037	12,903	4,493	54,070	3,500	57,570
1973	3,747	14,742	5,108	62,595	4,162	66,757
1974	4,513	17,622	6,281	74,717	5,369	80,085
1975	5,336	21,088	7,689	86,971	6,591	93,562
1976ᴾ	6,166	24,808	8,918	99,669	7,943	107,612
1975						
January	412.5	1,623.4	574.7	6,587.5	502.2	7,089.7
February	414.7	1,639.2	577.9	6,613.0	503.3	7,116.3
March	427.4	1,663.5	588.6	6,759.3	513.6	7,272.9
April	428.9	1,695.3	592.7	6,886.1	521.8	7,407.9
May	440.1	1,747.8	627.0	7,211.0	546.2	7,757.2
June	449.0	1,774.3	664.1	7,434.3	562.4	7,996.6
July	458.4	1,767.5	691.2	7,496.1	568.0	8,064.1
August	451.4	1,622.6	665.4	7,365.2	556.4	7,921.7
September	457.2	1,941.3	678.2	7,732.8	585.3	8,318.2
October	461.6	1,867.8	649.2	7,666.2	580.5	8,246.7
November	462.7	1,870.1	669.1	7,615.2	575.6	8,190.7
December	471.9	1,874.5	711.1	7,604.1	575.2	8,179.2
1976ᴾ						
January	475.3	1,914.2	662.4	7,548.0	602.8	8,150.7
February	484.4	1,933.3	681.0	7,666.9	612.2	8,279.1
March	502.1	1,939.1	681.8	7,805.3	621.6	8,426.9
April	506.5	1,943.3	703.0	7,933.8	631.8	8,565.6
May	508.2	2,239.8	727.1	8,425.0	673.9	9,098.9
June	521.5	2,201.4	767.8	8,651.7	691.7	9,343.3
July	524.8	1,879.4	787.8	8,372.4	667.2	9,039.6
August	519.3	1,884.2	775.6	8,345.7	665.3	9,011.0
September	525.4	2,100.6	770.0	8,572.4	681.7	9,254.2
October	530.1	2,309.8	776.8	8,810.3	700.9	9,511.2
November	529.9	2,161.8	765.1	8,669.3	690.8	9,360.1
December	538.4	2,299.5	819.4	8,862.3	702.9	9,565.2

Table based on the 1960 Standard Industrial Classification. Figure not adjusted for seasonality.
[1]Excludes military pay and allowances.
[2]Includes fishing and trapping.

8.21 Summary of selected working conditions of non-office and office employees in manufacturing and all industries, 1975 and 1976

Item		1975		1976	
		Manu-facturing industries	All industries[1]	Manu-facturing industries	All industries[1]
		Coverage			
NON-OFFICE EMPLOYEES					
Reporting establishments	*No.*	6,302	15,937ʳ	6,204	15,956
Employees	"	776,924	2,151,622	771,899	2,198,975
OFFICE EMPLOYEES					
Reporting establishments	*No.*	6,245	17,281	6,165	17,497
Employees	"	258,701	1,165,679	262,669	1,252,230
		Percentage of non-office employees			
STANDARD WEEKLY HOURS					
35 hours or less		3	3	3	3
Over 35 and under 37½ hours		2	4	1	4
37½ hours		3	9	3	10
Over 37½ and under 40 hours		1	6	1	7
40 hours		80	65	81	64
Over 40 and under 44 hours		5	3	5	3
44 hours and over		6	5	5	5
VACATIONS WITH PAY[2]					
Three weeks		93	89	94	90
After: 1 year or less		2	17	3	14
2 years		1	4	1	4
3 years		10	10	11	15
4 years		3	6	4	7
5 years		45	35ʳ	48	35
6-9 years		22	11	18	9
10 years		8	5	7	4
11 years or more		1	1	1	—
Four weeks		82	83	84	85
After: 9 years or less		3	12	4	14
10 years		11	15ʳ	12	18
11-14 years		13	18	19	19
15 years		37	28	36	26
16-19 years		10	5	7	3
20 years		7	4	6	3
21 years or more		1	1	—	—
Five weeks		56	54ʳ	59	58
After: 24 years or less		31	31	36	36
25 years		23	18ʳ	21	15
30 years		—	6ʳ	1	1
Six weeks[3]		20	18	23	21
After: 24 years or less		2	4	3	4
25 years		2	2	3	3
30 years		7	7ʳ	9	7
31 years or more		1	1ʳ	1	—
Paid holidays (statutory, public, etc.)					
8 days or less		11	11	9	9
9 days		18	13	15	11
10 days		31	34	28	32
11 days		22	26	27	32
12 days or more		17	14	19	14
		Percentage of office employees			
STANDARD WEEKLY HOURS					
Under 35 hours		1	6	2	3
35 hours		22	23	22	27
Over 35 and under 37½ hours		11	12	9	15
37½ hours		40	43	41	40
Over 37½ and under 40 hours		4	1	4	1
40 hours and over		22	13	23	13
VACATIONS WITH PAY[2]					
Three weeks		97	97	97	98
After: 1 years or less		6	28	6	31
2 years		2	8	2	10
3 years		12	11	13	15
4 years		3	5	3	5
5 years		47	35	52	30
6-9 years		21	6	17	5
10 years		6	3	5	2
Four weeks		90	92	91	93
After: 9 years or less		4	6	5	9
10 years		18	22	19	24
11-14 years		12	14	16	16
15 years		40	43	38	35
16-19 years		9	3	7	2
20 years or more		7	4	5	6

8.21   Summary of selected working conditions of non-office and office employees in manufacturing and all industries, 1975 and 1976 (concluded)

Item	1975		1976	
	Manu-facturing industries	All industries[1]	Manu-facturing industries	All industries[1]
	Percentage of office employees (concluded)			
VACATIONS WITH PAY[2] (concluded)				
Five weeks[3]	65	59	67	67
After: 24 years or less	38	29	41	34
25 years	24	15	24	19
30 years	—	14	1	1
Six weeks[3]	19	13	21	14
After: 24 years or less	2	1	2	3
25 years	3	3	3	3
30 years	9	5	11	5
31 years or more	2	1	2	1
Paid holidays (statutory, public, etc.)				
8 days or less	6	5	6	4
9 days	17	12	14	10
10 days	32	35	29	32
11 days	28	33	31	39
12 days	6	5	9	5
13 days or more	10	9	11	10

[1]Includes all major industries except agriculture, fishing, hunting, trapping, construction, and the non-logging part of forestry.
[2]Legislation in all jurisdictions in Canada entitle employees to 2 weeks annual vacation with pay, generally after 1 year of employment.
[3]Includes other provisions.

8.22   Trusteed pension funds, income, expenditures and assets, 1974-76

Item	1974	1975	1976
TRUST ARRANGEMENTS	*No.*	*No.*	*No.*
(a) Corporate trustees	2,821	2,776	2,705
(b) Individual trustees	736	736	731
(c) Combinations of (a) and (b)	95	84	82
(d) Pension fund societies	28	26	25
Total trusteed funds	3,680	3,622	3,543
INCOME	*$'000,000*	*$'000,000*	*$'000,000*
Total contributions	2,128	2,727	3,392
Employer	1,417	1,873	2,259
Employee	711	854	1,133
Investment income	1,164	1,326	1,639
Net profit on sale of securities	49	45	59
Other	20	12	14
Total income	3,361	4,110	5,104
EXPENDITURES			
Pension payments out of funds	754	900	1,037
Cost of pension purchased	26	29	35
Cash withdrawals	201	175	248
Administration costs	22	27	32
Net loss on sale of securities	99	122	93
Other expenditures	68	37	9
Total expenditures	1,170	1,290	1,454
ASSETS (book value)			
Investment in pooled funds	1,196	1,206	1,461
Investment in mutual funds	41	30	27
Investment in segregated funds of insurance companies	200	248	355
Bonds	8,537	10,146	11,863
Bonds of or guaranteed by Government of Canada	366	405	620
Bonds of or guaranteed by provincial governments	4,550	5,310	6,380
Bonds of Canadian municipal governments, school boards, etc.	797	876	966
Other Canadian	2,813	3,539	3,887
Non-Canadian	11	16	10
Stocks	4,773	5,313	6,213
Canadian, common	4,069	4,504	5,239
Canadian, preferred	96	99	79
Non-Canadian, common	601	704	894
Non-Canadian, preferred	7	6	1
Mortgages	1,936	2,471	3,349
Insured residential (NHA)	1,072	1,345	1,961
Conventional	864	1,126	1,388
Real estate and lease-backs	53	96	144
Miscellaneous			
Cash on hand and in chartered banks	302	391	398
Guaranteed investment certificates	126	133	172
Short-term investments	755	872	790
Accrued interest and dividends receivable	169	178	225
Accounts receivable	194	123	232
Other assets	2	3	5
Total assets	18,284	21,210	25,234

8.23   Unemployment insurance statistics, 1972-76, and by month 1976

Year, month and end of period	Activity				
	Insured population[1]	Claims data ('000)		Benefits data	
	'000	Claimants for UIC benefits (end of period)[1,2]	Initial and renewal claims received	Number of weeks '000	Average weekly payment $
1972	7,845	804	2,470	30,462	61.79
1973	8,264	828	2,238	29,537	68.45
1974	8,617	828	2,410	28,461	74.89
1975	8,951	1,049	2,857	37,327	84.64
1976	9,249	1,006	2,678	36,189	92.89
1976					
January	8,950	1,179	277	3,746	91.63
February	8,997	1,196	189	3,713	92.52
March	9,036	1,173	201	4,295	92.88
April	9,060	1,103	193	3,409	92.37
May	9,243	1,020	184	3,082	91.77
June	9,464	902	195	2,797	90.87
July	9,670	892	212	2,408	90.92
August	9,630	860	171	2,682	91.73
September	9,245	849	231	2,345	93.20
October	9,261	875	226	2,190	94.56
November	9,225	956	292	2,735	95.47
December	9,209	1,068	304	2,787	97.65

	Benefits data (concluded)					
	Benefits paid ($'000)					
	Regular	Sickness	Maternity	Retirement	Fishing	Total[3]
1972	1,764,031	58,854	36,431	2,440	20,403	1,871,802
1973	1,850,930	80,179	66,750	3,691	20,297	2,004,212
1974	1,924,543	98,321	81,708	4,164	22,675	2,119,213
1975	2,907,715	110,990	102,161	5,836	23,622	3,144,022
1976	3,019,686	129,804	139,624	18,048	28,881	3,342,246
1976						
January	314,180	9,761	9,713	1,842	5,349	341,825
February	313,263	9,761	9,047	2,474	6,178	341,909
March	364,153	11,717	11,432	2,272	6,142	397,096
April	287,336	9,981	10,635	1,554	2,767	313,676
May	257,729	9,458	11,073	1,411	966	281,135
June	229,328	9,497	11,790	1,273	76	252,549
July	193,672	10,692	11,913	1,395	20	217,053
August	218,778	11,822	12,876	1,694	11	244,561
September	191,721	11,922	12,595	1,191	9	216,856
October	180,498	10,953	12,682	1,030	15	205,288
November	230,494	12,338	13,819	1,057	1,098	259,450
December	238,534	11,900	12,049	855	6,250	270,848

[1] Annual figures are annual averages.
[2] Persons who have applied for or are in receipt of unemployment insurance benefits at end of month.
[3] Figures are adjusted for cancellation of warrants and collection of overpayments; total includes ordinary, seasonal and fishing benefits.

8.24   Fatal occupational injuries and illnesses[1], by industry, 1974-76

Industry	Number			Percentage of total		
	1974[r]	1975[r]	1976	1974[r]	1975[r]	1976
Agriculture	33	13	16	2.3	1.1	1.7
Forestry	85	71	58	6.0	6.1	6.3
Fishing and trapping	11	27	26	0.8	2.3	2.8
Mining, quarrying and oil wells	203	158	143	14.4	13.5	15.4
Manufacturing	305	222	161	21.6	19.0	17.4
Construction	232	217	167	16.4	18.6	18.0
Transportation, communication and other utilities	254	216	197	18.0	18.5	21.3
Trade	119	74	52	8.4	6.3	5.6
Finance, insurance and real estate	7	3	7	0.5	0.3	0.8
Service	101	83	52	7.1	7.1	5.6
Public administration	63	84	47	4.5	7.2	5.1
Total	1,413	1,168	926	100.0	100.0	100.0

[1] The Canada Department of Labour compiles statistics of all fatal industrial accidents; worker compensation statistics (Table 8.25) include only those accidents covered by legislation.

8.25   Compensation claims and payments made, for occupational injuries and illnesses, 1975 and 1976

Year and province	Compensation claims					Worker compensation payments[2] $'000
	Medical aid only[1]	Non-fatal disabling injury and illness	Fatal injury and illness	Total disabling injury and illness	Total injuries and illnesses	
1975						
Newfoundland	5,700	6,260	30	6,290	11,990	7,127
Prince Edward Island	1,374	1,453	3	1,456	2,830	1,079
Nova Scotia	18,746	11,850	24	11,874	30,620	18,111
New Brunswick	18,381	8,576	48	8,624	27,005	10,541
Quebec	136,505	147,109	241	147,350	283,855	184,359
Ontario	223,954	143,762	221	143,983	367,937	274,477
Manitoba	17,693	18,440	41	18,481	36,174	15,907
Saskatchewan	15,432	14,145	45	14,190	29,622	18,642
Alberta	48,794	34,576	124	34,700	83,494	44,760
British Columbia	57,326	54,284	180	54,464	111,790	93,444
Total, 1975	543,905	440,455	957	441,412	985,317	668,447
1976P						
Newfoundland	7,298	6,578	42	6,620	13,918	7,513
Prince Edward Island	1,352	1,386	2	1,388	2,740	1,582
Nova Scotia	17,410	11,524	35	11,559	28,969	21,153
New Brunswick	16,685	9,515	34	9,549	26,234	13,486
Quebec	126,074	156,344	266	156,610	282,684	218,101
Ontario	243,460	154,317	203	154,520	397,980	336,506
Manitoba	16,760	18,871	35	18,906	35,666	18,642
Saskatchewan	19,103	15,519	40	15,559	34,662	21,050
Alberta	58,695	41,363	125	41,488	100,183	56,237
British Columbia	63,644	56,110	150	56,260	119,904	100,729
Total, 1976P	570,481	471,527	932	472,459	1,042,940	794,999

[1]Injuries requiring medical treatment but not causing disability for a sufficient period to qualify for compensation; the period varies among provinces.
[2]Includes, except where noted otherwise, payments to compensate loss of earnings, medical aid payments, cost of rehabilitation and hospitalization (not including capital expenditures) and pensions paid (not pensions awarded) for temporary and permanent disabilities.

8.26   Union membership in Canada, 1968-77

Year	Members '000	Union membership as percentage of civilian labour force	Union membership as percentage of non-agricultural paid workers
1968	2,010	26.6	33.1
1969	2,075	26.3	32.5
1970	2,173	27.2	33.6
1971	2,231	26.8	33.6
1972	2,388	27.8	34.6
1973	2,591	29.2	36.1
1974	2,732	29.4	35.8
1975	2,884	29.8	36.9
1976	3,042	30.6	37.3
1977	3,149	31.0	38.2

8.27   Union membership, by type of union and affiliation, as at January 1976 and 1977

Year, type and affiliation	Unions No.	Membership	
		No.	%
1976			
International unions	90	1,508,078	49.6
AFL-CIO/CLC	72	1,266,179	41.7
CLC only	5	152,824	5.0
AFL-CIO only	6	3,852	0.1
Unaffiliated unions	7	85,223	2.8
National unions	106	1,424,965	46.8
CLC	33	734,196	24.1
CNTU	9	165,685	5.4
CSD	3	26,101	0.9
CCU	9	21,248	0.7
Unaffiliated unions	52	477,735	15.7
Directly chartered local unions	244	29,460	1.0
CLC	97	14,728	0.5
CNTU	4	662	[1]
CSD	143	14,070	0.5
Independent local organizations	139	79,769	2.6
Total	579	3,042,272	100.0

8.27 Union membership, by type of union and affiliation, as at January 1976 and 1977 (concluded)

Year, type and affiliation	Unions No.	Membership No.	%
1977			
International unions	89	1,544,717	49.0
AFL-CIO/CLC	71	1,278,834	40.6
CLC only	5	166,359	5.3
AFL-CIO only	6	3,857	0.1
Unaffiliated unions	7	95,667	3.0
National unions	103	1,492,093	47.4
CLC	25	702,046	22.3
CNTU	9	172,201	5.5
CSD	3	25,593	0.8
CCU	9	20,822	0.7
Unaffiliated unions	57	571,431	18.1
Directly chartered local unions	259	30,883	1.0
CLC	113	16,300	0.5
CNTU	3	513	[1]
CSD	143	14,070	0.5
Independent local organizations	141	81,520	2.6
Total	592	3,149,213	100.0

[1]Less than 0.1%.

8.28 Average annual compound percentage increases in base rates, in major collective agreement settlements, 1975 and 1976

Year and quarter	Manufacturing	Non-manufacturing[1]	All industries[1]
1975			
1st quarter	15.0	18.6	18.3
2nd ``	15.7	19.3	18.8
3rd ``	12.6	19.0	17.1
4th ``	14.0	14.7	14.5
Annual	13.9	17.8	17.0
1976			
1st quarter	11.8	15.3	14.2
2nd ``	9.7	11.2	11.0
3rd ``	9.5	9.7	9.7
4th ``	6.1	9.2	8.4
Annual	8.9	10.8	10.5

[1]Excluding construction.

8.29 Year-over-year percentage and cents-per-hour increases in base rates under major collective agreements, by month, 1975 and 1976

Year and month	Collective agreement[1] Manufacturing %	¢	Non-manufacturing[2] %	¢	All industries[2] %	¢
1975						
January	15.5	41.0	15.7	57.4	14.7	54.1
February	10.2	39.9	15.0	55.4	14.0	52.1
March	9.9	39.3	15.3	56.7	14.2	53.0
April	9.6	38.2	14.7	55.8	13.7	52.1
May	9.7	39.1	14.6	55.9	13.6	52.3
June	10.1	40.7	15.7	60.1	14.4	55.6
July	9.9	40.0	14.6	57.2	13.4	52.8
August	9.0	37.4	14.8	58.4	13.5	53.6
September	9.0	36.9	14.7	58.8	13.3	53.6
October	8.8	37.3	14.8	59.6	13.3	54.1
November	10.7	45.6	14.8	60.0	13.6	55.5
December	10.5	45.1	15.3	62.4	13.9	57.0
1976						
January	10.4	45.1	13.0	55.1	12.4	52.5
February	10.6	45.9	13.1	55.8	12.7	54.0
March	10.6	46.3	13.4	57.4	13.0	55.5
April	10.5	45.9	13.6	59.3	13.1	56.7
May	10.5	46.1	13.7	59.9	13.2	57.7
June	11.8	52.1	13.1	58.1	13.0	57.2
July	12.1	53.8	13.8	61.9	13.7	61.2
August	10.9	49.2	14.2	64.6	13.9	62.5
September	10.1	46.3	15.0	68.6	14.3	65.2
October	9.8	45.4	15.9	73.8	15.0	69.2
November	7.9	37.6	16.1	74.7	14.9	68.8
December	9.0	42.9	15.3	72.0	14.4	67.5

[1]Based on all major collective agreements covering 500 or more employees in force except those in the construction industry. In the case of a retroactive increase, the increase is shown in the month of the settlement rather than in the month to which it was retroactive. [2]Excluding construction.

8.30 Strikes and lockouts, by industry and jurisdiction, 1976 with totals for 1972-76

Industry and jurisdiction	Strikes and lockouts beginning during year	Strikes and lockouts in existence during year		
		Strikes and lockouts	Workers involved	Duration in man-days
INDUSTRY				
Agriculture	—	—	—	—
Forestry	3	4	784	36,320
Fishing and trapping	1	1	350	350
Mining	45	49	24,930	579,430
Manufacturing	374	457	166,534	4,493,260
Construction	75	76	135,668	2,856,370
Transportation and utilities	137	142	52,065	622,630
Trade	82	93	8,518	199,550
Finance	6	8	168	13,110
Service	137	147	148,840	1,298,490
Public adminis- tration	58	59	22,883	62,680
Various industries[1]	3	3	1,010,200	1,447,700
JURISDICTION				
Newfoundland	46	51	8,110	130,560
Prince Edward Island	4	4	515	8,030
Nova Scotia	22	25	7,550	196,680
New Brunswick	34	41	17,232	242,640
Quebec	314	357	448,542	6,465,650
Ontario	239	279	109,504	1,671,090
Manitoba	25	30	8,935	98,190
Saskatchewan	57	57	20,239	139,900
Alberta	26	27	7,532	106,910
British Columbia	81	93	83,149	1,490,680
Federal public service	12	12	9,097	12,490
Federal industries	60	62	20,535	217,070
"Day of protest" (Oct. 14)	1	1	830,000	830,000
Total 1976	921	1,039	1,570,940	11,609,890
1975	1,103	1,171	506,443	10,908,810
1974	1,170	1,216	592,220	9,255,120
1973	677	724	348,470	5,776,080
1972	556	598	706,474	7,753,530

[1]Includes multi-industry strikes and lockouts and the "day of protest" on October 14, 1976.

8.31 Estimated composition of total employee compensation, in Canada, 1976

Item	All industry			
	Average expenditure per employee $	Percentage of total compensation	Percentage of gross payroll	Composition of total compensation[1] $
Direct payments to employees				
Pay for time worked				
Basic pay for regular work	10,855	75.5	82.1	100.00
Commissions, incentive bonus	252	1.7	1.9	2.33
Overtime (straight-time pay)	271	1.9	2.1	2.50
Premium pay				
Overtime, including holiday work	139	1.0	1.1	1.28
Shift work	40	0.3	0.3	0.37
Other	38	0.3	0.3	0.35
Total, pay for hours worked	11,595	80.6	87.7	106.82
Paid absence				
Paid holidays	489	3.4	3.7	4.50
Vacation pay	715	5.0	5.4	6.59
Sick leave pay	142	1.0	1.1	1.31
Personal or other pay	23	0.2	0.2	0.21
Total, paid absence	1,369	9.5	10.4	12.61
Miscellaneous direct payments				
Bonuses—Christmas, year- end, etc.	57	0.4	0.4	0.53
Severance pay	23	0.2	0.2	0.21
Taxable benefits				
Provincial medicare	71	0.5	0.5	0.65
Other benefits	56	0.4	0.4	0.52
Other payments	50	0.4	0.4	0.46
Total, miscellaneous direct payments	257	1.8	1.9	2.37
Total (gross payroll)	13,221	91.9	100.0	121.80

8.31 Estimated composition of total employee compensation, in Canada, 1976 (concluded)

Item	All industry			
	Average expenditure per employee $	Percentage of total compensation	Percentage of gross payroll	Composition of total compensation[1] $
Employer contributions to employee				
Welfare and benefit plans				
Worker compensation	157	1.1	1.2	1.45
Unemployment insurance	175	1.2	1.3	1.61
Canada or Quebec pension plan	130	0.9	1.0	1.20
Private pension plans	489	3.4	3.7	4.50
Quebec Health Insurance Board	38	0.3	0.3	0.35
Private life and health plans	149	1.0	1.1	1.37
Supplementary unemployment benefit	4	—	—	0.04
Other plans or funds	20	0.1	0.2	0.18
Total, employer contributions	1,162	8.0	8.8	10.70
Total, compensation	14,383	100.0	108.8	132.50

[1]Basic pay for regular work $100.00.

8.32   Average wage and salary rates in selected occupations for certain metropolitan areas and cities, and for Canada, Oct. 1, 1975 and 1976[1]

Occupation	Halifax-Dartmouth, NS		Saint John, NB		Montreal, Que.		Ottawa-Hull Ont., Que.		Toronto, Ont.	
	1975	1976	1975	1976	1975	1976	1975	1976	1975	1976
MAINTENANCE TRADES	$ an hr	$ an hr	$ an hr	$ an hr	$ an hr	$ an hr	$ an hr	$ an hr	$ an hr	$ an hr
Carpenter	5.52	6.21	5.14	6.40	5.43	6.19	5.90	6.35	5.91	6.41
Electrician	6.01	6.78	6.40	7.63	5.99	6.75	6.64	7.32	6.47	7.15
Machinist	6.01	6.34	5.81	6.83	5.86	6.59	6.30	7.01	6.21	6.74
Millwright	5.55	6.63	6.46	7.95	5.79	6.57	6.24	7.24	6.27	6.99
Pipefitter	6.40	7.01	6.46	7.89	6.15	6.96	6.40	7.36	6.59	7.08
Tool and die maker	5.51	—	5.39	—	5.80	6.45	6.28	6.01	6.40	6.97
Welder	5.85	6.50	6.46	7.61	5.86	6.57	5.87	7.01	6.00	6.55
SERVICE OCCUPATIONS										
Truck driver, light and heavy	4.37	4.96	4.72	5.30	5.03	5.69	5.13	5.88	5.31	5.76
Trucker, power	5.49	5.74	4.84	5.70	5.18	5.64	5.10	5.92	5.16	5.66
General labourer	4.32	4.75	4.52	5.06	4.34	4.95	4.51	4.69	4.53	5.20
OFFICE OCCUPATIONS, MALE	$ a wk	$ a wk	$ a wk	$ a wk	$ a wk	$ a wk	$ a wk	$ a wk	$ a wk	$ a wk
Bookkeeper, senior	204	235	223	234	224	247	225	252	233	255
Clerk, general office, intermediate	156	176	181	196	165	190	158	176	171	188
Clerk, general office, senior	196	215	216	246	212	238	194	215	216	238
Clerk, order	158	195	193	193	186	208	200	218	196	215
Draughtsman, intermediate	216	235	254	290	215	236	220	247	231	253
Draughtsman, senior	251	285	292	338	268	303	259	286	289	317
OFFICE OCCUPATIONS, FEMALE										
Clerk, general office, intermediate	144	160	142	158	147	170	156	174	156	173
Bookkeeping, billing and calculating machine operator, senior	144	162	128	153	130	167	160	172	148	174
Secretary, senior	166	180	171	190	176	202	187	197	136	205
Stenographer, junior	132	145	128	151	133	153	133	151	146	162
Stenographer, senior	149	166	140	165	150	176	157	174	163	182
Telephone operator	120	136	127	146	130	152	141	155	144	157
Typist, junior	116	129	118	137	118	138	128	143	132	145
Typist, senior	132	147	123	146	133	155	135	150	147	162

8.32   Average wage and salary rates in selected occupations for certain metropolitan areas and cities, and for Canada, Oct. 1, 1975 and 1976¹ (concluded)

Occupation	Winnipeg, Man.		Regina, Sask.		Edmonton, Alta.		Vancouver, BC		Canada	
	1975	1976	1975	1976	1975	1976	1975	1976	1975	1976
	$ an hr	$ an hr	$ an hr	$ an hr	$ an hr	$ an hr	$ an hr	$ an hr	$ an hr	$ an hr
MAINTENANCE TRADES										
Carpenter	5.47	6.36	5.58	6.88	6.38	6.83	7.44	8.59	5.03	6.66
Electrician	6.22	6.85	6.64	7.65	7.42	7.57	7.67	8.85	6.52	7.32
Machinist	5.99	6.69	6.70	7.60	6.55	7.67	7.47	8.63	6.13	6.90
Millwright	6.10	6.66	6.44	7.08	7.13	7.90	7.52	8.75	6.38	7.21
Pipefitter	6.12	6.76	6.31	8.06	7.32	7.69	7.53	8.44	6.47	7.32
Tool and die maker	5.93	6.59	—	—	7.60	7.10	7.66	8.34	6.28	6.97
Welder	5.97	6.59	6.00	6.90	6.85	7.67	7.37	8.68	6.29	7.15
SERVICE OCCUPATIONS										
Truck driver, light and heavy	4.94	5.34	4.85	5.43	5.33	5.89	6.53	7.12	5.18	5.77
Trucker, power	4.75	5.09	4.55	5.34	5.18	6.03	6.73	7.64	5.33	6.01
General labourer	4.53	4.98	3.77	5.00	5.23	5.57	5.67	6.42	4.63	5.27
	$ a wk	$ a wk	$ a wk	$ a wk	$ a wk	$ a wk	$ a wk	$ a wk	$ a wk	$ a wk
OFFICE OCCUPATIONS, MALE										
Bookkeeper, senior	208	251	215	255	244	282	255	286	226	251
Clerk, general office, intermediate	167	186	158	175	189	200	185	198	169	191
Clerk, general office, senior	204	231	203	210	215	240	241	258	212	237
Clerk, order	172	193	191	215	192	213	222	241	193	216
Draughtsman, intermediate	217	235	215	253	245	271	246	274	224	249
Draughtsman, senior	277	293	260	269	282	299	297	328	273	302
OFFICE OCCUPATIONS, FEMALE										
Clerk, general office, intermediate	145	161	143	165	161	178	167	188	153	174
Bookkeeping, billing and calculating machine operator, senior	129	157	139	184	144	190	163	192	155	172
Secretary, senior	171	207	177	204	189	206	200	222	181	201
Stenographer, junior	136	151	131	155	149	167	155	180	137	160
Stenographer, senior	158	181	147	176	181	198	177	199	158	182
Telephone operator	128	142	127	147	145	155	157	172	137	154
Typist, junior	122	135	113	144	141	158	142	159	129	146
Typist, senior	143	160	138	170	166	181	157	179	144	164

¹The rates cover all major industries except agriculture, fishing, hunting, trapping, construction, and the non-logging part of forestry.

Sources

8.1 - 8.5 Labour Force Survey Division, Census and Household Surveys Field, Statistics Canada.
8.6 - 8.10 Census Characteristics Division, Census and Household Surveys Field, Statistics Canada.
8.11 - 8.20 Labour Division, General Statistics, Statistics Canada.
8.21 Labour Data Branch, Canada Department of Labour.
8.22 - 8.23 Labour Division, General Statistics, Statistics Canada.
8.24 - 8.25 Occupational Safety and Health Branch, Canada Department of Labour.
8.26 - 8.30 Labour Data Branch, Canada Department of Labour.
8.31 Labour Division, General Statistics, Statistics Canada.
8.32 Labour Data Branch, Canada Department of Labour.

Scientific research and development

Chapter 9

Tables

Scientific research and development

Chapter 9

Science in Canada 9.1

Development of Canada's natural resources and industry has involved the federal government in scientific activities since the establishment of Canada in 1867. The two main areas of investigation have been the natural sciences, including engineering, and the social sciences, including the humanities.

Natural sciences 1977-78 9.1.1

In such natural sciences as biology, chemistry, physics, astronomy, geology and oceanography, data are collected on the expenditures and manpower devoted to research and development (R&D) and related scientific activities (RSA). Although research and development is the central element, related scientific activities precede, complement and extend this work. RSA includes data collection, information testing and standardization, feasibility studies and educational support.

Federal expenditures in the natural sciences were expected to reach $1,374.5 million in 1977-78, an increase of 11% over 1976-77, and representing a little more than 3% of the total 1977-78 budget. R&D accounted for 71% of this, 56% for intramural work done in federal establishments and laboratories. RSA expenditures for 1977-78 were estimated at $393.2 million. Of this, scientific data collection represented 53%, scientific information 15%, feasibility studies 14%, testing and standardization 6% and education support 5%.

Three federal government sectors were expected to account for approximately half the natural science expenditures for 1977-78: the Department of Fisheries and the Environment $313.8 million; National Research Council of Canada $276.5 million; and the Department of Agriculture $130.3 million.

Federal support of social sciences 9.1.2

Social sciences include all disciplines involving the study of human actions and conditions, except the performing arts, and the social, economic and institutional mechanisms affecting them as well as such applied social science fields as anthropology, economics, human geography, business administration, communications, criminology and industrial relations.

Federal expenditures on the social sciences were expected to reach $461.0 million in 1977-78, an increase of 10% over 1976-77. R&D accounted for $147.8 million and RSA $313.0 million. It was anticipated that the federal government would perform 72% of the scientific activities. Universities and non-profit institutions would receive 13% of total expenditures, foreign performers 6% and business enterprises 4%.

Three sectors of the federal government were expected to account for about 45% of social science expenditures for 1977-78: Statistics Canada $148.0 million; Canada Council $31.6 million; and Department of National Health and Welfare $27.4 million. Social science expenditures were 1.1% of the total federal budget (1.1% in 1976-77).

Government, business and university sectors 9.1.3

The federal government is a major source of funds for scientific activities carried out in universities and business enterprises. It is also responsible for major programs in space, nuclear energy, natural resource identification and development, agriculture and economic and social data collection.

The business enterprise sector uses science to develop new products or processes for industry. The university sector trains manpower for all three sectors and carries out fundamental research which may have no immediate application in the other two.

Main federal government concerns centre on energy shortages, development of ocean and northern resources, environmental changes and the productivity of Canadian industry. In federal spending, natural sciences were expected to account for 75% of total expenditures on science in 1977-78. In the natural sciences in 1977-78, 71% of spending was earmarked for R&D while only 30% of the social science expenditures was designated for this activity.

Real growth of scientific activities is undoubtedly less than would appear, since expenditure data do not show the effects of inflation. No completely acceptable method of deflating scientific expenditures has yet been devised. If expenditures are deflated by the implicit price index of the GNE, 1976-77 expenditures were as follows: total, natural sciences, $1,260.9 million in current dollars ($786.1 million in 1971 dollars); total, social sciences, $439.6 million ($274.1 million).

9.1.4   Science policy

In 1966 the federal government established the Science Council of Canada, a Crown corporation charged with assessing independently Canada's scientific and technological resources, requirements and potential and making recommendations by publication of reports. The council is concerned with both R&D and with application of science and technology to social and economic problems. It draws its membership from industry, universities and government, and its views are independent of those of the internal government structure.

The council has published several reports of its own and others based on commissioned studies from consultants. Topics include energy conservation, technology transfer in construction and a case study of offshore petroleum exploration. The council recommended that Canada focus its scientific and technological effort through major programs designed to help solve social and economic problems. These include a space program, water resources management and development, transportation, urban development, computer applications and scientific and technological aid to developing countries.

In 1967, a special Senate committee on science policy was formed to consider and report on the scientific policy of the federal government. The first report, published in 1970, describes what the committee considered major deficiencies in the policy. The second, in 1972, contains specific recommendations on targets and strategies for the 1970s. The third volume recommended changes in federal structures concerned with science and technology. In 1977 the committee published its fourth and final report, concerned mainly with changes in science policy since publication of the first report and with Canadian activities in futures research, surveying human activity as it may develop in the medium- and long-term future.

9.2   Federal agencies

Information on federal government expenditures on scientific activities is secured by annual surveys carried out co-operatively by the Ministry of State for Science and Technology, the Treasury Board Secretariat and Statistics Canada. Each survey covers the costs of scientific programs for the preceding fiscal year and estimates for the following two years. (*Science statistics bulletin,* Vol. 2, No. 1, Statistics Canada Catalogue 13-003.)

Twenty-seven departments and agencies reported natural science expenditures with eight planning to spend over $50 million in 1977-78. The major funder of natural science R&D was the National Research Council with 25% of current expenditures. The Department of Fisheries and the Environment was the major funder of RSA with 44%.

Table 9.5 shows the expenditures for natural science activities by department or agency and performer. In 1977-78 approximately 61%, or $844.1 million, of federal government expenditures in the natural sciences was for work done within its own establishments. An estimated full-time equivalent of 23,342 federal employees was engaged in natural science activities, with 15,327 in R&D work.

Most payments to extramural performers for natural science activities go to Canadian industry (53% of 1977-78 extramural expenditures) and to Canadian universities (35%). Support of industrial R&D is provided principally through special programs designed to develop a research capacity in Canadian industry. Support of R&D in Canadian universities and related institutions is provided primarily through grants programs. Expenditures for natural science activities are shown in Tables 9.3, 9.5 and 9.6, and for social science activities in Tables 9.7, 9.8, 9.10 and 9.11.

Most federally-supported social science activities (72%) are performed intra-murally. Data collection is the major activity ($164.5 million planned for 1977-78) followed by R&D ($127.3 million). Fifty departments and agencies devote resources to social science activities, 16 with annual expenditures of over $5 million.

Three areas of the federal government account for approximately half the social science expenditures. For 1977-78, Statistics Canada was allocated $148.0 million, Canada Council $31.6 million and National Health and Welfare $27.4 million.

Department of Fisheries and the Environment 9.2.1

The Department of Fisheries and the Environment is the major funder of scientific activities. Expenditures for 1977-78 were set at $313.8 million in natural sciences and $14.6 million in social sciences.

A major performer of R&D is the Fisheries and Marine Service, operating nine research establishments across Canada with headquarters in Ottawa and research vessels on both coasts. Research concerns use and conservation of freshwater fisheries and marine resources. The service also conducts development activities in support of industries that depend on fishery resources. The Ocean and Aquatic Affairs Directorate conducts oceanographic research and surveys and charts coastal and inland navigable waters.

The Atmospheric Environment Service performs such basic research as studies of atmospheric electricity and applied research to support weather forecasting and data collection. Work is done on the climates of Canada and the application of meteorological information to such other activities as pollution research. The service collects large quantities of meteorological data, develops and tests meteorological instruments and operates the National Library of Meteorology at Toronto.

The Environmental Management Service consists of four main elements. The Lands Directorate is concerned with land classification, land inventory and land-use planning. Scientific activities consist of data collection and information services. The Inland Waters Directorate conducts research on the scientific aspects of the behaviour of water, on improved methods of water and waste water treatment and on the development of water treatment technology. Much scientific activity is conducted at the Canada Centre for Inland Waters in Burlington, Ont. The Canadian Forestry Service carries out most research into the protection and utilization of forest resources and improvement of tree growth. It operates regional laboratories, field stations and experimental areas across Canada. The Canadian Wildlife Service is responsible for research on the protection and preservation of wildlife.

The Environmental Protection Service is responsible for developing and enforcing environmental protection regulations and controls. It also informs other federal departments administering legislation under which environmental regulations are developed.

National Research Council of Canada 9.2.2

The National Research Council (NRC) is the principal federal agency with responsibility for scientific activities. Created in 1917 to provide Canada with qualified scientists and to promote research, its operations cover all aspects of scientific effort through three programs: engineering and natural science research, scientific and technical information, and grants and scholarships in aid of research.

The engineering and natural science research program accounts for most of the council's intramural research. Its six activities are: basic and exploratory engineering and scientific research; research on long-term problems of national concern; research in direct support of industrial innovation and development; research to provide technological support of social objectives; national facilities; and research and services related to standards.

The engineering laboratories consist of the divisions of building research, mechanical engineering, radio and electrical engineering, and a national aeronautical establishment; the physical/chemical sciences laboratories include divisions of physics and chemistry and an institute of astrophysics; biological sciences laboratories include a division of biological sciences, a regional laboratory in Saskatoon and another in Halifax.

Research is conducted on long-term problems relating to energy, transportation, food, building and construction. The study of energy-related problems covers a wide range of projects, including energy conservation, wind turbines, laser fusion research, separation of isotopes and power handling at extreme low temperatures.

Examples of other research are: legume seed research on the Prairies, expected to lead to commercial production in Canada of high protein foods derived from such legume crops as field peas and fava beans; research in mechanical engineering to establish design, installation and operation of ship hulls, oil platforms and terminal structures in an Arctic environment; and field and laboratory investigations into problems relating to northern construction and technology.

In direct support of industrial innovation and development, intramural research is performed for industrial companies on request, contract or collaboration basis.

Research to provide technological support of such social objectives as public safety, environment, health and education is conducted in many divisions. The biological sciences division is studying anaerobic bacterial digestion of organic wastes from food processing and sewage for production of methane and reduction of pollutant content. The divisions of physics, biological sciences, mechanical engineering and building research and the aeronautical establishment are co-operating in a joint study of environmental and physiological noise problems.

The National Research Council supports research efforts in industry, universities and other government departments through provision of national facilities. Examples are a large-scale towing tank operated by the Marine Dynamics and Ship Laboratory, a wind-tunnel complex at the aeronautical establishment and the Algonquin Radio Observatory.

Research and services related to standards include all work in support of standards, codes and specifications in the national or international public domain. The physics division is responsible for maintaining physical standards. An environmental secretariat assembles criteria for authorities responsible for setting environmental standards. A building research division provides support to NRC associate committees responsible for national building and fire codes.

In the field of scientific and technical information NRC is responsible for operation of the Canada Institute for Scientific and Technical Information, publication of journals of research and the development of a network of scientific and technical information services.

The NRC has a program to assist industry to become more competitive and innovative by promoting formation of R&D teams in industry. A research grant provides money for an applied research project conceived by a company with an end product or process in view. To be eligible, companies must be incorporated in Canada, undertake to do most of the proposed research in Canada, exploit results through Canadian operations and have access to export markets for the product. NRC pays direct salaries of scientists, engineers and technicians. The company provides laboratory space, equipment and consumable supplies and pays overhead costs. Estimated grants were $16.6 million for 1977-78.

A program of scholarships and grants for research remained with the NRC pending establishment of a research council for natural sciences and engineering in 1978. The program promotes and supports research in Canadian universities and the provision of qualified manpower in natural sciences and engineering. The program has three principal objectives: to support excellence in research for creation of new knowledge in the natural sciences and engineering, to promote and support development of research in selected fields of regional and national importance, and to assist in development of qualified manpower. The program includes grants for development, training and development of highly qualified manpower, and national and international activities. Expenditures for this program were planned at $98.0 million in 1977-78.

Statistics Canada   9.2.3

Statistics Canada is the major performer of activities in the social sciences with expenditures estimated at $148.0 million in 1977-78.

Statistics Canada provides statistical information needed to understand the Canadian economy and society. This is used to develop and monitor economic and social policies and programs of virtually all levels of government, as well as to support research work and decision-making. For example, the unemployment statistics and the Consumer Price Index are key indicators of the economic health of the nation. The agency's program consists almost totally of related scientific activities in the social sciences and represents nearly a third of total federal spending on social sciences RSA.

One of Statistics Canada's key jobs is conducting the census at five- and 10-year intervals on population and housing; the latest 10-year census was in 1971 and the five-year census in 1976. The bureau also undertakes a comprehensive census of agriculture at the same time and regularly surveys social and economic changes under more than 20 broad headings.

The growth of Statistics Canada, both in personnel and in the complexity of statistical activity, has paralleled Canada's development as a modern industrial state. The agency's staff includes the largest single body of social scientists in one organization in Canada. Several hundred additional persons are employed part-time on special surveys and censuses.

Statistical collection covers every area of Canada. Most Statistics Canada information is available to the public through publications. For users requiring information in a more sophisticated form there is an increasing output on microfilm, computer tapes and special tabulations.

Department of Agriculture   9.2.4

The agriculture department's research program has the largest budget devoted entirely to research — expected expenditures of $107.0 million in 1977-78. Research, conducted at the Central Experimental Farm in Ottawa and at regional laboratories across Canada, involves all elements of the food chain, soils, crops, animals, plant and animal products and diseases, in addition to problems of food processing and storage. Other programs cover production and marketing.

Rising labour and energy costs have made agriculture increasingly dependent on technology. This, combined with Canada's size, wide range of climatic and soil conditions and correspondingly wide range of agricultural crops and animals, makes it essential to co-ordinate research.

The research branch performs most of the department's research studies and supports a number of programs. One is aimed at economical conversion of cellulose and carbohydrate waste materials to feed-stock for ruminants and other animal species; others are concerned with extraction and processing of proteins derived from plants, models for predicting crop production on the basis of soil and weather information, and a control system for wild oats.

The branch is pursuing food-related research programs intended to improve the genetic characteristics of crops and livestock. One byproduct of such studies is that branch scientists have developed techniques for sexing and for the successful transfer of cattle embryos. These techniques will increase the international exchange of superior gene pools. Research into the causes of early pregnancy failure, mechanisms of transmission of diseases which reduce reproduction, and artificial control of the female reproductive cycle may lead to improvements in the productivity of Canadian breeding stock.

Emphasis of energy research is on energy efficiency of agricultural production systems and conversion of animal wastes to usable energy forms. Environmental programs continue to emphasize research into use of biotic agents for controlling agricultural pests and studies of the nature and effect of toxicants arising from infestations, additives, chemical control agents or inadvertent contamination.

The federal government, co-operating with provincial governments, supports an active soil survey program. Surveys have shown that land available for agriculture and food production is limited. Only about 5% of Canada's total land area, or about 49 million hectares (120 million acres), is improved farmland and it is estimated that no more than 16 million hectares (40 million acres) of land, most of it marginal, remains to be brought into use. This has given added impetus to land-use research.

The economics branch is expanding its activities in production economics, new technologies and management systems for farm use, appraisal of agricultural research resources and studies of energy use in agriculture. The production-marketing branch helps producer and industry groups, universities and provincial agencies on development and adaptation of new crops and varieties for commercial production. The animal health branch is intensifying research on diagnostic procedures for animal diseases, development of wildlife rabies vaccine and studies of disease prevention in high density cow-calf operations.

Agricultural research is conducted through a network of federal, provincial, university and industrial organizations. About 50% of the work is performed in federal laboratories. The agriculture department, as the focus of this federal involvement, has played an active role in developing the research infrastructure and in establishing co-operative research programs. Through its research agreement program, the department awards about $1.3 million a year to scientists at Canadian universities. It contributes to the provinces for expansion of veterinary science teaching facilities at the universities of Guelph, Montreal and Saskatchewan.

9.2.5 Atomic Energy of Canada Limited

Atomic Energy of Canada Limited (AECL), with intramural R&D expenditure of about $63 million in 1977-78, is a Crown corporation responsible for nuclear research and utilization. The main R&D centres are Chalk River Nuclear Laboratories, Chalk River, Ont. and Whiteshell Nuclear Research Establishment at Pinawa, Man. These laboratories carry out a full range of activities: underlying research in physics, chemistry, materials science and radiation biology; research and development on advanced nuclear reactors and other nuclear power systems; and research and development to improve current models of nuclear power plants. Three other groups, covering power projects, heavy water projects and commercial products, are responsible for utilization. They also carry out development work related to commercial objectives.

The corporation's prime responsibility is to develop nuclear energy technology to meet Canadian requirements. Its objective is to make available by the year 2000 about 80 000 MW of nuclear capacity (one and one-half times Canada's present total electric capacity). It also produces radioisotopes and develops associated products such as radiation processing equipment and radiotherapy instruments for use in medicine and industry.

Applied research and development activities, mainly performed at Chalk River, Ont., and Pinawa, Man., are carried out on power reactor systems, nuclear fuel, environmental protection and radioactive waste management, heavy water production, radiation equipment and radioactive isotopes. Applied work is supported by basic research in physics, chemistry and materials science. There is close collaboration with utilities and industry since this program provides the technological base for the largest industrial program ever initiated, developed and put into industrial practice by Canadians. About 30,000 people are employed in the Canadian nuclear industry.

R&D ranges from work at the laboratory bench to experiments using multi-million dollar research reactors and associated facilities. Much of the nuclear power activity involves the CANDU (CANada-Deuterium-Uranium) pressurized heavy water system; as well, work is conducted in support of heavy water plants. Particular attention is paid to developing reliability through sound design and good maintenance so that high capacity factors already achieved (87% for the Pickering, Ont., generating station in 1976) will continue. A slowly increasing percentage of the work is devoted to the development of new fuel cycles to ensure nuclear fuel supplies adequate for centuries. To protect people and the natural environment from effects of radiation, about 10% of the research effort is devoted to radioactive waste management, health physics, environmental research and biology research.

Department of Energy, Mines and Resources   9.2.6

The Department of Energy, Mines and Resources (EMR) promotes discovery, development, use and conservation of the country's mineral and energy resources. The earth sciences program provides the basic geodetic survey and topographical mapping of Canada. It conducts geological research and surveys to provide data on earth materials and terrain, to assess geological and terrain factors affecting use of these materials, and to develop techniques for monitoring characteristics of earth materials and terrain features. The program also conducts geophysical, seismic, gravity and magnetic studies of the earth's crust and interior, as well as research and field surveys in the area of the Arctic continental shelf. Through a centre for remote sensing the department is involved in development of facilities and techniques for production and use of data gathered by satellites and high-flying aircraft.

The program on mineral and energy resources includes research on the technology of mining, extraction, metallurgy, processing and use of metals and alloys; processing and use of fossil fuels; and minerals and mineral processing. Studies are conducted on pollution from thermal, metallurgical and mining processes and development of prevention and abatement techniques. Geological research and surveys are an important part of this program's activities, including research on the geological history of the earth, development of geological instruments and methods and surveys to describe and interpret the bedrock geology and to provide information to facilitate the discovery of mineral deposits.

Research activities in geoscience, minerals and energy technology provide the base needed for minerals and energy policy development. Research programs are directed in part to developing new techniques in exploration and resource analysis for land and offshore areas. Other research improves methods of understanding mineral and hydrocarbon occurrences, encouraging exploration and resource assessment.

One research program is directed toward a better understanding of Canadian earthquakes and seismic risk and hazard. Investigations of gravity and magnetic fields and the geothermal conditions of the earth are carried out in aid of navigation, transportation, communications, surveying and geophysical prospecting.

Some of the R&D activities bear directly on issues of environmental protection. For example, studies of permafrost not only provide information that will help in planning northern development but also provide data for assessing the environmental impact.

The geodetic framework, responsibility of a surveys and mapping branch, consists of tens of thousands of precisely located points and is essential to all other surveys and studies as well as for most large engineering projects. Multi-purpose topographic mapping is the foundation of a host of activities such as resource development, transportation, communications, urban and rural administration, education, defence and recreation.

The growing technology of sensing from satellites and aircraft is used by the remote sensing centre to provide data for such activities as ice reconnaissance, crop forecasting and forest fire prevention.

Department of National Defence   9.2.7

R&D projects for the defence department are varied and often have important applications in other areas. Many relate to defence of Canada's frontiers, especially the North, involving such problems as human and machine adaptation to extreme cold. Testing and standardization activities are conducted primarily by test and evaluation establishments of the Canadian Armed Forces.

The main responsibility for science and technology rests with a research and development branch. It is establishing a broad technology base to be maintained in six of its own research establishments across Canada, in industry and in other departments and agencies. In line with the contracting-out policy, external resources are used where practical but in many vital defence areas the technology base is maintained in-house. Outside contractors are used to provide advice on military training or operations and on human performance in the military environment. They are also used for equipment-related activities, ranging from demonstrations of feasibility, through development to performance evaluation and engineering tests.

One program conducted by the Defence Research Establishments (DRE) is a technical program on acoustic detection of submarines. It is carried out jointly by the DRE Atlantic and the DRE Pacific in the three oceans bordering Canada, using research ships and scientific equipment developed mainly by DRE. The program aims to improve military ability for reconnaissance, location and surveillance of submarines in waters of Canadian interest. It is concerned with all aspects of underwater acoustic propagation: natural and man-made interfering noise, signal processing and analysis, and transducer technology. International co-operation and information exchange are important parts of the program.

9.2.8 Department of Industry, Trade and Commerce

A major function of the department is to assist product and process development and to increase productivity in Canadian industry through greater use of research and application of advanced technology. Its main tool is financial assistance programs. Total scientific expenditures of the department were estimated at $83.0 million for 1977-78, $81.1 million for natural science activities.

In 1977 several of the development incentive programs were consolidated into one, the Enterprise Development Program (EDP). It provides financial support for product development, pre-production design and engineering, productivity studies and studies to determine market feasibility and strategies. The EDP normally finances up to 50% of eligible costs of specific innovation projects. It assists firms when projects appear commercially viable but represent a significant burden on the firm's resources. There was provision for up to $26.7 million for this program in 1977-78.

The department also has a program on industrial energy conservation research and development providing assistance for R&D aimed at new energy-conserving technologies; $1.5 million was included for it in the 1977-78 federal estimates.

The Industrial Research and Development Incentives Act (IRDIA), providing support for industrial R&D since 1966, was discontinued as of December 31, 1975. Since IRDIA grants were for past R&D, the department accepts applications for expenditures made by industry before the end of 1975. Total spending for this program reached $275.5 million as of March 1977. Grants authorized for 1977-78 were estimated at $45.8 million.

The department assists in establishment and maintenance of industrial research institutes at universities. Those which have been supported under the program are in Nova Scotia Technical College, École Polytechnique de Montréal, Ryerson Polytechnical Institute, and the universities of McGill, British Columbia, Quebec, Western Ontario, Alberta, Sherbrooke, Guelph, Dalhousie and Carleton.

9.2.9 Department of National Health and Welfare

The department provides substantial support for R&D and other activities in health sciences. A health resources fund was established in 1966 to help construction of teaching and research facilities at universities, hospitals and other institutions engaged in health research and training. The department provides capital grants covering up to 50% of the cost of approved projects.

The department provides funds for R&D in welfare-related subjects, drug abuse, fitness and sports participation, research related to health needs, health planning and the utilization of health services and health care facilities. R&D projects are supported primarily in Canadian universities and non-profit institutions. Grants are provided to provincial governments through an income security and social assistance program.

9.2.10 Medical Research Council

The Medical Research Council (MRC) supports research and development in health sciences (excluding public health) in Canadian universities and affiliated institutions. Research is supported primarily in the faculties of medicine, dentistry and pharmacy but projects in other areas relevant to health problems are considered. Research funds are distributed through three main programs: grants for research, direct personnel support and special programs. Estimated 1977-78 payments of the MRC to Canadian universities totalled $54.2 million.

The major portion of MRC expenditures are for research grants. There are two main types: operating grants and major equipment grants. These are intended to cover normal direct costs of research. To encourage maximum use of facilities major equipment grants are normally made to heads of departments or divisions where equipment will be located. Wherever possible, highly specialized equipment is provided for regional or national use — an example is the high resolution mass spectrograph facility at McMaster University. Operating grants represent the bulk of grants program expenditures. Normally made to a principal investigator to support his own research, grants are not intended to cover the entire costs of a project; space and basic facilities must be provided by the institution. The scientific merit of applications is assessed by the council's grants committees, comprised of working scientists assisted by external reviewers.

Special programs of the MRC are designed to promote development of medical research in Canada. This includes financing of MRC groups for research in especially productive areas — for example, a transplant research group at the University of Alberta. Development grants help universities recruit highly qualified investigators for full-time positions in areas (geographic or subject) needing development. To encourage collaboration and exchange of information the council offers visiting professorships, awards to visiting scientists and support for scientific symposia. General research grants are made to deans of medicine, dentistry and pharmacy for use at their discretion in support of research in their schools.

Canadian International Development Agency (CIDA) 9.2.11

The main objective of CIDA is to support efforts of developing countries toward economic growth and evolution of their social systems in ways that will produce a wide distribution of benefits of development to the populations of these countries, enhance the quality of life and improve the capacity of all population sectors to participate in national development.

Expenditures by CIDA include grants to international research organizations, scholarships to foreign students for study in Canada and assistance to Canadian scholars for studies related to international development.

Expenditures for scientific activities in 1977-78 were expected to total $83.6 million, $62.5 million for natural science activities. Funds to the foreign sector for both natural and social sciences were estimated at $24.0 million. The agency participates in an international consultative group on agricultural research, which co-ordinates support for such agencies as the International Maize and Wheat Improvement Centre and the International Rice Research Institute. CIDA contributes over $7 million — about 9% of the international total — in such grants.

CIDA also supports an African international laboratory for research on animal diseases in a search for immunological solutions to trypanosomiasis (spread by tse-tse flies) and East Coast fever (spread by ticks). Control of these diseases in Africa will increase pastureland available for ranching and other animal agriculture.

International Development Research Centre 9.2.12

This centre helps developing regions to build up research capabilities, innovative skills and institutions to solve their problems. It is a public corporation funded by the federal government through a CIDA grant. The main approach to its goal is through support of specific projects.

Research to improve food production and nutrition has been a prime concern. Other projects have studied modernization and its consequences, especially in rural communities in developing countries. Environmental health, disease prevention and health care, and the many variables that influence the size of families, have been focal points of research. The centre also collects and disseminates information about development.

A division of agriculture and nutrition supports research on crops, farming systems and reforestation in arid and semi-arid lands. This includes multiple cropping, growing more than one crop in the same year on the same land, and inter-cropping with two or more types of crop at a time. To develop new animal feed, experiments are conducted on using parts of plants or trees not generally considered edible, and byproducts such as the fleshy pulp of coffee and waste products of sugar cane. Other research is carried out on fish farming, shellfish culture, fish preservation and processing, post-harvest technology to protect the crop after it has been harvested, and the needs of the rural family.

Another division focuses on the effect of modernization and change, especially on rural peoples, on strategies for harmonious urban and rural development, formation of appropriate science and technology policies, delivery systems for primary education, and on formulation of population policies. The division also administers a scholarship program to increase the number of persons trained in development problems.

The main research interests of the division for population and health sciences are environmental health and disease prevention, fertility regulation, and rural health care delivery in developing countries. An example is support for research into biological control of diseases spread by insects. The centre approved a one-year grant of $500,000 to a special program, co-ordinated through the World Health Organization, to develop new tools for prevention, diagnosis and treatment of tropical parasitic diseases. This will help countries where the diseases are endemic by providing training in biomedical sciences and various forms of institutional support. The focus initially will be on major human parasitic infections in tropical zones.

9.2.13 Department of Communications

The department expected to spend $19.5 million for natural science activities in 1977-78. It undertakes scientific and technical research both directly, at its Communications Research Centre (CRC) near Ottawa, and indirectly, through a program of industry and university contracts for specific research projects.

The CRC has had wide experience in defence communications, high frequency transmission, the ionosphere and radar. The department is re-orienting research efforts to relate them more closely to the public telecommunications sector — to telephone, telegraph, broadcasting, cable distribution, data networks and mobile communications.

The department is also re-examining its radio research programs to ensure that research is carried out in line with its responsibility to manage the radio frequency spectrum.

Preliminary consultations began with industry to identify priorities for research into urban communications. A long-range program was outlined to look at possible effects on communications of such factors as energy shortages, conservation, employment and inflation.

A joint project of the department and the RCMP to develop a computer terminal for use in police cars continued. The terminal, including a video screen and typewriter keyboard mounted near the dashboard, would put mobile police officers in instant communication with a nationwide computer information system.

In the North, hunters, trappers and others in small remote communities often need reliable, portable communications systems. The department is conducting research on the practicability of a combined short-range relay system and a longer range, high-frequency radio system for providing reliable low-cost trail communications. Another project is aimed at developing techniques for integrating high-frequency radio transmissions in the North with existing communications and satellite networks.

Microwaves are used extensively in both terrestrial and satellite communications. A research program is under way to study effects of rain, turbulence and other atmospheric conditions on microwave systems.

The location of satellite earth stations is an important part of satellite communications planning; through a contract with Teleglobe Canada, the department has been studying site diversity.

The departmental radar research laboratory investigates new uses of radar such as in remote sensing of the environment; studies the application of new technology to radar systems; helps users specify and select new radar equipment; and investigates operating problems of radar systems.

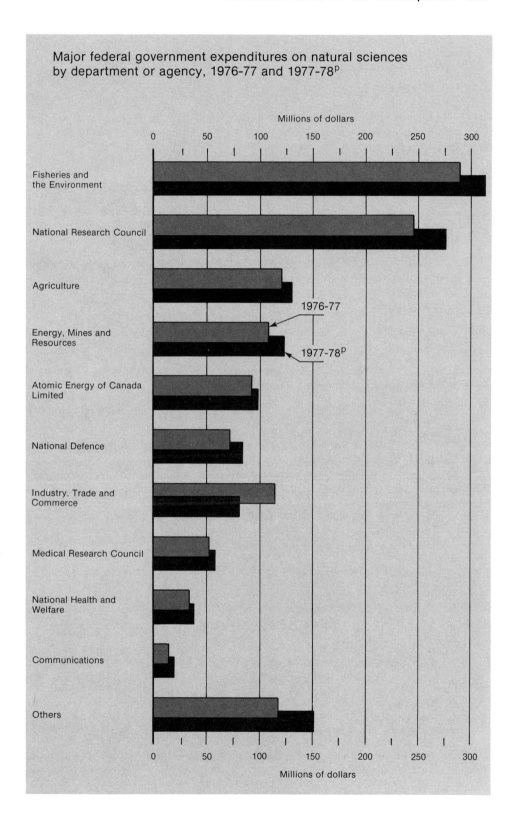

Major federal government expenditures on natural sciences by department or agency, 1976-77 and 1977-78ᴾ

In efforts to improve radio communication, the departmental radio communications laboratory continued experiments using the ionosphere, which deflects radio waves, to communicate over long distances.

9.2.14   The Canada Council

The council expected to spend $31.6 million in 1977-78 in support of activities in the social sciences, $23.5 million in the university sector. Funds for R&D amounted to $13.9 million. The council carries out its work mainly through a broad program of fellowships and grants. It shares with the Department of External Affairs responsibility for Canada's cultural relations with other countries, and administers the Canadian commission for UNESCO and special programs financed by private donations.

Research funds are channelled through five programs: grants to university faculty and other scholars for free research in the social sciences; the Killam grants (senior research scholarships and special postdoctoral research scholarships to support scholars of exceptional ability in significant research); leave fellowships for university faculty who wish to engage in some form of creative scholarship, research or study; research fellowships to permit younger scholars to undertake full-time research; and, as part of the cultural exchange program administered for the Department of External Affairs, grants to Canadian scholars for research in France.

Canada Council support for scientific information activities includes publication grants to specialized journals and block grants to the Humanities Research Council of Canada and the Social Science Research Council of Canada for publication of scholarly manuscripts and for support of attendance at annual meetings of Canadian learned societies.

In 1977-78 council expenditures for education support was planned at $11.7 million. This included doctoral fellowships for students in the social sciences who have completed at least one year of graduate study beyond the honours BA or its equivalent; grants to Canadian universities and organizations to support prominent visiting scholars from other countries; and grants to foreign students for advanced study in Canada.

9.2.15   Ministry of State for Science and Technology

The ministry is responsible for policies for the optimum development and application of science and technology in Canada. It reviews and assesses scientific and technological activities and programs in many federal departments and encourages co-operation among federal and provincial governments, public and private organizations and with other nations.

The government branch has overall responsibility for projects with a direct impact upon government policy and activities in science and technology. Its three divisional areas concern government projects, international programs and program review and assessment.

Government projects division, in consultation with other departments, furthers policy development in the oceans, space and northern science technology and reviews research and development in forestry.

The importance of Canada's offshore natural resources has caused MOSST to become involved in policy development and implementation on ocean matters. In particular, the division has devoted attention to the government's goal of achieving excellence in operating on and below ice-covered waters.

The review and assessment division interacts with departments on program planning and advises Treasury Board on requests by departments and agencies for financial and manpower resources. Criteria have been developed and advice provided on the decentralization of science and technology facilities, transportation R&D, on whether to make or buy scientific equipment and effects of cost increases on budgets of granting councils. Usefulness of special indicators to measure scientific activity is being examined as an aid to management of science and technology resources.

The industry branch identifies scientific and technological implications of policies and programs affecting the industrial sector and aims at a co-ordinated approach to R&D assistance and promotion. It makes proposals involving industrial R&D and science and technology which cut across departmental lines or which are outside the province of other departments and agencies. The branch also forecasts potential impact of scientific and technological advances upon Canadian society and environment.

A university branch advises government on general policy affecting federal support of university research and is a link between the government and the university community.

R&D in industry   9.3

Industry, performing about 44% of all Canadian R&D, is the largest performing sector. This is 13% more than the combined total of provincial and federal governments and 19% more than the university and private non-profit sectors.

Canadian industrial R&D effort, however, falls well behind most other industrialized countries when R&D expenditures are compared to gross domestic product (GDP). In fact, the ratio of Canada's GDP to R&D is only 30% to 50% of the ratios for the Federal Republic of Germany, Sweden, the United States, France and Japan.

Most industrial R&D is performed by a small number of firms. In recent years, the 25 leading performers have accounted for more than 50% of Canadian R&D spending; the first 200 performers spent approximately 88% of all such money.

Not only is R&D spending concentrated in a few companies but it is concentrated regionally. Ontario and Quebec account for 85% of current R&D spending. Another 13% is spent in British Columbia and Alberta, and the remaining 2% in other provinces.

Industry's total R&D spending increased from $132 million in 1959 to $950.2 million in 1977. But in terms of constant dollars, although R&D spending increased rapidly during the early 1960s it has since slowed to an almost static level.

Canadian industry finances much of its own R&D. In 1975 reporting companies provided 73% of R&D funds; 11% came from the federal government; 9% from other Canadian sources; and 7% from foreign sources. Foreign sources are mainly parent or affiliated companies. The proportion of funds coming from each of these sources has not changed much in recent years. Most federal government financial support goes to aircraft and electrical products industries.

All industries do not have the same need for R&D. Some, like electrical products, compete largely through new products based on R&D. Others, such as food and beverages, rely more on advertising and style than on R&D. Furthermore, subsidiary companies may rely on foreign parents for most R&D requirements. For example, Chrysler, Ford and General Motors in the United States together spent $2,462.6 million on R&D in 1976, four times as much as all Canadian industry. The different R&D intensity of industries is illustrated by the following percentages of non-government financed R&D to manufacturing value-added for 1974: electrical products 4.6%; petroleum and coal products 2.8%; machinery 2.5%; chemical and chemical products 2.1%; primary metals 1.4%; transportation equipment 0.8%; paper and allied products 0.6%; food and beverages 0.3%; and metal fabricated products 0.1%.

Provincial agencies   9.4

Economic planning   9.4.1

Nova Scotia set up a voluntary planning unit, an organization representing non-government elements of the community, in 1963 to involve the private sector in economic and social development.

The organization has: sector committees representing grass roots elements of producers, private business, labour and government in agriculture, construction, fisheries, forestry, mining, tourism, transportation and secondary manufacturing; advisory councils in consumer affairs, education, energy and labour-management affairs; a provincial planning board made up of sector and council chairmen plus other representatives of business, labour and government; and a small professional staff which provides administrative and technical support to volunteer groups.

The unit facilitates identification of problems by the private sector and relates appropriate private and public resources in an attempt to resolve these problems; and involves the private sector in analysis of government planning proposals prior to final approval.

Through this agency government has a contact with major elements of the private sector and the private sector has a forum for discussing problems and a direct channel to government for submitting co-ordinated views on development planning.

A major activity of 1976-77 was a detailed examination of implications of spruce budworm infestation in Nova Scotia. Another was assistance to the provincial government in developing new taxing arrangements for forest and agricultural land.

Activities include an examination of barriers to economic growth in Nova Scotia; an assessment of economic options open through 1990 with a focus on the developing energy crisis; an examination of implications in Canada's new 200-mile fishing limit; the monitoring of an inventory of manpower skills for the construction industry; and a review of manpower needs and training programs in all economic sectors.

Quebec Planning and Development Office. This agency has a mandate to prepare plans, programs and projects for social, economic, and territorial development which take regional characteristics into account and encourage better use of resources. It advises the government on policies and programs of various departments and promotes their integration. It co-ordinates research, studies, surveys, and inventories of other government departments and agencies and acts as a liaison office among agencies when several are involved in implementing a project.

Established by an act of the Quebec legislature in 1969, the office replaced the planning board of 1968 and the economic advisory council of 1961-68.

The Ontario Economic Council, established by legislation in 1968, represents a cross-section of informed people who pool knowledge and experience on social and economic questions, commission research and make policy recommendations to the public and private sectors. Its 21 members represent business, industry, finance, labour, agriculture and universities. Each serves without compensation for a term of one, two or three years.

Essentially, the council operates as an independent advisory body publishing its findings. Reports published in 1977 cover pension policy, day care, industrial research and development, education, transport policy, housing and land use policy and preventive medicine.

Until recently research activities have concentrated on six main areas — health, urban development, education, social security, national independence and Northern Ontario development. Two new areas of major interest are intergovernmental relations and government regulation.

9.4.2 Provincial research councils

Eight provinces have established research councils or foundations, with responsibility for assisting firms with technical problems and aiding development of provincial natural resources.

Nova Scotia has a Crown corporation with control vested in a board of directors appointed by the province. Its objective is to assist in the economic development of Nova Scotia by promoting, stimulating and encouraging effective utilization of science and technology by industry and government and to undertake, singly or in conjunction with others, such research, development, surveys, investigations and operations as may be appropriate.

The corporation's laboratories in Dartmouth were built on a site donated by the province. The staff of 91 includes 39 engineers and scientists and 37 technicians. The corporation's six scientific and technical divisions provide a strong multidisciplinary capability.

A geophysics division carries out gravity, seismic, magnetic and other surveys on land and at sea. A chemistry division stresses R&D related to minerals and other natural resources. Services are available to industry and government in inorganic and food chemistry, pollution control and chemical engineering.

A research division provides a service using mathematical techniques of systems analysis. Divisions on engineering physics and ocean technology emphasize ocean-oriented electronic and mechanical engineering. Another division provides technical information on materials, equipment and processes and engineering assistance to manufacturing industries. The biology group researches distribution, growth, conservation and utilization of commercial seaweed. It also does microbiological research related to water pollution and treatment of industrial waste waters.

New Brunswick set up a council on research and productivity in 1962. It is governed by an independent group of citizens from management, labour and the professions, appointed for three-year terms. Capital investment was provided by the federal government. Most council operations are carried out on a cost-recovery basis under contract with industry, trade associations and national and international agencies. The council maintains a centre for engineering and problem-solving, industrial research and development.

A Quebec industrial research centre was created in 1969. Its primary mission is to contribute to economic development of Quebec through encouragement of innovation in manufacturing enterprises, in particular those of small and medium size. It collaborates closely with such industries in research and development covering various fields of applied science and directed toward creation of new processes and products. In addition to laboratory activities, it provides an industrial and technological information service available to all Quebec enterprises.

Ontario's research foundation was established in 1928 as an independent corporation. Its board of governors is drawn from the industrial, commercial and scientific communities. The organization was financed initially by an endowment fund provided by industrial and commercial corporations through the Canadian Manufacturers Association and a matching grant from the provincial government. Most income is derived from contract research undertaken mainly for industry. Total income in 1977 was about $11 million. Since 1967 the Ontario government has provided an annual grant, with the amount related directly to foundation income from Canadian industry. The foundation is concerned primarily with the development of industry through application of science and technology. At the request of the various levels of government, it undertakes work relative to federal and provincial needs. Foundation activities are not restricted to Ontario; work is undertaken for any organization in Canada on an equal basis.

The foundation, with a staff of about 320, has provided large and small companies with a variety of services ranging from short-term investigations, market research and feasibility studies through product and process development to long-range scientific investigations. All projects are conducted confidentially, including business, technical or proprietary information revealed by clients or prospective clients. Patents resulting from contract research and development studies are assigned to the client.

Manitoba's research council consists of seven members and three advisory committees whose members represent natural-resource-based industry, manufacturing, labour, the universities and government. Its main purpose is to assist Manitoba industry to improve its market position by developing a more scientifically based production capability. The council maintains an office in Winnipeg and is establishing a small laboratory at Portage la Prairie. Permanent staff members are provided by the provincial government or hired by the council. The work is financed by provincial government appropriations and by contracts with the provincial and federal governments, and fees and service charges may be levied. The council promotes or carries out research and development related to natural resources and industrial operations. Much research sponsored by the council is aimed at establishing Manitoba as a centre of excellence in food products, electronics, materials research and health care products.

Through a technical assistance centre, industries are encouraged to use new technological developments. The centre is staffed by engineers and scientists with extensive industrial experience. During 1976-77 the centre received more than 150 inquiries for assistance and provided technical data on material selection and properties.

In Saskatchewan, the research council was set up in 1947 under an act of the legislature. It carries out research in natural and management sciences with the aim of improving the provincial economy. At first the council carried out its research programs at the University of Saskatchewan by means of grants to staff members and scholarships to graduate students. The 1947 act was amended in 1954 to empower the council to acquire property, employ staff and conduct its own financial affairs. Laboratory buildings were built on the university campus in 1958 and extended in 1963. The present program places emphasis on consulting and technical assistance to industry and provincial government departments, and research in metallic and industrial minerals, water, the environment, slurry pipeline transportation and selected aspects of agriculture. A large part of the program is carried out by a full-time staff of about 140 but some council research is still promoted by grants to university staff. Members of the council are representatives of the Saskatchewan government, the university and industry.

The Alberta government set up a research council in co-operation with the University of Alberta in 1921 to promote mineral development. Natural resources studies still receive considerable attention but strong emphasis is placed on research related to establishing new industries in the province, to transportation and to environmental problems. The principal areas of activity are fossil fuels development and utilization, mineral resource evaluation and research, groundwater and soils investigation, chemical product and process development, technical and economic evaluations, microbiology, technical assistance to industry, gasoline and oil testing, pipeline transportation, highway research, river engineering, environmental studies and hail research.

The council is directed by a 15-member board, representing the government, the universities and industry. Advisory committees of specialists drawn from these sectors review research projects. The council is financed by provincial government appropriations and through research contracts with private industry and government agencies. Main council laboratories and offices are on the University of Alberta campus, with a pilot plant and laboratory facility east of Edmonton and subsidiary offices and laboratories in other parts of the city. The full-time staff comprises approximately 400.

In British Columbia BC Research is a non-profit industrial research society with offices and laboratories at Vancouver. Its activities enable even the smallest firms to improve their competitive position in Canadian and world markets by the use of up-to-date scientific knowledge. The agency does contract research for clients on a confidential basis, initiates in-house research programs to promote and use the resources of the province, and provides a free technical information service in collaboration with the National Research Council of Canada. It is active in applied biology, chemistry, engineering — physics, ocean engineering, operations research, industrial engineering — and social impact and economic studies.

Sources
9.1 - 9.3 Science Statistics Centre, Education, Science and Culture Division, Statistics Canada.
9.4 Supplied by respective provincial departments and agencies.

Tables

9.1 Total expenditure on natural science R&D, by source of funds, 1967-77 (million dollars)

Year	General government[1]	Industry	University	Private non-profit	Foreign	Total
1967r	431	273	120	9	18	853
1968r	474	280	116	11	17	900
1969r	499	325	132	11	18	985
1970r	525	334	146	13	23	1,040
1971r	573	367	140	20	31	1,130
1972r	606	358	136	21	33	1,153
1973r	666	390	135	19	39	1,249
1974r	729	491	157	28	43	1,448
1975	759	570	193	29	53	1,603
1976	851	617	208	31	58	1,766
1977	925	668	225	34	64	1,916

[1]Federal and provincial governments and provincial research councils.

9.2 Total expenditure on natural science R&D, by sector of performance, 1967-77 (million dollars)

Year	General government[1]	Industry	University and private non-profit	Canada total
1967r	310	336	207	853
1968r	335	342	222	900
1969r	339	395	251	985
1970r	352	416	272	1,040
1971r	380	468	283	1,130
1972r	407	460	287	1,153
1973r	444	504	301	1,249
1974r	490	611	347	1,448
1975	510	692	401	1,603
1976	552	781	433	1,766
1977	602	846	468	1,916

[1]Federal and provincial governments and provincial research councils.

9.3 Federal government expenditure on the natural sciences, by department or agency and scientific activity, 1976-77 and 1977-78 (million dollars)

Department or agency	Current R&D	Current related scientific activities					Administration of extra-mural programs	Capital	Total
		Data collection	Scientific information	Testing and standardization	Feasibility studies	Education support			
1976-77									
Fisheries and the Environment	104.6	128.1	14.1	1.9	1.5	1.5	0.3	37.3	289.3
National Research Council	202.7	0.4	13.2	7.8	0.3	9.0	5.2	7.2	245.8
Agriculture	104.2	0.9	4.1	0.1	—	—	—	11.4	120.7
Atomic Energy of Canada Limited	83.1	0.9	1.9	—	—	—	—	6.4	92.3
Energy, Mines and Resources	60.6	34.7	3.9	0.9	1.7	—	1.0	5.4	108.2
National Defence	63.9	—	1.6	0.2	1.9	—	1.0	3.3	71.9
Industry, Trade and Commerce	110.1	—	—	—	—	1.1	3.3	—	114.5
Medical Research Council	49.5	0.2	0.2	—	—	1.0	1.3	—	52.2
National Health and Welfare	17.7	6.5	0.8	7.1	0.4	0.1	0.4	1.0	34.1
Communications	13.1	—	0.2	—	0.3	—	—	1.2	14.8
Others	40.3	17.5	12.1	3.4	30.1	5.0	7.0	1.6	117.1
Total	849.8	189.2	52.1	21.4	36.2	17.7	19.5	74.9	1,260.9

9.3 Federal government expenditure on the natural sciences, by department or agency and scientific activity, 1976-77 and 1977-78 (million dollars) (concluded)

Department or agency	Current R&D	Current related scientific activities					Adminis-tration of extra-mural programs	Capital	Total
		Data collec-tion	Scien-tific infor-mation	Testing and stan-dardiz-ation	Feasi-bility studies	Educa-tion support			
1977-78P									
Fisheries and the Environment	114.2	140.8	15.2	2.1	1.5	1.3	0.3	38.4	313.8
National Research Council	228.0	0.5	15.4	8.8	0.6	9.6	6.3	7.3	276.5
Agriculture	116.5	0.9	4.2	0.1	—	—	—	8.4	130.3
Atomic Energy of Canada Limited	90.4	0.9	2.0	—	—	—	—	5.6	98.9
Energy, Mines and Resources	70.4	38.2	4.5	1.0	1.8	—	1.1	6.0	123.0
National Defence	71.7	—	1.7	0.2	2.1	—	1.2	7.0	83.8
Industry, Trade and Commerce	76.3	—	—	—	—	1.1	3.7	—	81.1
Medical Research Council	54.9	0.2	0.3	—	—	1.3	1.5	—	58.2
National Health and Welfare	19.3	8.7	1.1	7.7	0.4	0.1	0.4	1.4	38.2
Communications	14.2	—	0.2	—	0.3	—	—	4.8	19.5
Others	49.8	16.8	14.2	3.9	49.7	5.6	7.1	3.4	151.2
Total	905.7	207.0	58.8	23.8	56.4	19.0	21.6	82.3	1,374.5

9.4 Federal government employees engaged in activities in the natural sciences, by department or agency and category, 1976-77 and 1977-78 (full-time equivalent)

Department or agency	Intramural R&D			Intramural related scientific activities			Adminis-tration of extra-mural programs	Total
	Scien-tific and profes-sional	Tech-nical	Other	Scien-tific and profes-sional	Tech-nical	Other		
1976-77								
Fisheries and the Environment	1,098.5	1,020.0	750.5	763.0	1,801.0	1,386.0	12.8	6,831.8
Agriculture	966.0	1,051.0	1,884.0	42.5	55.0	97.0	—	4,095.5
National Research Council	785.0	896.0	692.0	199.0	117.0	263.0	159.0	3,111.0
Energy, Mines and Resources	628.0	381.0	230.5	271.0	338.0	385.4	55.0	2,288.9
National Defence	446.0	559.0	835.0	87.0	2.0	116.0	55.0	2,100.0
Atomic Energy of Canada Limited	598.0	794.0	853.0	28.0	22.0	47.0	—	2,342.0
National Health and Welfare	162.0	80.0	23.0	203.0	292.0	100.5	15.5	876.0
Communications	171.0	85.0	8.0	4.0	7.0	47.0	—	322.0
Consumer and Corporate Affairs	—	—	—	199.0	27.0	201.0	—	427.0
Others	129.5	89.2	50.4	42.0	67.5	89.0	396.4	864.0
Total	4,984.0	4,955.2	5,326.4	1,838.5	2,728.5	2,731.9	693.7	23,258.2
1977-78P								
Fisheries and the Environment	1,124.5	1,026.0	822.5	762.0	1,765.0	1,423.0	12.8	6,935.8
Agriculture	963.0	1,047.0	1,864.0	42.5	55.0	97.0	—	4,068.5
National Research Council	794.0	892.0	689.0	201.0	116.0	260.0	170.0	3,122.0
Energy, Mines and Resources	622.0	377.0	243.5	264.0	331.0	375.4	58.1	2,271.0
National Defence	447.0	559.0	836.0	88.0	2.0	117.0	55.0	2,104.0
Atomic Energy of Canada Limited	597.0	777.0	849.0	27.0	22.0	47.0	—	2,319.0
National Health and Welfare	165.0	93.0	16.0	207.0	297.0	105.5	16.5	900.0
Communications	173.0	83.0	8.0	2.0	10.0	48.0	—	324.0
Consumer and Corporate Affairs	—	—	—	200.0	28.0	201.0	—	429.0
Others	126.1	83.8	49.8	42.0	70.5	87.0	410.0	869.2
Total	5,011.6	4,937.8	5,377.8	1,835.5	2,696.5	2,760.9	722.4	23,342.5

9.5   Federal government expenditure on the natural sciences, by department or agency and performer, 1976-77 and 1977-78 (million dollars)

Department or agency	Federal govern- ment	Canadian industry	Canadian universities	Canadian non-profit institutions	Other Canadian performers	Foreign	Total
1976-77							
Fisheries and the Environment	273.3	7.9	3.8	1.4	2.5	0.4	289.3
National Research Council	105.6	42.0	89.8	1.3	0.7	6.4	245.8
Agriculture	116.2	0.8	3.1	—	0.5	0.1	120.7
Atomic Energy of Canada Limited	64.4	27.0	0.5	—	0.4	—	92.3
Energy, Mines and Resources	84.6	14.8	1.3	0.1	7.4	—	108.2
National Defence	54.9	14.2	1.2	0.1	—	1.5	71.9
Industry, Trade and Commerce	3.3	105.9	0.9	3.8	0.6	—	114.5
Medical Research Council	1.4	—	48.6	—	—	2.2	52.2
National Health and Welfare	23.3	0.3	9.2	1.2	0.1	—	34.1
Communications	10.5	3.6	0.6	—	—	0.1	14.8
Canadian International Development Agency	0.4	28.2	2.0	2.8	—	3.8	37.2
Transport	9.3	12.0	1.8	0.1	0.4	—	23.6
Consumer and Corporate Affairs	9.8	—	—	—	—	—	9.8
Regional Economic Expansion	—	1.0	—	—	—	—	1.0
Atomic Energy Control Board	—	0.2	0.4	—	—	—	0.6
Others	16.3	10.6	2.5	0.7	1.7	13.1	44.9
Total	773.3	268.5	165.7	11.5	14.3	27.6	1,260.9
1977-78P							
Fisheries and the Environment	295.4	10.0	3.9	1.5	2.6	0.4	313.8
National Research Council	118.1	50.3	102.3	1.4	0.7	3.7	276.5
Agriculture	124.9	1.0	3.6	0.1	0.6	0.1	130.3
Atomic Energy of Canada Limited	63.8	34.1	0.6	—	0.4	—	98.9
Energy, Mines and Resources	90.6	18.0	2.0	0.1	12.2	0.1	123.0
National Defence	63.0	17.1	1.4	0.2	—	2.1	83.8
Industry, Trade and Commerce	3.7	73.8	0.9	2.1	0.6	—	81.1
Medical Research Council	1.5	—	54.2	0.1	—	2.4	58.2
National Health and Welfare	26.0	0.5	9.5	1.9	0.3	—	38.2
Communications	14.6	4.2	0.6	—	—	0.1	19.5
Canadian International Development Agency	0.5	47.3	2.2	3.0	—	9.5	62.5
Transport	11.9	11.9	1.9	0.2	0.5	—	26.4
Consumer and Corporate Affairs	10.7	—	—	—	—	—	10.7
Regional Economic Expansion	—	1.2	—	—	—	—	1.2
Atomic Energy Control Board	0.1	0.2	0.3	—	—	—	0.6
Others	19.3	12.7	2.4	0.5	2.7	12.2	49.8
Total	844.1	282.3	185.8	11.1	20.6	30.6	1,374.5

9.6   Federal government expenditure on the natural sciences, by activity and performer, 1976-77 and 1977-78 (million dollars)

Scientific activity	Federal government	Canadian industry	Canadian universities	Canadian non-profit institutions	Other Canadian performers	Foreign	Total
1976-77							
R&D	*503.5*	*217.0*	*151.7*	*7.1*	*12.2*	*24.8*	*916.3*
Current expenditure	452.1	217.0	151.7	7.1	12.2	24.8	864.9
Administration of extramural programs	15.0	—	—	—	—	—	15.0
Capital expenditure	51.4	—	—	—	—	—	51.4
Related scientific activities	*269.8*	*51.4*	*14.0*	*4.4*	*2.2*	*2.8*	*344.6*
Current expenditure	246.4	51.4	14.0	4.4	2.2	2.8	321.2
Scientific data collection	166.6	20.4	1.4	0.3	0.5	—	189.2
Scientific information	47.7	0.6	0.1	0.9	1.0	1.8	52.1
Testing and standard- ization	20.5	0.5	0.2	—	0.2	—	21.4
Feasibility studies	5.6	29.7	0.4	0.1	0.1	0.3	36.2
Education support	1.4	0.3	11.9	3.1	0.3	0.7	17.7
Administration of extra- mural programs	4.5	—	—	—	—	—	4.5
Capital expenditure	23.5	—	—	—	—	—	23.5
Total	773.3	268.5	165.7	11.5	14.3	27.6	1,260.9

9.6 Federal government expenditure on the natural sciences, by activity and performer, 1976-77 and 1977-78 (million dollars) (concluded)

Scientific activity	Federal government	Canadian industry	Canadian universities	Canadian non-profit institutions	Other Canadian performers	Foreign	Total
1977-78P							
R&D	*548.5*	*210.9*	*170.2*	*6.1*	*18.0*	*27.6*	*981.3*
Current expenditure	489.2	210.9	170.2	6.1	18.0	27.6	922.0
Administration of extramural programs	16.4	–	–	–	–	–	16.4
Capital expenditure	59.3	–	–	–	–	–	59.3
Related scientific activities	*295.6*	*71.4*	*15.6*	*5.0*	*2.6*	*3.0*	*393.2*
Current expenditure	272.6	71.4	15.6	5.0	2.6	3.0	370.2
Scientific data collection	183.6	20.5	1.6	0.5	0.7	–	207.0
Scientific information	53.9	0.7	0.2	0.9	1.2	1.9	58.8
Testing and standard-ization	22.6	0.6	0.2	0.1	0.3	–	23.8
Feasibility studies	6.1	49.3	0.5	0.1	0.2	0.3	56.4
Education support	1.2	0.3	13.1	3.3	0.3	0.8	19.0
Administration of extra-mural programs	5.2	–	–	–	–	–	5.2
Capital expenditure	23.0	–	–	–	–	–	23.0
Total	844.1	282.3	185.8	11.1	20.6	30.6	1,374.5

9.7 Federal government expenditure on the social sciences, by department or agency and scientific activity, 1976-77 and 1977-78 (million dollars)

Department or agency	Current R&D	Current related scientific activities					Adminis-tration of extra-mural programs	Capital	Total
		Data collec-tion	Infor-mation services	Economic and feasi-bility studies	Operations and policy studies	Educa-tion support			
1976-77									
Statistics Canada	3.6	160.2	7.2	1.1	1.5	–	–	1.0	174.6[1]
Canada Council	13.0	–	3.0	–	0.3	10.8	1.8	–	28.9
National Health and Welfare	16.8	1.2	1.0	1.0	1.2	1.1	1.2	–	23.5
International Develop-ment Research Centre	6.3	–	3.4	–	1.2	0.4	6.4	–	17.7
Canadian International Development Agency	12.4	–	–	–	0.6	4.7	0.4	–	18.1
National Library	–	–	14.0	–	–	–	–	–	14.0
Manpower and Immigration	3.7	2.5	0.2	0.7	4.2	0.1	0.1	–	11.5
Treasury Board	0.1	1.2	0.4	–	9.6	–	–	–	11.3
Indian Affairs and Northern Development	5.2	0.4	0.3	0.2	1.0	–	0.4	2.5	10.0
Urban Affairs	3.1	–	0.3	2.1	1.6	0.1	0.1	0.1	7.4
Others	43.8	23.0	21.0	4.5	22.2	2.9	4.5	0.7	122.6
Total	108.0	188.5	50.8	9.6	43.4	20.1	14.9	4.3	439.6
1977-78P									
Statistics Canada	4.3	132.2	7.9	1.3	1.6	–	–	0.7	148.0
Canada Council	13.9	–	3.7	–	0.3	11.7	2.0	–	31.6
National Health and Welfare	19.0	1.6	1.6	1.2	1.6	1.2	1.2	–	27.4
International Develop-ment Research Centre	6.9	–	3.7	–	1.2	0.4	7.0	–	19.2
Canadian International Development Agency	14.8	–	–	–	0.7	5.1	0.5	–	21.1
National Library	–	–	15.4	–	–	–	–	–	15.4
Employment and Immigration[2]	4.0	2.6	0.3	0.8	4.4	0.1	0.1	–	12.3
Treasury Board	0.1	1.4	0.5	–	11.2	–	–	–	13.2
Indian Affairs and Northern Development	5.4	0.5	0.3	0.3	1.2	0.1	0.5	2.7	11.0
Urban Affairs	1.6	0.2	2.0	2.4	2.9	0.4	0.1	–	9.6
Others	57.3	26.0	24.5	6.4	28.7	3.6	4.5	1.2	152.2
Total	127.3	164.5	59.9	12.4	53.8	22.6	15.9	4.6	461.0

[1]Includes additional expenditures for the 1976 Census.
[2]Employment and Immigration as of Aug. 15, 1977.

9.8   Federal government employees engaged in activities in the social sciences, by department or agency and category, 1976-77 and 1977-78 (full-time equivalent)

Department or agency	Intramural R&D			Intramural related scientific activities			Adminis-tration of extramural programs	Total
	Scientific and profes-sional	Technical	Other	Scientific and profes-sional	Technical	Other		
1976-77								
Statistics Canada	78.0	43.0	105.0	790.0	649.0	4,732.0	—	6,397.0
National Library	—	—	—	176.0	32.0	282.0	—	490.0
Manpower and Immigration	40.0	—	48.0	122.0	—	135.0	2.0	347.0
National Health and Welfare	96.5	15.0	36.0	19.5	18.5	24.5	35.5	245.5
Treasury Board	—	—	—	39.0	4.0	246.0	—	289.0
International Development Research Centre	—	—	—	31.0	—	21.0	175.0	227.0
Indian Affairs and Northern Development	73.5	25.0	11.0	25.5	6.0	7.0	20.7	168.7
Canada Council	—	—	—	4.0	—	4.0	87.8	95.8
Urban Affairs	23.5	2.5	31.0	16.0	1.5	20.5	7.0	102.0
Canadian International Development Agency	—	—	—	5.0	2.0	9.0	19.0	35.0
Others	722.9	103.7	336.1	612.1	320.1	1,134.7	127.3	3,356.9
Total	1,034.4	189.2	567.1	1,840.1	1,033.1	6,615.7	474.3	11,753.9
1977-78P								
Statistics Canada	82.0	49.0	78.0	776.0	722.0	3,749.0	—	5,456.0
National Library	—	—	—	174.0	37.0	279.0	—	490.0
Employment and Immigration[1]	37.0	—	43.0	120.0	—	127.0	2.0	329.0
National Health and Welfare	91.5	16.0	32.0	24.5	18.5	35.5	39.5	257.5
Treasury Board	—	—	—	33.0	9.0	239.0	—	281.0
International Development Research Centre	—	—	—	31.0	—	22.0	178.0	231.0
Indian Affairs and Northern Development	73.5	25.0	11.0	27.5	8.0	8.5	20.7	174.2
Canada Council	—	—	—	4.0	—	4.0	88.8	96.8
Urban Affairs	7.6	1.0	10.0	15.4	2.0	25.4	6.0	67.4
Canadian International Development Agency	—	—	—	5.0	2.0	9.0	19.0	35.0
Others	697.9	106.7	335.1	651.1	317.6	1,094.7	112.9	3,316.0
Total	989.5	197.7	509.1	1,861.5	1,116.1	5,593.1	466.9	10,733.9

[1]Employment and Immigration as of Aug. 15, 1977.

9.9   Federal government expenditure on the social sciences, by department or agency and performer, 1976-77 and 1977-78 (million dollars)

Department or agency	Federal government	Canadian business enterprises	Canadian universities	Canadian non-profit institutions	Other Canadian performers	Foreign	Total
1976-77							
Statistics Canada	174.6	—	—	—	—	—	174.6
Canada Council	2.0	—	21.5	2.3	—	3.1	28.9
National Health and Welfare	8.5	0.1	5.3	4.7	4.5	0.4	23.5
International Develop-ment Research Centre	7.9	0.3	0.1	0.1	0.5	8.8	17.7
Canadian International Development Agency	1.1	—	2.1	2.7	—	12.2	18.1
National Library	14.0	—	—	—	—	—	14.0
Manpower and Immigration	10.0	1.1	0.1	—	0.3	—	11.5
Treasury Board	11.2	—	—	0.1	—	—	11.3
Indian Affairs and Northern Development	7.9	0.6	0.4	0.4	0.7	—	10.0
Urban Affairs	3.3	1.3	1.1	0.5	1.2	—	7.4
Fisheries and the Environment	10.7	1.1	0.3	—	—	—	12.1
Transport	3.9	2.0	0.1	—	1.7	—	7.7
Public Archives	8.2	—	—	—	—	—	8.2
Finance	6.4	—	—	—	—	—	6.4
Central Mortgage and Housing Corporation	3.3	0.4	1.0	0.3	0.6	0.1	5.7
Others	65.2	5.7	4.1	2.0	5.0	0.5	82.5
Total	338.2	12.6	36.1	13.1	14.5	25.1	439.6

9.9   Federal government expenditure on the social sciences, by department or agency and performer, 1976-77 and 1977-78 (million dollars) (concluded)

Department or agency	Federal government	Canadian business enterprises	Canadian universities	Canadian non-profit institutions	Other Canadian performers	Foreign	Total
1977-78P							
Statistics Canada	148.0	—	—	—	—	—	148.0
Canada Council	2.2	—	23.5	2.6	—	3.3	31.6
National Health and Welfare	9.4	0.2	7.0	5.3	4.9	0.6	27.4
International Development Research Centre	8.6	0.3	0.1	0.1	0.5	9.6	19.2
Canadian International Development Agency	1.2	—	2.5	2.9	—	14.5	21.1
National Library	15.4	—	—	—	—	—	15.4
Employment and Immigration[1]	10.8	1.1	—	—	0.4	—	12.3
Treasury Board	13.1	—	—	0.1	—	—	13.2
Indian Affairs and Northern Development	8.7	0.6	0.5	0.4	0.8	—	11.0
Urban Affairs	2.3	1.7	2.7	0.2	2.5	0.2	9.6
Fisheries and the Environment	11.8	1.3	0.8	0.7	—	—	14.6
Transport	4.8	6.8	1.2	0.2	1.0	—	14.0
Public Archives	9.5	—	—	—	—	—	9.5
Finance	7.7	—	—	—	—	—	7.7
Central Mortgage and Housing Corporation	3.5	0.7	1.5	0.4	1.3	0.1	7.5
Others	77.0	6.4	5.0	2.4	7.3	0.5	98.9
Total	334.2	19.1	44.8	15.3	18.7	28.9	461.0

[1]Employment and Immigration as of Aug. 15, 1977.

9.10   Federal government expenditure on the social sciences, by activity and performer, 1976-77 and 1977-78 (million dollars)

Scientific activity	Federal government	Canadian business enterprises	Canadian universities	Canadian non-profit institutions	Other Canadian performers	Foreign	Total
1976-77							
R&D	59.3	6.1	20.3	6.2	9.8	18.3	120.0
Current expenditures	56.6	6.1	20.3	6.2	9.8	18.3	117.3
Administration of extramural programs	9.3	—	—	—	—	—	9.3
Capital expenditures	2.7	—	—	—	—	—	2.7
Related scientific activities	278.9	6.5	15.8	6.9	4.7	6.7	319.5
Current expenditures	277.3	6.5	15.8	6.9	4.7	6.7	317.9
Education support	0.1	—	12.5	3.0	0.9	3.6	20.1
Data collection	184.1	2.5	0.8	0.4	0.7	—	188.5
Information services	42.8	1.0	1.0	2.8	1.2	1.9	50.8
Economic and feasibility studies	6.2	1.3	0.7	0.4	1.0	—	9.6
Operation and policy studies	38.6	1.8	0.7	0.3	0.9	1.2	43.4
Administration of extramural programs	5.6	—	—	—	—	—	5.6
Capital expenditures	1.6	—	—	—	—	—	1.6
Total	338.2	12.6	36.1	13.1	14.5	25.1	439.6
1977-78P							
R&D	67.3	9.6	23.1	7.2	11.0	21.7	140.0
Current expenditures	64.3	9.6	23.1	7.2	11.0	21.7	137.0
Administration of extramural programs	9.7	—	—	—	—	—	9.7
Capital expenditures	3.0	—	—	—	—	—	3.0
Related scientific activities	266.9	9.4	21.7	8.1	7.7	7.3	321.0
Current expenditures	265.4	9.4	21.7	8.1	7.7	7.3	319.5
Education support	0.2	—	14.3	3.3	0.9	4.0	22.6
Data collection	158.5	2.8	1.0	0.5	1.7	—	164.5
Information services	49.1	0.9	3.0	3.1	1.8	2.0	59.9
Economic and feasibility studies	6.8	2.3	1.6	0.1	1.6	—	12.4
Operation and policy studies	44.6	3.5	1.8	1.1	1.7	1.2	53.8
Administration of extramural programs	6.2	—	—	—	—	—	6.2
Capital expenditures	1.5	—	—	—	—	—	1.5
Total	334.2	19.1	44.8	15.3	18.7	28.9	461.0

9.11   R&D expenditure of Canadian industrial firms, 1968-77 (million dollars)

Year	In Canada					Payments outside Canada	Total
	Intramural			Extramural	Total[1]		
	Current	Capital	Total				
1968[r]	306.2	36.2	342.2	16.2	345.9	36.6	382.5
1969[r]	345.0	49.7	394.7	22.8	398.6	38.8	437.4
1970[r]	365.5	50.4	415.9	30.5	419.4	45.4	464.8
1971[r]	404.2	63.3	467.5	31.5	470.5	52.5	523.0
1972[r]	412.4	47.1	459.5	34.6	462.7	58.3	521.0
1973[r]	460.6	43.3	504.0	44.3	507.3	70.1	577.4
1974[r]	532.3	78.5	610.9	52.9	614.4	80.9	695.3
1975	622.9	69.3	692.2	59.6	699.9	87.1	787.0
1976	702.1	78.9	781.1	64.6	791.3	91.3	882.6
1977	763.8	82.5	846.4	70.5	859.1	91.1	950.2

[1]To avoid double counting, certain transfers from one respondent to another have been subtracted from the sum of all Canadian intramural and extramural expenditures. Such transfers would be entered once as intramural and once as extramural.

9.12   Industrial R&D expenditure, 1974 and 1975

Industry group	Intramural R&D expenditure			Source of funds			
	Current $'000,000	Capital $'000,000	Total $'000,000	Reporting company		Federal government	
				$'000,000	% of total	$'000,000	% of total
1974							
Mines and wells	23.4	5.2	28.6	22.6	79	1.8	6
Chemical-based	104.9	22.9	127.8	109.7	86	7.3	6
Wood-based	25.3	1.8	27.1	15.2	56	2.6	10
Metals	48.0	7.6	55.6	48.6	87	2.7	5
Machinery and transportation equipment	105.8	4.6	110.4	54.8	50	39.5	36
Electrical	134.5	19.6	154.1	102.1	66	19.6	13
Other manufacturing	7.4	1.5	8.9	6.5	73	1.0	11
Other industries	41.8	10.4	52.2	32.9	63	9.7	19
Total	491.1	73.6	564.7	392.4	69	84.2	15
1975							
Mines and wells	37.5	3.3	40.8	26.2	64	3.2	8
Chemical-based	137.3	16.3	153.5	137.1	89	8.2	5
Wood-based	28.1	2.4	30.5	21.2	70	2.1	7
Metals	63.0	9.8	72.8	59.5	82	2.7	4
Machinery and transportation equipment	127.4	10.2	137.6	81.7	59	32.5	24
Electrical	155.7	10.4	166.1	112.0	67	21.9	13
Other manufacturing	10.0	1.1	11.1	9.9	89	0.9	8
Other industries	63.9	15.9	79.8	60.0	75	6.3	8
Total	622.9	69.3	692.2	507.6	73	77.6	11

Sources

9.1 - 9.12 Education, Science and Culture Division, Institutional and Public Finance Statistics Branch, Statistics Canada.

Renewable resources   Chapter 10

Tables

Renewable resources   Chapter 10

Forestry   10.1

Canada is a major exporter of forest products. Exports of wood, wood products and paper in 1976 amounted to $6,556 million which was 18% of the value of all commodity exports. Paper and paperboard constituted 36% of all forest products exports; newsprint alone accounted for 30%.

The forests of Canada are largely coniferous and make up 35% of the total land area; of this forest, less than 5% is reserved — parks and reserves where, by legislation, wood production is not primary. In 1975, 114 million cubic metres of roundwood was cut. The harvesting and processing of this timber generated work for 281,000 persons with $3,336 million in salaries and wages. The value added by processing beyond the raw materials stage amounted to $6,285 million which was 8.6% of the value added of all goods-producing industries.

British Columbia, Ontario and Quebec are the most important timber-producing provinces. In 1975 British Columbia sawmills produced 64% of all lumber in Canada and most of the sulphate pulp and softwood plywood while Ontario and Quebec produced most of the groundwood pulp and hardwood plywood.

There is a growing awareness of the importance of the forest in such areas as recreation, wildlife habitat and stream flow regulation. Recognition of these values is fostering a broader and more realistic concept of forestry.

Forest resources   10.1.1

Forest regions   10.1.1.1

Forests cover a vast area in the north temperate climatic zone but wide variations in physiographic, soil and climatic conditions cause marked differences in their character; hence, eight fairly well-defined forest regions can be recognized. By far the largest of these is the boreal region which represents 82% of the total forested area. The Great Lakes–St. Lawrence region covers 6.5% and the subalpine region 3.7%. The montane, coast, and Acadian regions each account for approximately 2% while the remaining Columbia and deciduous regions each represent less than 1%.

Boreal forest region. This region comprises the greater part of the forested area of Canada. It forms a continuous belt from Newfoundland and the coast of Labrador westward to the Rocky Mountains and northwestward to Alaska. White spruce and black spruce are characteristic species; other prominent conifers are tamarack, which ranges generally throughout, balsam fir and jack pine in the eastern and central portions, and alpine fir and lodgepole pine in the western and northwestern parts. Although the boreal forests are primarily coniferous there is a general admixture of deciduous trees such as white birch and poplar; these are important in the central and south-central portions, particularly along the edge of the prairie. In turn, the proportion of spruce and larch increases to the north and, with the more rigorous climate, the close forest gives way to an open lichen-woodland which finally changes into tundra. In the eastern section, along the southern border of the region, there is a considerable intermixture of species from the Great Lakes–St. Lawrence forest, such as eastern white pine, red pine, yellow birch, sugar maple, black ash and eastern white cedar.

Great Lakes–St. Lawrence forest region. Extending inland from the edges of the Great Lakes and the St. Lawrence River lies a forest of a mixed nature characterized by eastern white pine, red pine, eastern hemlock and yellow birch. With these are associated certain dominant broad-leaved species common to the deciduous forest region, including sugar maple, red maple, red oak, basswood and white elm. Other species with wide ranges are the eastern white cedar and largetooth aspen and, to a lesser extent, beech, white oak, butternut and white ash. Boreal species such as white spruce,

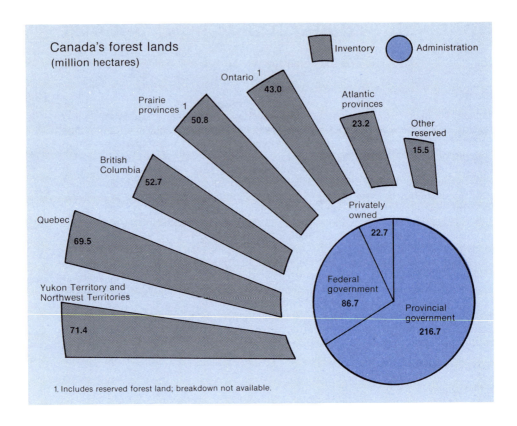

Canada's forest lands
(million hectares)

Inventory Administration

Ontario [1] 43.0

Prairie provinces [1] 50.8

Atlantic provinces 23.2

Other reserved 15.5

British Columbia 52.7

Quebec 69.5

Yukon Territory and Northwest Territories 71.4

Privately owned 22.7

Federal government 86.7

Provincial government 216.7

1. Includes reserved forest land; breakdown not available.

black spruce, balsam fir, jack pine, poplar and white birch are intermixed, and red spruce is abundant in certain central and eastern portions. This region extends in a westward direction into southeastern Manitoba but does not include the area north of Lake Superior.

Subalpine forest region. This is a coniferous forest located on the mountain uplands of Alberta and British Columbia, from the Rocky Mountain range through the interior of British Columbia to the Pacific Coast inlets. The characteristic species are Engelmann spruce, alpine fir and lodgepole pine. There is a close relationship between the subalpine and the boreal forest regions, which also has black spruce, white spruce and trembling aspen. There is also some penetration of interior Douglas fir from the montane forest, and western hemlock, western red cedar and amabilis fir from the coastal forests. Other species are western larch, whitebark pine, limber pine and, on the Coast Mountains, yellow cypress and mountain hemlock.

Montane forest region. The region occupies a large part of the interior uplands of British Columbia, as well as a part of the Kootenay Valley and a small area on the east side of the Rocky Mountains. It is a northern extension of the typical forest of much of the western mountain system in the United States, and comes in contact with the coast, Columbia, and subalpine forest regions. Ponderosa pine is a characteristic species of the southern portions. Douglas fir is found throughout, but more particularly in the central and southern parts; lodgepole pine and trembling aspen are generally present, the latter particularly well represented in the north-central portions. Engelmann spruce and alpine fir from the subalpine forest region, together with white birch, are found in the northern parts. White spruce, although primarily boreal in affinity, also grows here. Extensive prairie communities of bunch grasses and herbs are found in many river valleys.

Coast forest region. This region is part of the Pacific Coast forest of North America. Essentially coniferous, it consists principally of western red cedar and western hemlock, with Sitka spruce abundant in the north and Douglas fir in the south. Amabilis fir and yellow cypress are represented throughout the region and, with mountain hemlock and alpine fir, are common at the higher altitudes. Western white pine is found in southern parts, while western yew is in widely scattered groups. Deciduous trees, such as black cottonwood, red alder and bigleaf maple, have a limited distribution. Arbutus, a broad-leaved evergreen, and Garry oak grow only on the southeast coast of Vancouver Island, the adjacent islands and mainland. Both species are predominently southward in the United States.

Acadian forest region. This forest covers the greater part of the Maritime provinces and is closely related to the Great Lakes–St. Lawrence forest and, to a lesser extent, the boreal forest. Red spruce is a characteristic though not exclusive species, and associated with it are balsam fir, yellow birch and sugar maple, with some red pine, eastern white pine, jack pine and eastern hemlock. Beech was formerly more important than at present, but beech bark disease has drastically reduced its numbers in Nova Scotia, Prince Edward Island and southern New Brunswick. Other abundant species are white spruce, black spruce, red oak, white elm, black ash, red maple, white birch, grey birch and poplars. Eastern white cedar, although present in New Brunswick, is extremely rare elsewhere.

Columbia forest region. A large part of the Kootenay Valley, the upper valleys of the Thompson and Fraser rivers and the Quesnel Lake area of British Columbia contain a coniferous forest, called the Columbia forest region, which closely resembles the coast forest region. Western red cedar and western hemlock are the characteristic species in this interior wet belt. Associated trees are the interior Douglas fir which has general distribution and, in southern parts, western white pine, western larch, grand fir and western yew. Engelmann spruce from the subalpine forest region is important in the upper Fraser Valley and is found to some extent at the upper levels of the forest in the rest of the region. At lower elevations in the west and in parts of the Kootenay Valley, the forest merges with the montane forest region and in a few places borders directly on grassland.

Deciduous forest region. A small portion of the deciduous forest, widespread in the United States, extends into southwestern Ontario between lakes Huron, Erie and Ontario. Here, with the deciduous trees common to the Great Lakes–St. Lawrence forest region, such as sugar maple, beech, white elm, basswood, red ash, white oak and butternut, are scattered a number of other deciduous species which have their northern limits in this locality. Among these are the tulip tree, cucumber tree, pawpaw, red mulberry, Kentucky coffee tree, redbud, black gum, blue ash, sassafras, mockernut hickory, pignut hickory, black oak and pin oak. In addition, black walnut, sycamore and swamp white oak are confined largely to this region. Conifers are few but there is scattered distribution of eastern white pine, tamarack, eastern red cedar and eastern hemlock.

Grasslands. Although not a forest region, the prairies of Manitoba, Saskatchewan and Alberta support several species of trees in great numbers. Trembling aspen forms groves around wet depressions and continuous dense stands along the northern boundary. Several other species of poplar are usually found along rivers and in moist locations, along with willows and some white spruce. There are sporadic stands of white birch, Manitoba maple, bur oak and ash. In British Columbia, where grasslands are confined to deep valleys and low areas of the interior, there are scattered numbers of ponderosa pine, birch, poplar, spruce and mountain alder.

Forest land   10.1.1.2

Inventories of Canadian forest resources are made periodically by provincial forest authorities and, with their co-operation, the Canadian Forestry Service of the fisheries and the environment department compiles national statistics.

The 1973 national forest inventory reported an area of 3.2 million square kilometres of forest land (Table 10.1). Of this total, 131 140 km² are reserved by legislation for primary uses other than timber production. Currently, about 75% of the non-reserved forest land of Canada has been inventoried in the sense of gathering statistically reliable information on area and forest cover.

Provincial Crown forest land constitutes 69% of the non-reserved forest land of Canada, leaving 23% under federal jurisdiction and 8% in private ownership. Although precise use of private forest land is a matter of speculation, individual studies and limited statistics suggest that timber production still predominates despite a tendency to convert some of this land to recreational use.

The estimates of volumes of timber, given by province in Table 10.1, are also subject to constant revision as more accurate and complete inventories are compiled. The volumes reported in the 1973 national forest inventory are somewhat larger than those reported previously due to updating of inventories in some provinces. The estimates, however, are low because timber volumes for Labrador, the Yukon Territory and Northwest Territories are not available and because British Columbia has adopted procedures whereby data on volume of mature timber only were compiled.

10.1.1.3   Land use

The lands directorate of the fisheries and the environment department is responsible for investigating national aspects of land use in terms of management, research, planning and environmental concerns.

In support of resource management, the directorate operates a number of mapping programs. The largest is the Canada Land Inventory (CLI). Under federal-provincial agreement, all settled lands of Canada have been classified according to their capabilities for agriculture, forestry, recreation, wildlife, sportfish and present land use (circa 1967). These data, widely used for land-use planning at the regional level, have been placed in a computer system known as the Canada Geographic Information System (CGIS), enabling the production of statistics on land capability at the national level. In response to the need for other mapping techniques for those areas not covered by the CLI program, a biophysical land classification system has been developed and applied in the James Bay area. Further development of ecological (biophysical) land classification methodologies through the application of satellite imagery and high-altitude aerial photography is a major area of concentration. A map series on northern land use information, in co-operation with the Indian and northern affairs department, provides information regarding current land-use patterns and activities at a reconnaissance scale of 1:250,000 for the Yukon Territory and the Mackenzie District of the Northwest Territories.

National research programs have focused on trends and factors affecting land use. A map folio is being produced to define, locate and describe Canada's most important lands for agriculture, forestry, recreation, wildlife, urban growth and energy, in the form of a national perspective.

The directorate provides a secretarial service and plays an active role in a committee on ecological (biophysical) land classification. This federal-provincial committee is developing a national ecological land classification system. The directorate provides the leadership for an interdepartmental task force on land-use policy. The task force is preparing a paper with respect to a federal policy on land use in Canada. It will deal with land-use problems, the role of the federal government in land use and opportunities for the federal government to help resolve the problems.

The directorate plays the lead federal role in the federal-provincial agreement to conduct baseline environmental studies in Northern Quebec. This agreement, commonly known as the James Bay environmental studies, was extended to March 1979. The directorate is also involved in the negotiation of native land claims in various parts of Canada.

10.1.2   Forest depletion

The average annual forest utilization by cutting is shown in Table 10.2. The primary sources of Canada's current wood production are the areas of non-reserved Crown forest land allocated to wood production and private forest land. These two ownerships constitute 163.8 million hectares. On a volume basis, it was estimated in 1976 that the annual allowable cut to maintain productive forests was 255 million cubic metres. From 1966-75, the average annual total of wood harvested amounted to about 121 million m³, approaching half the allowable cut. In addition to cutting, extensive forest depletion is caused by fire, insects, diseases and natural mortality but no reliable estimates of these losses, either physical or economic, are available. A total of 10,358 forest fires occurred across Canada in 1976, destroying 2.2 million hectares of forest land (Table 10.3).

A large surplus of timber exists in Canada although there are shortages in some regions and species which could be overcome by increased silvicultural and management techniques. In addition, greater utilization of individual trees and of certain species could extend the resource.

Forest administration 10.1.3

Federal forestry programs 10.1.3.1

The federal government is responsible through several departments and agencies for protection and administration of forest resources in the Yukon Territory and Northwest Territories and on other federal lands such as national parks, Indian reserves, military areas and forest experiment stations. In addition, there are important federal responsibilities with respect to the nation's forest resources as a whole. These responsibilities, which relate to forestry research and development and the provision of information and technical services, are defined and established by the Forestry Development and Research Act (1966) and the Department of the Environment Act (1970).

The primary federal organization concerned with forestry is the Canadian Forestry Service of Fisheries and Environment Canada. Its headquarters organization consists of a director general with five branch directors covering policy analysis and development, forest protection, production and environmental forestry, forestry relations and technology transfer and forest utilization.

To promote improved management of forest resources and better forest products, the Canadian Forestry Service, through publications, workshops and seminars, disseminates technical information to forest resource and wood processing managers. Forest advisory services are provided for federal departments and international agencies, and assessments are also made of operational trials and treatments conducted by resource organizations and industry. In the forest products field, special testing services (unavailable commercially) are provided both for government and industry. The service also helps develop codes and specifications for forest products.

Through its publications, press releases, films, displays, visitor centres and demonstration areas, the service seeks to increase public awareness and understanding of forest values and to enlist support in the protection and wise use of forest resources.

Provincial forestry programs 10.1.3.2

All provincial forest land with the exception of minor portions in national parks, federal forest experiment stations, military areas and Indian reserves (except in Newfoundland) is administered by the respective provincial governments. The forestry program of each is outlined below.

Newfoundland. The forest resources of this province are geographically separated by the Strait of Belle Isle into two distinct regions — the island of Newfoundland and Labrador on the mainland. A forest inventory of Labrador, completed in 1975, was conducted on lands south of 56°N. The inventoried area disclosed a total of 258 012 square kilometres of which 55 374 km² were productive forest area; the total volume of black spruce and balsam fir in Labrador was 320.7 million cubic metres. A forest inventory of the island of Newfoundland showed that of a total area of 111 445 km², over 37 863 km² were classified as productive forest. This area supports a total gross volume of 281.8 million m³ of softwoods and hardwoods. The principal commercial species of trees are black spruce and balsam fir. White pine, white spruce, and white and yellow birch are of lesser commercial importance.

On the island 60% of the productive forest lands has been licensed, leased, or is owned by the pulp and paper industry while 37% remains under the direct jurisdiction of the province. Tenure of the remaining 3% is varied and includes federal and provincial parks.

Responsibility and authority over Crown forests in the province are vested in the forestry branch of the forestry and agriculture department. The branch employs more than 200 professional, technical and support staff. In four regions of the province 19 management units undertake the operational field work of forest protection, timber surveys, permits, enforcement, scaling, silviculture and forest management. Headquarters at St. John's is responsible for planning and program development. A new forest policy of increased utilization on a sustained yield basis, backed by legislation and intensified forest management, is being implemented following a study of all aspects of forests in the province.

A forest management inventory of the province is being conducted and is expected to be completed within five years. The legislation requires every owner of 121 hectares or more of forest land either to submit a plan for certification utilizing the annual sustainable yield of such land or to pay a high tax on the basis of unmanaged land. Limit holders are also being assessed an annual tax for managed land on the basis of area held.

The province's forest resource is primarily used to produce newsprint, linerboard and lumber. Two newsprint mills, one at Grand Falls and the other at Corner Brook, have a combined production capacity of approximately 1 860 tonnes a day. There is also a growing sawmill industry producing about half of the province's lumber requirements. This proportion is increasing and it is estimated that about 70% of provincial needs can be produced from the mix and extent of timber resources available. The total forest industry contributes about $173 million annually to the gross provincial product.

Forest research is principally carried out by the Canadian Forestry Service of the federal fisheries and environment department. Post high school education in forestry is available at Memorial University and at the College of Trades and Technology (CTT). Memorial University offers a three-year diploma course in forestry and is affiliated with the forestry faculty of the University of New Brunswick. At the CTT students may obtain a diploma after successfully completing a two-year forest technology course.

Prince Edward Island. About 45% of the 5 656 km² of land area is tree-covered. The wooded areas consist of scattered patches with a greater concentration in the eastern and western sections. All woodland is privately owned except some 129 km² of provincially owned forest land.

The forestry branch of the agriculture and forestry department administers all forestry matters in the province — reforestation, protection, extension and woodlot improvement. The reforestation program is not only concentrating its efforts on the genetic improvement of the commercially important tree species still existing in the province, but also on those endangered species that have practically disappeared due to shipbuilding and overcutting.

Nova Scotia. Of Nova Scotia's land area of 52 841 km², 44 442 km² are classed as forest and 75% of the forest land is considered suitable for regular harvesting. Although 91% of the forest land in Canada is held by the Crown in the right of the federal and provincial governments, only 24% is so held in Nova Scotia. Of the private woodlands, 71% are in parcels of up to 405 ha.

Provincial Crown lands are administered by the lands and forests department through a staff of foresters and rangers. Extension personnel assist owners of small private woodlands. The department administers the Lands and Forests Act as it pertains to all lands and is responsible for forest fire suppression. Forest fire detection is facilitated through 35 observation towers and an aerial patrol service with two helicopters and six fixed-wing aircraft. In 1976, 541 fires burned 17 535 ha of forest. The largest fire covered 13 365 ha. Fire suppression crews and rangers with equipment are stationed throughout the province.

The forest industry is important to the economy contributing about $150 million to the gross provincial product annually. In 1976 there were in operation some 438 sawmills of various types and sizes, one hardboard mill, two newsprint mills, one groundwood pulp mill and one chemical pulp mill. Roundwood production was 2.5 million m³, of which 2.4 million m³ was domestic pulpwood, 85 000 m³ was peeled pulpwood for export, and 13 601 m³ was poles, piling and pit props. Sawn products accounted for 449 970 m³ (volume-in-product), of which 432 990 m³ was lumber. Chip production totalled 311 300 m³, of which 299 980 m³ derived from sawmill residues and 10 075 m³ came from a recent development in Nova Scotia, whole-tree chipping.

A small reforestation program, active since the 1930s, has been greatly expanded in the 1970s. Experimental work on container planting, direct seeding, soil capability and site preparation continues, and efforts are being made to improve seed sources. Total softwood inventory as of October 1976 was 10.4 million seedlings and transplants, and 2.5 million trees were planted.

Timber, pulpwood and Christmas trees are sold through public tender, and cutting on Crown lands is done on recommendation of resource managers of the lands and forests department. Management cruises, regeneration studies and experimental cuttings are conducted on Crown lands and a program of operating these lands under long-term, integrated-use management plans is under way. During 1975-76, 1 299 ha of unleased Crown forest were thinned and improved, bringing the total area of Crown silvicultural treatments to 19 845 ha since 1965. Thirty-seven kilometres of new Crown land access road were added to the existing 579 km.

The provincial forest inventory, a continuous system designed to operate on a 10-year cycle, commenced its second cycle in 1971. Aerial colour photography, begun on Cape Breton Island in 1969, is being extended to the rest of the province. Remeasurement of a system of 1,765 randomly located sample plots every five years provides continuing data on growth, harvest rates and mortality.

Forest research is carried on by federal government agencies and the Nova Scotia Research Foundation. Investigations cover stand improvement, tree nutrition, cutting methods, and insect and disease activities. Extension projects include fire prevention, a province-wide motion picture program, distribution of information on forest and wildlife conservation, promotion of the Christmas tree industry, a hunter safety program, woodlot improvement, preparation of material for the mass media, and technical assistance to sawmill operators.

New Brunswick. Of New Brunswick's 72 092 km² approximately 87% is classed as forest land suitable for regular harvest. About 46% of the forest land is owned by the Crown, administered and managed by the natural resources department through its five forest regions and four support branches. The department has taken over administration of forest extension programs for privately owned woodlots.

The forest industry is of prime importance to the economy of New Brunswick, directly contributing over $220 million in value-added from primary forestry and forest-related industries and directly employing nearly 14,000 people. The total volume of standing timber is estimated at 580 million m³; coniferous species make up 70% and deciduous species the remainder. Approximately 8.5 million m³ of timber are currently harvested annually with 70% of the harvest being cut as pulpwood.

A large-scale silvicultural program has been initiated by the natural resources department and funded under a federal-provincial agreement. In 1977, approximately 17 million seedlings were planted on Crown lands with a planned increase to 30 million seedlings by 1980.

To evaluate new methods of timber allocation a pilot area has been selected in northeast New Brunswick. A forest management licence for approximately 4 050 km² has been cancelled and replaced by a long-term guarantee to provide annually, to the former licensee, a specified volume of standing timber for harvesting.

New Brunswick carries out an aerial spraying program to protect balsam fir and spruce from the spruce budworm. It has been carried out since 1952 by a Crown corporation sponsored by the provincial government and several of the major forest products companies.

New Brunswick does not maintain a forest research organization but co-operates with the Canadian Forestry Service and the University of New Brunswick in its research program. The university offers undergraduate and graduate courses in forestry leading to BScF and MScF degrees. It is also responsible for the administration of the Maritime Forest Ranger School in conjunction with the governments of New Brunswick and Nova Scotia and with private industry.

Quebec. Forests with economic potential cover 684 000 km², about 45% of the total area of the province. This forest cover stretches northward to an irregular line near 52°N in the east and west and 53°N in the centre of the province. Private forests cover an area of 70 000 km². Public forests cover 614 600 km² of which 492 000 km² are productive and under management plans. Public forests carry a volume of almost 3 804 million m³ of standing timber of various species; private forests contain 470 million m³. Coniferous species make up 75% of the total volume. Private forests account for about 20% of the annual cut, about 5.7 million m³. Forests account for about 25% of the gross provincial product.

The lands and forests department controls the development and use of woodlands, and undertakes conservation measures. Principal management controls are: the annual inventory of some 78 000 km² of forest land; study and regulation of silvicultural practices for this area and the zoning of the land for its best use; and restoration of lands destined for forestation by replanting or by proper treatment. To achieve this, Quebec maintains some 100 million plants in nursery stock. Regulations governing the use of the forests cover operational control, the issuing of permits for establishment of mills and cutting permits, measurement of wood harvested on Crown land, aid to development of private forests, and building and maintenance of forest roads. Through regional conservation groups this branch is responsible for forest protection against insects, fire and fungus attack.

Ontario. Forested land in Ontario amounts to 803 852 km², of which 425 981 km² are classified as forest land bearing or capable of bearing commercial timber suitable for regular long-term harvest. About 90% of the productive forested land is owned by the Crown, administered and managed by the provincial ministry of natural resources through three main programs: lands and waters, outdoor recreation and resources.

The forest resources branch is responsible for the regeneration, tending and improvement of the forests under The Woodlands Improvement Act and the promoting of forestry on privately owned lands. The branch operates 10 nurseries with a current production target of about 73.5 million trees. Complementing this are up-to-date tree improvement and nursery soil management programs. The branch, directly or indirectly, supervises all planting projects on Crown lands but regeneration agreements have been signed with all major licensees under which the latter assume responsibility for planting projects.

During 1976, 46 million nursery-produced trees were planted on about 24 160 ha of Crown and agreement lands, and tubed seedlings were planted on about 2 005 ha. Other silvicultural treatments included the direct seeding of 27 039 ha, treatment for natural regeneration on 18 572 ha and stand improvement (cleaning, spraying, thinning and pruning) on 38 325 ha. In all, 110 108 ha of Crown and agreement lands were silviculturally treated in 1976 to promote regeneration or to improve the forests. Owners of private lands may purchase planting stock for forestry purposes from government nurseries at nominal prices and may also receive free professional advice on any forestry matter. In 1976 (spring and fall), planting stock furnished for private lands totalled 15.8 million units. Under The Woodlands Improvement Act it is possible to have planting and improvement work carried out completely under government direction and mainly at public expense. Since 1966, the program has provided assistance for 97 917 ha of privately owned land.

Ontario has enabling legislation to permit municipalities and conservation authorities to place abandoned and submarginal agricultural lands to which they have acquired title under agreement with the ministry, which undertakes to plant and manage the properties for a specified period. Over 106 634 ha under such agreements are managed intensively and the older plantations are receiving regular thinnings. The trees removed are in demand for pulpwood, posts, poles and sawlogs, making the undertakings financially attractive. In addition, properties near population centres have acquired value as recreational areas. Forest pest problems in 1976 were dominated by the spruce budworm which infested almost 14.8 million hectares, but spraying operations to control this insect were limited to 40 875 ha in high-value local areas. Smaller acreages on Crown lands and lands managed under agreement were also treated for white pine weevil, pine and spruce sawflies, white grubs, white pine blister rust, annosus root rot and mice.

The forest research section provides scientific and technical knowledge for the management of forest lands and is more specifically oriented toward attaining production targets. Various disciplines including tree ecology and physiology, site and fertilization, tree genetics and breeding, mensuration, silviculture, equipment design and development are used to solve problems in tree improvement, stock production, regeneration and forest tending. The results of province-wide research are published in journals and reports. Research headquarters is at Maple and there are four field stations at Thunder Bay, Sault Ste Marie, Dorset and Midhurst.

The timber sales branch co-ordinates and supervises preparation of management plans for Crown management units and approves the plans prepared for company management units. Forest inventory requirements and priorities for such plans are determined by the branch. As of March 31, 1977, 179 plans (88 Crown units, 26 company units and 60 agreement forests) were completed or under way for about 533 540 km². The planning of access roads crucial to proper management is also part of the branch's responsibilities. The branch arranges for the allocation, disposition and measurement of Crown timber through Crown land licensing, timber sales and wood scaling. During 1976, some 440 Crown timber licences covering an area of 255 133 km² were effected. The Crown land harvest amounted to 13.1 million m³. Ontario's primary wood-using industries are licensed and their performance is monitored. In 1976, there were 761 primary wood-using plants in Ontario. The branch is also responsible for promotion of new industrial development and growth of the forest industry. Information is collected and analyzed on production, transportation and utilization of timber.

The air service and fire management branch is responsible for the 518 000 km² area under organized forest protection which is divided into eight regions and 38 districts. In 11 additional administrative districts, south of this area in the highly developed agricultural counties, municipalities are responsible for fire control. The vast inaccessible areas to the north of the fire districts, totalling over 295 000 km², do not support significant stands of merchantable timber and, except for the protection of private property and human life, are not normally protected. Within the fire regions, agreements were in effect in 1976-77 with 209 municipalities for prevention and control of forest fires. An agreement was also in effect with the federal government for fire protection of 392 127 ha of Indian lands.

Organized forest fire detection is accomplished primarily by aerial patrols with limited backup detection provided by several lookout towers in areas of high value such as Algonquin Park. Public reporting of forest fires is an important part of the program. The basic fire-fighting strike force comprises 135 trained five-man fire crews and 39 fire-bombing aircraft. The natural resources ministry owns 49 aircraft, most of which can drop either long- or short-term retardant on fires. Rented helicopters are also used. The communications system includes a network of ground stations, radiotelephones, fireline radios, aircraft radios, portable aircraft radiotelephones, Telex and facsimile.

Manitoba. The administration of provincial Crown forest lands in Manitoba is the responsibility of the renewable resources division of the renewable resources and transportation services department. The renewable resources division contains the lands and forests branch, the fish and wildlife branch, the planning and development branch and management services. All forestry activities occur in the forestry section of the lands and forests branch. The one exception occurs in the planning and development branch where long-term forest resource utilization and development plans are developed by a senior forest planner.

The forestry branch contains three major forestry programs: forest inventory, forest protection and forest management. These three functional areas administer the various forestry acts and regulations. Policy guidelines, projects and procedures are established in conjunction with regional forestry staff assistance who are subsequently responsible for implementation and delivery of the program. The head office of the three functional programs is also responsible for co-ordination, evaluation and effectiveness of the programs once they have been implemented.

The forest management office is responsible for the licensing, allocation of timber sales and permits, statistical data and collection of royalties on timber harvest. Forest management is also responsible for control measures relating to propagation, improvement and management of forests. The forest inventory section determines the extent of Manitoba's forests and how much may be harvested on a sustained annual basis. The forest protection section co-ordinates fire, insect and disease control activities.

A provincial forest nursery is maintained to supply stock for reforestation programs and a dynamic tree improvement program has been initiated to ensure that future supplies of seedlings will be of the highest possible quality. Seedlings are supplied to farmers for worklots and to commercial Christmas tree producers. An average of 3 million seedlings are planted annually in reforestation projects on Crown lands. Conventional planting programs are being reduced in some areas and reforestation of cutover lands is being achieved through scarification and seeding. Forest improvement consists of thinning, cleaning and chemical spraying to remove undesirable species and encourage growth of preferred trees on plantations and in natural stands. Forest inventories cover about 26 806 km² annually and, on the basis of these inventories, working plans with annual allowable cuts are made.

Forest management licences may be granted for periods of up to 20 years and are renewable. Timber sales may be from one year upward and timber permits for periods of up to one year. Three pulp and paper mills and one large sawmill provide the backbone for Manitoba's primary forest industry. A dozen intermediate-sized sawmill operations augment the production of the four larger mills. Numerous small sawmills and timber harvesting operations provide the balance of production.

There are 332 477 km² under forest protection with zones of priority in less accessible areas. Fires are detected through a comprehensive network of lookout towers and a highly efficient aircraft detection system and supporting ground patrols. Approximately 233 000 km² are covered by aerial patrols.

Public education in fire prevention and forest conservation is carried out through radio, television, newspapers, pamphlets, signs, films and tours.

Saskatchewan. The forests of Saskatchewan cover 352 000 km² of which 115 000 km² are productive and suitable for harvest.

The forestry branch of the tourism and renewable resources department consists of four sections — management, wood products and operations, inventory and silviculture — and develops and evaluates forest policies which are carried out by regional authorities. The province is divided into seven resource administration regions and further divided into resource officer districts. The Northern Saskatchewan department administers the northern forested area. Forest protection is also the responsibility of this department, and is effected by a network of 75 lookout towers, supplemented by patrol aircraft during high hazard periods. A VHF communication system is operated in towers, vehicles, aircraft and bush camps for detection and suppression of forest fires. Helicopters and fixed-wing aircraft capable of water-dropping provide aerial support. Six land-based Tracker aircraft were purchased, equipped to drop long-term fire retardants, and operate from four new airports constructed in 1977.

Alberta. The 383 751 km² of forest lands in Alberta include 276 494 km² capable of producing forest crops. The Alberta forest service of the energy and natural resources department through its five branches (program support, timber management, forest protection, land use and reforestation and reclamation) is responsible for their administration. Jurisdiction is decentralized into 10 forests, each responsible for the forest area within its boundaries. Each forest is under the control of a superintendent supported by specialists in timber management, fire, land use, construction and communications. These forests are further subdivided into ranger districts under a district forest officer responsible to the superintendent.

The timber management branch is responsible for the timber quota system, management and annual operating plans for leased and licensed Crown lands, forest management plans and disposal of Crown timber. The branch carries on silvicultural programs, processes applications, takes inventories of forest resources, inspects cutting areas to ensure proper logging practices and collects dues and fees.

The forest protection branch is in charge of all phases of protection including prevention, detection and suppression of fires. This branch includes specialists such as a meteorologist and a telecommunications officer; an aircraft dispatch section assists in the overall protection program.

The forest land use branch is responsible for planning and supervising land-use practices in the forested area including grazing, recreation and watershed management, particularly on the east slopes of the Rocky Mountains containing the headwaters of the North and South Saskatchewan rivers. The reforestation and reclamation branch plans and carries out reforestation and reclamation projects on provincial forest lands and operates the new Pine Ridge Forest Nursery, which will produce 20 million tree seedlings a year beginning in 1981 for use by industry and government in restocking cutover and burned over lands in Alberta.

Basic research in the forestry program is generally carried out by the federal forestry service, largely through a federal research laboratory in Edmonton.

British Columbia. Over 544 920 km² or 60% of British Columbia's area, is classified as forest land. This includes over 7.82 million m³ of mature merchantable timber, most of it coniferous species. Of this, 95% is publicly owned and managed by the British Columbia forest service. For management purposes, the province is divided into six forest districts with headquarters at Vancouver, Kamloops, Nelson, Williams Lake, Prince Rupert and Prince George. Further decentralization of authority is effected by subdivision into 99 ranger districts, each managed by a forest ranger who supervises harvesting, reforestation, silviculture and environmental protection. Fourteen directional, servicing or policy-forming divisions constitute the head office of the forest service at Victoria.

Efforts continue to bring the province's forest resources under sustained-yield management even though with an annual cut (1976) of 69.5 million m³ the total inventory would appear sufficient to support current needs in perpetuity. Sustained-yield administration has resulted in a greater proportion of the annual forest harvest coming from the interior of the province; in 1976 the wet belt forests on the coast accounted for about 46.3% of the total forest cut and the interior for 53.7%. Almost all interior forest is publicly owned, with most of the privately owned, leased or licensed forests on the coast. There are several systems of timber disposal. The tree farm licence is a contract between the government and a company or individual whereby the latter manages an area of forest land, including any privately held forest land, on a sustained-yield basis. Tree farm licences are subject to re-examination for renewal every 21 years. Public sustained-yield units are areas within which the forest service manages the Crown timber on a sustained-yield basis. Within the public sustained-yield units, recognized established logging operators can apply for timber sale licences or timber sale harvesting licences which entitle them to log at a given rate per year.

Forest fire prevention and suppression are vital aspects of planned sustained-yield management. Contracted air tankers, fire-spotter aircraft and helicopters are employed during the fire season in order to achieve early discovery and attack on forest fires.

Close liaison with the federal forestry service provides detailed information on insect and fungal enemies of the forest and on fire research.

To achieve an efficient administration of multiple use of Crown forest lands, the forest service, in conjunction with other government ministries, has developed the integrated use concept. The forest service recognizes that inevitably some forest lands will be withdrawn from timber production to accommodate other users. These losses must be offset by increased production on remaining areas.

10.1.4   Statistics of the forest industries

The forests of Canada provide raw materials for several large primary industries. Much of the output of the primary forest industries is exported; the sawmill industry and the pulp and paper industry, especially, contribute substantially to the value of the export trade of Canada and thereby provide an important part of the foreign exchange necessary to pay for imports. Statistics of manufacturing activity and total activity of the wood industries and the paper and allied industries are given in Chapter 17.

10.1.4.1   Logging industry

Tables 10.4 and 10.5 give the estimated quantities of wood cut in Canada, by province and by type of product, for 1973-76. The total volume of wood cut increased from 115 million m³ in 1975 to 140 million m³ in 1976.

10.1.4.2   Wood industries

The standard industrial classification subdivides the wood industries group as follows: sawmills and planing mills, shingle mills, veneer and plywood mills, sash, door and other millwork plants, hardwood flooring mills, manufacturers of prefabricated buildings, manufacturers of kitchen cabinets, wooden box factories, the coffin and casket industry, the wood preservation industry, the wood handles and turning industry, particleboard, and miscellaneous wood industries.

The sawmills and planing mills, the shingle mills, the veneer and plywood mills and the particleboard plants (the latter are included in the miscellaneous wood industries group) use mainly roundwood as a raw material and sometimes are called primary wood industries. The secondary wood industries further manufacture part of the production of the primary wood industries into a great variety of products. However, most of the production of the primary wood industries is not further processed.

Sawmill and planing mill industry. Lumber is by far the most important single commodity in this industry and British Columbia is the most important province in this field. The total value of shipments of establishments classified to this industry in 1975 amounted to $1,996.9 million of which lumber accounted for $1,473.3 million. In addition to this lumber, a small amount is produced by establishments classified to other industries bringing total lumber production in Canada in 1975 to 27 304 million m³.

Shingle mill industry. Most of the shingles and shakes produced in Canada are from British Columbia mills. Considerable quantities are produced by establishments classified to other industries and by individuals intermittently operating one or two shingle machines or producing shingles by hand; although no adequate measure of this production is available it is known to contribute significantly to the total.

Veneer and plywood industry. The production of hardwood veneer and plywood in Canada is confined largely to the eastern provinces and the production of softwood veneer and plywood almost entirely to British Columbia. For the latter, Douglas fir is most commonly used because of availability of its large-diameter logs from which large sheets of clear veneer can be obtained. Of the hardwoods, birch is by far the most important species. Although most of the raw materials are of Canadian origin, some decorative woods are imported, particularly walnut.

Most of the production of softwood veneers is further manufactured into softwood plywood by Canadian mills. Some hardwood veneers are also shipped to other veneer and plywood mills in Canada for further manufacture or to other industries such as the furniture industry for veneering purposes but a significant portion is exported.

10.1.4.3   Paper and allied industries

The standard industrial classification subdivides the paper and allied industries group into the pulp and paper industry, the asphalt roofing manufacturers, the paper box and bag manufacturers, and other paper converters.

Pulp and paper industry. This industry is by far the most important of the group. Part of its production is consumed in Canada or serves as raw material for the paper-using or secondary paper and allied industries. A great part of it is exported, particularly newsprint and various types of pulp, most of it to the United States. Some plants included in the pulp and paper industry classification also convert basic paper and paperboard into more highly manufactured papers, paper goods and boards but their output represents only a small part of Canada's total production of converted papers and boards. Table 10.9 gives shipment and production figures for pulp for 1974-77 and Table 10.10 gives shipments of basic paper and paperboard for 1973-75. Table 10.11 shows exports of pulp and newsprint for 1973-76.

Asphalt roofing manufacturers. These establishments produce composition roofing and sheathing, consisting of paper felt saturated with asphalt or tar and, in some cases, coated with a mineral surfacing. Their total shipments in 1975 were valued at $121.3 million.

Paper box and bag industries. These industries include manufacturers of folding cartons and set-up boxes, of corrugated boxes and of paper bags. Their total shipments in 1975 amounted, respectively, to $307.2 million, $511.4 million and $352.3 million.

Other paper converters. This group produces a host of paper products, among them envelopes, waxed paper, clay-coated and enamelled paper and board, aluminum foil laminated with paper or board, paper cups and food trays, facial tissues, sanitary napkins, paper towelling and napkins and toilet tissue. The total value of manufacturing shipments of this industry in 1975 amounted to $717.3 million.

Fisheries   10.2

Canada has co-operated with other nations to conserve high-seas fisheries resources through joint research projects and international agreements and took further action to protect and manage the fisheries in its coastal areas by extending its coastal fisheries jurisdiction to 200 nautical miles, effective January 1, 1977. Several bilateral agreements have been concluded with foreign countries to allow them to continue to fish within Canada's extended jurisdiction for stocks surplus to Canada's harvesting capacity and to provide a smooth transition to the new regime of fisheries management off the Canadian coasts.

The federal government has full legislative jurisdiction over the coastal and inland fisheries of Canada and all laws for the protection, conservation and development of these fisheries resources are enacted by Parliament. The management of fisheries is shared with the provincial governments to which certain administrative responsibilities have been delegated.

The federal fisheries and the environment department controls all fisheries, both marine and freshwater, in Newfoundland, Prince Edward Island, Nova Scotia, New Brunswick, the Yukon Territory and Northwest Territories. In Ontario, Manitoba, Saskatchewan and Alberta all fisheries are managed by the provincial governments. In Quebec, the provincial government manages both marine and freshwater fisheries but the inspection of fish and fishery products produced for sale outside the province is carried out by the federal department, as in all other provinces. In British Columbia, the fisheries for marine and anadromous (fish that migrate to the sea from fresh water) species are managed by the federal department, but the provincial government manages freshwater fisheries. In the national parks fisheries are managed by the Canadian Wildlife Service. In most instances, licences for sport fishing are distributed by the respective provincial or territorial governments which retain all revenues so collected.

Close contact with provincial authorities is maintained through fisheries and marine service regional offices. Co-ordination and discussion between federal and provincial fisheries managers on policies, programs and matters of mutual concern are facilitated through several federal-provincial committees.

Federal government activities   10.2.1

The work of the federal government in the conservation, development and general regulation of the nation's coastal and freshwater fisheries is carried out by the fisheries and marine service.

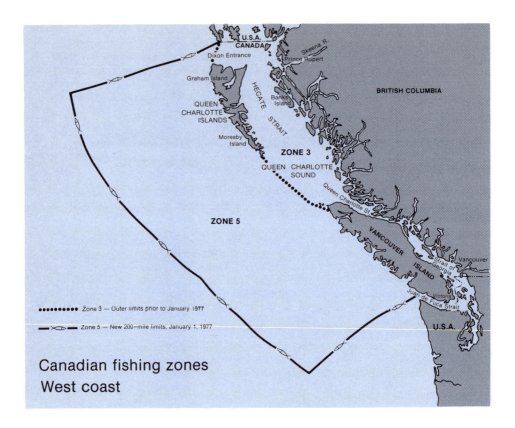

Canadian fishing zones
West coast

The fisheries and marine service, a major component of the fisheries and the environment department, undertakes a broad range of responsibilities. These include: management of Canada's ocean and some inland fisheries; fisheries and oceanographic research contributing to the understanding, management and optimum use of renewable aquatic resources and marine and fresh waters; hydrographic surveying and charting of navigable coastal and inland waters; administration of small craft harbours; environmental impact studies affecting coastal and inland waters; and research in support of international agreements relating to fisheries management and marine environmental quality.

Functions of the fisheries and marine service are grouped under three major units: fisheries management, ocean and aquatic sciences and small craft harbours. The service carries on most of its programs at regional and field locations. Regional headquarters for fisheries management are located at Vancouver, BC; Winnipeg, Man.; Quebec, Que.; Halifax, NS; and St. John's, Nfld.; and for ocean and aquatic sciences at Patricia Bay, BC; Burlington, Ont.; and Dartmouth, NS. Research institutes and laboratories are located at a number of centres across Canada.

Several appointed public corporations and boards are involved in activities closely aligned with those of the fisheries and marine service, including the Fisheries Prices Support Board, the Canadian Saltfish Corporation and the Freshwater Fish Marketing Corporation.

International fisheries. Many injurious effects on aquatic resources are results of historical practice, insufficient knowledge, multiple uses of water, social and economic conditions, and national and international competition. Problems under national control are corrected as conditions warrant but many resources are shared with other nations and must be managed jointly.

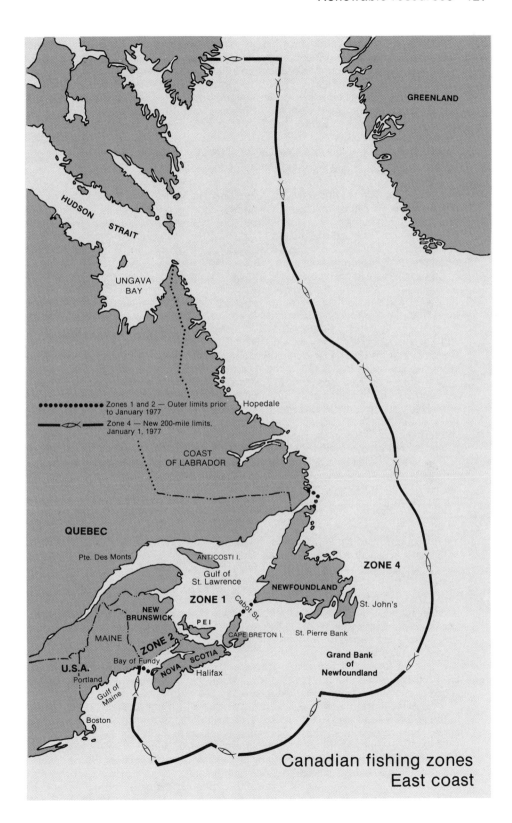

GREENLAND

HUDSON

STRAIT

UNGAVA
BAY

●●●●●●●●●●●●● Zones 1 and 2 — Outer limits prior   Hopedale
               to January 1977
─────⊲⊳───── Zone 4 — New 200-mile limits,
               January 1, 1977

COAST
OF LABRADOR

QUEBEC

Pte. Des Monts

ANTICOSTI I.

Gulf of
St. Lawrence

ZONE 4

NEWFOUNDLAND

St. John's

ZONE 1

NEW
BRUNSWICK

Cabot St.

P E I

MAINE

CAPE BRETON I.

St. Pierre Bank

ZONE 2

SCOTIA

Bay of Fundy

U.S.A.

NOVA

Halifax

Grand Bank
of
Newfoundland

Portland

Gulf of
Maine

Boston

Canadian fishing zones
East coast

Canada co-operates with many nations in obtaining scientific data and formulating policies for development and conservation of fisheries through membership in 10 international fisheries commissions and one international council. These international organizations are established under the terms of formal conventions. Canadian representatives are appointed by order-in-council and include officials of the fisheries and the environment department and members of the fishing industry.

Canada maintains membership in the fisheries committee of the Food and Agriculture Organization of the United Nations and in the Codex Alimentarius Commission which is concerned with world food quality standards.

10.2.2 Provincial government activities

Newfoundland. The provincial fisheries department promotes development in all sectors of the province's fishing industry. Experiments and demonstrations are conducted on new designs of fishing gear and modification of existing types, construction of multi-purpose fishing craft and exploration of potential fishing grounds to increase catching efficiency and landings. Subsidies are also paid to fishermen for fishing vessels and certain types of inshore fishing gear.

Loans are made to processors for the establishment and expansion of fish processing plants and for deep sea draggers. Aid to fishermen to build modern vessels capable of a greater variety of fishing operations and larger production is provided by loans from the Newfoundland Fisheries Loan Board. The Fishing and Coasting Vessels Rebuilding and Repairs (Bounties) Act authorizes financial assistance in maintaining and prolonging the life of the existing fleet. The Coasting Vessels (Bounties) Act and the Fishing Ships (Bounties) Act authorize the granting, for locally built ships over 12 years of age, of a bounty of 35% of the approved cost of the work provided the vessel is over 10 gross registered tons. A small boat bounty program provides a bounty of 35% on the approved cost of fishing boats measuring in length from 6.08 to 10.7 metres or over, providing that they do not exceed 10 tons gross. Loans are available to fishermen to build new boats, purchase used boats, acquire new engines, buy certain approved types of mechanical and electronic fishing equipment and convert boats from one type of fishing operation to another.

In terms of direct employment generated, fisheries continue to outrank all other resource sectors. In 1976 approximately 15,000 fishermen and 7,000 plant workers were engaged in the industry. Total landings of all fish species amounted to 338 million kilograms, with a landed value of $63 million and a market value of approximately $160 million.

The inland waters of Newfoundland, although they provide excellent sport fishing, are not commercially exploited. Lakes and ponds actually remain under the authority of the tourism department but, under federal-provincial agreement, these waters, including rivers and streams, are under federal control in matters of conservation and guardianship.

Prince Edward Island. The Prince Edward Island fishing industry ranks third in the island economy. The landed value of the 1976 catch was in excess of $12.6 million and after processing, its value exceeded $30 million. The industry involves 3,000 fishermen and helpers and between 700 and 800 people in the fish processing industry at 24 processing facilities.

The PEI fisheries department, with six divisions, supplements the activities of the federal fisheries and marine service and is responsible for administration of programs aimed at upgrading the industry and increasing returns to those engaged in it.

An active program in the aquaculture division is aimed at diversifying opportunities in the fisheries sector through increased production, improved quality and broadening of the resource base in shellfish, salmonids, seaplants and lobsters, with emphasis in 1977-78 placed on oyster industry development, shellfish culture, seaplant seeding and salmonid culture. The resource harvesting division conducts exploratory fishing projects and resource assessment studies to locate and quantify new and existing species, and conducts gear technology studies relating to harvesting methods and equipment. In 1977-78, projects included scallop and clam resource development, groundfish catch expansion, seaplant resource development, pelagic species development and lobster resource development.

The product handling division is responsible for port development programs designed to improve fish handling, processing and holding techniques and to ensure better fish quality. Water systems, offloading ramps, fishermen's bait sheds, weather shelters and unloading systems and wet fish storage facilities were to be provided at additional ports in 1977-78. The processing and quality control division administers the PEI Fish Inspection Act and Regulations, enforces quality standards, assists the processing industry to improve methods and productivity and promotes new product development.

The economics and statistics division provides technical assistance to the fishing industry in financial management, feasibility analysis and statistical studies. The extension division administers information programs, conducts field demonstrations, prepares fishermen's training programs and conducts technical upgrading programs.

Loans are made to fishermen and the fishing industry through a Crown corporation established in 1969 which is empowered to grant credit in the sectors of fisheries, industry, tourism and agriculture. Provincial responsibilities concerning freshwater fisheries are discharged by the environmental conservation services division of the environment department.

Nova Scotia. The fishing industry in Nova Scotia is of major importance to the province's economy. Landed value of fish in 1976 was approximately $102 million while the market value was in the order of $235 million. Fish products account for more than 30% of Nova Scotia's exports. Over 11,500 fishermen and 4,500 plant workers are directly employed in the industry and 191 fish processing plants are in operation.

The Nova Scotia fisheries department is engaged in almost all aspects of the fishing industry and has significant input into some of the policies and programs legislated and administered by the federal government.

The primary thrusts of the department are in industrial and resource development, training, and field service. The industrial development division is responsible for programs related to fishing vessels, gear and equipment, harbours and wharves, port facilities and processing plants. It deals with equipment and facilities involved in catching, handling, processing and marketing fish and fish products, and provides technical assistance and direction to fishermen and processors, as well as incentive financial assistance to encourage improvement and new technology aimed at greater productivity. Loans are available to fishermen and processors through the Nova Scotia Resources Development Board.

The resource utilization division is involved in a broad range of projects directed toward making the greatest use of all fishery resources, processing these to the highest level, and marketing the products at a good price. It includes efforts in aquaculture aimed at producing fishery resources in a controlled environment. Activity is also directed to development of under-utilized and unexploited species, recovery and use of fish presently discarded, and encouragement of production of more food fish and less fish meal.

The training and field services division is responsible for extensive training programs for commercial fishermen. It operates the fisheries training centre in Pictou as well as courses held in fishing communities throughout the province. The division also provides a field service consisting of nine fisheries representatives who act in liaison with fishermen and all segments of the fishing industry.

New Brunswick. Commercial fishing is one of the most important industries of New Brunswick, employing about 5,000 fishermen with annual earnings of $25.5 million and 7,000 plant workers. The annual marketed value of all fish and shellfish products is about $120 million; 73% of exported production goes to the US. New Brunswick's commercial fisheries, both tidal and inland, are under the jurisdiction of the federal government, while angling in Crown waters is the responsibility of the provincial natural resources department. The New Brunswick government plays a major role in resource assessment and development, fisheries training, financial assistance to the industry and long-term planning. The fisheries department has a highly qualified staff grouped into six branches: administration, inspection and marketing, planning, development, training and fishermen's loan board. Close liaison is maintained with other federal and provincial departments and agencies concerned with the fishing industry.

The administrative branch is responsible for budgeting, accounting and personnel administrations within the department. It provides highly qualified staff for departmental duties and services.

An inspection and marketing branch administers the New Brunswick Fish Inspection Act and regulations. To avoid duplication of personnel, arrangements have been made with the federal fisheries and environment department for inspection duties. The branch promotes expansion and modernization of existing fish processing plants and establishment of new plants. It carries out a program of product development to increase the added value of the catch. Another program is aimed at promoting fish consumption within the province. The inspection branch studies existing and potential markets in collaboration with other government agencies. Other services for the fishing industry include grants to provide ice facilities and bait sheds.

The research and development branch carries out programs in co-operation with the federal fisheries and environment department. Technical and financial assistance is given the provincial department for projects undertaken toward modernizing fishing methods, experimenting with new types of fish-catching equipment and demonstrating its operation to fishermen, and exploring and developing unexploited or under-exploited species of molluscs, crustaceans, fishes and seaweeds. This work has resulted in the establishment of snow crab, shrimp, Irish moss, tuna, sea urchin and eel fisheries.

The fisheries training branch provides technical training to fishermen and plant personnel; training is also provided for managers and supervisors. The branch operates a school of fisheries at Caraquet in northeastern New Brunswick where, in the 1976-77 academic year, 162 fishermen received training. The marine emergency duty centre is a step forward in the field of technical training in the fisheries sector. This facility includes a two-storey mock-up of a fishing vessel used for fire fighting and a separate building with classroom facilities to improve training.

A planning and co-ordination branch strives to formalize planning of government efforts in fisheries, aquatic resources and fisheries development. To improve programming, the branch has taken over the setting up of a mechanism for industry-government consultation.

The Fishermen's Loan Board of New Brunswick, a provincial corporation established in 1946, now operates under the Fishermen's Loan Act of 1952 and the regulations of November 1, 1963. The board consists of two major branches, the loans administration branch and the boat building branch. The former includes the credit section which investigates and secures loans with insurance, while the accounting section collects repayments. The latter consists of the vessel inspection, evaluation and contracts section. The board's main function is to make loans to the fishing industry for modernizing and developing the fishing fleet. It provides financial assistance at moderate interest rates to fishermen and processing firms and corporations to build modern fishing vessels, finance major repairs, and purchase engines and equipment.

Since the board's inception, it has granted 3,736 loans to New Brunswick fishermen for a total of over $60.8 million. Outstanding loans amounted to $29.6 million as of March 1977. Loans are repayable within five to eight years on most small inshore fishing vessels. Repayment schedules on large trawlers may extend to 15 years based on the gross proceeds of the catch. Others are on a 15-year annual instalment contract. Most new fishing vessels being built for fishermen and processing firms in the province are financed by the board.

Sport fishing contributes substantially to the economy of the province. Great Atlantic salmon rivers like the Miramichi, the Restigouche and the Saint John are known around the world for their prolific production of this majestic game fish and attract many thousands of tourists each year. Anglers catch as many as 50,000 salmon a year in the Miramichi system alone. Many other species are sought by both residents and non-residents in hundreds of streams, rivers and lakes. Tuna sport fishing has become an interesting venture for tourists in northeastern New Brunswick.

Quebec. In 1975 Quebec fishermen landed 53 million kg of fish and shellfish in the vast reservoir formed by the St. Lawrence River, gulf and estuary. The landed value to the fishermen was $14.5 million and the market value of the produce was $29.0 million.

The industry is of prime importance on a regional basis. It is the backbone of the economy of the Magdalen Islands and the lower North Shore and is a major activity in the Gaspé peninsula. Overall there are 6,460 commercial fishermen, including full-time coastal fishermen, sea-going helpers and officers and crew operating the seiners, long-liners and draggers. Some 30 processing plants employ about 1,300 workers. In this sector, commercial fishing has a multiplier effect on employment and incomes. Fishermen and shipowners build and repair their fishing vessels within the region, thus giving employment to shipyards. Local labour is also used for building and maintaining the various marine installations necessary for docking, safety and discharge of cargo, for operating ice-making plants, and in freezer and storage operations.

In 1975 redfish (33.1%), cod (25.8%) and herring (17.7%) made up 76.6% of the total catch. In terms of value, the proportions were: cod 29.3%, lobster 20.9%, redfish 13.9% and shrimp 10.3%.

The Quebec sea-going fishing fleet includes wooden or steel-hulled vessels of between 15 and 450 net registered tons; 3,728 craft of all types are engaged in the coastal fishery. The government has tried to modernize the ocean-going fleet through grants and construction loans for the building of a 40 m steel seiner and prototype 20 m container-seiner also steel-hulled, as well as seven wooden long-liners and draggers.

Government aid to the commercial fishery consists of loans for building or refitting of vessels, grants toward acquisition of coastal craft and fishing gear, and a wide range of technical assistance. The commercial fisheries branch allocated $1.1 million during the year in grants for boat-building, the purchase of fishing gear, collection of catch from coastal fishermen, land-based teams, marketing assistance and marine insurance. Interest-free loans amounting to $700,000 were approved for construction and repair of fishing vessels.

The main objectives of the commercial fisheries branch under the Canada–Quebec Agreement of 1968 (renegotiated in 1971) were a more efficient use of funds from the private as well as government sectors and concentration of fisheries in centres with well-equipped port facilities. Under the terms of the initial agreement, $4.8 million was budgeted for facilities related to ocean-going fisheries, and this amount was increased to $10 million in the agreement as renewed. The program had been expected to reach its peak in 1975 but due to increased costs it became necessary to make a supplementary agreement in the amount of $14 million to be spread over the 1974-78 period to continue work already begun. These facilities will eventually be completed under a new agreement, and talks are already under way.

In the Gaspé region, five production centres have been set aside, three to be developed as industrial fisheries complexes at Rivière-au-Renard, Paspébiac and Grande-Rivière for specialized production, with secondary production centres at Newport and Sandy Beach. In the Magdalen Islands, two centres have been set aside — an industrial complex at Cap-aux-Meules and a secondary production centre at Havre-Aubert. Landing points will supplement these centres, providing coastal fishermen with unloading and storage facilities. They will not have processing plants on site but will be linked by a fish transportation system to the nearest production centres. The commercial fisheries branch hopes to have landing points completed at Gascons and Les Méchins in the Gaspé region and Millerand and Étang-du-Nord in the Magdalen Islands before the present agreement expires.

Fishing in Quebec's inland waters falls under the jurisdiction of the tourism, fish and game department. To maintain the high standard of the sport in Quebec, the department carries out various development projects as well as wildlife research. Through its fisheries service, the department also rears several species of fish for restocking Quebec's lakes, rivers and streams to protect aquatic life.

Excellent fishing may be found in all provincial parks and reserves. Gaspé and Laurentide parks are renowned for trout fishing and the waters of Chibougamau Reserve and La Vérendrye Park abound in pickerel, pike and lake trout. Fifteen salmon rivers are open to anglers — the Petit Saguenay, Cap Chat, Ste-Anne, St. Jean,

Matapédia, Dartmouth, Port Daniel, Petite Cascapédia, Restigouche, Loutre, Jupiter 12, Jupiter 30, York, Chaloupe and Saumon.

A wildlife council, consisting of members drawn from those involved in the field, submits recommendations to the government concerning legislation required to maintain satisfactory fishing conditions or to deal with other problems created by the constant evolution of modern life and its effects on wildlife.

Ontario. Ontario's fishery resources are administered by the fisheries branch of the natural resources ministry under the authority of the federal Fisheries Act, the Ontario Fishery Regulations and the Ontario Game and Fish Act.

The commercial freshwater fishing industry in Ontario has a capital value of over $18 million and produced an annual yield of 19 million kg of fish in 1976 for which fishermen received $13 million in sales. In addition, nearly 4.5 million kg of bait fish were caught. Subsequent handling and processing of fish result in a contribution of about $25 million to the provincial economy. The widely scattered industry, centred chiefly on the Great Lakes, provides employment for about 2,200 commercial food-fish fishermen and 2,500 bait-fish fishermen; many more are employed indirectly. Approximately 900 are engaged in fish handling and processing. The species harvested commercially include yellow perch, smelt, whitefish, pickerel, pike, lake trout, herring, chub, carp, white perch, sturgeon, white bass, bullhead, catfish, eel, goldeye, sunfish, burbot, freshwater drum, rock bass, crappie, sauger and suckers. Slightly under 90% of all fish landed in Ontario are harvested from the Great Lakes. More than 350 smaller inland lakes, mainly in northwestern Ontario, are commercially fished.

Fishing methods and equipment have been modernized during the past few years and include the use of diesel-driven steel-hull tugs with depth sounding devices, radar and ship-to-ship and ship-to-shore communications. Modern icing facilities and transportation methods are in use as well as new types of fishing gear. Programs to develop more efficient and economical fishing and processing techniques have resulted in efficient bulk-handling techniques for smelt and a viable fish-meal plant which produces a marketable product from fish-processing wastes and fish unsuitable for food. Trawling on Lake Erie has proved efficient in harvesting smelt year-round. Most Ontario fishermen are organized into local associations mainly represented by a provincial council of commercial fisheries.

Ontario has an estimated freshwater area of approximately 177 388 square kilometres. Excellent angling opportunities are available for such prized fish as brook, rainbow and lake trout, yellow pickerel (walleye), smallmouth and largemouth bass, northern pike, and maskinonge. Quantities of hatchery-reared coho and chinook salmon are released annually in the western basin of Lake Ontario and provide good fishing during late summer and fall. A wide selection of ice-angling equipment including snowmobile rentals is available and seasons have been extended in many parts of the province for certain species of fish.

Revenue from the sale of angling licences in 1976 was $5.5 million. Prices and numbers sold vary greatly according to licence type. Canadian residents bought 24,909 licences at $4.00; non-residents bought 454,212 seasonal licences at $10.75 and 159,839 at $6.00. Total expenditures in Ontario related to resident and non-resident angling were estimated to be over $400 million in 1976. The management of this resource is administered by a field staff of conservation officers, biologists and technicians.

Ontario operates 14 fish hatcheries and rearing stations, notably for brook, rainbow and lake trout, splake, smallmouth and largemouth bass and maskinonge. The basic aim of the hatcheries is the economic production of high-quality species to sustain and rehabilitate recreational and commercial fishing. Studies are conducted on the improvement of transportation and planting techniques, including the use of aircraft and trucks, to improve survival and returns to the angler. The marking of hatchery fish by removal of a single fin is providing valuable information on survival of fish stocks and angler success; 180 fish sanctuaries provide protection during spawning. Research programs are directed toward specific fisheries management problems in the Great Lakes and in the smaller inland waters.

Manitoba. Manitoba's interior location belies the importance of its fisheries resources which stem from an abundance of fresh water in about 104 000 km² of lakes and streams covering 16% of the province.

In the year ended March 31, 1977, the commercial fishery produced 8.7 million kg of fish. The value to the fishermen increased from $5.9 million in 1975-76 to $7.6 million in 1976-77. Summer catch represented 68% of the value of the yearly catch. Lake Winnipeg contributed 3.6 million kg (41%), followed by the northern waters with 2.5 million kg (29%), other southern lakes with 1.1 million kg (12%), Lake Manitoba with 0.9 million kg (10%), and Lake Winnipegosis with 0.6 million kg (8%). In 1976-77, whitefish contributed 2.5 million kg, pike 1.7 million kg, walleye (pickerel) 2.6 million kg and sauger 1.3 million kg. A miscellany of species contributed 0.6 million kg. All of the commercial catch is marketed by the Freshwater Fish Marketing Corporation, a federal Crown agency, and is exported mainly to the United States. Gill-nets are the main fishing gear. About 1,704 fishermen were licensed during open-water fishing and 1,858 in winter fishing. During 1976-77, there were 2,953 individuals licensed.

Administration of both sport and commercial fisheries is controlled by the minister responsible for renewable resources and transportation services. The following are identifiable components of fisheries administration in Manitoba: program management, planning and economics, research, monitoring, extension, stocking, development, acts and regulations.

The sport fishery is an important use of the fishery resource, with walleye, pike, perch and several kinds of trout the principal sport species. In 1976-77, 189,337 angling licences were sold, 150,027 of them purchased by Canadian residents.

Saskatchewan. Fisheries resources are administered by the fisheries and wildlife branch of the tourism and renewable resources department and by the resource development branch of the Northern Saskatchewan department. The latter, with headquarters in La Ronge, administers the northern commercial fishery and the former, with head office in Prince Albert, the southern commercial fishery and the provincial sport fishery.

During 1976, 2,307 commercial fishing licences were issued to fish 215 lakes. The harvest of 4.9 million kg was worth $2.7 million to the fishermen. The industry, although widely scattered, is centred chiefly in the northern half of the province; about 70% of the production came from northern waters. In order of market value, the species composition of the catch was whitefish, walleye, lake trout, pike and tullibee.

One shallow saline lake in southern Saskatchewan produced 16 000 kg of brine shrimp and brine shrimp eggs. These are processed for sale to fish hobbyists. In 1976, 405 000 kg of buffalofish, a sucker species, and carp were harvested from the Qu'Appelle drainage, and 22 000 kg of bait fish were harvested by 32 commercial bait fishermen.

Interest in aquaculture remained stable in 1976 with the licensing of 2,466 aquaculture enterprises to raise rainbow trout. The majority of operations were intended for the private use of the owner. About 893,000 rainbow fingerlings stocked in the spring of 1976 resulted in an estimated 131 000 kg harvest.

In 1976, 196,529 angling licences were sold. Northern pike, walleye, lake trout, perch, arctic grayling, rainbow trout and goldeye continued to be the principal species taken. A continuous program of inventory of sport fishing stocks is maintained to provide up-to-date information for management purposes. During 1976, 233 waters were examined. Expansion of the exotic-species program continued with about 100 lakes and streams having established populations of trout and salmon to date.

The provincial hatchery at Fort Qu'Appelle reared 25.3 million fish of seven species for distribution in 159 waters in 1976. Rainbow trout was the species most widely distributed, being stocked in 50 waters. Walleye was stocked in 44 waters, brook trout in 38, brown trout in 15, northern pike in five, perch in four and whitefish in three.

The limnological and fisheries research program is designed to provide information on the productivity of water bodies, the abundance and relationship of fish species, and to investigate and assess factors affecting fish populations. This information is subsequently used to develop fishery management policies and programs. Angler and commercial catch data are collected to improve management of the fishery resource.

Alberta. Commercial and sport fishing are administered by a fish and wildlife division of the recreation, parks and wildlife department, under the authority of the Fisheries Act (Canada) and the Fish Marketing Act (Alberta).

Production of commercial fish from Alberta's 16 796 km² of fresh water for the fiscal year ended March 31, 1977 was 1.9 million kg, very close to the 1975-76 total. The landed value of the catch was $1.1 million and the market value was $2.1 million. Lake whitefish was the most valuable species caught commercially accounting for 60% of the total landings and 76% of the total value. Tullibee had the second highest landings followed by pike, walleye, suckers, ling, perch and lake trout.

All fishing licence sales increased in 1976-77 with 257,636 angling licences sold of which 254,254 were to resident and non-resident Canadians and 3,382 to non-resident non-Canadians. In addition 2,512 trophy lake licences, 581 spear fishing licences, 1,457 private, 45 commercial and 11 restricted game fish farm licences were purchased. In 1976-77, 210 lakes were stocked with almost 6.1 million fish: 75.3% rainbow trout, 7.5% perch, 6.8% lake trout, 5.2% brook trout, 4.9% walleye, 0.2% cutthroat trout and 0.04% lake whitefish.

British Columbia. The fisheries department, formed in 1947, was replaced in 1957 by the recreation and conservation department; the marine resources branch is the provincial organization concerned with marine commercial fisheries. Jurisdiction over the fisheries resources of British Columbia rests with the federal authority. The province administers non-tidal fisheries although the regulations covering them are made under federal order-in-council on the advice of the province.

The provincial Fisheries Act provides for taxation of fisheries and, under civil and property rights, for regulation and control of the various fish processing plants under a system of licensing. The commercial harvesting of oysters and marine aquatic plants is regulated by provincial permits and licences. Provision is made for arbitration of disputes regarding fish prices that may arise between fishermen and operators of licensed plants. Administration of the act involves collection of revenue and supervision of plant operations.

Regulation of net fishing in non-tidal waters, including commercial fishing and authority for regulation of the game fisheries in non-tidal waters, is vested in the fish and wildlife branch which operates a number of trout hatcheries and egg-taking stations for restocking purposes.

The marine resources branch co-operates closely with the fisheries and marine service of Canada. Biological research into those species of shellfish over which the province has control, principally oysters and marine plants, is conducted by this branch and the federal fisheries and marine service at the Pacific Biological Station, Nanaimo, BC, under agreement with the federal and provincial authorities.

10.2.3   Statistics of the fishing industry

The waters off the Pacific and Atlantic coasts of Canada rank among the most productive fishing grounds in the world and provide a livelihood to some 50,000 sea fishermen. Inland waters support another 7,500 fishermen, while an additional 14,000 persons are employed in fish processing plants.

10.2.3.1   Fish landings

Fish landings declined by 0.6% in 1975 to slightly less than 877 731 tonnes compared to 882 398 t in 1974. However, the demand for fishery products generated sufficiently higher prices to more than offset the declining catch. The result was a 2% increase in the gross earnings of fishermen to $291 million in 1975 and a 2.6% increase in the marketed value of this catch at $713 million (Tables 10.13 and 10.14).

Atlantic Coast landings were up 1.3% to 706 878 t in 1975 while the landed value increased by 11.8% to $183.6 million. Groundfish account for about 40.7% of the total value of the landings. The quantity of groundfish landed increased by 0.5% while the two other major groups, pelagic and estuarial and molluscs and crustaceans, increased by 0.8% and 16.1% respectively. Landings of scallops (the second most valuable shellfish species) increased in quantity to 8 356 t from 6 370 t in 1974, and increased 38.4% in value to $25.7 million (Table 10.15).

Pacific Coast landings dropped from 135 057 t in 1974 to 126 660 t in 1975, most of this in the salmon fishery (26 343 t). This decline of 6.2% represents a reduction in fishermen's incomes of about $21 million.

In 1975, landings of halibut, the mainstay of the Pacific groundfish fishery, increased by 1 762 t (52.1%) and 86.1% in value. Total landings, however, were 5 146 t, less than 50% of the previous 10-year average landings.

Landings of salmon, the most important species to the Pacific fishery, dropped to 34 551 t in 1975 from 60 893 t in 1974, a decline of 43%. This represented a $27 million loss to Pacific Coast salmon fishermen. A shift in the 1975 salmon catch breakdown saw pink become the principal species at 29% of the catch in volume terms followed by coho at 20% and spring at 19%. In value terms, however, coho represented 26.4% closely followed by spring (25.9%) and pink (17.4%).

Products and marketing   10.2.3.2

In 1975 the marketed value of processed product on the Atlantic Coast was $483 million (an increase of 14.4%) and on the Pacific Coast $167 million (a decrease of 24.2%). Atlantic Coast production of frozen fillets and blocks rose 16.8% to $141 million in 1975 with flatfish representing 32% of the total, being followed by redfish and cod with 28% and 20% respectively.

On the Pacific Coast, salmon, as a group, although showing a decline of 911,512 cases (21.8 kilograms each), were again the most valuable species in Canada in 1975 with all types of products having a marketed value of $100 million. Canned salmon pack was valued at $43 million in 1975, a decline of $71 million as compared to 1974.

The fur industry   10.3

The value of the 1976-77 Canadian production of raw furs amounted to $72.1 million, made up of $47.8 million (66%) from wildlife pelts and $24.3 million (34%) from farm pelts. The $47.8 million total was a record high, due mainly to higher values for most types of pelts. Production in 1975-76 amounted to $53.9 million.

Fur trapping. Prices for almost all kinds of Canadian wild furs have been on the increase and in 1976-77 pelt values were substantially above historic levels. The higher returns have encouraged trappers to work their traplines to full advantage, resulting in increased production of many species, especially the long-haired types such as fox, raccoon and coyote. Lynx is also high on the list of popular furs; however, in 1976-77 this cyclic species was approaching the lower end of its period of abundance and the numbers taken reflected this decline.

With the encouragement provided by recent strong price levels, production of many of the fur-bearer species approached optimum levels. This has not been the case for many years. Throughout the 1950s and 1960s raw fur prices failed to keep pace with the general price rise and there was little incentive for trappers to work their traplines to maximum potential. As a result a good percentage of fur bearers went unharvested.

Fur farming. Mink are raised in all provinces except Newfoundland. In 1976 the principal producers, in order of importance, were Ontario, British Columbia, Nova Scotia, Quebec and Alberta (Table 10.20).

With minor fluctuations, mink pelt production in Canada has declined since the peak year of 1967 when the output was 1,967,323 pelts. Lower returns in the face of higher production costs have been responsible for this decline. Many mink farmers ceased operations and the number of mink farms declined from 1,359 in 1967 to 397 in 1976. In earlier years beginners in the mink business got started through the acquisition of a small number of breeding animals and built up from that point. Now entry into the business on a scale that would hold the promise of some return on investment within a reasonable time involves a high outlay of capital. This is a limiting factor in attracting newcomers to the industry.

In 1976, 2,130 fox pelts were produced on 81 farms across the country; this is 9% above the 1975 output of 1,962 pelts from 55 farms. The increase in production was attributed to the improved market for all the long-haired furs; values for ranched fox pelts have risen sharply in the past decade, and the 1976 average price of $233.47 a pelt was the highest recorded since 1919 when the average was $201.74. Encouraged by the upturn, producers are expanding their operations and the demand for breeding animals is stronger than for many years.

Fur marketing. The bulk of Canada's fur production is sold at public auction through five fur auction firms in Montreal, North Bay, Winnipeg, Regina and Vancouver. At the auctions, furs are purchased through competitive bidding by buyers who may be purchasing for their own account or for firms in Canada or abroad. Canadian furs are usually sold in the raw or undressed state, facilitating entry into the many countries which maintain tariffs on imports of dressed furs.

In 1976-77 exports of raw furs amounted to $59.6 million, 13% above the 1975-76 exports valued at $52.8 million. Imports for 1976-77 totalled $82.4 million, 10% above the $74.9 million of furs imported in the previous year. The increase in imports was due not only to a healthy fur retail business in Canada, but also to requirements occasioned through growing exports of fur garments. In 1976 exports of fur garments amounted to $40.8 million, the highest value on record for this class of export.

The export of fur fashion garments on an important scale is a fairly new development on the Canadian fur scene. Historically, Canadian exports of furs have consisted mainly of undressed pelts from fur farms and the trapline. There are fairly definite limits to which this type of export can be developed. The production of wildlife pelts is relatively limited and not likely to be increased to any meaningful extent. In addition, in view of the highly competitive world fur farming situation, it is not practicable to visualize continuous increase in the production and export of ranch-raised furs.

In the fur manufacturing industry no such limits apply. Other factors, however, are present, principally import tariffs and competition from fur manufacturers in the importing countries. A high degree of efficiency in design and manufacture is required by Canada to compete, and there is a growing export group among Canadian fur manufacturers which is extending the horizons of this formerly largely domestic industry.

10.4 Wildlife

Wildlife is an important renewable natural resource. The original inhabitants of what is now Canada depended on it for food and clothing and still do in some remote areas. The coming of the Europeans brought development of the fur trade which guided the course of exploration and settlement. When the country was being developed, a number of mammals and birds became seriously depleted or extinct. As settlement progressed, wildlife habitat was reduced by cutting and burning of forests, pollution of streams, industrial and urban development, drainage of wetlands, building of dams, and other changes in the land.

Today, the arctic and alpine tundra, a major vegetational region, has begun to show serious man-made changes. The adjacent sub-arctic and sub-alpine non-commercial forests have been affected principally by increased human travel which has brought an increase in the number of forest fires, although the great forests farther south retain much of their original character despite exploitation. Arable lands, originally forest or grassland, have completely changed but they have, in some cases, become more suitable than the original wilderness for some forms of wildlife.

Canada is known for its varied and abundant wildlife. It maintains most of the world's stock of woodland caribou, mountain sheep, wolves, grizzly bears and wolverines. For a long time, certain species were protected from man and predator. Now, because of better understanding of how nature works, it has been recognized that many factors cause fluctuations in wildlife numbers, and hunting seasons and bag limits are based to a greater extent on environment.

In 1885, the Rocky Mountain Park (now Banff National Park) was established in Alberta, preserving an area of over 6 475 square kilometres in its natural state; in 1887, the continent's first bird sanctuary was established at Last Mountain Lake in Saskatchewan; in 1893 when wood bison faced extinction, laws were passed to protect them and a nucleus herd of plains bison was established at Wainwright, Alta. in 1907. These were among the early attempts at wildlife conservation in Canada.

As a natural resource, wildlife within each province comes under the jurisdiction of the provincial government. The federal government is responsible for wildlife on federal land and for research and management of migratory birds.

The Canadian Wildlife Service   10.4.1

The Canadian Wildlife Service (CWS) began as an agency to administer the Migratory Birds Convention Act passed in 1917. It was expanded in 1947 to meet the need for scientific research in wildlife management and is now a part of the environmental management service of the fisheries and the environment department.

The CWS conducts scientific research into wildlife problems in the Northwest Territories, the Yukon Territory and the national parks. Research projects in various areas of Western and Northern Canada continue on both polar and grizzly bear populations. Caribou and muskox in Northern Canada are species of concern and the CWS is conducting long-term studies of both species in co-operative programs with the Northwest Territories fish and wildlife service.

The CWS also carries out research in the national parks. Studies in limnology, ornithology, mammalogy and general ecosystem relationships are in progress. Long-term studies on wolf and grizzly bear ecology have just begun and a biophysical inventory of the mountain parks is continuing in Jasper and Banff national parks. A bison-livestock interaction study is proceeding in and around Wood Buffalo National Park. Shorter duration projects are defined each year and undertaken for Parks Canada according to its priorities.

The Convention on International Trade in Endangered Species of Wild Fauna and Flora was signed by Canada in July 1974. The CWS was designated the scientific and management authority for the convention in Canada. The Canada Wildlife Act, passed by Parliament in 1973, provides the federal government and the CWS with a legislative basis for undertaking joint federal-provincial management programs. Under the act, the CWS has initiated a rare and endangered species program. Continuing studies on the wood bison, whooping crane and peregrine falcon are to be augmented with new projects on other species. The International Agreement on the Conservation of Polar Bears came into effect on May 26, 1976. Canada was the first of the five signatories to ratify it. As administrator of the Migratory Birds Convention Act the CWS, in consultation with provincial wildlife agencies, recommends annual revisions of the regulations which govern open seasons, bag limits and hunting practices. The RCMP with CWS and provincial co-operation enforces the act and regulations.

The loss of wetlands to drainage and filling for agricultural and other purposes poses a serious threat to waterfowl. To counteract this the CWS in co-operation with provincial agencies began a major program in 1967 to preserve wetlands by purchase or long-term lease. Since then, 19 000 hectares have been bought for $9 million. The CWS also has charge of 80 bird sanctuaries covering 115 000 square kilometres.

The CWS conducts two annual surveys of waterfowl hunters, selected from the 471,500 holders of the Canada migratory game bird hunting permits, to obtain estimates of the species and age of the major waterfowl species taken by hunters. Other continuing projects related to migratory game birds include a national goose harvest survey, annual surveys of crop damage in the Prairie provinces and of waterfowl populations and habitat conditions in Western Canada and a program to reduce hazards caused by birds flying near airports. Bird-banding provides valuable information on migration and biology of birds, and is especially useful in waterfowl management. CWS headquarters in Ottawa keeps sets of continental banding records and controls the activities of banders operating in Canada.

Special attention is given to species greatly reduced in number or in danger of extinction. The program in which 21 young were raised from whooping crane eggs taken from the breeding grounds and incubated at the Patuxent Wildlife Research Center in Maryland is continuing. Eventually, progeny from these chicks will be released into the wild but only after a sufficiently large supply of breeding birds has been developed. In 1977, eight chicks were produced and by July the total wild population was estimated to be 79 birds.

Research continues on the effects of toxic chemicals on wildlife at various sites across the country. In Alberta, a study continues on the effect of herbicides on wildlife habitat. Field work on the relation between chemical contamination of the lower Great Lakes and the breeding success of fish-eating birds was continued under a Canada–United States Great Lakes water quality agreement. Research is proceeding on the effects of differing habitat quality and chemical contamination on the reproductive success of loons in eastern and northeastern Ontario. Research on the effects of different pesticides and spraying procedures on forest song birds was intensified in 1977.

Studies continued into the health of game and fur-bearing animals and rodents in Northern Canada and into parasitism in these mammals as well as in birds. Measures were taken to control anthrax among bison in Wood Buffalo National Park and in the Northwest Territories; no outbreaks occurred in 1977.

Under the interpretation program, the CWS operates four wildlife centres across Canada. Wye Marsh Wildlife Centre at Midland, Ont. interprets the northern hardwood biotic region; Cap Tourmente and Percé wildlife centres, both in Quebec, focus on the habitat of the greater snow geese and the natural and human history of the Atlantic gulf coast, respectively; and Creston Valley Wildlife Centre highlights the Columbia biotic region. Construction is under way on an interpretation centre in Saskatchewan, which will focus on the prairie grassland biotic region.

Research on both consumptive and non-consumptive use of the wildlife resource is a growing concern. The CWS has participated in several projects to shed light on the role of wildlife in the social and economic spheres in Canada.

The CWS has been participating in the Canada Land Inventory, a federal-provincial program to gather information on how land in the settled parts of Canada is being used, and how best it could be used for agriculture, forestry, recreation and wildlife.

10.4.2 Provincial wildlife management

Newfoundland. The functions of the wildlife division are: to preserve all indigenous species from extinction; to provide other species where suitable unused habitat exists, bearing in mind the real and aesthetic values of wildlife that are important to man; to maintain all species in the greatest number possible, consistent with the habitat needs of the species and without serious conflict with the other resource needs; and to provide and regulate the harvest surplus of wildlife populations.

The division manages game populations through changes in the hunting regulations. Research is conducted mainly on caribou and moose, but ptarmigan, arctic hare, martens, otter, mink, muskrat, ospreys and bald eagles are also being studied. Management surveys are conducted on all game species and some fur bearers. Transplant programs are carried out on the two rare animals — arctic hare and pine marten — to try and re-establish them throughout the island portion of the province.

Prince Edward Island. The fish and wildlife division of the environment department has full or partial responsibility for research and management of all wildlife on Prince Edward Island. All non-migratory wildlife is the responsibility of the province while management responsibilities for fish and migratory birds are shared with the federal government. A prime responsibility is the continual monitoring of game populations to assist in setting seasons and bag limits. Attempts are being made to establish a viable population of pheasants by the introduction of a new species. Beaver transplants to vacant habitat are being carried on to increase their range. Studies on black duck behaviour and red fox population are nearing completion.

Habitat improvement is of prime importance for all forms of wildlife. Fishery management consists largely of building fish ladders to facilitate fish passage and other stream improvement measures such as stream bed stabilization. Efforts are continuing to establish an early run salmon stock on the Morell River.

Nova Scotia. The wildlife division of the lands and forests department is concerned primarily with the maintenance of a stable and healthy environment for the purpose of ensuring optimum populations of vertebrate wildlife.

Inventories are conducted on an annual basis to monitor the population of important game and non-game species. Some other activities include: assistance with the preparation of integrated resource management plans for Crown lands, fur-bearer research, marshland management in co-operation with Ducks Unlimited (Canada), environmental impact studies and resources education programs including mandatory hunter training.

Among other concerns are co-operative programs with the Trappers Association of Nova Scotia and Nova Scotia Wildlife Federation, providing technical information to federal and provincial agencies involved with programs affecting land use and water quality; designation of unique wildlife habitats to ensure long-term management and protection; maintenance of a modest "put and take" trout fishery; biological assessment of lakes and streams; co-operative wildlife research with Acadia University; and the preparation and updating of legislation.

New Brunswick. The wildlife resources of New Brunswick are the responsibility of the fish and wildlife branch of the natural resources department. Orders-in-council issued under the New Brunswick Game Act provide a means of controlling bag limits and hunting pressure in the utilization of surpluses of wildlife species.

Biological surveys of game animals are carried out to determine the condition of populations. Principal game populations managed are: moose, white-tailed deer, black bear, beaver, muskrat and woodcock. Research and active management programs to integrate forestry practices with deer winter habitat requirements will continue to be the main thrust of deer management.

A trapper information and education program was initiated in 1976 and has received full support from the New Brunswick Trappers Association. Its primary objectives include the personal involvement of resident trappers in the wise use and management of the fur resource in the province. The current demand for long-haired furs such as bobcat, fox and fisher has diverted trapping pressure from beaver, otter, mink and muskrat.

Sport fishing contributes substantially to the economy. Atlantic salmon anglers fished 113,609 days to catch 52,243 salmon and grilse during 1976. The Miramichi River system accounted for 73% (38,318) of the salmon and grilse angler catch during this period. However, more angler-days are spent fishing for brook trout in New Brunswick. In 1976, 1,250,000 angler-days were spent participating in the inland salmon and trout sport fishery.

Quebec. The maintenance, improvement and protection of wildlife within Quebec is the responsibility of the wildlife branch of the tourism, fish and game department. The branch comprises three divisions. The wildlife management and operations division, with its regional services, is responsible for the management of all wildlife and particularly of species of interest to hunters and fishermen. Biologists are assigned to nine administrative regions and their work covers the inventory and study of animal populations as well as the improvement of the populations and their habitats. The fish hatchery service operates six hatcheries, inspects commercial hatcheries and controls imports of eggs and salmonids. The wildlife research branch conducts projects to improve the basic knowledge of fish and wildlife in order to help wildlife managers. The wildlife protection branch enforces fishing and hunting regulations and informs the public of the scope and importance of such regulations.

Ontario. Wildlife management in Ontario is administered by the fish and wildlife division of the natural resources ministry. Objectives are to provide and encourage a continuous supply of recreational and economic opportunities and to develop public awareness of relevant ecological principles. The wildlife branch is responsible for wildlife management, distributed among the main office, eight regional and 49 district offices.

The deer herd in Ontario has declined during recent decades, manifested by a southward retraction of 320 kilometres in the northern limit of the range of white-tailed deer and reduced deer populations throughout the remaining range north of agricultural Southern Ontario. The primary cause of the decline has been growth of the forest following the reduction of logging and fires since the 1930s. Several severe winters reduced the deer herd to a size compatible with or below the carrying capacity of the range. The number of hunters has declined more slowly than the number of deer resulting in relatively more pressure on the remaining herd.

The management program has been aimed at increasing the amount of food available in the summer range and in winter yards, maintaining suitable winter cover, and reducing hunting pressure in problem areas.

Moose management is concentrated on population and harvest inventory and evaluation of the effects of various timber harvesting practices on moose range. The number of moose hunters has been increasing and more intensive management measures for moose are being formulated.

In upland game and waterfowl management, effort is directed to the maintenance and improvement of habitat. Management is carried out on areas under agreement between landowners and the province and on provincial wildlife areas. These areas have helped increase opportunities for nature study and hunting in southern areas and some of them ensure preservation of the wetland habitats important to a great variety of wildlife, especially waterfowl. Waterfowl banding, production surveys and harvest inventory assist in developing waterfowl management programs. Improved monitors of upland game abundance are being developed for management purposes.

The major effort in fur management is directed toward beaver, with aerial censuses of beaver colonies and specimen collections by trappers. Monthly summaries of all fur bearers taken by each trapper are prepared. The harvest of beaver, marten and fisher is controlled by quota. About three-quarters of the fur harvested is auctioned through the Ontario Trappers' Association Fur Sales Service in North Bay. A concerted effort to develop more humane traps was started in 1973 in co-operation with the trappers' association. Workshops on humane trapping, pelt preparation, animal biology and management practices have continued to upgrade trappers' skills and knowledge.

Manitoba. The Manitoba renewable resources and transportation services department is responsible for programs designed to maximize the recreational and economic benefits of wildlife resources while preserving the ecological diversity of native species. Authority provided by provincial legislation (The Wildlife Act, the Predator Control Act and regulations) allows for legal protection and management of the 26 mammal, 160 bird, five reptile and three amphibian species. The federal Migratory Birds Convention Act deals with the protection of migratory game birds, migratory insectivorous birds and other migratory birds.

Wildlife authorities manage wildlife, game bird, goose and fur-bearing animal refuges and 50 wildlife management areas. Distribution of hunting and trapping pressure by setting definite seasons and bag limits is one management tool used. A quota system is used for moose, elk and caribou. Wildlife habitat development projects were undertaken successfully in 1977.

A five-year federal-provincial program designed to revitalize Manitoba's primary wild fur industry became effective in April 1975. Returns to trappers for 1976-77 were a record high. In 1977 Grant's Lake managed hunting area continued in its fourth year of operation and for the third successive year a managed waterfowl hunt was held on private land within Oak Hammock managed hunting area.

Saskatchewan. The fisheries and wildlife branch of the tourism and renewable resources department administers and manages the province's wildlife resources. The legislative authority is provided through the Game and Fur Acts.

Wildlife management programs are aimed at maintaining and enhancing wildlife populations for 100,000 hunters and an even greater number of non-consumptive users. Because consumptive demands exceed the supply of most big game species, hunting has been restricted, and hunting licences allocated by a computer draw. A new moose management program is currently being tested in an attempt to alleviate some problems created by restricted seasons.

Habitat loss continues to be a major problem and initiatives are being undertaken to arrest this loss. A new habitat protection and development division has been created. Aquatic habitat protection guidelines are currently being applied throughout the forested area of the province. Funds from hunting licences have permitted acquisition of more than 16 000 hectares of prime wildlife habitat. Attempts are under way to improve hunter-landowner relations with season manipulation and good hunting habits. A special program is in place to reduce grain crop losses caused by waterfowl.

Fur management stresses conservation, utilization and development of the fur resources. Training sessions update trappers on humane trapping techniques and quality pelt preparation.

Current fisheries studies evaluate reasons for fish population fluctuations and investigate productivity of selected waters. Aquatic wildlife projects are currently under way on land use as it affects wildlife. Specific projects on the basic ecological requirements of moose, elk and game birds are in progress. Special fisheries and wildlife studies have been completed in connection with a comprehensive Qu'Appelle Valley land-use evaluation and development plan. Implementation of recommendations is under way.

Alberta. The management of the fish and wildlife resources of Alberta is under the jurisdiction of the fish and wildlife division of the recreation, parks and wildlife department.

The fisheries branch is responsible for the maintenance and enhancement of fish populations and habitat. Fisheries management staff at seven centres administer the fisheries resource in their areas by conducting surveys, setting catch limits and monitoring land-use developments.

Fisheries habitat protection staff continued to review applications associated with industry development projects while habitat development staff had five major projects approved for improvement of fish habitat. The fisheries research section attempted to develop a trout stocking formula for pothole fisheries; investigated the causes and possible remedies to reduce mortality of trout during transportation for stocking; and developed a method of fish tagging for aerial monitoring of fish migration. The fish culture section produced 5.34 million trout for stocking in province waters while an aquaculture specialist and a commercial fisheries co-ordinator provided liaison between the government and the private sector.

Wildlife populations are managed for aesthetic, recreational and economic purposes. To ensure sustained optimum yields and harvests, the following methods are used: determination of population inventories and production, delineation and modification of limiting factors through habitat protection and development, intensive enforcement and public education.

Research and management efforts continued on ungulates, waterfowl and upland birds. Mapping of key habitat areas for ungulates and fur bearers is now centring on the eastern slopes of the Rockies as a result of the government's recently adopted zoning policy. As part of a fur-bearer management program, a questionnaire has been sent to 3,000 registered Alberta trappers. A computer analysis of trapline harvests from 1970 was also conducted. In 1976 the first annual game bird survey questionnaire was mailed to a random selection of Alberta hunters to determine harvest and recreational opportunity provided by the hunting of pheasant, Hungarian partridge, and ruffed, spruce, blue, sage and sharp-tailed grouse. Results indicated approximately 307,000 hunter-days in the 1976 season. Hunter-days in 1976 totalled 362,000 for moose, elk and deer.

British Columbia. The fish and wildlife branch of the recreation and conservation ministry is responsible for the protection, enhancement and use of wildlife and freshwater fish resources of British Columbia. Administrative and technical headquarters are in Victoria; seven regional headquarters in the main centres of population, 59 district offices, three fish hatcheries and a number of permanent field stations operate throughout the province. The branch licenses hunters and anglers and enforces closed seasons, bag limits and other measures. It licenses and regulates trapping of fur-bearing animals, commercial propagation of game birds and fish, and activities of big-game guides. The branch enhances the abundance and health of desirable species of animals by the acquisition of key areas of range for big game and waterfowl and by the stocking of lakes.

The branch objective is to contribute to the economy of British Columbia through wise management of game resources and non-tidal fisheries, paying attention to such matters as pollution and integrated use of lands for forestry, agriculture, transportation, mining and wildlife. The branch conducts programs of education and information to make the public aware of the value of wildlife resources and of the principles of wise management.

10.4.3   Territorial wildlife management

Yukon Territory. The Yukon game branch of the tourism, conservation and information department manages the Yukon's wildlife resource. It administers and enforces game and fur export ordinances and assists federal agencies in enforcement of the Migratory Birds Convention Act, Canada Wildlife Act, International Agreement on the Conservation of Polar Bears, International Agreement of Trade in Endangered Species, the Game Export Act and the Freshwater Fishery Regulations. With headquarters in Whitehorse, it has five regional offices in Dawson City, Mayo, Ross River, Watson Lake and Haines Junction.

The branch promotes judicious use of big game species, upland game birds and sport fish for residents and non-residents, licensing hunters and anglers and enforcing closed seasons, bag limits and other regulatory measures. It licenses and regulates trapping of fur-bearing animals and activities of outfitters and guides. To increase knowledge about wildlife species and provide the basis for proper management, it conducts and supports biological research and public educational programs.

Northwest Territories. The fish and wildlife service manages the wildlife resources of the Northwest Territories, and provides opportunities for native peoples to follow their traditional pursuits of hunting, trapping and fishing. It has headquarters in Yellowknife, four regional offices at Fort Smith, Inuvik, Frobisher Bay, and Rankin Inlet, and 25 area offices throughout the territories.

Wildlife management is carried out mainly by harvest monitoring and control. Harvest quotas are allocated by management zones. Management studies are conducted primarily to establish the abundance, productivity and seasonal distribution of large mammals, including the polar bear.

Trapping is encouraged through a series of programs designed to assist native peoples to return to the land. Included are trappers' incentive grants (a fur subsidy program based on a percentage of the season's harvest), fur marketing service, and an outpost camp program which provides financial assistance to groups who wish to move back to the land and live off the natural resources available through hunting and trapping.

The service is responsible for administration of sports fishing licences. Under permit from the federal fisheries and marine service, fish and wildlife officers monitor commercial fisheries and the testing of lakes and rivers to determine the viability of commercial operations to supply local domestic markets.

The fish and wildlife service is involved in environmental management through participation in various federal-territorial advisory committees. Close liaison is maintained with hunter and trapper associations as a link between the resource-dependent residents and the companies involved in exploration and development of non-renewable resources.

A Northwest Territories game advisory council advises the commissioner of the Northwest Territories on matters pertaining to wildlife policy and legislation. All members of the council are northern residents and represent native hunters and trappers and the outdoor recreation industry.

Sources

10.1 - 10.1.2 Information Services Directorate, Department of Fisheries and the Environment.

10.1.3 Information Services Directorate, Department of Fisheries and the Environment; provincial returns from respective provincial government departments.

10.1.4 Manufacturing and Primary Industries Division, Industry Statistics Branch, Statistics Canada.

10.2 - 10.2.1 Information Services Directorate, Department of Fisheries and the Environment.

10.2.2 Supplied by the respective provincial government departments.

10.2.3 Manufacturing and Primary Industries Division, Industry Statistics Branch, Statistics Canada.

10.3 Agriculture Division, Institutions and Agriculture Statistics Field, Statistics Canada.

10.4 - 10.4.1 Information Services Directorate, Department of Fisheries and the Environment.

10.4.2 Supplied by the respective provincial government departments.

10.4.3 Supplied by the respective territorial government departments.

Tables

10.1 1973 National Forest Inventory

Province or territory	Forest land km²		Non-reserved forest land tenure km²			Volume[3] Million cubic metres		
	Total[1]	Non-reserved[2]	Crown provincial	Crown federal	Privately owned	Soft-woods	Hard-woods	Total
Newfoundland	127 494	124 790	119 813	- -	4 977	220	35	255
Prince Edward Island	2 505	2 489	121	4	2 364	4	2	6
Nova Scotia	44 443	42 517	10 364	85	32 068	178	76	254
New Brunswick	63 107	61 841	27 976	215	33 650	415	164	579
Quebec	696 059	695 083	622 021	752	72 310	2 746	948	3 694
Ontario	432 233	430 137	384 051	4 152	41 934	2 588	1 650	4 238
Manitoba	135 474	131 893	128 194	1 028	2 671	345	101	446
Saskatchewan	128 198	122 301	116 753	1 530	4 018	293	199	492
Alberta	276 569	246 238	236 926	1 793	7 519	953	569	1 522
British Columbia	544 923	527 457	500 253	1 603	25 601	7 373	205	7 578[4]
Yukon Territory	232 361	210 309	- -	210 309	- -	- -	..	..
Northwest Territories	547 116	504 017	- -	504 017	- -	- -	..	..
Canada	3 230 482	3 099 072	2 146 472	725 488	227 112	15 115	3 949	19 064

[1]Land capable of producing stands of trees 10 cm dbh and larger on 10% or more of the area.
[2]Excludes land in parks, game refuges, water conservation areas and nature preserves, where, by legislation, wood production is not primary.
[3]Non-reserved inventoried areas only; excludes Labrador, the Yukon Territory and Northwest Territories for which no data are available.
[4]Mature timber only.

10.2 Forest utilization, 10-year average 1966-75

Item	Usable wood '000,000 m³	Percentage of total depletion
Products utilized		
Logs and bolts[1]	75.8	62.5
Domestic use	75.0	61.9
Exported	0.8	0.6
Pulpwood	40.0	33.1
Domestic use	38.1	31.4
Exported	2.0	1.7
Fuelwood (incl. wood for charcoal)	4.1	3.4
Other products	1.3	1.0
Total utilization	121.3	100.0

[1]Includes some wood used in pulp manufacture.

10.3 Forest fire losses, 1976

Province or territory	Number of fires	Area burned ha
Newfoundland	348	196 461
Prince Edward Island	30	115
Nova Scotia	541	17 535
New Brunswick	411	4 616
Quebec	1,108	465 551
Ontario	3,985	544 108
Manitoba	1,128	65 095
Saskatchewan	636	90 985
Alberta	774	23 032
British Columbia	888	56 947
Yukon Territory	112	52 807
Northwest Territories	313	641 832
Other federal lands	84	125
Total	10,358	2 159 209

10.4   Volume of wood cut, by province, 1973-76 (thousand cubic metres)

Province or territory	1973	1974	1975	1976
Newfoundland	2 952	3 211	2 452	2 345
Prince Edward Island	172	190	167	164
Nova Scotia	3 596	3 998	3 540	3 455
New Brunswick	8 868	8 781	6 906	7 479
Quebec	29 350	32 712	28 407	29 062
Ontario	18 446	18 868	14 215	17 878
Manitoba	1 828	2 101	2 022	1 742
Saskatchewan	2 718	2 778	2 313	2 866
Alberta	5 599	5 057	4 964	5 627
British Columbia	70 137	60 086	50 077	69 528
Yukon Territory and Northwest Territories	140	147	198	127
Canada	143 806	137 929	115 263	140 275

10.5   Volume of wood cut, by type of product, 1973-76 (thousand cubic metres)

Type of product	1973	1974	1975	1976
Logs and bolts	96 887	85 295	73 543	98 279
Pulpwood	42 378	48 034	37 063	36 836
Fuelwood	3 462	3 512	3 765	3 900
Poles and piling	106	72	[1]	[1]
Round mining timber	55	55	[2]	[2]
Fence posts	427	462	[2]	[2]
Miscellaneous roundwood	491	499	891	1 260
Total	143 806	137 929	115 263	140 275

[1]Included with logs and bolts.
[2]Included with miscellaneous roundwood.

10.6   Lumber production and shipments and value of all shipments of the sawmill and planing mill industry, by province, 1974 and 1975

Year and province or territory	Lumber Production m^3	Quantity shipped[1] m^3	Value of shipments[1] $'000	Value of shipments[1] of goods of own manufacture $'000
1974				
Newfoundland	25 370	[2]	[2]	3,557
Prince Edward Island	..	—	—	[2]
Nova Scotia	346 662	309 041	22,640	29,076
New Brunswick	692 172	661 300	46,433	78,795
Quebec	4 989 730	5 028 111	292,155	380,876
Ontario	2 585 533	2 293 955	152,841	198,957
Manitoba	238 902	227 191	11,565	17,418
Saskatchewan	280 920	320 872	14,635	16,368
Alberta	1 157 095	1 198 890	57,197	67,612
British Columbia	20 294 146	20 929 968	1,278,239	1,536,297
Yukon Territory and Northwest Territories	3 061	[2]	..	..
Canada	30 613 591	30 998 997	1,877,695	2,329,835
1975				
Newfoundland	17 669	16 614	1,143	3,447
Prince Edward Island	..	..	..	[2]
Nova Scotia	291 779	294 828	22,020	29,098
New Brunswick	635 947	660 448	44,314	74,859
Quebec	4 451 762	4 719 245	263,370	364,544
Ontario	1 895 564	1 825 193	117,256	180,044
Manitoba	136 745	146 270	9,928	17,655
Saskatchewan	293 388	354 455	17,245	19,708
Alberta	714 714	1 016 872	48,706	62,839
British Columbia	16 149 839	16 878 807	948,858	1,243,731
Yukon Territory and Northwest Territories	7 122	7 122	410	[2]
Canada	24 594 529	25 919 854	1,473,250	1,996,856

[1]Shipment figures contain some duplication because sales of lumber from one sawmill to another are reported as shipments by both establishments.
[2]Confidential.

10.7 Lumber shipments[1] of the sawmill and planing mill industry, by species, 1973-75

Kind of wood	1973		1974		1975	
	Quantity m^3	Value $'000	Quantity m^3	Value $'000	Quantity m^3	Value $'000
Spruce	16 518 095	985,036	15 518 959	822,771	13 702 856	694,316
Douglas fir	3 706 637	247,481	3 107 443	202,576	2 510 441	150,158
Hemlock	6 072 052	402,424	5 135 203	335,491	3 548 926	216,489
Cedar (red and white)	2 375 712	207,474	2 131 114	191,313	1 945 449	158,613
White pine	620 443	48,754	553 542	47,976	423 167	36,105
Jack pine	1 327 402	73,512	1 170 986	62,951	930 845	49,594
Maple	518 547	43,186	504 788	42,500	364 559	29,067
Yellow birch	314 928	26,186	245 894	22,267	219 407	18,658
Lodgepole pine	1 700 570	93,619	1 589 868	71,479	1 477 804	65,144
Balsam fir	416 276	27,051	315 867	19,898	262 902	16,245
Other	1 120 623	72,433	725 331	58,473	533 502	38,861
Total	34 691 285	2,227,156	30 998 995	1,877,695	25 919 858	1,473,250

[1]See footnote 1, Table 10.6.

10.8 Veneer and plywood shipments, by type, 1973-75

Type	1973		1974		1975	
	Quantity m^3	Value $'000	Quantity m^3	Value $'000	Quantity m^3	Value $'000
Veneer						
Softwoods	607 492[1]	37,063	488 770[1]	32,439	159 315[1]	30,439
Hardwoods	143 597[2]	45,419	118 462[2]	44,006	73 597[2]	35,589
Softwood plywood	2 148 902[3]	289,666	1 842 081[3]	262,244	207 777[3]	280,104
Hardwood plywood	249 551[4]	59,334	218 176[4]	57,610	28 405[4]	50,105

[1]Basis 3.175 mm.
[2]Surface measure.
[3]Basis 9.525 mm unsanded.
[4]Basis 6.35 mm sanded two sides.

10.9 Pulp shipments and production, 1974-77

Item		1974	1975	1976P	1977P
Mill shipments of pulp[1]	'000 t	7 603	5 650	6 594	5 338
	$'000	2,205,290	1,982,617	..	..
Groundwood pulp	'000 t	272	221	339	298
	$'000	49,714	42,908	..	..
Chemical pulps	'000 t	7 331	5 429	6 255	5 039
	$'000	2,155,576	1,939,709	..	..
Pulp production[2]	'000 t	19 677	15 113	18 048	14 852
Quebec	''	6 352	5 198	..	..
Ontario	''	3 877	2 540	..	..
British Columbia	''	5 267	3 998	..	..
Other provinces[3]	''	4 181	3 377	..	..

[1]Includes screenings.
[2]The differences between these figures and the quantities of mill shipments represent the amounts of pulp further manufactured by the reporting companies.
[3]Prince Edward Island is the only province in which there is no production.

10.10 Shipments of basic paper and paperboard, by type and by province, 1973-75

Type and province		1973	1974	1975
TYPE				
Newsprint paper	'000 t	8 373	8 761	7 046
	$'000	1,339,412	1,854,917	1,847,343
Book and writing paper	'000 t	891	1 032	680
	$'000	321,394	495,976	346,490
Wrapping paper	'000 t	601	616	408
	$'000	148,210	224,601	166,504
Paperboard	'000 t	2 062	2 142	1 504
	$'000	378,355	550,271	412,441
All other papers	'000 t	287	302	254
	$'000	64,909	100,197	88,693
Total	'000 t	12 214	12 853	9 892
	$'000	2,252,280	3,225,962	2,861,471
PROVINCE				
Quebec	'000 t	5 056	5 534	4 676
	$'000	938,560	1,381,144	1,335,388
Ontario	'000 t	3 114	3 273	2 053
	$'000	665,202	916,770	645,841
British Columbia	'000 t	1 990	1 930	1 437
	$'000	326,324	446,792	420,619
Other provinces[1]	'000 t	2 053	2 116	1 726
	$'000	322,194	481,256	459,623

[1]Prince Edward Island is the only province in which there is no production.

10.11 Exports of pulp and newsprint to Britain, United States and all countries, 1973-76

Commodity and year	Britain		United States		All countries	
	Quantity _t_	Value _$'000_	Quantity _t_	Value _$'000_	Quantity _t_	Value _$'000_
Pulp						
1973	348 109	61,506	3 370 362	609,227	5 912 310	1,054,166
1974	398 001	108,812	3 561 127	1,060,380	6 424 386	1,861,235
1975	403 787	145,412	2 654 164	991,879	4 963 550	1,817,998
1976	480 168	169,965	3 217 807	1,165,432	6 113 748	2,173,319
Newsprint						
1973	449 183	76,604	6 264 307	1,067,833	7 616 781	1,285,928
1974	448 188	107,443	6 401 788	1,352,758	7 891 882	1,721,768
1975	340 266	102,416	5 103 920	1,357,892	6 348 654	1,741,990
1976	432 769	130,055	5 675 372	1,595,477	6 997 267	1,997,371

10.12 Persons employed in the primary fishing industry, by province, 1974 and 1975

Province or territory	Sea fisheries		Inland fisheries	
	1974	1975	1974	1975
Newfoundland	..	..	...	...
Prince Edward Island	..	..	...	...
Nova Scotia	..	..	...	...
New Brunswick	..	..	...	...
Quebec	5,703	6,470	..	694
Ontario	...	...	2,208	2,220
Manitoba	...	...	1,685[1]	1,688
Saskatchewan	...	...	785[1]	1,651
Alberta	...	...	304[1]	1,279
British Columbia	11,906	12,578	...	...
Yukon Territory	...	...	55	91
Northwest Territories	...	...	112	136
Canada	..	..	..	..

[1]Fishermen who sold fish to the Freshwater Fish Marketing Corporation.

10.13 Value of all fishery products, by province, 1974 and 1975 (thousand dollars)

Province or territory	1974	1975
Newfoundland	114,612	120,753
Prince Edward Island	20,402	27,760
Nova Scotia	183,644	208,574
New Brunswick	100,581	116,274
Quebec	29,836	30,609
Ontario	19,310	22,104
Manitoba } Saskatchewan } Alberta } Northwest Territories } British Columbia and	19,217	22,072
Yukon Territory[1]	220,559	167,099
Total[2]	685,416	713,338

[1]Includes landings by Canadian fishermen in United States ports.
[2]The sum of provincial totals differs from Canada total as intershipments between provinces have been removed from the Atlantic Coast.

10.14 Landings of sea and inland fish and other sea products, by province, 1974 and 1975

Province or territory	1974		1975	
	Quantity _t_	Value _$'000_	Quantity _t_	Value _$'000_
Newfoundland	234 481	42,098	240 532	45,728
Prince Edward Island	16 177	8,921	14 541	12,410
Nova Scotia	283 277	78,480	281 743	91,010
New Brunswick	113 626	21,525	123 122	25,515
Quebec	53 863	14,051	53 575	15,433
Ontario	24 940	10,023	20 576	11,052
Manitoba	11 352	5,147	11 851	5,961
Saskatchewan	5 923	1,971	4 723	2,238
Alberta	2 364	931	1 946	975
British Columbia[1]	135 058	100,976	123 929	79,681
Yukon Territory	4	..	28	12
Northwest Territories	1 333	738	1 165	677
Canada	882 398	284,861	877 731	290,692
Sea fish	832 711	265,218	832 156	263,130
Inland fish	49 687	19,643	45 575	27,562

[1]Includes halibut landed in United States ports.

10.15   Landings of the chief commercial fish, 1974 and 1975

Area and species	1974		1975	
	Quantity t	Value $'000	Quantity t	Value $'000
ATLANTIC COAST				
Groundfish	*380 438*	*73,848*	*382 356*	*74,740*
Catfish	3 064	427	2 221	315
Cod	130 579	32,126	119 862	30,599
Flounder and sole	105 631	17,792	33 450	5,661
Haddock	12 315	5,736	15 942	7,119
Hake	11 302	1,883	10 244	1,851
Halibut	1 114	1,911	950	1,994
Pollock	20 963	3,251	21 833	3,655
Redfish	87 693	9,480	102 915	11,044
Other	7 777	1,242	74 939	12,502
Pelagic and estuarial	*275 525*	*25,723*	*277 750*	*27,722*
Alewives	5 692	460	4 500	461
Herring[1]	225 599	13,445	241 906	13,798
Mackerel	16 637	1,804	12 752	1,641
Salmon	2 215	3,657	2 212	3,617
Smelts	1 901	664	1 856	504
Swordfish	1	- -	16	9
Other	23 480	5,693	14 508	7,692
Molluscs and crustaceans	*39 060*	*64,035*	*45 386*	*80,987*
Clams	2 569	747	2 448	760
Lobsters	14 237	37,963	17 488	48,378
Oysters	1 249	508	1 633	739
Scallops	6 370	18,572	8 356	25,708
Other	14 635	6,245	15 461	5,402
Other[2]	2 622	650	1 386	153
Total, Atlantic Coast	697 645	164,256	706 878	183,602
PACIFIC COAST				
Groundfish	*16 422*	*9,876*	*21 097*	*15,120*
Cod (gray)	7 072	1,921	8 308	2,231
Flounder and sole	3 318	1,048	4 539	1,322
Halibut[3]	3 384	5,440	5 146	10,125
Lingcod	1 762	1,117	1 946	848
Sablefish	339	231	629	467
Other	547	119	529	127
Pelagic and estuarial	*111 451*	*87,831*	*99 347*	*61,226*
Herring	44 669	12,043	59 638	13,267
Salmon	60 893	73,998	34 551	46,913
Chum	12 468	10,680	5 380	7,206
Coho	9 319	13,834	7 091	12,401
Pink	11 061	5,791	9 961	6,900
Sockeye	21 276	29,841	5 635	8,184
Spring	6 720	13,803	6 436	12,172
Other	49	49	48	50
Other	5 889	1,790	5 158	1,046
Molluscs and crustaceans	*7 184*	*3,269*	*6 216*	*3,335*
Clams	1 110	383	1 173	333
Crabs	1 135	1,317	1 140	1,435
Oysters	3 676	880	3 036	883
Shrimps and prawns	1 202	624	784	534
Other	61	65	83	150
Total, Pacific Coast	135 057	100,976	126 660	79,681
INLAND				
Freshwater fish	*46 374*	*19,232*	*40 627*	*21,250*
Bass	1 070	407	1 170	709
Catfish	463	276	571	364
Herring, lake (cisco) and tullibee	1 522	873	2 231	1,302
Perch	6 305	4,350	4 625	4,545
Pickerel (yellow)	3 694	3,421	3 766	3,764
Pike	3 517	913	3 118	886
Saugers	1 800	1,252	1 931	1,414
Smelts	7 667	963	7 862	1,202
Sturgeon	71	124	78	153
Trout	642	403	577	420
Whitefish	8 110	4,680	7 723	5,270
Other	11 513	1,570	6 975	1,221
Other[4]	3 338	454	2 341	679
Total, Inland	49 712	19,686	42 968	21,929
Total	882 414	284,918	876 506	285,212

[1]Includes sardines.
[2]Includes livers and scales; excludes seaweeds and "Other" (seals, whales, oils).
[3]Includes landings by Canadian fishermen at United States ports.
[4]Sea fish caught inland.

10.16   Market value of all fishery products, by area and species, 1974 and 1975 (thousand dollars)

Area and species	1974	1975	Area and species	1974	1975
ATLANTIC COAST			PACIFIC COAST		
Groundfish	*197,119*	*215,354*	Groundfish	*16,306*	*24,081*
Catfish	1,476	1,316	Cod (gray)	4,531	5,255
Cod	70,115	61,571	Flounder and sole	2,217ʳ	2,850
Flounder and sole	41,358	46,047	Halibut²	6,996	12,873
Haddock	10,832	14,948	Lingcod	1,766	1,857
Hake	3,939	3,563	Sablefish	492	981
Halibut	3,029	3,928	Other	304ʳ	265
Pollock	8,586	11,161			
Redfish	29,732	47,122	Pelagic and estuarial	*198,110*	*136,473*
Other	28,052	25,698	Herring	29,815	34,867
			Salmon	165,841	99,748
Pelagic and estuarial	*93,572*	*105,655*	Chum	22,350	11,588
Alewives	1,197	1,485	Coho	24,375	19,783
Herring (includes sardines)	63,386	63,064	Pink	22,215	21,040
Mackerel	4,972	3,383	Sockeye	65,221	19,538
Salmon	5,547	7,061	Spring	18,864	20,270
Smelts	1,385	1,283	Other	12,816	7,529
Tuna	14,409	27,013	Other	2,454	1,858
Other	2,676	2,366			
			Molluscs and crustaceans	*5,585*	*5,408*
Molluscs and crustaceans	*114,812*	*144,861*	Clams	795	794
Clams	1,957	1,829	Crabs	2,255	2,295
Crabs	15,658	12,499	Oysters	1,064	1,090
Lobsters	59,053	68,282	Shrimps and prawns	1,217	1,010
Oysters	833	1,119	Other	254	219
Scallops	32,775	53,670			
Other	4,536	7,462	Other sea products	451	1,056
			Total, Pacific Coast	220,452	167,018
Other sea products	19,193	17,615	Total, Inland	40,268	45,110
Total, Atlantic Coast¹	424,696	483,485	Total	685,416	695,613

¹Excludes duplication.
²Includes halibut landed by Canadian fishermen in United States ports.

10.17   Pacific Coast production of canned salmon, 1974 and 1975

Kind	1974		1975	
	Quantity *21.8-kg cases*	Value *$'000*	Quantity *21.8-kg cases*	Value *$'000*
Chum	230,634	14,227	37,093	2,412
Coho	160,051	12,962	61,081	5,341
Pink	307,192	21,720	240,593	17,055
Sockeye	709,181	64,009	164,055	17,562
Spring	20,279	1,293	13,952	846
Steelhead	1,546	101	597	36
Total	1,428,883	114,312	517,371	43,252

10.18   Atlantic Coast production of frozen fillets and fish blocks, 1974 and 1975

Area and species	1974		1975	
	Quantity *t*	Value *$'000*	Quantity *t*	Value *$'000*
NEWFOUNDLAND	*45 919*	*65,479*	*48 359*	*72,232*
Cod	15 225	22,776	11 536	15,693
Haddock	97	145	56	104
Redfish	6 754	7,173	11 408	17,209
Flatfish	19 023	32,143	18 524	34,659
Other	4 820	3,242	6 835	4,567
MARITIMES¹	*42 500*	*46,132*	*48 898*	*59,464*
Cod	8 649	11,960	8 030	11,221
Haddock	2 386	4,361	3 869	7,212
Redfish	11 304	11,378	11 324	16,015
Flatfish	4 578	7,262	4 712	8,747
Other	15 583	11,171	20 963	16,269
QUEBEC	*8 300*	*9,557*	*7 416*	*9,768*
Cod	1 526	2,738	1 359	1,714
Redfish	5 757	5,310	4 578	6,024
Flatfish	848	1,388	1 033	1,736
Other	169	121	446	294
TOTAL, ATLANTIC COAST¹	*96 719*	*121,168*	*104 673*	*141,464*
Cod	25 400	37,474	20 925	28,628
Haddock	2 483	4,506	3 925	7,316
Redfish	23 815	23,861	27 310	39,248
Flatfish	24 449	40,793	24 269	45,142
Other	20 572	14,534	28 244	21,130

¹Data slightly overstated due to interprovincial shipments included in the Maritimes.

10.19   Pelts of wildlife fur-bearing animals taken, by kind, years ended June 30, 1975-76 and 1976-77

Kind	1975-76 fur season			1976-77 fur season		
	Pelts No.	Total value $	Average value $	Pelts No.	Total value $	Average value $
Badger	5,124	156,441	30.53	6,834	261,713	38.30
Bear						
White	406	192,700	474.63	530	310,165	585.22
Black or brown	3,531	154,523	43.76	3,402	149,444	43.93
Grizzly	8	1,520	190.00	6	1,350	225.00
Beaver	334,924	6,723,401	20.07	404,625	9,836,998	24.31
Cougar	58	9,570	165.00	—	—	—
Coyote	61,779	3,150,383	50.99	65,819	3,933,303	59.76
Ermine (weasel)	76,199	68,113	0.89	102,998	106,210	1.03
Fisher	8,698	702,997	80.82	9,664	921,795	95.38
Fox						
Blue	116	6,599	56.89	467	18,253	39.09
Cross and red	55,064	2,555,659	46.41	52,914	3,049,971	57.64
Silver	583	26,738	45.86	868	46,176	53.20
White	26,797	724,678	27.04	36,375	1,299,359	35.72
Not specified	10,125	559,508	55.26	11,674	604,130	51.75
Lynx	13,162	2,845,416	216.18	15,132	3,317,503	219.24
Marten	53,108	910,787	17.15	102,632	2,044,210	19.92
Mink	69,901	1,106,189	15.82	116,537	2,292,316	19.67
Muskrat	2,102,016	7,412,311	3.53	2,554,879	10,719,316	4.20
Otter	16,005	1,156,679	72.27	19,932	1,376,188	69.04
Rabbit	865	131	0.15	1,547	356	0.23
Raccoon	79,253	1,513,926	19.10	99,339	2,212,625	22.27
Seals						
Fur, North Pacific[1]	6,609	232,067	35.11	5,181	330,185	63.73
Hair	161,082	2,907,054	18.05	170,625	3,148,221	18.45
Skunk	747	1,102	1.48	1,256	2,794	2.22
Squirrel	445,507	320,128	0.72	823,621	644,809	0.78
Wildcat	3,103	295,069	95.09	3,459	320,216	92.57
Wolf	4,879	300,667	61.62	6,150	388,569	63.18
Wolverine	871	133,497	153.27	925	168,897	182.59
Total	3,540,520	34,167,853	...	4,617,391	47,505,072	...

[1]Commonly known as Alaska fur seal; value figures are the net returns to the Canadian government for pelts sold.

10.20   Value of fur pelts produced by province, 1974-77 (dollars)

Province	Value of wildlife pelts produced			Value of mink pelts produced on fur farms[1]		
	1974-75	1975-76	1976-77	1974	1975	1976
Newfoundland	1,741,204	1,639,825	1,902,187	—	—	—
Prince Edward Island	69,682	104,861	125,934	122,539	160,902	228,309
Nova Scotia	340,361	501,906	828,587	1,808,225	2,293,853	3,106,370
New Brunswick	465,559	582,412	702,451	128,873	229,027	286,317
Quebec	4,106,794	4,169,987	5,667,929	1,462,009	2,001,145	2,440,972
Ontario	6,141,910	7,976,545	11,076,084	7,017,241	8,593,362	10,782,325
Manitoba	2,554,766	4,316,986	5,582,231	975,962	989,475	1,320,182
Saskatchewan	2,006,580	4,376,192	6,759,941	379,684	257,659	251,904
Alberta	2,829,234	4,967,840	7,376,561	1,455,478	1,576,768r	1,464,627
British Columbia	1,349,773	1,832,071	2,095,691	3,079,308	3,280,234	3,938,203
Yukon Territory	403,543	367,677	430,104	—	—	—
Northwest Territories	2,081,640	2,742,484	4,317,187	—	—	—
Canada	25,070,358[2]	34,167,853[2]	47,505,072[2]	16,429,319	19,382,425r	23,819,209

[1]Fur farm data are for calendar years.
[2]Includes hair seal from the Maritime provinces and Alaska fur seal.

10.21   Exports and imports of furs, by kind, years ended June 30, 1975-76 and 1976-77 (thousand dollars)

Kind of fur	1975-76 fur season			1976-77 fur season		
	Britain	United States	All countries	Britain	United States	All countries
EXPORTS						
Undressed						
Beaver	1,059	759	6,540	1,130	763	8,970
Chinchilla	—	179	190	5	186	197
Ermine (weasel)	54	4	123	195	3	208
Fisher	18	382	707	11	592	949
Fox, all types	1,849	1,686	6,847	2,126	1,863	7,589
Lynx	57	2,019	2,850	—	2,723	3,865
Marten	106	744	955	171	1,497	2,029
Mink	624	7,362	12,723	752	9,704	16,173
Muskrat	3,159	510	5,209	5,692	486	9,830
Otter	—	30	72	5	129	265
Rabbit	—	42	79	—	34	62
Seal	1,061	97	4,866	906	10	4,612
Squirrel	283	—	325	561	—	569
Wolf	1,213	676	3,702	1,067	1,167	4,312
Other	1,314	1,009	7,602	1,586	915	8,863

10.21 Exports and imports of furs, by kind, years ended June 30, 1975-76 and 1976-77 (thousand dollars) (concluded)

Kind of fur	1975-76 fur season			1976-77 fur season		
	Britain	United States	All countries	Britain	United States	All countries
EXPORTS (concluded)						
Dressed						
Mink	—	164	1,262	1	232	1,760
Raccoon	—	22	216	—	—	285
Fur plates, mats	—	—	121	3	38	140
Other	12	1,663	2,497	30	1,327	2,590
Fur goods apparel	2,151	4,724	24,773	2,166	4,670	28,309
Total	12,960	22,072	81,659	16,407	26,339	101,577
IMPORTS						
Undressed						
China and Jap mink	382	43	969	81	—	98
Fox	1,772	4,233	10,456	2,052	4,938	10,929
Kolinsky	150	—	387	46	—	93
Mink	4,145	7,010	27,291	3,097	7,185	28,001
Muskrat	29	6,660	6,693	—	6,608	6,608
Persian lamb	28	36	140	36	4	239
Rabbit	—	136	163	—	29	29
Raccoon	31	20,029	20,081	—	22,672	22,710
Other	286	7,245	8,682	103	12,675	13,655
Dressed						
Hatters' furs	—	156	469	—	209	558
Mink	51	3,847	3,968	882	5,194	6,119
Seal	30	981	1,130	11	868	926
Sheep and lamb	466	731	2,208	916	2,530	4,588
Fur plates, mats	118	570	2,775	171	594	2,883
Other	254	2,128	3,335	389	5,150	6,729
Fur goods apparel	183	1,350	3,311	278	1,291	5,693
Total	7,925	55,155	92,058	8,062	69,947	109,858

Sources

10.1 - 10.3 Information Services Directorate, Department of Fisheries and the Environment.
10.4 - 10.18 Manufacturing and Primary Industries Division, Industry Statistics Branch, Statistics Canada.
10.19 - 10.21 Agriculture Division, Institutions and Agriculture Statistics Field, Statistics Canada.

Agriculture

Tables

Agriculture   Chapter 11

Agriculture in Canada   11.1

Trends and highlights   11.1.1

Statistics Canada estimated national realized net farm income for 1977 at $3.3 billion, down 11% from 1976. Cash receipts were estimated at $10.1 billion and operating and depreciation charges at $7.8 billion. The forecast for 1978 was for a further decline of 6% in net farm income.

In 1977, the agriculture minister and a committee of Cabinet ministers with jurisdictions related to food launched a series of meetings with national food-related associations, provincial governments and interested individuals. The aim of the meetings was to develop a national food strategy.

Several new programs were introduced. New grants were announced to expand feed storage in areas of the country where supply is inadequate. Grants were made to commercial grain storage facilities and to farmers increasing feed storage on the farm. More than $42 million was committed by the federal government for these programs.

To help farmers increase marketing efficiency, the government implemented the Advance Payments for Crops Act. Several groups took advantage of the program including the Ontario Soya-Bean Growers Marketing Board, BC Tree Fruits Ltd., Scotia Gold Co-op Ltd. of Kentville, NS, Quinte Fruit Growers Ltd. of Trenton, Ont. and the Fédération des producteurs de pommes du Québec. Under this act, the federal government guarantees, and pays interest on, loans made to producer organizations whose members require advance payments for storable crops.

The Agricultural Stabilization Board made support payments in 1977 on slaughter cattle, corn, greenhouse tomatoes, greenhouse cucumbers, summer pears, British Columbia cherries, apricots, prune plums and early table potatoes. Support payments for the 1976 sugar beet crop and 1977 beef cow-calf production would be made in 1978. Under the Agricultural Stabilization Act, certain commodities are supported at not less than 90% of the five-year average market price, indexed to reflect changes in cash costs of production.

The Agricultural Products Board agreed to purchase surplus Ontario freestone peaches and Ontario Keiffer pears in 1977 to ensure that the year's large crops of the two fruits would be processed. Through a new crop development fund, the agriculture department supported 27 projects in 1977 covering a wide range of research, from the feasibility of peanut production in Southern Ontario to growing sorghum in southwestern Prairie regions and German grape varieties in the Okanagan Valley. Under a fresh fruit and vegetable storage construction financial assistance program, the department provided $1.4 million for new or improved storage facilities.

Three provinces signed amendments to the small farm development agreement program in 1977. British Columbia, Saskatchewan and Prince Edward Island all agreed to renew the program in which the federal government assigns farm management advisers to the provincial ministries of agriculture to work with small-scale family farms. Expenditures under the program in 1977-78 were expected to be more than $1 million in Saskatchewan and about $200,000 each in BC and PEI.

Major policy changes were announced in brucellosis control and in pesticide importation regulations. The country was to be divided into three regions, according to the level of brucellosis infestation, and new regulations would be announced to control cattle movement between regions. In the new pesticide importation policy, all pesticides sold, distributed and used in Canada would have to be registered and individual farmers could no longer import chemicals for use on their own farms.

In June the Cabinet approved in principle the creation of a national broiler agency. At year end, negotiations were still under way with the provinces toward implementing the program.

When drought struck Western Canada in spring and early summer, farmers took advantage of cultivation and irrigation practices researched by Agriculture Canada scientists to save what might have been a disastrous crop.

The 1977-78 dairy policy enabled the federal government to write off $152 million in export sales debt. In all, the government contributed $477 million toward the dairy program through the Canadian Dairy Commission.

New crop varieties licensed by the department included Coulter, a durum wheat; Saturn, a sunflower variety; Candle, a new rapeseed variety; and Norstar wheat.

During 1977 a new $22 million agriculture centre at Lethbridge, Alta. was officially opened. The centre, with a floor area of 25 084 square metres, is shared by research and food production and marketing staff of the federal department and regional offices of the provincial agriculture department.

11.1.2   Agricultural regions

Climate, soil conditions and geography have combined to form several distinct farming regions in Canada. A harsh northern climate restricts most agriculture to the southern portion of the country and nearly all farms lie within 483 kilometres of the southern border. In the Atlantic provinces and Central Canada farming is limited to coastal regions and river valleys, and soils vary in depth and fertility. In the Prairie region soil is fertile but rain is light. Farming is limited to high plateaus and river valleys in the western mountainous region.

Farming is an important business in Canada. About 68.3 million hectares (168.8 million acres) in 10 provinces are cultivated; 44.1 million hectares (109.0 million acres) are improved land. In 1976, farm cash receipts exceeded $9.9 billion and agricultural exports exceeded $3.9 billion.

There are four main types of farms in Canada. Livestock farms include those specializing in feedlot finishing of cattle, large-scale feeding of hogs bought as weanlings, dairying, poultry production for meat and eggs, and breeding and raising livestock. Grain farms produce such crops as wheat, oats, flax and rapeseed. Special crop farms produce vegetables, fruits, potatoes or other root crops, tobacco or forest products. Other farms combine livestock and grain production. Although each region has its specialties, none is limited to one type of farming.

The Atlantic region. This area includes Newfoundland, Prince Edward Island, Nova Scotia, New Brunswick and the Gaspé district of Quebec. It is hilly, with a general covering of relatively fertile soil developed under forest cover. The climate is modified by the sea, but also affected by cold currents from the coast of Labrador and winds from the north. Precipitation averages 760 to 1 400 millimetres annually. Mixed farming is general and forage crops support a healthy livestock industry. Some small farmers combine fishing or lumbering with farming.

Nova Scotia's main agricultural areas surround the Bay of Fundy and Northumberland Strait where they are protected from Atlantic gales. Dairying and poultry production are common and beef farming is increasing. The Annapolis Valley is famous for fruit, particularly apples. New Brunswick produces potatoes and livestock in the Saint John River Valley and there is mixed farming in the northwest. More than a third of the commercial farms in the province are dairy farms.

Farming is the principal occupation on Prince Edward Island. Potatoes are the leading crop but the fertile land also supports mixed grains, dairying and other livestock enterprises. Small fruits and vegetables are produced.

In Newfoundland agriculture is of only local importance because of rough terrain. Bogland offers some potential for reclaiming and vegetable farming.

The central region. This lowland area bordering the St. Lawrence River includes the Ottawa Valley and extends through Southern Ontario to Lake Huron. Fertile soils, mostly formed by glacial drift and lake sediment developed under deciduous forest cover, and a mild climate modified by the Great Lakes and the St. Lawrence River, account for varied farming. Precipitation averages 760 to 1 140 mm a year. It is the most densely populated part of the country, providing large markets for farm produce.

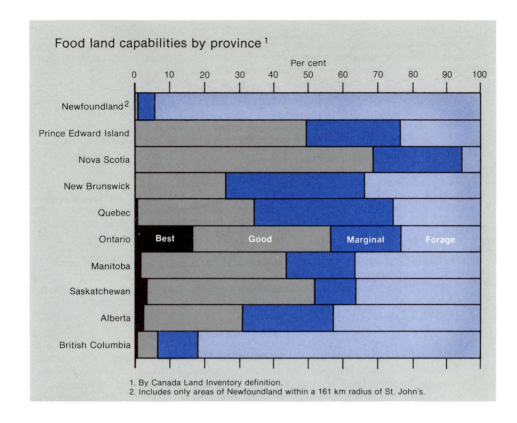

Food land capabilities by province[1]

Per cent

1. By Canada Land Inventory definition.
2. Includes only areas of Newfoundland within a 161 km radius of St. John's.

Well over half the commercial farms of Quebec are now dairy farms, a change from the traditional small mixed farming of old Quebec. Fairly large butter and cheese industries rely on these farms. Livestock farms, specializing in beef cattle, hogs or sheep, and mixed farms are common, and poultry and egg production is increasing. Forage crops account for the largest cultivation and oats and corn for feed are produced. Fruits, particularly apples, and vegetables are becoming prime crops, and sugar beets and flue-cured tobacco are grown and processed.

Ontario has specialized crops in more southerly regions, but also by far the largest number of commercial livestock farms and is second to Quebec in dairy farms. Forage crops are the largest cultivated crops; others are corn, mixed grains, winter wheat, oats and barley.

Dairy farms are concentrated in Middlesex, Oxford and Perth counties in southwestern Ontario, in the Bruce Peninsula and in the eastern counties. Beef is a specialty in Lake Huron and Georgian Bay areas. Sheep, poultry and hog production is widespread. Ontario is a major producer of apples and the Niagara Peninsula accounts for most of Canada's tender tree fruits and grapes. Vegetables are grown near most large centres. Maple syrup is a major sideline for farmers in Ontario and Quebec.

The Prairie region. Manitoba, Saskatchewan and Alberta contain 75% of the farmland in Canada. Precipitation that averages only 330 to 510 mm a year and a climate of bitter winters and short hot summers favour the production of high quality hard red spring wheat, by far the largest single crop in all three provinces. Rangeland and pasture also support a large number of cattle and livestock rearing in general is a major industry.

Manitoba has the highest rainfall of the three provinces and an average of 100 frost-free days, resulting in more varied farming. Wheat and other grains predominate but rapeseed is also grown, and there is mixed farming with an emphasis on livestock. Vegetables, sugar beets and sunflowers are grown south of Winnipeg and processed locally. Dairy farms are common around Winnipeg; hog production and sheep farms are widespread and beef cattle are raised in the southwest.

Saskatchewan grows about two-thirds of all Canada's wheat and large quantities of other grains, aided by light spring rainfall and long sunny days. Rapeseed is a popular crop and irrigation assists vegetable and forage crops. Mixed farming is common in the north where rainfall is higher, and turkey farming as well as egg and broiler chicken production are increasing. Hogs and beef cattle are gaining in importance.

Alberta is second to Saskatchewan in grain production but has more beef cattle than any other province. These are concentrated in large ranches in the south and in the Rocky Mountain foothills. Cattle-feeding operations are expanding and Alberta is a leading producer of hogs and sheep. Irrigation in the south aids in producing canning crops, sugar beets and forage crops. Dairy and poultry products are prominent in the mixed-farm economy. In the northwest the Peace River district produces grain and livestock.

The Pacific region. The most westerly region, British Columbia, is occupied largely by mountains and forests. Only 2% of the area is agricultural. There is no single regional climate: the Pacific Coast has mild temperatures and high rainfall; the interior has moderate temperatures with parts as dry as the Prairies; and the central interior, although a little cooler, has fairly high precipitation. Farms tend to be small and highly productive and are concentrated in the south-central mainland and southern Vancouver Island.

Livestock and dairying account for the greatest part of BC's agricultural production. Hogs and beef cattle are raised on many farms, beef particularly in the central and southern interior areas. Dairying and poultry meat and egg production are concentrated in the lower Fraser Valley where the population is large. Mixed farming is scattered throughout BC.

British Columbia is Canada's largest producer of apples. The Okanagan Valley is also noted for tree fruits such as peaches, plums and cherries. Raspberries and strawberries are grown in the Fraser Valley and on Vancouver Island along with other horticultural crops — apricots, grapes, tomatoes, sweet corn and potatoes. The processing industry is well developed. Vancouver Island's mild climate also produces flowering bulbs.

The northern region. The agricultural region north of latitude 55° consists of parts of northern British Columbia, the Yukon Territory and the Mackenzie River Valley in the Northwest Territories. Agricultural settlement in the area is not encouraged by the harsh climate and small population. Precipitation varies from light in the northern Yukon to heavy on the mountainous coast of BC. Frosts can occur in any month, but crops grown on northern slopes escape some damage. The North is estimated to have 1.2 million hectares (3 million acres) of potentially arable land and large expanses of grazing land, but there are probably fewer than 30 commercial farms in the region. Dairy products, beef cattle, forage crops, feed grains and vegetables are produced for the small local market.

11.1.3   Farm ownership and labour

Most farms in Canada are owned by the farmers who operate them, but as individual farms increase in size more land is being rented. By 1976, 31% of Canadian farmers rented some of the land they farmed; 6% rented all their land. Payment is usually cash or a share of crops or receipts.

Farm families provide most of the labour required on farms, although experienced workers are often employed on dairy farms and seasonal workers are required for harvests. In the West, operators of combine harvesters often move their machinery with the harvest, starting in the US and moving into Canada later in the season. Potato harvesters follow the same pattern in the East.

Transportation   11.1.4

Railways have been the traditional method of transporting agricultural products to large markets and ports. The Prairie provinces in particular rely on trains to move wheat and livestock to Canadian markets and to elevators in Vancouver, Churchill and Thunder Bay for shipment abroad. Bulky products such as sugar beets are usually shipped by rail.

Many products are now shipped by road. Although railways have retained their importance on the Prairies, many branch lines have been abandoned in other areas and most farmers now ship their produce at least part way in their own trucks. Eggs, poultry, cream, fruits and vegetables go to local markets by road and milk is generally collected at farms by tank trucks. Commercial farms and co-operatives use trucks for marketing and distributing agricultural products and in delivering supplies to farms.

Water routes supplement these means. The Great Lakes have long been used to ship grain from Thunder Bay to Eastern Canada and since the opening of the St. Lawrence Seaway in 1959 the lakes have been open to ocean-going vessels. Churchill is another seasonal port for Prairie grains. Vancouver and Halifax are year-round ports.

The Hall Commission on grain handling and transportation in 1977 recommended action to be taken on 10 139 kilometres (6,300 miles) of Prairie branch lines. The Prairie Rail Action Committee (PRAC) was subsequently appointed by the transport minister to examine recommendations regarding 3 702 km (2,300 miles) of line recommended for transfer to the Prairie Rail Authority as well as other proposals. As the result of government action some 2 916 km (1,812 miles) received permanent protection, 3 737 km (2,322 miles) are protected until the end of 1978 and 3 486 km (2,166 miles) were passed to the Canadian Transport Commission for abandonment proceedings. PRAC was asked to advise the government on: further transfers of branch lines to the permanent network guaranteed at least until the year 2000; abandonment of branch lines at a date to be specified by PRAC; priorities for branch line upgrading work; and the examination of the Prairie Rail Authority concept put forward by the Grain Handling and Transportation Commission.

Marketing and supplies   11.1.5

Marketing of Canada's farm products is a blend of private trading, public sales and auctions, and sales under contract and through co-operatives or marketing boards. Methods vary with type of product, region and preference of producers.

Canada's principal livestock markets are at Montreal, Toronto, Winnipeg, Calgary and Edmonton, but other outlets vary from large stockyards to country collection points. Most cattle and calves are marketed by auction at public stockyards; the remainder go directly to packing plants or are exported. Most hogs, sheep and lambs are sold directly to packing houses; sales of hogs are usually handled by marketing boards.

Egg sales in Canada are regulated by a Canadian egg marketing agency, and a turkey marketing agency serves turkey producers. Chickens raised for meat are marketed through provincial marketing boards which have authority to allocate producer quotas, set producer prices and collect levies.

Marketing fluid milk is a provincial responsibility and quality, prices and deliveries are regulated by provincial marketing agencies which estimate market requirements and assign producers a share of the market. A marketing plan allocates producers a share of the Canadian market for milk used for manufacturing. It is in effect in all provinces except Newfoundland. Market shares under this plan are administered by provincial marketing agencies under the direction of the Canadian Dairy Commission.

Most grain marketed in Canada is grown in the Prairie provinces. The Canadian Wheat Board is responsible for various aspects of marketing wheat, oats, barley, rye, flax and rapeseed in Western Canada. In Ontario all wheat grown is sold through the Ontario Wheat Producers' Marketing Board.

Fruit and vegetables are distributed through fresh and frozen food markets, canneries and other processors. Most produce is grown under a contract or a pre-arranged marketing scheme; marketing boards, producers' associations and co-operatives are common. Tobacco is controlled by marketing boards in Ontario and

Quebec, soybeans by a board in Ontario and sugar beets by contracts with refineries in Quebec, Manitoba and Alberta.

Farmers' co-operatives are usually organized to handle or market producers' crops or livestock, to supply goods and services needed in farming, or both. Co-operative pooling arrangements for farm products guarantee farmers cash advances on deliveries whether products are sold immediately or not.

Marketing of seed in Canada is carried on by private seed companies, farmer-owned co-operatives and seed growers. Seed grades are established by federal regulation. Pedigree seed is produced by members of the Canadian Seed Growers' Association under conditions that ensure purity of the variety.

Farm machinery, building materials, fertilizers, agricultural chemicals and other supplies are obtained through commercial and co-operative outlets. Statistics on farm implement and equipment sales appear in Chapter 18, Merchandising and trade, and on manufacturing of agricultural implements in Chapter 17, Manufacturing.

11.2 Federal government services

11.2.1 Canada Department of Agriculture

Responsibilities of the department cover three broad areas: research, promotional and regulatory services and assistance programs. Research aims at solving practical farm problems by applying fundamental scientific research to soil management, agricultural engineering, and crop and animal production. Promotional and regulatory services attempt to control and eradicate crop and livestock pests and register chemicals and other materials used for these purposes. Also included are inspection and grading of agricultural products and setting up crop and livestock improvement policies. Assistance programs include price stabilization, compensation, and income security in the event of crop failure.

11.2.2 Government and the grains industry

Government's interest and involvement in the grains industry predates Confederation and is a record of policies relating to land use and settlement; transportation; grain elevators, storage, handling and forwarding; marketing methods and opportunities; income security; and the many ramifications of international competition and the search for international co-operation in the sale of grain. The federal government's role in the grains industry is carried out by the agriculture department, the industry, trade and commerce department and two semi-autonomous bodies which report to Parliament through federal ministers: the Canadian Grain Commission and the Canadian Wheat Board.

Three other agencies also play integral roles: the Canadian International Grains Institute, the Canada Grains Council and a special advisory group on grains, the Grains Group. The grains institute contributes to the maintenance and expansion of markets for Canadian grains and oilseeds and their products in Canada and abroad. The grains council provides a forum for co-ordination, consultation and consensus on industry recommendations to government. The special advisory group is charged with co-ordinating, reviewing and recommending federal policies on grains.

Grains Group. In 1970 the minister responsible for the Canadian Wheat Board organized the special advisory group made up of policy advisers representing the departments of agriculture, industry, trade and commerce, and transport. The group examines problems of the grains industry in production, transportation and handling, and marketing. It co-ordinates, reviews and recommends federal policies for these areas. Policies subsequently adopted are implemented through government departments or other agencies concerned with the grains industry.

Production. The Canada Department of Agriculture conducts a research program in plant breeding and production methods to improve varieties, yields and quality of grains for which there is a domestic and export demand. An innovation has been the provision each March, well in advance of spring planting, of information on initial prices to be guaranteed farmers for new crops of wheat, oats and barley, and on minimum deliveries to be accepted by the Canadian Wheat Board during the crop year. These are announced by the minister responsible for the wheat board.

Marketing. To broaden assistance provided for sales and market development of grains, oilseeds and products, pertinent services of the industry, trade and commerce department are consolidated in the grain marketing office. Regular contact is maintained with the Canadian Wheat Board, other agencies and organizations concerned with grain marketing, trade commissioners abroad and the private trade sector. A program of trade promotion that includes participation in missions and trade fairs abroad is also maintained.

The grains and oilseeds marketing incentives program provides help in the form of cost or risk sharing to projects designed to increase sales of grains, oilseeds and their derivatives. Normally these are projects which would not be realized without incentives. Canadian companies, agencies, industry associations, universities, institutes and similar bodies are qualified applicants. Projects cover various fields such as grain handling, storage, processing, market testing of products, developing new products or processes, feeding trials and demonstrations and feasibility studies related to expansion of exports.

With the co-operation of the processing industry, federal and provincial governments and universities, a $5 million pilot plant in Saskatoon was opened in 1977. The POS Pilot Plant Corporation (protein, oil and starch) is a non-profit corporation directed by subscribing members.

Credit. Canada has been selling grain on credit since 1952. The original program provided for grain sales on terms of up to three years at commercial rates of interest. In 1968 the government approved a broadened and improved program for the sale of Canadian grain on credit to improve its competitive position in export markets. The new program allowed exporters to respond quickly to export opportunities in developing countries and on more favourable credit terms in some circumstances.

All credit sales are now on terms of three years or less, financed under the Canadian Wheat Board Act with a government guarantee. Credit sales of other than the western wheat, oats and barley which are marketed by the wheat board, and sales on terms of more than three years, are insured for government accounts under the Export Development Act on terms of up to three years.

Food aid. The Canadian food aid program has expanded from $2 million in 1962-63 to $249 million in 1977-78. Since 1963 food aid under bilateral and multilateral aid programs has been administered by the Canadian International Development Agency. Most of the food consists of wheat and wheat products, but rapeseed and rapeseed oil are also included. In the past about 80% of Canada's food aid was extended to foreign governments under bilateral programs, with the remaining 20% going through multilateral channels, mainly the world food program. Over the last 10 years more than 84 different countries have received food aid from Canada. Regular contributions of flour are also made to the United Nations Relief and Works Agency.

At the world food conference in Rome in November 1974, Canada pledged the allocation of one million tonnes of grain annually for three years ending in 1978, with 400 000 t of this amount to be channelled through the world food program. A commitment to increase substantially other types of food aid was also made.

The Canadian Grain Commission

11.2.3

A Canadian grain commission was established by the Canada Grain Act in April 1971, replacing the board of grain commissioners for Canada, established in 1912. It is composed of a chief commissioner and two commissioners and is under the jurisdiction of the federal agriculture department, with headquarters at Winnipeg and offices across Canada, the largest in Vancouver, Thunder Bay and Montreal.

The commission administers the Canada Grain Act, including inspection, weighing and storage of grain; fixes maximum tariffs for charges by licensed elevators; establishes grain grading standards; and operates the Canadian government elevators at Moose

Jaw, Saskatoon, Calgary, Edmonton, Lethbridge and Prince Rupert. All elevator operators in Western Canada and in Eastern Canada handling western-grown grain for export, as well as grain dealers in Western Canada, must be licensed by the commission and must file security by bond or otherwise as a guarantee for performance of all obligations imposed upon them by the Canada Grain Act or its regulations. On a fee basis, the commission provides official inspection, grading and weighing of grain, as well as registration of terminal elevator and eastern elevator receipts. The economics and statistics division of the commission is the basic source of information on grain handled through the Canadian licensed elevator system. The commission is also responsible for administering the Grain Futures Act which provides for the supervision of grain futures trading.

The commission's grain research laboratory conducts surveys of the quality of each year's grain crops and of grain moving through the Canadian elevator system. It provides information on quality of varieties and grades of grain to the inspection division, collaborates with plant breeders in studies on new grain varieties and undertakes basic research in relation to quality characteristics of cereal grains and oilseeds.

The commission's assistant commissioners — one in Alberta, two in Saskatchewan, one in Manitoba and one in Ontario — investigate complaints of producers and inspect licensed elevators. All grain elevators, equipment and stocks of grain may be inspected at any time.

The commission sets up western and eastern grain standards committees which participate in establishing grain grades and grade specifications and recommend standard and export standard samples for various grades. It also appoints grain appeal tribunals to hear appeals against grading of grain by the commission's inspectors; decisions of these tribunals are final.

11.2.4 The Canadian Wheat Board

The Canadian Wheat Board was set up under the Canadian Wheat Board Act of 1935 to market in an orderly manner, in interprovincial and export trade, grain grown in Canada. The wheat board became the sole marketing agency for Prairie wheat, oats and barley sold interprovincially or internationally. With the introduction of a new domestic feed grains policy in August 1974, marketing of feed grains for domestic use was removed from exclusive wheat board jurisdiction and these grains are now traded on the open market. The wheat board remains the sole purchaser and seller of feed grains for export. Other crops, such as rye, rapeseed, flaxseed, buckwheat and mustard are marketed by the private grain trade.

Sale of Prairie-grown wheat, oats and barley is carried out through sales negotiated directly by the wheat board, or through grain exporting companies acting as its agents.

Delivery of the kinds, grades and quantities of grain needed by customers is essential to the board's marketing program. This is accomplished in two stages. The first involves delivery of grain by the producer from his farm to the local country elevator under a quota system for the kind and grade of grain required to meet market commitments, and to allocate delivery opportunities equitably among all grain producers. The second stage involves movement of grain from country elevators to large terminals in Eastern Canada, at Thunder Bay, Churchill, and the West Coast by the railways. Shipping of grain from Thunder Bay to eastern positions is done largely by lake vessels. Extensive planning and a high degree of co-ordination within the grain handling and transportation industry are required. The wheat board, which co-ordinates the entire movement, programs rail shipments from country elevators to terminals on a weekly basis in accordance with sales requirements.

The producer selling to the wheat board receives payment in two stages. An initial payment price is established by order-in-council before the start of a crop year; this price, less handling costs at the local elevator and transportation costs to Thunder Bay or Vancouver, is the initial price received by the producer and is, in effect, a guaranteed floor price. If the wheat board, in selling the grain, does not realize this price and the necessary marketing costs, the deficit is borne by the federal treasury; after the end of the crop year when the board has sold all the grain or otherwise disposed of it in accordance with the Canadian Wheat Board Act, the board, if authorized by order-in-council, makes a final payment to producers.

Since implementation of the new domestic feed grains policy, a producer delivering feed grains to a country elevator has the option of selling the grain to the wheat board or on the open market. In the latter case he will, on delivery, receive a payment representing the final price in contrast to the Canadian Wheat Board system of initial and final payments. As a result of a modification effective in August 1976 in the feed grain policy, the wheat board offers feed grains to the domestic market at a corn competitive price.

The Prairie Grain Advance Payments Act, administered by the wheat board, provides that producers may receive through their elevator agents interest-free cash advances on farm-stored grain in accordance with a prescribed formula. The purpose is to make cash available to producers pending delivery of their grain under the quotas established. An advance of up to $45,000 (depending on the number of producers involved in the operation) may be issued to multi-farm farms, such as partnerships, co-operative and corporate farms. The maximum total advance prescribed by regulation may not exceed $15,000 for any individual for the crop year. The act also contains provisions for special advance payments to maximums of $7,500 for unharvested grain and $1,500 for drying of grain.

Two-Price Wheat Act. To mitigate the effects of sharp price fluctuations on domestic wheat consumers, the federal government implemented a two-price system for wheat in September 1973. The system provides a guaranteed price to the domestic miller of $3.25 for 36.4 cubic decimetres of bread wheat used for domestic consumption. Under the Two-Price Wheat Act, given royal assent on June 19, 1975, the government makes up the difference to farmers between the pegged domestic price and the export price on all sales into the domestic market, to a maximum payment of $1.75 for 36.4 cubic decimetres. A similar system is in place for durum wheats.

The grains and special crops division of Agriculture Canada's food production and marketing branch is responsible for administering this program and has supervised distribution of about $348 million from its inception to the end of 1977. Payments are distributed directly to farmers in Quebec and the Maritimes, while in Ontario and the Prairies, monthly payments are made to the Ontario Wheat Producers' Marketing Board and the Canadian Wheat Board, for distribution among farmers through the price pooling scheme operated by each organization.

The program will remain in effect until 1980, providing farmers with long-range price assurance and market stability.

The Canadian International Grains Institute was incorporated in July 1972. It operates in affiliation with the Canadian Wheat Board and the Canadian Grain Commission and financial responsibility is shared by the federal government and the wheat board. It is designed to help maintain and enlarge markets at home and abroad for Canadian grains, oilseeds and their products, and offers instructional programs to selected participants from countries purchasing these commodities and to Canadians associated with the grain industry. Courses are offered in grain handling, transportation, marketing, flour milling, bread baking and macaroni manufacturing, and lectures and practical training are given in analytical methods used in processing and using grains and oilseeds. Located in the Canadian Grain Commission Building in Winnipeg the institute includes classrooms, conference rooms, offices, library, laboratories, an 8.16 tonne, 24-hour-capacity flour mill and a pilot bakery.

The Canada Grains Council was established in 1969 to improve co-ordination within the industry on recommendations to government. Its principal aim is to co-ordinate activities directed at increasing Canada's share of world markets for grains and grain products and effecting their efficient use in Canada. Membership is open to all non-governmental organizations and associations whose members are engaged in grain production, processing, handling, transportation or marketing.

Administrative costs of the council are shared by federal government and industry members. The council currently has 29 member organizations representing thousands of individuals. At least two general meetings are held each year; the board of directors meets about 10 times a year. The council is served by a small secretariat.

The Western Grain Stabilization Act became effective in April 1976. Its objective is to protect producers against a large unexpected decline in either world grain prices or in sales of Canadian grain, increases in the cash costs of producing that grain or in any combination of those factors. The support given will prevent the net cash flow, the difference between total receipts from the production and sale of cereals and oilseeds and the cash costs of production, in each calendar year, from falling below the average of net cash flow in the previous five calendar years.

Under this voluntary program, participating grain producers contribute a levy of 2% of their grain sales up to a maximum of $25,000 a year to the western grain stabilization fund. The federal government contributes an equal amount to double the participating farmers' contributions. Detailed literature on the program is available from the Western Grain Stabilization Administration in Winnipeg.

11.2.5 Federal farm assistance programs

Changes in the past few decades have dictated the need for a different approach to some problems Large-scale mechanization and, in some segments of the industry, automation have reduced manpower requirements significantly; the number of farms has declined but the size and efficiency of farms have increased; marketing and income problems have taken different forms; and a decline in some rural communities has occurred together with problems of regional disparity. Legislation enacted to meet these situations provides price support, dairy market and producer income stabilization, crop insurance, feed grain assistance, credit facilities, marketing assistance, and other forms of assistance to meet emergency or long-term conditions. These measures are administered by the federal agriculture department or by organizations responsible to the agriculture minister except for the Farm Improvement Loans Act (administered by the finance department), Prairie Grain Advance Payments Act (industry, trade and commerce) and the Agriculture and Rural Development Act (ARDA) and Prairie Farm Rehabilitation Act (PFRA) programs (regional economic expansion).

The Agricultural Stabilization Board, established in 1958 by the Agricultural Stabilization Act and amended in July 1975, is empowered to stabilize prices of agricultural products to help the industry get fair returns for labour and investment, and to maintain a fair relationship between prices received by farmers and their costs of goods and services.

The act provides that the board shall take action to stabilize prices of agricultural commodities at prescribed price levels. These commodities are slaughter cattle, hogs, sheep, industrial milk, industrial cream, corn, soybeans, and oats and barley produced outside designated areas defined in the Canadian Wheat Board Act. The prescribed price of a named commodity is calculated at 90% of the five-year average of market price, or at such higher percentage as the Governor-in-Council may determine, indexed to reflect cash cost of production in that year as compared to the preceding five years. The Governor-in-Council may also designate other commodities for support on a similar basis. The board may stabilize the price of any product by offer to purchase, by making deficiency payments or other authorized payments for the benefit of producers. Stabilizing prices by means of assistance payments has helped the agricultural industry balance production and demand.

Since the inception of the act the cost of stabilization programs has totalled over $2 billion. The board maintains a revolving fund of $250 million; losses incurred are made up by parliamentary appropriations. An advisory committee, named by the agriculture minister and composed of farmers or representatives of farm organizations, advises the board and the minister on matters relating to stabilization.

The Agricultural Products Board was established in 1951 to administer contracts with other countries for purchase or sale of agricultural products and to perform other commodity operations as Canadian needs may dictate. The board's activities have included purchasing surplus Canadian commodities with resulting improvement in producer prices. Some of these commodities have been processed, packaged and delivered to the world food program as part of Canada's commitment to the Food and Agriculture Organization of the United Nations.

The Crop Insurance Act passed in 1959 (RSC 1970, c.C-36), permits the federal government to help the provinces in making all-risk crop insurance available to farmers on a shared-cost basis under federal-provincial agreements. Crop insurance can protect the farmer against unforeseen losses by spreading their impact over a number of years. The initiative for establishing crop insurance rests with the provinces and programs are developed to meet provincial requirements.

The federal government contributes a portion of premium costs or administration costs and shares the risk by providing loans or reinsurance when indemnities greatly exceed premiums and reserves. Farmers pay 50% of total premiums required to make the programs self-sustaining. The remainder is contributed by the federal government where the province elects to pay all administrative costs, or the province may elect to share the remaining premium and administrative costs equally with the federal government.

In the 1977-78 crop year, 112,000 farmers purchased some $1.5 billion in crop insurance coverage. Premiums totalled $148 million (including government contributions). The number of farmers participating increased by 16% over 1976-77 while coverage increased by 28%.

There was widespread threat of drought over most of the Prairies in late 1976 and early 1977, but timely rains during the 1977 growing season alleviated this situation except in Southern Alberta and southwestern Saskatchewan. Some $100 million would be paid out in indemnities on the 1977 crops with wet weather during harvest the major cause of loss in all provinces.

The Canadian Livestock Feed Board, established by the Livestock Feed Assistance Act (1966), is a Crown agency reporting to Parliament through the agriculture minister. It has four main objectives: to ensure that feed grain is available to meet the needs of livestock feeders; that adequate storage space in Eastern Canada is available for feed grain to meet the needs of livestock feeders; that the price of feed grain in Eastern Canada and in British Columbia remains reasonably stable; and that there is fair equalization of feed grain prices in Eastern Canada and in BC.

The board may make payments related to the cost of feed grain storage and transportation, the latter payments having been made since 1941. Since April 1967, the freight subsidy has been administered by the livestock feed board. Initially, it was applied only to feed grains produced in the Prairie provinces and designated for domestic livestock consumption in Eastern Canada and British Columbia. Subsequently it was extended to the movement of Ontario corn and wheat into the Atlantic provinces and Quebec.

The feed freight assistance program underwent substantial changes as part of the domestic feed grain policy. These changes, which became effective in August 1976, included reductions of $6.61 a tonne in rates of assistance to Ontario and Western Quebec (as far east as Montreal), with lesser reductions east of there. Rates of assistance to Eastern Quebec and the Atlantic provinces remain unchanged. Expenditures under the program were reduced from recent levels of about $20 million a year to about $10 million. Effective 1977, for a period of up to five years, the board may make payments against carrying charges for feed grains stored at feed mills in Eastern Canada and BC. The purpose of this federal program is to encourage expansion of grain storage in grain deficient areas.

The Farm Credit Corporation. This corporation (FCC) is responsible for the administration of the Farm Credit Act and the Farm Syndicates Credit Act and is an agent of the agriculture department in administering the land transfer plan of the small farm development program. Responsibility for lending decisions and operations is decentralized into seven branch offices, one for the Atlantic region and one for each of the other provinces. Field officers work out of 108 offices across Canada.

The Farm Credit Act, designed to meet long-term mortgage credit needs of Canadian farmers, provides three types of mortgage loans. Borrowers must be of legal age to enter into a mortgage agreement and loans are made only to Canadian citizens or those with landed immigrant status. All loans are repayable on an amortized basis within a period not exceeding 30 years. Funds for lending under the act are borrowed from the finance minister. In the fiscal year ending March 31, 1977 there were 4,465 loans for a total of $301.4 million.

The Farm Syndicates Credit Act authorizes the corporation to make loans to syndicates of three or more farmers for machinery, equipment or buildings. Loans can be made to syndicates to a maximum of $100,000 or $15,000 per qualifying member, whichever is the lesser. Loans are repayable over a period not exceeding 15 years for building and permanently installed equipment and seven years for mobile machinery. In 1976-77, the corporation lent $1.3 million to syndicates, representing 67 loans.

The small farm development program came into effect in September 1972. Under the land transfer plan of this program the corporation makes grants to owners of small farms who wish to sell for retirement or any other reason. During 1976-77 it approved $2.3 million in vendor assistance grants to 736 farmers, and a total of $568,600 was approved under the special credit provisions to help 32 farmers expand their operations.

The Farm Improvement Loans Act, administered by the finance department, is designed to facilitate credit by way of loans made by chartered banks and other lenders toward improvement or development of a farm. It includes purchase of implements and livestock; purchase and installation of agricultural equipment or a farm electrical system; major repairs or overhaul of agricultural implements and equipment; fencing or works for drainage on a farm; construction, repair or alteration of farm buildings including the family dwelling; and purchase of additional farmland. Credit is provided on security related to the purchase or project and on terms suited to the individual borrower.

The legislation has been continued through extensions since 1945, usually for three-year periods, the latest from July 1, 1977 to June 30, 1980. Maximum repayment period for land purchase is 15 years, and for all other purposes 10 years. The maximum loan or amount that may be outstanding to a borrower at any one time is $75,000. From inception of the program to December 31, 1977, loans amounting to about $3,656 million were made. During the same period, payments were made to the banks under a guarantee provision in respect of 5,833 claims amounting to $7.6 million, representing a loss ratio of 0.2%. In the first six months of 1977, 9,480 loans for $67.2 million were made and 61 claims under the guarantee were made for a value of $134,915.

11.3   Provincial government services

11.3.1   Departments of agriculture

Newfoundland. Government agricultural services in Newfoundland are provided by the forestry and agriculture department. Principal branches are: agriculture, lands and forestry. Programs are carried to the public by a regional services branch. Three regional supervisors, with agricultural representatives, each serve the public in a specified area known as an agriculture management unit.

Departmental policies in support of the agriculture industry include: a land clearing grant for private farmers, a capital assistance grant for purchase of buildings and equipment, subsidized provision of agricultural limestone, bonus payments for retention of quality breeding stock, grants-in-aid for construction of vegetable storage facilities, a subsidized regional pasture program, subsidized crop and livestock insurance programs, a subsidized veterinary services program, grants to agricultural societies, and technical information and farm management services.

Departmental assistance is also given under a provincial farm development loan board, a Newfoundland marketing board and a Newfoundland farm products corporation. The department produces and sells to farmers swine breeding stock which is free of specific diseases and which results from a controlled breeding program.

Prince Edward Island. The agriculture and forestry department is composed of the following branches and sections: a field services branch with a farm management section, district offices section, farm development section, livestock section, and crops and engineering section; a forestry branch, with a forest nursery division and Bunbury nursery division; a technical services branch; and a veterinary and dairy branch. The office of management services provides support to the four branches through its administration, information and economics, marketing and statistics sections. Management services also co-ordinates program planning and evaluation.

Work of the department is aimed at serving the needs of farm families, especially in efforts to stabilize and increase farm incomes and improve farm management ability. Programs range from 4-H soil-testing and developing individual plans for the family farm development program to crop insurance promotion, working with commodity groups and providing modern production recommendations.

Nova Scotia. The agriculture and marketing department directs the government's agricultural program. The department is concerned primarily with administration, extension, economics, horticulture and biology, livestock services, market development, soils and crops, and formal agricultural education through the Nova Scotia Agricultural College in Truro. The department is particularly interested in encouraging rural people to help themselves through such organizations as the Nova Scotia Federation of Agriculture, the Nova Scotia Fruit Growers' Association and other commodity-oriented groups.

New Brunswick. Provincial government agricultural policy and programs are directed by the agriculture and rural development department. Branches are concerned with administration, extension, livestock and poultry, veterinary services, communications and marketing, plant industry, agricultural engineering, home economics, credit unions and co-operatives, and planning and development. The province also has a farm adjustment board, farm products marketing commission, dairy products commission, and forest products commission.

Quebec. The Quebec agriculture department promotes agricultural development by providing technical and other assistance to help farmers modernize facilities, improve production and increase their standard of living. The department is responsible for boards and commissions administering programs concerning farm credit, agricultural product marketing, crop insurance, stabilization of farm income, encouraging new initiatives in food production and a provincial sugar refinery. The department's four branches deal with production, marketing, administration, and research and education.

Through the production branch, farmers in Quebec may call upon the expertise provided by 12 regional offices, five laboratories, and a young farmers' section, with staff specialized in information, technical training, and regional and resource development. An engineering directorate is responsible for mechanized operations, farm buildings and machines, and agricultural water services. Veterinary services run the contributory animal health insurance plan as well as the provincial animal pathology laboratories at Quebec and St-Hyacinthe. Also available are animal husbandry services, a swine insemination centre in St-Lambert, and an artificial insemination centre in St-Hyacinthe. Plant production services operate a Manicouagan potato breeding centre, a pilot blueberry processing plant at Normandin, a St-Bruno blueberry freezing plant, a St-Norbert maple products centre, Deschambault beekeeping centres, a soil laboratory at La Pocatière, and a plant laboratory in St-Hyacinthe.

The marketing sector serves both the farmer and the consumer. A marketing branch includes marketing, economic studies, and technical assistance to the food industry. A food inspection branch deals with crop products, dairy products and meat products, provides quality control and fraud suppression services, and food analysis at laboratories at Quebec and St-Hyacinthe.

The research and education branch includes an institute of agricultural and food technology at St-Hyacinthe, an institute of agricultural technology at La Pocatière, and research services for crop protection, soils and research stations.

A Quebec agricultural research and services council promotes and co-ordinates agricultural research in the province and determines research priorities in close co-operation with a plant products board, an animal husbandry board, and a food products board.

The administration branch is responsible for administrative and financial services, including subsidies, compensation, and land grants as well as computer services.

Ontario. The agriculture and food ministry conducts a variety of programs to develop a sound agricultural industry. Most assistance is given through self-help programs which benefit the individual farmer. The ministry administers 55 legislative acts and has 54 county and district offices.

Under a federal-provincial rural development agreement, 1975-79, the province shares equally with the federal government the cost of certain rural development programs. They cover farm enlargement and adjustment, rural resource development and assistance to rural industries to increase employment for rural people.

Agricultural manpower services branch, in co-operation with the colleges and universities ministry and the federal employment and immigration commission, helps to identify, develop and implement agricultural manpower training programs. Under a federal-provincial agricultural manpower agreement, money is provided to Ontario fruit, vegetable and tobacco growers for building and renovating accommodation for seasonal farm workers. The branch helps recruit and place full-time agricultural workers for Ontario farmers and co-operates with other countries in providing international programs for young agriculturists.

The soils and crops branch conducts extension programs in soil management and crop production. The branch has four sections — field crops, horticultural crops, seeds and weeds, and pest control. The branch also administers the Weed Control Act and the Grain Elevator Storage Act.

In the veterinary services branch, a laboratory section with six laboratories provides diagnostic, investigational, consultation and extension services and administers the Fur Farms Act. A meat inspection section administers the Meat Inspection Act. A regulatory and communicable diseases section administers acts, policies and programs concerned with disease control, animal care and sale of livestock medicines.

A livestock branch supervises livestock improvement programs and administers provincial laws relating to livestock. Programs include dairy herd improvement; beef cattle, sheep and swine performance testing; ram premium policy; federal-provincial sheep transportation assistance; testing of feed samples, a computerized ration formulation program; warble fly control and Northern Ontario livestock assistance. The branch makes grants to regional livestock clubs that hold sales and livestock shows, and sponsors exhibits of livestock outside the province. A staff of specialists provides feeding and management advice to livestock producers.

An Ontario stockyards board, operating under the federal Livestock and Livestock Products Act, provides a marketing service for Ontario livestock producers and protects their bargaining power.

A crop insurance commission, a branch of the ministry, offers contributory insurance against weather, insect and disease damage to winter wheat, spring grain, hay, corn (both silage and grain); soybeans, white beans; tomatoes, green peas, green and wax beans, lima beans and sweet corn for processing; red beets, apples, peaches, grapes, plums, sweet and sour cherries, pears, set onions, seed onions, coloured beans, new seeding and flue-cured tobacco, black tobacco, a burley tobacco, seed corn and flax. The total cost of administration is paid by the Ontario government and 50% of the premium is paid by the federal government.

The province has a milk commission, an administrative tribunal to which dairy producers, processors and others may appeal. The commission co-operates with Ontario's milk marketing board, cream producers' marketing board, and dairy council in dairy policy planning and development.

The milk industry branch is responsible for all regulative and administrative work required under the Milk Act, Oleomargarine Act, and Edible Oil Products Act, and regulations under the Farm Products Grades and Sales Act. The branch administers milk quality, fluid milk, milk products, plant record audit, farm inspections, and central milk testing programs.

A farm products inspection branch inspects fruit and vegetables for grade, and promotes improved methods of disease control, grading, packaging, marketing, handling, storing and transportation. The farm products marketing board administers 21 producer commodity boards covering some 44 commodities with a total market value of approximately $1 billion annually.

The province's food council is responsible for finding methods to improve domestic and export market opportunities for Ontario agriculture and food products. An Ontario food terminal, operating under the Ontario Food Terminal Act, offers farmers services of one of the largest volume wholesale fruit and vegetable markets in Canada.

Research and education are administered by the education and research and special services division. An advisory agricultural research institute reviews research programs and recommends priorities. The provincial entomologist reports on insect control programs and the provincial apiarist reports on the bee and honey industry.

The extension branch, through agricultural representatives at 54 county and district offices, relays information on research development and advice on farm management to farmers. Agricultural engineers work throughout the province. Northern Ontario assistance policies are also administered by the branch, which assists 4-H clubs and a junior farmers' association of Ontario.

The home economics branch conducts extension programs for rural adult groups and for young people's 4-H homemaking clubs on foods, nutrition, clothing, textiles, home furnishings, home crafts, and family and community life.

The information branch distributes publications, news releases, radio tapes and television film clips. A film library distributes more than 2,000 films annually. The market information service provides up-to-date commodity quotations to the media and individual producers daily using radio and audio-tape facilities.

The agricultural and horticultural societies branch offers advice and financial aid to agricultural and horticultural societies and plowmen's associations and manages an international plowing match and farm machinery show. The economics branch researches marketing, policy, production, land use and dairying, and works with Statistics Canada to collect and publish statistics on farm production and marketing.

The food land development branch provides an agricultural perspective to land-use planning. Staff contribute to and comment on official plans, amendments to plans and subdivision applications, and project plans for hydro, highways, pipelines and other facilities. Interim management of government-owned agricultural lands is accomplished through a land-lease program. The branch administers the Drainage Act and the Tile Drainage Act to provide loans and grants for draining agricultural lands. Staff also provide policy recommendations on alternative land-use programs.

Manitoba. The agriculture department serves through four divisions: marketing and production; rural development; regional; and management and operations. These divisions include the following branches.

An animal industry branch encourages improvement and efficient production of all classes of livestock, and helps to improve the quality of dairy products at producer and processor levels through inspection, consultation, education and laboratory quality control.

A soils and crops branch encourages development, production and improvement of cereal, forage and special crops and horticulture, and promotes policies that encourage good field crop husbandry, soil conservation, land development and weed control. An economics branch deals with educational and development programs in farm management and agricultural economics, carries out special studies and supervises farm diversification. A marketing branch conducts market development research and analysis to establish long-term markets for farm products. A veterinary services branch operates a diagnostic laboratory for animal diseases, administers the Veterinary Services District Act and the Veterinary Scholarship Fund Act, and works in co-operation with practising veterinarians.

A technical services branch provides programs in agricultural engineering, entomology and beekeeping and offers technical help to rural residents installing modern farm water systems. A community and family programs branch carries out education and development programs in 4-H and youth, agricultural manpower, community affairs, rural counselling and resource analysis. A communications branch provides press, radio and television services to mass media outlets and produces and distributes a wide range of instructional materials. The regional division includes five regions with 38 district offices, each staffed with agricultural representatives. The major role of this division is extension of educational programs and advice in agriculture and rural development.

The Manitoba Marketing Board supervises the operation of producer marketing boards responsible for the orderly marketing of hogs, milk, vegetables, eggs, broiler chickens, root crops, turkeys and honey. An agricultural products marketing commission helps market regulated agricultural products outside the control of producer boards.

Saskatchewan. Saskatchewan Agriculture has three main divisions: production and marketing, farm resources development, and extension and rural development. The department includes support services and a planning and research secretariat.

The production and marketing division administers legislation designed to improve production, handling, processing and marketing of specific farm products. It includes the following branches: plant industry, animal industry, veterinary services, and marketing and economics; and a milk control board, a crop insurance corporation, a hog marketing commission and sheep and wool marketing commission.

The extension and rural development division serves a co-ordinating function for extension programs and activities to help farm families develop and maintain viable farm units. It includes three branches. The regional extension services branch consists of about 100 field staff working throughout the province. The family farm improvement branch provides technical advice and services relating to farmstead mechanization. The newly formed irrigation branch, with headquarters at Outlook, is assisting development of a South Saskatchewan River irrigation project.

The extension and rural development division co-ordinates activities of FarmStart Corporation which administers a credit and grant program for persons establishing or expanding livestock operations.

The farm resources development division is primarily responsible for the development of land and water resources for agricultural uses through a lands branch and conservation and land improvement branch. The division is responsible for construction work for the South Saskatchewan River irrigation project and for development work in community pastures. The conservation and land improvement branch provides engineering services to the department in water management, including flood control, drainage and irrigation and is responsible for implementing Qu'Appelle conveyance and flood protection projects. The lands branch administers over 2.8 million hectares (7.0 million acres) of provincial lands for agricultural use. More than 12,000 farmers and ranchers lease land from the lands branch as full units or as additions to their private holdings. The remaining 607 028 hectares (1.5 million acres) are in provincial and co-operative pastures providing grazing for over 150,000 head of cattle belonging to more than 5,500 farmers.

Saskatchewan's land bank commission, agricultural implements board, and farm ownership board, are included in the farm resources development division. The land bank commission provides an alternative for farmers not wishing to commit themselves immediately to a heavy investment in land. It also provides Saskatchewan land-owners with a continuing sales opportunity for land, enables new farmers to start farming independent of substantial family assistance and permits farmers with insufficient land to add to their holding without raising large sums of money for capital investment. Major activities of the agricultural implements board include registration of implement distributors, licensing and inspection of retail vendors, and investigating complaints regarding warranties and repair parts availability. The farm ownership board deals mainly with farmland ownership by non-residents and non-agricultural corporations.

Alberta. The agriculture department activities are co-ordinated by an executive committee, while a secretariat, in consultation with agribusiness, farm organizations and researchers, advises the department on planning and research.

The office of the Farmers' Advocate ensures protection of the rights of individual farmers. The office investigates problems and complaints of farmers not relating to the provincial government and its agencies.

Marketing programs and activities are carried out by a marketing and economic services group and an international marketing group, with a view to expanding domestic and foreign markets for Alberta's farm products and to encourage increased food and agricultural processing in Alberta.

An economic services division provides economic, statistical, business and market analysis to facilitate decision making by producers and industry, and to encourage efficient use of resources.

A marketing services division supports commodity groups concerned with marketing, conducts consumer education and food promotional programs and assists in the expansion or creation of new processing facilities and development of new food products.

The international marketing group of Alberta Agriculture assists Alberta exporters of agricultural commodities, processed food and feed products and technical services. International marketing is the provincial government link between Alberta's agricultural industry and world markets. Market development services and programs are designed to supplement and complement those of the federal government.

Development divisions of the department are responsible for programs designed to advise producers, ensure survival of family farms and promote the interests of rural communities. An extension division operates through 64 district offices co-ordinated by six regional directors, with a staff of regional specialists in livestock, plant industry, engineering and home economics. Farm development division branches deal with engineering, home design and agricultural services. Programs in a home economics and 4-H division include home management, nutrition, family living, and 4-H youth training and leadership development. An irrigation division services the development and upgrading of irrigation projects and farms. An Alberta agricultural development corporation guarantees or makes loans for development of agricultural enterprises.

A plant industry division provides extension, research and resource assistance relating to crop improvement and protection; pest control, weeds, soils and fertilizers; horticulture, apiculture and special projects, and has a horticultural research centre at Brooks. A tree nursery at Oliver supplies trees for farm planting and reforestation.

An animal industry division administers legislation, policies and programs related to beef cattle, swine, sheep, horse and poultry industries. This involves general extension and many specific programs such as record of performance, artificial insemination, semen evaluation, feeder associations, warble control, brand registration, brand inspection, stray animals, research projects, cost studies in poultry and a broad range of industry licensing.

The dairy division administers legislation, policies and programs relating to the dairy industry. The testing, grading and purchasing of milk and cream by all dairy plants are regulated. Standards are set for construction, equipment, sanitation, and quality relating to milk production on farms and in dairy plants. A computerized dairy herd improvement program provides a guide to producers on which to base breeding, feeding and culling programs. A specialized extension service is provided in all areas of dairying. A dairy control board administers quota allocations for both fluid milk and manufacturing milk producers, and also a provincial fluid milk pooling system and fluid milk pricing.

A veterinary services division provides diagnostic laboratories for animal diseases and conducts investigations of disease conditions. It provides lecture service for the

University of Alberta and for other groups and promotes policies aimed at reducing losses by means of disease control, stockyard inspection and swine health programs.

British Columbia. The agriculture ministry comprises five divisions, each designed to serve various requirements of the industry. In addition to its Victoria headquarters, the ministry maintains 19 district offices, one laboratory each for dairy, entomology and veterinary sectors, and testing stations for beef cattle and poultry.

During the fiscal year 1977-78 the ministry intensified several programs aimed at implementation of production cost efficiencies at all levels of the food chain. A significant feature was the introduction of computerized least-cost ration formulations in the dairy and beef industries. Associated with the same objectives was expansion of co-ordinated resource management projects on Crown rangelands and a continuation of several market promotions, a number established in conjunction with a federal program.

11.3.2 Agricultural education

All of the provinces of Central and Western Canada have agricultural colleges associated with universities giving undergraduate and postgraduate courses in agricultural science and home economics. Ontario, Quebec and Saskatchewan have degree-granting veterinary colleges. In addition, all of these provinces have agricultural colleges, schools of agriculture or diploma courses offering basic training to young people intending to return to farms or interested in employment in businesses allied with agriculture.

Alberta has three agricultural colleges, Fairview, Olds and Lakeland (Vermilion campus), offering a broad range of diploma programs. Gradual expansion of the module approach in courses has increased and students may enter credit programs at a greater variety of times and locations. Also available are non-credit short courses focusing on specific agricultural activities.

A recently introduced green certificate program provides on-the-job and classroom training for farm hands and farm managers. The program is a joint project of farmers, Alberta government departments of agriculture and advanced education and manpower, the three agricultural colleges and Lethbridge Community College.

Several more unusual college programs such as turfgrass management and floriculture attract students from other provinces. The colleges participate in interprovincial and international agricultural education under exchange and world youth programs, providing orientation sessions and receiving groups from many countries. In recent years the colleges have expanded their curricula to meet both growing manpower needs of the business and industrial sectors and widely diversified interests of rural communities.

A number of public and private colleges in Alberta offer one or two years of university-transfer courses applicable toward degree programs in agriculture and veterinary medicine.

In Saskatchewan, the college of agriculture, college of veterinary medicine, and school of agriculture are all located at the University of Saskatchewan, Saskatoon. The colleges of agriculture, veterinary medicine, and home economics are degree-granting institutions providing undergraduate and postgraduate training. The school of agriculture is a two-year diploma course providing basic training for young people wishing to return to farms or seeking employment with related industries. A two-year farm machinery mechanics course is offered at Kelsey Institute, Saskatoon for trainees for the farm machinery service industry.

In Manitoba the major formal agricultural education is centred in the faculty of agriculture, University of Manitoba. The university offers a four-year course leading to a Bachelor of Science in agriculture and a two-year course leading to a diploma in agriculture. The faculty of agriculture also has an extensive program for graduate studies in agricultural sciences.

In Quebec agricultural science is taught at McGill and Laval universities. The education department offers a course in farm management and operation at two CEGEPs, and 15 school boards offer vocational training in agriculture in secondary schools. The Quebec agriculture department also operates two institutes of agricultural technology. The education, research and special services division of the Ontario agriculture and food ministry offers five diploma-course programs at the Ontario Agricultural College, University of Guelph, and at the colleges of agricultural technology at Centralia, Kemptville, New Liskeard and Ridgetown. In the Atlantic provinces, agricultural education is centred in the Nova Scotia Agricultural College at Truro, NS. This college provides the first two years of a four-year program in agricultural science, the first two years in agricultural engineering with the final two years provided by other faculties in Eastern Canada. The college offers several technical programs associated with farming and agribusiness and a variety of vocational courses designed to update farmers and other industry personnel.

Yearly statistics of agriculture   11.4

Collection, compilation and publication of statistics relating to agriculture are the responsibility of Statistics Canada. Valuable information is obtained through the censuses, partial-coverage mailed questionnaire surveys, and probability surveys, and from the administrative records of government operations.

Statistics Canada collects and publishes primary and secondary statistics of agriculture annually and monthly. Primary statistics relate mainly to reporting crop conditions, crop and livestock estimates, wages of farm labour and prices received by farmers for their products. Secondary statistics relate to farm income and expenditure, per capita food consumption, marketing of grain and livestock, dairying, milling and sugar industries and cold storage holdings. By collecting annual and monthly statistics, the federal agriculture department and various provincial departments, as well as such agencies as the Canadian Grain Commission and the Canadian Wheat Board, contribute statistical data and aid directly in Statistics Canada survey work. Thousands of farmers throughout Canada send in reports voluntarily and dealers and processors also provide much valuable data. The figures in this section do not include estimates for Newfoundland; agriculture plays a minor part in Newfoundland's economy and commercial production of most agricultural products is small. Subsection details are given for the most recent year available with earlier comparisons; figures for the latest year are subject to revision and many of those given for earlier years have been revised since the publication of the *Canada Year Book 1976-77.*

Farm income   11.4.1

Cash receipts from farming operations. Estimates of cash receipts from farming operations include cash revenue from the sale of farm products, Canadian Wheat Board participation payments on previous years' grain crops, cash advances on farm-stored grains and deferred income from the sale of grain in Western Canada, deficiency payments made by the Agricultural Stabilization Board and supplementary payments. Cash receipts from the sale of farm products include returns from all sales of agricultural products except those associated with direct inter-farm transfers. The prices used to value all products sold are prices to farmers at the farm level; they include any subsidies, bonuses and premiums that can be attributed to specific products but do not include storage, transportation, processing and handling charges which are not actually received by farmers.

Total cash receipts from farming operations for 1976 are estimated at $9,975 million, 0.5% below the revised 1975 value of $10,028 million. Increased receipts from the sale of livestock and livestock products did not offset lower cash returns from the sale of field crops. Contributing to the 3.5% increase in livestock cash receipts were increases from cattle, calves, sheep and lambs, poultry and eggs.

The major factors behind the decline in crops receipts were lower Canadian Wheat Board and Ontario Winter Wheat Producers' Marketing Board payments in 1976 amounting to $468 million as compared to the record high of $1,004 million received by farmers in 1975. The effect of this decline on total cash receipts was partially offset by lower deferments of grain receipts into the following year and an increase in cash advances on farm stored grains.

Farm net income. Two different estimates of farm net income from farming operations are prepared by Statistics Canada. *Realized net income* is obtained by adding together farm cash receipts from farming operations, supplementary payments and the value of income in kind, and deducting farm operating expenses and depreciation charges. This estimate of farm net income therefore represents the amount of income from farming that operators have left for family living, personal taxes and investment. The second estimate is referred to as *total net income,* and is obtained by adjusting realized net income to take into account changes occurring in inventories of livestock and stocks of field crops on farms between the beginning and end of the year. This estimate is used in calculating the contribution of agriculture to the income component of the system of national accounts and for making comparisons with net income of non-farm business enterprises (Table 11.3).

It is estimated that in Canada, excluding Newfoundland, realized net income of farm operators from farming operations amounted to $3,740.8 million in 1976, a 13.4% decrease from the revised 1975 amount of $4,320.2 million.

Contributing to this decrease was an estimated 0.5% increase in realized gross income to $11,004.2 million in 1976 from the 1975 figure of $10,953.6 million. This small increase in realized gross income was caused by decreases in both cash receipts and supplementary payments while income in kind increased by 11.3%.

Farm operators incurred operating expenses and depreciation charges totalling an estimated $7,263.3 million, 9.5% above the 1975 value of $6,633.4 million. Although expenditures on most farm input items were above the 1975 levels, some leading contributors to the increase in expenses were fuel and lubricants, electricity and telephone, property taxes and interest on farm business debt.

Total farm net income from farming operations (realized net income adjusted for inventory changes) is estimated at $4,040.3 million for 1976. The value of inventory change during 1976 was plus $299.5 million compared to plus $225.4 million in 1975. This expansion in inventories was due partly to a large increase in farm stocks of crop products which was partially offset by a decrease in livestock and poultry stocks.

11.4.2 Field crops

The bulk of the Canadian grains and oilseeds (excluding corn) is grown in the three Prairie provinces and the Peace River block of British Columbia. Wheat is the most important product and is produced largely for human consumption. Oats and barley are grown primarily for use as livestock feed. Of the oilseeds, rapeseed yields edible oil and flaxseed is crushed to produce linseed oil for industry; both these crops also produce meal for livestock feed.

Prairie production of wheat usually amounts to about three times domestic consumption, so this is an export-oriented industry. The same may be said of rapeseed and flaxseed. The coarse grains on the other hand do not enter into international trade to the same extent but large quantities do leave the Prairie provinces to be used as feed in Central and Eastern Canada and British Columbia.

There are approximately 160,000 grain producers in Western Canada (Canadian Wheat Board permit holders, 1976) and the crop is sold on world markets. A high quality information system covering annual production, stocks on hand and details on movement and location of supplies is essential to the smooth functioning of the trade. It is customary, for instance, to commit supplies for delivery before harvesting the crop from which such supplies will be drawn. Statistics Canada, in co-operation with the Canadian Grain Commission, the grain trade, the provincial departments of agriculture and the Canadian farmers, plays a leading role in providing this service.

Canada's 1977 wheat crop, estimated at 19.65 million tonnes, was 20.1% below the 1976 crop of 23.59 million tonnes and 15.0% above the 1975 crop of 17.08 million tonnes. The average yield per hectare at 1.943 tonnes (0.786 tonnes an acre) was 7.9% less than the 1976 yield of 2.096 tonnes (0.848 tonnes an acre) and 7.8% more than the 1975 yield of 1.802 tonnes (0.729 tonnes an acre). The average protein content of the 1977 crop of hard spring wheat was 13.1%, higher than the 1976 level of 12.8% and the 1975 level of 13.0%.

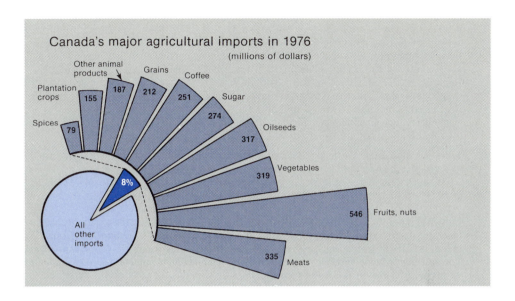

Canada's major agricultural imports in 1976
(millions of dollars)

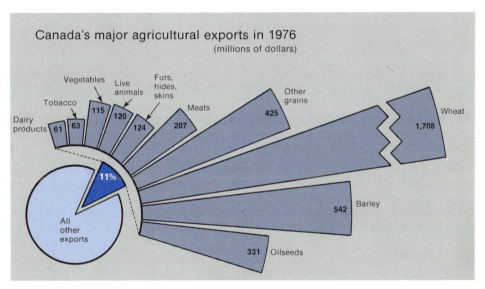

Canada's major agricultural exports in 1976
(millions of dollars)

Areas of cultivation, yields and prices of the principal field crops for the years 1972-77, with averages for 1967-71 are shown in Table 11.37; areas and production of field crops by province for 1976 and 1977 in Table 11.4 and areas and production of grain in the Prairie provinces for the years 1973-77 in Table 11.5. Table 11.6 shows stocks of Canadian grain on hand in Canada and in the United States on July 31 for the years 1973-77 with averages for the 10-year period 1967-76.

Livestock and poultry

11.4.3

Total number of cattle and calves in Canada, excluding Newfoundland, at July 1, 1977 was estimated at 14.6 million head, compared to 15.2 million head at July 1, 1976. Total cattle and calves in Newfoundland at June 1, 1976 according to the census, was 7,061 head. Milk cows two years and over were estimated at 1.9 million head, down 1.0% from July 1, 1976. Beef cows two years and over were estimated at 4.2 million head, down from 4.3 million head at July 1, 1976.

Exports of cattle and calves in 1976 totalled 477,794, up from 223,603 in 1975. Imports, at 199,247 head were up from 129,501. Beef exports, cold dressed carcass weight equivalent, increased from 20 million kilograms in 1975 to 59 million kg in 1976. Beef imports increased from 89 million kg in 1975 to 144 million kg in 1976.

Agriculture Canada reported that the weighted average price of A1 and A2 steers over 453.6 kg at Toronto for 1976 was $92.35.

Livestock slaughter. Cattle slaughtered at federally inspected packing plants amounted to 3,761,419 head in 1977, up 2.3% from 3,676,284 head in 1976.

Calves slaughtered at federally inspected packing plants amounted to 645,591 head in 1977, down 1.5% from 655,443 head in 1976.

The July 1, 1977 estimate for total number of pigs in Canada excluding Newfoundland was 6.2 million, up from 5.8 million in 1976. Pigs slaughtered in federally inspected and approved plants in 1977 numbered 8.0 million, up from 7.5 million in 1976 as reported by Agriculture Canada. The weighted average price at Toronto (for index 100 hogs, dressed) in 1976 was $1,413/tonne compared to $1,482/ tonne in 1975 and $1,038/tonne for the 1971-75 average.

The number of sheep and lambs on farms at July 1, 1977 was estimated at 532,500, down from 562,600 at July 1, 1976. (In Newfoundland on July 1, 1976 there were 9,159 sheep.) Sheep and lambs slaughtered in federally inspected packing plants in 1977 totalled 132,585 compared to 187,674 in 1976 and 186,566 in 1975. Imports of live animals increased from 57,601 in 1975 to 66,807 in 1976. Imports of mutton and lamb decreased from 20.2 million kg in 1975 to 17.4 million kg.

Wool. Estimates of production of shorn wool in 1976 at 1.2 million kilograms were 12.0% lower than in 1975. Average farm price per kilogram was 81.3 cents in 1976 compared to 66.1 cents a kg in 1975.

Poultry and eggs. Estimated number of laying hens on farms at July 1, 1976 was 22.3 million compared to 23.6 million at June 1, 1975. Production and consumption of poultry meat are shown in Table 11.8.

Table 11.18 shows production and value of farm eggs by province. Egg production totalled 455.6 million dozen in 1977, compared to 437.9 million dozen in 1976. The rate of lay per 100 layers rose to 22,139 from 21,975 in 1976 and the farm selling price of eggs averaged 65.0 cents a dozen compared with 64.0 cents a dozen in 1976. The Atlantic provinces produced 7.9% of all eggs in 1977, Quebec 16.3%, Ontario 38.8%, the Prairie provinces 24.1% and British Columbia 12.8%.

11.4.4 Dairying

Previous to 1975, milk production figures were derived by converting dairy products back to their milk equivalent. Beginning in 1975, published figures represent milk production sold off farms to first receivers, usually the marketing boards in each province. Therefore, data on milk production for the years 1975 and 1976 are not comparable to earlier years.

Total farm cash receipts have been adjusted for levies collected by provincial authorities in all provinces on manufacturing milk and cream shipments. Levies by category of milk are not available and therefore farm cash receipts may not add.

Number of dairy cows in Canada at July 1, 1977 was 1,975,000 head, the lowest in recent times. In 1977 milk production stood at 7 743.2 million kilograms compared to 7 685.2 million kg in 1976. Production is concentrated in Central Canada with Quebec and Ontario together accounting for 74.6% of Canadian production in 1977.

Production of creamery butter in 1977 was 113.3 million kg, compared with the 1976 output of 114.1 million kg. Quebec accounted for 51.8% of butter production and Ontario 29.3%.

Total production of factory cheese for 1977 was 134.3 million kg, some 7.1% above production in 1976; Quebec accounted for 43.8% of the output and Ontario 42.7%.

Total production of concentrated whole milk products, which includes condensed milk, evaporated milk, whole milk powder, partly skimmed evaporated milk and others, decreased 1.6% from 1976 while production of concentrated milk byproducts, which includes condensed skim milk, evaporated skim milk, skim milk powder, buttermilk powder, whey powder, casein and others, increased 31.9%.

Horticultural crops   11.4.5

Fruits and vegetables. Fresh and processed fruits and vegetables account for more than 40% in quantity of all food consumed in Canada. There are over 30 fruit and vegetable crops grown commercially in Canada with an annual farm value of almost $600 million.

The most important fruit grown in Canada is still the apple. Commercial apple orchards are found in Nova Scotia, New Brunswick, Southern Quebec, much of Ontario, and the interior of British Columbia, particularly in the Okanagan Valley. Tender tree fruits — pears, peaches, cherries, plums — are also grown in Ontario, with the most important concentrations in the Niagara Peninsula and in Essex County. These fruits, as well as apricots, are also grown on a large scale in the southern Okanagan Valley of British Columbia.

Strawberries and raspberries are cultivated commercially in the Maritimes, Quebec, Ontario and British Columbia. British Columbia fruit growers also produce small quantities of loganberries commercially on the lower mainland and on Vancouver Island. Grapes are grown in the Niagara district of Ontario and on a smaller scale in British Columbia. The native blueberry is found wild over large areas in Canada and is harvested in commercial quantities in the Atlantic provinces, Quebec and Ontario. A cultivated crop is grown in British Columbia. Cranberries are found in volume in Newfoundland, Prince Edward Island, Nova Scotia and British Columbia. Table 11.13 shows estimated commercial production and farm value of fruit grown in 1974-77.

Potatoes are the most important vegetable produced in Canada. The Maritime provinces are recognized as comprising the major growing region of the country. About one-third of the potato crop is processed by Canadian processors each year. In most years trade in potatoes with other countries is significant. In 1976, 131.4 million kilograms of fresh potatoes were exported and an additional 100.1 million kg were exported for seed.

The mushroom has been gaining importance in the Canadian diet. In 1976 the quantity of fresh mushrooms sold reached a record peak of 12.7 million kg while the quantity processed dropped to 7.1 million kg. For several years the amount of mushrooms processed in Canada has been decreasing, primarily because of increased competition from canned imports.

Considerable proportions of fruit and vegetable crops are canned, frozen or otherwise processed each season. Peas, corn, beans and tomatoes are the main vegetables processed. Many of the vegetables for processing are grown by farmers under contract to processors; however, the proportion of vegetables grown under this system is decreasing.

Imports of fruits and vegetables increased 16% between 1975 and 1976, and exports increased by about 13%. The United States was again Canada's major trading partner.

Honey. Honey is produced commercially in all provinces except Newfoundland. Alberta is consistently the largest producer followed by Ontario, Manitoba and Saskatchewan. In 1977, 27.7 million kilograms of honey was produced, a modest increase from the 1976 crop. In 1976 an average of 0.85 kg of honey was available for each Canadian.

In recent years the amount of honey sold directly to consumers has been increasing. Beekeepers' co-operatives are active in marketing in several provinces. Processors still buy huge quantities of honey and pasteurize it to facilitate storage, shipment, and uniformity of quality.

In 1977 approximately 8.97 million kg of honey was exported. More than half went to the United States but the Federal Republic of Germany was also a major buyer.

Maple products. Maple syrup is produced commercially in Nova Scotia, New Brunswick, Quebec and Ontario. The bulk of the crop comes from the Eastern Townships of Quebec, a district famous as the centre of the maple products industry. Virtually all maple products exported go to the United States with the largest proportion moving as syrup. Much of the syrup sold in Canada is marketed directly to the consumer from the producer; however, a considerable amount of both sugar and syrup is sold each year to processing firms. Production and value of maple sugar, syrup and taffy, by province, are shown in Table 11.18.

Greenhouse operations. The total area operated under glass and plastic in 1975 and 1976 amounted to 3.36 million square metres and 3.54 million m², respectively, while the total value of growers' sales stood at $129.9 million in 1975 and $146.2 million in 1976.

Nursery trades industry. In 1976 this industry had a total revenue of $129 million. Approximately 43% of this represented grower sales of traditional nursery stock and 34% was earned by supplying the demand for contracted services.

Tobacco. Total production decreased from 106.0 million kilograms in 1975 to 81.5 million kg in 1976. The average value per kilogram advanced from $2.06 in 1975 to $2.23 in 1976 (Tables 11.19 and 11.20). For information on tobacco products see Table 11.21.

11.4.6 Prices of farm products

The index of farm prices of agricultural products (Table 11.22) was designed to measure changes occurring in the average prices farmers receive at the farm from the sale of farm products. In comparing current index numbers with those prior to August 1975, the following points should be considered. Prices of all western grains used in the construction of the index prior to that date are final prices; all later figures are adjusted initial prices only for wheat, oats and barley. Any subsequent participation payments will be added to prices currently used and the index revised upward accordingly. Average cash prices of major Canadian grains are given in Table 11.23 and yearly average prices of Canadian livestock in Table 11.24.

11.4.7 Food consumption

Food consumption data shown in Table 11.25 represent supplies of food available for consumption by Canadians in 1975 and 1976. All calculations are made at the retail level of distribution except for meat which is compiled at the wholesale level. Amounts of food actually consumed would be lower than that shown because of losses and waste occurring after the products reach the consumer.

All basic foods are classified under 17 main commodity groups. The total for each group is computed using a common denominator, for example: milk solids (dry weight) for the dairy products group; fat content for fats and oils; and fresh equivalent for fruits. Most foods are included in their basic form, such as flour, fat, sugar, rather than in more highly manufactured forms.

In 1976 the trend was to healthier eating patterns. Less butter was available for consumption as consumers switched to less saturated fats such as margarine. More fresh fruits and vegetables were demanded by consumers and, as a result, imports to Canada increased during the winter. Although the use of canned fruit and vegetable products remained relatively stable, consumption of frozen products declined. Egg consumption declined about four eggs per person for the second year in a row. Grain and yogurt consumption increased.

11.5 1976 Census of Agriculture

This section presents a limited amount of information from the 1976 Census of Agriculture; details are contained in volumes 11, 12 and 13 of the 1976 Census of Canada. A list of the census reports released is available on request.

Number of census-farms. For census year 1976 a census-farm was defined as a holding of 0.4 hectares (one acre) or more with sales of agricultural products during 1975 of

$1,200 or more. This definition was changed from that used for census year 1971 when a sales figure of $50 was used instead of $1,200. Census data for 1971 are given here to correspond with the 1976 definition of a census-farm. The total number of census-farms in Canada in 1976, at 300,118, was almost the same as in 1971 when there were 299,868 census-farms (Table 11.27).

Farm areas. Total area of census-farms in 1976 was 67.2 million hectares (166.0 million acres), a 3.4% increase from the 64.9 million hectares (160.5 million acres) recorded in 1971 (Table 11.28). Nova Scotia, New Brunswick and Quebec reported decreases in farm areas. All other provinces reported increases. For Canada as a whole, the 43.7 million hectares (108.0 million acres) of improved land for 1976 increased 4.1% from the area recorded in 1971. The area of unimproved land increased 2.2% to 23.5 million hectares (58.0 million acres) in 1976 from 23.0 million hectares (56.7 million acres) in 1971. The area under crops, increasing by 4.8%, accounted for the majority of this increase.

Economic classification of census-farms. All census-farms in 1976, except institutional farms, such as experimental farms, community pastures and Indian reserves (but including Hutterite colonies), were classified as one of the four major historically comparable economic classes with a total of 10 breakdowns (Table 11.29). Economic classification of census-farms presents a measure of the productive size of the holding. The method of collecting sales information in the 1976 Census differed from previous censuses. Historically, the sales were reported by commodity. In 1976, however, the operator was asked to indicate one of several ranges which corresponded to his total sales of agricultural products during 1975.

Type of farm. Since the sales information collected in 1976 was not detailed by commodity as in the past, the farm typing scheme was based on potential value of sales by commodities. Potential value of sales was imputed from the 1976 physical inventories reported for the census-farm. With the exception of farms classified as institutional, all census-farms in 1976 with $2,500 or more of agricultural sales were classified as one of 10 major product types. Major differences in types compared with previous censuses are the deletion of the forestry type of farm and the division of the cattle-hog-sheep type into cattle and hogs. The small number of sheep farms is included in miscellaneous speciality. A farm was classified in a particular product type if 51.0% or more of the potential sales of this holding were obtained from this class of products.

Size of census-farms. In 1976, 43% of census-farms in Canada contained less than 97.1 hectares (240 acres) compared with 42% in 1971. This relatively small change in the proportion of such holdings suggests that the trend toward consolidation of farms into larger holdings may have moderated. However, the wide variation in distribution of farms by size between provinces continues. The proportion of farms under 97.1 hectares (240 acres) in the Atlantic provinces ranged from 50.2% in New Brunswick to 88.4% in Newfoundland; in Quebec 69.0%, Ontario 74.1%, Manitoba 20.7%, Saskatchewan 10.0%, Alberta 21.7% and British Columbia 70.6%.

Age of census-farm operators. The proportion of census-farm operators under 35 years of age in 1976 was approximately 19% of the total. Operators in the age groups 35-54 were 50% and those in the age groups 55 and over were 31% of the total. The corresponding percentages for 1971, at 15%, 53% and 32%, respectively, indicate a trend to younger farm operators.

Farm machinery. Table 11.36 indicates that between 1971 and 1976 forage crop harvesters increased by 25.9%, farm trucks by 25.5%, swathers by 13.1%, tractors by 12.0%, pick-up balers by 7.5%, automobiles by 5.4% and grain combines by 4.7%.

Product and marketing controls   11.6

Governments in Canada at both federal and provincial levels have from the earliest years been concerned with the importance of encouraging and assisting in the development of a productive and efficient agricultural sector. To this end numerous measures have been enacted over the years. Originally, emphasis was on production increases and control and eradication of pests and diseases. Gradually, however, with rising production and increasing specialization on the part of farmers, problems in marketing began to emerge.

To ensure quality, inspection and grading procedures and standards were established, but the periodic collapse of prices caused by bumper crops and intensified by the general inability of large numbers of producers to bargain on an equitable basis with far fewer buyers has been a much more difficult part of the marketing problem.

The first attempt to provide bargaining power to producers was the organization of voluntary marketing co-operatives. All provinces eventually passed legislation for incorporation of these co-operatives, and most of them also provided additional assistance in various forms. Federally, introduction of the Agricultural Products Co-operative Marketing Act provided for financial guarantees to producers willing to market their crops on a pooling-of-returns basis. More information on co-operative organizations is given in Chapter 18.

Although much co-operative marketing was initially successful, it was found that the voluntary aspect represented a serious weakness. Many members dropped out in good times to make their own deals. A type of marketing organization was needed with the legal power to control the output of all producers of a certain product in a certain area, and as a result marketing control legislation was adopted providing for various types of boards, agencies and commissions.

11.6.1 Quality standards

The federal and provincial departments of agriculture co-operate in establishing and enforcing quality standards for various foods. Some control over size and types of containers used is exercised by the Canada Department of Agriculture, and the Department of Consumer and Corporate Affairs enforces regulations pertaining to weights and measures.

Standards related to health and sanitation in food handling are developed and enforced at all three levels of government. Examples of provincial and municipal action include laws pertaining to pasteurization of milk, inspection of slaughter houses and sanitary standards in restaurants. At the federal level, inspection by the health of animals branch of the agriculture department of all meat carcasses that enter into interprovincial trade is required; the foods and drugs directorates of the health and welfare department have wide responsibility for food composition standards; and the consumer and corporate affairs department has jurisdiction over advertising.

11.6.2 Marketing controls

The Agricultural Products Co-operative Marketing Act (RSC 1970, c.A-6) was passed in 1939 as a result of a federal government decision to assist orderly marketing by encouraging establishment of pools that would give the producer the maximum sales return for his product, less a margin for handling expenses agreed upon in advance. All agricultural products except wheat produced in the Canadian Wheat Board area are eligible for marketing assistance under this act.

The purpose of this act is to aid farmers in pooling returns from sale of their products by guaranteeing initial payments and thus assisting in the orderly marketing of the product. The government may undertake to guarantee a certain minimum initial payment to the producer at the time of delivery of the product, including a margin for handling; sales returns are made to the producer on a co-operative plan. Amount of the initial payment is set at the discretion of the minister taking into account current and estimated market prices. For 1976 crops, agreements were made for marketing winter wheat and beans in Ontario, apples, strawberries and rutabagas in Prince Edward Island.

The Canadian Dairy Commission, established in 1966, was the first national marketing agency to be established since creation of the Canadian Wheat Board in 1935. The commission has the power to stabilize the market by offering to purchase major dairy products, butter and skim milk powder, at fixed prices and to package, process, store, ship, insure, import, export or sell or otherwise dispose of any dairy product purchased by it. The commission may also pay subsidies to producers of manufacturing milk and cream. These payments, up to $266 million in 1977-78, supplement returns to producers and permit market prices to be kept at reasonable levels. Each producer is eligible for subsidy on shipments covered by his market share quota. The commission, indirectly, pools returns to producers from products sold on the domestic and export markets through an export equalization fund. Money for this is collected by levies from producers in all provinces except Newfoundland under a market-sharing quota program and remitted to the commission. The funds are used to equalize export prices with domestic prices for products exported below domestic prices over a multiple-year period.

A comprehensive milk marketing plan, to balance demand and supply and equalize export assistance, was agreed to by the Canadian Dairy Commission and the milk marketing agencies of Ontario and Quebec in January 1971, establishing a market-sharing quota (MSQ) system for industrial milk and cream and that portion of milk, shipped by fluid producers, which is used for manufacturing purposes. Cream shippers in Quebec, Ontario and Prince Edward Island entered the plan in 1971. Producers in Alberta, Manitoba and Saskatchewan came under the program in 1972, British Columbia in 1973, and Nova Scotia and New Brunswick in 1974. All manufacturing milk and cream sold in Canada now come under the market-sharing program. The arrangement provides that each producer receives returns related to the target support price for manufacturing shipments up to his market share. The target support price is achieved through the offer-to-purchase program which stabilizes markets, plus direct payments to producers. Producer returns for deliveries over market share are related to world prices for surplus dairy products.

Producer marketing boards were introduced during the 1930s to give agricultural producers legal authority under certain conditions to control marketing of their produce. The Natural Products Marketing Act of 1934 attempted to provide this power at the federal level but the courts ruled that the subject was outside federal jurisdiction. The subsequently introduced Natural Products Marketing (British Columbia) Act, 1936 was found to be within the powers of provincial governments and it has since been used as a model for marketing board legislation as it evolved in all provinces.

The basic feature which enables marketing boards to control marketing is the compulsory aspect. A new board usually has to be first approved by a majority vote of the producers of the affected product. If it is approved, all producers of the product in the designated area, other than those who may be exempted as below a specified minimum production level, are required by law to market their produce under authority of the board. Depending on a board's objectives and the type of product, its powers and duties may only involve negotiating a minimum price or may include production or marketing quotas, designated times and places for marketing, or such other powers which may be considered necessary to ensure an orderly and equitable market.

The powers of a producer marketing board provided by provincial legislation are necessarily limited to trade within the province. Under the Agricultural Products Marketing Act (RSC 1970, c.A-7), the federal government may delegate to a marketing board powers with respect to interprovincial and export trade similar to those it holds under provincial authority with respect to intraprovincial trade. This act also gives the Governor-in-Council the right to authorize a provincial marketing board to impose and collect levies from persons engaged in production and marketing of commodities controlled by it for the purposes of the board including creation of reserves and equalization of returns.

The federal Farm Products Marketing Agencies Act passed in January 1972 is the enabling legislation for the creation of national marketing agencies or boards. National agencies may be set up, when producers and provincial authorities desire it, for any agricultural commodities which, owing to widespread production in Canada or for other reasons, cannot be effectively marketed in an orderly manner under the jurisdiction of individual provincial boards.

The National Farm Products Marketing Council (NFPMC) was established by the Farm Products Marketing Agencies Act in 1972 to advise the agriculture minister on matters pertaining to establishment of marketing agencies. It reviews their operations, assists them in promoting more effective marketing of farm products, and co-ordinates related activities of provincial governments as well as efforts of producers to establish marketing plans. Membership of the NFPMC includes representatives of consumer, labour and business interests.

The first national agency formed under the act, the Canadian Egg Marketing Agency, commenced operation in June 1973 and the second, the Canadian Turkey Marketing Agency, in March 1974. These federal agencies work in conjunction with provincial egg and turkey boards; they do not deal directly with producers.

Operation of these agencies has led to a great deal of interest on the part of producers of other commodities to establish such organizations. The NFPMC has been approached by representatives of producers of corn, tobacco, pregnant mares' urine and various fruits and vegetables. Proposals to establish a national marketing agency were received from the Canadian broiler council, representing broiler chicken producers. These various groups have indicated support for the concept of supply management being applied at the national level.

During 1975-76 there were 108 provincially authorized marketing boards operating in Canada, including the milk control boards which have a lesser degree of producer control than the others, as well as the earlier mentioned federal boards. Boards are established in all provinces, led by Quebec with 25 and Ontario with 21. An estimated 64% of 1975 farm cash income was received from sales made under the jurisdiction of marketing boards. A variety of agricultural commodities were sold under marketing boards, including grains, hogs, milk, fruit, potatoes and other vegetables, tobacco, poultry, eggs, wood, soybeans, honey, maple products and pulpwood. At year end 1976, the federal government had delegated authority to 79 provincial boards to control marketing of their products in interprovincial and export trade.

Sources

11.1 Information Division, Canada Department of Agriculture; Grains Group, Department of Industry, Trade and Commerce.

11.2 Information Division, Canada Department of Agriculture; Grains Group, Department of Industry, Trade and Commerce; Agriculture Stabilization Board; Crop Insurance Division, Canada Department of Agriculture; Canadian Livestock Feed Board; Farm Credit Corporation; Guaranteed Loans Administration, Department of Finance.

11.3 Supplied by respective provincial government departments.

11.4 Agriculture Division, Institutions and Agriculture Statistics Field, Statistics Canada; Food, Beverages and Textiles Section, Manufacturing and Primary Industries Division, Statistics Canada.

11.5 Census of Agriculture Division, Institutions and Agriculture Statistics Field, Statistics Canada.

11.6 Marketing and Trade Division, Canada Department of Agriculture; Canadian Dairy Commission; Information Division, Canada Department of Agriculture.

Tables

It should be noted that figures shown for the latest year are subject to revision, and that some figures for earlier years have been revised. Figures for Newfoundland are not included, as agricultural activity there is of minor importance and production small.

11.1   Cash receipts from farming operations (excluding supplementary payments), by province, 1972-77 (thousand dollars)

Province	1972	1973r	1974r	1975r	1976	1977p
Prince Edward Island	44,840	73,304	84,693	83,427	104,869	90,603
Nova Scotia	73,572	96,499	103,537	116,236	124,029	129,139
New Brunswick	64,955	95,822	103,233	100,043	113,967	106,550
Quebec	776,355	980,080	1,149,617	1,342,037	1,359,766	1,428,458
Ontario	1,622,681	1,992,585	2,486,908	2,649,785	2,769,932	2,855,037
Manitoba	487,812	619,429	825,371	935,004	897,160	892,061
Saskatchewan	1,202,828	1,467,146	2,039,831	2,468,789	2,286,186	2,107,236
Alberta	917,353	1,201,211	1,686,475	1,876,103	1,847,633	1,968,235
British Columbia	246,750	335,230	389,199	426,765	471,484	493,573
Total	5,437,146	6,861,306	8,868,864	9,998,189	9,975,026	10,070,892

11.2   Cash receipts from farming operations, by commodity or other source, 1973-77 (thousand dollars)

Item	1973	1974	1975	1976	1977p
Wheat	870,069	1,548,421	1,698,703	1,645,298	1,666,563
Wheat, Canadian Wheat Board payments	331,182	506,112	826,920	412,214	131,029
Oats	30,652	55,307	60,018	71,570	58,668
Oats, Canadian Wheat Board payments	12,861	—	27,809	11,665	13,956
Barley	229,475	468,566	465,701	462,079	334,253
Barley, Canadian Wheat Board payments	100,800	96,069	149,479	44,107	102,280
Canadian Wheat Board cash advances	29,947	48,105	19,866	102,689	125,699
Canadian Wheat Board cash advance repayments	-23,476	-36,684	-33,765	-44,247	-120,025
Deferred grain receipts	-305,246	-625,816	-695,450	-546,270	-434,034
Liquidation of deferred grain receipts	—	305,246	625,816	695,450	546,270
Rye	20,476	24,923	27,033	27,626	21,738
Flaxseed	119,062	135,435	81,340	78,779	90,402
Rapeseed	247,008	337,896	261,197	224,738	445,812
Soybeans	50,348	78,511	44,925	77,077	78,404
Corn	99,155	164,156	152,979	157,640	157,595
Sugar beets	22,293	43,799	39,919	36,275	31,282
Potatoes	159,969	211,475	164,805	213,165	203,502
Fruits	140,849	140,232	130,497	136,284	159,005
Vegetables	161,490	191,677	242,593	244,539	243,620
Floriculture and nursery	92,344	117,759	143,365	146,456	145,256
Tobacco	142,807	207,681	198,188	210,025	182,629
Other crops	172,574	149,719	170,186	163,110	198,832
Total, cash receipts from crops	2,704,639	4,168,589	4,802,124	4,570,269	4,382,736
Cattle and calves	1,479,511	1,680,581	1,817,975	1,963,335	2,057,426
Pigs	825,494	778,092	886,471	840,166	838,840
Sheep and lambs	10,673	12,867	13,466	13,689	12,201
Dairy products	849,455	1,095,903	1,343,286	1,319,180	1,419,472
Poultry	437,939	472,150	419,353	479,547	495,826
Eggs	243,785	269,095	260,165	282,508	281,170
Other livestock and products	73,678	75,730	77,224	85,697	89,890
Total, cash receipts from livestock and products	3,920,535	4,384,418	4,817,940	4,984,122	5,194,825
Forest and maple products	44,613	43,174	45,965	53,657	58,162
Dairy supplementary payments	131,022	221,059	259,770	258,868	269,449
Deficiency payments	60,497	36,041	22,596	24,791	48,704
Provincial income stabilization program	—	15,583	49,794	83,319	117,016
Total, cash receipts (excl. supplementary payments)	6,861,306	8,868,864	9,998,189	9,975,026	10,070,892
Supplementary payments	11,508	57,475	30,290	—	—
Total, cash receipts	6,872,814	8,926,339	10,028,479	9,975,026	10,070,892

11.3   Net income of farm operators from farming operations[1], by item and by province, 1973-76 (thousand dollars)

Item and province	1973[r]	1974[r]	1975[r]	1976
ITEM				
1. Cash receipts from farming operations	6,861,306	8,868,864	9,998,189	9,975,026
2. Income in kind	677,525	827,265	925,075	1,029,149
3. Supplementary payments	11,508	57,475	30,290	—
4. Realized gross income (items 1+2+3)	7,550,339	9,753,604	10,953,554	11,004,175
5. Operating and depreciation charges	4,777,964	5,827,446	6,633,354	7,263,348
6. Realized net income (items 4-5)	2,772,375	3,926,158	4,320,200	3,740,827
7. Value of inventory changes	515,406	-91,307	225,382	299,501
8. Total gross income (items 4+7)	8,065,745	9,662,297	11,178,936	11,303,676
Total, net income (8-5)	3,287,781	3,834,851	4,545,582	4,040,328
PROVINCE				
Prince Edward Island	30,222	56,558	21,652	60,299
Nova Scotia	31,244	25,139	27,225	30,565
New Brunswick	42,612	56,138	21,762	49,712
Quebec	367,477	397,808	474,799	388,298
Ontario	717,647	878,548	999,222	813,528
Manitoba	374,875	340,458	431,720	359,162
Saskatchewan	902,886	1,139,834	1,504,091	1,423,999
Alberta	671,735	790,799	893,013	722,652
British Columbia	149,083	159,569	172,098	192,113

[1]Includes estimated value of farm homes, supplementary payments made under the provisions of the Prairie Farm Assistance Act and payments under the Western Grain Producers' Acreage Payment Regulations.

11.4   Harvested area and production of field crops, by province, 1976 and 1977

Field crop and province	Area				Total production			
	1976		1977		1976		1977	
	'000 hectares	'000 acres	'000 hectares	'000 acres	'000 tonnes		'000 tonnes	
						'000 bu		'000 bu
WHEAT	*11 252*	*27,804*	*10 114*	*25,002*	*23 587.0*	*866,672*	*19 650.7*	*722,036*
Prince Edward Island	7	16	6	14	20.3	745	15.4	564
Nova Scotia	2	4	2	5	4.4	162	4.7	171
New Brunswick	2	6	3	8	5.7	209	7.2	265
Quebec	32	79	36	90	60.4	2,220	74.5	2,736
Ontario								
Winter	209	516	239	590	667.9	24,542	822.5	30,220
Spring	7	18	8	20	16.2	594	18.5	680
Manitoba	1 541	3,800	1 290	3,200	2 803.2	103,000	2 748.7	101,000
Saskatchewan	7 160	17,700	6 557	16,200	14 995.8	551,000	12 709.6	467,000
Alberta	2 266	5,600	1 943	4,800	4 953.2	182,000	3 184.3	117,000
British Columbia	26	65	30	75	59.9	2,200	65.3	2,400
OATS	*2 410*	*5,954*	*2 131*	*5,268*	*4 831.3*	*313,268*	*4 302.9*	*279,011*
Prince Edward Island	19	47	17	43	48.3	3,132	34.5	2,240
Nova Scotia	8	19	6	15	15.5	1,002	10.2	660
New Brunswick	20	48	16	39	32.0	2,073	20.7	1,345
Quebec	225	557	203	501	336.7	21,834	327.6	21,242
Ontario	221	545	182	450	345.9	22,427	330.4	21,424
Manitoba	506	1,250	425	1,050	940.8	61,000	894.5	58,000
Saskatchewan	647	1,600	627	1,550	1 388.0	90,000	1 233.8	80,000
Alberta	728	1,800	627	1,550	1 634.7	106,000	1 388.0	90,000
British Columbia	36	88	28	70	89.4	5,800	63.2	4,100
BARLEY	*4 354*	*10,758*	*4 649*	*11,489*	*10 513.3*	*482,866*	*11 515.5*	*528,906*
Prince Edward Island	11	26	11	28	33.5	1,540	29.6	1,358
Nova Scotia	2	4	2	4	4.8	220	4.0	186
New Brunswick	3	7	3	6	5.9	270	4.3	199
Quebec	19	47	20	50	38.7	1,777	43.2	1,985
Ontario	150	371	129	320	343.1	15,759	337.0	15,478
Manitoba	647	1,600	770	1,900	1 458.8	67,000	2 046.6	94,000
Saskatchewan	1 194	2,950	1 497	3,700	2 961.1	136,000	3 592.5	165,000
Alberta	2 246	5,550	2 145	5,300	5 486.7	252,000	5 290.7	243,000
British Columbia	82	202	73	180	180.7	8,300	167.6	7,700
FALL RYE	*233*	*576*	*221*	*547*	*416.4*	*16,390*	*364.5*	*14,349*
Quebec	5	12	4	10	7.5	296	6.3	248
Ontario	23	57	17	41	45.6	1,794	32.3	1,271
Manitoba	36	90	36	90	67.3	2,650	68.6	2,700
Saskatchewan	101	250	91	225	170.2	6,700	138.4	5,450
Alberta	65	160	71	175	119.4	4,700	114.3	4,500
British Columbia	3	7	2	6	6.4	250	4.6	180
SPRING RYE	*17*	*43*	*21*	*52*	*24.2*	*950*	*28.0*	*1,100*
Manitoba	1	2	1	2	1.3	50	1.3	50
Saskatchewan	7	17	10	25	10.2	400	14.0	550
Alberta	10	24	10	25	12.7	500	12.7	500
ALL RYE	*250*	*619*	*242*	*599*	*440.6*	*17,340*	*392.5*	*15,449*
Quebec	5	12	4	10	7.5	296	6.3	248
Ontario	23	57	17	41	45.6	1,794	32.3	1,271
Manitoba	37	92	37	92	68.6	2,700	69.9	2,750
Saskatchewan	108	267	101	250	180.4	7,100	152.4	6,000
Alberta	75	184	81	200	132.1	5,200	127.0	5,000
British Columbia	3	7	2	6	6.4	250	4.6	180

11.4   Harvested area and production of field crops, by province, 1976 and 1977 (concluded)

Field crop and province	Area 1976 '000 hectares	Area 1976 '000 acres	Area 1977 '000 hectares	Area 1977 '000 acres	Total production 1976 '000 tonnes	Total production 1976	Total production 1977 '000 tonnes	Total production 1977
						'000 bu		'000 bu
PEAS	*24*	*61*	*34*	*86*	*43.3*	*1,595*	*55.6*	*2,046*
Quebec	—	—	—	—	—	—	—	—
Ontario	—	—	—	—	—	—	—	—
Manitoba	15	38	22	55	26.9	990	36.7	1,350
Saskatchewan	6	15	9	23	10.3	380	12.8	470
Alberta	3	7	3	7	5.4	200	5.4	200
British Columbia	¹	1	¹	1	—	25	—	26
BEANS	*61*	*150*	*65*	*160*	*82.5*	*1,818*	*39.9*	*880*
Quebec	—	—	—	—	—	—	—	—
Ontario	61	150	65	160	82.5	1,818	39.9	880
SOYBEANS	*153*	*378*	*202*	*500*	*250.4*	*9,200*	*517.1*	*19,000*
Ontario	153	378	202	500	250.4	9,200	517.1	19,000
BUCKWHEAT	*28*	*67*	*43*	*107*	*19.7*	*909*	*43.9*	*2,016*
New Brunswick	—	—	—	—	—	—	—	—
Quebec	5	12	5	13	5.1	236	5.7	264
Ontario	5	11	6	14	5.9	273	5.5	252
Manitoba	18	44	32	80	8.7	400	32.7	1,500
MIXED GRAINS	*645*	*1,593*	*624*	*1,543*	*1 568.7*	*76,852*	*1 607.2*	*78,737*
Prince Edward Island	34	83	32	78	111.9	5,481	87.1	4,267
Nova Scotia	2	6	2	5	6.6	325	5.0	243
New Brunswick	3	6	2	5	5.7	280	3.7	180
Quebec	43	107	40	99	94.8	4,644	91.3	4,475
Ontario	334	826	328	810	815.9	39,972	900.6	44,122
Manitoba	69	170	69	170	140.8	6,900	163.3	8,000
Saskatchewan	49	120	40	100	112.3	5,500	85.7	4,200
Alberta	109	270	109	270	275.6	13,500	265.4	13,000
British Columbia	2	5	2	5	5.1	250	5.1	250
FLAXSEED	*323*	*800*	*576*	*1,420*	*276.9*	*10,900*	*609.7*	*24,000*
Manitoba	212	525	304	750	160.0	6,300	320.1	12,600
Saskatchewan	81	200	223	550	86.4	3,400	241.3	9,500
Alberta	30	75	49	120	30.5	1,200	48.3	1,900
RAPESEED	*720*	*1,778*	*1 348*	*3,330*	*836.9*	*36,900*	*1 775.8*	*78,300*
Manitoba	101	250	182	450	102.1	4,500	258.5	11,400
Saskatchewan	304	750	567	1,400	387.8	17,100	782.4	34,500
Alberta	304	750	567	1,400	335.7	14,800	703.1	31,000
British Columbia	11	28	32	80	11.3	500	31.8	1,400
						'000 lb.		'000 lb.
SUNFLOWER SEED	*20*	*50*	*67*	*165*	*24.0*	*53,000*	*79.4*	*175,000*
Manitoba	20	50	67	165	24.0	53,000	79.4	175,000
MUSTARD SEED	*35*	*87*	*73*	*182*	*35.2*	*77,800*	*79.3*	*175,000*
Manitoba	7	18	16	40	6.5	14,400	16.3	36,000
Saskatchewan	19	47	40	100	19.0	42,000	47.6	105,000
Alberta	9	22	17	42	9.7	21,400	15.4	34,000
						'000 bu		'000 bu
SHELLED CORN	*709*	*1,752*	*730*	*1,801*	*3 771.0*	*148,457*	*4 302.9*	*169,394*
Quebec	63	155	68	167	312.6	12,307	346.6	13,644
Ontario	639	1,580	652	1,610	3 429.2	135,000	3 916.9	154,200
Manitoba	7	17	10	24	29.2	1,150	39.4	1,550
POTATOES	*105*	*263*	*111*	*276*	*2 339.9*	*51,585*	*2 487.6*	*54,841*
Prince Edward Island	21	52	22	55	565.0	12,457	540.4	11,913
Nova Scotia	2	4	2	4	25.9	570	28.3	623
New Brunswick	22	56	23	57	505.4	11,143	503.2	11,093
Quebec	17	43	18	45	323.4	7,129	364.3	8,032
Ontario	18	46	19	48	429.4	9,466	488.1	10,760
Manitoba	14	34	15	37	190.5	4,200	249.5	5,500
Saskatchewan	1	2	1	2	20.0	440	19.0	420
Alberta	7	16	7	17	163.3	3,600	167.8	3,700
British Columbia	4	11	4	11	117.0	2,580	127.0	2,800
						'000 tons		'000 tons
TAME HAY	*5 731*	*14,162*	*5 621*	*13,890*	*24 972.0*	*27,527*	*24 735.4*	*27,266*
Prince Edward Island	56	138	58	144	235.0	259	243.1	268
Nova Scotia	72	178	74	183	339.3	374	326.6	360
New Brunswick	72	179	74	183	305.7	337	316.6	349
Quebec	1 119	2,765	1 078	2,665	4 840.7	5,336	4 496.9	4,957
Ontario	1 154	2,852	1 111	2,745	6 550.8	7,221	6 742.2	7,432
Manitoba	567	1,400	567	1,400	2 358.7	2,600	2 358.7	2,600
Saskatchewan	890	2,200	870	2,150	3 084.4	3,400	2 721.6	3,000
Alberta	1 538	3,800	1 518	3,750	5 624.5	6,200	5 806.0	6,400
British Columbia	263	650	271	670	1 632.9	1,800	1 723.7	1,900
FODDER CORN	*488*	*1,204*	*492*	*1,215*	*14 423.3*	*15,899*	*15 118.3*	*16,665*
Prince Edward Island	5	12	4	10	157.9	174	82.6	91
Nova Scotia	5	12	5	13	167.8	185	116.1	128
New Brunswick	2	6	3	7	75.3	83	59.0	65
Quebec	104	256	104	258	3 309.4	3,648	3 480.0	3,836
Ontario	350	864	350	865	10 125.1	11,161	10 673.0	11,765
Manitoba	13	33	16	39	254.0	280	317.5	350
British Columbia	8	21	9	23	333.8	368	390.1	430
SUGAR BEETS	*33*	*80*	*26*	*64*	*1 172.5*	*1,292*	*1 008.2*	*1,111*
Quebec	3	8	2	4	120.4	133	68.3	75
Manitoba	13	31	10	25	349.9	386	399.2	440
Alberta	17	41	14	35	702.2	774	540.7	596

¹Less than 500 hectares.

11.5 Harvested area and production of grain in the Prairie provinces, 1973-77

Grain	1973	1974	1975	1976	1977
HARVESTED AREA *('000 ha)*					
Wheat	9 348	8 701	9 227	10 964	22 752
Oats	2 145	1 942	1 882	1 881	3 964
Barley	4 593	4 532	4 209	4 087	9 907
Rye	236	321	294	219	381
Flaxseed	587	587	567	323	277
Rapeseed	1 275	1 255	1 720	709	826
PRODUCTION *('000 t)*					
Wheat	15 622	12 655	16 329	9 789	18 643
Oats	4 210	3 054	3 547	1 679	3 516
Barley	9 667	8 252	8 905	4 412	10 930
Rye	327	442	470	219	349
Flaxseed	493	351	445	576	610
Rapeseed	1 207	1 143	1 724	1 316	1 744
ACREAGES *('000 acres)*					
Wheat	23,100	21,500	22,800	27,100	24,200
Oats	5,300	4,800	4,650	4,650	4,150
Barley	11,350	11,200	10,400	10,100	10,900
Rye	582	792	727	543	542
Flaxseed	1,450	1,450	1,400	800	1,420
Rapeseed	3,150	3,100	4,250	1,750	3,250
PRODUCTION *('000 bu)*					
Wheat	574,000	465,000	600,000	836,000	685,000
Oats	273,000	198,000	230,000	257,000	228,000
Barley	444,000	379,000	409,000	455,000	502,000
Rye	12,865	17,410	18,500	15,000	13,750
Flaxseed	19,400	13,800	17,500	10,900	24,000
Rapeseed	53,200	50,400	76,000	36,400	76,900

11.6 Carryover of Canadian grain, 10-year average 1967-76 and crop years ended July 31, 1973-77 (thousand tonnes)

Grain and year	Total in Canada and United States	Total in Canada	In commercial storage in Canada	On farms in Canada	Prairie provinces On farms	In primary elevators
Wheat						
Av. 1967-76	15 369.4	15 369.4	8 897.0	6 472.4	6 373.8	5 063.9
1973	9 944.6	9 944.6	6 814.8	3 129.8	2 993.7	3 830.6
1974	10 089.0	10 089.0	7 884.5	2 204.5	2 041.2	4 026.1
1975	8 037.5	8 037.5	6 404.6	1 632.9	1 496.9	2 707.3
1976	7 979.3	7 979.3	6 400.8	1 578.5	1 496.8	2 713.7
1977	13 318.1	13 318.1	6 160.4	7 157.7	7 076.0	2 526.3
Oats						
Av. 1967-76	1 584.8	1 584.8	371.9	1 212.9	979.3	228.1
1973	1 228.8	1 228.8	226.4	1 002.4	802.0	127.6
1974	1 193.4	1 193.4	298.9	894.5	694.0	169.3
1975	1 132.1	1 132.1	391.8	740.3	539.8	260.3
1976	1 231.1	1 231.1	429.1	802.0	616.8	257.0
1977	1 327.9	1 327.9	294.7	1 033.2	848.2	162.7
Barley						
Av. 1967-76	3 705.9	3 705.9	1 943.6	1 762.3	1 635.1	1 121.9
1973[r]	4 202.7	4 202.7	2 286.7	1 916.0	1 785.3	1 414.6
1974	4 537.6	4 537.6	3 100.6	1 437.0	1 306.3	1 849.6
1975	4 104.3	4 104.3	2 993.9	1 110.4	979.8	1 526.4[r]
1976	2 763.5	2 763.5	1 674.9	1 088.6	979.7	771.4
1977	3 218.3	3 218.3	2 086.1	1 132.2	1 045.1	1 334.7
Rye						
Av. 1967-76	281.6	281.6	189.1	92.4	92.4	102.0
1973	261.7	261.7	221.1	40.6	40.6	120.5
1974	267.0	267.0	203.5	63.5	63.5	107.9
1975	339.5	339.5	250.6	88.9	88.9	150.2
1976	312.0	312.0	223.1	88.9	88.9	149.0
1977	347.2	347.2	288.7	58.4	58.4	175.1
Flaxseed						
Av. 1967-76	275.0	275.0	210.7	64.3	64.3	98.1
1973	194.9	194.9	179.7	15.2	15.2	71.5
1974	200.9	200.9	162.8	38.1	38.1	95.0
1975	218.6	218.6	155.1	63.5	63.5	88.6
1976	380.6	380.6	329.8	50.8	50.8	192.9
1977	211.9	211.9	186.5	25.4	25.4	115.1
Rapeseed						
Av. 1967-76	397.7	397.7	301.8	95.9	95.7	173.1
1973	469.0	469.0	459.9	9.1	9.1	198.7
1974	280.9	280.9	246.9	34.0	34.0	134.6
1975	399.9	399.9	309.2	90.7	90.7	173.6
1976	1 041.4	1 041.4	689.9	351.5	349.2	480.9
1977	199.2	199.2	174.3	24.9	24.9	81.4

11.7 Livestock slaughtered at federally inspected establishments, 1971-77

Year	Cattle	Calves	Sheep	Pigs
1971	2,786,908	464,240	205,082	9,742,759
1972	2,878,591	402,370	214,769	9,357,143
1973	2,878,016	291,524	234,206	8,721,921
1974	2,975,833	392,811	185,077	8,939,335
1975	3,337,687	682,094	186,566	7,656,334
1976	3,676,284	655,443	187,674	7,493,245
1977	3,761,419	645,591	132,585	8,007,313

11.8 Production and domestic disappearance of poultry meat[1], 1975 and 1976

Year and item	Net production t	Total supply t	Domestic disappearance t	Per capita consumption kg
1975				
Fowl and chickens	315 838	343 795	331 925	14.56
Turkeys	85 988	110 561	97 917	4.31
Geese	955	1 043	649	0.03
Ducks	3 757	4 886	4 629	0.20
Total	406 538	460 285	435 120	19.10
1976				
Fowl and chickens	357 686	392 332	373 123	16.10
Turkeys	94 353	113 126	97 160	4.20
Geese	1 433	1 436	1 210	0.05
Ducks	4 359	5 805	5 333	0.23
Total	457 831	512 699	476 826	20.58

[1]Eviscerated weight.

11.9 Production and utilization of milk, by province, 1975-77

Province and year		Farm sales of milk and cream			Whole milk used on farms		Total milk production
		Fluid t	Industrial		Farm home consumed t	Fed to livestock t	t
			Milk t	Cream[1] t			
Prince Edward Island	1975	12 599	50 689	16 354	2 896	3 787	86 325
	1976	13 995	54 949	14 093	2 962	4 709	90 708
	1977	15 442	55 502	12 695	2 792	4 278	90 709
Nova Scotia	1975	96 689	39 369	5 951	3 382	4 026	149 417
	1976	99 645	46 307	5 890	3 942	4 769	160 553
	1977	100 645	48 728	5 541	3 502	3 822	162 238
New Brunswick	1975	63 120	17 435	16 034	2 827	3 616	103 032
	1976	65 364	24 062	13 992	2 759	3 777	109 954
	1977	66 646	23 421	11 950	2 327	3 363	107 707
Quebec	1975	540 111	2 374 462	16 753	37 805	102 431	3 071 562
	1976	556 617	2 245 559	9 535	39 094	159 503	3 010 308
	1977	568 870	2 314 491	7 621	36 173	125 497	3 052 652
Ontario	1975	978 582	1 485 144	106 164	37 855	135 629	2 743 374
	1976	997 027	1 384 649	98 303	38 287	199 469	2 717 735
	1977	1 002 510	1 418 280	95 034	36 488	169 291	2 721 603
Manitoba	1975	106 898	129 224	63 757	13 468	20 078	333 425
	1976	110 993	126 618	51 444	13 199	26 699	328 953
	1977	114 438	128 110	49 609	11 544	24 430	328 131
Saskatchewan	1975	95 741	33 114	63 362	19 874	20 217	232 308
	1976	91 011	47 501	52 809	21 592	21 704	234 617
	1977	99 923	51 691	48 733	18 201	15 323	233 871
Alberta	1975	183 523	204 487	126 870	23 015	41 000	578 895
	1976	195 875	203 395	108 850	23 576	49 594	581 290
	1977	207 366	215 058	98 913	23 568	39 791	584 696
British Columbia	1975	266 599	149 524	2 027	7 241	20 653	446 044
	1976	269 424	141 100	1 613	7 835	31 145	451 117
	1977	276 774	144 773	1 414	8 110	30 560	461 631
Total	1975	2 343 862	4 483 448	417 272	148 363	351 437	7 744 382
	1976	2 399 951	4 274 140	356 529	153 246	501 369	7 685 235
	1977	2 452 614	4 400 054	331 510	142 705	416 355	7 743 238

[1]Farm separated cream expressed in terms of milk equivalent.

11.10   Cash receipts from milk and cream, sold off farms, by province, 1975-77 (thousand dollars)

Province and year		Farm sales of milk and cream				Supplementary payments	Total cash receipts
		Fluid purposes	Industrial purposes		Total		
			Delivered as milk	Delivered as cream			
Prince Edward Island	1975	3,052	6,767	1,373	11,192	3,553	14,745
	1976	3,525	7,276	1,207	12,008	3,961	15,969
	1977	4,062	7,400	1,198	12,660	4,001	16,661
Nova Scotia	1975	25,737	6,946	477	33,160	2,300	35,460
	1976	28,052	8,096	503	36,651	2,903	39,554
	1977	29,302	8,673	499	38,474	2,977	41,451
New Brunswick	1975	14,500	3,430	1,257	19,187	1,777	20,964
	1976	17,715	3,495	1,130	22,340	2,041	24,381
	1977	18,578	3,206	1,053	22,837	1,894	24,731
Quebec	1975	138,893	380,857	1,280	521,030	125,473	646,503
	1976	146,779	345,657	808	493,244	127,711	620,955
	1977	153,784	373,642	680	528,106	133,979	662,085
Ontario	1975	245,869	236,052	7,993	489,914	82,990	572,904
	1976	256,840	217,108	7,911	481,859	84,398	566,257
	1977	270,134	233,570	8,340	512,044	86,681	598,725
Manitoba	1975	26,465	18,898	4,779	50,142	9,931	60,073
	1976	28,162	19,927	3,998	52,087	9,742	61,829
	1977	29,841	19,604	4,123	53,568	10,099	63,667
Saskatchewan	1975	21,928	4,922	4,955	31,805	5,611	37,416
	1976	22,241	6,208	4,295	32,744	5,806	38,550
	1977	25,364	7,116	4,754	37,234	6,004	43,238
Alberta	1975	45,228	32,482	10,193	87,903	16,916	104,819
	1976	51,659	31,739	9,265	92,663	17,257	109,920
	1977	55,661	35,331	8,835	99,827	16,676	116,503
British Columbia	1975	72,614	30,170	131	102,915	7,883	110,798
	1976	79,239	28,374	109	107,722	7,894	115,616
	1977	87,352	30,431	107	117,890	8,216	126,106
Total	1975	594,286	720,524	32,438	1,347,248	256,434	1,603,682
	1976	634,212	667,880	29,226	1,331,318	261,713	1,593,031
	1977	674,078	718,973	29,589	1,422,640	270,527	1,693,167

11.11   Production of butter and cheese, by province, 1975-77 (tonnes)

Province and year		Butter				Cheese Factory[1]
		Creamery	Farm	Whey	Total	
Prince Edward Island	1975[r]	1 413	—	52	1 465	[2]
	1976	1 662	—	45	1 707	[2]
	1977	1 689	—	52	1 741	[2]
Nova Scotia	1975	533	—	—	533	2 896[3]
	1976	561	—	—	561	3 059[3]
	1977	580	—	—	580	3 524[3]
New Brunswick	1975	1 072	—	—	1 072	[2]
	1976	903	—	—	903	[2]
	1977	923	—	—	923	[2]
Quebec	1975	65 628	—	1 903	67 531	53 273
	1976	58 524	—	2 192	60 716	56 187
	1977	58 780	—	1 622	60 402	58 768
Ontario	1975[r]	39 165	—	2 166	41 331	52 203
	1976	33 686	—	2 228	35 914	52 917
	1977	32 472	—	1 777	34 249	57 300
Manitoba	1975	4 746[r]	—	—	4 746[r]	4 627
	1976	4 055	—	—	4 055[r]	4 616
	1977	4 131	—	—	4 131	5 929
Saskatchewan	1975	4 038	—	—	4 038	552
	1976	3 750	—	—	3 750	768
	1977	3 723	—	—	3 723	1 542
Alberta	1975	9 338	—	—	9 338	2 861[r]
	1976	8 282	—	—	8 282	4 609
	1977	8 417	—	—	8 417	5 473
British Columbia	1975	2 726	—	—	2 726	1 579
	1976	2 666	—	—	2 666	1 494
	1977	2 548	—	—	2 548	1 790
Total	1975[r]	128 659	—	4 121	132 780	120 658
	1976	114 089	—	4 465	118 554	125 464
	1977	113 263	—	3 451	116 714	134 326

[1]Factory-made cheese includes cheddar and other cheese made from whole milk and cream. Amounts for other cheese are included in Quebec, Ontario and Alberta figures, but, as fewer than three firms reported in the other provinces, data cannot be included except in the Canada total.
[2]Included with Nova Scotia.
[3]Includes Prince Edward Island and New Brunswick.

11.12   Domestic disappearance of dairy products, 1975-77

Product	1975r Total t	Per capita[1] kg	1976 Total t	Per capita[1] kg	1977 Total t	Per capita[1] kg
Milk and cream	*2 500 760*	*110.03*	*2 574 732*	*111.82*	*2 610 579*	*111.96*
Milk	2 150 528	94.62	1 215 211	96.21	2 255 668	96.74
Cream as milk	350 232	15.41	359 521	15.61	354 911	15.22
Butter	*123 895*	*5.45*	*121 473*	*5.28*	*108 995*	*4.67*
Creamery	119 815	5.27	116 979	5.08	105 713	4.53
Farm	—	—	—	—	—	—
Whey	4 080	0.18	4 494	0.20	3 282	0.14
Cheese	*162 099*	*7.13*	*164 123*	*7.13*	*167 452*	*7.27*
Cheddar	46 002	2.02	38 271	1.66	34 383	1.47
Process	58 085	2.56	61 802	2.69	63 938	6.04
Other	58 012	2.55	64 050	2.78	69 131	3.06
Concentrated whole milk products[2]	*110 365*	*4.86*	*110 368*	*4.79*	*122 326*	*5.25*
Evaporated	91 168	4.01	91 174	3.96	98 284	4.22
Condensed	7 572	0.33	7 008	0.30	8 615	0.37
Powdered	837	0.04	1 272	0.05	2 738	0.12
Concentrated milk byproducts[3]	*108 802*	*4.79*	*121 705*	*5.28*	*160 565*	*6.89*
Evaporated	10 057	0.44	7 278	0.32	17 868	0.77
Condensed	876	0.04	701	0.03	1 868	0.08
Powdered	62 638	2.76	65 095	2.82	79 513	3.41
All dairy products in terms of milk						
Butter	2 803 665	123.36	2 737 306	118.88	2 473 694	106.09
Cheese	1 610 306	70.85	1 621 494	70.42	1 675 390	71.86
Concentrated	244 917	10.78	246 443	10.70	285 511	12.24
Total[4]	7 733 242	340.27	7 755 145	336.91	7 631 888	327.32

[1]Includes Newfoundland for all manufactured dairy products.
[2]Includes, in addition to the items listed, partly skimmed evaporated milk, whole milk powder of less than 26% fat, formula milks, evaporated milk of 2% fat, and concentrated liquid milk.
[3]Includes, in addition to the items listed, powdered buttermilk, sugar of milk, casein, powdered whey, special formula skim milk products and concentrated liquid skim milk. Since the quantities used for human consumption and livestock feeding cannot be separated, per capita figures include both.
[4]Includes ice cream mix in terms of milk.

11.13   Estimated commercial production and farm value of fruit, 1974-77

Kind of fruit and year	Quantity/Weight t	Farm value $'000	Kind of fruit and year	Quantity/Weight t	Farm value $'000
Apples			Peaches		
1974	406 335r	53,973	1974	51 079	14,553
1975r	460 422	41,978	1975r	59 201	15,555
1976	408 908	60,774	1976	49 008	15,989
1977P	402 440	..	1977P	46 428	..
Apricots			Pears		
1974	2 885	937	1974	37 844r	6,798
1975r	3 790	952	1975	38 218	6,428r
1976	2 800	623	1976	29 279	4,292
1977P	..	..	1977P	40 991	..
Cherries (sour)			Plums and prunes		
1974	7 671	3,756	1974	6 426	1,699
1975	7 653	2,739r	1975r	10 528	2,577
1976	4 851	2,856	1976	6 366	1,750
1977P	7 786	..	1977P	7 640	..
Cherries (sweet)			Raspberries		
1974	8 477	4,347	1974	6 834	5,318
1975r	12 525	5,855	1975r	6 873	4,891
1976	8 722	4,134	1976	6 648	5,633
1977P	10 541	..	1977P	7 699	..
Strawberries			Grapes		
1974	14 725	11,045	1974	72 430	18,563
1975	16 590	13,649	1975	76 774	20,689
1976	18 311	15,851	1976	80 674	20,382
1977P	20 127	..	1977P	65 138	..
Cranberries			Blueberries		
1974	4 270	1,039	1974	8 843	3,946
1975	5 742	1,594	1975r	14 390	8,178
1976	6 498	2,296	1976	11 312	8,704
1977P	6 720	..	1977P	15 328	..

11.14   Estimated commercial area and production of vegetables, 1975-77

| Vegetables | 1975[r] | | | | 1976 | | | | 1977[1] | |
| | Area | | Production | | Area | | Production | | Area | |
	ha	acres	t	'000 lb.	ha	acres	t	'000 lb.	ha	acres
Asparagus	1 562	3,860	2 801	6,175	1 658	4,096	2 636	5,811	1 571	3,883
Beans	9 054	22,372	42 972	94,738	7 576	18,721	36 230	79,873	..	..
Beets	1 049	2,591	19 857	43,778	941	2,326	19 008	41,906	944	2,333
Cabbage	4 119	10,178	93 004	205,038	4 172	10,309	104 792	231,028	4 354	10,760
Carrots	6 566	16,224	222 491	490,508	5 813	14,365	177 616	391,576	6 095	15,061
Cauliflower	1 437	3,551	18 368	40,494	1 250	3,088	17 752	39,136	1 338	3,307
Celery	455	1,125	25 456	56,120	515	1,273	25 923	57,151	493	1,218
Corn	31 333	77,425	274 417	604,986	30 657	75,755	263 723	581,409	..	..
Cucumbers	3 801	9,393	63 098	139,108	3 694	9,129	57 465	126,688	3 168	7,830
Lettuce	1 969	4,866	32 082	70,729	1 881	4,648	31 782	70,068	1 996	4,934
Onions	3 352	8,282	100 040	220,551	3 716	9,182	86 568	190,849	3 777	9,333
Parsnips	174	430	2 690	5,930	176	434	2 613	5,761	171	422
Peas	27 043	66,825	78 242	172,494	22 138	54,703	62 120	136,951	..	..
Rutabagas	3 783	9,347	97 646	215,272	3 410	8,426	107 182	236,295	3 356	8,294
Spinach	358	884	2 461	5,425	337	833	3 055	6,735	356	880
Tomatoes	11 728	28,981	382 296	842,819	11 526	28,482	408 218	899,968	..	..

[1]Production not available for 1977.

11.15   Honey production, by province, and total value, 1974-77, with 10-year average for 1965-74 and 1966-75

| Province | | Average | | 1974 | 1975 | 1976 | 1977 |
		1965-74	1966-75				
Prince Edward Island	t	19	20	11	31	34	45
Nova Scotia	''	116	119	129	174	160	171
New Brunswick	''	61	66	55	88	98	91
Quebec	''	1 324	1 449	1 394	2 327	1 884	2 634
Ontario	''	3 636	3 583	2 447	3 915	3 269	3 606
Manitoba	''	3 664	3 760	3 892	3 654	5 479	5 180
Saskatchewan	''	3 230	3 239	3 233	2 946[r]	4 028	3 774
Alberta	''	8 328	8 049	8 137	6 307	9 290	9 730
British Columbia	''	1 399	1 368	1 469	1 613	1 202	2 516
Total production	t	21 777	21 653	20 767	21 055[r]	25 444	27 747
Total value	$'000	12,053	13,473	21,135	22,965[r]	25,085	30,616

11.16   Harvested area, yield, production and value of sugar beets, 1973-77

| Year | Sugar beets | | | | | | | | |
| | Harvested area | | Yield per ha (acre) | | Total production | | Average price per | | Total farm value |
	ha	acres	kg	tons	t	tons	t $	ton $	$'000
1973	27 778	68,640	32 468	14.48	901 889	994,162	43.15	39.15	38,917
1974[r]	27 336	67,548	27 495	12.27	751 596	828,493	55.47	50.32	41,690
1975	32 166	79,482	29 304	13.07	942 600[r]	1,039,039[r]	..	..	..
1976	33 100	80,020	35 423	16.15	1 172 500	1,292,371	..	..	..
1977	25 500	63,674	39 537	17.45	1 008 200	1,111,268	..	..	..

11.17   Production and value of maple sugar and maple syrup, by province, 1974-77, with 5-year average for 1969-73

| Province and year | Maple sugar | | Maple syrup | | Total value, sugar and syrup $'000 |
	Quantity kg	Value $'000	Quantity kL	Value $'000	
Nova Scotia					
Av. 1969-73	6 000	13	20	35	48
1974	4 000	12	18	45	57
1975	4 000	14	14	40	54
1976	2 000	9	18	55	64
1977	5 000	24	36	108	132
New Brunswick					
Av. 1969-73	9 000	21	40	72	93
1974	9 000	29	27	70	99
1975	9 000	35	41	116	151
1976	12 000	50	41	120	170
1977	10 000	48	36	127	175

11.17   Production and value of maple sugar and maple syrup, by province, 1974-77, with 5-year average for 1969-73 (concluded)

Province and year	Maple sugar		Maple syrup		Total value, sugar and syrup $'000
	Quantity kg	Value $'000	Quantity kL	Value $'000	
Quebec					
Av. 1969-73	129 000	225	7 595	9,697	9,922
1974	117 000	299	7 269	11,028	11,327
1975	136 000	366	5 587	9,562	9,928
1976	159 000	486	7 283	13,985	14,471
1977	170 000	594	7 669	15,732	16,326
Ontario					
Av. 1969-73	6 000	17	750	1,215	1,232
1974	7 000	28	646	1,486	1,514
1975	4 000	20	546	1,437	1,457
1976	6 000	33	605	1,741	1,774
1977	4 000	24	664	2,056	2,080
Total					
Av. 1969-73	150 000	276	8 405	11,019	11,295
1974	137 000	368	7 960	12,629	12,997
1975	153 000	435	6 188	11,155	11,590
1976	179 000	578	7 947	15,901	16,479
1977	189 000	690	8 405	18,023	18,713

11.18   Production, and value of farm eggs, by province, 1976 and 1977

Province	1976				1977			
	Average number of layers '000	Average production per 100 layers No.	Egg production '000 doz	Total value (sold and used) $'000	Average number of layers '000	Average production per 100 layers No.	Egg production '000 doz	Total value (sold and used) $'000
Newfoundland	374	21,861	6,810	5,608	394	21,728	7,135	5,926
Prince Edward Island	142	21,075	2,496	1,833	151	20,912	2,625	1,950
Nova Scotia	928	21,590	16,744	13,074	942	22,328	17,531	13,505
New Brunswick	507	20,719	8,758	6,935	510	21,514	9,141	7,264
Quebec	3,933	22,896	75,035	54,106	3,928	22,648	74,105	54,859
Ontario	8,368	23,736	165,506	108,431	9,032	23,511	176,878	116,159
Manitoba	2,560	22,895	48,871	28,651	2,616	23,063	50,270	29,847
Saskatchewan	1,136	19,685	18,635	12,758	1,192	19,647	19,522	13,589
Alberta	2,287	20,988	39,964	27,571	2,289	20,992	40,032	27,851
British Columbia	2,720	24,304	55,096	38,892	2,798	25,046	58,402	40,454
Total	22,955	21,975	437,915	297,859	23,852	22,139	455,641	311,404

11.19   Harvested area, production and value of the commercial crop of leaf tobacco, by province, 1974-76

Year	Quebec			Ontario			Other provinces		
	Harvested area ha	Production t	Value $'000	Harvested area ha	Production t	Value $'000	Harvested area ha	Production t	Value $'000
1974	3 597	4 992	9,432	44 591	109 281	216,786	1 526	2 203	4,376
1975	3 749	6 705	13,072	36 182	96 553	199,724	1 679	2 795	5,701
1976	3 753	4 909	10,312	[1]	73 657	165,194	1 758	2 930	6,492
	Harvested area acres	Production '000 lb.	Value $'000	Harvested area acres	Production '000 lb.	Value $'000	Harvested area acres	Production '000 lb.	Value $'000
1974	8,889	11,006	9,432	110,186	240,924	216,786	3,771	4,857	4,376
1975	9,265	14,783	13,072	89,408	212,862	199,724	4,149	6,162	5,701
1976	9,274	10,824	10,312	[1]	162,388	165,194	4,345	6,460	6,492

[1]Commencing with the 1976 crop year, producers of flue-cured tobacco in Ontario changed from an area harvested basis to weight delivered formula; thus these data are no longer available.

11.20 Harvested area, production and value of the commercial crop of leaf tobacco, by main type, 1974-76

Type of tobacco and year		Harvested area		Average yield per ha (acre)		Total production		Average farm price		Gross farm value $'000
		ha	acres	kg	lb.	t	'000 lb.	kg $	lb. ¢	
Flue-cured	1974	48 987	121,047	2 336	2,084	114 439	252,294	1.987	90.1	227,317
	1975	40 262	99,490	2 571	2,294	103 523	228,230	2.069	93.8	214,182
	1976	¹	¹	¹	¹	79 214	174,638	2.246	101.9	177,908
Burley	1974	379	937	2 401	2,143	910	2,007	1.623	73.6	1,477
	1975	463	1,145	2 259	2,015	1 046	2,307	1.790	81.1	1,872
	1976	526	1,300	2 148	1,917	1 130	2,492	1.862	84.4	2,104
Cigar leaf	1974	494	1,220	1 330	1,187	657	1,448	1.447	65.7	951
	1975	556	1,375	1 644	1,466	914	2,016	1.531	69.4	1,399
	1976	574	1,418	1 202	1,073	690	1,522	1.617	73.3	1,116
Total²	1974	50 119	123,846	2 324	2,073	116 477	256,787	1.980	89.8	230,594
	1975	41 611	102,822	2 549	2,274	106 053	233,807	2.060	93.5	218,497
	1976	¹	¹	¹	¹	81 496	179,672	2.233	101.3	181,998

¹Commencing with the 1976 crop year, producers of flue-cured tobacco in Ontario changed from an area harvested basis to weight del:.ered formula; thus these data are no longer available.
²Inci..des other types not specified.

11.21 Production and disposition of tobacco products, 1974-76

Item and year		Total production	Sales		
			In Canada¹	Ship/air stores embassies/Canada²	For export — bulk shipments, including Canadian missions abroad²
Cigarettes ('000)	1974	59,603,356	57,122,801	200,524	570,795
	1975	58,260,128	57,755,795	200,842	671,627
	1976	61,559,245	60,744,885	249,494	658,977
Cigars ('000)	1974	595,716	590,993	1,833	10,787
	1975	475,749	494,222	1,759	8,560
	1976	560,397	507,445	1,819	5,306
Manufactured tobacco Fine cut³ (kg)	1974	6 640 918	6 712 454	449	4 076
	1975	6 653 180	6 765 029	354	7 029
	1976	6 774 553	6 582 420	419	8 105
Pipe tobacco⁴ (kg)	1974	556 780	609 485	1 766	10 528
	1975	383 203	374 792	644	8 964
	1976	408 971	390 559	556	14 884
Other⁵ (kg)	1974	594 711	600 478	9	—
	1975	554 050	579 906	—	—
	1976	586 020	534 787	—	—

¹Includes samples and goods invoiced to wholesalers, retailers, and institutions which are subject to excise duty, less returned goods credited to same.
²Excise duty exempt.
³Includes tobacco, intended for cigarettes.
⁴Includes tobacco, intended for pipe smoking.
⁵Other tobacco, plug, snuff, chewing and twist.

11.22 Average index¹ of farm prices of agricultural products, by province, 1972-77 (1961 = 100)

Province	1972	1973ʳ	1974ʳ	1975ʳ	1976ʳ	1977
Prince Edward Island	153.3	266.6	297.7	265.1	314.7	282.8
Nova Scotia	132.9	178.0	209.9	215.6	228.2	229.4
New Brunswick	140.5	228.1	261.2	235.2	296.5	277.8
Quebec	154.7	198.9	229.0	260.1	261.0	264.9
Ontario	143.4	188.4	209.5	225.2	226.2	233.2
Manitoba	119.7	189.3	235.3	226.6	209.8	196.5
Saskatchewan	112.0	189.5	250.1	230.0	203.0	183.8
Alberta	129.6	197.0	236.8	230.2	210.4	200.3
British Columbia	134.9	169.2	202.5	205.6	211.0	215.8
Total	132.9	191.9	229.6	231.6	222.6	217.6

¹A description of this index, its coverage and the methods used will be found in Statistics Canada *Quarterly bulletin of agricultural statistics* (Catalogue 21-003) for July-September 1969.

11.23   Average cash prices a metric tonne of major Canadian grains, crop years ended July 31, 1973-77 (basis, in store Thunder Bay)

Year	Averages in dollars per tonne						
	Wheat		Oats[1]	Barley[1]	Rye[2]	Flaxseed[2]	Rapeseed[2]
	No. 1 N.	1 C. W. Red Spring 14%[1]	2 C. W.	3 C.W. — Six-Row	2 C. W.	1 C. W.	1 Canada
1973	...	96.50	70.84	81.12	62.45	190.05	160.50
1974	...	201.86[3]	113.15	141.23[4]	116.09	399.29	280.26
1975	...	193.41	121.90	162.99[4]	103.14	375.67	318.90
1976	...	172.10	121.63	151.14	105.34	274.10	226.76
1977	...	123.86	109.12	141.97	94.57	276.00	288.93

[1]Canadian Wheat Board daily fixed prices.
[2]Winnipeg Commodity Exchange daily closing cash quotations.
[3]1 C. W. Red Spring 13.5%.
[4]2 C. W. Six-Row.

11.24   Weighted average prices per 100 kg of Canadian livestock at public stockyards, 1964-76 (dollars)

Item	Average prices									
	1964-73	1969-73	1974	1975	1976	1964-73	1969-73	1974	1975	1976
	Toronto					Calgary				
A1,2 steers over 454 kg	69.45	79.98	108.84	103.60	92.35	65.06	75.33	105.65	96.03	86.42
D1,2 cows	48.83	57.19	63.25	50.77	57.43	46.16	55.49	57.45	46.96	50.51
Feeder steers over 363 kg	69.20	82.01	98.00	87.10	87.70	66.62	80.18	96.01	84.77	80.16
Choice and good veal calves	85.34	96.92	113.56	85.12	92.11	75.53	94.93	66.34	59.79	60.14
Index 100 hogs, dressed	76.26	81.90	110.87	148.19	141.32	67.00	72.18	100.35	144.20	133.51
Good lambs	68.01	77.03	107.23	129.06	117.77	52.76	59.35	81.35	88.07	—
	Winnipeg					Edmonton				
A1,2 steers over 454 kg	66.82	77.12	107.50	96.56	86.53	64.07	74.30	100.75	91.07	82.12
D1,2 cows	48.63	58.20	59.79	46.89	50.53	44.33	53.20	52.03	41.78	46.01
Feeder steers over 363 kg	67.51	81.26	89.55	77.98	79.41	66.71	80.67	91.12	81.13	78.70
Choice and good veal calves	94.18	112.70	154.98	97.53	103.70	79.68	99.60	73.28	63.69	69.44
Index 100 hogs, dressed	70.79	76.19	101.68	137.90	130.14	67.11	72.69	98.90	143.21	131.99
Good lambs	58.14	67.48	99.08	125.16	88.67	52.23	59.15	80.93	89.07	86.77

11.25   Per capita supplies of food moving into consumption 1975 and 1976 with average for 1970-74

Kind of food and weight base		kg per capita per annum		
		Average 1970-74	1975	1976
CEREALS	retail wt	68.84	72.33	75.14
Wheat flour	"	58.83	61.55	64.15
Rye flour	"	0.38	0.39	0.41
Oatmeal and rolled oats	"	1.58	1.45	1.55
Pot and pearl barley	"	0.05	0.04	0.05
Corn flour and meal	"	2.16	2.10	1.99
Buckwheat flour	"	0.01	0.01	0.01
Rice	"	2.65	2.45	2.63
Breakfast food	"	3.18	4.34	4.35
SUGARS AND SYRUPS	sugar content	49.25	41.04	43.20
Sugar	retail wt	45.82	39.61	41.85
Maple sugar	"	0.19	0.15	0.18
Honey	"	0.81	0.87	0.85
Other	"	3.69	0.82	0.71
PULSES AND NUTS	retail wt	5.74	7.34	5.14
Dry beans	"	1.63	1.33	0.83
Baked canned beans	"	..	2.73	..
Dry peas	"	0.83	1.21	0.67
Peanuts	"	2.48	3.99	2.65
Tree nuts	"	0.80	0.81	0.99
OILS AND FATS	fat content	19.36	19.91	10.32
Margarine	retail wt	4.42	5.25	5.48
Shortening and shortening oils	"	7.39	7.89	7.73
Salad oils	"	3.00	3.54	4.08
Butter	"	6.65	5.25	5.05
FRUIT	fresh equiv.	118.74	130.45	138.67
Fresh	retail wt	53.84	58.75	65.71
Canned	net wt canned	13.05	11.63	12.42
Frozen	retail wt	1.38	1.27	1.11
Juice	net wt canned	17.06	20.68	20.73
Tomatoes, fresh	retail wt	5.30	5.12	5.36
canned	net wt canned	2.97	2.47	3.00
Tomato juice	"	3.78	4.04	3.68
pulp, paste and purée	"	1.17	0.80	..
ketchup	"	2.01	2.34	..
Citrus fruit, fresh	retail wt	13.13	14.46	17.86
juice	net wt canned	7.40	10.05	10.26

11.25 Per capita supplies of food moving into consumption 1975 and 1976 with average for 1970-74 (continued)

Kind of food and weight base		kg per capita per annum		
		Average 1970-74	1975	1976
FRUIT (concluded)				
Apples, fresh	retail wt	11.11	11.72	13.39
canned	net wt canned	0.12	0.16	0.01
juice	"	2.50	4.25	3.55
frozen	retail wt	0.15	0.15	0.14
sauce	net wt canned	0.60	0.45	0.49
pie filling	"	0.21	0.20	0.20
Apricots, fresh	retail wt	0.11	0.11	0.10
canned	net wt canned	0.20	0.15	0.11
Bananas, fresh	retail wt	9.04	9.30	10.31
Blueberries, fresh	"	..	0.43	0.28
canned	net wt canned	..	0.04	- -
frozen	retail wt	..	..	..
Cherries, fresh	"	0.53	0.72	0.84
canned	net wt canned	0.12	0.16	0.09
frozen	retail wt	0.27	0.19	0.18
Cranberries, fresh	"	..	0.31	0.34
Melons, fresh	"	..	3.88	3.83
Peaches, fresh	"	2.30	2.54	2.39
canned	net wt canned	1.57	1.50	1.30
frozen	retail wt	..	—	—
Pears, fresh	"	1.62	1.79	1.84
canned	net wt canned	0.83	0.64	0.71
Pineapples, fresh	retail wt	0.18	0.30	0.31
canned	net wt canned	..	1.12	1.08
juice	"	0.43	0.41	0.36
Plums, fresh	retail wt	0.87	1.01	0.93
canned	net wt canned	0.13	0.10	0.09
Raspberries, fresh	retail wt	..	0.02	0.07
canned	net wt canned	0.04	0.03	0.03
frozen	retail wt	0.20	0.24	0.19
Strawberries, fresh	"	0.83	0.85	1.11
canned	net wt canned	0.04	0.04	0.03
frozen	retail wt	0.65	0.72	0.59
Grapes, fresh	"	4.54	5.41	5.84
Unspecified, fresh	"	..	0.78	..
canned	net wt canned	1.94	1.26	1.20
frozen	retail wt	0.08	..	..
juice	net wt canned	2.60	1.75	2.68
jams, jellies, marmalade	processed wt	2.16	1.94	2.45
VEGETABLES[1]	fresh equiv.	53.43	56.60	58.61
Fresh	retail wt	34.91	43.26	46.91
Canned	net wt canned	8.52	6.96	6.98
Frozen	retail wt	2.89	2.59	1.57
Cabbage, fresh	"	4.64	5.76	6.55
Lettuce	"	6.43	8.26	8.85
Spinach, fresh	"	0.26	0.28	0.35
Carrots, fresh	"	6.47	7.85	7.35
canned	net wt canned	0.48	0.33	0.19
frozen	retail wt	0.33	0.51	0.44
Beans, fresh	"	0.48	0.49	0.59
canned	net wt canned	1.67	1.47	1.34
frozen	retail wt	0.36	0.32	0.33
Peas, fresh	"	0.11	0.05	0.04
canned	net wt canned	2.16	1.86	..
frozen	retail wt	1.10	1.11	..
Beets, fresh	"	0.25	0.21	..
canned	net wt canned	0.36	0.28	0.32
Cauliflower, fresh	retail wt	0.86	0.98	1.12
Celery, fresh	"	3.18	3.58	3.86
Corn, fresh	"	1.65	1.97	2.47
canned	net wt canned	2.26	2.31	1.89
frozen	retail wt	0.34	0.20	0.23
Cucumbers, fresh	"	1.26	1.32	1.79
Onions, not processed	"	5.48	6.11	6.26
Asparagus, fresh	"	0.16	0.29	0.26
canned	net wt canned	0.22	0.20	0.24
frozen	retail wt	..	..	..
Rutabagas, fresh	"	2.04	2.33	2.67
Broccoli, fresh	"	..	0.63	0.72
frozen	"	..	0.11	0.14
Brussels sprouts, fresh	"	..	0.10	0.12
frozen	"	..	0.09	0.10
Unspecified, fresh	"	0.54	1.15	1.26
canned	net wt canned	1.33	0.46	0.47
frozen	retail wt	0.51	0.10	0.10
MUSHROOMS	fresh equiv.	1.41	1.33	1.28
Fresh	retail wt	0.42	0.52	0.57
Canned	net wt canned	0.75	0.97	0.85
POTATOES	fresh equiv.	70.70	70.30	63.48
White	"	70.46	70.02	63.08
Sweet	"	0.24	0.28	0.40

11.25   Per capita supplies of food moving into consumption 1975 and 1976 with average for 1970-74 (concluded)

Kind of food and weight base		kg per capita per annum		
		Average 1970-74	1975	1976
MEAT	carcass wt	73.47	74.73	79.10
Pork	"	27.38	23.11	24.07
Beef	"	41.02	46.32	50.09
Veal	"	1.71	2.43	2.22
Mutton and lamb	"	1.63	1.29	1.07
Offal	"	1.73	1.58	1.65
Canned meat[2]	net wt canned	..	0.29	0.31
EGGS	fresh equiv.	13.90	12.70	12.45
POULTRY[3]	eviscerated wt	20.62	19.28	20.60
Chicken	"	14.13	13.42	14.83
Fowl	"	1.55	1.34	1.29
Turkey	"	4.68	4.29	4.20
Duck	"	0.18	0.20	0.23
Goose	"	0.08	0.03	0.05
FISH	edible wt	5.59	5.80	7.30
Fish and shellfish fresh and frozen[4]	"	3.29	3.80	4.80
Fish, cured (smoked, salted, pickled)	"	0.35	0.30	0.60
Fish and shellfish, canned	"	1.95	1.70	1.90
MILK AND CHEESE	milk solids	28.60	36.33	37.84
Cheddar cheese	retail wt	1.98	3.55	3.53
Process cheese	"	2.33	2.54	2.64
Other cheese	"	1.99	2.62	2.87
Cottage cheese	"	1.01	1.02	1.05
Evaporated whole milk	"	5.19	3.99	3.90
Condensed whole milk	"	0.37	0.33	0.30
Powdered whole milk and cream	"	0.05	0.04	0.05
Miscellaneous milk products[5]	"	0.07	0.47	0.58
Powdered skim milk[6]	"	2.43	2.74	2.76
buttermilk	"	0.17	0.13	0.11
whey	"	1.11	0.96	1.67
Miscellaneous byproducts[7]	"	0.84	0.39	0.23
Fluid whole milk[8]	"	124.07	116.03	117.87
Milk in ice cream	"	18.14	18.62	18.32
BEVERAGES				
Tea	tea leaf equiv.	1.10	1.10	1.15
Coffee	green beans	4.15	4.31	4.38
Cocoa	"	1.62	1.32	1.42

[1]Includes pickles, relishes, vegetables used in soups.
[2]Per capita consumption not comparable with previous years.
[3]Excludes Newfoundland.
[4]Excludes herring, fresh and frozen, and all fish used for bait.
[5]Includes formula milk, concentrated liquid milk and malted milk.
[6]Part of this product is used for animal feeds.
[7]Includes evaporated and condensed skim milk, condensed buttermilk, sugar of milk, formula skim milk products and concentrated liquid skim milk.
[8]Includes cream expressed as milk.

11.26   Supply, distribution and disappearance of meats, 1971 and 1974-76

Item		1971[1]	1974	1975	1976
BEEF					
Animals slaughtered	'000	3,372.9	3,629.3	4,069.9	4,376.1
Estimated dressed weight	t	851 654	906 779	993 763	1 087 019
On hand, Jan. 1	"	19 719	23 721	20 040	22 560
Imports for consumption	"	69 471	81 413	86 621	143 170
Total supply	"	940 844	1 011 913	1 100 424	1 252 749
Exports	"	51 544	26 064	20 325	58 547
Used for canning	"	..	..	..	..
On hand, Dec. 31	"	16 042	20 040	21 117	35 010
Domestic disappearance	"	873 258	965 809	1 058 982	1 159 192
Per capita disappearance	kg	40.5	43.0	46.4	50.1
VEAL					
Animals slaughtered	'000	838.2	615.8	1,008.8	973.6
Estimated dressed weight	t	44 686	35 115	55 462	52 055
On hand, Jan. 1	"	2 411	2 011	1 565	1 521
Imports for consumption	"	2	2	2	—
Total supply	"	47 097	37 126	57 027	53 576
Exports	"	2	2	2	—
Used for canning	"	..	..	..	..
On hand, Dec. 31	"	1 527	1 565	1 515	2 151
Domestic disappearance	"	45 570	35 561	55 512	51 425
Per capita disappearance	kg	2.1	1.6	2.4	2.2

11.26   Supply, distribution and disappearance of meats, 1971 and 1974-76 (concluded)

Item		1971[1]	1974	1975	1976
MUTTON AND LAMB					
Animals slaughtered	'000	422.8	424.3	423.5	409.9
Estimated dressed weight	t	8 250	8 240	8 205	7 913
On hand, Jan. 1	''	5 107	3 281	3 091	2 007
Imports for consumption	''	23 934	17 417	20 250	17 390
Total supply	''	37 291	28 938	31 546	27 310
Exports	''	42	57	85	125
Used for canning	''	..	..	..	..
On hand, Dec. 31	''	5 020	3 091	1 870	2 496
Domestic disappearance	''	32 229	25 790	29 591	24 689
Per capita disappearance	kg	1.5	1.1	1.3	1.1
PORK					
Animals slaughtered	'000	11,904.0	10,289.3	8,358.3	8,617.2
Estimated dressed weight[3]	t	706 359	611 093	494 608	511 918
On hand, Jan. 1	''	12 192	15 020	10 384	7 850
Imports for consumption	''	7 690	36 374	44 760	88 880
Total supply	''	726 241	662 487	549 752	608 648
Exports	''	44 074	41 809	40 681	39 165
Used for canning	''	..	..	..	..
On hand, Dec. 31	''	13 140	10 384	7 526	12 440
Domestic disappearance	''	669 027	610 294	501 545	557 043
Per capita disappearance	kg	31.0	27.2	22.0	24.1
OFFAL					
Estimated production	t	60 508	58 823	59 363	62 817
On hand, Jan. 1	''	3 656	4 637	4 233	4 423
Imports for consumption	''	4 052	4 037	2 913	2 881
Total supply	''	68 216	67 497	66 569	70 121
Exports	''	20 620	25 503	26 973	27 149
Used for canning	''				
On hand, Dec. 31	''	4 099	4 233	4 178	4 845
Domestic disappearance	''	43 497	37 761	35 418	38 127
Per capita disappearance	kg	2.0	1.7	1.5	1.6

[1]Intercensal revisions.
[2]Included with beef.
[3]Trimmed of larding fat.

11.27   Number of census-farms, by province, censuses 1971[1] and 1976

Province or territory	1971	1976
Newfoundland	402	398
Prince Edward Island	3,462	3,054
Nova Scotia	3,534	3,441
New Brunswick	3,486	3,244
Quebec	48,207	43,097
Ontario	75,645	76,983
Manitoba	29,585	29,963
Saskatchewan	71,319	69,578
Alberta	53,205	57,310
British Columbia	11,014	13,033
Yukon Territory and Northwest Territories	9	17
Canada	299,868	300,118

[1]Based on 1976 census-farm definition.

11.28   Use of agricultural land, by province, censuses 1971¹ and 1976 (hectares, with acres in parentheses)

	Newfoundland 1971	Newfoundland 1976	Prince Edward Island 1971	Prince Edward Island 1976	Nova Scotia 1971	Nova Scotia 1976
Improved land	6 112 (15,104)	9 755 (24,105)	179 540 (443,654)	194 147 (479,749)	130 497 (322,464)	147 853 (365,353)
Under crops²	2 705 (6,683)	3 518 (8,694)	130 405 (322,239)	151 841 (375,208)	83 739 (206,924)	100 068 (247,273)
Pasture (improved)	2 689 (6,645)	5 488 (13,561)	39 911 (98,623)	36 367 (89,866)	36 231 (89,530)	37 217 (91,965)
Summerfallow	166 (410)	128 (316)	3 222 (7,962)	2 192 (5,417)	2 132 (5,269)	2 647 (6,542)
Other	553 (1,366)	621 (1,534)	6 001 (14,830)	3 747 (9,258)	8 393 (20,741)	7 921 (19,573)
Unimproved land	15 042 (37,170)	19 668 (48,601)	90 713 (224,156)	83 902 (207,327)	277 566 (685,880)	252 396 (623,684)
Woodland	3 315 (8,191)	6 800 (16,803)	70 022 (173,029)	68 529 (169,339)	222 828 (550,621)	206 150 (509,408)
Other	11 727 (28,979)	12 868 (31,798)	20 690 (51,127)	15 373 (37,988)	54 737 (135,259)	46 246 (114,276)
Total	21 155 (52,274)	29 423 (72,706)	270 253 (667,810)	278 050 (687,076)	408 062 (1,008,344)	400 249 (989,037)

	New Brunswick 1971	New Brunswick 1976	Quebec 1971	Quebec 1976	Ontario 1971	Ontario 1976
Improved land	165 711 (409,482)	171 844 (424,635)	2 292 925 (5,665,943)	2 245 347 (5,548,374)	4 043 807 (9,992,467)	4 333 292 (10,707,799)
Under crops²	113 465 (280,379)	126 546 (312,702)	1 566 883 (3,871,854)	1 737 809 (4,294,219)	2 971 846 (7,343,593)	3 414 815 (8,438,192)
Pasture (improved)	37 919 (93,700)	36 225 (89,515)	600 894 (1,484,842)	439 488 (1,085,998)	844 808 (2,087,567)	710 235 (1,755,028)
Summerfallow	2 949 (7,288)	1 636 (4,042)	27 757 (68,588)	18 235 (45,059)	81 752 (202,014)	69 327 (171,310)
Other	11 378 (28,115)	7 437 (18,376)	97 391 (240,659)	49 816 (123,098)	145 401 (359,293)	138 916 (343,269)
Unimproved land	259 827 (642,047)	230 484 (569,539)	1 447 507 (3,576,867)	1 408 787 (3,481,188)	1 698 877 (4,198,016)	1 633 523 (4,036,525)
Woodland	209 553 (517,816)	190 311 (470,270)	1 041 715 (2,574,134)	1 021 485 (2,524,145)	775 287 (1,915,775)	760 984 (1,880,432)
Other	50 275 (124,231)	40 173 (99,269)	405 792 (1,002,733)	387 301 (957,043)	923 590 (2,282,241)	872 540 (2,156,093)
Total	425 539 (1,051,529)	402 328 (994,174)	3 740 432 (9,242,810)	3 654 134 (9,029,562)	5 742 684 (14,190,483)	5 966 816 (14,744,324)

	Manitoba 1971	Manitoba 1976	Saskatchewan 1971	Saskatchewan 1976	Alberta 1971	Alberta 1976
Improved land	4 982 862 (12,312,921)	5 181 499 (12,803,765)	18 462 689 (45,622,303)	18 895 957 (46,692,931)	11 085 839 (27,393,709)	11 790 925 (29,136,013)
Under crops²	3 565 502 (8,810,549)	3 829 500 (9,462,902)	10 876 547 (26,876,536)	10 589 287 (26,166,700)	7 056 627 (17,437,308)	7 614 120 (18,814,902)
Pasture (improved)	277 492 (685,697)	307 067 (758,779)	778 557 (1,923,857)	900 779 (2,225,875)	1 062 955 (2,626,619)	1 310 083 (3,237,287)
Summerfallow	1 031 155 (2,548,040)	925 666 (2,287,371)	6 586 578 (16,275,790)	7 176 505 (17,733,533)	2 740 527 (6,771,991)	2 610 729 (6,451,254)
Other	108 713 (268,635)	119 266 (294,713)	221 007 (546,120)	229 385 (566,823)	225 730 (557,791)	255 992 (632,570)
Unimproved land	2 315 522 (5,721,780)	2 429 496 (6,003,416)	7 337 359 (18,131,010)	7 536 671 (18,623,523)	8 049 842 (19,891,595)	8 249 057 (20,383,865)
Woodland	343 181 (848,020)	362 881 (896,699)	374 228 (924,737)	407 091 (1,005,945)	558 355 (1,379,726)	683 094 (1,687,963)
Other	1 972 340 (4,873,760)	2 066 615 (5,106,717)	6 963 131 (17,206,273)	7 129 580 (17,617,578)	7 491 487 (18,511,869)	7 565 962 (18,695,902)
Total	7 298 384 (18,034,701)	7 610 995 (18,807,181)	25 800 048 (63,753,313)	26 432 628 (65,316,454)	19 135 682 (47,285,304)	20 039 981 (49,519,878)

	British Columbia 1971	British Columbia 1976	Yukon Territory and Northwest Territories 1971	Yukon Territory and Northwest Territories 1976	Canada 1971	Canada 1976
Improved land	627 887 (1,551,544)	736 237 (1,819,281)	490 (1,211)	576 (1,422)	41 978 362 (103,730,802)	43 707 432 (108,003,427)
Under crops²	398 260 (984,121)	473 803 (1,170,794)	88 (219)	460 (1,137)	26 766 069 (66,140,405)	28 041 767 (69,292,723)
Pasture (improved)	139 207 (343,989)	161 543 (399,182)	382 (943)	65 (161)	3 821 046 (9,442,012)	3 944 558 (9,747,217)
Summerfallow	62 243 (153,807)	66 296 (163,821)	13 (32)	21 (51)	10 538 495 (26,041,191)	10 873 382 (26,868,716)
Other	28 177 (69,627)	34 594 (85,484)	7 (17)	30 (73)	852 751 (2,107,194)	847 724 (2,094,771)
Unimproved land	1 464 465 (3,618,773)	1 615 566 (3,992,150)	195 (481)	1 220 (3,015)	22 956 914 (56,727,775)	23 460 771 (57,972,833)
Woodland	278 617 (688,477)	275 166 (679,950)	— (—)	18 (44)	3 877 101 (9,580,526)	3 982 510 (9,840,998)
Other	1 185 849 (2,930,296)	1 340 400 (3,312,200)	195 (481)	1 202 (2,971)	19 079 813 (47,147,249)	19 478 261 (48,131,835)
Total	2 092 353 (5,170,317)	2 351 802 (5,811,431)	685 (1,692)	1 796 (4,437)	64 935 275 (160,458,577)	67 168 202 (165,976,260)

¹Based on 1976 census-farm definition.   ²Includes field, vegetable, fruit and nursery cropland.

11.29 Economic classification of census-farms[1], by province, censuses 1971 and 1976 (number)

Economic class	Province or territory					
	Newfoundland		Prince Edward Island		Nova Scotia	
	1971	1976	1971	1976	1971	1976
Value of products sold of						
$100,000 or over		42		149		244
75,000-$99,999	24	13	115	92	213	106
50,000- 74,999		23		132		207
35,000- 49,999	11	15	88	170	122	189
25,000- 34,999	33	14	133	197	161	179
15,000- 24,999	38	28	288	368	383	250
10,000- 14,999	38	30	385	359	321	244
5,000- 9,999	57	55	909	543	621	480
2,500- 4,999	81	53	862	514	747	620
1,200- 2,499	95	95	674	523	946	907
Institutional farms	25	30	8	7	20	15
Total	402	398	3,462	3,054	3,534	3,441

Economic class	New Brunswick		Quebec		Ontario	
	1971	1976	1971	1976	1971	1976
Value of products sold of						
$100,000 or over		147		953		4,517
75,000-$99,999	138	100	902	732	4,603	3,189
50,000- 74,999		171		2,268		6,649
35,000- 49,999	132	195	837	3,907	4,041	6,746
25,000- 34,999	180	208	1,625	5,266	5,698	6,563
15,000- 24,999	344	309	5,094	7,440	11,532	8,953
10,000- 14,999	366	261	6,898	5,268	9,950	7,431
5,000- 9,999	657	501	14,257	6,252	16,527	11,524
2,500- 4,999	786	597	11,319	5,184	13,316	12,041
1,200- 2,499	865	738	7,172	5,752	9,894	9,289
Institutional farms	18	17	103	75	84	81
Total	3,486	3,244	48,207	43,097	75,645	76,983

Economic class	Manitoba		Saskatchewan		Alberta	
	1971	1976	1971	1976	1971	1976
Value of products sold of						
$100,000 or over		973		1,695		2,692
75,000-$99,999	606	731	737	2,050	2,236	1,743
50,000- 74,999		1,980		6,020		3,989
35,000- 49,999	531	2,533	895	8,310	1,797	4,598
25,000- 34,999	926	3,344	1,919	9,989	2,657	5,633
15,000- 24,999	3,138	5,277	7,556	13,745	7,292	8,631
10,000- 14,999	4,263	3,887	11,496	9,497	8,007	7,220
5,000- 9,999	8,984	4,912	23,840	9,586	14,246	9,892
2,500- 4,999	6,888	3,628	16,487	5,366	10,298	7,499
1,200- 2,499	4,212	2,646	8,122	3,036	6,494	5,242
Institutional farms	37	52	267	284	178	171
Total	29,585	29,963	71,319	69,578	53,205	57,310

Economic class	British Columbia		Yukon Territory and Northwest Territories		Canada	
	1971	1976	1971	1976	1971	1976
Value of products sold of						
$100,000 or over		937		—		12,349
75,000-$99,999	869	433	—	—	10,443	9,189
50,000- 74,999		681		—		22,120
35,000- 49,999	572	624	—	1	9,026	27,288
25,000- 34,999	728	628	—	—	14,060	32,021
15,000- 24,999	1,204	1,128	—	—	36,869	46,129
10,000- 14,999	1,070	1,166	—	—	42,794	35,363
5,000- 9,999	2,015	2,044	—	2	82,113	45,791
2,500- 4,999	2,167	2,369	3	3	62,954	37,874
1,200- 2,499	2,353	2,985	6	10	40,833	31,223
Institutional farms	36	38	—	1	776	771
Total	11,014	13,033	9	17	299,868	300,118

[1]Based on 1976 census-farm definition; farms with sales of $1,200 or over.

11.30 Census-farms with sales of $2,500 or more, classified by type of farm and by province, Census 1976 (number)

Type of farm	Province or territory					
	Nfld.	PEI	NS	NB	Que.	Ont.
Dairy	60	846	912	838	24,072	15,617
Cattle	12	334	518	333	2,511	18,343
Hogs	17	216	151	95	2,728	4,461
Poultry	46	19	134	103	1,103	1,655
Wheat	—	13	4	11	143	975
Small grains (excl. wheat farms)	—	126	40	65	1,674	13,443
Field crops, other than small grains	47	471	83	662	1,053	1,782
Fruits and vegetables	42	30	355	164	1,811	3,595
Miscellaneous specialty	15	21	155	66	884	2,437
Mixed	34	448	167	152	1,291	5,305
Livestock combination	2	319	80	80	459	3,658
Field crops combination	—	77	3	15	127	169
Other combinations	32	52	84	57	705	1,478
Total	273	2,524	2,519	2,489	37,270	67,613
	Man.	Sask.	Alta.	BC	YT and NWT	Canada
Dairy	1,639	570	1,933	1,437	—	47,924
Cattle	5,186	7,713	19,505	3,136	1	57,592
Hogs	821	501	1,176	116	—	10,282
Poultry	285	116	307	564	—	4,332
Wheat	7,407	43,817	8,500	206	—	61,076
Small grains (excl. wheat farms)	9,127	10,213	14,719	870	—	50,277
Field crops, other than small grains	275	44	524	220	2	5,163
Fruits and vegetables	64	25	37	2,153	—	8,276
Miscellaneous specialty	274	173	662	813	1	5,501
Mixed	2,187	3,086	4,534	495	2	17,701
Livestock combination	1,465	1,954	3,105	185	—	11,307
Field crops combination	160	24	365	65	—	1,005
Other combinations	562	1,108	1,064	245	2	5,389
Total	27,265	66,258	51,897	10,010	6	268,124

11.31 Lake shipments of Canadian grain from Thunder Bay, navigation seasons 1975 and 1976 (tonnes)

Year and item	To Canadian ports	To US ports	To overseas ports	Total shipments
1975				
Wheat	8 748 689	—	112 674	8 861 363
Oats	271 949	—	112 196	384 145
Barley	2 213 191	128 657	92 271	2 434 119
Rye	136 883	13 962	82 456	233 301
Flaxseed	19 379	—	119 558	138 937
Rapeseed	18 966	—	35 560	54 526
Mustard seed	—	—	—	—
Total	11 409 057	142 619	554 715	12 106 391
1976				
Wheat	7 885 415	21 612	93 048	8 000 075
Oats	274 858	—	311 468	586 326
Barley	2 470 036	82 880	104 109	2 657 025
Rye	49 179	12 357	57 449	118 985
Flaxseed	36 746	—	102 070	138 816
Rapeseed	—	—	55 702	55 702
Mustard seed	—	—	—	—
Total	10 716 234	116 849	723 846	11 556 929

11.32 Supply and disposition of Canadian grain, crop years ended July 31, 1976 and 1977 (thousand tonnes)

Item	Wheat	Oats	Barley	Rye	Flaxseed	Rapeseed
CROP YEAR 1975-76						
Carryover, Aug. 1, 1975	8 036.8	1 132.0	4 104.2	340.4	218.5	399.2
Production in 1975	17 077.9	4 466.3	9 519.0	523.2	444.5	1 839.3
Imports	—	—	—	—	—	—
Total, supply	25 114.7	5 598.3	13 623.2	863.6	663.0	2 238.5
Exports[1]	12 285.2	280.7	4 341.5	299.7	195.6	682.7
Domestic use[2]	4 849.8	4 086.9	6 518.7	251.5	86.4	514.8
Total, disposition	25 114.7	5 598.3	13 623.2	863.6	663.0	2 238.5
Carryover, July 31, 1976	7 979.7	1 230.7	2 763.0	312.4	381.0	1 041.0

11.32 Supply and disposition of Canadian grain, crop years ended July 31, 1976 and 1977 (thousand tonnes) (concluded)

Item	Wheat	Oats	Barley	Rye	Flaxseed	Rapeseed
CROP YEAR 1976-77						
Carryover, Aug. 1, 1976	7 979.7	1 230.7	2 763.0	312.4	381.0	1 041.0
Production in 1976	23 587.9	4 831.8	10 514.0	439.4	276.9	836.9
Imports	—	—	—	—	—	—
Total, supply	31 567.6	6 062.5	13 277.0	751.8	657.9	1 877.9
Exports[1]	13 447.3	493.5	3 799.3	167.6	332.8	1 018.3
Domestic use[2]	4 754.6	4 235.0	6 203.0	243.9	116.8	653.2
Total, disposition	31 567.6	6 062.5	13 277.0	751.8	657.9	1 877.9
Carryover, July 31, 1977	13 365.7	1 334.0	3 274.6	340.3	208.3	206.4

[1]Includes seed wheat, wheat flour in terms of wheat; seed oats, rolled oats and oatmeal in terms of oats; and malt in terms of barley.
[2]Includes human food, seed requirements, industrial use, loss in handling and animal feed.

11.33 Licensed grain storage capacity and grain in store, crop year 1976-77

Grain storage position	Licensed storage capacity	Canadian grain[1] in licensed storage			Proportion of licensed storage capacity occupied		
	Aug. 1, 1976 '000 t	July 31, 1976 '000 t	Mar. 31, 1977 '000 t	July 31, 1977 '000 t	July 31, 1976 %	Mar. 31, 1977 %	July 31, 1977 %
Primary elevators	9 629	4 565	5 567	4 473	47.4	57.8	46.5
Process elevators	576	214	188	205	37.1	32.6	35.6
Terminal	3 901	1 677	2 162	1 593	43.0	55.4	40.8
Other	3 412	3 292	2 369	2 966	96.5	69.4	86.9
Total	17 519	9 748	10 286	9 237	55.6	58.7	52.7

[1]Wheat, oats, barley, rye, flaxseed and rapeseed.

11.34 Wheat milled for flour, and production and exports of wheat flour, 10-year average 1967-76 and crop years ended July 31, 1974-77

Crop year	Wheat milled for flour '000 t	Wheat flour production '000 t	Wheat flour exports	
			Amount '000 t	Production %
Av. 1966-67 — 1975-76	2 394	1 758	478	27.2
1973-74	2 304	1 695	371	21.9
1974-75	2 419	1 770	370	20.9
1975-76	2 550	1 840	469	25.5
1976-77	2 523	1 864	532P	28.6

11.35 Selected farm machinery, by province, censuses 1971[1] and 1976 (number)

Type of machine		Province or territory					
		Nfld.	PEI	NS	NB	Que.	Ont.
Automobiles	1971	205	3,329	2,813	2,994	40,733	80,901
	1976	255	3,293	3,013	2,925	39,838	86,380
Motor trucks	1971	387	2,695	3,393	3,256	18,112	59,293
	1976	411	3,369	3,916	3,991	21,713	73,374
Tractors	1971	328	5,046	4,959	5,345	71,463	148,057
	1976	429	5,366	5,631	5,862	80,017	165,623
Grain combines	1971	—	1,180	292	776	5,373	23,787
	1976	1	1,248	310	759	5,925	24,914
Swathers	1971	6	92	71	111	3,583	7,505
	1976	6	121	170	263	6,719	11,598
Pick-up hay balers	1971	63	1,816	1,913	1,978	29,746	36,089
	1976	108	1,834	2,005	1,953	30,012	37,481
Forage crop harvesters	1971	8	143	278	199	4,708	12,768
	1976	13	353	368	324	5,979	15,674

11.35 Selected farm machinery, by province, censuses 1971[1] and 1976 (number) (concluded)

Type of machine		Province or territory					
		Man.	Sask.	Alta.	BC	YT and NWT	Canada
Automobiles	1971	28,212	59,980	46,032	11,080	1	276,280
	1976	28,895	61,365	52,205	13,042	12	291,223
Motor trucks	1971	37,918	112,861	86,193	12,504	14	336,626
	1976	48,323	140,684	109,694	17,106	17	422,598
Tractors	1971	58,710	127,466	101,914	17,114	12	540,414
	1976	65,176	139,487	116,316	21,352	25	605,284
Grain combines	1971	21,854	59,975	39,910	1,428	—	154,575
	1976	23,174	61,126	42,689	1,704	2	161,853
Swathers	1971	22,406	59,825	38,989	1,764	—	134,352
	1976	23,526	62,544	43,944	3,119	2	152,012
Pick-up hay balers	1971	12,806	28,158	27,009	3,845	3	143,426
	1976	13,950	30,860	31,093	4,927	4	154,227
Forage crop harvesters	1971	1,312	2,601	3,780	1,786	1	27,584
	1976	1,569	3,275	5,099	2,062	1	34,717

[1]Based on 1976 census-farm definition; farms with sales of $1,200 or over.

11.36 Age of census-farm operators, by province, censuses 1971[1] and 1976 (number)

Age of operator	Province or territory					
	Newfoundland		Prince Edward Island		Nova Scotia	
	1971	1976	1971	1976	1971	1976
Under 25 years	7	14	68	52	37	46
25-34 "	39	36	448	402	349	443
35-44 "	87	97	758	677	696	696
45-54 "	146	125	947	820	1,056	951
55-59 "	57	61	521	381	551	457
60-64 "	38	43	357	354	407	422
65-69 "	16	17	203	196	228	220
70 years and over	12	5	160	172	210	206
Total	402	398	3,462	3,054	3,534	3,441
	New Brunswick		Quebec		Ontario	
	1971	1976	1971	1976	1971	1976
Under 25 years	48	69	934	1,062	1,257	1,537
25-34 "	325	403	7,069	7,109	9,527	10,853
35-44 "	735	613	12,130	10,435	18,029	17,367
45-54 "	1,128	922	15,229	13,247	21,675	21,865
55-59 "	543	455	6,166	5,301	9,438	9,130
60-64 "	388	401	3,931	3,349	7,177	7,231
65-69 "	168	225	1,681	1,492	4,754	4,520
70 years and over	151	156	1,067	1,102	3,788	4,480
Total	3,486	3,244	48,207	43,097	75,645	76,983
	Manitoba		Saskatchewan		Alberta	
	1971	1976	1971	1976	1971	1976
Under 25 years	782	1,532	2,270	4,772	1,175	2,058
25-34 "	3,837	4,926	8,932	11,103	7,379	9,204
35-44 "	6,633	6,025	15,690	13,436	12,921	13,614
45-54 "	8,844	8,054	21,107	18,111	15,653	15,711
55-59 "	4,076	3,772	9,434	8,725	6,348	6,607
60-64 "	2,907	3,028	6,965	6,795	4,853	4,960
65-69 "	1,519	1,552	3,907	3,694	2,936	2,832
70 years and over	987	1,074	3,014	2,942	1,940	2,324
Total	29,585	29,963	71,319	69,578	53,205	57,310
	British Columbia		Yukon Territory and Northwest Territories		Canada	
	1971	1976	1971	1976	1971	1976
Under 25 years	127	193	—	—	6,705	11,335
25-34 "	1,176	1,641	1	—	39,082	46,120
35-44 "	2,878	3,395	3	5	70,560	66,360
45-54 "	3,281	3,972	3	4	89,069	83,782
55-59 "	1,411	1,431	—	2	38,545	36,322
60-64 "	1,058	1,140	1	1	28,082	27,724
65-69 "	642	665	1	3	16,055	15,416
70 years and over	441	596	—	2	11,770	13,059
Total	11,014	13,033	9	17	299,868	300,118

[1]See footnote 1, Table 11.35.

11.37   Harvested area, yields and prices of principal field crops, 1972-77, with average for 1967-71

Crop and year	Area '000 ha	'000 acres	Yield kg per ha	bu per acre	Production '000 t	'000 bu	Average price $ per t	$ per bu	Total value[1] $'000
Wheat									
Av. 1967-71	9 421	23,280	1 603	23.8	15 106	555,048	51.41	1.40	776,621
1972	8 640	21,350	1 680	25.0	14 514	533,288	68.44	1.86	993,349
1973	9 575	23,661	1 688	25.1	16 159	593,738	164.29	4.47	2,654,826
1974	8 934	22,077	1 488	22.1	13 295	488,513	154.62	4.21	2,055,608
1975	9 479	23,423	1 802	26.8	17 078	627,515	..	..	..
1976	11 252	27,804	2 096	31.2	23 587	866,672	..	..	..
1977	10 114	25,002	1 943	28.9	19 651	722,036	..	..	..
Oats									
Av. 1967-71	2 899	7,164	1 841	48.3	5 336	345,984	39.59	0.61	211,255
1972	2 470	6,104	1 874	49.2	4 630	300,208	57.61	0.89	266,733
1973	2 711	6,698	1 859	48.8	5 041	326,880	102.18	1.58	515,076
1974	2 471	6,106	1 590	41.7	3 929	254,745	101.33	1.56	398,120
1975	2 411	5,958	1 853	48.6	4 467	289,619	..	..	..
1976	2 410	5,954	2 005	52.6	4 831	313,268	..	..	..
1977	2 131	5,268	2 019	53.0	4 303	279,011	..	..	..
Barley									
Av. 1967-71	4 064	10,042	2 100	39.0	8 535	392,023	34.06	0.74	290,696
1972	5 062	12,509	2 229	41.4	11 285	518,316	57.26	1.25	646,184
1973	4 839	11,958	2 113	39.3	10 224	469,570	114.63	2.50	1,171,970
1974	4 775	11,800	1 843	34.3	8 802	404,286	101.39	2.21	892,447
1975	4 468	11,041	2 131	39.6	9 520	437,251	..	..	..
1976	4 354	10,758	2 415	44.9	10 513	482,866	..	..	..
1977	4 649	11,489	2 477	46.0	11 516	528,906	..	..	..
Rye									
Av. 1967-71	324	800	1 269	20.2	411	16,193	36.24	0.92	14,894
1972	257	634	1 339	21.3	344	13,524	56.75	1.44	19,521
1973	257	634	1 412	22.5	363	14,282	100.34	2.55	36,423
1974	341	843	1 408	22.4	480	18,914	87.98	2.23	42,232
1975	320	790	1 634	26.1	523	20,585	..	..	..
1976	250	619	1 764	28.0	441	17,340	..	..	..
1977	242	599	1 620	25.8	392	15,449	..	..	..
Mixed grains									
Av. 1967-71	739	1,826	2 529	50.1	1 869	91,559	41.51	0.85	77,576
1972	836	2,065	2 547	50.5	2 129	104,285	50.46	1.03	107,437
1973	810	2,002	2 444	48.5	1 980	97,013	94.82	1.94	187,735
1974	733	1,811	2 248	44.6	1 648	80,754	102.49	2.09	168,909
1975	743	1,835	2 467	48.9	1 833	89,807	..	..	..
1976	645	1,593	2 433	48.2	1 569	76,852	..	..	..
1977	624	1,543	2 575	51.0	1 607	78,737	..	..	..
Flaxseed									
Av. 1967-71	807	1,994	802	12.8	647	25,489	96.67	2.45	62,545
1972	535	1,321	836	13.3	447	17,617	158.53	4.02	70,863
1973	587	1,450	840	13.4	493	19,400	366.53	9.31	180,697
1974	587	1,450	598	9.5	351	13,800	373.05	9.49	130,940
1975	567	1,400	785	12.5	445	17,500	..	..	..
1976	323	800	858	13.6	277	10,900	..	..	..
1977	576	1,420	1 059	16.9	610	24,000	..	..	..

11.37 Harvested area, yields and prices of principal field crops, 1972-77, with average for 1967-71 (concluded)

Crop and year	Area '000 ha	Area '000 acres	Yield kg per ha	Yield	Production '000 t	Production '000 bu	Average price $ per t	Average price	Total value¹ $'000
Rapeseed				*bu per acre*		*'000 bu*		*$ per bu*	
Av. 1967-71	1 136	2,808	977	17.4	1 110	48,940	96.07	2.18	106,636
1972	1 323	3,270	983	17.5	1 300	57,300	139.30	3.16	181,086
1973	1 275	3,150	947	16.9	1 207	53,200	251.93	5.72	304,078
1974	1 279	3,160	909	16.2	1 163	51,300	311.51	7.06	362,288
1975	1 748	4,320	1 001	17.8	1 749	77,100	..	..	..
1976	720	1,778	1 162	20.8	837	36,900	..	..	..
1977	1 348	3,330	1 318	23.5	1 776	78,300	..	..	..
Corn for grain									
Av. 1967-71	443	1,094	5 156	82.2	2 284	89,932	50.03	1.27	114,269
1972	537	1,327	4 708	75.0	2 528	99,538	64.91	1.65	164,100
1973	530	1,310	5 289	84.2	2 803	110,365	104.80	2.66	293,751
1974	591	1,460	4 360	69.5	2 577	101,440	121.73	3.09	313,705
1975	635	1,569	5 740	91.5	3 645	143,493	..	..	..
1976	709	1,752	5 319	84.7	3 771	148,457	..	..	..
1977	730	1,801	5 895	94.1	4 303	169,394	..	..	..
Potatoes			*cwt*			*'000 cwt*		*$ per cwt*	
Av. 1967-71	121	299	19 149	170.8	2 317	51,075	40.19	1.82	93,125
1972	99	244	20 111	179.8	1 991	43,886	75.06	3.65	149,443
1973	106	261	20 358	182.6	2 158	47,586	114.29	5.47	246,636
1974	115	283	21 748	195.1	2 501	55,146	60.83	2.76	152,147
1975	105	260	20 905	185.9	2 195	43,390	..	..	..
1976	105	263	22 286	196.3	2 340	51,585	..	..	..
1977	111	276	22 414	199.0	2 488	54,841	..	..	..
Tame hay			*ton*			*'000 tons*		*$ per ton*	
Av. 1967-71	5 020	12,406	4 452	1.99	22 350	24,636	20.68	18.76	462,163
1972	5 042	12,459	4 179	1.86	21 073	23,229	25.50	23.14	537,446
1973	5 200	12,850	4 492	2.00	23 358	25,748	32.02	29.04	747,838
1974	5 274	13,033	4 369	1.95	23 044	25,402	47.92	43.47	1,104,256
1975	5 267	13,014	4 467	1.99	23 526	25,933	..	..	..
1976	5 731	14,162	4 357	1.94	24 972	27,527	..	..	..
1977	5 621	13,890	4 400	1.96	24 735	27,266	..	..	..

¹Gross value of farm production; does not represent cash income from sales.

Sources

11.1-11.18 Agriculture Division, Institutions and Agriculture Statistics Field, Statistics Canada.
11.19-11.21 Food, Beverages and Textiles Section, Manufacturing and Primary Industries Division, Statistics Canada.
11.22-11.26 Agriculture Division, Institutions and Agriculture Statistics Field, Statistics Canada.
11.27-11.30, 11.35-11.36 Census of Agriculture Division, Institutions and Agriculture Statistics Field, Statistics Canada.
11.31-11.34, 11.37 Agriculture Division, Institutions and Agriculture Statistics Field, Statistics Canada.

Mines and minerals

Chapter 12

Tables

Mines and minerals

Chapter 12

Canada's mineral industry

12.1

Canada leads the world in mineral exports and ranks third in mineral production behind the United States and the Soviet Union. The mineral industry has been a major factor in Canada's economic development and is still the main force in the northward advance of Canada's frontiers of population and economic activity.

The overall demand for Canadian non-fuel minerals in 1976 was up from 1975. Market conditions, however, varied widely among commodities. The value of mineral production in 1976 increased 15.4% compared with a 13.6% increase in 1975.

Canada's mineral production in 1976 was valued at $15,393 million compared with $13,347 million in 1975. Shipments of metals increased by 9%, non-metals 22%, fuels 20% and structural materials 6% during 1976.

Canada produces about 60 different minerals from domestic deposits. The 10 leading minerals comprised 84% of the total output by value in 1976 compared with 82% in 1975 and 83% in 1974. The 1976 value for the 10 leading minerals totalled $12,954 million. Individual values were: petroleum $4,128 million, natural gas $2,467 million, iron ore $1,241 million, nickel $1,232 million, copper $1,126 million, zinc $862 million, natural gas byproducts $794 million, asbestos $446 million, cement $339 million and sand and gravel $321 million. The first four accounted for 59% of the total value of mineral production in 1976 compared to 57% in 1975 (Tables 12.1 - 12.5).

Canada produces many minerals needed for modern economies although a few, such as manganese, chromium, bauxite and tin, are imported.

Export sales

12.1.1

The strength of the industry is based on export sales. About 82.0% of total mineral production was exported with crude minerals comprising 67.2% of total mineral exports.

Exports of crude and fabricated non-fuel mineral products brought several periods of sustained expansion in the economy in the past and have been a major factor in increased export trade. In the first nine months of 1976 these exports were valued at $5,200 million or 18.8% of Canada's total exports of all products. This proportion is typical of the past decade and has been maintained despite the sharp increase in Canada's automobile trade with the United States in the late 1960s. Increased US demand was the main factor in the increase in Canada's exports. Major consumers of Canada's exports of mineral products were: United States 55%, European Economic Community (EEC) 22%, and Japan 8%. Comparable percentages for 1975 were: United States 68%, EEC 15% and Japan 9%. Exports to EEC increased markedly although the United States remained Canada's most important export customer.

Sectors of production

12.1.2

Mineral production is divided into four sectors: metallics, non-metallics, mineral fuels and structural materials. The contribution of each of these groups to the total value of production in 1976 was as follows (1975 figures in brackets): mineral fuels 51.9% (49.8%), metallics 34.0% (36.0%), non-metallics 7.4% (7.0%) and structural materials 6.6% (7.2%). Value of mineral fuels production increased with the continued rise in export sales. Structural materials are sold mainly in the domestic market where demand is more stable.

Prices of most minerals, especially non-ferrous metals, showed great strength in part of 1974, but in general declined in 1975 and 1976. This was partly due to inflation and partly a result of an economic expansion that had been greater than anticipated in the US, Japan and Europe, followed by recessionary conditions in late 1974 and 1975.

Leading minerals

12.1.3

Petroleum, natural gas, iron ore, copper, nickel and zinc together contribute three-quarters of total Canadian mineral output value. Petroleum and natural gas production and refining is Canada's largest mineral industry. Domestic production and exports are small in the context of the world industry but are of great significance to Canada. The industry's growth in the past two decades has been of particular importance because of its effect on the balance of payments, as a source of revenue to the several levels of government, and for its impact on engineering and construction.

In 1976 total production of crude oil, gas and gas byproducts was valued at $7,389 million, an increase of 21.8% over the 1975 value of $6,067 million. Crude oil production is concentrated in Alberta, with Saskatchewan in second place and minor production elsewhere. The pattern of crude oil distribution in Canada reflects a national oil policy, allocating markets west of the Ottawa Valley to Canada's mid-continent producers while Quebec and Atlantic markets were supplied by overseas oil. Canada has produced oil almost equivalent to its total domestic needs but has imported oil in Eastern Canada from overseas and exported western oil to US markets. The possibility of depletion has been of concern and has affected amounts made available for export. Alberta oilfields are producing at close to capacity and the region's economic reserves of oil will last 13 years at current depletion rates. Canada's North is the focus of optimism for large-scale oil finds.

Natural gas is an important domestic product and an increasingly important export product. Generally gas and oil are found together. In Canada, western provinces have the major proven reserves of gas. The value of gas and gas byproducts produced in 1976 was $3,261 million compared with oil at $4,128 million.

Canada's gas reserves are sufficient for 21 years but known reserves of commercial gas declined for the first time in 1972. This does not include discoveries of gas in the Arctic because there is as yet no economical method of transporting it to southern markets. Sales of natural gas and gas byproducts showed an increase of 42% over the previous year. Existing proven reserves of gas are sufficient to meet normal domestic market growth and to continue meeting present export commitments only in the short term. However, if long-term domestic market growth and current exports to the United States are to be maintained, additional reserves of natural gas must be found in the next decade.

Iron ore production rose in 1976 to 56.9 million tonnes. Production was valued at $1,241 million, an increase of 35.2% compared to 1975. Of the 44.7 million tonnes exported in 1976, the United States received 24.5 million tonnes, EEC 14.2 million tonnes and Japan 5.1 million tonnes. Projects currently under way in the Quebec–Labrador area will increase both production and pelletizing capacities. Production capacity is expected to increase from 47 million tonnes of iron ore products in 1970 to approximately 79 million tonnes by 1978. Newfoundland, Quebec, Ontario and British Columbia are the only producers of iron ore.

Nickel ranked fourth among Canadian minerals produced in 1976. World oversupply, which led to accumulation of large stockpiles by Canadian producers during 1971 and 1972, eased in 1973 as demand increased. Canada is the world's leading producer of nickel.

Copper was fifth by output value in 1976. Production of recoverable copper from Canadian mines dropped to 747 135 tonnes, a decrease of 2% from 1975. Copper remained in oversupply in the world but a better balance between supply and demand was achieved. Copper is produced in all provinces except Prince Edward Island and Alberta. British Columbia accounted for 36.6%, Ontario for 34.6% and Quebec 16.1% of 1976 production.

Zinc production in 1976 declined nominally to 1.04 million tonnes from 1.06 million tonnes. Value also declined as a result, but Canada remained the world's leading producer of zinc.

12.1.4 Growth of the industry

In 1976 mineral investment (including both capital and repair expenditures) in mineral fuels was $2.5 billion, 32% higher than 1975, compared with non-metal mines at $562.9 million, up 6.2%, and metal mines at $1,394.0 million, up 13.8%. Similarly, in mineral manufacturing, investment in non-metallics at $390.6 million was 6.9% higher than

1975, compared with petroleum and coal products at $488.5 million, down 16.3%, and primary metals at $1,301.2 million, down 9.9%.

Overall capital expenditures increased in the minerals sector and the mineral fuels sector (excluding coal). Investment in iron mines increased to $485.6 million from $393.1 million in 1974. Capital expenditures in the mineral fuels sector (excluding coal) increased to $2,270.1 million in 1976, an increase of 32.3% over 1975, with more emphasis placed on exploration in areas where new finds of gas and oil have been reported.

The volume index of mineral production, which measures the mining industry's absolute growth, based on the revised index of 1971 = 100, increased to 110.4 from 109.3 in 1975.

Alberta accounted for 45.4% and Ontario 16.9% of the Canadian output value of minerals in 1976. Quebec accounted for 9.9%, British Columbia 9.2%, Saskatchewan 5.9%, Newfoundland and Labrador 4.7%, Manitoba 3.1%, New Brunswick 1.7%, Yukon Territory 1.4%, Northwest Territories and Nova Scotia 0.8% each and Prince Edward Island for a minimal amount. Tables 12.6 - 12.8 show mineral production and value by province.

Provincial summary   12.2

Newfoundland. Mineral production in Newfoundland and Labrador was valued at $756 million compared to $551 million in 1975, an increase of 37.2%. Iron ore production increased 23.8% to 28.0 million tonnes. Production of lead increased 77.3% in 1976 and zinc production rose 41.7%. Asbestos production increased 48.3%. Fluorspar concentrate production decreased 50% in 1976 to 72 500 tonnes valued at $2.2 million.

Prince Edward Island. Sand and gravel is the only mining product of this province. Production, valued at $1.7 million, decreased 4.9% in value from 1975.

Nova Scotia. Total mining production increased 15.6% in value to $117.2 million in 1976. The quantity of coal produced was 1.99 million tonnes, up from 1.66 million tonnes in 1975. Production of non-metallics increased to $32.9 million from $28.8 million in 1975, with gypsum accounting for $13.8 million or 42%.

New Brunswick. Mineral production increased to $255 million in 1976 from $232 million in 1975. Metal products, mined mainly around Bathurst, represented 85.5% of total mineral output. Zinc, lead and copper were the principal minerals produced. Coal production dropped to 290 000 tonnes from 418 000 tonnes in 1975.

Quebec. Total mineral output was $1,521 million, an increase of 22.7% over 1975. Metallics accounted for 50.3% of production, non-metallics 28.9% and structural materials 20.8%. Iron ore, at 21.3% of total mineral output, copper 11.9% and zinc 6.7%, were the major metallics produced. Asbestos was the major non-metallic, accounting for 22.6% of the total. Titanium dioxide, a non-metallic, is produced only in Quebec and has firm world markets; production was valued at $74.4 million in 1976 compared to $55.8 million in 1975.

Ontario produced minerals valued at more than $2,594 million in 1976, mostly metallics. The value of metallics increased by 10.5% from the previous year. Nickel made up 38.3% of provincial output, copper 15.1%, zinc 10.3%, iron ore 10.2% and precious metals 7.9%. The range of minerals is more diverse in Ontario than in any other province. Output of fuels is relatively small. The principal non-metals — salt, nepheline syenite, asbestos, gypsum, quartz and sulphur — are produced in relatively small quantities. Structural materials produced increased in value to $353 million from $333 million in 1975.

Manitoba. Mineral production in 1976 was valued at $478 million, a decrease of 9.7% from 1975. Metallic minerals accounted for 81.0% of the total, with nickel 49.9%, copper 17.9% and zinc 10.7%. Crude petroleum contributed 6.9% of the provincial total. Manitoba produced 19.3% of Canada's nickel, a decrease of 6.9% from the total value for 1975.

Saskatchewan produces mainly mineral fuels and non-metals. Crude petroleum and potash represented 48.0% and 39.8% of 1976 mineral production. Metallics constituted only 2.8% of the total. Renewed interest in uranium resulted in increased production and in 1976 Saskatchewan accounted for 35.8% of total Canadian production.

Alberta. Mineral production was valued at almost $7,000 million in 1976 with crude petroleum, natural gas and natural gas byproducts representing 94.5% of the total. Sulphur, produced as a byproduct in processing natural gas, represented 1.5% of mineral production. Alberta produced 85.5% of Canada's petroleum and 93.3% of Canada's natural gas in 1976. Coal production accounted for 3.2% of provincial mineral production. Structural materials made up most of the remainder.

British Columbia. Mineral output increased 9.6% to $1,421 million in 1976. Metallics comprised 50.0% and mineral fuels 39.9% of all mineral production with copper accounting for 29.0%, zinc 6.3%, molybdenum 6.6% and lead 3.1%. Coal represented 21.6%, natural gas 9.2% and crude petroleum 8.0% of the total. Production of copper in all forms was increased with mine production value up 13.5% from 1975. Coal production decreased to 7.7 million tonnes in 1976. Asbestos was the leading non-metallic mineral produced.

Northwest Territories. The value of mineral production in 1976 increased to $213 million from $206 million in 1975. Metallic minerals accounted for almost all of the total. Zinc comprised 56.2%, lead 12.6%, gold 10.8% and silver 7.0% of total mineral output. Crude oil and natural gas are of considerable potential value.

Yukon Territory. The value of production decreased to $131 million compared to $230 million in 1975. Zinc made up 32.6% of total production, asbestos 26.2%, lead 14.5%, copper 12.6% and silver 10.2%. Output is not large by national standards but is increasing rapidly.

12.3   Metals

12.3.1   Iron ore

Iron ore shipments in 1976 were 56.9 million tonnes compared with the previous year's 44.9 million tonnes. These figures include about 0.45 million tonnes of byproduct iron ore. The value of the shipments including byproduct ore was $1.2 billion compared with $918 million for 1975. The increase in shipments was due mainly to improvements in the operations at Wabush Mines, of the Iron Ore Co. of Canada (IOC) and to new production from the Mount Wright operation in Quebec, owned by Quebec Cartier Mining Co., a subsidiary of United States Steel Corp.

Iron ore was produced by 16 mining companies at 18 locations, with nine operations in Ontario, four in Quebec, two in Labrador, two in British Columbia and one in Quebec–Labrador.

Shipments by province in 1976 were as follows: Newfoundland 27.9 million tonnes; Quebec 17.7 million tonnes; Ontario 10.3 million tonnes and British Columbia 809 000 tonnes. IOC is the largest Canadian producer with 25.1 million tonnes followed by Quebec Cartier Mining at 14.1 million tonnes and Wabush Mines at 5.5 million tonnes.

Exports increased from 36.0 million tonnes to 44.7 million tonnes in 1976. Exports were expected to increase considerably in the following two years with the Mount Wright development reaching full production of 18.0 million tonnes a year in 1978, all for export.

Imports decreased from 4.8 million tonnes in 1975 to 3.0 million tonnes in 1976. This was due to increased production from Wabush Mines in which the Steel Company of Canada Ltd. (STELCO) and Dominion Foundries and Steel Ltd. (DOFASCO) have an interest. Imports, however, were expected to increase during the next three years as new taconite deposits in the US, in which STELCO, DOFASCO and Algoma Steel Corp. have taken equity participation, are fully developed and production reaches capacity.

Starting in 1977 it was expected that DOFASCO would receive about 0.6 million tonnes of pellets a year from its terms of investment in the Eveleth mine in Minnesota.

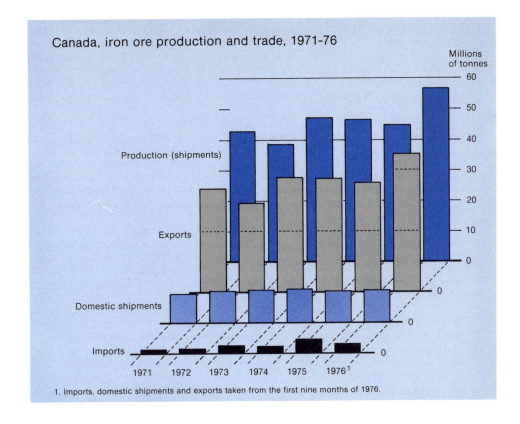

Canada, iron ore production and trade, 1971-76

Millions of tonnes

Production (shipments)

Exports

Domestic shipments

Imports

1971   1972   1973   1974   1975   1976[1]

1. Imports, domestic shipments and exports taken from the first nine months of 1976.

In 1979 STELCO will start receiving some 3.5 million tonnes of pellets a year from its equity participation in the Erie, Eveleth and Hibbing mines in Minnesota and the Tilden mine in Michigan, while Algoma will receive some 3.1 million tonnes of pellets a year from its investment in the Tilden mine.

Quebec Cartier continued to develop the Mount Wright project in Quebec. Production in 1976 was some 8 million tonnes. Included in the development is construction of a concentrator with a minimum production capacity of 18 million tonnes of iron ore concentrate (66% iron) with provision for an ultimate annual capacity of 24 million tonnes. Maximum capacity was expected to be reached in 1978. Two-thirds of production would be exported to Europe and one-third to US Steel. Quebec Cartier has 8.23% ownership in a group involved in the Fire Lake development in Quebec. Other participants include Sidbec, a provincial steel agency, with 50.1% of the ownership, and British Steel Corp. with 41.67%. The new company formed to develop Fire Lake is called Sidbec-Normines Ltd. Crude ore from Fire Lake is transported by rail to the Lac Jeannine concentrator where 6 million tonnes a year of concentrate are expected to be produced from both the Fire Lake mine and the Lac Jeannine mine, where ore reserves neared depletion in 1977. At Port Cartier a pellet plant was under construction to use concentrate from the Lac Jeannine concentrator. When completed the plant was expected to produce 3 million tonnes a year of pellets with an iron content of 65% and a silica content greater than 6% and 3 million tonnes a year of pellets grading 68% iron and less than 2% silica. The former product is for use in standard blast furnaces, the latter for use as direct reduction feed mainly for Sidbec's operations at Contrecoeur.

Most design problems have been solved at IOC's Sept-Îles, Que., pellet plant and it was expected full production would be attained in 1978. The pellet plant is designed for an annual production of 6.0 million tonnes; in 1976 production reached 3.7 million tonnes. The plant treats a blend of 85% blue ore and 15% yellow ore from the Schefferville area.

La Société de Développement de la Baie James (SDBJ) continued drilling on its property at Lac Albanel, Que. Reserves stood at over one billion tonnes of magnetite ore grading 31% iron. This would permit production of 9 million tonnes a year of pellets and concentrates for a period of 30 years. SDBJ intends to start production between 1985 and 1990.

In Ontario, areas that attracted recent attention were Lake St. Joseph, Geraldton–Nakina and Bending Lake. Because of their location, Algoma is the steel company in Canada most likely to show interest in their development. In 1976, however, nothing was done on the Lake St. Joseph property and not much was expected to be done for the next couple of years due to escalation in construction costs, less growth than expected in the steel industry in North America and investments by Algoma in the Tilden mine in the United States. Metallurgical work continued on the Geraldton property but at a reduced rate.

Jointly with Algoma, Steep Rock Iron Mines Ltd. studied the possibility of developing the Bending Lake iron ore deposit, 25 kilometres northwest of Atikokan. The ore at Bending Lake consists of magnetite grading 20% iron and reserves are sufficient to sustain a mining operation for 20 years. The ore would be mined by open-pit methods, concentrated at Bending Lake and shipped by rail or pipeline to Atikokan for pelletizing.

In 1976 a direct-reduction plant of STELCO at the Griffith mine operated from January to March. The plant then closed because of operating problems. During that period it operated at 80% capacity and produced almost 30 000 tonnes of sponge iron. The plant reopened later but was shut down completely in August. The plant has a rated capacity of 400 000 tonnes of sponge iron a year using a mixture of coal and iron pellets to produce sponge iron.

A direct-reduction plant of Sudbury Metals at Falconbridge, Ont., owned by Allis-Chalmers Corp. and National Steel Corp., officially opened May 12, 1976. But the plant was shut down in October following an explosion that occurred while it was not in operation. The plant has a designed capacity of 275 000 tonnes of sponge iron a year. It uses INCO oxide pellets as feed material and oil and gas as a reductant.

In British Columbia, Texada Mines Ltd. ceased operations at its iron ore mine near Gillies Bay, Texada Island, on December 17, 1976, when ore reserves were exhausted. Texada Mines started production in 1952 and at the end of 1976 had produced 10.5 million tonnes of concentrates. All of Texada's production was sold to Japanese consumers. The mine closure resulted in the loss of 180 jobs.

Other developments in the iron ore industry in Canada were as follows: the transport department investigated the possibility of increasing tolls on the St. Lawrence Seaway and Welland Canal; metallurgical work was conducted by the Canada Centre for Mineral and Energy Technology (CANMET) laboratories in Ottawa on iron material from the Peace River area of Alberta. The metallurgy of this material is complex as it has a high content of silica, phosphorus and sulphur; Combustion Engineering Ltd. investigated the possibility of using coal in pelletizing plants as a substitute for oil and gas; Campbell Chibougamau Mines Ltd. started a preliminary market study for production of pellets from an iron ore deposit located near Chibougamau, Que. The pellets to be produced would have a high titanic content of 1.0%; Canadian National and Canadian Pacific railways were studying the possibility of using hydroelectric energy rather than oil in transporting iron ore from the Quebec–Labrador region.

12.3.2 Nickel

Canadian production of nickel in 1976 amounted to 262 492 tonnes valued at $1.23 billion. World production of nickel increased 4.9%; Canadian production decreased 1.0% because of lower production rates instituted in 1975. Consumption of nickel in the non-communist world was about 496 700 tonnes compared with about 408 600 tonnes in 1975, when one of the sharpest declines in consumption in the history of the industry was recorded. Producer stocks rose to about three times normal at the end of the year.

At Sudbury, Ont., the International Nickel Co. of Canada, Ltd. (INCO) continued development of the Clarabelle open-pit extension and the Levack East mine, where production is expected to begin in 1984. In Manitoba, development work continued at the three operating mines, Birchtree, Pipe and Thompson. INCO started construction of a new rolling mill in the Sudbury district for direct rolling of metal powders to make coinage strip. The plant was scheduled to be in operation in 1978.

Six companies mined nickel ores in Canada during 1976. Falconbridge Nickel Mines Ltd., the second largest producer, continued development work at the Lockerby mine, scheduled to be producing at capacity in 1978. The Thierry deposit, near Pickle Crow, Ont., of Union Minière Explorations and Mining Co. Ltd. started production in August 1976. Three nickel producers ceased mining operations during the year.

Copper 12.3.3

Canadian mine production of recoverable copper amounted to 747 135 tonnes valued at $1,126.2 million in 1976 (Table 12.9). Canada produced 9.4% of the world's newly mined copper and ranked as the fourth largest producer. World mine production of copper increased 9% from 1975. Canadian exports of copper concentrates decreased 9% while exports of refined copper increased 12%. Domestic consumption of copper fell by 25% to the lowest level since 1964.

Copper and nickel-copper ores were smelted at five locations in Canada at the end of 1976. INCO continued to operate an oxygen flash smelter at Copper Cliff, Ont. Falconbridge operated a smelter at Falconbridge, Ont., treating nickel-copper concentrates. Ores and concentrates from most mines in the Atlantic provinces, Quebec and Ontario were processed at the Noranda smelter of Noranda Mines Ltd. or at the Murdochville smelter of Gaspé Copper Mines Ltd., both in Quebec, where major expansion programs have been completed. At Murdochville, smelter production was 67 000 tonnes of anode copper in 1976. A 270 000 tonnes a year sulphuric acid plant has been built and some acid was used to leach copper from low-grade oxide ores from the Copper Mountain mine. The expanded facilities encountered serious and lengthy start-up problems in 1974 and 1975. At Noranda, operation of a new continuous smelting process reactor, capable of producing 50 000 tonnes a year of blister copper in one furnace directly from concentrates, improved during the year. A run of 180 days was achieved without shutdown for refractory repairs. Operation of the reactor began early in 1973. A shortage of concentrates was experienced at Noranda in 1976 and production fell to 208 000 tonnes of anode copper compared with peak production of 244 000 tonnes in 1974. Hudson Bay Mining and Smelting Co. Ltd. operates a smelter at Flin Flon, Man. and produces anode copper which is refined at the Montreal refinery of Canadian Copper Refiners Ltd.

Falconbridge reactivated its smelter modernization program. Expenditures in 1976 were $32 million of an estimated total cost of $97 million. Start-up of the new facilities was scheduled for 1978.

Afton Mines Ltd. continued to build a new copper smelter at Kamloops, BC. Start-up of the project was scheduled for 1977. The smelter is to produce 22 000 tonnes of blister annually to be exported under long-term contract to the United Kingdom.

A new Canadian hydrometallurgical process to produce copper from concentrates underwent pilot plant testing in 1976. The process was developed by Sherritt Gordon Mines Ltd. and Cominco Ltd. with financial assistance from the federal government. The test demonstrated the commercial viability of the process, said to be environmentally clean and applicable to a wide range of copper sulphide concentrates. The process produces high purity copper and sulphur in elemental form.

Electrolytic copper refineries were operated by INCO at Copper Cliff, Ont., and by Canadian Copper Refiners at Montreal East, Que. INCO's copper refining capacity at Copper Cliff was 192 000 tonnes a year. Copper is recovered in part as a byproduct from the refining of nickel. Canadian Copper Refiners has a capacity of 435 000 tonnes of refined copper a year, making it the world's largest copper refinery.

As a result of depressed copper consumption in Japan and higher treatment charges sought by Japanese smelters, the shift of Canadian concentrate sales away from Japan continued in 1976. The displaced concentrates were processed at smelters in North

America. Japan, however, continued to be the most important market for Canadian copper concentrates, consuming 70% of all concentrate exports.

Further erosion of mine production capacity occurred, particularly in Eastern Canada, with decisions to close a number of small mines in Quebec. Development of new copper mines took place at a reduced pace.

In 1976 copper production in Canada increased by 2% to 747135 tonnes. Production in the Atlantic provinces and Manitoba decreased by 12%. In Quebec production increased by 2% and in British Columbia by 6%. In Ontario production was unchanged from 1975.

Union Minière opened a new mine at Pickle Lake, Ont., in 1976 at a cost of $104 million. The mine is expected to produce 13000 tonnes of copper in concentrate annually.

Copper production in Newfoundland in 1976 came from two mines and totalled 6764 tonnes valued at $10.2 million. In New Brunswick copper production from four mines was 9678 tonnes valued at $14.6 million. In Quebec production increased to 120411 tonnes valued at $181.5 million from 117556 tonnes valued at $165.2 million in 1975. About 20 mines were operating in Quebec in 1976, the main production centres being Rouyn–Noranda, Val d'Or, Matagami, Chibougamau, Murdochville and Stratford Centre.

Copper was produced at about 30 mines in Ontario in 1976, the main operations being the nickel-copper mines of the Sudbury district, copper-zinc and copper mines near Timmins, and copper-zinc mines near Manitouwadge. Ontario producer shipments amounted to 258981 tonnes valued at $390.4 million compared to 257778 tonnes valued at $361.4 million in 1975.

Production in Manitoba and Saskatchewan was 66283 tonnes valued at $99.9 million. The major producer was Hudson Bay Mining which produced copper in Manitoba's Flin Flon and Snow Lake areas. Sherritt Gordon Mines, Ltd. at Lynn Lake, Man., Fox Lake, Sask., and Ruttan, Man., and INCO at Thompson, Man., were the other main producers.

Production of copper in British Columbia in 1976 amounted to 273541 tonnes valued at $412.3 million compared to 258518 tonnes valued at $363.3 million in 1975. Most production comes from large open-pit mines. Production in the Yukon Territory increased substantially in 1976 due to higher shipments of concentrates from Whitehorse Copper Mines Ltd.

12.3.4 Lead and zinc

Canadian production of lead in 1976 was 259083 tonnes valued at $129 million, a decrease of 25.8% in volume and 17.1% in value compared to 1975 (Table 12.12). Output of refined lead was 175720 tonnes, a 2.5% increase from 1975.

Production of zinc in 1976 was 1.0 million tonnes valued at $862 million. Production decreased 1.2% in value and 1.5% in volume from 1975 (Table 12.13). Output of refined zinc was 472316 tonnes in 1976, up 10.6%.

Exports of refined lead in 1976 increased to 129311 tonnes, up 6.2% from 1975. Exports of refined zinc increased to 350487 tonnes, up 41.7% from 1975. Exports of lead in ores and concentrates declined due to strikes at several major producers.

In the Atlantic provinces lead production increased 21.1% to 77994 tonnes and zinc production decreased 9.7% to 134733 tonnes in 1976. Brunswick Mining and Smelting Corp. Ltd. in New Brunswick was the largest area producer.

In Quebec zinc production decreased 1.9% to 122934 tonnes. Mining companies associated with Noranda Mines Ltd. produced most of the zinc in Quebec. Lemoine Mines Ltd. near Chibougamau completed construction at its zinc-copper mine late in 1975 and began production early in 1976. Only 814 tonnes of lead was produced in Quebec in 1976.

In Ontario zinc production decreased 4.8% to 320030 tonnes. Texasgulf Canada Ltd., a wholly owned subsidiary of Texasgulf, Inc., operates Canada's largest zinc mine at Timmins and was the largest zinc producer in the province. Lead production in Ontario increased slightly to 3379 tonnes.

In Manitoba and Saskatchewan zinc production remained unchanged at 70 079 tonnes. Hudson Bay Mining and Smelting and Sherritt Gordon Mines Ltd. were the only zinc producers in the two provinces.

In British Columbia zinc production increased 13.6% to 113 266 tonnes and lead production increased 25.5% to 88 633 tonnes. The Ruth Vermont mine of Consolidated Columbia River Mines Ltd. operated for part of the year but was placed in receivership at the end of 1976. Northair Mines Ltd. began production at its Brandywine Falls silver-gold-lead-zinc property in May 1976. Lead production was expected to be at the rate of 1 000 tonnes a year. Production at Cominco Ltd.'s Sullivan and H.B. mines was improved during the year because of mining higher grade ore and better recoveries in the mills.

In the Yukon Territory lead production decreased 79.0% to 38 254 tonnes and zinc production decreased 78.9% to 51 723 tonnes. Cyprus Anvil Mining Corp., Canada's largest lead producer, suffered a series of strikes in 1976 which severely restricted output. Kerr Addison Mines Ltd. and Canadian Natural Resources Ltd., which jointly own the Grum lead-zinc-silver deposit near Faro, undertook a major feasibility study of this deposit, which contains over 26 million tonnes of ore, but had not made a production decision.

In the Northwest Territories lead production decreased 36.1% to 53 679 tonnes and zinc production increased 11.9% to 144 307 tonnes. Pine Point Mines Ltd., the major producer of lead and zinc, produced a lesser tonnage of lower grade ore and suffered technical problems in the mill which hampered production. Discussions continued with the federal government concerning development of the Polaris property of Arvik Mines Ltd. on Little Cornwallis Island. The Polaris orebody contains an estimated 22.7 million tonnes of ore grading 18% zinc-lead. Nanisivik Mines Ltd. commenced production at its zinc-lead-silver property on Baffin Island in October 1976. Production at capacity is expected to be 6 000 tonnes of lead a year and 60 000 tonnes of zinc a year. Ore reserves are estimated at 6.3 million tonnes grading 15.5% combined lead-zinc.

There were four zinc refineries in operation at the end of 1976 — Canadian Electrolytic Zinc Ltd. in Quebec, Hudson Bay Mining and Smelting in Manitoba, Texasgulf Canada in Ontario and Cominco in British Columbia. Combined metal refinery capacity at year end 1976 totalled 636 000 tonnes a year.

The lead refinery of Cominco at Trail, BC, with a capacity of 154 000 tonnes annually, and that of Brunswick Mining and Smelting Corp. Ltd. at Belledune, NB, with a capacity of 72 000 tonnes, were Canada's only producers of primary lead metal.

Gold   12.3.5

The most significant event for the gold mining industry in 1976 was the start of gold sales by the International Monetary Fund (IMF) to dispose of 777.6 million grams of gold from its official reserves over a four-year period. The IMF program also includes restitution of an equal quantity of gold, in four equal parts, over the four-year period to member nations at the official gold price of 35 Special Drawing Rights (SDR) an ounce of gold in proportion to their quotas in the fund, with first distribution made in January 1977.

The opening gold price on the London gold market and the high for 1976 was US$4.51 a gram. An announcement on January 8 by the IMF that an agreement had been reached on the format of the gold auctions and that an immediate start on sales should be made resulted in a downward trend in the price. At the first auction June 2, 1976, the accepted gold price bid was US$4.05 a gram, near the open market price. Subsequently the price dropped sharply to a 1976 low of US$3.31 a gram on August 31. The price recovered substantially when it was recognized that strong industrial demand had developed and the market could absorb the gold offered by the IMF auctions. The closing gold price for 1976 was US$4.32 a gram. Average gold price for 1976, based on the afternoon fixing price on the London market, was US$4.01 a gram.

Gold production in Canada in 1976 was 52.4 million grams valued at $207.8 million compared with 51.4 million grams in 1975 valued at $270.8 million. Volume of production increased 1.9%, the first increase in Canadian gold output since 1960. At the end of the year, 22 lode gold mines were in operation. One lode mine began operations in 1976, one closed and another opened in the first part of the year but was forced to close later because of financial problems.

Lode gold mines accounted for 71.8% of the total gold produced in Canada in 1976 compared with 73.0% in 1975. Gold recovered as a byproduct from base-metal mining accounted for 27.4% and placer mining 0.8%. Ontario continued to be the leading gold-producing province, accounting for 43.5% of the total, followed by Quebec with 28.1%, Northwest Territories with 11.2% and British Columbia with 10.5%. Canada ranked third in world gold production, well behind South Africa and the Soviet Union.

All gold produced in the Atlantic provinces in 1976 was recovered as a byproduct of base-metal mining. Gold production totalled 230 000 grams compared with 509 661 grams in 1975.

Gold production in Quebec in 1976 amounted to 14.7 million grams compared with 14.2 million grams in 1975. Ore reserves at the Horne mine of Noranda Mines Ltd. were exhausted and the mine closed in July 1976. The Horne mine came into production in 1927 and had been a major producer of byproduct gold. Lower byproduct production from this mine was offset by increased gold production from most other base-metal mines. All lode gold mines recorded increased output.

Gold production in Ontario in 1976 was 22.8 million grams compared with 23.5 million grams in 1975. Gold produced from lode gold mines accounted for 91.9% of the provincial total. In December 1975, Rengold Mines Ltd. began operations at its leased property near Missanabie, the former Renabie mine, but was forced to close in June 1976 because of financial difficulties. Bulora Corp. Ltd. exhausted its ore reserves at its Red Lake district gold mine, the former Madsen Red Lake Mine, and closed in 1976. The Ross mine of Hollinger Mines Ltd. at Holtyre and Hollingers main mine in Timmins were sold to Pamour Porcupine Mines Ltd.

Virtually all gold produced in the Prairie provinces was recovered as a byproduct from base-metal ores. Production in 1976 was 2.0 million grams, a slight increase.

The major portion of gold produced in British Columbia in 1976 was recovered as a byproduct of base-metal mines, mainly from treatment of copper ores. Northair Mines began production in May 1976 at its gold-silver-lead-zinc mine about 113 kilometres north of Vancouver. Concentrator capacity is 278 tonnes a day. This is the first new metal mine to come into production in British Columbia since 1972. Some placer gold was recovered from the Cariboo and Atlin districts. Total gold production in 1976 was 5.5 million grams compared with 4.9 million grams in 1975.

Gold production in the Yukon Territory in 1976 was 965 000 grams compared with 997 986 grams in 1975. Gold was recovered from placer mines and base-metal operations.

Gold produced in the Northwest Territories was recovered from lode gold mines near Yellowknife. Production in 1976 was 5.8 million grams compared with 5.5 million grams in 1975. Cominco completed sinking of a new shaft to a depth of 1 768 metres and was installing shaft facilities. Mill capacity increased from 410 to 590 tonnes a day.

12.3.6 Silver

Canada's mine production of silver in 1976, 1.3 million kilograms, was 37 090 kilograms more than 1975. Canada in 1976 was the world's third largest mine producer of silver, surpassed by the USSR and Mexico. Other major producers were Peru and the US.

Mine production of silver from base-metal ores in New Brunswick, Ontario, British Columbia and the Northwest Territories accounted for most of the increase. Silver output in Quebec, recovered mainly from base-metal ores, increased 8.5% in 1976 from 1975.

Ontario was the leading silver-producing province with output in 1976 accounting for almost 40% of Canadian mine production. The largest producer in Canada was Texasgulf Canada Ltd., which recovered over 323 943 kilograms of silver in copper, lead and zinc concentrates at its Kidd Creek mine near Timmins, Ont.

In the Prairie region much of the silver came from eight base-metal mines operated by Hudson Bay Mining near Flin Flon and Snow Lake, Man. Most of the remainder was derived from the Fox and Ruttan copper-zinc mines operated by Sherritt Gordon Mines Ltd. at Lynn Lake and Ruttan, Man.

Base-metal ores continued to be the main source of British Columbia's mine production of silver. Cominco, the province's major silver producer, recovered silver from the lead-zinc-silver ores of its Sullivan mine in southeastern British Columbia and from purchased ores and concentrates. Byproduct silver output from the Sullivan mine was considerably higher in 1976 than in 1975 because of higher grade and greater tonnages of ores produced.

Silver production in 1976 in the Northwest Territories was substantially higher than in 1975 because of greater output by both Echo Bay Mines Ltd. and Terra Mining and Exploration Ltd. Echo Bay and Terra Mining, which operate silver-copper properties near Port Radium on the east shore of Great Bear Lake, were the principal producers.

A decline of 53% in silver production in 1976 over 1975 in the Yukon Territory resulted mainly from lower byproduct output at the lead-zinc-silver mine of Cyprus Anvil Mining Corp. at Faro because of strikes.

Base-metal ores continued to be the main source of Canadian silver output, accounting for over 96% of total mine production in 1976. Most of the remaining 4% came from silver-cobalt ores mined in the Cobalt district of Northern Ontario and the balance was byproduct recovery from lode and placer gold ores.

Canadian silver production was valued at about $175.1 million in 1976 (Table 12.4). The $3.7 million decrease from 1975 resulted from slightly lower prices. The price in Canada fluctuated in 1976 between a low of $122.7 a kilogram and a high of $159.1 a kilogram. Reported industrial consumption of silver in 1976 was 202.2 tonnes compared with 317.3 tonnes in 1975. Additional quantities of 255.3 tonnes in 1976 and 311.9 tonnes in 1975 were used by the Royal Canadian Mint for silver coins commemorating the 1976 Olympic Games.

In 1976 refined silver was produced at six Canadian primary refineries, the largest being Canadian Copper Refiners Ltd. at Montreal East, Que. It recovered 699 859 kilograms from the treatment of anode and blister copper. The silver refinery of Cominco at Trail, BC, was the second largest producer, recovering 293 306 kilograms of byproduct silver in the processing of lead and zinc ores and concentrates. Other producers of refined silver were INCO at Copper Cliff, Ont., from nickel-copper concentrates, Canadian Smelting and Refining (1974) Ltd. at Cobalt, Ont., mainly from silver-cobalt ores and concentrates produced by the Cobalt area mines, and the mint at Ottawa, from gold bullion. At Belledune, NB, Brunswick Mining and Smelting Corp. recovered byproduct silver bullion from lead concentrates treated in a blast furnace.

Molybdenum 12.3.7

Canadian shipments of molybdenum in 1976 were 14.4 million kilograms valued at $91.9 million. Over 95% of Canadian molybdenum is produced in British Columbia. Quebec is the only other producing province. Canada is the second largest producer in the world, accounting for some 20% of world production.

Prior to 1969, most molybdenum in Canada was produced from primary sources. Since 1969, molybdenum has been produced as a byproduct or a coproduct with copper from large low-grade copper-molybdenum deposits in British Columbia and these deposits have become an important source of supply. In 1976, byproduct and coproduct molybdenum accounted for approximately 45% of Canadian production.

There are two primary producers of molybdenum in Canada — Endako Mines Division of Canex Placer Ltd. and Brynnor Mines Ltd. — both in British Columbia. Endako is the largest producer, accounting for approximately half Canada's production. In 1976, molybdenum was recovered as a byproduct or coproduct of copper at three mines in British Columbia, Brenda Mines Ltd., Lornex Mining Corp. Ltd. and Utah Mines Ltd., and from one mine in Quebec, Gaspé Copper Mines Ltd. Brenda is the second largest producer in Canada, accounting for approximately 25% of molybdenum production.

Two properties were under active consideration in 1976. The first was the extension of the Boss Mountain deposit of Brynnor Mines. A high-grade portion of the deposit is being mined by Brynnor; however, this high-grade area will be depleted within five years. A feasibility study was being done on opening a new mine on the lower-grade part of the orebody during 1977. The second is the possible reopening of a former producer,

British Columbia Molybdenum Ltd., by Climax Molybdenum Corp. of British Columbia Ltd. The mine, which last produced in 1972, was bought by Climax in 1973 from Kennecott Copper Corp. Ltd. In its last full year of production, some 2.3 million kilograms of molybdenum in concentrate was produced.

12.3.8   Platinum group metals

Production of these metals in 1976 was 13.4 million grams valued at $48.8 million compared with 12.4 million grams in 1975 valued at $56.5 million. Canada produces platinum metals as a byproduct of nickel refining. When nickel matte is electrolytically refined, the platinum group metals — platinum, palladium, rhodium, ruthenium, iridium and osmium — concentrate in the residue. The residue or sludge is upgraded and sent to refineries in Britain and the US for recovery of the platinum metals. Canada ranked third in world platinum metals production in 1976 behind South Africa and the USSR.

A slower-than-expected recovery in the industrial world's economy in 1976 resulted in a comparatively stable price pattern for all platinum group metals except rhodium. The price of rhodium increased substantially because of its expected use as a catalyst to control automotive emission of nitrogen oxide pollutants. Producer price for platinum at the end of 1976 was US$5.21 to $5.53 a gram and for palladium US$1.77 to $1.93 a gram.

12.3.9   Cobalt

Canadian shipments of cobalt amounted to 1.37 million kilograms valued at $11.8 million in 1976 compared with 1.34 million kilograms valued at $11.6 million in 1975. Cobalt is recovered principally as a byproduct of nickel-copper ores; there is also a small amount of cobalt recovered from silver-cobalt ores.

Canada's leading producer, INCO, recovers cobalt as an oxide at its nickel refinery at Thompson, Man., and as both oxide and metal at its Port Colborne, Ont., nickel refinery. INCO also produces cobalt oxide and metals at its refinery in Clydach, Wales, from nickel matte produced in Canada. The Clydach refinery also processes some crude oxides produced in Canada into upgraded salts and metal. Falconbridge Nickel recovers cobalt, from nickel matte produced in Canada, at its cobalt refinery in Kristiansand, Norway.

Sherritt Gordon Mines recovers cobalt metal powder from nickel end-solutions at its hydrometallurgical refinery at Fort Saskatchewan, Alta. The refinery treats nickel-copper concentrates from its Lynn Lake mine operation in Manitoba and also treats nickel matte purchased from several nickel operations in Western Australia. Sherritt Gordon closed its Lynn Lake operation in 1976. As a result any future cobalt production from the refinery at Fort Saskatchewan will be from purchased concentrates or from refining concentrates.

12.3.10   Columbium (niobium) and tantalum

Canadian shipments of columbium as columbium pentoxide were 1.66 million kilograms valued at $6.94 million in 1976 compared with 1.66 million kilograms valued at $6.85 million in 1975.

St. Lawrence Columbium and Metals Corp., with a mine, mill and concentrator near Oka, Que., is Canada's only producer of columbium and has one of only two mines in the world that produce columbium in pyrochlore concentrates as a primary product; the other larger operation is in Brazil.

Niobec Inc., a joint venture of Teck Corp., Copperfields Mining Corp. and Quebec Mining Exploration Co., continued development of its St-Honoré pyrochlore deposit, some 13 kilometres north of Chicoutimi, Que. The mill has been designed for an initial capacity of 1 361 tonnes of ore a day with provision for rapid expansion to 1 814 tonnes a day if demand warrants it. Niobec began production early in 1976.

There is only one producer of tantalum concentrates in Canada, Tantalum Mining Corp. of Canada Ltd. (Tanco), from its mine and mill at Bernic Lake, Man. In 1976, Tanco produced 127 813 kilograms of tantalum pentoxide in concentrate, down from 181 009 kilograms in 1975. The lower production was a direct result of a three-month strike at the mine in 1976. Tanco conducted a $250,000 underground drilling and exploration program in 1976. An additional 160 000 tonnes of reserves were discovered, an amount only sufficient to maintain Tanco's estimated mine life at six to seven years. However, some 300 000 tonnes of material were found and if prices continue the strong upward trend of recent years, this material could be worth mining. Tanco is the world's single largest mine source of tantalum.

Tungsten   12.3.11

There is only one producer of tungsten concentrates in Canada, Canada Tungsten Mining Corp. Ltd. from its mine in the Northwest Territories. In 1976, Canada Tungsten produced some 2.2 million kilograms of tungsten trioxide in concentrate, an increase of almost 60% over 1975. The large increase was the result of two factors: a resolution of metallurgical problems which plagued Canada Tungsten in 1975 (tungsten recovery rates in the mill rose from 71.1% in 1975 to 81.6% in 1976); and an improved market, which allowed Canada Tungsten to operate at 97.8% of rated capacity in 1976 as opposed to 92.7% in 1975. Canada Tungsten announced that it will double its mining and milling capacity by 1979. The expansion program will cost some $10 million.

Brunswick Tin Mines Ltd. completed metallurgical testing on ores from its tungsten-molybdenum-bismuth orebody near St. Andrews, NB, in 1976. Test results were satisfactory and Brunswick Tin was looking for a partner to participate in development of the orebody. Amax Exploration, Inc., wholly owned subsidiary of Amax Inc., reported that it had identified a scheelite deposit in the MacMillan Ross area of the Northwest Territories. With possible reserves of 30 million tonnes averaging 0.9% tungsten, this could be the largest known single deposit in the world. While Amax Exploration completed several studies on the feasibility of developing the deposit, there was no indication when development could begin. One discovery was announced during the year by Cordilleran Engineering Ltd. Cordilleran delineated an area of extensive tungsten mineralization on the Yukon Territory–British Columbia border. Cordilleran and several other companies planned more detailed exploration work in 1977.

Cadmium   12.3.12

Cadmium production in 1976 was 1 292.0 tonnes valued at $7.5 million compared to 1 191.7 tonnes valued at $9.0 million in 1975. Most zinc ores in Canada contain recoverable cadmium in quantities varying from 0.001% to 0.067% and zinc concentrates contain up to 0.7% cadmium. The largest mine production comes from Kidd Creek mine of Texasgulf Canada near Timmins, Ont., followed by the Geco mine of Noranda Mines at Manitouwadge, Ont. Other important producers are Cominco Ltd. in British Columbia, Hudson Bay Mining and Smelting in Saskatchewan and Manitoba, the Noranda group of companies in Ontario, Quebec and New Brunswick, Pine Point Mines Ltd. in the Northwest Territories and Anvil Mining Corp. in the Yukon Territory.

Cadmium is recovered as a byproduct from the smelting and refining of zinc ores and concentrates. Metallic cadmium is recovered as a byproduct at the electrolytic zinc plants of Cominco at Trail, BC, Hudson Bay Mining and Smelting at Flin Flon, Man., Canadian Electrolytic Zinc Limited at Valleyfield, Que., and Texasgulf Canada near Timmins, Ont. In 1976 metallic cadmium produced in Canada totalled 1 342.3 tonnes compared to 1 142.5 tonnes in 1975.

Selenium and tellurium   12.3.13

Production of selenium in 1976 increased to 260 000 kilograms valued at $9.1 million from 182 385 kilograms valued at $7.4 million in 1975. Production of tellurium increased to 24 000 kilograms valued at $529,000 from 19 854 kilograms valued at $414,074 in 1975. Selenium and tellurium are recovered from anode muds resulting from the electrolytic refining of copper at the plants of Canadian Copper Refiners at Montreal East, Que., and INCO at Copper Cliff, Ont.

Magnesium   12.3.14

Canadian production of magnesium was 5 858 tonnes valued at $12.2 million. Production was up from 3 826 tonnes in 1975, but well below the 9 650 tonnes reached in 1969. World production of primary magnesium in 1976 is estimated at 244 200 tonnes compared with 258 300 in 1975. The United States produced almost half the world output. Exports of Canadian magnesium metal have entered the US duty-free under a Canada–US defence production sharing program but this program has recently operated on a reduced scale. The US duty on magnesium ingots and further-processed products has been reduced progressively in accordance with negotiations under the General Agreement on Tariffs and Trade. However, only in certain high-purity items can the Canadian product find a market in the US. Exports of Canadian magnesium ingots face a 20% tariff when entering the US domestic market whereas the comparable Canadian tariff is 5%.

The only Canadian producer of primary magnesium, Chromasco Corp. Ltd., has operated a mine and smelter at Haley, Ont., 80.5 kilometres west of Ottawa, since 1942.

12.4 Industrial minerals

12.4.1 Asbestos
Canadian shipments of asbestos fibre were 1.54 million tonnes valued at $445 million in 1976, up from 1.06 million tonnes valued at $267 million in 1975. Production returned to near-normal levels in 1976 following supply disruptions of the previous year. All Canadian production consists of chrysotile and in 1976 approximately 80% of this fibre came from Quebec, 7% from the Yukon Territory, 6% from Newfoundland, 5% from British Columbia and 2% from Ontario.

Canada is the world's largest exporter of asbestos, shipping approximately 95% of its production to more than 70 countries. The United States is the largest market, accounting for about 40% of Canadian exports, followed by Japan, the Federal Republic of Germany, Britain and France. These five countries consumed about 70% of Canadian exports, which totalled about 1.5 million tonnes in 1976.

World demand for asbestos fibre is expected to remain strong for several years, largely based on expanded needs in developing countries. Most companies are allocating large capital expenditures to environmental improvements to comply with stricter regulations both inside and outside of the work place.

United Asbestos Inc. near Matachewan, Ont., operated below capacity in 1976. Run-in problems, accentuated by a need to meet new environmental control regulations, meant inability to realize full capacity operation of 90 000 tonnes a year. Total asbestos production from Ontario, including that of Hedman Mines Ltd., was 27 000 tonnes.

At Cassiar, BC, Cassiar Asbestos Corp. Ltd. completed a waste-removal program and a footwall stabilization program to control the amount of groundwater in the mine. Fibre production was approximately 71 000 tonnes in 1976.

In Quebec, Abitibi Asbestos Mining Co. Ltd., a subsidiary of Brinco Ltd. situated 84 kilometres north of Amos, continued evaluation of its property. Feasibility studies were well advanced with ore reserves estimated at 90 million tonnes averaging 3.5% asbestos fibre. Similarly, Rio Algom Mines Ltd. continued feasibility studies on a deposit owned by McAdam Mining Corp. Ltd. about 32 kilometres east of Chibougamau, Que. In the Yukon Territory 103 000 tonnes of fibre were shipped from Cassiar's Clinton Creek mine. Production was expected to cease in 1978 when ore reserves were exhausted.

Advocate Mines Ltd., Newfoundland's only asbestos producer, produced nearly 90 000 tonnes of fibre in 1976.

12.4.2 Clay and clay products
Shipments of clay and clay products from domestic sources in 1976 were valued at $92 million, up from the 1975 figure of $86 million. Deposits of clay for use in the manufacture of papers, refractories, high quality whitewares and stoneware products are scarce in Canada. Consequently, many of these products, as well as china clay (kaolin), fire clay, ball clay and stoneware clay, are largely imported. In Canada, common clays and shales, being higher in alkalis and lower in alumina than the other clays, are used to manufacture brick and tile products.

Potash 12.4.3

Canadian shipments, all from Saskatchewan, amounted to 5.1 million tonnes of potassium dioxide equivalent in 1976 compared with 4.7 million tonnes shipped in 1975 (Table 12.18). Installed production capacity was 12.9 million tonnes of potassium chloride. In 1975 the industry operated at 64% capacity. During 1976 the Saskatchewan government continued a program of acquiring potash mines through the Crown corporation Potash Corp. of Saskatchewan.

About 95% of the world's potash output is used for fertilizers, the balance being used for industrial purposes.

In New Brunswick, Potash Co. of America suspended exploratory drilling and began development plans on a potash lease granted in 1973. A lease was issued to the International Minerals and Chemical Corp. (Canada) for exploration and development of potash and salt on a 200 square kilometre tract near Salt Springs. Of 10 holes drilled, seven made intersections in potash.

Salt 12.4.4

Canadian shipments of salt amounted to about 6.0 million tonnes valued at almost $76 million in 1976. About 70% of the total was rock salt used principally for snow and ice control on streets and highways and for chemical manufacturing. The remainder is fine vacuum salt and salt as brine used for producing caustic soda and chlorine.

There are three rock salt mines, one in Nova Scotia and two in Ontario. Salt is also produced as a byproduct of potash mining in Saskatchewan. The two companies drilling for potash in New Brunswick were also exploring for salt, and would continue development. Fine salt evaporator plants and brining operations are located in Nova Scotia, Ontario, Manitoba, Saskatchewan and Alberta.

The Quebec government through Seleine Inc., a subsidiary of Quebec Mining Exploration Co., advanced its plans to develop a salt mine on Grosse-Île in the Magdalen Islands. Total capital costs for the mine and an associated port were forecast to exceed $45 million. The viability of the project depends on federal funding of the required port near the mine. Quebec Mining expects initial production at the rate of about 900 000 tonnes a year to begin in 1980.

Sulphur 12.4.5

Canadian sulphur shipments in all forms in 1976 amounted to 4.6 million tonnes valued at $79 million (Table 12.20). Shipments decreased 4% in volume and 22% in value compared to 1975. Reduced volume reflected the world economic downturn which began in late 1974.

Canadian sulphur is obtained from three sources: sour natural gas and petroleum, including the tar sands, which produce elemental sulphur; smelter gases which produce sulphuric acid; and pyrite concentrates used in the manufacture of sulphuric acid. Small amounts of elemental sulphur are recovered as a byproduct of electrolytic refining of nickel sulphide matte and a small quantity of liquid sulphur dioxide is produced from pyrites and smelter gases. In Canada 83% of sulphur shipments in 1976 were in elemental form, nearly all from sour natural gas.

Canadian production of sulphur in all forms peaked in 1973 at 8.1 million tonnes, 7.4 million tonnes in elemental form. In 1976, total output was estimated at 7.1 million tonnes, the 12% decline reflecting reduced output from sour natural gas in Western Canada. Since 1968 Canada has been the world's largest exporter of elemental sulphur.

Gypsum 12.4.6

In 1976 Canadian production of crude gypsum decreased to 5.6 million tonnes from 5.7 million tonnes in 1975, most of it exported to the Eastern US. Exports were mainly from Nova Scotia and Newfoundland quarries operated by Canadian subsidiaries of US gypsum products manufacturers.

Nine companies produced crude gypsum in Canada in 1976 at 14 locations, while five manufactured gypsum products at 18 locations. Production of gypsum in Canada is closely related to the building construction industry, particularly residential building in both Canada and the Eastern US.

12.4.7   Nepheline syenite

Nepheline syenite was produced from two operations on Blue Mountain, 40 kilometres northeast of Peterborough, Ont. In 1976 production was estimated at 541 000 tonnes. Although a 16% increase over 1975 (Table 12.22), this production level represents a return to pre-slump output. The value of shipments in 1976 was $10.8 million, up 23% from 1975. Exports accounted for 77% of total shipments. Sales to the US, representing 97% of Canada's total exports, increased 17%. Nepheline syenite is preferred to feldspar as a source of essential alumina and alkalis in glass manufacture. Other uses include the manufacture of ceramics, enamels, paints, papers, plastics and foam rubber. Canada is the world's largest producer of nepheline syenite.

12.4.8   Structural materials

The value of all construction undertaken in Canada in 1976 was roughly $31.7 billion, an increase of 12.8% over 1975. Production of structural materials, including cement, sand and gravel, stone, clay and clay products and lime, was valued at $1.1 billion in 1976, representing 15% of the total value of mineral production in Canada.

Canadian production of cement in 1976 was 9.8 million tonnes, a reduction reflecting less construction in place and fewer concrete-intensive projects. Cement was produced in all provinces except Prince Edward Island with Ontario and Quebec accounting for 70% of the Canadian total. Cement production capacity in Canada at the end of 1976 was about 15 million tonnes a year, excluding the capacity of five clinker grinding plants, two of them (belonging to Canada Cement Lafarge Ltd.) former fully integrated cement plants. During 1976, capacity changes indicated a net reduction of 77 000 tonnes a year despite the addition of 596 000 tonnes a year by St. Marys Cement Ltd. at its St. Marys, Ont., plant, where a four-stage suspension preheater and new kiln were added during conversion from a wet to a dry process. The rehabilitation and conversion of the Canada Cement Lafarge Ltd. Montreal East plant was slowed because of market conditions, with the result that at year end the clinker-producing capacity was non-existent and the plant could be used for grinding only.

In October 1976 St. Lawrence Cement Co. announced acquisition of Ciment Indépendant Inc. including the cement plant at Joliette, Que., a construction division, four ready-mix plants and two crushed-stone operations. Early in 1977 the company completed the purchase of all assets of Universal Atlas Cement Division's plant at Hudson, NY, for $8.2 million. Universal Atlas Cement is a division of United States Steel Corp.

Production of sand and gravel in 1976 was 248 million tonnes valued at $321 million (Table 12.24). Sand and gravel must be quarried, screened, washed, stockpiled and transported in large volume to compensate for the low unit value received. Transportation and handling often double the plant cost, making it economically desirable to establish plants close to major consuming centres. Urban expansion has greatly accelerated the demand for sand and gravel and many pits and quarries have been over-run by growing communities. Sand and gravel are used as fill, as granular base course and finish course in highway construction and as aggregate in concrete and asphalt.

Production of stone in 1976 was about 88 million tonnes valued at $209.6 million (Table 12.25). Dimension stone, for use as building and ornamental stone, accounts for about 1% of total production. Crushed stone for use as aggregate in concrete and asphalt, as railroad ballast and road metal accounts for about 80% and the remainder is used in the metallurgical, chemical and allied industries.

12.5   Mineral fuels

12.5.1   Coal

For production figures see Tables 12.4 and 12.8. For an outline of the industry see Chapter 13, Energy.

Oil and natural gas

Canadian production of crude oil and natural gas liquids in 1976 declined for the third consecutive year to 86 936 million cubic metres. Crude oil output, including synthetic crude oil from the Athabasca tar sands amounted to 77.8 million cubic metres or 213 000 cubic metres a day, about 6% less than in 1975 (Table 12.27). Gas plant production of natural gas liquids totalled 18 million cubic metres or 49 000 cubic metres a day. Natural gas production dropped slightly in 1976 to 86 858 million cubic metres or 244 million cubic metres a day (Table 12.28).

At the end of 1976 Canada's proven liquid hydrocarbon reserves, which include conventional crude oil and natural gas liquids (propane, butane and pentanes plus), amounted to 1.24 billion cubic metres. This is comprised of 1.0 billion cubic metres of crude oil and 0.246 billion cubic metres of natural gas liquids. These estimates do not include oil in the Athabasca sands. At the 1976 annual production level of 92.6 million cubic metres, the life index (ratio of reserves to production) for conventional crude oil and natural gas liquids increased to 13.8 years. This increase was due to reduced production rates, not an increase in proven reserves. The reserve position of most provinces declined. Alberta was the most notable as total reserves, including natural gas liquids, dropped by 66 million cubic metres. According to the Canadian Petroleum Association (CPA), proven remaining marketable reserves of natural gas increased by about 36 830 million cubic metres to a total of 1.6 trillion cubic metres in 1976. Using the 1976 level of production, the life index for natural gas increased to 21.85 years. In compiling its reserve estimates, the CPA assumed that Mackenzie Delta gas would be brought to market via the same pipeline system as Alaskan gas and so could be categorized as proven. This is not the case for gas from Arctic islands; a minimum reserve base is required before it can be considered within economic reach. Therefore gas reserves that have been found there and in offshore areas are classified as probable rather than proven. Alberta with 1.3 trillion cubic metres of marketable gas reserves accounted for 78% of Canadian reserves at the end of 1976.

Canadian refinery capacity increased substantially in 1976, mainly with the expansion of the Irving Oil Ltd., Saint John, NB, refinery to 39 747 cubic metres a day. By 1978 Canadian refinery capacity will have increased to almost 397 500 cubic metres a day from the 331 000 cubic metres at the start of 1976.

Alberta. During 1976 crude oil production in Alberta declined by 18 000 cubic metres a day to 174 000 cubic metres a day and accounted for 84% of total Canadian crude oil production. Of this amount, synthetic crude oil production from the Athabasca tar sands averaged 7 519 cubic metres a day in 1976.

Both exploratory and development drilling footages increased slightly in Alberta in 1976, partly due to provincial incentive programs but primarily because of substantial increases in field prices for both oil and gas. Drilling statistics show that, in 1976, the total number of wells drilled in all categories amounted to 5,042, for a total length of 4.87 million metres, 1.22 million metres more than in 1975. Despite increased efforts, no large oil discoveries were made.

The shallow gas-bearing formations of both northern and southern regions continued to be the principal target for explorers in Alberta and several discoveries were recorded. The foothills were again a prime exploratory target. Several follow-up wells were drilled on the discoveries made in 1975. Early indications are that recoverable reserves of natural gas from the area exceed 28 328 million cubic metres. Successful gas completions increased by 1,271 wells to 3,193 in 1976.

According to an appraisal of Alberta's oil sands completed in 1973 by the province's resources conservation board, there is an ultimate in-place reserve of crude bitumen of 159 billion cubic metres of which 39.7 billion cubic metres are recoverable by known methods of technology. The bulk of recoverable reserves are located in the Athabasca deposit with the remainder distributed between the Cold Lake, Peace River, Wabasca and Buffalo Head Hills deposits. Of the 39.7 billion cubic metres of recoverable synthetic crude oil, only 4.2 billion cubic metres are amenable to open-cast mining methods and all of this is located in the Athabasca deposit. The remaining 35.5 billion cubic metres are expected to be eventually recovered by recovery techniques still in the experimental stage of development. At present Great Canadian Oil Sands Ltd., at 8 000 cubic metres a day, is the only oil sands producer. It has been in operation since 1967. Construction of the Syncrude Canada Ltd. project was about 50% complete by the end of 1976. Predicted operation at production of 20 000 cubic metres a day was scheduled to start in 1978. Proposals put forward by two major oil companies were withdrawn because of large capital outlays required for construction and the risk of not receiving a satisfactory return on investment.

Saskatchewan. Saskatchewan's crude oil production declined by 5% to 8.9 million cubic metres in 1976, accounting for 12% of the Canadian total. Marketable natural gas production, at 1 902 million cubic metres, accounted for 1.9% of total Canadian production. Total drilling in Saskatchewan amounted to 229 670 metres compared with 197 878 metres in 1975. No significant oil or gas discoveries were made.

British Columbia. In 1976 production of crude oil in British Columbia increased by 4% to 6 519 cubic metres a day and represented 3% of total national production. Net withdrawals of natural gas in the province totalled 11 601 million cubic metres, down 4% from 1975.

In 1976 exploratory drilling at 147 966 metres was up 63 349 metres from 1975 and development drilling increased by 91 254 metres to 135 124 metres. No significant oil discoveries were made but several significant gas discoveries were recorded.

Manitoba. Production of crude oil in Manitoba amounted to 626 000 cubic metres in 1976, down slightly from 1975. All fields were producing at maximum capability. There was no natural gas production and no oil discoveries were made. Drilling in the province increased by 50% to 13 515 metres.

Yukon Territory, Northwest Territories and Arctic islands. Crude oil production north of the 60th parallel is confined to the Norman Wells field in the Northwest Territories. Oil from this field is processed in a small local refinery which serves local markets. Natural gas production declined to 54 million cubic metres in 1976, because of a substantial reduction in production from the Pointed Mountain field, principal producing gas field in the Northwest Territories.

There were 27 wells drilled in Northern Canada in 1976 for a total of 83 807 metres compared with 44 wells and 113 218 metres in 1975. Some significant discoveries were made despite the decline in drilling. In the Mackenzie Delta, Gulf Oil Canada Ltd. drilled several successful gas discoveries in the Parson's Lake area and considerably extended the as-yet-undefined limits of the field. One well tested gas of 0.58 million cubic metres a day. In the Beaufort Sea, Dome Petroleum Ltd. used three specially designed drill ships. The Dome discovery well, drilled 56 kilometres offshore, encountered natural gas at 2 987 metres, and would be further explored in 1977. In the islands, Panarctic Oil Ltd. made two significant gas discoveries in April 1976. These wells substantially enlarged previous estimates of the size of the Hecla field on the east coast of Melville Island, now over 40 kilometres long containing an estimated 99 108 million cubic metres of natural gas.

Eastern Canada. Aggregate drilling in Ontario increased in 1976 by 5% to 73 272 metres. Exploratory drilling accounted for 48% of the total, down 2% from the previous year. No noteworthy discoveries were made. In Lake Erie, one gas discovery was recorded by Consumers Gas Co.

Offshore from the East Coast 10 wells were drilled for a total of 22 793 metres in 1976, compared to nine wells and 26 314 metres in 1975. All wells were in the exploratory category. Drilling commenced in this region in 1966. A total of 136 wells have since been drilled, from which eight significant discoveries of oil and gas have been made. The two most important of these were drilled on the Labrador Shelf, the Bjarni H-81 in 1973 and the Gudrid in 1974. Two more potential gas discoveries were made on the Labrador Shelf in 1975, the Eastcan Snorri J-90 was drilled to depth in 1976 and is considered to be commercial at the present time, however, the other well (Karlsefni H-13) is not considered to be commercial. In 1976, Petro-Canada revitalized exploration on the Grand Banks and Scotian Shelf with exploration agreements with three separate major oil firms. Petro-Canada's programs will cost in the order of $20 million. Shell Oil Ltd. engaged in a $24 to $48 million exploration program on Sable Island, along with Mobil Oil Canada Ltd., Texas Eastern Exploration of Canada and Texaco Exploration Canada Ltd. While the prospects of the Scotian Shelf do not appear to be significant on a world basis, they may be sufficient to help the Atlantic provinces reduce dependence on imported oil.

Uranium 12.5.3

Canadian uranium shipments in 1976 totalled 6 058 tonnes of uranium oxide, a 10% increase over 1975 (Table 12.16). Of these shipments 64% came from three operations in Ontario — those of Denison Mines Ltd. and Rio Algom Mines Ltd., both near Elliot Lake, and Madawaska Mines Ltd., which began production near Bancroft late in the year. The remainder came from two operations in Northern Saskatchewan, those of Eldorado Nuclear Ltd. near Uranium City and Gulf Minerals Canada Ltd. at Rabbit Lake, which began production late in 1975. Canadian production was higher due mainly to an increased number of producers. A lack of experienced miners continued to plague the industry despite substantial training and house-building programs established by most producers. Eldorado, Denison and Rio Algom continued with major programs to expand mining and milling operations. In addition, early in 1976 Agnew Lake Mines decided to proceed with full-scale production of its underground mining leaching program near Espanola, Ont., and Amok Ltd. made known its plans for the development of its Cluff Lake deposits in the Carswell Dome area of Saskatchewan.

In response to a rise in uranium prices, increased uranium exploration was evident in virtually all provinces and territories. A significant discovery was made in mid-1975 at Key Lake, Sask., as the result of a drilling program carried out by Uranerz Exploration and Mining Ltd. jointly with Inexco Mining Co. and the Saskatchewan Mining Development Corp. A joint federal-provincial uranium reconnaissance program continued in 1976. The main objective of the program is to provide high-quality data to indicate areas where there is the greatest probability of discovering new deposits. During 1976 the program covered 555 000 square kilometres using airborne gamma spectrometry and regional geochemical surveys. In total this program is expected to take 10 years and cost some $30 million.

Late in December 1976 the federal government announced further strengthening of the safeguard requirements which apply to the export of Canadian nuclear reactors and uranium. The Saskatchewan government modified its initial uranium royalty proposal made in November 1975, after hearing from the companies involved. A uranium resource group, established in the federal energy department to assess Canada's uranium resources, published its second annual report in June 1976. It projected that Canadian uranium production would increase to 10 000 tonnes in 1980 and 11 540 tonnes in 1984.

During 1976 the Atomic Energy Control Board (AECB) announced that it had approved only one uranium export contract — for delivery of some 230 tonnes by Agnew Lake Mines Ltd. to a South Korean utility. Agnew Lake also had sales to United States and Swedish utilities pending AECB approval at year end. Approved contract and export commitments totalled some 84 600 tonnes as of June 1976.

Manufactured metals 12.6

Aluminum 12.6.1

Canadian primary aluminum output decreased to 633 428 tonnes in 1976 from 887 023 tonnes produced in 1975. Labour strikes at the Quebec smelters of the Aluminum Company of Canada Ltd. (ALCAN) caused the decline. Canada produced 6.3% of the non-communist world's primary aluminum and ranks as the fifth largest producer. World production of primary aluminum increased marginally by less than 1.0%. Canadian exports of aluminum in 1976, mainly ingot form (including some fabricated materials), were 535 707 tonnes, a slight increase from 533 739 tonnes exported in 1975.

The value of the 1976 exports was $496.2 million compared with $464.4 million in 1975, an increase of 6.9%, resulting from higher prices.

Two companies operate primary aluminum smelters in Canada, ALCAN and Canadian Reynolds Metals Co. Ltd. During 1976, ALCAN commissioned a second high-speed cold-rolling mill at its Kingston, Ont., works, increasing rolling capacity to 135 000 tonnes a year. The company also expanded its continuous casting and rolling capabilities at Jonquière, Que. Canadian Reynolds operates a smelter at Baie-Comeau, Que. Smelter production was 138 400 tonnes, an increase of 15% from 119 500 tonnes in 1975. Canadian Reynolds also began continuous strip casting for production of aluminum siding at its expanded Cap-de-la-Madeleine, Que., plant.

Canada's aluminum industry is totally dependent on the import of bauxite and alumina. No economic deposits of bauxite are found in Canada. Bauxite is imported for the production of alumina. This aluminum oxide is an intermediate product which is reduced in an electric furnace to aluminum metal. Approximately 4.5 tonnes of bauxite are refined to 2.0 tonnes of alumina, which in turn are smelted to obtain one tonne of aluminum. ALCAN's refinery at Arvida, Que., the only alumina refinery in Canada, supplies the company's four smelters with alumina. The process consumes from 7 to 8 kWh per 454 grams of aluminum produced. This high consumption of electric power explains the proximity of Canada's aluminum smelters to large hydroelectric power sources.

Canada's major import sources for bauxite ore are Suriname, Guinea, United States, Guyana, Australia and the People's Republic of China. Suppliers of alumina are Australia, United States, Federal Republic of Germany, Jamaica, France, and Netherlands Antilles. Export markets for Canadian primary aluminum are the United States, People's Republic of China, United Kingdom, Hong Kong, Turkey and Brazil.

The countries on which Canada relies for bauxite, with the exception of the US and the People's Republic of China, are part of a minerals producers group known as the International Bauxite Association (IBA). The IBA was formally instituted at a meeting in March 1974 by Australia, Guinea, Guyana, Jamaica, Sierra Leone, Suriname and Yugoslavia. In November 1974, the Dominican Republic, Ghana and Haiti became members and Indonesia joined in 1975. The IBA seeks an improved situation in trade stability or prices by virtue of co-ordinated group action. Canada, a pre-eminent producer of aluminum, is vitally interested in the activities and policies of the IBA.

12.6.2 Iron and steel

Crude steel production increased marginally to 13.1 million tonnes in 1976 from 13.0 million tonnes in 1975. Steel shipments from plants increased by 3.6% to about 9.8 million tonnes, principally due to export sales since disposition of rolled steel products to domestic markets declined by 2% to 8.5 million tonnes. Consumption fell fractionally to 9.8 million tonnes.

Although overall demand for steel products was poor the consumer goods sector was an exception, staying buoyant throughout the year, particularly due to strong sales of North American automobiles. So demand for flat-rolled products showed a strong gain over 1975. But capital goods and construction markets remained depressed. Demand for structurals, rods and bars was poor. This indicated a lack of investment and reflected the uncertainty of many manufacturers concerning prospects over the next few years for performance of the Canadian economy.

Variation in demand levels for different steel products resulted in variable performances by Canada's three big steel companies in 1976. Dominion Foundries and Steel Ltd. (DOFASCO), which produces mainly flat-rolled steel products, experienced a profitable year in response to demand by the automobile industry and, to a lesser extent, the appliances industry. By contrast, Algoma Steel Corp. Ltd., which produces significant tonnages of structural steel and related products, struggled under adverse market conditions pervasive in the capital goods sector. The Steel Company of Canada Ltd. (STELCO), which produces the largest range of steel products in Canada, had a reasonable year, with high demand for flat-rolled products more than compensating for slack demand for long products. Activity of many regional steel producers remained at depressed levels.

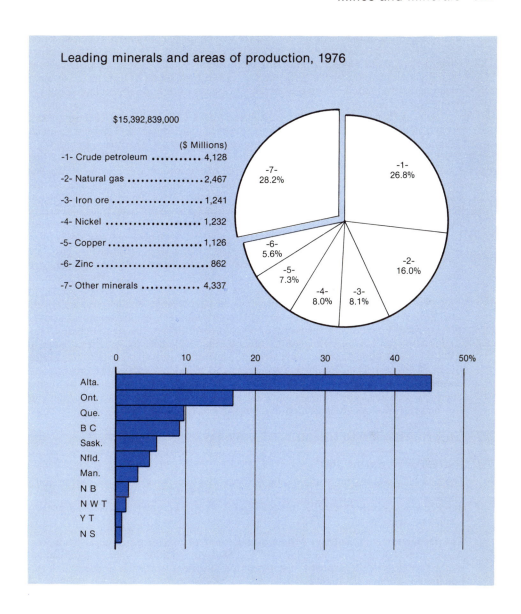

Leading minerals and areas of production, 1976

$15,392,839,000

($ Millions)
-1- Crude petroleum 4,128
-2- Natural gas 2,467
-3- Iron ore 1,241
-4- Nickel 1,232
-5- Copper 1,126
-6- Zinc 862
-7- Other minerals 4,337

Basic oxygen production amounted to 7.9 million tonnes, up from 7.3 million tonnes in 1975. This type of steelmaking accounted for roughly 60% of total steel production. Electric steel and steel castings production at 2.1 million tonnes was down nearly 400 000 tonnes. Basic open-hearth production was down by 100 000 tonnes to 3.0 million tonnes.

Average steel furnace capacity utilization declined marginally to 75.8% during the year. This compares with 76.6% in 1975 and 81.9% in 1974. Some of the 1976 experience was due to some new capacity not being available for the entire year and to capacity affected by strikes. However, it also reflected measures to reduce operating levels because of sluggishness in many steel product markets through much of the year.

Many companies continued with expansion programs during the year. STELCO, the largest Canadian steel producer, continued building a new steel complex at Nanticoke, Ont., on Lake Erie in 1976. Initial capacity will be 1.2 million tonnes with start-up scheduled for April 1980. The output of slabs will be trucked to Hamilton for rolling into finished steel products. DOFASCO's new basic oxygen plant continued under construction at its Hamilton site with completion set for 1978. Capacity will be approximately 910 000 tonnes. Algoma was in the final stages of a major investment program that has raised effective steelmaking capacity from under 2.7 million tonnes to 3.9 million tonnes annually.

Among the smaller producers Sidbec-Dosco Ltd. continued a major plant expansion at Contrecoeur, Que., designed to more than double steel capacity to about 1.5 million tonnes by the end of 1977. Interprovincial Steel and Pipe Corp. Ltd. (IPSCO) of Regina, the largest steel and pipe producer in Canada with capacity of about 500 000 tonnes a year, carried out investigations on expansion of capacity in 1976. The motivating factor was the potential market for large diameter pipe to be used in any northern pipeline to transport natural gas. The Atlas Steels Division of Rio Algom Mines Ltd., Canada's largest producer of stainless and specialty steels, continued a major program of conversion and expansion of steel producing operations. At Welland, Ont., a new melt shop was due for completion in 1977. Sydney Steel Corp., in Sydney, NS, continued to encounter difficulties with production and markets in 1976. The provincial Crown corporation investigated a rehabilitation program to improve production capability.

Exports of steel by Canadian producers increased 39% to 1.8 million tonnes in 1976. Value of exports increased approximately 12% to $655 million. Approximately two-thirds of total export trade went to the United States. Shipments were up sharply to EEC countries as exports jumped from 33 000 tonnes in 1975 to 237 700 tonnes in 1976.

Imports fell 18% to 1.3 million tonnes although value declined only fractionally to $598 million. Although imports were down, Japan increased shipments to Canada by over 30% to some 434 000 tonnes. Toward the end of 1976 imports were increasing and orders for forward delivery in early 1977 were said to be mounting. Concern was being shown by Canadian producers as many of these offshore quotations appeared to be at low price levels, possibly indicating dumping.

12.7   Government aid to the mineral industry

12.7.1   Federal government aid

The federal government helps mining by providing detailed geological, geophysical, topographical, geodetic, geographical and marine data; technical information concerning the processing of ores, industrial minerals and fuels on a commercial scale; and certain tax incentives.

The Department of Energy, Mines and Resources. This federal department was created on October 1, 1966 (RSC 1970, c.E-6). In addition to its administrative establishment, the department has three main sectors — science and technology, mineral development and energy policy.

The science sector has divisions for mineral and energy technology, geological surveys, surveys and mapping, earth physics, a polar continental shelf project, a centre for remote sensing, and an explosives branch.

The centre for mineral and energy technology, a large laboratory and pilot-plant complex, conducts research into methods of extracting and processing minerals and fuels. Emphasis is placed on recovery techniques for ores and minerals with low-grade impurities or complex mineral composition. Fuels research includes evaluation of fossil fuels and development of refining methods for the low-grade, high-sulphur petroleum of the Athabasca oil sands. One project is aimed at lowering waste rock production and costs by improving wall design of open-pit mines. Research is also being conducted to improve burning qualities of coal. In the related area of the extraction of metal by heat, research is concentrated on development of a shaft electric furnace for smelting iron ore. In mineral sciences, the centre carries out physical, chemical, crystallographic and magnetic studies to determine characteristics important to extraction and processing methods. The centre also produces standard reference ores and metals needed by mining and metallurgical companies. In metals research, in addition to improving techniques for metal forming, attention is focused on ensuring structural soundness of metal pipelines for the Arctic. Other programs are directed toward reduction of pollution and conversion of mineral waste into useful materials such as fillers and ceramics.

The centre is assisted by a national advisory committee on mining and metallurgical research, comprising representatives from industry, government and universities.

The geological survey maps and studies the geology of Canada. Its activities support two departmental programs — mineral and energy resources and earth sciences. A principal aim of the former is to ascertain available mineral and energy resource potential and the survey is active in estimating the amount and distribution of mineral and fuel resources. This is done by providing a systematic geological framework, by defining settings favourable to mineral and fuel occurrences and by appraising foreign resources. The earth sciences program is concerned with use and conservation of resources and preservation of the environment. The survey provides information on land resources and terrain performance derived from geological, geomorphic, geophysical, geotechnical and related studies of earth and rock materials, land forms and associated dynamic processes.

Each year the survey sends about 100 parties into various parts of Canada. The results of its studies are published in memoirs, bulletins, papers, maps and scientific technical journals. Its headquarters is in Ottawa and of several regional offices the largest are an institute of petroleum geology in Calgary and an Atlantic geoscience centre in Dartmouth, NS. The former studies geology of Canada's western and northern sedimentary basins and the latter investigates bottom morphology and structure of the continental shelves and the floors of the open ocean. A smaller group of geologists on the West Coast is developing similar marine geology studies.

The earth physics branch carries out geophysical work of interest to the mineral industry. It collects and publishes maps and charts on the geomagnetic field in Canada. Most of this information is obtained from airborne geomagnetic surveys which have ranged over all of Canada and as far as Scandinavia. The branch maintains a network of 11 permanent magnetic observatories, including an automatic magnetic observatory system at Yellowknife, NWT, which opened in 1974. It also operates a network of 33 seismic stations to study the earth's interior and assess seismic risk. In gravity research, another means of studying the composition of the earth's crust, the branch maps variations in gravity on a regional basis including the Arctic and the continental shelves. The results are available in a new gravity map of Canada on a scale of 1:5,000,000 or 50.7 kilometres to the centimetre for easy comparison with new geological and tectonic maps of Canada on a similar scale. Geothermal studies in mines and deep boreholes provide information to the mineral industry on underground thermal conditions, including permafrost.

No mineral development is possible without accurate, large-scale topographical maps. The mapping branch has completed topographical mapping of the country at the medium scale of 1:250,000, or about 2.5 kilometres to the centimetre. About 40% of the larger-scale mapping at 1:50,000 has been completed of more settled areas and those of greater economic importance. Also available for selected areas are maps at other scales. Another branch function is establishment of a basic network of survey control points across Canada that provide precise figures of latitude, longitude and elevation above sea level. In addition to topographic maps, the branch produces various multicoloured maps for other government agencies, aeronautical charts and the National Atlas of Canada, describing physical, economic and social geography. The branch's air photo library has on file over 4 million aerial photographs of Canada, both black and white and colour, taken over the last half-century from aircraft and more recently from space satellites.

The explosives branch administers the Canada Explosives Act, which controls the manufacture, authorization, storage, sale, importation and transportation by road of explosives.

The mineral development sector is responsible for research programs and policies in the field of non-renewable resources. It conducts fundamental and applied resource-engineering-economic research and field investigation into non-renewable resource problems on a total industry basis, in a regional, national and international context. The work covers all aspects of the mineral industry from resource to consumption. The sector publishes reports and advises government departments and agencies on policy. Current activities include regional studies of the mineral economy in Canada; assessment of mineral projects for which federal support has been requested; resource and reserve studies in a number of mineral commodities; and the safeguarding of Canadian mineral interests through participation in the work of international agencies. The sector administered the Emergency Gold Mining Assistance Act to aid communities largely dependent upon the gold mines until the act expired in June 1976. In collaboration with the Canadian International Development Agency and with the support of industry, the sector sets up training courses for mineral scientists, technologists and economists brought to Canada under various aid programs, and advises on mineral projects undertaken by Canada as an aid to developing countries. The sector publishes extensively and maintains a listing of about 16,000 mineral showings and deposits in Canada that may be consulted by the public.

The energy policy sector is primarily a policy-making group with a direct impact on the mining industry. It assesses individual projects and developments relative to each energy source and in terms of their inter-relationships. It appraises trends in oil and gas exploration and production, transportation, processing and marketing in Canada and abroad, and provides information to federal government agencies, industry and the public on oil and gas developments. In the field of uranium, the sector continues to co-ordinate matters in the areas of stockpiling, establishment of enrichment facilities and export. With respect to coal, it provides research and development grants and advises on production expansion rates in the light of profitability and projected demand. The sector also administers federal interests in offshore mineral resources as well as in federally owned mineral rights in the provinces.

Tax incentives to the mining industry. Although mineral industry enterprises are subject to federal income tax, certain benefits granted to them under the Income Tax Act serve as incentives to exploration and development. An outline of tax amendments and regulations affecting mining in general appears in previous editions of the *Canada Year Book*. Some general information is included in the sub-section on corporation income tax in Chapter 20 of this edition. The most up-to-date information on income tax allowances which apply to the mining industry may be obtained from Revenue Canada, Taxation and appropriate provincial tax offices.

12.7.2   Provincial government aid

Newfoundland. The mines and energy department provides the following services in support of exploration and mining activities: programs of geological and geochemical surveying and of mineral assessment to encourage development of mineral resources; inspection of exploration work and mining operations; control of removal of beach sand and gravel as a conservation measure; identification of mineral rock specimens; technical advice; co-operation with the geological survey and other federal government agencies; and publication of data for educational and informational purposes. Geological, geochemical and geophysical reports and maps, and compilations of general data pertaining to specific areas are available at nominal cost. Other information from unclassified files is available to interested parties. Prospectors' or miners' permits are issued and mining claims are recorded.

Nova Scotia. The mines department carries out safety inspection of mines, quarries and allied processing plants, development sites and storage facilities for explosives. Diamond drills are available to exploration and producing companies on a contract basis and industry is assisted in surface and subsurface development and construction projects. The department also carries out mine rescue and first aid training and administers all matters relating to mineral rights.

A resources and geological services division carries out geological, geochemical and geophysical surveys and studies of occurrences of specific minerals. It publishes the results in annual and specific reports, including maps. An analytical and ore dressing section, affiliated with Nova Scotia Technical College, provides an important service to both the department and industry.

New Brunswick. The natural resources department has two branches providing services to the mining industry, a mines branch and mineral resources branch.

The mines branch administers safety regulations. Mines and associated plants are inspected regularly and mine rescue training is conducted. Mineral statistics reports and reviews of mining operations are prepared.

The mineral resources branch provides the mineral (including petroleum) and construction industries with basic geological, geochemical and geophysical data to assist in the discovery, development and utilization of the province's mineral resources. The administration of Crown-owned mineral, petroleum-gas, bituminous shale and granular aggregate resources is the responsibility of this branch. This includes issuing prospecting licences, recording mining claims and issuing mining licences and leases. Regional offices and core libraries are maintained at Sussex, Fredericton and Bathurst. Reports and maps pertaining to exploration work filed for assessment credit are kept in these offices and are available to the public.

Quebec. A mines branch of the natural resources department implements the Mining Act and the Mining Duties Act. The branch incorporates three directorates: geology, mining domain and mineral economics and development. The geology directorate offers six services: geological exploration, mineral deposits, geotechnics, technical documentation, cartography and technical revision. The directorate undertakes the geological study of the province and produces detailed area reports and geoscientific maps. The geotechnics service deals with developmental problems of particular environments and their geological setting.

The mining domain directorate includes a mining titles service, mines inspection service and engineering service. The directorate controls the mining rights granted on Crown lands, registers mining claims and issues development permits or special permits governing sale or rental of lands for mining purposes. It ensures that holders of mining rights carry out development work prescribed. Mine inspectors verify that working conditions in mines, quarries and mills conform to safety standards. The directorate is also responsible for carrying out whatever engineering work is required to open up new mining areas or operations, including building of access roads and mining townsites.

The economics and development directorate is concerned with optimum use of Quebec's mineral resources in line with both development and conservation. Three services are now being organized: development, economic evaluation and promotion. The directorate's aim is to identify and promote projects that lend themselves to quick and tangible results in development of mineral resources. It arranges studies on marketing, financing, transportation, development and exploitation techniques, profitability and other criteria affecting mineral resource development. Included is a mineral research centre for metallurgical research and development.

Ontario. A mines division of the natural resources ministry promotes and regulates use of available supplies of minerals by resource products industries. The division has four branches.

A mineral resources branch, essentially a planning group, ensures the orderly development and use of non-renewable resources.

The geological branch encourages exploration through geological, geophysical and geochemical surveys of the province, publication of maps and reports on mineral occurrences and education of prospectors and others involved in mineral exploration. It is also responsible for a mineral exploration assistance program designed to stimulate exploration in the Red Lake, Geraldton–Beardmore, Kirkland Lake and Co-balt–Gowganda areas. Atikokan and Eastern Ontario areas were included in 1974 and 1975. One-third of the cost of exploration to a maximum of $33,333 is provided to small- and medium-sized companies to explore in these areas.

The research branch is composed of the assay laboratory in Toronto and the Temiskaming laboratory in Cobalt. The assay laboratory provides assay and analytical services and conducts investigations to aid in discovery and development of mineral deposits. It serves the mining industry and the public at large. The Temiskaming testing laboratory operates a bulk sampling and assay laboratory to assist producers in marketing their silver-cobalt ores.

The mines engineering branch administers a part of the Mining Act requiring regular examination of all operating mines, quarries, sand and gravel pits and certain metallurgical works to ensure the health and safety of employees and the public. Regional geologists and mines engineers provide advice or support. The lands administration and surveys and mapping branches handle recording of mining claims, assessment work and preparation of title to mining lands.

Manitoba. The resources division of the Manitoba mines, resources and environmental management department offers the following services: recording staking and acquisition of Crown mineral rights; compiling assessment information and inspecting mineral rights dispositions; compiling geological data on mineral occurrences; issuing reports and maps covering geological, geophysical and geochemical surveys; operating an analytical and assay laboratory to help evaluation of mineral occurrences and classification of rocks and minerals; giving engineering approval of mining works and inspecting mining operations with regard to the health and safety of employees; controlling in-plant environmental and safety regulations; training mine rescue crews and inspecting mine rescue facilities; inspecting oil well drilling sites. A mineral disposition regulation provides an option for provincial participation in mineral exploration and development. The exploration operations branch explores for, investigates and assesses mineral occurrences in the province.

Saskatchewan. Changes in the organization of the mineral resources department in 1976 resulted in the engineering and mineral lands branches being combined to form a mines branch. The branch is responsible for: the disposition, development and conservation of all Crown minerals, except fuels; reclamation of land distributed by mining operations; environmental aspects of mining operations; and engineering studies of long-term effects of mining.

The geological survey branch carries out fundamental geoscientific activities pertinent to the evaluation of the province's mineral potential.

A petroleum and natural gas branch is responsible for resource conservation, engineering and environmental management functions with regard to oil, natural gas and pipelines.

The potash management branch is responsible for development, production, disposition and conservation of potash resources in Saskatchewan and taxation related to these resources.

The government announced in November 1975 a policy of acquiring at least 50% ownership in the province's potash industry. The Potash Corporation of Saskatchewan Act, 1976, established a corporation that may own and operate potash mines and manufacture and deal in all kinds of fertilizers. It also may enter into joint or other ventures relating to the industry.

Officers of the mines safety branch of the labour department make regular examinations of all mines to ensure proper conditions for health and safety. Safety education is also part of the work of this branch.

Alberta. A new energy resources department was formed in 1975 out of the predecessor departments of lands and forests and mines and minerals. The new department is responsible for energy resources and renewable resources. Energy resources include a Syncrude equity division, economic planning and systems division, finance division and energy resources division. The Syncrude division manages Alberta's 10% equity participation in the Syncrude project, negotiation of agreements on behalf of the province and the provision of management liaison.

The economic planning division provides analysis and interpretation of economic data for all forms of energy for consideration in formulation of revenue management policies.

The finance division manages collection and verification of petroleum, natural gas, coal and other royalties. It also collects lease rentals, mineral taxes, revenue from sales of Crown reserves, incidental revenues and provides accounting and budget services.

Energy resources is responsible for advice in the management of mineral resources, assists in policy formulation and is responsible for administration and management of

Crown minerals. Included are a minerals subdivision, earth sciences branch, interdepartmental affairs office and senior technical advisers in oil, gas and coal.

British Columbia. The British Columbia mines and petroleum resources department assists the mining industry through two established branches, mineral resources and petroleum resources.

Inspectors of the mineral branch are stationed throughout the province. They inspect coal mines, metal mines and quarries, examine prospects, mining properties, roads and trails, and carry out special investigations under the Mineral Act. Environmental control inspectors conduct surveys on dust, ventilation and noise and recommend improvements in environmental conditions. Other inspectors administer roads and trails and prospectors grub-stake programs, and reclamation sections of provincial mining statutes.

A geological division carries out a variety of geological studies and publishes data. It assesses the mineral potential of land; collects, stores and disseminates geological statistical data; and records the exploration and mining activities of the industry. An inventory of mineral deposits is under way to establish a quantitative appraisal. The division offers a number of free assays for prospectors, identifies rocks and minerals, and conducts lectures in prospecting. The mining titles division administers laws concerning acquisition of rights to minerals and coal. It provides information, including maps, on mineral claims and placer leases and their ownership and also data on the ownership, location and status of coal licences and leases.

A petroleum resources branch administers the Petroleum and Natural Gas Act and related regulations. Every well location must be approved by the branch before drilling begins. All drilling and production operations are inspected frequently for compliance with regulations governing facilities and practices, plugging of abandoned wells, surface restoration of well sites, procedures for well-testing and measurement, disposal of produced water, fire protection and general conservation. Complaints of property damage are investigated. Comprehensive records of all drilling and producing operations are published or made available for study. Samples of bit cuttings as well as all core from every well drilled are retained for study, and detailed reservoir engineering and geological studies are carried out. Estimates of reserves of oil and natural gas are made annually. Crown-owned oil and natural gas rights are evaluated prior to disposition by public tender.

There was a change of government in the province in December 1975 and many laws affecting the mining industry were under review in 1976. Mineral tax and royalty acts were amended and new legislation introduced. Amendments were also made to the Mineral Act.

Mining legislation   12.8

Federal and departmental jurisdictions   12.8.1

Mineral rights vested in the Crown in right of Canada include those in the Yukon Territory and Northwest Territories and offshore within the limits of Canada's continental margins, as well as those underlying certain federal lands in the provinces.

The Supreme Court of Canada in its opinion of November 1967 stated that, as between Canada and the province of British Columbia, Canada has proprietary rights in and legislative jurisdiction over "lands, including the mineral and other natural resources, of the seabed and subsoil seaward from the ordinary low-water mark on the coast of the mainland and the several islands of British Columbia, outside the harbours, bays, estuaries and other similar inland waters, to the outer limit of the territorial sea of Canada, as defined in the Territorial Sea and Fishing Zones Act. . . ." The court also said the federal government has legislative jurisdiction "in respect of the mineral and other natural resources of the seabed and subsoil beyond that part of the territorial sea of Canada. . . to a depth of 200 metres or, beyond that limit, to where the depth of the superjacent waters admits of the exploitation of the mineral and other natural resources of the said areas. . . ."

The energy, mines and resources department, through a resource management and conservation branch, is responsible for administration and enforcement of legislation and regulations relating to mineral resources off Canada's coasts, in the Hudson Bay and Hudson Strait regions, and for federally owned mineral rights that become available for development in the provinces. The Indian affairs and northern development department is responsible for mineral rights in the Yukon Territory and Northwest Territories and in Canada's Arctic offshore regions.

Mineral rights of Indian reserves in the provinces are also vested in the Crown in right of Canada and are administered by the Indian affairs department in consultation with Indian band councils. The rights to a reserve may be taken up only after the band has approved development through a referendum vote. The minerals are then administered under special oil and gas or mining regulations. The Indian oil and gas regulations allow disposal of rights by public tender in the form of permit or lease parcels. The mining regulations provide for disposal based on terms negotiated with the Indian band council. The councils thus assume a share of responsibility in management of their mineral resources. Officers of the department are advisers to the Indian councils on mineral matters and are responsible for administration and enforcement of relevant regulations.

12.8.2   Federal mining laws and regulations

Mining exploration and development is carried out in the Yukon Territory in accordance with the provisions of the Yukon Quartz Mining Act and the Yukon Placer Mining Act. In the Northwest Territories, including Arctic coastal waters, operations are governed by the Canada Mining Regulations. Regulations for dredging, coal mining and quarrying are common to both territories. In the Yukon Territory, mining rights may be acquired by staking claims. A one-year lease may be obtained to prospect for the purposes of placer mining, renewable for two additional one-year periods; a 21-year lease, renewable for a like period, may be obtained under the quartz mining act.

Under the Canada Mining Regulations, a prospector must be licensed. Staked claims must be converted to lease or relinquished within 10 years. In certain areas, a system of exploration over large areas is allowed by permit. Any individual 18 years of age or any joint-stock company incorporated or licensed to do business in Canada may hold a prospector's licence. No lease is granted to an individual unless the minister of the department involved is satisfied that the applicant is a Canadian citizen and will be the beneficial owner of the interest acquired. No lease is granted to a corporation unless it is incorporated in Canada and unless the minister is satisfied that at least 50% of the issued shares are owned by Canadian citizens or that the shares are listed on a recognized Canadian stock exchange. Any new mine beginning production after the mining regulations came into force in 1961 is not required to pay royalties for 36 months.

An exploration assistance fund for petroleum and other minerals in the territories was established in 1966. Assistance to a single applicant is limited in aggregate to $50,000, but not exceeding 40% of the approved cost of an exploration program. Assistance is available only to Canadian citizens or companies incorporated in Canada. It is designed to encourage investment from Canadian sources not previously attracted to investment in northern exploration operations.

12.8.3   Provincial mining laws and regulations

In general, all Crown mineral lands within provincial boundaries (with the exception of those in Indian reserves, national parks and other lands under federal jurisdiction) are administered by provincial governments. The exception is Quebec where mining rights on federal lands are administered by the province.

The granting of land in any province except Ontario no longer automatically carries with it mining rights upon or under such land. In Ontario, mineral rights are expressly reserved if they are not to be included. In Nova Scotia, no mineral rights belong to the owner of the land except those pertaining to gypsum, agricultural limestone and building materials, and deposits of either limestone or building materials may be declared to be minerals. Such declaration is based on economic value or the public interest. In such case, the initial privilege of acquiring the declared minerals lies with the owner of the surface rights who must then conform with the requirements of the mining act. In Newfoundland, mineral and quarry rights are expressly reserved. Some early grants in British Columbia, Alberta, Saskatchewan, Manitoba, New Brunswick, Quebec and Newfoundland also included certain mineral rights. Otherwise, mining rights must be obtained separately by lease or grant from the provincial authority. Mining activities may be classified as placer, general minerals (or veined minerals and bedded minerals), fuels (coal, petroleum and gas) and quarrying. Provincial mining regulations under these divisions are summarized in the following paragraphs.

In most provinces where placer deposits occur, regulations define the size of placer holdings, the terms under which they may be acquired and the royalties to be paid.

General minerals are sometimes described as quartz, lode, or minerals in place. The most elaborate laws and regulations apply in this division. In all provinces except Alberta, Saskatchewan and Manitoba, a prospector's or miner's licence, valid for one year, must be obtained to search for mineral deposits, the licence being general in some areas but limited in others; a claim of promising ground of a specified size may then be staked. In British Columbia a licence is required only for staking and any number of dispositions may be staked under one licence. A claim must be recorded within a time limit and payment of recording fees made, except in Quebec where no fees are required. Work to a specified value per year must be performed on the claim for a period up to 10 years except in Quebec, where a development licence may be renewed on a yearly basis; in Manitoba and Saskatchewan no work commitment is required in the first year of the claim. The maximum life of a prospecting licence in Nova Scotia is six years continuous from the original date of issue, after which the operator is expected to go to lease with a productive deposit. In Quebec and Nova Scotia a specified cost of work must be performed; any excess amount expended may be applied to subsequent renewals of the development licence. The taxation applied most frequently is a percentage of net profits of producing mines or royalties. In Saskatchewan, subsurface mineral regulations covering non-metallics stipulate the size and type of dispositions that may be made, the required expenditures for work to maintain the disposition in good standing, provide for fees, rentals and royalties, and set out generally the rights and obligations of the holder.

Coal, petroleum and natural gas. In provinces where coal occurs, the size of holdings is laid down together with conditions of work and rental under which they may be held. In Quebec, the search for petroleum and natural gas may be carried out under an exploration licence followed by an operating lease; the exploration licence covers a period of five years and an area of not over 24 281 hectares, whereas the operating lease extends over a 20-year period and an area not less than 202 hectares or more than 2 023 hectares. In Nova Scotia, mining rights to certain minerals, including petroleum, occurring under differing conditions may be held by different licensees. Provision is sometimes made for royalties. Acts or regulations govern methods of production. In the search for petroleum and natural gas, an exploration permit or reservation is usually required; however, in Saskatchewan, Alberta and British Columbia leases usually follow the exploration reservation whether or not any discovery of oil or gas is made. In Alberta, exploration costs are applicable in part on the first year's lease rental, in Manitoba they may be applied to the lease rental for a period up to three years, in British Columbia credit is given for up to 24 months rental, and in Saskatchewan credit is given for up to three years rental, having regard to the amount of excess credit established. In other provinces, discovery of oil or gas is usually prerequisite to obtaining a lease or grant of a limited area, subject to carrying out drilling obligations and paying a rental, a fee, or a royalty on production.

Quarrying regulations define the size of holdings and the terms of lease or grant. In Nova Scotia, sand deposits of a quality suitable for uses other than building purposes and limestone deposits of metallurgical grade belong to the Crown; gypsum deposits belong to the owner of the property. In New Brunswick, quarriable substances (ordinary stone, building and construction stone, sand, gravel, peat and peat moss) are vested in the owner of the land in or on which they lie; the minister with the approval of the

Cabinet may designate a shore area lying outside Crown land to be subject to the act; and no person shall take or remove or cause to be taken or removed more than 0.383 cubic metres of a quarriable substance from Crown land or a designated shore area without obtaining a permit or lease. On Quebec public lands and on those granted to individuals after January 1, 1966, the stone, sand and gravel, like other building materials, belong to the Crown; quarries located on land granted to individuals prior to 1966 remain in the possession of the owners of the surface; the right to exploit all building materials except sand and gravel may be acquired by ordinary staking-out and the right to work sand and gravel beds is set by regulation. In Saskatchewan, sand and gravel on the surface and all sand and gravel obtainable by stripping off the overburden or other surface operation belong to the owner of the surface of the land. In Alberta, sand, gravel, clay and marl recovered by excavating from the surface belong to the owner of the surface of the land. British Columbia, Manitoba and Saskatchewan have made provision for participation by the Crown, at the option of the Crown, in all future mineral development. Such participation may be by way of association, joint venture or otherwise, usually through a Crown corporation. Copies of mining legislation including regulations and other details may be obtained from provincial authorities concerned.

Sources

12.1 - 12.7.1 Minerals and Metals Division, Mineral Development Sector, Department of Energy, Mines and Resources.

12.7.2 Resources and Development Division, Mineral Development Sector, Department of Energy, Mines and Resources.

12.8 - 12.8.3 Mining Industry Financial and Corporate Analysis Division, Mineral Development Sector, Department of Energy, Mines and Resources.

Tables

12.1 Value of mineral production, 1886-1976

Year	Total value $'000	Value per capita $	Year	Total value $'000	Value per capita $	Year	Total value $'000	Value per capita $
1886	10,221	2.23	1925	226,583	24.38	1965	3,714,861	189.11
1890	16,763	3.51	1930	279,874	27.42	1970	5,722,059	268.68
1895	20,506	4.08	1935	312,344	28.84	1971	5,962,692	276.46r
1900	64,421	12.15	1940	529,825	46.55	1972	6,408,026	293.93r
1905	69,079	11.51	1945	498,755	41.31	1973	8,369,515	379.69r
1910	106,824	15.29	1950[1]	1,045,450	76.24	1974r	11,751,445	525.46
1915	137,109	17.18	1955	1,795,311	114.37	1975	13,346,724	588.04
1920	227,860	26.63	1960	2,492,510	139.48	1976p	15,392,839	669.47

[1]Value of Newfoundland production included from 1950.

12.2 Value of mineral production, by class, 1968-76 (thousand dollars)

Year	Metallics	Non-metallics	Fuels	Structural materials	Total
1968	2,492,600	446,922	1,343,163	439,563	4,722,249
1969	2,377,523	450,189	1,465,400	441,172	4,734,284
1970	3,073,344	480,538	1,717,731	450,446	5,722,059
1971	2,940,287	500,827	2,014,410	507,168	5,962,692
1972	2,955,655	513,488	2,367,554	571,329	6,408,026
1973	3,850,072	614,523	3,227,142	677,778	8,369,515
1974r	4,820,675	895,891	5,201,723	833,156	11,751,445
1975	4,793,853	939,180	6,653,355	960,336	13,346,724
1976p	5,241,151	1,142,516	7,993,404	1,015,768	15,392,839

12.3 Quantity indexes of production of the principal mining industries, 1967-76 (1971=100)

Mining industry	1967	1968	1969	1970	1971	1972r	1973r	1974r	1975r	1976
Metal mines	84.3	88.5	83.1	99.2	100.0	97.8	110.3	110.5	102.8	108.3
Placer gold and gold quartz	132.6	120.4	116.9	104.0	100.0	90.9	81.1	69.9	71.2	74.6
Iron	77.0	92.6	81.1	102.5	100.0	89.3	114.9	112.9	112.5	139.4
Miscellaneous	83.9	84.8	81.6	97.7	100.0	100.3	110.5	111.9	101.9	102.2
Non-metal mines (except coal)	79.1	87.4	92.1	96.1	100.0	100.8	108.5	124.6	103.6	118.1
Asbestos	85.9	92.9	90.9	100.1	100.0	102.0	104.6	110.4	71.9	104.9
Mineral fuels	66.7	72.7	79.9	92.2	100.0	118.3	134.0	127.9	118.3	111.8
Coal	69.2	67.2	66.8	86.0	100.0	148.3	160.6	158.4	201.7	193.8
Crude oil and natural gas	66.5	73.3	81.2	92.8	100.0	115.8	131.8	125.4	111.5	105.1
Total, mines (incl. milling) quarries and oil wells	77.7	83.3	83.9	95.8	100.0	106.5	118.9	118.3	109.3	110.4

12.4　Quantity and value of mineral production, 1975 and 1976

Mineral		1975		1976	
		Quantity '000	Value $'000	Quantity '000	Value $'000
METALLICS		...	*4,793,853*	...	*5,241,151*
Antimony	kg	...	7,369	...	7,270
Bismuth	"	157	2,647	154	2,491
Cadmium	"	1 192	8,967	1 292	7,462
Calcium	"	428	1,005	558	1,521
Cobalt	"	1 354	12,548	1 373	11,769
Columbium (Cb₂O₅)	"	1 662	6,854	1 656	6,935
Copper	"	733 826	1,030,502	747 135	1,126,156
Gold	g	51 433	270,830	52 444	207,796
Indium	"	6 967			
Iron ore	t	44 893	918,065	56 902	1,241,263
Iron, remelt	"	...	80,753		65,086
Lead	kg	349 133	155,973	259 083	129,388
Magnesium	"	3 826	8,788	5 858	12,248
Mercury	"	414		—	—
Molybdenum	"	13 027	71,201	14 416	91,873
Nickel	"	242 180	1,100,523	262 492	1,232,143
Platinum group	g	12 417	56,493	13 375	48,790
Selenium	kg	182	7,362	260	9,134
Silver	"	1 235	178,864	1 272	175,128
Tantalum (Ta₂O₅)	"	...			
Tellurium	"	20	414	24	529
Tin	"	319	2,366	275	1,873
Tungsten (WO₃)	"	1 478			
Uranium (U₃O₈)	"	5 517	...	6 058	...
Yttrium (Y₂O₃)	"	...			
Zinc	"	1 055 151	872,328	1 039 688	862,296
NON-METALLICS		...	*939,180*	...	*1,142,516*
Asbestos	t	1 056	267,246	1 549	445,523
Barite	"	...	2,306	...	1,860
Diatomite	"	...			
Feldspar	"	—	—	—	...
Fluorspar	"	—	—	—	2,246
Gemstones	kg	110	414	...	414
Gypsum	t	5 719	20,304	5 663	22,906
Helium	m³	...			
Magnesitic dolomite and brucite	t	...	5,358	...	5,116
Nepheline syenite	"	468	8,869	541	10,828
Nitrogen	m³	...			
Peat moss	t	361	22,273	363	22,500
Potash (K₂O)	"	4 673	358,570	5 126	361,442
Pyrite, pyrrhotite	"	21	127	31	240
Quartz	"	2 492	13,112	2 376	13,895
Salt	"	5 123	59,714	5 752	75,691
Soapstone, talc, pyrophyllite	"	66	1,538	65	1,774
Sodium sulphate	"	472	22,049	490	24,878
Sulphur, in smelter gas	"	695	9,641	781	15,454
Sulphur, elemental	"	4 079	91,847	3 781	63,339
Titanium dioxide, etc.	"	...	55,812	...	74,410
FUELS		...	*6,653,355*	...	*7,993,404*
Coal	t	25 259	586,423	25 311	604,000
Natural gas	m³	87 485 758	1,520,661	86 858 171	2,466,621
Natural gas byproducts	"	17 835	782,337	16 543	794,325
Oil, crude	"	83 001	3,763,934	77 843	4,128,458
STRUCTURAL MATERIALS		...	*960,336*	...	*1,015,768*
Clay products		...	85,977	...	92,110
Cement	t	9 965	320,173	9 850	339,159
Lime	"	1 602	46,907	1 825	54,099
Sand and gravel	"	247 155	305,181	247 660	320,800
Stone	"	88 921	202,099	87 180	209,600
Total		...	13,346,724	...	15,392,839

12.5   Percentage of the total value contributed by principal minerals, 1967-76

Mineral	1967	1968	1969	1970	1971	1972	1973	1974	1975	1976
METALLICS[1]	*52.2*	*52.8*	*50.2*	*53.7*	*49.3*	*46.1*	*46.0*	*41.2*	*35.9*	*34.1*
Copper	13.3	12.9	12.4	13.6	12.7	12.5	13.8	12.0	7.7	7.3
Gold	2.6	2.2	2.0	1.5	1.3	1.9	2.3	2.3	2.0	1.4
Iron ore	10.7	11.3	9.6	10.3	9.3	7.6	7.2	6.2	6.9	8.1
Lead	2.0	1.9	2.0	2.2	1.8	1.8	1.5	1.1	1.2	0.8
Molybdenum	0.9	0.8	1.1	1.0	0.6	0.7	0.6	0.5	0.5	0.6
Nickel	10.6	11.2	10.2	14.5	13.4	11.2	9.7	8.3	8.2	8.0
Platinum group	0.8	1.0	0.7	0.8	0.7	0.5	0.5	0.6	0.4	0.3
Silver	1.4	2.2	1.8	1.4	1.2	1.2	1.4	1.7	1.3	1.1
Uranium	1.2	1.1	1.1	..	..	..	..	..	..	..
Zinc	7.3	6.9	7.8	7.0	7.0	7.5	7.8	7.4	6.5	5.6
NON-METALLICS[1]	*9.3*	*9.5*	*9.5*	*8.4*	*8.4*	*8.0*	*7.3*	*7.7*	*7.0*	*7.4*
Asbestos	3.7	3.9	4.1	3.6	3.4	3.2	2.8	2.6	2.0	2.9
Gypsum	0.3	0.3	0.3	0.2	0.3	0.3	0.3	0.2	0.2	0.1
Nepheline syenite	0.1	0.1	0.1	0.1	0.1	0.1	0.1	0.1	0.1	0.1
Potash	1.5	1.4	1.5	1.9	2.3	2.1	2.1	2.6	2.7	2.3
Quartz	0.1	0.1	0.1	0.1	0.1	0.2	0.1	0.1	0.1	0.1
Salt	0.6	0.7	0.6	0.6	0.7	0.6	0.6	0.5	0.5	0.5
Sodium sulphate	0.2	0.2	0.2	0.1	0.1	0.1	0.1	0.1	0.2	0.2
Sulphur, in smelter gas	0.2	0.2	0.2	0.1	0.1	0.1	0.1	0.1	0.1	0.1
Sulphur, elemental	1.6	1.7	1.3	0.5	0.4	0.3	0.3	0.6	0.7	0.4
Titanium dioxide, etc.	0.5	0.6	0.6	0.6	0.7	0.6	0.6	0.4	0.4	0.5
FUELS[1]	*28.2*	*28.4*	*31.0*	*30.0*	*33.8*	*37.0*	*38.6*	*44.4*	*49.9*	*51.9*
Coal	1.3	1.1	1.1	1.5	2.0	2.4	2.1	2.6	4.4	3.9
Natural gas	4.5	4.8	5.5	5.5	5.7	6.2	5.4	6.2	11.4	16.0
Oil, crude	19.8	19.8	21.4	20.2	22.7	24.5	26.8	30.1	28.2	26.8
STRUCTURAL MATERIALS	*10.3*	*9.3*	*9.3*	*7.9*	*8.5*	*8.9*	*8.1*	*6.7*	*7.2*	*6.6*
Clay products	1.0	1.0	1.0	0.9	0.8	0.8	0.7	0.6	0.6	0.6
Cement	3.3	3.1	3.4	2.7	3.1	3.3	2.9	2.3	2.4	2.2
Lime	0.4	0.4	0.4	0.4	0.4	0.4	0.4	0.4	0.4	0.3
Sand and gravel	3.3	2.8	2.6	2.3	2.6	2.8	2.6	2.0	2.3	2.1
Stone	2.3	2.0	1.9	1.5	1.6	1.6	1.5	1.4	1.5	1.4
Total	100.0	100.0	100.0	100.0	100.0	100.0	100.0	100.0	100.0	100.0

[1]Includes minor items not specified.

12.6   Value of mineral production, by province, 1967-76 (thousand dollars)

Year	Province or territory						
	Newfound-land (incl. Labrador)	Prince Edward Island	Nova Scotia	New Brunswick	Quebec	Ontario	Manitoba
1967	266,365	1,775	52,544	90,440	741,436	1,194,549	184,679
1968	309,712	977	56,940	88,452	725,078	1,355,629	209,626
1969	256,936	452	58,562	94,593	717,156	1,222,172	246,275
1970	353,261	640	58,159	104,791	803,286	1,593,039	332,214
1971	343,431	978	60,083	107,233	766,473	1,554,777	330,060
1972	290,659	1,097	57,522	120,171	785,962	1,535,683	323,292
1973	374,418	1,680	60,808	164,178	935,530	1,854,695	414,013
1974	448,473	1,454	80,251	213,519	1,192,440	2,429,530	486,249
1975	550,879	1,786	101,399	231,628	1,239,929	2,350,006	529,618
1976P	756,007	1,700	117,201	255,057	1,521,321	2,594,042	478,120

Year	Saskat-chewan	Alberta	British Columbia	Yukon Territory	Northwest Territories	Canada
1967	361,824	974,366	379,555	14,991	118,283	4,380,805
1968	357,082	1,092,444	389,307	21,366	115,636	4,722,249
1969	344,625	1,205,308	433,633	35,403	119,171	4,734,284
1970	379,190	1,395,994	490,158	77,512	133,814	5,722,059
1971	409,956	1,640,508	540,527	93,111	115,555	5,962,692
1972	409,889	1,978,750	677,883	106,781	120,337	6,408,026
1973	509,773	2,760,227	978,037	150,667	165,489	8,369,515
1974	790,330	4,518,383	1,155,787	171,538	223,050	11,711,004
1975	861,606	5,746,571	1,296,802	230,151	206,349	13,346,724
1976P	908,554	6,995,572	1,421,096	131,069	213,100	15,392,839

12.7   Value of metallics, non-metallics, fuels and structural materials produced, by province, 1975 and 1976 (thousand dollars)

Year and province or territory	Metallics	Non-metallics	Fuels	Structural materials	Total
1975					
Newfoundland (incl. Labrador)	512,262	21,005	—	17,612	550,879
Prince Edward Island	—	—	—	1,786	1,786
Nova Scotia	1	28,882	44,586	27,930	101,399
New Brunswick	197,231	6,253	7,206	20,938	231,628
Quebec	668,296	253,462	8	318,163	1,239,929
Ontario	1,948,966	56,186	11,554	333,300	2,350,006
Manitoba	448,137	4,804	31,445	45,232	529,618
Saskatchewan	18,622	385,955	433,766	23,263	861,606
Alberta	40	100,802	5,569,399	76,330	5,746,571
British Columbia	622,208	49,010	529,802	95,782	1,296,802
Yukon Territory	196,020	32,821	1,310	—	230,151
Northwest Territories	182,070	—	24,279	—	206,349
Canada, 1975	4,793,853	939,180	6,653,355	960,336	13,346,724
1976P					
Newfoundland (incl. Labrador)	699,919	38,699	—	17,389	756,007
Prince Edward Island	—	—	—	1,700	1,700
Nova Scotia	—	32,927	54,500	29,774	117,201
New Brunswick	217,945	6,758	6,383	23,971	255,057
Quebec	765,699	439,174	2	316,446	1,521,321
Ontario	2,153,488	75,742	11,788	353,024	2,594,042
Manitoba	387,330	4,366	32,995	53,429	478,120
Saskatchewan	25,116	392,357	462,612	28,469	908,554
Alberta	—	72,955	6,829,549	93,068	6,995,572
British Columbia	710,487	45,078	567,033	98,498	1,421,096
Yukon Territory	96,009	34,460	600	—	131,069
Northwest Territories	185,158	—	27,942	—	213,100
Canada, 1976P	5,241,151	1,142,516	7,993,404	1,015,768	15,392,839

12.8   Detailed mineral production, by province, 1975 and 1976P (thousands)

Mineral		Province or territory					
		Newfoundland (incl. Labrador)		Nova Scotia		New Brunswick	
		1975	1976	1975	1976	1975	1976
METALLICS	$	512,262	699,919	1	—	197,231	217,945
Antimony	kg	—	—	—	—		
	$	—	—	—	—	5,901	5,900
Bismuth	kg	—	—	—	—	134	131
	$	—	—	—	—	2,327	2,122
Cadmium	kg	5	5	—	—	24	24
	$	37	26	—	—	182	139
Calcium	kg	—	—	—	—	—	—
	$	—	—	—	—	—	—
Cobalt	kg	—	—	—	—	—	—
	$	—	—	—	—	—	—
Columbium (Cb₂O₅)	kg	—	—	—	—	—	—
	$	—	—	—	—	—	—
Copper	kg	7 500	6 764	—	—	11 212	9 678
	$	10,541	10,194	—	—	15,758	14,587
Gold	g	404	405	- -	—	105	125
	$	2,128	1,596	1	—	555	460
Indium	g	—	—	—	—	—	—
	$	—	—	—	—	—	—
Iron ore	t	22 586	27 970	—	—	—	—
	$	468,600	643,455	—	—	—	—
Iron, remelt	t	—	—	—	—	—	—
	$	—	—	—	—	—	—
Lead	kg	5 219	9 253	—	—	59 092	61 687
	$	2,331	4,621	—	—	26,399	30,808
Magnesium	kg	—	—	—	—	—	—
	$	—	—	—	—	—	—
Mercury	kg	—	—	—	—	—	—
	$	—	—	—	—	—	—
Molybdenum	kg	—	—	—	—	—	—
	$	—	—	—	—	—	—
Nickel	kg	—	—	—	—	—	—
	$	—	—	—	—	—	—
Platinum group	g	—	—	—	—	—	—
	$	—	—	—	—	—	—

12.8 Detailed mineral production, by province, 1975 and 1976[p] (thousands) (continued)

Mineral		Newfound-land (incl. Labrador)		Nova Scotia		New Brunswick	
		1975	1976	1975	1976	1975	1976
METALLICS (concluded)							
Selenium	*kg*	—	—	—	—	—	—
	$	—	—	—	—	—	—
Silver	*kg*	14	16	—	—	157	156
	$	2,005	2,194	—	—	22,752	21,497
Tantalum (Ta_2O_5)	*kg*	—	—	—	—	—	—
	$	—	—	—	—	—	—
Tellurium	*kg*	—	—	—	—	—	—
	$	—	—	—	—	—	—
Tin	*kg*	—	—	—	—	—	—
	$	—	—	—	—	—	—
Tungsten (WO_3)	*kg*	—	—	—	—	—	—
	$	—	—	—	—	—	—
Uranium (U_3O_8)	*kg*	—	—	—	—	—	—
	$	—	—	—	—	—	—
Yttrium (Y_2O_3)	*kg*	—	—	—	—	—	—
	$	—	—	—	—	—	—
Zinc	*kg*	32 198	45 616	—	—	149 210	171 733
	$	26,619	37,833	—	—	123,357	142,432
NON-METALLICS	*$*	*21,005*	*38,699*	*28,882*	*32,927*	*6,253*	*6,758*
Asbestos	*t*	58	86	—	—	—	—
	$	18,139	33,383	—	—	—	—
Barite	*t*	—	—	..	..	—	—
	$	—	—	941	760	—	—
Diatomite	*t*	—	—	—	—	—	—
	$	—	—	—	—	—	—
Feldspar	*t*	—	—	—	—	—	—
	$	—	—	—	—	—	—
Fluorspar	*t*	—	—	—	—	—	—
	$	—	2,246	—	—	—	—
Gemstones	*kg*	—	—	—	—	—	—
	$	—	—	—	—	—	—
Gypsum	*t*	583	558	3 895	3 874	53	42
	$	2,315	2,436	12,806	13,804	209	177
Helium	*m³*	—	—	—	—	—	—
	$	—	—	—	—	—	—
Magnesitic dolomite and brucite	*t*	—	—	—	—	—	—
	$	—	—	—	—	—	—
Nepheline syenite	*t*	—	—	—	—	—	—
	$	—	—	—	—	—	—
Nitrogen	*m³*	—	—	—	—	—	—
	$	—	—	—	—	—	—
Peat moss	*t*	—	—	7	7	82	87
	$	—	—	495	500	5,458	5,800
Potash (K_2O)	*t*	—	—	—	—	—	—
	$	—	—	—	—	—	—
Pyrite, pyrrhotite	*t*	—	—	—	—	—	—
	$	—	—	—	—	—	—
Quartz	*t*	—	—	..	..	—	—
	$	160	218	130	231	—	—
Salt	*t*	—	—	768	902	—	—
	$	—	—	14,510	17,632	—	—
Soapstone, talc, pyrophyllite	*t*	—	—	—	—	—	—
	$	391	416	—	—	—	—
Sodium sulphate	*t*	—	—	—	—	—	—
	$	—	—	—	—	—	—
Sulphur, in smelter gas	*t*	—	—	—	—	42	39
	$	—	—	—	—	586	781
Sulphur, elemental	*t*	—	—	—	—	—	—
	$	—	—	—	—	—	—
Titanium dioxide, etc.	*t*	—	—	—	—	—	—
	$	—	—	—	—	—	—
FUELS	*$*	—	—	*44,586*	*54,500*	*7,206*	*6,383*
Coal	*t*	—	—	1 657	1 996	418	290
	$	—	—	44,586	54,500	7,100	6,300
Natural gas	*m³*	—	—	—	—	2 577	2 775
	$	—	—	—	—	55	59
Natural gas byproducts	*m³*	—	—	—	—	—	—
	$	—	—	—	—	—	—
Oil, crude	*m³*	—	—	—	—	1	1
	$	—	—	—	—	51	24
STRUCTURAL MATERIALS	*$*	*17,612*	*17,389*	*27,930*	*29,774*	*20,938*	*23,971*
Clay products	*$*	457	475	3,155	3,915	1,310	2,464
Cement	*t*	..	..	..	..	..	..
	$	4,678	5,014	6,094	7,059	5,839	8,967
Lime	*t*	—	—	—	—	..	..
	$	—	—	—	—	1,303	1,440
Sand and gravel	*t*	6 237	5 080	8 906	8 618	3 834	3 538
	$	9,587	9,200	14,044	14,400	4,238	4,100
Stone	*t*	877	816	1 581	1 451	3 241	2 540
	$	2,889	2,700	4,637	4,400	8,248	7,000
Total 1975	*$*	550,879	...	101,399	...	231,628	...
Total 1976	*$*	...	756,007	...	117,201	...	255,057

12.8 Detailed mineral production, by province, 1975 and 1976ᵖ (thousands) (continued)

Mineral		Province or territory					
		Quebec		Ontario		Manitoba	
		1975	1976	1975	1976	1975	1976
METALLICS	$	668,296	765,699	1,948,966	2,153,488	448,137	387,330
Antimony	kg	—	—	—	—	—	—
	$	—	—	—	—	—	—
Bismuth	kg	1	- -	3	—	—	—
	$	16	5	42	—	—	—
Cadmium	kg	124	109	666	690	44	34
	$	932	629	5,012	3,992	331	192
Calcium	kg	—	—	428	558	—	—
	$	—	—	1,005	1,521	—	—
Cobalt	kg	—	—	1 088	1 139	266	234
	$	—	—	10,278	9,679	2,270	2,090
Columbium (Cb₂O₅)	kg	1 662	1 656	—	—	—	—
	$	6,854	6,935	—	—	—	—
Copper	kg	117 556	120 411	257 778	258 981	64 495	56 687
	$	165,221	181,500	361,432	390,361	90,646	85,444
Gold	g	14 155	14 743	23 488	22 830	1 497	1 431
	$	74,538	58,460	123,679	90,414	7,881	5,674
Indium	g	—	—	—	—	—	—
	$	—	—	—	—	—	—
Iron ore	t	11 501	17 754	9 504	10 369	—	—
	$	215,155	324,607	219,024	264,111	—	—
Iron, remelt	t	..	..	—	—	—	—
	$	80,753	65,086	—	—	—	—
Lead	kg	1 644	923	6 192	6 379	119	275
	$	735	460	2,766	3,186	53	137
Magnesium	kg	—	—	3 826	5 858	—	—
	$	—	—	8,788	12,248	—	—
Mercury	kg	—	—	—	—	—	—
	$	—	—	—	—	—	—
Molybdenum	kg	—	461	—	—	—	—
	$	—	2,946	—	—	—	—
Nickel	kg	—	—	179 095	211 584	63 085	50 908
	$	—	—	811,329	993,704	289,194	238,439
Platinum group	g	—	—	12 417	13 375	—	—
	$	—	—	56,493	48,790	—	—
Selenium	kg	125	159	49	49	6	34
	$	5,030	6,986	1,972	1,951	260	130
Silver	kg	106	115	464	486	31	28
	$	15,308	15,794	67,176	67,000	4,513	3,817
Tantalum (Ta₂O₅)	kg	—	—	—	—	..	..
	$	—	—	—	—	..	..
Tellurium	kg	10	15	7	6	2	2
	$	213	331	144	138	41	40
Tin	kg	—	—	287	166	—	—
	$	—	—	2,165	967	—	—
Tungsten (WO₃)	kg	—	—	—	—	—	—
	$	—	—	—	—	—	—
Uranium (U₃O₈)	kg	—	—	4 794	3 892	—	—
	$	—	—	..	..	—	—
Yttrium (Y₂O₃)	kg	—	—	..	..	—	—
	$	—	—	..	..	—	—
Zinc	kg	125 241	122 934	335 852	320 030	64 045	61 935
	$	103,541	101,960	277,660	265,426	52,948	51,367
NON-METALLICS	$	253,462	439,174	56,186	75,742	4,804	4,366
Asbestos	t	802	1 263	15	26	—	—
	$	176,942	343,164	1,494	3,797	—	—
Barite	t	—	—	—	—	—	—
	$	—	—	—	—	—	—
Diatomite	t	—	—	—	—	—	—
	$	—	—	—	—	—	—
Feldspar	t	—	—	—	—	—	—
	$	—	—	—	—	—	—
Fluorspar	t	—	—	—	—	—	—
	$	—	—	—	—	—	—
Gemstones	kg	—	—	—	—	—	—
	$	—	—	—	—	—	—
Gypsum	t	—	—	631	571	83	83
	$	—	—	2,936	2,658	286	203
Helium	m³	—	—	—	—	—	—
	$	—	—	—	—	—	—
Magnesitic dolomite and brucite	t	..	..	—	—	—	—
	$	5,358	5,116	—	—	—	—
Nepheline syenite	t	—	—	468	541	—	—
	$	—	—	8,869	10,828	—	—
Nitrogen	m³	—	—	—	—	—	—
	$	—	—	—	—	—	—
Peat moss	t	156	151	10	11	32	33
	$	7,998	7,800	875	870	2,498	2,470
Potash (K₂O)	t	—	—	—	—	—	—
	$	—	—	—	—	—	—
Pyrite, pyrrhotite	t	21	31	—	—	—	—
	$	127	240	—	—	—	—
Quartz	t	548	523	1 096	1 155	618	442
	$	5,334	5,075	4,358	5,400	1,865	1,517
Salt	t	—	—	3 763	4 246	27	28
	$	—	—	32,259	44,272	150	161
Soapstone, talc, pyrophyllite	t	..	..	..	..	—	—
	$	523	794	624	564	—	—

12.8   Detailed mineral production, by province, 1975 and 1976ᴾ (thousands) (continued)

Mineral		Quebec 1975	Quebec 1976	Ontario 1975	Ontario 1976	Manitoba 1975	Manitoba 1976
NON-METALLICS (concluded)							
Sodium sulphate	t	—	—	—	—	—	—
	$	—	—	—	—	—	—
Sulphur, in smelter gas	t	99	130	342	370	—	—
	$	1,368	2,575	4,749	7,318	—	—
Sulphur, elemental	t	—	—	1	2	- -	1
	$	—	—	21	35	5	15
Titanium dioxide, etc.	t	...	...	—	—	—	—
	$	55,812	74,410	—	—	—	—
FUELS	$	8	2	11,554	11,788	31,445	32,995
Coal	t	—	—	—	—	—	—
	$	—	—	—	—	—	—
Natural gas	m³	1 416	255	309 645	141 584	—	—
	$	8	2	6,508	5,760	—	—
Natural gas byproducts	m³	—	—	—	—	—	—
	$	—	—	—	—	—	—
Oil, crude	m³	—	—	112	100	702	626
	$	—	—	5,046	6,028	31,445	32,995
STRUCTURAL MATERIALS	$	318,163	316,446	333,300	353,024	45,232	53,429
Clay products	$	16,468	14,243	44,769	50,926	1,386	1,318
Cement	t	3 293	2 683	3 800	3 764	457	586
	$	101,789	88,733	110,708	116,162	17,056	22,606
Lime	t	581	622	769	916	...	...
	$	19,009	20,570	19,994	24,236	2,246	2,805
Sand and gravel	t	82 039	83 007	69 705	68 039	16 417	17 418
	$	70,488	75,900	95,579	95,200	22,934	25,200
Stone	t	50 763	51 710	27 526	26 853	477	454
	$	110,410	117,000	62,250	66,500	1,610	1,500
Total 1975	$	1,239,929	...	2,350,006	...	529,619	...
Total 1976	$	...	1,521,321	...	2,594,042	...	478,120

Mineral		Saskatchewan 1975	Saskatchewan 1976	Alberta 1975	Alberta 1976	British Columbia 1975	British Columbia 1976
METALLICS	$	18,622	25,116	40	—	622,208	710,487
Antimony	kg	—	—	—	—	364	375
	$	—	—	—	—	1,468	1,370
Bismuth	kg	—	—	—	—	19	23
	$	—	—	—	—	262	364
Cadmium	kg	6	14	—	—	321	413
	$	42	82	—	—	2,415	2,390
Calcium	kg	—	—	—	—	—	—
	$	—	—	—	—	—	—
Cobalt	kg	—	—	—	—	—	—
	$	—	—	—	—	—	—
Columbium (Cb₂O₅)	kg	—	—	—	—	—	—
	$	—	—	—	—	—	—
Copper	kg	7 905	9 596	—	—	258 518	273 541
	$	11,110	14,463	—	—	363,338	412,308
Gold	g	454	591	8	—	4 863	5 506
	$	2,394	2,342	40	—	25,605	21,820
Indium	g	—	—	—	—	6 967	...
	$	—	—	—	—	...	...
Iron ore	t	—	—	—	—	1 302	809
	$	—	—	—	—	15,285	9,090
Iron, remelt	t	—	—	—	—	—	—
	$	—	—	—	—	—	—
Lead	kg	—	—	—	—	70 612	88 633
	$	—	—	—	—	31,546	44,264
Magnesium	kg	—	—	—	—	—	—
	$	—	—	—	—	—	—
Mercury	kg	—	—	—	—	414	—
	$	—	—	—	—	...	—
Molybdenum	kg	—	—	—	—	13 027	13 955
	$	—	—	—	—	71,201	88,927
Nickel	kg	—	—	—	—	—	—
	$	—	—	—	—	—	—
Platinum group	g	—	—	—	—	—	—
	$	—	—	—	—	—	—
Selenium	kg	2	18	—	—	—	—
	$	100	67	—	—	—	—
Silver	kg	8	10	- -	—	197	255
	$	1,208	1,387	- -	—	28,487	35,108
Tantalum (Ta₂O₅)	kg	—	—	—	—	—	—
	$	—	—	—	—	—	—
Tellurium	kg	1	1	—	—	—	—
	$	16	20	—	—	—	—
Tin	kg	—	—	—	—	32	109
	$	—	—	—	—	201	906
Tungsten (WO₃)	kg	—	—	—	—	—	—
	$	—	—	—	—	—	—
Uranium (U₃O₈)	kg	723	2 166	—	—	—	—
	$	...	...	—	—	—	—

12.8 Detailed mineral production, by province, 1975 and 1976ᴾ (thousands) (continued)

Mineral		Province or territory					
		Saskatchewan		Alberta		British Columbia	
		1975	1976	1975	1976	1975	1976
METALLICS (concluded)							
Yttrium (Y₂O₃)	kg	—	—	—	—	—	—
	$	—	—	—	—	—	—
Zinc	kg	4 539	8 144	—	—	99 669	113 266
	$	3,752	6,755	—	—	82,400	93,940
NON-METALLICS	$	385,955	392,357	100,802	72,955	49,010	45,078
Asbestos	t	—	—	—	—	77	71
	$	—	—	—	—	37,850	30,719
Barite	t	—	—	—	—	...	...
	$	—	—	—	—	1,365	1,100
Diatomite	t	—	—	—	—	...	...
	$	—	—	—	—	...	...
Feldspar	t	—	—	—	—	—	—
	$	—	—	—	—	—	—
Fluorspar	t	—	—	—	—	—	—
	$	—	—	—	—	—	—
Gemstones	kg	—	—	—	—	110	...
	$	—	—	—	—	414	414
Gypsum	t	—	—	—	—	474	535
	$	—	—	—	—	1,752	3,628
Helium	m³	—	...	—	—	—	—
	$	—	...	—	—	—	—
Magnesitic dolomite and brucite	t	—	—	—	—	—	—
	$	—	—	—	—	—	—
Nepheline syenite	t	—	—	—	—	—	—
	$	—	—	—	—	—	—
Nitrogen	m³	—	...	—	—	—	—
	$	—	...	—	—	—	—
Peat moss	t	5	7	19	19	50	48
	$	323	400	1,060	1,100	3,565	3,560
Potash (K₂O)	t	4 673	5 126	—	—	—	—
	$	358,570	361,442	—	—	—	—
Pyrite, pyrrhotite	t	—	—	—	—	—	—
	$	—	—	—	—	—	—
Quartz	t	109	102	...	...	18	31
	$	169	168	840	1,125	256	161
Salt	t	277	281	288	295	—	—
	$	7,509	7,833	5,286	5,793	—	—
Soapstone, talc, pyrophyllite	t	—	—	—	—	—	—
	$	—	—	—	—	—	—
Sodium sulphate	t	...	...	...	...	—	—
	$	19,184	22,221	2,865	2,657	—	—
Sulphur, in smelter gas	t	—	—	—	—	212	242
	$	—	—	—	—	2,938	4,780
Sulphur, elemental	t	12	15	4 027	3 720	39	43
	$	200	293	90,751	62,280	870	716
Titanium dioxide, etc.	t	—	—	—	—	—	—
	$	—	—	—	—	—	—
FUELS	$	433,766	462,612	5,569,399	6,829,549	529,802	567,033
Coal	t	3 549	4 627	10 055	10 687	9 580	7 711
	$	9,634	12,900	183,015	223,800	342,088	306,500
Natural gas	m³	1 738 796	1 608 397	73 335 204	73 652 127	11 168 874	10 498 755
	$	10,449	8,250	1,405,247	2,302,235	77,342	130,137
Natural gas byproducts	m³	154	135	17 307	16 055	374	353
	$	6,381	5,787	760,501	772,414	15,455	16,124
Oil, crude	m³	9 408	8 824	70 332	65 799	2 286	2 337
	$	407,302	435,675	3,220,636	3,531,100	94,917	114,272
STRUCTURAL MATERIALS	$	23,263	28,469	76,330	93,068	95,782	98,498
Clay products	$	2,730	3,098	8,530	8,727	7,172	6,944
Cement	t	223	327	877	1 048	888	919
	$	9,860	15,171	29,347	36,948	34,801	38,499
Lime	t	—	—	110	132	36	34
	$	—	—	3,342	4,093	1,013	955
Sand and gravel	t	8 313	7 439	20 453	23 133	30 322	30 572
	$	10,672	10,200	33,952	42,100	41,900	42,800
Stone	t	—	—	235	181	4 221	3 175
	$	—	—	1,159	1,200	10,896	9,300
Total 1975	$	861,606		5,746,571		1,296,802	
Total 1976	$		908,554		6,995,572		1,421,096

12.8 Detailed mineral production, by province, 1975 and 1976ᵖ (thousands) (continued)

Mineral		Yukon Territory 1975	1976	Northwest Territories 1975	1976	Canada 1975	1976
METALLICS	$	*196,020*	*96,009*	*182,070*	*185,158*	*4,793,853*	*5,241,151*
Antimony	kg	—	—	—	—	..	..
	$	—	—	—	—	7,369	7,270
Bismuth	kg	—	—	—	—	157	154
	$	—	—	—	—	2,647	2,491
Cadmium	kg	2	3	- -	—	1 192	1 292
	$	15	12	1	—	8,967	7,462
Calcium	kg	—	—	—	—	428	558
	$	—	—	—	—	1,005	1,521
Cobalt	kg	—	—	—	—	1 354	1 373
	$	—	—	—	—	12,548	11,769
Columbium (Cb₂O₅)	kg	—	—	—	—	1 662	1 656
	$	—	—	—	—	6,854	6,935
Copper	kg	8 487	11 039	375	438	733 826	747 135
	$	11,929	16,639	527	660	1,030,502	1,126,156
Gold	g	998	965	5 461	5 848	51 433	52 444
	$	5,255	3,910	28,754	23,120	270,830	207,796
Indium	g	—	—	—	—	6 967	..
	$						
Iron ore	t	—	—	—	—	44 893	56 902
	$	—	—	—	—	918,065	1,241,263
Iron, remelt	t	—	—	—	—	..	..
	$	—	—	—	—	80,753	65,086
Lead	kg	122 864	38 254	83 391	53 679	349 133	259 083
	$	54,889	19,104	37,254	26,808	155,973	129,388
Magnesium	kg	—	—	—	—	3 826	5 858
	$	—	—	—	—	8,788	12,248
Mercury	kg	—	—	—	—	414	—
	$						
Molybdenum	kg	—	—	—	—	13 027	14 416
	$	—	—	—	—	71,201	91,873
Nickel	kg	—	—	—	—	242 180	262 492
	$	—	—	—	—	1,100,523	1,232,143
Platinum group	g	—	—	—	—	12 417	13 375
	$	—	—	—	—	56,493	48,790
Selenium	kg	—	—	—	—	182	260
	$	—	—	—	—	7,362	9,134
Silver	kg	197	98	61	108	1 235	1 272
	$	28,531	13,446	8,884	14,885	178,864	175,128
Tantalum (Ta₂O₅)	kg	—	—	—	—	..	..
	$	—	—	—	—	..	..
Tellurium	kg	—	—	—	—	20	24
	$	—	—	—	—	414	529
Tin	kg	—	—	—	—	319	275
	$	—	—	—	—	2,366	1,873
Tungsten (WO₃)	kg	—	—	1 478	—	1 478	..
	$	—	—	..	—	..	
Uranium (U₃O₈)	kg	—	—	—	—	5 517	6 058
	$	—	—	—	—	..	..
Yttrium (Y₂O₃)	kg	—	—	—	—	..	..
	$	—	—	—	—	..	..
Zinc	kg	115 395	51 723	129 002	144 307	1 055 151	1 039 688
	$	95,401	42,898	106,650	119,685	872,328	862,296
NON-METALLICS	$	*32,821*	*34,460*	—	—	*939,180*	*1,142,516*
Asbestos	t	104	103	—	—	1 056	1 549
	$	32,821	34,460	—	—	267,246	445,523
Barite	t	—	—	—	—	..	..
	$	—	—	—	—	2,306	1,860
Diatomite	t	—	—	—	—	..	..
	$	—	—	—	—	..	..
Feldspar	t	—	—	—	—	—	—
	$	—	—	—	—	—	—
Fluorspar	t	—	—	—	—	—	..
	$	—	—	—	—	—	2,246
Gemstones	kg	—	—	—	—	110	..
	$	—	—	—	—	414	414
Gypsum	t	—	—	—	—	5 719	5 663
	$	—	—	—	—	20,304	22,906
Helium	m³	—	—	—	—	..	..
	$	—	—	—	—	..	..
Magnesitic dolomite and brucite	t	—	—	—	—	..	..
	$	—	—	—	—	5,358	5,116
Nepheline syenite	t	—	—	—	—	468	541
	$	—	—	—	—	8,869	10,828
Nitrogen	m³	—	—	—	—	..	..
	$	—	—	—	—	..	..
Peat moss	t	—	—	—	—	361	363
	$	—	—	—	—	22,273	22,500
Potash (K₂O)	t	—	—	—	—	4 673	5 126
	$	—	—	—	—	358,570	361,442
Pyrite, pyrrhotite	t	—	—	—	—	21	31
	$	—	—	—	—	127	240
Quartz	t	—	—	—	—	2 492	2 376
	$	—	—	—	—	13,112	13,895
Salt	t	—	—	—	—	5 123	5 752
	$	—	—	—	—	59,714	75,691

12.8   Detailed mineral production, by province, 1975 and 1976p (thousands) (concluded)

Mineral		Province or territory					
		Yukon Territory		Northwest Territories		Canada	
		1975	1976	1975	1976	1975	1976
NON-METALLICS (concluded)							
Soapstone, talc, pyrophyllite	t	—	—	—	—	66	65
	$	—	—	—	—	1,538	1,774
Sodium sulphate	t	—	—	—	—	472	490
	$	—	—	—	—	22,049	24,878
Sulphur, in smelter gas	t	—	—	—	—	695	781
	$	—	—	—	—	9,641	15,454
Sulphur, elemental	t	—	—	—	—	4 079	3 781
	$	—	—	—	—	91,847	63,339
Titanium dioxide, etc.	t	—	—	—	—	...	...
	$	—	—	—	—	55,812	74,410
FUELS	$	*1,310*	*600*	*24,279*	*27,942*	*6,653,355*	*7,993,404*
Coal	t	—	—	—	—	25 259	25 311
	$	—	—	—	—	586,423	604,000
Natural gas	m³	53 009	28 317	876 237	925 961	87 485 758	86 858 171
	$	1,310	600	19,742	19,578	1,520,661	2,466,621
Natural gas byproducts	m³	—	—	—	—	17 835	16 543
	$	—	—	—	—	782,337	794,325
Oil, crude	m³	—	—	160	156	83 001	77 843
	$	—	—	4,537	8,364	3,763,934	4,128,458
STRUCTURAL MATERIALS	$					*960,336*	*1,015,768*
Clay products	$	—	—	—	—	85,977	92,110
Cement	t	—	—	—	—	9 965	9 850
	$	—	—	—	—	320,172	339,159
Lime	t	—	—	—	—	1 602	1 825
	$	—	—	—	—	46,907	54,099
Sand and gravel	t	—	—	—	—	247 155	247 660
	$	—	—	—	—	305,181	320,800
Stone	t	—	—	—	—	88 921	87 180
	$	—	—	—	—	202,099	209,600
Total 1975	$	230,150	...	206,349	...	13,346,724[1]	...
Total 1976	$	...	131,069	...	213,100	...	15,392,839[2]

[1]Includes 928 864 tonnes of sand and gravel valued at $1,786,565 produced in Prince Edward Island.
[2]Includes 816 000 tonnes of sand and gravel valued at $1,700,000 produced in Prince Edward Island.

12.9   Producers' shipments of copper, by province, and total value, 1970-76

Year	Province or territory					
	New-foundland	Nova Scotia	New Brunswick	Quebec	Ontario	Manitoba
	t	t	t	t	t	t
1970	13 783	24	7 277	156 618	267 703	43 460
1971	12 682	15	9 313	167 669	274 306r	50 135
1972	8 630	—	9 354r	160 056	262 832	54 279
1973	7 844	3	9 353	143 191	260 656	64 712
1974	5 654	5	11 369	144 234	283 897	71 083
1975	7 500	—	11 212	117 556	257 778	64 495
1976p	6 764	—	9 678	120 411	258 981	56 687

Year	Saskat-chewan	British Columbia	Yukon Territory	Northwest Territories	Canada	
					Shipments	Value
	t	t	t	t	t	$'000
1970	17 666	96 000	7 149	599	610 279	779,242
1971	10 111	127 287	2 328	625	654 471r	760,016
1972	11 382	211 834r	793	514	719 674r	806,427
1973	9 275	317 605r	10 517	787	823 943r	1,157,507
1974	7 987	287 549	9 111	492	821 381	1,402,571
1975	7 905	258 518	8 487	375	733 826	1,030,502
1976p	9 596	273 541	11 039	438	747 135	1,126,156

12.10   Producers' shipments of nickel, by province, and total value, 1970-76

Year	Quebec	Ontario	Manitoba	Saskat-chewan	British Columbia	Yukon Territory	Canada	
	t	*t*	*t*	*t*	*t*	*t*	Quantity *t*	Value *$'000*
1970	727	203 441	71 777	—	1 546	—	277 491	830,167
1971	679	195 729	69 461	—	1 154	—	267 023	800,064
1972	277	171 846	60 080	—	1 470	1 276	234 949	717,485
1973	328	178 395	67 660	—	1 119	1 545r	249 047r	813,101
1974	—	209 051r	59 331	—	689	—	269 071r	974,594
1975	—	179 095	63 085	—	—	—	242 180	1,100,523
1976P	—	211 584	50 908	—	—	—	262 492	1,232,143

12.11   Iron ore shipments and production of pig iron and steel ingots and castings, 1970-76

Year	Iron ore shipments						Production of pig iron *'000 t*	Production of steel ingots and castings *'000 t*
	Newfound-land (incl. Labrador) *'000 t*	Quebec *'000 t*	Ontario *'000 t*	British Columbia *'000 t*	Canada			
					Quantity *'000 t*	Value *$'000*		
1970r	21 373	13 651	10 731	1 704	47 459	588,631	8 227	11 113
1971	19 846	11 219	10 141	1 751	42 957	555,136	7 834	11 047
1972	16 395	10 537	10 664	1 139	38 735	489,023	8 488	11 854
1973	22 133	12 674	11 271	1 420	47 498	606,106	9 547	13 304
1974r	22 027	12 545	10 906	1 307	46 785	724,150	9 452	13 568
1975	22 586	11 501	9 504	1 302	44 893	918,065	9 228	12 944
1976P	27 970	17 754	10 369	809	56 902	1,241,263	9 801	13 326

12.12   Producers' shipments of lead from Canadian ores, by province, and total value, 1970-76

Year	Newfoundland *t*	Nova Scotia *t*	New Brunswick *t*	Quebec *t*	Ontario *t*
1970	16 084	1 178	56 858	1 959	10 850
1971	12 230	376	59 334	587	8 088
1972	11 069	—	41 268	1 520	9 621
1973	7 660	264	39 926	1 226	10 429
1974	14 052	197	47 829	958	9 173
1975	5 219	—	59 092	1 644	6 192
1976P	9 253	—	61 687	923	6 379

Year	Manitoba *t*	British Columbia *t*	Yukon Territory *t*	Northwest Territories *t*	Canada	
					Quantity *t*	Value *$'000*
1970	458	97 449	59 725	108 502	353 063	123,138
1971	182	112 458	98 582	76 035	367 872	109,488
1972	178	88 519	101 116	81 846	335 137	113,990
1973	58	84 892	106 831	90 668	341 954	121,676
1974	40	55 253	90 242	76 525	294 269	134,330
1975	119	70 612	122 864	83 391	349 133	155 973
1976P	275	88 633	38 254	53 679	259 083	129,388

12.13   Producers' shipments of zinc, by province, and total value, 1970-76

Year	Newfoundland *t*	Nova Scotia *t*	New Brunswick *t*	Quebec *t*	Ontario *t*	Manitoba *t*
1970	27 137	—	146 142	186 000	308 662	35 800
1971	18 899	—	146 523	158 230	331 780	22 667
1972	24 115	—	158 336	148 092	365 950	41 374
1973	7 888	—	174 690	140 849	414 006	60 233
1974	19 092	—	149 809	125 447	435 502	62 484
1975	32 198	—	149 210	125 241	335 852	64 045
1976P	45 616	—	171 733	122 934	320 030	61 935

Year	Saskatchewan *t*	British Columbia *t*	Yukon Territory *t*	Northwest Territories *t*	Canada	
					Quantity *t*	Value *$'000*
1970	19 807	125 006	70 745	216 416	1 135 715	398,859
1971	7 844	138 551	105 748	203 497	1 133 739	418,161
1972	15 082	121 721	107 604	154 103	1 136 377	477,783
1973	12 178	137 381	114 905	164 450	1 226 580	652,944
1974	5 902	77 734	79 151	171 886	1 127 007	867,135
1975	4 539	99 669	115 395	129 002	1 055 151	872,328
1976P	8 144	113 266	51 723	144 307	1 039 688	862,296

12.14   Producers' shipments of gold, by province, and total value, 1970-76

Year	New-foundland g	Nova Scotia g	New Brunswick g	Quebec g	Ontario g	Manitoba g	Saskat-chewan g
1970	211 846	—	159 250	21 866 211	36 143 546	1 077 487	1 396 204
1971	228 331	—	131 754	20 118 942	35 270 938	935 064	807 446
1972	437 595	1 306	99 687	16 785 582	31 703 867	1 182 927	949 496
1973	446 179	—	161 800	14 888 612	28 686 830	1 493 682	825 082
1974	360 956	—	133 621	13 702 917	24 917 151	1 634 612	471 404
1975	404 096	93	105 472	14 155 441	23 487 729	1 496 730	454 608
1976P	405 000	—	125 000	14 743 000	22 830 000	1 431 000	591 000

Year	Alberta g	British Columbia g	Yukon Territory g	Northwest Territories g	Canada Quantity g	Value $'000
1970	4 728	3 147 579	555 570	10 352 606	74 915 027	88,057
1971	2 457	2 781 055	450 161	9 590 415	70 316 563	79,903
1972	93	3 799 570	126 871	9 563 666	64 650 660	119,742
1973	5 443	5 883 067	648 974	7 747 098	60 786 767	190,376
1974	3 017	5 041 283	823 371	5 737 565	52 825 897	263,794
1975	7 651	4 862 656	997 986	5 460 651	51 433 113	270,830
1976P	—	5 506 000	965 000	5 848 000	52 444 000	207,796

12.15   Producers' shipments of silver, by province, and total value, 1970-76

Year	Average price per kg (Canadian funds) $	New-foundland kg	Nova Scotia kg	New Brunswick kg	Quebec kg	Ontario kg	Manitoba kg
1970	59.48	24 678	2 229	142 390	132 562	618 226	20 552
1971	50.16	17 530	1 720	157 310	136 171	581 064	21 595
1972	53.69	17 820	—	121 505	110 667	609 245	25 143
1973	81.21	17 820	710	110 998	94 897	610 170	33 678
1974	148.83	17 284	784r	138 860	92 388	555 272	39 421
1975	144.87	13 841	—	157 049	105 662	463 694	31 153
1976P	137.71	15 925	—	156 109	114 679	486 552	27 714

Year		Saskat-chewan kg	British Columbia kg	Yukon Territory kg	Northwest Territories kg	Canada Quantity kg	Value $'000
1970	59.48	15 301	202 525	131 901	85 990	1 376 354[1]	81,864
1971	50.16	7 426	238 694	178 774	91 209	1 431 493[1]	71,797
1972	53.69	11 958	215 424	155 174	126 257	1 393 193	74,803
1973	81.21	14 253	236 990	188 922	168 592	1 477 030[1]	119,954
1974	148.83	7 005	181 706	180 082	118 728	1 331 531[1]	198,166
1975	144.87	8 339	196 638	196 943	61 319	1 234 642[1]	178,864
1976P	137.71	10 078	254 956	97 634	108 085	1 271 732	175,128

[1]Includes relatively small quantities produced in Alberta.

12.16   Quantity and value of producers' shipments of uranium (U₃O₈), by province, 1970-76

Year	Ontario Quantity t	Value $'000	Saskatchewan Quantity t	Value $'000	Canada Quantity t	Value $'000
1970	3 029	..	695	..	3 724	..
1971	3 180	..	546	..	3 726	..
1972	3 823	..	606	..	4 429	..
1973	3 681	..	636	..	4 317	..
1974	3 830	..	521	..	4 351	..
1975	4 794	..	723	..	5 517	..
1976P	3 892	..	2 166	..	6 058	..

12.17 Quantity and value of producers' shipments of asbestos, 1970-76

Year	Quantity '000 t	Value $'000
1970	1 508	208,147
1971	1 483	203,999
1972	1 530	206,089
1973	1 690	234,323
1974	1 644	302,013
1975	1 056	267,246
1976P	1 549	445,523

12.18 Producers' shipments of potash, 1970-76

Year	K₂O eq. '000 t	Value $'000
1970	3 103	108,695
1971	3 629	134,955
1972	3 494	135,513
1973	4 453	176,876
1974	5 776	308,925
1975	4 673	358,570
1976P	5 126	361,442

12.19 Producers' shipments of salt, by province, and total value, 1970-76

Year	Nova Scotia '000 t	Ontario '000 t	Manitoba '000 t	Saskat- chewan '000 t	Alberta '000 t	Canada Quantity '000 t	Value $'000
1970	621	3 772	26	184	258	4 861	36,098
1971	806	3 785	24	190	222	5 027	40,111
1972	738	3 663	28	228	258	4 915	40,144
1973	682	3 776	33	259	299	5 049	49,631
1974	783	4 043	29	269	323	5 447	60,619
1975	768	3 763	27	277	288	5 123	59,714
1976P	902	4 246	28	281	295	5 752	75,691

12.20 Quantity and value of sulphur produced from smelter gases and in pyrite and pyrrhotite shipments, and of elemental sulphur sales, 1970-76

Year	Sulphur in smelter gases Quantity[2] '000 t	Value $'000	Producers' shipments pyrite and pyrrhotite Gross weight[3] '000 t	Sulphur content '000 t	Value $'000	Sales of elemental sulphur[1] Quantity '000 t	Value $'000
1970	640	7,433	329	160	1,699	3 219	28,354
1971	561	4,632	288	141	1,162	2 857	21,300
1972	616	5,118	114	54	456	3 299	19,588
1973	687	10,070	24	12e	173	4 168	23,816
1974	663	9,813	49	24e	347	5 033	68,556
1975	695	9,641	21	11	127	4 079	91,847
1976P	781	15,454	31	16	240	3 781	63,339

[1]Recovered from sour natural gas and nickel sulphide ores.
[2]Includes sulphur in acid made from roasting zinc sulphide concentrates at Arvida and Port Maitland.
[3]Excludes pyrite and pyrrhotite used to produce iron residues or sinter.

12.21 Producers' shipments of gypsum, by province, and total value, 1970-76

Year	New- foundland '000 t	Nova Scotia '000 t	New Brunswick '000 t	Ontario '000 t	Manitoba '000 t	British Columbia '000 t	Canada Quantity '000 t	Value $'000
1970	446	4 332	66	487	156	245	5 732	14,199
1971	509	4 436	70	634	118	313	6 080	15,083
1972	667	5 442	68	659	160	352	7 348	19,336
1973	734	5 605	83	685	172	331	7 610	21,067
1974	504	5 351	79	702	188	400	7 224	22,437
1975	583	3 895	53	631	83	474	5 719	20,304
1976P	558	3 874	42	571	83	535	5 663	22,906

12.22 Production and exports of nepheline syenite, 1970-76

Year	Production Quantity '000 t	Value $'000	Exports Quantity '000 t	Value $'000
1970	442	5,801	347	5,063
1971	469	6,206	372	5,333
1972	507	5,902	401	5,789
1973	516	7,860	408r	6,138r
1974	560	9,179	455r	8,023r
1975	468	8,869	356	7,125
1976P	541	10,828	415	8,280

12.23   Producers' shipments and value, imports, exports and apparent consumption of cement, 1970-76

Year	Shipments (sold or used)		Imports '000 t	Exports[1] '000 t	Apparent consumption[2] '000 t
	'000 t	$'000			
1970	7 208	155,740	88	514	6 783
1971	8 234	183,374	51	806	7 479
1972	9 107	210,685	39	1 178	7 968
1973	10 225	241,945	117	1 279	9 063
1974[r]	10 585	281,958	251	1 148	9 688
1975	9 965	320,172	421	935	9 451
1976[p]	9 850	339,159	315	920	9 245

[1]Standard portland cement.
[2]Shipments plus imports less exports.

12.24   Producers' shipments of sand and gravel, by province, and total value, 1970-76

Year	New-foundland '000 t	Prince Edward Island '000 t	Nova Scotia '000 t	New Brunswick '000 t	Quebec '000 t	Ontario '000 t
1970	3 933	750	6 520	6 244	33 380	75 185
1971	5 048	1 410	5 447	4 522	37 743	70 426
1972	4 929	1 432	8 978	6 859	40 817	69 291
1973	5 866	1 481	10 295	8 666	46 759	73 090
1974[r]	6 144	884	10 503	7 485	60 248	72 561
1975	6 237	929	8 906	3 834	82 039	69 705
1976[p]	5 080	816	8 618	3 538	83 007	68 039

Year	Manitoba '000 t	Saskatchewan '000 t	Alberta '000 t	British Columbia '000 t	Canada Quantity '000 t	Value $'000
1970	13 544	8 131	14 553	21 606	183 846	133,558
1971	15 145	10 270	16 945	26 538	193 494	152,628
1972	13 393	7 722	18 648	32 225	204 294	178,100
1973	11 596	6 202	16 880	30 959	211 794	213,437
1974	17 272	10 741	22 410	31 048	239 296[r]	236,985[r]
1975	16 417	8 313	20 453	30 322	247 155	305,181
1976[p]	17 418	7 439	23 133	30 572	247 660	320,800

12.25   Producers' shipments of stone[1], by province, and total value, 1970-76

Year	New-foundland '000 t	Nova Scotia '000 t	New Brunswick '000 t	Quebec '000 t	Ontario '000 t
1970	165	1 081	1 199	27 346	25 105
1971	185	1 491	1 298	34 033	25 618
1972	185	950	1 725	37 543	28 206
1973	357	738	2 500	43 686	32 240
1974	617	1 511	2 833[r]	50 014[r]	31 261
1975	877	1 581	3 241	50 763	27 526
1976[p]	816	1 451	2 540	51 710	26 853

Year	Manitoba '000 t	Alberta '000 t	British Columbia '000 t	Canada Quantity '000 t	Value $'000
1970	1 155	151	3 058	59 260	87,976
1971	918	167	2 982	66 692	96,537
1972	553	178	3 419	72 759	103,326
1973	567	146	3 475	83 709	128,693
1974	2 223	163	4 211	92 833[r]	177,207[r]
1975	477	235	4 221	88 921	202,099
1976[p]	454	181	3 175	87 180	209,600

[1]Excludes limestone used in Canadian lime and cement industries.

12.26   Value (total sales) of producers' shipments of clay products made from domestic clays, by province, 1970-76 (thousand dollars)

Year	Province or territory				
	New-foundland	Nova Scotia	New Brunswick	Quebec	Ontario
1970	37	2,816	940	8,160	28,649
1971	80	1,844	627	6,565	30,538
1972	257	1,684	668	8,300	30,484
1973	260	2,101	840	9,725	34,601
1974r	436	742	1,244	12,194	37,969
1975	457	3,155	1,310	16,468	44,769
1976p	475	3,915	2,464	14,243	50,926
	Manitoba	Saskat-chewan	Alberta	British Columbia	Canada
1970	346	1,819	4,657	4,367	51,791
1971	469	1,140	4,031	4,900	50,194
1972	667	1,758	4,438	4,301	52,557
1973	1,257	2,014	4,782	5,590	61,170
1974r	1,366	2,406	5,964	4,732	67,053
1975	1,386	2,730	8,530	7,172	85,977
1976p	1,318	3,098	8,727	6,944	92,110

12.27   Quantity and value of production[1] of crude oil, by province, 1970-76

Year	New Brunswick		Ontario		Manitoba		Saskatchewan	
	Quantity '000 m³	Value $'000	Quantity '000 m³	Value $'000	Quantity '000 m³	Value $'000	Quantity '000 m³	Value $'000
1970	2	14	167	2,840	939	14,858	14 227	199,770
1971	2	13	152	2,727	891	15,413	14 064	217,829
1972	1	12	140	2,499	836	14,588	13 798	214,057
1973	2	14	129	2,866	808	17,148	13 663	264,057
1974	1	11	117	4,342	755	27,164	11 757	397,835
1975	1	51	112	5,046	702	31,445	9 408	407,302
1976p	1	24	100	6,028	626	32,995	8 824	435,675
	Alberta		British Columbia		Northwest Territories		Canada	
	Quantity '000 m³	Value $'000	Quantity '000 m³	Value $'000	Quantity '000 m³	Value $'000	Quantity '000 m³	Value $'000
1970	53 802	876,887	4 051	60,943	135	1,142	73 323	1,156,454
1971	59 064	1,055,769	4 016	63,984	150	1,208	78 339	1,356,943
1972	70 625	1,272,903	3 806	63,710	141	1,059	89 347	1,568,828
1973	86 129	1,894,724	3 389	65,643	153	2,240	104 273	2,246,692
1974	81 948	2,985,549	3 012	103,501	152	3,167	97 742	3,521,569
1975	70 332	3,220,636	2 286	94,917	160	4,537	83 001	3,763,934
1976p	65 799	3,531,100	2 337	114,272	156	8,364	77 843	4,128,458

[1]Gross production of crude oil and condensate, less returned to formation.

12.28   Natural gas production[1], by province and total value, 1970-76

Year	New Brunswick		Quebec		Ontario		Saskatchewan		Alberta	
	Quantity '000 m³	Value $'000	Quantity '000 m³	Value $'000	Quantity '000 m³	Value $'000	Quantity '000 m³	Value $'000	Quantity '000 m³	Value $'000
1970r	3 714	108	4 696	25	483 196	6,488	1 772 467	7,332	52 966 869	265,912
1971r	2 976	91	4 819	25	460 424	6,333	2 015 185	8,952	58 537 937	290,672
1972r	2 750	57	5 291	26	350 424	4,768	1 951 359	8,932	67 548 151	338,709
1973r	2 294	31	5 588	28	269 806	3,678	1 865 715	9,044	71 857 999	388,696
1974r	2 492	44	5 182	27	213 424	3,248	1 745 337	9,001	71 854 432	645,138
1975	2 577	55	1 416	8	309 645	6,508	1 738 796	10,449	73 335 204	1,405,247
1976p	2 775	59	255	2	141 584	5,760	1 608 397	8,250	73 652 127	2,302,235
	British Columbia		Yukon Territory		Northwest Territories		Canada			
	Quantity '000 m³	Value $'000	Quantity '000 m³	Value $'000	Quantity '000 m³	Value $'000	Quantity '000 m³	Value $'000		
1970r	9 247 286	35,200	—	—	2 320	35	64 480 548	315,100		
1971r	9 710 098	36,269	24 610	90	8 427	117	70 764 476	342,549		
1972r	12 236 623	43 043	73 611	279	333 987	1,372	82 502 196	397,186		
1973r	13 287 773	46,052	—	—	1 044 127	4,324	88 333 302	451,853		
1974r	11 491 233	60,581	32 225	190	894 812	5,337	86 239 137	723,766		
1975	11 168 874	77,342	53 009	1,310	876 237	19,742	87 485 758	1,520,661		
1976p	10 498 755	130,137	28 317	600	925 961	19,578	86 858 171	2,466,621		

[1]Gross production, less field-flared and waste and re-injected.

Sources
12.1 - 12.28 Minerals and Metals Division, Mineral Development Sector, Department of Energy, Mines and Resources.

Energy

Chapter 13

Tables

Energy

Chapter 13

Canada's energy concerns

13.1

The international scene

13.1.1

Increasing world concern surrounds the adequacy of future energy supply prompted by growth in energy demand of recent years and dependence on Middle East oil.

The need for concerted action by major industrial nations was emphasized at a meeting of the International Energy Agency (IEA) chaired by the Canadian minister of energy, mines and resources in October 1977. It was concluded that "as early as the 1980s the world will not have sufficient oil and other forms of energy available." To support the need for strong measures, Canada was committed to limiting imported oil in 1985 to the lesser of one-third of oil requirements or 127 190 cubic metres (800,000 barrels) a day.

One objective adopted for the 19-member group of IEA countries was to hold total oil imports to not more than 4.1 million m³ (26 million bbl) a day in 1985 and to reinforce national energy policies to meet this objective. Twelve principles were established as guidelines to implement national measures.

Early in 1976 the IEA adopted a program of long-term co-operation to promote energy conservation, accelerate development of alternative sources, promote research and development and reduce legislative and administrative obstacles. Canada participates in this program, as well as in many of the co-operative research and development agreements. In 1978 Canada was expected to participate in the first full-scale test of the IEA emergency oil-sharing program.

In international energy affairs, membership in such international bodies as the United Nations, the Organization for Economic Co-operation and Development, the North American Treaty Organization as well as participation in conferences such as the 1976-77 conference on international economic co-operation, co-chaired by Canada, and the international nuclear fuel cycle evaluation, provide the forum for Canadian activities.

A conference on international economic co-operation, in which Canada played an active role in the energy and development commissions, ended in mid-1977 with mixed results. No agreement was reached in the important areas of debt relief assistance to developing countries, protection of purchasing power for oil-producing developing countries, stabilization of commodity prices and measures for industrial assistance to developing countries. However, establishment of an international energy co-operation and development program was agreed on to ease the transition for all countries to a new energy economy as world supplies of conventional hydrocarbons from petroleum become depleted.

Discussion and consultation on emerging world energy problems continue in many international forums, such as the UN, where Canada actively promotes international co-operation in the transition to new sources, stabilizing supplies and prices, implementing conservation measures and facilitating the financing of exploration and research and development.

A list of countries from which Canada imports oil, by volume and by value, appears in Table 13.1.

Energy research and development

13.1.2

An interdepartmental task force on energy R&D was established in 1974 to develop, implement and co-ordinate a federal program of energy R&D. The office of energy research and development was set up at that time in the energy, mines and resources department to serve as a secretariat; the report of the task force *Science and technology for Canada's energy needs* was published in 1975. It recommended major increases in federally-supported energy R&D, to be implemented in co-operation with provincial governments and industry, under the leadership of a new permanent interdepartmental panel on energy R&D.

This panel, supported by the energy research and development office, established new priorities for R&D and obtained increases in funding of $10 million each in both 1976-77 and 1977-78, raising the total for energy R&D in 1977-78 to $138 million. The latest increase for 1977-78 was apportioned as follows: renewable energy $4.4 million, energy conservation $3.7 million, fossil fuels $1.5 million, and transportation and transmission $0.4 million. Additional energy R&D funding was anticipated for environmental studies, biomass energy, health and safety programs for energy industries, tidal energy, fossil fuels resource assessment and socio-economic energy research.

A breakdown of total expenditures on energy research and development reflects supplementary funding of $15.0 million, announced in 1978, bringing total annual federal expenditure to $144.5 million for 1978-79 (Table 13.2).

13.1.3   Renewable energy sources

Government activity in renewable energy resources increased significantly in 1977. Federal spending rose from $4.4 million in 1976-77 to $7.4 million in 1977-78. Of the $13.5 million allotted for 1978-79, half was earmarked for solar energy research and development. Within the federal energy, mines and resources department a renewable energy resources branch was formed to develop government policies and programs in this area.

Research studies on the potential of solar and biomass energy, as well as investigations into the labour, economic, and industrial impact of these new technologies, are being carried out. A number of technical and general publications are available and a national advisory committee on conservation and renewable energy has been set up to provide direct public advice on these matters and related policy issues.

The National Research Council of Canada (NRC) directs research and development activities in renewable energy, and its new solar energy research group emphasized development of solar heating systems for multiple-unit residential quarters in 1977. Other NRC activities include testing the performance of solar collectors, and data monitoring and collection.

Prospects for solar energy. About 32% of the total energy demand in Canada is for low-grade heat below 100°C. In principle, almost a third of total demand could be met from heat generated by solar energy, but in practice many obstacles to the widespread use of solar space heating still exist, not the least of which is high initial cost. Other difficulties include the lack of sturdy and long-lasting commercially available solar panels adapted to the Canadian climate and the non-existence of any significant solar manufacturing industry in Canada; the absence of adequate consumer protection standards or legal guarantees to the right to sunlight; and unfavourable municipal and property tax structures. In the light of these institutional rather than purely technical difficulties, it is estimated that solar space and domestic water heating will contribute only between 5% and 7% of the total energy budget by the year 2000. At the end of 1977, there were between 100 and 150 solar heated houses in Canada in various stages of completion.

Biomass energy. Another form of renewable energy appears to hold greater promise in the short term than solar heating — energy from biomass. In particular, wastes and residues from forest industries, if put to effective use, could make the entire industry energy self-sufficient. The key technological step is the wood gasifier, to convert wood particles to a gas that burns with a higher flame temperature than wood itself. Other uses of biomass energy forms include manufacture of methanol from wood products and efficient use of municipal and agricultural wastes.

Wind energy. An NRC-designed vertical-axis windmill in the Magdalen Islands was successfully commissioned in 1977. Built by Dominion Aluminium Fabricating Co. and operated by Hydro-Québec, it has a peak operating capacity of 230 kilowatts (kW), and is intended to supplement the local electrical grid, now powered entirely by diesel generators. Use of wind to generate electricity appears to be competitive with conventional energy sources in some parts of Canada such as the Atlantic provinces, the coastal regions of Hudson Bay and southern Alberta — areas where winds are sufficiently strong and constant to make operation of windmills economically feasible or where electricity is being generated by expensive diesel fuel, or both.

Other renewable energy technologies. Other renewable methods of converting energy include harnessing the tidal power in the Bay of Fundy. The cost of producing electricity from this source seems to be approaching the cost-competitive line of other methods of electrogeneration. In the West, geothermal energy still awaits proof of technical feasibility.

Energy from all conventional sources is discussed in detail later in this chapter.

Energy conservation   13.2

A policy paper on energy conservation — *Energy conservation in Canada: programs and perspectives* — published by energy, mines and resources, outlined measures to increase efficiency of energy use to reduce the growth rate of energy consumption in an environmentally and socially acceptable way. Given the right combination of circumstances, the average annual growth in energy use could be in the range of 3.5% to 2.0% by 1990. A preliminary estimate based on consumption in the first nine months of 1977 indicates an increase in energy consumption of 3.3% over 1976.

If, in 1990, the mix of energy sources were the same as that prevailing in 1975, estimates indicate that primary energy savings could amount to: petroleum, 1.29 quads equivalent to the annual output of five Syncrude-size oil sands plants, or about 95 390 cubic metres (600,000 barrels) a day of crude oil; 0.51 quads of natural gas equivalent to about 20% of total Canadian production in 1975; as well as 0.79 quads of electric power, equivalent to the annual output of 13 Pickering-size nuclear plants; and 0.21 quads of coal equivalent to 7.7 million tonnes (8.5 million short tons) of bituminous coal. [Note: 1 quad = 1 quadrillion Btus (10^{15} Btus) = 172 million bbl of crude oil]

This is not a forecast of energy demand since many economic, social and technological changes may occur by the 1990s. Moreover, conservation savings in some sectors cannot be quantified with any accuracy.

In 1975 an industrial energy conservation program was developed: there are now 12 task forces representing specific sectors of industry. Their role is to set energy efficiency targets, exchange information on conservation opportunities and discuss approaches to the entire conservation program with the federal government.

In the public sector, the federal government's internal energy conservation program set a target to reduce energy use for 1976-77 by 10% from the previous year, holding that level for the next 10 years. From preliminary data, it appears an energy saving of about 9% was achieved, valued at more than $25 million a year.

In January 1977, the federal government announced special assistance programs for Prince Edward Island and Nova Scotia, because of their high dependence on foreign oil for electricity generation. This program included grants for householders to insulate, energy audit buses for industry, and industry consultation and grant programs. Based on the success of the insulation grant program, a $1.4 billion national home insulation grant program began September 1, 1977. In 1977, a special R&D fund of $1.5 million was created to improve energy efficiency in industrial processes. The energy bus program was made available to all provinces.

In June 1977 a draft code for energy conservation in new buildings was released for public comment and most provinces agreed to adopt energy standards at least as high as those contained in the new draft code by early 1978.

A number of important steps have been taken by the federal government to conserve energy. Fleet average performance standards for new cars for 1980 and 1985 have been introduced to ensure that total gasoline consumption in Canada in 1985 will be below the level of 1976 even though more cars likely will be on the roads. Provincial governments have been urged to adopt a 90 kilometres per hour speed limit on most highways and to impose higher registration fees for heavy cars. A surtax of $100 on automobile air conditioners has been imposed. Graduated weight or fuel economy taxes for cars and station wagons ranging from $30 to $300 are in effect for cars over 2 007 kilograms and increases are being considered. Auto makers and dealers have been urged to display automobile fuel economy ratings under a voluntary program. The program may become mandatory. The federal government collects a 10-cent a gallon excise tax on gasoline, partly to encourage thrift in its use. Federal sales taxes have been removed on insulation and on energy saving equipment and rapid depreciation allowances are available for similar equipment. A program of energy labelling for appliances is to be initiated.

In April 1976 the federal government issued a blueprint to manage Canada's energy future. The document outlined nine policy elements and five major related targets to deal with energy problems over the next 10 to 15 years with the general aim of reducing vulnerability to arbitrary changes in price or supply of imported energy by using domestic resources to the greatest possible extent, and by protecting against interruptions in imported supplies.

The policy areas include appropriate energy pricing, conservation, increased exploration and development, better resource information, substituting domestic energy for expensive imported energy, new or improved transportation and transmission systems, emergency preparedness, increased research and development, and greater Canadian content and participation.

The five major energy-related targets in the energy strategy document included: moving domestic oil prices toward international levels and moving domestic prices for natural gas to an appropriate competitive relationship with oil by 1980; reducing the average rate of growth of energy use to less than 3.5% a year by 1986; reducing net dependence on imported oil in 1985 to one-third of total oil demand; maintaining self-reliance in natural gas until such time as northern resources can be brought to market under acceptable conditions; and doubling exploration and development in frontier areas under acceptable social and environmental conditions.

13.3   Energy supply and demand

Canada's energy needs are met by oil, natural gas, coal, uranium and electricity. In terms of primary energy consumption, the share of oil as an energy source is 46%, that of natural gas 19% and coal 8%; of the total, 7% is used to produce electricity. About 27% of energy needs are met by hydro and nuclear power. Although nuclear power accounts for little more than 2% of total supply, it will become an increasingly important source of electric power. Hydroelectricity and thermal generation of electricity from coal, while remaining significant, will decline in relative importance as nuclear power development increases and the use of natural gas and oil is gradually phased out. By the end of the century, probably no more than 50% of primary energy consumption will be met by natural gas and oil.

The relative importance of energy sources, in terms of Canada's trade, is shown in Table 13.3. There was a marked change in the export-import balance in the period 1966-76, from a deficit of $107.3 million in the value of energy in 1966, on a trade balance basis, to a surplus of $862.6 million in 1976. A downward trend again became apparent in 1974 as crude oil exports declined. In 1976, with further reductions in crude oil exports, oil imports rose and the decline continued.

Canada's primary energy demand increased at an average annual rate of 5.1% over the period 1960-76, while energy use per capita grew annually by 3.4%. Higher prices, increasing attention to conservation measures and other policies should reduce per capita growth during the remainder of this decade.

Growth in oil use and related supply trends since 1965 are illustrated in Table 13.4. Production of crude oil and gas liquids increased by about 90% in the 10-year period to 1975; declines in production and exports, evident in 1975, continued in 1976. The most notable trend in 1975-76 was the decline in the export-import surplus from 6.4 megalitres (ML) a day to a deficit of 17.3 ML a day.

The natural gas supply and demand situation is illustrated in Table 13.5. In the 10-year period to 1975, production of marketable pipeline gas and domestic demand increased by more than 130%. In 1976 exports remained virtually static while domestic demand continued to grow, but at a slower rate than in previous years. With no new export approvals since 1970, and none planned except in emergencies, domestic demand growth will relate directly to the ability of the industry to develop new markets to be supplied from present producing areas, pending the opening of new sources in frontier areas.

More than 1,500 companies are involved in the Canadian petroleum industry, excluding a further 5,000 independent gas station operations; however, the top 30 oil producers accounted for 85.8% of Canadian oil production and the top 30 gas producers for 57.3% of gas production in 1976. This degree of concentration has diminished during the last decade as smaller companies move from exploration to production.

The federal government has taken a number of steps to boost supplies of oil and natural gas. It has made large direct investments in oil and gas projects including an investment of $300 million in Syncrude of Canada Ltd., which was to begin producing oil from the Athabasca tar sands in 1978. The project was granted the world price for oil. Development of new technology to tap Western Canada's heavy oil and oil sands resources is a high priority of research and includes a $96.0 million joint research fund with Alberta, and a $16.1 million heavy oils program undertaken with Saskatchewan.

Federal guarantees were provided for construction of a pipeline to move Alberta oil from Sarnia to Montreal, lessening Quebec's dependence on foreign oil. About 73% of the energy demand of that province is met by oil.

By 1977 it appeared the main future source of Alberta oil would be the costly oil sands and heavy oils. The price of this oil will be very much higher than the field price of $73.90 per cubic metre as of January 1, 1978. By world standards Canadian prices remain low.

By agreement with provincial governments, a schedule of increases has been established to bring the field price of oil to $86.48/m³ by January 1979. The schedule is subject to review. In May 1975 the minister of energy, mines and resources announced increased export prices of natural gas to $49.44 a thousand cubic metres in August, to $56.50/thousand m³ in September 1976 and to $68.51/thousand m³ in January 1977.

For the domestic market, the federal government raised the price of natural gas, after consultation with the provinces, from approximately $28.96/thousand m³ to $44.14/thousand m³ at Toronto November 1, 1975. This was increased to $49.62/thousand m³ in July 1976 and to $53.15/thousand m³ in January 1977.

Announcement of the scheduled price increases brought a sharp improvement in natural gas prospects. Federal funding and encouragement have been given to exploration in the Arctic and preliminary investigations are under way on eventual transportation of Arctic gas to southern markets. These include the possibility of an eventual Canadian link with the Alaska Highway natural gas pipeline, an application for a pipeline from the Arctic islands, participation by Petro-Canada, the national oil company, in efforts to ship liquefied Arctic gas to the Atlantic Coast by tanker, and the expansion of Quebec and Maritime access to western natural gas.

Oil and natural gas   13.4

Production   13.4.1

Production of Canadian crude oil and equivalent in 1976 declined 8.4% or an average of 21 megalitres a day from the 1975 production of 251 ML a day. Exports of crude oil and equivalent to the United States dropped 39.4 ML a day while domestic demand increased 9.4 ML a day.

In Alberta, production of conventional crude oil was down 17 ML a day, synthetic crude oil increased 1 ML a day and pentanes plus decreased 3 ML a day for a total decline of 19 ML a day or 8.8%. Saskatchewan crude oil production decreased 7.7% or 2 ML a day.

British Columbia crude oil and equivalent production remained steady at 7 ML a day while Manitoba also maintained its level of production at 2 ML a day.

13.4.2   Exploration and development

Exploratory drilling achieved record levels in 1976. Some 1,900 wells were drilled during the year, up 29% from 1,472 a year earlier. Details of drilling activity appear in Tables 13.7 and 13.8.

The high level of exploration in the West was more than maintained during the first half of 1977, resulting in a record number of discoveries. Among these was a new oil-producing venue in the West Pembina area — the first significant oil discovery in Western Canada since 1965.

It was estimated that more than 6,000 wells were drilled in Canada in 1977, up from a total of 5,682 in 1976. Alberta was the major contributor to the total, accounting for 87% of total depth drilled and 90% of well completions. Development drilling increased 32% to 2.9 million metres and exploratory drilling 34% to 1.96 million metres in the province.

The upward trend in natural gas development that began in 1975 continued in 1976 and accelerated during the first half of 1977. Completions rose by 60% over 1975 to 3,375 wells, 3,193 of which were drilled in Alberta where development of large, low-yield shallow gas reservoirs of southeastern and northern areas continued. Province-wide, about 200 single well gas discoveries were made in 1976, several of which may attain major field status. Significant deeper zone discoveries continued into 1977.

Among the latter was a Devonian Leduc reef discovery in the Pinto area, about 32 kilometres west of the Berland River field. In the Pass Creek area, several significant gas discoveries have been recorded in the past two years and this is considered one of the most promising exploration areas. To the northwest, in the Elmworth area, exploration companies have had some success in evaluating low-permeability gas reservoirs parallelling the eastern edge of the foothills for considerable distances in northwestern Alberta and northeastern British Columbia. Economics, technological development and availability of natural gas markets will likely dictate the pace of development of these low-yield resources.

In British Columbia, both depth and number of wells drilled increased substantially. Exploration companies completed the year with 175 wells drilled, including 86 potential gas producers. In Manitoba and Saskatchewan, drilling was up slightly from 1975 levels while activity declined in the Arctic.

Northern regions. The number of wells drilled and onshore geophysical activity declined in 1976 compared to 1975, but marine seismic work again showed an increase. Drilling depths decreased by 19%; however, total exploration costs increased by 12% to $240 million. Expenditures for 1977 increased a further estimated 20% although the number of wells declined 10%.

In the Mackenzie Delta–Beaufort Sea region, there were two discoveries in 1976, the oil and gas find of Sun et al Garry P-04 (actually made in 1975 but finalized in 1976) and the gas find of Imperial Netserk F-40. First exploratory drilling took place in the deeper waters of the Beaufort Sea north of the delta during 1976, with two wells drilled using ice-strengthened drill-ships. An additional three wells were drilled in the Beaufort Sea by Dome Petroleum in 1977, resulting in two gas discoveries (Ukalerk C-50 and Kopanoar M-13) and one oil and gas find (Nektoralik K-59).

Delineation drilling resulted in six successful completions in the Parsons Lake gas field, including a new-pool discovery of oil and gas, the Gulf-Mobil Kamik D-48, east of the main field. A single successful delineation well was also drilled in the Niglintgak field. In 1977, four delineation wells were drilled at Parsons Lake, including three gas successes and one unsuccessful step-out to the Kamik oil find of 1976. Single successful delineation wells were also drilled in the Taglu and Kumak gas fields.

In the Arctic islands, Panarctic et al Jackson Bay G-16A, drilling from a reinforced ice platform west of Ellef Ringnes Island, made a gas discovery similar to previous ones in this area. No discoveries of oil or gas were made in the islands region in 1977.

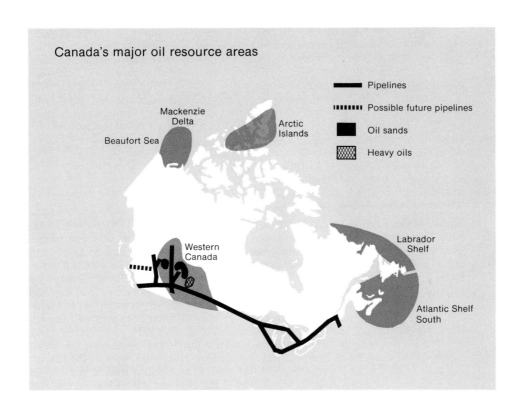

Canada's major oil resource areas

Pipelines

Possible future pipelines

Oil sands

Heavy oils

Mackenzie Delta

Beaufort Sea

Arctic Islands

Western Canada

Labrador Shelf

Atlantic Shelf South

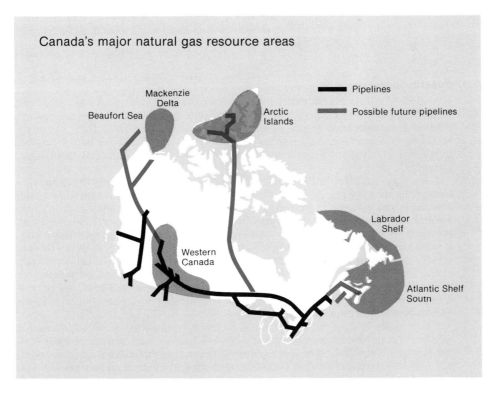

Canada's major natural gas resource areas

Pipelines

Possible future pipelines

Mackenzie Delta

Beaufort Sea

Arctic Islands

Western Canada

Labrador Shelf

Atlantic Shelf Soutn

Two successful delineation wells and one unsuccessful were drilled in the offshore extension of the Hecla gas field west of Melville Island, all from reinforced shore-fast ice platforms. One well was a deeper-pool discovery, encountering gas pay in sandstone below the main Borden Island reservoir. Delineation drilling continued in 1977 with one unsuccessful well at Hecla and one unsuccessful step-out to the Drake Point gas field offshore east of Melville Island. In July 1977, Panarctic reported proven reserves in both fields combined at 260 billion cubic metres.

The Bent Horn F-72A oil discovery of 1975 on Cameron Island was successfully offset in 1976 by the A-02 well 1.6 kilometres to the southwest, testing at 865 m³ of oil a day. Two further delineation wells drilled in 1977 were abandoned.

In the southern Northwest Territories, one dry exploratory well was drilled in 1976. In 1977, the Columbia Gas et al Kotaneelee YTH-38 well, a step-out to the North Beaver River discovery of 1964 in the southeast Yukon Foothills belt, tested gas at 600 000 m³ a day absolute open flow.

Eastern offshore region. Off the East Coast, 11 wells were drilled in 1976 compared with 10 in 1975 although the depth decreased; five wells were drilled off Labrador, where most industry interest has now concentrated. Entry of Petro-Canada into the East Coast exploration play early in the year was largely responsible for the six wells drilled off Nova Scotia in 1976. No discoveries were made off the East Coast in 1976, but production testing confirmed hydrocarbons at the Snorri J-90 well, drilled by the Total Eastcan group in 1975. The well gave flow rates of 275 000 m³ a day of gas and 37 m³ a day of condensate, the third wet gas discovery by the group on their Labrador Shelf permit block. No drilling took place off Newfoundland–Labrador in 1977.

The only drilling in 1977 was carried out on the Scotian Shelf with two wells completed, both involving Petro-Canada. The company also began work on the first of five wells in the shallow waters around Sable Island, using a jack-up type drilling-unit to delineate two discoveries made by Mobil in 1972 and 1973 and test undrilled prospects. Results of this program, scheduled for completion in 1979, will probably determine the commercial viability of oil and gas production in this region of the shelf.

Exploration expenditures off the East Coast for 1976 amounted to $62 million, down 22% from 1975, with 1977 costs estimated at about $23.5 million, a further 62% decrease, reflecting continuing inactivity off Labrador.

No exploration took place off the West Coast or in Hudson Bay in 1976 or 1977.

13.4.3 Federal incentives

Other federal initiatives contributed to the increase in exploration. A bill was introduced to create a Petroleum Corporations Monitoring Act. If passed, the act would require petroleum companies to report aspects of their cash flow resulting from increased revenues thus enabling the government to ensure that a significant portion of the increase is invested in exploration and development of petroleum and natural gas resources.

A strong incentive to exploration has been provided by a revised tax structure. Companies investing new oil or gas revenue in exploration R&D pay taxes at a lower rate than those which do not. In fact, oil and gas exploration and development expenditures amount to at least 70% of industry cash flows. Since 1975, both federal and provincial tax changes have encouraged exploration investment. Activity has been especially heavy in Alberta where additional supplies of natural gas have been located. A special federal tax allowance on wells costing $5 million or more has been introduced. This tax benefit can be applied to any source of income, making it available to investors outside the oil industry. The three-year moratorium on the proposed progressive incremental royalty for new oil and gas discoveries in the frontier regions has been extended until 1982 to encourage the search for new supplies.

Companies holding rights to potential oil and natural gas areas will, in future, be required to undertake certain exploratory work within a reasonable time. New legislation and regulations govern land tenure, royalties and Canadian-content guidelines for about 526 million hectares of land in Canada's northern and offshore oil and gas frontier regions. As an interim measure, these lands will be made available, on a selective basis, from time to time, for oil and gas exploration and development. In addition, applicants for leases on 12.5 million hectares of land have been given three land-tenure options.

Petro-Canada has been given first option for exploration agreements in frontier and offshore oil and gas areas over the next seven years to build up its holdings. Canadian content in frontier oil and gas exploration and development will be strengthened by Petro-Canada's option to acquire up to 25% working interest on lands where no significant discoveries have been made. This option is based on the amount of Canadian ownership in exploration ventures and cannot be exercised if more than 35% is already Canadian-owned.

Panarctic Oils Ltd., in which the federal government has invested $92 million, has had major successes in finding natural gas in the Arctic.

The full effect of recent exploration activity on reserves cannot yet be assessed.

Reserves   13.4.4

At the end of 1977 Canada's proven liquid hydrocarbon reserves, including conventional crude oil and natural gas liquids, amounted to 1.26 billion cubic metres made up of 0.95 billion m^3 of crude oil and 0.30 billion m^3 of natural gas liquids. These estimates do not include oil in the Athabasca bituminous sands. At the 1977 annual production level of 93.9 million m^3 the life index (reserves-to-production ratio) for conventional crude oil and natural gas liquids was 13.4 years, the same as in 1976.

The reserve position of most provinces declined except in Alberta where total reserves including natural gas liquids increased by 21 million m^3. The rise is not because of an increase in proven reserves of crude oil, but rather because ethane reserves have been included for the first time. The Canadian Petroleum Association (CPA) estimated Alberta's remaining recoverable reserves of crude oil at 0.81 billion m^3 and natural gas liquids at 0.29 billion m^3. Together, these represent about 88% of Canada's proven reserves. Saskatchewan's reserves of liquid hydrocarbons declined from 102 million m^3 to 97 million m^3 and accounted for 7.8% of the national total.

Natural gas liquids from the recently discovered, but as yet unproduced, gas fields in the Mackenzie Delta are included in the estimates but oil from the frontier regions is not, because discovered reserves of crude oil in the territories are negligible and currently well beyond economic reach.

At the end of 1977 the CPA estimated Canada's proven reserves of marketable gas at 1 684 057 million m^3, 42 022 million m^3 more than in 1976. Using the 1977 level of production of 23 422 million m^3 the life index increased to 22.95 years in 1977 from 21.85 years in 1976. Gross additions to reserves amounted to 102 965 million m^3, including 52 451 million m^3 due to extensions to existing fields, 9 939 million m^3 to new discoveries and 40 574 million m^3 to previously-estimated field reserves. Almost all of the increase was accounted for by increases in reserves in Alberta and the territories. Gross additions of marketable gas in Alberta amounted to 77 296 million m^3, most of it from extensions to existing fields. Gas reserves in the territories, which include the Mackenzie Delta and the Arctic islands, increased 15 508 million m^3, primarily by revisions for fields in the Arctic islands.

Alberta, with 1 310 035 million m^3 of marketable gas reserves, accounted for 78.0% of Canadian reserves at the end of 1977, British Columbia 11.2% and the territories 9.0%.

According to an appraisal of Alberta's oil sands completed in 1973 by Alberta's energy resources conservation board, ultimate recoverable reserves of synthetic crude oil from all of Alberta's bituminous deposits amount to 39.7 billion m^3. Of this, approximately 4.2 billion m^3 is considered recoverable by open-pit mining methods similar to those now in use at the Great Canadian Oil Sands Ltd. plant near Fort McMurray. Most of the oil from the deeper formations will only be recoverable by on-site thermal or other techniques still being developed.

In addition to these known resources, the Geological Survey of Canada estimated in 1975 that between 2.5 and 5.4 billion m^3 of combined crude oil and natural gas liquids remain to be discovered at the high (90%) and low (10%) probabilities respectively, about 80% of which will be found in frontier areas. For natural gas, between 4.1 and 8.3

trillion m³ may exist at the high and low probability levels respectively, about 90% of it in frontier areas. Details on how these estimates were made can be found in the Geological Survey Report EP 77-1 *Oil and natural gas resources of Canada 1976.*

13.5 Oil refining, gas processing

The rate of growth of petroleum product demand in recent years has been reduced because of a combination of factors — a slowdown in economic activity, higher product prices and energy conservation efforts. Surplus refining capacity has resulted, mostly in the East. Net sales of petroleum products were 96 million cubic metres in 1976, up 3.6% from 93 million a year earlier, a sharp reversal from the annual growth pattern of 5.4% over the past decade.

Table 13.9 gives details of oil refinery capacity in Canada for 1977, showing scheduled completion dates for new facilities. At the end of 1977, there were 38 refineries in Canada with total capacity of 390 800 m³ per day.

Net sales of natural gas in Canada increased 1 500 million m³ to an estimated 65 848 million m³ and exports were up slightly.

Natural gas processing capacity at the end of 1976 was 473.4 million m³/d, only 7 million m³/d more than in 1975. This small increase reflects the fact that no new major plants came on stream during 1976 although a record number of smaller ones were commissioned.

With major new gas reserve discoveries in 1976 and 1977, gas processing capacity will probably increase substantially in the near future. Plant output includes pipeline gas, propane, butanes, pentanes plus and sulphur.

Refinery expansions during the next several years are expected to centre on completion of projects already under way. These include the Texaco Canada Ltd. 15 103 m³/d Nanticoke refinery on the north shore of Lake Erie (1978 completion) and a Petrosar Ltd. petrochemical refinery in Sarnia, scheduled to operate in late 1977. This plant, when fully operational, will produce 5 644 m³/d of petrochemicals and 18 280 m³/d of petroleum products. Expansions under way include the Consumers Co-operative Refineries Ltd. plant in Regina and Chevron Standard Ltd. refinery in British Columbia.

Some expansion of existing refineries is already in progress. In 1977 Canada had 38 operating refineries with a total capacity at year end of more than 390 000 m³/d, compared to 37 refineries in 1976 with a total capacity of more than 357 000 m³/d. Refinery runs in 1976 were about 286 170 m³/d, about the same as in 1975.

In 1976 Canadian refineries yielded an average 36% of motor gasoline, 31% of middle distillates including light heating oil, diesel oil and jet fuel and about 19% of heavy fuel oil. Other products included liquefied petroleum gas, petrochemical feedstocks, aviation gasoline, asphalt, coke and lubricating oil. To meet the high yields of light products most refineries are equipped with catalytic crackers, and total installed cracking capacity in 1976 was equivalent to about 29% of crude distillation capacity.

Catalytic reforming amounted to about 18% of crude capacity. This process upgrades gasoline quality and also delivers aromatic petrochemical feedstocks. To meet the need for high quality low-sulphur distillates, hydrogen-treating plants have been installed totalling 36% of crude feed and it is common practice to hydrosulphurize most or all gas, oil and light distillates. Seven hydrocracking units have been installed in Canada capable of treating 5% of crude feed. This new process is used to upgrade heavy fuels to motor gasoline and middle distillates.

At Sarnia, Ont. three refineries are integrated with nine petrochemical companies. The oil refineries supply petroleum gases, naphtha and aromatics. The chemical companies convert them to a large number of intermediate and final products. Western Canadian natural gas is also piped into this complex. The intermediate products include ethylene, propylene, butadiene, aromatics and ethylene oxide. Final products include carbon black, synthetic rubbers, detergent alkylates, polyethylene, polystyrene, polyvinylchloride, ammonia, fertilizers, petroleum additives and many others. Many products are sold back to the refineries for blending into fuel products. Fuels are piped directly to the petrochemical plants for process heat and power requirements. Montreal and Edmonton are also major petrochemical centres but plants are distributed widely across Canada.

In the past, location and size of Canada's refineries was determined by the tendency to install them close to centres of consumption. Thus, approximately 57% of total capacity is in the populous regions of Southern Ontario and Quebec. Ontario has two main refining centres, in Sarnia and southwest of Toronto; Quebec has the largest refining centre, in Montreal, as well as a refinery in Quebec City. British Columbia has seven refineries, most close to Vancouver.

More recently the size of individual refineries is being increased for economies of scale, particularly in Alberta, Saskatchewan and Manitoba. Many small refineries have been phased out and replaced by two large refineries in Edmonton, close to the main sources of crude. They will confine the area subject to any environmental risk. Saskatchewan will lose one small refinery, but one of the two remaining will be expanded. Environmental control and conservation equipment to meet new standards is being installed.

A third factor influencing refinery location has been proximity to deepwater ports where crude input is received by tanker. The economies obtained with huge tankers have stimulated construction of large refineries in the Atlantic provinces, specifically at Saint John, NB and Point Tupper, NS. These are located in areas of relatively low population density so that a major proportion of their output is either shipped inland or re-exported. Changes in international markets had a major impact on the export refineries in 1976, resulting in a marked decrease in product exports. Production of Canadian refineries is closely in balance with total market demand, although there is some interchange of individual products to and from the United States. Both exports and imports were down from 1975.

Transportation   13.6

Natural gas   13.6.1

The authorization of large-volume gas removal from British Columbia and Alberta, beginning in the mid-1950s, led to construction of the first major gas transmission lines in Canada. Today, the complete system serves major Canadian centres from Vancouver to Montreal and transports gas to the international border for US markets from California to New England.

Most Canadian natural gas must be processed before it can be marketed. Gathering lines take raw gas from producing wells to a collection point on a transmission system or to the inlet of a gas processing plant. Main transmission systems receive marketable gas from field gathering lines or plants and transport it through trunk lines to Canadian distribution companies or to interconnected US transmission pipelines at the international border. Distribution systems serve the ultimate customers in the centres of population.

Gas pipeline constructed in 1976 showed a marked increase over 1975 as 9 446 kilometres of pipeline were added to gas transmission, distributing and gathering systems, compared with 6 430 km in 1975. By the end of 1976 total cumulative gas pipeline length was 133 376 km.

Gas transmission and distribution lines accounted for the bulk of the increase as construction in these categories reached record proportions. Much of the gathering system construction was in Alberta where a record number of new gas fields were brought into production in 1976. Gas gathering lines increased 3 772 km and distributing pipelines systems were enlarged 3 766 km.

Hearings into the feasibility of building a natural gas pipeline from the mainland Arctic regions, initiated by the National Energy Board (NEB) in October 1975 continued in 1976.

Mr. Justice Thomas Berger was appointed to examine the socio-economic impact of a Mackenzie Valley pipeline, and Kenneth Lysyk to conduct an inquiry into the Alaska

Highway pipeline. The government subsequently accepted the finding of the NEB, favouring the Alaska Highway route supported by the Lysyk report. This pipeline would transport only US gas from the north slope of Alaska along the Alaska pipeline right-of-way to Fairbanks, then follow southeast through Alaska across the Yukon Territory to Dawson City to be joined later with a 76-centimetre pipeline linking the Mackenzie Delta to Whitehorse, northeastern BC, and northwestern Alberta. This system will join existing pipelines in BC and Alberta for distribution in US and, at a later date, Canadian markets. The route will generally parallel the Alaska–Canada Highway south of Whitehorse.

The Canadian pipeline legislation was adopted in early 1978. Foothills Pipelines (Yukon) Ltd. will be the parent company for the Canadian portion (about 823 km). The Foothills consortium includes Westcoast Transmission Co. Ltd., Alberta Gas Trunk Line Co. Ltd., TransCanada Pipelines Ltd. and Alberta Natural Gas Ltd.

The Polar Gas Project, which proposes construction of a 120-cm island-hopping pipeline to deliver natural gas to southern markets from the Arctic islands, came closer to realization in 1976 with the discovery of substantial new gas reserves in that area. The problem is not so much in constructing the line, but rather in establishing gas reserves large enough to sustain the economic operation of the line long enough to justify the capital expenditures. Current industry estimates place this "threshold" at a minimum of 707 750 million cubic metres. There are strong indications that about half of this reserve requirement has already been found.

If sufficient reserves can not be found to justify building such a high-cost pipeline, other transportation such as ice-strengthened liquefied natural gas tankers may be considered as an alternative. Although this is a relatively expensive method of transporting gas, it involves a smaller initial capital cost than pipelines and a more rapid rate of return on investment. In 1976, Petro-Canada contributed $7 million to Polar Gas, partly in support of engineering and economic studies to identify preferred ways of transporting gas from the Arctic.

Founding participants of the group were Panarctic Oils, TransCanada Pipelines, Canadian Pacific Investments, Tenneco Texas Eastern and Pacific Lighting. Polar Gas filed an application with the NEB late in 1977 to build a pipeline.

The TransCanada system, Canada's longest pipeline, begins at the Alberta border near Burstall, Sask., where it receives gas bought in Alberta from Alberta Gas Trunk Line. It receives gas from four Saskatchewan locations, then passes south of Regina to a point south of Winnipeg where it branches into two lines. The original line goes east to Thunder Bay, North Bay and south to Toronto. At Toronto this line again divides with the westward branch serving the Hamilton area as well as delivering gas to the US at Niagara Falls; the eastward branch follows the Lake Ontario shore and the St. Lawrence River to Montreal before terminating at Philipsburg, Que., on the international border. The largest gas pipeline construction project in 1976 was carried out by TransCanada Pipelines Ltd. on its main line between Toronto and Montreal with the completion of 77 km of a 61-cm line and in addition 27 km of a 40-cm lateral to Ottawa.

Alberta Gas Trunk Line Co. Ltd. transports most of Alberta's export gas from producing fields to the provincial boundaries for delivery to interprovincial carriers. Its two main segments are the Foothills Division and the Plains Division. The former transports gas for the Alberta Natural Gas, Alberta and Southern, and Westcoast Transmission systems; the latter for TransCanada and Consolidated pipelines. In the northwest of the province a smaller system, the Northern Division, delivers gas to the main Westcoast Transmission trunk line.

The Westcoast Transmission Co. Ltd. large-diameter line extends from Fort Nelson in the northeastern corner of BC to Sumas on the Canada–US border, near Vancouver. The system includes a number of lateral lines gathering gas from producing areas in BC, western Alberta and the Pointed Mountain field in the Northwest Territories. In addition to serving Vancouver and communities along its route, Westcoast delivers gas to Pacific Northern Gas Ltd., a distribution company serving communities and industries along an 805-km route between the Westcoast main line at Summit Lake and the Pacific Coast communities of Prince Rupert and Kitimat. It also supplies Inland Natural Gas Co. which operates an extensive distribution system serving communities in southern and central BC. Westcoast's export sales are made to the El Paso Natural Gas Co. for distribution in the Pacific northwest region of the US.

Transportation of oil

Canadian oil moves to markets through an intricate network of oil pipelines extending from the producing fields west to Sumas, BC, near Vancouver, and east to the Niagara area of Ontario. This network serves Canadian refineries in British Columbia, Alberta, Saskatchewan, Manitoba and Ontario, and US markets in the Puget Sound, mid-west, Chicago and upper New York state areas. At year end 1976, the length of the entire pipeline system was 32 703 km.

Prime components are the trunk lines of Interprovincial Pipe Line Ltd., Canada's largest oil pipeline, and the Trans Mountain Oil Pipe Line Co. Both lines start in Edmonton and are fed by a network of gathering lines transporting oil to the main trunk lines at that point. Outside Alberta, the Interprovincial pipeline receives and transports Saskatchewan and Manitoba crude oil.

Trans Mountain operates a pipeline system which carries crude and natural gas liquid from Edmonton and other points in Alberta and British Columbia to Burnaby, BC, and a subsidiary operates branch lines to refineries in the state of Washington.

The other prime mover of oil from Alberta, the Aurora pipeline, with a length of only 1.6 kilometres within Canada, receives crude oil and equivalent from the Rangeland gathering system and moves it to Billings, Montana, both for refining and further shipment to points in the US mid-west.

The oil embargo of the winter of 1973, coupled with frequent price increases of offshore oil, led the federal government to decide on a policy of an all-Canadian coast-to-coast pipeline network for security of supply, self-reliance in oil and oil products and to further economic development throughout the country.

In May 1975, the government approved the company's application to extend the Interprovincial system from Sarnia to Montreal to provide consumers in eastern Ontario and western Quebec with access to more secure domestic supplies of Canadian crude oil. Pipeline construction at 872 km increased in 1976 largely owing to completion of this Sarnia–Montreal link. Elsewhere pipeline construction continued the decline that began in 1973. The lack of new oil discoveries and regulated cutbacks in crude oil production were responsible for this decrease in activity.

Interprovincial Pipe Line Ltd.'s 76-cm oil pipeline from Sarnia to Montreal was completed in June 1976 and was the only large-diameter project finished during the year. The line will eventually have a capacity of 55 643 cubic metres a day and initial throughput is 39 745 m³/d. Fully powered with 16 pumping stations, capacity of the line could approach 109 696 m³/d if necessary, and flow in the line can be reversed.

In product pipelines, construction commenced on the 321-km, 30.4-cm $300 million natural gas liquids pipeline from Edmonton to Sarnia via the US. After receiving approval from the NEB to export ethylene and ethane to the US, Dome Petroleum Ltd., one of the principals in the system, began work on 12 river crossings in 1976. The 11 925 m³/d project had been planned for more than five years and was to be completed early in 1978. Canadian portions of the system are called the Cochin pipeline and in the US the Dome segment. The system primarily will carry ethane, ethylene and propane from plants near Edmonton and Red Deer, Alta., crossing the border near Sherwood, North Dakota, and again at Windsor, Ont. A spur line will supply Columbia Gas System's synthetic natural gas plant at Green Springs, Ohio, with 6 360 m³/d of gas liquids.

In Ontario, Trans-Northern Pipe Line Co. has completed looping a 22-km stretch of its products pipeline south of Ottawa. It is also planning to reactivate two pumping stations to move more products from Toronto area refineries to the Ottawa market.

In October 1976, Alberta Oil Sands Pipeline Ltd. received approval from Alberta's energy resources conservation board to build an oil line for transmission of synthetic crude oil from the Mildred Lake extraction plant of Syncrude Canada Ltd. to the International Pipe Line Ltd. terminal in Edmonton. The proposed line will comprise some 434 km of 58-cm pipe and 11 km of 53-cm pipe with four pumping stations along the route. Completion of the line was to be timed to coincide with the start-up of the Syncrude plant scheduled for 1978.

13.7   Coal

Forecasts of Canadian coal production in the next few years indicate that total output will continue to increase. In the immediate future the thermal sector will experience the largest gains as Alberta, Saskatchewan, Ontario, New Brunswick and Nova Scotia have plans or work under way to expand the use of Canadian coals to produce electricity.

If all projects now under consideration go ahead, coal consumption by utilities would approximately double by 1985. New large-scale coal movements from British Columbia, Alberta and Saskatchewan to Ontario, projected for 1979, indicate Canadian thermal coals can be competitive with imported coals. The export thermal sector has levelled off at about 10 or 11 million tonnes and is unlikely to increase substantially in the immediate future.

Total coal production in Canada in 1977 approached 29 million t. Imports approached 15.5 million t for a total supply of 44.5 million t. On the demand side, consumption of thermal coal approximated 21 million t while that for coking coal was about 7 million and the industrial-commercial sector used 2 million. Exports were expected to total about 12.5 million t.

Canadian production of coal in 1976 was 25.5 million t valued at $607 million (Table 13.10), up slightly from 1975 (25.3 million t). Output increased in Nova Scotia, Alberta and Saskatchewan but decreased in New Brunswick and British Columbia. Western Canadian production reached 23.2 million t, while output from Nova Scotia and New Brunswick mines totalled 2.3 million t.

About 11.9 million t of coal were exported in 1976 (Table 13.11), with British Columbia and Alberta contributing 95% and Nova Scotia making up most of the other 5%. Japan received 10.6 million t in 1976; the remaining 1.2 million t went to 11 other countries. Exports in 1977 were forecast at 12.5 million t. Imports from the US in 1976 were 14.5 million t, down from 15.2 million in 1975.

In 1976-77 the general world economic slowdown led to lower steel production and a consequent reduced demand for coking coal. Japan's steel industry, major consumer of Canadian coking coals, operated below capacity and was not expected to revert to full production until the 1980s. As a result, exports of coking coal, 10.0 million t in 1974, remained stable during 1975 (10.8 million t) and 1976 (10.6 million t) and were expected to change very little in 1977. Canadian coking coal producers are now actively seeking other smaller Asian as well as Latin American markets.

Demand for thermal coal has increased as several provinces have either expanded their use or made commitments to use coal to meet growing energy requirements. Nova Scotia, New Brunswick, Ontario, Manitoba, Saskatchewan and Alberta consumed 19.1 million t of coal to generate electricity in 1976 (Table 13.12) and 1977 figures were expected to be 20% higher.

Domestic coal, mainly sub-bituminous in Alberta and lignite in Saskatchewan, supplied about 11 million t to Western Canada power stations. Bituminous coal is used in small quantities for thermal power generation in New Brunswick and Nova Scotia and imported by Ontario Hydro in large amounts from the United States. Demand for coal by other general industrial and commercial users reached 1.7 million t in 1976.

13.7.1   Production areas

British Columbia. Coal mining is centred southeast in the Crowsnest Pass region.

Kaiser Resources Ltd., with two operating mines in the Crowsnest coalfield, produced about 5.4 million tonnes of clean coal in 1976. Work was under way on a new hydraulic panel scheduled for production in 1979. Work on the proposed new Hosmer-Wheeler hydraulic mine south of Kaiser's operations continued during 1976 but in 1977 a delay in development of this mine was announced pending further sales negotiations.

The Fording Coal Ltd. mine near Elkford, about 64 kilometres north of Sparwood, BC produced approximately 1.6 million t of clean coal in 1976, all of it shipped to Japan.

The company was evaluating a potential hydraulic mine at Eagle Mountain in anticipation of expanded output.

Byron Creek Collieries produced approximately 343 000 t of clean coal in 1976 for both domestic and export markets. Production will be expanded to meet increased demands from Ontario Hydro for western bituminous coal over the next few years.

Studies continued in 1976 in several areas of BC on potential metallurgical and thermal coal developments. In northeastern BC, marketing, transportation, socio-economic and feasibility studies were under way at several locations. Prospects for exporting coking coal in the early 1980s depend on world demand and competitiveness in world markets. In southeastern BC, European, Japanese and Canadian interests continued feasibility studies on new and existing properties. BC Hydro and Power Authority conducted studies into the potential use of its Hat Creek lignite deposits near Ashcroft for future power generation.

In mid-1977, the province announced a policy to form the basis for current and future development of coal resources. The new policy outlined a licence system incorporating competitive bidding; committed BC's coal resources to development consistent with other provincial objectives; retained the existing $1.36 a tonne royalty; tied BC's export coal prices to world prices; and reserved BC's coal for provincial, other Canadian and finally, foreign needs.

Alberta. In terms of volume, Alberta is Canada's leading coal-producing province, mining bituminous, sub-bituminous, and semi-anthracite coals with the sub-bituminous coal used primarily to generate electricity. Most bituminous coal is now exported to Japan for coke making, although some is destined for shipment to Ontario for electric power generation in the near future. Bituminous production reached 4.6 million t in 1976 and sub-bituminous, 6.4 million, representing increases of 12% and 8% respectively over 1975 levels.

Alberta continued to expand its sub-bituminous industry in 1976 and 1977 to meet increasing demand for electrical energy and satisfy some of Saskatchewan's electricity needs. On the Prairies, thermal electric plants are generally located close to coal mines to facilitate low cost power generation. At Wabamun Lake, west of Edmonton, Calgary Power Ltd. operates two power plants using coal from two surface mines. Other coal-fired power plants include the Drumheller, Battle River and Grande Cache stations.

In 1976 and 1977 new coal-fired thermal electric generating stations to meet Alberta's power needs in the 1980s were proposed. Calgary Power Ltd. and Alberta Power Ltd. took the first steps to obtain government approval for their proposed Keephills and Sheerness generating stations, for operation in 1984 and 1985.

Four mines produce coking coal in Alberta. In 1976 the largest operator, McIntyre Mines Ltd. near Grande Cache, produced 1.9 million t of clean coking coal and 580 000 t of middlings. Most of the coking coal was sold to the Japanese steel industry while middlings coal went to Alberta Power Ltd. for power generation at Grande Cache.

Luscar Ltd.'s Cardinal River mine produced approximately 1.7 million t of clean coal in 1976, all for the Japanese steel industry. Development of two new pits began near existing operations.

Canmore Mines Ltd. produced 91 000 t of semi-anthracite coal in 1976 while Coleman Collieries Ltd. extracted 955 000 t of clean coal from its underground and open-pit mines. Feasibility studies continued at Coleman's Tent Mountain No. 5 mine to supplement dwindling capacity at existing underground operations.

Several mines produced sub-bituminous coal in the plains region of the province for power generation. Output at Calgary Power Ltd.'s Highvale mine west of Edmonton reached 2.6 million t in 1976; the nearby Whitewood mine produced 2.1 million. Both supply on-site power stations. Southeast of Edmonton, Manalta Coal Ltd.'s Vesta mine and Forestburg Collieries Ltd.'s Diplomat mine processed 533 000 and 773 000 t, respectively, for power generation and industrial markets. Output at Manalta's Roselyn mine northeast of Calgary reached 362 000 t in 1976, most marketed to Saskatchewan Power's generating station at Saskatoon.

In 1976, the Alberta Cabinet approved Luscar Sterco Ltd.'s Coal Valley project to supply bituminous coal to Ontario Hydro over a 15-year period beginning in 1978. Coal will move by rail to Thunder Bay, through the new coal terminal now under construction, and then by freighter to Lake Erie ports.

Saskatchewan. In 1976, five lignite mines in Southern Saskatchewan supplied about 4.7 million t of lignite coal. The Manitoba and Saskatchewan Coal Co. Ltd.'s Boundary Dam mine produced 1.5 million t while Manalta's Utility mine produced 1.2 million t, all for Saskatchewan Power's Boundary Dam power station. Other production included 763 000 t from the M&S mine at Bienfait and 1.1 million from Manalta's Klimax mine, both serving power generation and industrial markets, and 19 000 t from the new Saskatchewan Power Corp. Souris Valley mine.

Lignite production in Saskatchewan expanded in 1976 to meet exceptional requirements in Manitoba. However, Manitoba's coal demands were expected to decline in 1978 given normal weather and precipitation conditions. Saskatchewan is building a new thermal power station south of Moose Jaw, and this, along with commitments to supply Ontario Hydro with lignite coal for its new Thunder Bay generating units, will require an increase in production over the next few years.

New Brunswick. In 1976, N.B. Coal Ltd., a provincial Crown company, produced a total of 296 000 t of clean coal from the Minto coalfield, most of it for the provincial electric utility company; the rest went to industrial and commercial markets.

Nova Scotia. Production of clean coal in Nova Scotia reached approximately 2.0 million t in 1976. Most production came from three mines of Cape Breton Development Corp. (DEVCO) — the Lingan mine, No. 26 Colliery and the Prince mine. About 80% of Nova Scotia's 1976 production was destined for thermal markets. Output at the new Prince mine totalled 143 000 t of clean coal with ultimate capacity rated at approximately 635 000 t. Nova Scotia continued to sell coal to Europe, but made the first of a series of shipments to The Steel Co. of Canada Ltd. at Hamilton, Ont. in 1976. A total of 2.3 million t is scheduled for delivery over five years. Plans call for increased production to meet higher domestic demand for thermal coal.

13.7.2 Federal incentives

An assessment of Canada's coal resources and reserves in 1976 was published by energy, mines and resources in 1977 — the first such estimate ever published. Canada's coal reserves are defined as that portion of coal resources that has been reasonably well delineated and can be produced with current technology and delivered at competitive market prices. Current reserves of recoverable coal are estimated at 717.0 million tonnes of coking coal and 5.2 billion t of thermal coal. These estimates are considered conservative because they do not include reserves of several companies and utilities. For 1976, resources of immediate interest were estimated at 31.9 billion t of measured resources, 14.6 billion of indicated resources and 181.5 billion of inferred resources.

The report outlines the department's intentions for implementing a national coal inventory program and a national coal data system in conjunction with the provinces and industry. The inventory program ultimately will result in the determination and compilation of data and information on the quantity, quality, mineability and economics of Canadian coal and provide a meaningful estimate of coal reserves in terms of cost and availability. A comprehensive coal policy, recognizing the vital provincial role, is being prepared. Upgrading of transportation facilities at the Lakehead will facilitate movement of western coal to Ontario. The federal government is financing research into methods of converting coal into gaseous and liquid fuels. Eight research agreements have been signed. Funding is to be increased to $1.25 million. In Nova Scotia electrical costs are high because of reliance on imported oil and the federal government has contributed $9.2 million to help that province develop other fuels, especially coal, to replace oil.

13.8 Uranium and nuclear energy

Canada's uranium industry continued to expand in 1976 and 1977 to meet increasing domestic and export commitments. Shipments of uranium from Canadian producers have risen from a low of 2 847 tonnes of elemental uranium (U) in 1968 to 5 627 t in

1976 and 5 953 t in 1977. Recent increases were due in part to expanded output from the three established producers, and in part to three new producers, the first starting in late 1975 and the most recent in 1977. Further increases in production can be expected as a result of continued expansion and the probable development of several new operations. Based on known deposits, Canada could be producing some 12 500 tonnes U/year by 1985.

A sellers market for uranium during 1974 and 1975 saw prices in the $104 per kilogram range early in 1976. Exploration activity significantly expanded in Canada with expenditures of some $40 to $50 million. Preliminary figures indicate expenditures of $60 to $70 million in 1977. Activity was reported in every province and the territories, but attention focused on northern Saskatchewan, where two significant discoveries were made at Key Lake — one in late 1975 and the other in mid-1976. Estimates of recoverable uranium resources continued to increase.

A uranium resource appraisal group was established within the energy, mines and resources department in September 1974 to audit Canada's uranium resources annually. The group completed its third annual assessment early in 1977 under two price categories, up to $88/kg ($40/lb.) U_3O_8 and from $88 to $132/kg ($40 to $60/lb.) U_3O_8.

Resources recoverable for up to $132/kg ($60/lb.) U_3O_8 were estimated at: measured 83 000 tonnes elemental uranium; indicated 99 000 t; and inferred 307 000 t. The following conversion factors have been used to calculate these reserves: $88/kg ($40/lb.) U_3O_8 = $104/kg U; 1.0 tonne U (elemental uranium) = 1.2999 short tons U_3O_8.

After deducting 1976 production of 4 850 tonnes U from 1975 resources in these categories, resources have increased by some 14.3% over those reported in 1975.

In January 1977 the Canadian government embargoed exports of Canadian nuclear materials and technology to all countries which had not completed negotiations of revised nuclear safeguard agreements with Canada. These revisions were a result of Canada's two-stage effort, announced in 1974 and 1976, to strengthen export requirements in the interests of nuclear safeguards. Shipments of Canadian uranium to a number of countries were affected, principally Japan, countries of the European Economic Community, Switzerland and the United States. At year end 1977 agreements were in place with all but Japan and Switzerland, although negotiations with Japan were concluded early in 1978.

A number of inquiries under way in 1976 and 1977 could affect the shape of Canada's future uranium and nuclear power industry. Among these were the Saskatchewan government's Cluff Lake inquiry board, examining the environmental, health and safety aspects of Amok Ltd.'s Cluff Lake uranium project, as well as the overall implications of further uranium developments in Saskatchewan; an Ontario government environmental assessment board's hearings on environmental impact of expansion of uranium operations in the Elliot Lake area; and the Porter Commission on electric power planning in Ontario. Ontario Hydro's commitment to nuclear power forms a major part of the Porter inquiry.

In August 1977 the energy, mines and resources department published a report on management of Canada's nuclear wastes. It concluded: "The country needs a consolidated plan for the management of radioactive wastes now: a piecemeal, hesitant approach to this challenge will not be in the national interest." The report recommended the following targets: 1978, declare a national plan to deal with nuclear wastes and acceleration of R&D programs; 1983, choose at least two hard-rock sites in Ontario to be developed for geological disposal; 1985, have shafts sunk and testing under way in the hard-rock sites; 1988, start construction of irradiated-fuel handling facilities at one site; 1990, start test disposal of immobilized irradiated fuel and immobilized reactor wastes; 1995 to 2000, have an operating repository capable of receiving Canada's annual output of irradiated fuel.

In November 1977 the minister of energy, mines and resources tabled a proposed new act for nuclear control and administration to replace the 30-year-old Atomic Energy Control Act. Its main objective is separation of responsibilities for health, safety, security and environmental matters from those dealing with commercial and promotional aspects of nuclear energy. The former responsibilities would be administered by the Atomic Energy Control Board (AECB) to be renamed the Nuclear Control Board (NCB), which would report to Parliament through the minister of state for science and technology. The minister of energy, mines and resources would retain responsibility for, and administer, all commercial and promotional matters.

Domestic uranium requirements, estimated at 560 tonnes U in 1977, are expected to rise to some 2 400 and 4 200 t a year by 1985 and 1990, respectively.

In mid-1977 nuclear generating capacity operating in Canada exceeded 4 000 megawatts, 94% of it in Ontario. Additional capacity totalling 7 900 MW was either under construction or committed, and scheduled for operation by 1986, some 84% to be in Ontario.

Early in 1978, the Ontario government signed an agreement with uranium producers at Elliot Lake to supply the projected uranium requirements of Ontario Hydro into the next century at a cost of $7 billion.

13.9   Electric power

13.9.1   Electric power development

Total installed generating capacity increased by 11.0% in 1976 to 68 088 megawatts with additions totalling 6 736 MW (2 194 MW hydro, 3 386 MW fossil-fired steam, 342 MW combustion turbines, 14 MW diesel and 800 MW nuclear capacity). Generating units that became operational in 1976 but were not yet available for normal commercial service are included.

Load growth in terms of energy in 1976 increased 7.1% over 1975. On a national basis, electrical energy consumption totalled 284.1 TWh (1 terawatt hour = 10^9 kilowatt hours), distributed across the country in the ratio of approximately 33% in each of Quebec and Ontario, 13% in British Columbia, 4% to 6% in each of Alberta and Manitoba, and 2% to 3% in each of Newfoundland/Labrador (excluding Churchill Falls), New Brunswick, Nova Scotia and Saskatchewan, with Prince Edward Island, the Yukon and Northwest Territories each accounting for less than two-tenths of 1.0% of the total. Growth rates varied considerably across the country — 10.1% in the Yukon Territory to +15.5% in Newfoundland. On a nationwide basis, total residential consumption grew 12.7%, commercial 7.8% and industrial 1.7%.

The national growth rate of approximately 7.0% (−0.3% in 1975) represents a return to usual levels following a partial recovery of demand in the industrial sector which, in 1975, decreased 11.4%. Total generation for the year (293.4 TWh) increased 7.6%. Hydro represented 72.6% of the total in 1976 (74.2% in 1975); nuclear energy 5.6% (4.3% in 1975); and conventional thermal 21.8% (21.5% in 1975). Coal comprised 61.7% of thermal or 13.4% of the total and oil 20.3% of thermal or 4.4% of the total.

While this growth could be viewed as a return to the long-term growth trend, such a conclusion must be treated with caution. Price increases and conservation policies have undoubtedly led to some decrease in consumption, but substitution of electricity for other energy forms, either because of relative price changes or for security of supply, may cause some increases in demand. Time lags in demonstrating full customer response to these factors is inevitable in both cases. In the short run however economic activity seems to be the most likely explanation of variations in demand occurring in 1975 and 1976.

Net export of electrical energy in 1976 was 9.3 TWh or 3.2% of net generation, up 26.1% from 7.4 TWh (2.7% in 1975). The change reflected some recovery in economic activity. [Gross national expenditure (GNE) in 1971 dollars showed an increase of 4.6% in 1976 after a rise of only 0.6% in 1975.]

13.9.2   Generating capacity

Power generating capability measures available generating resources of all hydro and thermal facilities at the time of the one-hour firm peak load for each reporting company and is not equal to the installed capacity of such generating facilities.

Hydroelectric power generation 13.9.3

Hydroelectric generation forms a significant though decreasing part in Canada's electrical development. By the end of 1976, the hydro portion of the country's total generating capacity had fallen to 58% from over 90% 20 years earlier.

In view of Canada's extensive water resources, many undeveloped sites would seem to be potential sources of hydroelectric power, but all are not economically viable. Only a fraction of the sites with a theoretical power potential can actually be developed competitively. Before a site can be termed a source of potential power, a detailed analysis of such factors as cost, geography, geology and ecology must be performed. Until such a study is completed on a national scale, estimates of Canada's undeveloped hydropower resources (estimated to exceed 60 000 megawatts), may be misleading.

The maximum economic installation at a power site can be determined only by careful consideration of all conditions and circumstances pertinent to its individual development. For a number of reasons, it is normal practice to install units with a combined capacity in excess of the available continuous power at the flow available 50% of the time, and frequently in excess of the power available at the arithmetical mean flow. Excess capacity may be installed for use at peak-load periods, to take advantage of periods of high flow, or to facilitate plant or system maintenance. In some instances, storage dams have been built after initial development to smooth out fluctuations in river flows. In other cases, deficiencies in power output during periods of low flow have been offset by auxiliary power supplied from thermal plants or by interconnection with other plants operating under different load conditions or located on rivers with different flow characteristics. The extent to which installed capacity exceeds available continuous power at various rates of flow depends on factors that govern the system or plant operation, and this varies widely from one area to another.

Distribution of installed hydroelectric generating capacity given in Table 13.13 shows that substantial amounts of water power have been developed in all provinces and territories except Prince Edward Island. As natural-resource development proceeds, the fortunate incidence of water power near mineral, forest and other resources becomes increasingly apparent. The vast hydro potential of northern rivers may well prove to be a prime factor in the eventual realization of the natural wealth of the Canadian North.

Water-power resources of Nova Scotia and New Brunswick, although small compared to other provinces, are a valuable source of energy and make a substantial contribution to the economies of the two provinces. Numerous rivers provide moderate size power sites within economic transmission distance of principal cities and towns. Others are advantageously situated for development of timber and mineral resources. These provinces have, however, turned to thermal generation, initially coal-fired, with a subsequent shift to oil. Construction of a nuclear power plant in New Brunswick is under way and there are indications of a possible return to coal as a fuel source for new installations.

Thermal power generation 13.9.4

Immense water-power resources and the rapid pace of development have tended to overshadow the contribution being made by thermal energy to Canada's power economy. From a modest 133 megawatts of generating capacity installed at the end of 1900, Canada's installed hydro capacity rose to 39 475 MW by the end of 1976 and thermal capacity to 28 613 MW (Table 13.13).

The same table shows that thermal generation is predominant in Prince Edward Island and Nova Scotia. By the end of 1971, the Yukon Territory had joined the Northwest Territories, Alberta, Saskatchewan and Ontario in having more than half their total capacity thermal-electric. Thermal generation is expected to become increasingly predominant in Ontario.

Over 90% of all thermal power generating equipment in Canada is driven by steam turbines fired by coal, oil, gas or, in the case of nuclear equipment, uranium. The magnitude of loads carried by steam plants combined with the economies of scale has led to the installation of steam units with capacities as high as 540 MW, and units in the 800-MW size range were committed for as early as 1976. Additions of these larger units

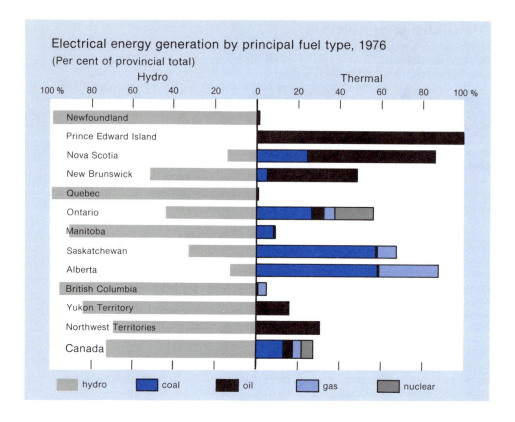

Electrical energy generation by principal fuel type, 1976
(Per cent of provincial total)

Hydro / Thermal

Newfoundland
Prince Edward Island
Nova Scotia
New Brunswick
Quebec
Ontario
Manitoba
Saskatchewan
Alberta
British Columbia
Yukon Territory
Northwest Territories
Canada

hydro   coal   oil   gas   nuclear

are only possible where systems are large enough to accommodate them. Additional types of thermal generation are provided by gas turbine and internal combustion equipment; their flexibility makes them particularly suitable for meeting power loads in smaller centres, especially in isolated areas. Gas turbines are frequently used for peak loads because of their rapid start-up capability and low capital cost.

After World War II, industrial expansion and rapidly growing residential and agricultural development placed extremely heavy demands on power generating facilities which were impossible to satisfy from hydro sources alone. An extensive program of thermal plant construction began in the early 1950s; by 1956 thermal capacity represented 15% of the total. Since then, the annual installed capacity has averaged 56% hydroelectric with the rest in thermal generation. At the end of 1976 thermal capacity accounted for 42% of Canada's installed capacity.

Thermal plants accounted for only 21.8% of total generation in 1976 (Table 13.15) because much of the capacity installed is operated for peak-load duty only, with hydroelectric capacity providing base-load generation. This pattern will change as additional nuclear-fuelled plants which can operate economically at high capacity for base-load purposes are introduced.

Details of the type of fuel used, by province, appear in Table 13.12. The table shows a significant increase in coal consumption to generate electricity in 1976. More than 19.0 million tonnes were used, valued at $322.0 million in the latest year, up from 16.6 million t valued at $214.6 million in 1975. Oil showed only a small volume increase but higher oil prices pushed the dollar value to more than $155.0 million from $89.9 million a year earlier. The volume of natural gas used was virtually unchanged, but higher prices again increased the cost of fuel for this type of generation to nearly $112.0 million from $87.2 million in 1975.

Nuclear thermal power

13.9.5

Commercial electric power generated from a nuclear reactor in Canada dates back to 1962 when the 20-MW nuclear power demonstration station at Rolphton, Ont., forerunner of a series of large nuclear stations, fed power for the first time into a distribution system in Ontario.

Atomic Energy of Canada Limited (AECL), a federal Crown company incorporated in 1952, has concentrated on development of the CANDU power reactor using heavy water (deuterium oxide) as a moderator for slowing the neutrons released by nuclear fission. The high neutron economy obtained by using this moderator with neutron-transparent core materials (zirconium alloys) means Canada's abundant resources of natural uranium may be used as fuel. The CANDU system is sufficiently flexible that enriched uranium, plutonium recovered from spent fuel, or thorium may be incorporated into its fuel system.

Production of heavy water has been a critical item in the Canadian nuclear power program. The first 726 tonne-a-year production plant at Ontario Hydro's Bruce nuclear power development on Lake Huron went into operation in 1973 and is producing at over 80% of its design capacity. Ontario Hydro started the first of two additional plants at the Bruce site and completion was scheduled for 1978. In Nova Scotia rehabilitation of the Glace Bay plant continued; start-up was scheduled for 1976. Operation of the Port Hawkesbury plant was considerably improved after modifications. Ownership of the plant was transferred to AECL in 1975. There will be a two-year delay in construction of the new 726-tonne-a-year La Prade plant owned by AECL at the Gentilly site in Quebec. The first 363-tonne unit is expected to be operating by 1981.

At Douglas Point, on the shore of Lake Huron, the country's first full-scale nuclear power station went into operation in 1966. The station, built with the co-operation of Ontario Hydro, houses a 220-MW CANDU reactor. Experience gained in the design and operation of the Rolphton and Douglas Point reactors has led to development of larger units. Construction of the 2 160-MW Pickering nuclear station near Toronto is now complete; two of the station's four units came on line in 1971 and the other units in 1972 and 1973. Work on the Bruce nuclear station for Ontario Hydro is proceeding with four 800-MW units planned for installation from 1976 to 1979. In addition, a duplicate of the Pickering station, at the Bruce site, has been committed and Hydro-Québec and New Brunswick Electric Power Commission have started construction of 600-MW CANDU stations at Gentilly and at Point Lepreau.

A further step in development of the CANDU reactor is the use of boiling light water instead of pressurized heavy water as the coolant. The initial Gentilly nuclear power station (Gentilly 1) uses boiling light water; this station came into service in 1971 with 266 MW of nuclear-electric capacity.

Load demand and energy use

13.9.6

Firm power peak load is the measure of the maximum average net kilowatt demand of one-hour duration from all loads, including commercial, residential, farm and industrial consumers as well as the line losses. Such load demand increased at the rate of 7.5% a year from 1963 to 1973 and 7.4% a year from 1969 to 1973; this rate of increase is expected to remain unchanged through 1980. As a result of the rapid increase in generating capability and the somewhat slower but steady increase in peak loads, together with the slight reduction in deliveries of firm power to the US, the indicated reserve on net capability in the 1961-75 period increased each year except 1961, 1963, 1964, 1966 and 1972. The reserve ratio as a percentage of firm power peak load reached a high of 28.2% in 1960, fell to 13.7% in 1968, but is expected to rise to 20.0% in 1980.

As indicated in Table 13.16, total electrical energy consumed in Canada during 1976 showed residential loads up 12.7% from a year earlier. Commercial loads climbed 7.8% in the period while industrial demand crept ahead 1.7%. Industrial accounted for the largest segment of total demand, followed by commercial and residential, which have steadily increased their share of the total since the 1950s.

While availability of electric energy at reasonable cost is an important element in Canada's industrial growth, in only a few industries is the cost of electric power a key element in economic competitiveness. Energy distribution for industry can be subdivided roughly as follows: mineral industry (including smelting and refining), one-third, pulp and paper industry, one-quarter, chemical manufacturing, one-tenth, and the rest to all other industrial categories.

The growth among non-industrial consumers results from greater reliance on facilities powered by electricity. Increasing quantities of electricity are required to meet rapidly escalating demands for heating, cooling, lighting, transportation, elevators, electrical appliances and farm machinery. The shift of population from rural areas to cities and towns, where electrical demand is greatest, has contributed to this growth.

Details of the regional pattern of electric energy use can be seen in Table 13.16. In 1976 more than two-thirds of total available electric power was consumed in Ontario and Quebec with all other regions accounting for the remaining one-third. The share of total consumption by these other regions has, however, been rising (combined total of 26% in 1960 but 33% in 1976).

Among the more significant percentage increases in total energy demand were a 15.5% rise in Newfoundland because of higher consumption in the residential and commercial sector; a 12.4% gain in New Brunswick, where residential and commercial load increases offset a decline in industrial consumption; and an 11.9% gain in British Columbia reflecting higher demand in all sectors, including an 18.9% increase in industrial use.

13.9.7   Electric power transmission

Loads handled by small, widely scattered generating systems in the early days of the electric power industry did not warrant the expense of interconnecting power systems. However, increased demand for dependable electric power and improved techniques resulting in lower power transmission costs have led to a reappraisal of the benefits of integrating power systems for better reliability of service and greater flexibility of operation. Most of Canada's generating stations today are components of large, integrated, and often interconnected, power systems operated by power utilities.

Research has developed techniques enabling power producers to use hydroelectric sites once considered as beyond economic transmission distances. Most noticeable is the progressive stepping up of transmission line voltages. A number of transmission lines are designed for operation at 500 kilovolts (kV) and 735 kV. A 924-kilometre 500-kV line carries power from the Peace River to the lower mainland of British Columbia. In Ontario, a 700-km 500-kV line brings power from hydro plants in the James Bay watershed to Toronto. In 1965 Hydro-Québec became the first utility in the world to transmit electric power at 735 kV over the 604-km line linking Quebec's Manicouagan–Outardes hydro complex with urban demand centres at Quebec City and Montreal. By the end of 1971, the initial program for 1 976 km of the 735-kV line was completed, and three 735-kV circuits connecting the Churchill Falls generation station to the Hydro-Québec grid have also been added.

Most power is transmitted as alternating current but there are three applications of high-voltage direct-current (HVDC). A 260-kV HVDC line links the BC mainland and Vancouver Island. It has a capacity of 312 MW and includes 34 km of undersea cable; it is a monopolar system using the ground as a return path for current. Capacity was doubled in 1976 to 624 MW. A 450-kV HVDC system was placed in service in 1973 linking the Kettle generation station on the Nelson River to Winnipeg where two 893-km lines have been completed and converter equipment with a capacity of 1 620 MW is in service. The planned ultimate rating of this system is 3 200 MW. Another application designed to provide a non-synchronous tie between the power systems of New Brunswick and Quebec is a 320-MW back-to-back HVDC system at Eel River, NB, using solid state thyristor valves instead of the mercury arc valves used for the earlier HVDC systems in BC and Manitoba.

Interconnections of 66 kV and 138 kV already exist between British Columbia and Alberta and a 230-kV tie is being planned. Saskatchewan, Manitoba, Ontario and portions of the Quebec system are interconnected and, through the Ontario Hydro system, are linked with systems in the northeastern US. Quebec, New Brunswick and

Nova Scotia systems are interconnected. The first major international tie connecting regions of the Maritimes in Canada with the US was the completion of a 345-kV link between the New Brunswick and Maine systems. British Columbia has an international tie with the Pacific northwest (500 kV) and a 230-kV link between Manitoba and the US was completed in 1970.

The search for economies in transmission systems has led to changes in materials used and in tower erection and cable-stringing methods. Guyed V-shaped and Y-shaped transmission towers are being used instead of self-supporting towers where the terrain is suitable, and erection costs are being reduced by using helicopters to transport tower sections to the site.

Electric utilities   13.9.8

Federal regulation of electric utilities regarding export of electric power and construction of transmission lines for such exports falls within the jurisdiction of the NEB.

Power is generated in Canada by publicly and privately operated utilities and by industrial establishments (Table 13.17). Most of the total power generated in 1975 came from public utilities, followed by privately owned utilities and industrial establishments. However, ownership varies greatly in different areas of the country. Although Quebec power installations were at one time privately owned, most were transferred to public ownership in 1963. In Ontario almost all electric power has been produced by a publicly owned utility for over 60 years.

Determination of market prices and regulation of services is limited to competition with oil, gas and coal. Consequently, there is some regulation of electric utilities in all provinces. In all but two provinces major generation and main transmission of power is the responsibility of a provincial Crown corporation. Investor-owned electric utilities are prominent in Alberta, Newfoundland and Prince Edward Island and continue to play a significant role in Ontario and British Columbia. On a percentage basis, industrial generation has been declining steadily as purchasing power from utilities which can take advantage of larger unit sizes and operational flexibility becomes more attractive. Even when process steam is required for industry, it is sometimes advantageous to purchase both steam and power from the utility.

Aid to Atlantic provinces   13.9.9

The minister of energy, mines and resources and the premiers of New Brunswick, Nova Scotia and Prince Edward Island agreed on January 23, 1978 to accept a feasibility report and to incorporate the Maritime Energy Corporation (MEC). When established MEC would undertake the following functions: financing construction and operation of regional electrical generation and transmission projects; co-ordination and direction (dispatching) of day-to-day system operation; provision of regional planning staff for generation and bulk transmission facilities; co-ordination of expanded facilities and agreements for expansion of external transmission lines for import and export; carrying out R&D for Maritime utilities.

Certain measures to encourage electric power development projects in the four Atlantic provinces were announced by the federal energy minister in April 1977: Improvement of electrical supply in Atlantic Canada is being promoted by loans for nuclear power in New Brunswick, financial aid for the submarine cable between New Brunswick and PEI, an interconnection between New Brunswick and Nova Scotia, and offers of aid for an inventory of Newfoundland energy resources and for a transmission system to carry power from the proposed Gull Island hydroelectric project.

In addition to federal-provincial home insulation programs, the federal government has contributed $9.2 million to help Nova Scotia develop other fuels, especially coal, to replace oil.

Other federal initiatives. A loan program has been set up to finance a $14 million program aimed at strengthening regional transmission links between New Brunswick and Nova Scotia to increase energy flow among the three Maritime provinces. Funding has been assured for a study and demonstration project for electrical load management to reduce capital requirements for power generation by reducing peak loads. This will be a detailed study in conjunction with one or more power utilities designed to analyze and apply load management experience in other jurisdictions and to expand on the findings of an analysis by the regional economic expansion department with the co-operation of Nova Scotia Power.

13.10 Provincial activities, 1976

13.10.1 Newfoundland

About 99% of the province's electrical energy is hydraulic in origin and the rest oil-fired thermal. With the addition of Churchill Falls' production the province's hydro generation (38.8 terawatt hours) exceeded that of Ontario (38.3 TWh) and British Columbia (36.7 TWh). Churchill Falls power supplied under contract to Quebec totalled 32.1 TWh (82.8% of total hydro production for the province).

Energy demand from the Newfoundland and Labrador system increased 15.5%. Consumption by residential and commercial customers grew by more than 15% and 11% respectively over the previous year. Statistics for industrial consumption are still incomplete but preliminary data indicate a substantial increase reflecting a return to normal demand following depressed sales resulting from industrial disputes in energy-intensive industries in 1975.

Major additions in 1976 were confined to combustion turbine installations of 53.8 megawatts at Stephenville and 25 MW at Green Hill near Grand Bank. Growth in demand over the next several years is expected to average about 8% a year and to meet this increased load, combustion turbines will be added at St. John's (Hardwoods Station — 54 MW, 1977), Flower's Cove (15 MW, 1978) and Port aux Basques (25 MW, 1980). An additional 154-MW hydro unit at Bay d'Espoir was scheduled for service in 1977 and a third 150-MW oil-fired unit at the Holyrood thermal station was scheduled for 1979.

Transmission developments included completion of a second 105-kilometre 230-kilovolt circuit from Bay d'Espoir to a terminal station near Grand Falls and construction of a 274-km 138-kV line from the Churchill Falls plant to Goose Bay thus replacing the thermal generation station by lower cost hydro supply.

13.10.2 Prince Edward Island

Electric power generation is entirely dependent on oil fuel, and rapid increases in oil prices have led to very high electricity costs. The submarine cable interconnection with New Brunswick scheduled for service in 1977 should relieve the province's dependence on local small-scale oil-based generation by providing access to power supply from larger, more efficient fossil-fuelled plants and from nuclear generation on the mainland.

The Prince Edward Island/New Brunswick interconnection facility, owned by PEI, involves an investment of some $36 million, $27 million of it financed by the federal government in the form of an $18 million grant and a $9 million long-term loan. This 138-kV interconnection will comprise two three-phase cables with a firm rating of 100 MW (100 MW each cable plus 100% standby) linking terminals near Cape Tormentine, NB and Borden, PEI.

Provincial load grew 6.3% over 1975. Energy demand increased 6.9% for residential customers and 8.4% for commercial.

13.10.3 Nova Scotia

To reduce dependency on expensive imported oil, the Nova Scotia Power Corporation has intensified its search for alternative sources of energy. Confirmation of adequate coal reserves under an exploration program carried out in 1977 will lead to an ultimate capacity of 600 MW at the Lingan thermal station and will be a major step in the displacement of offshore oil. Studies are in hand on conversion of existing oil-fired stations to coal, and the introduction of coal/oil slurry technique to maximize the use of coal in boilers designed for oil firing.

Provincial load growth was 6.6%. Increases in energy demand were recorded in residential (6.0%), commercial (8.8%) and industrial (10.2%) categories. About 86% of the province's electrical energy was produced by thermal generation and 62% of this (58% of total supply) was oil-fired.

In 1976, Nova Scotia Power completed two major additions to its generating system. Four 30-MW combustion turbine units were commissioned at the Burnside Industrial Park in Dartmouth to provide peaking capacity, and a 150-MW oil-fired unit was added to increase the capacity of the Tufts Cove station to 350 MW.

Construction began on the first two 150-MW coal-fired units at Lingan in Cape Breton. The decision to proceed with this project followed negotiations with the Cape Breton Development Corporation (DEVCO) for the supply of coal. The first Lingan unit is scheduled for operation in 1979 and the second in 1981.

The last substantial hydro site in the province at Wreck Cove has been developed to provide 200 MW of peaking capacity with a 100-MW unit commissioned in 1977 and another scheduled for 1978.

New Brunswick   13.10.4

The provincial load grew 12.4% over 1975. A decrease of 13.8% in industrial demand was offset by increases in residential and commercial demand in excess of 22% and 35% respectively. About 88% of the provincial electrical energy supply (excluding exports) was produced within the province with the rest imported from Quebec. Internal generation was 52% hydro, 5% coal-fired thermal and 43% oil-fired thermal (representing 33%, 3% and 27% respectively of the net provincial supply, excluding exports).

Net exports to Nova Scotia of 373 gigawatt hours (GWh) and the US (2 434 GWh) together represented 42.0% of internal generation and, respectively, 10.0% and 65.1% of the net import from Quebec. On an energy balance basis, less than 25.0% of power imported from Quebec was retained within the province, but such an analysis ignores the important timing aspect of export/import flows as well as peaking and reserve capacity benefits.

The New Brunswick Electric Power Commission has undertaken an expansion program to raise the installed generating capacity in 1980 by 1 370 megawatts, an increase of almost 70% over the 1976 level of 1 973 MW. Additions will include hydro, oil- and coal-fired thermal and nuclear power plants. Capacity of the Mactaquac hydro station on the Saint John River was increased to 638 MW with the installation of the fifth and sixth units in 1978; a 200-MW unit, capable of burning oil or coal (from the redeveloped Minto field), is to be added in 1979 to the Dalhousie thermal station which now has a single 150-MW oil unit.

The first nuclear power station in the Maritimes is under construction at Point Lepreau, west of Saint John, on the north shore of the Bay of Fundy. Initial operation of the first 630-MW CANDU unit is expected by 1980. Provision is being made for addition of a second unit at some future date to help reduce the province's dependence on high cost fuels for production of electricity.

In 1976 the first two of three 320-MW units were commissioned at the Coleson Cove oil-fired station. The final unit was expected to be in commercial service in 1977. Of the total 960-MW capacity, 400 MW has been committed to export to US utilities in the state of Maine for a 10-year period.

Planning for reinforcement of the provincial power grid is well under way. The existing system will eventually have an overlay at 345 kilovolts for reinforcement of major north-south transmission, and to connect the nuclear unit in southern New Brunswick with load centres. The first stage of this planned expansion is a 345-kV transmission line between Coleson Cove and the substation at Salisbury serving Moncton. This is also the terminal for the reinforced Nova Scotia/New Brunswick interconnection. A third such interconnection went into service in December 1976, operating initially at 138 kV but scheduled to expand to 345 kV in 1978 when the new Coleson Cove/Salisbury 345-kV circuit is completed.

A small windmill-powered generator was placed in service at the Sackville district office in the eastern part of the province and a solar-assisted heat pump heating system is being installed in a new district office building in Shediac.

13.10.5   Quebec

Over 99% of Quebec's electrical energy supply is generated hydraulically. Total electrical consumption in the province grew 5% compared with 1975. A 9.7% decline in industrial consumption was offset by increases of 22.8% in domestic and 11.0% in commercial categories. Consumption exceeded generation by some 16.5 terawatt hours, the deficiency being supplied by the Churchill Falls station in Labrador. Supply in excess of provincial demand was exported to New Brunswick, Ontario and the United States and in total was equivalent to about 48.6% of the 32.1 TWh-supply imported from Churchill Falls.

Electricity's share of the energy market in Quebec is expected to increase because of its availability and certainty of supply compared with that of oil, as well as the multiple use aspect of electric power. It is forecast that by 1990 there will be significant substitution of electricity for other forms in home heating.

To conserve non-renewable energy sources, the Hydro-Québec research institute has undertaken several projects related to development of alternative sources of generation (wind, solar, thermal, nuclear fusion) and new storage forms for energy.

Hydro-Québec's expansion program is designed to meet an average annual growth rate of 7.8% over the next 15 years. A deficit in peak-load capacity is forecast for 1979 and will be alleviated, in part, through purchase of 200 megawatts of peak-load capacity from the New Brunswick Power Commission. An additional 240 MW of peaking capacity will be provided by installation in 1979 of a four-unit combustion turbine station (La Citière) near the Hertel substation.

In 1976, the remaining five (197.2 MW) hydraulic units at Hydro-Québec's Manicouagan 3 station were placed in service, raising total capacity to 1 183.2 MW. The only addition to thermal capacity was a 53.3-MW combustion turbine unit at Cadillac in the Abitibi system. Because the Abitibi system will remain isolated from the integrated system until 1979, installation of two similar units was proposed for 1977.

Construction of the Manicouagan–Outardes hydro complex continued with completion scheduled for 1978 of three 151.3-MW units at the Outardes II station. Outardes II will be Hydro-Québec's third development on the Outardes River and will replace the present 50-MW development at Chute-aux-Outardes. Construction continued on the 685-MW single unit CANDU nuclear station, Gentilly II, near Trois-Rivières. This plant is expected to go into service in 1979.

Initial power supplies from the James Bay generation complex should become available in 1980. The La Grande River is being developed in the first phase at four sites with total maximum capacity of 10 190 MW. The first station, LG-2, will contain 16 units, each of 333 MW for a total ultimate capacity of 5 328 MW; six units are scheduled for service in 1980, six more units in 1981 and the remaining four in 1982. First power from LG-3 is expected in 1982 and from LG-1 in 1983. The LG-3 station will consist of 10 units of 192 MW each, for a total capacity of 1 920 MW. The LG-1 station will be a 910-MW development in 10 units. At the LG-4 site, seven 254-MW units are planned for service in 1984 with the eighth and final unit to follow in 1985 for a total capacity of 2 032 MW.

In conjunction with Ontario Hydro, the possibility of improving the present minimal interconnection between the Quebec and Ontario systems is being studied. Hydro-Québec has been authorized by the NEB to export power and off-peak energy to the Power Authority of the State of New York during the summer months (April to October) under a 13-year licence. This arrangement allows Quebec and PASNY to take advantage of the seasonal diversity of their respective peak demands. The Hydro-Québec system will be connected to the PASNY system by a 765-kV transmission line extending from the Châteauguay substation near Beauharnois to the Marcy substation, near Utica, NY. It was scheduled for service by mid-1978.

13.10.6   Ontario

Total electrical energy made available for use in the province in 1976 was 7.6% above 1975. Energy demand by sector showed increases in residential (5.6%), commercial (4.1%), and industrial (6.0%) categories. Ontario Hydro reported a December peak load of 15 896 megawatts, 9.5% above 1975. Generation totalled 87.2 terawatt hours of which 43.9% was from hydro, 18.8% from nuclear units and 37.3% from fossil fuels (70.8% coal, 13.8% gas, 15.4% oil — representing, respectively, 26.4%, 5.1% and 5.7% of total generation).

Ontario imported 13.2 TWh, representing 13.8% of provincial energy needs from other provinces. Net exports to the US of 4.2 TWh (4.8% of total generation) and exports to other provinces (0.4 TWh) resulted in a net import of 8.7 TWh — about 9.1% of electrical energy consumed within the province.

Ontario Hydro added a sixth 500-MW unit at its Nanticoke coal-fired station. Two 573.75-MW units were installed at the Lennox oil-fuelled station in eastern Ontario raising the total capacity of that station to 2 295 MW, the first major oil-fuelled power station in the Ontario Hydro system. Increases in hydro generation were limited to a 37.05-MW unit at Arnprior and a 24-MW unit at Andrew's Falls.

At the Bruce generating station, an 800-MW nuclear unit came into operation and was scheduled for commercial service in early 1977. A 12.16-MW gas turbine unit was commissioned at the same station and two 12.16-MW gas turbine units became operational at the nearby Bruce heavy water plant during the year; two more similar units were planned for 1977.

An extensive program of nuclear generation is expected to add 10 960 MW of new capacity in the period 1977-87. This program consists of four four-unit stations, three employing 800-MW units and one, Pickering B, with 540-MW units. At Bruce A, one 800-MW unit will be added each year from 1977 through 1979. Other nuclear additions now scheduled are Pickering B, 2 160 MW, 1981-83; Bruce B, 3 200 MW, 1982-85; and Darlington, 3 200 MW, 1984-87.

In northwestern Ontario, an extension to the coal-fired Thunder Bay plant is expected to add 300 MW in two units in 1980. A four-unit, 800-MW coal-fuelled generating station at Atikokan is projected for 1983-84. Both the Thunder Bay addition and the Atikokan stations are being designed to use western coal.

Fossil-fuelled additions will include the seventh and eighth 500-MW units at the Nanticoke station in 1977. An oil-fired station at Wesleyville, near Port Hope, similar in design to the Lennox station, is tentatively scheduled for service in 1981-83.

The only hydro unit currently scheduled by Ontario Hydro is a second 37.05-MW unit at Arnprior to be installed in 1977. Great Lakes Power Co. is proposing to add two 7.5-MW hydraulic units in 1981 at the St. Mary's station.

Manitoba   13.10.7

A federal loan of $193 million, in addition to an earlier $244 million provided for the Nelson River development, will assist in movement of power from Manitoba's Nelson River sites.

Work is proceeding on schedule at the Long Spruce hydro station despite extensive remedial work to rectify damage caused by a fire in one unit. Long Spruce, the second major development on the Nelson River, will have an ultimate capacity of 980 megawatts over 10 units by 1980. The first two units were scheduled for 1977 to be followed by four units in 1978 and the remaining four in 1979.

The Churchill River diversion scheme is almost complete. Up to 850 000 cubic decimetres of water per second can now be diverted from the Churchill to the Nelson.

The next major hydro development is now in progress at Limestone, downstream from Long Spruce. The first stage cofferdam for this 1 100-MW station is under construction and the first three 110-MW units are scheduled for service in the fall of 1984 to be followed by four units in 1985 and the remaining three in 1986.

Beyond Limestone plans are tentative, but a 10-unit station (1 080 MW) site at Conowapa downstream from Limestone is being considered with first generation projected for 1987.

A capability is being developed to begin the addition of nuclear generation in Manitoba during the mid-1980s. A decision on whether or not to proceed will depend on comparison with the economics of installing additional hydro capacity on the Nelson River system.

The capacity of the Nelson River HVDC transmission system has been increased by addition of the final two converter units for Bipole I, scheduled for service by mid-1977, to double capacity to 1 620 MW. To provide additional capacity for Long Spruce generation and later for Limestone, further additions to converter capacity comprising Bipole II will be added in stages in a new converter station at Henday near the Limestone generating station, and in an extension of the Dorsey terminal near Winnipeg; 900 MW was scheduled for 1978 and the final 900 MW for 1983-84. With these additions, the full capability of the two ±450-volt DC transmission circuits between the Nelson River generating sites and Winnipeg will be used.

Manitoba Hydro was granted a licence by the NEB to construct a second international 230-kilovolt transmission circuit, extending from the Ridgeway substation near Winnipeg to the US (Minnesota) border near Sprague, Man. Manitoba Hydro will export interruptible energy and short-term power to Minnesota Power and Light over this line, which was brought into service late in 1976, and will derive important system support benefits in the event of loss of supply from the north.

Manitoba's total load increased 1.9% from 1975 because of growth in the southern system load. Residential and farm demand showed an increase of 10.5%, most of it because of an increase in the number of customers. Commercial consumption increased 6.1%. The apparent 5.3% decrease in industrial demand was due to a decline in mining activity.

In-province generation was 14.0 terawatt hours, down 5.5% from that of 1975. Hydro generation of 12.7 TWh represented 91.0% of total generation compared with 14.3 TWh (96.8% of total) in 1975. The drop in hydro production was due to low water flows and was partially compensated by thermal production which increased from 0.5 TWh in 1975 to 1.3 TWh in 1976.

13.10.8 Saskatchewan

There were no additions to generating capacity in Saskatchewan in 1976; 67.2% of power generated in the province came from thermal power stations and 32.8% from hydro.

Total load increased 4.7% over 1975. This is lower than expected, and was due to zero growth in potash production, a decline in pipeline pumping and a mild winter. However, a 3.5% decline in industrial use was more than offset by increases in other categories. Sales to residential and farm customers increased 5.3%, due in part to a 3.3% increase in number of customers, and commercial consumption increased 4.1%.

Saskatchewan Power Corp. forecasts an annual load growth over the next five years of 6.6% for net system energy requirements anticipating industrial expansion, increased use of electricity for space heating and normal growth in customer sales categories.

Future plans for new generation include a 300-megawatt lignite-fired unit at Boundary Dam in 1977 to be followed in 1979 by the first 300-MW unit at the Poplar River lignite-fired plant near Coronach in south-central Saskatchewan. Transmission developments linked to new generation will include the construction of several 230-kilovolt lines.

13.10.9 Alberta

AEC Power Ltd. (a subsidiary of Alberta Energy Corporation and Calgary Power Ltd.) is building a 260-megawatt thermal generating station to supply power and process heat to the nearby Syncrude oil sands mining and refining project. Load growth in 1976 was 6.8%. Energy demand by sector showed increases over 1975 in residential (6.4%), commercial (9.0%) and industrial (8.5%) categories. More than 87% of electrical supply was generated in coal (58%) or gas (29%) thermal plants.

In 1976, Calgary Power completed four major construction phases at the Sundance coal thermal station on Lake Wabamun: a 486-hectare cooling pond commissioned late in 1975; conversion of the ash disposal system from a slurry to a dry-haul system enabling collection of both bottom ash and fly ash; and construction of the 750-MW (two 375-MW units) addition was completed with the No. 3 unit commissioned in 1976 and No. 4 in service in early 1977. Construction is in progress for two additional 375-MW units (Nos. 5 and 6) scheduled for commissioning in 1978 and 1980. Demand for coal from Calgary Power's mining operations increased 11% over 1975; a further 22% increase was expected in 1977 to fuel the Sundance and Wabamun thermal plants.

British Columbia 13.10.10

British Columbia derived over 95% of its electricity needs from hydro generation in 1976. Current planning indicates that hydro development will continue through the year 2000 but thermal generation will become important by the mid-1980s. Coal, an abundant resource in BC, will become the principal fuel source with the advent of larger thermal stations. BC Hydro is currently planning the development of a substantial lignite deposit at Hat Creek in central BC and is tentatively planning a 2 000-megawatt thermal station at that site.

The first two of four 435-MW units scheduled for installation in the underground power station at the Mica Dam on the Columbia River were commissioned late in the year. Units 3 and 4 were scheduled for service in 1977 with two additional units to be installed after 1978 as required.

Installation of a 53.9-MW gas turbine unit at the Keogh station near Port Hardy was the only addition to thermal capacity in 1976.

On the Peach River, construction of the Site I hydro project downstream of the Portage Mountain station continued on schedule. This plant is to have four 175-MW units, with first power expected in 1979. Units 3 and 4 are scheduled for service in 1980.

Construction is under way at the Seven Mile hydro site on the Pend-d'Oreille River. The diversion tunnel was completed and by year end construction of the main access road was nearing completion. Three of the four planned 202-MW units are scheduled for service in 1980. The fourth will be installed later when needed.

Plans for another major hydro development on the Columbia River at Revelstoke are well advanced. A licence was granted late in 1976 and tenders were called in January 1977 for the diversion tunnel and access roads. This station will have an initial installation of four 450-MW units, with first power scheduled for 1983; two additional units may be added at a later date.

Major transmission developments in 1976 included completion of BC Hydro's second 500-kilovolt receiving station, the Meridian substation near Port Coquitlam, to provide delivery of power from Mica to the 230-kV network in the Vancouver area. The two-line 500-kV Mica transmission facility will be completed by 1978. West Kootenay Power and Light Co. added 45 kilometres of 230-kV transmission line between Kelowna and Vernon. This line was connected for initial use at 138 kV and was to be brought to full power (230 kV) in 1977 when terminal facilities were completed. At year end, work was close to completion on an expansion of the HVDC underwater link between the mainland and Vancouver Island.

Total electrical energy demand in BC increased by 11.9% from 1975, apparently chiefly because of increasing industrial demand (18.9%) and to a lesser extent increases in both residential (9.1%) and commercial (7.7%) consumption.

Some 8.6% of total generation in BC (3.3 terawatt hours) was exported. More than 86.0% of exports went to the US through existing interconnections with the Bonneville Power Authority; the remaining 1.1% of generation was transferred to Alberta via the existing 138-kV Crowsnest interconnection.

The territories 13.11

In the Northwest Territories, electric energy use increased approximately 2.9%. In the Yukon Territory, energy consumption decreased 10.1% as a result of a 24.8% decrease in industrial (mining) consumption caused by labour unrest; this reduction was partially offset by increases in residential (14.5%) and commercial (2.9%) categories.

Major additions to generating capacity in the Northwest Territories included the commissioning of two 5-megawatt hydro units at Snare Forks. Diesel installations by the Northern Canada Power Commission added 13.5 MW of new capacity, 95% of it in the Northwest Territories. Future hydro developments in the Yukon Territory and Northwest Territories have been suspended pending a reassessment of short-term needs. In 1977, Northern Canada Power expects to have two 100-kilowatt packaged gas turbine units available for emergency use at smaller plants over its entire service area. A 2 500-kW gas turbine unit will also be available at larger plants. The addition of 2 500 kW of diesel capacity is scheduled for the Northwest Territories during 1977.

Transmission construction in 1976 included 16 kilometres of 115.0-kV line to connect the Snare Forks plant to the Snare system and completion of a portion of the 34.5-kilovolt loop at Yellowknife. In the Yukon Territory, a 34.5-kV line to Marsh Lake was completed and two reactors were added to the Whitehorse system.

13.12 Electric power statistics

Electric power statistics (Tables 13.17 and 13.18) are based on reports of all electric utilities and all industrial establishments generating energy, regardless of whether any is sold, and therefore show the total production and distribution of electric energy in Canada. Utilities are defined as companies, commissions, municipalities or individuals whose primary function is to sell most of the electric energy that they have either generated or bought. Industrial establishments are defined as companies or individuals generating electricity mainly for use in their own plants.

13.13 Financing energy self-reliance

An energy, mines and resources department analysis of the Canadian energy industry's ability to raise sufficient capital over the next 12 years indicates that some financing problems for specific energy projects are possible, but fears of widespread difficulties are groundless. A summary of the analysis *Financing energy self-reliance,* was released in March 1978. The study is a background paper to the 1976 report *Energy strategy for Canada; policies for self-reliance.*

Two analyses project energy expenditures of $181 billion (in 1975 dollars) but in different energy supply mixes. The first, optimistic about frontier oil and gas supplies, is based on three frontier pipelines. The second projects only one frontier pipeline but places more reliance on synthetic crude oil from the oil sands and heavy oils of Western Canada and on electricity.

13.13.1 Petroleum financing

Total petroleum investments including oil sands and refinery expenditures will range from $42 billion in the second analysis to $49 billion in the first. Major petroleum firms will probably generate cash flows sufficient to finance all their projected investments and the strong cash flow position should enable them to borrow if necessary to enhance the economics of oil sands or heavy oil projects. Smaller firms, primarily Canadian, are not likely to enjoy the same availability of funds. Canadian content rules may give them opportunities, but funds may restrict their access to tar sands and frontier development.

The petroleum industry faces new challenges in unconventional oil, gas and frontier development. These may require innovative approaches to financing. The industry will likely step up its diversification into non-petroleum areas, especially if frontier or tertiary recovery are less profitable than hoped. The system of monitoring cash flows and capital expenditures already in place would provide adequate warning to government should this process impede national energy objectives.

13.13.2 Paying for pipelines

Capital investments for pipelines vary from $17 billion in the second analysis to $28 billion in the first, with most of the difference accounted for by construction of the two additional frontier pipelines. This industry can be studied in two distinct segments in future: first is the conventional pipeline industry in southern Canada; the second involves construction and operation of frontier pipelines. Financial conditions in each sector are so different that each should be considered separately.

Conventional pipelines. Capital expenditures in this sector will be confined to replacement and expansion of existing lines, with the possible addition of new lines such as the proposed Quebec and Maritime system. These can be financed by normal methods and should present no difficulties.

Frontier pipelines. The existing pipeline industry is expected to contribute significant equity toward the completion of frontier transportation systems. However, about 75% of the investment will be financed by debt, probably on a project financing basis. Domestic and foreign capital markets can absorb the debt if the terms are right, although some compensation to debt holders for risks may be needed — for example, some participation in profits.

Electric power costs   13.13.3

Electrical power generation represents the largest investment ranging from $90 billion in the first and $110 billion in the second analysis. The main problem seems to be that for the next 15 years emphasis on remote hydro and nuclear growth will transform the risk characteristics of utility debt offerings. Most new capacity will be highly capital intensive (large-unit nuclear or hydro projects). The larger risks associated with such projects may require greater sharing among utilities and governments. Since most Canadian utilities are Crown corporations, equity is unlikely to be forthcoming from the private sector. There should be an increased role for innovative financing in the electrical sector over the medium term.

Costs of efficient-size generating units are very high relative to the financial capacity of some provinces and utilities. Moreover, such additions to capacity are very large in relation to existing load, so that errors in forecasting load growth can lead to under use of capacity, with potentially serious financial effects. All of these risks are likely to decrease in importance over time. With experience, nuclear reactor construction costs should drop — a tendency that may be reinforced by technological breakthroughs. As costs fall, financial capacity of the provinces and utilities should grow so that the relative importance of additional financial constraints in building a new nuclear plant will decline. As the system grows, the percentage increase in capacity due to the addition of a single efficient-size nuclear station will be smaller, thus minimizing the penalty from errors in demand forecast.

It would appear therefore that some provinces and utilities may need federal help to adjust to a structural change in the electrical sector. However, once the adjustment is made, the need for this aid should diminish. The next 10 to 15 years will be crucial.

Sources
13.1 - 13.13 Energy Policy Co-ordination, Department of Energy, Mines and Resources.

Tables

13.1 Oil imports, by country, 1976 and 1977

Country	Volume		Value	
	1976 '000 m^3	1977 '000 m^3	1976 $'000	1977 $'000
Algeria	794	88	65,335	8,501
Ecuador	130	442	9,398	37,088
Egypt, Arab Republic of	128	364	9,064	32,080
French Africa, n.e.s.	125	—	9,034	—
Gabon	741	187	53,745	14,067
Iran	9 534	6 592	694,416	535,063
Iraq	1 829	1 335	132,585	109,331
Kuwait	313	233	22,435	20,052
Libya	1 386	—	107,314	—
Mexico	42	55	3,146	4,878
Netherlands Antilles	—	359	—	29,017
Nigeria	1 907	424	153,113	37,390
Saudi Arabia	6 702	8 872	481,475	711,710
Trinidad and Tobago	117	224	8,023	17,808
Union of Soviet Socialist Republics	163	270	12,398	23,783
United Arab Emirates	833	165	61,785	14,176
United Kingdom	—	92	—	7,283
United States	119	3 103	9,840	283,875
Venezuela	16 395	15 199	1,245,396	1,317,540
Yemen	2 771	77	201,583	5,853
Total, all countries	44 029	38 081	3,280,085	3,209,495

13.2 Federal energy R&D expenditures, 1976-77 to 1978-79[e]

Item	1976-77		1977-78		1978-79	
	$'000	%	$'000	%	$'000	%
Renewable energy	4,437	4	7,437	6	13,564	9
Energy conservation	8,025	7	11,396	9	16,716	12
Fossil fuels	12,326	10	13,826	10	15,226	11
Transportation and transmission of energy	5,218	4	6,182	5	7,432	5
Nuclear energy	90,028	75	90,028	69	90,288	62
Co-ordination and monitoring	160	- -	1,025	1	1,238	1
Total	120,194	100	129,894	100	144,464	100

13.3 Trade in energy, 1966 and 1976 (million dollars)

Item	1966	1976
Petroleum		
Exports	348.4	2,485.5
Imports	451.6	3,446.4
Balance	-103.2	-960.9
Natural gas		
Exports	108.8	1,616.5
Imports	17.6	8.8
Balance	91.2	1,607.7
Petroleum and natural gas		
Exports	457.2	4,102.0
Imports	469.2	3,455.2
Balance	-12.0	646.8
Coal and coke		
Exports	14.6	566.2
Imports	153.3[1]	570.2[1]
Balance	-138.7	-4.0

13.3   Trade in energy, 1966 and 1976 (million dollars) (concluded)

Item	1966	1976
Electric energy		
Exports	16.2	161.7
Imports	10.2	9.2
Balance	6.0	152.5
Radioactive ores		
Exports	36.4	67.3
Imports	—	—
Balance	36.4	67.3
Total		
Exports	524.4	4,897.2
Imports	631.7	4,034.6
Balance	-107.3	862.6

[1]Includes rail freight charge from the mine to the US border.

13.4   Petroleum, supply and demand, 1965, 1975 and 1976 (megalitres a day)

Item	1965	1975	1976
Supply			
Production	*146.2*	*275.7*	*254.0*
Crude oil	139.1	250.6	229.4
Gas plant, liquefied petroleum gas	7.1	25.1	24.6
Imports	*88.5*	*136.6*	*120.9*
Crude oil	62.8	129.9	114.4
Products	25.7	6.7	6.5
Total, supply	234.7	412.3	374.9
Demand			
Domestic demand	*186.4*	*273.5*	*282.9*
Motor gasoline	56.9	93.9	97.6
Diesel fuel	15.8	29.9	31.8
Light fuel oil	34.7	44.2	46.1
Heavy fuel oil	39.2	42.2	50.7
Other (including refinery use and loss)	39.8	63.3	56.7
Exports	*48.1*	*143.0*	*103.6*
Crude oil	46.9	112.5	74.7
Products	1.2	30.5	28.9
Total, demand	234.5	416.5	386.5
Inventory change	0.2	-4.2	-11.6

13.5   Natural gas, supply and demand, 1965, 1975 and 1976 (gigalitres)

Item	1965	1975	1976
Supply			
Net production	40 844.4	97 529.4	98 070.3
Marketable pipeline gas[1]	29 141.0	69 266.2	69 614.5
Imports	501.2	288.8	116.1
Total, supply	29 642.2	69 555.0	69 730.6
Demand			
Domestic demand	*16 225.6*	*37 511.5*	*38 819.8*
Residential	5 303.8	8 472.4	8 857.6
Commercial	8 061.8	8 095.8	8 350.7
Industrial	2 860.0	20 943.3	21 611.5
Other (including pipeline losses)	1 614.1	3 618.9	3 262.1
Exports	11 459.9	26 810.5	27 003.1
Total, demand	29 299.6	67 940.9	69 085.0
Storage and line pack increase	342.6	1 614.1	645.6

[1]After deduction of field and plant use/loss, processing shrinkage.

13.6   Crude oil and equivalent production, by province, 1973-76 (megalitres a day)

Item and province	1973	1974	1975	1976	Percentage change 1975-76
Crude oil	*277*	*260*	*220*	*201*	*-8.6*
Alberta	228	217	185	168	-9.2
Saskatchewan	37	32	26	24	-7.7
British Columbia	9	8	6	6	—
Manitoba	2	2	2	2	—
Other	1	1	1	1	—
Pentanes plus/condensate	*27*	*26*	*24*	*21*	*-12.5*
Alberta	26	25	23	20	-13.0
Saskatchewan	- -	- -	- -	- -	
British Columbia	1	1	1	1	—
Synthetic crude oil					
Canada — Alberta	8	7	7	8	+14.3
Total	312	293	251	230	-8.4
Alberta	262	249	215	196	-8.8
Saskatchewan	37	32	26	24	-7.7
British Columbia	10	9	7	7	—
Manitoba	2	2	2	2	—
Other	1	1	1	1	—

13.7   Wells drilled, by type, region and depth, selected years, 1955-76

Region and type	1955 No.	1955 m	1960 No.	1960 m	1975 No.	1975 m	1976 No.	1976 m
Western Canada								
Alberta	1,620	2 569 687	1,692	3 192 547	3,646	3 649 170	5,042	4 871 708
New field wildcats	307	540 709	338	663 641	350	413 714	336	475 244
Other exploratory	105	133 180	223	356 945	847	1 043 549	1,235	1 489 092
Development	1,208	1 895 798	1,131	2 171 961	2,449	2 191 907	3,471	2 907 372
British Columbia	36	63 104	143	229 608	80	128 488	175	283 091
New field wildcats	34	59 135	60	111 501	12	29 656	6	16 305
Other exploratory	2	3 969	11	16 992	35	54 962	77	131 662
Development	—	—	72	101 115	33	43 870	92	135 124
Saskatchewan	912	986 113	602	720 449	267	197 878	257	229 670
New field wildcats	312	360 495	113	142 801	55	47 788	51	56 876
Other exploratory	50	54 715	28	30 237	52	41 442	89	73 650
Development	550	570 903	461	547 411	160	108 648	117	99 144
Manitoba	361	257 688	67	44 790	7	6 687	16	13 514
New field wildcats	59	53 131	10	9 298	4	3 628	10	7 400
Other exploratory	10	7 236	3	1 942	—	—	3	3 011
Development	292	197 321	54	33 550	3	3 059	3	3 103
Yukon Territory and Northwest Territories	6	3 739	32	32 299	42	113 218	31	86 154
New field wildcats	6	3 739	32	32 299	30	78 805	20	56 569
Other exploratory	—	—	—	—	2	7 569	2	3 885
Development	—	—	—	—	10	26 844	9	25 700
Total, Western Canada	*2,935*	*3 880 331*	*2,536*	*4 219 693*	*4,042*	*4 095 441*	*5,521*	*5 484 137*
New field wildcats	718	1 017 209	553	959 540	451	573 591	423	612 394
Other exploratory	167	199 100	265	406 116	936	1 147 522	1,406	1 701 300
Development	2,050	2 664 022	1,718	2 854 037	2,655	2 374 328	3,692	3 170 443
Eastern Canada								
Ontario	387	145 077	307	123 877	138	65 701	144	73 272
New field wildcats	64	34 213	39	20 846	49	28 788	40	26 459
Other exploratory	57	28 205	55	33 479	16	8 302	15	8 862
Development	266	82 659	213	69 552	73	28 611	89	37 951
Quebec	9	3 117	6	1 380	3	3 307	4	8 543
New field wildcats	9	3 117	5	1 307	3	3 307	4	8 543
Other exploratory	—	—	—	—	—	—	—	—
Development	—	—	1	73	—	—	—	—
Atlantic provinces	9	7 906	3	6 969	7	17 134	2	3 271
New field wildcats	2	1 462	3	6 969	7	17 134	2	3 271
Other exploratory	—	—	—	—	—	—	—	—
Development	7	6 444	—	—	—	—	—	—
East Coast offshore	—	—	—	—	10	32 753	11	31 787
New field wildcats	—	—	—	—	10	32 753	11	31 787
Other exploratory	—	—	—	—	—	—	—	—
Total, Eastern Canada	*405*	*156 100*	*316*	*132 226*	*158*	*118 895*	*161*	*116 873*
New field wildcats	75	38 792	47	29 122	69	81 982	57	70 060
Other exploratory	57	28 205	55	33 479	16	8 302	15	8 862
Development	273	89 103	214	69 625	73	28 611	89	37 951
Canada	3,340	4 036 430	2,852	4 351 919	4,200	4 214 336	5,682	5 601 010
New field wildcats	793	1 056 001	600	988 662	520	655 573	480	682 454
Other exploratory	224	227 305	320	439 595	952	1 155 824	1,421	1 710 162
Development	2,323	2 753 125	1,932	2 923 662	2,728	2 402 939	3,781	3 208 394

13.8   Wells drilled, by type and region, 1975 and 1976

Region	Oil		Gas		Dry		Total	
	1975	1976P	1975r	1976P	1975r	1976P	1975r	1976P
Western Canada	*782*	*722*	*2,044*	*3,306*	*1,226*	*1,423*	*4,042*	*5,521*
Alberta	670	550	1,922	3,193	1,064	1,229	3,646	5,042
Saskatchewan	105	154	85	16	77	87	267	257
British Columbia	2	13	31	86	47	76	80	175
Manitoba	2	3	—	—	5	13	7	16
Yukon Territory and Northwest Territories	3	2	6	11	33	18	42	31
Eastern Canada	*4*	*3*	*69*	*69*	*85*	*89*	*158*	*161*
Ontario	4	3	68	67	66	74	138	144
Quebec	—	—	—	2	3	2	3	4
Atlantic provinces	—	—	—	—	7	2	7	2
East Coast offshore	—	—	1	—	9	11	10	11
Total	786	725	2,113	3,375	1,311	1,512	4,200	5,682

13.9   Oil refining, by province, 1976 and 1977

Year and province or territory	Existing refineriesP			New refineries planned or under construction	
	No.	Capacity '000 m³/d	% of total	Capacity '000 m³/d	Scheduled for completion
1976					
Newfoundland	2	18.1	5.0	—	—
Nova Scotia	3	28.9	8.1	—	—
New Brunswick	1	39.7	11.2	—	—
Quebec	7	102.7	28.8	—	—
Ontario	7	87.4	24.5	18.2	1978
Manitoba	1	4.8	1.4	—	—
Saskatchewan	2	6.4	1.8	—	—
Alberta	6	42.9	11.9	—	—
British Columbia	7	26.0	7.2	—	—
Northwest Territories	1	0.5	0.1	—	—
Total	37	357.4	100.0	18.2	1978
1977					
Newfoundland	2	18.1	4.6	—	—
Nova Scotia	3	28.9	7.4	—	—
New Brunswick	1	39.7	10.2	—	—
Quebec	7	102.7	26.3	—	—
Ontario	8	105.6	27.0	15.1	1978
Manitoba	1	4.8	1.2	—	—
Saskatchewan	2	14.3	3.7	—	—
Alberta	6	50.1	12.8	—	—
British Columbia	7	26.1	6.7	—	—
Northwest Territories	1	0.5	0.1	—	—
Total	38	390.8	100.0	15.1	1978

13.10   Coal production[1], by type and province, 1975-77

Type and province	1975		1976		1977P	
	'000 t	$'000	'000 t	$'000	'000 t	$'000
Bituminous	*15 752*	*552,797*	*14 388*	*568,312*	*15 331*	*623,550*
Nova Scotia	1 656	44,586	2 000	57,755	2 132	80,500
New Brunswick	418	7,100	296	5,881	290	6,450
Alberta	4 097	159,023	4 583	206,919	4 218	192,750
British Columbia	9 580	342,088	7 509	297,757	8 691	343,850
Sub-bituminous Alberta	5 958	23,992	6 410	23,248	7 575	26,150
Lignite Saskatchewan	3 548	9,634	4 678	15,537	5 488	20,800
Total	25 259	586,423	25 476	607,100	28 394	670,500

[1]Includes production of clean coal and shipments of raw coal from the mine.

13.11   Coal, supply and demand, 1965, 1975 and 1976 (thousand tonnes)

Item	1965	1975r	1976
Supply			
Net production	10 512	25 259	25 476
Imports	14 765	15 819	14 622
Total, supply	25 277	41 078	40 098
Demand			
Consumption	*23 262*	*25 513*	*28 220*
Electrical utilities	6 985	16 538	19 045
Metallurgical use	5 326	7 294	7 389
General industry	10 951	1 681	1 786
Exports	1 237	11 431	11 857
Total, demand	24 499	36 944	40 077
Inventory change	778	4 134	21

13.12   Fuel used by electrical utilities to generate power, by province, 1975 and 1976

Year and province or territory	Coal Quantity *t*	Coal Value $	Petroleum fuels Quantity *kL*	Petroleum fuels Value $	Gas Quantity *kL*	Gas Value $
1975						
Newfoundland	—	—	144 846	8,736,912	—	—
Prince Edward Island	—	—	166 410	8,393,227	—	—
Nova Scotia	570 269	11,963,453	1 181 010	27,667,145	—	—
New Brunswick	247 216	4,380,515	423 641	18,295,625	—	—
Quebec	—	—	45 636	3,776,126	—	—
Ontario	6 834 225	172,193,581	172 230	9,363,344	1 655 999	54,542,795
Manitoba	322 684	2,207,576	18 902	1,757,183	822	24,419
Saskatchewan	3 251 157	10,321,342	13 270	559,238	226 386	5,189,605
Alberta	5 345 119	13,549,943	12 947	777,880	1 583 970	17,902,102
British Columbia	—	—	54 007	4,747,755	563 276	9,525,516
Yukon Territory	—	—	20 143	1,898,965	—	—
Northwest Territories	—	—	42 856	3,885,115	—	—
Canada	16 570 670	214,616,410	2 295 898	89,858,515	4 030 453	87,184,437
1976						
Newfoundland	—	—	124 292	7,185,920	—	—
Prince Edward Island	—	—	173 952	9,572,145	—	—
Nova Scotia	726 968	23,040,823	1 221 138	41,530,383	—	—
New Brunswick	208 179	4,487,719	634 728	32,892,360	—	—
Quebec	—	—	73 641	4,655,162	—	—
Ontario	7 682 973	251,412,189	800 336	44,698,984	1 412 722	71,912,806
Manitoba	989 644	9,938,485	20 688	2,255,654	2 903	140,770
Saskatchewan	3 537 223	14,034,952	4 092	354,692	320 722	12,134,135
Alberta	5 994 588	19,093,008	6 858	618,381	3 010 695	26,753,017
British Columbia	—	—	50 258	4,839,905	32 693	956,220
Yukon Territory	—	—	8 152	938,378	—	—
Northwest Territories	—	—	52 509	5,538,412	—	—
Canada	19 139 575	322,007,176	3 170 644	155,080,376	4 779 735	111,896,948

13.13   Installed generating capacity, as at Dec. 31, 1975 and 1976 (megawatts)

Year and province or territory	Steam Conventional	Steam Nuclear	Gas turbine	Internal combustion	Total thermal	Hydro	Total
1975							
Newfoundland (incl. Labrador)	355	—	36	66	456	6 206	6 662
Prince Edward Island	70	—	41	7	118	—	118
Nova Scotia	1 012	—	55	6	1 073	160	1 233
New Brunswick	621	—	23	9	653	680	1 333
Quebec	675	266	—	74	1 014	13 831	14 845
Ontario	8 917	2 400	509	26	11 853	7 008	18 861
Manitoba	447	—	28	23	498	2 475	2 973
Saskatchewan	1 085	—	159	33	1 277	567	1 844
Alberta	2 641	—	195	37	2 874	718	3 592
British Columbia	1 274	—	293	135	1 703	5 353	7 056
Yukon Territory	—	—	—	40	40	56	96
Northwest Territories	1	—	2	86	89	35	124
Canada, 1975	17 098	2 666	1 340	544	21 648	37 090	58 738
Net additions, 1975	802	—	134	10	947	311	1 258
Percentage increase, over 1974	4.9	—	11.1	1.9	4.6	0.8	2.2

13.13 Installed generating capacity, as at Dec. 31, 1975 and 1976 (megawatts) (concluded)

Year and province or territory	Steam		Gas turbine	Internal combustion	Total thermal	Hydro	Total
	Conventional	Nuclear					
1976P							
Newfoundland (incl. Labrador)	355	—	114	72	541	6 206	6 747
Prince Edward Island	71	—	41	7	118	—	118
Nova Scotia	1 162	—	205	6	1 373	160	1 533
New Brunswick	1 261	—	23	9	1 293	680	1 972
Quebec	667	266	53	79	1 065	15 012	16 077
Ontario	12 893	3 200	551	22	16 665	7 050	23 715
Manitoba	447	—	28	20	495	2 475	2 970
Saskatchewan	1 085	—	157	24	1 266	567	1 833
Alberta	3 538	—	196	48	3 781	718	4 500
British Columbia	1 370	—	350	137	1 857	6 500	8 357
Yukon Territory	—	—	—	45	45	58	103
Northwest Territories	1	—	2	111	113	49	162
Canada, 1976	22 847	3 466	1 720	580	28 613	39 475	68 088
Net additions, 1976	3 386	800	342	14	4 543	2 194	6 736
Percentage increase, over 1975	14.8	30.0	24.8	2.5	18.9	5.9	11.0
Per cent of total capacity, 1976	33.6	5.1	2.5	0.8	42.0	58.0	100.0

13.14 Capability and firm power peak-load requirements, actual 1966 and 1973-76, and forecast 1977-81 (megawatts)

Item	Actual					Forecast				
	1966	1973	1974	1975	1976	1977	1978	1979	1980	1981
NET GENERATING CAPABILITY										
Hydroelectric	21 459	34 807	36 624	37 318	38 543	40 263	41 520	42 274	45 451	47 527
Steam, conventional	6 634	15 161	13 694	16 484	18 884	22 298	23 120	24 156	24 812	25 809
nuclear	—	2 284	1 775	2 284	2 284	3 713	4 279	5 615	6 249	6 857
Internal combustion	257	375	393	410	406	420	425	433	428	439
Gas turbine	583	1 180	1 156	1 437	1 783	2 034	2 037	2 287	2 615	2 766
Total, net generating capability	28 933	53 807	53 642	57 933	61 900	68 728	71 381	74 765	79 555	83 398
Receipts of firm power from United States	100	1	2	1	51	7	138	81	28	37
Deliveries of firm power to United States	87	416	394	228	656	633	407	480	543	500
Total, net capability	28 946	53 392	53 250	57 706	61 295	68 102	71 112	74 366	79 040	82 935
PEAK LOADS										
Firm power peak loads within Canada	25 921	42 699	42 528	45 995	49 399	53 155	56 709	60 725	64 999	69 368
Indicated shortages	—	—	—	192	138	—	—	—	—	—
Total, indicated peak loads within Canada	25 921	42 699	42 528	46 187	49 537	53 155	56 709	60 725	64 999	69 368
Indicated reserve	3 025	10 693	10 722	11 519	11 758	14 947	14 403	13 641	14 041	13 567

13.15 Electric energy generation, by fuel type, 1976

Province or territory	Hydro		Thermal										Nuclear		Total generation	
	GWh	%	Coal		Oil		Gas		Total				GWh	%	GWh	%
			GWh	%	GWh	%	GWh	%	GWh	%						
Newfoundland[1]	38 774	99.0	—	—	417	1.0	—	—	417	1.0			—	—	39 190	13.4
Prince Edward Island	—	—	—	—	445	100.0	—	—	445	100.0			—	—	445	0.2
Nova Scotia	792	14.0	1 369	24.2	3 501	61.8	—	—	4 870	86.0			—	—	5 662	1.9
New Brunswick	3 409	51.6	355	5.4	2 843	43.0	—	—	3 198	48.4			—	—	6 607	2.3
Quebec	77 440	99.6	—	—	272	0.4	—	—	272	0.4			—	—	77 712	26.5
Ontario	38 292	43.9	23 021	26.4	5 007	5.7	4 487	5.1	32 515	37.3			16 430	18.8	87 237	29.7
Manitoba	12 729	90.9	1 200	8.6	75	0.5	—	—	1 275	9.1			—	—	14 004	4.8
Saskatchewan	2 463	32.8	4 299	57.2	51	0.7	702	9.3	5 052	67.2			—	—	7 515	2.6
Alberta	1 951	12.4	9 155	58.0	41	0.3	4 633	29.4	13 829	87.6			—	—	15 779	5.4
British Columbia	36 689	95.2	—	—	133	0.3	1 720	4.5	1 854	4.8			—	—	38 543	13.1
Yukon Territory	258	84.0	—	—	49	16.0	—	—	49	16.0			—	—	307	0.1
Northwest Territories	254	69.4	—	—	112	30.6	—	—	112	30.6			—	—	366	0.1
Canada	213 049	72.6	39 399	13.4	12 947	4.4	11 542	3.9	63 888	21.8			16 430	5.6	293 367	100.0

[1]Including export to Quebec of 32 105 GWh hydroelectric energy.

13.16 Electric energy consumed in Canada, by sector demand and province, 1975 and 1976 (megawatt hours)

Item and year	Nfld.	PEI	NS	NB	Que.	Ont.	Man.	Sask.	Alta.	BC	YT	NWT	Canada
Residential													
1975	1 152 010	194 517	1 682 828	1 633 103	19 737 427	19 281 752	3 417 004	2 234 916	3 408 274	6 741 120	63 068	29 901	59 575 920
1976	1 333 995	207 943	1 784 557	2 005 373	24 239 314	20 369 717	3 774 616	2 353 666	3 628 098	7 354 844	72 237	43 138	67 167 498
Percentage change	15.8	6.9	6.0	22.8	22.8	5.6	10.5	5.3	6.4	9.1	14.5	44.3	12.7
Commercial													
1975	902 467	184 174	1 411 968	1 603 602	20 086 898	28 086 866	3 600 952	2 104 360	6 831 302	7 762 024	89 446	64 973	72 729 032
1976	1 005 883	199 704	1 535 721	2 169 078	22 293 899	29 235 769	3 822 032	2 190 428	7 446 493	8 361 736	92 051	53 001	78 405 795
Percentage change	11.5	8.4	8.8	35.3	11.0	4.1	6.1	4.1	9.0	7.7	2.9	-18.4	7.8
Industrial													
1975	3 983 109	—	1 793 227	2 537 195	39 957 080	24 915 666	3 526 133	1 915 255	3 096 821	14 919 062	161 653	174 466	95 635 464
1976		—	1 976 871	2 186 966	36 091 813	26 418 352	3 338 354	1 847 900	3 358 822	17 732 919	121 487	179 486	97 236 079
Percentage change			10.2	-13.8	-9.7	6.0	-5.3	-3.5	8.5	18.9	-24.8	2.9	1.7
Total¹													
1975	6 135 169	418 644	5 662 279	6 710 708	89 754 706	89 144 720	12 010 756	7 050 165	15 074 112	32 575 423	341 197	356 086	265 233 965
1976	7 085 651	444 989	6 035 598	7 540 170	94 218 713	95 934 601	12 235 912	7 378 443	16 092 511	36 438 557	306 626	366 435	284 078 206
Percentage change	15.5	6.3	6.6	12.4	5.0	7.6	1.9	4.7	6.8	11.9	-10.1	2.9	7.1

¹Includes allowance for losses, cyclical billing and other adjustments.

13.17   Electric power generated by utilities and industrial establishments, 1975 (kilowatt hours)

Province or territory	Utilities			Industrial establishments	Total
	Public	Investor owned	Total		
Newfoundland	2 750	32 660	35 410	394	35 803
Prince Edward Island	4	417	421	—	421
Nova Scotia	5 023	—	5 023	475	5 498
New Brunswick	4 037	100	4 136	541	4 677
Quebec	54 523	3 625	58 148	17 959	76 108
Ontario	73 466	1 529	74 995	3 564	78 558
Manitoba	14 742	—	14 742	76	14 818
Saskatchewan	6 281	575	6 857	233	7 090
Alberta	3 395	10 980	14 376	724	15 100
British Columbia	22 670	869	23 538	11 003	34 542
Yukon Territory	296	26	322	30	352
Northwest Territories	356	25	381	44	425
Canada	187 543	50 805	238 348	35 044	273 392

13.18   Capital investment by electrical utilities, 1965-77 (million dollars)

Year	Construction				Machinery equipment	Total
	Generation	Transmission distribution	Other structures	Total		
1965	397	331	28	756	193	948
1966	468	296	23	786	356	1,142
1967	561	325	29	916	382	1,298
1968	493	332	64	889	443	1,332
1969	478	305	72	856	484	1,340
1970	581	449	28	1,057	554	1,610
1971	572	472	36	1,079	668	1,747
1972	636	449	50	1,135	619	1,754
1973	926	502	71	1,499	695	2,194
1974	1,049	598	53	1,700	1,054	2,753
1975	1,446	929	90	2,465	1,436	3,901
1976e	..	..	..	2,926	1,436	4,362
1977e	..	..	..	3,695	1,682	5,377

Sources

13.1 - 13.18 Manufacturing and Primary Industries Division, Industry Statistics Branch, Statistics Canada; and Energy Policy Sector, Department of Energy, Mines and Resources.

Housing
and construction

Chapter 14

Tables

Housing and construction

Chapter 14

Housing starts in Canada in 1977 declined to 245,724 units from the 1976 record level of 273,203. Housing completions, however, increased from 236,249 to 251,789 units. Although the number of dwellings subsidized by direct loans under the National Housing Act (NHA) was lower than in 1976, those financed by private lenders under NHA mortgage insurance agreements, and with additional funds from assisted home-ownership and assisted rental programs, more than doubled in 1977.

The federal government and housing
14.1

Although the federal government entered the housing field in 1918 when it made money available to the provinces for re-lending to municipalities, the first general piece of federal housing legislation was the Dominion Housing Act passed in 1935. This was followed by national housing acts in 1938 and 1944. The present National Housing Act, defined as an act to promote the construction of new houses, the repair and modernization of existing houses and the improvement of housing and living conditions, was passed in 1954.

In general the federal government, through successive housing acts, has attempted to stimulate and supplement the market for housing rather than assume direct responsibilities that belong to other levels of government or that could be borne more effectively by private enterprise. The aim has been to increase the flow of mortgage money and to encourage lenders to make loans on more favourable terms to prospective home-owners. Almost half the country's 7.5 million dwelling units have been built since the first covering legislation. About one-third of these were financed under the housing acts.

All provinces have complementary legislation providing for joint federal-provincial housing and land assembly projects, and most have enacted housing legislation.

Ministry of State for Urban Affairs (MSUA)
14.1.1

Since its creation in 1971, MSUA has been identifying and analyzing settlement and urbanization problems of federal concern and developing policies to improve the quality of life in Canadian cities.

Under the British North America Act, responsibility for Canada's municipalities and matters of local concern rests with the provinces and their municipal governments. But federal government policies, programs and projects affect the pattern, quality of life and economic base in Canadian settlements. Results must be beneficial to urban areas, and federal initiatives should take into account objectives of the provinces and their local governments.

MSUA works with other federal departments to incorporate and integrate urban concerns into federal policies and programs. Urban policy units provide MSUA with expertise and interdepartmental policy liaison, and collect and analyze information to define Canada's urban issues and their relevance to the federal government. Directorates in this work include human environment, settlement patterns, natural environment and resources, urban economy, metropolitan community development, non-metropolitan community development and urban networks.

Ministry operations units co-operate with the provinces and municipalities so that federal programs can respond to their objectives. Through intergovernmental and interdepartmental liaison MSUA learns about provincial and municipal urban development plans and in certain cases encourages and financially supports initiatives; it also provides financial help to organizations and institutions concerned with the future of urban Canada.

MSUA's directorate for international affairs expands the ministry's policy development and implementation by co-operating with other countries and international organizations on matters related to human settlements research, expertise and technology.

A recent reorganization brought the ministry into a closer relationship with Central Mortgage and Housing Corporation. Both retain separate responsibilities, but share expertise in urban-related matters. One deputy minister is responsible for both.

The ministry's operating expenditures in 1976-77 were $10.8 million. Contributions supporting urban development projects initiated by other levels of government and institutions in urban management and development totalled $2.2 million.

14.1.2 Central Mortgage and Housing Corporation (CMHC)

This corporation is the federal agency which administers the National Housing Act. Recent years have seen significant shifts in emphasis in the legislation which CMHC administers. Once primarily concerned with making mortgage loans and insuring loans made by approved NHA lenders, CMHC has become responsible for a growing number of socially oriented housing programs. Amendments to the National Housing Act have been designed to facilitate the supply of housing and to make it possible for more Canadians, particularly low- and moderate-income Canadians, to own or rent accommodation according to their needs. Through assisted home ownership and non-profit corporations and co-operatives there is an increasing range of choices by which those who need housing may obtain it.

Special programs are available to provincial and municipal governments to help deal with the varied impacts of urbanization. The neighbourhood improvement program, residential rehabilitation assistance program, new communities program and land assembly program are aimed at responding to changing local and regional needs and conditions.

CMHC is also concerned with developing new and innovative solutions to Canadian housing problems. On its own account and in co-operation with other governments and the private sector, CMHC seeks new ways of creating housing and housing forms, using land and servicing it, approaching planning and dealing with social, economic, physical and technological problems associated with housing and communities.

14.2 Government assistance

Social housing. The largest shortfall in loan commitments under CMHC's 1977 capital budget was for construction and acquisition of housing for low- and moderate-income earners. Commitments for public housing, to be rented according to an approved rent-to-income scale, and financed by CMHC loans to the provinces and joint investments by CMHC and the provinces, totalled only $199 million in 1977 compared to a budget allocation of $353 million. These commitments related to 7,547 dwelling units in public housing projects.

The emphasis in NHA assistance for Canadians with low and modest incomes has shifted in recent years toward non-profit and co-operative housing and away from public housing. Non-profit organizations can provide and operate homes for low-income families, the elderly, or special groups such as the handicapped, and can be constituted exclusively for charitable purposes or be provincially or municipally owned. Besides NHA loans at below market interest rates and capital contributions of 10% of the project's cost, start-up funds are available which help defray expenditures prior to the loan application. Also, land can be leased from CMHC. Its funding of community resource organizations enables the corporation to provide technical aid in project planning and management.

There are two types of co-operative housing. Non-profit co-operatives are organized with a view to collective ownership and management, while housing constructed through a building co-operative remains in individual ownership. All types of NHA assistance provided for non-profit organizations are available to building co-operatives. Under these programs commitments were made for 6,174 dwelling units and

1,188 hostel beds in 1977. Commitments were also made to rent 2,776 dwellings from private landlords, for occupancy on the same rent-to-income basis.

By year end 172,220 occupied units were subsidized under federal-provincial agreements, with rents on a scale graduated according to the tenant's income. The federal government's share was $141 million, up from $117 million in 1976.

Rural and Native Housing Program. The objectives of this program are to provide adequate housing in rural communities with populations of 2,500 or less, to give those eligible the opportunity to become involved in the entire housing process and create job opportunities, and to encourage the development and use of house designs that meet rural needs. Since the program was initiated in 1975 with a five-year target of 50,000 completions, over 15,000 units have been started or rehabilitated. In 1977 the total was 7,563 units, 70% of them rehabilitated dwellings. Under the NHA, $3 million was made available in sustaining grants for associations formed to organize or assist community and native groups, and in provision of training programs and secondment of technical experts.

Market housing. Three principal forms of NHA assistance are available for housing to be sold or rented on the private market. These are the provision of insurance on mortgage loans made by approved lenders; of direct CMHC mortgage loans on terms similar to those from approved lenders; and of CMHC incentives through the assisted home-ownership and assisted rental programs. These incentives are available whether a project is financed by an approved lender or by CMHC.

During 1977 direct lending for market housing by CMHC declined even further from the low level of 1976 and at $40 million for 1,127 units was below the amount allocated in the corporation's capital budget. This reduction was due to the availability of private mortgage lending with NHA mortgage insurance.

NHA-insured mortgage loans made by private lenders were worth $6 billion in 1977, an increase of 38% from the previous year. With these loans, 114,414 new dwelling units and 62,014 existing units were financed, nearly double the total for 1976. A large part of the NHA-insured mortgage lending on new housing, financed by private lenders, was accompanied by aid from the assisted home-ownership and assisted rental housing programs.

Assisted Home-Ownership Program (AHOP). The program encourages approved lenders and the building industry to make moderately-priced housing available for sale. To promote home-ownership, CMHC provides loans and grants to qualified purchasers of designated units.

To be designated for AHOP assistance, housing must be built within maximum sale price limits imposed by CMHC. These limits vary from market to market as a reflection of different production costs, and were increased in a number of areas in 1977, partly because of higher insulation requirements. In 1977, some 32,090 dwellings were designated for AHOP assistance, a decrease of 20% from the previous year, while actual commitments for such assistance amounted to 31,743 units, well below the budgeted provision of 43,390 units.

Assisted Rental Program (ARP). The program stimulates the production of moderately-priced housing for rent. Projects are mostly privately financed through approved lenders and have to be built within size and price levels determined by CMHC.

Assistance is available to builders when the cost of constructing and operating a project exceeds the rents that can be charged. This was a strong incentive to production in 1977, as were substantial tax advantages provided by the federal government through capital cost allowances. Help is provided through a second mortgage loan. Its approval is conditional upon the builder entering into an operating agreement with CMHC on rent levels.

Commitments for ARP assistance were made for 60,125 dwelling units, substantially higher than the total of 23,102 units in 1976. An extra budgetary allocation was made available in 1977 due to the large number of applications for assistance and the need for new rental construction.

14.2.1   Land and municipal infrastructure

An adequate supply of serviced land for residential development is a major objective of the federal government, as are the stabilization or reduction of serviced land prices, the elimination of water and soil pollution, and the promotion of high standards of community environment.

In 1977, under NHA land assembly provisions, CMHC made direct loan commitments or approved joint loan investments with the provincial governments of $44 million. This low level of investment reflected the emphasis put on development of existing land holdings rather than on acquisition of additional land. Sanitary sewage collection and treatment, storm sewers and water supply all qualify for federal assistance, identical to that for land assembly except that 25% of the loan is forgiven. Grants are also provided to help the preparation of regional sewerage plans and to offset excessive per capita costs of sewer installation in smaller communities and areas where difficult topography makes installation costs high. Most of the sewage treatment facilities in Canada in recent years have been financed with the assistance of CMHC; the long-term goal is to achieve a national standard of sewage treatment by 1985. Direct NHA sewage and water treatment loans amounting to $247 million were granted to municipalities, compared to a budget allocation of $281 million and the previous year's total of $302 million.

Grants provided under the NHA for development of regional sewage and water plans increased from $489,000 for 12 projects in 1976 to $950,000 for 25 projects in 1977.

Under a municipal incentives grants program designed to be in effect for the three-year period 1976-78, grants became available to encourage municipalities to develop modest housing, to make more economical use of land through increased density, and to help municipalities offset higher expenditures associated with medium-density development. The program was operational in all provinces in 1977.

To qualify for a $1,000 grant, a unit does not have to be financed under the NHA but must be priced within the AHOP and ARP limits established by CMHC. Total value of these grants in 1977 was $35.5 million for 35,500 dwelling units.

14.2.2   Neighbourhood improvement and residential rehabilitation

NIP and RRAP. The principal sources of NHA funding for neighbourhood and residential improvement are the Neighbourhood Improvement Program (NIP) and the Residential Rehabilitation Assistance Program (RRAP).

NIP encourages and supports efforts of municipalities to revitalize older residential neighbourhoods which are occupied predominantly by people of low and moderate incomes. Projects are planned and implemented by the municipalities and neighbourhood residents. NIP operates under annual agreements between CMHC and the provinces, which designate the municipalities to be eligible for assistance, and they in turn select eligible neighbourhoods. In 1977 agreements were signed with most provinces and 148 new NIP areas were selected, bringing the total since the inception of the program in 1974 to 478. NIP loan commitments during 1977 totalled $15.4 million, little changed from 1976, while grants increased from $49.4 million to $54.3 million.

These grants have been used for providing social and recreational amenities such as community and daycare centres, parks and playgrounds, improvement of water and sewer services and upgrading of sidewalks and street lighting. Since NIP emphasizes a comprehensive approach to rehabilitation and conservation, loans available to home-owners and landlords under the residential rehabilitation assistance program to improve and repair old and substandard dwellings play a major role in NIP areas. The NHA loans granted under this program are in part forgivable depending on income. In 1977, RRAP loans were approved for $69.4 million, nearly all allocated for the year. In 1976, loans totalling $61 million were granted with a prospective $42 million in loan forgiveness.

Home Improvement Loan Program. Under the home improvement loan provision of the NHA, chartered banks and approved credit instalment agencies are authorized to make loans for home improvements at favourable interest rates. These loans are guaranteed by CMHC in return for an insurance fee. In 1977, loans were approved for $18.1 million, relating to 2,474 loans.

Insulation programs. During 1977, under federal government direction, CMHC established two new programs for energy conservation in residential dwellings. The home insulation program was instituted in February in Prince Edward Island and Nova Scotia, two provinces particularly affected by rising energy costs. In September, the Canadian Home Insulation Program (CHIP) was initiated; this program applied to older housing in all areas of Canada except Prince Edward Island and Nova Scotia, although it was not until December 1977 that Quebec and Alberta agreed to participate. CMHC administers a special fund of $40.3 million established for CHIP. In 1977 some 109,800 applications for grants were approved and $14.4 million advanced.

Research and development   14.2.3

Policy development work undertaken in 1977 related to issues associated with ongoing CMHC programs and also to alternative means of achieving housing objectives nationally. The existing social programs of the NHA and NIP received special attention.

The corporation's research program continued to study technological innovations in the housing field. Support was also given to investigations into land market problems, new ways of meeting housing needs in rural and northern areas, special requirements of the handicapped, and the problems of the mobile housing industry.

The professional standards and technology sector took over the work of the former development group of the corporation in 1977 in developing technological innovations for housing. This group analyzes the problems of standards relating to community development and housing design to ensure their consistent and effective application under the NHA. It also helped develop standards for solar heating equipment and plans for research and development of solar energy for space and hot water heating.

Work relating to waste disposal and treatment continued on a demonstration plan, known as the Canadian water energy loop (CANWEL), in a Toronto apartment building. This is a project funded by CMHC to determine the feasibility of converting domestic sewage into usable water, and turning solid waste to energy. Toward the end of 1977, a municipal experimental CANWEL plant at Vaudreuil, Que. was completed and began trial operations.

Demonstration projects   14.2.4

CMHC plays a major role in developing and demonstrating innovative solutions to community and housing problems. The objective of the corporation's demonstration program is to plan and have built attractive communities and reasonably priced homes in a variety of urban settings. It is intended that such accommodation will be built by the private sector, including non-profit and co-operative groups.

The projects test and evaluate various designs and plans, and examine alternative means of financing, tenure, service and other aspects of development. CMHC co-operates closely with provincial and local authorities within whose jurisdiction the projects are situated. A Maryfield demonstration community in Charlottetown, PEI is a medium-density development of affordable houses designed to conserve land and energy in contrast to the sprawl of bungalows on large lots. A Woodroffe demonstration community in Ottawa, Ont. provides a mix of housing types and preserves the traditional advantages of suburban living while increasing land use efficiency and improving access to shops and other services. LeBreton Flats, a new community in the Ottawa inner city, was designed to house families of mixed incomes in eight different pilot units, each incorporating a solar heating system for hot water and supplementary space heating. A Fournier demonstration community in Hull, Que. is designed to revitalize a neighbourhood by building some 850 new housing units, a community shopping and recreation area, and a park and open spaces on the shoreline around the development. A Revelstoke, BC demonstration community where both home-ownership and rental accommodation will be available includes such other innovations as a road system that minimizes snow clearance for the municipality and a technique for reducing heating requirements.

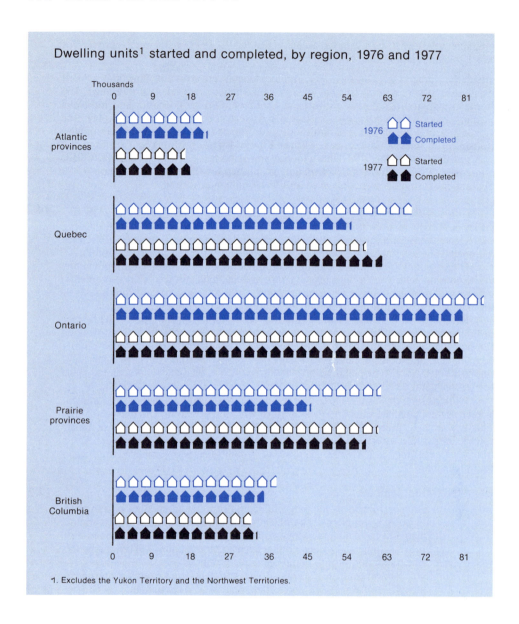

Dwelling units[1] started and completed, by region, 1976 and 1977

1. Excludes the Yukon Territory and the Northwest Territories.

14.3 Financial operations

A 1977 capital budget of $1.8 billion was approved by the government for CMHC loans and investments under the National Housing Act. During the year, commitments amounted to only $1.4 billion; this shortfall was caused partly by the lack of response to some NHA programs and partly by the deliberate action by the corporation in restraining the growth in undisbursed commitments.

14.3.1 Loans and investments

Funds advanced against loans and investment commitments in 1977 amounted to $1.1 billion, the greater part related to commitments from previous years. After repayments, this investment in loans and joint federal-provincial projects accounted for most of the increase of $644 million in the corporation's assets in 1977, to $9.8 billion at year end.

The corporation borrows its investment funds from the Bank of Canada. In 1977, these borrowings rose to $9.7 billion, an increase of 6.6% from 1976.

Grants, contributions and subsidies   14.3.2

Acting as the agent of the federal government, as distinct from its role as a financial institution, the corporation during 1977 advanced $524 million in subsidies, grants and contribution provisions of the NHA. This was 62% more than the $323 million in 1976. These funds provide financing for public housing subsidies, grants to non-profit corporations, co-operatives and home-owners, and for the forgiveness of interest payments under AHOP and ARP. The corporation's net income for 1977, after tax, was $10.8 million, up from $7.1 million in the preceding year.

Census and survey data on housing   14.4

Since 1941 decennial censuses of Canada have provided a comprehensive inventory of the nation's dwelling stock in a complete housing census taken in conjunction with the censuses of population and agriculture. Detailed information covering the 1941-71 period may be found in the relevant census volumes and reports. Summary data from the 1971 and 1976 censuses included here relate to a selection of the housing characteristics for which data were collected. More detailed information, including cross-classifications of the data, may be obtained from the user inquiry service of census field, Statistics Canada. Much of the present data was derived from the annual survey of household facilities and equipment carried out by Statistics Canada in conjunction with the May 1976 labour force survey.

Dwellings and housing growth rates   14.4.1

The 1976 Census recorded a total of 7.17 million private occupied dwellings in Canada. (A dwelling, for census purposes, is a structurally separate set of living quarters with a private entrance either from outside the building or from a common hall or stairway inside.) This total represented an 18.8% increase in dwellings since the 1971 Census. It is apparent that, despite slower population growth resulting from declining birth rates and lower immigration, the need for dwellings has continued to increase at a slightly higher rate than was observed in 1966-71. These trends in housing growth rates, including comparisons from 1966 to 1976 for such characteristics as type of dwelling and tenure, are summarized in Table 14.3.

Dwelling types, tenure and size   14.4.2

Dwelling types. Single detached homes continued to be the predominant type of housing in Canada in 1976, although their relative numbers have gradually declined in favour of multiple-type dwellings. Ten years earlier, at the 1966 Census, 62.4% of all dwellings were single detached but this percentage gradually dropped to 59.5% in 1971 and 55.7% in 1976. In the 1966-76 period single detached homes increased by 23.4%, whereas multiple-type units — single attached (double and row houses), apartments and movable dwellings — grew at the significantly higher rate of 63.1%.

Table 14.4 shows the distribution of the two broad dwelling-type classes in 1976 by province and by metropolitan area. Saskatchewan had the largest proportion of single detached homes, 77.1% of its occupied dwellings. Almost 60% of Quebec's dwellings were multiple-type units and only 40% were single detached, by far the lowest proportion among the provinces. The distribution within the major metropolitan centres reflected these provincial ratios in general terms, except that in most cases the proportion of multiple-unit dwelling types was considerably higher than for the province as a whole. This was particularly evident in Montreal and Toronto, where only 24.2% and 39.8%, respectively, of all occupied dwellings were single detached homes.

Tenure. Home-ownership increased 21.8% between 1971 and 1976 and the number of rented dwellings increased 14.1%. The faster growth rate in home-ownership in recent years reversed the trend observed in 1966-71 when the number of rented dwellings increased 25.5% compared with the 11.2% increase in owned dwellings.

As in the case of dwelling types, there was considerable variation among provinces in the proportions of owned dwellings in 1976, ranging from a low of 50.4% in Quebec to a high of 80.6% in Newfoundland. Table 14.5 shows that the increase in owner occupancy in the 1971-76 period was characteristic of all provinces, and at a fairly uniform rate.

There was a significant difference in the proportion of home-ownership between urban and rural areas. While 55.6% of private dwellings in urban areas were owner-occupied, the proportion in rural areas was 84.2%. Table 14.5 also shows that, in general, percentage of home-ownership varied inversely with size of urban communities. In urban areas of 500,000 population and over, for example, 48.2% of private dwellings were owner-occupied, compared with a range from 58.4% to 70.6% in smaller urban areas. In rural areas over 90% of dwellings in farming communities were owner-occupied, but a slightly lower percentage of home-ownership (82.4%) was reported for non-farm communities.

Dwelling size. The average size of Canadian dwellings in the period 1971-76 remained virtually constant at 5.4 rooms, although as Table 14.6 shows, the average number of rooms per dwelling declined in all provinces east of Ontario and in Manitoba. In the nation as a whole, Prince Edward Island had the highest average in 1976 at 5.89 rooms per dwelling, and Manitoba the lowest at 5.06.

14.4.3   Period of construction and length of occupancy

Period of construction. Figures from the 1971 Census indicate that 28.8% of the occupied housing stock was built after 1960. Table 14.7 shows the percentage distribution in 1971 of period of construction by province and by census metropolitan area. There were significant variations from province to province in the proportion of new dwellings. The lowest percentages of dwellings built after 1960 were found in the Atlantic provinces, Prince Edward Island having the lowest at 19%. Newfoundland was an exception, however, its 28.6% being close to the national average. Figures above the national average were found only in Alberta, British Columbia, and the Yukon Territory and Northwest Territories, which reported, respectively, 34.5%, 35.3%, and 58.4% of dwellings built after 1960. Values for census metropolitan areas ranged from 19.3% for Windsor to 40.9% for Edmonton.

Length of occupancy. The 1971 Census data on length of occupancy of household heads, displayed in Table 14.7, indicate the mobility of Canadians. Of all household heads in 1971, 66.8% had lived for 10 years or less in the dwelling in which they were enumerated, and 17.8% for less than one year — little changed from the 68.6% and 15.3%, respectively, in 1961.

The pattern of provincial variation for length of occupancy was similar to that for period of construction. The proportion of household heads occupying their present dwelling for 10 years or less was lowest in the Maritimes, ranging from 50.2% in Prince Edward Island to 55.0% in New Brunswick. The highest percentages were 70.6% in Alberta, 74.1% in British Columbia, and 90.7% in the Yukon Territory and Northwest Territories. For census metropolitan areas the range extended from 59.5% for Windsor to 76.1% for Calgary and 76.1% for Edmonton. Fully 25.7% of household heads in Calgary had occupied their dwellings for less than one year.

14.4.4   Heating fuel

In view of world energy shortages, data on home heating fuels are of particular interest. The 1976 survey data show that 47.5% of occupied Canadian dwellings were heated principally by oil or other liquid fuels, while 36.7% used gas. The major change since 1961 was a strong increase in the proportion of dwellings heated by gas, from 18.8% to 36.7%. This was offset by a correspondingly large decrease, from 10.6% to just 0.2%, in the proportion of dwellings using coal or coke. The category of other fuels declined from 14.3% to 2.4%, largely as a result of an increase from 0.7% to 13.1% in the proportion of dwellings which were electrically heated and a corresponding decrease in the proportion of dwellings heated by wood.

Table 14.8 gives the percentage distribution of dwellings by principal heating fuel, by province and by metropolitan area. There was a sharp difference between Quebec and Ontario in the proportions of dwellings heated principally by oil and by gas. In Quebec and the Atlantic provinces the proportion using gas as fuel was never higher than 6.9% (Quebec), while the proportion using oil was 70.5% or higher in all cases. In Ontario and the western provinces the proportion using gas was never lower than 45.1% (Ontario and British Columbia) and went as high as 89.7% for Alberta, while the proportion using oil was never above 41.8% and was as low as 4.6% in Alberta.

Household facilities and equipment   14.4.5

Survey data. The annual survey of household facilities and equipment provides an inventory to measure advances in living standards and to provide data for market research. The 1976 survey covered items such as plumbing and sanitary facilities, heating equipment, and accessories such as refrigerators, freezers, dishwashers, clothes dryers and television sets. Only the data on the first of these subjects, that is, the incidence of homes with running water, bath and toilet facilities, are shown by province in Table 14.9.

Continuing the rising trend in recent decades, there was again a marked improvement in the number of dwellings equipped with plumbing and sanitary facilities during the 1961-76 period. Dwellings with running water increased from 89.1% of all dwellings in 1961 to 99.0% in 1976. Similarly, households reporting a bath or shower for their exclusive use advanced from 77.1% to 97.6%, and households with exclusive use of a flush toilet from 79.0% to 98.1%.

Annual estimates. Table 14.10 presents some summary statistics from the 1976 sample survey on household facilities and equipment. About 36,000 households, chosen by area sampling methods, were included. Unlike decennial censuses, the sample survey cannot produce data for the smaller localities and areas, but much of the information shown in Table 14.10 for Canada is available also for individual provinces and selected metropolitan areas. This table shows that the percentage of households having colour TVs has risen from 44.5% in 1974 to 53.4% in 1975 and 60.6% in 1976. Households with black and white TVs decreased correspondingly over the same three-year period to 62.4% in 1976 from 67.8% a year earlier and 73.3% two years earlier. Households with dishwashers increased to 18.6% in 1976 from 12.9% in 1974; and those with two or more cars increased 24.4% in 1976 from 23.0% a year earlier and 21.6% in 1974.

In May 1976, more than one household in 10 reported air conditioning facilities: 670,000 households owned window-type air conditioners and 254,000 households were equipped with central unit air conditioners.

Construction   14.5

Value of construction work   14.5.1

The data on the construction industry represent the estimated total value of all new and repair construction performed by contractors and by the labour forces of utility, manufacturing, mining and logging firms, government departments, home-owner builders and other persons or firms not primarily engaged in the construction industry. Table 14.11 shows the value of new and repair construction work and Table 14.12 the value of such work performed by contractors and others during the period 1973-77, preliminary expenditures for 1976 and intentions for 1977. Table 14.21 gives estimates of total expenditures in Canada on each type of construction for which information is available.

Principal statistics of the construction industry for 1975-77 are shown by province and for contractors, utilities, governments and others in Table 14.13. The statistics given for Canada as a whole may be considered as relatively accurate but those for individual provinces and by class of builder are approximations only. All estimates given for cost of materials used are based on ratios of this item to total value of work performed, derived from annual surveys of construction work performed by contractors and others and applied to the total value-of-work figures. Estimates of labour content are similarly based but, in addition, are adjusted to include working owners and partners and their withdrawals.

Chapter 21 includes detailed price index numbers of construction and capital goods, which measure price changes in residential and non-residential building materials and changes in construction wage rates; price indexes of highway construction which show annual costs to provincial governments in contracts awarded for highway construction as a percentage of prices paid in 1971; and price indexes of electrical utility construction (distribution systems, transmission lines, transformer stations) which provide an estimate of the impact of price change on the cost of materials, labour and equipment used in constructing and equipping such utilities.

14.5.2   Building permits issued

The estimated value of proposed construction is indicated by the value of building permits issued. Figures of building permits are collected from approximately 2,000 municipalities across the country and are available for individual municipalities, metropolitan areas, provinces and economic areas in Quebec, Ontario and Manitoba.

The total value of permits issued for building construction in 1976 was $12.2 billion, a figure 15.1% higher than in 1975. Residential construction value increased by 20% and overall non-residential construction value also increased by 5.7%, reflecting an increase of 15.3% in the industrial, 13.1% in the commercial, and a 13% decrease in the institutional and government construction sectors.

Permit values rose in all provinces except Nova Scotia and Ontario which showed a slight decrease. The value of building permits issued in each province in 1975 and 1976 is given in Table 14.14, in 50 municipalities in Table 14.15 and in 22 metropolitan areas in Table 14.16. These metropolitan areas made up 70% of the 1976 total for Canada.

14.6   Capital expenditures

Capital spending in Canada by all sectors of the economy during 1977 was expected to reach about $46.5 billion, an increase of 10.5% over the 1976 level of about $42.1 billion. These estimates were in current dollars without any adjustment for price increase and reflected the intended outlays by respondents between May and July 1977. The survey covered business establishments, educational and other institutions and governments at all levels.

Intended capital expenditures on new construction in 1977 were estimated at $29.7 billion, an increase of 9.9% over the 1976 total of $27.1 billion. The two main elements of construction are shown at $10.3 billion ($9.8 billion in 1976) for residential and $19.5 billion ($17.3 billion) for non-residential. The increase for housing was 5.2% and for non-residential construction, 12.5%.

Acquisition of new machinery and equipment during 1977 was expected to amount to $16.8 billion, 11.8% above the 1976 total of $15 billion. Rates of increase were 6.0% for 1976 and 17.0% in 1975.

Table 14.17 shows the trend in capital spending over the years 1968-77 in both current and constant (1971) dollars. Table 14.18 summarizes capital and repair expenditures by economic sector and Table 14.19 contains details of the manufacturing, mining and utilities sectors for 1975-77. A summary of capital expenditures in Table 14.20, representing gross additions to the capital stock of each province and territory, reflects economic activity in the area and employment and income-giving effects in other regions. For example, spending millions of dollars on plant and equipment in Western Canada may generate considerable activity in machinery industries in Ontario and Quebec as well as construction activity in the western provinces.

Sources
14.1 - 14.3 Editorial Services, Central Mortgage and Housing Corporation.
14.4 Census Characteristics Division, Census Content and Analysis Branch, Statistics Canada.
14.5 - 14.6 Construction Division, Industry Statistics Branch, Statistics Canada.

Tables

14.1   Dwelling units[1] started and completed, by type of financing, 1975-77 and by region, 1976 and 1977

Year and region	Dwelling units started					Dwelling units completed
	National Housing Act		Conventional institutional loans	All other financing	Total	
	CMHC	Approved lenders loans				
1975	41,800	47,132	66,905	75,619	231,456	216,964
1976	24,087	93,883	71,776	83,457	273,203	236,249
1977	17,819	102,462	71,700	53,743	245,724	251,789
1976						
Atlantic provinces	3,632	4,089	6,339	6,733	20,793	21,340
Quebec	2,766	27,615	15,067	23,300	68,748	54,301
Ontario	8,863	37,050	24,413	14,356	84,682	80,302
Prairie provinces	5,276	16,690	15,713	23,574	61,253	45,396
British Columbia	3,550	8,439	10,244	15,494	37,727	34,910
1977						
Atlantic provinces	1,996	5,169	5,870	3,311	16,346	17,778
Quebec	3,074	23,660	13,214	17,632	57,580	61,979
Ontario	5,149	40,262	28,686	5,033	79,130	80,717
Prairie provinces	4,798	21,251	15,597	18,664	60,310	58,084
British Columbia	2,802	12,120	8,333	9,103	32,358	33,231

[1]Excludes the Yukon Territory and Northwest Territories.

14.2   Dwelling units started in metropolitan areas, large urban centres and urban agglomerations, 1976 and 1977

Area	Dwelling units started					
	1976[1] Total	1977[2]				
		Total	Single detached	Semi-detached and duplex	Row	Apartment; other
METROPOLITAN AREAS	166,729	156,820	51,987	13,568	21,751	69,514
Calgary	11,360	13,190	3,638	1,174	3,331	5,047
Chicoutimi–Jonquière	964	1,262	863	62	55	282
Edmonton	12,370	12,206	3,955	526	2,486	5,239
Halifax	3,499	4,277	1,219	160	246	2,652
Hamilton	5,490	3,956	1,624	810	809	713
Kitchener	3,926	3,466	1,345	369	612	1,140
London	3,318	3,973	1,389	274	664	1,646
Montreal	37,531	27,193	9,350	993	495	16,355
Oshawa	3,500	2,672	842	458	684	688
Ottawa–Hull	7,059	7,429	1,450	1,164	2,535	2,280
Quebec	5,427	8,456	4,880	512	41	3,023
Regina	3,070	3,497	1,696	150	216	1,435
Saint John	4,167	2,341	1,295	502	116	428
St. Catharines–Niagara	1,732	516	425	31	—	60
St. John's	1,386	1,012	403	32	41	536
Saskatoon	2,965	3,976	1,409	102	—	2,465
Sudbury	1,058	1,265	678	290	152	145
Thunder Bay	1,491	1,620	670	152	245	553
Toronto	26,555	27,918	5,472	4,086	6,081	12,279
Vancouver	16,702	15,257	5,365	650	2,160	7,082
Victoria	4,439	3,166	699	246	61	2,160
Windsor	2,002	1,819	896	26	37	860
Winnipeg	6,718	6,353	2,424	799	684	2,446
LARGE URBAN CENTRES AND URBAN AGGLOMERATIONS	15,477	13,846	6,563	1,373	682	5,228
Brantford	1,526	1,026	465	334	109	118
Guelph	1,159	943	162	76	—	705
Kamloops	708	526	317	144	29	36
Kelowna	1,392	713	353	24	4	332
Kingston	1,481	1,154	481	232	19	422
Moncton	904	406	378	6	—	22
North Bay	446	514	229	206	6	73
Peterborough	970	649	223	6	84	336

14.2   Dwelling units started in metropolitan areas, large urban centres and urban agglomerations, 1976 and 1977 (concluded)

Area	Dwelling units started					
	1976[1]	1977[2]				
	Total	Total	Single de-tached	Semi-de-tached and duplex	Row	Apart-ment; other
LARGE URBAN CENTRES AND URBAN AGGLOMERATIONS (concluded)						
Prince George	1,488	977	885	17	—	75
Sarnia	1,288	1,500	421	142	315	622
Sault Ste Marie	747	749	243	150	21	335
Shawinigan	275	365	266	2	—	97
Sherbrooke	728	1,796	633	18	6	1,139
St-Jean	831	654	245	8	—	401
Sydney/Sydney Mines	753	765	539	—	80	146
Trois-Rivières	781	1,109	723	8	9	369
Other areas	90,997	75,058	49,853	3,432	4,188	17,585
Canada[3]	273,203	245,724	108,403	18,373	26,621	92,327

[1]1976 data based on the 1971 Census.
[2]1977 data based on the 1976 Census.
[3]Excludes the Yukon Territory and Northwest Territories.

14.3   Summary of housing characteristics, censuses of 1966, 1971 and 1976

Item		1966	1971	1976	Percentage increase	
					1966-71	1971-76
Total occupied private dwellings	No.	5,180,475	6,034,510	7,166,095	16.5	18.8
	%	100.0	100.0	100.0	...	...
TYPE OF DWELLING						
Single detached	No.	3,234,125	3,591,770	3,991,540	11.1	11.1
	%	62.4	59.5	55.7	...	...
Single attached	No.	401,755	679,590[1]	587,180	69.2	-13.6
	%	7.8	11.3	8.2	...	...
Apartment and duplex	No.	1,516,420	1,699,045[1]	2,412,660	12.0	42.0
	%	29.3	28.2	33.7	...	...
Mobile	No.	28,180	64,105	174,710	127.5	172.5
	%	0.5	1.1	2.4	...	...
TENURE						
Owned	No.	3,269,970	3,636,925	4,431,235	11.2	21.8
	%	63.1	60.3	61.8	...	...
Rented	No.	1,910,505	2,397,580	2,734,860	25.5	14.1
	%	36.9	39.7	38.2	...	...

[1]In 1971 the "single attached" included some "apartment"; consequently, there should be more dwellings in "apartment and duplex" and less dwellings in "single attached" than the figures shown in this table.

14.4   Type of dwelling, by province and by census metropolitan area, 1976

Province and census metropolitan area	Total occupied private dwellings[1]	Single detached	Multiple-unit types[2]	Single detached %	Multiple-unit types[2] %
PROVINCE					
Newfoundland	131,665	95,925	31,455	72.9	23.9
Prince Edward Island	32,930	24,315	7,000	73.8	21.3
Nova Scotia	243,100	162,550	66,570	66.9	27.4
New Brunswick	190,435	125,830	52,585	66.1	27.6
Quebec	1,894,110	745,595	1,120,630	39.4	59.2
Ontario	2,634,620	1,494,465	1,117,365	56.7	42.4
Manitoba	328,005	219,950	100,140	67.1	30.5
Saskatchewan	291,155	224,510	55,755	77.1	19.2
Alberta	575,280	372,420	174,610	64.7	30.4
British Columbia	828,290	516,485	268,690	62.4	32.4
Yukon Territory	6,495	3,425	2,165	52.7	33.4
Northwest Territories	10,020	6,070	2,865	60.6	28.7
Canada	7,166,095	3,991,540	2,999,840	55.7	41.9

14.4   Type of dwelling, by province and by census metropolitan area, 1976 (concluded)

Province and census metropolitan area	Total occupied private dwellings[1]	Single detached	Multiple-unit types[2]	Single detached %	Multiple-unit types[2] %
METROPOLITAN AREA					
Calgary, Alta.	155,155	90,765	62,475	58.5	40.3
Chicoutimi, Que.	33,850	16,165	17,080	47.8	50.5
Edmonton, Alta.	179,635	100,345	77,350	55.9	43.1
Halifax, NS	81,845	39,335	39,030	48.1	47.7
Hamilton, Ont.	172,510	101,470	70,805	58.8	41.0
Kitchener, Ont.	87,880	47,305	40,455	53.8	46.0
London, Ont.	91,770	51,505	39,830	56.1	43.4
Montreal, Que.	924,635	223,365	698,750	24.2	75.6
Oshawa, Ont.	41,445	24,935	16,440	60.2	39.7
Ottawa-Hull, Ont., Que.	225,105	94,105	129,445	41.8	57.5
Quebec, Que.	164,600	60,065	103,180	36.5	62.7
Regina, Sask.	49,790	33,310	16,195	66.9	32.5
Saint John, NB	34,065	14,780	17,635	43.4	51.8
St. Catharines, Ont.	97,395	67,860	29,205	69.7	30.0
St. John's, Nfld.	36,800	18,475	17,690	50.2	48.1
Saskatoon, Sask.	44,800	28,315	16,045	63.2	35.8
Sudbury, Ont.	45,710	26,080	19,030	57.1	41.6
Thunder Bay, Ont.	37,270	26,240	10,735	70.4	28.8
Toronto, Ont.	909,530	361,560	547,435	39.8	60.2
Vancouver, BC	407,560	231,915	171,080	56.9	42.0
Victoria, BC	81,005	46,995	32,680	58.0	40.3
Windsor, Ont.	80,190	53,705	25,565	67.0	31.9
Winnipeg, Man.	197,305	115,400	81,090	58.5	41.1

[1]Includes mobile homes and other movable dwellings.
[2]Includes double and row houses, apartments, duplexes and dwellings attached to non-residential structures.

14.5   Owned and rented dwellings, by province and type of locality, censuses of 1971 and 1976

Province or territory and type of locality	1971 Owned	1971 Rented	Percentage Owned	Percentage Rented	1976 Owned	1976 Rented	Percentage Owned	Percentage Rented
PROVINCE								
Newfoundland	88,335	22,110	80.0	20.0	106,180	25,485	80.6	19.4
Prince Edward Island	20,725	7,155	74.3	25.7	25,225	7,700	76.6	23.4
Nova Scotia	147,705	59,800	71.2	28.8	176,055	67,040	72.4	27.6
New Brunswick	109,450	48,185	69.4	30.6	136,795	53,640	71.8	28.2
Quebec	761,340	843,450	47.4	52.6	953,960	940,155	50.4	49.6
Ontario	1,400,340	825,145	62.9	37.1	1,676,250	958,370	63.6	36.4
Manitoba	190,585	97,790	66.1	33.9	217,685	110,320	66.4	33.6
Saskatchewan	194,535	73,035	72.7	27.3	219,925	71,230	75.5	24.5
Alberta	296,705	167,910	63.9	36.1	372,825	202,455	64.8	35.2
British Columbia	422,785	244,765	63.3	36.7	540,635	287,655	65.3	34.7
Yukon Territory and Northwest Territories	4,425	8,240	35.0	65.0	5,705	10,820	34.5	65.5
Canada	3,636,925	2,397,585	60.3	39.7	4,431,235	2,734,860	61.8	38.2
TYPE OF LOCALITY								
Urban	2,572,885	2,164,535	54.3	45.7	3,123,330	2,489,720	55.6	44.4
500,000 and over	956,765	1,118,550	46.1	53.9	1,367,720	1,462,950	48.2	51.8
100,000-499,999	556,375	428,770	56.5	43.5	631,220	403,305	61.0	39.0
30,000- 99,999	304,450	230,365	56.9	43.1	338,915	241,440	58.4	41.6
5,000- 29,999	449,685	248,740	64.4	35.6	469,225	250,200	65.2	34.8
Under 5,000	305,610	138,105	68.9	31.1	316,245	131,825	70.6	29.4
Rural	1,064,045	233,050	82.0	18.0	1,307,905	245,145	84.2	15.8
Non-farm	758,830	210,830	78.3	21.7	1,071,475	229,200	82.4	17.6
Farm	305,210	22,215	93.2	6.8	236,425	15,945	93.7	6.3

14.6   Average number of rooms per dwelling, by province, 1971 and 1976

Province	Average number of rooms per dwelling 1971[1]	Average number of rooms per dwelling 1976[2]
Newfoundland	5.8	5.66
Prince Edward Island	6.1	5.89
Nova Scotia	5.7	5.51
New Brunswick	5.7	5.68
Quebec	5.2	5.19
Ontario	5.6	5.59
Manitoba	5.2	5.06
Saskatchewan	5.3	5.49
Alberta	5.4	5.40
British Columbia	5.2	5.25
Total	5.4	5.41

[1]1971 Census.
[2]Household facilities and equipment survey, May 1976.

14.7 Period of construction and length of occupancy, by province and by census metropolitan area, 1971 (percentage)

Province and census metropolitan area	Period of construction				Length of occupancy					
	Before 1946	1946-60	1961-71[1]	Total	Less than 1 year	1-2 years	3-5 years	6-10 years	More than 10 years	Total
PROVINCE										
Newfoundland	35.0	36.4	28.6	100.0	11.1	11.7	13.5	16.1	47.6	100.0
Prince Edward Island	62.2	18.7	19.0	100.0	11.7	11.4	12.3	14.8	49.7	100.0
Nova Scotia	53.2	25.6	21.1	100.0	14.4	13.1	12.7	14.1	45.7	100.0
New Brunswick	50.4	27.9	21.7	100.0	14.1	13.0	13.3	14.6	45.0	100.0
Quebec	37.5	33.9	28.6	100.0	17.2	17.5	15.5	17.9	32.0	100.0
Ontario	39.8	32.2	28.0	100.0	17.5	17.0	16.2	16.3	32.9	100.0
Manitoba	42.6	32.2	25.3	100.0	17.8	14.6	14.6	16.2	36.8	100.0
Saskatchewan	42.3	32.6	25.1	100.0	14.7	12.6	14.2	18.1	40.4	100.0
Alberta	26.8	38.7	34.5	100.0	20.6	16.7	16.5	16.8	29.4	100.0
British Columbia	29.6	35.1	35.3	100.0	22.5	18.3	17.6	15.7	25.9	100.0
Yukon Territory and Northwest Territories	9.5	32.2	58.4	100.0	34.9	26.4	17.9	11.5	9.2	100.0
Canada	38.0	33.2	28.8	100.0	17.8	16.6	15.8	16.6	33.2	100.0
METROPOLITAN AREA										
Calgary, Alta.	19.7	40.6	39.7	100.0	25.7	18.1	17.3	15.0	23.9	100.0
Chicoutimi—Jonquière, Que.	36.8	40.4	22.8	100.0	15.8	15.5	14.0	15.4	39.2	100.0
Edmonton, Alta.	18.0	41.2	40.9	100.0	23.6	18.6	17.3	16.6	23.8	100.0
Halifax, NS	34.8	32.8	32.4	100.0	22.7	17.5	16.1	15.4	28.2	100.0
Hamilton, Ont.	37.5	34.1	28.4	100.0	16.7	17.0	16.9	16.6	32.8	100.0
Kitchener, Ont.	32.9	29.9	37.2	100.0	21.7	18.4	15.4	15.2	29.2	100.0
London, Ont.	39.4	29.7	30.9	100.0	21.2	17.2	16.5	14.6	30.5	100.0
Montreal, Que.	31.1	36.4	32.4	100.0	19.5	19.9	16.5	19.0	25.1	100.0
Ottawa—Hull, Ont., Que.	28.7	33.9	37.4	100.0	21.5	19.1	17.4	17.2	24.9	100.0
Quebec, Que.	35.2	29.6	35.2	100.0	18.9	17.3	15.9	17.4	30.5	100.0
Regina, Sask.	30.5	35.9	33.7	100.0	21.5	16.9	15.7	18.6	27.3	100.0
Saint John, NB	53.7	23.9	22.4	100.0	16.1	16.0	16.7	17.4	33.8	100.0
St. Catharines—Niagara, Ont.	41.8	35.8	22.4	100.0	14.9	14.3	14.8	15.9	40.1	100.0
St. John's, Nfld.	36.2	32.2	31.6	100.0	14.6	14.6	15.6	17.7	37.5	100.0
Saskatoon, Sask.	27.6	35.0	37.3	100.0	24.3	17.5	15.7	16.7	25.8	100.0
Sudbury, Ont.	30.6	41.4	28.1	100.0	19.6	18.1	15.7	15.3	31.2	100.0
Thunder Bay, Ont.	44.4	35.3	20.2	100.0	14.6	14.1	15.2	16.9	39.2	100.0
Toronto, Ont.	30.7	35.3	34.0	100.0	19.1	19.4	17.5	17.2	26.8	100.0
Vancouver, BC	30.8	35.5	33.7	100.0	22.1	17.9	17.3	15.6	27.1	100.0
Victoria, BC	36.2	32.9	30.9	100.0	22.0	18.1	17.3	16.6	26.0	100.0
Windsor, Ont.	49.1	31.6	19.3	100.0	16.9	14.0	13.5	15.1	40.5	100.0
Winnipeg, Man.	41.5	33.4	25.1	100.0	20.2	16.1	15.1	15.5	33.1	100.0

[1]Includes the first five months only of 1971.

14.8 Percentage distribution of principal fuels used for home heating, by province and by selected census metropolitan area, 1976[1]

Province and census metropolitan area	Total	Oil or other liquid fuel	Piped gas	Coal or coke	Other[2]	Electricity
PROVINCE						
Newfoundland	100.0	70.8	. . .	- -	- -	24.6
Prince Edward Island	100.0	90.6	. . .	- -	- -	. . .
Nova Scotia	100.0	83.7	. . .	2.2	4.4	8.8
New Brunswick	100.0	81.4	. . .	- -	4.6	13.4
Quebec	100.0	70.5	6.9	- -	2.5	20.0
Ontario	100.0	41.8	45.1	- -	1.4	11.6
Manitoba	100.0	22.0	62.7	- -	1.3	12.7
Saskatchewan	100.0	23.3	72.0	- -	- -	- -
Alberta	100.0	4.6	89.7	1.3	3.1	- -
British Columbia	100.0	37.8	45.1	- -	2.3	14.5
Canada	100.0	47.5	36.7	0.2	2.4	13.1
METROPOLITAN AREA						
Calgary, Alta.	100.0	. . .	98.4	. . .	. . .	. . .
Edmonton, Alta.	100.0	. . .	97.9	. . .	. . .	. . .
Halifax, NS	100.0	92.1	. . .	. . .	. . .	7.6
Hamilton, Ont.	100.0	41.1	50.7	. . .	. . .	7.7
Kitchener—Waterloo, Ont.	100.0	45.5	39.6	. . .	. . .	14.9
London, Ont.	100.0	16.5	76.0	. . .	. . .	7.1
Montreal, Que.	100.0	70.2	12.3	. . .	. . .	16.8
Ottawa—Hull, Ont., Que.	100.0	61.1	22.8	. . .	. . .	15.8
Quebec—Lévis, Que.	100.0	79.8	. . .	. . .	. . .	19.6
St. Catharines—Niagara, Ont.	100.0	18.8	76.3	. . .	. . .	4.6
Toronto, Ont.	100.0	36.2	55.6	. . .	. . .	7.9
Vancouver, BC	100.0	30.1	63.1	. . .	. . .	6.0
Victoria, BC	100.0	73.8	. . .	. . .	. . .	24.7
Windsor, Ont.	100.0	6.7	77.3	. . .	. . .	16.0
Winnipeg, Man.	100.0	9.9	81.5	. . .	. . .	8.5

[1]Household facilities and equipment survey, May 1976.
[2]Includes bottled gas, wood and sawdust.

14.9   Percentage of dwellings with specified facilities, by province, 1976[1]

Province	Percentage of dwellings with		
	Running water	Bath or shower (exclusive use)	Flush toilet
Newfoundland	93.8	86.9	89.2
Prince Edward Island	90.6	90.6	90.6
Nova Scotia	96.9	90.3	92.5
New Brunswick	97.7	93.6	94.8
Quebec	99.8	98.3	99.6
Ontario	99.5	98.9	99.0
Manitoba	96.5	96.5	95.9
Saskatchewan	95.7	94.6	93.1
Alberta	98.2	97.6	97.4
British Columbia	99.5	98.8	99.0
Canada	99.0	97.6	98.1

[1]Household facilities and equipment survey, May 1976.

14.10   Annual estimates of household facilities and equipment, 1976

Item	Estimated households 1976 (May) '000	Percentage of households		
		1976 (May)	1975 (April)	1974 (April)
Total households	6,918	100.0	100.0	100.0
Principal heating facilities				
Furnaces	5,451	78.8	79.5	80.8
Oil	2,937	42.5	45.1	46.9
Gas	2,470	35.7	33.6	33.1
Wood, coal and other	44	0.6	0.6	0.7
Other equipment	1,467	21.2	20.5	19.2
Oil	350	5.1	6.4	6.6
Gas	113	1.6	2.1	2.4
Wood, coal and other	97	1.4	1.1	1.3
Electricity	907	13.1	10.5	8.6
Cooking fuel				
Electricity	6,013	86.9	85.1	83.6
Piped gas	565	8.2	9.2	10.0
Bottled gas	106	1.5	1.8	1.8
Wood or coal	93	1.3	1.5	1.8
Kerosene, oil or other	141	2.0	2.4	2.5
Fuel used for piped hot water supply				
Electricity	3,393	49.0	49.4	49.2
Gas	2,396	34.6	33.8	32.6
Oil	909	13.1	13.1	13.4
Coal, wood and other	29	0.4	0.3	0.5
No hot water supply	191	2.8	3.3	4.3
Refrigerators and home freezers				
Electric refrigerators	6,861	99.2	99.3	98.9
Home freezers	3,012	43.5	41.8	39.8
Washing machines				
Automatic	3,870	55.9	52.1	49.0
Other electric	1,400	20.2	24.8	28.6
Clothes dryers	3,782	54.7	51.6	48.3
Telephones	6,677	96.5	96.4	95.8
Radios				
All types, except car	6,817	98.5	98.3	98.2
FM receivers	5,293	76.5	75.7	71.7
TV sets				
All types	6,684	96.6	96.8	96.4
Black and white	4,318	62.4	67.8	73.3
Colour	4,193	60.6	53.4	44.5
Record players	5,294	76.5	75.7	74.5
Automobiles	5,491	79.4	78.9	78.0
One automobile	3,803	55.0	55.9	56.4
Two or more automobiles	1,688	24.4	23.0	21.6
Miscellaneous				
Window-type air conditioners	670	9.7	9.2	7.6
Central-unit air conditioners	254	3.7	3.2	2.6
Automatic dishwashers	1,289	18.6	15.2	12.9
Adult-sized bicycles	2,708	39.1	33.8	30.4
Snowmobiles	678	9.8	10.0	9.3
Floor polishers	2,834	41.0	[1]	48.3
Snow blowers	542	7.8	[1]	6.5
Lawn mowers — power	3,543	51.2	[1]	49.2
Boats	975	14.1	[1]	13.5
Overnight camping equipment	1,628	23.5	[1]	21.4
Vacation homes	476	6.9	[1]	7.3
Snow skis	1,406	20.3	[2]	[2]

[1]Data not collected in 1975.
[2]New item in 1976.

14.11 Value of new and repair construction work performed, 1972-77[1]

Year	New $'000,000	Repair $'000,000	Total $'000,000	Total construction as percentage of gross national expenditure
1972	14,469	2,820	17,289	16.5[r]
1973	16,954	3,220	20,174	16.5[r]
1974[r]	20,772	3,921	24,693	17.1
1975[r]	24,056	4,320	28,376	17.6
1976	27,063	4,710	31,773	17.2
1977	29,155	5,194	34,349	..

[1]Actual expenditures 1972-75; preliminary actual 1976; intentions 1977.

14.12 Value of construction work performed, by contractors and others, 1973-77[1] (million dollars)

Item	1973	1974[r]	1975[r]	1976	1977
Contract construction	15,904	19,618	22,592	25,165	26,810
New	13,883	17,071	19,801	22,102	23,383
Repair	2,021	2,547	2,791	3,063	3,427
Other construction[2]	4,270	5,075	5,784	6,608	7,539
New	3,071	3,701	4,255	4,961	5,772
Repair	1,199	1,374	1,529	1,647	1,767
Total, construction	20,174	24,693	28,376	31,773	34,349
New	16,954	20,772	24,056	27,063	29,155
Repair	3,220	3,921	4,320	4,710	5,194

[1]Actual expenditures 1973-75; preliminary actual 1976; intentions 1977.
[2]Work done by the labour forces of utilities, government departments and other employers not primarily engaged in the construction industry.

14.13 Labour content, cost of materials and value of work performed in construction, by province and by employer, 1975-77[1]

Province and employer	Year	Labour content No.	Labour content Value $'000	Cost of materials used $'000	Value of work performed $'000
PROVINCE					
Newfoundland	1975	17,528	219,267	239,373	617,083
	1976	16,601	248,285	274,695	712,511
	1977	16,488	266,808	290,431	750,201
Prince Edward Island	1975	3,016	36,893	46,495	107,535
	1976	2,531	32,732	41,266	95,270
	1977	2,375	33,059	41,231	95,941
Nova Scotia	1975	22,352	276,217	306,462	764,168
	1976	21,309	307,974	345,537	852,091
	1977	21,673	341,236	382,296	949,166
New Brunswick	1975	22,709	297,766	321,388	828,972
	1976	20,891	284,441	312,096	785,694
	1977	20,225	298,262	322,833	812,194
Quebec	1975	159,583	2,475,136	2,742,868	7,110,980
	1976	148,518	2,548,862	2,892,611	7,469,455
	1977	147,445	2,754,235	3,101,953	8,188,591
Ontario	1975	217,335	3,183,499	3,567,106	8,989,399
	1976	213,165	3,455,564	3,908,134	9,831,206
	1977	203,538	3,556,384	4,033,135	10,155,404
Manitoba	1975	24,810	378,084	362,742	1,012,643
	1976	26,285	457,531	448,035	1,232,524
	1977	25,885	486,904	480,188	1,311,563
Saskatchewan	1975	29,513	411,244	449,587	1,130,553
	1976	32,169	497,235	554,038	1,381,405
	1977	33,177	557,724	612,765	1,533,244
Alberta	1975	88,531	1,331,781	1,485,484	3,920,862
	1976	120,231	1,692,414	1,956,596	5,052,957
	1977	127,319	1,939,891	2,226,027	5,756,893
British Columbia, the Yukon Territory and Northwest Territories	1975	77,248	1,324,799	1,583,147	3,894,149
	1976	79,391	1,488,493	1,769,913	4,360,065
	1977	79,788	1,627,736	1,948,225	4,796,128
Canada	1975	662,625	9,934,686	11,104,652	28,376,344
	1976	681,091	11,013,531	12,502,921	31,773,178
	1977	677,913	11,862,239	13,439,084	34,349,325

14.13   Labour content, cost of materials and value of work performed in construction, by province and by employer, 1975-77[1] (concluded)

Province and employer	Year	Labour content		Cost of materials used $'000	Value of work performed $'000
		No.	Value $'000		
EMPLOYER					
Contractors	1975	525,722	7,979,338	8,874,159	22,591,878
	1976	534,773	8,749,868	9,952,588	25,165,288
	1977	526,204	9,296,093	10,526,676	26,809,749
Utilities	1975	61,794	970,617	1,176,839	3,035,988
	1976	63,130	1,081,593	1,299,868	3,347,421
	1977	69,025	1,288,891	1,563,794	4,029,577
Governments	1975	34,002	434,115	348,412	1,233,677
	1976	38,212	529,963	411,679	1,466,459
	1977	38,253	571,384	441,818	1,582,534
Miscellaneous	1975	41,107	550,616	705,242	1,514,801
	1976	44,976	652,107	838,786	1,794,010
	1977	44,431	705,871	906,796	1,927,465

[1] Actual expenditures 1975; preliminary actual 1976; intentions 1977.

14.14   Value of building permits issued, by province, 1975 and 1976 with totals for 1972-76 (thousand dollars)

Province or territory and year		Residential construction			Non-residential construction			Total
		New	Improve-ments	Total	Indus-trial	Commer-cial	Institutional and government	
Newfoundland	1975	26,474	3,027	29,501	4,864	17,565	12,838	64,768
	1976	70,221	5,892	76,113	1,435	19,176	9,919	106,643
Prince Edward Island	1975	33,529	2,444	35,973	4,508	7,726	22,249	70,456
	1976	25,420	3,161	28,581	6,920	6,632	6,481	48,614
Nova Scotia	1975	183,179	15,934	199,113	16,458	75,798	49,547	340,916
	1976	176,236	18,649	194,885	34,112	44,945	35,802	309,744
New Brunswick	1975	102,948	9,217	112,165	15,094	50,931	53,607	231,797
	1976	119,861	13,954	133,815	14,066	62,032	41,427	251,340
Quebec	1975	1,044,023	85,282	1,129,305	208,180	432,078	261,889	2,031,452
	1976	1,443,501	121,161	1,564,662	217,893	702,578	318,235	2,803,368
Ontario	1975	2,227,173	235,226	2,462,399	434,829	918,508	493,205	4,308,441
	1976	2,290,399	272,964	2,563,363	473,498	685,839	351,622	4,074,322
Manitoba	1975	157,033	12,238	169,271	15,164	70,395	36,322	291,152
	1976	239,882	17,737	257,619	35,377	101,565	37,971	432,532
Saskatchewan	1975	227,692	12,085	239,777	13,128	85,700	91,580	430,185
	1976	295,950	16,556	312,506	19,421	109,087	75,400	516,414
Alberta	1975	729,517	41,385	770,902	85,905	280,130	134,460	1,271,397
	1976	1,153,171	67,855	1,221,026	73,238	407,485	129,357	1,831,106
British Columbia	1975	902,120	69,314	971,434	76,868	307,290	184,255	1,539,847
	1976	1,029,777	85,517	1,115,294	130,938	398,699	160,843	1,805,774
Yukon Territory	1975	—	—	—	—	—	—	—
	1976	19	9	28	5	—	—	33
Northwest Territories	1975	8,828	272	9,100	828	5,362	1,834	17,124
	1976	7,485	438	7,923	3,112	7,475	931	19,441
Total	1972	3,480,725	157,686	3,638,411	520,479	1,411,599	893,119	6,463,608
	1973	4,552,220	210,862	4,763,082	853,754	1,970,340	971,762	8,558,938
	1974	4,299,034	276,676	4,575,710	1,315,898	2,292,665	1,095,802	9,280,075
	1975	5,642,516	486,424	6,128,940	875,826	2,251,483	1,341,786	10,598,035
	1976	6,851,922	623,893	7,475,815	10,010,015	2,545,513	1,167,988	12,199,331

14.15   Estimated value of proposed construction as indicated by building permits issued in 50 municipalities, 1975 and 1976 (thousand dollars)

Province and municipality	1975	1976	Province and municipality	1975	1976
NEWFOUNDLAND			Oshawa	48,936	77,811
St. John's	39,272	49,473	Ottawa	190,315	165,780
			Scarborough (borough)	282,352	224,322
PRINCE EDWARD ISLAND			Thunder Bay	46,679	54,424
Charlottetown	2,742	3,923	Toronto	538,134	261,790
			Windsor	56,509	66,613
NOVA SCOTIA			York (borough)	15,490	34,147
Halifax	79,723	128,609	York North (borough)	212,009	226,637
NEW BRUNSWICK			MANITOBA		
Fredericton	34,443	42,305	Fort Garry		
Moncton	25,478	21,970	St. Boniface ⎫		
Saint John	74,013	65,625	St. James ⎬ ²	220,639	347,520
			Winnipeg ⎭		
QUEBEC			SASKATCHEWAN		
LaSalle	17,933	18,369	Moose Jaw	19,360	26,369
Montreal	287,534	404,474	Prince Albert	17,354	36,517
Quebec	48,325	102,162	Regina	150,939	169,858
Ste-Foy	40,476	92,634	Saskatoon	127,839	128,853
St-Laurent	43,275	100,901			
Sept-Iles	21,439	13,279	ALBERTA		
Sherbrooke	27,000	28,551	Calgary	391,313	447,817
Trois-Rivières	25,723	18,553	Edmonton ⎫ ³	303,076	482,711
			Jasper Place ⎭		
ONTARIO			Lethbridge	43,874	60,244
Brampton	85,185¹	96,876¹	Medicine Hat	33,834	46,700
Burlington	63,520	52,649	Red Deer	25,858	59,680
Etobicoke (borough)	100,573	86,852			
Hamilton	148,186	104,591	BRITISH COLUMBIA		
Kitchener	72,855	86,820	Burnaby District	105,549	105,563
London	105,287	97,928	Richmond Township	103,648	101,005
London Township	2,980	3,019	Surrey District	90,128	98,884
Mississauga	229,317	332,938	Vancouver	152,282	242,684
Nepean Township	31,330	48,841	Victoria	56,392	75,645

¹Includes townships of Chinguacousy and Toronto Gore and part of Mississauga.
²Metropolitan Corporation of Greater Winnipeg.
³Jasper Place included with Edmonton following annexation.

14.16   Estimated value of building permits issued in metropolitan areas, 1975 and 1976 (thousand dollars)

Metropolitan area	1975	1976	Metropolitan area	1975	1976
Calgary	391,313	447,817	Saint John	74,789	65,625
Chicoutimi–Jonquière	56,264	55,627	St. Catharines–Niagara	142,934	150,257
Edmonton	394,968	618,533	St. John's¹	39,279	49,437
Halifax	170,593	128,609	Saskatoon	127,839	128,853
Hamilton	277,126	209,470	Sudbury	57,477	56,293
Kitchener	148,196	170,087	Thunder Bay	49,590	58,657
London	129,102	125,292	Toronto	1,835,372	1,566,326
Montreal	1,109,303	1,427,874	Vancouver	702,950	846,855
Ottawa–Hull	436,385	598,614	Victoria	177,476	201,660
Quebec	195,075	306,768	Windsor	82,446	92,250
Regina	150,939	169,858	Winnipeg	220,639	347,520

¹Although this is a metropolitan area, only St. John's proper is included in the building permits survey.

14.17   Capital expenditures¹ on construction and on machinery and equipment, in current and constant (1971) dollars, 1968-77

Year	Capital expenditures ($'000,000)						Capital expenditures as percentage of gross national expenditure²	
	Construction		Machinery and equipment		Total			
	Current dollars	Constant 1971 dollars	Current dollars	Constant 1971 dollars	Current dollars	Constant 1971 dollars	Current dollars	Constant 1971 dollars
1968	9,909	11,486	5,546	6,142	15,455	17,628	21.7	21.9
1969	10,824	11,918	6,103	6,580	16,927	18,498	21.6	21.9
1970	11,319	11,962	6,479	6,673	17,798	18,635	21.0	21.4
1971	13,274	13,274	6,910	6,910	20,184	20,184	21.8	21.8
1972ʳ	14,470	13,716	7,748	7,559	22,218	21,275	21.6	21.3
1973ʳ	16,953	14,602	9,665	8,999	26,618	23,601	22.2	22.1
1974ʳ	20,771	15,084	12,111	9,798	32,882	24,882	23.2	22.6
1975ʳ	24,054	15,409	14,162	9,931	38,216	25,340	24.3	22.8
1976ʳ	27,061	15,661	15,011	9,981	42,072	25,632	23.4	22.1
1977	29,727	—	16,782	—	46,509	—	—	—

¹Actual expenditures 1968-75; preliminary actual 1976 and mid-year intentions 1977.
²The percentage is calculated by dividing "Gross Fixed Capital Formation", as defined by the National Income and Expenditure Accounts, by the total "Gross National Expenditure".

14.18 Summary of capital and repair expenditures, by economic sector, 1975-77[1] (million dollars)

Type of enterprise and year		Capital			Repair			Capital and repair		
		Construction	Machinery and equipment	Total	Construction	Machinery and equipment	Total	Construction	Machinery and equipment	Total
Agriculture and fishing	1975	479.6	1,968.2	2,447.8	190.2	392.3	582.5	669.8	2,360.5	3,030.3
	1976	507.1	2,323.0	2,830.1	203.3	434.2	637.5	710.4	2,757.2	3,467.6
	1977	508.3	2,283.4	2,791.7	219.2	489.0	708.2	727.5	2,772.4	3,499.9
Forestry	1975	90.9	104.6	195.5	33.2	130.4	163.6	124.1	235.0	359.1
	1976	93.7	91.3	185.0	39.6	136.5	176.1	133.3	227.8	361.1
	1977	95.1	104.8	199.9	41.8	145.8	187.6	136.9	250.6	387.5
Mining, quarrying and oil wells	1975	1,968.1	643.9	2,612.0	302.7	699.5	1,002.2	2,270.8	1,343.4	3,614.2
	1976	2,395.0	1,052.7	3,447.7	287.3	770.8	1,058.1	2,682.3	1,823.5	4,505.8
	1977	2,956.8	1,295.0	4,251.8	322.1	825.8	1,147.9	3,278.9	2,120.8	5,399.7
Construction industry	1975	80.2	488.2	568.4	19.4	430.4	449.8	99.6	918.6	1,018.2
	1976	89.8	545.7	635.5	21.5	481.2	502.7	111.3	1,026.9	1,138.2
	1977	97.2	590.7	687.9	23.4	521.0	544.4	120.6	1,111.7	1,232.3
Manufacturing	1975	1,568.5	3,952.9	5,521.4	381.0	2,064.1	2,445.1	1,949.5	6,017.0	7,966.5
	1976	1,468.8	3,803.0	5,271.8	397.5	2,221.0	2,618.5	1,866.3	6,024.0	7,890.3
	1977	1,740.7	4,612.5	6,353.2	430.9	2,408.5	2,839.4	2,171.6	7,021.0	9,192.6
Utilities	1975	4,402.9	3,629.9	8,032.8	697.4	1,531.6	2,229.0	5,100.3	5,161.5	10,261.8
	1976	4,626.8	3,737.8	8,364.6	755.6	1,646.1	2,401.7	5,382.4	5,383.9	10,766.3
	1977	5,673.5	4,179.2	9,852.7	823.1	1,788.2	2,611.3	6,496.6	5,967.4	12,464.0
Trade (wholesale and retail)	1975	350.9	644.7	995.6	97.6	136.4	234.0	448.5	781.1	1,229.6
	1976	352.6	672.1	1,024.7	95.2	136.9	232.1	447.8	809.0	1,256.8
	1977	376.2	693.8	1,070.0	101.7	146.8	248.5	477.9	840.6	1,318.5
Finance, insurance and real estate	1975	1,559.9	195.0	1,754.9	93.8	36.4	130.2	1,653.7	231.4	1,885.1
	1976	1,455.4	290.6	1,746.0	110.1	34.8	144.9	1,565.5	325.4	1,890.9
	1977	1,549.3	257.0	1,806.3	116.3	38.3	154.6	1,665.6	295.3	1,960.9
Commercial services	1975	857.2	1,608.1	2,465.3	39.5	187.2	226.7	896.7	1,795.3	2,692.0
	1976	597.4	1,559.7	2,157.1	35.3	234.7	270.0	632.7	1,794.4	2,427.1
	1977	291.1	1,770.0	2,061.1	38.3	263.4	301.7	329.4	2,033.4	2,362.8
Institutions	1975	1,190.6	293.4	1,484.0	163.8	60.3	224.1	1,354.4	353.7	1,708.1
	1976	1,119.1	302.9	1,422.0	175.5	64.2	239.7	1,294.6	367.1	1,661.7
	1977	1,100.7	311.3	1,412.0	182.5	66.0	248.5	1,283.2	377.3	1,660.5
Government departments	1975	4,391.8	632.8	5,024.6	725.8	166.6	892.4	5,117.6	799.4	5,917.0
	1976	4,591.7	632.5	5,224.2	774.4	145.9	920.3	5,366.1	778.4	6,144.5
	1977	5,064.9	683.8	5,748.7	838.7	153.5	992.2	5,903.6	837.3	6,740.9
Housing	1975	7,113.9	—	7,113.9	1,575.6	—	1,575.6	8,689.5	—	8,689.5
	1976	9,763.9	—	9,763.9	1,813.7	—	1,813.7	11,577.6	—	11,577.6
	1977	10,273.5	—	10,273.5	2,055.5	—	2,055.5	12,329.0	—	12,329.0
Total	1975	24,054.5	14,161.7	38,216.2	4,320.0	5,835.2	10,155.2	28,374.5	19,996.9	48,371.4
	1976	27,061.3	15,011.3	42,072.6	4,709.0	6,306.3	11,015.3	31,770.3	21,317.6	53,087.9
	1977	29,727.3	16,781.5	46,508.8	5,193.5	6,846.3	12,039.8	34,920.8	23,627.8	58,548.6

[1]Actual expenditures 1975; preliminary actual 1976; intentions 1977.

14.19 Capital and repair expenditures for certain economic sectors, 1975-77[1] (million dollars)

Type of enterprise and year		Capital			Repair			Capital and repair		
		Construction	Machinery and equipment	Total	Construction	Machinery and equipment	Total	Construction	Machinery and equipment	Total
MANUFACTURING										
Food and beverages	1975	118.2	280.9	399.1	34.6	173.1	207.7	152.8	454.0	606.8
	1976	115.5	305.8	421.3	34.8	158.6	193.4	150.3	464.4	614.7
	1977	163.8	357.7	521.5	34.5	170.8	205.3	198.3	528.5	726.8
Tobacco products	1975	5.7	15.6	21.3	2.6	12.2	14.8	8.3	27.8	36.1
	1976	3.3	13.1	16.4	3.7	11.1	14.8	7.0	24.2	31.2
	1977	3.1	21.6	24.7	3.8	11.9	15.7	6.9	33.5	40.4
Rubber	1975	23.0	100.4	123.4	5.6	54.9	60.5	28.6	155.3	183.9
	1976	13.2	75.3	88.5	4.9	57.3	62.2	18.1	132.6	150.7
	1977	15.1	92.5	107.6	5.2	59.5	64.7	20.3	152.0	172.3
Leather	1975	2.7	10.6	13.3	1.3	7.8	9.1	4.0	18.4	22.4
	1976	2.9	9.1	12.0	1.6	8.3	9.9	4.5	17.4	21.9
	1977	2.9	9.8	12.7	1.7	8.4	10.1	4.6	18.2	22.8
Textile	1975	31.6	131.6	163.2	8.3	47.9	56.2	39.9	179.5	219.4
	1976	15.5	88.4	103.9	8.4	52.3	60.7	23.9	140.7	164.6
	1977	19.4	68.8	88.2	8.7	46.4	55.1	28.1	115.2	143.3
Knitting mills	1975	0.3	10.7	11.0	0.8	4.6	5.4	1.1	15.3	16.4
	1976	0.8	7.4	8.2	0.8	3.0	3.8	1.6	10.4	12.0
	1977	0.7	7.7	8.4	0.7	2.9	3.6	1.4	10.6	12.0
Clothing	1975	8.5	14.6	23.1	2.0	5.7	7.7	10.5	20.3	30.8
	1976	2.9	14.3	17.2	1.8	5.5	7.3	4.7	19.8	24.5
	1977	5.7	12.8	18.5	1.6	5.5	7.1	7.3	18.3	25.6

14.19 Capital and repair expenditures for certain economic sectors, 1975-77[1] (million dollars) (continued)

Type of enterprise and year		Capital			Repair			Capital and repair		
		Con-struc-tion	Ma-chinery and equip-ment	Total	Con-struc-tion	Ma-chinery and equip-ment	Total	Con-struc-tion	Ma-chinery and equip-ment	Total
MANUFACTURING (concluded)										
Wood	1975	73.6	192.8	266.4	16.6	125.9	142.5	90.2	318.7	408.9
	1976	64.9	167.1	232.0	18.1	138.6	156.7	83.0	305.7	388.7
	1977	79.7	189.2	268.9	17.4	150.3	167.7	97.1	339.5	436.6
Furniture and fixtures	1975	9.2	20.6	29.8	3.3	8.8	12.1	12.5	29.4	41.9
	1976	8.4	11.6	20.0	2.5	7.7	10.2	10.9	19.3	30.2
	1977	5.9	12.4	18.3	2.6	7.6	10.2	8.5	20.0	28.5
Paper and allied industries	1975	110.1	467.2	577.3	29.1	348.3	377.4	139.2	815.5	954.7
	1976	155.7	564.6	720.3	37.2	432.6	469.8	192.9	997.2	1,190.1
	1977	182.8	687.1	869.9	41.8	473.6	515.4	224.6	1,160.7	1,385.3
Printing, publishing and allied industries	1975	18.4	77.0	95.4	6.1	21.2	27.3	24.5	98.2	122.7
	1976	16.2	72.8	89.0	6.0	20.5	26.5	22.2	93.3	115.5
	1977	13.2	83.3	96.5	5.7	21.8	27.5	18.9	105.1	124.0
Primary metals	1975	200.5	614.4	814.9	65.8	563.4	629.2	266.3	1,177.8	1,444.1
	1976	147.3	467.1	614.4	60.9	625.9	686.8	208.2	1,093.0	1,301.2
	1977	151.8	618.4	770.2	69.2	687.6	756.8	221.0	1,306.0	1,527.0
Metal fabricating	1975	50.3	160.1	210.4	15.5	92.7	108.2	65.8	252.8	318.6
	1976	45.3	134.6	179.9	15.0	82.0	97.0	60.3	216.6	276.9
	1977	44.2	152.8	197.0	14.5	89.5	104.0	58.7	242.3	301.0
Machinery	1975	49.1	101.7	150.8	15.4	35.5	50.9	64.5	137.2	201.7
	1976	30.5	88.9	119.4	16.1	40.7	56.8	46.6	129.6	176.2
	1977	38.8	105.9	144.7	15.8	42.9	58.7	54.6	148.8	203.4
Transportation equipment	1975	76.5	198.3	274.8	19.0	114.1	133.1	95.5	312.4	407.9
	1976	62.5	125.2	187.7	25.7	127.3	153.0	88.2	252.5	340.7
	1977	79.0	269.0	348.0	25.2	131.2	156.4	104.2	400.2	504.4
Electrical products	1975	28.3	108.6	136.9	12.1	52.9	65.0	40.4	161.5	201.9
	1976	34.4	111.8	146.2	12.7	52.6	65.3	47.1	164.4	211.5
	1977	23.4	131.6	155.0	12.7	58.2	70.9	36.1	189.8	225.9
Non-metallic mineral products	1975	41.1	158.0	199.1	14.4	151.8	166.2	55.5	309.8	365.3
	1976	46.8	192.9	239.7	13.3	137.6	150.9	60.1	330.5	390.6
	1977	48.3	219.1	267.4	14.4	148.7	163.1	62.7	367.8	430.5
Petroleum and coal products	1975	337.5	112.9	450.4	96.1	37.0	133.1	433.6	149.9	583.5
	1976	265.2	90.0	355.2	98.8	34.5	133.3	364.0	124.5	488.5
	1977	306.3	106.1	412.4	117.0	42.0	159.0	423.3	148.1	571.4
Chemical and chemical products[2]	1975	374.7	599.8	974.5	28.5	186.1	214.6	403.2	785.9	1,189.1
	1976	425.9	684.1	1,110.0	30.5	206.6	237.1	456.4	890.7	1,347.1
	1977	549.1	794.6	1,343.7	33.5	230.7	264.2	582.6	1,025.3	1,607.9
Miscellaneous	1975	9.2	30.1	39.3	3.9	20.2	24.1	13.1	50.3	63.4
	1976	11.6	31.1	42.7	4.7	18.3	23.0	16.3	49.4	65.7
	1977	7.5	33.8	41.3	4.9	19.0	23.9	12.4	52.8	65.2
Capital items charged to operating expenses	1975	—	547.0	547.0	—	—	—	—	547.0	547.0
	1976	—	547.8	547.8	—	—	—	—	547.8	547.8
	1977	—	638.3	638.3	—	—	—	—	638.3	638.3
Total, manufacturing	1975	1,568.5	3,952.9	5,521.4	381.0	2,064.1	2,445.1	1,949.5	6,017.0	7,966.5
	1976	1,468.8	3,803.0	5,271.8	397.5	2,221.0	2,618.5	1,866.3	6,024.0	7,890.3
	1977	1,740.7	4,612.5	6,353.2	430.9	2,408.5	2,839.4	2,171.6	7,021.0	9,192.6
MINING[3] Gold	1975	20.7	5.3	26.0	0.8	10.9	11.7	21.5	16.2	37.7
	1976	17.0	4.4	21.4	0.8	9.8	10.6	17.8	14.2	32.0
	1977	18.0	4.0	22.0	0.7	9.9	10.6	18.7	13.9	32.6
Iron	1975	225.8	83.8	309.6	20.5	191.3	211.8	246.3	275.1	521.4
	1976	..	..	..	..	..	..	..	..	..
	1977	..	..	..	..	..	..	..	..	..
Copper-gold-silver	1975	109.2	51.8	161.0	10.3	125.1	135.4	119.5	176.9	296.4
	1976	121.0	76.4	197.4	11.5	126.1	137.6	132.5	202.5	335.0
	1977	112.3	88.7	201.0	11.5	139.5	151.0	123.8	228.2	352.0
Silver-lead-zinc	1975	55.3	29.3	84.6	7.2	23.7	30.9	62.5	53.0	115.5
	1976	43.4	21.5	64.9	10.2	28.5	38.7	53.6	50.0	103.6
	1977	52.9	41.4	94.3	10.5	30.7	41.2	63.4	72.1	135.5
Other metal mines[4]	1975	88.6	45.1	133.7	24.9	95.7	120.6	113.5	140.8	254.3
	1976	306.0	256.4	562.4	44.3	317.0	361.3	350.3	573.4	923.7
	1977	270.0	359.0	629.0	47.9	333.0	380.9	317.9	692.0	1,009.9
Asbestos	1975	42.5	19.3	61.8	4.1	46.0	50.1	46.6	65.3	111.9
	1976	65.3	28.4	93.7	6.5	68.9	75.4	71.8	97.3	169.1
	1977	60.9	47.3	108.2	6.7	72.5	79.2	67.6	119.8	187.4
Other non-metal mines[5]	1975	70.3	190.3	260.6	19.7	138.3	158.0	90.0	328.6	418.6
	1976	62.1	175.7	237.8	17.3	138.7	156.0	79.4	314.4	393.8
	1977	124.7	211.9	336.6	17.8	153.2	171.0	142.5	365.1	507.6
Petroleum and gas[6]	1975	1,355.7	219.0	1,574.7	215.2	68.5	283.7	1,570.9	287.5	1,858.4
	1976	1,780.2	489.9	2,270.1	196.7	81.8	278.5	1,976.9	571.7	2,548.6
	1977	2,318.0	542.7	2,860.7	227.0	87.0	314.0	2,545.0	629.7	3,174.7

14.19 Capital and repair expenditures for certain economic sectors, 1975-77[1] (million dollars) (concluded)

Type of enterprise and year		Capital			Repair			Capital and repair		
		Construction	Machinery and equipment	Total	Construction	Machinery and equipment	Total	Construction	Machinery and equipment	Total
MINING[3] (concluded)										
Total, mining	1975	1,968.1	643.9	2,612.0	302.7	699.5	1,002.2	2,270.8	1,343.4	3,614.2
	1976	2,395.0	1,052.7	3,447.7	287.3	770.8	1,058.1	2,682.3	1,823.5	4,505.8
	1977	2,956.8	1,295.0	4,251.8	322.1	825.8	1,147.9	3,278.9	2,120.8	5,399.7
UTILITIES										
Transportation										
Air transport	1975	29.8	234.3	264.1	5.0	160.7	165.7	34.8	395.0	429.8
	1976	31.4	65.6	97.0	7.7	171.0	178.7	39.1	236.6	275.7
	1977	30.5	68.3	98.8	7.8	180.6	188.4	38.3	248.9	287.2
Railway transport	1975	370.9	322.6	693.5	327.8	440.1	767.9	698.7	762.7	1,461.4
	1976	326.9	334.8	661.7	344.8	431.6	776.4	671.7	766.4	1,438.1
	1977	335.2	258.6	593.8	378.2	481.9	860.1	713.4	740.5	1,453.9
Water transport and services	1975	59.0	89.3	148.3	19.0	57.7	76.7	78.0	147.0	225.0
	1976	29.7	84.2	113.9	15.9	56.4	72.3	45.6	140.6	186.2
	1977	56.7	98.0	154.7	17.7	58.1	75.8	74.4	156.1	230.5
Motor transport	1975	25.5	136.2	161.7	11.7	161.6	173.3	37.2	297.8	335.0
	1976	32.0	139.7	171.7	11.1	190.8	201.9	43.1	330.5	373.6
	1977	33.8	153.6	187.4	10.7	198.5	209.2	44.5	352.1	396.6
Urban transit systems	1975	160.5	115.7	276.3	19.5	58.6	78.1	180.0	174.3	354.3
	1976	162.9	131.4	294.3	25.3	67.5	92.8	188.2	198.9	387.1
	1977	213.1	165.6	378.7	29.4	69.8	99.2	242.5	235.4	477.9
Pipelines	1975	321.2	40.7	361.9	20.5	9.7	30.2	341.7	50.4	392.1
	1976	292.6	49.1	341.7	21.7	10.9	32.6	314.3	60.0	374.3
	1977	324.4	57.2	381.6	21.1	11.6	32.7	345.5	68.8	414.3
Capital items charged to operating expenses	1975	—	23.8	23.8	—	—	—	—	23.8	23.8
	1976	—	22.5	22.5	—	—	—	—	22.5	22.5
	1977	—	23.4	23.4	—	—	—	—	23.4	23.4
Total, transportation	1975	966.9	962.6	1,929.5	403.5	888.4	1,291.9	1,370.4	1,851.0	3,221.4
	1976	875.5	827.3	1,702.8	426.5	928.2	1,354.7	1,302.0	1,755.5	3,057.5
	1977	993.7	824.7	1,818.4	464.9	1,000.5	1,465.4	1,458.6	1,825.2	3,283.8
Communication										
Broadcasting[7]	1975	51.4	101.6	153.0	5.8	9.5	15.3	57.2	111.1	168.3
	1976	62.6	85.3	147.9	5.4	9.9	15.3	68.0	95.2	163.2
	1977	55.6	86.2	141.8	5.0	10.4	15.4	60.6	96.6	157.2
Telephone and telegraph	1975	533.9	1,158.5	1,692.4	92.0	430.6	522.6	625.9	1,589.1	2,215.0
	1976	564.9	1,265.0	1,829.9	116.9	478.4	595.3	681.8	1,743.4	2,425.2
	1977	624.3	1,376.5	2,000.8	128.4	519.5	647.9	752.7	1,896.0	2,648.7
Capital items charged to operating expenses	1975	—	22.1	22.1	—	—	—	—	22.1	22.1
	1976	—	23.9	23.9	—	—	—	—	23.9	23.9
	1977	—	25.9	25.9	—	—	—	—	25.9	25.9
Total, communication	1975	585.3	1,282.2	1,867.5	97.8	440.1	537.9	683.1	1,722.3	2,405.4
	1976	627.5	1,374.2	2,001.7	122.3	488.3	610.6	749.8	1,862.5	2,612.3
	1977	679.9	1,488.6	2,168.5	133.4	529.9	663.3	813.3	2,018.5	2,831.8
Miscellaneous utilities										
Grain elevators	1975	18.2	16.6	34.8	6.4	6.7	13.1	24.6	23.3	47.9
	1976	18.1	28.7	46.8	10.0	9.3	19.3	28.1	38.0	66.1
	1977	39.0	39.9	78.9	9.1	9.3	18.4	48.1	49.2	97.3
Electric power	1975	2,660.8	1,296.3	3,957.1	163.2	178.2	341.4	2,824.0	1,474.5	4,298.5
	1976	2,926.3	1,436.2	4,362.5	168.7	203.3	372.0	3,095.0	1,639.5	4,734.5
	1977	3,758.7	1,741.4	5,500.1	185.1	229.6	414.7	3,943.8	1,971.0	5,914.8
Gas distribution	1975	150.2	42.5	192.7	21.4	10.3	31.7	171.6	52.8	224.4
	1976	146.9	42.0	188.9	21.2	8.4	29.6	168.1	50.4	218.5
	1977	171.3	45.3	216.6	22.4	9.8	32.2	193.7	55.1	248.8
Other utilities[8]	1975	21.5	9.3	30.8	5.1	7.9	13.0	26.6	17.2	43.8
	1976	32.5	6.9	39.4	6.9	8.6	15.5	39.4	15.5	54.9
	1977	30.9	12.1	43.0	8.2	9.1	17.3	39.1	21.2	60.3
Capital items charged to operating expenses	1975	—	20.4	20.4	—	—	—	—	20.4	20.4
	1976	—	22.5	22.5	—	—	—	—	22.5	22.5
	1977	—	27.2	27.2	—	—	—	—	27.2	27.2
Total, miscellaneous utilities	1975	2,850.7	1,385.1	4,235.8	196.1	203.1	399.2	3,046.8	1,588.2	4,635.0
	1976	3,123.8	1,536.3	4,660.1	206.8	229.6	436.4	3,330.6	1,765.9	5,096.5
	1977	3,999.9	1,865.9	5,865.8	224.8	257.8	482.6	4,224.7	2,123.7	6,348.4
Total, utilities	1975	4,402.9	3,629.9	8,032.8	697.4	1,531.6	2,229.0	5,100.3	5,161.5	10,261.8
	1976	4,626.8	3,737.8	8,364.6	755.6	1,646.1	2,401.7	5,382.4	5,383.9	10,766.3
	1977	5,673.5	4,179.2	9,852.7	823.1	1,788.2	2,611.3	6,496.6	5,967.4	12,464.0

[1]Actual expenditures 1975; preliminary actual 1976; intentions 1977.
[2]Includes expenditures for heavy water plants.
[3]Capital construction expenditures include on-property exploration and development but exclude outside or general exploration.
[4]Includes capital and repair expenditures for metal and non-metal exploration companies.
[5]Includes coal mines, gypsum, salt, potash and miscellaneous non-metal mines, and quarrying.
[6]Includes expenditures on facilities related to petroleum and gas wells and extraction of petroleum from shales or sands, natural gas processing plants and contract drilling for petroleum and gas.
[7]Includes community antenna television and satellite communication systems.
[8]Includes toll highways and bridges, warehousing, water systems of private and provincial enterprises and other utilities.

14.20 Capital and repair expenditures, by province, 1975-77[1],[2] (million dollars)

Province or territory and year		Capital			Repair			Capital and repair		
		Con-struc-tion	Ma-chinery and equip-ment	Total	Con-struc-tion	Ma-chinery and equip-ment	Total	Con-struc-tion	Ma-chinery and equip-ment	Total
Newfoundland	1975	527.7	186.5	714.2	88.9	165.8	254.7	616.6	352.3	968.9
	1976	619.9	185.4	805.3	92.6	173.4	266.0	712.5	358.8	1,071.3
	1977	571.2	177.2	748.4	100.2	187.3	287.5	671.4	364.5	1,035.9
Prince Edward Island	1975	82.9	34.8	117.7	24.5	12.5	37.0	107.4	47.3	154.7
	1976	74.4	32.0	106.4	21.0	13.8	34.8	95.4	45.8	141.2
	1977	83.4	35.1	118.5	21.9	14.3	36.2	105.3	49.4	154.7
Nova Scotia	1975	636.5	285.1	921.6	127.9	144.5	272.4	764.4	429.6	1,194.0
	1976	718.1	271.7	989.8	133.5	140.9	274.4	851.6	412.6	1,264.2
	1977	808.7	304.2	1,112.9	145.4	140.0	285.4	954.1	444.2	1,398.3
New Brunswick	1975	727.1	432.2	1,159.3	101.4	134.6	236.0	828.5	566.8	1,395.3
	1976	669.2	437.7	1,106.9	115.4	145.4	260.8	784.6	583.1	1,367.7
	1977	695.8	469.6	1,165.4	122.6	163.2	285.8	818.4	632.8	1,451.2
Quebec	1975	6,119.2	3,051.2	9,170.4	992.1	1,280.3	2,272.4	7,111.3	4,331.5	11,442.8
	1976	6,384.5	3,052.5	9,437.0	1,085.0	1,363.4	2,448.4	7,469.5	4,415.9	11,885.4
	1977	7,228.1	3,554.0	10,782.1	1,199.9	1,484.9	2,684.8	8,428.0	5,038.9	13,466.9
Ontario	1975	7,492.5	5,427.8	12,920.3	1,496.0	2,138.5	3,634.5	8,988.5	7,566.3	16,554.8
	1976	8,150.1	5,556.8	13,706.9	1,681.0	2,390.8	4,071.8	9,831.1	7,947.6	17,778.7
	1977	8,498.1	6,057.4	14,555.5	1,837.8	2,595.0	4,432.8	10,335.9	8,652.4	18,988.3
Manitoba	1975	823.1	698.0	1,521.1	189.2	262.7	451.9	1,012.3	960.7	1,973.0
	1976	1,020.8	750.4	1,771.2	212.4	237.6	450.0	1,233.2	988.0	2,221.2
	1977	1,087.2	768.9	1,856.1	233.4	253.3	486.7	1,320.6	1,022.2	2,342.8
Saskatchewan	1975	922.9	856.9	1,779.8	207.5	273.7	481.2	1,130.4	1,130.6	2,261.0
	1976	1,160.0	985.4	2,145.4	220.5	299.7	520.2	1,380.5	1,285.1	2,665.6
	1977	1,280.3	1,015.4	2,295.7	245.3	333.4	578.7	1,525.6	1,348.8	2,874.4
Alberta	1975	3,395.3	1,611.7	5,007.0	525.3	520.4	1,045.7	3,920.6	2,132.1	6,052.7
	1976	4,497.2	2,099.6	6,596.8	556.2	554.8	1,111.0	5,053.4	2,654.4	7,707.8
	1977	5,307.4	2,477.4	7,784.8	641.4	614.4	1,255.8	5,948.8	3,091.8	9,040.6
British Columbia	1975	2,996.9	1,426.1	4,423.0	543.1	857.5	1,400.6	3,540.0	2,283.6	5,823.6
	1976	3,380.7	1,473.2	4,853.9	563.3	942.5	1,505.8	3,944.0	2,415.7	6,359.7
	1977	3,703.0	1,835.2	5,538.2	611.9	1,011.6	1,623.5	4,314.9	2,846.8	7,161.7
Yukon Territory and Northwest Territories	1975	330.4	151.4	481.8	24.1	44.7	68.8	354.5	196.1	550.6
	1976	386.4	166.6	553.0	28.1	44.0	72.1	414.5	210.6	625.1
	1977	464.1	87.1	551.2	33.7	48.9	82.6	497.8	136.0	633.8
Canada	1975	24,054.5	14,161.7	38,216.2	4,320.0	5,835.2	10,155.2	28,374.5	19,996.9	48,371.4
	1976	27,061.3	15,011.3	42,072.6	4,709.0	6,306.3	11,015.3	31,770.3	21,317.6	53,087.9
	1977	29,727.3	16,781.5	46,508.8	5,193.5	6,846.3	12,039.8	34,920.8	23,627.8	58,548.6

[1] Actual expenditures 1975; preliminary actual 1976; intentions 1977.
[2] Capital expenditures on machinery and equipment include an estimate for "capital items charged to operating expenses", in the manufacturing, utilities and trade totals.

14.21   Value of construction work performed, by type of structure, 1975-77¹ (thousand dollars)

Type of structure	1975			1976			1977		
	New	Repair	Total	New	Repair	Total	New	Repair	Total
BUILDING CONSTRUCTION									
Residential	7,113,857	1,575,805	8,689,662	9,763,811	1,813,859	11,577,670	9,807,914	2,055,296	11,863,210
Industrial	1,146,232	363,446	1,509,678	1,075,584	379,295	1,454,879	1,090,387	411,052	1,501,439
Factories, plants, workshops and food canneries	957,316	277,648	1,234,964	871,635	283,121	1,154,756	901,854	306,703	1,208,557
Mine and mine mill buildings	133,742	50,088	183,830	156,399	58,638	215,037	119,508	63,262	182,770
Railway stations and roadway buildings	33,758	21,493	55,251	30,822	22,920	53,742	46,152	25,174	71,326
Railway shops, engine houses, water and fuel stations	21,416	14,217	35,633	16,728	14,616	31,344	22,873	15,913	38,786
Commercial	3,378,962	353,503	3,732,465	2,970,534	357,187	3,327,721	2,879,433	377,521	3,256,954
Warehouses, storehouses and refrigerated storage	320,002	37,731	357,733	293,102	40,073	333,175	292,802	42,686	335,488
Grain elevators	17,517	5,485	23,002	22,060	8,690	30,750	36,322	9,514	45,836
Hotels, clubs, restaurants, cafeterias and tourist cabins	234,127	30,591	264,718	171,723	26,102	197,825	151,318	26,306	177,624
Office buildings	1,357,821	132,779	1,490,600	1,395,895	142,003	1,537,898	1,495,758	151,651	1,647,409
Stores, retail and wholesale	537,916	78,739	616,655	476,334	72,330	548,664	463,835	75,387	539,222
Garages and service stations	148,190	46,769	194,959	135,136	46,396	181,532	172,399	51,312	223,711
Theatres, arenas, amusement and recreational buildings	761,214	19,195	780,409	474,710	20,506	495,216	265,703	19,542	285,245
Laundries and dry-cleaning establishments	2,175	2,214	4,389	1,574	1,087	2,661	1,296	1,123	2,419
Institutional	1,376,134	184,768	1,560,902	1,323,637	187,774	1,511,411	1,362,505	202,207	1,564,712
Schools and other education buildings	803,348	98,771	902,119	715,862	103,788	819,650	736,940	107,602	844,542
Churches and other religious buildings	27,984	11,083	39,067	31,698	10,107	41,805	23,824	9,911	33,735
Hospitals, sanatoria, clinics and first-aid stations	337,845	50,205	388,050	359,956	54,860	414,816	374,671	58,254	432,925
Other	206,957	24,709	231,666	216,121	19,019	235,140	227,070	26,440	253,510
Other building	908,566	208,142	1,116,708	925,232	237,197	1,162,429	1,017,169	260,210	1,277,379
Farm buildings (excl. dwellings)	316,098	124,422	440,520	340,750	133,836	474,586	367,617	143,274	510,891
Broadcasting, radio and television, relay and booster stations, and telephone exchanges	133,316	18,323	151,639	126,719	18,604	145,323	158,068	21,027	179,095
Aircraft hangars	30,131	7,287	37,418	32,452	8,801	41,253	28,767	9,591	38,358
Passenger terminals, bus, boat, air and other	136,349	3,251	139,600	150,751	3,437	154,188	182,210	4,091	186,301
Armouries, barracks and drill halls	13,410	11,306	24,716	13,475	11,870	25,345	14,674	13,189	27,863
Bunkhouses, dormitories, camp cookeries, bush depots and camps	16,745	8,413	25,158	16,923	9,994	26,917	12,846	10,777	23,623
Laboratories	54,878	5,786	60,664	65,253	5,890	71,143	70,743	6,277	77,020
Other	207,639	29,354	236,993	178,909	44,765	223,674	182,244	51,984	234,228
Total, building construction	13,923,751	2,685,664	16,609,415	16,058,798	2,975,312	19,034,110	16,157,408	3,306,286	19,463,694
ENGINEERING CONSTRUCTION									
Marine	142,208	38,445	180,653	129,355	40,421	169,776	179,134	43,149	222,283
Docks, wharves, piers and breakwaters	86,014	20,907	106,921	62,088	20,498	82,586	98,263	22,404	120,667
Retaining walls, embankments and riprapping	10,716	1,269	11,985	13,546	864	14,410	12,285	952	13,237
Canals and waterways	6,849	3,264	10,113	10,229	3,937	14,166	21,138	4,656	25,794
Dredging and pile driving	7,807	7,528	15,335	7,776	7,504	15,280	12,655	9,610	22,265
Dyke construction	17,546	7,450	17,996	19,174	354	19,528	15,617	9,312	15,929
Logging booms	315	826	1,141	545	718	1,263	1,386	794	2,180
Other	12,961	4,201	17,162	15,997	6,546	22,543	17,790	4,421	22,211

Road, highway and airport runways	*1,964,287*	*417,659*	*2,381,946*	*2,117,521*	*496,748*	*2,614,269*	*2,225,696*	*510,694*	*2,736,390*
Highway, road and street construction (incl. grading, scraping, oiling, filling)	1,863,080	404,049	2,267,129	1,999,828	482,667	2,482,495	2,097,829	497,445	2,595,274
Parking lots	19,525	5,620	25,145	17,267	6,325	23,592	19,180	5,056	24,236
Sidewalks and paths	38,742	5,611	44,353	40,267	5,262	45,529	50,388	5,410	55,798
Runways, landing fields and tarmac	42,940	2,379	45,319	60,159	2,494	62,653	58,299	2,783	61,082
Waterworks and sewage systems	*1,123,314*	*118,028*	*1,241,342*	*1,198,490*	*117,326*	*1,315,816*	*1,376,209*	*132,401*	*1,508,610*
Tile drains, drainage ditches and storm sewers	216,419	37,300	253,719	215,639	29,058	244,697	244,630	35,953	280,583
Water mains, hydrants and services	269,846	41,546	311,392	326,263	46,318	372,581	401,336	50,551	451,887
Sewage systems and connections	557,865	26,228	584,093	539,036	27,694	566,730	603,404	30,213	633,617
Water pumping stations and filtration plants	69,462	10,324	79,786	106,110	11,670	117,780	111,341	12,980	124,321
Water storage tanks	9,722	2,630	12,352	11,442	2,586	14,028	15,498	2,704	18,202
Dams and irrigation	*111,505*	*25,970*	*137,475*	*89,173*	*24,331*	*113,504*	*100,645*	*27,571*	*128,216*
Dams and reservoirs	42,104	9,294	51,398	28,035	6,863	34,898	33,837	8,343	42,180
Irrigation and land reclamation projects	69,401	16,676	86,077	61,138	17,468	78,606	66,808	19,228	86,036
Electric power	*2,639,704*	*185,460*	*2,825,164*	*2,908,946*	*189,211*	*3,098,157*	*3,651,458*	*209,438*	*3,860,896*
Electric power generating plants, including water conveying and controlling structures	1,708,065	52,347	1,760,412	1,906,418	56,369	1,962,787	2,437,062	61,975	2,499,037
Electric transformer stations	5,096	1,175	6,271	5,472	1,117	6,589	5,351	1,179	6,530
Power transmission and distribution lines, and trolley wires	898,604	121,869	1,020,473	974,647	126,917	1,101,564	1,179,040	138,846	1,317,886
Street lighting	27,939	10,069	38,008	22,409	4,808	27,217	30,005	7,438	37,443
Railway, telephone and telegraph	*739,569*	*359,558*	*1,099,127*	*788,885*	*402,894*	*1,191,779*	*829,370*	*149,365*	*1,268,735*
Railway tracks and roadbeds	272,285	228,636	500,921	249,821	244,312	494,133	248,796	266,891	515,687
Signals and interlockers	21,609	31,026	52,635	18,151	33,877	52,028	32,594	37,279	69,873
Telegraph, telephone and cablevision lines, and underground and marine cables	445,675	99,896	545,571	520,913	124,705	645,618	547,980	135,195	683,175
Gas and oil facilities	*1,528,787*	*321,027*	*1,849,814*	*2,033,939*	*300,129*	*2,334,068*	*2,595,493*	*347,782*	*2,943,275*
Gas mains and services	120,038	20,362	140,400	130,311	19,370	149,681	137,817	20,887	158,704
Pumping stations, oil	19,190	6,446	25,636	38,177	4,644	42,821	13,817	6,016	19,833
Pumping stations, gas	7,413	2,851	10,264	6,119	3,857	9,976	17,627	3,904	21,531
Oil storage tanks	54,146	5,192	59,338	31,129	6,107	37,236	13,738	7,226	20,964
Gas storage tanks	13,638	933	14,571	17,103	671	17,774	8,484	730	9,214
Oil pipelines	150,977	9,604	160,581	114,834	1,645	126,479	66,324	9,285	75,609
Gas pipelines	51,895	3,306	55,201	54,952	2,653	57,605	90,374	3,142	93,516
Oil and gas wells	870,447	121,814	992,261	1,310,185	111,150	1,421,335	1,677,138	123,073	1,800,211
Oil refinery — processing units	141,124	82,325	223,449	206,460	81,480	287,940	392,477	96,628	489,105
Natural gas processing plants	99,919	68,194	168,113	124,669	58,552	183,221	177,697	76,891	254,588
Other engineering	*1,882,680*	*168,728*	*2,051,408*	*1,737,985*	*163,714*	*1,901,699*	*2,039,903*	*177,323*	*2,217,226*
Bridges, trestles, culverts, overpasses and viaducts	290,732	37,975	328,707	221,266	37,823	259,089	278,484	42,987	321,471
Tunnels and subways	63,715	3,167	66,882	66,042	1,139	67,181	66,030	1,014	67,044
Incinerators	3,583	1,322	4,905	3,239	1,132	4,371	4,500	1,340	5,840
Park systems, landscaping and sodding	54,836	20,489	75,325	82,402	16,031	98,433	96,193	16,547	112,740
Swimming pools, tennis courts and outdoor recreation facilities	45,837	4,031	49,868	26,337	1,772	28,109	25,514	2,165	27,679
Mine shafts and other below surface workings	321,317	5,316	326,633	383,494	7,303	390,797	418,236	6,138	424,374
Fences, snowsheds, signs and guard rails	68,394	36,509	104,903	67,200	35,546	102,746	73,967	41,313	115,280
Other engineering	1,034,266	59,919	1,094,185	888,005	62,968	950,973	1,076,979	65,819	1,142,798
Total, engineering construction	10,132,054	1,634,875	11,766,929	11,004,294	1,734,774	12,739,068	12,997,908	1,887,723	14,885,631
Total, all construction	24,055,805	4,320,539	28,376,344	27,063,092	4,710,086	31,773,178	29,155,316	5,194,009	34,349,325

Actual expenditures 1975; preliminary actual 1976; intentions 1977.

Sources
14.1 - 14.2 Central Mortgage and Housing Corporation.
14.3 - 14.10 Housing and Families Group, Content and Analysis Branch, and Consumer Income and Expenditure
   Division, Household Statistics Branch, Census and Household Surveys Field, Statistics Canada.
14.11 - 14.21 Construction Division, Industry Statistics Branch, Statistics Canada.

Transportation

Chapter 15

Tables

Transportation   Chapter 15

Transportation services   15.1

New water transport surveys   15.1.1
Because of Canada's size, geography and dependence on trade, water transportation has
always occupied a dominant role within the overall economic system. Historically, the
earliest industries of the country developed because of convenient access to water
transportation. To the present day, water transport has continued to be a relatively cheap
and easy means of moving raw materials and consumer goods. Section 15.4 provides a
broad view of water transport in Canada.

In the early 1970s the transportation and communications division of Statistics
Canada developed a comprehensive and accurate survey to replace the annual Water
Transportation Report which for more than 30 years had surveyed water transportation
in Canada. The Water Transportation Data Sheet, 1974, established a new statistical
benchmark for the water transport industry. Statistics for the years 1974 and 1975 based
on the data sheet and a revised survey indicate a sharply higher level of water transport
activity than was represented by the previous survey method.

Water transportation in 1975 generated revenues of $1.231 billion for 560
Canadian domiciled for-hire, private and sightseeing carriers, compared with 1974
revenues of $1.215 billion for 587 carriers. The largest portion of these revenues in
1975, $1.128 billion, was generated by 485 for-hire carriers representing the for-hire
water transportation industry; in 1974 there were 512 for-hire carriers which generated
$1.142 billion. The water transport operations of 50 private carriers accounted for $100.6
million in 1975 compared with $71 million by 54 private carriers the previous year.
Sightseeing undertakings contributed the balance of the total revenue.

The 560 carriers in 1975 employed 20,082 vessel crew who earned wages totalling
$278.9 million. By comparison, in 1974 the 587 carriers employed 20,054 crew whose
total wages were $244.4 million.

Government promotion and regulation   15.1.2
The federal government plays a twofold role in developing transportation services —
promotional, to ensure the growth and development of transportation appropriate to
need; and regulatory, including economic regulation of rates and services and the
application of technical regulations to meet safety requirements. The first category has
included building of canals from the time of Confederation to the construction of the St.
Lawrence Seaway, underwriting railway development and branch-line extension,
establishing Air Canada, investing in airports and aeronautical installations and building
the Trans-Canada Highway.

Transport Canada (the department of transport) is a corporate structure of
administrations and Crown corporations with varying degrees of autonomy, and
separate agencies for development and economic regulation. As outlined in Appendix 1
of this edition, it includes a department headquarters staff and three major operating
administrations: the Canadian marine transportation administration, the Canadian air
transportation administration and the Canadian surface transportation administration.
The Arctic transportation directorate works with other bodies in dealing with
transportation problems in the North. The Transportation Research and Development
Centre provides the national focus for changing technology and development in
transportation. The Canadian Transport Commission performs an independent
regulatory role. Air Canada, Canadian National Railways and the Northern Transportation Co. Ltd. are autonomous Crown corporations which report to Parliament through
the minister of transport.

The transport department and the various Crown agencies have jurisdiction over
canals, harbours, shipping, civil aviation and interprovincial and international railways.

Jurisdiction over for-hire extraprovincial and international highway transport also rests with the federal government but these powers are exercised by the provincial highway transport boards as provided for in the federal Motor Vehicle Transport Act of 1954 (RSC 1970, c.M-14), except for the Roadcruiser service operated by Canadian National Railways in Newfoundland, which was exempted by an order-in-council in July 1976. This bus service is now under the jurisdiction of the motor vehicle transport committee of the Canadian Transport Commission, and is subject to provisions of the National Transportation Act (RSC 1970, c.N-17).

Railway regulation was developed in a period when railways enjoyed a virtual transportation monopoly. Measures to protect the public against excessive charges, unjust discrimination and other objectionable monopoly practices, together with measures to ensure safe operation, have subjected railways to the most comprehensive regulation of any Canadian industry. However, the rapid growth of road, air and pipeline services has ended the near-monopoly and forced the railways into a highly competitive situation.

The National Transportation Act (RSC 1970, c.N-17) defines a national transportation policy for Canada with a view to achieving maximum efficiency in all available modes of transportation at the lowest cost. The act established the Canadian Transport Commission (CTC) to carry out functions formerly performed by the Board of Transport Commissioners for Canada, the Air Transport Board and the Canadian Maritime Commission. It created a framework within which the CTC might regulate interprovincial and international motor transport and transportation by pipeline of commodities other than oil and gas.

The Canadian Transport Commission has established several committees, five of which are railway transport, air transport, water transport, commodity pipeline transport and motor vehicle transport. The commission is a court of record. Its decisions are binding within its jurisdiction and may be reviewed only by appeal to the Supreme Court of Canada on a question of law or jurisdiction, or by the Governor-in-Council. However, a party to a licence application under the Aeronautics Act or the Transport Act may appeal to the transport minister.

The commission has jurisdiction under several acts, including the Railway Act, the Aeronautics Act and the Transport Act, over transportation by rail, air and inland water.

Under the Railway Act the commission has jurisdiction over construction, maintenance and operation of railways that are subject to the legislative authority of Parliament, including matters of engineering, location of lines, crossings and crossing protection, safety of train operation, operating rules, investigation of accidents, accommodation for traffic and facilities for service, abandonment of operation and uniformity of railway accounting. The commission regulates tolls for the use of international bridges and tunnels.

Except for certain statutory rates, and subject to certain powers to deal with rates that the commission finds to be contrary to the public interest, the railways are free under the statutes to establish freight rates in accordance with market forces. However, such rates must be compensatory, as defined in the Railway Act, and the commission may prescribe tolls for captive shippers if existing tolls unduly favour the railways.

The commission is responsible for the economic regulation of commercial air services in Canada and is required to advise the transport minister on matters relating to civil aviation. The regulatory function relates to Canadian air services within Canada and abroad and to foreign air services operating into and out of Canada. It is because of this responsibility that the CTC participates in bilateral negotiations for the exchange of traffic rights. The commission is responsible for licensing commercial air services and regulating the licensees. It issues regulations dealing with, among other things, the classification of air carriers and commercial air services, licences, tariffs, service schedules and statistical reporting.

The CTC takes an active part in the work of international organizations and conferences related to economic matters of air transport. Of greatest significance in 1977 was the special air transport conference convened by the International Civil Aviation Organization in Montreal and the 22nd session of assembly of the organization.

As provided by the Transport Act, the commission grants licences for ships to transport goods and passengers between ports or places in Canada on the Great Lakes and on the Mackenzie and Yukon rivers. Provisions of the act do not apply to the transport of goods in bulk on waters other than the Mackenzie River. The commission must determine that the service is required. Tariffs of tolls must be filed and the commission has regulatory powers over such tolls.

The commission, under the Pilotage Act, is empowered to investigate objections to proposed tariffs of pilotage charges, to hold public hearings, and to make its recommendations to the appropriate pilotage authority. Under the Shipping Conferences Exemption Act, ocean carriers which are members of a shipping conference are required to file with the commission copies of their agreements, arrangements, contracts, patronage contracts and tariffs. These documents are made available for public inspection.

The commission is also authorized, under the St. Lawrence Seaway Authority Act, to consider any complaint of unjust discrimination in existing tariffs and to report its findings to the authority.

In Canada the coasting laws restrict the operation of ships from one point to another to Canadian or British ships, depending upon the area. To enable a ship of any foreign country to engage in such operations, application has to be made to the national revenue minister to obtain a waiver of the coasting laws. The commission is then called upon to advise the minister as to the availability of a suitable Canadian ship.

Rail transport 15.2

Canadian railway transport is dominated by two transcontinental systems, supplemented by a number of regional railways. The government-owned Canadian National Railway system is the country's largest public utility and operates the longest trackage in Canada. It serves all 10 provinces as well as the Great Slave Lake area of the Northwest Territories. It also operates a highway transport service, a fleet of coastal steamships, a chain of large hotels and resorts, a telecommunications service, and as an autonomous subsidiary a scheduled Canadian and international air service, Air Canada. Canadian Pacific Ltd. is a joint-stock corporation operating a railway service in eight provinces. Similar to and competitive with the Canadian National Railway system, it is a multi-transport organization with a fleet of inland and ocean-going vessels as well as coastal vessels, a fleet of trucks, a chain of year-round and resort hotels, a telecommunications service, and a domestic and foreign airline service. Through a major subsidiary, Canadian Pacific Investments Ltd., interests are held in mining (for example, Cominco Ltd.), oil and gas, forest products, real estate and related operations, hotel and food services, and steel production and associated services. The British Columbia Railway operates over a 2 045-kilometre route from North Vancouver to Fort Nelson in northeastern British Columbia. The Northern Alberta Railway, jointly owned by CP and CN, serves the area north of Edmonton with a 1 484-km system. Northern Ontario is served by the provincially owned Ontario Northland Railways with a 925-km system stretching from North Bay to Moosonee, and by the privately owned Algoma Central Railway operating over 500 km of line between Sault Ste Marie and Hearst.

In addition, a US–Canada passenger service inaugurated by the National Railroad Passenger Corp., AMTRAK, is operated between Seattle, Wash. and Vancouver, BC and between Montreal, Que. and Washington, DC via New York City, Springfield, Mass. and resort areas in Vermont.

The largest contributors to Canada's total 1976 railway revenue were Canadian National (54.5%) and Canadian Pacific (36%). The Quebec North Shore and Labrador Railway, built to transport ore and concentrates from the iron mines of the Schefferville and Wabush areas of Quebec and Labrador to water transportation facilities on the St. Lawrence River, accounted for 2% of the revenues. Other individual railways contributing 1% or more of the total revenue were the British Columbia Railway (1.7%) and the Ontario Northland Railways (1.1%).

In recent years, railways have faced strong competition from highway and air transport. Still indispensable for carrying bulk commodities, railways are necessary to the development of natural resources in isolated areas. Only pipelines have competed with them by providing an alternate economical means of transporting the products of oil and gas fields for long distances overland.

The rapid growth of containerization has made the integration of railway, highway, shipping and other modes of transport of growing importance. Canada's two major railways are already involved in several forms of transportation. They have evolved from a virtual monopoly, through a highly competitive stage to co-operation and co-ordination with other modes of transport.

15.2.1   Government aid

In the 19th century governments promoted railway building. Private developers received assistance in land grants, cash payments, loans or purchase of shares. Debenture issues of the Canadian National Railway system, except those for rolling-stock, are guaranteed by the federal government. Provincial governments had guaranteed the bonds of some lines that were later incorporated in the CNR system. As these mature or are called, they are paid off by the CNR in large measure through funds raised by issuing new bonds guaranteed by the federal government. At December 31, 1976 railway bonds guaranteed by the Government of Canada amounted to $574 million.

The National Transportation Act provided for normal railway subsidy payments of $110 million for 1967, declining by $14 million a year, the last payment being $12 million for 1974, and allows railways to file claims and receive specific payments for losses incurred on branch lines and passenger-train services. Total payments of $255.7 million for 1975 represented specific payments to the two major railways, exceeding and replacing their shares of the normal subsidy. Claims for 1976 had to be filed by June 30, 1977.

Truckers receive federal assistance through freight rate subsidies similar to the subsidies to railways provided under the Maritime Freight Rate Act. Since 1969 the Atlantic Region Freight Assistance Act has allowed subsidies on goods moved from Nova Scotia, Prince Edward Island, New Brunswick, the island of Newfoundland, and Quebec south of the St. Lawrence River and east of highway 23 to points in Canada outside that territory. In October 1970 assistance was authorized for goods moved by highway transport within that area as well. In April 1974 selective assistance for specified commodities moving by railway or highway transport to points in Canada outside the territories was authorized at a level of 50% of that portion of the freight rate earned within the particular territory.

15.2.2   Rail transport statistics

Trackage and rolling-stock. Table 15.2 illustrates the historical development of first main track from 28 416 kilometres in 1900 to 70 471 km in 1976. It also presents statistics on main and other types of track by province and territory and that operated by Canadian carriers in the US for the years 1972-76.

Table 15.3 gives freight and passenger equipment in operation in 1975 and 1976. Freight carrying capabilities of the railways are steadily being improved with larger, more efficient cars and locomotives and modernized handling and terminal services. Each year hundreds of units, particularly freight cars, are converted and modified for specific types of traffic or are replaced by special-purpose equipment for particular hauling jobs. Because of the fuel consumption efficiency of the railways and higher fuel costs, there is a trend to greater freight movement by rail. Container and piggyback traffic has also increased.

Revenue freight. Total freight carried by all common carrier railways (including national loadings and receipts from US connections) in 1975 and 1976 is shown in Table 15.4 under the commodity structure adopted in 1970 based on Statistics Canada's standard commodity classification. Despite some loss of continuity with previous data, the new commodity breakdown permits improved comparisons with other series (such as water transport, imports and exports) which are also based on this classification.

Capital structure and finance. Tables 15.5 - 15.8 give information on capital investment in road and equipment, and on operating revenues, expenses and net income of all common carrier railways operating in Canada, except those of the Cartier Railway which are not available. In transportation statistics a distinction is made between expenditures and expenses. In the following data, the term expenses refers to the cost of furnishing rail transportation service and of associated operations, including maintenance and depreciation of the plant used in such service.

The capital structure of the Canadian National Railway system is presented in Table 15.6 and financial details of operations in both Canada and the United States in Table 15.7. Revenues and expenses include those of express and commercial communications and highway transport (rail) operations. Tax accruals and rents are charged to operating expenses.

Total operating revenues and expenses of common carrier railways operating in Canada (except the Cartier Railway) continued to rise, both reaching peak levels in 1976; increases over 1975 amounted to 16.8% and 9.8%, respectively (as calculated from Table 15.8). A net operating income of $116.6 million was recorded in 1976.

Road transport   15.3

The federal government establishes motor vehicle safety standards, while registration of motor vehicles and regulation of motor vehicle traffic lie within the jurisdiction of the provincial and territorial governments.

Federal safety regulations   15.3.1

The Motor Vehicle Safety Act, in effect since January 1971, establishes mandatory safety standards for new motor vehicles to ensure minimum standards of vehicle safety and environment protection. The standards are for the safe design, construction and functioning of new motor vehicles in order to save lives and prevent injuries. The legislation, administered by the transport department, applies to all new motor vehicles manufactured in or imported into Canada. It requires that all such vehicles and their components meet motor vehicle safety regulations at the point of manufacture or importation and obliges manufacturers to issue notices of safety defects. The safety of vehicles in use is a provincial responsibility.

The road and motor vehicle safety branch of the transport department administers the Motor Vehicle Safety Act and the Motor Vehicle Tire Safety Act, and enforces regulations pertaining to them. It joined with the provinces in a five-year co-operative program aimed at reducing by 15% the fatality rate on Canadian roads by 1979. In 1976, for the first time in 10 years, the fatality rate dropped markedly by 30% from the all-time 1973 high. The department is also constructing a motor vehicle test centre at Blainville, Que., and conducting research into cost-effective measures to improve traffic safety. The branch works closely with other federal government departments, the provinces and international organizations on specific road safety projects.

Safety regulations currently include 43 standards for the design and performance of passenger cars, trucks, buses, motorcycles, competition motorcycles, minibikes and trailers; six standards limiting motor vehicle exhaust, evaporative and noise emissions; and 11 standards for snowmobiles. These standards are reviewed and revised regularly to keep pace with engineering or technical advances. The regulations require all Canadian motor vehicle manufacturers or distributors to apply the national safety mark, accompanied by a label certifying compliance with all applicable federal motor vehicle safety standards, to every classified vehicle produced after January 1, 1971. Vehicles imported for sale or private use must also be certified.

The Motor Vehicle Tire Safety Act, adopted by Parliament in April 1976, provides authority for the enforcement of minimum safety standards for certain motor vehicle tires manufactured in or imported into Canada.

Motor vehicle and traffic regulations   15.3.2

Motor vehicle and traffic regulations in force in 1977 are outlined in Table 15.14. This table includes requirements in all provinces and territories for a driver's licence for

Kilometre Guide	Calgary	Charlottetown	Edmonton	Fredericton	Halifax	Montreal	Ottawa	Quebec	Regina	St. John's	Saskatoon	Thunder Bay	Toronto	Vancouver	Victoria	Whitehorse	Winnipeg	Yellowknife
Calgary	●	4917	299	4558	5042	3743	3553	4014	764	6183	620	2050	3434	1057	1123	2385	1336	1811
Charlottetown	4917	●	4949	359	232	1184	1374	945	4163	1294	4421	2878	1724	5985	6051	7034	3592	6460
Edmonton	299	4949	●	4598	5082	3764	3574	4035	785	6212	528	2071	3455	1244	1310	2086	1357	1511
Fredericton	4558	359	4598	●	346	834	1024	586	3813	1622	4070	2527	1373	5634	5700	6684	3241	6109
Halifax	5042	232	5082	346	●	1318	1508	912	4297	1349	4554	3011	1857	6119	6185	7168	3726	6593
Montreal	3743	1184	3764	834	1318	●	190	270	2979	2448	3236	1693	539	4801	4867	5850	2408	5275
Ottawa	3553	1374	3574	1024	1508	190	●	460	2789	2638	3046	1503	399	4611	4677	5660	2218	5086
Quebec	4014	945	4035	586	912	270	460	●	3249	2208	3507	1963	810	5071	5137	6120	2678	5546
Regina	764	4163	785	3813	4297	2979	2789	3249	●	5427	257	1286	2670	1822	1888	2871	571	2297
St. John's	6183	1294	6212	1622	1349	2448	2638	2208	5427	●	5684	4141	2987	7248	7314	8298	4855	7723
Saskatoon	620	4421	528	4070	4554	3236	3046	3507	257	5684	●	1543	2927	1677	1743	2614	829	2039
Thunder Bay	2050	2878	2071	2527	3011	1693	1503	1963	1286	4141	1543	●	1384	3108	3174	4157	715	3582
Toronto	3434	1724	3455	1373	1857	539	399	810	2670	2987	2927	1384	●	4492	4558	5528	2099	4966
Vancouver	1057	5985	1244	5634	6119	4801	4611	5071	1822	7248	1677	3108	4492	●	66	2697	2232	2411
Victoria	1123	6051	1310	5700	6185	4867	4677	5137	1888	7314	1743	3174	4558	66	●	2763	2298	2477
Whitehorse	2385	7034	2086	6684	7168	5850	5660	6120	2871	8298	2614	4157	5528	2697	2763	●	3524	2704
Winnipeg	1336	3592	1357	3241	3726	2408	2218	2678	571	4855	829	715	2099	2232	2298	3524	●	2868
Yellowknife	1811	6460	1511	6109	6593	5275	5086	5546	2297	7723	2039	3582	4966	2411	2477	2704	2868	●

Official highway distances

different types of vehicles, times of renewal of licences, types of motor vehicle insurance and speed limits.

Registration plates. Motor vehicles and trailers are usually registered annually with the payment of specified fees. Most motor vehicles carry a registration plate on the front and one on the rear; trailers carry one on the rear. In Saskatchewan and Alberta passenger cars, vehicles licensed for Drive-ur-self service and trucks carry two plates. Vehicles operated by dealers, motorcycles and off-highway vehicles have one rear licence plate. In Quebec a single licence plate is displayed on motor vehicles. In some provinces, multi-year plates are issued and validated annually by stickers. In some provinces registration plates stay with the vehicle when it is sold, but in others the owner retains them.

Safety responsibility legislation. Each province has enacted safety responsibility legislation. In general, laws provide for the automatic suspension of the driver's licence and motor vehicle registration of a person convicted of a serious offence (impaired driving, driving under suspension, dangerous driving) or a person whose uninsured vehicle is involved directly or indirectly in an accident resulting in damage of a specified amount or injury or death to any person.

Insurance and unsatisfied judgment fund. In 1977 most provinces and the two territories had a compulsory insurance plan in effect (Table 15.14). Other provinces had enacted legislation providing for the establishment of a fund, frequently called an unsatisfied judgment fund (in Ontario and Alberta, the Motor Vehicle Accident Claims Fund). Judgments awarded for damages arising out of motor vehicle accidents which cannot be collected by the ordinary process of law were paid out of this fund. In Newfoundland, Prince Edward Island, Nova Scotia and Quebec the fund was maintained by insurance companies. In provinces where insurance was not compulsory, the funds were obtained by collecting an annual fee from the registered owner of every uninsured motor vehicle or from every person to whom a driver's licence was issued.

Additional provincial details. Other details of motor vehicle and traffic regulations particular to individual provinces in 1977 follow.

In Newfoundland, one year's experience as a licensed driver is required to drive commercial classes other than cars or light trucks. Accidents resulting in personal injury or death, or property damage in excess of $200 must be reported to a police officer. All motor vehicles must display two licence plates, with the registration year ending March 31 of any year.

In Nova Scotia all vehicle registrations expire December 31 of each year; a three-month extension is granted for passenger vehicles and one month for commercial vehicles. All vehicles are required to be inspected annually. It is an offence to operate a motor vehicle without financial responsibility insurance. Driver files are continuously reviewed, with a demerit point system and a medical advisory committee on driver licensing used to make drivers improve or remove the unsafe driver from the highways.

In New Brunswick a person must be age 14 or over to operate a moped, but 16 to operate a motorcycle.

In Quebec, use of seat belts is mandatory for persons occupying the front seat of a pleasure vehicle. The period during which a learner-driver's permit is valid is five months. The owner of a vehicle makes his contribution to the automobile insurance plan (coverage against bodily injury provided by the province) when he registers his vehicle. At the same time, the bureau makes certain that the owner is insured against damage to property (this insurance is through private companies).

In Ontario, exemption from registration is granted for six months for non-residents from other provinces and three for residents of another country or state. A certificate of mechanical fitness is required before a vehicle sold second-hand can be issued a permit for operation. Accidents resulting in personal injury or property damage in excess of $400 must be reported to a police officer.

In Manitoba all motor vehicles and trailers must be registered; semi-trailers must also be registered and display an identification number plate. Multi-year plates are issued and validated annually by stickers. Passenger car registration fees are based on weight. Vehicle inspection is compulsory by random selection and at time of sale of used vehicles. Mopeds may be driven with any licence class. The driver's licence is also a certificate of insurance for which an insurance premium is payable. Non-residents, or new residents, holding a valid licence may drive up to 90 days.

In Saskatchewan all motor vehicles including trailers and semi-trailers must be registered. Registrations may be transferred from one vehicle to another by the owner. Proof of ownership must be provided on application. Out-of-province vehicles may be operated for 90 days without requiring Saskatchewan registration. Students may operate out-of-province vehicles while attending a school in Saskatchewan. Out-of-province drivers involved in an accident in Saskatchewan are required to provide proof of financial responsibility; failure to do so will result in the vehicle being impounded. All accidents resulting in damage in excess of $200 or personal injury must be reported to a police officer immediately.

Alberta motor vehicles and trailers are registered annually with the payment of specified fees and providing proof of vehicle insurance. Passenger car registration plates remain with the owner when the vehicle is sold. Regulations permit non-residents temporarily in Alberta to operate vehicles currently registered in their home province or in the United States for six months; the period is extended to a school year for out-of-province students whose vehicles carry non-resident student stickers.

In British Columbia, seat belts must be used at all times in all passenger vehicles of 1965 or later, and commercial vehicles of 1973 or later. The driver is responsible for ensuring that passengers between ages 6 and 15 are properly buckled up; persons age 16 and older are responsible for themselves. In BC 60-70% of all motor vehicles are required to be mechanically inspected.

In Yukon Territory, registration plates remain with the owner of the vehicle and may be transferred to a vehicle of the same type upon payment of a prescribed fee. All vehicles in the Yukon Territory must be registered; an exemption is given to tourists who do not remain more than 90 days. Safety regulations require all vehicles to meet certain standards.

15.3.3 Road transport statistics

Roads and streets. At the end of 1974 Canada had 295 807 kilometres of highways and roads under federal or provincial jurisdictions and 564 912 km of roads and streets under local government jurisdiction (Table 15.9). Most is in the more populated sections. Roads built by logging, pulp and paper, and mining companies provide some access to remote communities but large areas of most provinces and the territories are still sparsely settled and are virtually without roads.

Table 15.10 presents expenditure data for all roads and streets in 1974 and 1975. In 1975 total expenditures equalled $3,804 million, an increase of 22% over the previous year. Construction expenditures increased 24% and maintenance and administration costs rose 19.2%.

Road motor vehicles. Registrations continue to increase yearly, a record of 11.4 million being reached in 1975. Of that total, 8.9 million were passenger cars. Registration by province is given in Table 15.11 and types of vehicles registered by province in Table 15.12.

Taxation of motive fuels, motor vehicles, garages, drivers and chauffeurs is a source of provincial revenue. In every province licences or permits are required for motor vehicles, trailers, operators or drivers, paid chauffeurs, dealers, garages and gasoline and service stations. The more important sources from which revenue from motor vehicles is derived are shown in Table 15.13.

Motive fuels for motor vehicle use are taxable at the point of sale. To estimate the amount of fuel sold for road motor vehicles, tax-exempt sales to the federal government and other consumers, exports and sales on which tax refunds are paid are eliminated from gross sales. As shown in Table 15.15, consumption of gasoline, which was taxed at road-use rates and which is used almost entirely for automotive purposes, rose 1.9% in 1976 and net sales of diesel oil were up 8.4%.

Statistics of intercity bus companies for 1975 and 1976 are shown in Table 15.16. Table 15.17 presents summary statistics of the Canadian urban transit industry, and Table 15.18 of the motor carriers (freight).

15.4 Water transport

The Canada Shipping Act (RSC 1970, c.S-9), is the most significant statute dealing with shipping. Other legislative measures include the Pilotage Act, the Arctic Waters Pollution Prevention Act and the Navigable Waters Protection Act. Under the Canada Shipping Act, the Arctic Waters Pollution Prevention Act and their amendments, the federal government has complete jurisdiction over the regulation of shipping in Canadian-controlled waters.

15.4.1 Shipping

Except in the case of the coasting trade, all Canadian waterways, including canals, lakes and rivers, are open on equal terms to the shipping of all countries, and Canadian shipping must compete with foreign flag shipping.

The carriage of goods and persons from one Canadian port to another, commonly referred to as the coasting trade, is restricted to ships registered in Canada within the region from Havre-Saint-Pierre on the St. Lawrence River, upstream to the head of the Great Lakes. Elsewhere in Canada the coasting trade is restricted to ships registered and owned in a Commonwealth country.

Canadian registry. Part I of the Canada Shipping Act sets out the sizes, types and ownership of vessels to be registered. As at December 31, 1976, there were 31,593 ships constituting 4,374,923 gross tons (12 388 404 cubic metres) registered in Canada. This represents an increase over the previous calendar year of 305 ships.

Shipping traffic. Table 15.22 shows the number and tonnage of all vessels (except those of less than 15 registered net tons (equivalent to 42 m³), Canadian naval vessels and fishing vessels) entering Canadian customs and non-customs ports.

Freight movement through large ports takes different forms, including cargoes for or from foreign countries and cargoes loaded and unloaded in coastwise shipping, that is, domestic freight moving between Canadian ports. Table 15.23 presents data by province on cargoes loaded and unloaded from vessels in international and coastwise shipping. In 1976 a total of 279 million tonnes were loaded and unloaded at the principal Canadian ports, compared with 274.9 million tonnes in 1975. In-transit movement in vessels that pass through harbours without loading or unloading and movements from one point to another within harbours are also numerous in many ports.

Shipping statistics, which cover traffic in and out of both customs and non-customs ports, do not include freight in transit or freight moved from one point to another within the harbour. Table 15.24 shows the principal commodities loaded and unloaded in international and coastwise shipping at 20 ports handling large cargo volumes in 1976. These ports handled 83.9% of all Canada's international shipping and 64.5% of the coastwise trade. The specific commodities shown are those transported in volume and often in bulk form.

Ports and harbours   15.4.2

The ports and harbours of Canada comprise 25 large deep-water ports and about 650 smaller ports and multi-purpose government wharves on the East and West coasts, along the St. Lawrence Seaway and Great Lakes, in the Arctic, and on interior lakes and rivers.

Administration of Canadian ports is generally under Transport Canada's Canadian Marine Transportation Administration (CMTA). Canada's harbours are subdivided into National Harbours Board ports, harbour commission ports, public harbours and government wharves. About 2,000 fishing harbours and facilities for recreational boating are administered by the fisheries and the environment department.

The National Harbours Board, a Crown corporation, is responsible for administering the Jacques Cartier and Champlain bridges at Montreal, the grain elevators at Prescott and Port Colborne, Ont. and port facilities such as wharves and piers, transit sheds and grain elevators at the harbours of St. John's, Nfld.; Halifax, NS; Saint John and Belledune, NB; Sept-Îles, Chicoutimi, Baie-des-Ha! Ha!, Quebec, Trois-Rivières and Montreal, Que.; Churchill, Man.; and Vancouver and Prince Rupert, BC. The number of vessels and the tonnage handled at these ports in 1976 and 1977 are reported in Table 15.25.

Value before depreciation of fixed assets administered by the board was $680 million at December 31, 1976, increased from $436 million at December 31, 1973. These figures include expenditures for developing berthing and terminal facilities, grain elevators and harbour buildings and equipment. The net book value after deducting accumulated depreciation was $510 million. During 1976, the board had capital expenditures of $19.1 million. Included in this was $6.2 million at Vancouver, $2.5 million at Prince Rupert and $2.2 million in Saint John.

Canada's major multi-purpose harbours are administered by harbour commissions. These are corporate federal bodies, operating semi-autonomously under the general supervision of the department. The harbour commissions include municipal as well as federal appointees, and are responsible for general administration, operation and maintenance as well as for close liaison with the department and with the provincial, regional and local interests they serve.

More than 300 public harbours are directly administered by CMTA. Most harbourmasters and wharfingers at these ports are fees-of-office employees appointed by the minister of transport.

Many of the government wharves for which CMTA is responsible are located within public harbours and are used for commercial traffic including auto and truck ferries. Some major interprovincial federal ferry terminals are administered by the Canadian Surface Transportation Administration. Provincial governments administer intraprovincial ferry wharves.

Transport Canada is responsible for planning and providing adequate public port facilities to serve commercial interests and for improving or phasing out facilities in response to economic growth or changes in traffic patterns resulting from new industries, new types of ships and new developments in cargo handling. Specialized deep-water terminals for bulk commodities, particularly coal and oil, are also provided when needed under long-term full cost-recovery agreements with individual shippers. These often complement related development programs sponsored by the regional economic expansion department.

The department establishes and collects fees from users of port facilities, and all rates assessed by ports under federal jurisdiction are subject to departmental approval. Harbour dues, cargo rates, wharfage, berthage and other charges on goods and vessels are subject to some regional and local variation.

In addition to public facilities, there are extensive wharf and associated cargo handling facilities owned by private companies, usually for handling coal, iron ore, petroleum, grain and pulpwood.

The continuing trend to larger ships has resulted in increased investment in ports for facilities farther from shore, channel dredging, larger turning basins and more complex systems of aids to navigation and traffic control. Also, environmental considerations often require expensive terminal construction.

Increasing use of containers has brought significant changes in cargo routing and handling. Container ships travel at high speeds and port turnaround time is critical. Port facilities have to be more efficient and specialized; they include special ramps for roll-on roll-off vessels; large container cranes which can handle 20 or more 14-tonne containers an hour; special container packing facilities; large open storage areas for containers, automobiles, lumber and bulk products like coal; and rail and truck loading and unloading facilities.

15.4.3   The St. Lawrence Seaway

The St. Lawrence Seaway Authority, constituted as a corporation by act of Parliament in 1951, undertook the construction (and subsequent maintenance and operation) of Canadian facilities between Montreal and Lake Erie to allow navigation by vessels of 8.23 metres draft. At the same time, construction of similar facilities in the International Rapids section of the St. Lawrence River was undertaken by the Saint Lawrence Seaway Development Corporation of the United States. The seaway was opened to commercial traffic on April 1, 1959 and officially inaugurated on June 26, 1959. A portion of the third Welland Canal and the Canadian lock at Sault Ste Marie are also under the seaway authority's jurisdiction for operation and maintenance.

Seaway traffic. Tables 15.26 and 15.27 give combined traffic statistics for the St. Lawrence and Welland canals in 1975 and 1976. Duplicate transits are eliminated so that the figures show actual shipments through the St. Lawrence Seaway.

In 1976, 3,454 ships carrying about 28.5 million tonnes of cargo moved upbound through the seaway and 3,478 vessels carrying 36.6 million tonnes moved downbound. Ocean-going ships carried 20.4% of the total cargoes and lakers 79.6%. Of the total tonnage carried upbound in 1976, 23.3 million tonnes were domestic cargo, 5.2 million tonnes were foreign traffic; downbound, 28.5 million tonnes were domestic freight and 8.1 million tonnes were carried to and from foreign ports.

On the Montreal–Lake Ontario section, upbound traffic amounted to 26.2 million tonnes in 1976 and downbound traffic to 23.2 million tonnes, an increase of 13.2% over 1975. Almost 71.2% of the former was accounted for by iron ore shipped from St. Lawrence ports to Hamilton and Lake Erie and the downbound traffic consisted largely of overseas shipments of grain. There were 73 more upbound transits and 82 more downbound transits in 1976 than in 1975, indicating a slight increase in the number of vessels using this portion of the seaway. Bulk commodities made up 91.7% of the total traffic through the section in 1976, the principal commodities through the St. Lawrence canals being iron ore, wheat, corn, barley, manufactured iron and steel, and coke. Traffic patterns show that 29.6% of the total movement was between Canadian ports, 43.3% between Canadian and United States ports, and 27% consisted of foreign trade to and from Canada and the United States. The small remainder was traffic between ports in the United States.

There were 5,892 transits through the Welland Canal in 1976, with a cargo volume of 23.1 million tonnes upbound and 35.1 million tonnes downbound; bulk cargo accounted for 94.1% of the traffic. Although many vessels pass through both the St. Lawrence and Welland canals on through trips, there is a substantial amount of local traffic between Great Lakes ports which involves only the Welland Canal. These movements are largely of iron ore, grain and coal. The Welland Canal traffic was 9.1 million tonnes greater than that reported for the Montreal–Lake Ontario section.

Income of the St. Lawrence Seaway Authority for the 12-month period ended March 31, 1977 amounted to $27.4 million, made up of toll revenue of $22.7 million assessed for transits through the seaway locks between Montreal and Lake Erie and sundry revenues (rentals, wharfage, bridge revenue) of $4.7 million. Total expenses (excluding depreciation and interest) for the 12-month period ended March 31, 1977 amounted to $32.6 million, of which operation and maintenance expenses amounted to $22.8 million and regional and headquarters administration expenses $9.8 million (Table 15.28).

Canadian Coast Guard 15.4.4

The Canadian Coast Guard, a component of CMTA, is headed by a commissioner. A headquarters organization and five regional offices have the following objectives: the support of waterborne commerce through facilities and services which promote the safe and efficient movement of marine traffic; the provision of the framework necessary to permit the marine industry to conduct its affairs in an orderly and efficient manner; the support of the objectives of other government departments and agencies as they apply to the marine field; the promotion of continuous improvement, innovation, growth or phase-out of various types of marine transportation and the associated ancillary services; and the recovery of financial costs from the users or other beneficiaries of facilities and services provided by the coast guard.

Each of the five Canadian Coast Guard regions has a number of field offices. The commissioner of the coast guard also has a functional responsibility for the Pacific, Great Lakes, Laurentian and Atlantic Pilotage authorities. Each authority is a Crown corporation established under the Pilotage Act.

The aids and waterways branch. Within the branch the marine aids division is responsible for policies and standards for all shore-based and floating aids to navigation. It provides policies and standards for marine traffic control systems and routing schemes and administers the Navigable Waters Protection Act. The waterways development division is responsible for planning, research and development on navigable waterways, hydraulic model activities, hydraulic engineering expertise and water resources management.

Ship safety branch is concerned with the safety of life and property at sea and the protection of the environment from detrimental effects of ships through the following divisions. The Board of Steamship Inspection employs steamship inspectors; develops standards for the design and construction of ships, their machinery, fittings and equipment, and translates these standards into statutes and regulations to be followed by builders and operators of vessels; and establishes procedures to ensure that these standards are met. Nautical services division develops standards for loading, unloading and stowing of cargo, quantities of cargo, work practices, number and qualifications of personnel, discipline aboard ship, and navigating and operating procedures, including traffic routing; operates the registry of ships and administers the licensing of small vessels; reviews measurement of ships; and protects the interests of the owners of wrecked ships and their cargo and the interest of the Crown in unclaimed wrecks. Air cushion vehicles division prepares domestic regulations for the design, construction, maintenance and operation of air cushion vehicles and ensures that domestic regulations concerning air cushion vehicle crews and maintenance engineers are implemented.

Fleet systems branch is responsible for approximately 160 vessels of different sizes performing various functions. These vessels are all attached to coast guard regions according to the type of work and workload in each region. Headquarters provides policies, standards and guidelines, and is operationally responsible for the ships in the Arctic in the summer months. The Canadian Coast Guard College at Point Edward, NS trains navigation and marine engineering officers for the Canadian Coast Guard fleet.

Telecommunications and electronics branch provides expert knowledge in communications and computers, quality assurance, guidance systems and ship electronics and avionics to Canadian Coast Guard line operations. It also provides ship-to-shore radio communications for ships using Canadian waters from 45 radio stations which also broadcast weather reports and notices of danger to navigation.

Pilotage branch establishes national technical standards and conducts research to ensure that standards are maintained. It prescribes standards for health, uniform financial reporting procedures and procedures for hearings held by the four regional pilotage authorities, and recommends the establishment of compulsory pilotage areas.

The Pilotage Act, in effect since 1971, established four regional pilotage authorities as Crown corporations with wide powers to set up, operate and maintain, in the interest of safety, an efficient pilotage service within their region. They are the Atlantic pilotage authority, the Laurentian pilotage authority, the Great Lakes pilotage authority and the Pacific pilotage authority.

Legislation branch is responsible for certain legislative activities and for participation in Law of the Sea matters. It is also involved in amendments to the maritime pollution claims fund and in the development of a maritime code which is expected to replace the existing Canada Shipping Act.

The senior marine officer, emergencies, is responsible for developing contingency plans; developing, testing, acquiring and allocating certain pollution clean-up equipment; advice and assistance at the time of a pollution incident; and certain wartime activities.

Regional marine emergency officers take over at the time of an incident and ensure that appropriate contingency plans are followed. They develop regional contingency plans and are members of provincial and municipal committees dealing with oil pollution matters.

The senior marine officer, casualty investigations, is responsible for developing accident investigation procedures and carrying out fact-finding and preliminary inquiries according to the provisions of the Canada Shipping Act.

The coast guard fleet performs many duties. During winter, East Coast icebreaking operations are co-ordinated by an ice operations office in Sydney, NS. During summer, the Arctic operations of the fleet are co-ordinated jointly by Ottawa headquarters and an ice operations office at Frobisher Bay. The coast guard fleet has one cable-laying vessel stationed at St. John's, Nfld. under year-round contract. Two rotational weather-ships off the BC coast provide meteorological data. Buoy vessels and helicopters carry out aids to navigation work; smaller craft are also used for this purpose where possible. The coast guard operates a number of specialized vessels for search and rescue work which, during distress situations, are controlled by a national defence department co-ordinator. These vessels include cutters, lifeboats and hovercraft. The cutters are normally deployed on patrol when not involved in search and rescue work.

The Laurentian coast guard region is responsible for channel maintenance in the St. Lawrence River downstream from Montreal and in the Saguenay River, including sweeping and administration of dredging contracts.

Civil aviation 15.5

Administration and policy 15.5.1

Administration. Civil aviation in Canada is under the jurisdiction of the federal government and is administered under the authority of the Aeronautics Act and the National Transportation Act. The first part of the Aeronautics Act deals with the technical side of civil aviation, including matters of aircraft registration, licensing of personnel, establishing and maintaining airports and facilities for air navigation, air traffic control, accident investigation and the safe operation of aircraft; it is administered by the Canadian Air Transportation Administration. The second part deals with the economic aspects of commercial air services and gives the Canadian Transport Commission certain regulatory functions in commercial air services. The third part deals with matters of internal administration.

Federal civil aviation policy. Domestically, the role of the regional carriers and their relationship with the main line carriers (Air Canada and Canadian Pacific Air Lines) remained basically unchanged from 1966. Regional carriers provided regular route operations into the North and operated local or regional routes to supplement the domestic main line operations of Air Canada and CP Air. They had greater scope in developing routes and services. In 1969 the transport minister defined more precisely the regions in which each of the five regional carriers would be permitted to supplement, or authorized to replace, main line operations.

The relative roles of Air Canada and CP Air in the domestic sphere were defined in the transcontinental policy of 1967. These were based on a formula that would maintain Air Canada's pre-eminence on transcontinental services, on the assumption that the carrier might from time to time be called on to perform special services not necessarily in its commercial best interests. In 1973 their international role was defined. The policy assigned specific areas and countries to Air Canada and CP Air which they would serve once bilateral agreements had been satisfactorily concluded. This division was aimed at assisting the airlines in long-range planning for both passenger and cargo services.

The development of air policy continues, with a new focus on the question of third level or local air carriers, whose scope of operation is increasing rapidly.

Canada's position in aviation as well as its geographical location makes it imperative to co-operate with other nations engaged in international civil aviation. Canada therefore played a major part in the establishment of the International Civil Aviation Organization (ICAO) which has its headquarters in Montreal. By the end of 1977, Canada had bilateral agreements with 32 other countries.

Airports. Transport Canada owns 160 airports in Canada including such major international airports as Vancouver, Calgary, Edmonton, Winnipeg, Toronto, Montreal (2), Halifax and Gander. Of these, 90 are operated by the department and the rest by municipalities and other organizations. Municipal airports, served by scheduled air service, are eligible to receive an operating subsidy from the department, which also assists in the construction of smaller community airports through capital grants.

Air traffic control. The primary functions of air traffic control in Transport Canada are to prevent collisions between aircraft operating within controlled airspace and between aircraft and obstructions in the manoeuvering area of controlled airports; and to expedite and maintain a safe, orderly flow of air traffic. These functions are carried out by controllers in airport control towers, terminal control units and area control centres.

The airspace reservation co-ordination office provides reserved airspace for specified operations within controlled airspace and information to other pilots concerning these reservations and military activity areas in controlled and uncontrolled airspace. The office, in Ottawa, provides the service in all Canadian airspace and in the Gander oceanic control area.

Telecommunications and electronics. The Canadian Air Transportation Administration (CATA) also provides telecommunications electronics and flight service to other components of the department, to other departments and agencies and to civil aviation users in Canada. The branch prepares specifications, designs telecommunications and electronics systems, and procures electronic equipment and systems employed in civil aviation in Canada, and also maintains this highly complex electronic equipment.

Radio operators are employed at 116 air radio stations in Canada. They are responsible for pre-flight weather briefings, flight planning, monitoring of sophisticated aids to navigation, broadcast services, and airport advisory services to aircraft. Over a year, this activity involves approximately 900,000 flight plans, 1.5 million air-ground communications and more than 1.3 million landings and take-offs of aircraft at Canadian airports that have no control towers.

Airworthiness. To comply with a 1970 CATA policy decision, the airworthiness division of civil aeronautics is responsible for validating the airworthiness certification of all foreign and domestic manufactured aircraft and components prior to issuing a type approval or a certificate of airworthiness; and for supervising manufacturers and repair organizations for compliance to Canadian airworthiness standards.

15.5.2 Commercial air services

The Canadian flag carriers operating international and domestic air routes are Air Canada and CP Air, which together in 1976 earned 68% of the total operating revenues of Canadian commercial air carriers. The five regional carriers (Eastern Provincial Airways, Nordair, Quebecair, Pacific Western Airlines and Transair) earned 13% of the total operating revenues. The remaining 19% was earned by some 600 smaller airlines, many of them operating in areas of Canada which are relatively inaccessible by surface transport. On international routes, the Canadian flag carriers are authorized to provide scheduled services to Europe, the Soviet Union, Asia Minor, Japan and Hong Kong, Mexico and South America, Morocco, the Caribbean, Australia and the United States (including Hawaii); 37 foreign airlines have scheduled services between Canada and other countries.

The Canadian Transport Commission (air transport committee), in its directory of Canadian commercial air services, classifies commercial air carriers in two major groups, domestic and international.

Domestic air carriers, operating wholly within Canada, are divided into seven classes. Scheduled carriers provide public transportation of persons, goods or mail to designated points according to a service schedule, at a toll per unit. Regular specific point carriers, to the extent that facilities are available, provide public transportation to points according to a service pattern, at a toll per unit. Specific point carriers provide public transportation, serving points consistent with traffic requirements and operating conditions, at a toll per unit. Charter carriers offer public transportation from a base specified in the licence, at a toll per kilometre or per hour for the charter of the entire aircraft, or at such other tolls as may be permitted by the air transport committee. Contract carriers do not offer public transportation but carry persons or goods solely under contract. Flying clubs incorporated as non-profit organizations provide flying training and recreational flying. Specialty carriers operate for purposes not provided by other classes such as aerial photography and survey, aerial distribution (crop dusting, seeding), aerial inspection, reconnaissance and advertising, aerial control (fire control, fire-fighting, fog dispersal), aerial construction and air ambulance and mercy services.

International air carriers operate between points in Canada and points in any other country. They constitute two more classes of carrier. International scheduled carriers provide public transportation serving points according to a service schedule at a toll per unit. International carriers which are domestic and foreign air carriers operate specific point or contract commercial service.

15.5.2.1 Canada's international flag carriers

Air Canada, a Crown corporation incorporated in 1937 as Trans-Canada Air Lines, maintains passenger, mail and commodity services over a network extending to 59 destinations in Canada, the United States, Ireland, the British Isles, Europe, Bermuda and the Caribbean. Despite a substantial operating profit, the airline posted a net loss in

1976 due to an air transport industry shut-down in mid-year. The company's 40th year of service was otherwise marked by a period of financial progress, increased productivity and service improvements.

Operating revenues passed the billion dollar mark for the first time, increasing to $1.057 billion while operating expenses reached $1.018 billion, up 11%. Air Canada again carried more than 10 million passengers during the year.

At December 31, 1976 the airline's fleet consisted of 118 aircraft: six Boeing 747s, 10 Lockheed L-1011s, 35 DC-8s, 14 Boeing 727s and 53 DC-9s. Two L-1011s were leased for the peak summer traffic periods only.

Canadian Pacific Air Lines Ltd. (CP Air), a private airline, was established in 1942 by integrating 10 air carrier bushline companies and has since developed into a major international flag carrier. In 1976 CP Air carried 2.5 million revenue passengers. Operating revenues for the year reached $350 million.

CP Air's network radiates from the company's headquarters in Vancouver to Japan, Hong Kong, the Netherlands, Hawaii, Fiji, Australia, Portugal, Spain, Italy, Greece, Israel, Mexico, Peru, Chile and Argentina. There are regular West Coast flights between Vancouver, San Francisco and Los Angeles. Within Canada CP Air's transcontinental services link Vancouver, Edmonton, Calgary, Winnipeg, Toronto, Ottawa and Montreal; the company also operates interior services in British Columbia, Alberta and the Yukon Territory. CP Air operates 25 aircraft: four Boeing 747s, 12 Douglas DC-8s, seven Boeing 737s and two Boeing 727s. On order in 1978 were two Douglas DC-10s (with options for two more) and two Boeing 737s (with options for one more).

Regional airlines

15.5.2.2

Eastern Provincial Airways (1963) Ltd. is the regional carrier for the Atlantic provinces. In 1976 it carried 593,000 revenue passengers and 4 616 tonnes of freight. Operating revenues were $38.7 million, 8% higher than 1975 revenues of $35.9 million. Scheduled services were operated to Charlottetown, PEI; Moncton–Chatham–Charlo–Fredericton and Saint John, NB; Sydney and Halifax, NS; Deer Lake–Stephenville–Gander and St. John's, Nfld.; Goose Bay–Wabush (Labrador City) and Churchill Falls in Labrador; and Montreal and the Magdalen Islands in Quebec.

The company's fleet at the end of 1975 consisted of seven Boeing 737s, and two Hawker-Siddeley 748s.

Nordair Ltée-Ltd., with its head office in Montreal, was established in 1957 by the merger of Mont Laurier Aviation and Boreal Airways. Since its formation Nordair has expanded steadily; it operates scheduled services in Quebec, Ontario and the Northwest Territories, as well as extensive domestic and international charter flights throughout Canada and from Eastern Canada to the southern United States, the Caribbean and since 1977 Mexico.

Scheduled services operate between Montreal, Ottawa, Hamilton, Windsor and Pittsburgh. Other scheduled services are operated between Montreal, Val-d'Or, Fort George, Matagami, La Grande, Chibougamau, Great Whale River and Fort Chimo, Que. and Frobisher Bay and Resolute Bay, NWT. In September 1977 Nordair filed an application with the Canadian Transport Commission to provide jet service between Toronto, Sault Ste Marie, Thunder Bay, Dryden and Winnipeg. In 1976 Nordair carried more than 550,000 passengers and in 1977 more than 580,000.

Nordair's charter flights accommodate inclusive tour travel and group travel. Under contract with the United States Air Force, Nordair provides air services between the DEW-line sites along the Arctic Coast and, under contract with the federal supply and services department, operates ice reconnaissance services for the federal environment department. In 1978 Nordair's fleet was composed of one Douglas DC-8, six Boeing 737s, two Lockheed L-188s, and three Fairchild F-227s. A Boeing 737 was ordered for delivery in June 1979.

Pacific Western Airlines Ltd., with executive offices in Calgary, operates scheduled passenger and cargo services over 22 531 unduplicated route-kilometres in Western and northwestern Canada. Main line scheduled services operate from Vancouver to Victoria, Comox, Powell River, Campbell River, Port Hardy and Sandspit on the Pacific Coast. Service is also provided between Vancouver and Seattle. In the interior and northern region of BC, flights are scheduled from Vancouver to Penticton, Kelowna, Cranbrook, Castlegar and Kamloops through to Calgary and Edmonton, Alta.; in the northern interior, service is provided from Vancouver to Smithers, Dawson Creek, Williams Lake, Quesnel and Prince George. From Edmonton, scheduled flights operate to Cambridge Bay, Fort Simpson, Fort Smith, Hay River, Inuvik, Norman Wells, Resolute and Yellowknife in the Northwest Territories and to Fort Chipewyan, Fort McMurray, High Level and Peace River, Alta. and Uranium City, Sask. The only no-reservations AirBus service in Canada operates between Calgary and Edmonton. Pacific Western also operates extensive international passenger and cargo charters.

In 1976 there were 2.19 million passengers carried. The company's fleet of 24 aircraft includes one Boeing 727, 13 Boeing 737s, two Boeing 707s, three Convair 640s and three Lockheed Hercules freighters (cargo only).

Quebecair, with its head office at Montreal International Airport, Dorval, offers scheduled services in Quebec and Labrador. The company was founded in 1946 under the name Le Syndicat d'Aviation de Rimouski. The name was changed to Rimouski Airlines in 1947 and to Quebecair in 1953 when it was amalgamated with Gulf Aviation. During 1965 Quebecair acquired Northern Wings Ltd. and Northern Wings Helicopters, in 1967 A. Fecteau Transport Ltée and in 1974 Air Gaspé Inc.

Quebecair operates scheduled services and the subsidiaries handle flights by light aircraft, charter and contract services. Scheduled services are operated over 9 656 km serving a number of localities in nine economic regions of Quebec and Labrador. Points linked are Montreal, Quebec City, La Malbaie (Charlevoix), Baie-Comeau (Hauterive), Churchill Falls (Twin Falls), Gagnon, Wabush (Labrador City), Mingan, Mont-Joli–Rimouski, Saguenay (Bagotville), Schefferville, Sept-Îles, La Grande 2 (James Bay), Gaspé, Îles-de-la-Madeleine, Rouyn–Noranda, Bonaventure, Charlo, Roberval, Port Meunier, Ste-Anne-des-Monts, Senneterre, Val-d'Or, Chibougamau, Matagami, Blanc Sablon, Saint-Paul, Old Fort Bay, Saint-Augustin, La Tabatière, Tête à la Baleine, Harrington Harbour, Gethsemani, Kégaska, Natashquan, Aguanish, Baie Johan Beetz and Havre-Saint-Pierre. Quebecair also operates group charters within Canada and to the US, Caribbean, Bermuda, Mexico, South America and Europe using jet aircraft.

Revenue passengers transported by Quebecair in 1977 numbered 588,900 on scheduled services and 308,000 passengers on international charter services.

Quebecair's large and varied fleet of aircraft enables it to meet the diverse requirements of today's charter market. The combined fleet of Quebecair and its subsidiaries totals 63 units: two Boeing 707s, one Boeing 727, three BAC 1-11 jets, four F-27 turbo-props, four Douglas DC-3s, 11 DHC-3 Otters, three Cessna 180s, 14 DHC-6 Beavers, three Cessna 185s, three Beechcraft 99s, one Hawker-Siddeley 748 and 14 helicopters of various types.

Transair Ltd. This company was formed in November 1969 through the merger of Transair Ltd. and Midwest Airlines Ltd., both of Winnipeg. With headquarters at the Winnipeg International Airport, the company operates scheduled services in Manitoba, the NWT, the Yukon Territory and northwestern Ontario as far east as Toronto. Points linked are Gillam, The Pas, Flin Flon, Lynn Lake, Thompson, Churchill, Yellowknife, Whitehorse, Dryden, Thunder Bay, Sault Ste Marie and Toronto. Several other points in the NWT are also served by flights from Churchill. Midwest Airlines Ltd., a wholly owned subsidiary of Transair, operates in Manitoba's inland region. Charter flights operate from Canada to Florida, Hawaii, Mexico and the Carribbean.

The Transair fleet consists of three Boeing 737s, two Fokker F-28s, two YS-11s, 13 Bell Jet Ranger helicopters and one Bell 204 heavy lift helicopter. In 1975 Transair carried 409,000 passengers, compared with 484,000 passengers in 1974.

15.5.2.3 Commonwealth and foreign scheduled commercial air services

At the end of 1976 there were 37 foreign air carriers holding valid Canadian operating certificates and licences issued for international scheduled commercial air services into

Canada: Aeroflot (USSR), Aeronaves dé Mexico S.A., Air France, Air Jamaica (1968) Ltd., Alitalia-Linee Aeree Italiane, Allegheny Airlines Inc., American Airlines Inc., British Airways, British West Indian Airways, Caribbean Airlines, Czechoslovak Airlines, Delta Airlines Inc., Eastern Air Lines, El Al Israel Airlines Ltd., Empresa-Consolidada Cubana de Aviacion, Finnair, Frontier Airlines Inc., Hughes Air West (a division of Hughes Air Corporation), Iberia Air Lines of Spain, Irish International Airlines, Japan Air Lines Company Ltd., KLM Royal Dutch Airlines, Lot Polish Airlines, Lufthansa German Airlines, North Central Airlines Inc., Northwest Airlines Inc., Olympic Airways S.A., Qantas Airways Limited, Royal Air Maroc, Sabena Belgian World Airlines, Scandinavian Airlines System, Seaboard World Airlines Inc., Swissair, Transportes Aeroes Portugueses S.A.R.L., United Air Lines Inc., Western Air Lines Inc. and Wien Air Alaska Inc.

Civil aviation statistics   15.5.3

Airport activity. The upward trend in air traffic activity continued in 1976 but at a slower rate than in previous years. The 60 major airports operating during the year handled nearly 6.5 million aircraft landings and take-offs compared with nearly 4.9 million by 53 airports in 1971, reflecting a 32.5% increase. A decade ago, traffic handled by the 33 towers at that time amounted to 3.3 million movements. At the three national defence department airports where civilian passenger traffic is allowed, 115,744 aircraft movements were recorded. The 136 smaller airports without control tower facilities, which report daily traffic counts, registered 2,146,676 movements.

Toronto International airport continued to be the leader in itinerant activity with 235,998 movements, followed by Vancouver International with 211,102 and Montreal International (Dorval) with 157,711. The increase reported for Vancouver International was almost double the one reported by Toronto while the 30,149 decrease in the total for Montreal International (Dorval) reflected the rerouting of international traffic to Mirabel International airport.

Light aircraft weighing under 1 814 kilograms continued to account for the largest share of the itinerant activity in 1976. Heavy airline aircraft such as the Boeing 707 and 747, DC-8, DC-10 and the Tristar accounted for 150,764 movements. Piston engine aircraft contributed the major share (60.9%) of overall itinerant traffic. Jet aircraft accounted for 25.9% and other aircraft such as turbo-props, helicopters and gliders for the remainder.

There were 284,055 international movements recorded in 1976, an increase of 21,576 or a 8.2% gain over 1975. The international airports at Toronto, Montreal, Vancouver and Mirabel in that order, were responsible for 58.4% of the international total.

In 1976 Canada's busiest airport in terms of overall traffic was Edmonton Municipal, with a total of 279,867 movements, followed closely by Saint Hubert, Que. with 265,396. Both these satellite airports reported high "local" counts, characterized by light aircraft traffic largely of a training or recreational nature.

Commercial air services. Tables 15.30 and 15.31 provide statistics on commercial air services of Canadian airlines with gross annual flying revenues exceeding $150,000 and of scheduled foreign airlines. The data for Canadian airlines refer to both domestic and international operations. Figures for the scheduled foreign airlines pertain only to the hours and distance flown over Canadian territory, excluding passengers and goods in transit through Canada. Table 15.31 contains comparative data for domestic and international traffic in 1975 and 1976.

Urban transportation   15.6

Almost 60% of all transportation activity in Canada takes place in urban areas, where approximately 75% of the population lives. About 80% of all urban travel takes place in private automobiles. Growing adverse public reaction to further road building and increasing concern over energy, air pollution and congestion generated by private cars has led to a new emphasis on public transit, including buses, subways and streetcars.

Although prime responsibility for urban transportation lies with the provincial and municipal governments, the federal government has been reviewing its transportation policies in this area and has taken some initiatives in the urban transit field. Transport Canada has established an urban transportation research branch to develop and demonstrate improvements to traffic management and public transport.

The demand for adequate transport facilities in urban areas has placed a heavy financial burden on municipalities. Provincial cost-sharing programs which have assisted in meeting the capital and operating costs of transportation systems in urban areas have until recently been strongly oriented to freeways and roads. Several provinces are now shifting the emphasis from highway construction toward transit planning and construction.

Newfoundland does not have a current program related to urban transportation problems although considerable work and planning have been carried out to improve access to, from and across St. John's. The St. John's harbour arterial road was expected to be completed in 1979 and work on a cross-town arterial road, started in 1977, continued in 1978. The city is served by a bus system subsidized by the provincial government at $4 per capita of the city population. No federal financial assistance is received but discussions are ongoing at the federal level related to future urban transportation requirements.

Nova Scotia. In early 1978 the province was carrying out railway relocation studies in two major urban areas, Truro–Colchester and the city of Dartmouth. These studies, undertaken as tri-level ventures, were designed to introduce solutions to auto-rail conflict problems and core area development.

In June 1977, the municipal affairs minister announced that the Nova Scotia government would provide $12 million for urban transit over the next five years, including 50% of the capital cost of transit equipment, a per capita grant up to $3 toward shared operating deficits and 100% of the cost of approved demonstration projects and transit studies. The federal urban transportation assistance program announced in October 1977 would add $8.3 million to the program over the next five years for a total of $20.3 million, the most ambitious transportation program in the Atlantic region.

New Brunswick. The NB six cities public transit study, completed in 1976 at a cost of $150,000, reviewed existing systems and recommended plans for each city: Saint John, Moncton, Fredericton, Bathurst, Edmundston and Campbellton. Funding was by the federal regional economic expansion department and the province. Early in 1978 the province implemented a policy of grants to the municipalities for capital projects, where these projects also received federal funds under Transport Canada's urban transportation assistance program. Urban transit deficits in Saint John, Moncton and Fredericton were paid by the municipalities with help from the province's program of unconditional grants to municipalities.

Quebec is developing an integrated multi-mode urban transportation policy. A program of aid for public transit was introduced in 1975 to encourage an alternative to individual transportation. The Quebec government, through its transport department, would pay the full cost of studies on setting up or improving public transit systems and would subsidize 30% of acquisition or improvement costs to transit corporations or inter-municipal groups, for vehicles manufactured in Quebec. To obtain these on the best possible terms, the department consolidated bus purchases for transit commissions over five years and prepared a tender call for the purchase of 1,200 urban transport buses. It subsidized operating deficits of public transit systems at rates of 45% to 55% depending on the utilization rate of each system. In medium-density areas where the quality of service must be upgraded, municipal corporations could be set up. Where such a transit corporation took over a system, the Quebec transport department could pay up to 33% of takeover costs.

On this basis government subsidies, which totalled $85.6 million in 1976, exceeded $125 million in 1977, including $3 million for transportation of the handicapped. To increase the viability of urban transit, the government promoted the integration of school buses into the public transit system of more than 40 municipalities. Five transit commissions in Quebec received most of the government assistance including the Montreal urban community transit commission, the Quebec urban community transit commission, the Laval transit commission, the Outaouais regional transportation commission and the Montreal South Shore transit commission. Together they serve more than 3 million people. The Montreal transit authority received $67 million in government grants in 1976, $2.3 million of it for the purchase of buses and $30 million toward repayment of the debt incurred for the subway system.

All transit commissions have revised their routes and effected various improvements such as reserved bus lanes and high-speed routes (Quebec), off-road loading bays and express services (Outaouais) and métrobuses (Montreal). They began or continued a program of installing bus-passenger shelters and inaugurating reduced rates for senior citizens. All were also studying or implementing a monthly pass system and published route timetables to inform users of services available.

In addition to transport department studies regarding service to Mirabel and intermodal transit on the Montreal South Shore, two studies on points of origin and destination were carried out by the Montreal urban community and Laval transit authorities. The South Shore authority took part in a pilot study with the federal and provincial governments on sharing operating costs and deficits with various municipalities served by a transit organization. The Quebec urban community completed a study into standardizing services.

The department and representatives of the transit authorities of Montreal, Laval and the South Shore have worked together on the Montreal transportation committee. The recommendations made by this committee would serve as the basis for development of a transportation policy for the entire Montreal metropolitan region. This policy was aimed at co-ordinating all transit systems and grappling with the integration of plans for the Montreal region: the projected express system to serve Mirabel and other parts of the metropolis (TRAMM), train service linking the suburbs with the downtown core, now provided by both CN and CP, future extensions of the subway, and the transportation situation on the South Shore. Published in 1977, the report of the committee sets out various methods of retaining the diverse clientele and of encouraging others to use public transit rather than private cars.

Ontario. The provincial government has an urban transportation subsidy program that encourages the upgrading and use of public transit in cities and towns. The program is designed to make public transit more attractive and convenient, providing a balanced way to move people in a "people-oriented" society. Under the transit subsidy program, from 1971 to 1976 the province paid 50% of a municipality's operating deficit, 75% subsidy on capital expenditures and 75% for transit studies. In 1976, operating subsidies totalled $45.5 million and capital subsidies $23.5 million. In 1977 a new method of computing operating transit subsidies was adopted. It was based on population and financial targets and aimed at creating incentives toward more efficient transit systems. It would provide special assistance to major new transit services aimed at offsetting the usual low ridership during the first years of operation. By the end of 1977 approximately 60 municipalities were operating public transit systems.

Through the transportation and communications ministry, Ontario subsidizes and administers demonstration projects. The province provides 100% funding for these projects over a period of time. Then the municipality has the option of taking the project over under a normal subsidy. The province also backs new concepts that make public transit more attractive and efficient.

The Toronto Area Transit Operating Authority (TATOA) was created in 1974 to provide transit systems crossing regional boundaries of metropolitan Toronto and adjacent municipalities of Peel and York. In 1977, Halton and Hamilton–Wentworth, previously associate members, gained full membership status in TATOA as well as the regional municipality of Durham. While the regional governments retained full responsibility for transit within their boundaries, TATOA planned to improve inter-regional connections by co-ordinating facilities, equipment, personnel training, service schedules and fare structures.

TATOA became responsible for Ontario's commuter services, the GO Trains and the GO Bus service, both inter-regional operations. GO Buses were operated by Gray Coach lines and Travelways under contract to TATOA.

In 1976 GO Bus operations were altered to intrude less into the city core by routing to convenient suburban subway terminals and GO Transit rail heads. In the Newmarket corridor, GO services were realigned to provide service to the Toronto Transit Commission (TTC) Finch subway terminal. GO also assumed responsibility for local bus service between the Finch subway station and Richmond Hill. This was formerly supplied by TTC under contract to the local municipalities of Markham, Richmond Hill and Vaughan with significant operating deficits. The change brought about an estimated annual saving of $250,000 in transit operation in the area.

TATOA sought other opportunities to combine its service with local services and make better passenger connections between GO (inter-regional) and the TTC, Mississauga transit, Markham transit and other regional systems.

The original GO Train, operated for TATOA by Canadian National Railways, has been providing commuter service at capacity along Toronto's Lakeshore route. To provide extra capacity to satisfy rush hour demand, 80 specially designed double-decker coaches were ordered and were expected to begin service on this line in October 1978, each having 75% more seats than a GO coach. Use of the new coaches was expected to make equipment available on proposed lines servicing Richmond Hill in 1978 and Streetsville in 1980.

TATOA has been assigned the task of co-ordinating interior modifications of the Union Station and reconstruction of the Bathurst Street junction, opening the way for expanded commuter rail operations into downtown Toronto. It has initiated a pilot telephone information project to centralize public transit information in the Markham, Vaughan and Richmond Hill areas.

Manitoba. In late 1977 a feasibility study was completed at a cost of about $350,000 on development of an 11.3-kilometre transportation corridor along an existing railway right-of-way from the Winnipeg city centre to the University of Manitoba. The study recommended the establishment of a bus-way using the corridor.

The province provided about $800,000 for innovative transit programs in Winnipeg, including a dial-a-bus system in the southern part of the city, a downtown shuttle service (DASH) operating during business hours, suburban feeder services in several areas, a bus shelter design program, and a preferential signing and control system for buses at a number of major intersections. Also included for the first time was $225,000 for a Handi-Transit system for disabled persons. The system operates on regularly-established routes and also makes casual pickups of disabled persons whenever possible.

Through the highways department, the province provided direct operating grants of $8.31 million to Winnipeg, $292,955 to Brandon, $42,111 to Flin Flon and $15,000 to Thompson to help cover operating deficits. It also provided $6.13 million to Winnipeg and $148,000 to Brandon to assist in transit bus purchases.

Saskatchewan. Telebus offers door-to-door service within individual zones of Regina and door-to-connection with scheduled line service for trips between zones. Most of the vehicles are small, carrying 16-24 passengers, but during peak hours standard 42-passenger buses are added. About one-third of Regina receives full service and the entire area receives at least part-time service.

In 1978 the transit assistance for cities program continued to provide cities 50% of the cost of approved rolling stock, 75% of the cost of demonstration projects and studies, 75% of transit facility construction costs, and a subsidy of three cents a passenger. Ten cities participated in the 1977-78 fiscal year with expenditures of nearly $2.1 million.

A program of urban assistance for transportation of the handicapped, begun in 1975, provided 75% of cost of approved rolling stock; 50% of incurred operating deficit; 75% of transit system facility construction costs; 75% of costs for transit studies and demonstration projects. Six cities participated in the 1977-78 fiscal year with expenditures over $600,000.

Alberta committed approximately $400 million in urban transportation assistance to its cities between 1974 and 1984. The Alberta transportation department initiated a number of six-year programs in 1974, providing assistance to Alberta cities. Included were: research assistance totalling $9.6 million for studies on public transportation, inter-urban transportation and demonstration projects, among others; public transit capital assistance totalling nearly $97.56 million; deficit subsidies for operating public transit systems, expected to total approximately $22 million over six years; and railway relocation study assistance funds, added in 1975, equal to 50% of the federal government's contributions to approved railway relocation study projects.

A new assistance program initiated in 1977 and expected to continue until 1984 is providing, as the initial input of funds, a minimum of $160 million to assist Alberta cities in planning and constructing one major continuous roadway through each city. The first two projects under construction in Edmonton and Calgary, the first cities eligible, were scheduled to be completed by 1984.

Studies assisted by the research program include a major surface transportation noise attenuation study at a cost of over $400,000 involving evaluations of social, psychological and technological effects and ways to reduce roadway noise. Another major project assisted was an in-service evaluation of articulated buses by the cities of Edmonton and Calgary. This project, the first of its kind in North America, was expected to cost in excess of $300,000 over the two-year term of the tests. In all, some 30 projects, including most of Alberta's cities, were assisted.

British Columbia. The transit services division of the municipal affairs and housing ministry provides mobility for BC residents and a cost-effective and environmentally sound way of transporting people in larger urban areas, by providing bus and other transit services in conjunction with local governments, in areas not served by BC Hydro. It provides for future growth of travel in major metropolitan areas and aids in community development by planning and implementing various forms of advanced transit. These include the Burrard Inlet Ferries as well as proposed light rapid transit and commuter rail services. It aids in the development of government policy on public transportation by providing technical data and analysis on needs, ways of providing for needs and on general industry development.

The division emphasizes: the inter-relationship between transit and urban land use; use of an interdisciplinary team for administering transit service, including engineers, economists, town planners, management specialists, geographers and architects; and multi-modal approach to public transit, employing diesel buses, trolley buses, taxi services, light rapid transit, commuter rail and marine passenger services.

In 1977 the Burrard Inlet ferries began operation. This water-borne rapid transit system linked the Lonsdale corridor of North Vancouver with the downtown area of Vancouver. With vessels that resemble floating subway cars, it enabled people to make the 1.75 nautical mile crossing in 10 minutes, and used the first automated self-service fare system in Canada.

The Sea-Bus, with a capital cost of $36.9 million, was designed to provide an alternative to new bridge expenditures for handling peak-hour traffic. It can move the increasing peak transit loads at a lower operating cost than buses. In addition, having a major transit hub in the Lonsdale Quay is expected to help revitalize the lower end of Lonsdale. Sea-Buses dock at the Granville waterfront station on the Vancouver side, a transport facility based around the Canadian Pacific Railway station. As part of its Sea-Bus program, the ministry helped refurbish this historic building and construct a pedestrian connection from the station to Granville Plaza and Granville Mall.

The Sea-Buses are fully co-ordinated with the bus service. Where the ferry offered a travel time advantage, bus services to downtown Vancouver were re-routed to the North Vancouver terminal, with an annual saving of $1.6 million. Where buses had the advantage, they continued in use and some new bus routes were installed. A ridership target was a short term goal for evaluating the Sea-Buses' early performance: 7,500 patrons were expected for a typical weekday. In winter actual levels range from 8,000 to 10,000 and in summer 12,000 to 15,000, many of them tourists.

There was no major expansion of the metropolitan bus system operated by BC Hydro, but Richmond bus services were reorganized. Ongoing adjustments were made in the capital region by BC Hydro and by Vancouver Island Transit Ltd., resulting in improved service.

To establish small city transit services a municipality passes a transit enabling bylaw and petitions the province for service. Once approved, the transit services division staff work with local planners to design routes and set service standards. In situations where local government does not operate buses, a purchase-of-service contract is put to open tender and awarded to the lowest qualified bidder. This firm is then provided with buses from the provincial fleet under a lease that controls their use and maintenance. Planning, timetables, bus stop signs, publicity and marketing are provided by the transit services division in conjunction with the municipality. Thus each city has considerable local input, while buses, bus stop signs and timetables throughout the province have identical colours and designs.

Three major service expansions took place in 1977. The bus systems in Trail, Maple Ridge, Penticton and Kamloops were studied, service was improved and productivity increased.

Transit service opened in Kelowna in July 1977, with expansion in bus services in the built-up area of the city. A new bus system in Prince Rupert began operation in August 1977, with three new buses from the provincial fleet. Providing a uniform fleet of vehicles for operations around British Columbia permits economies of volume purchasing, ensures standardization of mechanical elements and offers a consistent quality of service to the public. Transit services division provides technical backup to the operators of small city systems and maintains an inventory of essential parts. Field checks assure that vehicles are well maintained. No new vehicles were ordered in 1977 by the ministry, but BC Hydro acquired 30 diesel buses. The fleet owned by both the ministry and BC Hydro in December 1977 totalled 1,308 units.

Sources

15.1.1 Transportation and Communications Division, Industry Statistics Branch, Statistics Canada.
15.1.2 Canadian Transport Commission, Public Affairs, Department of Transport.
15.2 Transportation and Communications Division, Industry Statistics Branch, Statistics Canada.
15.2.1 Canadian Transport Commission.
15.2.2 Transportation and Communications Division, Industry Statistics Branch, Statistics Canada.
15.3.1 Public Affairs, Department of Transport.
15.3.2 Supplied by respective provincial and territorial governments.
15.3.3 Transportation and Communications Division, Industry Statistics Branch, Statistics Canada.
15.4 - 15.4.1 Public Affairs, Department of Transport; Transportation and Communications Division, Industry Statistics Branch, Statistics Canada.
15.4.2 Public Affairs, Department of Transport; National Harbours Board.
15.4.3 The St. Lawrence Seaway Authority.
15.4.4 - 15.5.1 Public Affairs, Department of Transport.
15.5.2 Canadian Transport Commission and the respective airlines.
15.5.3 Transportation and Communications Division, Industry Statistics Branch, Statistics Canada.
15.6 Public Affairs, Department of Transport and the respective provincial governments.

Tables

15.1   Net subsidies paid for maintenance of coastal and inland shipping services, 1975-76 and contract subsidies 1976-77 (dollars)

Service	Net subsidy[1] 1975-76	Contract subsidy[2] 1976-77
WEST COAST		
Gold River—Zeballos, BC	107,315	107,000
Vancouver—northern BC ports	2,540,247	2,186,625
Vancouver Island—Ahousat freight service	67,100	73,200
Vancouver Island—Kyuquot freight service	75,500	83,846
Central BC coast—Coast Ferries Ltd.	—	195,598
EAST COAST		
St. Barbe, Nfld.—Blanc-Sablon, Que.	225,000	214,941
Burnside—St. Brendan's, Nfld.	64,956	82,475
Carmanville—Fogo Island, Nfld.	139,545	173,081
Cobb's Arm—Change Islands, Nfld.	58,000	86,100
Dalhousie, NB—Miguasha, Que.	5,036	—
Grand Manan and the mainland, NB	259,000	259,000
Greenspond—Badger's Quay, Nfld.	46,909	55,750
Montreal—Quebec—Rimouski—north shore ports, Que.	1,748,000	2,102,586
Grindstone (Îles-de-la-Madeleine), Que.—Souris, PEI	1,109,430	1,188,282
Pelee Island and the mainland, Ont.	122,958	142,265
Pictou, NS—Charlottetown, PEI—Grindstone, Que.	159,240	—
Portugal Cove—Bell Island, Nfld.	993,424	993,107
Prince Edward Island—Newfoundland	231,588	—
Wood Islands, PEI—Caribou, NS	2,407,728	3,601,951
Montreal, Que.—Corner Brook—St. John's, Nfld.	3,810,073	3,675,893
Little Bay Islands—St. Patrick, Nfld.	205,946	209,886
Total	14,376,995	15,431,586

[1]Net amount paid after recapture.
[2]Negotiated amount subject to recapture.

15.2   Railway track kilometres operated, 1900-76

First main[1] track Year	km	Area and type of track	1972	1973	1974	1975	1976
1900	28 416	First main					
1905	32 971	Newfoundland	1 519	1 519	1 519	1 519	1 493
1910	39 801	Prince Edward Island	409	409	409	409	409
1915	56 137	Nova Scotia	2 007	2 007	2 007	2 010	2 010
1920	62 451	New Brunswick	2 680	2 680	2 680	2 680	2 678
1925	64 937	Quebec	8 568	8 705	8 705	8 687	8 674
1930	67 668	Ontario	15 894	15 894	15 847	15 789	15 783
1935	69 067	Manitoba	7 633	7 633	7 633	7 398	7 397
1940	68 502	Saskatchewan	13 784	13 784	13 778	13 668	13 599
1945	68 159	Alberta	10 023	10 021	10 021	9 915	9 788
1950[2]	69 168	British Columbia	7 488	7 686	7 702	7 702	7 702
1955	69 916	Yukon Territory	93	93	93	93	93
1960	70 858	Northwest Territories	208	208	208	208	208
1965	69 454	United States	546	546	637	637	637
1970	70 784						
1971	71 057	Total, first main	70 852	71 185	71 239	70 715	70 471
1972	70 851						
1973	71 185	All other	25 769	25 772	25 719	25 917	25 848
1974	71 239						
1975	70 716						
1976	70 471	Total	96 621	96 957	96 958	96 632	96 319

[1]Defined as a single track extending the entire distance between terminals, upon which the length of the road is based.
[2]Newfoundland included from 1950.

15.3   Railway rolling-stock in service as at Dec. 31, 1975 and 1976

Type	1975	1976	Type	1975	1976
Locomotives	*3,977*	*4,008*	Freight cars	*193,197*	*193,401*
Steam	—	—	Automobile	2,776	3,541
Diesel-electric	3,963	3,994	Ballast	2,199	2,670
Electric	14	14	Box	92,669	88,644
			Flat	25,722	26,305
Passenger cars	*1,936*	*1,855*	Gondola	21,370	21,377
Turbo train			Hopper	29,287	31,801
Power unit cars	6	6	Ore	7,731	8,236
Coach	15	15	Refrigerator	5,016	4,898
Parlour	6	6	Stock	2,359	2,093
Self-propelled cars	117	116	Tank	379	325
Coach	713	737	Other	3,689	3,511
Combination	32	35			
Dining	87	80	Privately owned cars	*22,000*	*25,446*
Parlour	116	118	Tank	14,699	14,233
Sleeping	344	306	Other	7,301	11,213
Baggage, postal and express	495	431			
Other	5	5			

15.4   Commodities[1] hauled as revenue freight by railways, 1975 and 1976 (thousand tonnes)

Commodity	1975	1976	Commodity	1975	1976
LIVE ANIMALS	*218*	*139*	Gypsum	3 639	3 665
			Limestone	3 747	3 691
Cattle	206	130	Other crude non-metallic minerals	13 319	12 281
Other live animals	12	9	Waste materials	891	789
FOOD, FEED, BEVERAGES			FABRICATED MATERIALS,		
AND TOBACCO	*31 122*	*30 634*	INEDIBLE	*58 305*	*60 893*
Meat, fresh or frozen	287	247	Lumber	5 888	7 183
Other animal products	201	207	Other wood fabricated materials	2 011	2 184
Barley	5 537	5 926	Wood pulp and other pulp	4 464	5 096
Wheat	15 226	14 063	Newsprint	4 325	4 248
Other grains	2 254	2 404	Other paper and paperboard	2 790	3 083
Milled cereals and cereal products	1 803	1 880	Chemicals	4 925	5 614
Fruits and fruit preparations	536	627	Potash	7 300	7 864
Vegetables and vegetable preparations	1 195	1 254	Other fertilizers	2 101	1 797
Sugar	410	383	Petroleum and coal products	12 559	12 171
Other food and food preparations	946	794	Metals and primary metal products	6 317	5 637
Animal feed	2 264	2 424	Cement	1 731	1 836
Beverages	415	383	Other fabricated materials	3 894	4 180
Tobacco and tobacco products	48	42			
			END PRODUCTS, INEDIBLE	*9 719*	*10 024*
CRUDE MATERIALS, INEDIBLE	*117 349*	*125 310*	Road motor vehicles and parts	5 432r	6 149
Crude animal and vegetable materials	1 642	1 617	Other end products	4 287r	3 875
Pulpwood (logs and chips)	7 629	10 723			
Other crude wood materials	2 315	1 991	SPECIAL TYPES OF TRAFFIC	*7 923*	*10 789*
Textile fibres	85	87	Piggyback (trailer and containers)[2]	3 930	6 641
Iron ore	49 004	57 838	Freight forwarder	1 804	1 810
Nickel-copper ore	4 087	4 912	Other special traffic	2 189	2 338
Bauxite ore and alumina	2 380	1 139			
Other metallic ores	7 203	6 589	NON-CARLOAD SHIPMENTS[3]	*1 344*	*1 005*
Scrap metal, slags and drosses	2 140	1 995			
Coal	18 875	17 645	Total	225 980	238 794
Crude oil and bituminous substances	393	348			

[1]In this table duplications are eliminated, for example, freight that is interlined between two or more Canadian railways is counted only once. The statistics do not cover United States operations of Canadian railways except for the Canadian Pacific Railway line through Maine, US, and certain other short mileages which are deemed to be an integral part of the Canadian railway system. Sections of United States railways operating into Canada are regarded as Canadian railways and are included. Freight carried by the Cartier Railway is included in this table; however, financial data for this railway are not available for inclusion in the financial tables.
[2]Excludes traffic moved in railway-operated containers and trailers.
[3]Includes express-rated traffic.

15.5   Capital invested in railway road and equipment property, 1972-76 (thousand dollars)

Investment	1972	1973	1974	1975	1976
Road	176,137	203,405	284,628	359,926	345,736
Equipment	Cr. 53,720	30,571	77,321	174,650	117,932
General	4,405	8,426	5,789	Cr. 6,983	10,728
Undistributed[1]	Cr. 2,637	49,250	Cr. 13,755	29,890	14,879
CNR non-rail property	5,379	14,989	20,363	18,595	19,312
CPR " "	Cr. 16,189	27,003	Cr. 46,929	Cr. 1,459	Cr. 21,596
Other " "	8,172	7,258	12,811	12,754	17,163
Total	124,185	291,652	353,983	557,483	489,275
Cumulative investment to Dec. 31	8,585,977	{ 8,877,629 } { 8,848,751[2] }	9,202,734	9,760,217	10,249,492

[1]Credit entries in this table result when the annual "write-offs" are greater than the annual investment in any category.
[2]Revised to reflect restatement of data by two railways.

15.6 Capital structure of the Canadian National Railway System as at Dec. 31, 1972-76 (thousand dollars)

Year	Shareholders' capital		Funded debt held by public		Government loans and appropriations— active assets in public accounts	Total
	Government of Canada shareholders' account	Capital stock held by public	Guaranteed by federal and provincial governments	Other		
1972	2,023,540	4,345	809,532	2,024	1,082,453	3,921,894
1973	2,023,540	4,345	803,474	2,024	1,088,898	3,922,281
1974	2,162,550	4,345	596,229	2,024	1,292,574	4,057,722
1975	2,224,606	4,345	582,888	2,024	1,438,071	4,251,934
1976	1,496,789	4,345	657,055	2,024	1,405,142	3,565,355

15.7 Total operating revenues, operating expenses, net revenue, fixed charges and deficits of the Canadian National Railway System (Canadian and United States operations), 1972-76 (thousand dollars)

Year	Total operating revenue	Total operating expenses	Income before fixed charges	Total fixed charges	Net income or deficit	Cash deficit or surplus[1]
1972	1,334,047	1,293,081	76,476	90,177	Dr. 13,701	Dr. 17,822
1973	1,482,507	1,440,001	76,168	96,240	'' 20,072	'' 21,324
1974	1,817,106	1,782,456	86,990	124,016	'' 37,025	'' 37,733
1975	1,916,778	1,986,426	Dr. 22,046	138,535	'' 160,581	'' 16,368
1976	2,274,396	2,151,548	165,740	153,983	Cr. 11,757	Cr. 11,764

[1]Contributed by or paid to the Government of Canada.

15.8 Railway operating revenues and expenses (Canadian operations), 1971-76

Item and year	Total revenues $'000	Total expenses (before fixed charges) $'000	Ratio of expenses to revenues %	Per km of first main track			Freight revenue per freight-train km $
				Revenues $	Expenses $	Net revenues $	
All railways							
1971	1,805,661	1,698,206	92.07	25,521	24,002	1,519	15.26
1972	1,940,594	1,842,575	93.24	27,389	26,006	1,383	15.74
1973	2,122,988	2,032,984	93.84	29,823	28,559	1,264	17.46
1974	2,568,994	2,512,922	96.24	36,062	35,275	787	19.32
1975	2,733,811	2,801,967	101.31	38,659	39,623	Dr. 964	20.59
1976	3,192,485	3,075,928	94.67	45,302	43,648	1,654	23.81
CNR							
1971	923,012	888,808	96.26	24,475	23,568	907	14.81
1972	1,017,510	988,813	97.13	27,100	26,336	764	15.98
1973	1,118,767	1,096,184	97.98	29,796	29,195	601	18.27
1974	1,385,730	1,371,929	99.00	36,946	36,578	368	20.67
1975	1,462,646	1,555,308	106.34	39,201	41,685	Dr. 2,484	22.43
1976	1,739,033	1,682,953	96.78	46,798	45,289	1,509	22.64
CPR							
1971	659,912	613,921	89.45	25,475	23,700	1,775	14.80
1972	711,168	653,184	87.55	27,462	25,223	2,239	15.91
1973	772,859	711,316	87.57	29,853	27,476	2,377	16.17
1974	935,179	867,495	88.53	35,996	33,391	2,605	17.71
1975	1,007,306	940,306	90.20	39,192	36,585	2,607	21.79
1976	1,148,746	1,060,896	87.90	44,810	41,383	3,427	25.96

15.9   Road and street kilometres classified by type and by province as at Dec. 31, 1973 and 1974

Year, province or territory and jurisdiction	Surfaced				Earth	Total
	Rigid pavement	Flexible pavement	Gravel	Other		
1973						
FEDERAL AND PROVINCIAL JURISDICTION	*2 938*	*114 289*	*143 420*	*10*	*27 600*	*288 258*
Newfoundland	—	3 420	5 890	—	291	9 601
Prince Edward Island	19	2 906	1 403	—	956	5 285
Nova Scotia	21	8 636	16 588	10	19	25 273
New Brunswick	—	8 913	11 154	—	3	20 070
Quebec	576	27 433	28 495	—	8 130	64 634
Ontario	1 777	15 725	10 179	—	332	28 012
Manitoba	512	7 232	11 829	—	97	19 669
Saskatchewan	—	14 600	7 858	—	2 927	25 386
Alberta	3	9 920	21 792	—	6 394	38 109
British Columbia	31	15 403	22 365	—	8 385	46 185
Yukon Territory and Northwest Territories	—	101	5 866	—	66	6 033
MUNICIPAL JURISDICTION	*9 785*	*89 856*	*315 282*	*304*	*143 249*	*558 476*
Newfoundland	10	1 035	1 558	2	171	2 775
Prince Edward Island	18	140	19	—	2	179
Nova Scotia	126	1 535	303	—	2	1 965
New Brunswick	113	1 751	406	—	60	2 329
Quebec	3 436	19 748	16 377	135	3 454	43 150
Ontario	2 984	40 914	76 946	85	5 844	126 773
Manitoba	1 988	1 255	40 105	14	19 663	63 025
Saskatchewan	266	3 302	88 042	14	81 911	173 536
Alberta	594	9 284	87 183	13	31 876	128 950
British Columbia	253	10 815	4 149	40	259	15 516
Yukon Territory and Northwest Territories	—	76	195	—	10	280
1974						
FEDERAL AND PROVINCIAL JURISDICTION	*1 804*	*125 487*	*142 852*	*2*	*25 663*	*295 807*
Newfoundland	—	3 830	5 597	—	304	9 732
Prince Edward Island	19	3 030	1 313	—	950	5 312
Nova Scotia	10	9 310	16 948	—	19	26 287
New Brunswick	—	9 418	10 797	—	11	20 226
Quebec	629	30 054	29 692	2	8 066	68 444
Ontario	560	18 451	9 226	—	232	28 469
Manitoba	558	8 122	11 011	—	146	19 838
Saskatchewan	—	15 220	7 633	—	2 583	25 436
Alberta	3	11 074	21 913	—	4 381	37 371
British Columbia	24	16 859	22 626	—	8 884	48 393
Yukon Territory and Northwest Territories	—	117	6 095	—	87	6 299
MUNICIPAL JURISDICTION	*10 203*	*93 068*	*311 648*	*296*	*149 696*	*564 912*
Newfoundland	10	1 125	1 548	6	193	2 882
Prince Edward Island	18	151	6	—	8	183
Nova Scotia	126	1 553	293	6	2	1 979
New Brunswick	77	1 819	515	—	55	2 466
Quebec	2 662	18 707	14 671	122	3 180	39 342
Ontario	4 260	41 917	75 119	90	5 845	127 232
Manitoba	1 954	1 065	38 502	2	18 387	59 909
Saskatchewan	254	3 251	89 439	13	84 830	177 787
Alberta	594	12 237	87 309	19	36 926	137 086
British Columbia	249	11 175	4 081	37	261	15 804
Yukon Territory and Northwest Territories	—	68	164	—	10	241

15.10   Construction, maintenance and administration expenditure on roads and streets, by province, years ended Mar. 31, 1974 and 1975 (thousand dollars)

Item and province or territory	Construction		Maintenance and administration		Total expenditure	
	1974	1975	1974	1975	1974	1975
EXPENDITURE ON PROVINCIAL, FEDERAL AND OTHER UTILITY ROADS[1, 2]	*1,233,265*	*1,514,888*	*569,026*	*704,649*	*1,802,291*	*2,219,537*
Newfoundland	42,267	49,570	25,322	32,977	67,589	82,547
Prince Edward Island	8,227	6,974	9,225	9,983	17,452	16,957
Nova Scotia	48,768	55,546	40,831	41,702	89,599	97,248
New Brunswick	51,089	84,082	32,114	39,190	83,203	123,272
Quebec	449,097	539,467	174,079	203,254	623,176	742,721
Ontario	336,912	369,545	121,685	144,472	458,597	514,017
Manitoba	29,377	36,115	30,374	44,407	59,751	80,522
Saskatchewan	47,674	65,029	25,908	33,920	73,582	98,949
Alberta	81,236	157,453	43,138	53,579	124,374	211,032
British Columbia	99,632	126,241	51,410	85,909	151,042	212,150
Yukon Territory and Northwest Territories	38,987	24,867	14,940	15,257	53,927	40,124

15.10   Construction, maintenance and administration expenditure on roads and streets, by province, years ended Mar. 31, 1974 and 1975 (thousand dollars) (concluded)

Item and province or territory	Construction		Maintenance and administration		Total expenditure	
	1974	1975	1974	1975	1974	1975
EXPENDITURE ON MUNICIPAL ROADS[2,3]	*603,373*	*761,764*	*712,414*	*822,696*	*1,315,787*	*1,584,460*
Newfoundland	4,569	7,176	7,264	7,732	11,833	14,908
Prince Edward Island	355	747	977	1,262	1,332	2,009
Nova Scotia	8,569	13,292	11,106	12,805	19,675	26,097
New Brunswick	6,824	19,728	9,904	14,850	16,728	34,578
Quebec	139,340	163,230	179,733	203,403	319,073	366,633
Ontario	262,927	309,229	321,518	340,866	584,445	650,095
Manitoba	18,434	42,263	29,099	35,706	47,533	77,969
Saskatchewan	28,429	33,938	42,998	54,899	71,427	88,837
Alberta	66,522	99,598	65,057	97,031	131,579	196,629
British Columbia	66,279	69,587	43,650	52,498	109,929	122,085
Yukon Territory and Northwest Territories	1,125	2,976	1,108	1,644	2,233	4,620

[1]Includes small amounts paid by private companies and other organizations in connection with railway grade crossings and overpasses.
[2]Provincial and federal subsidies to municipalities amounted to $388 million in 1974 and $437 million in 1975 and should be added to provincial and federal expenditures and subtracted from municipal expenditures to arrive at net expenditures for the respective levels of government.
[3]Fiscal year for municipalities ends the previous Dec. 31.

15.11   Motor vehicles registered for road use, by province, 1971-75

Province or territory	1971[1]	1972	1973	1974	1975
Newfoundland	129,200	140,650	153,585	163,975	170,612
Prince Edward Island	42,691	45,430	49,141	53,332	55,459
Nova Scotia	310,383	304,028	325,871	346,392	345,198
New Brunswick	216,710	235,108	256,042	274,173	288,658
Quebec	2,279,722	2,370,405	2,556,260	2,799,352	2,702,272
Ontario	3,209,862	3,382,497	3,583,379	3,891,603	4,079,108
Manitoba	419,314	428,360	471,507	508,751	535,808
Saskatchewan	464,924	496,214	523,557	568,918	613,269
Alberta	813,395	864,397	933,673	1,035,562	1,073,020
British Columbia	1,115,028	1,191,953	1,281,917	1,333,277	1,554,081
Yukon Territory	11,796	11,232	10,663	13,620	13,723
Northwest Territories	9,111	11,158	12,845	13,048	11,435
Canada	9,022,136	9,481,432	10,158,440	11,002,003	11,442,643

[1]Includes a small number of snowmobiles.

15.12   Types of motor vehicles registered, by province, 1975 with totals for 1971-74

Year, province or territory	Passenger cars[1]	Trucks and buses[2]	Motorcycles and mopeds	Other[3]	Total
Newfoundland	127,300	35,800	2,867	4,645	170,612
Prince Edward Island	40,661	13,104	1,694	[4]	55,459
Nova Scotia	262,187	73,978	8,550	483	345,198
New Brunswick	218,919	55,614	9,656	4,469	288,658
Quebec	2,188,895	295,627	184,869	32,881	2,702,272
Ontario	3,404,000	597,806	77,302	[4]	4,079,108
Manitoba	395,098	131,001	9,314	395	535,808
Saskatchewan	348,855	245,645	9,465	9,304	613,269
Alberta	715,713	327,589	29,718	[4]	1,073,020
British Columbia[5]	1,156,964	370,958	26,159	[4]	1,554,081
Yukon Territory	6,912	4,885	483	1,443	13,723
Northwest Territories	4,803	5,797	797	38	11,435
Canada 1975	8,870,307	2,157,804	360,874	53,658	11,442,643
1974	8,472,224	2,027,565	321,167	181,047[6]	11,002,003
1973	7,866,084	1,843,307	287,820	161,229[6]	10,158,440
1972	7,407,275	1,681,977	248,501	143,679[6]	9,481,432
1971	6,967,247	1,556,557	198,867	299,465[6,7]	9,022,136[7]

[1]Includes taxis and rent-a-car.
[2]Includes other types of motor vehicles, in certain provinces or territories, while certain classes of trucks and/or buses have been included under passenger cars in 5 provinces.
[3]Includes ambulances, fire trucks and some government vehicles.
[4]Included in trucks and buses or in passenger cars.
[5]Estimated, due to new licence program.
[6]Figures not complete; farm tractors (where registered) included from 1971 to 1974.
[7]Includes some licensed snowmobiles for 1971.

15.13 Provincial revenue from the registration and operation of motor vehicles, by province, for the licence year 1975 (dollars)

Province or territory	Motor vehicle licences and fees[1]	Chauffeur and driver licences	Public service vehicle fees[2]	Motive fuel taxes	Other[3]	Total	Commission allowed gasoline agents[4]
Newfoundland	6,542,754	689,689	237,907	34,279,964	1,434,950	43,185,264	134,819
Prince Edward Island	2,093,808	139,388	88,904	8,219,308	323,591	10,864,999	84,260
Nova Scotia	17,282,697	1,146,256	431,411	56,038,671	1,823,452	76,722,487	387,231
New Brunswick	14,539,947	646,450	[5]	46,216,319	1,272,642	62,675,358	249,450
Quebec	130,550,185	9,861,779	8,220,630	419,771,055	7,821,418	576,225,067	2,161,712
Ontario	189,152,204	9,849,539	8,917,226	577,986,806	27,607,883	813,513,658	[6]
Manitoba	12,599,122	1,634,507	4,655,120	58,205,912	217,728	77,312,389	288,961
Saskatchewan	19,435,889	1,002,402	[5]	45,329,084	2,887,692	68,655,067	760,111
Alberta	34,592,239	1,984,752	250,386	82,429,202	4,552,784	123,809,363	990,491
British Columbia	46,043,434	1,972,877	769,425	170,910,080	137,789	219,833,605	1,431,378
Yukon Territory	786,859	43,381	214,771	3,384,029	491,206	4,920,246	—
Northwest Territories	419,102	30,846	114,754	2,626,736	165,631	3,357,069	—
Canada	474,038,240	29,001,866	23,900,534	1,505,397,166	48,736,766	2,081,074,572	6,488,413

[1]Includes passenger cars, motor trucks and buses, motorcycles, other motor vehicles, trailers and transfer of motor vehicle ownership.
[2]Includes passenger and freight.
[3]Includes gasoline or service station licences, garage licences, fines for infractions of motor vehicle act and other miscellaneous revenue.
[4]Deducted from gross tax collections to obtain motive fuel taxes.
[5]Included with motor vehicle licences and fees.
[6]Commission payments discontinued, effective May 1, 1972.

15.14 Motor vehicle and traffic regulations, 1977

Province or territory	Age required for operator's licence	Type of licence[1]	Licence renewal	Motor vehicle protection	Speed limits[2]
Newfoundland	16 (motorcycle) 17 (automobile or truck)	Classified[1]	Every three years on the licensee's birth date	Insurance compulsory (minimum of $75,000 public liability)	90km/h (day) 80km/h (night)
Prince Edward Island	16	Chauffeur Operator	Every three years at the end of the licensee's birth month	Insurance compulsory Judgement recovery fund and third party liability	88km/h (day) 80km/h (night)
Nova Scotia	16	Chauffeur Operator	Every three years at the end of the licensee's birth month	Unsatisfied judgement fund	80km/h
New Brunswick	16[3,4]	Operator	Every two years at the end of the licensee's birth month	Unsatisfied judgement fund	80km/h (unless otherwise posted)
Quebec	16 (driving course compulsory) 18 (5 months learner permit or driving course)	Classified[1]	Every two years; odd-numbered years for those born in odd-numbered years; and even years for those born in even years	Insurance compulsory	90km/h
Ontario	16	Classified[1]	Every three years on the licensee's birth date	Motor vehicle accident claims fund	100km/h (unless otherwise posted)
Manitoba	16[3]	Classified[1]	Annually at the end of the licensee's birth month	Insurance compulsory	100km/h (unless otherwise posted)
Saskatchewan	16[3]	Classified[1]	Annually at the end of the licensee's birth month	Insurance compulsory	80km/h (unless otherwise posted)
Alberta	16[3]	Chauffeur Operator	Every five years expiring on the licensee's birth date[5]	Motor vehicle accident claims fund	100km/h (day) 90km/h (night)
British Columbia	16	Operator Special tests for taxi and tractor trailer drivers	Every five years expiring on the licensee's birth date	Insurance compulsory	80km/h
Yukon Territory	16[3]	Chauffeur Operator	Every three years on the licensee's birth date[5,6]	Insurance compulsory	97 km/h (unless otherwise posted)
Northwest Territories	16 (operator) 18 (chauffeur)	Chauffeur Operator	Annually on March 31	Insurance compulsory	90km/h (unless otherwise posted)

[1]Classified driver licensing is part of a Canada-wide program to match the driver's skills, experience and responsibilities with the type of vehicle being driven for example, automobiles, motorcycles, buses or heavy trucks and tractor trailers. Several provinces have adopted this plan, discontinuing the chauffeur licence and issuing a driver's licence in a certain class. [2]Slower speeds are required in cities, towns and villages. [3]Age 18 required for certain types of vehicles. [4]Out-of-province driver must be 18 years old. [5]Annually, for applicants, 70 years of age and over, for certain classes of licence. [6]Annually, where a medical report is required.

15.15   Sales of motive fuels, by province, 1972-76 (thousand litres)

Province or territory	1972[r]	1973[r]	1974[r]	1975	1976
Gasoline					
Newfoundland	426 274	480 735	534 685	540 285	553 000
Prince Edward					
Island	136 453	153 947	160 291	167 394	169 190
Nova Scotia	899 843	983 942	1 043 161	1 087 903	1 109 306
New Brunswick	783 370	843 704	884 366	947 878	967 933
Quebec	6 871 852	7 446 381	7 556 244	8 176 727	8 233 337
Ontario	10 421 423	11 200 639	11 514 614	11 898 816	11 977 129
Manitoba	1 133 619	1 212 172	1 272 738	1 320 216	1 355 354
Saskatchewan	1 102 812	1 180 710	1 249 168	1 347 108	1 408 966
Alberta	2 318 206	2 528 560	2 740 597	2 938 584	3 167 612
British Columbia	2 780 355	3 087 186	3 306 360	3 316 892	3 413 862
Yukon Territory	36 377	40 852	43 491	47 688	46 582
Northwest Territories	34 667	32 838	33 557	34 553	35 359
Total, net sales	26 945 253	29 191 666	30 339 271	31 824 044	32 437 630
Sales exempt from tax or taxed at lesser rates	2 296 000	2 212 000	2 563 000	2 690 000	2 477 260
Total, gross sales	29 241 000	31 404 000	32 902 000	34 514 000	34 914 890
Diesel oil					
Total, net sales	2 722 932	3 293 907	3 868 779	4 058 578	4 400 418
Liquefied petroleum gases					
Total, net sales	..	..	32 674	31 601	32 477

15.16   Canadian intercity bus industry, 1975 and 1976

Year and item		Class 1 and 2 ($500,000 and over)	Class 3 ($100,000-499,999)	Class 4, 5 and unknown (under $100,000)	Total, all classes
1975					
Establishments reporting	No.	19	10	25	54
Total operating revenue	$'000	160,331	2,535	823	163,689
Number of employees (including working owners)	No.	4,953	144	62	5,159
Equipment operated					
Highway buses	"	1,235	54	33	1,322
Urban and suburban buses	"	373	15	9	397
School buses	"	65	18	3	86
Other equipment	"	1	7	6	14
Total, equipment	"	1,674	94	51	1,819
Fare passengers carried	'000	33,493	1,443	..	34,936
Total vehicle kilometres travelled	'000	173 172	4 056	..	177 227
1976					
Establishments reporting	No.	18	7	33	58
Total operating revenue	$'000	184,224	2,345	3,012	189,581
Number of employees (including working owners)	No.	5,280	118	159	5,557
Equipment operated					
Highway buses	"	1,396	55	80	1,531
Urban and suburban buses	"	429	—	15	444
School buses	"	62	15	26	103
Other equipment	"	1	6	3	10
Total, equipment	"	1,888	76	124	2,088
Fare passengers carried	'000	31,952	1,217	..	33,169
Total vehicle kilometres travelled	'000	177 305	4 886	..	182 191

15.17   The Canadian Urban Transit Industry, 1975 and 1976

Year and item		Class 1 and 2 ($500,000 and over)	Class 3 ($100,000-499,999)	Class 4, 5 and unknown (under $100,000)	Total, all classes
1975					
Establishments reporting	No.	32	23	9	64
Total operating revenue	$'000	506,264	12,916	1,375	520,555
Number of employees (including working owners)	No.	25,836	827	68	26,731
Equipment operated					
Highway buses	"	18	26	1	45
Urban and suburban buses	"	7,532	378	34	7,944
School buses	"	359	57	5	421
Other equipment	"	1,910	1	2	1,913
Total, equipment	"	9,819	462	42	10,323
Fare passengers carried	'000	1,118,032	27,804	. .	1,145,836
Total, vehicle kilometres	'000	493 375	18 937	. .	512 312
1976					
Establishments reporting	No.	39	22	14	75
Total operating revenue	$'000	653,179	13,807	2,213	669,199
Number of employees (including working owners)	No.	27,456	745	101	28,302
Equipment operated					
Highway buses	"	18	8	3	29
Urban and suburban buses	"	8,071	405	73	8,549
School buses	"	231	46	4	281
Other equipment	"	1,861	—	5	1,866
Total, equipment	"	10,181	459	85	10,725
Fare passengers carried	'000	1,195,279	23,674	. .	1,218,953
Total, vehicle kilometres	'000	547 315	15 833	. .	563 148

15.18   Commodities transported by motor carriers, by mass, 1975 (thousand tonnes)

Commodity	1975	Commodity	1975
LIVE ANIMALS	*1252*	Crude oil and bituminous substances	593
		Other crude non-metallic minerals	18 195
Cattle	714	Waste materials	1 099
Other live animals	538		
		FABRICATED MATERIALS, INEDIBLE	*42 630*
FOOD, FEED, BEVERAGES AND TOBACCO	*16 638*	Lumber and sawn timber	3 570
		Other wood fabricated materials	646
Meat, fresh or frozen	487	Wood pulp and other pulp	386
Other animal products	2 790	Newsprint	932
Grains	1 773	Other paper and paperboard	516
Milled cereals and cereal products	412	Chemicals and related products	3 854
Fruits and fruit preparations	312	Fertilizers and fertilizer materials	416
Vegetables and vegetable preparations	1 822	Petroleum and coal products	12 330
Sugar and sugar preparations	445	Metals and primary metal products	9 044
Other food and food preparations	4 220	Cement and concrete basic products	5 913
Animal feed	1 318	Other fabricated materials	5 023
Beverages	2 679		
Tobacco and tobacco products	380	END PRODUCTS, INEDIBLE	*10 755*
		Road motor vehicles and parts	2 625
CRUDE MATERIALS, INEDIBLE	*28 667*	Other end products	8 130
Crude animal and vegetable materials	723	CONTAINERS AND CLOSURES	*2 052*
Pulpwood	1 672		
Other crude wood materials	3 672	GENERAL FREIGHT	*3 455*
Textile fibres	18		
Iron ore	140	Total	105 449
Other metallic ores	2 532		
Coal	23		

1975 For-hire trucking survey (Catalogue 53-224) — a survey of the intercity movements of commodities by motor carriers with revenue of at least $100,000 annually.

15.19　Canadian for-hire trucking industry, excluding household-goods movers, by province, 1975 and 1976[1]

Item		Province or territory					
		Atlantic provinces		Quebec		Ontario	
		1975	1976P	1975	1976P	1975	1976P
Establishments reporting	No.	218	212	650	637	739	703
Total operating revenues	$'000	132,342	148,870	547,767	618,588	1,051,532	1,158,240
Total operating expenses	"	126,615	141,884	530,637	594,802	994,814	1,112,455
Net operating revenues	"	5,727	6,986	17,130	23,785	56,718	45,784
Average number of employees (including working owners)	No.	4,336	4,292	20,583	20,306	34,197	34,081
Equipment:							
Trucks	"	1,412	1,393	7,818	7,263	7,144	6,440
Tractors	"	1,145	1,205	7,381	7,680	12,881	12,959
Semi-trailers	"	2,270	2,505	11,413	12,981	25,342	26,372
Full-trailers	"	239	229	2,826	2,480	2,178	2,600
Other equipment	"	148	113	1,082	944	1,506	1,266
Total, equipment	"	5,214	5,445	30,520	31,348	49,051	49,637
		Manitoba[2]		Saskatchewan[2]		Alberta	
		1975	1976P	1975	1976P	1975	1976P
Establishments reporting	No.	101	98	84	84	369	369
Total operating revenues	$'000	156,345	185,876	28,588	35,526	322,615	354,980
Total operating expenses	"	148,108	177,644	26,936	34,136	307,417	336,686
Net operating revenues	"	8,237	8,231	1,652	1,390	15,198	18,294
Average number of employees (including working owners)	No.	4,781	4,710	933	973	7,981	8,167
Equipment:							
Trucks	"	1,057	920	234	203	1,915	1,538
Tractors	"	1,659	1,506	437	468	2,852	3,115
Semi-trailers	"	3,969	4,396	528	719	6,406	6,839
Full-trailers	"	134	56	32	27	378	473
Other equipment	"	663	207	43	27	278	429
Total, equipment	"	7,482	7,085	1,274	1,444	11,829	12,394
		British Columbia		Yukon Territory and Northwest Territories		Canada	
		1975	1976P	1975	1976P	1975	1976P
Establishments reporting	No.	407	356	10	8	2,578	2,467
Total operating revenues	$'000	329,085	356,712	4,516	5,027	2,572,790	2,863,822
Total operating expenses	"	320,075	344,642	4,600	4,610	2,459,201	2,746,863
Net operating revenues	"	9,010	12,069	(84)	416	113,589	116,959
Average number of employees (including working owners)	No.	9,939	9,762	101	87	82,851	82,378
Equipment:							
Trucks	"	2,941	2,430	39	31	22,560	20,218
Tractors	"	2,908	2,840	38	34	29,301	29,807
Semi-trailers	"	4,867	5,040	53	60	54,848	58,912
Full-trailers	"	398	468	9	1	6,194	6,334
Other equipment	"	544	592	2	2	4,266	3,580
Total, equipment	"	11,658	11,370	141	128	117,169	118,851

[1]Revenue classes 1, 2 and 3 only.
[2]Class I Saskatchewan domiciled carriers were grouped with Manitoba to meet confidentiality requirements.

15.20　Canadian for-hire trucking industry, excluding household-goods movers, by revenue class, 1975 and 1976

Year and item		Class 1 ($2,000,000 and over)	Class 2 ($500,000- 1,999,999)	Class 3 ($100,000- 499,999)	Total, all classes
1975					
Establishments reporting	No.	183	537	1,858	2,578
Total operating revenue	$'000,000	1,532	574	466	2,573
Equipment operated					
Straight trucks	No.	8,916	5,167	8,477	22,560
Truck tractors	"	16,693	7,298	5,310	29,301
Trailers (semi- and full-)	"	40,421	13,325	7,296	61,042
Other equipment	"	1,994	1,205	1,067	4,266
Total, equipment	"	68,024	26,995	22,150	117,169

15.20 Canadian for-hire trucking industry, excluding household-goods movers, by revenue class, 1975 and 1976 (concluded)

Year and item		Class 1 ($2,000,000 and over)	Class 2 ($500,000-1,999,999)	Class 3 ($100,000-499,999)	Total, all classes
1976P					
Establishments reporting	*No.*	197	546	1,724	2,467
Total operating revenue	*$'000,000*	1,767	628	467	2,863
Equipment operated					
Straight trucks	*No.*	8,329	4,983	6,906	20,218
Truck tractors	``	17,173	7,446	5,188	29,807
Trailers (semi- and full-)	``	44,726	13,632	6,888	65,246
Other equipment	``	1,530	1,278	772	3,580
Total, equipment	``	71,758	27,339	19,754	118,851

15.21 Canadian for-hire trucking industry[1], excluding household-goods movers, by major type of service, 1975 and 1976

Item and year		General freight	Bulk liquids	Dump (sand, gravel, snow)	Forest products	Other commodities	Total
1975							
Establishments operating	*No.*	1,074	207	356	292	649	2,578
Total operating revenue	*$'000*	1,629,596	192,228	126,756	89,599	534,611	2,572,790
Total operating expenses	``	1,560,151	175,938	119,952	87,678	515,481	2,459,201
Net operating revenue	``	69,445	16,290	6,804	1,921	19,130	113,589
Average number of employees (including working owners)	*No.*	56,366	4,949	3,575	2,820	15,141	82,851
Equipment operated							
Trucks	``	15,441	998	1,912	819	3,390	22,560
Tractors	``	18,158	1,953	923	1,099	7,168	29,301
Semi-trailers	``	38,447	3,625	897	1,168	10,711	54,848
Full-trailers Other equipment	`` }	6,806	261	633	603	2,157	10,460
Total, equipment	``	78,852	6,837	4,365	3,689	23,426	117,169
1976P							
Establishments operating	*No.*	1,021	210	274	295	667	2,467
Total operating revenue	*$'000*	1,795,740	156,481	106,203	123,726	681,670	2,863,822
Total operating expenses	``	1,723,326	146,155	102,083	119,296	656,000	2,746,863
Net operating revenue	``	72,413	10,326	4,120	4,429	25,669	116,959
Average number of employees (including working owners)	*No.*	55,727	3,665	2,887	3,230	16,869	82,378
Equipment operated							
Trucks	``	14,203	850	1,374	728	3,063	20,218
Tractors	``	18,221	1,626	987	1,300	7,673	29,807
Semi-trailers	``	40,949	1,998	901	1,550	13,514	58,912
Full-trailers Other equipment	`` }	6,016	637	617	599	2,045	9,914
Total, equipment	``	79,389	5,111	3,879	4,177	26,295	118,851

[1]Revenue classes 1, 2 and 3 only.

15.22 Vessels entered at Canadian ports, 1973-76

Year	In international seaborne shipping		In coastwise shipping		Total	
	Vessels	Net registered tons[1]	Vessels	Net registered tons[1]	Vessels	Net registered tons[1]
1973	23,436	121,466,828	58,456	90,480,407	81,892	211,947,235
1974	20,992	112,832,551	53,368	86,095,620	74,360	198,928,171
1975	20,225	115,591,697	46,867	83,731,925	67,092	199,323,622
1976	21,898	124,070,721	41,581	83,173,354	63,479	207,244,075

[1]The capacity of the spaces within the hull, and the enclosed spaces above the deck, available for cargo and passengers; excluding spaces used for the accommodation of officers and crew, navigation, propelling machinery and fuel. A registered ton is equivalent to 100 cu ft and it is expected that this internationally recognized measure, like the nautical mile and the knot, will continue in use for some considerable time.

15.23   Cargoes loaded and unloaded at principal Canadian ports from vessels in international seaborne and coastwise shipping, by province, 1976 with total for 1975 (tonnes)

Province and port	International		Coastwise		Total 1976	Total 1975
	Loaded	Unloaded	Loaded	Unloaded		
NEWFOUNDLAND	*1 270 607*	*865 713*	*1 466 733*	*2 873 718*	*6 476 771*	*11 329 097*
St. John's	13 494	11 480	104 147	719 852	848 973	908 193
Stephenville	498 052	12 740	140 069	257 693	908 554	1 037 451
Holyrood	1 443	345 655	228 468	157 422	732 988	638 532
Corner Brook	217 599	35 699	12 578	387 386	653 262	559 499
Come By Chance	169 525	95 450	493 037	9 992	768 004	5 760 351
Port aux Basques	1 833	511	61 113	400 879	464 336	571 725
PRINCE EDWARD ISLAND	*158 435*	*5 937*	*66 013*	*589 946*	*820 331*	*799 227*
Charlottetown	26 517	2 761	65 672	523 444	618 394	733 903
NOVA SCOTIA	*6 809 388*	*8 432 632*	*4 444 427*	*1 814 707*	*21 501 154*	*22 780 932*
Halifax	2 638 250	5 136 660	2 001 230	698 911	10 475 051	10 652 827
Port Hawkesbury	1 675 493	3 218 261	1 386 892	66 709	6 347 355	7 002 039
Hantsport	1 093 924	—	—	—	1 093 924	941 937
Sydney	620 062	46 257	104 178	738 653	1 509 150	2 308 286
Little Narrows	553 248	—	277 049	—	830 297	680 958
North Sydney	10 961	716	366 317	67 507	445 501	565 558
NEW BRUNSWICK	*2 977 986*	*6 214 774*	*1 919 501*	*782 990*	*11 895 251*	*11 882 984*
Saint John	2 047 787	6 043 984	1 905 792	91 588	10 089 010	9 843 568
Dalhousie	474 372	—	—	134 327	608 699	646 371
Newcastle	268 820	985	—	196 859	466 664	405 719
QUEBEC	*59 614 738*	*14 211 330*	*11 561 517*	*20 689 157*	*106 076 742*	*99 851 300*
Sept-Îles–Pointe-Noire	26 563 996	243 161	3 245 413	1 094 185	31 146 755	27 392 587
Montreal	4 086 607	3 501 687	3 971 713	4 006 241	15 566 248	16 903 517
Port-Cartier	17 814 565	1 614 001	4 691	2 253 207	21 686 464	15 991 752
Quebec	3 150 039	4 260 239	961 003	3 789 227	12 160 508	11 336 189
Baie-Comeau	3 262 126	1 100 491	45 750	2 588 199	6 996 566	6 653 310
Sorel	2 030 281	464 113	11 554	2 842 845	5 348 793	6 085 926
Port-Alfred	142 501	1 138 270	12 379	319 717	1 612 867	3 632 550
Trois-Rivières	976 186	380 916	33 465	1 642 009	3 032 576	2 577 697
Havre-Saint-Pierre	270 585	—	2 135 327	21 757	2 427 669	2 691 974
Contrecoeur	754 956	1 411 845	215 797	346 629	2 729 227	3 175 161
Chicoutimi	1 089	—	—	612 410	613 499	613 818
Forestville	—	—	592 102	32 262	624 364	544 545
ONTARIO	*9 888 363*	*22 469 757*	*23 305 279*	*16 042 060*	*71 705 459*	*73 084 929*
Thunder Bay	3 155 135	106 012	14 233 808	814 483	18 309 438	18 168 934
Hamilton	306 029	6 648 590	299 277	4 834 816	12 088 712	12 945 858
Sarnia	1 082 435	3 163 348	2 582 324	399 392	7 227 499	8 246 695
Sault Ste Marie	157 510	3 955 779	208 191	1 464 395	5 785 875	5 380 254
Toronto	180 511	1 007 153	175 401	1 447 314	2 810 379	2 709 909
Clarkson	262 517	113 985	283 013	1 781 614	2 441 129	2 871 599
Windsor–Walkerville	831 076	638 437	853 009	490 273	2 812 795	1 761 716
Port Colborne	878 048	103 630	121 581	635 139	1 738 398	1 984 802
Colborne	—	—	1 838 026	—	1 838 026	2 319 006
Picton	961 958	59 921	210 845	43 591	1 276 315	957 292
Goderich	684 517	10 945	689 394	185 054	1 569 910	1 658 353
Lakeview (Port Credit)	—	1 755 241	283 702	3 765	2 042 708	1 726 839
Depot Harbour	647 283	—	—	—	647 283	666 607
Little Current	412 015	3 774	—	99 742	515 531	503 495
Midland	—	—	28 678	765 700	794 378	958 784
Prescott	—	98 265	83 910	330 184	512 359	490 852
Nanticoke	—	3 269 412	—	105 566	3 374 978	3 575 337
Badgeley Island	70 257	—	374 235	—	444 492	415 842
Parry Sound	—	26 722	1 471	443 637	471 830	504 708
MANITOBA	*719 700*	*28 630*	*20 091*		*768 421*	*638 525*
Churchill	719 700	28 630	20 091	—	768 421	638 525
BRITISH COLUMBIA	*33 375 921*	*4 240 279*	*11 097 876*	*11 003 863*	*59 717 939*	*54 500 965*
Vancouver[1]	23 916 215	2 687 690	2 307 088	3 561 410	32 472 403	32 224 399
New Westminster	1 024 105	491 295	154 040	928 737	2 598 177	2 250 877
Nanaimo	871 630	13 088	77 773	474 274	1 436 765	1 151 959
Duncan Bay–Campbell River	416 160	103 892	141 700	718 189	1 379 941	1 046 568
Britannia Beach	28 444	—	732 155	118 434	879 033	826 099
Victoria	732 764	209 698	1 634 833	344 723	2 922 018	1 793 132
Powell River	239 110	58 885	325 209	485 988	1 109 192	1 100 386
Crofton	882 482	10 950	16 166	280 495	1 190 093	874 985
Kitimat	354 320	525 241	152 060	32 503	1 064 124	844 671
Prince Rupert	752 441	11 204	92 155	111 925	967 725	962 255
Port Alberni	641 084	22 964	13 550	195 348	872 946	893 142
Tasu	903 228	—	—	285	903 513	962 770
Andys Bay	—	—	5 661	549 668	555 329	415 784
Squamish	322 434	23 297	48 990	89 966	484 687	209 679
Vanguard	—	—	201 694	511 596	713 290	472 746
Gold River	142 733	16 250	38 555	159 145	356 683	262 803
Sooke	416 236	—	40 470	13 395	470 101	315 185
Howe Sound	—	10 886	69 763	558 446	639 095	469 420
NORTHWEST TERRITORIES	—	*5 987*	*730*	*85 725*	*92 442*	*97 862*
Total	114 815 138	56 475 039	53 882 166	53 882 166	279 054 508	274 965 821

[1]Includes Roberts Bank.

15.24 Principal commodities in water-borne cargo loaded and unloaded at ports handling large tonnage in 1976 (tonnes)

Port and commodity	International		Coastwise		Total
	Loaded	Unloaded	Loaded	Unloaded	
VANCOUVER[1]	23 916 215	2 687 690	2 307 088	3 561 410	32 472 403
Coal, bituminous	9 249 353	–	87 577	–	9 336 930
Wheat	3 415 304	–	2 177	–	3 417 481
Sand and gravel	–	759 518	15 334	1 863 289	2 638 141
Sulphur in ores	2 125 235	–	8 647	–	2 133 882
Lumber and timber	1 912 844	18 456	75 789	129 465	2 136 554
Potash	1 475 944	–	–	–	1 475 944
Logs	115 819	18 144	17 345	972 414	1 123 722
Barley	1 760 565	–	–	–	1 760 565
Fuel oil	114 485	37 980	847 548	86	1 000 099
Pulpwood	150 476	–	502 326	18 135	670 937
Rapeseed	547 149	–	–	–	547 149
Containerized freight	412 196	418 043	–	–	830 239
Salt	–	285 501	42 393	–	327 894
Gasoline	21 388	–	331 465	–	352 853
Pulp	554 685	–	7 117	107 653	669 455
Cement	51 376	31 421	8 126	94 891	185 814
Newsprint	14 559	–	635	223 185	238 379
Inorganic chemicals	8 396	30 210	84 046	290	122 942
Limestone	82 463	–	3 402	123 813	209 678
Flaxseed	92 379	–	–	–	92 379
Asbestos	57 181	–	–	–	57 181
Waste and scrap	1 594	24	61 583	7 575	70 776
Building paper and board	2 866	112	25 383	3 655	32 016
Petroleum and coal products	234 809	–	454	12	235 275
Concentrated and complete feeds	182 086	–	82	–	182 168
Phosphate rock	–	575 494	–	–	575 494
Other commodities not listed	1 333 065	512 789	185 659	16 948	2 048 461
SEPT-ÎLES–POINTE-NOIRE	26 563 996	243 161	3 245 413	1 094 185	31 146 755
Iron ore and concentrates	26 560 440	–	3 233 686	–	29 794 126
Fuel oil	–	–	2 372	900 102	902 474
Bentonite	–	196 865	–	–	196 865
Other commodities not listed	3 556	46 296	9 355	194 083	253 290
PORT-CARTIER	17 814 565	1 614 001	4 691	2 253 207	21 686 464
Iron ore and concentrates	14 321 246	–	–	–	14 321 246
Wheat	1 913 144	392 789	–	1 606 335	3 912 268
Corn	1 142 124	1 058 104	–	43 345	2 243 573
Barley	204 498	21 745	–	221 293	447 536
Fuel oil	–	–	–	266 386	266 386
Other commodities not listed	233 554	141 362	4 691	115 849	495 456
THUNDER BAY	3 155 135	106 012	14 233 808	814 483	18 309 438
Wheat	115 981	–	8 400 021	–	8 516 002
Iron ore and concentrates	1 690 690	–	2 472 365	–	4 163 055
Barley	206 369	–	2 560 528	–	2 766 897
Oats	365 753	–	285 132	–	650 885
Rapeseed	59 757	–	–	–	59 757
Flaxseed	100 860	–	38 761	–	139 621
Fuel oil	–	–	–	256 264	256 264
Gasoline	–	–	–	158 941	158 941
Hulls, screenings and chaff	–	–	49 542	–	49 542
Newsprint	159 987	–	–	–	159 987
Rye	72 166	–	56 995	–	129 161
Concentrated and complete feeds	237 703	–	–	–	237 703
Other commodities not listed	145 868	106 012	370 464	399 278	1 021 622
MONTREAL	4 086 607	3 501 687	3 971 713	4 006 241	15 566 248
Fuel oil	715 225	652 653	2 634 009	412 759	4 414 646
Wheat	1 112 201	51 018	–	1 809 565	2 972 784
Containerized freight	942 264	702 773	–	–	1 645 037
Gasoline	18 448	–	819 901	177 345	1 015 694
Crude petroleum	–	–	–	81 606	81 606
Salt	4 672	318 879	25	281 613	605 189
Cement	46 608	–	9 132	6	55 746
Gypsum	–	–	–	435 693	435 693
Raw sugar	–	383 995	–	–	383 995
Barley	65 415	15 135	–	303 899	384 449
Corn	74 689	197 916	11 548	37 804	321 957
Organic chemicals	123 676	20 291	4 438	1 758	150 163
Coal, bituminous	–	198 528	19 836	19 770	238 134
Petroleum and coal products	3 727	15 527	16 950	269 949	306 153
Structural shapes	14 498	77 627	31 370	90	123 585
Lubricating oil and grease	941	36	105 056	142	106 175
Machinery	25 856	22 477	30 486	4 446	83 265
Manganese ore	–	183 154	–	–	183 154
Plate and sheet steel	44 808	35 535	2 789	102	83 234
Molasses, crude	–	98 059	–	–	98 059
Bars and rods, steel	25 362	36 452	4 296	–	66 110
Other commodities not listed	868 220	491 632	281 879	169 693	1 811 424
QUEBEC	3 150 039	4 260 239	961 003	3 789 227	12 160 508
Crude petroleum	–	3 553 566	58 033	127 713	3 739 312
Fuel oil	228 615	–	664 274	671 278	1 564 167
Wheat	1 255 868	8 437	–	1 319 990	2 584 295
Barley	155 344	–	–	336 119	491 463
Containerized freight	361 108	275 255	–	–	636 363
Gasoline	17 755	–	102 657	466 958	587 370
Zinc ore and concentrates	538 765	–	–	–	538 765

15.24 Principal commodities in water-borne cargo loaded and unloaded at ports handling large tonnage in 1976 (tonnes) (continued)

Port and commodity	International		Coastwise		Total
	Loaded	Unloaded	Loaded	Unloaded	
QUEBEC (concluded)					
Pulpwood	—	—	—	592 101	592 101
Corn	76 097	172 611	1 493	72 051	322 252
Alumina and bauxite ores	—	49 841	33 432	—	83 273
Newsprint	176 801	—	—	—	176 801
Salt	—	78 649	—	121 742	200 391
Asbestos	5 248	—	—	—	5 248
Other commodities not listed	334 439	121 880	101 113	81 276	638 708
HAMILTON	*306 029*	*6 648 590*	*299 277*	*4 834 816*	*12 088 712*
Iron ore and concentrates	—	1 736 028	—	4 237 738	5 973 766
Coal, bituminous	—	4 396 621	—	277 809	4 674 430
Fuel oil	—	—	509	155 348	155 857
Plate and sheet steel	205 594	1 126	159 197	19 269	385 186
Sand and gravel	—	145 543	—	64	145 607
Soybeans	—	127 484	—	—	127 484
Gasoline	—	—	—	13 764	13 764
Other commodities not listed	100 436	241 789	139 570	130 824	612 619
HALIFAX	*2 638 250*	*5 136 660*	*2 001 230*	*698 911*	*10 475 051*
Crude petroleum	—	4 495 445	—	—	4 495 445
Gypsum	1 441 757	—	136 912	—	1 578 669
Fuel oil	57 097	—	1 308 859	429 373	1 795 329
Containerized freight	504 806	528 357	—	—	1 033 163
Gasoline	102 301	—	538 283	141 089	781 673
Wheat	269 771	—	—	56 791	326 562
Other commodities not listed	262 517	112 857	17 176	71 659	464 209
SAINT JOHN	*2 047 787*	*6 043 843*	*1 905 792*	*91 588*	*10 089 010*
Crude petroleum	—	5 266 211	—	—	5 266 211
Fuel oil	57 270	15 537	1 403 298	44 092	1 520 197
Gasoline	82 283	—	455 642	47 496	585 421
Wheat	439 034	—	—	—	439 034
Containerized freight	550 912	381 277	—	—	932 189
Raw sugar	—	220 616	—	—	220 616
Pulp	315 913	499	—	—	316 412
Wheat flour	128 837	—	—	—	128 837
Newsprint	143 105	1 361	—	—	144 466
Other commodities not listed	330 434	158 343	46 852	—	535 629
SARNIA	*1 082 435*	*3 163 348*	*2 582 324*	*399 392*	*7 227 499*
Coal, bituminous	—	1 821 703	—	77 670	1 899 373
Fuel oil	916 063	—	1 302 200	63 180	2 281 443
Gasoline	81 364	—	861 473	6 047	948 884
Limestone	—	1 232 123	—	—	1 232 123
Wheat	—	—	87 837	96 653	184 490
Petroleum coal products	—	23 133	25 350	10 638	59 121
Inorganic chemicals	5 719	—	183 548	—	189 267
Lubricating oil and grease	550	—	18 397	92 666	111 613
Organic chemicals	60 897	—	1 758	3 221	65 876
Other commodities not listed	17 842	86 389	101 762	49 318	255 311
BAIE-COMEAU	*3 262 126*	*1 100 491*	*45 750*	*2 588 199*	*6 996 566*
Wheat	1 049 571	210 236	—	961 845	2 221 652
Barley	1 306 906	—	—	1 244 577	2 551 483
Newsprint	289 753	—	9 754	—	299 507
Alumina and bauxite ores	—	270 328	—	—	270 328
Fuel oil	—	—	—	162 979	162 979
Corn	371 418	256 905	—	168 724	797 047
Soybeans	222 973	220 772	—	—	443 745
Aluminum	19 702	—	34 473	18	54 193
Other commodities not listed	1 803	142 251	1 523	50 056	195 633
PORT HAWKESBURY	*1 675 493*	*3 218 261*	*1 386 892*	*66 709*	*6 347 355*
Crude petroleum	930 504	3 212 308	—	—	4 142 812
Fuel oil	84 550	—	1 110 877	61 228	1 256 655
Gypsum	564 475	—	—	—	564 475
Petroleum and coal products	—	—	77 972	5 481	83 453
Gasoline	462	—	198 044	—	198 506
Other commodities not listed	95 502	5 953	—	—	101 455
SAULT STE MARIE	*157 510*	*3 955 779*	*208 191*	*1 464 395*	*5 785 875*
Coal, bituminous	—	2 273 527	—	—	2 273 527
Iron ore and concentrates	—	1 020 566	—	1 080 045	2 100 611
Limestone	—	659 465	—	—	659 465
Fuel oil	—	—	—	221 892	221 892
Plate and sheet steel	9 678	—	111 662	—	121 340
Gasoline	—	—	—	128 338	128 338
Other commodities not listed	147 832	2 221	96 529	34 120	280 702
SOREL	*2 030 281*	*464 113*	*11 554*	*2 842 845*	*5 348 793*
Titanium ore	—	—	—	2 134 600	2 134 600
Wheat	670 023	71 138	—	586 598	1 327 759
Slag, drosses, byproducts	717 410	—	—	—	717 410
Pig iron	431 207	—	7 649	—	438 856
Coal	—	220 914	—	—	220 914
Barley	87 911	23 101	—	102 975	213 987
Corn	84 600	84 063	—	4 824	173 487
Fuel oil	—	—	—	8 123	8 123
Other commodities not listed	39 131	64 897	3 905	5 724	113 657

15.24 Principal commodities in water-borne cargo loaded and unloaded at ports handling large tonnage in 1976 (tonnes) (concluded)

Port and commodity	International		Coastwise		Total
	Loaded	Unloaded	Loaded	Unloaded	
NANTICOKE	—	*3 269 412*	—	*105 566*	*3 374 978*
Coal, bituminous	—	3 269 412	—	105 566	3 374 978
Other commodities not listed	—	—	—	—	—
TROIS-RIVIÈRES	*976 186*	*380 916*	*33 465*	*1 642 009*	*3 032 576*
Wheat	540 254	23 658	24 628	543 415	1 131 955
Fuel oil	—	27 177	5 046	598 892	631 115
Barley	74 871	—	1 642	99 904	176 417
Pulpwood	—	—	—	205 002	205 002
Newsprint	184 306	—	—	—	184 306
Salt	—	11 793	—	76 565	88 358
Other commodities not listed	176 755	318 288	2 149	118 232	615 424
VICTORIA	*732 764*	*209 698*	*1 634 833*	*344 723*	*2 922 018*
Wheat	169 659	—	—	2 177	171 836
Logs	6 845	70 322	40 969	6 840	124 976
Sand and gravel	—	31 026	1 447 092	2 790	1 480 908
Lumber and timber	405 110	3 904	72 709	3 719	485 442
Cement	82 287	21 319	—	33 935	137 541
Gasoline	—	—	—	95 657	95 657
Fuel oil	—	32 261	155	139 888	172 304
Other commodities not listed	68 864	50 867	73 908	59 716	253 355
WINDSOR	*831 076*	*638 437*	*853 009*	*490 273*	*2 812 795*
Salt	727 001	290	726 125	—	1 453 416
Gasoline	—	—	—	192 847	192 847
Fuel oil	23 847	17 794	6 643	132 262	180 546
Plate and sheet steel	911	3 378	120	90 834	95 243
Limestone	—	381 519	—	—	381 519
Other commodities not listed	79 318	235 455	120 121	74 331	509 225
TORONTO	*180 511*	*1 007 153*	*175 401*	*1 447 314*	*2 810 379*
Fuel oil	—	—	123 888	133 147	257 035
Coal, bituminous	—	44 662	—	—	44 662
Cement	1	—	—	418 132	418 133
Wheat	—	—	24 603	284 832	309 435
Salt	—	104 115	—	203 611	307 726
Raw sugar	—	182 812	—	—	182 812
Soybeans	—	319 657	—	11 665	331 322
Barley	—	—	—	85 682	85 682
Other commodities not listed	180 510	355 908	26 911	310 244	873 573
NEW WESTMINSTER	*1 024 105*	*491 295*	*154 040*	*928 737*	*2 598 177*
Sand and gravel	—	75 070	9 072	361 858	446 000
Logs	8 571	—	6 895	288 212	303 678
Pulpwood	130 933	—	95 494	2 716	229 143
Lumber and timber	177 551	23	10 795	6 622	194 991
Cement	—	—	11 045	172 720	183 765
Pulp	291 023	3 017	1 361	6 278	301 679
Other commodities not listed	416 028	413 186	19 378	90 331	938 923

[1]Includes Roberts Bank.

15.25 Vessels and tonnage handled by harbours administered by the National Harbours Board[1], 1976 and 1977

Port or elevator	Year	Vessel arrivals		Cargo handled *tonnes*	Grain elevator shipments *bu[3]*
		No.	Gross registered *tons[2]*		
St. John's, Nfld.	1976	1,279	2,972,000	934 627	—
	1977	1,274	2,786,000	924 382	—
Halifax	1976	1,802	14,781,980	10 944 512	17,262,141
	1977	2,109	17,767,000	14 027 017	19,054,842
Saint John	1976	1,807	16,830,034	5 498 659	16,287,563
	1977	1,848	21,310,465	8 396 388	18,622,603
Belledune, NB	1976	30	306,987	265 804	—
	1977	30	269,485	232 034	—
Sept-Îles	1976	1,484	22,311,000	31 957 349	—
	1977	1,514	22,021,000	32 447 271	—
Chicoutimi	1976	130	505,034	609 459	—
	1977	126	500,403	641 120	—
Baie-des-Ha! Ha!	1976	287	1,638,265	1 688 026	—
	1977	393	3,584,816	4 241 185	—
Quebec	1976	1,429	11,934,000	12 749 341	80,024,677
	1977	1,535	13,922,000	15 316 278	113,815,033
Trois-Rivières	1976	881	4,845,000	3 223 212	34,847,962
	1977	507	3,605,000	2 663 254	29,467,202

15.25   Vessels and tonnage handled by harbours administered by the National Harbours Board[1], 1976 and 1977 (concluded)

Port or elevator	Year	Vessel arrivals		Cargo handled *tonnes*	Grain elevator shipments *bu[3]*
		No.	Gross registered *tons[2]*		
Montreal	1976	4,041	33,702,748	19 188 337	109,311,829
	1977	4,363	39,367,464	19 808 891	117,667,441
Prescott	1976	76	594,000	547 051	12,577,278
	1977	99	793,000	512 065	11,874,095
Port Colborne	1976	20	126,969	146 368	4,780,459
	1977	35	232,270	295 644	11,621,793
Churchill	1976	50	564,350	768 274	30,026,013
	1977	61	606,793	815 951	28,801,593
Vancouver	1976	22,300	81,037,879	37 078 365	229,836,347
	1977	22,746	87,720,947	42 736 810	266,590,748
Prince Rupert	1976	1,434	3,704,000	1 425 684	18,392,483
	1977	1,406	3,053,000	1 137 783	19,261,304
Total	1976	37,050	195,854,246	127 025 068	553,346,752
	1977	38,046	217,539,643	144 196 075	636,776,654

[1]National Harbours Board data may differ in some instances from data in Tables 15.23 and 15.24, due to some differences in physical definitions of the ports, and to the use in some cases of different source documents.
[2]The capacity of the spaces within the hull, and the enclosed spaces above the deck, available for cargo and passengers; including spaces used for the accommodation of officers and crew, navigation, propelling machinery and fuel. A registered ton is equivalent to 100 cu ft and it is expected that this internationally recognized measure, like the nautical mile and the knot, will continue in use for some considerable time.
[3]Metric conversion will be available for 1978.

15.26   Summary statistics of St. Lawrence Seaway traffic[1], 1975 and 1976

Item	Upbound				Downbound			
	1975r		1976		1975r		1976	
	No. of transits	Cargo *t*	No. of transits	Cargo *t*	No. of transits	Cargo *t*	No. of transits	Cargo *t*
Type of vessel								
Ocean								
Cargo	805	3 672 303	876	5 036 933	820	7 383 860	883	7 790 140
Tanker	31	218 067	37	216 988	34	214 967	39	281 080
Inland								
Cargo	1,947	15 740 426	1,947	21 232 592	1,932	29 422 510	1,921	27 188 668
Tug and barge	80	515 951	75	732 735	77	170 551	72	50 324
Tanker	343	1 322 886	226	814 694	330	991 373	227	999 976
Coastal								
Cargo	62	398 965	57	364 077	44	299 864	63	237 804
Tug and barge	25	74 547	10	27 964	31	51 553	14	23 277
Tanker	19	92 343	19	113 020	22	117 875	17	65 883
Non-cargo								
Tug and barge	141	—	104	—	130	—	103	—
All other[2]	106	—	103	—	120	—	139	—
Total	3,559	22 035 488	3,454	28 539 003	3,540	38 652 553	3,478	36 637 152
Type of cargo								
Bulk	1,439	19 262 014	1,626	25 231 193	2,511	37 562 184	2,406	35 333 932
General	379	2 417 977	443	2 941 939	138	313 174	167	641 790
Mixed	94	355 497	92	365 871	180	777 195	177	661 430
Passengers	9	—	—	—	9	—	—	—
In ballast								
Ocean	274	—	216	—	42	—	69	—
Laker	1,084	—	839	—	388	—	398	—
Coastal	34	—	31	—	23	—	19	—
Other	246	—	207	—	249	—	242	—
Type of traffic								
Domestic								
Canada to Canada	1,275	5 853 247	985	6 008 984	1,531	15 330 791	1,390	14 492 761
Canada to United States	1,301	12 052 712	1,444	17 079 451	33	239 182	10	84 131
United States to Canada	51	98 194	17	32 560	1,022	15 237 926	1,031	13 742 710
United States to United States	97	140 964	90	164 087	100	245 827	121	246 329
Foreign								
Canada								
Import	180	508 525	180	493 277	—	—	—	—
Export	—	—	—	—	196	1 204 470	221	1 419 545
United States								
Import	655	3 381 846	738	4 760 644	—	—	—	—
Export	—	—	—	—	658	6 394 357	705	6 651 676

[1]Combined traffic of the Montreal — Lake Ontario section and the Welland Canal, with duplications eliminated.
[2]Includes naval vessels.

15.27 St. Lawrence Seaway traffic¹ classified by type of cargo, 1975 and 1976

Commodity	1975ʳ Cargo t	% of total	1976 Cargo t	% of total
AGRICULTURAL PRODUCTS	*21 448 166*	*35.3*	*20 292 998*	*31.1*
Wheat	12 320 454	20.3	9 206 497	14.1
Corn	3 124 864	5.2	4 285 272	6.6
Rye	237 715	0.4	126 790	0.2
Oats	461 767	0.8	660 965	1.0
Barley	2 459 828	4.0	3 130 767	4.8
Flour, wheat	13 400	—	27 627	—
Flour, edible, other	21 516	—	13 375	—
Soybeans	1 558 826	2.6	1 488 033	2.3
Soybean oil, cake and meal	111 704	0.2	87 714	0.1
Beans and peas	124 256	0.2	168 345	0.3
Malt	71 728	0.1	85 974	0.1
Flaxseed	138 467	0.2	143 087	0.2
Other agricultural products	803 641	1.3	868 552	1.4
ANIMAL PRODUCTS	*140 012*	*0.2*	*240 768*	*0.4*
Packing house products, edible	6 716	—	51 204	0.1
Hides, skins and pelts	41 563	0.1	38 623	0.1
Other animal products	91 733	0.1	150 941	0.2
MINERAL PRODUCTS	*30 012 719*	*49.5*	*35 000 314*	*53.7*
Bituminous coal	7 900 120	13.0	6 913 630	10.6
Coke	863 032	1.4	1 825 936	2.8
Iron ore	17 238 672	28.4	22 334 896	34.3
Aluminum ore and concentrates	26 394	—	47 486	0.1
Clay and bentonite	172 897	0.3	250 936	0.4
Gravel and sand	270 181	0.5	294 392	0.5
Stone, ground or crushed	1 134 387	1.9	993 982	1.5
Stone, rough	3 974	—	—	—
Petroleum, crude	165 422	0.3	3 243	—
Salt	1 356 131	2.2	1 514 765	2.3
Phosphate rock	62 745	0.1	36 247	—
Sulphur	31 577	0.1	43 742	0.1
Other mineral products	787 187	1.3	741 059	1.1
FOREST PRODUCTS	*109 800*	*0.2*	*75 030*	*0.1*
MANUFACTURES AND MISCELLANEOUS	*8 693 135*	*14.3*	*9 297 845*	*14.3*
Gasoline	316 519	0.5	301 652	0.5
Fuel oil	2 331 857	3.8	1 860 807	2.8
Lubricating oils and greases	147 001	0.2	191 035	0.3
Petroleum products, other	131 497	0.2	247 091	0.4
Rubber, crude, natural and synthetic	23 877	—	12 928	—
Chemicals	383 043	0.6	452 830	0.7
Sodium products	134 761	0.2	117 190	0.2
Tar, pitch and creosote	60 892	0.1	47 571	0.1
Pig iron	123 642	0.2	124 922	0.2
Iron and steel, bars, rods, slabs	558 665	1.0	275 406	0.4
Iron and steel, nails, wire	25 128	—	50 499	0.1
Iron and steel, manufactured	1 668 319	2.9	2 715 318	4.2
Machinery and machines	82 826	0.1	96 470	0.1
Cement	205 420	0.3	165 592	0.2
Wood pulp	52 374	0.1	70 594	0.1
Newsprint	63 022	0.1	43 813	0.1
Syrup and molasses	17 921	—	39 788	0.1
Sugar	287 676	0.5	318 120	0.5
Food products	132 092	0.2	100 632	0.2
Scrap iron and steel	599 280	1.0	463 654	0.7
Other manufactures and miscellaneous	1 347 323	2.3	1 601 933	2.4
PACKAGE FREIGHT	*284 209*	*0.5*	*269 200*	*0.4*
Domestic	284 209	0.5	269 200	0.4
Foreign	—	—	—	—
Total	60 688 041	100.0	65 176 155	100.0

¹Combined traffic of the Montreal — Lake Ontario section and the Welland Canal, with duplications eliminated.

15.28 St. Lawrence Seaway Authority expenditure, years ended Mar. 31, 1976 and 1977 (dollars)

Item	1976¹	1977
Administration		
Headquarters	6,421,843	5,543,028
Regional	4,966,314	4,217,735
Operation and maintenance		
Salaries and benefits	19,380,609	17,257,884
Maintenance materials and services	5,951,156	3,874,936
Other operation and maintenance expenses	2,071,025	1,720,231
Total	38,790,947	32,613,814

¹15-month period ended Mar. 31, 1976.

15.29   Aircraft movements by class of operation at airports with Department of Transport air traffic control towers, 1971 and 1973-76

Operation	1971	1973	1974	1975	1976
Local operations[1]	2,736,404	2,667,345	3,153,170	3,404,782	3,448,856
Itinerant operations[2]	1,999,938	2,586,625	2,539,541	2,993,399	3,038,271
Total, movements	4,895,376	5,253,970	5,692,711	6,398,181	6,487,127
Number of towers	53	56	57	60	60

[1]Landing or take-off by aircraft that remain at all times within the tower control zone.
[2]Landing or take-off by aircraft that enter or leave the tower control zone.

15.30   Summary statistics[1] of commercial air services, 1971 and 1973-76

Item		1971	1973	1974	1975	1976
Canadian carriers, revenue traffic only						
Unit toll transportation[2]						
Departures	'000	428	524	562	578	591
Hours flown	"	506	590	629	641	630
Kilometres flown	"	254 215	294 182	320 374	331 761	326 777
Passengers carried	"	11,082	15,316	17,225	17,697	17,931
Passenger-kilometres	"	15 387 669	21 881 755	24 771 485	25 189 055	26 169 729
Cargo and excess baggage						
tonne-kilometres	"	416 463	493 376	526 209	541 769	572 034
Mail tonne-kilometres	"	71 102	87 076	94 225	94 301	114 998
Cargo and excess baggage	'000 t	154	179	191	199	202
Mail carried	"	35	44	48	45	56
Bulk transportation[3]						
Departures	'000	439	445	494	472	508
Hours flown	"	545	634	678	705	700
Kilometres flown	"	94 465	113 872	120 205	127 193	133 215
Passengers carried	"	1,402	1,790	2,011	2,414	2,653
Passenger-kilometres	"	3 123 922	3 995 971	4 372 142	6 322 821	6 595 275
Goods tonne-kilometres	"	70 244	118 989	119 316	137 798	93 174
Freight carried	'000 t	112	147	142	153	129
Other flying services[4]						
Hours flown	'000	182	233	252	268	256
Canadian carriers, all services						
Revenue traffic						
Departures	'000	867	969	1,056	1,050	1,099
Hours flown	"	1,233	1,457	1,559	1,614	1,586
Kilometres flown	"	348 680	408 054	440 579	458 954	459 994
Passengers carried	"	12,484	17,106	19,237	20,111	20,585
Passenger-kilometres	"	18 511 590	25 877 725	29 143 627	31 511 876	32 765 005
Goods tonne-kilometres	"	557 809	699 440	739 752	773 868	780 205
Goods carried	'000 t	301	369	381	397	387
Non-revenue traffic						
Hours flown	'000	40	55	55	53	46
Passenger-kilometres	"	..	..	..	..	..
Goods tonne-kilometres	"	..	..	..	..	..
Fuel consumed	'000 l	2 363 062	2 916 399	3 242 871	3 429 275	3 372 837
Oil consumed	"	1 591	1 609	1 705	1 837	1 741
Average employees	'000	30	34	39	40	40
Salaries and wages paid	$ '000	304,209	413,341	511,660	603,691	669,971
Operating revenues	"	884,404	1,214,069	1,552,644	1,833,207	1,991,338
Operating expenses	"	827,794	1,133,584	1,481,678	1,766,705	1,935,883
Canadian and scheduled foreign carriers, all services						
Hours flown	'000	1,265	1,493	1,590	1,649	1,623
Kilometres flown	"	367 668	429 196	462 348	480 135	481 133
Passengers carried	"	15,723	21,707	24,257	25,245	26,234
Goods carried	'000 t	362	460	473	490	494

This table includes data for Levels I-IV carriers only.
[1]Although most figures in this table have been taken from the audited reports of commercial air carriers, some preliminary figures have been used.
[2]Transportation of passengers or goods at a toll per unit.
[3]Transportation of passengers or goods at a toll per kilometre or per hour for the entire aircraft.
[4]Comprises activities such as flying training, aerial photography, and aerial advertising.

15.31 Comparative statistics of domestic and international air traffic, 1975 and 1976

Year and item		Canadian airlines		Scheduled foreign airlines		Total
		Domestic services	International services	United States[1]	Other foreign[1]	
1975						
Unit toll transportation[2], revenue traffic only						
Departures	'000	518	60	..	..	..
Hours flown	"	483	158	16	18	675
Kilometres flown	"	225 554	105 951	9 247	11 166	351 918
Passengers carried	"	13,787	3,909	3,592	1,339	22,627
Passenger-kilometres	"	14 103 050	11 086 005	463 856	1 171 023	26 823 934
Goods tonne-kilometres	"	304 125	331 945	5 362	44 691	686 123
Goods carried	'000 kg	166 055	77 931	37 967	53 738	335 691
Bulk transportation[3], revenue traffic only						
Departures	'000	453	19	..	..	..
Hours flown	"	639	66	- -	1	706
Kilometres flown	"	82 324	45 124	312	455	128 215
Passengers carried	"	974	1,441	57	146	2,618
Passenger-kilometres	"	208 780	6 114 093	14 204	83 152	6 420 229
Goods tonne-kilometres	"	69 480	68 328	39	472	138 319
Freight carried	'000 kg	141 116	12 412	40	472	154 040
1976						
Unit toll transportation[2], revenue traffic only						
Departures	'000	532	59	..	..	..
Hours flown	"	477	153	17	19	666
Kilometres flown	"	223 675	103 104	9 782	10 699	347 260
Passengers carried	"	13,815	4,116	4,016	1,439	23,386
Passenger-kilometres	"	14 256 944	11 912 786	505 857	1 296 945	27 972 532
Goods tonne-kilometres	"	324 761	362 271	6 627	52 995	746 654
Goods carried	'000 kg	176 277	81 671	45 221	60 689	363 858
Bulk transportation[3], revenue traffic only						
Departures	'000	487	21	..	..	..
Hours flown	"	631	69	- -	1	701
Kilometres flown	"	84 964	48 253	171	486	133 874
Passengers carried	"	1,065	1,589	46	149	2,849
Passenger-kilometres	"	231 485	6 363 790	11 022	96 118	6 702 415
Goods tonne-kilometres	"	59 336	33 837	- -	1 784	94 957
Freight carried	'000 kg	123 336	5 464	- -	669	129 469

[1] Hours and kilometres flown are those flown only over Canada.
[2] Transportation of passengers or goods at a toll per unit.
[3] Transportation of passengers or goods at a toll per kilometre or per hour for the entire aircraft.

Sources

15.1 Water Transportation Assistance Directorate, Financial Planning and Control, Department of Transport.

15.2 - 15.13, 15.15 - 15.24 Transportation and Communications Division, Industry Statistics Branch, Statistics Canada.

15.14 Respective provincial government departments.

15.25 National Harbours Board.

15.26 - 15.28 St. Lawrence Seaway Authority.

15.29 - 15.31 Transportation and Communications Division, Industry Statistics Branch, Statistics Canada.

Communications

Chapter 16

Tables

Communications   Chapter 16

Telecommunications   16.1

The size, topography and climate of Canada have influenced the development of Canadian telecommunications. Vast networks of telephone, telegraph, radio and television facilities are necessary to provide efficient communication among Canadians and between Canada and the rest of the world. Canada possesses a mix of telecommunications systems — federal, provincial, municipal and investor-owned — whose operations are co-ordinated to carry messages to all parts of the country by land lines, microwave, tropospheric scatter, high-frequency radio and satellite communications systems and to other parts of the world by undersea cable and international satellites.

Canada's telecommunications carrier industry, with $14.5 billion invested in plant, is expanding at the rate of more than $2 billion a year. Investment for 1977 was $2 billion, a figure expected to reach $2.5 billion a year by 1980, increasing to $4 billion a year by 1985.

The Canadian Telecommunications Carriers Association (CTCA), established in 1972, provides the framework for co-operation on an industry-wide basis for major telecommunications carriers. The association consists of 21 telecommunications carrier organizations, each represented on the board of directors. It brings together in one organization the TransCanada Telephone System and its 10 members, the Canadian Independent Telephone Association, six other telephone companies, Canadian National and Canadian Pacific Telecommunications and Teleglobe Canada.

CTCA is active in the affairs of the Geneva-based International Telecommunication Union and attempts to secure, through the federal communications department, the compatibility of the Canadian telecommunications system with those of other countries.

Telecommunications media   16.1.1

Voice communications   16.1.1.1

Telephony. There are more than 14 million telephones and 22.5 million kilometres of circuits in Canada. Close to 12.5 million telephones are provided by the 10 member companies of the TransCanada Telephone System (TCTS): Alberta Government Telephones, British Columbia Telephone Company, Bell Canada, the Island Telephone Company Ltd., Manitoba Telephone System, Maritime Telegraph and Telephone Company Ltd., the New Brunswick Telephone Company Ltd., Newfoundland Telephone Company Ltd., Saskatchewan Telecommunications and Telesat Canada. Almost a million telephones are provided by edmonton telephones, Northern Telephone Ltd., Okanagan Telephone Company, Ontario Northland Communications, Québec-Téléphone, Télébec Ltée, Thunder Bay Telephone Department and Canadian National Telecommunications. There are also about 1,000 smaller companies.

Canadian National Telecommunications provides telephone service for residents in the Yukon Territory and Northwest Territories, parts of Newfoundland and in northern sections of British Columbia.

Each Canadian telecommunications company is responsible for service in its own territory and for integrating its facilities with those of all other telephone organizations.

Collectively, these companies operate the world's longest microwave system and have access to Canada's domestic satellite system — a telecommunications blend which carries phone conversations, radio and television programs and computer data coast to coast. Through the integrated North American network and Teleglobe Canada's intercontinental connections, the system can reach nearly all the world's 425 million telephones.

Each year Canadians place some 5.4 million telephone calls to countries outside North America. In September 1976 TCTS and Teleglobe Canada introduced direct dialing from Vancouver to the United Kingdom, the Federal Republic of Germany, Japan, Hong Kong, Australia, New Zealand and the Philippines. Over the next two years, the service will be expanded to include most major Canadian cities. On December 3, 1977, direct dialing was introduced from a number of exchanges in Montreal and Quebec City to 27 foreign countries and, the same day, all Canadian subscribers were able to dial direct to the Bahamas, Bermuda, the Cayman Islands and Jamaica.

For a basic monthly charge most telephone users can place as many calls as they wish in a defined area and talk as long as they like. With the expansion of major cities and the merging of small towns creating larger communities, most telephone companies have introduced extended area service which enables customers to place calls in a much wider area without paying long distance rates. For this service the customer pays a slightly higher fee, based on the number of telephones within his extended area.

Ownership and regulation of Canada's telecommunications carriers varies. The majority of telephones in Canada are owned and operated by investor-owned companies such as Bell Canada, British Columbia Telephone Company, Québec-Téléphone, Maritime Telegraph and Telephone Company Ltd. and New Brunswick Telephone Company Ltd. The Island Telephone Company Ltd., Newfoundland Telephone Company Ltd., Northern Telephone Ltd., Télébec Ltée and Okanagan Telephone Company are subsidiaries of investor-owned telephone companies.

Three major investor-owned systems, Bell Canada, British Columbia Telephones and CP Telecommunications, are federally regulated by the Canadian Radio-television and Telecommunications Commission; the other investor-owned systems are regulated by provincial agencies.

Alberta Government Telephones, Manitoba Telephone System and Saskatchewan Telecommunications are provincially owned corporations. Ontario Northland Communications, a division of Ontario Northland Transportation Commission, a provincially owned corporation, provides telephone and telegraph services in the northeastern part of Ontario.

Thunder Bay Telephone Department and the Edmonton telephones system are the country's two largest municipal systems. Many smaller telephone companies are grouped in the Canadian Independent Telephone Association. Canadian National and Teleglobe Canada are federal Crown corporations.

16.1.1.2 Record communications

Public message service. Canada's public message-telegram-service is provided by CNCP Telecommunications. A joint venture of the telecommunications divisions of the Canadian National and Canadian Pacific railways, CNCP offers public message service in all provinces and territories. Messages can be forwarded or received from any point in Canada or throughout the world via cable and satellite facilities of Teleglobe Canada.

There has been a gradual decline in public message volumes and a corresponding growth in Telex and Teletypewriter Exchange Service (TWX). But the service continues on behalf of users who are not on Telex or TWX. Most of them file messages at telegraph offices via telephone and Telex. Few telegrams are filed at the counter.

Telex and TWX. Each year some 3.6 million Telex and TWX messages and one million telegrams to overseas points are switched through the facilities of Teleglobe Canada; the total worldwide complex provides access to more than 900,000 TWX and Telex subscribers.

Telex, the first North American dial-and-type teleprinter service, was introduced in Canada in 1956. It has since grown to more than 35,000 customers and 160 exchanges. It interconnects with 55,000 Telex subscribers and 40,000 TWX subscribers in the United States, and with Telex network throughout the world.

TWX has some 5,000 subscribers in Canada who have the capability to reach TWX users in the United States and, through an agreement between TCTS and Western Union Telegraph Co., with US Telex users. TWX subscribers connect with overseas customers through International Telex, provided by Teleglobe Canada.

Telex and TWX are now considered universal services, available to some 200 countries and territories, almost half of which are linked to Teleglobe Canada's computer-controlled exchange permitting subscriber-to-subscriber dialing without operator assistance. Computerized switching integrates the Canadian domestic Telex and TWX networks with the overseas network, handling more than 4,000 messages an hour.

Data communications   16.1.1.3

Member companies of the TransCanada Telephone System and CNCP Telecommunications offer a wide selection of the data communications services necessary in a modern industrial country.

The telecommunications carriers provide a range of terminals for transmission and reception. There are printer terminals that can be used for computer access, cathode ray tube terminals that display information on a screen and a variety of more specialized machines. Customers may also use their own terminal equipment.

A number of different transmission systems may be used. Many customers have private-line networks linking scattered locations. Others employ pay-as-you-use data transmission services. Transmission speeds vary from less than 100 words a minute to the equivalent of 50,000.

Introduction of digital transmission networks in early 1973 provided the first nationwide commercial digital systems in the world. Digital transmission permits reduced costs by more efficient use of existing circuits and ensures improved accuracy, vital in high-speed data transfer. Introduction of packet switching and digital circuit switching systems in 1977 was another major development.

Provision of data communications in Canada is undertaken competitively by two major national carriers, CNCP Telecommunications and the TransCanada Telephone System. Data communication between Canada and points outside North America is provided through Teleglobe Canada. In co-operation with the British post office, Teleglobe inaugurated a Canada–UK data link in January 1976. Work is under way to expand the service to other countries with the aim of introducing a public data network in conjunction with domestic carriers and foreign telecommunications administrations.

The network   16.1.1.4

Three microwave routes and a satellite system form the backbone of Canada's telecommunications network. Two of the routes belong to the TransCanada Telephone System, the third to CNCP Telecommunications. Canada's first coast-to-coast microwave system, completed in 1958 by TCTS, and extending almost 6 400 kilometres carries the bulk of network traffic. Telesat Canada provides additional facilities throughout Canada over satellite communications, and Teleglobe Canada uses Intelsat satellites as well as undersea cables to provide the global connection.

Telesat Canada launched Anik I, the world's first domestic commercial communications satellite on November 9, 1972. A back-up, Anik II, was launched in April 1973, and another, Anik III, in May 1975.

Initial commercial service to Telesat customers began during January 1973 through a network of earth stations strategically located across Canada. Basically, satellite communication is a long microwave link; transmission is comparable to that of existing microwave systems with the added capability of transmitting to areas which had not been well served by more conventional means.

The Anik series provides television distribution to all parts of Canada and improved telephone communications to the North, and supplements existing microwave systems. The Anik generation of satellites has a projected seven-year life cycle.

Satellites used by Telesat and Teleglobe Canada are stationed about 35 900 kilometres above the earth. Although Anik is exclusively a Canadian domestic system, other satellites in the Intelsat international system and a vast network of undersea cables make it possible for Canadians to communicate with virtually all countries in the world.

Satellite transmission began with the launching of Telstar in 1962, 10 years after the first telephone and multi-purpose submarine cable in the world was laid across the Atlantic by Teleglobe Canada and three other carriers. This first cable, with 80 circuits and the most recent, with 1,840 circuits, still help meet growing demand for overseas telecommunications.

16.1.1.5 Telecommunications in the North

Anik is the Inuit word for brother. Anik I opened a new era of telecommunications in the North providing reliability, flexibility and new services, including television broadcasting, to remote communities not served by surface facilities. Northern communication is accomplished mainly by microwave and tropospheric scatter systems and high-frequency radio. Both methods, as well as land line facilities, are still used.

Telecommunications services in the North are operated mainly by CN Telecommunications and Bell Canada. British Columbia Telephone Company provides telecommunications services along the West Coast to Alaska.

CN Telecommunications covers an area that runs north through British Columbia from Fort St. John and includes all of the Yukon Territory and Northwest Territories west of longitude 102°. Bell Canada serves the eastern half of the Northwest Territories up to and including Grise Fjord in the Arctic Circle and all Northern Quebec. Newfoundland Telephones operates in Labrador while Ontario Northland Communications serves northeastern Ontario.

Throughout the North, CNT and Bell Canada automatic telephone exchanges are connected to the Canadian networks, through them to the North American networks and through Teleglobe Canada to overseas networks. Microwave tropospheric scatter and the domestic satellite systems are used to penetrate the heart of the Arctic and connect to the North American continental telecommunications network. It is also possible to communicate within the coverage area through high-frequency equipment with mining camps, oil and gas exploration sites, construction camps and outposts.

16.1.2 Telephone and telegraph statistics

Telephone statistics. In 1976 Canada had an estimated 821 telephone systems compared to 860 in 1975; of these, 806 filed returns with Statistics Canada compared to 850 in 1975 (Table 16.1). Although the number of co-operative systems declined from 737 in 1975 to 720 in 1976, growth in the telephone industry was particularly evident in the large telephone companies. The largest incorporated telephone company, Bell Canada, operates in Ontario, Quebec and the Northwest Territories. In 1976 it owned and operated 8.3 million of approximately 13.9 million telephones in Canada. The BC Telephone Company, also owned by shareholders, operated 1.4 million of the total telephones in 1976.

Table 16.2 shows the distribution of telephones by province in 1976. Of the 1976 total, 70.3% or 9.8 million were residential telephones and 4.1 million were business telephones. Alberta had the most telephones per 100 population with 65.2, followed by Ontario at 63.7 and British Columbia at 61.9. As Table 16.3 shows, each Canadian averaged 953 calls in 1976.

Table 16.4 shows capitalization, revenue and expenditure of telephone companies plus the number of employees, salaries and wages paid for 1971-76. Provincial figures for 1975 and 1976 are given in Table 16.5.

Telecommunications statistics. Nine telecommunications companies operated in Canada during 1976. This was the fourth year of commercial operations of Telesat Canada, which added over $29 million in revenue to the operation of commercial telecommunications carriers. The operating revenues of telecommunications companies increased from $259.1 million in 1975 to $278.3 million in 1976 or 7.4% while expenses for the same period increased from $193.8 million to $213.7 million or 10.3% (Table 16.6). The property and equipment for these nine telecommunications companies increased by $38.1 million to $977.7 million in 1976, from $939.6 million in 1975. These figures include investment in property and equipment by Telesat Canada, which in 1976 was reported at $155.1 million.

16.1.3 Federal regulations and services

The Department of Communications. The department, established in April 1969, is responsible for ensuring that all Canadians obtain the best possible access to an expanding range of communications services. This involves not only technological research and planning, but also exploration of social, human and economic issues which result from changing patterns of communications. The department protects Canadian interests in the realm of international telecommunications and manages the radio frequency spectrum to permit development and growth of radio communications.

The duties, powers and functions of the communications minister include all matters relating to telecommunications over which the Parliament of Canada has jurisdiction, not by law assigned to any other department, branch or agency of the federal government. The general development and use of communication undertakings, facilities, systems and services for Canada also come under the minister's jurisdiction. The department has four sectors: policy, space program, research and services.

The policy sector recommends international and national telecommunications policies and proposes legislation for the government's consideration. This sector co-ordinates federal-provincial relations and is the point for contacts with Teleglobe Canada and the Canadian Radio-television and Telecommunications Commission. It provides technological and socio-economic forecasts, identifies areas needing research and development, and carries out strategic planning.

The department's field organization, with 48 offices in the Atlantic, Quebec, Ontario, Central and Pacific regions, is primarily concerned with management of the electromagnetic spectrum.

The space sector comprises all space-related communications activities. These include Canada's experimental satellite launched in 1976; relations with Telesat Canada and other agencies concerned with space; development of new space systems and applications; and planning and international functions in communications technology.

The research sector carries out research and development in communications, both in-house and through university and industrial contracts. It is concerned with new communications systems and services, the use of the radio frequency spectrum and providing scientific advice to help formulate departmental policy and develop new programs. The department's principal research facility is a research centre near Ottawa.

A telecommunications regulatory service sets technical standards for broadcasting facilities and equipment, issues technical certificates and radio operating licences and manages the radio frequency spectrum. A government telecommunications agency provides consulting and centralized telecommunications services for the government.

The Canadian Radio-television and Telecommunications Commission Act, put into effect April 1, 1976, transferred regulatory jurisdiction over certain telecommunications common carriers previously exercised by the Canadian Transport Commission to the Canadian Radio-television and Telecommunications Commission (CRTC). Telephone and telegraph companies incorporated under federal legislation are now subject to the jurisdiction of the CRTC.

Radiocommunications in Canada, except for matters covered by the Broadcasting Act, are regulated under the Radio Act and regulations and the Canada Shipping Act and ship station radio regulations. The Radio Act and regulations provide for licensing of radio stations performing terrestrial radio services, and licensing earth and space stations engaged in space radiocommunication services. Radiocommunications in Canada are administered in accordance with an international telecommunication convention and its radio regulations, an international civil aviation convention, and an international convention for the safety of life at sea. A number of Canada–United States conventions and agreements are also in effect: a convention for the promotion of safety on the Great Lakes by means of radio; a convention relating to the operation, by citizens of either country, of certain radio equipment or stations in the other country; an agreement relating to the co-ordination and use of some radio frequencies; television and FM agreements; and an agreement relating to the operation in either country of radiotelephone stations licensed in the citizens radio service of the United States and the general radio service of Canada. Canada is also a party to a North American regional broadcasting agreement.

The Canadian Radio-television and Telecommunications Commission (CRTC). Under the Canadian Radio-television and Telecommunications Commission Act the CRTC also rules on all federally-regulated broadcast undertakings. The CRTC issues

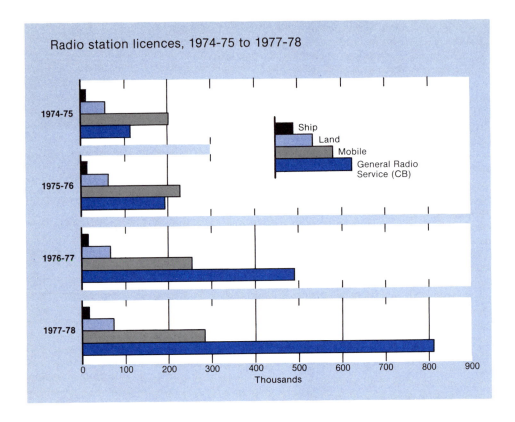

Radio station licences, 1974-75 to 1977-78

broadcasting licences under the Broadcasting Act of 1968 after the minister of communications certifies that the applicant has satisfied the requirements of the Radio Act and regulations, and has been or will be issued a technical construction and operating certificate. Broadcasting undertakings include radio (AM and FM) and television broadcasting stations, community antenna television (CATV) systems, and network operations. Availability of technical facilities for broadcasting is subject to terms of Canada–US agreements, covering television and FM . The CRTC also regulates tariffs of telephone and telegraph companies in Canada.

International telegraph and telephone communications are subject to the International Telecommunication Convention and its regulations or regional agreements, or both. Overseas cables landed in Canada are subject to external submarine cables regulations under the Telegraphs Act.

Licensing and regulating of radiocommunications. Licensing is the federal government's method of maintaining control over radiocommunications in Canada. Under the Radio Act, radio stations (other than those used in broadcasting undertakings) using any form of Hertzian wave transmission, including television and radar, must be licensed by the communications department unless exempted by regulation. The general radio regulations provide for six classes of radio station licence: coast, land, mobile, ship, earth and space. Various categories of service may be authorized under each class, such as public commercial service, private commercial service, amateur and experimental.

The number of radio station licences in force in the year ended March 31, 1976 was 515,222 compared with 395,614 for 1975. These figures include stations operated by federal, provincial and municipal government departments and agencies, stations on ships and aircraft registered in Canada and stations in land vehicles operated for both public and private purposes, but they do not include stations in the broadcasting service. Total licences in force as of March 31, 1977 were 893,781, an increase of 378,559.

A large part of the increase in radio station licences over 1976 was due to citizen's band radio, officially known in Canada as General Radio Service (GRS). More than 330,000 Canadians hold GRS licences and the figure is increasing rapidly.

Radio standards are drawn up in consultation with the electronics industry, organizations, associations and the public, taking into account technical factors affecting frequency spectrum utilization, reliability of apparatus, and compatibility under conditions of service. The communications department develops standard specifications and tests apparatus for compliance with the standards.

Licensing involves assigning specific frequencies to each station. Bands of frequencies are allocated for various types of services, often on a shared non-interference basis. Frequency selection, compatibility evaluation, domestic registration and notification to the International Frequency Registration Board (IFRB) of the International Telecommunications Union at Geneva are carried out to ensure efficient use of the spectrum. Assignments are made in keeping with international and domestic statutes and regulations, regional agreements and domestic policies. The IFRB is notified of frequency assignments for technical examination and for inclusion in a master international frequency register so that Canadian assignments will receive international recognition and be given protection from interference by foreign stations.

Enforcement activities of the communications department include technical inspection of all radio stations including monitoring and measurement of radiated signals to ensure compliance with regulations and conditions of licensing; location and suppression of radio interference; technical examination of candidates for the various classes of certificates of proficiency in radio which must be held by operators of radio stations; and direction of prosecutions in the courts. These functions are carried out through personnel located at five regional offices, 42 district offices, 10 fixed monitoring stations, eight mobile monitoring vehicles and 13 regional spectrum observation centres.

International services   16.1.4

Teleglobe Canada is the link between domestic telephone and telecommunications carriers and almost every country in a global communications system. Its mandate is to establish, maintain and operate Canada's external telecommunications services and co-ordinate their use with the services of other countries.

Canadians now may telephone around the world almost as easily as they call across town. Businessmen contact overseas clients rapidly. Over vast distances, television viewers receive live satellite coverage of major events.

Radio and television   16.2

Broadcasting, like other communication systems in Canada, has evolved to meet the needs of a comparatively small population concentrated along the southern fringe of a vast national territory. Approximately 90% of the population has several choices of both radio and television service available from Canadian sources. Only 2%, by reason of distance or topography, is without at least one radio and one television service. This 2% is being reduced by developments in technology and projects such as a current program of the Canadian Broadcasting Corporation (CBC) to provide radio and TV services in all communities of 500 or more people.

These general statistics reflect the development in the distribution of radio and television program services. A first objective has been to establish east-west networks to link up communities from Atlantic to Pacific across Canada's southern border. This was a principal concern in radio during the late 1920s and 1930s, and an urgent concern between 1952 when television broadcasting began in Canada and 1959 when basic television networks were completed between British Columbia and Newfoundland.

Service to Canada's North has awaited availability of resources and development of technology. Until recently, service was provided by adaptations of resources normally used to serve southern Canada or by improvisations such as CBC's scheme for circulating recorded TV programs to low-powered, self-contained television transmitters, or "frontier packages", which were installed in remote or isolated communities.

Satellite technology is the key to one major aspect of the northern problem of large distance and small population. CBC's Northern Television Service (NTS) moved to live satellite transmission in 1973. NTS stations north of the 60th parallel (31 in 1977) obtain their programming by satellite mainly from the CBC transcontinental networks across southern Canada. A major objective is to develop a strong regional element in program service more relevant to the particular needs and interests of the people and communities of the North.

To underwrite its broadcasting services, Canada combines the resources of public ownership and commercial operations. Hundreds of private business firms operate everything from cable television systems in small communities to major broadcasting stations in metropolitan centres. The CBC (Société Radio-Canada or simply Radio-Canada in French) as the national publicly owned system is the main element of the public sector and the largest broadcasting enterprise in the country. At the other end of the public ownership scale are small community antenna systems or rebroadcasting transmitters operated by community associations in remote areas. In between are the regional, educational television services operated by provincial governments in Quebec and Ontario and the Alberta education department's radio operation.

The country has an equally complicated combination of conventional or originating broadcasting stations (379 AM radio stations, 134 FM radio, and 101 television), and repeater or rebroadcasting stations (335 AM radio rebroadcasters, 144 FM radio, and 733 television), plus a cable television industry made up of 427 individual systems or undertakings serving more than half Canada's households.

According to current statistics, there is at least one radio in service in Canada for each Canadian. About 98% of the country's households are equipped with radio, almost seven out of 10 with more than one receiver, and there are 6.7 million car radios in use.

Of the estimated 7.02 million households, 6.82 million have one or more television sets. An estimated 67.8% of households had colour television sets in May 1977. During 1977, households subscribing to cable television service passed the 50% mark and the industry's cable television service was available to 83% of Canadian households.

The CBC operates two nationwide television networks, one in English and one in French. There are two major commercially operated networks: the CTV network provides an English-language program service from coast-to-coast and the Réseau de télévision TVA provides French-language programming across Quebec. There are three regional television services which distribute programming from a basic originating station and several rebroadcasters: the privately owned Global Communications Ltd. in Southern Ontario; Radio-Quebec (Office de la radio-télévision du Québec) and TV Ontario (Ontario Educational Communications Authority), both educational and cultural systems operated by provincial authorities.

The CBC operates coast-to-coast AM radio networks in both French and English as well as FM radio networks in both languages that approach national distribution. There are no full-time AM or FM networks operated by private commercial interests although more than 100 private stations are affiliated with the English or French networks of the CBC. Many part-time regional networks of privately owned stations operate to present specific program services such as play-by-play accounts of major sporting events.

As a result of four major parliamentary examinations of broadcasting since the 1920s, Canadian law has come to regard publicly owned broadcasting, commercially based radio and TV, and cable television as "constituting a single system". The legislation from which this wording is taken, the Broadcasting Act which came into force in 1968, directs the Canadian Radio-television and Telecommunications Commission to regulate and supervise all aspects of the Canadian broadcasting system with a view to implementing certain policy objectives: effective ownership and control of broadcasting facilities in the hands of Canadians; a wide variety of programming which provides reasonable, balanced opportunity for the expression of differing views on matters of public concern; availability of service in English and French to all Canadians; and programming of high standard that makes use of predominantly Canadian creative and other resources.

These legislative objectives reflect the concerns of a country with a small population sharing a continent and a good deal of cultural background with a large and populous neighbour. Either directly or through cable television systems, more than 60% of English-speaking Canadians and more than 40% of French-speaking Canadians have access to United States stations and networks. As well, Canadian broadcast rights to popular programs produced in the US are relatively cheap, since the producer has already recovered costs from US sales. As a result, American programs make up a large proportion of material presented by Canadian television and radio stations.

The concern to have broadcasting service that will safeguard, enrich and strengthen the cultural, political, social and economic fabric of Canada as called for by the Broadcasting Act, is strongly represented by the publicly owned CBC. It also finds expression in CRTC regulations dealing with program content. The amount of broadcast time devoted to non-Canadian programming by television stations and networks is not to exceed 40% between the hours of 6:00 a.m. and midnight. For AM radio broadcasting, 30% of the musical compositions presented by stations and networks between 6:00 a.m. and midnight are to be Canadian according to the criteria set out in CRTC regulations. Individual FM radio station or network operators make specific commitments as to Canadian content which are conditions of their licences.

Cable television   16.2.1

Although relatively little attention had been paid by government or industry to the significance of cable television until the latter half of the 1960s, more than 10% of Canadian homes had become subscribers to CATV service by 1968. In 1969, the CRTC considered the capability of cable television technology to enlarge the coverage areas of US stations and networks in Canada and concluded that "the rapid acceleration of such a process throughout Canada would represent the most serious threat to Canadian broadcasting since 1932 before Parliament decided to vote the first Broadcasting Act." The regulations for cable television issued by the commission in 1975 reflect this concern: local and regional stations are given precedence over distant stations in the order of priority that CATV must use in assigning the distribution channels available on any given cable television system; and there is provision for the substitution of the signals of a local or regional station for the signals of a distant station of lower priority when an identical program is being transmitted during the same period.

Pay television. The CRTC policy on cable television issued in December 1975 said that it was premature to introduce a comprehensive pay television service in Canada at that time. In June 1976 the commission invited submissions on pay television about "the form and function of an organization, institution or agency to assemble, produce and acquire programming for distribution to licensed broadcasting undertakings for pay television on a national or regional basis in English and in French."

Such a structure would have to meet three objectives for pay television set out by the communications minister: It must provide a range of programming which would not duplicate that offered by broadcasters and must do so without siphoning programs from the broadcast system. It must ensure the production of high-quality Canadian programs that Canadians would watch. It must ensure that programs would be produced in Canada for international sale.

In May 1977, the CRTC held public hearings to examine and discuss the material it had received in more than 100 briefs about pay television. The commission was expected to issue its comments on the subject early in 1978.

Canadian Broadcasting Corporation (CBC)   16.2.2

Facilities and coverage. The CBC operates two national television networks, English and French; four radio networks, AM and FM in English and French; a special medium and shortwave radio service in the North including native language programs; and an international shortwave and transcription service. In 1977 the CBC owned some 412 radio outlets (full stations or rebroadcasting transmitters), and its radio network service was also carried on 112 privately owned outlets. CBC-owned television stations or rebroadcasters totalled about 296, and the CBC television networks also included 245 privately owned affiliates or rebroadcasters. The corporation has production centres in Toronto (English), Montreal (French), and in several cities across the country.

CBC AM radio networks are within reach of 99% of the Canadian population and CBC television networks cover 98%. Remaining unserved locations are gradually being provided with radio and TV transmitters under the Accelerated Coverage Plan (ACP), a six-year program approved by the federal government in 1974. The ACP will involve nearly 700 engineering projects by the time it is completed. By late 1977, applications had been submitted to the CRTC for 250 ACP projects and licences have been granted for 174.

The demand for a unifying reflection of Canadian society through the CBC was brought into focus with the 1976 Quebec election and the constitutional issues arising from it. For the CBC this meant increased emphasis on continuing responsibilities such as fair and thorough news coverage, the reflection of Canada's cultural diversity and special co-operative efforts between the French and English networks.

General programming. CBC radio and television continued to offer varied program schedules in news, current affairs, music, drama, sports, religion, science, children's programs, consumer interests and light entertainment. Interest in greater regional development was expressed both outside and inside the CBC. Plans were announced to increase regional resources and production capacity, to enrich regional news and current affairs capabilities in particular, and to start on a five-year plan of gradually increased regional production in drama, music and variety. The radio services celebrated the 40th anniversary of the CBC, and continued the development of new programs and schedules for the AM and FM-stereo networks. The Northern Radio Service combined network programming with local and regional broadcasts in English, French and 10 native languages and dialects. One of the goals for CBC television, moving into its 25th anniversary year, was to maintain attractive Canadian programs in competition with other sources of television.

All CBC networks supported the work of Canadian artists and performers through presentation of Canadian drama, literature, music and films. Renewed attention was also given to regional talent development and intraprovincial programming. CBC programs or performers won more than 60 awards in Canadian and international competitions. Selected programs from English and French CBC television networks were made available for post-broadcast distribution to educational bodies through the National Film Board.

International activities. Radio Canada International (RCI), the CBC's overseas shortwave service with headquarters in Montreal, broadcasts daily in 11 languages and distributes free recorded programs for use by broadcasters throughout the world. In shortwave programming an attempt was made to draw firm lines between information and opinion, so as to avoid confusion for the listener. A central talks unit was created under the general supervision of the news department to commission commentaries and provide press reviews combining opinion from English- and French-language newspapers in Canada. Program personnel were regrouped under five target area desks: Eastern Europe, Western Europe, Africa, North America and Latin America.

Under agreement with the defence department, the CBC Armed Forces Service provides recorded and shortwave programs for Canadian forces radio stations in the Federal Republic of Germany, with staff seconded to manage the stations. The Armed Forces Service reports to Radio Canada International.

Finance. The CBC's total operating expenses for the fiscal year 1976-77 were $476.0 million, including $20.9 million for the 1976 Summer Olympics; $234.0 million for television, and $64.0 million for radio programs broadcast; and $59.2 million for television and $16.6 million for radio program distribution. This operating budget was provided by parliamentary appropriation of $389.0 million, general revenues of $86.0 million (including $82.0 million from commercials), $12.7 million from the Olympics Organizing Committee, with the balance covered by depreciation.

In constant dollars the increase in expenditures in 1976-77 was 10.1% over that of 1975-76. The largest part of the increase in operating expenditures went to the financing of price and wage increases. The balance was allocated to programming improvements and the operation of new facilities.

Statistics of the broadcasting industry   16.2.3

Statistics on radio and television broadcasting are obtained by Statistics Canada in co-operation with the Canadian Radio-television and Telecommunications Commission. In 1976 returns were received from 288 private radio reporting units and 65 television reporting units covering the operation of 402 private radio stations, 59 private originating television stations and the CTV network. Operating revenue of the broadcasting industry, including CBC, for the year amounted to $604.5 million, an increase of 19.5% over 1975. Of the total, radio accounted for $244.7 million or 40.5% and television for $359.8 million or 59.5%. Revenue from national and network time sales represented 54.3% of the total air time sales and local time sales were 45.7%. Operating expenses in 1976 at $800.8 million were 20.3% higher than in 1975. However, total operating revenue, plus the net cost of operating the CBC, which is financed from its parliamentary grant, exceeded these expenses, resulting in a net profit after depreciation and interest charges and other adjustment of $96.2 million for 1976 compared to $70.6 million in 1975.

In 1976 there were 24,680 employees engaged in the radio and television broadcasting industry, an increase of 1,184 or 5.0% over 1975. Salaries and wages paid by the industry totalled $413.7 million. After provision for income taxes, the final net profit of the private sector of the broadcasting industry in 1976 was $52.5 million compared with $34.2 million in 1975. The upswing in television broadcasting profits in 1976 was due in part to Global Communications Ltd. which in 1976 substantially reduced its losses of the previous two years while increasing its revenue 80.6% to $15.0 million; its expenses increased only 11.1%.

Statistics of the cable television industry. Table 16.7 presents financial statistics of the Canadian cable television industry. This industry, comprising 356 operating systems, reported an increase of 22.8% in total operating revenue for the year ended August 31, 1976, rising to $199.2 million from $162.3 million for the previous year. Subscription revenue from individual subscribers and multi-outlet contracts accounted for $181.4 million. Operating expenses before deducting interest and depreciation charges rose from $85.8 million to $108.1 million in 1976, resulting in net operating revenue of $91.1 million compared with $76.5 million in the previous year. After deducting interest, depreciation and making other adjustments, the industry achieved a net profit to August 31, 1976 of $36.0 million compared with $31.3 million earned in the previous year.

Postal service   16.3

The basic function of the Canadian Postal Service is to receive, convey and deliver postal matter. It maintains thousands of post offices and uses air, rail, road and water transportation facilities. Associated functions include sales of stamps and other articles of postage, registration of letters and other mail for dispatch, parcel insurance, accounting for COD articles and transaction of money-order business. Because of its transcontinental facilities, the post office assists other government departments with such tasks as selling hunting permits, collecting annuity payments, distributing income tax forms and public service employment application forms, and displaying official posters.

Post offices are established wherever the population warrants. In rural areas and small urban centres they transact all the functions of a city office. In larger urban areas, postal stations have functions similar to the main post office, including general delivery service, lock-box delivery and letter-carrier delivery. Canada's larger postal installations are semi- or fully-automated plants with optical character reading machines capable of reading printed or typed addresses; machines which automatically and at high speed cull, face and cancel stamps; letter sorting machines capable of handling 26,000 pieces of mail an hour; conveyors and chutes, parcel and bag sorting machines, wrapomatic parcel sealing machines, photo-electric counters and intercom systems. Outside some regular post office buildings there are stamp-vending machines and curbside mail boxes.

The operating service of Canada Post is organized into four regions divided into districts. The operating and support functions required to provide postal service are the responsibility of local postmasters who receive technical and administrative assistance from district and regional offices at strategic points.

Postal service is provided throughout Canada. The country's airmail system utilizes most transcontinental flights, supported by many branch and connecting lines, and links up with United States domestic and other international airmail systems. First-class domestic mail is carried by air between Canadian points whenever this expedites delivery. Air stage routes provide an all-class mail service to many northern areas which can be served only by air. There are over 74 030 km of airmail and air stage routes.

By the end of the fiscal year 1975-76 there were 8,506 postal facilities in operation in Canada. With the growth in urban density, letter-carrier service grew to provide more convenient service to city dwellers. By the end of March 1976, an additional 272,085 new points of call were added to the 5,236,843 already being served through 12,480 full and 469 partial letter-carrier routes from 279 post offices. Rural and suburban services were reduced slightly owing to the lessened demand. The number of rural routes decreased by 14 to 4,942 and suburban service was reduced by one to 44.

Revenue and expenditure of the post office for the year ended March 31, 1976 were $568.0 million and $768.3 million, respectively; gross revenue receipts were received mainly from postage, either in the form of postage stamps and stamped stationery, postage meter and postage register machine impressions, or in cash. During the year 34 million money orders were issued having a value of $1,089.0 million.

16.4   The press

Daily newspapers published in Canada in 1977 numbered 120, counting morning and evening editions separately. Combined circulation was about 5.2 million — about 82% in English and 18% in French (Table 16.8). Publishers' surveys show that each newspaper is read by an average of three persons.

Daily newspaper advertising net revenue in 1975 was $565.0 million and circulation revenue was $120.3 million. In 1977, there were 17 daily newspapers with a circulation in excess of 100,000, accounting for 59% of total circulation. There were 12 dailies published in French, 10 of them located in Quebec. Although the circulation of daily newspapers blankets the more populous areas well beyond publishing points, smaller cities and towns and rural areas are also served by 825 weekly newspapers catering to local interests. There were 274 newspapers and periodicals published in Canada by ethno-cultural groups.

About 27% of Canada's daily newspapers are privately owned or independent. There are three major newspaper chains in the country, Southam Press Ltd. (14 dailies), Thomson Newspapers Ltd. (35 dailies) and FP Publications Ltd. (nine dailies). Both Southam and Thomson Newspapers are publicly owned companies with shares traded on Canadian stock exchanges. Papers in the Thomson chain are concentrated in the smaller cities. Southam accounts for about 21% of total daily circulation, Thomson for 10% and FP for about 20%.

In addition to their own news-gathering staffs and facilities, Canadian newspapers subscribe to a number of syndicated agencies and wire services, the largest being The Canadian Press which is a co-operative agency owned and operated by Canadian dailies. Largely by teletype and wirephoto transmission, it provides its 110 member newspapers with world and Canadian news and also serves radio and television stations. CP has its own news-gathering staff and each member newspaper provides important local news for transmission to fellow members. Members share the cost in ratio to their circulations.

CP carries world news from Reuters (the British agency), from The Associated Press (the United States co-operative) and from Agence France-Presse (of France) and these agencies receive CP news on a reciprocal basis. CP maintains a French-language service in Quebec.

United Press International of Canada, the second major news wire service in Canada, is a private company and a part of United Press International World Service. It provides Canadian and international news and pictures to newspapers and TV and radio stations across Canada and is an outlet for Canadian news through United Press International facilities throughout the world. Certain foreign newspapers maintain bureaus in Ottawa and elsewhere in Canada to collect and interpret Canadian news.

Press statistics. Table 16.8 gives numbers and circulations of reporting English- and French-language newspapers, by province, for 1976 and 1977, estimated from *Canadian Advertising*. Circulation figures are given for daily English- and French-language newspapers only. Such circulation figures are relatively easy to obtain because, in their own interest, newspapers qualify for and subscribe to the Audit Bureau of Circulation. For these, ABC "net paid" figures have been used; "controlled" (free) distribution newspapers are not included. On the other hand, circulation data for foreign-language newspapers, weekly newspapers, weekend newspapers and magazines are incomplete. Ethno-cultural publications numbered 274 in 1977 (Table 16.9); 30 were Ukrainian, 27 Italian, 26 German, 19 Jewish, 15 Greek, 12 each by Dutch and Polish groups, 11 Chinese, 10 East Indian, nine Arabic and smaller numbers for people of 29 additional groups or national origins, as well as four inter-ethnic publications.

Ethnic Press Analysis Service. During 1976-77 an ethnic press analysis service, secretary of state department, received and analyzed over 225 ethnic newspapers and periodicals in more than 30 languages. This was carried out by a staff of 17 contract analysts. Information gathered was used to prepare a monthly review entitled the *Canadian Ethnic Press Review*. This publication, produced in a limited edition, is distributed to officers and libraries of various government departments and agencies. The service also carried on liaison activities with the Canada Ethnic Press Federation and its four affiliated press associations in Toronto, Montreal, Winnipeg and Vancouver.

Native Communications Program. A native communications program received Cabinet and Treasury Board approval in early 1974. Grants are provided to communications societies set up to serve the communications needs of all native people in a given area. In 1976-77, there were 11 such societies from the Yukon Territory to Nova Scotia receiving funds at a total cost of $1.6 million.

The communications societies are involved in a full range of media, including HF radio systems, trail radios, film, theatre, newspapers and radio programming.

Sources
16.1 - 16.1.1 Canadian Telecommunications Carriers Association.
16.1.2 Transportation and Communications Division, Industry Statistics Branch, Statistics Canada.
16.1.3 Information Services, Department of Communications.
16.1.4 Teleglobe Canada.
16.2 - 16.2.1 Information Services, Canadian Radio-television and Telecommunications Commission.
16.2.2 Audience Services, Canadian Broadcasting Corporation.
16.2.3 Transportation and Communications Division, Industry Statistics Branch, Statistics Canada.
16.3 Public Affairs Branch, Post Office Department.
16.4 The Canadian Press; Canadian Daily Newspaper Publishers Association; United Press International of Canada Ltd.; Ethnic Press Analysis Service, Department of the Secretary of State; Native Communications Program, Department of the Secretary of State.

Tables

16.1 Pole-line and wire length and number of telephones in use, 1971-76

Year	Systems reporting	Route length km	Length of wire km	Telephones in use			
				Business	Residential	Total	Per 100 population
1971	1,171	471 916	95 271 948	2,996,276	7,272,505	10,268,781	47.3
1972	1,170	487 631	101 382 574	3,183,076	7,804,065	10,987,141	50.0
1973	985	550 396	109 338 836	3,428,292	8,249,152	11,677,444	52.3
1974	904	564 880	117 695 690	3,691,581	8,762,750	12,454,331	55.0
1975	850	537 521	124 994 708	3,928,375	9,236,635	13,165,010	57.2
1976	806	582 821	135 443 220	4,126,554	9,758,501	13,885,055	59.6

16.2 Telephones in use, by province, 1976

Province or territory	Telephones						
	On private lines		On party lines		Extensions		Coin telephones
	Business	Residential	Business	Residential	Business	Residential	Business
Newfoundland	17,762	92,973	1,281	29,181	17,064	28,889	1,848
Prince Edward Island	4,802	19,025	406	12,272	3,725	7,373	347
Nova Scotia	35,612	191,317	1,526	46,277	31,752	67,009	3,474
New Brunswick	25,583	144,059	1,147	40,778	30,585	55,957	1,966
Quebec	289,867	1,667,580	7,990	288,315	254,376	579,486	26,351
Ontario	424,837	2,307,055	9,012	413,082	342,431	1,040,237	35,040
Manitoba	51,763	272,792	3,282	57,511	45,611	89,974	3,552
Saskatchewan	41,263	221,437	2,931	69,029	33,593	83,529	2,877
Alberta	112,297	517,532	3,719	74,023	94,908	228,596	8,342
British Columbia	139,131	608,333	2,917	222,424	101,152	266,388	8,519
Yukon Territory	2,000	3,466	167	1,529	1,306	874	142
Northwest Territories	3,733	7,183	62	1,059	2,159	1,862	205
Canada	1,148,650	6,052,752	34,440	1,255,480	958,662	2,450,174	92,663
	Private branch exchanges		WATS[1]	Centrex	Mobile	Total	Telephones per 100 population
	Business	Residential	Business	Business	Business		
Newfoundland	18,369	—	—	3,688	470	211,525	37.9
Prince Edward Island	4,312	—	—	—	63	52,325	42.9
Nova Scotia	29,415	—	1	9,506	355	416,244	49.8
New Brunswick	21,990	—	—	9,557	216	331,838	47.8
Quebec	359,926	46	2,722	140,910	963	3,618,532	57.6
Ontario	547,148	49	4,392	224,197	1,768	5,349,248	63.7
Manitoba	69,382	—	195	4,345	115	598,522	57.9
Saskatchewan	42,527	—	32	6,863	800	504,881	53.4
Alberta	149,046	—	—	14,690	13,658	1,216,811	65.2
British Columbia	170,139	—	633	31,768	3,423	1,554,827	61.9
Yukon Territory	1,565	—	—	514	—	11,563	55.1
Northwest Territories	2,469	—	—	—	7	18,739	48.0
Canada	1,416,288	95	7,975	446,038	21,838	13,885,055	59.6

[1] On wide area telephone service lines.

16.3 Local and long-distance calls, calls per capita and average calls per telephone, 1971-76

Year	Local calls '000	Long-distance calls '000	Total calls '000	Calls per capita	Average calls per telephone		
					Local	Long-distance	Total
1971	16,439,365	495,454	16,934,819	779	1,601	48	1,649
1972	17,776,963	571,944	18,348,907	835	1,618	52	1,670
1973	18,396,642	658,248	19,054,890	854	1,575	56	1,631
1974	19,936,758	764,248	20,701,006	914	1,601	61	1,662
1975	20,340,605	853,504	21,194,109	922	1,545	65	1,610
1976	21,301,349	917,812	22,219,161	953	1,534	66	1,600

16.4 Financial statistics of telephone systems, 1971-76

Year	Capital stock[1] $'000	Long-term debt $'000	Cost of plant $'000	Revenue $'000	Expenditure $'000	Full-time employees	Salaries and wages[2] $'000
1971	2,005,304	2,861,144	7,255,226	1,725,302	1,504,854	69,995	600,949
1972	2,067,681	3,065,290	7,960,368	1,924,840	1,673,433	72,671	681,187
1973	2,149,479	3,297,124	8,791,434	2,200,702	1,920,424	75,407	775,700
1974	2,308,008	3,764,305	10,039,662	2,514,907	2,234,221	81,225	921,007
1975	2,519,844	4,435,368	11,426,333	3,054,705	2,650,396	82,866	1,091,350
1976	2,709,137	4,988,387	12,936,322	3,485,404	3,112,719	83,864	1,269,868

[1]Includes premium on capital stock.
[2]Full-time and part-time.

16.5 Financial statistics of telephone systems, by province, 1975 and 1976

Year and province or territory	Capital stock[1] $'000	Cost of plant $'000	Revenue $'000	Expenditure $'000	Full-time employees	Salaries and wages[2] $'000
1975						
Newfoundland	40,396	169,423	44,730	39,511	1,220	13,604
Prince Edward Island	8,912	39,676	10,011	8,848	312	2,910
Nova Scotia	93,957	378,969	93,032	82,448	3,522	35,299
New Brunswick	73,711	296,665	78,479	68,788	2,660	30,728
Quebec[3]	1,864,102	6,724,631	1,899,693	1,585,210	45,199	590,504
Ontario	60,522	119,235	41,309	28,946	787	8,807
Manitoba	—	472,418	107,581	106,874	4,408	55,340
Saskatchewan	64,940	457,242	102,013	86,649	3,291	41,012
Alberta	2	1,310,100	296,039	296,480	10,128	147,435
British Columbia	313,302	1,457,974	381,818	346,642	11,339	165,711
Northwest Territories	—	—	—	—	—	—
Total	2,519,844	11,426,333	3,054,705	2,650,396	82,866	1,091,350
1976						
Newfoundland	49,597	192,159	53,727	46,945	1,256	16,022
Prince Edward Island	10,795	44,826	11,502	10,153	286	3,374
Nova Scotia	111,628	427,271	113,550	100,110	3,561	44,398
New Brunswick	91,124	341,560	96,181	82,413	2,638	34,871
Quebec[3]	1,949,553	7,520,675	2,061,482	1,813,355	44,567	681,768
Ontario	67,426	129,606	48,121	35,300	913	11,334
Manitoba	—	524,447	133,074	123,275	4,481	61,710
Saskatchewan	59,827	528,515	126,789	107,724	3,613	51,084
Alberta	2	1,535,012	379,105	372,098	10,831	173,504
British Columbia	369,185	1,692,251	461,873	421,346	11,718	191,803
Northwest Territories	—	—	—	—	—	—
Total	2,709,137	12,936,322	3,485,404	3,112,719	83,864	1,269,868

[1]Includes premium on capital stock.
[2]Full-time and part-time.
[3]Includes data of Bell Canada which operates in Quebec, Ontario and the Northwest Territories.

16.6 Summary statistics of Canadian telecommunications, 1971-76

Year	Operating revenues $'000	Operating expenses $'000	Net operating revenue $'000	Pole-line length km	Wire length km	Employees[1]	Telegrams '000	Cable-grams[2] '000	Money transfers $'000
1971	146,413	107,567	38,846	68 120	1 190 651	7,553	5,888	5,347	40,833
1972	163,190	115,308	47,882	64 803	1 225 588	7,323	5,052	6,457	49,594
1973	190,703	140,114	50,588	60 416	1 214 424	7,047	3,454	7,412	41,944
1974	230,078	172,554	57,524	57 594	1 203 894	7,163	3,743	7,292	55,305
1975	259,059	193,811	65,248	51 744	1 235 347	7,162	4,115	8,016	81,798
1976	278,311	213,749	64,562	49 085	1 309 636	6,973	2,747	9,295	63,033

[1]Excludes commission operators.
[2]Includes wireless messages and transatlantic Telex messages.

16.7   Operating and financial summary of the cable television industry, 1974-76

Item		1974	1975	1976
OPERATING REVENUE				
Direct subscribers	$	113,496,341	137,283,823	167,712,541
Indirect subscribers (apartments, etc.)	$	9,640,628	12,289,588	13,688,966
Installation (including reconnect)	$	8,271,154	9,160,805	11,261,226
Education services	$	56,575	60,116	41,379
Other	$	1,967,942	3,478,525	6,510,866
Total, operating revenue	$	133,432,640	162,272,857	199,214,978
OPERATING EXPENSES				
Program	$	4,593,991	6,247,383	10,169,015
Technical	$	26,988,704	34,865,947	45,229,334
Sales and promotion	$	6,679,056	7,695,424	9,342,781
Administrative and general	$	28,226,368	36,957,565	43,406,516
Depreciation	$	27,120,858	32,782,776	39,610,101
Interest expense	$	11,019,993	13,831,609	17,706,585
Other adjustments — addition to (or deduction from) income	$	(366,718)	1,413,654	2,296,605
Total, operating expenses	$	104,995,688	130,967,050	163,167,727
Net profit (loss) before income taxes	$	28,436,952	31,305,807	36,047,251
Salaries and other staff benefits	$	32,628,507	42,714,948	56,421,499
Number of employees, weekly average		3,691	4,084	5,157
SUBSCRIBERS				
Individual		1,986,307	2,235,621	2,496,948
Indirect (contract with apartment building owner)		574,480	625,316	646,367
Total, subscribers		2,560,787	2,860,937	3,143,315
HOUSEHOLDS SERVED				
Households in licensed area (including apartments, etc.)		4,365,580	4,600,328	4,985,304
Households offered service (cable passes by building)		4,044,559	4,318,060	4,706,402
Households in multiple dwellings, offered service (apartments)		1,090,627	1,158,179	1,020,818

16.8   Estimated numbers and circulations of reporting English-language and French-language newspapers, by province, 1976 and 1977

Province	1976			1977		
	Daily	Circulation[1]	Weekend	Daily	Circulation[1]	Weekend
	English-language newspapers					
Newfoundland	3	48,357	1	3	50,302	1
Prince Edward Island	3	31,216	—	3	31,922	—
Nova Scotia	6	168,949	—	6	173,780	—
New Brunswick	5	123,259	—	5	124,623	—
Quebec	3	295,453	1	3	305,745	1
Ontario	46	2,140,889	1	48	2,201,364	1
Manitoba	8	254,929	—	9	268,397	—
Saskatchewan	4	130,973	—	4	134,564	—
Alberta	8	383,836	—	8	405,952	—
British Columbia	18	557,672	—	19	568,353	—
Total	104	4,135,533	3	108	4,265,002	3
	French-language newspapers					
New Brunswick	1	14,296	—	1	14,984	—
Quebec	11	842,105	14	10	867,409	15
Ontario	1	45,146	—	1	46,457	—
Total	13	901,547	14	12	928,850	15

[1]Circulation not reported for all newspapers.

16.9   Ethnic newspapers and periodicals published in Canada, by province, 1977

Ethno-cultural group	NS	Que.	Ont.	Man.	Sask.	Alta.	BC	Total
Arabic	—	4	3	—	—	—	2	9
Armenian	—	1	2	—	—	—	—	3
Black	1	2	3	—	—	—	—	6
Bulgarian	—	—	1	—	—	—	—	1
Byelorussian	—	—	1	—	—	—	—	1
Chinese	—	—	3	—	—	—	8	11
Croatian	—	—	5	—	—	—	1	6
Czech	—	—	3	—	—	1	—	4
Danish	—	—	1	—	—	—	1	2
Dutch	—	1	8	1	—	—	2	12
East Indian	—	1	5	1	—	1	2	10
Estonian	—	—	2	—	—	—	—	2
Filipino	—	1	2	1	—	—	—	4
Finnish	—	—	3	—	—	—	1	4
German	—	—	9	13	1	—	3	26
Greek	—	6	4	—	—	—	5	15
Hungarian	—	—	5	—	—	—	—	5
Icelandic	—	—	—	2	—	—	—	2
Italian	—	8	16	—	—	1	2	27
Japanese	—	1	3	—	—	—	1	5
Jewish	—	9	5	3	—	1	1	19
Korean	—	—	4	—	—	—	2	6
Latvian	—	—	3	—	—	—	—	3
Lithuanian	—	1	2	—	—	—	—	3
Malayalam	—	—	1	—	—	—	—	1
Norwegian	—	—	—	—	—	—	1	1
Pakistani	—	1	1	—	—	—	1	3
Polish	—	—	10	1	—	1	—	12
Portuguese	—	3	3	—	—	—	2	8
Romanian	—	—	2	—	—	—	—	2
Russian	—	—	2	—	—	—	2	4
Serbian	—	—	4	—	—	—	—	4
Slovak	—	—	6	—	—	—	—	6
Slovenian	—	—	3	—	—	—	—	3
Spanish[1]	—	—	4	—	—	1	1	6
Swedish	—	—	1	—	—	—	1	2
Swiss	—	—	1	—	—	—	—	1
Ukrainian	—	—	17	8	2	2	1	30
Yugoslav	—	—	1	—	—	—	—	1
Inter-ethnic[2]	—	—	2	—	—	1	1	4
Total	1	39	151	30	3	9	41	274

[1] Includes newspapers and periodicals representing South and Central America as well as Spain.
[2] Includes more than one ethno-cultural group.

Sources
16.1 - 16.7 Transportation and Communications Division, Industry Statistics Branch, Statistics Canada.
16.8 Compiled by Canada Year Book Staff from *Canadian Advertising.*
16.9 Department of the Secretary of State.

Manufacturing

Chapter 17

Tables

Manufacturing   Chapter 17

Manufacturing industries   17.1

Shipments of the top 10   17.1.1

Preliminary estimates for total manufacturing shipments in 1977 for 172 industries in Canada were $109.8 billion. The top 10 industries alone accounted for $46.4 billion or 42.3% of the total value of these shipments. Most of these industries have important export markets. Those exporting over half their shipments, according to the 1974 destination of shipments survey, were: motor vehicle manufacturers (63%), pulp and paper mills (62%), motor vehicle parts and accessories manufacturers (58%), and sawmills and planing mills (53%).

The leading manufacturing industry in Canada in 1977, measured by the value of manufacturing shipments, was motor vehicle manufacturers at $8.4 billion. Prices and production both increased by 8% during the year. Shipments of passenger cars to the United States increased 15.5% to $3.9 billion and trucks increased 47.5% to $2 billion. Automotive products are significant components of the balance of trade between Canada and the US and have been growing as a percentage of all Canadian exports.

Petroleum refining was a close second ranking industry with $8.1 billion of shipments. Canada has one of the highest per capita consumptions of energy in the world. The climate, geography, industrial structure and high standard of living all contribute to the high demand. There have been substantial price increases in this industry in recent years in attempts to reach world market prices.

Pulp and paper mills was next highest with shipments of $6.5 billion. The industry's real domestic product rose 5% over 1976, a small increase despite the industry-wide strikes of 1975, carried into the first few months of 1976, which artificially deflated the production for those two years. Slaughtering and meat processing held its fourth ranking position with $4.3 billion of shipments. There has been growth in the number of establishments in this industry in most years of the past decade matched by steady increases in the index of real output. Prices for the last three years remained stable.

Four industries were in the $3 billion to $4 billion range. Iron and steel mills at $3.8 billion in shipments showed modest increases in both prices and production for 1977. Weak demand constricted the Canadian market while over-capacity among foreign producers led to increased competition in the export market. Motor vehicle parts and accessories manufacturers at $3.7 billion had a production increase of 6% while prices rose 10%. Automotive components shipped to the United States in 1977 increased 23% to $3.4 billion. The performance of this industry reflects that of motor vehicle manufacturers in the US where sales increased substantially in 1977.

Sawmills and planing mills with shipments of $3.6 billion had a record year in 1977 for lumber production which increased substantially over 1976 in spite of high increases in prices (16%). The gain was due mainly to continued strong demand for family dwellings in the US. Over half of this industry's products are exported; lumber exports to the United States alone jumped 55.5% in 1977, to almost $1.9 billion.

Output of dairy products at $3 billion has remained fairly stable during the past few years. A significant change over the past decade has been a steady drop in the number of establishments in the industry from 880 in 1970 to 519 in 1975. Advances in technology, such as refrigerated trucks and automation, and the production of a wide range of commodities by bigger plants has caused the disappearance of small, specialized establishments.

Much of the output from the miscellaneous machinery and equipment industry, the ninth largest, goes into capital formation. Capital expenditures for machinery and equipment rose 8% in 1977 with an expected increase to $4.6 billion in 1978. Shipments of $2.7 billion in this industry included such diverse products as elevators, bulldozer blades, snow blowers and fire-fighting equipment.

Shipments for the 10th largest industry, smelting and refining, were $2.2 billion in 1977. A major strike in a large smelting and refining company in 1976 coupled with production for inventory purposes in 1977 explains the huge increase in the industry's index of real domestic product over the two years (25%). Increased input costs escalated prices by 22%. The forecast for 1978 was unfavourable because of cutbacks and layoffs brought on by deflated markets.

17.1.2   Statistics on manufacturing

Manufacturing, as one of the most important sectors of the economy, in 1976 accounted for 21% of real domestic product, that is, the net output of goods and services produced in Canada in constant dollars. An annual census of manufactures carried out by Statistics Canada provides a data base used for publication and internal research purposes. The task of editing thousands of comprehensive questionnaires results in a time lag of more than a year before census results can be fully compiled and made public. A sample survey of the monthly shipments, inventories and orders provides estimates for some of the statistics and narrows the gap between the reporting period and publication date to 60 days. These are the data cited above for 1977.

Table 17.1 compares the value of shipments of goods of own manufacture, by province, for 1976 and 1977 (from the monthly survey) with data for 1975 and earlier censuses, and Table 17.2 makes similar comparisons for industry groups. Table 17.3 gives company data on profitability in various industry groups for the years 1974-76. Because these latter figures relate to companies and those derived from the census of manufactures relate to establishments (roughly speaking, plants), the two series are of limited comparability.

17.1.3   Destination of shipments

A survey covering the destination of manufacturing shipments was carried out in conjunction with the census of manufactures for 1974. This investigation produced a comprehensive picture of the flow of goods shipped between provinces and exported outside Canada. Such data have many uses ranging from market research to transportation studies.

Total manufacturing shipments reported for 1974 were $82.5 billion. Products valued at $40 billion were shipped to destinations within the province of manufacture. Goods shipped outside the province of origin accounted for $37 billion, 56% or $20.7 billion of this going to other provinces or territories and 44% or $16.3 billion going to other countries. Non-distribution of shipments by respondents to the census questionnaire accounted for $3.8 billion, while custom and repair work, unallocated by survey definition, measured $1.7 billion.

The Atlantic provinces and British Columbia showed a relatively low dependence on interprovincial markets and relatively high shipments of exports compared to the central regions of Canada (Table 17.20). Across Canada, the region of origin was consistently the source of greatest supply to itself (Table 17.21). The second greatest supplier for all regions, except itself, was Ontario; this province alone accounted for more than half of all Canadian shipments.

Results of such a survey are influenced by the fact that respondents may often not be aware of the ultimate destination of goods which they ship to middlemen or others. This accounts for most of the shortfall of shipments to other countries compared with apparently corresponding data from the external trade division of Statistics Canada on exports from Canada.

17.2   Federal assistance to manufacturing

The industry, trade and commerce department is responsible for stimulating the establishment, growth and efficiency of the manufacturing, processing and tourist industries in Canada, and also for developing export trade and external trade policies. It assists Canadian industries to initiate and take advantage of technological advances, improve products and services, increase productivity and expand domestic and foreign markets through a variety of programs and services. At each phase of the product

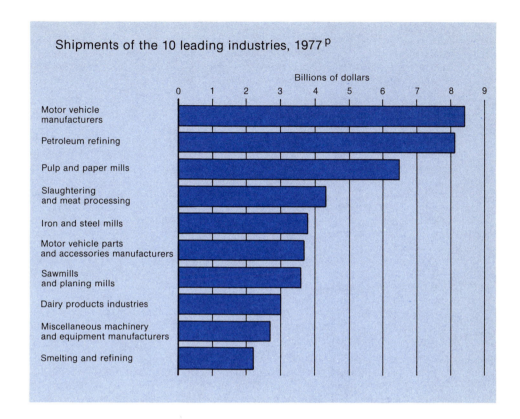

Shipments of the 10 leading industries, 1977 P

Billions of dollars

Motor vehicle manufacturers

Petroleum refining

Pulp and paper mills

Slaughtering and meat processing

Iron and steel mills

Motor vehicle parts and accessories manufacturers

Sawmills and planing mills

Dairy products industries

Miscellaneous machinery and equipment manufacturers

Smelting and refining

cycle — from research, development and design through production and marketing — the department can assist with information and funds.

Enterprise development   17.2.1

In April 1977 the Enterprise Development Program (EDP) replaced several innovative and adjustment assistance programs of the department to facilitate co-ordination among various forms of help, and make the programs more accessible, particularly to smaller and medium-sized Canadian businesses. The overall objective was to enhance growth in the manufacturing and processing sectors of the Canadian economy by providing assistance to selected firms to make them more productive and internationally competitive. The focus was on firms prepared to undertake, in relation to their resources, relatively high risk projects which were viable and promised attractive rates of return on the total investment.

EDP adopts a corporate approach to analysis of applicant firms to identify present and future requirements and to tailor a financing package that combines one or more forms of EDP assistance with other government assistance and private sector financing. This merchant banking flexibility has been described as investing in firms, not just supporting projects. A merchant bank is defined as a financial institution which endeavours to serve its clients by identifying, structuring and providing, or arranging for, all types of financing, and financial and management services required by a firm to realize its full potential.

The corporate/merchant banking approach is to examine the human, financial, physical and technological resources of a firm, the market opportunities and constraints and the firm's plans to exploit its market opportunities.

Decision-making is carried out at different levels. Mixed private sector and public sector boards provide pragmatic market-oriented decisions by drawing on the experience of prominent businessmen. Decentralized boards in the various regions of Canada, with delegated approval limits, make faster decisions possible, with awareness of regional business conditions a factor.

Forms of EDP assistance. Five types of grants are available to offset partially the cost of consultants to develop proposals for projects for assistance; to study market feasibility; to study productivity improvement projects which, while requiring no new technology, could involve some risk; to promote greater use of industrial design for mass-produced products; and for technological innovations which could lead to industrial growth and economic benefit to the firm and to the Canadian economy.

Loans or loan insurance for adjustment projects facilitate restructuring of firms by providing last resort financial assistance. The purpose is to help Canadian firms to meet international competition in both domestic and export markets. In some cases, the usual sources of financing are inadequate for smaller and medium-sized firms, and loan insurance (guarantees) can be provided. Direct loans may also be provided, but these are restricted to firms which have been injured by import competition. These loans and loan insurance may provide for plant expansion, equipment modernization or working capital. Because of risks associated with last resort financing, this type of assistance may be associated with one or more of the grants described above.

In addition, two forms of special purpose assistance are available: loans and grants to encourage restructuring of firms in footwear or tanning industries, and insurance on loans, leases and conditional sales agreements to air carriers in Canada and the United States to acquire de Havilland DHC-7 aircraft.

Industrial Research and Development Incentives Act. This legislation, enacted in March 1967, provided cash grants or equivalent tax credits to corporations that carried out research and development of benefit to Canada. As a result of restrictions in government spending, the payment of grants for work carried out after December 1975 was terminated, and the act was expected to be repealed in late 1978.

17.2.2 Automotive program

A Canada–United States agreement on automotive products in January 1965 provided for the removal of tariffs and other impediments to trade between the two countries in motor vehicles and original equipment parts. The basic objectives were creation of a broader market to permit benefits of specialization and scale, trade liberalization to enable both countries to participate in the North American market on an equitable basis, and development of conditions in which market forces would operate to attain economic patterns of investment, production and trade. As a result there were substantial increases in Canadian exports of vehicles and parts, employment in the industry, investment in new plants and expansion of existing facilities.

17.2.3 Machinery program

This program was introduced in January 1968 to increase efficiency in Canadian industry by enabling machinery users to acquire advanced capital equipment at the lowest possible cost while affording Canadian machinery producers tariff protection on what they manufacture. Canadian machinery producers are protected by a single statutory rate of duty which applies immediately when they are in a position to supply. This is particularly significant for Canadian producers of custom-engineered machinery.

The program covers general-purpose machinery, metalworking and woodworking machinery, construction and materials-handling equipment and various types of special industry machinery, such as pulp and paper and plastics industry machinery, and service industry equipment. The statutory rate of duty is 2½% British preferential and 15% most-favoured-nation. Since June 1971 the program has been extended to imports covering machinery for use in sawmills and logging.

The duty otherwise payable on machines, accessories, attachments, control equipment, tools and components may be remitted if such remission is in the public interest and the goods imported are not available from production in Canada. A machinery and equipment advisory board advises the minister of industry, trade and commerce on the eligibility of machinery for remission of duty. The board examines all tariff remission applications in respect of machinery and equipment or production tooling for the manufacture of original equipment, automotive parts and accessories. Final authority for granting remission lies with the Governor-in-Council.

Machinery producers may also apply for remission of duty on production parts and components which they cannot get in Canada. This is intended to stimulate Canadian machinery manufacturers to specialize their production and enable them to compete more effectively.

Industrial design   17.2.4

Design Canada, the administrative arm of the National Design Council and the branch of the department responsible for design in industry, manages a number of programs aimed at improving the products of Canadian secondary industry. Design Canada activities include co-funded programs of design assistance to industry; a design advisory service; design internship with industry; scholarships; design education advisory service and materials; product design case studies; audio-visual presentations and exhibits; and awards for design in industry.

Defence industry productivity   17.2.5

A program designed to enhance the technological competence of the Canadian defence industry in its export activities provides financial assistance to industrial firms for selected projects. Emphasis is placed on defence technology having civil export sales potential. Assistance may cover development of products for export purposes; acquisition of modern machine tools and other manufacturing equipment to meet exacting military standards; and assistance with pre-production expenses to establish manufacturing sources in Canada for export markets. Manufacturing equipment projects to be helped are selected on the basis that the machinery acquired will greatly increase productivity.

Shipbuilding industry assistance   17.2.6

A shipbuilding industry assistance program provides assistance to shipyards building or converting ships for domestic or export customers. Introduced in March 1975, this program supersedes the former ship construction subsidy regulations for domestic owners and shipbuilding temporary assistance program for export orders. Assistance is in two forms. An outright subsidy was introduced initially at 14% but revised to 20% of the approved cost of the eligible ship. An improvement grant of 3% is conditional upon the shipyard investing this and a matching amount for improved performance. The program encourages the use of Canadian materials, components and equipment when they are available at competitive prices.

Export market development   17.2.7

An export market development program is designed to help increase exports of Canadian goods and services. Canadian companies may obtain repayable contributions toward defraying approved expenses which would otherwise inhibit their attempts to earn a share of markets.

Section A, incentives for participation in capital projects abroad, is applicable anywhere outside Canada. The term capital projects is intended to describe facilities, systems and other projects requiring the provision of skilled services, engineering products and other capital goods. Section B, market identification and marketing adjustment, emphasizes manufactured goods but it can be more widely applied; it is applicable anywhere outside Canada and continental US. Section C, participation in trade fairs abroad, is not restricted as to markets, products or services; it is applicable anywhere outside Canada but participants in Canadian national stands at the same fair abroad are not eligible. Section D, incoming foreign buyers, also has no restrictions on markets, products or services; buyers from anywhere outside Canada and the continental US may be invited by a company to examine products and production in Canada or an agreed third location.

The department's contribution will normally be 50% of airfare and special and unusual costs and $70 a day toward personnel costs. If a company receiving assistance succeeds in obtaining the business sought, repayment of the department's contribution will be required, but no repayment is required if the company is unsuccessful.

17.2.8 Promotional projects

A program of trade fairs and missions was set up to promote the export of Canadian products and services. Its sponsored promotions are designed to meet particular requirements and include trade fairs abroad, trade missions, in-store promotions, travelling sample shows, incoming trade delegates and buyers programs, export-oriented training programs and, under the programs for export market development, incentives for participation in trade fairs abroad and incentives for foreign buyers.

17.2.9 Fashion design assistance program

Fashion/Canada, an incorporated body of representatives from federal and provincial governments and major associations of the Canadian fashion industry, administers this program. The designer development part of the program provides opportunities for students of fashion design and established fashion designers to acquire advanced training. Design and designer promotion publicizes good Canadian fashion design to attract domestic and foreign buyers and promotes Canadian fashion designers.

17.3 Provincial assistance to manufacturing

17.3.1 Newfoundland

The Newfoundland industrial development department assists prospective industry in determining desirable plant locations, preparing feasibility studies, and defining raw materials, transportation and labour costs as well as a variety of other economic data. The department also provides liaison with public and private sectors.

Financial assistance may be provided by the Newfoundland and Labrador Development Corporation through loans against the securities offered by the prospective enterprise, or by acquiring and holding shares or other securities of any company in the province, with the right of the enterprise to buy back these shares. The corporation also provides management advisory services.

The government may provide direct financial help based on cost-benefit analyses. Buildings and land may be provided on attractive terms. Industrial training facilities are available for specialized courses to meet requirements of incoming industry.

17.3.2 Prince Edward Island

Provincial assistance to manufacturers and processors is provided through Industrial Enterprises Inc. (IEI), an autonomous Crown corporation administered by an independent board of directors composed of businessmen. IEI provides money for new and existing manufacturing and processing industries. It identifies specific industrial opportunities, establishes their feasibility and provides management assistance in industrial engineering, marketing and finance.

The corporation makes loan capital, working capital and equity capital available. It constructs and rents serviced factory buildings and provides equipment leasing. It maintains industrial property and operates industrial parks at Charlottetown and Summerside, offering serviced lots of various sizes for sale to manufacturers and processors for warehouses and essential service businesses. Long-term financing is provided at attractive rates of interest.

IEI establishes contacts with venture capital groups in Canada and other countries who show an interest in provincial projects. Its internal consulting group provides management assistance to PEI companies.

17.3.3 Nova Scotia

Industrial Estates Ltd. (IEL) is a Nova Scotia Crown corporation created to promote and support the establishment and expansion of manufacturing industries in the province.

IEL can finance, at competitive interest rates, up to 100% of the cost of the land and buildings and up to 60% of the installed cost of production machinery of a new enterprise or a plant expansion. Financing of land and buildings over a 20-year period and of machinery over 10 years is customary. IEL can also design an incentives program to suit the needs of a project after careful evaluation of the economic impact of the project on Nova Scotia and Canada. IEL owns and operates the provincial industrial parks in Nova Scotia.

Mainland Investments Ltd. is a federal-provincial Crown company providing venture capital in the form of equity investments for mainland Nova Scotia.

The Nova Scotia Research Foundation conducts research into ocean technology, chemistry, biology and geophysics and offers advisory, technical and scientific services to industry and government.

The Nova Scotia Resources Development Board, affiliated with the Nova Scotia development department, provides term financing on the security of fixed assets for projects defined under the Industrial Loan Act, the Industrial Development Act and the Fishermen's Loan Act. It provides financing for tourism facilities, primary agriculture processing, fish plants, vessels and saw and planing mills.

The Nova Scotia development department has several other programs to help business and industry. A management development program subsidizes the salaries of NS graduates with MBA degrees who are hired by NS businesses. A marketing assistance program offers grants to assist Nova Scotia firms wishing to attend trade fairs and exhibits, conduct market identification investigations, attend market education courses, and host incoming buyers. A product design and development program provides grants to NS manufacturers for the design and development of products. A rural industry program offers capital grants to NS businesses to establish, expand or modernize their facilities outside the Halifax–Dartmouth city limits. An opportunity identification program assists new and established NS businesses to identify products likely to be commercially successful. An industrial malls program encourages the development of new, small businesses and industries by providing rental assistance in the first years of their existence, as well as advisory and some office services.

Other programs are offered by NS departments of agriculture and marketing, lands and forests, tourism, labour, fisheries and education which may be relevant to business and industry in certain sectors.

Municipal tax assistance is available for limited periods for new or expanding firms, on approval by the NS departments of municipal affairs and development.

The province co-operates closely with the Cape Breton Development Corp., a federal Crown corporation, and contributes financially to some of the industry-development projects sponsored by it.

New Brunswick 17.3.4

The commerce and development department has the major responsibility for the development of the manufacturing and processing sectors of the provincial economy. Its aims are to support and strengthen existing industries, attract new industry, increase the quality of employment, expand the tax base, maintain or improve social or environmental quality and alleviate regional disparities.

An industrial development branch is responsible for attracting new manufacturing and processing industries to New Brunswick. It analyzes and makes recommendations on all applications for financial assistance to industries wishing to locate in the province and to existing industries planning to expand. A regional development division is responsible for departmental liaison with the federal government on federal-provincial development agreements, capital expenditures in provincial industrial parks and with regional industrial development commissions. A commerce and industry services branch provides management, technical and product improvement services to industry; develops markets for manufactured or processed products; develops local processing of resources; and provides management, technical and financial services to industries in danger of failure. A planning branch evaluates cost-effectiveness of departmental programs and develops and modifies programs.

Three agencies are associated with the department and report to the commerce and development minister. The New Brunswick Industrial Development Board recommends financial assistance to manufacturers or processors, normally through a direct loan or loan guarantee. Terms and conditions are subject to individual negotiation but specifically require the applicant to provide reasonable equity and security in the form of a first charge on assets. Provincial Holdings Ltd. is a Crown corporation established to hold and administer the province's equity position in various companies. This agency is prepared to take an equity position in manufacturing industries wishing to locate in New Brunswick. The extent of the equity taken is negotiable and depends on various factors in a particular proposal. The Research and Productivity Council (RPC) provides a source of technical support services for New Brunswick industry. RPC carries out research and problem-solving on a cost-recovery basis for clients in Canada and abroad. An industrial engineering service and free technical information and assistance are made available to New Brunswick (and Prince Edward Island) companies by RPC in co-operation with the National Research Council.

17.3.5 Quebec

Quebec assistance to manufacturers is provided mainly through the Industrial Development Corp. The corporation's aims are to help transform Quebec's industrial structure through aid to high-technology industries, and to encourage existing industries to consolidate production facilities while adapting to modern techniques to improve their competitive position. Companies unable to obtain financial assistance at reasonable rates elsewhere may obtain aid if their operation contributes to the economic development of the province or any of its regions. Corporation assistance may be granted for capital investment for construction, purchasing, renovating or expanding plants, purchasing land, machinery, tools or equipment and facilities, purchasing or developing patents, improving the financial organization of the company or purchasing shares in manufacturing or commercial firms. Assistance may take various forms: loans at market interest rates; assumption of part of the costs of loans; repayment of part of the loans provided the company meets certain criteria; and purchase of part of a company's capital stock, provided the corporation at no time holds a majority of the capital stock or holds shares which bring its total investment in shares in such companies to more than 30% of its total assets.

In August 1977 two new programs were introduced by the industry and commerce department under the Act respecting Fiscal Incentives to Industrial Development (Bill 48). An industrial incentives fund was established to enable small and medium-sized companies to plan expansion, and encourage reinvestment of their profits. This continuing program enables companies with 200 or fewer employees, assets of less than $7.5 million and at least 50% of their gross revenue derived from production to deposit in the fund 50% of the provincial taxes otherwise payable. All sums deposited in the fund must be used within five and a half years following the taxation year to make an allowable expenditure related to manufacturing or processing operations, up to a maximum of 25%.

The second program provides manufacturing firms with a tax abatement to encourage regional economic development. It provides an abatement of 25% on the tax otherwise payable, up to a maximum of 25% of an allowable investment related to the operation of a manufacturing firm or $500,000 for the aggregate of the allowable investments. This applies to any corporation operating such a firm and making an investment of at least $50,000 prior to March 31, 1980, subject to certain economic and regional criteria.

A manufacturing or processing plant selling and delivering part of its Quebec production outside the province may receive an exemption from provincial sales tax on goods purchased for its own use, in proportion of its out-of-province sales to its total sales for the year. This measure also applies to provincial sales tax on gas and electricity used directly for manufacturing or processing. The manufacturer is also entitled to a partial reimbursement of sales tax paid on construction materials that he has used in his industrial buildings.

A manufacturing firm may receive total repayment of the tax paid on gasoline or diesel fuel if the fuel is used to operate machinery or as a raw material in the manufacture of certain products. Industrial machinery used in Quebec for manufacturing or processing is also exempt from provincial sales tax.

For a number of years the government of Quebec has sought easier access to foreign markets for Quebec businesses. The industry and commerce department has economic advisers attached to its delegations and Quebec offices abroad, provides financial and technical assistance to firms wishing to participate in industrial shows, organizes trade missions and provides information on export techniques and external trade. The Industrial Development Corp. also grants financial assistance to manufacturing and commercial operators exporting goods manufactured in Quebec.

In 1977 the Quebec national assembly authorized the creation of Quebec business development corporations. The aim is to provide risk capital for small and medium-sized manufacturing firms, as well as management assistance. Shareholders are entitled to tax abatements of up to 25% of the total investments, to a maximum of $25 per share.

A Quebec food crop programs corporation was set up to encourage and participate in the establishment, modernization, expansion, development, consolidation and grouping of food industries. It has two main methods of operation: participation as a co-partner in a business by purchasing risk capital, without at any time holding the majority of the capital stock, and selling its shares when the co-partners decide to repurchase them; and long-term loans at market interest rates (loans to shareholders). Its operations are restricted to the processing and marketing sectors.

The industry and commerce department offers companies technical assistance, mainly in the form of consultation services, negotiation of licence agreements, market studies and statistics. An industrial research centre also provides information and technical assistance services.

Ontario 17.3.6

Ontario northern and eastern development corporations are Crown agencies established by the Ontario government to provide services to business to stimulate industrial growth, economic development and employment opportunities. They report to the Ontario legislature through the industry and tourism minister. Their boards of directors represent the business and financial communities and organized labour.

Loan programs administered by the development corporations include an Ontario business incentive program to encourage industrial and economic development. Incentive loans are repayable, although initial repayment may be deferred. The loans may be interest free or at a rate lower than the Ontario Development Corp.'s prevailing rate of interest.

Term loan programs available to Canadian-owned companies in the province include small business loans for companies to expand their operations in manufacturing or services closely allied to manufacturing; venture capital loans to introduce new technology; loans for approved pollution control equipment; loans for tourist resort operators to upgrade facilities and establish new accommodation; export support loans to finance production and warehousing of goods for export against specific orders; and industrial mortgages and leasebacks to help establish or expand manufacturing facilities.

The Ontario Development Corp. administers Northam Industrial Park in Cobourg and Huron Industrial Park in Centralia, renting industrial space and housing. It also manages the sale and leasing of property in Sheridan Park near Toronto to companies engaged in industrial research and development.

The ministry of industry and tourism provides information on location and expansion of tourist operations, economic studies and other pertinent material; it advises tourist and service industry operators on ways to increase and improve their operations. An industry and trade division assists manufacturing companies and supporting service industries to maximize use of facilities, apply technology, establish new production facilities and find new business opportunities in both domestic and international markets. Technology, industrial design and development, product performance, energy management and loan programs are discussed in seminars with independent business people, industrial commissions and municipal councils. A manufacturing opportunity days program presents new products and processes, licensing opportunities, joint ventures from around the world and contracts for tendering.

Throughout the province 15 field offices meet local needs and problems and nine international field offices cover 35 countries; three are in the United States — Chicago, Los Angeles and New York; others are in Mexico City, Paris, Brussels, Frankfurt, London, Milan, Stockholm, Tokyo and Vienna. To aid the manufacturer, the international staff works in conjunction with industrial development officers in Toronto to arrange trade missions, business appointments, plant visits, incoming buyer missions, tourist incentive programs, and to provide consulting services with government and investment representatives. Some trade missions include visits to international trade fairs and exhibitions enabling executives from Ontario companies to see new products and manufacturing techniques. This program introduces Ontario companies to foreign concerns wishing to establish contacts for licensing, joint ventures or expansion in Canada.

17.3.7 Manitoba

The industry and commerce department encourages a planned expansion in the provincial economy through continued industrial and commercial development. The activities of the department encourage sales improvement, productivity improvement, new enterprise formation and existing enterprise expansion.

An industrial development branch assists companies or individual entrepreneurs who are considering new manufacturing plants and major expansions to existing plants. Professional service and advice are provided in such areas as feasibility and market studies. Site location reports are prepared that respond to client requirements for information on taxation, wage rates and labour availability, financial incentive programs, supply of raw materials and other operating factors.

A trade development branch (Manitrade) advises firms in developing and handling trade to export markets and acts as an export department or agent for small Manitoba companies.

The enterprise development group consists of nine branches and two associated agencies, all directly concerned with programs for small businesses. Its branches encourage small enterprise development and improvement, and new enterprise formation; provide management counselling; and advise on human resource management, and marketing, distribution and design of products. The two agencies are the Manitoba Research Council and the Manitoba Design Institute.

The Manitoba Development Corp. is a Crown corporation established in 1958. Its objectives are to assist businesses to increase productivity and to raise wage levels. It encourages exports and the use of locally developed products instead of imports.

The corporation provides money for businesses in the form of loans or bank loan guarantees. Technical and managerial help is made available in conjunction with the Manitoba industry and commerce department. The corporation also provides assistance in preparing applications for financing from other lenders and for federal and provincial grants.

17.3.8 Saskatchewan

The industry and commerce department has set up a development strategy aimed at providing jobs, increasing resource processing in the province and creating balanced economic growth. New programs have been introduced to encourage continuing development and innovation in Saskatchewan business. In 1976-77 the department was involved in the starting up and expansion of 60 industrial and commercial enterprises.

A small business interest abatement program offers grants to businesses to reduce costs on term financing used to establish new operations and expand or upgrade existing facilities. A small industry development program provides interest-free forgivable loans for manufacturing or processing industries to modernize, expand or build new facilities in the province. The amount covered by the loans varies according to the location of the industry. Financial aid for product design or quality improvement is available through a product development program. Product testing and prototype or special process development may also qualify under this plan.

Industries using management consultants can receive help through a management development program which offers grants up to 50% of consultation fees. This program provides financial assistance to industrial managers attending approved courses and the department also organizes and finances management seminars. A Saskatchewan mainstreet development program supplies consultant services and grants to help with storefront renovations in central business districts. The department provides counselling to business people through eight regional offices and two urban offices.

Jobs for the disadvantaged are provided through an economic development program for disadvantaged persons. By a federal-provincial special agreement under the Agricultural and Rural Development Act (ARDA), grants are made to projects employing native people.

An aid to trade program, through cost-sharing with the department, helps Saskatchewan manufacturers introduce new products and expand marketing territories. The department provides promotional opportunities for companies to expand their product market, and expertise for market analysis studies. It also produces a Saskatchewan foreign investment guide which outlines investment opportunities.

The Saskatchewan Economic Development Corp. (SEDCO), a Crown corporation, provides money for business enterprises. SEDCO was established in 1963 to provide loans for establishing or expanding manufacturing enterprises. In August 1972, its terms of reference were broadened to permit funding for virtually all types of businesses, including retail, wholesale and service enterprises. The most common form of SEDCO assistance is a first mortgage loan over a medium term. Security for such loans consists of pledges of land, buildings or equipment; support of individuals in the business is normally pledged as well. Repayment is designed to suit the income pattern of the enterprise, and may include step-payments, seasonal payments or similar arrangements. Terms vary from a few months to 20 years and amounts from a few thousands to many millions of dollars. The term is determined by the estimated life of security pledged and the earnings of the business. Equipment-based loans would be for 5 to 8 years, while building and equipment loans might be 8 to 12 years, and real estate alone as security would warrant a loan up to 20 years. Working capital loans would range from one month to two years.

In all cases, the corporation expects that owners of the borrowing company will have a reasonable equity in the enterprise. In certain instances, the corporation may take an equity investment in its own right if required to maintain a reasonable balance between debt and equity. The corporation also has industrial sites and buildings to lease to eligible businesses. Lease, lease-purchase or outright sale of such properties can be considered and, in certain circumstances, the corporation will construct a facility for sale or lease to a client.

Alberta   17.3.9

The Alberta Opportunity Co. (AOC) is a Crown agency created to promote economic growth by stimulating new businesses and aiding existing enterprises. AOC gives priority to Albertans and Alberta-owned enterprises, small businesses and centres of small population.

To qualify for assistance, a business may be a proprietorship, partnership, co-operative or corporate body, must operate for gain or profit, must be in Alberta and must provide assurance that any aid given will be used exclusively in Alberta. Eligible businesses include manufacturing, processing and assembly operations, service industries, commercial wholesale and retail trade, recreational facilities, tourist establishments, local development organizations, student business enterprises and new industries which are unique and valuable additions to the province. The program is not designed for finance companies, suppliers of residential accommodation other than tourist facilities, public utilities including power generation and distribution, or resource-based industries such as mining and oil and gas production.

Assistance may provide for establishing new businesses, acquiring fixed assets — land, buildings and equipment — expanding existing facilities, strengthening working capital, financing raw material or finished inventories for manufacturers, and research and development. Funds are made available directly or by guarantee in various forms.

Business counselling services may be provided by AOC and include management advice and guidance on financial, technical and marketing matters for small and intermediate-sized Alberta businesses which cannot afford to obtain this type of help elsewhere. Services are provided through the company's head office in Ponoka and branch offices in Calgary, Lethbridge, Grande Prairie, St. Paul, Medicine Hat, Edson and Edmonton.

17.3.10 British Columbia

The economic development ministry provides programs and services to government agencies, industries and the business community through two main branches. A policy planning and research branch formulates economic development flows for the province, co-ordinates implementation of strategies and provides information for many of the ministry's activities. The branch also implements programs originating from the Canada–British Columbia Industrial Development Subsidiary Agreement. A business and industrial development branch provides market development and promotional programs to assist existing businesses, encourage new industry and develop BC's exports. These include trade missions, trade shows, seminars, incoming buyers' programs, and market development.

A small business assistance program helps managers in analyzing and diagnosing business problems, recommending solutions and preparing financial proposals and applications for assistance through various aid programs. Through its technical assistance program, the branch shares companies' costs in undertaking market and financial feasibility studies, and studies designed to improve productivity. The aim of this program is to provide incentive to companies to expand their facilities, diversify their product lines, or enter into new business.

The ministry maintains liaison with the British Columbia Development Corp. This Crown corporation provides for acquisition and servicing of land suitable for industrial use in areas where serviced industrial land was not previously available, or where high land costs prohibited location of individual firms. It also provides financial assistance through loans to businesses wishing to expand existing operations or create new economic activity in the province. The ministry maintains a trade and industry office at British Columbia House in London, England.

17.4 Government aid and controls

17.4.1 Department of Consumer and Corporate Affairs

The federal consumer and corporate affairs department administers federal legislation and policies affecting business, and demonstrates that a competitive marketplace can benefit consumers, business people and investors. Four bureaus and a field operations service share responsibility for achieving the department's marketplace objectives.

The consumer affairs bureau co-ordinates government activities in the field of consumer affairs; it includes consumer services and consumer research branches and the consumer standards directorate. The corporate affairs bureau administers legislation and regulations pertaining to corporations; its branches are responsible for corporations, bankruptcy, securities and research. The intellectual property bureau administers laws pertaining to patents, copyright and industrial design, and trade marks, with a branch responsible for each of these fields. Canada's participation in international intellectual property organizations is the responsibility of the research and international affairs branch, and the role of informing Canadians of the services of the bureau is filled by a technical advisory services branch. The bureau of competition policy has branches specializing in resources, manufacturing, services and marketing practices. A research branch undertakes basic research projects. The Restrictive Trade Practices Commission is an independent administrative commission that reports directly to the minister.

The field operations service supervises the department's operations across Canada, staffing regional and district offices in Vancouver, Winnipeg, Toronto, Montreal and Halifax, and district and local offices in other cities. These offices ensure that laws and regulations administered by the department are uniformly applied and interpreted in all parts of the country. The field force includes consumer consultants and information officers, complaints officers, inspectors and specialists in the fields of bankruptcy and marketing practices such as misleading advertising.

Anti-combines legislation. Canadian anti-combines legislation seeks to eliminate restrictive trade practices in order to stimulate maximum production, distribution and employment through open competition. Legislative measures, including some formerly included in the Criminal Code, were amended in 1960 and consolidated into the Combines Investigation Act (RSC 1970, c.C-23). An act to amend this act was passed in December 1975 (SC 1974-75-76, c.76) and for the most part came into effect January 1, 1976, the remainder on July 1 of the same year.

In general terms, the Combines Investigation Act makes illegal the operation of combines that prevent, or lessen unduly, competition in production, manufacture, purchase, barter, sale, storage, rental, transportation or supply of a product of trade or commerce, or in the price of insurance.

Under the act it is illegal to participate in a merger or a monopoly that has operated, or is likely to operate to the detriment of the public, whether consumers, producers or others. Other sections of the act forbid misleading or deceptive advertising, either as to normal price or as to presumably factual statements describing goods or property offered for sale. The act also provides against double ticketing, pyramid selling, referral selling, bait and switch selling, and certain types of promotional contests. Resale price maintenance, price discrimination and "predatory price cutting" are also prohibited.

The assistant deputy minister for the bureau of competition policy, who is also the director of investigation and research, is responsible for investigating combines and other restrictive practices. The Restrictive Trade Practices Commission is responsible for appraising the evidence submitted to it by the director and the parties under investigation, and for making a report to the consumer and corporate affairs minister. When there are reasonable grounds for believing that a forbidden practice is engaged in, the director may obtain from the commission authorization to examine witnesses, search premises, or require written returns. After examining all the information available, if the director believes that it proves the existence of a forbidden practice, he submits a statement of the evidence to the commission and to the parties believed to be responsible for the practice. The commission then sets a time and place for a hearing at which both sides are represented. The commission prepares and submits a report to the minister; such reports are required to be published within 30 days. At the completion of an inquiry, the director may submit the evidence directly to the attorney general for prosecution without going to the commission.

The director may bring before the Restrictive Trade Practices Commission a broad range of business matters for review under civil procedures. The commission is empowered to issue appropriate remedial orders where serious anti-competitive effects are found.

Patents. Patents for inventions are issued under the provisions of the Patent Act (RSC 1970, c.P-4) and patent regulations have been proclaimed to carry into effect the objectives of the act. Applications for patents for inventions and requests for information about such patents should be addressed to the Commissioner of Patents, Bureau of Intellectual Property, Consumer and Corporate Affairs Canada.

On November 16, 1976, the patent office issued its one millionth patent. A bronze medallion was struck by the Royal Canadian Mint to commemorate the event.

By March 31, 1977, the complete office file of 1,007,800 issued patents was classified and organized into 339 main classes of technology which were further subdivided into 32,121 subclasses. These classes are constantly reviewed, and revised or extended as new technologies emerge and new combinations of known technologies are developed. During 1976-77 and the previous year, 20 classes consisting of 2,194 subclasses were completely revised; 1,911 new subclasses were established and 756 old subclasses were abolished in the partial revision of existing classes.

A search room and library are maintained by the patent office where the public may obtain information on Canadian and foreign patents. Printed copies of Canadian patents issued from January 1, 1948 are available at $1 each. During 1975-76 and 1976-77, the patent office handled an average of 675 requests daily for reference material and published weekly the *Patent Office Record* which contains a list of patents issued during the week covered, information about patent office services and information of concern to the patent profession.

Foreign patents may also be seen at the patent office library. British patents and their abridged specifications from 1617 to date and United States patents from 1845 to date are available, as well as many patents, indexes, journals and reports from Australia, India, Ireland, New Zealand, Pakistan, South Africa, Austria, Belgium, Colombia, Czechoslovakia, Egypt, France, Federal Republic of Germany, Italy, Japan, Mexico, the Netherlands, Norway, Sweden, Switzerland and Yugoslavia. A list of the foreign patents available is published in the *Patent Office Record.*

Copyright, industrial design and timber marks. Copyright protection is governed by the Copyright Act (RSC 1970, c.C-30) in force since 1924. Protection is automatic without any formality, although a system of voluntary registration is provided. Copyright exists in Canada in every original literary, dramatic, musical and artistic work and in contrivances by means of which sounds may be mechanically reproduced. The term for which the copyright exists is, except as otherwise expressly provided by this act, the life of the author and a period of 50 years after death.

The Industrial Design Act provides a maximum 10-year period of protection for shape, pattern, ornamentation and configuration applied to an article of manufacture, provided that the design is registered within one year of publication in Canada. Protection is granted if an examination does not reveal any other design already registered to be identical with or closely resembling the proposed design. The name of the proprietor, the letters Rd. (Registered) and the year of registration must appear upon the article to which the design applies.

People or companies floating timber on the inland waters of Ontario, Quebec and New Brunswick must, based on the Timber Marking Act, select a mark or marks and apply for their registration within one month after starting in this business.

Trade marks. The trade marks office, a branch of the intellectual property bureau, administers the Trade Marks Act (RSC 1970, c.T-10) which covers all legislation concerning the registration and use of trade marks and supersedes from July 1, 1954 former legislation enacted under the Unfair Competition Act, the Union Label Act and the Shop Cards Registration Act. Correspondence relating to an application for registration of a trade mark should be addressed to the Registrar of Trade Marks, Ottawa.

Applications are advertised in the *Trade Marks Journal,* a weekly publication that also gives particulars of every registration of a trade mark. The required fee payable on application for registration of a trade mark is $35 and for advertisement of an application, $25.

17.4.2 Trade standards

17.4.2.1 Standards Council of Canada

This council, with headquarters in Ottawa, is the national co-ordinating agency through which organizations concerned with voluntary standardization may co-operate in recognizing, establishing and improving standards in Canada. It enables these organizations to play a larger and more effective role in formulating and promoting the use of standards to meet the needs of the economy through a national standards system. Sponsored by the council, the system includes organizations involved in standards-writing, testing and certification. It encourages the development of national standards of Canada to meet both national and international responsibilities.

The objects of the council are to foster and promote voluntary standardization in fields relating to the construction, manufacture, production, quality performance and safety of buildings, structures, manufactured articles and products and other goods.

In the international field, the council appoints members and directs activities of the Canadian National Committee of the International Electro-Technical Commission and is the member body for Canada in the International Organization for Standardization. The council co-ordinates and integrates the national and international standards and oversees the accreditation of some 350 delegates to represent Canada at over 360 international technical committee meetings each year. The council's international standardization branch is located at Mississauga, Ont.

In January 1977 the council, in co-operation with the accredited standards writing organizations, established a standards information service, with a central information and referral service at its Ottawa offices.

Trade standards and regulations 17.4.2.2

In its consumer program, the consumer and corporate affairs department is responsible for administration of broad legislation affecting the business community. Policies and programming are determined by the consumer standards directorate, and field supervision by the field operations service.

Hazardous products. A product safety branch administers the Hazardous Products Act which deals with consumer goods. The act makes specific mention of products designed for household, garden, or personal use, for use in sports or recreational activities or for use by children. It also mentions without reference to end use, poisonous, toxic, flammable, explosive and corrosive products. The minister is empowered to establish mandatory standards for application in Canada. Compliance orders being enforced include the use of shatterproof glass in patio and shower doors, flammability standards for children's sleepwear and protective standards for hockey helmets. Regulations governing toys, rattles, cribs and portable car seats are designed to protect children. Other rigid specifications cover matches, charcoal, ceramics and electrical appliances.

General commodity field. The Consumer Packaging and Labelling Act and regulations administered by a consumer fraud protection branch are designed to give uniformity to packaging and labelling practices in Canada, reduce the possibilities of fraud and deception in packaging, and control the proliferation of packaged sizes. The legislation applies to most pre-packaged consumer products and came into effect in September 1975 for non-food items and in March 1976 for foods.

Regulations under the Textile Labelling Act, in effect since December 1972, require labels on all consumer textile articles. The label must include fibre names and percentages and the identification of the dealer. The regulations also deal with misrepresentation in both labelling and advertising. The textile care labelling system of coloured symbols recommending proper care for textile products is a voluntary program. The Canada standard size system for children's garments, developed by the Canadian Government Specifications Board in conjunction with the consumer and corporate affairs department, is administered under the National Trade Mark and True Labelling Act. This system is also voluntary, although dealers must register for a licence before claiming that the garment does, in fact, conform to the standard size and before attaching such a label to the product.

Control of marking of precious metal articles is maintained under the Precious Metals Marking Act. The regulations came into force in July 1973.

Food. In areas of health, grading, standards and composition, the Food and Drug Act, the Canadian Agricultural Products Standards Act and the Fish Inspection Act are generally applicable. The consumer and corporate affairs department is charged with administration of the economic fraud aspects in distribution. This responsibility relates mainly to labelling and advertising in any segment of the news media.

Advertising. Most legislation has particular requirements to ensure against misleading advertising. The deceptive marketing provisions of the Combines Investigation Act include general provisions against misleading advertising practices.

Measurement. The Weights and Measures Act prescribes the legal standards of weight and measure for use in Canada; it also ensures control of the types of all weighing and measuring devices used for commercial purposes, and provides for in-use surveillance directed toward the elimination of device-tampering and short-weight sales. A replacing act was passed by Parliament and new regulations were proclaimed in August 1974. The fundamental objectives of earlier legislation remain unchanged. The act is complementary to consumer packaging and labelling legislation.

The Electricity Inspection Act and the Gas Inspection Act control the approval before sale and the use of instruments used for billing of electricity and gas whether by meter or other type of device; they also provide for continual in-use inspection.

Corporations. The bureau of corporate affairs is concerned with much of the general legal framework that governs the orderly conduct of business under federal jurisdiction. The bureau is subdivided into branches for bankruptcy, corporations and corporate research.

The corporations branch administers the Canada Business Corporations Act, the Canada Corporations Act, the Canada Co-operatives Association Act and the Boards of Trade Act. The branch has a statutory duty to issue formal documents in connection with corporations created under other federal acts such as the Loan Companies Act, Trust Companies Act, the Canadian and British Insurance Companies Act, and the Railway Act.

All federal corporations other than those carrying on business as financial intermediaries must be incorporated under the Canada Business Corporations Act, proclaimed in December 1975. However, because that act does not repeal the old Canada Corporations Act until December 15, 1980, the branch is required to administer corporations subject to either act until that date. This policy of gradual implementation of the Canada Business Corporations Act was adopted to enable corporations to effect transition from the old to the new act with a minimum of pressure and inconvenience following a relatively simple continuance procedure. One part of the Canada Corporations Act continues to apply to all federal charitable and membership corporations.

Ancillary to its formal activities, the branch furnishes to the public copies of corporate documents and information about registered corporate names and trade marks. In 1976-77 the branch issued 17,000 documents compared to 23,000 in 1975-76; this drop reflects a decision to discontinue providing copies of financial statements which are published in monthly issues of the *Bureau of Corporate Affairs Bulletin*. There were also 32,895 registered corporate name and trade mark searches made, an increase of 11,038 from 1975-76.

By March 31, 1977 full automation of all name search services for information on registered corporate names and trade marks neared completion. The branch maintains a computerized file on approximately 500,000 corporate entities that exist in Canada (30,000 of which are federally incorporated) and a list of close to 150,000 registered trade marks.

Two programs begun in 1977 include plans for a national food strategy for Canada and energy conservation in use of large electrical household appliances. In conjunction with other federal departments and the provinces, the department conducts investigations to ensure that food strategy will provide Canadians with reasonably-priced, nutritious food while at the same time maintaining a strong, competitive food industry in Canada. With regard to energy conservation, the consumer and corporate affairs department takes the position that non-renewable energy sources will not last forever and that part of the answer lies in resource conservation and more efficient use of energy.

17.5 Bankruptcies and commercial failures

Two series of figures are included in this chapter which, although closely related in subject matter, cover different aspects of the field of bankruptcies and commercial failures. Table 17.13 is limited to the supervision, by the superintendent of bankruptcy, of the administration of bankrupt estates under the Bankruptcy Act (RSC 1970, c.B-3); it gives information on the amounts realized from the assets as established by debtors and indicates that values actually paid to creditors are invariably very much lower than such estimates alone would imply. It can therefore be assumed that this applies in even greater degree to the more extended fields covered in the second series (Tables 17.11 and 17.12) compiled by Statistics Canada, which is limited to bankruptcies and insolvencies made under federal legislation and includes business failures only.

The report issued annually by the superintendent of bankruptcy gives statistics and comments on various activities in the field of bankruptcy, such as prosecution for offences, issue of licences for trustees in bankruptcy, number of estates reported and closed during the year and costs of bankruptcy administration in Canada. These data are summarized in Table 17.13.

Returns under the Bankruptcy and Winding-up Acts. Statistics Canada data on bankruptcies and insolvencies cover only business failures coming under the federal Bankruptcy Act and the Winding-up Act. Table 17.11 gives yearly comparisons of liabilities — as estimated by debtors — for the main regions of the country. Table 17.12 shows the number of bankruptcies and insolvencies by industry and economic area for 1975 and 1976.

Administration of bankrupt estates. The Bankruptcy Act was revised in 1949 and amended in 1966. The amendments were instigated by exposures and suggestions of illegal and improper practices in connection with bankruptcy proceedings or administration. They do not constitute a complete revision of the Bankruptcy Act but were designed to provide, as an interim measure, remedies to the most urgent areas of complaints. They give the superintendent of bankruptcy direct and immediate authority in investigation and inquiry, and tighten the procedures and requirements in a number of areas, such as proposals which an insolvent person may make to his creditors. These amendments were intended to provide remedies in situations where abuses of the bankruptcy process are most likely to occur. The amendments also contain a part on the orderly payment of debts which may be brought into force in any province at the request of provincial authorities. Six areas have taken advantage of this part of the legislation: Alberta in April 1967, Manitoba in June 1967, Saskatchewan in April 1969, British Columbia in June 1970, Nova Scotia in July 1970 and Prince Edward Island in April 1971.

A small debtor program was instituted in June 1972. While it is not an amendment to the Bankruptcy Act, it authorizes federal employees who have been appointed as trustees to handle the estates of certain wage-earners who cannot obtain the services of a private trustee.

Sources
17.1 Manufacturing and Primary Industries Division, Industry Statistics Branch, Statistics Canada.
17.2 Information Services Branch, Department of Industry, Trade and Commerce.
17.3 Supplied by the respective provincial government departments.
17.4.1 Information and Public Relations, Department of Consumer and Corporate Affairs.
17.4.2 The Standards Council of Canada; Information and Public Relations, Department of Consumer and Corporate Affairs.
17.5 Superintendent of Bankruptcy, Department of Consumer and Corporate Affairs; Business Finance Division, General Statistics Branch, Statistics Canada.

Tables

17.1 Value of shipments of goods of own manufacture, by province, 1961 and 1974-77 (million dollars)

Province or territory	1961	1974	1975	1976P	1977P
Newfoundland	135.9	711.7	650.0	585.9	656.4
Prince Edward Island	30.6	94.1	108.6	..	..
Nova Scotia	381.4	1,696.1	1,819.1	1,978.8	2,225.1
New Brunswick	390.6	1,585.7	1,669.4	1,859.4	2,098.8
Quebec	7,022.2	22,396.8	23,966.5	25,791.6	28,145.3
Ontario	11,563.7	41,404.4	44,422.8	50,291.4	55,893.3
Manitoba	716.7	2,279.7	2,581.0	2,748.2	2,929.8
Saskatchewan	331.9	1,045.2	1,176.5	1,218.5	1,357.7
Alberta	935.5	3,821.3	4,726.5	5,273.4	6,118.4
British Columbia	1,927.0	7,411.1	7,326.5	8,718.6	10,229.3
Yukon Territory and Northwest Territories	3.4	9.0	13.6	..	..
Canada	23,439.0	82,455.1	88,460.4	98,597.3[1]	109,798.3[1]

[1]Includes Prince Edward Island, the Yukon Territory and Northwest Territories.

17.2 Value of shipments of goods of own manufacture, by industry group, 1974-77 (million dollars)

Industry group	1974	1975	1976P	1977P
Food and beverage industries	14,737.7	16,492.3	17,293.6	19,094.0
Tobacco products industries	704.9	831.5	855.0	892.8
Rubber and plastics products industries	1,833.5	1,955.8	2,302.4	2,530.0
Leather industries	570.1	619.2	706.6	710.9
Textile industries	2,477.8	2,439.0	2,772.7	3,000.7
Knitting mills	600.6	624.5	641.7	719.3
Clothing industries	2,076.6	2,306.6	2,569.2	2,763.7
Wood industries	3,991.1	3,802.6	4,901.1	5,991.3
Furniture and fixture industries	1,338.2	1,363.7	1,471.1	1,546.8
Paper and allied industries	7,677.4	7,131.6	8,162.7	8,905.2
Printing, publishing and allied industries	2,550.5	2,897.5	3,195.4	3,446.5
Primary metal industries	6,535.4	6,682.4	7,329.2	8,270.7
Metal fabricating industries (except machinery and transportation equipment industries)	5,834.0	6,216.7	6,779.6	7,358.5
Machinery industries (except electrical machinery)	3,137.8	3,731.6	3,892.2	3,990.1
Transportation equipment industries	10,173.9	11,193.0	12,847.9	15,123.0
Electrical products industries	4,344.9	4,599.3	5,007.3	5,150.9
Non-metallic mineral products industries	2,282.5	2,569.4	2,915.5	3,177.3
Petroleum and coal products industries	5,185.3	5,953.3	6,979.4	8,300.3
Chemical and chemical products industries	4,607.7	5,107.4	5,833.3	6,476.6
Miscellaneous manufacturing industries	1,794.9	1,943.0	2,141.3	2,349.5
All manufacturing industries	82,455.1	88,460.4	98,597.3	109,798.3

17.3 Net profit before taxes and extraordinary items, as a percentage of total revenue of corporations classified to the manufacturing industries, 1974-76

Industry group	1974	1975	1976
Food and beverage industries	4.5	4.7	4.2
Rubber industries	4.7	4.3	3.2
Textile industries[1]	5.4	2.7	2.2
Wood industries[2]	6.8	4.4	5.9
Paper and allied industries	14.9	7.9	6.0
Printing, publishing and allied industries	10.5	9.0	8.2
Primary metal industries	12.3	8.2	4.4
Metal fabricating industries	8.2	7.8	7.1
Machinery industries	9.0	9.1	8.5
Transportation equipment industries	4.9	3.9	4.5
Electrical products industries	7.5	8.9	6.0
Non-metallic mineral products industries	10.5	10.6	9.7
Petroleum and coal products industries	15.5	13.7	11.2
Chemical and chemical products industries	12.6	11.8	9.0
Miscellaneous manufacturing industries	8.6	8.3	8.3
All manufacturing industries	8.8	7.4	6.3

[1]Includes knitting mills and clothing industries.
[2]Includes furniture and fixture industries.

17.4   Summary statistics of manufactures, 1966-75

Year	Estab-lish-ments No.	Activity						

		Manufacturing activity						
		Production and related workers			Cost of fuel and electri-city[1] $'000	Cost of materials and supplies used $'000	Value of shipments of goods of own manu-facture $'000	Value added $'000
		Number	Man-hours paid '000	Wages $'000				
1966	33,377	1,172,943	2,498,012	5,575,206	731,726	20,642,695	37,303,455	16,351,740
1967	33,267	1,168,651	2,478,916	5,869,085	759,780	21,371,785	38,955,389	17,005,696
1968	32,643	1,160,226	2,458,791	6,278,429	808,764	23,090,970	42,061,555	18,332,204
1969	32,669	1,189,887	2,515,183	6,921,525	860,525	25,383,484	45,930,438	20,133,593
1970	31,928	1,167,063	2,450,058	7,232,256	903,264	25,699,999	46,380,935	20,047,801
1971	31,908	1,167,810	2,448,419	7,819,050	1,000,243	27,661,379	50,275,917	21,737,514
1972	31,553	1,213,106	2,547,609	8,763,104	1,078,916	31,137,946	56,190,740r	24,264,829r
1973	31,145	1,275,985	2,665,681	10,060,062	1,221,885	37,600,538	66,674,393	28,716,119
1974	31,535	1,300,792	2,713,436	11,637,073	1,623,617	47,499,791	82,455,109	35,084,752
1975	30,100	1,272,051	2,613,549	12,672,237	1,805,666	51,177,157	88,460,358	36,139,301

		Total activity						
		Working owners and partners		Total employees[2]		Cost of materials and supplies used and goods purchased for resale[3] $'000	Value of shipments and other revenue[4] $'000	Value added[5] $'000
		Number	With-drawals $'000	Number	Salaries and wages $'000			
1966	33,377	13,894	60,076	1,646,024	8,695,890	24,195,610	41,722,527	17,260,256
1967	33,267	13,377	59,187	1,652,827	9,254,190	25,546,764	44,143,808	18,049,639
1968	32,643	12,084	58,798	1,642,352	9,905,504	27,546,942	47,646,657	19,483,614
1969	32,669	11,583	59,128	1,675,332	10,848,341	30,347,637	52,130,615	21,456,276
1970	31,928	10,760	58,605	1,637,001	11,363,712	30,805,904	52,886,022	21,417,748
1971	31,908	10,286	60,939	1,628,404	12,129,897	33,462,590	57,479,421	23,187,881
1972	31,553	9,793	62,330	1,676,130	13,414,609	37,663,105	64,360,301r	25,981,742r
1973	31,145	8,981	..	1,751,066	15,220,033	45,697,053	76,689,795	30,766,506
1974	31,535	7,075	..	1,785,977	17,556,982	57,794,605	95,030,218	37,654,465
1975	30,100	6,977	..	1,741,545	19,160,724	62,381,833	102,178,371	38,715,600

[1]Cannot be reported separately for manufacturing and non-manufacturing activities but related substantially to manufacturing activity.
[2]Includes production and related workers, administrative and office employees, sales, distribution and other employees; excludes working owners and partners.
[3]Includes supplies used in both manufacturing and non-manufacturing activity.
[4]Includes shipments of goods of own manufacture, value of shipments of goods purchased for resale and other operational revenue.
[5]Value of total operational revenue less total cost of materials, supplies, fuel and electricity used and goods purchased for resale in the same condition; all adjusted for inventory changes where required.

17.5   Establishments in the manufacturing industries, by number employed and by province, 1975

Province or territory	Number employed									Total
	Under 5	5 to 9	10 to 19	20 to 49	50 to 99	100 to 199	200 to 499	500 to 999	1,000 or over	
Newfoundland	93	46	30	37	33	18	10	1	2	270
Prince Edward Island	38	24	24	19	8	3	1	—	—	117
Nova Scotia	191	130	123	118	62	29	24	8	4	689
New Brunswick	159	87	85	112	39	51	17	6	3	559
Quebec	2,274	1,482	1,615	1,932	930	626	375	88	53	9,375
Ontario	2,732	1,994	2,142	2,427	1,243	880	584	167	76	12,245
Manitoba	346	199	203	219	112	84	39	10	3	1,215
Saskatchewan	214	137	126	91	45	25	13	1	1	653
Alberta	497	391	351	298	139	94	39	11	1	1,821
British Columbia	1,029	575	492	521	228	152	105	19	10	3,131
Yukon Territory and Northwest Territories	9	5	6	5	—	—	—	—	—	25
Canada	7,582	5,070	5,197	5,779	2,839	1,962	1,207	311	153	30,100

17.6 Number of establishments in manufacturing industries, by industry group and employment size group, 1975

Industry group	Establishments with total employment of									Total
	Under 5	5 to 9	10 to 19	20 to 49	50 to 99	100 to 199	200 to 499	500 to 999	1,000 or over	
Food and beverage industries	1,327	906	820	795	401	267	169	42	13	4,740
Tobacco products industries	3	1	2	6	1	2	7	3	2	27
Rubber and plastics products industries	128	99	146	195	100	56	29	13	5	771
Leather industries	63	47	52	101	63	64	22	3	–	415
Textile industries	173	159	166	166	94	74	67	17	7	923
Knitting mills	18	31	40	80	58	53	23	2	1	306
Clothing industries	317	249	382	564	303	195	76	7	1	2,094
Wood industries	958	486	486	480	265	162	74	8	1	2,920
Furniture and fixture industries	842	301	267	296	120	95	36	1	1	1,959
Paper and allied industries	30	58	75	140	92	104	97	43	26	665
Printing, publishing and allied industries	1,372	794	642	468	191	81	50	13	7	3,618
Primary metal industries	23	20	50	95	47	57	45	21	22	380
Metal fabricating industries (except machinery and trans-portation equipment industries)	836	757	768	858	318	209	111	22	3	3,882
Machinery industries (except electrical machinery)	134	159	170	303	142	96	69	13	12	1,098
Transportation equipment industries	174	131	138	206	121	97	70	27	27	991
Electrical products industries	85	70	118	164	126	101	81	42	18	805
Non-metallic mineral products industries	239	201	282	245	103	78	41	8	2	1,199
Petroleum and coal products industries	14	14	9	20	9	18	13	3	1	101
Chemical and chemical products industries	153	175	160	244	135	84	75	18	2	1,046
Miscellaneous manufacturing industries	693	412	424	353	150	69	52	5	2	2,160

17.7 Establishments and shipments in the manufacturing industries, by shipments per establishment, 1975

Value group	Establish-ments No.	Value of shipments of goods of own manufacture $'000	Average per establishment $'000	Proportion of total shipments %
Up to $99,999	7,276	369,727	51	0.4
$ 100,000 - $ 199,999	4,082	586,699	144	0.7
200,000 - 499,999	5,377	1,766,184	328	2.0
500,000 - 999,999	3,774	2,706,750	717	3.1
1,000,000 - 4,999,999	6,482	14,603,844	2,253	16.5
5,000,000 - 9,999,999	1,531	10,722,928	7,004	12.1
10,000,000 - 24,999,999	1,013	15,739,425	15,537	17.8
25,000,000 - 49,999,999	312	10,708,075	34,321	12.1
50,000,000 and over	253	31,256,725	123,544	35.3
Total and average	30,100	88,460,358	2,939	100.0

17.8 Establishments in the manufacturing industries, by value of shipments of goods of own manufacture and by province, 1975

Province or territory	Up to $99,999	$100,000 to $199,999	$200,000 to $499,999	$500,000 to $999,999	$1,000,000 to $4,999,999	$5,000,000 to $9,999,999	$10,000,000 to $24,999,999	$25,000,000 to $49,999,999	$50,000,000 and over	Total
Nfld.	105	26	31	28	64		——— 13 ———		3	270
PEI	34	14	28		——— 41 ———		–		–	117
NS	206	98	126	81	123	29	12	6	8	689
NB	163	65	94	61	119	35	11	3	8	559
Que.	2,172	1,281	1,692	1,258	2,116	419	277	94	66	9,375
Ont.	2,599	1,613	2,206	1,557	2,805	691	507	147	120	12,245
Man.	340	163	200	148	250	67	32	9	6	1,215
Sask.	219	108	104	64	115	25	——— 13 ———		5	653
Alta.	481	263	348	224	328	84	55	23	15	1,821
BC	945	448	542	342	532	168	105	27	22	3,131
YT and NWT	12	3	6		——— 4 ———		–	–	–	25
Canada	7,276	4,082	5,377	3,774	6,482	1,531	1,013	312	253	30,100

17.9 Establishments and employment in the manufacturing industries, by number employed[1] per establishment, 1975

Size group	Estab-lishments No.	Employees No.	Working owners and partners No.	Proportion of total employment[1] %
Under 5 employed	7,582	13,437	4,804	1.0
5 - 9 "	5,070	32,716	1,477	2.0
10 - 19 "	5,197	71,398	493	4.1
20 - 49 "	5,779	182,136	171	10.4
50 - 99 "	2,839	198,907	21	11.4
100 - 199 "	1,962	274,642	5	15.7
200 - 499 "	1,207	362,824	6	20.8
500 - 999 "	311	211,088	—	12.1
1,000 or more "	153	305,716	—	17.5
Head offices	—	88,681	—	5.1
Total	30,100	1,741,545	6,977	100.0

[1]Includes working owners and partners.

17.10 Trends in domestic exports of manufactures, 1967-77 (million dollars)

Year	Fabricated materials	End products	Total manufactured goods[1]
1967	4,229.4	3,115.9	7,345.3
1968	4,855.1	4,351.5	9,206.6
1969	5,162.7	5,318.1	10,480.8
1970	5,866.4	5,551.0	11,417.4
1971	5,796.8	6,193.2	11,990.0
1972	6,578.2	7,136.2	13,714.4
1973	8,223.9	8,386.6	16,610.5
1974	10,695.7	9,236.8	19,932.5
1975	9,861.7	10,457.2	20,318.9
1976	12,323.1	13,095.9	25,419.0
1977	14,925.8	14,945.8	29,871.6

[1]These categories of exports are only approximately equivalent to exports of manufactured goods.

17.11 Estimated liabilities[1] of bankruptcies and insolvencies, 1972-76 (thousand dollars)

Year	Atlantic provinces	Quebec	Ontario	Prairie provinces	British Columbia	Total
1972	4,292	101,327	94,096	82,941	24,587	307,243
1973	37,578	131,062	86,734	17,396	23,931	296,701
1974	3,748	148,865	126,720	21,897	24,390	325,620
1975	2,496	167,524	102,076	21,037	32,163	325,295r
1976	839,922	150,461	137,379	26,112	31,419	1,185,293

[1]Estimated by debtors and therefore to be accepted with reservations.

17.12   Bankruptcies and insolvencies, by industry and economic area, 1975 and 1976

Industry	Atlantic provinces	Quebec	Ontario	Prairie provinces	British Columbia	Total	Estimated liabilities $'000
1975							
Primary industries	—	16	18	5	7	46	12,802
Manufacturing	2	110	100	13	16	241	76,173
Foods and beverages	1	11	4	2	1	19	4,593
Textiles	—	5	3	—	—	8	8,042
Clothing	—	13	1	—	1	15	3,019
Wood	—	21	13	2	5	41	13,047
Paper and allied industries	—	13	20	1	3	37	4,348
Primary and fabricated metal, machinery, transportation equipment, electrical products and non-metallic mineral products	1	25	34	7	3	70	34,511
Chemical	—	3	4	—	1	8	1,317
Other manufacturing industries	—	19	21	1	2	43	7,296
Construction	4	124	189	19	28	364	53,444
General contractors	4	50	74	6	12	146	25,756
Special trade contractors	—	74	115	13	16	218	27,688
Transportation, communication and other utilities	—	53	119	22	15	209	22,225
Trade	8	334	350	37	55	784	90,907
Food	3	34	32	1	11	81	7,378
General merchandise	1	7	19	1	1	29	10,281
Automotive products	1	60	78	14	10	163	6,790
Apparel and shoes	2	68	44	3	5	122	16,354
Hardware	—	17	5	2	2	26	1,921
Household furniture and appliances	1	35	49	7	5	97	8,466
Drugs	—	4	2	—	1	7	785
Other trades	—	109	121	9	20	259	38,932
Finance, insurance and real estate	4	10	28	2	5	49	23,542
Service	3	146	206	18	25	398	46,202ʳ
Education, health and welfare	—	7	8	2	1	18	8,655
Recreational	—	11	20	1	4	36	3,921ʳ
Business	—	23	39	3	4	69	11,714
Personal	1	11	26	—	4	42	1,712
Other services	2	94	113	12	12	233	20,200ʳ
Total, all industries	21	793	1,010	116	151	2,091	325,295ʳ
1976							
Primary industries	—	21	26	8	9	64	10,223
Manufacturing	4	113	102	14	15	248	911,340
Foods and beverages	—	9	5	1	3	18	1,646
Textiles	—	3	6	—	—	9	2,355
Clothing	—	20	8	1	—	29	6,820
Wood	1	21	22	4	5	53	30,057
Paper and allied industries	1	24	23	1	—	49	9,032
Primary and fabricated metal, machinery, transportation equipment, electrical products and non-metallic mineral products	—	20	28	5	7	60	22,731
Chemical	—	1	2	1	—	4	760
Other manufacturing industries	2	15	8	1	—	26	836,940
Construction	13	133	247	43	47	483	73,151
General contractors	9	46	73	14	18	160	48,707
Special trade contractors	4	87	174	29	29	323	24,443
Transportation, communication and other utilities	1	47	104	23	16	191	11,516
Trade	10	383	522	64	76	1,055	104,225
Food	2	56	46	5	8	117	12,119
General merchandise	—	9	22	2	1	34	1,456
Automotive products	2	71	98	17	21	209	15,259
Apparel and shoes	2	68	72	8	7	157	10,930
Hardware	1	7	13	2	3	26	1,999
Household furniture and appliances	1	39	69	6	7	122	10,474
Drugs	—	5	7	—	—	12	569
Other trades	2	128	195	24	29	378	51,419
Finance, insurance and real estate	1	9	28	4	8	50	18,194
Service	9	172	268	21	29	499	56,645
Education, health and welfare	2	9	16	1	2	30	4,596
Recreational	—	15	23	3	3	44	6,131
Business	2	22	53	2	4	83	13,098
Personal	—	18	29	2	2	51	2,910
Other services	5	108	147	13	18	291	29,909
Total, all industries	38	878	1,297	177	200	2,590	1,185,293

17.13   Summary statistics¹ of estates closed during 1974-76, under the Bankruptcy Act

Item		Province										
		Nfld.	PEI	NS	NB	Que.	Ont.	Man.	Sask.	Alta.	BC	Total
1974												
Bankrupt estates												
Estates closed	No.	—	1	19	6	1,969	3,344	90	59	127	720	6,335
Assets as estimated by debtors	$'000	—	500	692	347	23,953	12,375	1,369	312	1,256	3,106	43,910
Unsecured liabilities as estimated by debtors	..	—	737	11,913	346	87,577	85,164	4,438	2,297	3,730	21,255	217,457
Realization by trustee	..	—	234	406	24	11,268	9,988	717	253	394	2,550	25,834
Costs of administration	..	—	35	347	11	6,081	5,006	226	89	188	1,236	13,219
Costs as percentage of realization	%	—	15	85	46	54	50	32	35	48	48	51
Paid to unsecured creditors	$'000	—	199	59	13	5,187	4,982	491	164	206	1,314	12,615
Retained by secured creditors	..	—	397	541	553	22,961	20,719	818	523	808	5,694	53,014
Average percentage recovered by unsecured creditors	%	—	27	3	4	6	5	11	7	6	6	6
Proposals												
Proposals closed	No.	—	—	—	1	105	54	2	1	1	10	174
Unsecured liabilities as estimated by debtors	$'000	—	—	—	87	13,057	9,749	18	—	100	1,304	24,315
Paid to unsecured creditors	..	—	—	—	33	3,792	2,202	1	81	43	400	6,552
1975												
Bankrupt estates												
Estates closed	No.	5	—	12	15	1,781	2,434	232	49	652	667	5,847
Assets as estimated by debtors	$'000	1,340	—	1,851	50	19,519	13,815	1,916	1,860	5,895	4,915	51,161
Unsecured liabilities as estimated by debtors	..	1,721	—	3,648	347	553,157	91,599	7,024	16,798	16,860	22,085	713,239
Realization by trustee	..	1,048	—	864	35	10,859	8,873	660	350	2,650	2,527	27,866
Costs of administration	..	545	—	283	23	5,739	3,905	437	163	1,230	1,532	13,857
Costs as percentage of realization	%	52	—	33	66	53	44	66	47	46	61	50
Paid to unsecured creditors	$'000	503	—	581	12	5,120	4,968	223	187	1,420	995	14,009
Retained by secured creditors	..	1,547	—	245	61	36,649	754,501	2,478	1,223	12,049	7,422	816,175
Average percentage recovered by unsecured creditors	%	29	—	16	3	1	5	3	8	8	5	2
Proposals												
Proposals closed	No.	—	—	—	—	110	15	—	3	6	18	152
Unsecured liabilities as estimated by debtors	$'000	—	—	—	—	8,725	7,207	—	7,072	3,177	8,106	34,287
Paid to unsecured creditors	..	—	—	—	—	1,773	1,786	—	18	229	1,014	4,820
1976												
Bankrupt estates												
Estates closed	No.	—	—	15	4	2,184	3,550	296	109	392	802	7,352
Assets as estimated by debtors	$'000	—	—	254	410	1,015,633	16,972	852	26,180	2,381	3,627	1,066,309
Unsecured liabilities as estimated by debtors	..	—	—	3,026	349	1,366,244	105,629	5,621	64,416	7,175	32,193	1,584,653
Realization by trustee	..	—	—	296	15	9,274	13,813	981	488	1,334	4,060	30,261
Costs of administration	..	—	—	147	12	5,673	7,341	439	193	597	1,788	16,190
Costs as percentage of realization	%	—	—	50	80	61	53	45	40	45	44	54
Paid to unsecured creditors	$'000	—	—	149	3	3,601	6,472	542	295	737	2,272	14,071
Retained by secured creditors	..	—	—	9,845	11	821,215	389,640	158,841	49,892	1,660	9,091	1,440,195
Average percentage recovered by unsecured creditors	%	—	—	5	—	0.3	6	10	0.5	10	7	1
Proposals												
Proposals closed	No.	—	—	1	—	78	24	—	—	—	14	117
Unsecured liabilities as estimated by debtors	$'000	—	—	42	—	23,327	14,109	—	—	—	5,888	43,366
Paid to unsecured creditors	..	—	—	11	—	3,169	2,579	—	—	—	1,855	7,614

¹Excludes Small Debtor Program.

17.14 Summary statistics of manufactures, by industry group, 1975

Industry group	Establishments No.	Manufacturing activity							Total activity		
		Production and related workers			Cost of fuel and electricity $'000	Cost of materials and supplies used $'000	Value of shipments of goods of own manufacture $'000	Value added $'000	Total employees		Total value added $'000
		Number	Man-hours paid '000	Wages $'000					Number	Salaries and wages $'000	
Food and beverage industries	4,740	145,357	304,534	1,396,422	193,841	11,325,767	16,492,290	5,030,036	220,415	2,312,018	5,375,417
Tobacco products industries	27	6,540	12,218	71,806	3,555	506,963	831,522	363,352	9,686	117,332	366,845
Rubber and plastics products industries	771	39,473	80,067	355,806	34,722	972,106	1,955,825	964,082	52,963	531,854	1,030,971
Leather industries	415	23,440	47,222	158,242	4,659	308,751	619,191	309,247	26,834	200,819	316,474
Textile industries	923	56,450	116,687	435,869	51,515	1,344,492	2,439,005	1,067,877	71,050	614,644	1,099,789
Knitting mills	306	21,567	45,135	141,853	6,331	323,764	624,490	294,072	24,682	182,833	294,497
Clothing industries	2,094	89,347	177,302	573,274	7,856	1,201,314	2,306,619	1,105,674	100,528	722,130	1,143,188
Wood industries	2,920	81,588	167,763	846,138	79,405	2,098,283	3,802,635	1,654,542	97,717	1,070,932	1,689,596
Furniture and fixture industries	1,959	41,460	87,859	329,384	11,848	642,136	1,363,703	707,864	49,688	431,897	724,645
Paper and allied industries	665	95,794	181,293	1,054,795	418,836	3,346,680	7,131,614	3,397,527	127,342	1,553,080	3,469,803
Printing, publishing and allied industries	3,618	55,044	109,601	587,431	15,980	1,016,868	2,897,471	1,871,330	92,912	1,042,556	1,899,375
Primary metal industries	380	90,169	184,323	1,119,159	317,595	3,641,157	6,682,356	2,879,904	120,335	1,612,991	2,948,174
Metal fabricating industries (except machinery and transportation equipment industries)	3,882	117,106	245,259	1,241,513	61,289	3,056,957	6,216,654	3,149,631	150,899	1,727,946	3,278,286
Machinery industries (except electrical machinery)	1,098	63,699	132,518	698,998	27,411	1,999,276	3,731,625	1,765,710	92,290	1,073,613	2,046,762
Transportation equipment industries	991	120,833	256,333	1,446,395	80,239	7,636,055	11,193,031	3,516,350	159,642	2,033,079	4,208,537
Electrical products industries	805	82,711	170,818	776,409	34,312	2,189,898	4,599,292	2,317,885	125,868	1,364,138	2,631,778
Non-metallic mineral products industries	1,199	42,149	88,984	471,466	174,274	974,686	2,569,385	1,446,135	55,932	669,350	1,502,408
Petroleum and coal products industries	101	7,877	16,955	122,267	49,786	5,108,677	5,953,330	857,667	17,264	298,040	865,952
Chemical and chemical products industries	1,046	42,643	88,735	463,522	216,601	2,566,268	5,107,353	2,422,515	80,251	1,003,909	2,663,040
Miscellaneous manufacturing industries	2,160	48,804	99,942	381,486	15,611	917,059	1,942,966	1,017,902	65,247	597,561	1,160,062
Total	30,100	1,272,051	2,613,549	12,672,237	1,805,666	51,177,157	88,460,358	36,139,301	1,741,545	19,160,724	38,715,600

17.15 Summary statistics of the 40 leading industries, ranked according to value of shipments of goods of own manufacture, 1975

Industry	Establishments No.	Manufacturing activity — Production and related workers: Number	Man-hours paid '000	Wages $'000	Cost of fuel and electricity $'000	Cost of materials and supplies used $'000	Value of shipments of goods of own manufacture $'000	Value added $'000	Total activity — Total employees: Number	Salaries and wages $'000	Total value added $'000
1 Motor vehicle manufacturers	24	31,694	66,861	441,391	24,821	4,792,843	6,024,429	1,171,006	45,256	668,205	1,758,047
2 Petroleum refining	42	6,806	14,680	110,354	46,320	4,998,610	5,773,459	789,783	15,624	277,268	789,680
3 Pulp and paper mills	147	64,329	117,176	754,892	398,692	2,191,526	5,122,092	2,569,050	84,046	1,091,675	2,601,472
4 Slaughtering and meat processors	477	24,621	51,417	278,192	24,897	3,097,743	3,828,825	711,710	32,993	391,010	746,341
5 Iron and steel mills	46	42,169	86,419	554,499	139,982	1,797,393	3,147,692	1,348,021	54,003	761,004	1,364,022
6 Dairy products industry	519	13,780	29,499	141,454	36,111	2,026,277	2,612,789	576,693	27,988	304,905	631,318
7 Miscellaneous machinery and equipment manufacturers	873	43,573	91,030	465,092	17,040	1,223,676	2,432,778	1,227,889	62,221	706,982	1,377,083
8 Motor vehicle parts and accessories manufacturers	231	34,907	74,999	415,636	29,266	1,284,639	2,325,801	1,008,394	42,639	531,651	1,028,860
9 Sawmills and planing mills	1,368	40,788	84,899	462,430	51,179	1,113,446	1,996,855	849,323	49,156	577,555	850,833
10 Smelting and refining	28	22,932	45,613	287,742	141,134	564,944	1,550,615	844,537	35,577	491,840	886,405
11 Miscellaneous food processors	245	11,508	23,406	108,021	19,204	952,394	1,533,537	565,147	19,815	215,851	613,387
12 Communications equipment manufacturers	262	26,328	54,720	255,149	6,242	507,515	1,368,086	843,736	42,041	481,367	930,139
13 Commercial printing	2,078	32,955	65,856	331,727	8,571	565,429	1,353,952	783,782	43,485	478,481	797,013
14 Metal stamping and pressing industry	541	20,273	42,772	207,421	10,285	732,554	1,326,139	577,683	25,574	287,571	600,263
15 Feed industry	643	5,991	12,763	55,729	11,981	1,020,942	1,256,696	224,282	9,260	91,297	257,129
16 Publishing and printing	615	16,837	33,436	196,909	6,416	255,129	1,002,057	740,909	33,562	391,016	740,932
17 Rubber products industries	112	18,289	37,185	186,971	20,230	478,715	998,595	508,866	27,422	302,877	565,593
18 Plastics fabricating industry	659	21,184	42,882	168,834	14,492	493,391	957,229	455,215	25,541	228,977	465,377
19 Miscellaneous metal fabricating industries	484	18,488	38,371	178,484	11,591	459,504	927,996	469,117	23,810	251,352	487,023
20 Manufacturers of industrial chemicals (organic)	35	5,400	11,612	76,496	77,225	495,161	923,789	370,364	10,306	165,300	409,924
21 Manufacturers of electrical industrial equipment	193	17,594	36,558	174,818	7,820	402,480	911,538	501,347	29,293	334,499	580,675
22 Fabricated structural metal industry	161	13,870	29,180	182,918	6,803	391,084	879,952	478,004	19,101	262,721	510,994
23 Men's clothing factories	480	32,321	65,866	211,968	3,095	459,148	877,858	412,892	36,934	273,254	420,994
24 Wire and wire products manufacturers	276	13,559	28,160	160,979	11,010	504,369	875,081	382,697	17,614	203,405	394,785
25 Women's clothing factories	585	26,822	51,238	172,518	2,096	481,514	847,014	369,522	30,147	221,525	383,218
26 Bakeries	1,599	17,798	37,173	153,734	15,983	376,048	828,945	437,184	27,379	254,908	459,105
27 Manufacturers of industrial chemicals (inorganic)	91	6,117	12,635	81,370	91,065	353,759	822,774	415,203	9,904	138,009	439,259
28 Fruit and vegetable canners and preservers	208	11,726	24,469	91,103	9,276	500,074	791,096	309,572	15,607	137,541	334,244
29 Miscellaneous chemical industries	319	8,679	17,879	84,383	13,072	419,162	771,043	345,880	15,385	180,231	384,849
30 Agricultural implement industry	147	13,117	27,612	156,657	6,817	452,620	762,104	325,515	16,649	205,256	343,515
31 Cane and beet sugar processors	15	2,049	4,450	24,256	7,594	598,967	737,543	102,336	2,780	35,311	107,974
32 Soft drink manufacturers	288	6,432	13,318	61,479	8,995	398,346	732,529	325,602	13,808	146,268	359,683
33 Miscellaneous paper converters	217	11,711	23,755	107,006	7,416	388,539	717,334	319,978	17,059	178,359	345,181
34 Aircraft and aircraft parts manufacturers	101	13,864	29,400	152,747	6,976	311,921	707,789	425,538	22,289	273,425	438,036
35 Household furniture manufacturers	708	23,133	49,475	177,929	6,477	330,501	695,435	359,666	27,147	226,076	360,379
36 Breweries	44	7,011	14,735	100,301	12,153	221,141	693,656	460,351	11,652	172,441	476,230
37 Manufacturers of pharmaceuticals and medicines	134	6,707	13,587	60,009	5,471	246,244	654,447	418,402	14,793	171,946	457,167
38 Manufacturers of electric wire and cable	39	7,160	14,971	78,321	5,937	406,916	646,358	232,013	10,134	122,224	232,968
39 Ready-mix concrete manufacturers	356	7,703	16,531	93,656	16,353	309,132	596,026	271,824	9,541	122,860	282,597
40 Tobacco products manufacturers	17	5,471	10,132	63,745	2,597	243,134	581,870	338,860	8,357	106,195	341,844

17.16 Summary statistics of manufactures, by province, 1975

Province or territory	Establishments No.	Manufacturing activity — Production and related workers — Number	Man-hours paid '000	Wages $'000	Cost of fuel and electricity $'000	Cost of materials and supplies used $'000	Value of shipments of goods of own manufacture $'000	Value added $'000	Total activity — Total employees — Number	Salaries and wages $'000	Total value added $'000
Newfoundland	270	10,164	22,011	97,085	26,149	418,747	650,008	208,020	13,000	132,261	224,139
Prince Edward Island	117	1,817	3,817	13,090	2,234	74,173	108,562	34,132	2,353	18,107	36,741
Nova Scotia	689	27,895	57,564	247,063	51,332	1,139,514	1,819,094	675,705	37,365	362,252	700,019
New Brunswick	559	22,348	47,234	211,698	77,788	1,041,867	1,669,422	583,162	29,300	295,087	610,085
Quebec	9,375	394,333	813,095	3,520,808	537,647	13,662,494	23,966,501	9,958,016	532,932	5,382,265	10,458,512
Ontario	12,245	612,745	1,263,962	6,316,577	791,497	25,560,075	44,422,821	18,357,809	850,291	9,673,050	20,122,934
Manitoba	1,215	41,057	83,830	373,209	50,260	1,537,285	2,580,985	1,014,619	55,010	541,507	1,080,398
Saskatchewan	653	13,995	28,584	143,740	22,622	719,195	1,176,473	437,152	19,213	208,281	455,185
Alberta	1,821	45,781	94,635	491,988	65,828	3,131,381	4,726,466	1,581,096	64,678	741,073	1,638,347
British Columbia	3,131	101,701	198,363	1,254,927	180,058	3,884,032	7,326,464	3,284,674	137,138	1,804,129	3,383,285
Yukon Territory and Northwest Territories	25	215	454	2,051	251	8,394	13,561	4,915	265	2,711	5,956
Canada	30,100	1,272,051	2,613,549	12,672,237	1,805,666	51,177,157	88,460,358	36,139,301	1,741,545	19,160,724	38,715,600

17.17 Summary statistics of manufactures, by census metropolitan area, 1975

Census metropolitan area	Establishments No.	Manufacturing activity — Production and related workers — Number	Man-hours paid '000	Wages $'000	Cost of fuel and electricity $'000	Cost of materials and supplies used $'000	Value of shipments of goods of own manufacture $'000	Value added $'000	Total activity — Total employees — Number	Salaries and wages $'000	Total value added $'000
Calgary, Alta.	540	14,712	30,172	157,522	12,558	863,566	1,382,641	515,769	19,624	222,678	534,062
Chicoutimi-Jonquière, Que.	92	8,399	16,521	100,967	22,861	296,798	559,492	238,016	11,717	150,317	254,971
Edmonton, Alta.	624	18,091	37,590	200,727	28,963	1,362,818	2,003,202	645,473	25,133	293,462	671,269
Halifax, NS	129	5,299	11,002	52,304	5,632	424,191	602,636	174,036	7,394	77,533	176,805
Hamilton, Ont.	647	52,275	108,103	606,434	87,970	2,252,012	3,872,777	1,623,467	66,534	821,773	1,675,443
Kitchener, Ont.	516	37,818	77,233	360,513	23,043	1,053,819	1,962,626	903,302	47,999	488,002	934,865
London, Ont.	375	21,246	43,366	216,265	12,963	1,113,559	1,782,426	649,171	28,530	313,742	725,694
Montreal, Que.	5,128	214,148	442,013	1,914,680	164,758	7,877,132	13,580,866	5,629,752	280,883	2,819,692	5,876,232
Ottawa-Hull, Ont., Que.	342	13,636	27,705	137,121	27,991	413,950	840,911	401,630	19,929	221,777	429,810
Quebec, Que.	483	16,934	34,166	154,306	21,561	636,413	1,114,518	461,277	21,245	206,981	476,173
Regina, Sask.	135	4,214	8,668	49,307	6,966	260,813	427,721	158,147	5,764	68,733	165,771
Saint John, NB	75	5,586	12,113	64,148	19,677	548,590	736,190	181,499	7,086	84,635	185,949
St. Catharines-Niagara, Ont.	416	30,920	64,480	368,576	57,014	1,040,244	2,064,492	973,053	39,123	488,961	992,009
St. John's, Nfld.	75	2,002	4,283	19,356	2,328	57,399	113,454	54,468	2,803	28,322	59,624
Saskatoon, Sask.	140	3,975	8,138	37,898	3,466	209,225	328,083	113,113	5,256	53,548	118,113
Sudbury, Ont.	65	5,408	10,805	58,719	37,392	55,395	266,077	173,488	8,611	108,793	179,540
Thunder Bay, Ont.	95	5,809	9,817	57,963	16,596	169,081	346,717	162,574	7,472	77,478	167,395
Toronto, Ont.	5,894	240,949	499,846	2,391,524	163,699	9,616,264	16,958,575	7,219,667	327,819	3,584,706	7,730,886
Vancouver, BC	1,848	51,250	102,367	606,564	43,617	2,080,488	3,643,895	1,518,793	67,358	841,567	1,585,910
Victoria, BC	195	4,015	7,925	50,532	2,219	104,475	215,768	109,384	5,093	65,562	111,874
Windsor, Ont.	384	26,228	58,992	335,883	26,360	1,792,335	2,865,863	1,027,440	31,740	432,168	1,037,751
Winnipeg, Man.	850	33,198	67,772	294,645	23,673	1,229,438	2,040,167	795,209	43,272	415,470	840,046

17.18 Percentages of value of shipments of goods of own manufacture accounted for by the four leading enterprises in the 40 leading industries of Canada, ranked by 1974 shipments

Industry	Establishments No.	Enterprises No.	Value of shipments of goods of own manufacture $'000,000	Percentage of shipments accounted for by the four leading enterprises			
				1968	1970	1972	1974
Pulp and paper mills	147	68	5,703.2	35.9	36.2	34.5	34.0
Motor vehicle manufacturers	22	16	5,381.9	94.6	93.3	¹	90.1
Petroleum refining	43	17	5,057.2	78.2	79.0	73.7	67.8
Slaughtering and meat processors	487	433	3,579.0	55.4	54.8	54.0	50.2
Iron and steel mills	47	33	3,036.2	78.1	76.2	77.8	76.8
Sawmills and planing mills	1,530	1,414	2,329.8	22.1	20.9	18.2	18.8
Motor vehicle parts and accessories manufacturers	227	186	2,281.1	49.7	46.2	48.9	46.2
Dairy products industry	556	364	2,083.0	..	29.2	33.0	37.3
Miscellaneous machinery and equipment manufacturers	849	799	2,065.4	16.7	16.1	12.5	12.7
Smelting and refining	28	14	1,409.9	79.9	79.0	78.7	75.0
Miscellaneous food processors	256	205	1,381.7	33.8	33.7	35.2	39.5
Metal stamping and pressing industry	550	501	1,246.3	..	39.0	39.5	37.0
Communications equipment manufacturers	260	222	1,225.4	..	55.8	56.5	60.8
Feed industry	698	565	1,221.6	29.4	29.4	29.1	27.4
Commercial printing	2,208	2,127	1,214.7	15.9	17.1	19.5	20.9
Wire and wire products manufacturers	282	245	960.3	44.1	43.1	43.3	47.9
Plastics fabricating industry	674	619	958.4	20.8	16.2	13.3	11.3
Publishing and printing	614	531	878.8	35.4	37.6	42.5	48.7
Rubber products industries	109	90	875.2	..	¹	60.7	51.4
Manufacturers of industrial chemicals (organic)	34	24	859.8	..	60.5	59.9	61.6
Miscellaneous metal fabricating industries	501	473	858.1	18.9	14.6	15.1	13.4
Fabricated structural metal industry	163	140	832.8	45.1	44.0	38.0	44.0
Men's clothing industry	482	448	790.5	11.8	12.0	11.7	12.7
Manufacturers of electrical industrial equipment	184	139	764.5	58.0	55.6	51.1	50.7
Women's clothing factories	586	559	745.7	7.2	8.0	8.2	7.5
Bakeries	1,680	1,625	726.6	31.2	31.6	33.5	37.0
Miscellaneous chemical industries	328	259	724.2	..	36.4	32.3	33.2
Household furniture manufacturers	721	687	711.5	..	13.1	13.4	15.4
Fruit and vegetable canners and preservers	209	162	694.5	..	41.7	39.8	37.3
Miscellaneous paper converters	219	176	690.4	32.9	33.3	33.5	34.2
Manufacturers of electric wire and cable	38	19	674.2	77.4	83.0	79.2	80.2
Manufacturers of industrial chemicals (inorganic)	90	39	668.9	..	52.5	52.4	44.7
Cane and beet sugar processors	15	8	650.9	¹	¹	¹	92.0
Aircraft and aircraft parts manufacturers	95	91	620.2	76.9	72.0	¹	¹
Breweries	44	6	612.9	94.8	94.0	96.6	¹
Manufacturers of pharmaceuticals and medicines	136	120	579.8	28.0	29.6	27.8	25.6
Soft drink manufacturers	322	258	578.6	43.7	46.0	46.2	50.4
Fish products industry	348	264	576.4	40.1	39.2	42.5	44.5
Agricultural implement industry	143	140	571.9	67.4	70.6	65.3	¹
Corrugated box manufacturers	84	45	527.8	58.6	56.5	54.4	51.9

¹Confidential.

17.19 Selected statistics for the 100 largest manufacturing enterprises, by manufacturing value added, 1974

Enterprise group¹	Establishments²		Value of manufacturing shipments		Value added				Total employees		Production workers	
					Manufacturing activity		Total activity					
	No.	%	$'000,000	%	$'000,000	%	$'000,000	%	No.	%	No.	%
4 largest enterprises	148	0.5	6,389	7.7	2,527	7.2	2,808		98,090	5.5	72,684	5.6
8 "	236	0.8	11,160	13.5	4,251	12.1	5,018		172,070	9.6	124,369	9.6
12 "	326	1.0	14,325	17.4	5,710	16.3	6,503		230,213	12.9	169,558	13.0
16 "	490	1.6	17,904	21.7	6,896	19.7	7,790		284,589	15.9	204,707	15.7
20 "	682	2.2	20,572	24.9	7,846	22.4	8,796		322,252	18.0	230,005	17.7
25 "	814	2.6	22,327	27.1	8,779	25.0	9,747		358,163	20.1	255,415	19.6
50 "	1,261	4.0	30,499	37.0	12,188	34.7	13,292		480,787	26.9	337,967	26.0
100 "	1,835	5.8	39,733	48.2	16,224	46.2	17,517		649,793	36.4	450,951	34.7
All manufacturing enterprises	31,535	100.0	82,455	100.0	35,085	100.0	37,654		1,785,977	100.0	1,300,792	100.0

¹An enterprise consists of one company or a group of companies under a common control.
²Companies owning one or more establishments or plants. The above table does not include non-manufacturing establishments of enterprises.

17.20 Percentage distribution of shipments of manufacturing industries by destination and value, 1974

Region of origin	Region of destination									Total value $'000
	Atlantic provinces	Quebec	Ontario	Prairie provinces	British Columbia	Other countries	Unallocated	Custom and repair work	Total	
Atlantic provinces	41.1	9.0	7.4	0.9	0.5	34.3	4.7	1.9	100.0	4,087,619
Quebec	4.0	49.5	19.3	4.2	2.2	13.7	4.4	2.5	100.0	22,396,846
Ontario	2.9	11.2	50.3	5.9	3.0	20.3	4.4	1.8	100.0	41,404,361
Prairie provinces	1.2	6.0	6.3	63.0	8.3	8.4	3.8	2.4	100.0	7,146,162
British Columbia	0.7	1.6	3.3	7.9	39.6	37.6	6.5	2.4	100.0	7,411,103
Canada	4.7	20.2	31.7	10.3	6.4	19.8	4.6	2.1	100.0	82,455,110[1]

[1]Includes the Yukon Territory and Northwest Territories.

17.21 Sources of Canadian manufacturers' shipments destined to regions of Canada and to other countries, 1974[1]

Region of origin	Region of destination								
	Atlantic provinces	Quebec	Ontario	Prairie provinces	British Columbia	Other countries	Unallocated	Custom and repair work	Total
Atlantic provinces	43.2	2.2	1.2	0.4	0.4	8.6	5.1	4.5	5.0
Quebec	22.8	66.6	16.5	10.9	9.5	18.9	26.2	32.5	27.2
Ontario	30.7	27.9	79.6	28.9	23.3	51.7	48.4	42.9	50.2
Prairie provinces	2.1	2.6	1.7	52.9	11.2	3.7	7.2	9.7	8.7
British Columbia	1.2	0.7	0.9	6.9	55.7	17.1	12.7	10.3	9.0
Total	100.0	100.0	100.0	100.0	100.0	100.0	100.0	100.0	100.0
Canada ($'000)	3,890,252	16,669,149	26,175,270	8,512,101	5,274,004	16,286,634	3,784,259	1,749,739	82,455,110[2]

[1]Relative percentage participation of manufacturers of regions of Canada in apparent regional markets for Canadian manufacturers.
[2]Includes the Yukon Territory and Northwest Territories.

Sources
17.1 - 17.10 Manufacturing and Primary Industries Division, Industry Statistics Branch, Statistics Canada.
17.11 - 17.12 Business Finance Division, General Statistics Branch, Statistics Canada.
17.13 Superintendent of Bankruptcy, Department of Consumer and Corporate Affairs.
17.14 - 17.21 Manufacturing and Primary Industries Division, Industry Statistics Branch, Statistics Canada.

Merchandising and trade

Tables

Merchandising and trade

Chapter 18

Merchandising and service trades

18.1

This section deals with the distribution of goods and services from producer to consumer — principally through wholesale and retail channels and through service businesses — in what is generally known as the marketing process.

Merchandising industries include wholesaling and warehousing which exist in a variety of forms: wholesale merchants, agents and brokers, primary products dealers, manufacturers sales branches, petroleum bulk tank plants and truck distributors. Retailing encompasses all sales activities related to transmitting goods to consumers for household or personal use, both through traditional store locations and such facilities as direct selling and machine vending.

Statistics on merchandising and service industries are gathered through periodic business censuses as well as monthly, annual and occasional surveys. In recent years, considerable interest has been focused on the service trades, resulting in expanded statistical coverage.

Retail trade

18.1.1

Retail trade statistics are collected by Statistics Canada from monthly surveys of all retail chains (four or more stores in the same kind of business under one owner), and of a sample of independent retailers.

Table 18.1 shows retail trade by kind of business and by province from 1974 to 1977 and indicates the percentage changes during this period; retail sales rose from $44.8 billion to $61.6 billion, an increase of 37.6%. Above-average sales increases were recorded by sporting goods and accessories stores (66.7%), book and stationery stores (56.0%), and grocery, confectionery and sundries stores (54.2%). Among the smallest sales increases recorded were those in furniture, TV, radio and appliance stores (4.5%), general merchandise stores (12.0%) and specialty shoe stores (14.2%). All provinces showed retail sales increases of over 25.0%, with Alberta (57.6%) recording the largest increase, followed by New Brunswick (37.8%) and Prince Edward Island (37.4%); Manitoba, with an increase of 26.0%, recorded the smallest increase of all the provinces.

Chain and independent stores. A retail chain is an organization operating four or more retail stores in the same kind of business under the same legal ownership. All department stores are classified as chains even if they do not meet this definition. An independent retailer is one who operates one to three stores, even if he is a member of a voluntary group organization.

Table 18.2 provides information on the sales trends of chains and independent stores by kind of business in 1974 and 1977 and the percentage change during that period; retail sales through chain stores rose by 37.5% and those through independent stores by 37.6%. As in the past, combination store (groceries and meat) chains continued their sales increases (41.5%) at the expense of independent stores (27.6%). Although sporting goods and accessories store chains were less important in terms of total dollar volume than independents, from 1974 to 1977 they made more headway (216.9%) than independents (62.6%). Women's clothing stores swung from a category dominated by independent stores in 1974 to a group dominated by chain stores in 1977. Other increases in the strength of chains can be observed in general stores, men's clothing stores, family clothing stores, book and stationery stores and personal accessories stores. In each case, the sales increases of chains compared to those of independents were much higher in the 1974-77 period.

Table 18.3 illustrates the relative importance of chains by kind of business and the trends from 1974 to 1977. The percentages shown represent the chain store shares of the market, the balance being accounted for by independent stores. In 1977, chain stores accounted for 42.0% of the total market (and independent stores for 58.0%), unchanged from the 1974 distribution and compared to 41.8% in 1976. The largest changes occurred in general stores where chain stores increased their market share from 19.2% in 1974 to 28.0% in 1977 and in men's clothing stores where their market share increased from 24.4% in 1974 to 33.2% in 1977. The largest decline in the market share of chains occurred in household furniture stores, from 19.6% in 1974 to 15.5% in 1977, and in automotive parts and accessories stores, from 17.0% in 1974 to 13.5% in 1977.

Department stores. Department stores (Table 18.4) have shown one of the most consistent growth rates of all categories of retail trade. Their sales were exceeded only by combination stores (groceries and meat) and motor vehicle dealers. In 1977, department store sales reached $6,941 million, excluding catalogue sales, an increase of 36.9% from 1974. The market share of department stores for 1977 was 11.3%.

The largest increases for departments within such stores in 1977 were recorded by jewellery (58.1%), lingerie and women's sleepwear (52.6%), photographic equipment and supplies (52.2%) and hardware, paints and wallpaper (49.8%). Seven departments recorded sales of over $300.0 million, the highest of which were furniture and food and kindred products, each with sales of $367.8 million; followed by major appliances, $351.9 million; women's and misses' sportswear, $341.2 million; toiletries, cosmetics and drugs, $330.8 million; men's clothing, $325.3 million; and men's furnishings, $313.6 million. The lowest increase in sales was recorded in millinery (4.6%), which was also the department with the lowest sales, $16.0 million.

New motor vehicle sales. The largest homogeneous group of commodities sold through retail trade outlets is new motor vehicles. The statistical series based on the sales of this special category of consumer durable is a sensitive indicator of consumer attitudes and expectations regarding the national economy. In 1977, the $8,546 million in sales of new motor vehicles accounted for 72.7% of all retail sales by motor vehicle dealers ($11,750 million), 13.9% of total retail sales of $61,597 million, and 7.0% of total personal spending on consumer goods and services amounting to $121,955 million.

Statistics Canada obtains monthly new motor vehicle sales figures from both Canadian manufacturers and importers. They supply both unit and dollar sales. Users of the unit data however should be aware that they differ from data available from other sources, such as factory shipments and registrations, owing to variations in definition and treatment of new vehicles in relation to demonstrators, sales to the Canadian forces, semi-finished imports and sales of motors and chassis to coach body-builders.

The new motor vehicles referred to are passenger cars, trucks and buses sold by motor vehicle dealers. Excluded are all export sales and domestic sales of motorcycles, snowmobiles and other all-terrain vehicles. Passenger cars include not only private cars but taxis and car rental fleets and other passenger cars used for business and commercial purposes; commercial vehicles refer solely to trucks and buses. Vehicles manufactured overseas include only those imported (some by Canadian and US manufacturers) in a fully assembled state from countries other than the United States. When assembled on this continent some well-known foreign makes of cars and trucks are treated statistically as being Canadian- and US-made vehicles.

During 1977 record numbers of passenger cars and commercial vehicles were sold in Canada — 1,344,959 units valued at $8,546 million; this consisted of 991,398 passenger car units valued at $5,796 million and 353,561 commercial vehicle units valued at $2,750 million.

Passenger cars made in Canada and the US had their second highest sales year on record with unit sales of 797,752, a small advance over 1976 but short of the record of 835,679 unit sales in 1975. In the latest year the overall increase was mainly due to a 7.4% increase in sales over the year in Ontario where 40.3% of these vehicles were purchased. Total spending on these passenger cars rose 7.5% over the year from $4,523 million to $4,864 million in 1977.

The share of the domestic passenger car market supplied by imports in 1977 rose 3.3% from 1976 to 19.5%, with sales of 193,646 units for a value of $931 million. The bulk of these sales was imported from Japan, a rise of 32.8% during the year to 134,900 units, and fetched $567 million, 36.9% higher than in 1976.

Sales of commercial vehicles produced in North America, the source of most of these vehicles, 95.6% in 1977, increased 2.1% over the year to a record of 337,914 units valued at $2,673 million. Despite this overall expansion in sales during the year, declines in sales of these vehicles were recorded both in Manitoba and Saskatchewan in 1977 although these provinces are usually strong markets for new commercial vehicles. Sales of imported commercial vehicles amounted to 15,647 units, an increase of 12.2% over the 13,948 units sold in 1976, and receipts of $77 million were 18.9% higher than the $65 million in 1976.

Campus book stores. Retail trade statistics are collected annually from more than 200 book stores located on the campuses of universities and other post-secondary educational institutions. Owing to their location and the highly seasonal nature of their business, campus book stores are not included in the census of merchandising and services, nor are they included in the monthly estimates of retail trade. Since they are not considered retail outlets, a separate survey is conducted to provide data. In the 1976-77 academic year, as shown in Table 18.7, 210 campus book stores registered net sales of $82 million, an 11.8% increase over the previous year. Of the total dollar sales, 62.4% was accounted for by textbooks, 12.3% by trade books, 15.9% by stationery and supplies and 9.4% by sales of miscellaneous items.

Non-store retailing. Consumer goods, in addition to being sold in retail stores, often reach the household consumer through other channels. These channels bypass the retail outlet completely in moving from primary producer, manufacturer, importer, as wholesaler or specialized direct seller, to the household consumer. Statistics Canada conducts annual surveys of two distinct forms of non-store retailing: merchandise sales through vending machines and sales by manufacturers and distributors specializing in direct-sales methods such as catalogue and mail-order sales, door-to-door canvassing, and house parties.

Vending machine sales. This survey is designed to measure the value of merchandise sales made through automatic vending machines owned and operated by independent operators and subsidiaries or divisions of manufacturers and wholesalers of vended products. Excluded from coverage are the sales through many thousands of vending machines (carrying such commodities as cigarettes, beverages, confectionery) which are owned and operated by retail stores, restaurants and service stations; these sales statistics are usually inextricable from data collected in the course of other surveys.

During 1976, the 612 operators of 104,548 vending machines covered by this survey reported sales of $269.4 million, including $3.5 million from bulk confectionery machines (Table 18.8). These sales exceeded by 7.8% the sales of $250.0 million reported in 1975. Increased sales through the following principal types of machines were chiefly responsible for the overall advance in receipts between 1975 and 1976 (Table 18.9): cigarette machine sales which expanded 7.7% to gross $120.8 million; coffee machine sales which rose 8.3% to a total of $45.1 million and soft drink machine sales which increased 5.7% to account for $40.6 million. Notable gains were also recorded for packaged confectionery, pastry and snack food machines, in which receipts rose 11.4% to $27.1 million; fresh food dispensing machines, which recorded a 2.1% increase in sales to $15.8 million; packaged milk (and juice) machines, with receipts up 5.7% to $9.0 million; and hot canned food and soup machines which increased sales by 6.5% to reach $4.5 million in 1976.

Of the 82,430 full-size vending machines (excluding the small bulk confectionery machines) on location at year-end 1976, 37.3% were in industrial plants, 23.2% were in hotels, motels, taverns and restaurants, while 13.3% were in institutions such as hospitals, schools and colleges.

Direct selling. During 1976, Canadian householders spent $1,593.4 million on a wide variety of goods purchased directly through channels of distribution which bypass traditional retailing outlets (Table 18.10). Major product lines handled by direct-selling businesses include dairy products ($246.6 million), newspapers ($185.2 million), cosmetics and costume jewellery ($149.9 million), household electrical appliances, including vacuum cleaners ($139.0 million), and dinnerware, kitchenware and utensils ($92.0 million).

Door-to-door selling is the best known of the various channels of direct selling and accounted for 59.1%, or $942.1 million, of the $1,593.4 million (Tables 18.10 and 18.11) spent on direct sales in 1976. Sales made by mail are another mode of direct selling by which specialized retailers contact the household consumer. In 1976, mail-order business accounted for 16.1%, or $256.9 million, of direct sales. Commodities which rely heavily on this channel of distribution include magazines and phonograph records (100% in each case), books (66.7%), clothing (41.0%), pharmaceuticals and medicines (11.4%). It should be noted that these figures of mail-order purchases do not include data on foreign mail-order sales made to Canadians or the mail-order sales of Canadian department stores.

Other methods of direct selling which bypass the regular retail outlet and are included in the approximate $1.6 billion total sales figure are sales made from showrooms and premises of manufacturing companies and primary producers (which in 1976 accounted for 18.8%) and the miscellaneous sales made from temporary roadside stands and market stalls, at exhibitions and shows, and purchases of meals and alcoholic beverages on airlines, ferries and railways (6.0%).

18.1.2 Sales financing and consumer credit

Sales financing. Ancillary to the retailing industry are the financial institutions which facilitate consumer instalment purchases, particularly of the more expensive consumer durables such as automobiles and household appliances. Separate statistics have for many years been maintained by Statistics Canada on the retail instalment financing undertaken by the sales finance industry, especially their participation in the financing of automobile purchases. The firms in this industry include independent sales finance companies, the sales finance company subsidiaries of car, truck and farm implement manufacturers, and the sales financing business of consumer loan companies.

Not reported in these statistics are the instalment sales financing done by acceptance companies which are the subsidiaries of, or which are associated exclusively with, large retailing organizations. The sales financing activity of these companies is regarded as an extension of the merchandising function, and their statistics are included with the accounts receivable reported by department stores and other retail merchandising establishments. At year-end 1976 about a dozen such acceptance companies reported accounts receivable of $1,353.6 million for purchases of consumer goods through their associated retail outlets.

By year-end 1976 the sales finance industry, as delineated above, held outstanding balances of $3,392 million covering the retail instalment financing of both consumer goods ($1,134 million) and commercial and industrial goods ($2,258 million) (Table 18.12). During the course of the year, the industry augmented its purchases of new finance paper by $2,804 million, $1,074 million of consumer goods paper and $1,730 million of commercial and industrial finance paper.

Since 1970 the composition of the portfolios of sales finance companies has shifted from a preponderance of consumer goods paper to an emphasis on commercial and industrial goods financing. The latter class of paper now comprises 66.6% of all paper whereas in 1970 it amounted to 49.5%. The financing of passenger car sales still plays a significant role in the activity of sales finance companies. At year-end 1976 these companies held balances of $1,326 million for this class of finance paper (including balances on new passenger cars acquired for business use such as taxis and commercial fleets) amounting to 39.1% of their total holdings. In Table 18.12 these business purpose passenger cars are shown as commercial vehicles. The chartered banks have also increased their participation in passenger car financing over the years and now hold balances of $4,401 million.

Consumer credit. Estimates of total consumer indebtedness for 1977 and selected earlier years are shown in Table 18.13. These estimates are based on the outstanding balances recorded in the books of various financial institutions, retail merchandising establishments, public utilities and other credit-granting organizations. The consumer credit extended to individuals and families for non-commercial purposes can be in the form of cash advances, or the provision of goods and services on credit or through use of credit cards, and is generally repaid by regular instalments which include interest and other finance charges. Statistics on consumer indebtedness exclude fully-secured bank loans, home-improvement loans, and long-term indebtedness such as residential mortgages. Data are not available on certain other forms of consumer credit such as interpersonal loans, bills owed to dentists and other professional practitioners, and to clubs or other personal service establishments. A 1970 survey of families and unattached individuals showed that consumer debt accounted for 24% of all personal indebtedness, residential mortgages accounted for another 68%, and other miscellaneous debt accounted for the remaining 8% (Statistics Canada Catalogue 13-547).

By the end of 1977 the total amount of balances outstanding with the above-mentioned selected holders amounted to $31,234 million, an expansion of $3,496 million, up 12.6% over the level reached at the end of 1976. Chartered banks, with outstanding balances of $18,731 million, held 59.9% of total outstandings. Other major grantors of consumer credit were credit unions and caisses populaires with holdings of $4,568 million, 14.6% of the total; sales finance and consumer loan companies with holdings of $2,767 million, 8.8% of the total; and retail merchandising establishments with holdings of $2,628 million, 8.4% of the total outstanding.

In the present situation in the consumer credit market, the cash-lending institutions — chartered banks, credit unions and caisses populaires, consumer loan companies and life insurance companies' policy loans — account for the overwhelming share (87.6%) of consumers' credit needs. This is in marked contrast to the situation in earlier years when consumer credit requirements were mainly serviced by department stores and other retail establishments and sales financing companies, which arranged and financed instalment credit for household effects and other consumer durables including passenger cars.

Service trades   18.1.3

Service trades generally encompass those businesses, both commercial and non-commercial, which perform a service and in which the sale of goods constitutes only a minor function. Commercial service trades are classified generally into six principal groups: amusement and recreational services (such as movie theatres, bowling alleys, billiard parlours and health clubs); personal services (barber shops, beauty parlours, laundry and dry cleaning, laundromats and shoe repair shops); restaurant services (restaurants, take-out food shops, and other eating and drinking places); miscellaneous services (photographers, automobile and truck rentals and driving schools); services to business (lawyers, accountants, computer services, consultants, advertising agencies, media representatives); and accommodation services (hotels, motels and tourist camps). Non-commercial services include religious institutions, trade and professional associations, fraternal organizations and service clubs. Services related to education, health and finance are not included in this section. Automotive services, such as garages and other repair shops, are covered under retailing.

Traveller accommodation. Table 18.14 summarizes the major types of accommodation services in 1974 and 1975. Total accommodation receipts in 1975 amounted to $2,571.0 million, of which hotels accounted for the major share, 81.6%, with total receipts of $2,097.1 million. Receipts reported by motels totalled $301.8 million (11.7%) and the remaining $172.1 million (6.7%) was accounted for by tourist homes, tourist courts and cabins, outfitters and tent and trailer campgrounds. Total receipts include such source items as sales of rooms, food, alcoholic beverages, merchandise and other services provided by traveller accommodation business — telephone, valet, laundry and parking. A further breakdown of traveller accommodation data by province is in Table 18.15.

Food and beverage industry. A census-type survey of the food and beverage industry (restaurants, caterers and taverns) was carried out in 1976. It was the first such survey since 1971 and showed total receipts of $5,131.0 million, representing a 114.6% increase over the 1971 figure of $2,397.0 million. The survey included establishments primarily engaged in preparing and serving meals and beverages, such as regular restaurants, caterers, drive-in, take-out and industrial restaurants, and taverns. Excluded were establishments owned by and operated as an integral part of hotels, motels and other accommodation businesses; armed forces messes, private clubs, legion branches and eating and drinking places operated by establishments classified to an industrial sector other than the service trades. The provincial breakdown of food and beverage receipts for 1976 is given in Table 18.16.

Power laundries, dry-cleaning and dyeing plants. In 1975, 2,681 power laundries, dry-cleaning and dyeing plants operated in Canada with a revenue of $373.7 million, an increase of 16.1% from $321.8 million in 1974. Of these plants, 354 were laundries, with a revenue of $185.0 million, and 2,327 were dry-cleaning and dyeing plants with receipts totalling $188.7 million. In 1975 power laundries showed a 22.5% increase in revenue over 1974 while the revenue of dry-cleaning and dyeing plants increased 10.5%.

Motion picture exhibition, distribution and production. This industry consists of exhibitors who operate regular movie theatres and drive-in theatres, film distributors, and private firms and government agencies engaged in producing motion picture films.

In 1976 receipts from admissions were $224.0 million, of which $192.4 million were obtained by regular movie theatres and $31.6 million by drive-in theatres. Revenues obtained by regular theatres from other sources such as snack bars brought total receipts to $233.1 million. There were 1,129 regular theatres and 309 drive-in theatres in operation in Canada.

The average admission price was $2.34 (including taxes) in regular theatres and $2.42 in drive-in theatres. The average utilized seating capacity in Canada as a whole dropped from 16% in 1975 to 15% in 1976. Smaller centres continue to have higher capacity utilization than larger centres.

In 1976, 86 firms distributed films through 134 offices in Canada. Total receipts increased by 11.4% to $127.0 million in 1976 from $114.0 million in 1975. Revenue from the rental of films for theatrical use amounted to $78.6 million, representing 63.5% of total receipts. Revenue from the rental of films for television accounted for 29.3% and the remaining 7.2% came from the rental of film for other uses. In 1976, 2,580 new films were distributed, compared with 864 in 1975. New feature films numbered 690 and included 456 English, 187 French and 47 films in other languages. Of the 690 new feature films, 380 came from the US, 109 from France, 32 from Italy, 43 from Britain, 40 from Canada and 86 from other countries.

Motion picture production in 1976 was undertaken by 271 private firms reporting gross production revenue of $53.7 million. A total of 5,810 original motion picture films was reported, plus 1,000 revisions, making a total of 6,810 productions of which 6,515 were made by private firms and 295 by government agencies. These 6,810 productions included the following: commercials (3,396), long and short features (424), documentary and educational films (1,038), filmstrips (1,091) and other (861).

Advertising agencies. In 1976, 304 advertising agencies reported gross billings of $916.2 million. This only represents part of the total expenditure on advertising in the country since all advertising is not produced or placed by and through advertising agencies. Among expenditures not generally channelled through advertising agencies are classified advertisements in newspapers and a certain amount of catalogue and direct mail advertising. Of the total gross billings, which include media billings and production charges, $321.0 million was in print media (including newspapers, weekend roto magazines, consumer magazines, trade papers, yellow pages and farm publications), $394.7 million in television, $120.4 million in radio, $27.1 million for outdoor and transportation, $24.1 million for direct mail and $3.6 million for other media.

Total media billings amounted to $749.3 million; production charges accounted for a further $141.7 million and market research surveys and other services for $25.2

million, making a total of $916.2 million in gross billings, an overall increase of 27% over 1975.

A further comparison with 1975 shows that total advertising billings (media billings plus production charges) increased 27.8%; total media billings 31.4%; production costs 11.8%; and expenditures on market research surveys and other services 4.6%.

Computer service industry. In 1976 a survey of the computer service industry revealed that 516 companies in Canada provided services involving 506 computers of various capacities, 3,063 terminals and 3,309 access ports. Total operating revenue amounted to $1,048 million of which hardware sales and rentals accounted for $626 million, processing for $273 million, software for $74 million and equipment maintenance, education and other services for $75 million.

Of the total operating revenue, $195 million or 18.6% was generated from computer services provided to financial institutions, $243 million or 23.2% from forestry, mining and manufacturing, $181 million or 17.3% from all levels of government, $200 million or 19.1% from the wholesale, retail and service sectors, and $229 million or 21.8% from other businesses and institutions.

A further 502 companies in other industry groups also provided computer services as a secondary activity, producing revenues of $117 million, 53% of which was from processing, 29% from automated data processing hardware rentals and 18% from other computer services. These services were provided to financial institutions (31%), wholesale, retail and service sectors (33%) and the remaining 36% to businesses and institutions in other industry groups.

Wholesale trade   18.1.4

In the field of wholesale statistics a program of upgrading has been implemented which includes biennial coverage of the operations of wholesale merchants begun in 1973, and of agents and brokers, begun in 1974. As well as producing more up-to-date statistics on these two types of operation, the results, in the case of wholesale merchants, will be used as the base for a new sample of monthly sales and inventory estimates.

Wholesalers are primarily engaged in buying merchandise for resale to retailers; to industrial, commercial, institutional and professional users; to farmers for farm use; to other wholesalers; or act as agents in connection with such transactions. Businesses engaged in more than one activity, such as wholesaling and retailing or wholesaling and manufacturing, are considered to be primarily in wholesale trade if the greater part of their gross margin (the difference between the total sales and the cost of goods sold) is due to their wholesaling activity.

Wholesale trade statistics measure the total volume of Canadian wholesale trade, the total volume of trade (domestic and export sales) conducted by all wholesalers operating in Canada, whether they are Canadian-owned or subsidiaries of foreign companies. The total volume of trade measured by Statistics Canada cannot be equated with the value of goods passing through the wholesale sector of the economy because at times wholesale businesses sell to each other and thus the value of the same merchandise may be recorded more than once.

According to certain common characteristics, each wholesale establishment and location (wholesale outlet) is assigned to one of the following types of operation: primary product dealers (grain, livestock, raw furs, fish, leaf tobacco and pulpwood, including co-operative marketing associations); wholesale merchants (buying and selling goods on own account); agents and brokers (buying and selling goods for others on a commission basis); manufacturers sales branches (wholesale businesses owned by manufacturing firms for marketing their own products); or petroleum bulk tank plants and truck distributors (wholesale distribution of petroleum products).

Wholesale merchants account for about 60% of the total wholesale volume of trade and had estimated sales in 1976 of $49,988 million, up 10.2% from the 1975 volume of $45,377 million. Industrial goods trades accounted for $26,613 million of the 1976 total volume of trade while the remaining $23,374 million was in consumer goods trades, which showed an accelerated growth for the year of 11.4% compared with the industrial goods increase of 9.1%. Data for 1974-76 are given in Table 18.21.

Farm implement and equipment sales data are collected annually from manufacturers and importers active in the farm implement and equipment field. Dollar sales are reported at dealers' buying price before the deduction of dealers' cash discounts, value of trade-ins, volume or performance bonuses and export sales are excluded. In 1976 the dollar value of repair parts of $154.7 million was 7.4% greater than the $144.0 million reported for 1975.

Farm equipment sales reached a low point in 1970 but in subsequent years have shown a steady recovery, attaining a record level of $1,134.0 million in 1976. The two most important products were farm tractors with a sales volume of $449.3 million or 39.6% of total sales volume and harvesting machinery with sales of $225.5 million or 19.9% of all farm implement and equipment sales in 1976.

Construction machinery and equipment sales include sales by Canadian distributors, direct sales by manufacturers to end-users (at actual final selling price) and revenue derived from renting equipment to users. In 1976 new machinery entering the market (by outright sale, first lease or rental) was valued at $1,564.1 million, 2.3% below 1975 (Table 18.23). The sale of used machinery rose by 14.4% from $211.3 million in 1975 to $241.8 million in 1976. Rental income increased 4.4% from $118.3 million to $123.5 million. Of the $1,564.1 million, $566.1 million was accounted for by repair and consumable parts. The largest single item in terms of dollar sales was crawler-type tractors: 2,629 units entered the market for a value totalling $166.1 million. Sales of new equipment by distributors totalled $1,420.1 million, while sales by manufacturers amounted to $144.0 million.

Diesel and natural gas engines sold in Canada in 1976 totalled 19,025. This included 18,201 diesel engines, a decrease of 5.8% from the 19,328 sold in 1975 (Table 18.24) and 824 natural gas engines. In addition, 17,523 diesel engines were exported or re-exported (compared to 18,891 in 1975), and five natural gas engines were exported in 1976.

18.1.5   Co-operatives

Overall business volume of Canadian co-operatives passed the $6.0 billion mark in 1976 at $6,198.8, a gain of $657.0 million or about 12% over 1975. The marketing of farm and fish products accounted for more than half the increase but on a percentage basis marketings rose 11%, merchandise and supply sales 13% and service revenue 18% while miscellaneous income was almost unchanged. The year was generally one of growth with almost all revenue categories and regions benefitting from the upsurge. Despite a drop from the double-digit inflation of the previous two years, many co-operatives were severely affected by the cost-price squeeze.

Co-operatives covered here exclude recreational (such as community halls and rinks), financial (credit unions) and those run by native peoples. Those included are classified by their primary function into four main groups: marketing and purchasing (the largest), service, fishermen's and wholesales. The service group is frequently subdivided into service and production. Production co-operatives provide services directly related to agricultural production such as artificial breeding, or are directly involved in production such as co-operative farming. The first three groups are known as local co-operatives because they deal directly with individual members; the wholesale co-operatives perform wholesaling functions for the locals.

Asset value of local co-operatives expanded $242 million or approximately 10% in 1976 with about half of it going into property and equipment. Short- and long-term loans financed about three-fifths of the asset expansion while member equity, most heavily in the form of share capital, covered the rest. Member equity accounted for 39% of assets at year-end 1976, unchanged from 1975. Reporting associations registered a modest upturn for the second year in a row after years of decline and numbered almost 2,500. Co-operative membership continued to increase to about 2,350,000 at year end, a gain of about 240,000. Alberta continued as the province with the most co-operatives and members, while Saskatchewan, with its huge grain revenues, was again the leader in overall co-operative business volume.

Business volume of marketing and purchasing co-operatives rose in 1976 as did the number of associations and their shareholders or members. Farm marketings climbed $377 million or about 11% on the rebound from the modest growth of 1975. Whereas product categories were down in 1975 with the major exception of dairy products, all categories showed increases in 1976. Grain and seeds contributed about $200 million to the gain on the strength of a surge in Canadian wheat exports from the high reached in 1974 and 1975. Dairy product growth was again substantial although more modest than 1975 which saw the acquisition of additional large facilities from the private sector, notably in Quebec. In miscellaneous product marketings, gains were registered for maple products, tobacco, fur, lumber and wood, while honey, wool and other miscellaneous commodities were lower.

Sales of farm supplies and consumer goods by marketing and purchasing co-operatives rose $246 million or about 13% for the year, down from an inflation-fuelled increase of $361 million or 23% in 1975. This is particularly pertinent for food products which account for over one-quarter of supply sales and where the inflation rate declined to only about one-fifth that of the previous year. Sales of clothing and home furnishings showed a gain reflecting the continued expansion of co-operatives into the consumer goods field, particularly in urban areas. Feed sales were only slightly ahead, mostly due to a decline in prices rather than physical volume. Farm machinery sales were good in another year of high-level farm revenues, and would have been better but for poor spring weather on the Prairies. All other supply categories were also on the uptrend in line with larger farm product marketings and rising prices.

Control and sale of alcoholic beverages   18.1.6

The retail sale of alcoholic beverages in Canada is controlled by provincial and territorial government liquor control authorities. Alcoholic beverages are sold directly by most of these authorities to the consumer or to licensees for resale. However, in some provinces beer and wine are sold directly by breweries and wineries to consumers or to licensees for resale. During the year ended March 31, 1976, provincial government liquor authorities operated 1,536 retail stores and had 381 agencies in smaller centres.

Table 18.27 shows the value and volume of sales of alcoholic beverages in the years ended March 31, 1975 and 1976. The value does not always represent the final retail selling price of alcoholic beverages to the consumer because in some cases only the selling price to licensees is known. Volume of sales is a more realistic indicator of trends in consumption, but as a measure of personal consumption by Canadians it is subject to the same limitations as value sales and includes, in addition, purchases by non-residents.

Government revenue specifically related to alcoholic beverages and details of sales by value and volume for each province are given in Table 18.28. *The control and sale of alcoholic beverages in Canada* (Statistics Canada Catalogue 63-202) shows further detail as well as volume figures of production and warehousing transactions, the value and volume of imports and exports and the assets and liabilities of provincial liquor commissions.

International trade   18.2

Summary   18.2.1

There was a strong recovery for exports of about 15% in both 1976 and 1977 following the recession year of 1975 when exports increased only 2.5% from 1974 (Table 18.29). Imports grew less rapidly than exports, increasing 8% in 1976 and 12% in 1977.

The stronger growth in exports compared with imports was reflected in the trade balance which changed from a deficit of $1.4 billion in 1975 to surpluses of $0.7 billion in 1976 and $2.1 billion in 1977. The 1977 improvement in the trade balance coincided with a decline in 1977 of 7.8% in the value of the Canadian dollar against the US dollar. Trade balances are given on a customs basis, based on data tabulated from customs documents according to procedures and concepts explained in Section 18.2.5. Trade balances are also available on a balance-of-payments basis, reflecting a number of adjustments applied to the customs total to make them consistent with the concepts and definitions used in the System of National Accounts.

Import and export price indexes showed almost no increase in 1976 compared with 1975 (Table 18.37). Import prices increased less than 1% compared with 16% in 1975 and export prices less than 3% compared with 11% in 1975. Both indexes accelerated in 1977 but at a more moderate pace than the peak years of 1974 and 1975 for imports, and 1973 and 1974 for exports. Import prices were up 12% in 1977 compared with 1976, the depreciated Canadian dollar being a significant factor. Export prices were up 6%. A major contributor to the lower export price index increases in 1976 and 1977 was the weakness of food, feed, beverage and tobacco prices which were down 5% from the previous year in 1976 and down 10% in 1977, a considerable factor being the weakness of wheat prices, down 14% in 1976 and 23% in 1977.

Import volume grew less than 1.0% in 1977 compared with a 9.0% increase for domestic exports. Import volume was down from the previous year for all sections apart from end products which increased only 2.0%. Export volume increased for all sections apart from crude materials, down 4.0%, affected by weakness in world demand for metal ores and a reduction in crude petroleum exports to half the peak level reached in 1974.

Of 1977 import value, 61.9% was accounted for by end products (Table 18.30). Fabricated materials accounted for 16.7%, crude materials for 12.6%, and food, feed, beverages and tobacco for 7.7%. Largest commodity import groups were motor vehicles and parts 27.0%, up from 24.0% in 1975, industrial machinery 8.3%, crude petroleum 7.7%, down from 9.5% in 1975, chemicals 4.8% and agricultural machinery 3.2%.

End products, although only half as important for exports as for imports, accounted for slightly more than one-third of total export value in 1977, 34.5% compared with 32.2% in 1975 (Table 18.31). Fabricated materials also accounted for just over one-third of exports, 34.4% in 1977 up from 30.3% in 1975. Crude materials accounted for 20.4% and food, feed, beverages and tobacco for 10.1%.

In 1977 the most important export commodities were motor vehicles and parts 23.4%, up from 21.9% in 1975, lumber, wood pulp and newsprint 16.0%, crude and fabricated metals including ores and concentrates 14.5%, natural gas 4.7%, up from 3.4% in 1975, crude petroleum 4.0%, down from 6.2% in 1975 and wheat 4.2%, down from 6.2% in 1975.

18.2.2 Trade with the United States

The United States increased its share of Canadian trade in both directions, from 68.1% of imports in 1975 to 70.2% in 1977, and from 65.3% of exports in 1975 to 69.9% in 1977. Imports from the US increased 9.0% in 1976 and 15.0% in 1977 (Table 18.34) compared with 8.0% and 12.0% respectively for total imports. Exports increased even faster, gaining 19.0% over the previous year in 1976 and 20.0% in 1977, compared with 15.0% and 16.0% respectively for total exports.

After adjusting for conceptual differences, which normally add to the balance calculated from Canadian data, the reconciled trade surplus with the US measured $0.7 billion compared with a 1975 deficit of US$1.3 billion (Table 18.35).

An important part of Canada's trade with the US is the free passage of automotive products. The ratio of exports divided by imports has fallen steadily from a high of 112.5% in 1970 to 77.2% in 1975. It recovered to 88.6% in 1976 and 89.9% in 1977.

Automotive parts account for a much larger proportion of imports of automotive products than of exports because a high volume of parts are imported into Canada, assembled and exported in the form of complete vehicles. In this instance parts refers to components designed for motor vehicles or motor vehicle engines. Excluded are some general purpose components which may be used elsewhere than in a motor vehicle, such as tires, radios, batteries and generators. Parts accounted for 61% of automotive imports in 1977 compared with 37% of exports.

Exports of automotive goods have grown more rapidly than imports in both 1976 and 1977. Exports of vehicles were up 23% in 1976 and 25% in 1977 compared with increases of 6% and 19% respectively for imports. For automotive parts, exports were up 43% in 1976 and 22% in 1977 compared with gains of 19% and 24% respectively for

Canada's trade with principal trading areas, 1976 and 1977

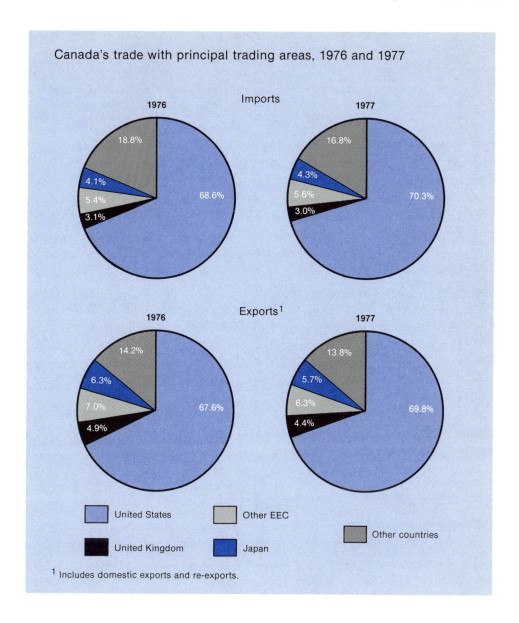

Imports

1976

18.8%

4.1%

5.4%

3.1%

68.6%

1977

16.8%

4.3%

5.6%

3.0%

70.3%

Exports[1]

1976

14.2%

6.3%

7.0%

4.9%

67.6%

1977

13.8%

5.7%

6.3%

4.4%

69.8%

United States

United Kingdom

Other EEC

Japan

Other countries

[1] Includes domestic exports and re-exports.

imports. Other year-to-year import changes significant in 1977 were end products which increased 15% (at the same rate as total imports from the US). Growth was particularly strong in the automotive sector, with the 1977 increase in total automotive imports from the US at 22%. This was counterbalanced by relatively low increases of 6% for machinery, 10% for other equipment and tools and no growth for personal and household goods. Fabricated material imports increased 13%, food, feed, beverages and tobacco 14%, and crude materials 28%, growth in crude petroleum imports being an important factor.

Increases in Canadian exports to the United States in 1977 compared with 1976 were between 22% and 30% for all sections apart from crude materials which increased only 2.6%. Crude material exports were affected by a 23.4% drop in crude petroleum exports counterbalanced by a 25.5% increase in natural gas exports.

End product exports to the United States increased 22% in 1977. Automotive exports increased 24%, industrial and agricultural machinery 25%, other equipment and tools 17%, apparel and footwear 24%. There was no increase in personal and household goods exports. Fabricated materials increased 30%. Sharp increases were shown by electricity of 133%, aluminum 66%, lumber 56% and iron and steel products 38%. Food, feed, beverages and tobacco increased 22%. Fruits and vegetables and preparations increased 63% and whisky 22%.

18.2.3   Trade with other countries

Japan's share of Canadian imports increased from 3.5% in 1975 to 4.3% in 1977. Canadian exports to Japan declined from 7.1% in 1973 to 5.7% in 1977.

Canadian trade with the United Kingdom continued to decline, from 6.0% of imports in 1967 to 3.0% in 1977, and from 10.3% of exports in 1967 to 4.4% in 1977. Canada's share of trade with other European Economic Community (EEC) countries has remained fairly stable over the last 10 years and in 1977 stood at 5.6% for imports and 6.3% for exports.

18.2.4   Trade in energy-related products

Total exports of all energy products, including crude petroleum, natural gas, coal, petroleum and coal products, electricity and radioactive ores, elements and isotopes, increased 7.6% in 1975, fell 1.5% in 1976, and increased 4.9% in 1977 to reach $5.7 billion. Imports increased 25.0% in 1975, fell 2.8% in 1976, and increased 3.7% in 1977 to reach $4.2 billion giving a net balance of $1.5 billion in 1977. The ratio of energy product exports divided by imports fell from 152.8% in 1974 to 131.5% in 1975 and has since recovered to 133.3% in 1976 and 134.8% in 1977.

After increasing 25.0% in 1975, crude petroleum imports decreased 0.8% in 1976 and 0.9% in 1977. Crude petroleum exports declined 11.0% in 1975, 25.0% in 1976 and 23.0% in 1977. Natural gas exports have compensated for much of the decline in crude petroleum exports. Combined exports of crude petroleum and natural gas increased 5.6% in 1975 and declined 5.8% in 1976 and 3.2% in 1977. Petroleum and coal products, electricity and radioactive ores, elements and isotopes have also compensated for declining crude oil exports. Imports and exports of coal and other crude bituminous substances have virtually balanced over the last four years.

18.2.5   Sources of statistics

Canada's external trade statistics are tabulated from copies of administrative documents collected by customs offices at ports across Canada. The Customs Act requires that each time goods are imported into or exported from Canada a document be filed with customs giving such descriptions of the goods and details of the transaction as are required for customs administration. It follows that the method of compilation of external trade statistics is determined and limited to some extent by customs regulations and procedures.

Statistics on trade in electricity and on exports of crude petroleum and natural gas cannot, for administrative reasons, be obtained from customs documents. They are instead collected by Statistics Canada.

Concepts and definitions used in the compilation of external trade statistics are published in *Summary of external trade* (Statistics Canada Catalogue 65-001). Among them are the following:

System of trade. Canadian statistics are tabulated according to the general system of trade. Thus imports include all goods which have crossed Canada's geographical boundary, whether they are entered through customs for immediate use in Canada or stored in bonded customs warehouses. Domestic exports include goods grown, extracted or manufactured in Canada (including goods of foreign origin which have been materially transformed in Canada). Re-exports are exports of goods of foreign origin which have not been materially transformed in Canada (including goods withdrawn for export from bonded customs warehouses).

Coverage. Merchandise trade includes only goods which add to or subtract from the stock of material resources in Canada as a result of their movement across the Canadian border.

Valuation. Exports are recorded at values which usually reflect the actual selling price. Most exports are valued at the place in Canada where they are loaded aboard a carrier for export (a mine, farm or factory) but a significant proportion of exports by water or air reflect values which include transportation charges to the port of export. Some overland shipments to the United States are recorded at a value which includes transportation charges to the ultimate destination.

Imports are generally recorded at the values established for customs duty purposes. Customs values are identical to selling prices for most arms-length transactions. However, customs values exceed company transfer prices for most transactions between affiliated firms. Import documents are required to show values which exclude all transportation charges. Some imports from the US are, however, purchased on a delivered basis and their prices reflect an allowance for transportation costs.

Trading partner attribution. Imports are attributed to the country from which the goods were first consigned directly to Canada, whether or not this is the country of origin. An exception is made in the case of goods of Central or South American origin consigned to Canada from the United States; such imports are credited to the country of origin.

Exports are attributed to the country which is the last known destination of the goods at the time of export. (Many primary products are shipped to entrepôt points, particularly in Europe, for re-export to the ultimate destination which is unknown when the goods leave Canada.) The country classification employed by Statistics Canada is designed for purposes of economic geography and therefore does not reflect the views or intentions of the federal government on international issues of recognition, sovereignty or jurisdiction.

Reconciliation. Canadian trade statistics rarely agree with the counterpart statistics of its trading partners. The major factors contributing to the discrepancies are differences in concepts and collection procedures. Conceptual differences are most common in statistical treatment of special categories of trade such as military supplies, government-financed gifts of commodities, postal and express shipments, tourist purchases, bunker and warehouse trade, in the definition of territorial areas, and in the system of crediting trade by countries. Differences in collection procedures lead to discrepancies in valuation, since the value of trade can be based on customs value, transaction value, or fair market value with or without the inclusion of transportation charges; in timing, since the definition of a statistical month or year can differ; and in the capture of trade data, since the documentation of export trade tends to be less closely monitored than import trade. The United States and Canada have agreed on concepts and definitions describing a framework within which it has been possible to reconcile differences in trade statistics published by the two countries.

Indexes of price and volume. The price indexes in Table 18.37 are current-weighted and are calculated from price relatives based on $1971 = 100$. The weights are trade quantities for the month, quarter or year to which the index refers and hence change from period to period. The volume index is derived by dividing a value index by the corresponding current-weighted price index. The resulting volume index is, therefore, weighted with fixed 1971 price weights. The price indicators selected are either commodity unit values calculated directly from the trade statistics or, particularly in the case of end products, price indexes obtained from other Canadian or foreign statistics.

An explanation of the methodology used to construct the indexes is contained in a reference paper entitled *The 1971-based price and volume indexes of Canada's external trade,* published in December 1976 as a supplement to *Summary of external trade* (Statistics Canada Catalogue 65-001).

Principal trading areas. Effective January 1976 the principal trading areas shown in some tables include new groupings which are defined as follows: other EEC — Belgium, Denmark, France, Federal Republic of Germany, Ireland, Italy, Luxembourg and the Netherlands (the UK is also a member of the EEC but is shown separately because of the importance of its trade with Canada); other OECD — Austria, Finland, Greece, Iceland, Norway, Portugal, Spain, Sweden, Switzerland, Turkey, Australia and New Zealand (the EEC countries, United States, Japan and Canada are also members of OECD); other America — this group includes all countries and territories of North and South America (other than the US and Canada) including Greenland, Bermuda and Puerto Rico.

18.3 Federal trade services

Canada's economy continues to be vitally dependent on international trade. Competition among industrial nations is intense and increased exports are not easy to achieve. A successful export trade can only be assured by combining good products, efficient production and aggressive, intelligent marketing with government support.

Federal government support is provided through the industry, trade and commerce department and the Export Development Corporation. The department assists Canadian industry throughout the complete cycle — from research, design and development through production to marketing of the finished product. The Export Development Corporation, a Crown agency which reports to Parliament through the minister of industry, trade and commerce, provides insurance, guarantees, loans and other financial facilities to help Canadian exporters.

18.3.1 Department of Industry, Trade and Commerce

The department has a number of units involved in international trade. An explanation of their role follows.

The office of general relations includes a general trade policy branch and a commodity trade policy branch, responsible, within the department, for formulating and implementing Canadian trade policy with particular reference to the activities of the General Agreement on Tariffs and Trade (GATT), the Organization for Economic Co-operation and Development (OECD), the United Nations Conference on Trade and Development (UNCTAD) and the trade aspects of domestic industrial and agricultural policies. It is also responsible for commodity trade policy questions generally and in particular the preparation and conduct of the negotiation of intergovernmental commodity arrangements and agreements.

The office of special import policy implements government policies relating to low-cost imports. It proposes action to be taken by government in the light of recommendations of the Textile and Clothing Board (with respect to imports of textiles and clothing) and of the Anti-dumping Tribunal (with respect to other low-cost products), as well as in other instances where low-cost imports have caused or threaten serious injury to domestic production. It conducts restraint negotiations with other governments and implements special measures of protection by means of import controls when required. It is also responsible for the conduct of international textile negotiations within GATT and participates in the work of the textiles surveillance body established under an arrangement regarding international trade and textiles. The office administers controls under the Export and Import Permits Act.

The export and import permits division is responsible for all matters relating directly or indirectly to commodity control measures under the authority of the Export and Import Permits Act and the United Nations Rhodesia Regulations. The purpose of the act is to ensure, by means of export controls, that there is an adequate supply and distribution in Canada of goods necessary for defence or other purposes; that no specified goods having a strategic nature will be made available to any destination wherein their use might be detrimental to the security of Canada; that an intergovernmental arrangement or commitment is implemented; that any action taken to promote the further processing in

Canada of a natural resource that is produced in Canada is not rendered ineffective by reason of the unrestricted export of that natural resource; and that the export of any raw or processed material that is produced in Canada in circumstances of surplus supply and depressed prices is limited or kept under surveillance.

By means of import controls, the act is also intended to ensure an adequate supply in Canada of goods that are scarce in world markets, or subject to governmental controls in the countries of origin or to allocation by intergovernmental arrangement; to implement action taken under certain other specified federal acts; and to implement an intergovernmental arrangement or commitment. Other functions of this division are to advise exporters and importers on interpretation and requirements of the export control, area control and import control lists and regulations; to study the economic implications of the act; and to review control lists and practices.

The international bureaus (European bureau, Pacific, Asia and Africa bureau, and Western Hemisphere bureau) are focal points on matters affecting Canada's trade and economic relations with other countries and areas. Bureau responsibilities include development of Canada's international trade strategy, market development programs for individual countries and areas and maintenance and improvement of access for Canadian products to export markets. The bureaus are centralized sources of information on Canada's trade with specific countries or regions and they provide a regional perspective for matters of both international trade relations and export trade development. They also provide information, advice and guidelines to government agencies and to the business community on foreign government trade and economic regulations and practices; maintain contact, normally through Canadian posts abroad, with foreign markets and foreign governments on matters pertaining to markets for Canadian exports; and provide advice to the department, to other Canadian government agencies and to the Canadian business community on export market problems and opportunities.

The trade commissioner service has 88 trade offices in 65 countries. Its primary role is to promote Canada's export trade and to represent and protect its commercial interests abroad. Accordingly, a trade commissioner performs a variety of tasks: to act as an export marketing consultant; to bring foreign buyers into contact with Canadian sellers; to help organize trade fairs and trade missions; to recommend modes of distribution and suitable agents; and to report on changes in tariffs, exchange controls and other matters affecting Canada's trade with the countries to which he is accredited. He initiates programs to develop new markets for Canadian products, responds to inquiries from Canadian firms and provides advice to the visiting Canadian exporter. He also acts on behalf of the foreign programs of a number of federal government departments and undertakes agricultural reporting at specified posts. For a Canadian firm wishing to develop a market in his territory, the trade commissioner can supply information on product use, if any, local production and import data, and prospective users or agents.

The scheduled return of trade commissioners for official tours of Canada helps Canadian firms interested in the export trade. Trade associations are informed in advance of these visits so that appointments may be arranged by businessmen wishing to meet trade commissioners, through the trade commissioner service, trade associations, or one of the department's regional offices.

The office of overseas projects stimulates, develops and sustains Canadian participation in export projects other than pure commodity or equipment transactions exclusive to the interests of one industry sector or the direct sales of engineering consultant or Canadian construction services. These services do not normally come to the office unless such transactions are part of an integrated response with other firms to an export opportunity. The office works with the appropriate bilateral and multilateral agencies to develop those financing accommodations needed to support exports of goods and services.

The responsiblity of the office includes identification, promotion and co-ordination of industry response to overseas project opportunities, and development of appropriate financial and risk-sharing facilities such projects demand. The office is the focus in the department for all capital and turnkey projects.

The grains marketing office is concerned with federal government activities in marketing assistance and industrial development for grain, oilseeds and their bulk derivatives. It contributes to overall grain production, transportation and marketing policy formulation and works closely with the Canadian Wheat Board on grain sales and promotion programs. Its continuing operational responsibility includes the institution and administration of programs designed to expand exports of grain, oilseeds and their products and to help stabilize the market. Among these are a grains and oilseeds marketing incentive program, grain credit facilities and Prairie grain advance payments. The office participates in the activities of international organizations concerned with grain and oilseeds such as the International Wheat Council and the Food and Agriculture Organization of the United Nations.

The transportation services branch is concerned with the transportation environment and with short- and long-term transportation problems that affect Canadian trade and industrial development. Continuing reviews are made of freight rates and services to shippers, and of regional, national and international transportation policies and measures that have an impact on Canadian trade. Assistance is provided to shippers in selecting appropriate transportation routes and modes at lowest possible freight costs. The branch participates in national and international organizations and conferences which are concerned with cargo movement, transport, distribution systems, simplification of documentation, facilitation of trade procedures, and international maritime development.

The office of international marketing consists of two branches: the promotional projects branch and the defence programs branch. The promotional projects branch administers a program through which projects are initiated, organized and implemented by the department. This includes participation in international trade fairs, solo shows and in-store promotions overseas; organization of technical seminars and trade missions abroad; and sponsoring of foreign visits to Canada to stimulate the sale of Canadian products in various export markets. The department provides promotional publicity in support of these projects.

In general, missions abroad are used for market investigation, evaluation and identification of technical market access problems; visiting missions are designed to invite foreign government or company representatives who can influence buying to inspect the industrial capacity and technical capabilities of Canadian firms and the products and services they can supply. Technical seminars acquaint potential buyers with Canadian expertise and technology in specific fields as a basis for joint ventures and sales of Canadian products and services. The trade visitors section of the program provides financial assistance at short notice to take advantage of foreign market opportunities by bringing foreign government trade representatives, buyers and export-oriented trainees to Canada.

The defence programs branch promotes defence export trade through marketing programs aimed at the sale of Canadian defence and defence-related high technology equipment to friendly countries and the establishment of arrangements with Canada's allies for co-operative industrial research, development and production in defence-related matters. A major activity is the Canada–United States program on defence development and production sharing which entails the joint development and reciprocal procurement of defence items.

18.3.1.1 Canadian Government Office of Tourism

The government's office of tourism is an agency of the industry, trade and commerce department. It is headed by an assistant deputy minister, tourism, who, through the deputy minister, advises the industry, trade and commerce minister on policy and operational matters relating to development and promotion of tourism in Canada. He represents federal interests in domestic and international tourism organizations.

A reorganization in March 1976 reflected changing conditions affecting tourism in Canada. The office was organized into two branches: marketing, and policy planning and industry relations.

The marketing branch helps to ensure a competitive travel product — market development and the actual marketing of the travel product which is Canada itself. This branch promotes travel to Canada from other countries and travel within Canada by Canadians and co-ordinates its activities with those of the provinces, territories and the private sector. The branch analyzes and identifies the market and uses sophisticated electronic and print advertising campaigns, direct mail, and a publicity and promotion program involving written material, displays, photographs and films.

The branch maintains 26 market development and promotional offices in the United States and seven overseas countries: Britain, France, Federal Republic of Germany, the Netherlands, Mexico, Australia and Japan. It also has travel trade programs to promote tours to and within Canada and to stimulate growth in the field of international congresses, incentive travel, conventions and corporate meetings.

The policy planning and industry relations branch is responsible for policy and development planning and research to ensure that the supply and demand sides of tourism have a balanced growth. The branch is responsible for co-ordinating liaison on tourism resources among other federal agencies, the provinces, territories and municipalities, and among private sector tourism-related organizations, both domestically and internationally. It also gathers and disseminates information on tourism to the travel industry, the media and the public and provides substantial marketing and administrative support for the tourism office.

Export Development Corporation (EDC)   18.3.2

EDC is a federally owned enterprise that operates on a commercially self-sustaining basis to help Canadian exporters meet international credit competition. It offers four types of assistance to exporters: export credit insurance, insuring Canadian firms against non-payment when Canadian goods and services are sold abroad; surety and related insurance for performance guarantees, insuring Canadian firms, surety companies and financial institutions against wrongful call by a foreign buyer, or non-performance of one or more members of a limited liability consortium; long-term export loans to foreign buyers of Canadian capital equipment and technical services; and foreign investment guarantees, insuring Canadians against loss of investments abroad by reason of political actions. EDC may also guarantee financial institutions against loss incurred in financing either the Canadian supplier or the foreign buyer.

Export credits insurance. Export credits insurance offers Canadian exporters protection against non-payment by foreign buyers for any reason beyond the control of either the exporter or the buyer. The main risks covered are: insolvency of or protracted default by the buyer; repudiation where the exporter is not at fault and where proceedings against the buyer would serve no purpose; transfer delays; loss through war or revolution; cancellation or non-renewal of an export permit and the imposition of restrictions on the export of goods not previously subject to restriction; and the incurring, as a result of interruption or diversion of a voyage, of additional handling, transport or other changes in respect to goods exported.

Classes of insurable transactions include: consumer goods and miscellaneous general commodities sold on short credit terms usual for the particular trade, and which normally range from documentary sight draft to a maximum of 180 days; capital goods such as heavy machinery sold on medium credit terms which may extend to a maximum of five years; services such as the supply of design, engineering, construction, technological and marketing services to a foreign customer; photogrammetric and geophysical surveys; and invisible exports such as the sale or licensing to a foreign customer of any right in a patent, trademark, or copyright, advertising fees, and fees to auditors and architectural consultants.

In the case of goods or services sold on short-term credit, a policy is issued which covers an exporter's entire export sales for one year. For goods or services sold on medium credit terms, specific policies are issued for each transaction.

There are two types of policies: a contracts policy which protects from the time an exporter receives the order until he is paid, and a shipments policy which protects him from the time of shipment only. The contracts policy is designed for the exporter who manufactures goods to particular specifications, or goods which are so marked or stamped that they are of no value except to the original buyer. EDC normally covers a maximum of 90% of the amount of the loss with the exporter required to retain the remaining 10%.

Aid in financing. To assist him in export financing, a policy-holder may ask EDC to assign the proceeds of any losses payable under a policy to a bank or other agent providing financing in respect of export sales. He may assign an individual bill or he may make a blanket assignment of all his foreign accounts receivable. As a further aid, EDC may issue 100% unconditional guarantees to chartered banks or other financial institutions agreeing to provide non-recourse supplier financing. Such guarantees are usually issued for insured sales of capital goods. EDC may also issue 100% unconditional guarantees to banks or other financial institutions agreeing to finance the manufacturing period of an insurable medium-term export credit sale.

Surety and related insurance. EDC recently introduced a surety and related insurance package for performance-type bonds. It may insure: the exporter against wrongful call of a performance bond up to 90%; a bank which issues such a guarantee to a foreign buyer on behalf of a Canadian exporter up to 100%; and a domestic surety company providing a performance bond to a foreign buyer or a back-up bond to a foreign surety company providing a bond to a foreign buyer. It may also insure against non-performance by one or more members of a limited liability consortium.

Long-term export loans. EDC makes loans to foreign purchasers of Canadian capital equipment and technical services, or guarantees private loans to such purchasers, at interest rates as internationally competitive as possible. Such loans are made when extended credit terms are required and when financing is not available through normal commercial sources. In most cases, the loans are made in conjunction with Canadian chartered banks or other financial institutions.

Some examples of capital equipment and services eligible for export financing by EDC include: conventional and nuclear power plants, electrification programs and transmission lines; equipment such as ships, aircraft, locomotives and rolling stock; equipment for telephone systems, microwave facilities and earth satellite stations as well as for wood, pulp and paper, chemical, mining, construction and metallurgical projects. Financing may be provided for service contracts related to appraisal and development of natural resources, primary and secondary industry projects and public utilities projects.

EDC loans are generally made to meet the requirements of specific transactions, usually on application made by the Canadian exporter who has developed the business on behalf of the foreign borrower. Under appropriate circumstances, EDC will arrange lines of credit to foreign countries. Such arrangements provide a general framework for future commercial dealings and serve to alert both the Canadian export community and potential buyers in a recipient country to the possibilities of doing business.

An export transaction financed or guaranteed must be one which justifies repayment terms normally in excess of five years. Repayment schedules vary according to industry practice. The project must be commercially viable as well as offer adequate security for the loan. The transaction must have the highest possible Canadian material and labour content and meet a minimum standard in this respect. Interest and fees levied by EDC generally reflect the need to meet those offered by foreign competition.

EDC generally provides loans up to 85-90% of the cost of the Canadian equipment, supplies and services. Canadian chartered banks and other financial institutions usually participate with EDC by taking 30-40% of the EDC loan. In addition, they often provide funds for down payment, pre-shipment, and local cost financing. They usually participate with EDC on a last-in–first-out basis and at their own risk.

Foreign investment guarantees. EDC can provide guarantees which protect Canadian businessmen investing abroad against loss due to the political events of expropriation, war or insurrection, or the inability to repatriate funds. The foreign investment guarantee can cover almost any right that the Canadian investor might acquire in a foreign enterprise, including equity, loans, management contracts, royalty and licensing agreements. Only new investments in countries receptive to foreign interests are accepted by EDC. The major criterion is that the investment maximizes the benefits to Canada and to the host country.

Investments may be in the form of cash, contribution in kind, or the issuing of a guarantee to another party investing in another country. The investment may be made directly in a foreign enterprise, or indirectly through a related company based in Canada, the host country, or even a third country.

Coverage under a guarantee can have a term of up to 15 years. It can be cancelled only by the investor, and not by EDC, as long as the conditions of the contract are maintained. The investor may elect to insure for one or more of the political risks and need cover only assets actually at risk.

The program calls for the investor to carry a percentage of the liability; the remainder is borne by EDC. This co-insurance requirement is extended to all contracts regardless of investor or country. The normal co-insurance to be carried by the investor is 15%.

Maximum liabilities. To implement its objective of promoting Canadian exports through insurance, loans and guarantees, EDC has authority to undertake maximum financial liabilities of almost $9 billion on a roll-over basis.

The ceiling for liabilities under contracts of export credits, insurance, guarantees and surety issued at EDC's risk is $2.5 billion. In addition, insurance and guarantees may be given at government risk when, having regard to any one transaction, the amount or term is considered to be excessive for EDC. A separate fund of $1.0 billion maximum outstanding liability is provided for this.

Financing long-term and, in exceptional cases, medium-term credit for major export sales of capital equipment and services may total $5,100 million. Within the overall ceiling, there are two authorities. One is for lending by EDC for its own account on approval by the board of directors to a limit on outstanding liabilities of $4,250 million. The other is for lending for the account of the Canadian government to a limit of $850 million. The latter facility applies to large export transactions or to other special situations considered by the government to be in the national interest. A ceiling of $250 million for liabilities under the foreign investment insurance program is provided.

Tariffs and trade agreements   18.4

Canadian tariff structure   18.4.1

Information relating to rates of duty, value for duty and anti-dumping duty is available from the national revenue department, customs and excise, which administers the Customs Act, customs tariff and the Anti-dumping Act. Details of the organization and functions of the Tariff Board will be found in Appendix 1.

The Canadian tariff consists, in the main, of four sets of tariff rates — British preferential, most-favoured-nation, general and general preferential.

British preferential tariff rates are applied to imported commodities from British Commonwealth countries, with the exception of Hong Kong, when conveyed without trans-shipment from a port of any British country enjoying the benefits of the British Commonwealth preferential tariff into a port of Canada. Some Commonwealth countries have trade agreements with Canada that provide for rates of duty, on certain specified goods, lower than the British preferential rates.

Most-favoured-nation rates are usually higher than the British preferential rates and lower than the general tariff rates. They are applied to commodities imported from countries with which Canada has trade agreements. These rates would apply to British countries when they are lower than the British preferential tariff rates. The most important trade agreement concerning the effective rates applied to goods imported from countries entitled to most-favoured-nation rates is the General Agreement on Tariffs and Trade (GATT).

General tariff rates are applied to goods imported from the few countries with which Canada has not made trade agreements.

The general preferential tariff came into effect on July 1, 1974, as a result of Canada's acceding to a generalized system of preferences designed to allow lower rates of duty on goods imported from developing countries. Generally, the rates are the lesser of the British preferential tariff or the most-favoured-nation tariff minus one-third.

Despite the numerous tariff items and the various rates of duty applicable to each item, there are numerous goods which are duty free under all four tariffs.

Valuation. In general, the Customs Act provides that the value for duty of imported goods shall be the fair market value of like goods as established in the home market of the exporter at the time when and place from which the goods are shipped directly to Canada when sold to purchasers located at that place with whom the vendor deals at arm's length and who are at the same or substantially the same trade level as the importer, and in the same or substantially the same quantities for home consumption in the ordinary course of trade under competitive conditions. In cases where like goods are not sold for home consumption and in a few special cases, other methods of determining the value for duty are employed. The value for duty ordinarily may not be less than the amount for which the goods were sold to the purchaser in Canada, exclusive of all charges after their shipment from the country of export.

Anti-dumping Act. Canada's Anti-dumping Act provides, in brief, that where goods are dumped (the export price is less than the normal value) and such dumping has caused, is causing, or is likely to cause material injury to the production of similar goods in Canada, or has materially retarded or is materially retarding the establishment of the production in Canada of similar goods as determined by the Anti-dumping Tribunal, there shall be levied, collected and paid an anti-dumping duty. This duty is in an amount equal to the margin of dumping of the entered goods.

Drawback. Drawback legislation is designed to remove the customs duty and sales tax included in the manufacturers' costs to enable them to compete more equitably both abroad and at home with foreign manufacturers. It does this by granting a drawback, in the case of Canadian exporters, of customs duty and sales taxes paid on imported parts or materials used in Canada in the manufacture of goods subsequently exported. In the case of certain strategic industries in Canada (aircraft, automobiles and other secondary manufacturers), their costs of plant equipment or key materials are also reduced in the same manner when the specified imported goods are used in eligible Canadian manufacturers. Other areas where drawbacks are payable include: ships stores; joint Canada–US projects; and imported goods exported or destroyed in Canada.

18.4.2 Tariff and trade arrangements

Canada's tariff arrangements with other countries fall into three main categories: trade agreements with a number of Commonwealth countries; the General Agreement on Tariffs and Trade (GATT); and other arrangements.

Canada signed the protocol of provisional application of the General Agreement on Tariffs and Trade on October 30, 1947 and brought the agreement into force on January 1, 1948. The agreement provides for scheduled tariff concessions and the exchange of most-favoured-nation treatment among the contracting parties, and lays down rules and regulations to govern the conduct of international trade. As at July 1976 there were 83 members and three provisional members. GATT is applied on a de facto basis also to a number of newly independent states pending decision as to their future commercial policies.

Trade relations between Canada and a number of other countries are governed by trade agreements of various kinds, by exchange of most-favoured-nation treatment under orders-in-council, by continuation to newly independent states of the same treatment originally negotiated with the countries previously responsible for their commercial relations and by even less formal arrangements.

Britain and Ireland were to terminate by 1977 the preferential tariffs extended to Canada. Phasing out these preferences began February 1, 1973, as a result of the accession by those countries to the European Economic Community (EEC).

Tariff and trade arrangements with Commonwealth countries as at September 1977   **18.4.2.1**

Australia. Trade agreement in force June 30, 1960, modified and continued by an exchange of letters, October 24, 1973. GATT effective January 1, 1948. (Bindings of rates of duty and margins of preference on specified products and exchange of tariff preferences.)

Bahamas. Relations are based on a Canada — British West Indies trade agreement and protocol thereto (see British West Indies). GATT de facto application. (Exchange of preferential tariff treatment.)

Bangladesh. GATT effective December 16, 1972. (Canada accords British preferential treatment. Bangladesh accords most-favoured-nation treatment to Canada.)

Barbados. Relations are based on a Canada — British West Indies trade agreement and protocol thereto (see British West Indies). GATT effective November 30, 1966. (Exchange of preferential tariff treatment.)

Botswana. GATT de facto application. (Canada accords British preferential tariff treatment.)

British West Indies — Belize (formerly British Honduras), Bermuda, Leeward Islands, Windward Islands. Canada — British West Indies trade agreement of 1925 together with the amending protocol signed July 8, 1966, was terminated December 31, 1973. Belize, Bermuda, the Leeward Islands and the Windward Islands participate in GATT. (Exchange of preferential tariff treatment.)

Cyprus. GATT effective August 16, 1960. (Exchange of British preferential tariff treatment.)

Fiji. Maintains de facto application of GATT. (Canada accords British preferential tariff treatment to Fiji. Fiji extends most-favoured-nation treatment to Canada.)

Gambia. GATT effective February 18, 1965. (Canada accords British preferential tariff treatment to Gambia. Gambia extends most-favoured-nation treatment to Canada.)

Ghana. GATT effective October 17, 1957. (Canada accords British preferential tariff treatment to Ghana. Ghana extends most-favoured-nation treatment to Canada.)

Grenada. Relations are based on a Canada — British West Indies trade agreement and protocol thereto (see British West Indies) which was terminated December 31, 1975. GATT de facto. (Exchange of preferential tariff treatment.)

Guyana. Relations are based on a Canada — British West Indies trade agreement and protocol thereto (see British West Indies) which was terminated December 31, 1975. GATT effective May 26, 1966. (Exchange of preferential tariff treatment.)

India. Since 1897 Canada has unilaterally accorded British preferential treatment without contractual obligation. GATT effective July 8, 1948. (Canada accords British preferential tariff treatment to India. India extends most-favoured-nation treatment to Canada.)

Jamaica. Relations are based on a Canada — British West Indies trade agreement and protocol thereto (see British West Indies) which was terminated December 31, 1975, GATT effective August 6, 1962. (Exchange of preferential tariff treatment.)

Kenya. GATT effective December 12, 1963. (Canada accords British preferential tariff treatment to Kenya. Kenya extends most-favoured-nation treatment to Canada.)

Lesotho. GATT de facto application. (Canada accords British preferential tariff treatment to Lesotho.)

Malawi. Malawi and Canada observe the terms of 1958 trade agreement between Canada and the former Federation of Rhodesia and Nyasaland. GATT effective July 6, 1964. (Exchange of preferential tariff treatment.)

Malaysia. GATT effective August 31, 1957. (Canada accords British preferential tariff treatment to Malaysia. Malaysia extends some preferential rates to Canada.)

Malta. Canada — United Kingdom trade agreement of 1937 was legal basis for exchange of British preferential tariff treatment. This agreement was terminated January 31, 1973. GATT effective September 21, 1964.

Mauritius. GATT effective March 12, 1968. (Exchange of British preferential tariff treatment.)

Nauru. Relations governed by trade agreement of 1937 between Canada and the United Kingdom. This agreement was terminated January 31, 1973. (Canada accords British perferential treatment to Nauru.)

New Zealand. Trade agreement in force May 24, 1932, modified by protocol of May 13, 1970 and exchange of notes July 26, 1973. GATT effective July 30, 1948. (Bindings of rates of duty on specified products and the exchange of tariff preferences.)

Nigeria. Relations governed by trade agreement of 1937 between Canada and the United Kingdom. GATT effective October 1, 1960. (Canada accords British preferential treatment to Nigeria. Nigeria extends most-favoured-nation treatment to Canada.)

Rhodesia. Canada does not recognize the present government of Rhodesia. (Trade embargo exists between Canada and Rhodesia with certain humanitarian exceptions.)

Sierra Leone. Relations governed by trade agreement of 1937 between Canada and the United Kingdom. GATT effective April 27, 1961. (Canada accords British preferential tariff treatment to Sierra Leone. Sierra Leone extends most-favoured-nation tariff treatment to Canada.)

Singapore. GATT effective August 9, 1965. (Canada and Singapore exchange British preferential tariff treatment.)

Sri Lanka. GATT effective July 29, 1948. (Canada accords British preferential tariff treatment to Sri Lanka. Sri Lanka extends most-favoured-nation tariff treatment to Canada.)

Swaziland. GATT de facto application. (Canada accords British preferential treatment to Swaziland.)

Tanzania, United Republic of. GATT effective for Tanganyika December 9, 1961 and extended to Zanzibar upon formation of United Republic, April 23, 1964. (Canada accords British preferential tariff treatment to Tanzania. Tanzania extends most-favoured-nation treatment to Canada.)

Tonga. GATT de facto application. (Exchange of British preferential tariff treatment.)

Trinidad and Tobago. Relations are based on a Canada — British West Indies trade agreement and protocol thereto (see British West Indies). GATT effective August 31, 1962. (Exchange of preferential tariff treatment.)

Uganda. GATT effective October 9, 1962. (Canada accords British preferential tariff treatment to Uganda. Uganda extends most-favoured-nation tariff treatment to Canada.)

United Kingdom. GATT effective January 1, 1948. (Exchange of preferential tariff rates with some minor reservations by Canada.)

Western Samoa. Relations are based on a Canada — New Zealand trade agreement. A general settlement to this agreement was declared in 1962. (Exchange of British preferential tariff treatment.)

Zambia. GATT de facto application. (Canada accords British preferential tariff treatment to Zambia. Zambia extends most-favoured-nation treatment to Canada.)

Tariff and trade arrangements with non-Commonwealth countries as at September 1977 18.4.2.2

Algeria. Franco — Canadian trade agreement of 1933 applied to Algeria. Algeria maintains de facto application of GATT. (Since the creation of Algeria as an independent state in 1962, Canada has continued to grant most-favoured-nation treatment.)

Argentina. GATT effective October 11, 1967. (Exchange of most-favoured-nation treatment.)

Austria. GATT effective October 19, 1951. (Exchange of most-favoured-nation treatment.)

Bahrain. Bahrain maintains de facto application of GATT. (Exchange of most-favoured-nation treatment.)

Belgium — Luxembourg. Convention of commerce with Belgium — Luxembourg Economic Union (including Belgian colonies) entered into effect October 22, 1924. GATT effective January 1, 1948. (Exchange of most-favoured-nation treatment.)

Benin, People's Republic of, (formerly Dahomey). Franco — Canadian trade agreement of 1933 applied to Dahomey. Admittance to GATT September 12, 1963. (Exchange of most-favoured-nation treatment.)

Bolivia. Order-in-council of July 20, 1935 accepted Article 15 of UK — Bolivia treaty of commerce. (Exchange of most-favoured-nation treatment.)

Brazil. Trade agreement in force April 16, 1943. GATT effective July 30, 1948. (Exchange of most-favoured-nation treatment.)

Bulgaria. Trade agreement effective January 7, 1974, provides for most-favoured-nation treatment and exception for British preferential tariffs. To be extended for yearly periods.

Burma. GATT effective July 29, 1948. (Exchange of most-favoured-nation treatment.)

Burundi. GATT effective July 1, 1962. (Exchange of most-favoured-nation treatment.)

Cameroon. Franco — Canadian trade agreement of 1933 applied to Cameroon. Admittance to GATT May 3, 1963. (Exchange of most-favoured-nation treatment.)

Central African Empire. Franco — Canadian trade agreement of 1933 applied to Central African Empire. Admittance to GATT May 3, 1963. (Exchange of most-favoured-nation treatment.)

Chad. Franco — Canadian trade agreement of 1933 applied to Chad. Admittance to GATT July 12, 1963. (Exchange of most-favoured-nation treatment.)

Chile. Trade agreement in force October 29, 1943. GATT effective March 16, 1948. (Exchange of most-favoured-nation treatment.)

China, People's Republic of. Canada — China trade agreement of October 13, 1973. (Exchange of most-favoured-nation treatment.)

Colombia. Treaty of commerce with United Kingdom of February 16, 1866 applied to Canada. Modified by protocol of August 20, 1912 and exchange of notes December 30, 1938. (Exchange of most-favoured-nation treatment.)

Congo. Franco — Canadian trade agreement of 1933 applied to Congo. Admittance to GATT May 3, 1963. (Exchange of most-favoured-nation treatment.)

Costa Rica. Modus vivendi in force January 20, 1951. (Exchange of most-favoured-nation treatment.)

Cuba. GATT effective January 1, 1948. (Exchange of most-favoured-nation treatment.)

Czechoslovakia. Convention of commerce in force November 14, 1928. GATT effective April 20, 1948. (Exchange of most-favoured-nation treatment.)

Denmark (including Greenland). Treaties of peace and commerce with United Kingdom of February 13, 1660 and July 11, 1670 apply to Canada. GATT effective May 28, 1950. (Exchange of most-favoured-nation treatment.)

Dominican Republic. Trade agreement in force January 22, 1941. GATT effective May 19, 1950. (Exchange of most-favoured-nation treatment, including scheduled concessions.)

Ecuador. Modus vivendi in force December 1, 1950. (Exchange of most-favoured-nation treatment.)

Egypt, Arab Republic of. Exchange of notes in force December 3, 1952. GATT effective May 9, 1970. (Exchange of most-favoured-nation treatment.)

El Salvador. Exchange of notes in force November 17, 1937. (Exchange of most-favoured-nation treatment.)

Equatorial Guinea. United Kingdom — Spain treaty of commerce of 1922 extended to Canada July 19, 1928. Canada — Spain trade agreement signed May 25, 1954. GATT de facto application. (Since the creation of Equatorial Guinea as an independent state in 1968, Canada has continued to grant most-favoured-nation treatment.)

Ethiopia. Exchange of notes effective June 3, 1955. (Exchange of most-favoured-nation treatment.)

Finland. Exchange of notes effective November 17, 1948. GATT effective May 25, 1950. (Exchange of most-favoured-nation treatment.)

France and French overseas territories. Franco — Canadian trade agreement in force June 10, 1933. Exchange of notes of May 17 and November 22, 1933 further modified by exchanges in 1934 and 1935. GATT effective January 1, 1948. (Exchange of most-favoured-nation treatment, including scheduled concessions.)

Gabon. Franco — Canadian trade agreement of 1933 applied to Gabon. Admittance to GATT May 3, 1963. (Exchange of most-favoured-nation treatment.)

Germany, Federal Republic of. GATT effective October 1, 1951. (Exchange of most-favoured-nation treatment.)

Greece. Modus vivendi by exchange of notes of July 24 and July 28, 1947. GATT effective March 1, 1950. (Exchange of most-favoured-nation treatment.)

Greenland. (See Denmark.)

Guatemala. Trade agreement in force January 14, 1939. (Exchange of most-favoured-nation treatment.)

Guinea. Franco — Canadian trade agreement of 1933 applied to Guinea. (Since the creation of Guinea as an independent state in 1958, Canada has continued to grant most-favoured-nation treatment.)

Haiti. Trade agreement in force December 8, 1938. GATT effective January 1, 1950. (Exchange of most-favoured-nation treatment.)

Honduras. Exchange of notes concerning a commercial modus vivendi in force July 18, 1956. (Exchange of most-favoured-nation treatment.)

Hong Kong. Trade agreement of 1937 between Canada and the United Kingdom applied to Hong Kong, was terminated January 31, 1973.

Hungary. Trade agreement between Canada and Hungary in force May 27, 1972. GATT effective September 9, 1973. (Exchange of most-favoured-nation treatment.)

Iceland. GATT effective April 21, 1968. (Exchange of most-favoured-nation treatment.)

Indonesia. Admittance to GATT February 24, 1950. (Exchange of most-favoured-nation treatment.)

Iran. United Kingdom — Persia commercial convention of 1903 modified by agreement March 21, 1960. (Canada grants most-favoured-nation tariff rates as long as Iran accords reciprocal treatment.)

Iraq. Special arrangement by order-in-council effective September 15, 1951. (Exchange of most-favoured-nation treatment.)

Ireland. Trade agreement of 1932 modified by exchange of letters on December 21, 1967, was terminated by Ireland on January 31, 1973. GATT effective December 22, 1967. (Exchange of most-favoured-nation treatment on non-preferential items.)

Israel. GATT effective July 5, 1962. (Exchange of most-favoured-nation treatment.)

Italy. Modus vivendi by exchange of notes in force April 28, 1947. GATT effective January 1, 1950. (Exchange of most-favoured-nation treatment.)

Ivory Coast. Franco — Canadian trade agreement of 1933 applied to Ivory Coast. Admittance to GATT December 31, 1963. (Exchange of most-favoured-nation treatment.)

Japan. Agreement on commerce effective June 7, 1954. GATT effective September 10, 1955. (Exchange of most-favoured-nation treatment.)

Khmer Republic (formerly Cambodia). Franco — Canadian trade agreement of 1933 applied to Cambodia. Maintains a de facto application of GATT in 1968. (Exchange of most-favoured-nation treatment.)

Korea. Canada — Korea trade agreement in force December 20, 1966. GATT effective April 14, 1967. (Exchange of most-favoured-nation treatment.)

Kuwait. GATT effective June 19, 1961. (Exchange of most-favoured-nation treatment.)

Laos. Franco — Canadian trade agreement of 1933 applied to Laos. (Since the creation of Laos as an independent state in 1955, Canada has continued to grant most-favoured-nation treatment.)

Lebanon. Special arrangement by order-in-council of November 19, 1946. (Canada grants most-favoured-nation tariff rates as long as Lebanon accords reciprocal treatment.)

Liberia. Special arrangement by order-in-council effective June 3, 1955. (Canada grants most-favoured-nation tariff rates as long as Liberia accords reciprocal treatment.)

Liechtenstein. (See Switzerland.)

Luxembourg. (See Belgium — Luxembourg.)

Madagascar, Democratic Republic of. Franco — Canadian trade agreement of 1933 applied to Madagascar. Admittance to GATT September 30, 1963. (Exchange of most-favoured-nation treatment.)

Maldives. GATT de facto application. (Canada accords British preferential tariff treatment to the Maldive Islands.)

Mali. Franco — Canadian trade agreement of 1933 applied to Mali. Mali maintains a de facto application of GATT. Since the creation of Mali as an independent state in 1960, Canada has continued to grant most-favoured-nation treatment.)

Mauritania. Franco — Canadian trade agreement of 1933 applied to Mauritania. Admittance to GATT September 30, 1963. (Exchange of most-favoured-nation treatment.)

Mexico. Trade agreement in force June 6, 1947. (Exchange of most-favoured-nation treatment.)

Morocco. Convention of 1856 on commerce and navigation and the general treaty between the United Kingdom and Morocco applied to Canada. (Since the creation of

Morocco as an independent state in 1956, Canada has continued to grant most-favoured-nation treatment.)

Netherlands. Convention of commerce of July 11, 1924 includes Netherlands Antilles and Suriname. GATT effective January 1, 1948. (Exchange of most-favoured-nation treatment.)

Nicaragua. Trade agreement in force December 19, 1946. GATT effective May 28, 1950. (Exchange of most-favoured-nation treatment.)

Niger. Franco — Canadian trade agreement of 1933 applied to Niger. Admittance to GATT December 31, 1963. (Exchange of most-favoured-nation treatment.)

Norway. Convention with United Kingdom of May 16, 1913 applied to Canada. GATT effective July 10, 1948. (Exchange of most-favoured-nation treatment.)

Pakistan. Canada unilaterally accords British preferential treatment without contractual obligation. GATT effective July 30, 1948. (Exchange of most-favoured-nation tariff treatment.)

Panama. Commercial agreement of 1928 between the United Kingdom and Panama applied to Canada. Treaty terminated March 21, 1941. (Exchange of most-favoured-nation treatment.)

Paraguay. Exchange of notes in force June 21, 1940. (Exchange of most-favoured-nation treatment.)

Peru. GATT effective October 7, 1951. (Exchange of most-favoured-nation treatment.)

Philippines. Trade agreement in force August 29, 1972. Granted provisional accession to GATT on August 9, 1973. (Exchange of most-favoured-nation treatment.)

Poland. Convention of commerce in force August 15, 1936. GATT effective October 18, 1967. (Exchange of most-favoured-nation treatment.)

Portugal, Portuguese adjacent islands and Portuguese overseas provinces. Trade agreement in force April 29, 1955. Accession to GATT May 6, 1962. (Exchange of most-favoured-nation treatment.)

Qatar. Qatar maintains de facto application of GATT. (Exchange of most-favoured-nation treatment.)

Romania. Trade agreement in force December 14, 1971. GATT effective November 14, 1971. (Exchange of most-favoured-nation treatment.)

Rwanda. GATT effective July 1, 1962. (Canada grants most-favoured-nation treatment.)

Senegal. Franco — Canadian trade agreement of 1933 applied to Senegal. GATT effective June 20, 1960. (Exchange of most-favoured-nation treatment.)

South Africa, Republic of. Trade agreement in force June 30, 1933. Exchange of notes August 2 and 31, 1935, effective retroactively from July 1, 1935. GATT effective June 13, 1948. (Exchange of British preferential rates on scheduled items. Exchange of most-favoured-nation treatment.)

Spain and Spanish possessions. Since August 1, 1928, Canada has adhered to UK — Spain treaty of commerce of October 31, 1922. GATT effective August 29, 1963. (Exchange of most-favoured-nation treatment.)

Sweden. UK — Sweden treaties of peace and commerce of March 18, 1826 applied to Canada. GATT effective April 30, 1950. (Exchange of most-favoured-nation treatment.)

Switzerland. UK — Switzerland treaty of friendship, commerce and reciprocal establishment of September 6, 1855 applied to Canada. By exchange of notes Liechtenstein included under terms of this agreement, effective August 2, 1947. GATT effective August 1, 1966. (Exchange of most-favoured-nation treatment.)

Syrian Arab Republic. Special arrangement by order-in-council of November 19, 1946. (Canada grants most-favoured-nation treatment tariff rates as long as Syria accords reciprocal treatment.)

Taiwan (Formosa). Special arrangement by order-in-council of April 6, 1948. (Canada grants most-favoured-nation treatment as long as Taiwan accords reciprocal treatment.)

Thailand. Modus vivendi effective April 22, 1969. (Exchange of most-favoured-nation treatment.)

Togo. Franco — Canadian trade agreement of 1933 applied to Togo. GATT effective March 20, 1964. (Exchange of most-favoured-nation treatment.)

Tunisia. Franco — Canadian trade agreement of 1933 applied to Tunisia. Trade agreement between Canada and Tunisia in force August 8, 1972. (Exchange of most-favoured-nation treatment.)

Turkey. Exchange of notes in force March 15, 1948. GATT effective October 17, 1951. (Exchange of most-favoured-nation treatment.)

Union of Soviet Socialist Republics. Trade agreement extended by several protocols, the last in force provisionally July 14, 1976, with effect from April 18, 1976. (Exchange of most-favoured-nation treatment and undertaking of USSR to purchase determined quantity of Canadian wheat.)

United States of America. Trade agreement of November 17, 1938 suspended as long as both countries continue to be contracting parties to GATT. GATT effective January 1, 1948. (Exchange of most-favoured-nation treatment.)

Upper Volta. Franco — Canadian trade agreement of 1933 applied to Upper Volta. Admittance to GATT May 3, 1963. (Exchange of most-favoured-nation treatment.)

Uruguay. Trade agreement in force May 15, 1940. GATT effective December 16, 1953. (Exchange of most-favoured-nation treatment.)

Vietnam, Republic of. Franco — Canadian trade agreement of 1933 applied to Vietnam. (Since the creation of Vietnam as an independent state, Canada has continued to accord most-favoured-nation rates.)

Yugoslavia. Canada acceded to Article 30 of United Kingdom — Yugoslavia treaty of 1927. Suspended April 15, 1941, resumed July 7, 1945. GATT effective August 25, 1966. (Exchange of most-favoured-nation treatment.)

Zaire. Convention of commerce between the Economic Union of Belgium and Luxembourg applied to Zaire. GATT effective September 11, 1971. (Exchange of most-favoured-nation treatment.)

Tariff preferences for specified countries 18.4.2.3

Canada implemented a system of tariff preferences for specified countries on July 1, 1974. Imports of most manufactured and semi-manufactured products from designated beneficiary countries will be subject to the lower of the British preferential tariff or the most-favoured-nation tariff, less one-third. The only notable product group to which the preference system does not apply is textiles.

Beneficiary countries: Afghanistan, Algeria, American Samoa, Angola, Antigua, Argentina, Ascension, Bahamas, Bahrain, Bangladesh, Barbados, Belize, Benin (formerly Dahomey), Bermuda, Bhutan, Bolivia, Botswana, Brazil, British Indian Ocean Territory, British Solomon Islands, British Virgin Islands, Brunei, Bulgaria, Burma, Burundi, Cameroon, Cape Verde Islands, Cayman Islands, Central African Empire, Chad, Chile, Christmas Island, Cocos Islands, Colombia, Comoro Archipelago, Congo, Cook Islands, Costa Rica, Cuba, Cyprus, Dominica, Dominican Republic, Ecuador, Egypt (Arab Republic of), El Salvador, Equatorial Guinea, Ethiopia, Falkland Islands, Fiji, French Polynesia, French Southern and Antarctic Territories, French Territory of the Afars and the Issas, Gabon, Gambia, Ghana,

Gibraltar, Gilbert and Ellice Islands, Greece, Grenada, Guam, Guatemala, Guinea, Guinea–Bissau, Guyana, Haiti, Honduras, Hong Kong, India, Indonesia, Iran, Iraq, Israel, Ivory Coast, Jamaica, Kenya, Khmer Republic (formerly Cambodia), Korea, Kuwait, Laos, Lebanon, Lesotho, Liberia, Madagascar (Democratic Republic of), Malawi, Malaysia, Maldives, Mali, Malta, Mauritania, Mauritius, Mexico, Montserrat, Morocco, Mozambique, Nauru, Nepal, Netherlands Antilles, New Caledonia and dependencies, Nicaragua, Niger, Nigeria, Norfolk Island, Pakistan, Panama, Papua New Guinea, Paraguay, Peru, Philippines, Pitcairn, Portugal, Portuguese adjacent islands, Portuguese overseas provinces, Qatar, Romania, Rwanda, St. Christopher — Nevis — Anguilla, St. Helena, St. Lucia, St. Pierre and Miquelon, St. Vincent, Sao Tomé and Principe, Senegal, Seychelles, Sierra Leone, Singapore, Somalia, Spanish North Africa, Sri Lanka, Sudan, Suriname (Republic of), Swaziland, Syrian Arab Republic, Tanzania (United Republic of), Thailand, Togo, Tonga, Trinidad and Tobago, Tristan da Cunha, Tunisia, Turkey, Turks and Caicos Islands, Uganda, United Arab Emirates, Upper Volta, Uruguay, Venezuela, Vietnam (Republic of), Virgin Islands of the United States, Western Samoa, Yemen Arab Republic, Yemen People's Democratic Republic, Yugoslavia, Zaire (Republic of), Zambia.

Sources

18.1 - 18.1.4 Merchandising and Services Division, Industry Statistics Branch, Statistics Canada.

18.1.5 Marketing and Trade Division, Policy and Economics Branch, Canada Department of Agriculture.

18.1.6 Public Finance Division, Institutional and Public Finance Statistics Branch, Statistics Canada.

18.2 External Trade Division, General Statistics Branch, Statistics Canada.

18.3 - 18.3.1 Information Services Branch, Department of Industry, Trade and Commerce.

18.3.2 Export Development Corporation.

18.4.1 Information Services, Department of National Revenue, Customs and Excise.

18.4.2 Special Projects, Department of National Revenue, Customs and Excise.

Tables

18.1 Retail trade, by kind of business and by province, 1974-77, percentage change 1974-77 and percentage distribution 1977

Kind of business and province	1974r $'000,000	1975r $'000,000	1976 $'000,000	1977 $'000,000	Percentage change 1974-77	Percentage distribution 1977
Kind of business						
Combination stores (groceries and meat)	8,342.4	9,728.4	10,434.4	11,452.9	+37.3	18.6
Grocery, confectionery and sundries stores	1,920.5	2,255.4	2,721.4	2,961.9	+54.2	4.8
All other food stores	870.3	897.1	1,033.0	1,045.1	+20.1	1.7
Department stores	5,055.1	5,786.0	6,509.6	6,939.9	+37.3	11.3
General merchandise stores	1,470.7	1,598.2	1,560.0	1,647.1	+12.0	2.7
General stores	887.2	995.5	1,083.6	1,202.6	+35.6	1.9
Variety stores	772.3	818.8	876.1	885.7	+14.7	1.4
Motor vehicle dealers	8,303.3	10,183.7	11,057.9	11,752.8	+41.5	19.1
Used car dealers	142.4	182.6	186.3	214.1	+50.4	0.4
Service stations	3,041.8	3,302.1	3,891.9	4,256.0	+39.9	6.9
Garages	553.3	554.8	704.0	814.9	+47.3	1.3
Automotive parts and accessories stores	918.6	1,088.0	1,141.2	1,189.3	+29.5	1.9
Men's clothing stores	606.4	663.5	761.1	755.6	+24.6	1.2
Women's clothing stores	745.0	862.7	1,005.9	1,093.9	+46.8	1.8
Family clothing stores	642.9	739.7	833.4	855.8	+33.1	1.4
Specialty shoe stores	51.6	48.9	55.3	58.9	+14.2	0.1
Family shoe stores	373.8	425.0	484.9	537.7	+43.9	0.9
Hardware stores	528.6	580.9	638.8	673.4	+27.4	1.1
Household furniture stores	694.5	803.2	870.4	930.0	+33.9	1.5
Household appliance stores	194.8	215.6	235.1	234.6	+20.4	0.4
Furniture, TV, radio and appliance stores	472.7	510.6	524.0	493.9	+4.5	0.8
Pharmacies, patent medicine and cosmetics stores	1,304.0	1,488.1	1,710.9	1,860.9	+42.7	3.0
Book and stationery stores	178.4	191.5	238.7	278.3	+56.0	0.5
Florists	161.9	170.6	204.2	227.6	+40.6	0.4
Jewellery stores	385.0	426.8	484.6	550.0	+42.9	0.9
Sporting goods and accessories stores	423.8	517.4	691.0	721.5	+66.7	1.2
Personal accessories stores	650.2	680.4	794.1	876.5	+34.8	1.4
All other stores	5,059.8	5,692.4	6,433.8	7,051.3	+39.4	11.4
Total	44,751.1	51,408.4	57,166.9	61,563.5	+37.6	100.0
Province or territory						
Newfoundland	843.3	972.0	1,048.3	1,142.2	+35.4	1.9
Prince Edward Island	208.2	240.6	262.1	286.1	+37.4	0.5
Nova Scotia	1,444.6	1,619.4	1,821.7	1,943.3	+34.5	3.2
New Brunswick	1,141.4	1,338.3	1,494.6	1,572.7	+37.8	2.5
Quebec	11,382.8	13,020.9	14,447.8	15,549.2	+36.6	25.2
Ontario	16,564.1	19,156.2	21,057.0	22,651.2	+36.8	36.8
Manitoba	1,988.6	2,191.9	2,407.9	2,506.4	+26.0	4.1
Saskatchewan	1,903.9	2,242.5	2,522.1	2,571.9	+35.1	4.2
Alberta	3,734.3	4,557.0	5,289.6	5,884.6	+57.6	9.5
British Columbia	5,428.6	5,938.6	6,674.7	7,298.3	+34.4	11.8
Yukon Territory and Northwest Territories	111.2	130.9	140.5	157.2	+41.4	0.3

18.2 Sales of chain and independent stores, by kind of business, 1974 and 1977 and percentage change 1974-77

Kind of business	Chain stores			Independent stores		
	1974 $'000,000	1977 $'000,000	Percentage change 1974-77	1974 $'000,000	1977 $'000,000	Percentage change 1974-77
Combination stores (groceries and meat)	5,801.4	8,211.7	+41.5	2,540.9	3,241.2	+27.6
Grocery, confectionery and sundries stores	334.1	442.9	+32.6	1,586.3	2,519.0	+58.8
All other food stores	88.3	98.6	+11.7	781.9	946.4	+21.0
Department stores	5,055.0	6,939.9	+37.3	—	—	—
General merchandise stores	1,173.4	1,261.7	+7.5	297.2	385.3	+29.6
General stores	170.7	337.0	+97.4	716.4	865.5	+20.8
Variety stores	584.5	674.8	+15.5	187.7	210.9	+12.4
Motor vehicle dealers	99.8	145.4	+45.7	8,203.4	11,607.3	+41.5
Used car dealers	—	—	—	142.3	214.1	+50.5
Service stations	584.6	850.8	+45.5	2,457.0	3,405.2	+38.6
Garages	—	—	—	553.2	814.9	+47.3
Automotive parts and accessories stores	156.6	162.7	+3.9	761.9	1,026.6	+34.7

18.2   Sales of chain and independent stores, by kind of business, 1974 and 1977 and percentage change 1974-77 (concluded)

Kind of business	Chain stores			Independent stores		
	1974 $'000,000	1977 $'000,000	Per-centage change 1974-77	1974 $'000,000	1977 $'000,000	Per-centage change 1974-77
Men's clothing stores	148.1	250.4	+69.1	458.2	505.2	+10.3
Women's clothing stores	346.8	581.4	+67.7	398.1	512.5	+28.7
Family clothing stores	259.5	396.9	+52.9	383.3	458.9	+19.7
Specialty shoe stores	21.8	24.9	+14.2	29.7	34.0	+14.5
Family shoe stores	203.4	316.9	+55.8	170.3	220.8	+29.7
Hardware stores	83.0	108.9	+31.2	445.5	564.4	+26.7
Household furniture stores	136.1	144.3	+6.0	558.3	785.6	+40.7
Household appliance stores	39.3	40.3	+2.5	155.4	194.3	+25.0
Furniture, TV, radio and appliance stores	135.3	125.1	-7.5	337.4	368.8	+9.3
Pharmacies, patent medicine and cosmetics stores	262.2	415.7	+58.5	1,041.7	1,445.1	+38.7
Book and stationery stores	63.3	115.3	+82.2	115.0	163.0	+41.7
Florists	9.2	10.9	+18.5	152.6	216.7	+42.0
Jewellery stores	159.9	229.3	+43.4	225.0	320.7	+42.5
Sporting goods and accessories stores	21.3	67.5	+216.9	402.3	654.0	+62.6
Personal accessories stores	134.3	219.1	+63.1	515.7	657.4	+27.5
All other stores	2,714.2	3,660.6	+34.9	2,345.5	3,390.6	+44.6
Total, all stores	18,787.3	25,834.0	+37.5	25,963.7	35,729.4	+37.6

18.3   Percentage market share of chain stores, by kind of business, 1974-77

Kind of business	1974	1975	1976	1977
Combination stores (groceries and meat)	69.5	69.2	71.0	71.5
Grocery, confectionery and sundries stores	17.4	16.7	14.6	15.0
All other food stores	10.2	9.7	9.1	9.5
Department stores	100.0	100.0	100.0	100.0
General merchandise stores	79.8	78.2	76.6	76.6
General stores	19.2	27.3	27.1	28.0
Variety stores	75.7	75.8	75.2	76.2
Motor vehicle dealers	1.2	1.2	1.2	1.2
Used car dealers	—	—	—	—
Service stations	19.2	21.3	20.3	20.0
Garages	—	—	—	—
Automotive parts and accessories stores	17.0	15.5	13.1	13.5
Men's clothing stores	24.4	27.0	28.9	33.2
Women's clothing stores	46.6	52.2	51.9	53.4
Family clothing stores	40.4	43.3	46.7	46.3
Specialty shoe stores	42.4	46.0	47.7	42.2
Family shoe stores	54.4	57.4	59.4	59.1
Hardware stores	15.7	15.6	16.7	16.2
Household furniture stores	19.6r	16.6r	20.0	15.5
Household appliance stores	20.2r	18.0r	18.0	17.2
Furniture, TV, radio and appliance stores	28.6r	29.6r	29.6	25.3
Pharmacies, patent medicine and cosmetics stores	20.1	22.4	22.4	22.4
Book and stationery stores	35.5	45.9	43.4	41.8
Florists	5.7	4.9	5.6	4.8
Jewellery stores	41.6	42.8	42.2	41.8
Sporting goods and accessories stores	5.0	4.0	7.9	9.6
Personal accessories stores	20.7	24.4	24.2	25.0
All other stores	53.6r	53.1r	53.2	52.7
Total, all stores	42.0r	41.8r	41.8	42.0

18.4   Department store sales by department, 1974-77

Department	Sales				
	1974 $'000,000	1975 $'000,000	1976 $'000,000	1977 $'000,000	Percentage change 1974-77
Women's, misses' and children's clothing					
Women's and misses' dresses, house-dresses, aprons and uniforms	107.7	117.2	133.5	141.9	+31.8
Women's and misses' coats and suits	106.5	122.2	137.6	140.5	+31.9
Women's and misses' sportswear	251.4	278.1	316.0	341.2	+35.7
Furs	20.5	19.4	24.7	24.8	+21.0
Infants' and children's wear and nursery equipment	134.7	160.1	178.0	189.8	+40.9
Girls' and teenage girls' wear	80.0	91.1	99.4	106.4	+33.0
Lingerie and women's sleepwear	82.5	98.1	115.4	125.9	+52.6
Intimate apparel	55.5	59.7	67.4	72.3	+30.3
Millinery	15.3	16.1	16.4	16.0	+4.6
Women's and girls' hosiery	53.7	57.5	63.4	68.2	+27.0

18.4   Department store sales by department, 1974-77 (concluded)

Department	Sales				
	1974 $'000,000	1975 $'000,000	1976 $'000,000	1977 $'000,000	Percentage change 1974-77
Women's and girls' gloves, mitts and accessories	79.6	89.9	104.9	115.6	+45.2
Women's, misses' and children's footwear	131.4	149.0	165.6	178.9	+36.1
Total, women's, misses' and children's clothing	1,118.8	1,258.4	1,422.1	1,521.4	+36.0
Men's and boys' clothing					
Men's clothing	242.6	279.3	303.8	325.3	+34.1
Men's furnishings	221.3	245.0	296.2	313.6	+41.7
Boys' clothing and furnishings	90.6	93.4	100.8	111.2	+22.7
Men's and boys' footwear	83.1	96.1	109.3	110.7	+33.2
Total, men's and boys' clothing	637.6	713.8	810.1	860.9	+35.0
Food and kindred products	255.5	330.9	344.6	367.8	+44.0
Toiletries, cosmetics and drugs	232.6	271.3	302.5	330.8	+42.2
Photographic equipment and supplies	84.1	106.8	121.3	128.0	+52.2
Piece goods	58.7	60.2	62.3	64.1	+9.2
Linens and domestics	129.4	148.9	172.2	191.2	+47.8
Smallwares and notions	54.7	59.0	69.9	71.5	+30.7
China and glassware	68.8	75.9	84.6	91.1	+32.4
Floor coverings	117.9	114.5	126.5	127.1	+7.8
Draperies, curtains and furniture covers	98.5	106.4	123.0	128.6	+30.6
Lamps, pictures, mirrors and all other home furnishings	55.5	59.3	68.3	74.2	+33.7
Furniture	282.9	289.8	336.9	367.8	+30.0
Major appliances	275.0	313.3	355.2	351.9	+28.0
Television, radio and music	232.5	245.5	283.8	290.1	+24.8
Housewares and small electrical appliances	172.5	210.4	243.4	256.8	+48.9
Hardware, paints and wallpaper	160.9	194.7	224.6	241.1	+49.8
Plumbing, heating and building materials	53.5	53.8	54.8	60.5	+13.1
Jewellery	115.6	136.9	167.6	182.8	+58.1
Toys and games	109.1	132.9	148.5	162.6	+49.0
Sporting goods and luggage	164.2	195.2	229.1	243.0	+48.0
Stationery, books and magazines	154.0	172.4	189.4	202.2	+31.3
Gasoline, oil, auto accessories, repairs and supplies	112.8	128.6	134.5	154.6	+37.1
Receipts from meals and lunches	113.7	135.0	151.4	164.3	+44.5
Receipts from repairs and services } All other departments	212.6	272.1	283.2	306.7	+44.3
Total, all departments	5,071.3	5,786.0	6,509.6	6,941.0	+36.9

18.5   Retail sales of new motor vehicles, 1970-77

Year	Passenger cars		Trucks and buses		Total	
	No.	$'000	No.	$'000	No.	$'000
1970	640,360	2,158,543	133,881	653,787	774,241	2,812,330
1971	780,762	2,737,516	159,570	815,535	940,332	3,553,051
1972	858,959	3,170,305	206,662	1,142,754	1,065,621	4,313,059
1973	970,828	3,835,173	255,870	1,535,201	1,226,698	5,370,374
1974	942,797	4,016,879	306,507	1,900,106	1,249,304	5,916,985
1975	989,280	5,018,402	327,349	2,242,606	1,316,629	7,261,008
1976	946,488	5,241,970	344,975	2,512,118	1,291,463	7,754,088
1977	991,398	5,795,552	353,561	2,750,341	1,344,959	8,545,893

18.6   Retail sales of new motor vehicles by type and source, 1970-77

Year	Passenger cars		Trucks and buses		Total	
	Canadian/US	Overseas	Canadian/US	Overseas	Canadian/US	Overseas
	Number					
1970	497,185	143,175	124,664	9,217	621,849	152,392
1971	592,319	188,443	147,001	12,569	739,320	201,012
1972	653,933	205,026	189,577	17,085	843,510	222,111
1973	782,914	187,914	235,449	20,421	1,018,363	208,335
1974	796,840	145,957	287,686	18,821	1,084,526	164,778
1975	835,679	153,601	310,590	16,759	1,146,269	170,360
1976	793,201	153,287	331,027	13,948	1,124,228	167,235
1977	797,752	193,646	337,914	15,647	1,135,666	209,293
	Thousand dollars					
1970	1,795,709	362,834	628,532	25,255	2,424,241	388,089
1971	2,225,121	512,395	779,544	35,991	3,004,665	548,386
1972	2,554,779	615,526	1,087,306	55,448	3,642,085	670,974
1973	3,197,173	638,000	1,466,448	68,753	4,663,621	706,753
1974	3,455,140	561,739	1,831,532	68,574	5,286,672	630,313
1975	4,350,220	668,182	2,174,855	67,751	6,525,075	735,933
1976	4,522,723	719,247	2,447,109	65,009	6,969,832	784,256
1977	4,864,157	931,395	2,673,007	77,264	7,537,234	1,008,659

18.7   Retail sales in campus book stores, academic years 1973-74 to 1976-77

Province and items sold	1973-74 $'000	1974-75 $'000	1975-76 $'000	1976-77 $'000	Percentage change 1975-76 to 1976-77
Province					
Atlantic region	4,148	4,762	5,522	5,997	+8.6
Nova Scotia	1,750	2,069	2,407	2,532	+5.2
New Brunswick	1,323	1,531	1,760	1,914	+8.8
Quebec	9,541	11,208	12,846	14,084	+9.6
Ontario	21,871	27,340	32,873	36,901	+12.3
Manitoba	2,772	3,505	3,925	4,531	+15.4
Saskatchewan	1,954	2,475	2,636	3,430	+30.1
Alberta	5,140	6,305ʳ	7,986	8,858	+10.9
British Columbia	4,542	5,711	7,542	8,211	+8.9
Total	49,968	61,306ʳ	73,330	82,012	+11.8
Items sold					
Textbooks[1]	33,108	39,847ʳ	45,305	51,147	+12.9
Trade books[2]	5,338	6,259	9,345	10,110	+8.2
Stationery and supplies	7,506	9,571	12,526	13,008	+3.8
Miscellaneous[3]	4,016	5,629	6,154	7,747	+25.9

[1]Includes all professional and educational books.
[2]Includes hard covers and paperbacks.
[3]Includes newspapers, magazines, periodicals and sundries.

18.8   Vending machine operators, 1964-76

Year	Firms No.	Annual change %	Machines[1] No.	Annual change %	Sales $'000	Annual change %
1964	651	−3.3	75,392	−3.9	78,561.8	+16.2
1965	764	+17.4	85,091	+12.9	89,815.4	+14.3
1966	769	+0.7	84,154	−1.1	107,539.6	+19.7
1967	790	+2.8	91,289	+8.5	119,650.9	+11.3
1968	791	+0.1	95,867	+5.0	127,058.6	+6.2
1969	..	..	100,948	+5.3	142,909.6	+12.5
1970	768	..	103,751	+2.8	156,822.1	+9.7
1971	697	−9.2	97,965	−5.6	162,249.1	+3.5
1972[2]	692	−0.7	106,758	+9.0	178,909.0	+10.3
1973	648	−6.4	104,253	−2.3	207,081.4	+15.7
1974	667	+2.9	106,278	+1.9	227,445.2	+9.8
1975	627	−6.0	110,287	+3.8	249,959.6	+9.9
1976	612	−2.4	104,548	−5.2	269,386.6	+7.8

[1]Maximum during the year; ovens, coin and bill changers are excluded.
[2]Beginning 1972, data of small operators excluded.

18.9   Sales through vending machines, distribution and percentage change, by selected type of machine, 1975 and 1976

Type of machine	1975 $'000	%	1976 $'000	%	Percentage change 1975-76
Cigarettes	112,212.4	44.9	120,835.2	44.9	+7.7
Beverages					
Coffee	41,655.8	16.7	45,124.9	16.8	+8.3
Soft drinks					
Can or bottle	18,330.5	7.3	21,136.7	7.8	+15.3
Disposable cups	20,059.5	8.0	19,435.8	7.2	−3.1
Packaged milk	8,551.9	3.4	9,036.9	3.4	+5.7
Other beverages	377.1	0.1	553.7	0.2	+46.8
Confections and foods					
Bulk confectionery	2,430.2	1.0	3,499.6	1.3	+44.0
Packaged confectionery	13,948.5	5.6	15,142.7	5.6	+8.6
Pastries	7,906.2	3.2	9,020.9	3.3	+14.1
Snack food	2,481.7	1.0	2,956.4	1.1	+19.1
Hot canned foods and soups	4,207.5	1.7	4,482.5	1.7	+6.5
Ice cream	891.6	0.3	928.8	0.3	+4.2
Fresh food (casseroles, hot dogs, sandwiches, salads)	15,436.6	6.2	15,766.0	5.9	+2.1
Other vending machines for food	793.3	0.3	923.9	0.3	+16.5
All other food and non-food	676.8	0.3	542.6	0.2	−19.8
Total	249,959.6	100.0	269,386.6	100.0	+7.8

18.10   Direct sales by commodity, 1973-76

Commodity	1973 $'000	1974 $'000	1975 $'000	1976 $'000	Percentage change 1975-76
Meat, fish and poultry	14,263	15,658	16,754	18,655	+11.3
Frozen food plans	29,352	30,701	27,220	21,580	−20.7
Dairy products	187,757	203,716	225,500ʳ	246,600	+9.4
Bakery products	45,533	46,763	47,900ʳ	53,500	+11.7
All other foods and beverages	35,712	42,584	43,685	44,105	+1.0
Canvas, awnings, sails and tents	6,032	6,967	8,305	8,201	−1.3
Clothing	9,878	11,362	12,123	15,516	+28.0
Fur goods	8,262	9,818	11,713	15,409	+31.6
Furniture, re-upholstery and repairs	42,204	51,000ʳ	55,000ʳ	60,500	+10.0
Books	72,022	82,889	90,909	120,010	+32.0
Newspapers	143,900ʳ	153,528ʳ	174,955ʳ	185,176	+5.8
Magazines	20,135	20,839	22,452	27,250	+21.4
Aluminum windows, doors, screens and awnings	21,354	24,109	27,220	31,632	+16.2
Dinnerware, kitchenware and utensils	38,731	49,806	67,981	92,007	+35.3
Sailboats and pleasure craft	12,528	12,568	15,718	16,412	+4.4
Household electrical appliances	84,750	103,519	120,929	139,015	+15.0
Pharmaceuticals and medicines	3,455	2,882	2,809	2,842	+1.2
Brushes, brooms, mops and household soaps and cleaners	26,099	31,331	36,115ʳ	38,110	+5.5
Cosmetics and costume jewellery	102,972	123,758	124,925	149,912	+20.0
Phonograph records	23,190	26,352	18,845	13,250	−29.7
Greenhouse flowers, nursery seeds and stocks	36,027	49,389ʳ	62,158ʳ	79,078	+27.2
Miscellaneous[1]	77,082	130,676	135,813ʳ	214,678	+58.1
Total, all commodities	1,041,238ʳ	1,230,215ʳ	1,349,029ʳ	1,593,438	+18.1

[1]Includes leather goods, textiles, stamps, coins and personal stationery and sales of merchandise to credit-card holders of gasoline oil companies.

18.11   Methods of distribution of direct sales, 1975 and 1976

Commodity	By door-to-door canvassing 1975 %	1976 %	By mail 1975 %	1976 %	From manu- facturers' premises 1975 %	1976 %	Through other channels[1] 1975 %	1976 %
Meat, fish and poultry	—	—	—	—	88.5	87.1	11.5	12.9
Frozen food plans	—	—	—	—	100.0	100.0	—	—
Dairy products	100.0	100.0	—	—	—	—	—	—
Bakery products	100.0	100.0	—	—	—	—	—	—
All other foods and beverages	54.8	54.2	—	—	1.3	1.9	43.9	43.9
Canvas, awnings, sails and tents	15.8	12.6	—	1.9	84.2	85.5	—	—
Clothing	15.6	14.4	34.1	41.0	50.3	44.6	—	—
Fur goods	—	—	—	—	100.0	100.0	—	—
Furniture, re-upholstery and repairs	—	1.3	—	—	100.0	98.7	—	—
Books	32.3	32.5	67.7	66.7	—	—	—	—
Newspapers	87.6ʳ	87.6	9.6ʳ	9.6	0.3ʳ	0.3	2.5ʳ	2.5
Magazines	—	—	100.0	100.0	—	—	—	—
Aluminum windows, doors, screens and awnings	57.2	51.3	—	—	42.8	48.7	—	—
Dinnerware, kitchenware and utensils	100.0	76.2	—	8.6	—	—	—	15.2
Sailboats and pleasure craft	—	—	—	—	100.0	100.0	—	—
Household electrical appliances	75.5	82.5	—	4.4	24.5	13.1	—	—
Pharmaceuticals and medicines	77.8	82.1	22.2	11.4	—	6.5	—	—
Brushes, brooms, mops and household soaps and cleaners	100.0	98.7	—	1.3	—	—	—	—
Cosmetics and costume jewellery	92.4	91.0	6.5	8.0	1.1	1.0	—	—
Phonograph records	—	—	100.0	100.0	—	—	—	—
Greenhouse flowers, nursery seeds and stocks	—	—	9.9ʳ	10.0	35.6ʳ	35.2	54.5ʳ	54.8
Miscellaneous	21.5	16.5	48.6	36.0	29.9	42.2	—	5.5
Total, all commodities	62.3ʳ	59.1	15.2ʳ	16.1	18.1ʳ	18.8	4.4ʳ	6.0

[1]Includes roadside stands, market stalls, shows, exhibitions and other display and demonstration venues.

18.12   Sales finance companies' new paper purchased and balances outstanding, by class of goods, 1973-76 (million dollars)

Class of goods	Paper purchased				Balances outstanding Dec. 31			
	1973	1974	1975	1976	1973	1974	1975	1976
Consumer goods	*1,080*	*1,127*	*1,041*	*1,074*	*1,150*	*1,168*	*1,156*	*1,134*
New passenger cars	515	548	546	534	609	617	638	630
Used passenger cars	186	199	185	192	200	203	201	204
Radio and television sets, household appliances, furniture and other consumer goods	379	380	310	348	341	348	317	300
Commercial and industrial goods	*1,383*	*1,453*	*1,560*	*1,730*	*1,529*	*1,870*	*2,080*	*2,258*
New commercial vehicles	784	838	871	899	826	1,049	1,122	1,201
Used commercial vehicles	111	120	119	121	115	134	151	151
Other commercial goods	488	495	570	710	588	687	807	906
Total	2,463	2,580	2,602	2,804	2,679	3,038	3,236	3,392

18.13   Consumer credit balances outstanding, selected holders and selected years, 1955-77 (million dollars)

Item	1955	1960	1965	1970	1975	1977e
Instalment financing by sales finance and consumer loan companies[1]	616	886	1,198	1,136	1,156	1,100
Cash loans						
Under $1,500	89	392	628	525	252	200
Over $1,500	173	100	348	1,190	1,504	1,400
Chartered banks	441	857	2,241	4,663	13,175	19,040
Quebec savings banks	2	6	16	22	58	90
Life insurance companies policy loans	250	344	411	759	1,149	1,270
Credit unions and caisses populaires	174	433	813	1,493	3,243	4,600
Department stores	226	368	575	720	1,232	1,300
Furniture, TV, radio and household appliance store loans	175	195	167	148	192	190
Other retail dealers	351	397	571	683	994	1,000
Other credit-card issuers	20	43	72	186	338	350
Public utility companies	..	..	116	181	295r	400
Trust and mortgage companies	..	..	..	..	199	360
Total	2,517	4,021	7,157	11,706	23,787r	31,300

[1] Data for years after 1970 show principal amount outstanding only, excluding unearned interest and other finance charges.

18.14   Summary statistics of major traveller accommodation groups, 1974 and 1975

Accommodation group	Locations No.	Rooms No.	Cabins and cottages No.	Tent trailer spaces No.	Total receipts $'000
1974					
Hotels	4,893	181,680	3,949	2,322	1,852,764
Motels	4,067	73,391	3,557	5,136	271,004
Tourist homes	393	2,883	33	44	2,589
Tourist courts and cabins	2,573	991	21,008	7,992	33,861
Outfitters	1,946	1,068	13,637	7,892	46,638
Tent and trailer campgrounds	3,057	395	3,581	284,029	64,123
Total	16,929	260,408	45,765	307,415	2,270,979
1975					
Hotels	4,834	188,506	3,814	2,491	2,097,148
Motels	4,066	74,680	3,237	5,216	301,770
Tourist homes	385	2,595	24	19	2,711
Tourist courts and cabins	2,531	1,020	20,733	8,017	36,256
Outfitters	1,937	1,332	13,478	8,273	51,026
Tent and trailer campgrounds	3,091	504	3,421	291,225	82,056
Total	16,844	268,637	44,707	315,241	2,570,967

18.15 Locations and receipts of major traveller accommodation groups, by province, 1974 and 1975

Province or territory	Hotels		Motels		Total receipts[1]	
	Locations No.	Receipts $'000	Locations No.	Receipts $'000	$'000	% distri- bution
1974						
Newfoundland	75	23,608	39	5,732	31,729	1.4
Prince Edward Island	26	5,238	58	3,299	10,147	0.4
Nova Scotia	102	34,178	159	12,139	49,208	2.2
New Brunswick	72	21,524	179	14,808	39,331	1.7
Quebec	1,778	407,712	738	48,669	479,828	21.1
Ontario	1,166	554,592	1,450	87,026	709,416	31.2
Manitoba	263	127,974	113	7,612	140,445	6.2
Saskatchewan	371	95,027	141	10,685	111,667	4.9
Alberta	454	240,717	317	25,785	272,084	12.0
British Columbia	525	319,398	843	51,950	390,661	17.2
Yukon Territory	41	10,756	20	[2]	14,327	0.6
Northwest Territories	20	12,040	10	[2]	15,001	0.7
Canada	4,893	1,852,764	4,067	271,004	2,270,979[3]	100.0
1975						
Newfoundland	81	28,768	37	6,808	38,211	1.5
Prince Edward Island	23	5,392	59	3,935	11,664	0.5
Nova Scotia	97	39,826	163	14,964	58,369	2.3
New Brunswick	72	23,662	171	16,441	43,597	1.7
Quebec	1,681	457,868	736	53,728	541,221	21.1
Ontario	1,170	615,070	1,436	94,134	785,225	30.7
Manitoba	261	136,407	113	8,628	151,548	5.9
Saskatchewan	390	112,365	153	13,430	132,301	5.2
Alberta	467	293,996	326	31,301	330,956	12.9
British Columbia	534	359,710	842	54,961	436,095	17.0
Yukon Territory	39	12,268	19	2,052	15,258	0.6
Northwest Territories	19	11,816	11	1,388	15,586	0.6
Canada	4,834	2,097,148	4,066	301,770	2,570,967[3]	100.0

[1]Includes tourist homes, tourist courts and cabins, outfitters, and tent and trailer campgrounds.
[2]Confidential.
[3]Components do not add to totals because there is no provincial breakdown for federal campgrounds.

18.16 Restaurant, caterer and tavern receipts, by province, 1976

Province	Receipts $'000	Establishments No.
Newfoundland	81,709	721
Prince Edward Island	23,399	157
Nova Scotia	142,680	910
New Brunswick	107,073	778
Quebec	1,560,769	10,827
Ontario	1,747,863	9,068
Manitoba	191,437	919
Saskatchewan	142,025	754
Alberta	449,585	1,575
British Columbia	677,217	3,028
Yukon Territory and Northwest Territories	7,586	50
Canada	5,131,343	28,787

18.17 Receipts, taxes and paid admissions of motion picture and drive-in theatres, 1969-76

Year	Motion picture theatres		Drive-in theatres	
	Admission receipts and taxes $'000	Paid admissions '000	Admission receipts and taxes $'000	Paid admissions '000
1969	109,848	78,918	16,691	11,308
1970	119,802	80,826	18,164	11,489
1972[1]	131,399	81,241	19,054	10,559
1973	139,541	77,438	22,171	11,581
1974	160,904	79,020	24,563	11,372
1975	195,545	84,161	31,256	12,843
1976	206,556	82,328	33,678	13,048

[1]No survey was conducted in 1971.

18.18   Billings of advertising agencies, 1973-76 (thousand dollars)

Type of medium or service	1973	1974	1975	1976
Media billings				
Print media	193,386	218,270	228,617	271,297
Television	192,171	229,950	248,167	341,848
Radio	63,045	73,185	77,361	111,125
Outdoor and transportation	13,923	14,450	16,204	25,008
Total, media billings	462,525	535,855	570,349	749,278
Production cost				
Print media	47,380	47,924	49,901	49,731
Television	38,512	38,687	40,892	52,867
Radio	5,651	8,136	6,916	9,270
Outdoor and transportation	2,433	1,776	2,380	2,093
Direct mail	16,889	21,004	20,132	24,107
Other	3,727	3,827	6,470	3,628
Total, production cost	114,592	121,354	126,691	141,696
Total, advertising billings	577,117	657,209	697,041	890,974
Research				
Market surveys	9,045	18,591	24,096	25,214
Total, gross billings	586,162	675,800	721,137	916,188

18.19   Revenue and expenditure of religious organizations, by province, 1974 and 1975

Province	Establishments		Revenue		Expenditure		Difference	
	1974 No.	1975 No.	1974 $'000	1975 $'000	1974 $'000	1975 $'000	1974 $'000	1975 $'000
Newfoundland	494	502	16,373	18,070	14,782	17,094	+1,591	+976
Prince Edward Island	190	187	5,626	6,139	5,079	5,294	+547	+845
Nova Scotia	1,140	1,149	31,159	35,690	28,357	32,795	+2,802	+2,895
New Brunswick	993	997	27,546	31,765	25,652	30,729	+1,894	+1,036
Quebec	3,048	3,053	140,139	154,300	131,102	143,966	+9,037	+10,334
Ontario	8,388	8,416	389,229	430,225	364,171	411,531	+25,058	+18,694
Manitoba	1,390	1,397	47,236	53,340	45,562	52,355	+1,674	+985
Saskatchewan	2,034	2,031	45,643	52,793	42,603	49,494	+3,040	+3,299
Alberta	2,184	2,196	82,693	93,749	73,698	88,667	+8,995	+5,082
British Columbia	2,056	2,066	84,281	92,942	79,998	91,868	+4,283	+1,074
Yukon Territory	38	39	759	751	660	708	+99	+43
Northwest Territories	56	56	626	952	670	1,030	−44	−78
Canada	22,011	22,089	871,308	970,716	812,334	925,531	+58,974	+45,185

18.20   Firms, total receipts and average funeral cost, by province, 1972 and 1976

Province	Firms No.		Total receipts $'000		Average funeral cost[1] (dollars)	
	1972	1976	1972	1976	1972	1976
Newfoundland	16	26	1,215	2,916	488	797
Prince Edward Island	19	15	616	997	499	794
Nova Scotia	68	64	4,556	6,821	576	906
New Brunswick	50	47	3,339	5,313	588	934
Quebec	363	381	32,811	55,814	794	1,119
Ontario	492	473	51,902	78,406	721	1,026
Manitoba	52	48	5,132	8,043	537	788
Saskatchewan	69	71	4,817	8,342	522	809
Alberta	52	63	6,783	11,915	530	827
British Columbia, Yukon Territory and Northwest Territories	75	71	8,289	13,398	469	636
Canada	1,256	1,259	119,460	191,965	665	956

[1]Includes cost of casket but excludes vaults and extra charges.

18.21   Sales of wholesale merchants, by kind of business, 1974-76

Kind of business	1974 $'000,000	1975 $'000,000	1976 $'000,000	Per-centage change 1975-76
Consumer goods trades	*18,866.1*	*20,987.0*ʳ	*23,374.0*	*+11.4*
Automotive parts and accessories	2,529.2	2,808.5	3,197.1	+13.8
Motor vehicles	917.2	972.8	941.5	−3.2
Drugs and drug sundries	892.6	982.9	1,023.8	+4.2
Clothing and furnishings	396.2	415.9	473.9	+14.0
Footwear	96.0	92.9	100.3	+8.0
Other textiles and clothing accessories	851.3	829.5ʳ	991.4	+19.5
Household electrical appliances	897.0	993.1	1,057.7	+6.5
Tobacco, confectionery and soft drinks	1,342.7	1,555.6	2,198.2	+41.3
Fresh fruits and vegetables	751.3	795.2	825.8	+3.8
Meat and dairy products	1,083.2	1,090.9	1,119.9	+2.7
Floor coverings	424.3	426.3	443.6	+4.1
Groceries and food specialties	5,804.0	6,693.4	7,394.4	+10.5
Hardware	903.8	956.1	1,080.1	+13.0
Other consumer goods	1,977.5	2,374.0	2,526.1	+6.4
Industrial goods trades	*24,344.1*	*24,390.1*	*26,613.4*	*+9.1*
Coal and coke	69.7	84.3	69.9	−17.0
Grain	4,267.3	4,278.2	4,440.1	+3.8
Electrical wiring supplies, construction materials, apparatus and equipment	798.2	856.8	863.6	+0.8
Other construction materials and supplies, including lumber	5,358.1	5,367.7	6,549.0	+22.0
Farm machinery	1,607.0	2,006.4	2,300.1	+14.6
Industrial and transportation equipment and supplies	3,888.5	4,496.9	4,753.7	+5.7
Commercial, institutional and service equipment and supplies	940.5	994.9	1,063.5	+6.9
Newsprint, paper and paper products	694.3	729.3	791.5	+8.5
Scientific and professional equipment and supplies	482.9	565.3	572.9	+1.3
Iron and steel	2,736.3	1,998.1	1,765.0	−11.7
Junk and scrap	1,024.2	614.7	671.5	+9.2
Other industrial goods	2,477.1	2,397.6	2,772.7	+15.6
Total, all trades	43,210.2	45,377.1ʳ	49,987.5	+10.2

18.22   Sales of farm implements and equipment, by province and by major group, 1973-76

Province and major group	1973 $'000	1974 $'000	1975 $'000	1976 $'000
Province				
Atlantic provinces	14,184	17,918	23,062	27,693
Quebec	67,343	84,821	108,559	125,785
Ontario	142,149	173,663	214,763	239,973
Manitoba	60,971	76,839	106,442	125,373
Saskatchewan	143,602	175,653	262,116	315,518
Alberta	130,073	163,604	226,734	268,054
British Columbia	15,544	21,199	24,623	31,689
Total	573,866	713,696	966,299	1,134,086
Major group				
Tractors — farm use	212,492	256,573	373,341	449,266
Ploughs	12,192	16,550	23,448	31,216
Tilling, cultivating and weeding machinery	40,769	56,414	80,530	97,700
Planting, seeding and fertilizing machinery	23,638	31,106	39,467	53,416
Haying machinery	41,489	49,444	64,987	72,079
Harvesting machinery	114,074	126,162	181,765	225,496
Machines for preparing crops for market or for use	33,436	43,166	50,399	43,491
Farm wagons, boxes and sleighs	23,901	29,101	35,639	32,010
Barn equipment	22,864	30,501	30,324	31,577
Farm dairy machinery and equipment	12,857	16,120	18,973	13,314
Spraying and dusting equipment	4,430	6,540	10,270	11,911
Pump and irrigation equipment and miscellaneous farm equipment	31,722	52,019	57,156	72,610

18.23   New construction machinery and equipment sales, by commodity group, 1975 and 1976

Commodity group	1975 Units No.	1975 Value $'000	1976 Units No.	1976 Value $'000
Tractors, crawler-type	2,968	182,476	2,629	166,130
Tractors, wheel-type	2,408	53,609	1,812	45,443
Skid-steer loaders	276	2,551	473	4,474
Front-end loaders, wheel-type	2,562	164,547	1,908	159,405
Attachments for tractors and front-end loaders	2,679	12,199	1,815	11,925
Scrapers	193	31,101	159	25,152
Off-highway haulers, heavy duty	373	73,254	254	59,739
Excavator/cranes, crawler mounted	849	97,250	822	105,926
Excavator/cranes, rubber tire mounted	1,194	76,259	1,068	53,861
Excavator/crane attachments	149	674	[1]	[1]
Tower and climbing cranes	16	1,129	[1]	[1]
Trenchers and ditchers	310	4,032	277	3,960
Graders, motor	890	52,235	722	50,779
Logging skidders	1,124	37,665	923	38,760
Compactors and rollers, vibratory, hand-guided	2,394	4,504	1,872	3,615
Compactors, vibratory	453	11,857	335	10,222
Compactors and rollers, static, self-propelled	264	4,211	170	3,790
Air compressors	1,108	18,676	828	14,806
Rock drills	1,133	17,161	787	18,292
Pumps, contractors' type	11,029	10,636	12,418	12,213
Contractors' tools, hand held	2,800	2,178	2,775	2,266
Concrete machinery	2,627	15,299	2,077	14,369
Asphalt equipment	359	14,943	290	16,427
Aggregate processing equipment	...	52,485	...	37,559
Forklift trucks	1,345	21,408	1,696	38,786
All other construction type machinery and attachments	...	90,797	...	99,188
All repair and consumable parts	...	547,394	...	566,074
Total	...	1,600,531	...	1,564,119

[1]Confidential.

18.24   Sales of diesel engines, by province, 1973-76

Province	Units[1] 1973	1974	1975	1976
Atlantic provinces	2,150	2,031	1,720	1,408
Quebec	4,337	5,456	4,243	3,291
Ontario	6,394	6,530	4,834	4,177
Manitoba	1,068	1,064	1,013	1,086
Saskatchewan	648	1,249	1,339	1,524
Alberta	2,300	3,798	2,692	3,083
British Columbia	3,890	4,387	3,487	3,632
Total	20,787	24,515	19,328	18,201

[1]Horsepower range 0-100 to 401 and over.

18.25   Summary statistics of co-operative marketing and purchasing associations, by province, 1972-76

Year and province		Associ- ations	Share- holders or members	Farm marketings $'000	Sales of merchan- dise $'000	Total business[1] $'000
1972		1,120	1,491,000	1,708,300	906,300	2,666,900
1973		1,116	1,527,000	2,176,100	1,178,600	3,415,700
1974		1,123	1,546,000	3,142,800	1,550,000	4,769,600
1975		1,144	1,633,000	3,363,400	1,910,900	5,362,200
1976		1,181	1,724,300	3,740,400	2,157,200	5,986,900
Newfoundland	1972	37	16,000	1,700	19,100	21,400
	1973	37	17,000	1,700	24,000	26,100
	1974	35	17,000	3,300	27,400	31,300
	1975	40	19,000	4,600	31,600	37,100
	1976	35	19,400	5,800	36,800	43,300
Prince Edward Island	1972	14	10,000	3,700	12,400	16,400
	1973	17	10,000	5,200	14,700	20,000
	1974	16	10,000	5,200	15,900	21,600
	1975	18	10,000	7,600	19,000	27,100
	1976	20	9,800	8,400	19,500	28,400
Nova Scotia	1972	81	34,000	52,800	37,300	92,000
	1973	84	35,000	72,700	45,300	120,000
	1974	83	34,000	89,800	58,700	151,000
	1975	82	36,000	107,400	65,800	176,900
	1976	81	36,800	112,000	72,000	188,100

18.25   Summary statistics of co-operative marketing and purchasing associations, by province, 1972-76 (concluded)

Year and province		Associ- ations	Share- holders or members	Farm marketings $'000	Sales of merchan- dise $'000	Total business[1] $'000
New Brunswick	1972	42	18,000	10,900	24,300	36,000
	1973	39	20,000	12,300	29,600	42,700
	1974	40	20,000	16,400	36,400	54,100
	1975	43	26,000	18,200	49,500	69,800
	1976	47	28,800	22,600	55,700	80,700
Quebec	1972	340	145,000	263,400	211,500	481,400
	1973	350	172,000	271,100	283,400	564,200
	1974	381	194,000	348,000	379,500	742,000
	1975	387	193,000	530,500	431,000	983,700
	1976	375	211,000	579,300	470,200	1,065,600
Ontario	1972	105	107,000	101,700	136,100	242,600
	1973	105	101,000	116,400	169,300	290,300
	1974	97	95,000	164,500	219,400	390,000
	1975	95	101,000	194,400	247,200	448,900
	1976	90	107,000	234,400	262,000	505,400
Manitoba	1972	74	182,000	59,600	68,800	146,100
	1973	73	185,000	88,600	83,600	192,300
	1974	71	190,000	126,300	109,900	258,300
	1975	79	199,000	124,300	142,100	286,700
	1976	79	199,300	77,600	167,600	268,000
Saskatchewan	1972	236	443,000	617,300	159,900	788,700
	1973	229	437,000	865,700	187,100	1,068,200
	1974	226	400,000	1,230,400	240,900	1,487,200
	1975	209	386,000	1,230,900	300,300	1,545,600
	1976	210	395,000	1,383,200	353,100	1,754,500
Alberta	1972	112	323,000	314,900	144,300	463,400
	1973	114	326,000	397,200	187,500	588,700
	1974	105	346,000	619,800	251,200	877,400
	1975	117	391,000	636,400	332,900	977,300
	1976	122	422,000	757,900	396,200	1,164,800
British Columbia	1972	75	65,000	138,900	59,300	202,200
	1973	63	67,000	165,700	75,500	244,800
	1974	64	76,000	186,600	102,100	294,600
	1975	69	85,000	213,000	126,400	347,000
	1976	67	92,000	238,400	138,200	379,800
Interprovincial	1972	4	148,000	143,300	33,100	176,600
	1973	5	157,000	179,500	78,500	258,400
	1974	5	164,000	352,500	108,600	462,000
	1975	5	187,000	296,100	165,200	462,100
	1976	5	203,200	320,800	185,900	508,300

[1]Includes service revenue and other income.

18.26   Sales of products handled by marketing and purchasing co-operatives, 1973-76 (thousand dollars)

Product	1973	1974	1975	1976
Marketing	2,176,100	3,142,800	3,363,400	3,740,400
Dairy products	514,000	643,300	916,300	1,001,200
Fruits and vegetables	58,500	80,400	76,700	86,000
Grains and seeds	1,106,400	1,951,200	1,921,500	2,127,300
Livestock and livestock products	363,200	329,600	293,800	334,500
Eggs and poultry	98,300	102,900	107,200	129,800
Miscellaneous	35,700	35,400	47,900	61,600
Purchasing	1,178,600	1,550,000	1,910,900	2,157,200
Food products	357,400	422,400	524,800	595,900
Clothing and home furnishings	35,500	44,500	56,400	66,300
Hardware	102,400	135,800	164,000	188,000
Petroleum products	166,200	217,100	269,000	338,600
Feed	236,700	342,100	360,800	363,500
Fertilizer and spray material	88,900	132,500	193,400	207,300
Machinery and equipment	77,600	91,200	138,000	159,100
Building material	59,600	81,500	98,500	126,900
Miscellaneous	54,300	82,900	106,000	111,600
Total	3,354,700	4,692,800	5,274,300	5,897,600

18.27   Value and volume of sales of alcoholic beverages, years ended Mar. 31, 1975 and 1976

Province or territory	Spirits		Wines		Beer		Total	
	1975	1976	1975	1976	1975	1976	1975	1976
	Value $'000							
Nfld.	29,288	32,157	3,811	4,096	44,085	51,342	77,184	87,595
PEI	9,161	10,929	1,114	1,294	6,264	7,438	16,539	19,661
NS	56,790	61,423	9,591	10,670	46,212	51,248	112,593	123,341
NB	36,157	42,745	6,514	7,634	37,943	44,579	80,614	94,958
Que.	264,019	299,597	108,051	128,214	297,490	369,898	669,560	797,709
Ont.	547,220	604,413	121,366	140,011	422,232	560,655	1,090,818	1,305,079
Man.	76,425	84,934	10,789	12,541	50,471	58,467	137,685	155,942
Sask.	70,047	71,156	7,070	7,207	44,334	52,110	121,451	130,473
Alta.	149,842	177,824	26,683	31,578	91,689	113,062	268,214	322,464
BC	207,963	248,913	54,142	70,338	126,575	128,593	388,680	447,844
YT	3,435	4,177	726	986	2,662	2,804	6,823	7,967
NWT	4,986	5,748	899	963	3,689	4,270	9,574	10,981
Canada	1,455,333	1,644,016	350,756	415,532	1,173,646	1,444,466	2,979,735	3,504,014
	Volume litres							
Nfld.	3 214	3 178	1 082	1 023	45 151	48 828	49 447	53 029
PEI	1 032	1 082	386	396	7 142	7 719	8 560	9 197
NS	6 323	6 210	3 596	3 332	56 907	55 811	66 826	65 353
NB	3 973	4 373	2 496	2 568	44 846	48 097	51 315	55 038
Que.	31 672	33 290	40 196	41 500	574 014	598 926	645 882	673 716
Ont.	68 240	69 476	43 250	46 246	709 699	718 918	821 189	834 640
Man.	9 201	9 306	4 874	5 092	80 882	83 433	94 957	97 831
Sask.	8 324	7 710	3 168	2 982	59 839	64 008	71 331	74 700
Alta.	16 716	18 134	10 783	11 756	131 266	139 544	158 765	169 434
BC	25 435	26 767	21 734	26 240	205 811	203 688	252 980	256 695
YT	323	368	200	200	2 896	2 037	3 419	2 605
NWT	441	455	278	218	3 091	3 355	3 810	4 028
Canada	174 894	180 349	132 043	141 553	1 921 544	1 974 364	2 228 481	2 296 266

18.28   Revenue of all governments[1] specifically derived from the control, taxation and sale of alcoholic beverages, years ended Mar. 31, 1972-76 (thousand dollars)

Government	1972	1973	1974	1975	1976
Government of Canada	471,936	499,819	554,177	613,709	640,696
Provincial and territorial governments					
Newfoundland	17,142	20,740	24,461	28,428	30,132
Prince Edward Island	4,510	5,182	5,874	6,698	7,793
Nova Scotia	28,269	33,648	37,529	42,618	46,906
New Brunswick	22,117	24,487	26,373	29,604	33,538
Quebec	142,618	151,997	164,920	184,798	204,851
Ontario	221,789	255,773	282,394	309,234	335,121
Manitoba	34,347	37,745	41,236	46,379	52,291
Saskatchewan	31,041	36,978	41,610	50,376	49,427
Alberta	64,493	73,799	84,204	94,750	107,338
British Columbia	85,419	97,484	108,870	120,643	150,274
Total, provincial governments	651,745	737,833	817,471	913,528	1,017,671
Yukon Territory	1,985	2,303	2,542	2,743	3,130
Northwest Territories	2,817	3,295	3,752	4,664	4,477
Total, provincial and territorial governments	656,547	743,431	823,765	920,935	1,025,278
Total, all governments	1,128,483	1,243,250	1,377,942	1,534,644	1,665,974

[1]Revenue of the Government of Canada comprises excise duties, excise taxes, import duties and certain fees and licences. Revenue of provinces and territories includes revenue collected directly by the provincial and territorial governments as well as revenue of liquor authorities but excludes revenue resulting from general retail sales taxation.

18.29   Total imports, exports and trade balance, Canada, 1957-77

Year	Imports		Exports		Trade balance	Ratio of exports to imports
	$'000,000	Percentage change over previous year	$'000,000	Percentage change over previous year	$'000,000	%
1957	5,488	−1.4	4,890	1.1r	−598r	89.1
1958	5,060	−7.8	4,899	0.1r	−161	96.8
1959	5,530	9.3	5,144	5.1r	−386	93.0
1960	5,495	−0.6	5,390	4.8	−105	98.2r
1961	5,781	5.2	5,903	9.5	122r	102.1
1962	6,294	8.9	6,357	7.7	63	101.0
1963	6,578	4.5	6,990	10.0r	412r	106.3
1964	7,488	13.8	8,303	18.8	815r	110.9
1965r	8,633	15.3	8,767	5.6	134	101.6r
1966r	10,072	16.7	10,325	17.8	253	102.5

18.29 Total imports, exports and trade balance, Canada, 1957-77 (concluded)

Year	Imports		Exports		Trade balance	Ratio of exports
	$'000,000	Percentage change over previous year	$'000,000	Percentage change over previous year	$'000,000	to imports %
1967	10,873	8.0r	11,420	10.6r	547	105.0
1968	12,360	13.7	13,679	19.8	1,319	110.7
1969	14,130	14.3	14,871	8.7	741	105.2
1970	13,952	−1.3	16,820	13.1	2,868	120.6
1971	15,617r	11.9	17,820	5.9	2,203r	114.1
1972	18,668r	19.5	20,150	13.1	1,482r	107.9
1973	23,325	24.9	25,421	26.2	2,096r	109.0
1974r	31,722	36.0	32,442	27.6	720	102.3
1975r	34,691	9.4	33,245	2.5	−1,446	95.8
1976	37,469	8.0	38,146	14.7	677	101.8
1977	42,068	12.3	44,197	15.9	2,129	105.1

18.30 Imports[1] into Canada from all countries, by section and selected commodities, 1973-77 and percentage of 1977 total (million dollars)

Section and commodity	1973	1974	1975	1976	1977	% of 1977 total
LIVE ANIMALS	137	112	75	109	52	0.1
FOOD, FEED, BEVERAGES AND TOBACCO	1,844	2,404	2,607	2,763	3,255	7.7
Meat and fish	339	314	338	546	534	1.3
Fruits and vegetables	609	693	774	861	1,039	2.5
Raw sugar	162	402	459	249	220	0.5
Coffee	124	132	169	250	423	1.0
CRUDE MATERIALS, INEDIBLE	2,018	4,073	5,086	5,085	5,287	12.6
Metal ores, concentrates and scrap	330	397	468	424	514	1.2
Coal	167	303	576	544	618	1.5
Crude petroleum	942	2,646	3,302	3,274	3,243	7.7
FABRICATED MATERIALS, INEDIBLE	4,282	6,482	5,944	6,218	7,010	16.7
Wood and paper	404	559	645	736	686	1.6
Textiles	659	817	740	841	890	2.1
Petroleum and coal products	215	374	276	220	301	0.7
Chemicals	1,023	1,537	1,476	1,682	2,010	4.8
Iron and steel	653	1,260	937	723	869	2.1
Non-ferrous metals	374	608	427	493	531	1.3
END PRODUCTS, INEDIBLE	14,797	18,362	20,654	22,789	26,042	61.9
Industrial machinery	2,126	2,720	3,208	3,221	3,489	8.3
Agricultural machinery and tractors	636	902	1,220	1,319	1,335	3.2
Motor vehicles	2,577	3,133	3,683	3,980	4,709	11.2
Motor vehicle parts	3,504	3,991	4,528	5,357	6,617	15.7
Communications equipment	813	957	853	1,096	1,252	3.0
Office machinery	497	609	659	736	798	1.9
Apparel	355	442	534	757	671	1.6
SPECIAL TRANSACTIONS — TRADE	247	289	325	505	421	1.0
Total, imports	23,325	31,722	34,691	37,469	42,068	100.0

[1]Includes other commodities not listed.

18.31 Domestic exports[1] from Canada to all countries, by section and selected commodities, 1973-77 and percentage of 1977 total (million dollars)

Section and commodity	1973	1974	1975	1976	1977	% of 1977 total
LIVE ANIMALS	145	90	83	134	151	0.3
FOOD, FEED, BEVERAGES AND TOBACCO	3,013	3,780	4,029	4,115	4,394	10.1
Meat and fish	638	525	579	719	909	2.1
Wheat	1,221	2,065	2,001	1,708	1,827	4.2
CRUDE MATERIALS, INEDIBLE	5,025	7,793	7,956	8,262	8,843	20.4
Metal ores, concentrates and scrap	2,000	2,376	2,233	2,501	2,722	6.3
Crude petroleum	1,482	3,420	3,052	2,287	1,751	4.0
Natural gas	351	494	1,092	1,616	2,028	4.7

18.31   Domestic exports[1] from Canada to all countries, by section and selected commodities, 1973-77 and percentage of 1977 total (million dollars) (concluded)

Section and commodity	1973	1974	1975	1976	1977	% of 1977 total
FABRICATED MATERIALS, INEDIBLE	8,224	10,696	9,862	12,149	14,926	34.4
Lumber	1,599	1,291	973	1,648	2,387	5.5
Wood pulp and similar pulp	1,082	1,889	1,831	2,177	2,156	5.0
Newsprint paper	1,288	1,726	1,744	1,998	2,381	5.5
Chemicals	700	985	1,025	1,366	1,740	4.0
Petroleum and coal products	312	611	638	559	649	1.5
Iron and steel	480	756	749	837	1,046	2.4
Non-ferrous metals	1,604	2,005	1,721	2,144	2,495	5.8
END PRODUCTS, INEDIBLE	8,387	9,237	10,457	12,539	14,946	34.5
Industrial machinery	554	764	928	883	1,164	2.7
Agricultural machinery and tractors	290	398	542	539	559	1.3
Motor vehicles	3,308	3,742	4,293	5,202	6,560	15.1
Motor vehicle parts	2,107	1,975	2,138	2,966	3,608	8.3
SPECIAL TRANSACTIONS — TRADE	45	80	79	130	68	0.2
Total, domestic exports	24,838	31,676	32,466	37,329	43,328	100.0
Total, re-exports	583	767	779	818	870	...
Total, exports	25,421	32,442	33,245	38,146	44,197	...

[1]Includes other commodities not listed.

18.32   Imports into Canada from all countries, from the United States and from the European Economic Community (EEC)[1], by section and commodity, 1976 and 1977 (thousand dollars)

Section and commodity	All countries		United States		EEC	
	1976	1977	1976	1977	1976	1977
LIVE ANIMALS	108,609	52,157	101,689	49,221	4,984	1,802
FOOD, FEED, BEVERAGES AND TOBACCO	2,762,670	3,255,431	1,447,480	1,657,002	297,606	402,881
Meat, fresh, chilled or frozen	329,662	281,393	209,227	195,000	467	253
Other meat and meat preparations	34,029	32,752	15,406	15,330	4,852	3,818
Fish and marine animals	182,042	219,412	100,406	119,579	9,025	10,406
Dairy produce, eggs and honey	75,236	80,291	23,631	26,681	34,530	35,356
Indian corn, shelled	95,091	61,424	95,090	61,403	—	15
Other cereals and cereal preparations	87,163	110,673	64,941	83,480	16,766	20,237
Bananas and plantains, fresh	59,134	66,632	117	1,824	4	2
Grapes, fresh	60,941	72,688	53,096	62,248	34	1
Oranges, mandarins and tangerines, fresh	60,101	71,933	43,912	54,316	—	18
Other fresh fruits and berries	130,282	138,821	117,996	126,794	1,194	365
Fruits, dried or dehydrated	34,857	46,406	18,209	24,587	95	310
Orange juice and concentrates	50,867	71,574	39,579	51,279	—	1
Other fruit juices and concentrates	16,528	24,168	12,208	16,832	690	1,094
Fruits and products, canned	54,266	61,495	28,672	36,155	2,360	2,850
Other fruits and fruit preparations	27,606	28,321	9,563	10,769	3,123	1,334
Nuts, except oil nuts	48,592	63,263	23,460	30,753	1,335	2,550
Tomatoes, fresh	49,275	58,427	39,343	39,248	—	—
Other fresh vegetables	174,679	211,002	166,050	196,385	322	562
Other vegetables and vegetable preparations	93,951	123,946	48,861	53,016	9,842	12,034
Raw sugar	249,296	219,868	4	1	1	—
Refined sugar, molasses and syrups	37,523	21,418	27,612	13,299	339	640
Sugar preparations and confectionery	55,739	67,368	22,702	24,266	25,460	33,909
Cocoa and chocolate	55,120	109,007	16,810	41,019	20,308	36,845
Coffee	250,477	423,444	62,534	121,601	16,120	24,930
Tea	35,870	73,289	2,364	3,118	8,616	23,564
Other foods and materials for foods	121,037	145,599	83,665	102,822	10,948	13,588
Oilseed cake and meal	70,248	90,690	70,219	90,686	—	—
Other fodder and feed	33,858	32,486	32,575	30,806	811	741
Distilled alcoholic beverages	73,993	87,386	5,429	6,101	51,611	64,116
Other beverages	94,154	138,244	3,694	6,063	72,686	107,709
Tobacco	21,053	22,012	10,107	11,543	6,067	5,635
CRUDE MATERIALS, INEDIBLE	5,085,039	5,287,274	1,399,976	1,796,942	84,851	99,830
Fur skins, undressed	75,446	80,733	46,485	54,173	18,057	16,914
Other crude animal products	43,738	46,171	34,280	35,512	3,122	2,107
Soybeans	81,136	98.953	81.130	98.942	—	3
Other oilseeds, oil nuts and oil kernels	45,028	48,168	33,645	45,750	539	458
Rubber and allied gums, natural	64,198	79,777	9,212	7,144	625	207
Other crude vegetable products	81,253	95,145	63,166	72,827	11,590	14,863
Crude wood materials	63,265	67,425	62,665	67,329	1	9
Wool and fine animal hair	34,149	36,080	4,742	3,714	16,218	16,486
Cotton	72,139	82,417	49,883	73,441	77	97
Man-made fibres	67,525	58,222	60,070	50,926	4,282	5,195
Other textile fibres	4,452	3,970	1,179	1,245	332	317

18.32 Imports into Canada from all countries, from the United States and from the European Economic Community (EEC)[1], by section and commodity, 1976 and 1977 (thousand dollars) (continued)

Section and commodity	All countries		United States		EEC	
	1976	1977	1976	1977	1976	1977
CRUDE MATERIALS, INEDIBLE (concluded)						
Iron ores and concentrates	81,514	76,274	75,802	64,596	–	59
Scrap iron and steel	48,307	29,746	48,299	29,710	7	20
Aluminum ores, concentrates and scrap	155,453	190,886	55,367	21,413	16,297	19,592
Other metals in ores, concentrates and scrap	139,206	217,079	40,781	82,280	6,871	9,167
Coal	544,312	617,677	544,312	615,851	–	–
Crude petroleum	3,273,927	3,243,187	9,840	283,880	–	7,283
Other crude bituminous substances	13,987	4,190	9,530	957	99	76
Abrasives, natural	12,422	13,599	9,543	10,729	2,670	2,657
Phosphate rock	54,472	55,012	54,254	54,000	–	–
Other crude non-metallic minerals	89,444	96,128	67,224	77,364	3,155	3,434
Other waste and scrap materials	39,664	46,435	38,565	45,159	908	887
FABRICATED MATERIALS, INEDIBLE	6,218,497	7,010,132	4,409,962	4,974,862	743,890	903,550
Leather and leather fabricated materials	103,345	86,158	67,694	54,069	17,922	16,461
Rubber fabricated materials	72,384	87,864	58,028	73,130	8,330	8,005
Lumber	164,155	171,300	155,280	160,792	63	62
Veneer	21,622	16,804	17,768	13,442	1,184	698
Plywood and wood building boards	104,012	80,602	60,013	30,893	338	251
Other wood fabricated materials	83,514	96,038	67,873	78,766	1,988	2,741
Wood pulp and similar pulp	32,889	34,719	25,923	30,204	845	859
Paper and paperboard	330,267	286,183	315,735	265,428	9,565	13,257
Cotton yarn and thread	24,998	26,594	12,443	12,355	2,847	2,795
Man-made fibre yarn and thread	107,366	113,013	71,690	76,696	18,446	18,403
Other yarn and thread	27,505	35,473	12,073	15,482	12,676	15,775
Cordage, twine and rope	19,811	26,741	5,639	7,394	2,617	2,687
Broad woven fabrics, wool and hair	27,224	30,501	906	787	11,154	16,348
Broad woven fabrics, cotton	147,398	136,199	66,289	63,711	7,716	7,283
Broad woven fabrics, man-made	96,964	97,454	41,680	40,680	21,778	23,451
Broad woven fabrics, mixed fibres	130,840	150,253	77,716	79,454	24,915	25,228
Other broad woven fabrics	18,779	17,407	1,295	1,272	1,523	2,051
Coated or impregnated fabrics	100,597	112,032	80,536	95,198	12,624	8,979
Other textile fabricated materials	139,146	144,007	84,475	95,670	23,497	21,905
Vegetable oils and fats, except essential oils	72,803	82,227	31,956	42,103	1,872	2,509
Other oils, fats, waxes, extracts and derivatives	60,362	65,325	50,850	56,170	6,533	5,487
Inorganic chemicals	190,988	248,034	151,976	190,898	26,381	35,888
Organic chemicals	413,397	493,284	284,266	348,312	87,880	96,166
Fertilizers and fertilizer materials	57,680	67,598	55,106	64,623	426	555
Synthetic and reclaimed rubber	75,758	81,957	62,314	63,325	6,728	11,282
Plastics materials, not shaped	302,658	381,875	276,022	326,746	19,470	46,682
Plastic film and sheet	109,720	124,776	94,299	108,244	9,977	10,229
Other plastics, basic shapes and forms	66,269	80,322	59,181	70,586	3,491	5,062
Dyestuffs, except dyeing extracts	46,061	52,302	15,987	18,422	24,325	27,892
Pigments, lakes and toners	31,765	34,159	21,436	20,968	8,144	10,694
Paints and related products	52,733	64,323	46,549	58,842	4,699	4,220
Other chemical products	334,628	381,418	293,850	333,672	32,835	38,161
Fuel oil	78,796	116,487	26,229	24,672	3,385	1,952
Lubricating oils and greases	46,386	56,852	39,673	48,085	672	2,416
Coke of petroleum and coal	54,143	88,317	50,189	80,042	3,953	8,275
Other petroleum and coal products	40,412	39,494	31,835	31,962	2,363	5,031
Bars and rods, steel	93,446	105,055	30,905	32,643	17,928	23,639
Plate, sheet and strip, steel	205,783	269,834	100,385	125,271	32,245	57,173
Structural shapes, steel and sheet piling	62,704	70,905	23,387	25,786	19,502	31,903
Pipes and tubes, iron and steel	130,854	169,307	78,833	103,521	13,950	12,394
Wire and wire rope, iron and steel	58,635	71,022	13,569	18,022	31,206	37,521
Other iron and steel and alloys	171,535	183,114	107,391	118,651	10,047	13,027
Aluminum, including alloys	160,688	205,289	143,721	184,617	14,350	17,346
Copper and alloys	69,224	88,630	48,366	55,979	9,876	22,659
Nickel and alloys	120,684	57,012	26,314	36,193	7,916	8,662
Precious metals, including alloys	48,201	58,946	38,698	50,439	5,416	3,625
Tin, including alloys	32,422	57,122	20,514	44,941	1,153	381
Other non-ferrous metals and alloys	62,158	64,183	35,315	42,629	11,057	6,068
Bolts, nuts and screws	113,499	152,993	99,802	134,211	3,733	3,872
Other basic hardware	136,453	148,154	107,779	116,694	11,671	10,920
Chains	25,264	31,343	13,755	17,326	5,515	5,724
Valves	114,506	130,179	77,799	92,352	26,212	24,630
Pipe fittings	70,914	76,197	53,235	57,571	9,342	9,446
Other metal fabricated basic products	180,624	196,114	157,563	164,647	10,182	11,712
Clay bricks, clay tiles and refractories	94,505	116,127	63,216	76,979	19,808	25,175
Sheet and plate glass	46,661	54,280	38,091	46,333	1,426	560
Other glass basic products	75,541	65,844	65,260	54,879	6,313	6,040
Abrasive basic products	40,212	51,119	33,480	41,358	4,010	5,845
Natural and synthetic gem stones	57,512	82,819	7,822	12,737	21,639	32,178
Other non-metallic mineral basic products	101,038	102,039	81,457	80,040	10,839	12,778
Electricity	9,195	14,718	9,195	14,718	–	–
Other fabricated materials, inedible	148,863	179,693	119,246	143,229	15,388	20,533

18.32 Imports into Canada from all countries, from the United States and from the European Economic Community (EEC)[1], by section and commodity, 1976 and 1977 (thousand dollars) (continued)

Section and commodity	All countries		United States		EEC	
	1976	1977	1976	1977	1976	1977
END PRODUCTS, INEDIBLE	22,789,295	26,042,096	18,048,547	20,800,923	1,978,012	2,159,495
Machinery	*4,539,637*	*4,823,901*	*3,728,873*	*3,948,114*	*541,731*	*585,722*
Engines and turbines, diesel and general purpose	69,398	85,174	41,812	58,552	26,993	26,090
Other engines and turbines, general purpose	127,042	125,397	86,868	92,498	26,392	22,962
Electric generators and motors	192,781	178,137	133,211	122,455	42,970	35,748
Bearings	106,570	121,981	76,201	86,697	13,787	15,416
Other mechanical power transmission equipment	85,542	103,717	71,476	85,101	10,718	16,267
Compressors, blowers and vacuum pumps	107,682	107,832	94,026	91,741	11,504	13,474
Pumps, except oil well pumps	77,535	90,085	63,575	72,311	7,965	10,360
Packaging machinery	62,816	70,406	51,434	56,933	9,466	10,601
Other general purpose industrial machinery	170,966	201,940	130,262	152,181	34,120	31,101
Conveyors and conveying systems	28,036	33,649	20,067	26,384	4,170	5,466
Elevators and escalators	16,141	12,117	13,460	10,772	2,483	1,197
Industrial trucks, tractors, trailers and stackers	86,800	100,790	74,211	87,953	6,543	8,366
Hoisting machinery	115,990	108,721	91,185	85,232	10,653	8,721
Other materials handling equipment	66,470	85,020	61,203	74,040	1,897	7,663
Drilling machinery and drill bits	182,206	212,135	166,125	184,419	5,564	8,720
Power shovels	201,490	167,525	188,110	157,914	7,178	6,448
Bulldozing and similar equipment	19,099	23,108	16,919	21,211	420	407
Front end loaders	143,836	131,141	129,768	118,323	3,885	3,334
Other excavating machinery	92,531	94,092	84,714	86,808	6,745	5,840
Mining, oil and gas machinery	165,757	162,055	125,375	136,796	28,968	17,252
Construction and maintenance machinery	139,971	153,104	129,496	140,395	7,580	9,566
Machine tools, metalworking	145,595	169,831	100,587	116,318	22,578	31,713
Welding apparatus and equipment	36,590	46,489	34,201	43,200	1,409	2,248
Rolling mill machinery	27,092	29,301	14,594	18,543	11,138	9,926
Other metalworking machinery	115,532	147,665	90,706	120,907	17,136	18,086
Pulp and paper industries machinery	70,303	80,568	50,654	57,645	10,812	7,342
Printing presses	47,339	57,079	34,280	39,345	11,283	14,227
Other printing machinery and equipment	68,942	80,063	58,226	66,578	8,695	10,217
Spinning, weaving and knitting machinery	31,750	35,436	17,323	19,921	9,755	10,528
Other textile industries machinery	54,686	55,326	35,035	35,007	15,181	15,114
Food, beverages and tobacco industries machinery	73,937	84,501	49,768	58,036	21,172	22,809
Plastics and chemical industry machinery	94,727	91,727	73,825	67,481	14,500	17,057
Other special industry machinery	195,762	242,845	147,644	176,664	33,243	45,570
Soil preparation, seeding and fertilizing machinery	117,323	121,136	109,319	111,859	2,910	4,433
Combine reaper-threshers	95,931	129,233	92,930	126,076	1,663	2,481
Other haying and harvesting machinery	128,384	118,869	123,839	114,903	3,043	3,024
Other agricultural machinery and equipment	142,625	162,045	133,807	152,101	7,166	8,081
Wheel tractors, new	438,948	425,049	362,586	340,246	55,978	74,586
Track-laying tractors and used tractors	131,185	124,106	110,729	108,008	3,820	2,046
Tractor engines and tractor parts	264,327	254,506	239,322	226,563	20,247	21,233
Transportation and Communication Equipment	*11,400,422*	*13,676,007*	*9,877,200*	*11,956,405*	*472,924*	*522,457*
Railway and street railway rolling stock	86,038	98,158	77,156	80,499	7,297	11,099
Sedans, new	2,444,324	2,985,709	2,030,360	2,506,363	174,506	207,223
Other passenger automobiles and chassis	359,198	370,607	290,941	318,424	24,825	14,661
Trucks, truck tractors and chassis	839,681	1,010,632	802,563	971,047	5,502	7,257
Other motor vehicles	336,548	342,249	273,713	246,077	6,270	6,114
Motor vehicle engines	507,922	545,322	447,027	491,605	1,941	1,356
Motor vehicle engine parts	460,868	691,095	446,677	673,160	7,880	8,239
Motor vehicle parts, except engines	4,387,804	5,380,755	4,256,767	5,244,508	59,111	65,034
Marine engines and parts	109,104	120,233	76,967	88,482	22,935	16,713
Ships, boats and parts, except engines	91,638	100,378	55,857	48,196	21,756	20,140
Aircraft, complete with engines	110,840	89,403	102,100	85,288	6,001	766
Aircraft engines and parts	132,895	161,220	114,661	137,753	17,966	22,776
Aircraft parts, except engines	168,965	187,444	161,270	181,205	4,990	4,685
Other transportation equipment	268,320	341,056	140,810	179,167	51,270	68,908
Telephone and telegraph equipment	95,284	112,370	78,461	98,765	7,088	6,234
Televisions, radio sets and phonographs	342,049	374,274	114,246	133,006	4,678	6,707
Electronic tubes and semi-conductors	98,040	112,340	78,316	91,102	7,153	8,583
Other telecommunication and related equipment	560,903	652,762	329,307	381,758	41,753	45,963
Other Equipment and Tools	*3,097,627*	*3,405,147*	*2,525,074*	*2,768,944*	*309,486*	*321,092*
Air conditioning and refrigeration equipment	194,062	202,244	169,442	173,196	11,809	15,131
Electric lighting fixtures and portable lamps	103,463	111,079	82,275	88,945	9,518	9,459
Switchgear and protective equipment	70,761	111,771	41,778	46,206	17,636	41,817
Industrial control equipment	45,966	56,787	41,358	51,168	2,665	3,289
Other electric lighting distribution equipment	164,249	151,806	106,234	121,301	42,418	15,254
Auxiliary electric equipment for engines	153,319	155,627	141,434	143,152	6,401	7,440

18.32   Imports into Canada from all countries, from the United States and from the European Economic Community (EEC)¹, by section and commodity, 1976 and 1977 (thousand dollars) (concluded)

Section and commodity	All countries		United States		EEC	
	1976	1977	1976	1977	1976	1977
END PRODUCTS, INEDIBLE (concluded)						
Electrical property measuring instruments	57,616	68,373	49,637	57,673	5,698	7,370
Miscellaneous measuring and controlling instruments	95,154	110,807	86,374	99,734	5,489	6,770
Medical and related equipment	101,092	115,442	90,302	101,315	6,728	8,812
Navigation equipment	22,759	17,718	21,039	16,032	920	1,056
Other measuring and laboratory equipment	256,268	299,213	197,609	241,405	36,289	33,460
Safety and sanitation equipment	83,151	98,879	76,686	84,025	4,799	7,303
Service industry equipment	69,527	77,985	60,559	69,159	6,927	6,531
Furniture and fixtures	187,986	214,194	129,902	149,924	30,806	35,409
Hand tools and cutlery	213,046	237,888	153,837	170,549	26,490	28,421
Electronic computers	428,355	552,819	408,532	521,701	13,284	20,932
Other office machines and equipment	308,015	245,192	202,285	149,755	49,967	36,823
Miscellaneous equipment and tools	542,837	577,325	465,789	483,706	31,641	35,814
Personal and Household Goods	*1,900,626*	*1,948,606*	*595,600*	*595,592*	*354,467*	*404,098*
Outerwear, except knitted	341,848	276,338	46,641	32,883	35,543	39,859
Outerwear, knitted	262,659	231,648	18,814	17,462	36,939	34,081
Other apparel and apparel accessories	152,580	163,075	45,616	45,935	24,242	26,633
Footwear	198,597	231,376	16,636	20,209	69,416	86,665
Watches, clocks, jewellery and silverware	148,814	172,018	59,889	55,670	34,523	40,223
Sporting and recreation equipment	147,420	161,266	79,234	78,531	21,743	22,847
Games, toys and children's vehicles	104,503	124,388	53,640	57,777	10,909	12,438
House furnishings	167,081	154,744	97,694	87,026	20,659	23,730
Kitchen utensils, cutlery and tableware	163,415	188,468	64,534	70,877	53,237	62,523
Other personal and household goods	213,710	245,285	112,903	129,221	47,256	55,099
Miscellaneous End Products	*1,850,982*	*2,188,436*	*1,321,800*	*1,531,868*	*299,403*	*326,126*
Medicinal and pharmaceutical products	173,176	220,190	81,037	101,954	57,600	53,939
Medical, ophthalmic and orthopedic supplies	146,115	192,800	106,511	138,700	26,925	34,825
Newspapers, magazines and periodicals	136,448	160,515	122,581	146,601	12,883	12,915
Books and pamphlets	222,847	247,379	174,734	193,793	44,547	48,927
Other printed matter	120,340	148,698	102,412	125,693	11,209	13,414
Stationers' and office supplies	89,188	106,425	65,532	78,929	14,627	16,434
Unexposed photographic film and plates	115,595	147,161	67,981	89,091	32,943	36,845
Other photographic goods	276,972	341,881	184,238	225,145	29,300	25,662
Containers and closures	133,964	159,114	120,041	144,129	8,387	9,126
Other end products, inedible	436,337	464,272	296,734	287,832	60,981	74,040
SPECIAL TRANSACTIONS — TRADE	504,708	421,092	328,985	273,060	71,013	70,775
Total, imports	37,468,819	42,068,183	25,736,640	29,552,010	3,180,355	3,638,333

In this table a dash indicates that either there was no trade or the amount was less than $500.
¹Includes the United Kingdom.

18.33   Domestic exports from Canada to all countries, to the United States and to the European Economic Community (EEC)¹, by section and commodity, 1976 and 1977 (thousand dollars)

Section and commodity	All countries		United States		EEC	
	1976	1977	1976	1977	1976	1977
LIVE ANIMALS	133,996	151,215	114,746	131,150	4,056	4,319
FOOD, FEED, BEVERAGES AND TOBACCO	4,114,610	4,394,085	925,450	1,128,709	883,101	875,479
Meat, fresh, chilled or frozen	188,310	208,542	78,374	73,132	20,718	21,552
Other meat and meat preparations	20,146	14,857	8,114	7,977	246	288
Fish, whole or dressed, fresh or frozen	112,576	158,375	52,260	61,714	39,050	54,699
Fish, fillets and blocks, fresh or frozen	174,610	245,528	161,063	206,380	11,199	37,052
Fish, preserved, except canned	58,441	64,877	27,155	26,166	3,992	7,045
Fish, canned	46,553	71,369	9,195	10,126	24,071	39,454
Shellfish	117,986	145,466	95,911	104,697	15,586	23,427
Dairy produce, eggs and honey	72,011	112,366	7,999	14,879	8,738	10,481
Barley	542,362	312,347	33,190	25,797	165,410	103,691
Wheat	1,707,822	1,826,772	2,991	4,230	365,752	372,157
Other cereals, unmilled	119,844	79,663	7,547	11,728	36,759	27,400
Wheat flour	122,726	114,727	203	67	152	13
Other cereals, milled	38,618	52,081	10,761	16,338	2,649	4,003
Cereal preparations	40,079	39,225	37,547	36,141	403	440
Fruits and fruit preparations	29,942	42,578	17,431	30,676	5,439	3,594
Vegetables and vegetable preparations	117,654	118,500	18,747	28,369	70,113	49,161
Sugar and sugar preparations	43,359	74,041	35,479	60,971	858	1,395
Other foods and materials for food	109,975	170,670	18,650	42,862	6,849	6,541
Oilseed cake and meal	18,310	31,230	785	2,649	17,484	23,996
Other feeds of vegetable origin	69,053	76,022	34,387	36,100	8,791	8,879
Other fodder and feed	44,252	56,584	14,610	17,407	20,377	22,271
Whisky	222,588	270,741	216,376	264,131	3,100	3,297
Other beverages	30,281	38,481	29,579	37,542	26	75
Tobacco	67,112	69,043	7,097	8,628	55,338	54,570

18.33   Domestic exports from Canada to all countries, to the United States and to the European Economic Community (EEC)[1], by section and commodity, 1976 and 1977 (thousand dollars) (continued)

Section and commodity	All countries 1976	All countries 1977	United States 1976	United States 1977	EEC 1976	EEC 1977
CRUDE MATERIALS, INEDIBLE	8,262,280	8,842,766	5,326,490	5,465,789	1,108,035	1,251,166
Raw hides and skins	68,805	89,151	18,206	19,795	10,764	13,296
Fur skins, undressed	56,504	69,421	17,726	20,861	22,262	29,359
Other crude animal products	22,241	20,421	17,856	16,491	1,856	1,677
Seeds for sowing	20,200	29,506	12,512	18,395	5,685	6,308
Flaxseed	66,278	93,538	9,729	12,595	26,976	49,878
Rapeseed	185,971	310,047	1,516	177	9,691	55,075
Other oilseeds, oil nuts and oil kernels	30,449	37,083	14,273	12,232	6,362	9,764
Other crude vegetable products	39,935	49,986	36,614	45,851	1,538	1,390
Pulpwood	11,611	12,081	7,428	8,673	4,183	2,611
Pulpwood chips	21,517	34,727	21,517	30,014	–	–
Other crude wood products	38,831	58,870	16,741	25,864	628	1,956
Textile and related fibres	28,142	52,353	11,383	13,995	8,406	3,285
Iron ores and concentrates	920,463	1,063,922	602,968	756,310	234,361	229,869
Scrap iron and steel	63,816	51,025	35,494	37,268	9,079	4,355
Aluminum ores, concentrates and scrap	30,121	42,933	24,043	33,220	1,366	1,330
Copper in ores, concentrates and scrap	342,705	310,618	73,944	44,100	12,413	21,952
Lead in ores, concentrates and scrap	34,887	53,349	7,150	8,106	5,675	10,905
Nickel in ores, concentrates and scrap	524,200	537,263	123,209	77,532	226,795	273,881
Precious metals in ores, concentrates and scrap	134,689	157,414	59,638	66,472	50,703	63,908
Zinc in ores, concentrates and scrap	241,457	205,514	20,613	23,943	155,667	121,005
Radioactive ores and concentrates	67,392	75,438	46,850	72,848	20,541	2,590
Other metals in ores, concentrates and scrap	140,970	224,320	25,778	43,655	69,704	96,554
Crude petroleum	2,286,675	1,750,637	2,286,675	1,750,637	–	–
Natural gas	1,616,490	2,028,053	1,616,490	2,028,053	–	–
Coal and other crude bituminous substances	560,878	651,148	505	55,028	15,323	25,636
Asbestos, unmanufactured	475,893	553,284	125,270	138,168	150,952	155,136
Sulphur	110,045	122,288	18,800	18,913	23,820	18,249
Other crude non-metallic minerals	92,831	126,102	53,956	63,901	32,458	49,945
Other waste and scrap materials	28,286	32,275	19,606	22,690	827	1,251
FABRICATED MATERIALS, INEDIBLE	12,148,886	14,925,834	8,376,828	10,854,093	1,976,509	2,020,470
Leather and leather fabricated materials	18,577	19,038	13,267	14,560	1,091	2,257
Lumber, softwood	1,609,994	2,338,629	1,195,353	1,869,774	197,868	205,520
Lumber, hardwood	37,686	48,451	21,083	23,977	15,127	22,566
Shingles and shakes	101,465	141,188	99,098	137,424	1,788	1,876
Other sawmill products	8,855	9,177	7,758	8,974	110	176
Veneer	53,608	64,116	44,906	55,175	6,904	6,359
Plywood	47,255	84,064	7,181	11,760	38,634	71,440
Other wood fabricated materials	52,358	99,273	40,974	87,868	9,035	8,436
Wood pulp and similar pulp	2,176,963	2,156,029	1,165,432	1,218,688	678,163	621,492
Newsprint paper	1,998,296	2,381,265	1,595,877	1,869,417	161,285	197,663
Other paper for printing	101,396	151,430	80,840	131,930	10,523	9,071
Paperboard	96,477	103,707	3,554	10,598	63,044	57,014
Other paper	140,744	172,701	70,623	96,856	30,996	32,277
Yarn, thread, cordage, twine and rope	13,061	22,166	5,082	10,574	3,011	3,431
Cotton broad woven fabrics	13,561	10,920	5,751	5,244	2,893	1,361
Other broad woven fabrics	32,411	32,462	3,495	5,017	15,668	14,946
Other textile fabricated materials	40,859	45,401	18,283	20,510	4,991	4,892
Oils, fats, waxes, extracts and derivatives	76,035	138,935	10,506	13,434	20,681	30,329
Chemical elements	97,628	110,692	33,356	39,741	56,694	51,015
Other inorganic chemicals	300,977	352,441	260,902	326,362	11,823	10,435
Organic chemicals	172,664	301,352	110,520	198,384	36,573	64,564
Fertilizers and fertilizer materials	547,209	659,969	439,719	541,454	10,647	4,595
Synthetic rubber and plastics materials	131,701	172,882	83,364	114,160	19,295	22,628
Plastics, basic shapes and forms	49,324	73,158	26,594	42,770	3,793	4,562
Other chemical products	66,712	69,014	41,427	43,680	8,920	9,804
Petroleum and coal products	558,788	649,390	472,675	585,034	32,109	8,651
Ferro-alloys	18,259	27,365	15,633	19,837	1,524	4,892
Primary iron and steel	93,638	141,118	61,677	86,152	24,555	39,419
Castings and forgings, steel	157,758	173,241	155,875	169,390	415	1,392
Bars and rods, steel	72,672	105,092	45,464	84,448	17,751	13,476
Plate, sheet and strip, steel	221,305	302,807	135,182	229,262	38,652	8,131
Railway track material	41,388	33,882	11,458	25,695	3	2
Other iron and steel and alloys	231,995	262,837	193,477	240,089	5,317	2,542
Aluminum, including alloys	466,569	769,448	322,653	536,275	20,595	23,027
Copper and alloys	522,583	522,036	170,401	196,751	261,744	241,270
Lead, including alloys	50,867	85,344	20,240	49,709	19,633	31,025
Nickel and alloys	443,592	436,937	326,710	309,478	65,105	88,298
Precious metals, including alloys	357,131	416,488	311,775	388,715	11,639	18,237
Zinc, including alloys	273,780	228,277	218,828	164,628	32,459	38,347
Other non-ferrous metals and alloys	29,353	36,769	18,952	22,038	8,461	11,029
Metal fabricated basic products	221,294	289,318	167,702	218,908	10,229	12,697
Abrasive basic products	79,460	90,917	70,773	80,068	4,952	4,933
Other non-metallic mineral basic products	114,014	162,672	84,736	135,832	7,242	10,377
Electricity	161,733	376,965	161,733	376,965	–	–
Other fabricated materials, inedible	46,887	56,471	26,341	36,489	4,568	4,012

18.33 Domestic exports from Canada to all countries, to the United States and to the European Economic Community (EEC)[1], by section and commodity, 1976 and 1977 (thousand dollars) (concluded)

Section and commodity	All countries 1976	All countries 1977	United States 1976	United States 1977	EEC 1976	EEC 1977
END PRODUCTS, INEDIBLE	12,539,223	14,945,785	10,299,520	12,552,963	496,217	495,543
Machinery	*1,422,273*	*1,723,118*	*1,010,136*	*1,260,395*	*79,935*	*82,479*
Engines and turbines, general purpose	42,798	53,981	27,353	30,358	1,757	890
Electric generators and motors	42,869	35,984	21,988	21,106	2,429	1,087
Other general purpose industrial machinery	167,688	236,911	100,079	129,758	8,530	10,355
Materials handling machinery and equipment	137,276	171,277	96,183	132,948	5,870	4,533
Drilling, excavating and mining machinery	118,145	181,384	49,322	106,173	11,936	9,324
Metalworking machinery	75,238	97,171	51,172	78,164	5,279	4,460
Woodworking machinery and equipment	54,618	59,213	28,078	32,944	15,075	16,206
Construction machinery and equipment	66,662	84,715	22,971	48,708	6,364	5,543
Plastics industry machinery and equipment	57,449	86,951	52,236	81,589	2,211	2,257
Pulp and paper industries machinery	36,638	48,188	19,577	24,821	1,425	735
Other special industry machinery	83,749	108,503	53,114	67,163	7,978	10,453
Soil preparation, seeding and fertilizing machinery	93,171	88,827	87,345	83,860	994	873
Combine reaper-threshers and parts	189,184	202,864	169,246	178,891	4,221	12,235
Other haying and harvesting machinery	71,186	70,729	63,075	64,548	4,233	1,711
Other agricultural machinery and equipment	80,227	82,043	73,099	76,210	1,083	1,143
Tractors	105,375	114,375	95,297	103,156	550	674
Transportation and Communication Equipment	*9,492,236*	*11,458,219*	*8,351,556*	*10,234,694*	*175,500*	*184,685*
Railway and street railway rolling stock	111,885	49,958	7,742	6,657	–	12
Passenger automobiles and chassis	3,637,845	4,285,318	3,415,738	3,948,142	6,329	13,782
Trucks, truck tractors and chassis	1,403,551	2,080,583	1,235,962	1,886,237	94	24
Other motor vehicles	160,534	194,220	119,505	127,878	2,269	1,230
Motor vehicle engines and parts	776,627	944,688	761,636	925,659	420	195
Motor vehicle parts, except engines	2,189,175	2,663,187	2,039,627	2,503,672	12,583	13,598
Ships, boats and parts	183,603	191,652	97,197	102,979	66,673	61,499
Aircraft, complete with engines	79,403	66,651	9,340	12,552	2,774	3,219
Aircraft, engines and parts	214,464	226,066	145,352	168,711	32,567	21,970
Aircraft parts, except engines	159,593	192,930	122,429	149,742	17,206	17,803
Other transportation equipment	179,902	158,004	171,164	149,991	635	1,020
Televisions and radio sets and phonographs	32,901	57,593	29,540	47,934	2,522	8,536
Other telecommunication and related equipment	362,753	347,370	196,325	204,539	31,427	41,798
Other Equipment and Tools	*832,147*	*932,464*	*496,944*	*583,009*	*122,711*	*126,734*
Heating and refrigeration equipment	37,434	44,218	20,863	26,596	2,647	3,025
Cooking equipment for food	11,548	13,601	2,510	5,089	4,450	5,208
Electric lighting and distribution equipment	106,438	126,108	53,000	65,209	10,379	12,045
Navigation equipment and parts	59,394	51,107	36,788	34,270	17,891	12,617
Other measuring, controlling, laboratory, medical and optical equipment	114,464	132,147	63,509	75,937	19,435	19,090
Hand tools and miscellaneous cutlery	30,457	38,750	11,028	14,881	2,541	3,883
Office machines and equipment	331,103	346,684	203,517	221,942	53,610	59,383
Other equipment and tools	141,309	179,849	105,728	139,086	11,757	11,483
Personal and Household Goods	*247,259*	*297,697*	*145,809*	*180,356*	*43,666*	*46,598*
Apparel and apparel accessories	109,087	126,309	57,570	70,006	27,312	27,852
Footwear	23,853	31,099	21,977	27,787	1,159	2,380
Toys, games, sporting and recreation equipment	45,624	54,562	31,511	37,169	5,125	5,994
Other personal and household goods	68,695	85,726	34,751	45,394	10,070	10,371
Miscellaneous End Products	*545,308*	*534,287*	*295,075*	*294,510*	*74,405*	*55,049*
Medicinal and pharmaceutical products	43,747	49,847	8,596	10,446	9,144	10,379
Medical, ophthalmic and orthopedic supplies	20,345	22,723	8,811	8,994	3,772	5,305
Printed matter	90,335	90,161	78,104	77,901	7,546	6,717
Photographic goods	59,914	62,083	50,217	50,078	5,353	6,565
Firearms, ammunition and ordnance	16,766	13,101	11,588	8,480	2,097	2,366
Containers and closures	37,598	58,435	29,104	48,188	1,405	2,207
Prefabricated buildings and structures	112,388	136,385	28,173	31,429	2,636	2,434
Other end products	164,215	101,552	80,482	58,995	42,452	19,075
SPECIAL TRANSACTIONS — TRADE	129,550	68,179	91,895	52,058	7,659	2,775
Total, domestic exports	37,328,545	43,327,864	25,134,929	30,184,762	4,475,577	4,649,752

In this table a dash indicates that either there was no trade or the amount was less than $500.
[1]Includes the United Kingdom.

18.34 Trade of Canada with principal trading areas, 1967-77

Item and year	United States		United Kingdom		Other EEC		Japan		Other countries	
	Value $'000,000	% of total	Value $'000,000	% of total	Value $'000,000	% of total	Value $'000,000	% of total	Value $'000,000	% of total
Imports										
1967	7,952	73.1	649	6.0	635	5.8	294	2.7	1,343r	12.4
1968	9,051	73.2	696	5.6	698	5.6	360	2.9	1,555r	12.6
1969	10,243	72.5	791	5.6	831	5.9	496	3.5	1,769r	12.5
1970	9,917	71.1	738	5.3	849	6.1	582	4.2	1,866	13.4
1971	10,951	70.1	837	5.4	984	6.3	803	5.1	2,043	13.1
1972	12,878	69.0	949r	5.1	1,215r	6.5	1,071	5.7	2,556	13.7
1973	16,502	70.7	1,005	4.3	1,477	6.3	1,011	4.3	3,330r	14.3
1974	21,387r	67.4	1,126	3.5r	1,920	6.1	1,430	4.5	5,859	18.5
1975	23,616r	68.1r	1,222	3.5	2,074	6.0	1,205	3.5	6,574r	19.0
1976	25,737	68.7	1,152	3.1	2,028	5.4	1,525	4.1	7,027	18.8
1977	29,552	70.2	1,281	3.0	2,357	5.6	1,802	4.3	7,076	16.8
Exports[1]										
1967	7,332	64.2	1,178	10.3	721	6.3	574	5.0	1,615	14.1
1968	9,285	67.9	1,226	9.0	789	5.8	608	4.4	1,771	12.9
1969	10,551	71.0r	1,113	7.5	887	6.0	626	4.2	1,694r	11.4
1970	10,900	64.8	1,501	8.9	1,242	7.4	813	4.8	2,364r	14.1
1971	12,025	67.5	1,395	7.8	1,145	6.4	831	4.7	2,424	13.6
1972	13,973r	69.3	1,385	6.9	1,178	5.8	965	4.8	2,649r	13.1
1973	17,129	67.4	1,604	6.3	1,581	6.2	1,814	7.1	3,293	13.0
1974	21,399r	66.0	1,929	5.9	2,175	6.7	2,231	6.9	4,708r	14.5
1975r	21,697	65.3	1,800	5.4	2,358	7.1	2,133	6.4	5,257	15.8
1976	25,796	67.6	1,868	4.9	2,664	7.0	2,389	6.3	5,429	14.2
1977	30,893	69.9	1,945	4.4	2,764	6.3	2,512	5.7	6,083	13.8

[1]Includes domestic exports and re-exports.

18.35 Measures of bilateral trade between Canada and the United States, 1971-76 (billions of US dollars)

Year	Published by Canada			Published by United States			Reconciled figures		
	Imports from US	Exports to US	Canadian surplus	Exports to Canada	Imports from Canada	Canadian surplus	From US to Canada	From Canada to US	Canadian surplus
1971	10.8	11.9	1.1	10.4	12.7	2.3	10.6	12.0	1.4
1972	13.0	14.1	1.1	12.4	14.9	2.5	12.6	14.2	1.5
1973	16.5	17.1	0.6	15.1	17.7	2.6	16.1	17.3	1.2
1974	21.7	21.7	—	19.9	22.3	2.3	21.1	22.1	0.9
1975	23.1	21.1	−1.9	21.7r	22.2	0.4	22.8r	21.4	−1.3
1976	25.9	25.9	—	24.1	26.2	2.1	25.5	26.2	0.7

18.36 Trade in energy-related products, 1966-77 (million dollars)

Item and year	Crude petroleum	Natural gas	Coal and other crude bituminous products	Petroleum and coal products	Electricity	Radioactive ores, elements and isotopes	Total
Domestic exports							
1966	322	109	14	29	16	44	534
1967	398	124	16	40	16	31	625
1968	446	154	18	50	14	36	718
1969	526	176	10	59	18	31	820
1970	649	206	30	88	34	34	1,041
1971	787	251	87	117	48	34	1,324
1972	1,008	307	107	210	68	70	1,770
1973	1,482	351	166	312	109	93	2,513
1974	3,420	494	319	611	175	99	5,118
1975	3,052	1,092	494	638	104	129	5,509
1976	2,287	1,616	561	559	162	243	5,428
1977	1,751	2,028	651	649	377	236	5,692
Imports							
1966	316	18	142	181	10	6	673
1967	323	20	145	193	10	4	695
1968	373	35	163	215	12	4	802
1969	393	16	115	224	9	5	762
1970	415	5	150	206	12	68	856
1971	541	7	151	213	11	21	944
1972	681	8	179	210	9	15	1,102
1973	942	8	167	215	6	24	1,362
1974	2,646	6	303	374	5	16	3,350
1975	3,302	8	577	276	13	13	4,189
1976	3,274	9	549	220	9	12	4,073
1977	3,243	1	623	301	15	41	4,223

18.36   Trade in energy-related products, 1966-77 (million dollars) (concluded)

Item and year	Crude petroleum	Natural gas	Coal and other crude bituminous products	Petroleum and coal products	Electricity	Radioactive ores, elements and isotopes	Total
Balance (Domestic exports minus imports)							
1966	6	91	−128	−152	6	38	−139
1967	75	104	−129	−153	6	27	−70
1968	73	119	−145	−165	2	32	−84
1969	133	160	−105	−165	9	26	58
1970	234	201	−120	−118	22	−34	185
1971	246	244	−64	−96	37	13	380
1972	327	299	−72	—	59	55	668
1973	540	343	−1	97	103	69	1,151
1974	774	488	16	237	170	83	1,768
1975	−250	1,084	−83	362	91	116	1,320
1976	−987	1,607	12	339	153	231	1,355
1977	−1,492	2,028	28	348	362	195	1,469

¹Less than $500,000.

18.37 Price and volume indexes of trade in Canada by section, 1974-77 (1971=100)

Item and year	Food, feed, beverages and tobacco		Crude materials, inedible		Fabricated materials, inedible		End products, inedible		All sections	
	Index	Percentage change from previous year	Index	Percentage change from previous year	Index	Percentage change from previous year	Index	Percentage change from previous year	Index	Percentage change from previous year
Current weighted price indexes										
Imports										
1974	162.8	25.3	285.5	121.3	148.0	33.7	115.8	9.9	135.9	23.4
1975r	172.0	5.7	350.1	22.6	162.6	9.9	135.6	17.1	157.4	15.8
1976	159.4	-7.3	358.9	2.5	162.9	0.2	139.1	2.6	158.2	0.5
1977	190.6	19.6	402.5	12.1	184.4	13.2	155.5	11.8	177.0	11.9
Domestic exports										
1974	221.7	50.7	209.9	72.6	160.0	31.1	114.7	9.8	156.3	32.6
1975r	223.8	0.9	244.5	16.5	183.1	14.4	127.5	11.2	173.1	10.7
1976	213.5	-4.6	256.6	4.9	191.2	4.4	134.1	5.2	177.5	2.5
1977	192.9	-9.7	285.3	11.2	211.8	10.8	143.9	7.3	188.7	6.3
Fixed weighted volume indexes										
Imports										
1974r	132.1	4.0	107.9	-8.9	139.4	13.1	161.2	13.0	149.5	10.2
1975r	135.6	2.6	109.9	1.9	116.4	-16.5	155.0	-3.8	141.1	-5.6
1976	155.0	14.3	107.2	-2.5	121.5	4.4	166.6	7.5	151.7	7.5
1977	152.8	-1.4	99.4	-7.3	121.1	-0.3	170.3	2.2	152.2	0.3
Domestic exports										
1974	83.4	-16.8r	113.8	-10.2r	115.3	-0.9	130.0	0.3	116.5	-3.8
1975r	88.0	5.5	99.7	-12.4	92.9	-19.4	132.5	1.9	107.8	-7.5
1976	94.2	7.0	98.5	-1.2	106.7	14.9	151.0	14.0	119.8	11.1
1977	111.4	18.3	94.8	-3.8	118.4	11.0	167.7	11.1	130.8	9.2

18.38   Value of total exports and imports, by geographic region and country, 1975-77 (thousand dollars)

Region and country[1]	Exports[2]			Imports[2]		
	1975r	1976	1977	1975r	1976	1977
WESTERN EUROPE						
United Kingdom	1,800,378 (3)	1,867,697 (3)	1,945,284 (3)	1,221,900 (2)	1,152,384 (4)	1,281,259 (4)
Gibraltar	12,121	3,160	14	—	14	1
Ireland	18,551	31,903	31,641	31,588	26,230	40,566
Malta	2,479	2,286	2,196	660	797	459
Austria	20,267	21,366	19,952	56,623	59,197	62,110
Belgium and Luxembourg	382,028 (8)	483,617 (7)	515,141 (7)	143,304	124,696	160,495
Denmark	28,644	32,839	49,091	78,258	76,146	81,036
Finland	20,546	17,160	20,246	28,904	34,529	38,452
France	351,336 (10)	403,747 (9)	369,785 (10)	487,471 (8)	438,990 (8)	521,977 (8)
Germany, Federal Republic of	611,486 (4)	709,536 (4)	778,052 (4)	795,133 (5)	815,502 (5)	963,396 (5)
Greece	36,531	35,801	35,449	16,976	29,396	31,450
Iceland	1,074	1,171	1,856	437	765	1,278
Italy	482,884 (6)	550,971 (5)	501,345 (8)	379,557 (9)	365,123 (9)	399,146 (9)
Netherlands	483,314 (5)	451,846 (8)	519,158 (6)	158,589	181,284	190,459
Norway	171,550	152,387	224,010	120,095	133,515	69,759
Portugal	18,229	19,037	34,751	28,570	22,387	26,346
Spain	116,722	128,868	133,084	101,921	105,729	114,037
Sweden	98,974	106,481	109,645	264,851	262,373	260,204
Switzerland	81,445	88,164	104,967	179,255	163,442	219,809
Total, Western Europe	4,738,558	5,108,036	5,395,667	4,094,091	3,992,499	4,462,238
EASTERN EUROPE						
Albania	10,487	145	324	22	—	27
Bulgaria	2,537	6,402	4,363	3,754	2,135	2,136
Czechoslovakia	10,072	17,808	15,040	46,445	40,695	43,081
German Democratic Republic	4,271	46,070	31,524	5,387	4,946	5,731
Hungary	6,889	6,035	8,191	15,036	15,979	20,457
Poland	116,640	126,612	148,743	40,811	44,956	46,743
Romania	62,331	38,786	12,564	19,238	24,255	22,122
Union of Soviet Socialist Republics	419,229 (7)	535,179 (6)	360,044	28,783	55,233	55,358
Yugoslavia	36,795	19,335	57,597	18,901	17,609	18,101
Total, Eastern Europe	669,253	796,373	638,389	178,377	205,808	213,756
MIDDLE EAST						
Bahrain	1,382	1,476	1,596	8	1	—
Cyprus	2,295	2,403	2,164	737	194	302
Qatar	1,542	4,161	3,351	6,411	1	1
United Arab Emirates	4,894	12,057	20,428	140,589	61,995	14,192
Egypt, Arab Republic of	6,672	35,367	50,808	335	10,299	33,840
Ethiopia	3,388	6,542	3,670	911	2,067	1,515
Iran	153,238	151,039	146,847	756,390 (6)	695,426 (6)	535,514 (7)
Iraq	67,679	54,769	58,886	133,956	133,630	111,316
Israel	65,578	60,593	52,200	28,213	38,396	42,549
Jordan	2,442	5,636	7,848	7	12	18
Kuwait	16,093	22,437	37,297	110,522	22,439	20,064
Lebanon	41,444	3,042	27,532	1,320	446	299
Libya	23,341	11,762	19,091	36,150	107,323	10
Saudi Arabia	38,680	107,418 (7)	109,841	745,961 (7)	481,614 (7)	712,281 (6)
Somalia	1,427	1,135	59	56	—	4

18.38   Value of total exports and imports, by geographic region and country, 1975-77 (thousand dollars) (continued)

Region and country[1]	Exports[2] 1975r	1976	1977	Imports[2] 1975r	1976	1977
MIDDLE EAST (concluded)						
Sudan	4,407	3,135	2,399	180	534	122
Syria	5,083	13,102	50,455	1,208	50	99
Turkey	46,307	65,602	52,163	3,382	5,890	7,048
Yemen	7,008	2,915	4,063	196,655	201,715	5,853
Total, Middle East	492,898	564,591	650,699	2,162,989	1,762,033	1,485,028
OTHER AFRICA						
Gambia	167	204	219	5	38	8
Ghana	19,749	19,669	24,907	5,776	4,100	4,671
Kenya	10,490	9,660	31,482	12,362	12,786	20,385
Malawi	2,000	1,818	3,307	248	115	388
Mauritius and Dependencies	1,414	674	697	—	1,512	—
Nigeria	41,747	32,831	32,144	63,057	—	6,483
Rhodesia	3	18	3	78,371	155,860	37,563
Sierra Leone	424	169	60	3,769	1,454	3,909
South Africa, Republic of	135,940	98,237	84,889	193,322	155,223	150,273
Tanzania	22,111	12,499	6,418	6,904	9,181	8,546
Uganda	251	327	645	2,547	1,846	2,547
Zambia	21,229	29,068	13,928	7	119	13
Commonwealth Africa, other	149	105	132	9,366	123	3,078
Algeria	101,278	94,931	186,463	1,704	65,420	47,822
Angola	1,007	935	3,432	479	1,000	—
Benin	2,099	1,652	693	52	—	41
Cameroon	7,853	2,985	1,469	3,861	5,369	2,058
French Africa, other	11,549	3,644	2,716	40	9,221	17,246
Gabon	803	2,450	1,536	25,828	61,676	22,686
Guinea	139	547	323	15,100	7,948	12,255
Ivory Coast	3,068	6,671	13,922	2,556	7,429	5,227
Liberia	2,624	3,183	3,721	589	1,284	59
Madagascar	731	1,260	1,041	1,520	2,780	—
Mauritania	2,715	809	4,612	14	—	1
Morocco	19,113	2,960	30,842	2,113	2,768	3,503
Mozambique	2,251	6,370	5,645	4,402	1,511	5,209
Portuguese Africa, other	1,388	324	1,201	10	278	—
Senegal	6,396	1,878	2,132	359	76	898
Spanish Africa	556	104	722	11	47	2
Togo	414	12,239	1,421	18	8	116
Tunisia	10,064	19,186	11,456	111	59	129
Zaire, Republic of	12,242	16,424	6,988	8,267	13,925	7,444
Total, other Africa	441,963	383,832	479,166	443,372	523,156	362,562
OTHER ASIA						
Bangladesh	85,766	37,689	59,324	5,030	8,586	7,101
Hong Kong	44,146	61,762	68,766	170,786	284,558	280,422
India	203,580	154,735	136,194	46,650	66,709	55,677
Malaysia	24,942	31,799	41,284	56,745	48,314	53,648
Pakistan	95,574	34,145	75,103	7,864	10,105	6,633
Singapore	37,128	32,882	39,402	46,624	77,626	93,509
Sri Lanka, Republic of	14,556	14,493	16,415	12,846	12,091	16,855

18.38   Value of total exports and imports, by geographic region and country, 1975-77 (thousand dollars) (continued)

Region and country[1]	Exports[2] 1975r	1976	1977	Imports[2] 1975r	1976	1977
OTHER ASIA (concluded)						
Afghanistan	774	1,520	3,100	117	208	531
Burma	362	4,250	6,181	18	21	7
China, People's Republic of	377,260 (9)	196,525	369,270	56,328	88,348	82,154
Indonesia	66,791	77,974	67,083	14,266	18,153	24,553
Japan	2,133,442 (2)	2,389,303 (2)	2,511,684 (2)	1,205,316 (3)	1,525,417 (2)	1,802,475 (2)
Khmer Republic — Laos	9	1	675	10		1
Korea, North	4,924	9,442	99	108	1,954	527
Korea, South	82,216	119,234	144,207	166,139	303,391	322,724
Philippines	60,043	52,625	76,147	22,430	31,365	39,397
Portuguese Asia	1	1	28	330	860	404
Taiwan	38,813	42,683	73,935	181,904	292,589	320,569
Thailand	22,556	38,563	54,114	6,065	9,086	12,081
Vietnam	4,832	171	11,132	228	27	117
Total, other Asia	3,297,714	3,299,798	3,754,143	1,999,803	2,779,408	3,119,383
OCEANIA						
Australia	253,414	370,491	411,237 (9)	344,756 (10)	339,272 (10)	353,473 (10)
Fiji	1,237	1,176	2,394	225	471	1,554
New Zealand	52,484	57,287	71,612	48,591	73,942	72,036
British Oceania, other	148	78	72	3		39
French Oceania	1,300	1,735	1,048	5	80	9
United States Oceania	2,080	1,141	1,071	—	65	37
Total, Oceania	310,663	431,907	487,434	393,581	413,829	427,148
SOUTH AMERICA						
Falkland Islands	7	165	805	—		—
Guyana	14,944	11,921	8,190	11,682	5,864	12,647
Argentina	58,634	47,744	106,507	13,160	21,048	22,809
Bolivia	5,559	3,820	4,142	5,336	4,623	11,222
Brazil	202,916	333,358	280,316	170,217	162,609	213,995
Chile	30,806	14,493	37,741	19,091	33,372	22,653
Colombia	38,873	59,989	61,319	32,239	41,657	63,659
Ecuador	22,776	26,552	20,684	21,148	30,308	68,077
French Guiana	23	90	156	—	—	191
Paraguay	405	350	445	1,232	2,528	4,034
Peru	82,544	50,734	48,349	11,412	15,365	37,526
Suriname	3,040	3,213	3,164	6,421	9,726	11,703
Uruguay	6,633	6,651	8,325	1,465	3,727	4,237
Venezuela	348,039	375,378 (10)	545,710 (5)	1,106,751 (4)	1,296,698 (3)	1,361,075 (3)
Total, South America	815,201	934,458	1,125,854	1,400,154	1,627,524	1,833,827
CENTRAL AMERICA AND ANTILLES						
Bahamas	14,417	15,821	13,298	23,880	11,163	10,943
Barbados	14,342	14,283	17,302	7,957	5,221	5,886
Belize	2,670	1,902	1,726	1,439	1,914	628
Bermuda	22,654	18,449	11,954	492	1,554	1,518
Jamaica	52,080	43,832	38,474	18,078	14,770	55,333

18.38   Value of total exports and imports, by geographic region and country, 1975-77 (thousand dollars) (concluded)

Region and country[1]	Exports[2]			Imports[2]		
	1975r	1976	1977	1975r	1976	1977
CENTRAL AMERICA AND ANTILLES (concluded)						
Leeward and Windward Islands	15,939	12,307	20,941	862	568	596
Trinidad and Tobago	32,007	38,981	53,256	25,083	21,344	39,878
Costa Rica	11,815	17,169	14,096	18,538	24,167	25,947
Cuba	228,944	259,804	185,067	81,456	60,312	45,375
Dominican Republic	28,345	22,741	25,760	24,305	29,014	24,704
El Salvador	8,190	10,416	13,523	8,069	9,746	14,802
French West Indies	1,611	2,036	1,243	17	126	66
Guatemala	11,191	21,817	16,557	19,475	17,056	23,305
Haiti	12,585	17,778	16,000	3,582	2,309	3,407
Honduras	8,140	13,229	9,003	11,755	17,405	18,696
Mexico	222,364	215,563	219,042	95,362	146,076	194,259
Netherlands Antilles	3,678	4,371	9,309	24,458	7,087	64,936
Nicaragua	4,045	4,754	9,178	6,061	13,753	14,436
Panama	17,763	19,155	20,239	5,880	5,411	12,933
Puerto Rico	52,656	59,493	60,778	24,730	38,318	53,885
Virgin Islands of the United States	1,776	1,358	2,468	24	33	89
Total, Central America and Antilles	767,214	815,259	759,214	401,503	427,349	611,622
NORTH AMERICA						
Greenland	2,535	2,088	2,021	284	448	515
St. Pierre and Miquelon	12,387 (1)	14,117	12,209	240	124	94
United States	21,697,083 (1)	25,795,873 (1)	30,892,636 (1)	23,616,318 (1)	25,736,640 (1)	29,552,010 (1)
Total, North America	21,712,004	25,812,079	30,906,866	23,616,842	25,737,212	29,552,619
Total, all countries	33,245,469	38,146,333	44,197,433	34,690,714	37,468,819	42,068,183

In this table a dash indicates that either there was no trade or the amount was less than $500.
[1] The country classification was designed for purposes of economic geography and does not reflect the views of the Government of Canada on international issues of recognition, sovereignty or jurisdiction.
[2] Figures in parentheses indicate rank of 10 leading countries, 1975-77.

Sources
18.1 - 18.24 Merchandising and Services Division, Industry Statistics Branch, Statistics Canada.
18.25 - 18.26 Marketing and Trade Division, Policy, Planning and Evaluation Branch, Canada Department of Agriculture.
18.27 - 18.28 Public Finance Division, Institutional and Public Finance Statistics Branch, Statistics Canada.
18.29 - 18.38 External Trade Division, General Statistics Branch, Statistics Canada.

Banking, finance and insurance

Chapter 19

Tables

Banking, finance and insurance

Chapter 19

Banking

19.1

The Bank of Canada

19.1.1

Canada's central bank, the Bank of Canada, began operations on March 11, 1935, under the terms of the Bank of Canada Act, 1934, which charged it with the responsibility for regulating "credit and currency in the best interests of the economic life of the nation" and conferred on it specific powers for discharging this responsibility. Through the exercise of these powers, the Bank of Canada influences the level of short-term interest rates and thus the growth of the money supply, narrowly defined as the public's holdings of chartered bank deposits and currency. Revisions to the act were made in 1936, 1938, 1954 and 1967, and are included in RSC 1970, c.B-2.

The provisions of the Bank of Canada Act enable the central bank to determine the total amount of cash reserves available to the chartered banks as a group and in that way to influence the level of short-term interest rates. The Bank Act, which regulates the chartered banks, requires that each chartered bank maintain a stipulated minimum average amount of cash reserves, calculated as a percentage of its Canadian dollar deposit liabilities, in the form of deposits at the Bank of Canada and holdings of Bank of Canada notes. The minimum cash reserve requirement, which came into effect under the legislation beginning February 1, 1968, is 12% of demand deposits and 4% of other deposits. The ability of the chartered banks as a group to expand their total assets and deposit liabilities is therefore limited by the total amount of cash reserves available. A decrease in cash reserves tends to cause short-term interest rates to rise, making it more costly for the public to hold non-interest-bearing deposits and currency. An increase in cash reserves would put downward pressure on interest rates and indirectly induce the public to hold more money. Control of interest rates thus provides some control over the growth of the money supply.

There are two methods by which the Bank of Canada can alter the level of cash reserves of the chartered banks. The technique employed more often is the transfer of government deposits between the central bank and the chartered banks. The second method is the purchase or sale of government securities.

The transfer of government deposits from the Bank of Canada to the chartered banks or the payment by the central bank for the securities purchased in the market adds to the cash reserves of the chartered banks as a group and puts them in a position to expand their assets and deposit liabilities. The more direct method of increasing bank reserves is the transfer of government deposits to the chartered banks. Such transfers, which the bank is authorized to make as the fiscal agent of the federal government, do not involve any immediate effect on security prices and yields in financial markets.

If the Bank of Canada wishes to decrease the reserves of the chartered banks, it may either transfer government deposits from accounts at the chartered banks to the government's account at the central bank or sell government securities in the market.

In using the powers at its disposal, the Bank of Canada attempts to bring about monetary conditions appropriate to both domestic and external conditions. The basic principle it relies upon to do this is that the money supply should grow along a path capable of accommodating the maximum rate of real growth consistent with a long run goal of price stability. In November 1975 the governor of the Bank of Canada announced an explicit target range for this monetary aggregate. It was expressed as a trend rate of increase of not less than 10% but well below 15%, measured from the average level of money holdings over the three months centred on May 1975. In August 1976 the target range was lowered to 8% — 12% a year measured from the three-month average level centres on March 1976; and in October 1977 the range was again lowered to 7% — 11% measured from the average level for the month of June 1977.

Steadiness in monetary growth tends to stabilize total spending in the economy. When the trend growth in national expenditure exceeds that of money holdings, for any appreciable period, interest rates tend to rise. Consequently, firms and individuals are induced to moderate their spending. On the other hand, if expenditure is sluggish compared to the growth of money supply, there is an incentive for business and consumer spending to expand.

The emphasis on the rate of growth of the narrowly defined money supply does not mean that other indicators are being ignored by the Bank of Canada. Indeed, many other factors provide useful information on the behaviour of the economy — among them are the movements in various economic series and monetary and credit totals.

The Bank of Canada leaves the allocation of bank and other forms of credit to the private sector of the economy. Each chartered bank is free to attempt to gain as large a share as possible of the total cash reserves available by competing for deposits and to decide what proportion of its funds to invest in particular kinds of securities and in loans to particular types of borrowers. The influence of the central bank — based in essence on its power to expand or contract chartered bank cash reserves as described above — operates through financial markets affecting relative rates of return on various assets in the economy. In this impersonal and indirect fashion monetary policy has its effect on total spending in the economy.

The Bank of Canada may buy or sell securities issued or guaranteed by Canada or any province, short-term securities issued by Britain, treasury bills or other obligations of the United States and certain types of short-term commercial paper. It may buy or sell gold, silver, nickel and bronze coin, or any other coin, and gold and silver bullion as well as foreign exchange and may accept non-interest-bearing deposits from the federal government, the government of any province, any chartered bank and any bank regulated by the Quebec Savings Bank Act. The Bank of Canada may open accounts in other central banks; accept deposits from other central banks, the International Monetary Fund, the International Bank for Reconstruction and Development, and any other official international financial organization; and pay interest on such deposits. The Bank of Canada does not accept deposits from individuals nor does it compete with the chartered banks in the commercial banking field. It acts as the fiscal agent for the federal government in payment of interest and principal and generally in respect of management of the public debt of Canada. The sole right to issue paper money for circulation is vested in the bank.

The central bank also may require the chartered banks to maintain, in addition to the legal minimum cash reserve requirement, a secondary reserve which the Bank of Canada may vary within certain limits. The secondary reserve, consisting of cash reserves in excess of the minimum requirement, treasury bills and day-to-day loans to investment dealers, cannot exceed 12%; effective February 1977, the required level was 5%. In the event the Bank of Canada wishes to introduce or increase the secondary reserve requirement, one month's notice to the chartered banks is required; the amount of any increase in the requirement cannot exceed 1.0% a month. In the case of a lowering of the secondary reserve requirement, however, the percentage change in any one month is not restricted.

The Bank of Canada may make loans or advances for periods not exceeding six months to chartered banks, or to banks to which the Quebec Savings Bank Act applies, on the pledge of certain classes of securities. Loans or advances may be made under certain conditions and for limited periods to the federal government or to any province. The bank must make public at all times the minimum rate at which it is prepared to make loans or advances; this rate is known as the bank rate. From November 1, 1956 until June 24, 1962, the bank rate was established weekly at a fixed margin of one-quarter of 1.0% above the latest weekly average tender rate for 91-day treasury bills. Bank rates since October 12, 1962 have been fixed from time to time and are given in Table 19.1. The rate at May 9, 1977 was 7.50% per annum.

On May 12, 1974 the Bank of Canada announced a change in practice concerning the maximum rate at which it would enter into purchase and sale agreements with money market dealers. The practice had been to set a purchase and resale agreement rate at one-quarter of 1.0% above the average 91-day treasury bill rate at the latest weekly tender, subject to a minimum of bank rate minus three-quarters of 1.0% and a maximum at the level of bank rate. Under the new practice the maximum rate is bank rate plus one-half of 1.0%.

Assets and liabilities of the Bank of Canada at December 31, 1974-76 are shown in Table 19.2. The bank is not required to maintain gold or foreign exchange reserves against its liabilities.

Prior to the 1967 amendment to the Bank of Canada Act, there was some uncertainty about the exact relationship between the central bank and the government. Changes in the Bank of Canada Act in 1967 were designed to clarify this matter. They provide for regular consultation between the governor of the Bank of Canada and the finance minister as well as for a formal procedure whereby, in the event of a disagreement between the government and the central bank which cannot be resolved, the government may, after consultation has taken place, issue a directive to the Bank of Canada as to the monetary policy that it is to follow. Any such directive must be in writing, must be in specific terms, and must be applicable for a specified period. It must be published immediately in the *Canada Gazette* and tabled in Parliament. The amendment makes it clear that the government must take ultimate responsibility for monetary policy and it provides a mechanism for that purpose, but the central bank still is responsible for day-to-day monetary policy and its execution.

The Bank of Canada is under the management of a board of directors composed of the governor, the deputy governor and 12 directors. The governor and deputy governor are appointed for terms of seven years by the directors, with the approval of the Governor-in-Council. Directors are appointed by the finance minister, with the approval of the Governor-in-Council, for terms of three years each. The deputy minister of finance is a member of the board but does not have the right to vote. There is an executive committee of the board composed of the governor, the deputy governor, two directors and the deputy minister of finance, who is without a vote; this committee has the same powers as the board except that its decisions must be submitted to the board at its next meeting. In addition to the deputy governor who is a member of the board, there may be one or more deputy governors appointed by the board of directors to perform such duties as are assigned by the board.

The head office of the Bank of Canada is in Ottawa. It has agencies in Halifax, Saint John, Montreal, Ottawa, Toronto, Winnipeg, Regina, Calgary and Vancouver and is represented by other institutions in St. John's and Charlottetown. In addition there are representatives of head office departments in Montreal, Toronto, Edmonton and Vancouver.

The Federal Business Development Bank was established by an act of Parliament in 1974 as a federal Crown corporation to succeed the Industrial Development Bank. Under the act, which came into force in October 1975, this bank aids establishment and development of business enterprises in Canada by providing financial and management services. It supplements such services available from other sources and it gives particular attention to the needs of smaller businesses.

It extends financial help in various forms to new or existing businesses of almost every type which are unable to obtain required financing from other sources on reasonable terms and conditions. To qualify for this financing, a business should have investment by other lenders and a continuing commitment to the business with reasonable expectation of success.

The bank's management counselling service can help small businesses improve their methods. This service, supplementing counselling services available from the private sector, makes available the experience of retired business persons.

To help improve management skills in small businesses, the bank conducts management training seminars in smaller communities across Canada. It publishes booklets on a wide range of topics pertaining to the management of small business and provides information about assistance programs for small business sponsored by the federal government and others.

The head office is in Montreal and there are five regional offices and 90 branch offices across Canada. Some 98% of the loans made by the bank are approved at the branch or regional offices.

19.1.2   Currency

How bank notes became the chief circulating medium in Canada prior to 1935 is described in the *Canada Year Book 1938* pp 900-905. Features of the development which then became permanent are outlined in the *Canada Year Book 1941* pp 809-810.

When the Bank of Canada began operations in 1935 it assumed liability for Dominion notes outstanding. These were gradually replaced in public circulation and partly replaced in cash reserves by the central bank's legal tender notes. Bank of Canada notes thus replaced chartered bank notes as the issue of the latter was reduced. Further restrictions introduced by the 1944 revision of the Bank Act cancelled the right of chartered banks to issue or reissue notes after January 1, 1945, and in January 1950 the chartered banks' liability for such of their notes issued for circulation in Canada as then remained outstanding was transferred to the Bank of Canada in return for payment of a like sum to the Bank of Canada.

Bank of Canada note liabilities for the years 1974-76 are given in Table 19.4. Note circulation in public hands as at December 31, 1976 amounted to $6,572.8 million, compared to $6,078.6 million in 1975 and $5,212.8 million in 1974. Bank of Canada statistics concerning currency and chartered bank deposits are given in Table 19.5.

19.1.3   Coinage

Under the Currency and Exchange Act (RSC 1970, c.C-39), gold coins may be issued in the denomination of $20 (nine-tenths fine or millesimal fineness 900); and subsidiary coins in denominations of $1, 50 cents, 25 cents, 10 cents (five-tenths fine or millesimal fineness 500, silver, or pure nickel), five cents (pure nickel), and one cent (bronze — copper, tin and zinc). Provision is made for the temporary alteration of composition in the event of a shortage of prescribed metals.

Table 19.6 gives figures for the value of Canadian coins in circulation. Receipts of gold bullion at the Royal Canadian Mint and bullion and coinage issued are given in Table 19.7.

The Ottawa Mint, established as a branch of the Royal Mint under the United Kingdom Coinage Act of 1870, was opened on January 2, 1908. On December 1, 1931, by an act of the Canadian Parliament it became the Royal Canadian Mint and operated as a branch of the finance department. The mint was established as a Crown corporation in 1969 by the Government Organization Act of 1969 to allow for a more industrial type of organization and for flexibility in producing coins of Canada and other countries; to buy, sell, melt, assay and refine gold and precious metals; and to produce medals, plaques and other devices. The mint reports to Parliament through the minister of supply and services.

On December 16, 1971, a decision was made by the Cabinet to locate a new plant for the production of coin for general circulation in Winnipeg. The plant was officially opened on April 30, 1976.

19.1.4   Chartered banks

Canada's commercial banking system consists of 11 private banks of which eight have been in operation for many years. Three began operation recently; one in July 1968, and two more in late 1976. Another, which began operations in January 1973, amalgamated in 1977 with one of the long-established banks. At the end of December 1977, these banks operated 7,324 banking offices in Canada and 280 abroad. Canadian chartered banks accept various types of deposit from the public including accounts payable on demand, both chequing and non-chequing, notice deposits and fixed-term deposits. In addition to holding a portfolio of securities, they make loans under a wide variety of conditions for commercial, industrial, agricultural and consumer purposes. They also deal in foreign exchange, receive and pay out bank notes, provide safekeeping facilities and perform various other services. For the most part, these operations are carried out by the extensive network of bank branches. Head offices of the banks confine their activities largely to general administration and policy functions, the management of investment portfolios and related matters. A detailed account of the branch banking system in Canada is given in the *Canada Year Book 1967* pp 1126-1128.

All banks operating in Canada are chartered by Parliament under the terms of the Bank Act. The act regulates certain internal aspects of bank operations such as the auditing of accounts, the issuing of stock, the setting aside of reserves and similar matters. In addition, the Bank Act regulates their relationship with the public, the government and the Bank of Canada.

The Bank Act has been revised at approximately 10-year intervals; the most recent revision was enacted by Parliament early in 1967 and came into effect May 1 of that year. In August 1976, the federal government published a white paper dealing with the next revision of the Bank Act. Because the proposals are still being discussed, legislation to extend the present Bank Act to March 31, 1979 was introduced in the House of Commons on December 13, 1977.

Chartered bank financial statistics for recent years are given in Tables 19.8 - 19.12; month-end data are available in the *Bank of Canada Review.*

Branches of chartered banks. Although there are fewer chartered banks now than at the beginning of the century, there has been a great increase in the number of branch banking offices. As a result of amalgamations, the number of banks declined from 34 in 1901 to 10 in 1931, and remained at that figure until the incorporation of a new bank — The Mercantile Bank of Canada — in 1953 brought the total to 11. Since then the amalgamation in 1955 of the Bank of Toronto and the Dominion Bank as the Toronto-Dominion Bank, the amalgamation of Barclays Bank (Canada) with the Imperial Bank of Canada in 1956 and the amalgamation of the Canadian Bank of Commerce and the Imperial Bank of Canada as the Canadian Imperial Bank of Commerce on June 1, 1961 reduced this number to eight. The Bank of British Columbia was granted a charter by Parliament in December 1966 and commenced operations in July 1968. The Unity Bank of Canada was granted a charter in November 1972, commenced operations in 1973, but amalgamated with the Provincial Bank of Canada in June 1977. The Canadian Commercial and Industrial Bank was granted a charter in June 1976 and commenced operations in September 1976. The Northland Bank received its charter in December 1975 and began operations in November 1976. The number of branches of the 11 chartered banks in each province at various periods between 1920 and 1977 is given in Table 19.13.

Branches of individual Canadian chartered banks by province as at December 31, 1976 and 1977 are given in Table 19.14. The Canadian banks also maintain about 280 offices abroad in more than 40 countries, providing important links in facilitating trade and handling international operations of all kinds.

Cheque payments. The value of cheques cashed in 50 clearing centres during 1977 reached a high of $2,718,308 million, an increase of 10.1% above the value of $2,469,599 million for 1976. All five geographic regions showed increases, with the Atlantic provinces showing an increase of 5.3%, Quebec 9.0%, Ontario 11.2%, the Prairie provinces 10.4% and British Columbia 7.0%. Payments in the two leading centres also reached new highs, Toronto advancing 12.1% and Montreal 11.2% over 1976.

Other banking institutions   19.1.5

In addition to the savings departments of the chartered banks and of trust and loan companies, there are provincial government financial institutions in Ontario and Alberta, and the Montreal City and District Savings Bank in Quebec, established under federal legislation and reporting monthly to the finance department. Co-operative credit unions also encourage savings and extend small loans to their members.

Province of Ontario Savings Office. The establishment of the Province of Ontario Savings Office was authorized by the provincial legislature at the 1921 session and the first branches were opened in March 1922. Interest at the rate of 6.25% per annum (as at January 1, 1978), compounded half-yearly, is paid on accounts; deposits are repayable

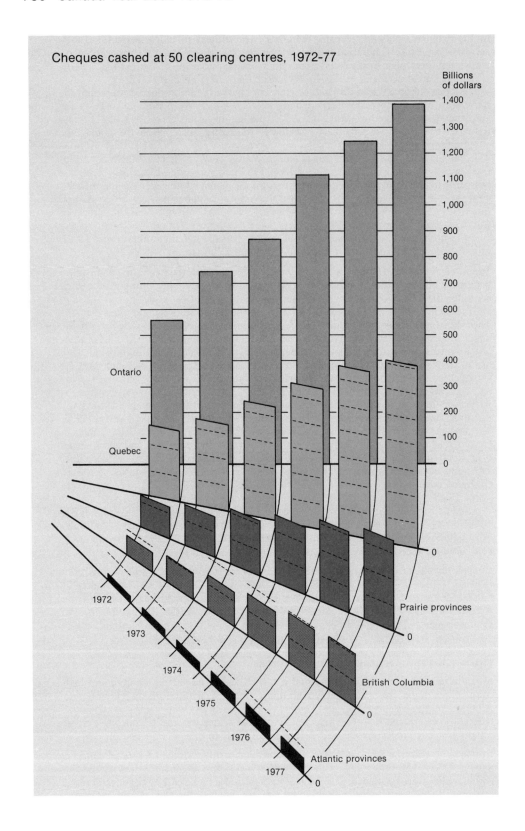

Cheques cashed at 50 clearing centres, 1972-77

on demand. Total deposits as at November 30, 1977 were $344 million and the number of depositors was approximately 81,800; 21 branches are in operation throughout the province.

Province of Alberta Treasury Branches. Established in 1938, this system operates 95 branches, one dependent branch, two sub-branches and 90 agencies throughout the province. As at March 31, 1977, deposits from customers totalled $1,052 million while loans to individuals, merchants, corporations and municipal bodies totalled $832.4 million. Profits for the year ended March 31, 1977, before allowances for reserves, were $12.8 million. Of this amount $6.1 million was transferred to the general revenue of the province. Financial services include current accounts; regular (chequing) and super (non-chequing) savings, interest-bearing accounts; capital savings, term deposits for terms ranging from one day to six years bearing competitive interest rates; agricultural loans, business loans, life-insured personal loans, mobile home financing, home improvement loans, small businessmen's loans, 20-year commercial and industrial mortgage loans, and 25-year residential mortgage loans. Treasury branches are also authorized lending agents for farm improvement loans and small business loans guaranteed by the federal government.

The Montreal City and District Savings Bank was founded in 1846 and has operated under a federal charter since 1871. On October 31, 1977 it had a paid-up capital and reserve of $34.0 million ($30.3 million in 1976), savings deposits of $1,169.8 million ($1,077.7 million) and total liabilities of $1,231.9 million ($1,123.9 million). Assets of a like amount included $262.1 million, ($258.6 million in 1976) consisting of federal, provincial, municipal and other securities.

Credit unions. The first credit union in Canada was founded in Lévis, Que., in 1900 to promote thrift by encouraging saving and to provide loans to members who could not get credit elsewhere or could get it only at high interest rates. For many years growth was slow; in 1911, when the first figures were available, assets amounted to $2 million and by 1940 they were only $25 million. However, since that time there has been a spectacular increase. The first credit union legislation was passed in Nova Scotia in 1932 followed by legislation in Manitoba and Saskatchewan in 1937 and in Ontario and British Columbia in 1938.

Credit unions are under provincial legislation. Almost all local offices in each province belong to central credit unions operating within the province. The number of chartered credit unions in Canada at the end of 1976 was 4,037. They reported a total membership of 7.7 million and assets of $15,077 million (Table 19.15). Quebec, with 4.1 million members and assets of $6,928 million, accounted for 53% of members and 46% of assets of all credit unions in Canada (Table 19.16).

Canadian credit unions in the 1960s and 1970s have continued their rapid growth. Outstanding loans of credit unions at year end increased 26.0% in 1976 over 1975 to reach $10,623 million. Assets at $15,077 million increased 22.3% and savings at $14,339 million increased 21.3% over 1975. Membership of 7.7 million represented 33.4% of the total population. Assets, liabilities and member equity in local credit unions are given in Table 19.17.

There were 20 central credit unions in 1976; these are organized as centralized banking entities to serve the needs of local credit union members, mainly by accepting deposits of surplus funds from them and providing a source of funds for them to borrow when they cannot meet the demand for local loans. Most centrals also admit co-operatives as members. Total assets of the centrals increased 21% to $3,155 million over 1975. The Credit Union National Association serves as the central organization for provincial centrals.

The centrals had combined total assets of $3,155 million at the end of 1976 compared to $2,601 million in 1975. Most funds are invested in securities and these are financed mainly by demand and term deposits from local credit union members. The combined total assets of local and central credit unions exceeded $18 billion at the end of 1976.

19.2 Other financial institutions

19.2.1 Trust and mortgage companies

Trust and mortgage companies are registered with either federal or provincial governments. They operate under the federal Loan Companies Act (RSC 1970, c.L-12) and the Trust Companies Act (RSC 1970, c.T-16), or under corresponding provincial legislation.

Trust companies operate as financial intermediaries in two areas: banking and fiduciary. Under the banking function, trust corporations can accept funds in exchange for their own credit instruments such as trust deposits and guaranteed investment certificates. This aspect of its business is often referred to as the guaranteed funds portion and differs little from the savings business of chartered banks.

Trust corporations are the only corporations in Canada with power to conduct fiduciary business. In this capacity they act as executors, trustees and administrators under wills or by appointment, as trustees under marriage or other settlements, as agents in the management of estates, as guardians of minor or incapable persons, as financial agents for municipalities and companies, as transfer agents and registrars for stock and bond issues, as trustees for bond issues and, where so appointed, as authorized trustees in bankruptcies.

Mortgage corporations may also accept deposits and may issue both short-term and long-term debentures. The investment of these funds is spelled out specifically in the acts under which most of the funds are invested in mortgages secured by real estate.

Trust and mortgage companies were established and grew rapidly under provincial legislation in the late 19th and early 20th centuries. Some companies were chartered by special acts of Parliament but it was not until 1914 that the federal government began to regulate trust and mortgage companies registered under its acts. In 1976 there were: 69 trust companies of which 20 were federally incorporated and eight federally supervised; 69 mortgage companies of which 18 came under federal jurisdiction. The federal superintendent of insurance regulates the federal companies and also, by arrangement with the provinces, trust and mortgage companies incorporated in Nova Scotia and trust companies incorporated in New Brunswick and Manitoba. Companies must be licensed by each province in which they wish to operate.

Although there may be some differences among the federal and provincial acts, broad lines of the legislation are common. In their intermediary business the companies have the power to borrow or, in the case of trust companies, to accept funds in guaranteed accounts subject to maximum permitted ratios of these funds to shareholder equity. The funds may be invested in specified assets which include: first mortgages secured by real property; government securities and the bonds and equity of corporations having established earnings records; loans on the security of such bonds and stocks; and unsecured personal loans. Trust and mortgage companies are not required to hold specified cash reserves, as are the chartered and savings banks, but there are broadly defined liquid asset requirements in a number of the acts.

In the 1920s trust and mortgage companies held about half the private mortgage business in Canada but their growth rate fell off sharply because of the effect of the depression and World War II on the mortgage business. Since then strong demand for mortgage financing has led to sustained rapid expansion.

At the end of 1976 total assets of trust companies in the Statistics Canada survey were $18,335 million compared with $14,559 million in 1975, an increase of 25%. Trust companies have been putting a high proportion of their funds into mortgages and 72% of their total assets were represented by mortgages at the end of 1976. The trust companies had $13,402 million in term deposits outstanding and $3,156 million in demand deposits at the end of 1976, accounting for 90% of total funds. About 20% of demand or savings deposits were in chequing accounts. There is considerable variety among the trust companies and a few have developed a substantial short-term business, raising funds by issuing certificates for terms as short as 30 days and also operating as lenders in the money market. But the main business of trust companies in their intermediary role is to channel savings into mortgages. In addition, trust companies, as at December 31, 1976, had $38 billion under administration in estate, trust and agency accounts. Summary statistics are given in Tables 19.18, 19.32 and 19.33.

Mortgage companies had total assets of $9,332 million at the end of 1976 compared with $8,017 million a year earlier. Their holdings of mortgages amounted to $7,548 million, or 81% of total assets. To finance their investments, these companies sold $5,525 million of term deposits and debentures and $804 million of demand deposits.

More complete and up-to-date financial information may be found in quarterly financial statements published by Statistics Canada and the Bank of Canada, the reports of the superintendent of insurance on loan and trust companies and the reports of provincial supervisory authorities.

Small loans companies 19.2.2

Small loans companies and money-lenders are subject to the Small Loans Act (RSC 1970, c.S-11). This act, first passed in 1939, sets maximum charges on personal cash loans not in excess of $1,500 and is administered by the federal department of insurance. Lenders not licensed under the act may not charge more than 1.0% a month. Those wishing to make small loans at higher rates must be licensed each year by the minister of finance under the Small Loans Act. The act allows maximum rates, including charges of every kind, of 2.0% a month on unpaid balances not exceeding $300, 1.0% a month on the portion of unpaid balances exceeding $300 but not exceeding $1,000 and one-half of 1.0% on any remainder of the balance exceeding $1,000. Loans in excess of $1,500 are not regulated and lenders operating entirely above this limit and the larger loans of licensed lenders are exempt from the act; nor does the act regulate charges for the instalment financing of sales. Prior to January 1, 1957, the act applied only to loans of $500 or less and the maximum interest charge allowed was 2.0% a month.

At the end of 1976, there were four small loans companies and 38 money-lenders licensed under the act. Small loans companies are incorporated federally; money-lenders include provincially incorporated companies. Many of the small loans companies and money-lenders are affiliated with other financial institutions, principally Canadian sales finance companies and US finance or loan companies. These affiliations reflect the close relationship between instalment financing and the consumer loan business.

Statistics Canada publishes quarterly balance sheets for sales finance and consumer loan companies as a whole and does not attempt to distinguish the two groups within the industry (see *Financial institutions*, Catalogue 61-006).

Annual figures of assets and liabilities given in Table 19.19 are from the department of insurance report. More complete data on the business of licensed lenders are given in the report on small loans companies and money-lenders, published annually by the superintendent of insurance.

Insurance 19.3

Insurance business is transacted in Canada by about 900 companies and societies. All are licensed or registered by provincial insurance authorities; at the end of 1976, 421 were also registered by the federal insurance department. Details of the classes of insurance each company or society is authorized to transact and statistical information may be found in the published reports of individual superintendents of insurance for the provinces. Financial statistics of the federally registered companies and fraternal benefit societies are published in the annual report of the federal superintendent.

Life insurance 19.3.1

Total life insurance in force in Canada at the end of 1976 amounted to $262,400 million of which about 93% was written by federally registered companies and fraternal benefit societies. Canadian companies reported an additional $56,828 million in force out of Canada at the end of 1976.

At the end of 1976, 153 companies were registered by the federal insurance department to transact life insurance (59 Canadian, 13 British and 81 foreign). There were also 42 registered fraternal benefit societies (16 Canadian and 26 foreign).

Table 19.20 gives figures since 1880 for amounts of new insurance effected and an analysis of amounts in force at the end of the year. Table 19.21 compares newly effected written business and total amounts in force for 1975 and 1976.

Net insurance premiums written in 1976 totalled $2,205 million compared to $2,004 million in 1975. Net insurance claims (death, disability and maturity) totalled $845 million in 1976 compared to $771 million in 1975. Table 19.22 gives a provincial analysis of the premium income in 1976 on a direct written basis only.

The major categories of assets and related liabilities of federally registered life insurance companies are given in Table 19.23. The major sources of income and selected expenditures are given in Table 19.24.

For registered fraternal benefit societies, certificates in force in Canada totalled $2,190 million at the end of 1976 compared to $1,959 million at the end of 1975. Premiums written in Canada totalled $37 million during 1976, of which $31 million was applicable to Canadian societies and $6 million to foreign societies. Canadian societies also reported $108 million in premiums written outside Canada.

19.3.2 Property and casualty insurance

Direct premiums written in Canada for property and casualty insurance, totalled $5,959 million in 1976, about 75% written by federally registered companies. The rest was written by other provincially licensed companies including a large number of parish, municipal, county and farmer mutuals, by Lloyd's and by provincial government insurance offices.

At the end of 1976, there were 348 companies (135 Canadian, 29 British and 184 foreign) registered by the federal insurance department to transact insurance other than life insurance.

For federally registered companies, premium income on a net basis totalled $4,386 million in 1976. See Table 19.25 (net premiums) and Table 19.26 (direct premiums) for further details.

Property insurance net premiums written in Canada during 1976 were $1,214 million. Net premiums earned in 1976 were $1,130 million and net claims were $673 million, a claims ratio of 60%. Net premiums for automobile insurance written in Canada during 1976 were $1,763 million. Net premiums earned in 1976 were $1,643 million and the net claims incurred were $1,157 million, a claims ratio of 70%.

Personal accident and sickness insurance net premiums written in Canada during 1976 were $980 million. Net premiums earned in 1976 were $962 million and net claims incurred were $752 million, a claims ratio of 78%. Net premiums for liability insurance written in Canada in 1976 were $231 million. Net premiums earned in 1976 were $218 million and net claims were $164 million, a claims ratio of 78%.

The major categories of assets and related liabilities of federally registered property and casualty insurance companies are given in Table 19.27.

Underwriting experience in Canada over the past 10 years has ranged from a loss of $54 million in 1964 to a gain of $51 million in 1967. The loss for 1976 was nearly $81 million (Table 19.28).

19.3.3 Fire losses

Fire losses in Canada reached $503.9 million in 1976, an increase of $40.1 million or 9.2% over losses reported in 1975. The total number of fires was 69,651, a decrease of 230 or 0.3% from 1975 (Tables 19.29 and 19.30). This represents an average daily loss of $1,380,532 from 190 fires. There were 829 deaths from fire in 1976, an increase of 7 or 0.8%. Of this total, 188 or 22.7% were children, a decrease of 7.3% over 1975.

19.3.4 Government insurance

19.3.4.1 Federal government insurance

Deposit insurance. The Canada Deposit Insurance Corporation was established in 1967 to provide, for persons having deposits with a member of the corporation, insurance against the loss of deposits up to a maximum of $20,000 for any one depositor.

Membership in the corporation is obligatory for chartered banks, Quebec savings banks and those federally incorporated loan and trust companies that accept deposits from the public. Provincially incorporated loan and trust companies that accept deposits from the public are eligible to apply for membership if they have the consent of the province of incorporation. The definition of deposits, set out in the general bylaw of the corporation, might be summarized as money received by a member institution that is repayable on demand or notice and money that is repayable on a fixed date not more than five years after the money is received. Deposits not payable in Canada or in Canadian currency are not insured.

Provincial government insurance 19.3.4.2

Manitoba. The Manitoba Public Insurance Corporation is a Crown corporation established under the Automobile Insurance Act. The act provides for establishment of a universal, compulsory automobile insurance plan and of other plans of automobile insurance within the province. The corporation started operations November 1, 1971. In mid-1975, the corporation began offering a wide range of non-compulsory general insurance coverages in competition with private insurance companies. Revenue for the plan comes from three sources — premiums on drivers' licences, premiums on vehicles, and a two-cents-a-gallon insurance premium on gasoline. Premiums are also based on such factors as year, make, model and use of the car, and rating territory, based on the address of the vehicle owner.

Saskatchewan. A Saskatchewan government insurance office, a Crown corporation established by the Saskatchewan Government Insurance Act, 1944, started business in May 1945. It provides all types of insurance other than sickness and life. The aim of the legislation is to provide residents with low-cost insurance designed for their needs. Rates are based on loss experience in Saskatchewan only and the surplus is invested, to the extent possible, within the province. Premium income for 1976 amounted to $43.2 million and earned deficit amounted to $722,000. The total amount made available to the Saskatchewan government finance office from 1945 to December 31, 1976 exceeded $10 million. Assets at the latter date were $120.8 million of which $64.2 million was invested in bonds and debentures issued by the province and by municipalities, hospitals and schools. Independent insurance agents, numbering 517, sell insurance throughout the province on behalf of the government insurance office.

The Automobile Accident Insurance Act, administered by the insurance office on behalf of the provincial government, provides a comprehensive automobile accident insurance plan. Premiums paid by motorists create a fund from which benefits are paid in the event of death, injury or damages sustained in automobile accidents. Any surplus over payments is used to increase benefits, reduce premiums or absorb deficits in periods of high accident frequency. The plan provides protection against loss arising out of a motorist's liability to pay for bodily injury or death of others and damage to property of others, up to a limit of $35,000, regardless of the number of claims arising from any one accident. Comprehensive coverage, including collision and upset, is also provided. From the inception of the act in 1946 to December 31, 1976, more than $420 million was paid in claims.

The insurance office, under contract with the Saskatchewan tourism and renewable resources department, offers insurance to farmers covering damage to crops by certain wildlife such as ducks, geese, sandhill cranes, deer, elk, bear and antelope.

Alberta. A variety of agencies in Alberta offer forms of prepaid protection corresponding to insurance, but the nature of the enabling legislation governing these plans emphasizes the fact that they do not constitute insurance. Because such exemptions are specifically provided by the insurance laws of the province, reference to these plans is necessary only to make it clear that they do not come within the scope of the Alberta Insurance Act. It should be noted that the Alberta Hail Insurance Act and the Alberta Crop Insurance Act are administered by the Alberta Hail and Crop Insurance Corporation and each contains a clause exempting its operations from the provisions of the Alberta Insurance Act.

Sources

19.1.1 - 19.1.2 Banking and Financial Analysis Department, Bank of Canada; Federal Business Development Bank.

19.1.3 Royal Canadian Mint.

19.1.4 Banking and Financial Analysis Department, Bank of Canada; The Canadian Bankers' Association; Business Finance Division, Business Statistics Field, Statistics Canada.

19.1.5 The Province of Ontario Savings Office; Treasury Branches of Alberta; The Montreal City and District Savings Bank; Business Finance Division, Business Statistics Field, Statistics Canada.

19.2.1 Business Finance Division, Business Statistics Field, Statistics Canada.

19.2.2 - 19.3.2 Special Services Division, Department of Insurance.

19.3.3 Dominion Fire Commissioner, Department of Public Works.

19.3.4 Canada Deposit Insurance Corporation; The Manitoba Public Insurance Corporation; Saskatchewan Government Insurance Office; Department of Consumer and Corporate Affairs, Government of Alberta.

Tables

19.1 Bank rates from Nov. 24, 1964 to May 9, 1977

Date of change	% per annum	Date of change	% per annum	Date of change	% per annum
Nov. 24, 1964	4.25	Mar. 3, 1969	7.00	Aug. 7, 1973	6.75
Dec. 6, 1965	4.75	June 11, 1969	7.50	Sept. 13, 1973	7.25
Nov. 14, 1966	5.25	July 16, 1969	8.00	Apr. 15, 1974	8.25
Jan. 30, 1967	5.00	May 12, 1970	7.50	May 13, 1974	8.75
Apr. 7, 1967	4.50	June 1, 1970	7.00	July 24, 1974	9.25
Sept. 27, 1967	5.00	Sept. 1, 1970	6.50	Nov. 18, 1974	8.75
Nov. 20, 1967	6.00	Nov. 12, 1970	6.00	Jan. 13, 1975	8.25
Jan. 22, 1968	7.00	Feb. 15, 1971	5.75	Sept. 3, 1975	9.00
Mar. 15, 1968	7.50	Feb. 24, 1971	5.25	Mar. 8, 1976	9.50
July 2, 1968	7.00	Oct. 25, 1971	4.75	Nov. 22, 1976	9.00
July 29, 1968	6.50	Apr. 9, 1973	5.25	Dec. 22, 1976	8.50
Sept. 3, 1968	6.00	May 14, 1973	5.75	Feb. 1, 1977	8.00
Dec. 18, 1968	6.50	June 11, 1973	6.25	May 9, 1977	7.50

19.2 Assets and liabilities of the Bank of Canada, as at Dec. 31, 1974-76 (million dollars)

Item	1974	1975	1976
Assets			
Foreign exchange	8.0	14.2	63.1
Advances to chartered and savings banks	8.0	—	23.0
Bills bought in open market, excluding treasury bills	139.7	44.1	104.8
Investments			
Treasury bills of Canada	1,590.3	2,081.4	2,085.6
Other securities issued or guaranteed by Canada maturing within three years	2,528.7	2,804.1	2,917.1
Other securities issued or guaranteed by Canada not maturing within three years	2,859.6	2,923.1	3,382.8
Bonds and debentures issued by Industrial Development Bank	892.0	1,029.5	858.4
Other securities	572.7	1,081.7	1,370.6
Industrial Development Bank capital stock	73.0	—	—
Bank premises	44.7	55.0	61.4
All other assets	467.4	462.9	976.4
Total, assets	9,184.1	10,495.9	11,843.2
Liabilities			
Capital paid up	5.0	5.0	5.0
Rest fund	25.0	25.0	25.0
Notes in circulation			
Held by chartered banks	1,077.6	1,204.4	1,240.2
All other	5,212.8	6,078.6	6,572.8
Deposits			
Government of Canada	16.7	26.6	32.5
Chartered banks	2,361.3	2,748.5	3,169.3
Other	101.3	63.3	123.5
Foreign currency liabilities	1.9	7.7	56.0
All other liabilities	382.5	336.7	618.8
Total, liabilities	9,184.1	10,495.9	11,843.2

19.3 Assets and liabilities of the Federal Business Development Bank, as at Dec. 31, 1976 and 1977

Item		1976r	1977
Assets			
Loans and investments[1]	$'000,000	1,372.5	1,450.3
Other assets	"	81.4	82.8
Total, assets	"	1,453.9	1,533.1
Liabilities			
Capital and reserves	$'000,000	162.2	175.8
Notes and debentures outstanding	"	1,246.4	1,306.9
Other liabilities	"	45.3	50.4
Total, liabilities	"	1,453.9	1,533.1
Accounts on books			
Amount outstanding	$'000,000	1,385.9	1,471.4
Customers on books	No.	31,541	33,093

[1]Net after allowance for doubtful accounts of $31 million in 1976 and $47 million in 1977.

19.4   Bank of Canada note liabilities, as at Dec. 31, 1974-76 (thousand dollars)

Denomination	1974	1975	1976
Bank of Canada notes			
$1	203,032	216,747	220,112
$2	143,047	161,901	166,723
$5	331,830	352,313	353,513
$10	1,057,131	1,135,514	1,142,928
$20	2,781,724	3,303,249	3,562,128
$25	46	46	46
$50	398,505	485,584	580,126
$100	1,236,165	1,451,646	1,587,884
$500	25	25	25
$1,000	126,084	163,141	186,683
Total	6,277,588	7,270,166	7,800,166
Note issues in process of retirement[1]	12,888	12,886	12,884
Total, Bank of Canada note liabilities	6,290,476	7,283,052	7,813,050
Held by:			
Chartered banks	1,077,645	1,204,448	1,240,219
Others	5,212,831	6,078,604	6,572,831

[1]Includes, in 1976, chartered banks' notes $8,132,675, Dominion of Canada notes $4,635,535, provincial notes $27,568 and defunct banks' notes $88,156; these amounts have changed little in recent years.

19.5   Canadian dollar currency and chartered bank deposits, as at Dec. 31, 1969-76 (million dollars)

Year	Currency outside banks			Chartered bank deposits				Total currency and chartered bank deposits[1]			
	Notes	Coin	Total	Personal savings deposits	Govern-ment of Canada deposits	Other deposits[1]	Total[1]	Total including govern-ment deposits	Held by general public		
									Including personal savings deposits	Excluding personal savings deposits	
1969	2,903	434	3,337	15,030	1,308	9,540	25,877r	29,214	27,906	12,876	
1970	3,106	461	3,568	16,615	1,257	10,972	28,845	32,412	31,155	14,540	
1971	3,506	488	3,993	17,783	2,239	14,572	34,594	38,587	36,348	18,565	
1972	4,056	518	4,574	19,949	2,407	16,892	39,248	43,822	41,415	21,466	
1973	4,620	589	5,209	24,604	2,361	19,220	46,186	51,395	49,034	24,429	
1974	5,213	656	5,868	29,789	4,682	21,784	56,255	62,124	57,442	27,652	
1975r	6,079	708	6,787	33,237	3,663	27,359	64,259	71,046	67,383	34,146	
1976	6,573	760	7,333	40,478	3,103	31,880	75,461	82,794	79,691	39,213	

[1]Less total float, (cheques and other items in transit).

19.6   Canadian coin[1] in circulation, as at Dec. 31, 1969-76

Year	Silver $'000	Nickel $'000	Tombac[2] $'000	Steel $'000	Bronze $'000	Total $'000	Per capita $
1969	316,715	117,199	549	3,444	43,004	480,911	22.62
1970	316,610	137,890	549	3,444	46,092	504,583	23.60
1971	317,033	159,151	549	3,443	49,297	529,473	24.42
1972	317,269	185,141	549	3,442	53,494	559,896	25.65
1973	325,981	243,246	549	3,441	58,259	631,476	28.58
1974	386,350	321,434	549	3,440	65,199	776,972	34.41
1975	444,548	382,334	549	3,440	71,490	902,361	39.58
1976	508,680	423,710	549	3,440	77,926	1,014,304	..

[1]The figures shown are of net issues of coin.
[2]Tombac, a copper-zinc alloy, was used to conserve nickel for war purposes; no coins of this metal have been issued since 1944.

19.7   Receipts of gold bullion at the Royal Canadian Mint and bullion and coinage issued, 1969-76

Year	Gold received '000 g	Gold bullion issued '000 g	Silver coin issued $'000	Nickel coin issued $'000	Bronze coin issued $'000
1969	66 779	64 975	—	41,741	3,301
1970	65 753	66 872	—	20,702	3,089
1971	62 518	62 487	556	21,277	3,207
1972	60 061	58 941	350	26,006	4,199
1973	45 909	46 126	8,804	58,128	4,768
1974	39 812	41 585	60,382	78,208	6,941
1975	33 561	32 845	58,203	60,939	6,295
1976	50 792	51 787	57,398	41,404	6,437

19.8   Statement of chartered bank assets and liabilities, as at Dec. 31, 1974-76 (thousand dollars)

Assets and liabilities	1974	1975	1976
Assets			
Gold coin and bullion	241,436	151,724	134,568
Other coin in Canada	53,867	69,291	65,681
Other coin outside Canada	1,305	1,954	1,853
Notes of and deposits with Bank of Canada	3,438,919	3,952,903	4,409,565
Government and bank notes other than Canadian	82,329	84,095	95,209
Deposits with banks in Canadian currency	562,260	499,081	615,478
Deposits with banks in currencies other than Canadian	14,885,099	15,468,113	19,330,256
Cheques and other items in transit (net)	2,639,562	2,360,024	1,355,499
Government of Canada treasury bills, at amortized value	3,702,842	3,434,189	4,141,269
Other Government of Canada issued or guaranteed securities maturing within three years, at amortized value	2,161,465	2,484,948	2,187,375
Government of Canada issued or guaranteed securities maturing after three years, at amortized value	2,199,996	1,814,666	2,259,325
Canadian provincial government issued or guaranteed securities, at amortized value	480,466	656,195	613,172
Canadian municipal and school corporation issued or guaranteed securities not exceeding market value	464,572	486,796	441,375
Other Canadian securities, not exceeding market value	2,097,336	2,236,988	2,958,093
Securities other than Canadian, not exceeding market value	637,089	506,899	524,676
Mortgages and hypothecs insured under the National Housing Act 1954	3,316,002	4,178,032	5,217,544
Day, call and short loans to investment dealers and brokers in Canadian currency, secured	1,372,160	1,373,057	1,571,359
Day, call and short loans to investment dealers and brokers in currencies other than Canadian, secured	525,897	427,429	453,740
Loans to Canadian provincial governments in Canadian currency	61,674	110,716ʳ	77,471
Loans to Canadian municipalities and school corporations in Canadian currency less provision for losses	1,455,530	1,795,280	1,924,387
Other loans in Canadian currency, less provision for losses	39,443,466	45,852,759ʳ	54,716,692
Other loans in currencies other than Canadian, less provision for losses	11,692,477	14,429,560	16,507,810
Bank premises at cost, less amounts written off	731,893	861,902	1,031,970
Securities of and loans to corporations controlled by the bank	379,207	409,109	540,527
Customers' liability under acceptances, guarantees and letters of credit, as *per contra*	4,287,685	4,645,998	5,075,809
Other assets	100,288	86,751	152,456
Total, assets	97,014,822	108,378,459	126,403,159
Liabilities			
Deposits by Government of Canada in Canadian currency	4,682,130	3,663,123	3,102,646
Deposits by Canadian provincial governments in Canadian currency	621,539	1,076,667	1,051,916
Deposits by banks in Canadian currency	924,805	1,285,360	1,108,851
Deposits by banks in currencies other than Canadian	15,196,536	16,268,270	20,750,774
Personal savings deposits payable after notice, in Canada, in Canadian currency	29,789,439	33,236,723	40,478,243
Other deposits payable after notice, in Canadian currency	11,209,857	13,357,248	17,657,561
Other deposits payable on demand, in Canadian currency	11,569,555	14,254,165	13,373,317
Other deposits in currencies other than Canadian	14,116,603	15,092,765	17,464,865
Advances from Bank of Canada, secured	8,000	—	23,000
Acceptances, guarantees and letters of credit	4,287,685	4,645,998	5,075,809
Other liabilities	554,294	682,584	713,273
Accumulated appropriations for losses	809,323	949,247	1,090,082
Debentures issued and outstanding	780,404	952,230	1,168,946
Capital paid up	354,500	379,290	410,925
Rest account	2,103,194	2,521,510	2,926,031
Undivided profits at latest financial year end	6,958	13,279	6,920
Total, liabilities	97,014,822	108,378,459	126,403,159

19.9   Canadian cash reserves, 1969-76 (million dollars)

Year	Cash reserves			Canadian dollar deposit liabilities	Average cash reserve ratio
	Bank of Canada deposits	Bank of Canada notes	Total		
1969	1,090	560	1,650	25,916	6.4
1970	1,112	587	1,699	27,066	6.3
1971	1,356	610	1,966	31,329	6.3
1972	1,615	686	2,301	36,951	6.2
1973	1,902	768	2,670	42,246	6.3
1974	2,106	888	2,993	49,814	6.0
1975	2,653	985	3,638	60,225	6.0
1976	2,991	1,071	4,063	69,642	5.8

Bank of Canada deposits are averages of the months in the year shown; the monthly levels are averages of the juridical days in that month. Bank of Canada notes and Canadian dollar deposits are also averages of the months in the year shown; the monthly levels in this case are averages of the four consecutive Wednesdays ending with the second last Wednesday in the previous month. Until June 1967 the required cash reserve ratio was 8% on both demand and notice deposits. For the next eight months the required minimum monthly average on demand deposits was increased by one-half of 1% per month and that on notice deposits was decreased by one-half of 1%. Since February 1968 the required ratios have been 12% for demand deposits and 4% for notice deposits as prescribed under the Bank Act.

19.10 Classification of chartered bank deposit liabilities payable to the public in Canada in Canadian currency, as at Apr. 30, 1976 and 1977 (number of accounts)

Deposit accounts of the public of:	1976			1977		
	Personal savings deposit accounts	Other deposit accounts of the public	Total deposit accounts of the public	Personal savings deposit accounts	Other deposit accounts of the public	Total deposit accounts of the public
Less than $100	8,503,981	154,867	11,164,305	8,742,476	154,152	11,500,079
$100 or over but less than $1,000	6,768,684	193,829	9,867,222	6,980,606	214,639	10,209,440
$1,000 or over but less than $10,000	5,832,258	186,445	7,220,516	6,631,781	215,751	8,134,007
$10,000 or over but less than $100,000	759,569	114,762	1,043,968	951,017	131,811	1,257,880
$100,000 or over	9,471	27,307	51,332	11,484	30,056	57,275
Total deposits	21,873,963	677,210	29,347,343	23,317,364	746,409	31,158,681

19.11 Classification of chartered bank loans in Canadian currency, as at Dec. 31, 1975 and 1976 (million dollars)

Class of loan	1975[r]	1976
General loans		
Personal	14,048.0	17,049.1
To individuals, fully secured by marketable bonds and stocks	829.6	834.9
Home improvement loans	43.8	37.4
To individuals, not elsewhere classified	13,174.6	16,176.7
Farmers		
Farm Improvement Loans Act	477.6	456.4
Other farm loans	2,240.0	2,883.0
Industry	8,532.7	9,791.4
Chemical and rubber products	553.2	530.0
Electrical apparatus and supplies	397.0	445.4
Foods, beverages and tobacco	1,048.3	1,151.8
Forest products	987.5	1,192.1
Furniture	144.2	189.0
Iron and steel products	1,252.5	1,549.2
Mining and mine products	993.9	975.9
Petroleum and products	1,330.2	1,782.1
Textiles, leather and clothing	607.8	644.0
Transportation equipment	462.8	450.6
Other products	755.4	881.3
Merchandisers	3,607.9	4,694.1
Construction contractors	1,513.5	2,065.5
Public utilities, transportation and communications	1,656.2	1,625.9
Other business	7,917.7	10,041.1
Religious, educational, health and welfare institutions	469.1	607.6
Total, general loans	40,462.8	49,214.2
Other loans		
Provincial governments	110.7	77.5
Municipal governments and school districts	1,795.3	1,924.4
Special	718.6	902.8
Other	369.0	492.2
Loans to finance the purchase of Canada Savings Bonds	495.2	524.5
Grain dealers and exporters	655.3	748.3
Instalment and other financial companies	743.4	428.0
Total, other loans	4,887.5	5,097.7
Total, loans in Canadian currency	45,350.3	54,311.9

19.12 Chartered bank revenues, expenses, shareholders' equity and accumulated appropriations for losses, as at Oct. 31, 1974-76 (million dollars)

Item	1974	1975	1976
For financial year ended Oct. 31			
Revenues			
Income from loans	6,807.6	7,664.4	8,605.7
Income from securities[1]	753.0	840.2	947.3
Other operating income	543.5	693.3	756.5
Total, revenues	8,104.0	9,197.9	10,309.5
Expenses			
Interest on deposits and bank debentures	5,269.8	5,519.0	6,198.4
Salaries, premiums, contributions and other staff benefits	1,148.7	1,430.5	1,699.9
Property expenses, including depreciation	314.6	377.4	452.1
Other operating expenses[2]	496.4	636.7	759.4
Total, expenses[3]	7,229.4	7,963.6	9,109.8

19.12   Chartered bank revenues, expenses, shareholders' equity and accumulated appropriations for losses, as at Oct. 31, 1974-76 (million dollars) (concluded)

Item	1974	1975	1976
Balance of revenue[3]	874.6	1,234.3	1,199.7
Less:			
Loss experience not included in other operating expenses	137.2	62.8	20.8
Appropriations for losses, net[4]	7.5	139.9	140.8
Income taxes	425.4	570.8	515.9
Leaving for dividends and shareholders' equity	304.4	460.8	522.2
Dividends	166.8	190.8	209.3
Total additions to shareholders' equity	196.4	448.0	455.1
From above operations	137.6	270.0	312.9
From issue of new shares including premiums	58.8	178.0	142.2
As at end of financial year			
Shareholders' equity			
Undivided profits	7.0	13.3	6.9
Rest account	2,060.0	2,476.5	2,922.9
Capital paid up	351.3	376.5	401.9
Total, shareholders' equity	2,418.3	2,866.3	3,331.7
Accumulated appropriations for losses	809.3	949.2	1,090.1

[1]Excludes realized profits and losses on securities held in investment account which are included in the item "Loss experience not included in other operating expenses".
[2]Includes provision for losses based on five-year average loss experience and taxes other than income taxes.
[3]Before provision for income taxes and appropriations for losses other than those included in "Other operating expenses".
[4]General and tax-paid appropriations for losses: net after any transfers out of accumulated appropriations for losses to undivided profits or rest account.

19.13   Branches[1] of chartered banks, by province, as at Dec. 31, 1920-77

Province or territory	1920	1930	1940	1950	1960	1970	1976	1977
Newfoundland	—	—	—	39	71	114	134	134
Prince Edward Island	41	28	25	23	27	30	32	32
Nova Scotia	169	138	134	144	173	202	231	229
New Brunswick	121	102	97	100	113	136	168	172
Quebec	1,150	1,183	1,083	1,164	1,427	1,524	1,569	1,598
Ontario	1,586	1,409	1,208	1,257	1,785	2,307	2,811	2,852
Manitoba	349	239	162	165	234	310	346	351
Saskatchewan	591	447	233	238	296	350	363	369
Alberta	424	304	172	246	394	521	654	684
British Columbia	242	229	192	294	514	684	844	864
Yukon Territory and Northwest Territories	3	4	5	9	17	21	37	39
Canada	4,676	4,083	3,311	3,679	5,051	6,199	7,189	7,324

[1]Figures include sub-agencies and sub-branches in Canada receiving deposits for the banks employing them.

19.14   Branches[1] of individual Canadian chartered banks, by province, as at Dec. 31, 1976 and 1977

Bank	Province or territory											
	Nfld.		PEI		NS		NB		Que.		Ont.	
	1976	1977	1976	1977	1976	1977	1976	1977	1976	1977	1976	1977
Bank of Montreal	35	35	3	3	35	34	28	27	217	215	490	489
The Bank of Nova Scotia	58	58	10	10	69	67	51	53	90	91	399	402
The Toronto-Dominion Bank	3	3	2	2	4	6	7	7	97	101	515	533
La Banque Provinciale du Canada	—	—	2	2	—	—	24	27	276	285	30	43
Canadian Imperial Bank of Commerce	18	18	9	9	37	37	26	26	208	214	756	768
The Royal Bank of Canada	20	20	6	6	85	84	31	31	224	231	576	585
Banque Canadienne Nationale	—	—	—	—	—	—	—	—	454	459	26	26
The Mercantile Bank of Canada	—	—	—	—	1	1	1	1	2	2	4	5
Bank of British Columbia	—	—	—	—	—	—	—	—	—	—	—	—
Unity Bank of Canada[2]	—	—	—	—	—	—	—	—	1	—	15	—
Canadian Commercial & Industrial Bank	—	—	—	—	—	—	—	—	—	—	—	1
Northland	—	—	—	—	—	—	—	—	—	—	—	—
Total	134	134	32	32	231	229	168	172	1,569	1,598	2,811	2,852

19.14   Branches[1] of individual Canadian chartered banks, by province, as at Dec. 31, 1976 and 1977 (concluded)

Bank	Province or territory											
	Man.		Sask.		Alta.		BC		YT and NWT		Canada	
	1976	1977	1976	1977	1976	1977	1976	1977	1976	1977	1976	1977
Bank of Montreal	71	70	63	64	126	124	170	170	6	6	1,244	1,237
The Bank of Nova Scotia	32	33	42	43	94	99	103	103	2	2	950	961
The Toronto-Dominion Bank	49	51	46	47	95	106	108	113	3	3	929	972
La Banque Provinciale du Canada	—	1	—	—	—	1	—	3	—	—	332	362
Canadian Imperial Bank of Commerce	88	89	108	109	191	198	237	241	22	22	1,700	1,731
The Royal Bank of Canada	98	99	104	104	138	143	194	203	4	6	1,480	1,512
Banque Canadienne Nationale	6	6	—	—	—	—	—	—	—	—	486	491
The Mercantile Bank of Canada	1	1	—	1	2	2	1	1	—	—	12	14
Bank of British Columbia	—	—	—	—	6	7	27	29	—	—	33	36
Unity Bank of Canada[2]	1	—	—	—	1	—	3	—	—	—	21	—
Canadian Commercial & Industrial Bank	—	—	—	—	1	2	1	1	—	—	2	4
Northland	—	1	—	1	—	2	—	—	—	—	—	4
Total	346	351	363	369	654	684	844	864	37	39	7,189	7,324

[1]Figures include sub-agencies and sub-branches in Canada for receiving deposits.
[2]Unity Bank of Canada amalgamated with La Banque Provinciale du Canada June 16, 1977.

19.15   Credit unions in Canada, 1969-76

Year	Credit unions chartered	Members	Assets $'000	Loans granted to members $'000
1969	4,769	5,002,722	4,064,065	1,525,655
1970	4,593	5,203,402	4,591,953	1,781,331
1971	4,441	5,623,994	5,587,728	1,828,888
1972	4,351	5,843,820	6,761,224	2,970,397
1973	4,256	6,382,054	8,465,786	3,765,767
1974	4,194	6,805,625	10,026,257	4,111,857
1975	4,117	7,268,552	12,331,379	4,983,082
1976	4,037	7,742,312	15,077,462	6,489,339

19.16   Summary statistics of credit unions, by province, 1975 and 1976

Year, province or territory	Credit unions chartered	Members	Assets $'000	Shares $'000	Deposits $'000	Loans granted to members $'000
1975						
Newfoundland	42	7,601	7,878	4,309	2,472	5,374
Prince Edward Island	13	18,095	16,202	5,748	5,977	9,289
Nova Scotia	126	132,595	123,666	59,445	52,366	75,086
New Brunswick	145	164,005	136,475	80,456	43,055	80,842
Quebec	1,611	3,921,406	5,745,824	566,047	4,861,294	1,770,199
Ontario	1,340	1,421,847	2,220,964	900,833	1,176,706	1,123,854
Manitoba	192	295,357	638,346	1,474	591,639	317,159
Saskatchewan	249	426,279	1,341,703	343,331	856,429	569,490
Alberta	212	284,729	543,442	113,448	376,466	357,732
British Columbia	183	595,644	1,554,269	201,992	1,221,727	672,587
Northwest Territories	4	1,444	2,610	101	2,060	1,470
Total	4,117	7,268,552	12,331,379	2,277,184	9,190,191	4,983,082
1976						
Newfoundland	36	8,565	10,302	5,418	4,040	6,951
Prince Edward Island	13	19,164	18,294	6,381	7,011	11,130
Nova Scotia	127	141,343	149,848	65,387	68,701	92,226
New Brunswick	138	178,671	176,150	98,149	61,335	111,497
Quebec	1,570	4,127,516	6,928,381	700,359	5,840,239	2,498,623
Ontario	1,329	1,502,165	2,692,371	1,026,601	1,499,736	1,303,437
Manitoba	192	317,765	782,616	1,578	724,134	372,084
Saskatchewan	247	458,284	1,550,410	358,426	1,031,665	667,767
Alberta	205	320,282	723,737	122,781	534,700	454,972
British Columbia	176	666,437	2,042,803	203,519	1,676,852	970,578
Northwest Territories	4	2,120	2,550	112	2,759	74
Total	4,037	7,742,312	15,077,462	2,588,711	11,451,172	6,489,339

19.17   Assets, liabilities and members' equity of local credit unions in Canada, 1974-76 (million dollars)

Item	1974	1975	1976	Item	1974	1975	1976
Assets				Fixed assets			
Cash and demand deposits				Land and buildings	180	204	228
On hand	168	192	207	Equipment and furniture	45	52	64
In banks	80	52	51	Stabilization fund deposits	19	28	36
In centrals	1,045	1,092	1,363	Other assets	110	152	191
Other	3	50	46	Total, assets	10,026	12,331	15,077
Investments				Liabilities			
Term deposits	563	1,158	1,305				
Government of Canada	28	33	26	Accounts payable			
Provincial governments	206	204	204	Interest	83	114	146
Municipal governments	442	411	396	Dividends	20	22	23
Shares in centrals	94	127	142	Other	38	44	49
Religious institutions				Loans payable			
Hospitals	145	194	252	Centrals	236	239	284
Other				Banks	8	9	6
Loans				Other	9	18	33
Cash loans				Deposits			
Personal	2,782	3,285	3,768	Ordinary	4,389	5,491	6,526
Farm	121	161	270	Term	2,884	3,699	4,925
Co-operatives and other enterprises	60	95	171	Other liabilities	51	59	97
Other	68	49	107				
Mortgage loans				Members' equity			
Dwellings	3,520	4,234	5,400	Share capital	1,988	2,277	2,589
Farm	171	208	247	Reserves	279	302	337
Co-operatives and other enterprises	140	320	558	Undivided earnings	41	57	62
Other	74	80	103	Total, liabilities and members' equity	10,026	12,331	15,077
Allowance for doubtful loans	38	50	58				

19.18   Revenues and expenses of trust and mortgage companies, 1974-76 (million dollars)

Item	Trust companies			Mortgage companies		
	1974	1975	1976	1974	1975	1976
Revenues						
Interest earned	1,017	1,202	1,551	552	670	821
Dividends	12	17	33	18	34	23
Fees and commissions	274	316	363	1	2	1
Other revenues	34	43	54	23	24	25
Total, revenues	1,337	1,578	2,001	594	730	870
Expenses						
Interest	864	1,004	1,300	432	511	644
Depreciation	8	9	11	2	2	3
Amortization	—	—	1	1	1	1
Income taxes	38	51	58	28	41	38
Other expenses	376	440	531	82	96	115
Total, expenses	1,286	1,504	1,901	545	651	801
Net profit	51	74	100	49	79	69

19.19   Assets and liabilities of small loans companies and money-lenders, 1975 and 1976 (thousand dollars)

Assets and liabilities	1975	1976
Assets		
Small loans balances	252,097	234,917
Balances, large loans and conditional sales agreements	1,160,500	1,117,640
Cash	15,671	8,973
Other	116,921	220,277
Total, assets	1,545,189	1,581,807
Liabilities		
Borrowed money	882,897	918,846
Unearned charges on large loans and conditional sales agreements	257,422	248,100
Reserves for losses	47,662	52,513
Paid-up capital	138,724	118,425
Surplus paid in by shareholders	21,354	42,991
Earned surplus	158,269	161,736
Other	38,861	39,196
Total, liabilities	1,545,189	1,581,807

19.20 Life insurance effected and in force in Canada by insurance companies under federal registration, 1880-1976 (million dollars)

Year	New insurance effected during year	Amounts in force Dec. 31			
		Canadian	British	Foreign	Total
1880	14	38	20	34	91
1900	68	267	39	124	431
1920	630	1,664	77	916	2,657
1940	590	4,609	146	2,221	6,975
1960	5,693	30,418	1,555	12,676	44,649
1970	12,915	76,775	5,727	28,615	111,116
1974	25,488	128,178	8,785	40,157	177,120
1975	32,526	151,974	10,476	45,629	208,079
1976	36,016	179,083	11,962	51,645	242,690

19.21 Amounts of ordinary[1] and group life insurance policies effected and in force in Canada by federally registered companies, 1975 and 1976 (million dollars)

Policies	Canadian		British		Foreign	
	1975	1976	1975	1976	1975	1976
Effected during year						
Ordinary[1]	10,749	12,783	1,436	1,748	4,620	4,983
Group	11,970	12,830	657	333	3,094	3,338
In force Dec. 31						
Ordinary[1]	61,907	69,917	7,464	8,548	22,025	24,629
Group	90,066	109,166	3,013	3,414	23,604	27,016

[1]Includes industrial policies.

19.22 Life insurance premiums (direct written), by province, 1976 (million dollars)

Province or territory	Ordinary[1]	Group	Total
Newfoundland	16	9	26
Prince Edward Island	5	2	7
Nova Scotia	49	18	67
New Brunswick	34	13	47
Quebec	444	184	628
Ontario	613	265	878
Manitoba	61	26	87
Saskatchewan	47	20	67
Alberta	114	47	160
British Columbia	143	64	207
Yukon Territory and Northwest Territories	2	1	3
Miscellaneous	18	2	20
Total	1,546	651	2,197

[1]Includes industrial policies.

19.23 Major assets and liabilities of federally registered life insurance companies as at Dec. 31, 1975 and 1976 (million dollars)

Assets and liabilities	Canadian[1]		British[2]		Foreign[2]	
	1975	1976	1975	1976	1975	1976
Assets						
Bonds	7,414	8,259	599	708	1,298	1,464
Stocks	1,820	1,991	280	286	7	8
Mortgages	8,162	8,758	517	584	1,561	1,644
Real estate	1,298	1,393	108	127	45	77
Policy loans	1,615	1,709	76	86	184	194
Other assets	920	1,017	126	136	165	161
Segregated	2,168	2,781	310	407	60	6
Total	23,397	25,908	2,016[3]	2,334[3]	3,320[3]	3,614[3]
Liabilities						
Actuarial reserves	16,580	18,122	1,510	1,675	2,635	2,775
Outstanding claims	271	272	12	12	53	57
Amounts on deposit	1,360	1,461	6	6	151	168
Other liabilities	1,719	1,893	56	56	173	175
Segregated	2,152	2,765	319	408	47	58
Total	22,082	24,513	1,903	2,157	3,059	3,233
Surplus or excess[4]	1,273	1,352	113	177	261	381
Capital stock	41	43	...	...	...	...

[1]Assets at book values, in and out of Canada (segregated funds at market values).
[2]Assets at market values, in Canada only.
[3]Includes assets under control of Chief Agent in Canada.
[4]Excess of assets over liabilities in Canada for British and foreign companies; for such companies, "capital stock" is not applicable in Canada.

19.24 Major items of income and expenditure of federally registered life insurance companies, 1976 (million dollars)

Income and expenditure	Canadian[1]	British[2]	Foreign[2]
Income			
Insurance premiums and annuity considerations	3,917	332	564
Investment income — regular funds	1,625	155	271
Net investment gain — segregated funds	293	25	2
Other items	108	14	40
Total income	5,943	526	877
Selected expenditure			
Claims incurred	1,825	133	303
Dividends to policyholders	375	37	91
Commissions and general expenses	897	70	162
Taxes, licences and fees	140	2	55

[1]Worldwide business of which $1,039 million was applicable to out-of-Canada business.
[2]Business in Canada only.

19.25 Property and casualty net premiums written and net claims incurred, by class of insurance and by incorporation of company, 1976 (million dollars)

Insurance class	Net premiums written				Net claims incurred
	Canadian	British	Foreign	Total	
Property[1]	648	83	483	1,214	673
Automobile	1,153	91	519	1,763	1,157
Liability	143	14	74	231	164
Accident and sickness	642	13	325	980	752
Other casualty[2]	82	12	65	159	73
Marine	12	6	21	39	24
Total	2,680	219	1,487	4,386	2,843

[1]Includes fire, personal property, real property, windstorm, earthquake, inland transportation, livestock, theft, forgery, plate glass.
[2]Includes hail, fidelity, surety, boiler and machinery, aircraft, credit, title, mortgage.

19.26 Property and casualty direct premiums written and claims incurred, by province and by category of company, 1976 (million dollars)

Province or territory	Premiums written				Claims incurred
	Companies federally registered	Companies provincially licensed	Lloyd's	Total	
Newfoundland	65	14	10	89	55
Prince Edward Island	20	2	—	22	12
Nova Scotia	155	1	4	160	107
New Brunswick	146	2	4	152	107
Quebec	1,244	385	98	1,727	1,106
Ontario	1,864	270	36	2,170	1,414
Manitoba	101	110	5	216	138
Saskatchewan	63	61	2	126	88
Alberta	448	69	17	534	296
British Columbia	323	396	23	742	447
Yukon Territory and Northwest Territories	19	1	1	21	14
Canada	4,448	1,311	200	5,959	3,784

19.27 Major assets and liabilities of federally registered property and casualty insurance companies, 1975 and 1976 (million dollars)

Assets and liabilities	Canadian[1]		British[2]		Foreign[2]	
	1975	1976	1975	1976	1975	1976
Assets						
Bonds	1,598	2,000	314	250	1,494	1,839
Stocks	546	719	90	73	75	76
Amounts due from agents and premiums receivable	287	362	35	39	133	133
Other	760	1,149	73	90	292	353
Total	3,191	4,230	512	452	1,994	2,401

19.27 Major assets and liabilities of federally registered property and casualty insurance companies, 1975 and 1976 (million dollars) (concluded)

Assets and liabilities	Canadian[1]		British[2]		Foreign[2]	
	1975	1976	1975	1976	1975	1976
Liabilities						
Unearned premiums	735	950	105	110	515	587
Unpaid claims	1,318	1,704	166	178	772	911
Other	440	677	21	28	191	210
Total	2,493	3,331	292	316	1,478	1,708
Surplus or excess[3]	285	573	220	136	516	693
Capital stock and amounts transferred	413	326	...	...	...	...

[1]Business in and out of Canada, investments on book value basis. Deduction, if any, for excess of market over book value in "Other" assets.
[2]Business in Canada only, investments on market value basis.
[3]Excess of assets over liabilities in Canada for British and foreign companies; for such companies, "capital stock" is not applicable in Canada.

19.28 Property and casualty insurance, underwriting results in Canada, 1976 with totals for 1971-76 (million dollars)

Registered companies	Underwriting income earned	Claims[1] incurred	Expenses incurred	Dividends to policyholders	Underwriting gain
Canadian[2]					
Property and casualty	1,918.4	1,246.7	679.2	18.1	-25.6
A and S branches[3]	571.8	475.3	92.1	40.1	-35.7
British	209.2	135.7	75.1	1.0	-2.2
Foreign					
Property and casualty	1,116.7	763.4	361.6	2.7	-11.0
A and S branches[3]	274.3	204.6	65.0	11.6	-6.8
Total, 1976	4,090.4	2,825.7	1,273.0	73.5	-81.3
1975	3,302.0	2,385.5	1,029.4	47.4	-160.3
1974	2,743.2	2,118.5	874.2	40.2	-289.7
1973	2,460.1	1,804.2	772.8	31.8	-148.7
1972	2,166.1	1,509.5	695.0	23.4	-61.8
1971	1,953.6	1,326.7	645.6	15.2	-33.9

[1]Includes adjustment expenses.
[2]Excludes transactions out of Canada.
[3]Accident and sickness branches of life insurance companies.

19.29 Fire losses[1], by type of property and cause of fire, 1975 and 1976

Type of property and reported cause of fire	1975		1976	
	Fires reported	Property loss $'000	Fires reported	Property loss $'000
Type of property				
Residential	39,750	166,795	35,565	191,406
Mercantile	4,625	83,068	2,374	59,877
Farm	5,127	31,406	1,779	27,097
Manufacturing	2,016	75,303	1,780	61,749
Institutional and assembly	2,458	42,549	2,993	77,979
Miscellaneous	15,905	64,694	25,160	85,786
Total	69,881	463,815	69,651	503,894
Reported cause				
Smokers' carelessness (including matches)	13,401	38,747	12,156	44,482
Stoves, furnaces, boilers and smoke pipes	5,435	36,521	5,624	38,703
Electrical wiring and appliances	11,079	76,055	14,396	99,668
Defective and overheated chimneys and flues	581	2,238	446	2,538
Hot ashes, coals and open fires	1,887	10,566	1,466	5,971
Petroleum and its products	2,865	12,463	2,154	14,480
Lights, other than electric	320	1,042	258	1,221
Lightning	3,007	4,454	938	6,356
Sparks on roofs	62	513	238	1,424
Exposure fires	1,713	14,751	1,743	14,644
Spontaneous ignition	587	9,316	737	1,246
Incendiarism	4,964	58,507	8,441	77,652
Miscellaneous known causes (explosions, fireworks, friction, hot grease or metal, steam or hot water pipes)	14,096	52,943	8,513	54,325
Unknown	9,884	145,699	12,541	141,185

[1]Excludes forest fires.

19.30   Fire losses[1], by province, 1973-76

Province or territory	Property loss		1975			1976		
	1973 $'000	1974 $'000	Fires reported	Property loss $'000	Loss per capita $	Fires reported	Property loss $'000	Loss per capita $
Newfoundland	7,312	12,774	685	15,835	28.42	464	7,658	21.85
Prince Edward Island	1,555	1,440	448	2,356	19.64	589	3,436	30.88
Nova Scotia	12,888	14,331	2,203	14,196	17.06	2,127	13,162	27.66
New Brunswick	9,592	15,388	1,032	19,620	28.51	1,206	15,442	22.25
Quebec	95,668	128,105	17,257	130,175	20.85	16,630	158,042	13.72
Ontario	114,772	128,899	23,913	131,552	15.79	23,109	143,102	15.74
Manitoba	15,147	21,903	5,649	22,177	21.57	7,225	28,578	17.04
Saskatchewan	6,711	7,934	2,452	21,729	23.23	2,892	13,392	28.16
Alberta	25,629	30,324	7,895	32,864	18.00	7,679	40,791	25.14
British Columbia	45,693	62,842	7,996	69,392	27.85	7,433	77,566	14.17
Yukon Territory and Northwest Territories	3,252	4,839	351	3,917	66.40	297	2,720	45.35
Canada	338,219	428,779	69,881	463,814	20.07	69,651	503,894	21.63

[1]Excludes forest fires.

19.31   Cheques cashed at 50 clearing centres, by province or region, 1976 and 1977 (million dollars)

Clearing centre	1976	1977	Clearing centre	1976	1977
ATLANTIC PROVINCES	57,869	60,948	London	27,302	29,287
			Niagara Falls	2,163	2,420
Charlottetown	2,100	2,149	Oshawa	19,007	19,635
Fredericton	5,788	6,450	Ottawa	39,130	46,651
Glace Bay	215	223	Peterborough	2,783	2,943
Halifax	20,437	22,622	St. Catharines	8,842	7,295
Moncton	5,286	3,425	Sarnia	4,169	5,442
Saint John	6,524	7,180	Sault Ste Marie	13,589	15,441
St. John's	15,787	16,931	Sudbury	3,399	3,763
Sydney	1,733	1,969	Thunder Bay	4,742	5,359
			Timmins	1,447	1,615
QUEBEC	651,094	709,854	Toronto	1,040,414	1,166,700
			Windsor	16,578	16,494
Chicoutimi	2,490	2,775			
Drummondville	1,388	1,458	PRAIRIE PROVINCES	317,917	351,085
Granby	1,706	1,868			
Montreal	548,487	609,659	Brandon	1,356	1,563
Quebec	87,697	83,741	Calgary	102,380	114,792
Saint-Hyacinthe	2,382	2,713	Edmonton	71,066	86,211
Shawinigan Falls	588	656	Lethbridge	3,155	3,453
Sherbrooke	3,455	3,780	Medicine Hat	1,413	1,486
Trois-Rivières	2,058	2,237	Moose Jaw	1,053	1,135
Valleyfield	843	968	Prince Albert	1,146	1,304
			Regina	25,442	27,277
ONTARIO	1,248,446	1,388,568	Saskatoon	7,123	8,408
			Winnipeg	103,784	105,457
Brantford	4,364	4,554			
Chatham	7,541	6,647	BRITISH COLUMBIA	194,272	207,853
Cornwall	1,791	1,857			
Guelph	3,203	3,315	Vancouver[1]	161,426	172,008
Hamilton	34,499	35,895	Victoria	32,846	35,845
Kingston	3,590	3,600			
Kitchener	9,893	9,655	Total	2,469,599	2,718,308

[1]Includes New Westminster.

19.32 Assets, liabilities and shareholders' equity of trust companies (company and guaranteed funds), 1974-76 (million dollars)

Item	1974	1975	1976	Item	1974	1975	1976
Assets				Liabilities			
Demand deposits, incl. cash and foreign currency	155	163	248	Demand and savings deposits			
Investments				Chequing	492	603	629
Investments in Canadian securities				Non-chequing	1,712	2,211	2,527
Federal	381	362	395				
Provincial	302	358	380	Term deposits			
Municipal	122	104	119	Under 1 year	1,695	1,351	1,468
Sales finance and commercial paper	318	250	318	1 to 5 years	7,421	9,037	11,786
Term deposits with chartered banks	658	948	1,347	Over 5 years	63	105	148
Other institutions	46	36	41				
Corporation bonds and debentures	436	467	428	Bank loans	23	21	33
Collateral loans	266	267	462				
				Short-term loans and notes payable	..	..	—
Mortgages							
Loans under NHA	1,582	1,717	1,942	Debts owing parent and affiliated companies	18	19	103
Conventional mortgage loans	7,264	8,824	11,227				
Investments in Canadian preferred and common shares	227	280	339	Interest, dividends, taxes and other payables	365	455	696
Investments in foreign securities	24	32	30				
Investments in subsidiary and affiliated companies	82	100	233	Shareholders' equity			
Interest, rents and other receivables	166	180	223	Capital paid up	308	366	474
Real estate and equipment	106	117	179	Investment reserves	35	38	23
Other assets	307	354	424	Reserve fund	212	226	198
				Retained earnings	98	127	250
Total, assets	12,442	14,559	18,335	Total, liabilities and shareholders' equity	12,442	14,559	18,335

19.33 Assets, liabilities and shareholders' equity of mortgage companies, 1974-76 (million dollars)

Item	1974	1975	1976	Item	1974	1975	1976
Assets				Liabilities			
Demand deposits, incl. cash and foreign currency	34	29	72	Demand and savings deposits			
Investments				Chequing	166	191	184
Investments in Canadian securities				Non-chequing	494	581	621
Federal	87	99	124	Term deposits			
Provincial	56	57	42	Under 1 year	188	156	223
Municipal	4	3	2	1 to 5 years	3,453	4,284	4,741
Sales finance and commercial paper	29	40	70	Over 5 years	492	535	561
Term deposits with chartered banks	212	226	284	Bank loans	65	83	51
Other institutions	3	2	4	Short-term loans and notes payable	360	383	474
Corporation bonds and debentures	68	65	91	Debts owing parent and affiliated companies	135	215	437
Collateral loans	35	42	85				
				Interest, dividends, taxes and other payables	241r	299r	356
Mortgages				Long-term debentures	562	590	929
Loans under NHA	688	767	914				
Conventional mortgage loans	4,821	5,793	6,634	Shareholders' equity			
Investments in Canadian preferred and common shares	112	141	151	Capital paid up	319	387	442
Investments in foreign securities	5	6	8	Investment reserves	21	26	32
Investments in subsidiary and affiliated companies	346	416	469	Reserve fund	113	129	126
Interest, rents and other receivables	81	88	115	Retained earnings	134	158	155
Real estate and equipment	62	60	61				
Other assets	100	183	206	Total, liabilities and shareholders' equity	6,743	8,017	9,332
Total, assets	6,743	8,017	9,332				

Sources

19.1 - 19.2, 19.4 - 19.5, 19.8 - 19.12 Bank of Canada.

19.3 Federal Business Development Bank.

19.6 - 19.7 Royal Canadian Mint.

19.13 - 19.14 The Canadian Bankers' Association.

19.15 - 19.18, 19.31 - 19.33 Financial Institutions Section, Business Finance Division, Business Statistics Field, Statistics Canada.

19.19 - 19.28 Special Services Division, Department of Insurance.

19.29 - 19.30 Dominion Fire Commissioner, Department of Public Works.

Government finance

Chapter 20

Tables

Government finance   Chapter 20

Consolidated finance statistics   20.1

Data on each level of government constitute the basis of the intergovernment consolidation which is presented for the years 1970 to 1976 in Table 20.1. The consolidation process integrates the separate levels of government to reveal the fiscal framework of the public sector viewed as an economic unit. As a result, the numerous intergovernmental transactions either as revenue or as expenditure are eliminated in order to obtain a measure of the collective impact of all government transactions upon the rest of the economy, in terms of services provided and taxes collected.

Federal government finance   20.2

General accounts   20.2.1

Tables 20.2 - 20.5 and 20.20 present financial statistics of the federal government prepared in accordance with the concepts published in *The Canadian system of government financial management statistics.* Financial statistics in Tables 20.7, 20.8 and 20.14 are extracted directly from the *Public Accounts of Canada.*

Tables 20.2 and 20.3 give details of gross general revenue and expenditure for the years ended March 31, 1974 to 1976. Revenue increased from $25,102 million to $34,703 million while expenditures rose from $24,277 million to $36,845 million.

Transfers from the federal government to provincial governments, territories and local governments for the year ended March 31, 1976 are shown in Table 20.20. Comparable figures for the two previous years are available in the *Canada Year Book 1976-77* pp 977-981.

Table 20.4 provides details of the assets and liabilities of the federal government as at March 31, 1974 to 1976. Table 20.5 analyzes gross bonded debt according to average interest rate, average term of issue and place of payment as at March 31, 1974 to 1976.

In addition to direct gross bonded debt, the federal government has assumed certain contingent liabilities. The major categories of this indirect or contingent debt are the guarantee of insured loans under the National Housing Act and the guaranteed bonds and debentures of Canadian National Railways. The remainder consists chiefly of guarantees of loans made by chartered banks to the Canadian Wheat Board, to farmers and to university students and of guarantees under the Export Development Act. Table 20.6 provides details of the contingent liabilities of the government as at March 31, 1976 and 1977.

Table 20.7 summarizes the public debt position during the period 1972-77 as to interest and amount outstanding. Details of unmatured debt and treasury bills outstanding and information on new security issues of the federal government may be found in the *Public Accounts of Canada.* They are summarized by standard classification in Statistics Canada publication *Federal government finance* (Catalogue 68-211).

Individual and corporation taxes   20.2.2

Statistics of income tax collections are gathered at the time the payments are made and are therefore up to date. Over 75% of individual taxpayers are wage- or salary-earners who have almost the whole of their tax liability deducted at the source by their employers. All other taxpayers are required to pay most of their estimated tax during the taxation year. Thus, the greater part of the tax is collected during the same year in which the related income is earned and only a limited residue remains to be collected when returns are filed. The collections for a given fiscal year include employer remittances of tax deductions, Canada Pension Plan contributions, unemployment insurance premiums and instalments, embracing portions of two or more taxation years, and year-end payments; they cannot therefore be closely related to the statistics for a given taxation year. As little information about a taxpayer is received when the payment is made and as a single cheque from one employer may frequently cover the tax payment of hundreds of employees, the payments cannot be statistically related to taxpayers by occupation or income. Descriptive classifications of taxpayers are available only from tax returns, but collection statistics, if interpreted with the current tax structure and the above factors in mind, indicate the trend of income in advance of final compilation of statistics. The statistics given in Table 20.8 pertain to revenue collections for fiscal years ended March 31, 1973-77.

Individual income tax statistics collected by the national revenue department are presented in Tables 20.9 - 20.11 on a calendar-year basis and are compiled from a sample of all returns received. Taxpayers and amounts of income and tax are shown for selected cities and by occupational class and income classes.

Statistics on the taxation of corporate income showing a reconciliation of income taxes to taxable income and book profits are published on an industry basis by Statistics Canada. Data for 1974 and 1975 are summarized for nine industrial divisions in Table 20.13. Income data are also available on a provincial basis, as shown in Table 20.12 for years 1971 to 1975.

20.2.3 Excise taxes

Excise taxes collected by the customs and excise branch are given for the years ended March 31, 1975 to 1977 in Table 20.14.

Gross excise duties collected for the year ended March 31, 1977 were: spirits $366 million; beer or malt liquor $184 million; tobacco, cigarettes and cigars $317 million, for a total of $867 million. A drawback of 99% of the duty may be granted when domestic spirits, testing not less than 50% over proof, are delivered in limited quantities for medicinal or research purposes to universities, scientific or research laboratories, public hospitals or health institutions in receipt of federal and provincial government aid.

20.3 Federal-provincial fiscal relations

Fiscal relations between the federal, provincial and territorial governments are governed either by an act of Parliament or by formal agreements. The British North America Act, 1867, the Public Utilities Income Tax Transfer Act, and the Federal-Provincial Fiscal Arrangements and Established Programs Financing Act, 1977 are the legislative measures under which fiscal transfers are paid by the federal government to the provinces. Payments under each of these acts are summarily described in this section.

20.3.1 British North America Act

Under this act, which forms the written constitution of the country, the federal government pays to the provinces statutory subsidies consisting of contributions toward the support of provincial governments. These include an allowance per head of population, allowances for interest on debt and other special amounts which were agreed upon under the arrangements subsequent to the enactment of the constitution in 1867. Total federal payments under this act amounted to $33.8 million in the fiscal year ended on March 31, 1977.

20.3.2 Public Utilities Income Tax Transfer Act

Pursuant to the Public Utilities Income Tax Transfer Act, the federal government turns over to the provinces 95% of the tax it collects from investor-owned public utilities on income which is attributable to the generation or distribution to the public of electrical energy, gas and steam. Payments amounted to $38.7 million in the fiscal year ended March 31, 1977.

20.3.3 Federal-provincial fiscal arrangements

Federal-provincial fiscal arrangements originated at the end of World War II. The first agreements were implemented for the years 1947 to 1952, pursuant to the Dominion-Provincial Tax Rental Agreement Act. The 1947 agreements started the series of five-year federal-provincial arrangements, each one modifying and broadening the terms and content of the preceding one. For instance, with the adoption in 1957 of the tax sharing arrangements, replacing the tax rental agreements in force since 1947, the federal government initiated an income tax abatement system in favour of the provinces. The 1957 formula, however, was modified by the 1962 agreements so that the provinces could establish their own income tax rates which could be higher or lower than the federal abatement. Further, the federal government offered to collect, together with its own income tax, any income tax that provinces levied.

In 1967 the equalization program was established pursuant to the enactment of the Federal-Provincial Fiscal Arrangements Act, 1967. New acts were passed in 1972 and 1977. These revisions did not modify the basic philosophy of redistributing part of the nation's wealth among the provinces. From its general revenue, the federal government compensates any province whose per capita revenue is below the national average because of a relative deficiency in the province's tax base. Thus, equalization payments are intended to ensure that all citizens are provided with comparable standards of public services throughout the country. The Federal-Provincial Fiscal Arrangements and Established Programs Financing Act, 1977 added provisions concerning the financing of established shared-cost programs, set out later in this section.

Fiscal equalization payments. According to the formula, known as the representative tax system, provincial revenue subject to equalization is divided into 29 revenue sources in the 1977 act, compared with 20 in the 1972 act, to reflect more accurately what provinces are taxing. For each revenue source, an economic revenue base is established uniformly for all provinces. To determine the amount of equalization to which a province is entitled, population of the province as a proportion of the population of all provinces and revenue base of the province as a proportion of that of all provinces are calculated for each of the 29 revenue sources. Where the former proportion is higher than the latter for any of the revenue sources, the province is said to have a fiscal capacity deficiency for that revenue source; if these proportions are reversed, the province is said to have a fiscal capacity excess. The total revenue of all provinces for each revenue source is multiplied by each province's respective fiscal capacity related to the appropriate revenue source and for any province the amount of equalization payable is the sum total of the deficiency products less the total of the excess products.

Since the beginning of this program in 1967, seven provinces have received equalization payments: Newfoundland, Prince Edward Island, Nova Scotia, New Brunswick, Quebec, Manitoba and Saskatchewan. Payments amounted to $1,921.6 million in the fiscal year ended March 31, 1976 compared with $549.6 million in the fiscal year 1967-68 when the program was initiated.

Tax collection agreements. Pursuant to the Federal-Provincial Fiscal Arrangements Act, 1962, the federal government undertook to collect for the provinces, with its own income tax, provincial personal and corporation income taxes provided that provinces' tax systems were uniform with the federal system. All provinces except Quebec signed the agreements for personal income tax, and all provinces except Quebec and Ontario for corporation income tax. This collection is made at no cost to the provinces except for a small fee for administration of special tax rebates implemented by some provinces.

The federal tax abatement system, introduced in 1957, was abandoned in 1972 and the federal rates of personal income tax were adjusted downward to take into account the previous abatements and modifications to the structure of the federal tax system. A new scale was established according to which it was estimated that a provincial tax rate of 30.5% of the new basic federal tax would produce the same revenue as did the 28% abatement granted under the 1967 arrangements.

Due to modifications brought by the 1977 act for financing established shared-cost programs, the enlarged personal income tax field available to the provinces would be equivalent to about 44% of basic federal tax. Provincial governments, however, are free to specify rates above or below 44% and so determine the impact of their income taxes. Section 20.6.2 gives provincial rates in 1977.

Provincial personal income tax revenue guarantee payments. The formula according to which the federal government guaranteed that the provinces would not suffer a loss of personal and corporation income tax revenue entailed by the 1971 revision of the Income Tax Act was completely modified by the 1977 act.

First, provincial revenue from corporation income tax was no longer subject to revenue guarantee payments. Second, the guarantee of provincial revenue from personal income tax was to be calculated for a given year in the five-year period April 1, 1977 to March 31, 1982, in relation to the immediately preceding year. For provinces which express their rates of personal income tax as a percentage of basic federal tax, the federal government would compensate any revenue losses they might incur as a result of policy changes which reduced the federal basic tax. However, such losses would have to exceed 1.0% of federal basic tax within a province before a guarantee payment could be made to that province. For provinces with their own personal income tax system — Quebec only in 1977 — a guarantee payment would be made if they made changes in their tax system similar to the federal changes during the same year.

Established programs financing. The 1977 act sets out provisions for financing established shared-cost programs, namely those for post-secondary education, hospital insurance, medical care and extended health care services. Through this new act all open-ended cost-sharing arrangements in the health care fields were terminated and cost-sharing provisions of the Hospital Insurance and Diagnostic Services Act and the Medical Care Act were replaced by new financing provisions.

The federal contributions under the new financing system take the form of a share of the field of income taxes, occupied so far by the federal government, and of cash payments. In the field of income taxes, the share of federal taxes transferred to the provinces is equal to 13.5% of basic federal tax and a 1.0% tax on corporation taxable income. These percentages include the former transfers of 4.357% of personal income tax and the 1.0% tax on corporation taxable income which were associated with the post-secondary education cost-sharing program. Therefore, the net additional tax transfer in favour of the provinces corresponds to 9.143% of the former basic federal tax. From 1977 onward, the latter was reduced to take into account such a transfer. As a result, there would be no increase in income tax to be paid by taxpayers if the provinces were to raise their rates to offset precisely the federal reduction.

In the case of Quebec, the tax change required consists only in a reduction of the special federal tax abatement granted to the residents of the province, from 24% to 16.5%, in order to express it in relation to the reduced basic federal tax with no loss to Quebec taxpayers. This special abatement is tantamount to the province's contracting out all the above-named shared-cost programs in 1964.

Cash payments are in four forms, as follows: (1) A basic per capita cash contribution equal to the amount obtained from the multiplication of the population of each province by an amount equivalent to 50% of the national average per capita contribution to the above-mentioned shared-cost programs in the base year, 1975-76, adjusted annually according to the rate of growth of the Canadian economy. (2) Transition payments to compensate for variations in the value of the tax transfers among the provinces to ensure that this value is at least equal to basic cash contributions. (3) Levelling adjustments to facilitate the transition to the new arrangements and to achieve equal per capita payments among provinces over a five-year period. Provinces below the national average will receive additional grants so as to reach this average in three years; provinces above the national average will be reduced to that average in five years. (4) A cash payment of $20 per capita, adjusted annually to take account of variations in the gross national product, in respect of some health care services formerly included in part in the Canada Assistance Plan, such as nursing home and adult residential care services. Other services are also included, namely intermediate care, converted mental hospitals, home care and ambulatory health care.

Alternative payments for standing programs. In 1964, the provinces were given an option to assume full financial and administrative responsibility for certain federal-provincial shared-cost programs in return for fiscal compensation. To this end, the Established Programs (Interim Arrangements) Act was enacted in April 1965; it was repealed through the 1977 act. Quebec alone took advantage of this legislation and contracted out all major shared-cost programs. Several amendments were made to the act between 1965 and 1972. As a result, the tax abatement granted to Quebec taxpayers, in respect of contracting out, was at the end of 1976 as follows: hospital insurance program 16%; special welfare program 5%; and youth allowances program 3%. However, the latter abatement has been fully recovered from Quebec since 1973 when federal youth allowances started to be paid to Quebec residents.

The new arrangements for federal income tax abatement in favour of the provinces required new calculations of the special abatement to Quebec related to contracting out. These calculations take into account the additional fiscal transfer of 13.5% granted to all provinces and the accompanying reduction in the basic federal tax. As a result, the revised tax abatement granted to Quebec taxpayers is 16.5% of the reduced federal basic tax commencing with the 1977 taxation year. This 16.5% abatement corresponds, in dollar value, to the former 24% abatement.

Provincial taxes and fees. According to the British North America Act, a government cannot levy taxes on another government. However, due to the growing complexities of the economic and commercial transactions of governments, the constitutional provisions for intergovernmental taxation have become increasingly difficult to observe, particularly when government purchases are made through suppliers in the private sector such as retailers and building contractors.

To remove, or at least minimize, the uncertainties and difficulties surrounding the paying of consumption taxes among governments, a set of indexes based on criteria applied to various types of expenditure has been devised and is incorporated in the 1977 federal-provincial fiscal arrangements. Under this act the federal government could enter into reciprocal taxation agreements with the provincial governments as of October 1977. Such agreements would run until March 31, 1981, with provisions for renewal. The terms of these agreements also apply to purchases by Crown corporations listed in parts of the Financial Administration Act and the Federal-Provincial Fiscal Arrangements and Established Programs Financing Act, 1977. As of February 1977, six provinces had agreed to enter into these reciprocal taxation agreements: Newfoundland, Prince Edward Island, Nova Scotia, New Brunswick, Quebec and Ontario.

With minor exceptions, consumption taxes levied by a level of government would be paid by the other level. Where federal taxes paid by one province exceeded that province's taxes paid by the federal government, the latter would compensate the difference through a grant in lieu of such taxes.

Provincial government finance   20.4

Because of variation from province to province in administrative structure and, to a lesser extent, in accounting and reporting practices, adjustments are made to financial data reported in public accounts to produce statistics comparable between different provinces and with those for the other levels of government. In 1972 the concepts and classifications of the national system of government financial statistics were redefined by Statistics Canada (see *The Canadian system of government financial management statistics*, Catalogue 68-506). Financial statistics for the years 1971 onward are compiled in accordance with these revisions and are not comparable with data for prior years published in earlier editions of the *Canada Year Book*.

Gross general revenue and expenditure for the year ended March 31, 1975 are given in Table 20.22, liabilities in Table 20.15, and liabilities of other governments and entities guaranteed by provincial and territorial governments in Table 20.16. More information on outstanding provincial bonds and debentures is in Table 20.17.

Local government finance   20.5

Local government taxation. In 1974, the latest year for which complete data are available, local government revenues from taxation rose 11.4% to $4,730 million, and the rate of collections declined to 98.77% from 99.99%. Taxes receivable expressed as a proportion of taxation revenue remained steady at 10.4%. Lower percentages of taxes receivable relative to taxation revenue were recorded in Newfoundland, Prince Edward Island, Nova Scotia, Quebec and the territories. Rates of collection increased slightly in 1974 compared with 1973 in Newfoundland, Nova Scotia, Quebec and the territories.

Local government revenue, expenditure and debt. General revenue of local government in 1974 increased 17.0% to $12,287 million over 1973, while general expenditure at $13,307 million showed an increase of 18.3%. Debenture and other long-term debt amounted to $11,913 million as at December 31, 1974 compared with $10,476 million at December 31, 1973. Details for revenue and expenditure are given in Tables 20.23 - 20.24. Preliminary data are also given for 1975 and 1976. Table 20.25 gives the direct debt of local government for 1974 and 1975.

20.6 Tax rates

Taxes are imposed in Canada by the three levels of government. The federal government has the right to raise money by any mode or system of taxation while provincial legislatures are restricted to direct taxation within the province. Municipalities derive their incorporation with its associated powers, fiscal and otherwise, provincially and are thus also limited to direct taxation.

A direct tax is generally recognized as one demanded from the very person who it is intended or desired should pay it. This concept has limited the provincial governments to the imposition of income tax, retail sales tax, succession duties and an assortment of other direct levies. In turn, municipalities acting under provincial legislation tax real estate, water consumption and places of business. The federal government levies taxes on income, excise taxes, excise and customs duties, and a sales tax.

Since 1941 a series of federal-provincial tax agreements has been concluded to promote the orderly imposition of direct taxes. The duration of each agreement was normally five years. Under earlier agreements, the participating provinces undertook — in return for compensation — not to use, or permit their municipalities to use, certain of the direct taxes. These were replaced by arrangements under which the federal personal and corporation income tax otherwise payable in all provinces and the estate tax otherwise payable in three provinces were abated by certain percentages to make room for provincial levies.

Federal tax amendments which became effective for the most part from 1972, included a new personal income tax rate structure which was not designed to be abated in the previous way. At the same time the federal estate tax was terminated. As a result, the arrangement under which federal taxes are abated has general application only for the corporation income tax. All provinces impose taxes on the income of individuals and corporations but only two provinces, Ontario and Quebec, impose taxes on property passing at death. The federal government has tax collection agreements under which it collects provincial personal income taxes for all provinces except Quebec and provincial corporation income taxes for all provinces except Ontario and Quebec. The provinces which impose succession duties also collect them.

20.6.1 Federal taxes

Individual income tax. The federal government has adopted a tax system in which taxpayers volunteer the facts about their incomes and calculate the taxes they must pay. Every individual who is resident in Canada is liable for the payment of income tax on all his income. A non-resident is liable for tax only on income from sources in Canada. Residence is the place where a person resides or where he maintains a dwelling ready at all times for his use. There are also statutory extensions of the meaning of resident to include a person who has been in Canada for an aggregate period of 183 days in a taxation year, a person who was during the year a member of the armed forces of Canada, an officer or servant of Canada or of any one of its provinces, or the spouse or dependent child of any such person. The extended meaning of resident also includes employees who go from Canada to work under certain international development assistance programs.

Canadian tax law uses the concepts of income and taxable income. Income means income from all sources inside or outside Canada and includes income for the year from businesses, property, offices and employments. Since January 1, 1972, it has also included half of any capital gains.

In computing income, an individual must include benefits from employment, fees, commissions, dividends, annuities, pension benefits, interest, alimony and maintenance payments. Also included are unemployment insurance benefits, family allowance payments, scholarships in excess of $500, benefits under a disability insurance plan to which his employer contributes and other miscellaneous items of income. A number of items are expressly excluded from income, including certain war service disability pensions, social assistance payments, compensation for an injury or death under provincial worker compensation acts and family income security payments.

Half of capital gains is included in income. Taxable capital gains are determined by deducting capital losses from capital gains and dividing by two. In the event that losses exceed capital gains, $2,000 of allowable capital losses may be deducted from other income. Capital gains or losses relate to disposition of property. Other gains or losses, for example, resulting from a lottery or gambling, are not included. The sale of personal property at a price not exceeding $1,000 and the sale of a home do not give rise to a capital gain or loss.

Certain amounts are deductible in computing income. These include contributions to a registered employees pension plan, premiums to a registered retirement savings plan, premiums under the unemployment insurance program, alimony payments and union dues. A taxpayer 18 years of age or over who does not own a house or whose spouse does not own one may deduct contributions up to $1,000 a year to a lifetime maximum of $10,000 to a registered home-ownership savings plan. The proceeds of such plans will be taxable when they are paid to the taxpayer unless they are applied by him to the purchase of a home. An employee may deduct 3% of salary or wages up to a maximum of $250 a year to cover expenses of earning his income. No receipts or details of actual expenditures are necessary to claim this deduction. Expenses of meals and lodging while away from home are deductible by employees who have to travel as they perform their work, such as employees who work on trains or who drive trucks. When a mother has her children cared for in order that she may work, she may deduct this expense subject to certain limitations. Expenses of moving to a new work location are deductible from income earned in the new location. Students attending universities, colleges or certain other certified educational institutions in Canada may deduct their tuition fees.

An individual carrying on a business may deduct business expenses. These include wages, rents, depreciation (called capital cost allowances), municipal taxes, interest on borrowed money, reserves for doubtful debts, contributions to pension plans or profit-sharing plans for his employees, and bad debts.

The amount of the guaranteed income supplement, which is a payment made to individuals over age 65 who have little or no income in addition to their old age pension, is also deductible. Individuals who have incurred business losses in other years may deduct these in computing taxable income. A taxpayer is able to deduct up to $1,000 of Canadian investment income from interest, dividends or capital gains. In addition, a taxpayer who is 65 or over is able to deduct up to $1,000 of his pension income including amounts he receives from pension plans and from annuities under registered retirement savings plans and deferred profit sharing plans. A taxpayer under 65 may deduct up to $1,000 of qualified pension income. This includes amounts received from a pension plan or as a consequence of the death of a spouse.

Having computed income, an individual calculates taxable income by subtracting certain exemptions and deductions. Before 1974 the levels of exemptions and deductions were fixed from time to time by Parliament. The introduction in the 1974 taxation year of a mechanism for indexing personal income tax results in automatic adjustments each year to reflect the inflation rate in the levels of exemptions and deductions. The adjusted personal exemptions and deductions for each year are based on such factors as married or single status, dependent children, other dependents, charitable donations, medical expenses, income of a spouse or children, age (if 65 or over) and certain disabilities. Details are provided in an annual tax guide which is sent to each taxpayer; copies are also available in post offices and district taxation offices.

The amount of tax is determined by applying a schedule of progressive rates to taxable income. The tax bracket limits are adjusted yearly by means of the indexing mechanism. Thus taxpayers are prevented from being pushed into higher marginal tax brackets in the absence of real growth in their income. The schedule of rates for the 1978 taxation year (as of January 1978) starts at 6% on the first $761 of taxable income and increases to 43% on taxable income in excess of $91,260. These rates were reduced in 1977 as part of the revised federal-provincial fiscal arrangements. The new arrangements contained a transfer of tax room to the provinces whereby federal tax rates were reduced and provincial tax rates were increased. The net effect of this transfer leaves the combined federal and provincial tax burden on individuals unchanged.

After all calculations are made, there is deducted from the tax otherwise payable an amount called the federal tax credit. In 1978, this is equal to the greater of $300 or 9% of tax payable to a maximum of $500. In addition, there is a child credit of up to $50 for each dependent child under age 18. The overall maximum for the tax credit and child credit is $500.

Individuals who reside in the Yukon Territory or the Northwest Territories or who reside outside Canada but are deemed to be residents in Canada for tax purposes (such as diplomats and others posted outside the country) must pay an additional tax (43% in 1978) of their tax otherwise payable. This tax is intended to correspond in an approximate way to the income tax imposed by the provinces on their residents.

To a large extent, individual income tax is payable as income is earned. Taxpayers on salary or wages have tax deducted from pay by the employer and in this way pay nearly 100% of their tax liability during the calendar year. The balance of the tax, if any, is payable at the time of filing the tax return. The deadline for individual income tax is April 30 for income of the previous calendar year. Individuals with more than 25% of income in a form not subject to tax deductions at source must pay tax by quarterly instalments. Returns of these individuals must be filed on or before April 30 of the following calendar year. Farmers and fishermen pay two-thirds of tax on or before December 31 each year and the remainder on or before April 30 of the following year. Table 20.18 shows the amount of personal income tax payable on various levels of income in 1978.

Canadian employers are required to deduct and remit to the government income tax from the amounts paid to their employees as wages and salaries. The government provides employers with deduction tables to guide them in calculating the amount of federal and provincial income taxes, Canada Pension Plan contributions and unemployment insurance premiums to be withheld.

Corporation income tax. The Income Tax Act levies a tax upon income from anywhere in the world of corporations resident in Canada and upon the income attributable to operations in Canada of non-resident corporations carrying on business in Canada. Half of capital gains must be included in income. In computing income, corporations may deduct operating expenses including municipal real estate taxes, reserves for doubtful debts, bad debts and interest on borrowed money.

Corporations may deduct over a period of years the capital cost of all depreciable property. The normal capital cost allowances are computed each year on the diminishing balance principle. Regulations established a number of classes of property and maximum rates. Typical rates include 5% for buildings, 20% for machinery and 30% for automobiles. Accelerated depreciation (full write-off in two years) is allowed on machinery and equipment acquired by manufacturers and processors after May 8, 1972 for use in Canada.

Current or capital expenditures on scientific research related to the business of the taxpayer may be written off for tax purposes in the year when incurred or any subsequent year.

A corporation whose principal business is mining, oil production or a related activity may deduct Canadian exploration expenses from income from any source in the

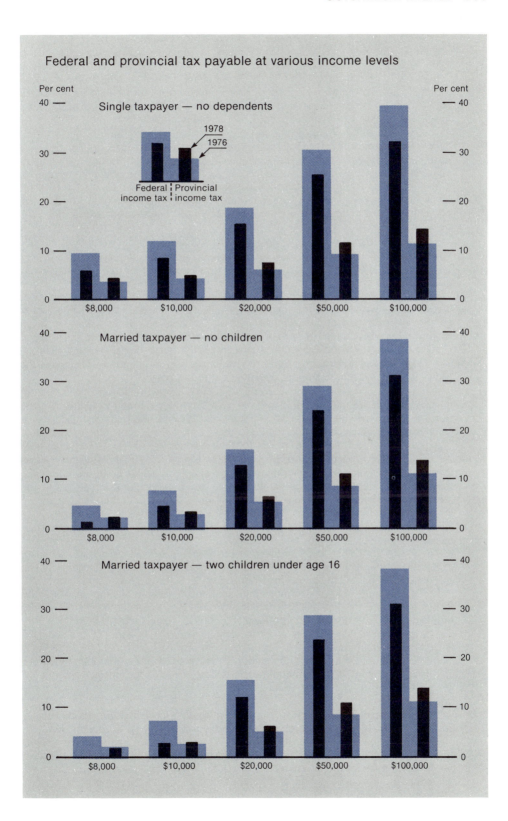

Federal and provincial tax payable at various income levels

Per cent

Single taxpayer — no dependents

1978
1976

Federal : Provincial
income tax : income tax

$8,000 $10,000 $20,000 $50,000 $100,000

Married taxpayer — no children

$8,000 $10,000 $20,000 $50,000 $100,000

Married taxpayer — two children under age 16

$8,000 $10,000 $20,000 $50,000 $100,000

year in which the expenses are incurred and any unused balance can be carried forward indefinitely. Corporations which do not meet the principal business test may deduct 100% of Canadian exploration expenses incurred between May 25, 1976 and July 1, 1979 in the year incurred. For such corporations Canadian exploration expenses incurred on or before May 25, 1976 must be amortized at 30% on a declining balance basis. For all corporations, the amount which may be deducted for Canadian development expenses may not exceed 30% of the unamortized balance.

Taxpayers with resource profits are entitled to a resource allowance equal to 25% of such resource profits before the deduction of interest expense, exploration and development expense and earned depletion. In addition to the resource allowance, a taxpayer with resource profits may deduct earned depletion in computing his income for a taxation year. The earned depletion deduction for a particular taxation year is the lesser of the earned depletion base (one-third of qualifying expenditures to date less previous claims) and 25% of the resource profits. Canadian exploration and development expenses are qualifying expenditures.

Provincial royalties and mining taxes are not deductible in computing taxable income for federal purposes.

Capital equipment and facilities for a new mine may be written off immediately against income from the mine. The assets eligible for this accelerated depreciation include buildings, mining machinery, processing facilities and such social capital as access roads, sewage plants, housing, schools, airports and docks. The accelerated write-off provision for new mines will also apply in the case of a major expansion of an existing mine where there has been at least a 25% increase in milling capacity. The list of eligible assets is the same as for new mines except that social capital does not qualify.

Taxpayers operating timber limits receive an annual cost allowance. The rate of the allowance is based on the amount of timber cut in the year.

In computing taxable income, corporations may deduct dividends received from other Canadian taxable corporations and also from certain non-resident affiliates. Business losses may be carried back one year and forward five years and deducted in computing taxable income. Corporations may also deduct donations to charitable organizations up to a maximum of 20% of their income.

The standard rate of corporation income tax is 46%. A special deduction reduces this rate on Canadian manufacturing and processing profits to 40%. In order to make room for provincial corporation income taxes (which range from 9% to 15%), the provinces have been granted an abatement of 10% of federal tax otherwise payable on income earned in a province.

A small business deduction reduces the standard federal rate of tax on certain business income to 25%. This rate is reduced to 20% on Canadian manufacturing and processing profits. This small business deduction is restricted to private Canadian corporations which are not controlled by a non-resident or by a Canadian public corporation. It applies only to income from an active business carried on in Canada and not to investment income. The maximum amount of taxable income on which the deduction may be calculated is $150,000 in any one year. A corporation is entitled only to this deduction until it has accumulated $750,000 of taxable income since 1971.

A corporation that qualifies as an investment corporation pays tax at a standard federal rate of approximately 29%. The investment income (other than dividends) of a private corporation is subject to the standard rate of federal tax (that is, 46% before provincial abatement) but an amount not exceeding 16⅔% of such income is refunded when dividends are paid to shareholders. The 10% abatement granted to the provinces similarly applies to all of the above special rates.

Dividends received by a private corporation from portfolio investments are subject to a special 25% tax but this is refunded when dividends are paid to shareholders.

A corporation may elect to pay a special 15% tax on its 1971 undistributed income on hand. Dividends received from this tax-paid undistributed income before 1979 are not included in the income of the receiving shareholder but the amount of the dividend will reduce the adjusted cost base of the shares for capital gains tax purposes. Dividends paid from the untaxed half of a private corporation's capital gains are also excluded from the income of the recipient shareholders but with no similar reduction in the adjusted cost base of the shares for capital gains tax purposes.

Special rules are provided for the taxation of special-purpose companies such as mutual fund corporations, life insurance companies, non-resident-owned investment companies and co-operatives.

A corporation may reduce its tax otherwise payable by a credit for taxes paid to foreign governments on foreign source income. This credit may not exceed the Canadian tax related to such income. A corporation may also deduct from its tax an amount equal to two-thirds of any logging tax paid to a province not exceeding 6⅔% of its income from logging operations in the province. (At present only Quebec and British Columbia impose logging taxes.) Corporations are required to pay their tax by monthly instalments throughout their taxation year.

Taxation of non-residents. An individual or corporation not resident in Canada is liable for Canadian income tax on income from employment or from carrying on business in Canada and on half of capital gains less losses on disposals of taxable Canadian property. The taxation of capital gains may be restricted by the provisions in tax treaties between Canada and other countries. The expression "carrying on business in Canada" includes producing, growing, packaging or improving any article in Canada and also soliciting orders or offering anything for sale in Canada through an agent or servant. However, this is usually modified by tax treaties so that an enterprise of the other country is taxed by Canada on its industrial and commercial profits only if it carries on business through a permanent establishment in this country.

The taxable income of non-resident individuals derived from employment or carrying on business or from capital gains in Canada is taxed under the same schedule of rates as are Canadian resident individuals.

Income earned by non-resident corporations carrying on business or from capital gains in Canada is taxed at the regular rates of corporation income tax. The distributable business earnings of a branch of a non-resident corporation are also subject to an additional tax often referred to as a branch tax. This tax applies to the branch earnings after taxes that are not reinvested in the business in Canada.

Certain specific items of income paid to non-residents from sources in Canada are subject to a 25% tax withheld at source by the Canadian payer. This non-resident withholding tax applies to interest (except interest on certain government and long-term corporate bonds and interest paid to certain exempt lenders), dividends, rents, royalties (including royalties from motion pictures and television films), management fees, income from a trust or estate, alimony, pension benefits (other than the old age pension and the Canada Pension Plan or Quebec Pension Plan benefits), payments from deferred income plans and the taxable portion of annuities.

The 25% rate of non-resident withholding tax may be modified by tax treaties. The rate of tax applicable to dividends is reduced by 5% in the case of dividends paid by a corporation that has a degree of Canadian ownership. Generally, a corporation is regarded as having a degree of Canadian ownership where 25% of its equity and voting shares are owned by Canadians or corporations controlled in Canada, or where the voting shares of the corporation are listed on a Canadian stock exchange and no more than 75% of its issued outstanding voting shares are owned by a non-resident alone or in combination with related persons.

Non-residents who receive from sources in Canada only the kinds of income subject to the non-resident withholding tax do not file returns to Canada. However, those who receive rents on real property, timber royalties, pension benefits or proceeds from deferred income plans may elect to file returns and be taxed at either personal or corporation rates.

Estate and gift taxes. The federal government formerly imposed an estate tax and a tax on gifts. They do not apply to deaths occurring after 1971 or gifts made after 1971.

Excise taxes. The Excise Tax Act levies a general sales tax and special excise taxes. These taxes are levied on goods imported into Canada as well as on goods produced in Canada. They are not levied on goods exported.

The general sales tax is 12%. It is levied on the manufacturer's sale price of goods produced or manufactured in Canada or on the duty-paid value of goods imported into Canada. Duty-paid value includes the amount of customs duties, if any. For alcoholic beverages and tobacco products the sale price for purposes of the sales tax includes excise duties levied under the excise act. The rate of sales tax on a long list of construction materials and equipment for buildings is 5%.

Some goods are exempt from sales tax. Drugs, electricity, fuels for lighting or heating, all clothing and footwear, foodstuffs and a comprehensive list of energy conservation, transportation and construction equipment are exempt. In addition articles and materials purchased by public hospitals and certain welfare institutions are not subject to sales tax. The products of farms, forests, mines and fisheries are, to a large extent, exempt as is most equipment used in farming and fishing. Machinery and equipment used directly in production, materials consumed or expended in production and equipment acquired by manufacturers or producers to prevent or reduce pollution to water, soil or air from their manufacturing operations are all exempt. A number of items are exempt when purchased by municipalities. These and other exemptions are set forth in the Excise Tax Act.

The Excise Tax Act also imposes a number of special excise taxes in addition to the sales tax. Where these are ad valorem taxes they are levied on the same price or duty-paid value as the general sales tax. Those levied as at December 31, 1977 are given in Table 20.19.

Excise duties. The excise act levies taxes (referred to as excise duties) upon alcohol, alcoholic beverages other than wines and tobacco products. These duties are not levied on imports but the customs tariff applies special duties to these products equivalent to the excise duties levied on the products manufactured in Canada. Exported goods are not subject to excise duties.

The duties on spirits are on a proof gallon basis. They do not apply to denatured alcohol intended for use in the arts and industries, or for fuel, light or power, or any mechanical purpose. Canadian brandy (a spirit distilled exclusively from juices of native fruits without the addition of sweetening materials) is subject to an excise duty. Excise duties are imposed on tobacco, cigars and cigarettes in addition to the special excise taxes.

Customs duties. Most goods imported into Canada are subject to customs duties at various rates as provided by tariff schedules. Customs duties which once were the chief source of revenue for the country have declined in importance as a source of revenue to the point where they now provide less than 10% of the total. Quite apart from its revenue aspects, however, the tariff still occupies an important place as an instrument of economic policy.

The Canadian tariff consists mainly of four sets of rates — general preferential, British preferential, most-favoured-nation and general. The general preferential rates apply to goods imported from designated developing countries. The British preferential rates are applied to imported commodities shipped directly to Canada from countries within the British Commonwealth. Rates lower than the British preferential are applied on certain goods imported from designated Commonwealth countries.

The most-favoured-nation rates apply to goods from countries that have been accorded tariff treatment more favourable than the general tariff but which are not entitled to the British or general preferential tariffs. Canada has most-favoured-nation arrangements with almost every country outside the Commonwealth. The most important is the General Agreement on Tariffs and Trade.

The general tariff applies to imports from countries not entitled to the British preferential, general preferential or most-favoured-nation treatment. Few countries are in this category and they are not significant in terms of trade coverage.

In all cases where the tariff applies there are provisions for drawbacks of duty on imports of materials used in the manufacture of products later exported. These drawbacks assist Canadian manufacturers to compete with foreign manufacturers of similar goods. There is a second class of drawbacks known as home consumption drawbacks. These apply to imported articles used in the production of specified classes of goods manufactured for home consumption.

Provincial taxes

All of Canada's provinces levy a wide variety of taxes, fees, licences and other forms of imposition. Among such levies, a relatively small number account for about 75% of total provincial revenue from own sources. Only the more important levies are briefly described here. Complete details may be found in *Principal taxes in Canada*, Statistics Canada Catalogue 68-201E.

Personal income tax. All provincial governments levy a tax on the income of individuals who reside within their boundaries and on the income earned by non-residents from sources within those boundaries. Rates of provincial individual income tax are expressed as percentages of basic federal tax, with the exception of Quebec which has its own system. The basic federal tax on which provinces apply their rates is the federal tax after the dividend tax credit but before any foreign tax credit and special federal tax reductions. In accordance with the new 1977 fiscal arrangements, as outlined in Section 20.3.3, new provincial rates were introduced in 1977 to take full advantage of the larger portion of the income tax field made available to provincial governments. These new rates are as follows: Newfoundland 58%; Prince Edward Island 50%; Nova Scotia 52.5%; New Brunswick 55.5%; Ontario 44%; Manitoba 56%; Saskatchewan 58.5%; Alberta 38.5%; and British Columbia 46%. Newfoundland increased its rates from 56.5% on July 1, 1977. The effective rate for the 1977 taxation year was 57.25%. For personal income tax, the federal government acts as collection agent for all provinces except Quebec.

In Quebec, provincial income tax is not related to basic federal tax but is levied at graduated rates which take into account the federal tax abatement granted to Quebec taxpayers pursuant to the province's contracting out all shared-cost programs in 1964. Due to the reduction in basic federal tax entailed by the established shared-cost programs financing measures introduced in the 1977 final arrangements, the former abatement of 24% granted to Quebec taxpayers had to be recalculated. As a result, this abatement was set at 16.5% so as to correspond to the former 24%. The rates are progressive, varying from 16% on taxable income between $2,000 and $9,000 to a maximum of 28% on income exceeding $60,000. The determination of taxable income is based on exemptions and deductions somewhat similar to those for the federal tax. Quebec does not participate in the federal tax collection agreements and therefore collects its own.

Ontario, Manitoba, Alberta and British Columbia have introduced tax credit schemes which are administered, at a small fee, through the federal tax collection machinery. Manitoba and Saskatchewan have introduced a surtax on provincial income tax payable in excess of a certain amount.

Corporation income tax. All provinces levy a tax on the taxable income of corporations. In provinces other than Quebec and Ontario, the provincial corporation income tax is imposed on the same basis as that established for federal corporation income tax purposes, and is collected by the federal government under tax collection agreements. In Quebec and Ontario, the determination of corporation taxable income follows closely, but not exactly, the federal rules and each collects its own levy. Corporate taxable income earned in a province is eligible for the 10% federal abatement to compensate corporations for provincial taxes payable. This 10% abatement does not apply to income earned in the Yukon Territory and Northwest Territories since they do not impose their own corporate income tax.

The rate that applies in Newfoundland is 14%; Nova Scotia 12%; Prince Edward Island 10%; Quebec 12%; and Alberta 11%. Five provinces introduced a preferential low tax rate for small business income. The dual corporate rates for these provinces are: New Brunswick 12%/9%; Ontario 12%/9%; Manitoba 15%/13%; Saskatchewan 14%/12%; and British Columbia 15%/12%.

Business taxes. Quebec, Ontario, Manitoba and British Columbia impose a tax on paid-up or utilized capital of corporations which have a permanent establishment within their boundaries. The rate is ⅕ of 1% in Quebec, Manitoba and British Columbia, and ³⁄₁₀ of 1% in Ontario. Certain types of companies such as banks, railway, express, trust and insurance companies are subject to special rules for computing taxable paid-up capital or special taxes, licences or fees applicable in such cases. Quebec has a place of business tax of $50 for companies whose paid-up capital exceeds $25,000 and $25 when below that amount.

Gift tax. In Quebec, Ontario and Manitoba a gift tax is levied and collected on the aggregate taxable value of gifts made in one year by a donor resident in a province as well as on a gift of real property situated within a province made by a donor who is not a resident in the province. The rates range from 15% on the first $25,000 to 50% on gifts in excess of $200,000. There are exemptions for gifts made to a spouse or charitable organization, deductions for gifts made to other recipients up to an aggregate annual amount, and credits for the tax levied by other jurisdictions on property situated outside the province.

Succession duties. Succession duty is levied on property of the deceased situated in a province regardless of where the deceased was domiciled at the time of death as well as on the dutiable value of property passing to the beneficiary who is a resident in a province. The rate depends on the net value of the whole estate wherever situated, the amount of property passing to the beneficiary, and the relationship of the beneficiary to the deceased. As of January 1977 only three provinces were still levying and collecting succession duties: Quebec, Ontario and Manitoba.

Provincial sales tax. All provinces except Alberta tax at a retail level a wide range of consumer goods and services purchased in or brought into the province. The tax is payable on the selling price of tangible personal property, defined to include certain services, purchased for own consumption or use and not for resale. Each provincial act, however, specifies a number of goods that are exempt. Exemptions include items related mainly to necessities of life and to material used in the farming or fishing industries. Provincial tax rates in 1977 were as follows: Newfoundland 10%; Prince Edward Island, Nova Scotia, New Brunswick and Quebec 8%; Ontario and British Columbia 7%; Manitoba and Saskatchewan 5%.

Gasoline and diesel fuel oil taxes. Each province and both territories impose a tax on the purchase of gasoline and diesel fuel by motorists and truckers and other fuel intended to generate motive power. A number of activities such as farming, fishing, mining or logging are either exempt from motive fuel taxation or are taxed at a preferred rate.

Tobacco taxes. A tax on consumers of tobacco products is levied in all provinces and both territories. Cigarettes are taxed on a unit basis at rates ranging from eight twenty-fifths of one cent per cigarette in Alberta and the Northwest Territories to one and two-tenths cents each in Newfoundland. The tax on cigars is calculated as a percentage or as an amount based on the final selling price. These taxes are usually collected at the wholesale level to facilitate collection and administration but may also be collected by retail dealers acting as collection agents of the province.

Amusement taxes and race track taxes. Each province with the exception of Newfoundland, Manitoba, Saskatchewan, Alberta and British Columbia has a tax on admission to places of entertainment. In Quebec, this tax is collected by municipalities which retain the proceeds even though the rate is determined by an act of the provincial government. In Manitoba and Saskatchewan the province does not levy the tax but has given the right to impose an admissions tax to its municipalities. In addition, all provinces levy a tax on all legal wagering on horse races in the province. The federal government also has a pari-mutuel-levy of six-tenths of 1% on monies wagered. This is for the supervision of race tracks.

Tax on premium income of insurance companies. All provinces and both territories impose a tax on the premium income of insurance companies. Ontario imposes a tax of 3% calculated on gross premiums and an additional tax of one-half of 1% on the premium income from property insurance. British Columbia levies a tax of 2% on gross premiums and 5% on the premiums paid to unlicensed insurers or reciprocal exchange. All other provinces tax premium income at the rate of 2%. Fire insurance comes under a separate act in each of the provinces and territories with the exception of Ontario and Manitoba.

Tax on logging operations. Quebec and British Columbia levy a tax on income from logging operations of individuals, partnerships, associations or corporations. The rate of taxation is 10% in Quebec and 15% in British Columbia on net income in excess of $10,000; if the net income is greater than $10,000 the whole amount is taxable with no basic exemption. In Quebec 33.3% of the tax is allowed as a deduction from provincial income tax. In British Columbia, as a result of the dual corporation income tax rate, the tax credit allowed from provincial corporation income tax is 44.4% of logging taxes paid by small businesses and 29.4% for the remainder. The federal income tax also allows a credit which is the lesser of two-thirds of the logging taxes paid to a province or 6⅔% of the logging income earned in a province.

Hospitalization and medical care programs. Ontario, Alberta, British Columbia and the Yukon Territory levy premiums and Quebec a flat rate personal income tax and a payroll tax, to help finance their hospitalization and medical care programs. The remaining provinces and territory finance the provincial share of their programs out of general revenue. (For details see Chapter 5, Health.)

Motor vehicle licences and fees. Each province levies a fee on the compulsory registration of a motor vehicle whereupon the vehicle is issued with licence plates. The fees vary from province to province and, in the case of passenger cars, may be assessed on the weight of the vehicle, the wheel base, the year of manufacture, the number of cylinders of the engine or at a flat rate for specified regions within a province or territory. The fees for commercial motor vehicles and trailers are based on the gross or curb weight for which the vehicle is registered, that is, the weight of the vehicle empty plus the load it is permitted to carry. Every operator or driver of a motor vehicle is required to register periodically and pay a fee for a driver's licence. The licences are valid for periods of from one to five years and the fees vary from $1 to $7 a year.

Land transfer taxes. Ontario levies a tax based on the price at which ownership of land is transferred. The tax for Canadian residents is 0.3% on the purchase up to $35,000 and 0.6% on anything in excess of that amount; for non-residents the tax is 20% of the purchase price. In addition, Ontario levies a tax of 20% on the increase in value on the sale of designated land (all real property except Canadian resource property). Quebec levies a 33% tax on the value of immovable property transferred to non-residents. Municipalities may levy duties on immovable property transferred to other than Canadian non-residents. In Alberta, a registration fee is charged proportional to the registered value of land; $5 for the first $1,000 and $1 for each additional $1,000 up to $25,000, and 50 cents per $1,000 in excess of that amount. There is also a fee charged on transfers of land based on the difference between the old registered price and the new registered price at the rate of .02% on any increase in the value of land up to $5,000 and .01% on any excess over $5,000. British Columbia and Saskatchewan do not have a land transfer tax but have an equivalent in land title fee which is based on land value.

Provincial property taxes. Most provinces levy, in varying degrees, real property taxes. In Prince Edward Island and New Brunswick, where services formerly carried out by municipal authorities were taken over by provincial governments, the real property tax field is shared by both provincial and municipal governments. The provincial governments levy a flat rate real property tax on a province-wide basis and each municipality has its own separate rate as required to meet its expenditure. All collections, however, are effected by the provinces which remit the municipal share to individual municipalities. Some provinces impose property taxes of limited application on land in unorganized areas not subject to a municipal rate. Nova Scotia imposes a property school tax on occupiers of land having more than 404.7 hectares (1,000 acres). A provincial property tax is levied in Ontario and Saskatchewan on the assessed value of real property in municipally unorganized territories where residents may enjoy provincial services. British Columbia's provincial property tax is levied on the assessed value of land and improvements in unorganized (non-municipal) areas at different rates between farmland and wild land.

20.6.3   Local taxes

For purposes of financial statistics local government is comprised of three principal categories — municipalities, local school authorities and special purpose authorities. Consequently, local taxes are levied by either one of these entities or by all of them depending upon the taxing powers granted to each of them by their respective provincial legislatures. For more than a century, the main source of revenue of local governments has been related to real properties within their jurisdictions. Various taxes have been gradually implemented to supplement the real property tax from which, however, they still derive the bulk of their revenue.

Local property tax. Municipalities throughout Canada levy taxes on real properties situated within their boundaries. Generally speaking, they set the rates and collect the proceeds of their own levy or, in addition, on behalf of other local governments in their area, particularly local school authorities. However, in most of Quebec outside the Montreal area and in the unorganized parts of Ontario, school boards levy and collect their own real property taxes.

The real property tax rate is generally expressed in mills (rate per $1,000 of the base) or as a rate per $100 of the base. This base is the assessed value of each property. Methods of determining assessed value vary widely not only among the provinces but also among municipalities within a province. However, for taxation purposes, it is generally referred to as fair market value which is considered to be a percentage of actual market value.

Business taxes. Among other taxes that municipalities levy, business taxes rank next to the real property tax as a producer of municipal revenue. Such taxes are levied directly on the tenant or the operator of a business. The bases on which business taxes are levied are very diversified among the provinces. The most common in use are: a percentage of the assessed value of real property, value of stock-in-trade, the assessed annual rental value of immovables and the area of premises occupied for business purposes.

Water charges. In general, municipalities recoup all, or part, of the cost of supplying water through special charges for water consumption. Such charges take various forms such as, for example, a water tax based on the rental value of the property occupied, or a charge based on the actual consumption of water.

Sources

20.1 - 20.2.1 Public Finance Division, Institutional and Public Finance Statistics Branch, Statistics Canada.

20.2.2 Statistics Section, Consulting and Statistics Division, Systems and Planning Branch, Department of National Revenue, Taxation; Business Finance Division, General Statistics Branch, Statistics Canada.

20.2.3 - 20.2.4 Public Finance Division, Institutional and Public Finance Statistics Branch, Statistics Canada.

20.3 Information Services Division, Department of Finance.

20.4 - 20.5 Public Finance Division, Institutional and Public Finance Statistics Branch, Statistics Canada.

20.6 - 20.6.1 Information Services Division, Department of Finance; Information Services, Department of National Revenue, Taxation.

20.6.2 - 20.6.3 Public Finance Division, Institutional and Public Finance Statistics Branch, Statistics Canada.

Tables

20.1 Consolidated government revenue and expenditure, after elimination of intergovernment transfers, fiscal years ended nearest Dec. 31, 1970-72 and 1974-76 (million dollars)

Source or function	1970r	1971r	1972r	1974	1975	1976e
Consolidated government revenue by source						
Income tax						
Individuals	9,147.7	10,194.5	12,007.3	17,326.0	19,137.9	22,443.2
Corporations	3,189.4	3,181.4	3,897.5	6,723.3	7,839.4	7,658.1
General sales tax	4,071.6	4,664.3	5,384.9	7,465.4	7,183.1	8,611.8
Real and personal property tax	3,210.9	3,424.4	3,707.8	4,353.5	5,051.8	5,859.5
Other tax revenue	6,970.3	7,500.1	8,153.5	12,793.1	13,973.3	16,003.0
Non-tax revenue	5,584.3	6,455.5	7,440.0	11,399.0	13,105.5	15,280.9
Total	32,174.2	35,420.2	40,591.0	60,060.3	66,291.0	75,856.5
Consolidated government expenditure by function						
General government	1,972.8	2,284.0	2,506.1	4,087.9	4,447.3	5,075.1
Protection of persons and property	3,078.7	3,374.4	3,650.0	4,809.3	5,717.3	6,853.2
Transportation and communications	3,246.6	3,682.9	4,084.2	6,013.1	6,783.6	7,148.3
Health	4,262.4	4,886.2	5,478.0	7,357.5	8,961.0	10,181.9
Social welfare	5,808.3	6,967.8	8,665.6	13,299.4	16,155.8	18,373.2
Education	5,993.7	6,538.5	6,953.0	8,792.4	10,653.6	12,600.8
Natural resources	537.8	629.4	720.3	2,077.3	2,756.5	2,706.3
Recreation and culture	584.2	759.8	910.8	1,432.3	1,797.1	1,889.2
Housing	296.1	509.6	427.6	544.3	928.9	1,015.3
Foreign affairs	289.1	311.5	385.4	584.1	747.7	797.4
Debt charges	2,617.7	3,069.4	3,374.9	4,695.3	5,729.7	6,678.6
Other expenditures	2,796.2	3,313.7	3,852.7	5,605.3	7,132.0	7,570.2
Total	31,483.6	36,327.2	41,008.6	59,298.2	71,810.5	80,889.5
Consolidated government revenue less consolidated government expenditure	+690.6	−907.0	−417.6	+762.1	−5,519.5	−5,033.0

20.2 Gross general revenue of the federal government, years ended Mar. 31, 1974-76 (million dollars)

Source	1974	1975	1976
Taxes			
Income			
Individuals	9,226	11,710	12,709
Corporations	3,710	4,836	5,748
On certain payments or credits to non-residents	324	428	481
General sales	3,590	3,866	3,515
Alcoholic beverages	460	502	549
Tobacco	611	625	647
Other commodities and services	20	32	85
Customs duties	1,385	1,809	1,887
Estate taxes	14	7	11
Social insurance levies[1]	1,017	1,670	1,949
Universal pension plan levies[2]	998	1,213	1,457
Oil export charge	287	1,669	1,063
Other	—	—	425
Total, taxes	21,642	28,367	30,526
Natural resources	14	21	28
Privileges, licences and permits	24	24	50
Sales of goods and services	965	941	969
Return on investments	1,568	1,904	2,177
Contributions to non-trustee public service pension plans	220	253	305
Postal receipts	586	611	561
Bullion and coinage	58	48	36
Fines and penalties	16	15	19
Miscellaneous	9	8	32
Total, gross general revenue from own sources	25,102	32,192	34,703
Specific purpose transfers from other levels of government	—	—	—
Total, gross general revenue	25,102	32,192	34,703

[1]Unemployment insurance.
[2]Canada Pension Plan.

20.3 Gross general expenditure of the federal government, years ended Mar. 31, 1974-76 (million dollars)

Function	1974	1975	1976
General government	1,382	1,710	1,873
Protection of persons and property[1]	2,602	2,890	3,397
Transportation and communications[2]	1,765	2,194	2,479
Health	1,951	2,296	2,781
Hospital care	1,068	1,310	1,745
Other	883	986	1,036
Social welfare	8,109	10,079	12,385
Universal pension plans	279	399	589
Old age security	3,035	3,445	3,934
Veterans' benefits	561	639	705
Unemployment insurance	2,159	2,490	3,328
Family and youth allowances	996	1,824	1,958
Assistance to disabled, handicapped, unemployed and other needy persons	814	955	1,449
Other	265	327	422
Education	919	1,039	1,178
Natural resources	570	1,582	1,981
Agriculture, trade and industry, and tourism	1,145	1,490	1,918
Environment	246	270	290
Recreation and culture	253	274	346
Labour, employment and immigration	331	344	444
Housing	138	212	338
Foreign affairs and international assistance	439	584	748
Supervision and development of regions and localities	144	152	142
Research establishments	302	326	503
General purpose transfers to other levels of government	1,883	2,696	2,688
Transfers to own enterprises	362	481	521
Debt charges	1,735	2,271	2,832
Other	1	1	1
Total, gross general expenditure	24,277	30,891	36,845

[1]Includes National Defence.
[2]Includes Post Office.

20.4 Assets and liabilities of the federal government, as at Mar. 31, 1974-76 (million dollars)

Item	1974	1975	1976	Item	1974	1975	1976
Assets				Liabilities			
Cash on hand or on deposit	440	2,684	1,819	Payables	8,794	9,848	11,384
Receivables	399	689	555	Loans and advances	—	—	—
Loans and advances	23,088	23,859	27,119	Treasury bills	4,905	5,630	6,495
Investments	15,757	18,820	22,177	Canada Savings Bonds	10,406	12,915	15,517
Other assets	1,166	1,726	1,808	Other bonds	13,860	14,541	15,684
				Other liabilities	6,209	6,535	7,606
Total, assets	40,850	47,778	53,478	Total, liabilities	44,174	49,469	56,686

20.5 Gross bonded debt of the federal government, average interest rate, term of issue and place of payment as at Mar. 31, 1974-76

Item		1974	1975	1976
Bonded debt	$'000	24,266,365	27,456,159	31,201,571
Average interest rate	%	6.78	7.33	7.66
Average term of issue	yr	11.68	10.97ʳ	10.84
Place of payment				
Canada	$'000	24,008,281	27,248,560	31,026,004
New York	"	258,084	207,599	175,567
Federal Republic of Germany	"	—	—	—

20.6 Contingent liabilities of the Government of Canada, years ended Mar. 31, 1976 and 1977

Item	1976		1977	
	Amount of guarantee $	Amount outstanding $	Amount of guarantee $	Amount outstanding $
Railway securities guaranteed as to principal and interest				
Canadian National 5% due May 15, 1977	70,884,500	70,884,500	68,774,500	68,774,500
Canadian National 4% due February 1, 1981	300,000,000	300,000,000	300,000,000	30,000,000
Canadian National 5¼% due January 1, 1985	82,017,000	82,017,000	79,667,000	79,667,000
Canadian National 5% due October 1, 1987	129,986,500	129,986,500	125,381,500	125,381,500
Total, railway securities	582,888,000[1]	582,888,000	573,823,000[1]	573,823,000
Other outstanding guarantees and contingent liabilities				
Loans made by lenders under Part IV of the National Housing Act, 1954 for home extension and improvements[2]	32,500,000	25,200,000	32,500,000	25,500,000
Insured loans made by approved lenders under the National Housing Act, 1954[2,3]	25,000,000,000	13,864,000,000	25,000,000,000	15,130,000,000
Liability for insurance and guarantees under the Export Development Act	1,750,000,000	839,940,000	1,750,000,000	1,037,149,000
Loans made by chartered banks and credit unions under the Farm Improvement Loans Act	343,000,000	106,500,000	370,800,000	124,200,000
Loans made by chartered banks and credit unions under the Fisheries Improvement Loans Act	10,900,000	8,700,000	10,800,000	8,500,000
Loans made by chartered banks and credit unions under the Small Businesses Loans Act	47,000,000	30,600,000	56,600,000	37,000,000
Loans made by chartered banks and credit unions under the Canada Student Loans Act[4]	1,009,300,000	567,000,000	1,166,300,000	640,800,000
Loans made by lenders under the Regional Development Incentives Act and the Regional Economic Expansion Act	19,024,000	14,177,000	26,915,000	21,816,000
Loans made by lenders under the Cape Breton Development Act	100,000,000	94,000,000	100,000,000	91,450,000
Loans made by lenders under the General Adjustment Assistance Program	250,000,000	61,765,285	250,000,000	68,418,301
Licensing agreement provisions in the sale of aircraft	17,600,000[5]	17,600,000	...	...
Canadian Commercial Corporation — disputed contract termination action	...	...	6,800,000	6,800,000
Loans to Indians or Indian bands, individuals, corporations or partnerships for the purpose of Indian Economic Development	40,180,323	31,845,622	42,705,427	29,323,823
Loans made by banks or other approved lenders for housing purposes as defined by the National Housing Act	...	...	20,463,972	20,163,265
Loans to Panarctic Oils Limited for its exploration program	...	...	12,000,000	12,000,000
Loans by Central Mortgage and Housing to Nanisivik Mines Ltd.	2,154,686	2,154,686	2,990,606	2,990,606
Loans made by banks to Canadair Limited to finance production of Lear Star 600 Aircraft	...	...	50,000,000	—
Agreement with Bombardier/MLW to purchase two LRC train sets in the event that Amtrak does not purchase the train sets under a lease purchase agreement	...	...	9,000,000	—
Canadian National Railways-bid bond to participate in a railroad project in Venezuela	...	...	7,500,000	—
Swine flu vaccine supplied by Glaxo (Canada) Ltd. to the Department of National Health and Welfare (1976)	...	...	1,800,000	1,800,000
Borrowing by Crown corporations guaranteed by the government				
Air Canada	...	...	23,565,100[6]	15,183,786
Petro-Canada	...	...	240,000,000	240,000,000
Canadian Wheat Board[7]	1,670,000,000	877,777,707	1,370,000,000	725,882,713
Total, contingent liabilities	30,874,547,009	17,124,148,300	31,124,563,105	18,812,800,494
Loans made by approved lending institutions under National Housing Act prior to 1954	Unstated	Indeterminate	Unstated	Indeterminate
Guarantees to owners of returns from moderate housing projects[8]	Unstated	Indeterminate	Unstated	Indeterminate

[1] Balances valued as at January 1, 1976 and 1977.
[2] As at December 31, 1975 and 1976.
[3] As reported by approved lenders as at December 31, 1975 and 1976.
[4] Includes contingent liability in respect of alternative payments to non-participating provinces.
[5] As of July 23, 1976 an out-of-court settlement was reached between the Government of Canada and Northrup Corporation Ltd. An amount of $9,000,000 was agreed upon by the interested parties in compensation for reconfiguration agreements, royalties and interests on the sale of CF-5 and NF-5 aircrafts to the Netherlands and damages related to the sale of CF-5 aircrafts to Venezuela. Out of that amount, $6,204,442 was paid by the Canadian Commercial Corporation, $2,695,558 by the Department of Supply and Services and $100,000 by the Department of National Defence through Canadair Ltd. their agency. Northrup Corporation Ltd. received payments in the amount of $8,196,024 as withholding taxes of $803,976 were deducted from the gross claim and remitted directly by the above entities to the Department of National Revenue.
[6] March 31, 1977 close rate. Converted from pounds sterling. One £=$1.8127 Canadian.
[7] Liability is subject to exchange rate in effect June 15, 1976 and 1977.
[8] As of December 31, 1976, Rental Guarantee Funds totalling $8,832,000 (December 31, 1975, $8,343,000) were held by the Central Mortgage and Housing Corporation for the purpose of settling claims. In 1975 the last of the rental contracts expired.

20.7 Government of Canada public debt and interest payments thereon, years ended Mar. 31, 1972-77

Year ended Mar. 31	Gross debt $'000,000	Net active assets $'000,000	Net debt $'000,000	Net debt per capita[1] $	Increase or decrease in net debt during year $'000,000	Interest paid on debt $'000,000	Interest paid per capita[2] $
1972	47,687	29,750	17,937	831.63	+615	1,964	89.95
1973	51,716	34,260	17,456	790.03	−481	2,105	96.43
1974	55,557	37,429	18,128	807.65	+672	2,549	115.36
1975	62,696	43,421	19,275	847.72	+1,147	3,164	138.78
1976	59,802	36,506	23,296	1,013.20	+4,022	3,908	169.97
1977	67,075	37,489	29,586	1,270.27	+6,290	4,661	200.11

[1]Based on the official estimates of population for June 1 of the year indicated.
[2]Based on the official estimates of population for June 1 of the year immediately preceding the one indicated.

20.8 Revenue collected (net of refunds) by the Department of National Revenue, Taxation, years ended Mar. 31, 1973-77 (thousand dollars)

Year ended Mar. 31	Income tax[1] Individual[2]	Corporation	Special refundable tax	Total	Estate tax and succession duties[3]	Total collections
1973	12,421,913	3,287,807	−840	15,708,880	71,594	15,780,474
1974	13,967,315	4,087,710	−396	18,054,629	39,117	18,093,746
1975	17,880,320	5,386,385	—	23,266,705	24,701	23,291,406
1976	20,013,553	6,610,695	—	26,624,248	—	26,624,248
1977	23,162,330	5,958,811	—	29,121,141	—	29,121,141

[1]Includes transfers to Old Age Security Fund.
[2]Includes non-resident withholding tax and Canada Pension Plan contributions by employers, employees and self-employed persons and unemployment insurance premiums.
[3]Includes federal estate taxes as well as succession duties and gift taxes collected on behalf of certain provinces.

20.9 Number of taxpayers and amounts of income and tax, by selected cities, 1974 and 1975

City and province	1974 Taxpayers	Total income assessed $'000,000	Federal tax payable $'000,000	1975 Taxpayers	Total income assessed $'000,000	Federal tax payable $'000,000
Brantford, Ont.	32,696	322.9	41.3	30,193	363.9	46.3
Calgary, Alta.	206,215	2,137.3	281.2	191,226	2,536.6	348.8
Dartmouth, NS	31,240	302.0	35.8	31,684	348.7	40.8
Edmonton, Alta.	244,569	2,499.0	324.6	226,305	2,947.9	403.9
Guelph, Ont.	29,840	298.3	37.5	29,148	354.8	45.3
Halifax, NS	65,876	632.8	76.7	66,601	730.4	87.4
Hamilton, Ont.	219,455	2,321.3	308.8	155,067	1,964.4	257.4
Hull, Que.	51,433	537.7	52.4	54,871	648.9	60.8
Kingston, Ont.	41,809	412.8	52.1	37,142	470.5	62.1
Kitchener — Waterloo, Ont.	84,910	828.7	104.7	78,996	942.7	118.2
London, Ont.	118,305	1,200.9	154.9	102,512	1,294.5	169.0
Moncton, NB	28,762	253.3	27.8	29,126	289.1	31.1
Montreal, Que.	798,436	8,680.8	856.4	790,462	9,832.1	948.0
New Westminster, BC	21,095	217.3	28.7	20,753	239.8	31.1
Niagara Falls, Ont.	30,864	294.2	34.5	24,810	292.6	35.6
Oakville, Ont.	30,920	377.6	56.7	28,413	427.4	64.9
Oshawa, Ont.	48,510	506.2	67.5	45,677	584.2	76.9
Ottawa, Ont.	225,569	2,503.5	344.7	207,993	2,787.8	378.8
Peterborough, Ont.	30,734	302.7	36.7	25,038	308.3	38.4
Quebec, Que.	147,666	1,567.7	152.1	146,934	1,783.9	164.2
Regina, Sask.	64,688	639.8	78.5	62,078	787.1	100.5
Saint John, NB	37,326	331.8	37.6	39,056	402.9	45.5
St. Catharines, Ont.	60,886	636.9	81.4	52,025	660.4	84.5
St. John's, Nfld.	44,397	406.5	49.3	44,807	476.7	58.4
Sarnia, Ont.	34,907	391.1	53.7	29,602	408.8	56.9
Saskatoon, Sask.	57,706	555.4	65.2	49,620	623.5	77.7
Sault Ste Marie, Ont.	33,482	352.1	45.7	26,852	357.7	47.4
Sherbrooke, Que.	32,374	323.4	29.1	32,273	362.0	31.4
Sudbury — Copper Cliff, Ont.	60,900	617.4	76.8	51,419	659.1	86.8
Sydney — Glace Bay, NS	40,025	333.5	34.3	41,975	390.5	38.5
Thunder Bay, Ont.	50,928	516.3	66.3	42,957	540.8	70.2
Toronto, Ont.	1,158,990	12,147.4	1,677.5	997,865	12,973.3	1,815.5
Trois-Rivières, Que.	21,362	216.0	19.6	20,031	237.8	22.0
Vancouver, BC	505,354	5,371.3	726.9	503,255	6,181.9	826.3
Victoria, BC	102,288	1,030.7	126.5	108,027	1,259.8	150.9
Windsor, Ont.	105,908	1,140.2	146.3	90,171	1,177.2	151.0
Winnipeg, Man.	262,891	2,487.8	295.8	270,955	2,883.5	331.9

20.10   Number of taxpayers and amounts of income and tax, by occupational class, 1974 and 1975

Occupational class	1974			1975		
	Taxpayers	Total income assessed $'000	Federal tax payable $'000	Taxpayers	Total income assessed $'000	Federal tax payable $'000
Employees	7,567,679	74,435,815	8,676,177	7,279,444	85,143,182	9,941,067
Farmers	187,233	2,421,318	287,841	162,223	2,619,786	322,666
Fishermen	17,532	141,697	14,008	15,293	122,754	9,870
Self-employed professionals						
Accountants	6,996	215,014	45,335	7,195	249,438	51,152
Medical doctors and surgeons	24,495	1,092,115	258,690	26,626	1,242,400	287,120
Dentists	5,652	200,789	45,108	5,851	239,135	54,521
Lawyers and notaries	10,658	455,639	111,471	11,697	499,823	116,736
Consulting engineers and architects	3,123	106,523	24,236	2,689	116,726	27,857
Entertainers and artists	6,652	60,581	6,720	6,657	71,519	7,435
Other professionals	18,777	283,820	46,647	18,091	325,439	53,072
Salesmen	24,455	316,880	42,285	24,974	354,231	44,157
Business proprietors	303,544	3,327,047	410,159	280,612	3,612,831	436,290
Investors	240,989	3,175,306	390,290	210,327	3,453,510	426,901
Property owners	55,450	669,889	90,783	47,380	733,730	102,444
Pensioners	249,209	1,691,691	108,080	150,569	1,469,928	94,267
All others	207,788	1,045,531	58,169	242,117	1,429,262	75,163
Total	8,930,232	89,639,655	10,615,999	8,491,745	101,683,693	12,050,715

20.11   Individual income tax statistics, by income class, 1974 and 1975

Income class based on total income	Taxpayers		Total income assessed		Federal tax payable		Average federal tax	
	1974	1975	1974 $'000	1975 $'000	1974 $'000	1975 $'000	1974 $	1975 $
Under $2,000	73,164	13,084	112,020	13,354	782	283	11	22
$2,000 and under $3,000	427,755	120,276	1,080,503	308,487	1,240	303	3	3
$3,000 " $5,000	1,381,666	722,346	5,622,715	3,058,688	158,234	58,115	115	80
$5,000 " $7,000	1,594,449	1,344,549	9,528,604	8,117,198	594,369	402,855	373	300
$7,000 " $10,000	1,985,792	1,982,519	16,779,778	16,759,854	1,519,358	1,302,047	765	657
$10,000 " $15,000	2,155,194	2,404,110	26,221,158	29,503,496	3,089,599	3,162,054	1,434	1,315
$15,000 " $25,000	1,037,703	1,508,031	18,957,754	27,812,730	2,796,929	3,791,757	2,695	2,514
$25,000 " $50,000	224,527	329,581	7,303,039	10,603,266	1,384,222	1,877,938	6,165	5,698
$50,000 and over	49,982	67,249	4,034,080	5,506,622	1,071,266	1,455,365	21,433	21,641
Total	8,930,232	8,491,745	89,639,651	101,683,693	10,615,999	12,050,715	1,189	1,419

20.12   Allocation of taxable income, by province, 1971-75 (million dollars)

Province or territory	1971	1972	1973r	1974	1975
Newfoundland	84.6	84.0	113.3	145.7	156.8
Prince Edward Island	15.0	16.0	19.6	26.9	31.1
Nova Scotia	145.0	159.2	188.8	260.4	319.5
New Brunswick	110.7	126.9	163.9	228.0	246.0
Quebec	1,629.8	1,863.2	2,307.2	3,244.5	3,402.5
Ontario	3,220.9	3,804.2	4,847.1	6,444.4	6,578.2
Manitoba	263.0	322.0	394.7	570.4	625.2
Saskatchewan	158.4	193.1	254.5	453.1	573.9
Alberta	653.6	796.3	1,129.5	2,221.4	3,486.7
British Columbia	793.4	934.7	1,428.2	1,703.8	1,718.8
Yukon Territory	5.0	6.0	8.1	20.4	40.0
Northwest Territories	10.6	17.0	41.7	85.6	55.2
Other	113.8	139.3	141.7	164.7	222.9
Canada	7,203.7	8,461.9	11,038.4	15,569.2	17,456.9

20.13 Income taxes by industrial division, 1974 and 1975 (million dollars)

Income taxes and year	Agri-culture, forestry, fishing	Mining	Manu-fac-turing	Con-struc-tion	Trans-portation, commun-ication, and other utilities	Whole-sale trade	Retail trade	Finance	Services	Total
Book profit before taxes[1]										
1974	119.2	2,622.6	8,050.4	717.6	1,317.6	2,019.0	1,017.9	2,271.8	737.4	18,873.3
1975	25.8	2,573.5	6,945.5	1,048.2	1,355.4	1,587.9	1,106.4	2,581.3	853.7	18,077.8
Taxable income										
1974	145.7	1,700.8	6,279.8	639.2	792.0	1,977.5	1,042.0	2,165.8	826.4	15,569.2
1975	149.7	2,529.4	6,201.1	947.3	917.5	1,786.3	1,170.9	2,770.6	979.2	17,456.9
Federal income taxes										
1974	31.3	531.0	1,900.7	161.6	297.7	649.1	285.4	737.5	244.3	4,838.6
1975	29.5	694.3	1,800.2	248.5	329.5	557.4	319.1	901.1	276.2	5,155.8
Provincial income taxes										
1974	15.5	155.3	713.0	71.3	91.7	231.8	117.4	231.3	92.5	1,719.9
1975	16.1	188.1	681.9	108.2	105.7	207.8	131.6	297.9	109.8	1,847.1
Total income taxes										
1974	46.8	686.4	2,613.6	232.9	389.4	881.0	402.8	968.9	336.8	6,558.5
1975	45.6	882.4	2,482.2	356.9	435.2	765.2	450.7	1,199.0	386.0	7,002.9

[1]After losses. Adjusted to exclude intercorporate dividends and net capital gains.

20.14 Excise taxes collected, by commodity, years ended Mar. 31, 1975-77 (million dollars)

Commodity	1975	1976	1977
Sales tax[1]	3,866	3,515	3,929
Other excise taxes			
Cigarettes, tobacco and cigars	367	370	405
Jewellery, watches and ornaments	21	32	30
Matches and lighters	2	3	3
Oil export charge	1,445	1,063	661
Oil export tax	224	—	—
Special excise tax (gasoline)	—	425	600
Wines	13	12	10
Sundry commodities	8	19	35
Interest and penalties	3	3	3
Less refunds		−1	−1
Total	5,949r	5,441	5,675

[1]Includes tax credited to the Old Age Security Fund.

20.15 Liabilities of provincial and territorial governments as at Mar. 31, 1975 and 1976 (thousand dollars)

Province or territory and year		Short-term bank loans and over-drafts	Payables	Loans and advances	Treasury bills	Savings bonds	Bonds and debentures	Other secu-rities	Deposits and other liabilities	Total
Newfoundland	1975	60,744	13,447	74,280	—	—	1,149,860	27,979	49,029	1,375,339
	1976	64,627	13,416	84,880	—	—	1,383,230	23,358	45,814	1,615,325
Prince Edward Island	1975	4,684	19,169	25,971	—	—	113,907	—	18,747	182,478
	1976	6,210	15,868	29,206	—	—	132,878	—	18,719	202,881
Nova Scotia	1975	16,680	84,937	108,085	—	—	1,180,859	1,500	48,405	1,440,466
	1976	52,372	105,746	156,874	—	—	1,245,142	—	45,889	1,606,023
New Brunswick	1975	1,674	102,907	40,511	7,892	—	686,142	—	34,421	873,547
	1976	1,699	130,726	47,826	10,605	—	784,756	—	29,319	1,004,931
Quebec	1975	87,265	955,817	775,565	103,813	289,455	3,326,165	12,038	153,403	5,703,521
	1976	126,858	1,513,229	894,069	91,513	556,877	3,959,620	—	277,779	7,419,945
Ontario	1975	—	47,531	262,375	—	—	9,421,267	—	322,263	10,053,436
	1976	—	42,186	355,414	325,000	—	11,511,035	—	378,708	12,612,343
Manitoba	1975	103,982	62,443	41,297	130,779	61,495	837,699	—	62,182	1,299,877
	1976	100,536	62,073	42,121	160,665	58,601	1,111,812	—	98,402	1,634,210
Saskatchewan	1975	55,620	28,465	17,256	29,143	7,978	847,431	—	970	986,863
	1976	47,685	18,607	15,552	27,880	5,751	949,903	—	70,892	1,136,270
Alberta	1975	35,825	177,201	36,262	101,894	—	1,522,105	—	73,177	1,946,464
	1976	28,033	210,741	56,425	49,936	—	1,702,376	—	87,337	2,134,848
British Columbia	1975	25,989	145,718	—	2,558	—	765,930	—	31,537	971,732
	1976	28,158	341,715	—	1,500	—	916,728	—	32,721	1,320,822
Yukon Territory	1975	—	3,091	52,753	—	—	—	—	3,092	58,936
	1976	—	4,251	50,474	—	—	—	—	3,762	58,487
Northwest Territories	1975	—	11,933	115,699	. —	—	—	—	3,711	131,343
	1976	—	12,691	118,865	—	—	—	—	2,387	133,943
Canada	1975	392,463	1,652,659	1,550,054	376,079	358,928	19,851,365	41,517	800,937	25,024,002
	1976	456,178	2,471,249	1,851,706	667,099	621,229	23,697,480	23,358	1,091,729	30,880,028

20.16   Liabilities guaranteed by provincial and territorial governments[1] as at Mar. 31, 1975 and 1976 (thousand dollars)

Province or territory and year		Bonds and debentures	Bank loans	Other	Total
Newfoundland	1975	149,806	176,207	120,561	446,574
	1976	222,987	180,891	142,420	546,298
Prince Edward Island	1975	8,568	4,992	843	14,403
	1976	7,965	5,469	832	14,266
Nova Scotia	1975	308,853	48,852	3,815	361,520
	1976	404,902	45,431	20,206	470,539
New Brunswick	1975	525,349	32,128	36,837	594,314
	1976	675,515	30,378	45,254	751,147
Quebec	1975	4,450,636	217,177	62,949	4,730,762
	1976	5,935,943	383,067	245,862	6,564,872
Ontario	1975	3,416,331	127,865	367,585	3,911,781
	1976	4,632,906	106,971	393,750	5,133,627
Manitoba	1975	1,390,611	–	2,230	1,392,841
	1976	1,596,774	–	10,347	1,607,121
Saskatchewan	1975	12,776	14,683	51,082	78,541
	1976	8,000	12,461	55,834	76,295
Alberta	1975	534,658	145,527	767,941	1,448,126
	1976	769,614	171,770	909,918	1,851,302
British Columbia	1975	3,446,916	8	34,393	3,481,317
	1976	4,123,932	–	42,097	4,166,029
Yukon Territory	1975	–	–	–	–
	1976	–	–	6,160	6,160
Northwest Territories	1975	–	103	4,790	4,893
	1976	–	95	11,636	11,731
Canada	1975	14,244,504	767,542	1,453,026	16,465,072
	1976	18,378,538	936,533	1,884,316	21,199,387

[1]Excludes liabilities of provincial government special funds guaranteed by provincial governments but considered as provincial government liabilities.

20.17   Bonds and debentures[1], by market, of provincial governments outstanding as at Mar. 31, 1975 and 1976 (thousand dollars)

Province and year		Domestic	Foreign Traditional United States	Europe	Other	International	Total
Newfoundland	1975	446,619	278,438	139,793	–	130,584	995,434
	1976	500,836	403,245	135,286	–	160,008	1,199,375
Prince Edward Island	1975	74,769	8,277	–	–	–	83,046
	1976	87,769	8,206	–	–	–	95,975
Nova Scotia	1975	375,400	422,821	20,028	–	50,915	869,164
	1976	338,400	467,796	19,611	–	50,315	876,122
New Brunswick	1975	209,303	209,707	–	–	31,543	450,553
	1976	194,135	281,305	–	–	30,543	505,983
Quebec	1975	2,249,124	740,280	105,817	32,682	122,203	3,250,106
	1976	2,919,285	976,108	77,961	32,682	115,075	4,121,111
Ontario	1975	1,102,562	1,733,546	61,253	–	–	2,897,361
	1976	1,517,969	2,262,981	56,445	–	–	3,837,395
Manitoba	1975	253,770	190,000	43,912	–	–	487,682
	1976	253,876	250,000	121,254	–	–	625,130
Saskatchewan	1975	322,861	114,181	9,103	–	–	446,145
	1976	379,253	102,541	–	–	–	481,794
Alberta	1975	588,069	122,404	–	–	81	710,554
	1976	497,739	116,171	–	–	34	613,944
British Columbia	1975	132,750	135,790	–	–	22,500	291,040
	1976	132,000	131,932	–	–	22,500	286,432
Total	1975	5,755,227	3,955,444	379,906	32,682	357,826	10,481,085
	1976	6,821,262	5,000,285	410,557	32,682	378,475	12,643,261

[1]Includes savings bonds.

20.18 Personal income tax payable on various levels of income, 1978 (dollars)

Status	Income[1]	Federal income tax[2]	Provincial income tax[3]
Single taxpayer — no dependents	2,000	—	—
	3,000	—	8
	4,000	—	54
	5,000	—	122
	8,000	480	343
	10,000	846	504
	15,000	1,896	966
	20,000	3,107	1,502
	50,000	12,750	5,830
	100,000	32,102	14,345
Married taxpayer — no children	4,000	—	—
	5,000	—	1
	8,000	94	173
	10,000	441	326
	15,000	1,425	759
	20,000	2,577	1,266
	50,000	11,984	5,493
	100,000	31,186	13,942
Married taxpayer — two children under age 16	4,000	—	—
	5,000	—	—
	8,000	—	150
	10,000	284	301
	15,000	1,261	731
	20,000	2,401	1,233
	50,000	11,874	5,445
	100,000	31,055	13,884

[1]The taxpayer is assumed to be under age 65 and to receive wage or salary income. Family allowances, at 1978 rates, are added to income where applicable. The taxpayer is assumed to take the standard deduction of $100 in respect of medical expenses and charitable contributions. In addition to personal exemptions, the employment expense deduction of 3% of wage and salary income to a maximum of $250, and social security contributions, calculated at 1978 rates, are deducted from income in computing taxable income.
[2]Federal income tax includes the tax cut of 9%, minimum $300, maximum $500, as well as the child credit ($50 per dependent child under age 18). The overall maximum for the tax cut and child credit is $500. The tax calculations represent the income tax provisions as of January 1978.
[3]Provincial income tax is calculated at the standard rate of 44% of federal basic tax. No account is taken of the various provincial tax reductions or credits.

20.19   Special excise taxes levied as at Dec. 31, 1976 and 1977

Item	Tax
Cigarettes	3¢ per 5 cigs.
Cigars	20½% ad valorem
Pipe tobacco, cut tobacco, snuff	90¢ per lb.
Jewellery, including articles of ivory, amber, shell, precious or semi-precious stones, clocks and watches[1], goldsmiths' and silversmiths' products, except gold-plated or silver-plated ware for the preparation or serving of food or drink	10% ad valorem
Lighters	10¢ per lighter
Playing cards	20¢ per pack
Slot machines — coin, disc or token-operated games or amusement devices	10% ad valorem
Matches	10% ad valorem
Tobacco, pipes, cigar and cigarette holders and cigarette rolling devices	10% ad valorem
Wines[2]	
Wines of all kinds containing not more than 7% absolute alcohol by volume	25¢ per gal
Non-sparkling wines containing more than 7% absolute alcohol by volume but not more than 40% proof spirit	50¢ per gal
Sparkling wines	$2.50 per gal
Wines (additional excise taxes)[3]	
Wines of all kinds containing not more than 7% absolute alcohol by volume	2½¢ per gal
Wines of all kinds containing more than 7% absolute alcohol by volume	5¢ per gal
Insurance premiums paid to British or foreign companies not authorized to transact business in Canada or to non-resident agents of authorized British or foreign companies	10% of net premium for property surety, fidelity and liability insurance. (Most other kinds of insurance are exempt.)
Air transportation tax on tickets purchased in or outside of Canada for transportation of persons	
(a) in the taxation area[4] (including travel in Canada)	8% ad valorem, maximum $8
(b) beginning in Canada and ending outside the taxation area	$8[5]
Automobiles, station wagons and vans designed for use as passenger vehicles[6]	$30 for the first 100 lb. in excess of the weight limit[7] $40 for the second 100 lb. in excess of the weight limit $50 for the third 100 lb. in excess of the weight limit $60 for each additional 100 lb. in excess of the weight limit
Motorcycles with engines that have a displacement of greater than 250 c.c.	5% ad valorem
Motors exceeding 20 hp (including drive assemblies) for boats other than motors for boats purchased for use by the Government of Canada and commercial fishermen, trappers and hunters	10% ad valorem
Aircraft but not including gliders or aircraft purchased or imported for use exclusively in the provision of such class or classes of air services as the Governor-in-Council may by regulation prescribe	10% ad valorem
Gasoline for personal use	10¢ per gal
Air conditioners designed for use in automobiles, station wagons, vans or trucks	$100 per unit

All the foregoing items, except insurance premiums, are also subject to the general sales tax of 12%. Cigarettes, cigars and tobacco are subject to additional taxes under the Excise Act (referred to as excise duties).
[1]Special excise tax only applies on the amount by which the sale price or the duty-paid value of the clock or watch exceeds $50.
[2]These taxes apply only to wines manufactured in Canada. The customs tariff on wines includes a levy on imported wines to correspond to the taxes on domestic production.
[3]These taxes apply to both domestic and imported wines.
[4]Includes Canada, the islands of St. Pierre and Miquelon, and the US except Hawaii.
[5]Reduced to $4 for a child under 12 travelling at a fare 50% or more below the applicable fare; nil if the fare is 90% below the applicable fare.
[6]Excludes ambulances, hearses, vehicles for police or firefighting.
[7]The weight limit is 4,425 lb. for automobiles and 5,000 lb. for station wagons and vans.

20.20   Transfers by the federal government to provincial governments, territories and local governments, year ended Mar. 31, 1976 (thousand dollars)

Payee and purpose	Nfld.	PEI	NS	NB	Que.	Ont.	Man.	Sask.	Alta.	BC	All provinces	YT	NWT	Canada
PROVINCIAL GOVERNMENTS AND TERRITORIES														
General purpose transfers														
Statutory subsidies	9,708	659	2,174	1,774	4,484	5,504	2,156	2,100	3,132	2,117	33,808	–	–	33,808
Federal corporation income tax on privately owned public utilities	1,773	475	–	–	2,465	7,027	1,184	59	16,538	2,274	31,795	363	109	32,267
Tax revenue guarantee payments	10,129	2,223	14,692	14,052	100,412	192,054	26,516	18,308	42,049	40,076	460,511	–	–	460,511
Equalization	191,314	49,048	281,262	202,504	999,204	–	150,468	47,843	–	–	1,921,643	–	–	1,921,643
Grants in lieu of taxes	307	541	–	1,400	–	–	882	280	1,858	816	2,757	167	290	3,214
Other	–	141	233	-289	3,326	4,461	–	–	–	1,759	12,958	25,204	119,880	158,042
Total, general purpose transfers	213,231	53,087	298,361	219,441	1,109,891	209,046	181,206	68,590	63,577	47,042	2,463,472	25,734	120,279	2,609,485
Specific purpose transfers														
General government														
Other	–	–	–	–	–	–	–	–	–	–	–	1,617	4,000	5,617
Protection of persons and property	221	66	499	405	3,548	5,096	724	686	1,249	1,774	14,268	8	337	14,613
Courts of law	174	44	436	351	3,169	4,602	649	617	1,132	1,617	12,791	8	314	13,113
Other (civil emergency measures)	47	22	63	54	379	494	75	69	117	157	1,477	–	23	1,500
Transportation and communications														
Road	500	362	–	3,390	14,188	14,147	10,935	8,874	11,646	3,749	67,791	–	–	67,791
Railway Grade Crossing Fund	475	35	740	–	3,428	4,483	–	–	2,080	1,249	12,490	–	–	12,490
Other	25	327	2,650	–	10,760	–	–	–	9,566	2,500	45,637	–	–	45,637
Rail	–	–	–	–	–	9,664	–	–	–	–	9,664	–	–	9,664
Health														
Hospital care	53,929	10,251	78,818	65,152	31,312	855,739	104,205	88,618	194,999	252,272	1,735,295	1,732	6,476	1,743,503
Hospital insurance and diagnostic services	53,929	10,251	78,818	65,152	31,312	855,739	104,205	88,618	194,999	252,272	1,735,295	1,625	3,573	1,740,493
Hospital care	–	–	–	–	–	–	–	–	–	–	–	107	2,903	3,010
Medicare	20,777	4,104	28,511	23,918	228,174	290,659	35,987	33,300	61,941	86,412	813,783	795	1,887	816,465
Medical Care Act	19,034	4,104	28,277	23,532	215,650	287,661	35,702	32,092	61,363	86,368	793,783	696	1,277	795,756
Health Resources Fund	1,743	–	234	386	12,524	2,998	285	1,208	578	44	20,000	–	–	20,000
Medicare — Indian and Inuit	–	–	–	–	–	–	–	–	–	–	–	99	610	709
Preventive services	–	2	92	3	–	–	104	33	22	–	256	–	–	256
Other														
Professional training	64	20	85	73	821	829	109	103	162	186	2,452	–	–	2,452
Total, health	74,770	14,377	107,506	89,146	260,307	1,147,227	140,405	122,054	257,124	338,870	2,551,786	2,527	8,363	2,562,676
Social welfare														
Assistance to disabled, handicapped, unemployed and other needy individuals	37,112	9,166	42,223	51,181	172,946	426,511	62,149	43,857	100,938	186,409	1,132,492	1,125	4,779	1,138,396
Old age assistance	-13	–	–	–	62	-8	-1	–	-1	–	39	–	–	39
Disabled persons allowances	203	107	931	771	28	11,620	1,797	1,822	3,241	1,022	21,542	137	56	21,735
Blind persons allowances	147	29	227	133	68	8	32	17	85	–	746	–	2	748
Canada Assistance Plan	36,775	9,030	41,065	50,277	172,788	414,891	60,321	42,018	97,613	185,387	1,110,165	988	4,721	1,115,874
Other	–	13	3,855	3,272	3,680	24,044	13,707	5,193	7,610	–	61,374	–	5	61,379
Total, social welfare	37,112	9,179	46,078	54,453	176,626	450,555	75,856	49,050	108,548	186,409	1,193,866	1,125	4,784	1,199,775
Education														
Indian and Inuit schools	–	–	–	–	–	–	–	–	–	174	174	–	–	174

20.20   Transfers by the federal government to provincial governments, territories and local governments, year ended Mar. 31, 1976 (thousand dollars) (continued)

Payee and purpose	Nfld.	PEI	NS	NB	Que.	Ont.	Man.	Sask.	Alta.	BC	All provinces	YT	NWT	Canada
PROVINCIAL GOVERNMENTS AND TERRITORIES (concluded)														
Specific purpose transfers (concluded)														
Education (concluded)														
Post-secondary	5,275	1,074	26,512	5,905	253,567	167,473	18,211	12,429	42,814	14,441	547,701	—	—	547,701
Post-secondary education	5,275	1,074	26,512	5,905	240,859	167,473	18,211	12,429	42,814	14,441	534,993	—	—	534,993
Canada Student Loans Act	—	—	—	—	12,708	—	—	—	—	—	12,708	—	—	12,708
Other	766	492	1,819	6,633	58,865	36,257	2,482	1,050	2,099	2,817	113,280	1,662	113	115,055
Total, education	6,041	1,566	28,331	12,538	312,432	203,730	20,693	13,479	44,913	17,432	661,155	1,662	113	662,930
Natural resources														
Forests														
Inventory of forest reserves	170	—	—	—	—	—	—	—	—	—	170	—	—	170
Mines	623	—	—	737	—	—	—	307	—	125	1,792	—	—	1,792
Oil and gas (Oil Export Tax Act)	—	—	—	—	—	—	1	21,023	175	7	21,206	—	—	21,206
Water power	—	—	—	—	2,377	—	—	1,000	—	4,418	7,795	—	—	7,795
Other	—	108	—	109	—	—	103	220	—	—	540	—	—	540
Total, natural resources	793	108	—	846	2,377	—	104	22,550	175	4,550	31,503	—	—	31,503
Agriculture, trade and industry, and tourism														
Agriculture	2,020	33,267	3,151	8,109	20,992	14,850	10,609	23,280	16,585	3,439	136,302	—	—	136,302
Agricultural and Rural Development Act	1,535	—	3,058	16	4,785	7,650	2,711	3,210	1,346	2,335	26,646	—	—	26,646
Land surveying and mapping	442	—	—	—	—	—	—	—	—	—	442	—	—	442
Rural area development	—	32,867	—	8,004	13,944	—	1,852	—	—	—	56,667	—	—	56,667
Canada Land Inventory	36	—	—	—	—	—	—	—	—	48	84	—	—	84
Rabies control	—	—	—	—	10	77	2	1	—	—	90	—	—	90
Crop insurance	7	400	93	89	2,008	5,672	4,418	19,739	14,794	1,056	48,276	—	—	48,276
Assistance re crop losses due to adverse weather	—	—	—	—	3	—	1,401	—	—	—	1,404	—	—	1,404
Research	—	—	—	—	242	1,451	—	—	—	—	1,693	—	—	1,693
Waterfowl crop depredation	—	—	—	—	—	—	225	330	445	—	1,000	—	—	1,000
Trade and industry	46,333	—	18,423	33,460	38,060	15,148	12,116	10,413	4,701	2,779	181,433	—	—	181,433
Tourism	54	23	—	—	64	95	327	34	—	75	672	12	—	684
Total, agriculture, trade and industry, and tourism	48,407	33,290	21,574	41,569	59,116	30,093	23,052	33,727	21,286	6,293	318,407	12	—	318,419
Environment														
Pest control operations	—	—	—	—	98	—	119	148	—	—	365	—	—	365
Recreation and culture	108	77	120	132	—	100	—	—	—	—	537	66	97	700
Labour, employment and immigration														
Labour and employment	858	2	2,221	2,886	16,786	10,687	2,402	8,144	4,252	12,451	60,689	—	57	60,746
Housing														
General assistance	168	—	—	—	—	—	—	—	—	—	168	140	—	308
Foreign affairs and international assistance	—	—	—	—	—	—	—	—	—	160	160	—	—	160
Supervision and development of regions and localities	4,457	2	4,472	1,155	7,414	—	1,164	—	121	—	18,785	—	—	18,785
Total, specific purpose transfers	173,435	58,921	210,909	206,520	850,515	1,864,012	275,454	258,712	449,314	571,688	4,919,480	7,157	17,751	4,944,388
Total, transfers to provincial governments and territories	386,666	112,008	509,270	425,961	1,960,406	2,073,058	456,660	327,302	512,891	618,730	7,382,952	32,891	138,030	7,553,873

20.20 Transfers by the federal government to provincial governments, territories and local governments, year ended Mar. 31, 1976 (thousand dollars) (concluded)

Payee and purpose	Nfld.	PEI	NS	NB	Que.	Ont.	Man.	Sask.	Alta.	BC	All provinces	YT	NWT	Canada
LOCAL GOVERNMENTS														
General purpose transfers														
Grants in lieu of taxes	749	1	5,179	—	15,154	35,540	5,704	2,114	4,684	8,189	77,314	293	429	78,036
Specific purpose transfers														
Transportation and communications	363	—	175	46	1,471	9,675	929	3,442	1,333	1,072	18,506	—	301	18,807
Air	363	—	175	46	725	1,741	929	334	709	730	5,752	—	301	6,053
Road	—	—	—	—	746	6,988	—	2,965	619	342	11,660	—	—	11,660
Rail	—	—	—	—	—	946	—	143	5	—	1,094	—	—	1,094
Social welfare														
Assistance to disabled, handicapped, unemployed and other needy individuals	4,000	1,016	248	1,665	15,697	6,020	299	219	479	2,678	32,321	64	24	32,409
Education														
Primary and secondary	—	—	—	—	1,326	102	696	—	742	—	2,866	—	—	2,866
Environment	709	177	1,883	1,984	11,807	24,944	2,301	89	3,623	6,205	53,722	—	—	53,722
Sewage collection and disposal	709	177	1,683	1,949	11,747	24,895	2,082	89	3,145	6,202	52,678	—	—	52,678
Other	—	—	200	35	60	49	219	—	478	3	1,044	—	—	1,044
Housing														
General assistance	1,467	27	294	364	5,568	9,418	1,580	81	505	914	20,218	—	—	20,218
Supervision and development of regions and localities	—	—	—	24	—	—	—	—	—	—	24	—	—	24
Total, specific purpose transfers	6,539	1,220	2,600	4,083	35,869	50,159	5,805	3,831	6,682	10,869	127,657	64	325	128,046
Total, transfers to local governments	7,288	1,221	7,779	4,083	51,023	85,699	11,509	5,945	11,366	19,058	204,971	357	754	206,082
Total, transfers to provincial governments, territories and local governments	393,954	113,229	517,049	430,044	2,011,429	2,158,757	468,169	333,247	524,257	637,788	7,587,923	33,248	138,784	7,759,955

20.21   Conditional grants and shared-cost programs as at Mar. 31, 1976 and 1977

Department and project	Provinces participating[1]		Federal contribution	
	1976	1977	1976 $'000	1977 $'000
AGRICULTURE				
Crop insurance	10	10	48,276	56,457
4-H Club assistance	10	10	177	181
Freight on livestock shipments to and from the Royal Agricultural Winter Fair, Toronto		8(PEI, Ont.)		114
Crop loss assistance	Que., Man.	8(Ont., Man.)	1,405	2,641
Contributions for rabies	Que., Ont., Man., Sask.	Que., Ont., Man., Sask.	90	84
Aid to universities — veterinary teaching	Que., Ont.	9(Ont.)	1,693	51
Other	7(PEI, Que., Ont.)	—	65	—
EMPLOYMENT AND IMMIGRATION				
Agricultural manpower	10	10	3,462	3,438
Manpower training research	PEI, NB, Ont., Sask.	PEI, NB, Ont., Sask.	158	31
ENERGY, MINES AND RESOURCES				
Aeromagnetic surveys	Nfld., Que., BC + NWT	Que., BC + NWT	1,219	1,174
BC – YT – NWT boundary and Sask. – NWT boundary	Sask., BC	BC	49	42
Mineral development programs	Nfld., Sask.	Sask.	710	276
Alberta iron processing and Peace River iron ore programs	Alta.	Alta.	129	42
Assistance to provinces for energy substitution and conservation programs		PEI, NS		33,765
Bay of Fundy tidal power study	NS, NB	NS, NB	142	824
Energy and energy-related research fund	Man., Sask.	Alta.		4,000
Non-renewable resource evaluation	Ont., Man., Sask.	Man.	228	182
Uranium reconnaissance program	Sask.	NB, Ont., Man., Sask., BC	357	667
Saskatchewan heavy oils program	Sask.	—	200	—
ENVIRONMENT				
Metropolitan Toronto and Upper Thames	Ont.	Ont.	1,000	6
Migratory birds crop depredation	Man., Sask., Alta.	Man., Sask., Alta.	141	826
Shore damage to property on Great Lakes	Ont.	Ont.	1,250	50
Industrial development	7(Man., Sask., Alta.)	8(Sask., Alta.)	4,368	652
Fraser River flood control	BC	BC	776	6,336
Environmental assessment	Que.	Que.	629	493
Delta project — Manitoba	Man.	Man.		463
Alberta oil sands environment research program	Alta.	Alta.		622
Flood risk mapping agreement	Nfld., Que.	NB, Que., Ont., Man.	756	1,000
Forest engineering research institute of Canada	Que.	Que.	225	183
Montreal area flooding	Que.	BC		514
Okanagan basin implementation		Man., Sask., Alta.		71
Prairie provinces water board	Man., Sask., Alta.	NB	72	72
Qu'Appelle valley agreement	Sask.	Sask.		—
Preservation of Atlantic salmon in Saint John River		NB	1,000	200
Solid waste disposal and management	Man., Sask., Alta.	NS, Ont.	338	731
Souris River study	Ont.	Man., Sask.	2,377	35
Southwestern Ontario dyking	Que.	Ont.	400	176
St. Lawrence water quality studies		Que.		1,454
Feasibility study of reclaiming marketable waste paper	Nfld., Man., BC	Ont.	76	434
Canada land inventory	Sask.		533	13
Churchill, Nelson Basin studies	Ont., Man., Sask.		853	—
Hydrometric agreements	Que.		15	—
Lac Seoul agreement	Man.		76	—
Lake Winnipeg, Nelson and Churchill rivers	BC		14	—
Prince Rupert environment assessment	Sask., Alta.		962	—
Rivière des Roches Weir	Ont.		1,660	—
Water quality, Great Lakes and lower lakes				—

20.21 Conditional grants and shared-cost programs as at Mar. 31, 1976 and 1977 (continued)

Department and project	Provinces participating[1] 1976	1977	Federal contribution 1976 $'000	1977 $'000
INDIAN AFFAIRS AND NORTHERN DEVELOPMENT				
Community development on and off reserve	Nfld.	Nfld.	4,664	4,500
Child care agreement	Man., Sask. + YT	Man., Sask. + YT	222	2,438
Maintenance of highway — Rocky Harbour to St. Pauls	Nfld.	Nfld.	218	213
Forest fire protection agreements	Ont.	Ont., Man., Sask.	58	166
Registered trapline fur agreement	Man.	Man.	220	440
Roads on and to reserves	Sask.	Man., Sask.	190	521
Natural resources agreements	Ont.	Ont.	200	312
Purchase of land	Nfld., PEI, BC	Nfld., NB, BC	3,381	2,311
Agricultural representative agreement	..	Sask.	..	130
Development services wildlife agreement	..	Man.	..	398
Forestry operations	Man.	Man.	15	25
Fredericton military compound	NB	NB	226	20
High-level fixed highway bridge across Chambly canal	Que.	Que.	487	163
Indian policing agreements	Ont.	Ont.	250	588
Maintenance of national historic parks	PEI, NB, BC	PEI, BC	1,003	800
Qu'Appelle corridor recreation and tourism planning	..	Sask.	..	89
Canada — Ontario — Rideau, Trent, Severn waterways	Ont.	—	46	—
Vocation and technical training	Ont.	—	128	—
INDUSTRY, TRADE AND COMMERCE				
Tourism	10 + YT	10 + YT and NWT	803	1,525
JUSTICE				
All programs included	10 + YT and NWT	10 + YT and NWT	12,884	13,487
NATIONAL DEFENCE				
Contributions to provinces and municipalities for civil defence purposes	10 + NWT	10 + YT and NWT	1,500	1,500
NATIONAL HEALTH AND WELFARE				
Health care programs				
Health Resources Fund Act	9(PEI)	8(PEI, Alta.)	20,000	24,050
Training of health personnel	10	10 + NWT	2,252	2,190
Medical Care Act	10 + YT and NWT	10 + YT and NWT	795,796	1,003,583
Hospital Insurance and Diagnostic Services Act	10 + YT and NWT	10 + YT and NWT	2,379,617[2]	2,734,191[3]
Income security and social assistance programs				
Old age security	Nfld., Que., Ont., Man., Alta.	Nfld., Que. Ont.	39[2]	17Cr.[3,4]
Blind Persons Allowance	9(BC) + NWT	9(BC)	794[2]	621[3]
Disabled Persons Allowance	9(BC) + NWT	9(BC)	1,153[2]	607[3]
Guaranteed income experimental projects	Man.	Man.	2,844[2]	3,120
Unemployment assistance	Alta. + NWT	Alta.	5[2]	76Cr.[3,4]
Canada Assistance Plan	10 + YT and NWT	10 + YT and NWT	1,364,690[2]	1,595,933[3]
Services to young offenders	NB, Ont.	NB. Ont.	16,967	15,797
Nursing home care	Ont., Man., Alta.	Ont., Man., Alta.	56,708	63,661
National welfare grants	NS, NB, Ont. + NWT	Ont., Sask. + NWT	35	31
Vocational rehabilitation of disabled persons	9(Que.) + YT and NWT	9(Que.) + YT and NWT	20,611	23,560
PUBLIC WORKS				
Maintenance cost of Perley Bridge — agreement that federal government pay 75%, Ontario 25%. Maintenance cost of Macdonald-Cartier Bridge — agreement that federal government pay 33⅓%, Ontario 66⅔%.	Ont.	Ont.	2,590	85

20.21 Conditional grants and shared-cost programs as at Mar. 31, 1976 and 1977 (concluded)

Department and project	Provinces participating[1] 1976	Provinces participating[1] 1977	Federal contribution 1976 $'000	Federal contribution 1977 $'000
REGIONAL ECONOMIC EXPANSION				
Fund for Rural Economic Development (FRED)				
Agricultural and Rural Development (ARDA)				
Special areas infrastructure and highways	10	10	280,576[5]	281,228[5]
General development agreements				
Interim planning agreements				
Miscellaneous agreements				
SECRETARY OF STATE				
Post-secondary education	10	10	534,993	648,700
Language texts for citizenship classes	Que., Ont., Man., Alta.	6(Nfld., PEI, NB, BC)	158	193
Bilingualism development	10 + YT and NWT	10 + YT and NWT	111,755	162,934
Citizenship and language instruction for immigrants	NS, Que., Ont., Man., Alta.	7(Nfld., PEI, NB)	1,130	3,930
TRANSPORT				
Contributions to assist in extending the network of highways and road facilities in the northern part of the province	Man., Sask., Alta., BC	Man., Sask., Alta., BC	10,000	10,735
Contributions to assist in upgrading the primary highway network	Man., Sask., Alta.	Man., Sask., Alta.	21,875	26,702
Contributions in the construction and operation of certain rail lines	..	BC	..	54,000
URBAN AFFAIRS				
National Capital Commission	Que., Ont.	Que., Ont.	19,141	6,130

[1]Provinces not participating are shown in parentheses.
[2]"Contracting out" has occurred. In order to be consistent, the total federal contribution has been included, although the contribution may have taken the form of a tax abatement and an operating cost adjustment payment or recovery.
[3]Includes the contribution to Quebec under the Established Programs (Interim Arrangements) Act, which may have taken the form of a tax abatement and an operating cost adjustment payment or recovery.
[4]Cr. indicates a recovery.
[5]Excludes amounts of $3,940 for 1976 and $6,548 for 1977, which were not allocated by province.

20.22 Gross general revenue and expenditure of provincial and territorial governments, year ended Mar. 31, 1975 (thousand dollars)

Source or function	Nfld.	PEI	NS	NB	Que.	Ont.	Man.	Sask.	Alta.	BC	YT	NWT	Canada
Gross general revenue by source													
Income tax													
Individuals	59,944	11,322	122,606	90,987	2,367,248	1,750,790	203,908	157,649	347,386	503,876	—	—	5,615,716
Corporations	21,470	3,045	30,639	25,350	422,029	742,305	77,221	46,993	275,565	242,975	—	—	1,887,592
General sales tax	96,558	15,843	102,987	90,177	1,049,902	1,568,829	142,640	122,797	—	405,809	—	—	3,595,542
Motive fuel tax	32,977	7,973	53,844	45,339	395,208	571,644	55,571	47,021	80,277	147,800	3,395	3,090	1,444,139
Health insurance premiums	—	—	—	—	—	548,095	—	—	59,558	89,210	765	—	697,628
Social insurance levies	7,910	1,223	11,837	13,073	185,033	282,991	18,752	18,581	45,701	89,257	—	—	674,358
Other provincial taxes	17,026	8,870	18,683	59,976	902,355	411,490	38,007	33,916	35,375	118,085	2,267	2,638	1,648,688
Natural resource revenue	6,389	360	3,737	10,369	69,130	210,584	36,818	292,827	1,406,859	339,045	152	141	2,376,411
Privileges, licences and permits	20,701	2,579	20,113	16,702	178,582	304,761	26,583	21,435	47,814	64,107	1,597	893	705,867
Liquor profits	14,328	3,998	38,072	27,218	137,950	228,554	41,337	40,037	89,756	119,063	1,810	3,699	745,822
Non-tax revenue from own sources	58,878	18,307	93,466	51,884	697,262	782,305	111,789	133,717	282,810	302,835	3,168	8,947	2,545,368
General purpose transfers from other levels of government	199,334	46,880	236,164	191,016	851,196	70,613	156,607	168,547	213,985	35,584	11,417	70,556	2,251,899
Specific purpose transfers from other levels of government	164,349	40,343	171,380	171,453	982,275	1,452,119	209,178	184,733	337,957	472,580	20,507	23,433	4,230,307
Total	699,864	160,743	903,528	793,544	8,238,170	8,925,080	1,118,411	1,268,253	3,223,043	2,930,226	45,078	113,397	28,419,337
Gross general expenditure by function													
General government	31,954	12,446	41,331	39,789	424,598	738,518	50,086	66,824	142,798	206,169	9,532	45,802	1,809,847
Protection of persons and property	19,826	3,327	25,198	20,063	295,341	393,239	53,909	42,493	95,123	122,162	3,198	3,175	1,077,054
Transportation and communications	83,409	16,924	97,727	119,117	737,798	765,855	85,212	119,224	226,432	346,278	14,257	3,867	2,616,100
Health	160,279	28,222	228,204	169,596	1,957,563	2,677,130	307,519	229,690	545,955	732,924	4,508	12,497	7,054,087
Social welfare	74,931	18,988	91,571	100,063	1,206,153	1,169,909	149,869	132,720	305,846	472,804	3,251	7,711	3,733,816
Education	194,484	42,931	224,463	219,445	2,029,431	2,256,520	258,358	213,887	574,714	511,115	10,470	30,636	6,572,454
Natural resources	35,099	2,252	18,611	13,980	137,479	107,795	28,998	20,580	81,903	111,892	440	1,628	560,657
Agriculture, trade and industry, and tourism	28,632	16,486	29,323	23,747	208,806	120,921	42,704	104,551	124,738	72,688	362	4,849	777,807
Housing	813	6,267	5,681	—	25,349	29,063	—	34,344	—	125,187	2,538	7,633	236,875
Debt charges	85,291	11,646	81,556	52,872	364,326	693,243	80,090	51,187	115,292	47,388	2,734	6,013	1,591,638
General purpose transfers to other levels of government	—	—	—	—	280,148	264,119	25,756	6,094	44,629	63,858	—	—	731,540
All other expenditures	71,281	5,537	59,951	52,376	299,750	438,759	71,307	55,940	106,018	94,887	4,922	15,147	1,275,875
Total	791,667	165,877	911,737	843,344	7,966,742	9,655,071	1,153,808	1,077,534	2,363,448	2,913,352	56,212	138,958	28,037,750

20.23   General revenue of local governments, fiscal years ended nearest Dec. 31, 1974-76 (thousand dollars)

Year and source	Nfld.	PEI	NS	NB	Que.	Ont.	Man.	Sask.	Alta.	BC	YT	NWT	Canada
1974													
Revenue from own sources	28,552	5,929	155,260	36,608	1,567,021	2,583,964	321,858	255,563	559,873	812,979	3,033	6,261	6,336,901
Taxes	19,173	3,261	113,730	22,988	1,137,480	2,022,551	232,001	190,919	352,973	631,332	1,302	2,669	4,730,379
Real property	12,865	3,040	90,750	22,977	960,721	1,722,520	206,191	169,190	305,571	583,023	1,267	2,439	4,080,554
Special assessments	485	78	3,274	11	76,069	45,547	8,034	7,264	16,468	28,502	35	76	185,843
Personal property	—	—	9,661	—	—	—	646	—	—	—	—	—	10,307
Corporations and business	4,123	142	7,719	—	75,517	244,917	15,234	11,774	30,934	19,686	—	154	410,200
Other	1,700	1	2,326	—	25,173	9,567	1,896	2,691	—	121	—	—	43,475
Grants in lieu of taxes	1,302	5	13,020	—	56,927	83,152	27,503	6,132	13,557	11,467	574	917	214,551
Federal government	292	—	4,516	—	18,844	32,743	5,935	1,353	3,831	3,930	270	395	72,109
Federal government enterprises	195	—	3,303	—	3,727	4,106	1,045	364	—	1,313	—	—	14,053
Provincial governments	224	—	766	—	14,109	10,867	12,767	627	9,400	2,377	304	522	51,963
Provincial government enterprises	104	—	4,430	—	4,514	25,411	4,208	2,638	181	3,263	—	—	44,568
Local government enterprises	—	—	—	—	4,896	10,025	—	860	145	584	—	—	15,962
Non-government organizations	487	5	5	—	10,837	—	3,548	290	—	—	—	—	15,896
Sales of goods and services	4,441	1,843	17,467	9,570	246,608	292,672	33,903	33,888	126,342	93,164	826	1,177	861,901
Water	2,771	816	6,996	6,463	163,913	128,104	15,331	13,412	30,464	36,306	552	861	405,989
Other	1,670	1,027	10,471	3,107	82,695	164,568	18,572	20,476	95,878	56,858	274	316	455,912
Rentals	226	104	1,113	814	9,028	12,982	1,276	1,345	9,883	26,475	29	127	63,402
Concessions and franchises	—	—	172	—	78	6,801	265	80	4,986	660	—	—	13,042
Licences and permits	293	54	776	616	10,058	24,614	3,557	1,992	8,456	17,295	137	165	67,946
Remittances from own enterprises	—	279	—	—	3,373	—	—	2,891	17,858	1,997	—	103	26,501
Interest	88	7	2,561	680	19,970	60,319	12,835	7,603	7,251	18,124	51	58	129,547
Interest and penalties on taxes	23	26	2,013	110	16,490	14,300	2,189	1,121	2,960	4,058	58	76	43,424
Fines	17	259	908	402	16,246	9,098	2,326	2,836	6,871	3,598	36	14	42,611
Miscellaneous	2,984	89	3,580	1,428	50,763	57,475	6,003	6,756	8,736	4,809	19	955	143,597
Transfers	26,423	27,586	194,178	47,818	1,639,123	2,455,902	234,426	246,618	596,492	473,556	1,310	6,854	5,950,286
General purpose	3,503	857	16,267	31,184	268,338	278,958	10,496	11,732	38,939	67,158	628	716	728,776
Provincial governments	3,503	857	16,267	31,184	268,338	278,958	10,496	11,732	38,939	67,158	628	716	728,776
Specific purpose	22,920	26,729	177,911	16,634	1,370,785	2,176,944	223,930	234,886	557,553	406,398	682	6,138	5,221,510
Federal government	7,270	558	6,497	7,796	22,970	33,477	2,828	1,586	8,003	9,802	34	641	101,462
Provincial governments	15,650	26,171	171,414	8,838	1,347,815	2,143,467	221,102	233,300	549,550	396,596	648	5,497	5,120,048
Total, general revenue	54,975	33,515	349,438	84,426	3,206,144	5,039,866	556,284	502,181	1,156,365	1,286,535	4,343	13,115	12,287,187
1975p													
Revenue from own sources	44,637	6,857	169,895	48,249	1,718,141	2,943,697	384,678	309,861	656,061	997,825	3,732	7,879	7,291,512
Taxes	31,156	3,993	130,751	32,459	1,254,594	2,333,156	287,572	228,052	420,122	799,456	1,566	4,029	5,526,906
Real property	22,997	3,741	102,261	32,459	1,043,567	1,952,154	257,459	198,687	363,032	745,107	1,522	3,830	4,726,816
Special assessments	1,510	77	5,493	—	116,409	58,491	8,953	10,782	17,871	32,012	44	78	251,720
Personal property	—	—	10,619	—	—	—	758	—	—	—	—	—	11,377
Corporations and business	4,709	175	9,128	—	82,970	319,909	17,125	12,369	39,147	21,969	—	121	507,622
Other	1,940	—	3,250	—	11,648	2,602	3,277	6,214	72	368	—	—	29,371
Grants in lieu of taxes	2,048	—	14,067	—	71,255	116,211	34,467	6,930	20,635	14,618	679	1,612	282,522
Federal government	633	—	4,861	—	18,689	38,441	6,209	1,639	4,918	4,422	279	677	80,768
Federal government enterprises	381	—	3,677	—	4,312	4,113	2,198	376	239	2,216	—	—	17,561
Provincial governments	45	—	942	—	16,760	25,214	5,301	640	7,711	2,109	49	935	59,996
Provincial government enterprises	217	—	4,567	—	4,438	37,955	8,278	2,783	3,361	4,929	339	—	66,540
Local government enterprises	—	—	—	—	9,826	10,143	1,979	1,045	2,820	43	12	—	25,856
Non-government organizations	772	20	20	—	17,230	345	10,502	447	1,586	899	—	—	31,801
Sales of goods and services	5,553	1,990	15,015	12,836	242,352	320,979	36,004	49,339	138,112	102,277	1,072	1,523	927,052
Water	3,264	637	7,397	7,290	170,371	139,868	13,307	15,770	35,123	40,645	740	1,184	435,596
Other	2,289	1,353	7,618	5,546	71,981	181,111	22,697	33,569	102,989	61,632	332	339	491,456

20.23 General revenue of local governments, fiscal years ended nearest Dec. 31, 1974-76 (thousand dollars) (concluded)

Year and source	Nfld.	PEI	NS	NB	Que.	Ont.	Man.	Sask.	Alta.	BC	YT	NWT	Canada
1975P (concluded)													
Rentals	480	112	1,068	805	750	15,631	1,456	2,029	11,615	29,708	29	211	63,894
Concessions and franchises			93	5		7,024	287	1,740	10,713	2,636	3	97	22,598
Licences and permits	523	57	1,064	708	28,705	25,218	3,576	3,385	11,351	17,118	157	136	92,089
Remittances from own enterprises		313		88	9,446		2,838	2,385	16,864	1,759			33,693
Interest	77	8	2,271	283	18,939	26,446	6,698	5,591	9,107	14,865	70	47	84,402
Interest and penalties on taxes	40	14	1,536		18,567	17,007	3,257	1,194	4,494	4,819	53	56	51,037
Fines	40	294	1,310	394	21,288	10,568	2,495	3,274	8,307	1,831	39	64	49,890
Miscellaneous	4,720	76	2,720	671	52,245	71,457	6,028	5,851	4,741	8,738	64	118	157,429
Transfers	32,400	34,716	243,317	48,084	1,902,459	2,808,866	214,246	270,854	689,769	557,111	5,006	8,382	6,815,210
General purpose	4,081	1,421	30,932	35,019	367,008	339,000	12,792	17,677	47,596	85,093	772	1,146	942,537
Provincial governments	4,081	1,421	30,932	35,019	367,008	339,000	12,792	17,677	47,596	85,093	772	1,146	942,537
Specific purpose	28,319	33,295	212,385	13,065	1,535,451	2,469,866	201,454	253,177	642,173	472,018	4,234	7,236	5,872,673
Federal government	5,748	2,438	6,703	5,284	16,300	41,646	1,394	6,099	4,939	10,666	50	94	101,361
Provincial governments	22,571	30,857	205,682	7,781	1,519,151	2,428,220	200,060	247,078	637,234	461,352	4,184	7,142	5,771,312
Total, general revenue	77,037	41,573	413,212	96,333	3,620,600	5,752,563	598,924	580,715	1,345,830	1,554,936	8,738	16,261	14,106,722
1976													
Revenue from own sources	54,117	7,643	189,092	62,123	2,050,387	3,444,009	463,016	351,443	791,016	1,158,429	4,449	11,036	8,586,760
Taxes	33,589	4,401	146,127	43,091	1,496,958	2,711,979	336,215	263,510	495,065	911,692	1,936	4,681	6,449,244
Real property	22,039	4,319	117,813	42,959	1,211,377	2,270,163	300,672	231,256	426,370	851,238	1,829	4,463	5,484,498
Special assessments	1,544	82	4,330	132	129,256	57,249	11,165	10,284	20,953	35,451	107	83	270,676
Personal property			13,571				665						14,236
Business	7,290		7,155		117,936	384,292	19,446	16,223	47,742	24,470		135	624,689
Other	2,716		3,258		38,389	275	4,267	5,747		493			55,145
Grants in lieu of taxes	3,426		15,845		90,629	139,723	41,123	8,923	23,823	18,488	797	1,502	344,279
Federal government	1,160		6,421		20,024	46,107	7,099	1,767	6,234	4,690	296	725	94,523
Federal government enterprises	365		3,422		5,790	5,409	2,973	594	326	2,322	53		21,254
Provincial governments	66		1,145		22,670	23,539	6,444	868	10,000	3,262	428	777	69,199
Provincial government enterprises	246		4,731		14,657	47,936	9,922	3,902	4,079	7,122	20		92,615
Local government enterprises					12,052	13,339	2,333	1,250	3,041				32,015
Non-government organizations	1,589		126		15,436	3,393	12,352	542	143	1,092			34,673
Sales of goods and services	8,893	2,308	15,979	16,006	300,076	382,002	47,531	49,785	167,546	122,172	1,272	1,767	1,115,337
Water	4,726	622	8,406	8,185	200,513	158,217	18,150	18,228	46,389	46,272	848	1,292	511,848
Other	4,167	1,686	7,573	7,821	99,563	223,785	29,381	31,557	121,157	75,900	424	475	603,488
Rentals	538	158	1,267	1,134	690	17,817	1,890	1,939	14,722	37,478	51	313	77,997
Concessions and franchises			148	6		9,104	331	2,269	14,250	3,565	4	220	29,937
Licences and permits	509	56	1,113	806	13,661	27,663	4,817	3,712	14,475	20,191	176	175	87,354
Remittances from own enterprises		300			10,363		4,500	2,378	25,255	1,776			44,572
Interest	157	5	2,534	239	21,683	44,230	9,932	6,994	13,972	22,147	65	66	122,024
Interest and penalties on taxes	265	24	1,756		22,544	19,495	3,987	1,235	5,360	5,912	78	66	60,722
Fines	39	285	1,793	387	31,226	12,498	2,980	4,333	9,280	2,679	51	75	65,626
Miscellaneous	6,661	106	2,530	454	62,557	79,498	9,710	6,365	7,268	12,329	19	2,171	189,668
Transfers	36,222	39,895	308,105	52,179	2,505,317	3,124,549	261,060	341,844	827,295	606,572	3,397	10,057	8,116,492
General purpose	6,187	1,561	40,594	35,434	371,709	361,622	17,435	20,285	51,339	100,214	777	2,386	1,009,543
Provincial governments	6,187	1,561	40,594	35,434	371,709	361,622	17,435	20,285	51,339	100,214	777	2,386	1,009,543
Specific purpose	30,035	38,334	267,511	16,745	2,133,608	2,762,927	243,625	321,559	775,956	506,358	2,620	7,671	7,106,949
Federal government	6,358	1,538	4,363	6,733	16,562	41,671	4,883	3,864	7,367	9,156	105	100	102,700
Provincial governments	23,677	36,796	263,148	10,012	2,117,046	2,721,256	238,742	317,695	768,589	497,202	2,515	7,571	7,004,249
Total, general revenue	90,339	47,538	497,197	114,302	4,555,704	6,568,558	724,076	693,287	1,618,311	1,765,001	7,846	21,093	16,703,252

20.24 General expenditure of local governments, fiscal years ended nearest Dec. 31, 1974-76 (thousand dollars)

Year and function	Nfld.	PEI	NS	NB	Que.	Ont.	Man.	Sask.	Alta.	BC	YT	NWT	Canada
1974													
General government	5,252	852	17,221	6,290	219,812	178,350	23,993	24,178	40,898	54,292	530	2,875	574,543
Executive and legislative	860	56	1,390	585	9,183	15,723	2,241	1,837	3,518	3,481	55	75	39,004
Administrative	2,972	669	13,320	5,103	210,629	145,642	17,189	20,243	34,450	41,671	413	2,673	494,974
Other	1,420	127	2,511	602	—	16,985	4,563	2,098	2,930	9,140	62	127	40,565
Protection of persons and property	2,523	1,298	24,548	18,299	283,098	405,861	40,721	29,958	82,287	112,361	695	579	1,002,228
Police services	749	876	10,003	9,067	182,528	228,676	20,281	16,357	42,379	49,323	72	19	566,224
Courts of law and correctional services	—	—	4,958	—	4,515	—	34	—	262	4,180	—	60	14,343
Fire-fighting services	1,499	411	8,540	8,217	83,827	145,081	14,692	11,209	33,980	45,774	524	372	354,126
Emergency measures	—	—	124	417	—	5,603	141	141	1,241	3,320	—	33	12,043
Regulatory services	62	3	514	367	—	21,326	247	753	2,811	7,242	59	65	33,449
Other	213	—	409	231	12,228	5,175	5,326	252	1,614	2,522	40	30	28,043
Transportation and communications	15,487	1,752	27,550	28,395	484,337	677,355	66,167	82,245	219,638	112,737	852	1,313	1,717,828
Common services	816	28	2,437	2,297	—	6,044	2,566	7,300	5,623	10,100	233	193	37,637
Road	14,405	1,724	24,822	25,816	382,083	615,344	62,118	74,507	212,000	99,587	619	1,098	1,514,123
Administration	159	—	877	533	41,798	14,371	1,784	9,241	4,277	3,447	22	—	76,509
Engineering	461	—	1,001	349	—	10,165	2,252	1,889	1,391	1,872	—	—	19,380
Roads and streets	10,892	1,423	17,480	19,428	242,541	491,109	43,563	51,176	169,623	77,216	362	884	1,125,697
Snow and ice removal	1,467	78	1,591	3,190	75,105	26,641	5,433	4,351	9,433	2,080	106	108	129,583
Bridges, subways, tunnels	3	—	32	1	16,138	3,302	3,841	3,316	14,746	1,906	16	62	41,643
Street lighting	927	205	2,102	1,622	3,396	23,165	2,251	3,053	5,699	6,895	71	34	51,038
Traffic services	135	18	1,455	464	3,105	17,083	229	1,027	5,448	4,432	36	10	35,488
Parking	23	—	108	229	—	14,613	582	431	541	1,270	—	—	17,454
Other	338	—	176	—	—	14,895	—	23	842	469	6	—	17,331
Public transit	—	—	—	256	102,254	18,172	1,266	438	2,015	907	—	—	121,806
Other	266	—	291	26	—	37,795	217	—	—	2,143	—	22	44,262
Environment	18,579	1,940	35,989	31,590	299,086	406,448	30,696	31,889	112,359	130,862	1,878	1,867	1,103,183
Water purification and supply	9,607	617	12,623	19,705	119,429	141,122	15,064	16,037	54,716	41,379	856	1,396	432,551
Sewage collection and disposal	6,574	942	18,452	9,395	131,223	199,588	7,675	9,344	37,964	68,119	616	217	490,109
Garbage and waste collection and disposal	2,346	269	4,762	2,490	45,819	61,427	7,690	6,459	17,658	20,177	406	248	169,751
Other	52	112	152	—	2,615	4,311	267	49	2,021	1,187	—	6	10,772
Health	24	6	38,470	16	3,155	264,468	40,060	65,425	161,638	16,660	38	30	589,990
Preventive services	14	6	367	16	3,069	69,675	3,405	2,374	3,820	8,980	38	30	91,794
Medical care	—	—	22	—	—	—	46	31	—	1,097	—	—	1,196
Hospital care	—	—	38,035	—	86	187,940	36,609	63,017	157,248	6,511	—	—	489,446
Other	10	—	46	—	—	6,853	—	3	570	72	—	—	7,554
Social welfare	—	23	27,040	—	13,796	366,052	5,259	4,241	16,899	66,212	—	—	499,522
Administration	—	—	2,094	—	13,796	10,200	1,316	146	2,596	5,977	—	—	36,125
Assistance	—	—	11,185	—	—	134,848	3,486	2,525	7,972	53,887	—	—	213,903
Services	—	—	12,695	—	—	216,600	373	1,105	5,023	3,949	—	—	239,745
Other	—	23	1,066	—	—	4,404	84	465	1,308	2,399	—	—	9,749
Housing — general assistance	2,322	201	2,066	5,429	34,670	53,937	7,526	2,542	7,280	20,549	44	1,489	138,055
Environmental planning and zoning	—	6	732	487	8,774	24,626	1,837	499	3,609	7,856	6	8	48,440
Community development	2,274	195	1,211	4,346	20,610	23,847	5,595	2,043	3,002	12,661	38	1,481	77,303
Other	48	—	123	596	5,286	5,464	94	—	669	32	—	—	12,312
Natural resources	—	—	55	—	3,175	52,234	1,441	1,412	2,295	6,078	32	955	63,113
Agriculture, trade and industry, and tourism	—	—	4,070	601	2,803	8,722	421	1,101	4,121	1,017	26	—	23,565
Agriculture	—	—	—	—	135	10	349	183	505	302	—	—	1,918
Trade and industry	—	—	3,891	287	2,668	8,711	106	505	3,595	350	26	—	20,196
Regional development commissions	—	—	31	—	—	—	—	—	62	79	—	—	424
Industrial parks and commissions	—	—	3,860	—	—	8,701	243	183	3,533	271	26	—	19,772
Tourism	—	—	179	314	372	—	72	128	21	365	—	—	1,451

20.24   General expenditure of local governments, fiscal years ended nearest Dec. 31, 1974-76 (thousand dollars) (continued)

Year and function	Nfld.	PEI	NS	NB	Que.	Ont.	Man.	Sask.	Alta.	BC	YT	NWT	Canada
1974 (concluded)													
Recreation and culture	5,602	785	10,252	9,771	166,819	353,272	37,853	23,483	120,296	129,592	269	818	858,812
Recreational facilities	5,158	745	6,794	7,911	100,694	249,919	30,692	16,109	108,409	107,531	265	759	634,986
Cultural facilities	440	26	3,118	1,696	40,136	97,359	5,898	7,055	10,774	18,616	4	42	185,164
Other	4	14	340	164	25,989	5,994	1,263	319	11,113	3,445	—	17	38,662
Education — primary and secondary	2,370	24,012	182,359	8,934	1,636,029	2,169,451	263,936	220,816	450,771	661,435	222	4,395	5,615,574
Fiscal services	7,520	2,477	24,527	8,335	323,088	410,297	50,398	28,990	92,641	135,037	161	590	1,084,721
Debt charges	5,985	2,306	18,173		316,798	262,018	35,569	18,339	66,699	97,626	155	549	832,558
Interest on short-term borrowing	896	85	3,319	1,117	15,054		1,155	4,913	3,325	9,437		87	51,076
Interest on long-term borrowing	4,729	2,115	14,391	7,046	272,152	250,255	25,468	13,328	61,082	86,984		389	738,094
Other	360	106	463		29,592	75	8,946	98	2,292	1,205		73	43,388
Transfers to reserves and allowances	517	171	5,423	463	6,290	92,731	8,350	9,342	18,443	37,202	6	41	179,170
Transfers to own enterprises	1,018	—	931	599	—	55,548	6,479	1,309	7,499	209	61	—	72,993
Other services	130	—	813	639	740	21,358		91	1,336	10,802	28	4	35,941
Total, general expenditures	59,809	33,346	394,960	110,019	3,467,805	5,367,805	568,471	515,270	1,312,459	1,457,634	4,582	14,915	13,307,075
1975													
General government	6,748	1,894	17,549	7,350	219,609	224,628	37,355	23,873	64,068	72,679	689	2,367	678,809
Executive and legislative	292	144	1,612	980	11,977	19,703	2,138	2,058	4,872	4,203	69	137	48,185
Administrative	4,687	1,544	14,249	5,556	137,068	187,815	32,983	17,571	53,625	58,097	569	2,068	515,832
Other	1,769	206	1,688	814	70,564	17,110	2,234	4,244	5,571	10,379	51	162	114,792
Protection of persons and property	3,160	1,658	26,184	22,346	371,591	508,104	49,065	35,182	101,451	135,856	813	1,053	1,256,463
Police services	420	1,057	11,669	11,311	243,328	290,141	23,177	19,425	54,972	62,418	17	117	718,052
Courts of law and correctional services		—	2,630	—	8,004	896	45	687	237	3,084	91	149	15,823
Fire-fighting services	2,110	589	10,715	9,697	107,882	170,158	16,989	12,509	38,935	57,007	584	568	427,743
Emergency measures	93	7	118	586		10,634	3,986	328	1,397	943	69	72	18,164
Regulatory services	65	—	654	476	12,377	25,942	2,649	954	3,996	9,607	52	89	44,501
Other	472	5	398	276		10,333	2,219	1,279	1,914	2,797	—	58	32,180
Transportation and communications	22,715	1,919	28,312	45,275	383,164	825,137	76,432	103,918	242,697	146,267	2,928	4,303	1,883,067
Common services	1,187	55	3,100	4,004		9,597	5,740	4,019	9,873	11,174	123	204	49,076
Road	20,371	1,864	25,180	40,596	381,063	814,577	69,283	99,355	231,836	131,893	2,805	3,872	1,822,695
Administration	229	44	1,057	917	36,321	32,819	486	1,557	4,219	3,468	44	89	82,464
Engineering	795	—	1,069	491		20,600		1,668	2,120	3,451	5	5	30,685
Roads and streets	15,320	1,317	15,800	31,980	210,468	583,492	48,585	83,613	190,973	104,728	2,295	3,361	1,291,932
Snow and ice removal	2,300	218	3,144	4,179	85,558	56,933	8,886	3,199	13,479	3,877	169	142	182,084
Bridges, subways, tunnels	12	—	693	38	9,885	36,067	1,926	3,120	3,426	1,611	160	—	56,938
Street lighting	1,088	247	2,305	2,045	27,653	29,181	4,075	3,456	7,736	7,797	87	116	85,786
Traffic services	48	38	995	562	7,580	33,749	2,424	1,599	7,830	4,749	40	42	59,656
Parking	28	—	79	351	3,598	15,666	227	500	859	1,229	5	19	22,556
Other	551	—	38	33		6,070	974	643	1,194	983	—	103	10,594
Public transit		—	—	105	32		45			214	—	—	396
Other	1,157	—	—	570	2,069	963	1,364	544	988	2,986	—	227	10,900
Environment	21,715		38,749	30,641	355,503	480,692	34,872	33,327	115,132	154,925		3,875	1,279,357
Water purification and supply	9,427	7,218	14,304	14,273	111,209	160,081	17,470	16,995	72,231	53,908	2,708	2,393	475,954
Sewage collection and disposal	8,950	2,559	19,047	13,638	170,499	208,076	8,660	9,737	27,207	72,523	1,104	725	544,586
Garbage and waste collection and disposal	3,300	4,135	5,339	2,730	70,680	100,567	8,490	6,357	14,507	27,473	1,389	661	240,669
Other	38	350	59	203	3,115	11,968	252	238	1,187	1,021	215	96	18,148
Health	61	174	34,209	203	3,819	286,782	41,439	75,519	175,114	12,862		61	630,121
Preventive services	20	14	641	26	3,729	80,433	1,658	2,561	4,945	9,545	38	50	103,653
Medical care		8	26	—	—	2,698	1,527	448	6	1,664	37	—	6,369
Hospital care		—	33,361	—	—	198,666	38,193	72,453	169,407	1,429	—	—	513,599
Other	41	6	181	177	90	4,985	61	57	756	224	1	11	6,500

20.24   General expenditure of local governments, fiscal years ended nearest Dec. 31, 1974-76 (thousand dollars) (concluded)

Year and function	Nfld.	PEI	NS	NB	Que.	Ont.	Man.	Sask.	Alta.	BC	YT	NWT	Canada
1975 (concluded)													
Social welfare	—	—	29,649	—	16,895	346,237	5,555	4,460	21,647	64,888	—	—	489,331
Administration	—	—	2,172	—	16,895	13,316	1,679	171	571	3,902	—	—	38,706
Assistance	—	—	11,725	—		171,790	3,762	2,313	9,011	48,953	—	—	247,554
Services	—	—	14,656	—		159,770	113	1,965	10,052	9,218	—	—	195,774
Other	—	—	1,096	—		1,361			2,013	2,815	—	—	7,297
Housing – general assistance	1,128	182	8,908	3,032	39,562	88,311	10,792	2,354	38,449	27,692	54	1,768	222,232
Environmental planning and zoning	15		1,049	595	14,470	31,790	2,524	447	4,815	9,981	13		65,699
Community development	1,035	182	7,490	2,338	21,068	52,083	8,234	1,693	30,919	16,734	27	1,727	143,530
Other	78		369	99	4,024	4,438	34	214	2,715	977	14	41	13,003
Natural resources	—	—	67	60	1,114	30,325	1,062	2,767	8,580	1,950	—	—	45,925
Agriculture, trade and industry, and tourism	—	3	4,708	916	7,392	14,284	673	1,450	5,222	1,686	10	—	36,344
Agriculture	—	3	4,426	490	4,348	11,844	527	332	4,511	1,288	—	—	27,769
Trade and industry	—	3	43	89	15	604	305		150	189	—	—	1,398
Regional development commissions	—	—	4,383	401	4,333	11,240	222	332	4,361	1,099	—	—	26,371
Industrial parks and commissions	—	—	282	426	3,044	2,440	146	1,118			10	—	8,575
Tourism	—	—							711	398		—	
Recreation and culture	10,151	853	12,741	15,292	229,493	389,951	37,921	26,082	105,048	128,423	2,434	1,721	960,110
Recreational facilities	9,624	698	7,783	11,409	175,065	274,440	27,119	18,967	92,125	106,129	2,317	1,671	727,347
Cultural facilities	461	122	4,403	3,410	30,028	103,806	8,767	6,668	10,745	18,502	10	39	186,961
Other	66	33	555	473	24,400	11,705	2,035	447	2,178	3,792	107	11	45,802
Education – primary and secondary	7,124	27,735	225,035		1,796,858	2,360,663	273,650	262,797	537,227	774,395	5,085		6,270,569
Fiscal services	10,297	2,681	26,578	10,526	429,807	464,229	46,106	29,852	116,462	135,586	427	962	1,273,513
Debt charges	8,828	2,552	20,074	10,004	356,816	276,746	30,507	18,027	82,000	108,980	275	937	915,746
Interest on short-term borrowing	906	220	2,239	1,028	9,579		1,474	406	4,985	7,028		777	38,506
Interest on long-term borrowing	7,404	2,321	15,619	8,849	312,044	264,837	27,693	14,314	72,632	100,540	273	794	827,320
Other	518	11	2,216	127	34,208	2,330	1,340	3,307	4,383	1,412	2	66	49,920
Transfers to reserves and allowances	330	129	5,273	522	14,991	83,448	7,026	9,914	21,599	26,023	152	25	169,432
Transfers to own enterprises	1,139		1,231		58,000	104,035	8,573	1,911	12,863	583			188,335
Other services	11	3	4,682	17	829	829	1,220	6,032	1,054	1,241	—	—	29,541
Total, general expenditure	83,110	44,160	457,371	135,658	3,869,259	6,020,172	616,142	607,613	1,532,151	1,658,450	10,101	21,195	15,055,382
1976													
General government	9,168	995	17,244	10,785	262,975	290,502	40,134	29,663	68,389	81,158	922	3,127	815,062
Executive and legislative	444	128	1,762	989	13,882	20,638	2,427	2,251	5,099	4,754	74	3,153	52,601
Administrative	6,966	723	13,954	8,833	163,560	255,736	34,228	22,021	55,339	63,589	690	2,845	628,484
Other	1,758	144	1,528	963	85,533	14,128	3,479	5,391	7,951	12,815	158	129	133,977
Protection of persons and property	5,211	1,839	28,874	26,710	434,956	580,980	52,535	43,967	116,745	157,693	999	1,221	1,451,730
Police services	377	1,203	15,168	12,312	293,008	332,713	27,300	26,453	63,703	72,567	999	43	844,847
Courts of law and correctional services			899		573	51		732	322	3,128	107	301	18,222
Fire-fighting services	3,793	611	11,169	12,429	115,873	189,259	19,205	13,768	42,482	66,624	738	656	476,607
Emergency measures	1	5	123	1,091	12,109	20,136	1,003	889	1,534	1,422		103	26,307
Regulatory services	141	12	845	637		32,851	3,561	1,318	4,380	10,848	93	97	54,783
Other	899	8	670	241	13,966	5,448	1,415	807	4,324	3,104	61	21	30,964

20.24   General expenditure of local governments, fiscal years ended nearest Dec. 31, 1974-76 (thousand dollars) (concluded)

Year and function	Nfld.	PEI	NS	NB	Que.	Ont.	Man.	Sask.	Alta.	BC	YT	NWT	Canada
1976 (concluded)													
Transportation and communications	26,244	2,213	31,124	47,697	473,360	811,847	94,180	129,709	281,841	172,438	1,680	4,126	2,076,459
Common services	1,704	65	3,325	5,314	—	10,694	5,979	4,741	10,009	12,334	97	546	54,808
Road	23,725	2,148	27,621	41,733	469,976	799,284	86,755	122,463	270,868	158,775	1,573	3,234	2,008,155
Administration	210	47	1,229	681	37,934	35,655	2,017	1,828	5,412	6,726	42	115	91,896
Engineering	840	—	1,148	662	—	21,484	1,060	2,686	2,321	4,616	—	144	34,961
Roads and streets	18,108	1,531	18,743	32,461	275,332	552,565	64,910	104,176	219,673	122,960	1,038	2,267	1,413,764
Snow and ice removal	2,144	194	2,657	4,698	103,821	70,768	7,413	3,807	13,860	5,571	139	287	215,359
Bridges, subways, tunnels	—	—	404	66	9,558	36,254	2,359	3,539	8,381	2,237	155	71	63,024
Street lighting	1,649	281	777	2,237	28,793	28,544	5,285	4,128	9,198	9,229	121	178	92,060
Traffic services	461	81	106	627	9,849	31,907	2,694	1,839	8,777	5,792	66	63	62,933
Parking	175	14	140	298	4,689	12,962	314	326	2,397	1,276	9	—	22,566
Other	138	—	—	3	—	9,145	703	134	849	368	3	109	11,592
Public transit	—	—	—	584	—	—	78	—	—	—	—	—	662
Other	815	—	66	66	3,384	1,869	1,368	2,505	964	1,329	10	346	12,834
Environment	23,803	4,207	29,227	44,366	471,927	514,283	44,343	45,909	118,511	164,340	4,631	5,541	1,471,088
Water purification and supply	9,337	1,951	12,945	20,024	168,968	177,905	22,547	23,748	62,050	58,793	2,438	2,457	563,163
Sewage collection and disposal	10,422	1,827	10,534	21,185	208,285	229,490	10,866	14,982	35,269	73,527	1,610	2,242	620,239
Garbage and waste collection and disposal	4,032	346	5,664	3,012	91,202	100,290	10,369	6,930	20,101	31,551	408	638	274,543
Other	12	83	84	145	3,472	6,598	561	249	1,091	469	175	204	13,143
Health	26	14	40,409	181	4,745	352,991	58,462	109,353	264,988	16,651	55	146	848,021
Preventive services	16	14	662	181	4,745	96,910	3,523	2,813	2,002	12,612	55	33	123,566
Medical care	—	—	37	—	—	4,860	1,488	476	—	2,583	—	—	9,444
Hospital care	—	—	39,596	—	—	246,881	53,401	106,018	262,288	1,427	—	—	709,692
Other	—	—	114	—	—	4,340	50	46	698	29	—	—	5,319
Social welfare	10	—	31,311	—	18,990	387,033	9,789	5,031	23,171	10,648	—	81	535,229
Administration	—	—	2,606	—	—	16,773	1,939	199	11,235	—	—	32	55,258
Assistance	—	—	15,694	—	18,990	168,824	7,691	2,963	626	3,516	—	—	245,054
Services	—	—	12,013	—	—	199,487	157	1,848	11,202	4,943	—	—	229,650
Other	—	—	998	—	—	1,949	2	21	108	2,189	—	—	5,267
Housing — general assistance	2,517	367	3,969	4,463	42,736	84,800	14,259	5,872	50,775	37,431	146	2,412	249,747
Environmental planning and zoning	11	5	1,453	762	16,937	36,950	2,471	1,406	5,849	10,562	68	15	76,489
Community development	2,419	357	1,471	3,648	19,332	36,125	8,831	2,975	39,968	25,664	78	2,355	143,223
Other	87	5	1,045	53	6,467	11,725	2,957	1,491	4,958	1,205	—	42	30,035
Natural resources	—	—	93	113	—	44,141	1,740	1,113	12,888	6,856	4	—	66,948
Agriculture, trade and industry, and tourism	—	209	1,312	858	8,830	19,825	1,138	586	12,008	5,592	36	—	50,394
Agriculture	—	3	825	519	5,553	18,223	537	407	10,089	4,529	—	—	40,685
Trade and industry	—	3	38	95	16	737	189	56	244	228	—	—	1,606
Regional development commissions	—	—	787	424	5,537	17,486	348	351	9,845	4,301	—	—	39,079
Industrial parks and commissions	—	—	487	339	3,277	1,602	601	179	1,919	1,063	—	—	9,709
Tourism	—	206	—	—	—	—	—	—	—	—	36	—	—
Recreation and culture	12,210	1,158	13,096	13,820	241,447	389,176	49,440	28,807	115,869	145,980	1,428	1,317	1,013,748
Recreational facilities	11,470	928	8,594	9,868	179,113	264,247	32,058	20,557	97,836	119,404	1,327	1,201	746,603
Cultural facilities	740	213	4,169	3,717	33,966	110,629	15,679	7,933	13,606	25,437	9	85	216,183
Other	—	17	333	235	28,368	14,300	1,703	317	4,427	1,139	92	31	50,962
Education — primary and secondary	6,220	34,766	297,518	4,701	2,309,268	2,832,215	314,398	323,370	619,833	994,711	—	—	7,737,000
Fiscal services	12,255	3,100	31,698	14,764	491,708	486,495	56,387	34,247	144,870	161,507	547	4,721	1,438,646
Debt charges	10,265	2,874	23,133	13,253	347,688	301,976	41,258	18,221	93,521	129,141	271	1,068	982,623
Interest on short-term borrowing	551	264	2,832	1,541	14,851	9,657	2,880	639	4,051	5,216	271	1,022	42,607
Interest on long-term borrowing	9,395	2,554	18,908	11,645	325,127	290,803	36,443	14,362	85,710	122,580	—	—	918,563
Other	319	56	1,393	67	7,710	1,516	1,935	3,220	3,760	1,345	—	46	21,453
Transfers to reserves and allowances	647	226	6,890	1,254	19,590	86,041	6,813	13,563	30,494	31,182	276	132	197,022
Transfers to own enterprises	1,343	—	1,675	257	124,430	98,478	8,316	2,463	20,855	1,184	—	—	259,001
Other services	20	5	4,889	6	9,755	432	390	5,777	688	172	2	46	22,136
Total, general expenditure	97,674	48,873	530,764	163,763	4,770,697	6,794,720	737,175	763,404	1,830,576	2,004,433	10,450	23,679	17,776,208

20.25 Direct debt of local governments, fiscal years ended nearest Dec. 31, 1974 and 1975 (thousand dollars)

Year and direct debt	Nfld.	PEI	NS	NB	Que.	Ont.	Man.	Sask.	Alta.	BC	YT	NWT	Canada
1974													
Long-term (debentured)	74,451	33,201	186,037	83,676	3,866,564[1]	4,095,971[2]	429,798	217,901	1,240,034	1,306,940	3,260	7,662	11,545,495
Less sinking funds	–	4,951	3,374	1,559	–	632,080	54,566	39,945	6,532	64,473	–	–	807,480
Net long-term (debentured)	74,451	28,250	182,663	82,117	3,866,564	3,463,891	375,232	177,956	1,233,502	1,242,467	3,260	7,662	10,738,015
Short-term borrowings	26,299	1,143	84,520	23,958	584,400	271,269	108,880	15,953	61,931	83,301	–	398	1,262,052
Accounts and other payables	22,871	5,622	47,818	42,648	471,588	259,265	45,810	36,077	152,188	115,305	662	2,784	1,202,638
Other liabilities	4,654	315	10,324	679	66,906	180,445	8,473	18,215	79,735	34,882	52	3,836	408,516
Total, direct debt less sinking funds	128,275	35,330	325,325	149,402	4,989,458	4,174,870	538,395	248,201	1,527,356	1,475,955	3,974	14,680	13,611,221
1975													
Long-term (debentured)	25,547	33,765	213,791	101,725	4,558,723[3]	4,260,757[2]	450,749	226,169	1,463,700	1,570,912	4,456	11,982	12,922,276
Less sinking funds	–	5,287	2,967	1,859	–	655,657	53,989	44,317	6,812	72,347	–	–	843,235
Net long-term (debentured)	25,547	28,478	210,824	99,866	4,558,723	3,605,100	396,760	181,852	1,456,888	1,498,565	4,456	11,982	12,079,041
Short-term borrowings	25,417	2,881	113,969	39,636	583,909	369,379	96,207	20,174	109,347	99,656	388	57	1,461,020
Accounts and other payables	105,498	6,105	85,062	52,530	670,801	587,001	57,748	59,520	187,378	145,010	1,026	5,294	1,962,973
Other liabilities	7,877	571	10,221	4,194	197,067	64,749	10,839	10,627	155,465	27,420	75	3,153	492,258
Total, direct debt less sinking funds	164,339	38,035	420,076	196,226	6,010,500	4,626,229	561,554	272,173	1,909,078	1,770,651	5,945	20,486	15,995,292

[1]Includes debenture debt of $993,969,000 for Quebec schools.
[2]Includes other long-term debt due to the Ministry of the Environment.
[3]Includes debenture debt of $1,139,741,000 for Quebec schools.

Sources
20.1 - 20.7 Public Finance Division, Institutional and Public Finance Statistics Branch, Statistics Canada.
20.8 -20.11Statistics Section, Statistical Services Division, Policy and Systems Branch, Department of National Revenue, Taxation.
20.12 - 20.13 Corporation Taxation Statistics, Business Finance Division, Statistics Canada.
20.14 - 20.17 Public Finance Division, Institutional and Public Finance Statistics Branch, Statistics Canada.
20.18 - 20.19 Tax Analysis and Commodity Tax Division, Department of Finance.
20.20 Public Finance Division, Institutional and Public Finance Statistics Branch, Statistics Canada.
20.21 Federal-Provincial Relations Division, Tax Policy and Federal-Provincial Relations Branch, Department of Finance.
20.22 - 20.25 Public Finance Division, Institutional and Public Finance Statistics Branch, Statistics Canada.

Selected economic indicators

Chapter 21

Tables

Selected economic indicators

Chapter 21

In this chapter various statistical statements and studies are presented, covering broad areas of Canadian economic activity. These are based on a Canadian system of national accounts which consists of national income and expenditure accounts, indexes of real domestic product, the balance of international payments and financial flows. The integrated aggregative economic accounts provide an inter-related framework for analysis of the Canadian economy and its relationship with other countries. In its broad outline, the Canadian national accounts system bears a close relationship to the international standard described in the United Nations publication *A system of national accounts.* There are also sections on price indexes and the Anti-Inflation Board.

National income and expenditure

21.1

National income and expenditure accounts provide accounting summaries for the nation and portray economic activity in terms of transactions taking place between major groups of transactors, namely, governments, corporate and government business enterprises, persons and unincorporated businesses and non-residents. By combining and summarizing these operations into their various classes, information may be obtained on the functioning of the economy which is of particular interest to governments concerned with problems of unemployment, taxation and prices, and to businessmen concerned with programs of investment and marketing.

Tables 21.1 - 21.9 are based on the revised historical series of the national income and expenditure accounts. Annual coverage since 1926 is available in Statistics Canada occasional publication *Income and expenditure accounts, 1926 to 1974,* Catalogue 13-531 and in the annual publication *System of national accounts — national income and expenditure accounts,* Catalogue 13-201.

National income. Net national income at factor cost measures the current earnings of Canadian factors of production (land, labour and capital) from productive activity. It includes wages and salaries, profits, interest, net rent and net income of farm and non-farm unincorporated business.

Gross national product (GNP), by totalling all costs arising in production, measures the market value of all final goods and services produced in the current period by Canadian factors of production. It is equal to national income plus net indirect taxes (indirect taxes less subsidies) plus capital consumption allowances and miscellaneous valuation adjustments.

Personal income is the sum of current receipts of income whether or not these receipts represent earnings from production. It includes transfer payments from government (such as family allowances, unemployment insurance benefits and war service gratuities) in addition to wages and salaries, net income of farm and non-farm unincorporated business, interest, dividends and net rental income of persons. It does not include undistributed profits of corporations and other elements of the national income not paid out to persons.

Gross national expenditure (GNE) measures the same aggregate as gross national product (total production of final goods and services at market prices) by tracing the disposition of production through final sales to persons, governments and business on capital account, including changes in inventories, and to non-residents (exports). Imports of goods and services, including payments of interest and dividends to non-residents, are deducted since the purpose is to measure only Canadian production.

21.1.1 Economic growth in 1977

Gross national product in 1977 was $207.7 billion, an increase of 9.3% from the level of the previous year; after adjustment for price changes, real GNP grew 2.6% compared with 4.9% in 1976. The major source of strength in 1977 was external demand. Real exports were 7.7% higher than in 1976, and with real imports increasing only 2.0%, real net exports (real exports less real imports) contributed more than one percentage point to the increase in real GNP. Real personal expenditure and real government current expenditure on goods and services both grew at rates well below their long-term averages, real gross fixed capital formation declined, and inventory investment fell substantially.

Wages, salaries, and supplementary labour income grew at about the same rate as total GNP in 1977, increasing 10.0% after a 15.2% rise in 1976. Corporation profits before taxes were up 11.4% compared to 1976, and after-tax profits increased 18.8%. Other non-wage income grew 9.1%, despite a 13.6% decline in accrued net income of farm operators.

Personal disposable income increased 10.5% in 1977 to a level of $139.3 billion, of which 10.7% was saved. Total personal expenditure rose 10.3%.

In real terms, personal expenditure increased 2.8% in 1977, with the largest percentage increase — 5.2% — in spending on services, reflecting higher net expenditure abroad and a large increase in spending on financial and legal services. Growth in spending on durable goods fell to 2.4% from 5.4% in 1976, with the largest decline in spending on household appliances. Expenditure on semi-durable goods increased 1.0%, as real expenditure on household furnishings was unchanged from 1976 and spending on clothing and footwear declined. Expenditure on non-durables was up only 0.7%, reflecting marginal growth in energy products and food, beverages and tobacco.

Gross fixed capital formation fell 8.6% in real terms in 1977, with all of the decline in business fixed investment. Business investment in residential construction fell 6.0%, and machinery and equipment investment dropped 1.7%. Because non-residential construction increased 3.7%, total business investment in plant and equipment was virtually unchanged from 1976.

After substantial accumulation in 1976, inventory investment swung to slight liquidation in real terms. Most of this swing was concentrated in the non-farm business sector, which grew by $122 million in 1977 after a very strong $1,068 million growth in 1976. On an industry basis, wholesalers liquidated inventories by $105 million in 1976. Retail trade recorded a moderate accumulation of $60 million in 1977, a decline of $411 million from $471 million in 1976. There was an accumulation of $95 million in manufacturing, approximately the same level as 1976. Farm inventories swung from an accumulation of $167 million in 1976 to a liquidation of $208 million in 1977.

The 10.0% rise in real exports of goods in 1977 was primarily the result of strong increases in exports of motor vehicles and parts and fabricated materials. Lumber and wheat exports were also strong in 1977. The weak 1.1% increase in real imports of goods was the result of widespread declines in imports of food, crude and fabricated materials, and industrial machinery, partially offsetting large increases in imports of passenger cars and motor vehicle parts.

The service account continued to deteriorate in 1977, but at a slower rate than in 1976. This was primarily a result of a reduction in the rate of growth of tourism abroad, offsetting accelerating interest and dividend payments to non-residents.

Total revenues of all levels of government combined (excluding intergovernmental transfers) increased 8.6% in 1977, a much slower rate of increase than in 1976. Most of the revenue components showed sharply lower rates of growth in 1977 than in 1976, reflecting the lower rate of growth in incomes.

Total government expenditure rose 11.2% for the year, just somewhat lower than the 1976 rate of 12.4%. Outlays on goods and services increased 10.0%, one of the smallest rates of change in several years, as wage and salary payments, which account for almost 60% of total government expenditure on goods and services, grew by 9.5%, after increases of about 20% in the previous two years. Transfer payments to persons increased 14.7%, a rate similar to that of 1976. With expenditures rising more rapidly

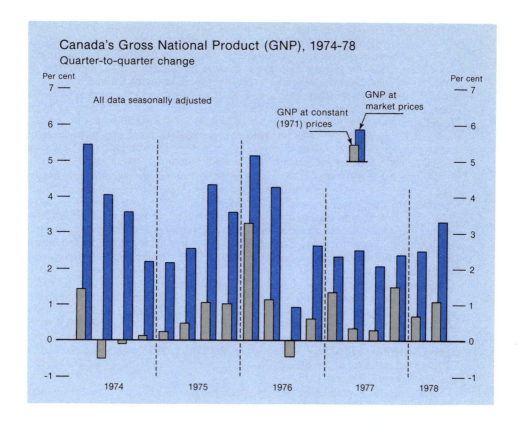

Canada's Gross National Product (GNP), 1974-78
Quarter-to-quarter change

than revenues, the deficit, on a national accounts basis for the government sector as a whole, rose from $3.4 billion in 1976 to $5.6 billion in 1977.

Industrial growth and change, 1949-77   21.2

In the three decades since 1949 there have been many changes in industrial output and industrial structure. In 1949 the output of the domestic economy (gross domestic product at factor cost in current dollars) was $15.3 billion; by 1977 this had increased more than twelvefold to $188.9 billion. But most of this change in output was due to changes in price and if the volume of production is expressed in terms of constant 1971 dollars the increase in output becomes less than a fourfold increase: from $27.8 billion in 1949 to $105.8 billion in 1977. This represents an average annual growth rate of 4.9%.

In the same period the number of persons employed increased from 5.0 million to 9.3 million for an average annual growth rate of 2.2%. Man-hours worked is a more sensitive measure of labour input and takes into account the decrease in the length of the average work week, the increase in vacations, work breaks and other time spent away from the job. Man-hours worked grew at a lesser rate than the number of persons employed for an average annual growth rate from 1949 to 1977 of 1.3%.

Productivity, measured by output per person employed and by output per man-hour, grew in this period at average annual rates of 2.5% and 3.3% respectively. Productivity is most often related to a single input, labour, but not all changes in productivity can be attributed solely to labour. While output per unit of labour input does reflect changes in education, skills, and effort of employed labour, it also reflects the contribution to output of capital investment, capacity utilization, changes in technology and process efficiency. Because of the difficulty in calculating the changes in these other inputs, measures of productivity are usually in terms of labour input.

The trend in output at the aggregate level obscures sharp fluctuations in growth both among industries and among regions. Even at the aggregate level, the pattern of output growth has shown considerable cyclical change. In analyzing the business cycle it has been customary to redefine output to exclude some components of output such as government and agriculture. However, the results of such analysis depend heavily on the definitions used and thus the number of business cycles identified during the last three decades may vary from four to seven or even eight, each one varying in duration and amplitude. Notwithstanding the insight given by more complex analyses, it is possible to generalize certain features of the trend in output in recent years.

From 1949 to 1959 the growth in aggregate real output averaged 5.0% although from 1954 to 1956 growth rates were recorded that were sharply higher than any others recorded during the entire post-war period. For example, the growth rate between 1954 and 1955 was 10.5% and between 1955 and 1956 it was 8.5%. The decade of the 1960s on the other hand was characterized by a high and sustained level of growth and the period remains unique for the rate of real output growth attained for its duration. From 1961 to 1969 the average annual rate of growth was 6.1%. In the middle of this period the growth of output peaked but the peak was sustained for several years.

Given that the 1960s showed unusual strength, it could be expected that the 1970s would see a return to lower growth rates. This has indeed been happening. Although the earlier years of the 1970s showed some strength particularly until 1973, the later years have brought exceptionally low rates of growth. Thus the 16 average annual growth rates of real output calculated from 1961 to 1977 show a declining trend. The growth rate from 1961 to 1977 of 5.2% is the highest; the 1962 to 1977 growth rate declines to 5.1% and the series continues the decline (with one exception) to the lowest growth rate of 2.8% between 1976 and 1977.

Since the 1950s production of both goods and services has moved in a generally cyclical manner. During the 1950s the amplitude of the cyclical movements in terms of the percentage deviation from trend was similar for both goods and services. Since the early 1960s, however, the production of services has shown considerably less cyclical volatility. Production of goods, on the other hand, has become more volatile with manufacturing being the principal contributor to cyclical instability. Production of goods other than manufactured goods, even including such volatile industries as agriculture and construction, has shown more stability than manufacturing since 1960.

In 1975 the sharp interruption in the growth of aggregate output pointed up a number of departures from the patterns of the 1950s and 1960s. The downturn in manufacturing was not only the most severe throughout the period under review but there were also unusual downturns in goods production other than manufacturing. Most notable of these were in metal mines, crude petroleum and natural gas, and electric power utilities. Downturns in these industries have been rare events in the past, and for electric power utilities unknown since 1945. The service industries displayed, in contrast to the 1950s and 1960s, sustained growth offsetting to some extent the downturn in goods production.

In 1977 real domestic product, as already noted, increased by only 2.8% following an increase of 4.6% in 1976; for the 1971-77 period the annual average growth rate was 4.0%. Service industries grew 3.3% and goods-producing industries by 2.0%. Average annual growth rates for the 1971-77 period were 4.7% for services and 2.8% for goods. Because the service industries include governmental and other non-commercial output it is sometimes useful to compare the commercial outputs of the goods and services industries. From 1961 to 1971 both the commercial goods and commercial services industries increased at nearly the same average annual rate of growth: 5.7% for goods and 5.8% for services. In the 1971 to 1977 period the growth rate of commercial goods production was about halved to 2.8%. But commercial services industries continued at nearly the same rate of growth, 5.7%.

There are a number of industries that provided considerable strength during the decade of the 1960s but which lagged in the 1970s. A notable example is the output from mines, quarries and oil wells which showed an average annual real growth rate of 6.2% from 1961 to 1971 but a rate of 1.4% from 1971 to 1977. Manufacturing output showed a growth rate of 6.3% from 1961 to 1971 but only 3.1% from 1971 to 1977. Among manufacturing industries the transportation equipment group showed strong growth in the 1960s, turning out a variety of products from locomotives to snowmobiles in addition to motor vehicles; the group as a whole showed a strong average annual growth rate of 12.3%. From 1971 to 1977 the growth rate for the group slumped to 4.5%, sustained at that level mainly by automobile and truck output. Among service industries, education showed enormous growth in the 1960s particularly in the post-secondary institutions. As a group, education and related services recorded an average annual real growth of 9.4% from 1961 to 1971 but from 1971 to 1977 this dropped to 1.4%. Air transport, which showed growth of 10.0% in the 1971-77 period, is down from the 15.0% annual real growth shown from 1961 to 1971.

The reasons why these industries grew rapidly in the 1960s and suffered a much diminished rate of growth in the 1970s are as diverse as the industries themselves. For most, a return to the higher rates of growth is not foreseeable so that, even while providing strength to a recovering economy, that strength is considerably muted as compared with the 1960s. For many other industries not mentioned in this brief summary, particularly those service industries that are continuing to provide strength in the 1970s, it is difficult to generalize the probable direction of their growth in output.

Aggregate productivity measures 21.2.1

The level of, and changes in, productivity have a vital influence on economic growth, overall cost structure, international competitiveness and, in the final analysis, on the quality of life. In the measurement of productivity, output is related to one or more kinds of inputs used in the production process.

The measures of productivity presented here relate output to a single input only, namely labour time. It must be emphasized that changes in output per unit of labour input cannot be attributed directly and solely to labour; such measures reflect not only changes in the skills and effort of the labour force but also the contribution of other productive resources with which labour works as well as the effectiveness with which all are combined and organized for the purpose of production. In other words, changes in technology, capital investment, capacity utilization, work flow, managerial skills and labour-management relations all have a bearing on movements in what is termed labour productivity. The measures of unit labour cost are the ratios of labour compensation to output. Unit labour cost can also be obtained as the ratio of average compensation to productivity; thus unit labour cost will increase when average compensation grows more rapidly than productivity.

Sources of data. The output components of the various indexes of output per unit of labour input and unit labour cost referred to here are the indexes of real domestic product (RDP) by industry. Developed within the conceptual framework of the system of national accounts, they measure in constant dollar terms the contribution of each component industry to total output.

The major sources for the employment and man-hour indexes were the monthly labour force and employment surveys and these were supplemented by data from such sources as the annual censuses of manufactures and mining and the decennial census of population. Since the data from these diverse sources varied considerably in their coverage, concepts and methods of compilation, care had to be exercised in selection, adaptation and combination of the data into aggregate measures of labour input which would be conceptually and statistically consistent, both internally and in relation to the output data. Labour force survey data were used for the paid worker estimates of agriculture and of fishing and trapping while those for manufacturing and mining were based on adjusted annual census data. Estimates for most of the remaining industry divisions were derived from employment survey data. Estimates of other than paid workers (own-account workers, employers and unpaid family workers) were derived mainly from the labour force survey. Estimates of average hours worked, needed for the indexes of output per man-hour, were also based on labour force survey data except in the case of manufacturing where man-hours data reported in the census of manufactures were also used. Labour compensation is the sum of wages, salaries, supplementary labour income and an imputed labour income for self-employed workers. For imputed labour income the average hourly income of paid workers is attributed to self-employed persons in the same industry division. Indexes of output per person employed, output per man-hour and unit labour cost for commercial industries and the major components are presented in Table 21.14.

Growth rates. Between 1961 and 1977, output per person employed in the commercial industries increased at an annual average rate of 2.8%. Output per man-hour increased more rapidly, 3.5% a year, reflecting reductions in the average work week. A review of the period since 1961 shows higher growth rates in labour productivity during the first decade than for the years since 1971. Output per man-hour increased at an average rate of 4.2% a year for the years 1961-71 and at an average rate of 1.8% a year for the years 1971-77. There is considerable variation in the year to year per cent change of output per man-hour for the years since 1971, with almost no growth in 1974 and 1975 and increases of 4.0% and 2.5% in 1976 and 1977 respectively.

Commercial goods-producing industries have recorded higher rate of growth in output per man-hour than have commercial service-producing industries, with the former increasing at an average rate of 4.4% a year and the latter at 2.5% a year for the years 1961-77. Output per man-hour in manufacturing increased 3.9% a year for the years 1961-77.

Unit labour cost for commercial industries increased at an average annual rate of 1.2% a year for the period 1961-66, 3.7% a year for the period 1966-71 and 10.3% a year for the period 1971-77. The increases in unit labour cost for manufacturing for the corresponding periods were 0.4%, 2.5% and 10.0%.

21.3 Price indexes

The price indexes provided here are classified into price indexes of goods and services paid by consumers at the retail level; industry and commodity price indexes at levels other than retail; purchase price indexes of selected capital goods; farm input price indexes; and securities price indexes.

21.3.1 Retail price indexes

This section describes price indexes currently available for commodities purchased by consumers at the retail level.

Consumer price index (CPI). The consumer price index measures the movement from month to month in retail prices of goods and services bought by a representative cross-section of the Canadian population living in urban centres with a population over 30,000. Since April 1973, the index has been based on the 1967 expenditure pattern of families ranging in size from two to six persons, with annual incomes from $4,000 to $12,000. The CPI is based on fixed weights; it measures the effect of changing prices on the cost of purchasing a given basket of goods and services containing about 300 items. In August 1975, the index was officially converted to a 1971 = 100 time reference base without altering the 1967 weighting pattern. Consumer expenditure patterns for 1974 were due to replace the 1967 weights in mid-1978.

Movements in the CPI up to the end of 1976 are contained in previous editions of the *Canada Year Book*. Based on changes in annual average indexes, the CPI advanced 8.0% in 1977, rising from a level of 148.9 in 1976 to 160.8. This increase was marginally above the 7.5% registered in 1976 but substantially lower than the annual increases of 10.9% and 10.8% observed in 1974 and 1975 respectively. On the basis of annual average index levels, the purchasing power of the 1971 consumer dollar declined from an average of 67 cents in 1976 to 62 cents in 1977.

Between December 1976 and December 1977 the All-Items CPI advanced 9.5% with all seven major components contributing to this change. The food index rose 15.4% and accounted for over two-fifths of the total advance while housing charges climbed

8.6% and were responsible for slightly less than one-third of the overall increase. Transportation charges, up 5.7%, contributed 9% to the total change while clothing prices, which increased by 7.5%, added a further 7%. The three remaining major components were responsible for adding an average of 4% each to the total change in the CPI. This 9.5% increase in the CPI was markedly higher than the moderate 5.8% increase registered between December 1975 and December 1976; the difference was primarily due to the 15.4% increase in food prices in 1977; in 1976 food prices declined 0.7%.

Another perspective of consumer price movements becomes available when the CPI is analyzed in terms of goods and services. Based on percentage changes in successive 12-month periods, the goods index, which had been declining at a faster rate relative to the services index for most of 1976, largely due to declining food prices, changed direction toward the end of that year and continued to move upward through 1977. The services index, on the other hand, continued a downward trend first observed in mid-1976. In the 12-month period December 1976 to December 1977, the goods index advanced 10.7% relative to a 7.5% increase registered by the service index. In the corresponding period December 1975 to December 1976, goods rose only 2.8% while services increased 11.1%.

For detail on movements in the CPI, see *Consumer price index,* Statistics Canada Catalogue 62-001, monthly or *Consumer prices and price indexes,* Statistics Canada Catalogue 62-010, quarterly. For additional information on methodology and weighting patterns, see *The consumer price index for Canada (1961 = 100) (revision based on 1967 expenditures),* Statistics Canada Catalogue 62-539.

Consumer price indexes for 14 selected regional cities are shown in Table 21.17. These indexes measure percentage changes in retail prices over time within the specified cities and should not be used to make comparisons of price levels between cities. On the basis of annual average index levels, consumer price increases between 1976 and 1977 ranged from 7.2% in Vancouver to 9.8% in Regina. Six of the 14 cities registered annual increases of less than 8% while six other cities showed increases of less than 9%. Between 1975 and 1976, annual average index level increases ranged from 6.7% in Montreal and Quebec City to 9.7% in Vancouver. For more detailed information on the movements of consumer price indexes for regional cities, see *Consumer price indexes for regional cities,* Statistics Canada Catalogue 62-009, monthly.

Table 21.18 provides data on the percentage changes in consumer price indexes based on annual averages for the period 1975 and 1976 for a selected group of countries. The data indicate that the CPI in Canada advanced faster than that of the United States in this period, but moved noticeably slower when compared with movements recorded for a number of industrialized European countries.

Intercity consumer price indexes. Table 21.37 provides indexes that compare levels of prices among 11 major Canadian cities. They express prices in each city as a percentage of the combined cities average which equals 100. The comparisons shown are those in effect as of June 1976 and September 1977 for six components of the CPI. The selected components in the table make up more than 60% of the average urban consumer's budget. For technical reasons, shelter costs (for both rented and owned facilities), clothing and restaurant meals are not included in the comparisons.

The retail prices used for the intercity comparisons are largely those routinely collected in each city for the production of the CPI. The exception is the component for food at home which is derived from data collected in a special survey undertaken in October 1975 and April 1977. Comparability between cities is ensured as far as possible by matching quotations for goods and services characterized by similar qualities and types of retail outlets. Since comparisons relate to prices that include sales and excise taxes, variations in the proportion of sales tax applied between provinces on largely non-food commodities may account for a large part of intercity price differentials.

Industry and commodity price indexes   21.3.2

Industry selling price indexes: manufacturing. Indexes of the selling prices of some 120 individual industries classified to manufacturing in the *Standard industrial classification* *manual,* Statistics Canada Catalogue 12-501 are produced and published monthly. In addition, indexes are available for major groups of manufacturing industries and for all manufacturing.

The indexes measure the change through time of prices received by manufacturers for their products. Prices reflected in the index are free on board from the manufacturing establishment, excluding taxes levied on manufacturers' sales. The items and weights in the current indexes are based on manufacturers' shipments in 1971. The composite index for manufacturing is presented in Table 21.19, for the years 1957-77.

The general wholesale index includes mainly manufacturers' prices but also incorporates prices of wholesalers and others. Prices are grouped according to a commodity classification scheme based on chief component material similarities. Indexes classified according to degree of manufacture are also available. In Table 21.20, the general wholesale index is presented for the period 1954-77.

Chemical and mineral process plant indexes. These are fixed-weighted indexes measuring the price movements of components representative of a processing installation bought by companies with a large processing activity, for example, chemicals, pulp and paper, petroleum refineries and mineral processing. The engineering and administration of the project, the construction of the buildings and the purchase and installation of the machinery and equipment required are all covered. Indexes for the period 1972-77 are given in Table 21.21.

21.3.3   Price indexes of selected capital goods

This section covers price indexes currently available for inputs into residential and non-residential building construction, completed new houses, highway construction, engineering construction, and machinery and equipment purchased by the construction and forestry industries.

Residential and non-residential building construction indexes. Price indexes of residential and non-residential building construction are fixed-weighted input indexes of materials and labour. They are presented in Tables 21.22 and 21.23 on the 1971 base. Indexes are available for the total inputs of material and labour for five regions and are also calculated for the inputs into individual trades in residential construction.

Since the building material prices in these indexes reflect price movements on purchasers' markets they contain federal sales tax changes. The wage rate component, from 1971 forward, is derived mainly from surveys of agreements conducted by the Canadian Construction Association for construction trades in various centres; these are base rates which reflect union scale or collective agreements. The combined indexes of materials and wage rates do not necessarily reflect changes in the price of construction output since they do not take account of changes in profit margins or in productivity but instead concentrate only on the prices of the inputs.

New housing price indexes. These are fixed-weighted indexes of builders' selling prices, measuring pure price change for work put in place and land for new residential construction. Monthly market selling prices agreed between builder and buyer in purchase contracts are reported monthly. In the price sample are major builders who construct 100 or more housing units a year (or possibly fewer units if comparable models can be priced through time). Prices cover dwelling structure, land and services to land if these are not provided separately by municipality; they exclude legal fees, provincial land transfer tax and similar costs to buyer for property acquisition. Price movements cover single unit houses in all cities and semi-detached and row condominiums in cities where this type of housing is widespread.

The indexes use a fixed-weighted formula. Weights reflect the relative importance of individual models in a firm's total output, and the relative importance of individual companies in the single family housing market. New firms brought into the sample are assigned weights reflecting their importance at time of entry. Quality changes not reflected in the index include model substitutions, changes in size and location of building lots, design and construction techniques and provision of such extra features as appliances. When actual cost data are not available quality changes are assumed to be proportional to differences in selling prices at a selected point in time. Indexes for 1972-77 with a base of 1971 = 100 are shown in Table 21.24. Indexes are also available on a 1976 base for Canada and 19 cities beginning in January 1975.

Highway construction indexes. These relate to prices paid by provincial governments in contracts awarded for highway construction. They are base-weighted indexes and measure the effect of price change on the cost of specified new highway construction projects represented by contracts of approximately $50,000 or more awarded by provincial governments. Indexes for the period 1956-76 are given in Table 21.25. Prices contained in the index are for units of construction work put in place by contractors. Also included are prices of materials usually supplied by the highways department.

Electrical utility construction indexes. The price indexes of electrical utility construction, which include those of distribution systems, transmission lines, transformer stations, hydroelectric and steam-electric generating stations, give an estimate of the impact of price change on the cost of materials, labour and equipment used in building and equipping electrical utilities. The index provides an estimate of how much it would cost to reproduce the base-period program of construction in another period using the same construction technology and assuming similar rates of profit and productivity. Price indexes for the years 1971-77 are presented in Table 21.26.

Price indexes of machinery and equipment. Table 21.27 shows base-weighted price indexes of machinery and equipment purchased by the construction and by the forestry industries. Prices used for the indexes are, for the most part, selling prices reported monthly by manufacturers, although in some cases distributors' prices are used. Prices of imported machinery and equipment are included in the index, represented either by commodity price indexes of the US government or by prices collected directly from foreign manufacturers. All prices have been adjusted as relevant to include duty, exchange and federal sales tax.

Farm input price indexes   21.3.4

Farm input price indexes measure, through time, changes in prices of commodities and services used as inputs into the agriculture industry. The weights for the indexes are based on a 1958 farm income and expenditure survey. The time base is 1961 = 100. Indexes for 41 series are published quarterly for Eastern, Western and all Canada. Annual averages for the total index are provided from 1968 to 1977 in Table 21.28.

Balance of international payments   21.4

The Canadian balance of international payments summarizes transactions between residents of Canada and those of the rest of the world. Current account transactions, which measure the flow of goods and services between Canada and other countries, are included, with minor adjustments, as a component of gross national expenditure. Capital account transactions between residents and non-residents are included in the financial flow accounts. A summary of the Canadian balance of international payments is provided in Table 21.29 for 1971-77. Table 21.30 contains some additional information on Canada's official international monetary assets for the period 1975-77 and Table 21.31 contains a detailed presentation of the Canadian balance of international payments for 1976-77.

Current account   21.4.1

During 1977 international transactions in goods, services and unilateral transfers led to a current account deficit of $4.2 billion. A surplus of $2.9 billion on merchandise trade was the second highest ever recorded but was more than offset by the continuing growth in the deficit on non-merchandise transactions which climbed to $7.1 billion.

Merchandise exports rose by more than 17% to $44.6 billion while merchandise imports increased by 13% to $41.7 billion, resulting in a trade surplus $1.8 billion higher than in 1976. Exports of automotive products accounted for the largest share of the overall increase in exports, rising by more than $2 billion to over $10 billion. Natural gas shipments rose sharply by over $450 million due mainly to price increases. Reflecting the maintenance of strength in the construction industry in the United States, exports of lumber continued the strong trend which began in 1976 and rose by over 40% to $2.4 billion. Newsprint sales increased by over $370 million but exports of woodpulp remained virtually static. In total, these commodities accounted for almost two-thirds of the growth in exports in 1977. Other significant increases were recorded for exports of aluminum and alloys, wheat, electricity, iron and steel, and chemicals. Exports of crude petroleum fell, despite price increases, by more than $400 million as the volume shipped dropped by almost one-third, in line with Canadian government policy to reduce Canada's crude petroleum exports.

Imports of automotive products increased by about $2 billion to account for over two-fifths of the growth of merchandise imports. Other increases occurred in a large number of commodities, the most notable being coffee, telecommunication equipment, iron and steel, chemicals and electronic computers. Reduced imports were recorded for wearing apparel, nickel and alloys, and office machinery.

Trade in crude petroleum yielded a deficit of nearly $1.5 billion compared to $1.1 billion in 1976. This was entirely due to the drop in the value of exports of crude petroleum as the level of imports was almost unchanged from 1976.

The deficit on non-merchandise transactions increased by more than one-third to $7.1 billion. Almost three-quarters of this increase was brought about by enlarged deficits on the interest and dividends and international travel accounts. Net payments of interest and dividends, consistently the largest component of the non-merchandise deficit, were more than $900 million higher than in 1976. Interest payments to non-residents surged ahead by almost 45% to nearly $2.7 billion due mainly to the servicing of the large foreign debt incurred by the provinces and their agencies. Interest receipts, which came mainly from earnings on Canadian monetary reserves, declined slightly again in 1977. Net payments of dividends at approximately $1 billion were 5% higher than 1976.

International travel transactions produced a deficit of over $1.6 billion, almost $500 million greater than in 1976. Canadians travelling abroad spent nearly $3.7 billion, an increase of $540 million. But non-resident travel expenditures in Canada rose by only 4% to $2 billion. In volume terms, Canadian travellers abroad increased by about 6% while non-resident travellers entering Canada dropped by approximately 2%.

21.4.2   Capital account

Capital movements during 1977 led to a net inflow of $2.8 billion comprising long-term inflows of nearly $4.4 billion and short-term outflows of $1.5 billion.

New issues of Canadian securities sold to non-residents provided an inflow of nearly $5.8 billion, the largest single component of the capital account. Although substantially lower than the corresponding inflow for 1976, which exceeded $9 billion, it was the second highest new issue inflow on record. Transactions in outstanding Canadian securities gave rise to offsetting movements, with a net inflow from trading in debt issues of $299 million and a net outflow from equity transactions of $91 million.

Foreign direct investment in Canada increased during 1977, resulting in a net capital inflow of $410 million following an outflow in 1976 due to repatriation of foreign-owned assets by Canadian government entities. Canadian direct investment abroad returned to the level of previous years, amounting to $790 million for the year, following a decline to $555 million in 1976.

Outflows due to loans and subscriptions by the federal government to foreign governments and international agencies increased substantially during 1977. At $522 million, the annual net outflow was about one-third higher than in 1976. Export credits directly or indirectly at the risk of the Canadian government also increased sharply, more than doubling in 1977 to lead to a net capital outflow of $532 million.

Short-term capital transactions resulted in a net outflow of $1.5 billion during the year. There were, however, substantial inflows in a number of categories. These included the Canadian chartered bank's foreign currency operations with non-residents, which produced a net inflow of nearly $1.4 billion. Canadian dollar deposits by non-residents also increased, giving rise to a net inflow of $225 million. In addition, there were inflows of funds through the Canadian money market, amounting to $422 million.

These inflows were more than offset, however, by changes in Canadian non-bank holdings of short-term funds abroad and by other short-term capital transactions. Canadian non-bank holdings of short-term funds abroad, mainly in the form of foreign bank deposits, gave rise to a net outflow of almost $500 million during the year, while other short-term capital transactions resulted in an outflow of nearly $3.2 billion. Other short-term capital transactions, the residual category of capital movements, contains the balancing item representing unidentified transactions in the balance of payments. The balancing item characteristically increases during periods of currency speculation, such as occurred during 1977 when the Canadian dollar came under consistent downward pressure. It was the largest component among other short-term capital transactions, consisting of estimated debit entries of approximately $2.3 billion in the year.

Official international monetary assets and liabilities 21.4.3

Excluding valuation adjustments, Canada's net official monetary assets declined by $1.4 billion in 1977. By the end of the year the level of Canada's official international reserves stood at $5 billion, equivalent to US$4.6 billion, the lowest level since 1970. Valuation adjustments raised the value of the reserves by $567 million. These adjustments represent the effect on the Canadian dollar value of the reserves of the depreciation of the Canadian dollar against other currencies and the Special Drawing Right (SDR), and the revaluation to market prices of gold transferred between the mint and the exchange fund account.

During the year there were decreases of $1.2 billion in holdings of foreign currencies (mainly US dollars), of $143 million in the Canadian position at the International Monetary Fund and of $80 million in holdings of SDRs, and an increase of $9 million in holdings of gold.

During the year, the Canadian dollar came under downward pressure in foreign exchange markets. It depreciated by US 7.72 cents during the year, closing at US 91.4 cents. The decline in the Canadian dollar started in the fourth quarter of 1976. It drifted downward throughout the year, interspersed by periods of stability and some slight appreciation, which in December more than offset the decline of the previous month. The Canadian dollar reached a low for the year of US 89.63 cents in October, the lowest level since September 1939, when it was US 89.29 cents. On a trade-weighted basis the Canadian dollar declined 9.4% during the year compared with 7.8% against the US dollar, reflecting the general depreciation of that currency against most other currencies. In response to pressure on the Canadian dollar, the government announced in October the establishment of a standby credit facility with the Canadian chartered banks under which the government could borrow up to US$1.5 billion. This facility was arranged to permit the exchange fund account to replenish its holdings of US dollars if and when circumstances required and was in addition to existing lines of credit, such as a swap facility for US$2 billion with the Federal Reserve System of the United States. By the end of the year no drawings had been made.

Foreign exchange 21.4.4

The dollar was established as the official currency of the united provinces of Canada on January 1, 1858, and extended to cover the new dominion by the Uniform Currency Act of 1870. The gold sovereign remained the standard for the Canadian dollar until 1910 when the currency was defined in terms of fine gold, making it the exact gold equivalent of the United States dollar. Both British and US gold coins, however, were legal tender in Canada during this period.

The 1870 act defined the Canadian dollar as 15/73 of the British gold sovereign, that is, the par rate of exchange between the dollar and the pound sterling was fixed at $4.866, making the Canadian currency the equivalent of the US dollar at parity. With only minor variations the value of the pound sterling in Canada remained at this level until 1914.

For a complete description of the fluctuations between Canadian and US dollars up to 1950 see the *Canada Year Book 1972* pp 1252-1254.

In September 1950, the finance minister announced that official fixed foreign exchange rates which had been in effect at varying levels since 1939 would be withdrawn in October and that the rate would henceforth be determined in the market for foreign exchange. This policy was carried out within the framework of exchange control until December 1951, when the foreign exchange control regulations were revoked by the Governor-in-Council, terminating the period of exchange control that had prevailed in Canada since 1939. The Foreign Exchange Control Act was repealed in 1952. In May 1962 the finance minister announced that the Canadian dollar was being stabilized at a fixed par value of 92.5 cents in terms of United States currency. This action was taken with the concurrence of the International Monetary Fund (IMF) and, in accordance with the articles of agreement of that organization, the Government of Canada undertook to maintain the Canadian exchange rate within a margin of 1% on either side of the established par value.

In May 1970, the federal government announced a decision not to maintain the exchange rate of the Canadian dollar within the 1% parity band prescribed by the IMF for the time being. The movements of the US dollar in Canadian funds from January 1970 to December 1977 are shown in Table 21.35 while Table 21.36 gives the value of the Canadian dollar in US funds and major overseas currencies. Details of Canada's official international reserves in US dollars are presented in Table 21.30.

21.5   Canada's international investment position

Canada has been among the world's largest importers of capital as the demand for real resources from abroad has been associated with a pattern of consistent current account deficits and net inflows of capital. This pattern was interrupted by sizable current account surpluses in 1970 and 1971 and then a much smaller one in 1973. In 1975 a current account deficit of $4.8 billion was registered followed by deficits of $4.2 billion in 1976 and 1977. In addition to capital inflows, which are a counterpart to net deficits on the current account, undistributed earnings accruing to non-residents have also been a significant factor. These two sources of funds have helped capital formation in Canada and stimulated production, earnings and employment.

Preliminary estimates produced on the basis of available data indicate that Canada's net indebtedness to other countries reached a book value of over $53.5 billion at the end of 1977. Canadian long-term investment abroad increased by some $3 to $27 billion. The major elements in this increase were outflows of long-term direct investment capital, reinvested earnings accruing to Canadians from their investments abroad and export credits. With the inclusion of short-term claims on non-residents, the total of Canada's external assets amounted to over $52.5 billion. Decreases in Canada's net official monetary assets and other short-term holdings of foreign exchange were more than offset by an increase in other short-term receivables.

Long-term foreign investment in Canada at $89 billion had increased by some $9 billion, reflecting predominantly an inflow of long-term portfolio capital and an increase in earnings accruing to non-residents. Net sales of government bonds were sizable though not as large as the record level registered in 1976. Other long-term liabilities including non-resident equity in Canada's assets abroad brought the total of long-term liabilities to about $94 billion. With the addition of various short-term claims Canada's external liabilities to non-residents exceeded $100 billion for the first time — some $106 billion at year-end 1977.

Canada's balance of international indebtedness rose by $7 billion to $42.3 billion at the end of 1975, the last year for which full estimates are available, almost double the growth of $3.6 billion in 1974. The balance of international indebtedness represents the amount by which Canada's outstanding liabilities or obligations to other countries exceed Canada's assets or claims against other countries. Following increases of 11% and 7% in 1974 and 1973 respectively, net indebtedness expanded sharply by 20%. At year-end 1975 Canada's gross liabilities, the amount owing to other countries in long-term and short-term forms, stood at $82.8 billion, an increase of 14% over 1974. This was offset partially by $40.5 billion in gross assets, the total amount owing to Canada by non-residents, which grew by 8% from $37.5 billion.

The term balance of international indebtedness is used here, in a balance of payments context, to include equity investment and contractual borrowing. It is determined by offsetting Canada's outstanding claims against other countries with the outstanding obligations of Canadians to other countries. The totals of international claims and obligations which have been acquired over the years, arising from capital transactions and other factors, constitute the international assets and liabilities determining a country's international investment position.

Canadian assets abroad   21.5.1

Among the assets, Canadian long-term investment abroad rose 14% in 1975 to $21.6 billion with direct investment rising 15% to $10.7 billion and miscellaneous investments, due largely to export credits, increasing 19% to $3.7 billion. Net official monetary assets which remained at $5.8 billion during the two previous years declined to $5.4 billion at year-end 1975 while short-term receivables rose 8% to $10.9 billion.

The book value of Canada's direct investment abroad at the end of 1975 amounted to nearly $10.7 billion, an increase of nearly $1.4 billion or about 15% over that recorded a year earlier. Net direct investment capital outflows abroad as recorded in the balance of payments accounted for some $795 million of this growth, with the reinvestment of earnings accounting for most of the balance.

An advance of $771 million boosted Canadian direct investment in the United States to nearly $5.7 billion, to maintain its share of total Canadian direct investment abroad at 53%. Over 75% of this growth occurred in the manufacturing and mining and petroleum industries although direct investment in merchandising at $257 million jumped 44% over 1974. A small increase of $40 million brought direct investment in all other North American countries, including those in the Caribbean, to $929 million.

Brazil, with nearly $1.1 billion of accumulated Canadian direct investment, was the second largest recipient country and represented nearly 90% of the $1.2 billion invested in South and Central America.

Investment in the United Kingdom, at more than $1 billion, rose $140 million, accounting for over half of the advance of $262 million registered for all countries in the European Economic Community. Offshore petroleum exploration in the United Kingdom sector of the North Sea contributed heavily to the increase of $39 million in mining and petroleum, which rose to $98 million at year-end 1975.

In the Afro-Asian countries Canadian direct investment grew 25% to $474 million. While this group represents only 4% of the direct investment total, investment in Asia (other than Japan) has grown significantly in recent years, with investment in mining and petroleum accounting for some 90% of the $81 million increase for 1975.

Canadian direct investment in manufacturing, over $5.3 billion, continued to represent half of total direct investment. Over three-fourths of the $628 million addition in manufacturing occurred in the non-ferrous metals, beverages, and iron and products sectors. Investment in mining and petroleum, and merchandising rose by 21% and 28% to nearly $2.4 billion and $485 million, respectively, while a 15% rise in the "other" category brought investment to over $1 billion. The utilities sector recorded a moderate gain of $80 million to nearly $1.5 billion.

External liabilities   21.5.2

At the end of 1975 Canadian gross external liabilities amounted to $82.8 billion. Long-term investment owned by non-residents comprised $68.6 billion or 83% of the total.

Direct investment, representing basically foreign capital (both long-term debt and equity) in Canadian enterprises attributable to the country of control, grew 10% to $39.8 billion. Direct investment continued to be the largest single component at 58%, down from 60% at year-end 1974.

During 1975 portfolio transactions in Canadian securities gave rise to record net capital inflows of $4.5 billion in the balance of payments. Foreign investment in government bonds increased 31% to reach $15.1 billion at the end of 1975 with provincial governments and their agencies the most active borrowers, accounting for nearly $12.4 billion, up almost $3 billion from 1974. The spate of new issues abroad was spurred by substantial capital requirements by provincial utilities to finance large-scale developments as well as by a lower rate of interest abroad. Foreign placements of bonds by municipal governments were also quite large, increasing 32% to a level of nearly $2.1 billion at the end of the year. Over 68% of the net growth in government bonds sold abroad was absorbed by US investors while 30% was provided by investors in overseas countries other than the United Kingdom.

Foreign investment in Canadian corporate portfolio investment rose 10%, the same rate as direct investment, to over $10.1 billion at year-end 1975. Miscellaneous investment — comprising such things as foreign investment in real estate, mortgages, private investment companies and assets administered for non-residents — increased marginally to nearly $3.6 billion.

Foreign long-term investment owned by US residents grew 13% to $52.9 billion at the end of 1975. Direct investment, the largest component, rose 11% to nearly $32.2 billion while holdings of government bonds advanced by 28% to nearly $11.1 billion. Of these claims by US investors, the capital stock of Canadian companies at nearly $26.2 billion comprised 49% while bonds and debentures, both government and corporate, represented 32%.

Long-term investment owned by investors in the United Kingdom rose 6%, the same rate as in 1974, to nearly $5.7 billion in 1975. There was a sizable gain of 19% in government bonds held while direct investment grew 5% to $3.7 billion. A breakdown of these liabilities to the United Kingdom by type of claim shows that about 60%, or $3.4 billion, was in the capital stock of Canadian companies while only 15% was invested in bonds and debentures.

Investment owned in all other countries increased 23% to $10 billion as investment in government bonds and other portfolio investment jumped 42% and 38% in the year to nearly $3.6 billion and $1.6 billion, respectively. A large number of new Canadian issues were sold in the Eurobond market in 1975, particularly in the latter part of the year. A number of factors such as the temporary removal of the 15% withholding tax on interest payments by corporations to non-residents on certain types of new issues, the high credit rating of the borrowers, and the high coupon rates offered made such issues highly marketable. Eurodollar rates in general were lower than Canadian bond yields and the registration requirements less costly to fulfil than those faced by borrowers in the United States. By type of claim, investment in bonds and debentures at nearly $4.4 billion comprised 44% of the total investment from this group of countries while investment in the capital stock of Canadian companies at nearly $3.6 billion, was 35%.

Long-term investment in manufacturing grew 10% to nearly $18.4 billion at the end of 1975. Within manufacturing the largest increases were registered by iron and products, chemicals and allied products and non-ferrous metals. About 90% of the growth in each of these three sectors was due to increased US investment. Investment in petroleum and natural gas and in mining and smelting both expanded 7% to $10.8 billion and $5.5 billion, respectively. Long-term investment in the financial industry and merchandising rose 16% and 13% to more than $7.5 billion and nearly $3 billion, respectively. A substantial portion of the increase in finance was registered by sales finance, insurance and real estate enterprises.

21.5.3 Foreign investment in Canadian industry

Foreign investment in Canadian industry is measured in terms of the proportion to long-term capital employed in selected industry groups of both foreign-owned long-term capital in those industries and total long-term capital employed in enterprises controlled by non-residents in those industries.

The estimated book value of total long-term capital, debt and equity employed in non-financial industries in Canada amounted to $129.7 billion at the end of 1974, an increase of 12% from 1973. Foreign-owned capital comprised 34% of this total, unchanged from the past three years. There was a shift, however, of one percentage point between investment owned in the United States, which fell to 27%, and investment owned in other countries, which rose to 7%. Foreign-controlled investment in Canadian non-financial industries declined to 33% from 34% in 1973. Between 1971 and 1974 there were successive declines of one percentage point annually in the proportion controlled by non-residents. Projections of the ratios of foreign control indicated, however, that they would remain at 33% for 1975 and 1976.

Financial activity in Canada, 1977   21.6

The total volume of funds raised by the domestic non-financial sectors of the Canadian economy increased substantially in 1977 with close to $45.7 billion raised, an increase of 11.9% over 1976. Despite the large increase in the demand for credit, recourse to foreign capital markets by Canadian borrowers was markedly reduced from 1976. In 1977 the foreign sector accounted for 10.6% of total funds supplied, down from 24% in 1976.

In the public sector, borrowing by the federal government was well above the pace recorded in 1977 and increased use was made of treasury bill financing. Net borrowing by the provincial government sector declined but substantial demands on capital markets were made and a large increase in the financial assets of the sector occurred. Borrowing by the associated enterprises of government, however, was well below the 1976 flow.

During 1977 there was a definite shift toward long-term borrowing on the part of private non-financial corporations. Long-term borrowing accounted for over 60% of the total of funds raised by this sector via conventional credit market channels. Equity financing, particularly floating-rate preferred shares, was a prominent feature of corporate finance in 1977.

Mortgage flows accounted for approximately one-third of the total volume of funds raised by the domestic non-financial sectors, up from 27% in 1976, while the flow of consumer credit declined both absolutely and as a proportion of total funds raised by the household sector. Direct lending assumed heightened importance in 1977 compared to 1976. Associated with the tendency toward long-term finance, the proportion of total funds supplied with the chartered banks as intermediaries declined while the relative importance of intermediation by other private domestic financial institutions increased.

An easing in demand for short-term credit, plus efforts of the monetary authorities to stimulate the growth of the money supply in the first half of the year, combined to put downward pressure on short-term interest rates, which dropped approximately 105 basis points over the first six months of the year. In the second half of the year short-term interest rates fluctuated, over a basis points range of 50 in the case of finance company paper, and 20 in the case of three-month treasury bills. In the United States, however, short-term interest rates moved upward as demand for short-term financing increased. These movements in short-term interest in Canada and the United States led to a rapid narrowing of the short-term uncovered interest rate differential between the two countries. Long-term interest rates fluctuated within fairly narrow bands in both countries; however, the movements in long-term interest rates were, on balance, such as to decrease the attractiveness of foreign capital markets to Canadian borrowers.

A summary of the financial market for 1976 and 1977 is presented in Table 21.38. More detailed data for individual sectors and summary matrices are available in the quarterly Statistics Canada publication *Financial flow accounts,* Catalogue 13-002.

The anti-inflation program   21.7

The anti-inflation program was implemented in the fall of 1975 by the federal government with co-operation of provincial governments. Prices had been rising by more than 10% per year for two consecutive years; demands for compensation increases were accelerating to more than 20% per year in an effort to make up for past losses and ensure against further losses in the future.

A four-point anti-inflation program was implemented, entailing: fiscal and monetary policies aimed at increasing total demand and production at a rate consistent with declining inflation; policies aimed at limiting growth of government expenditures and the rate of increase in public service employment; structural policies dealing with the special problems of energy, food and housing, to ensure a more efficient and competitive economy and to improve labour-management relations; a prices and incomes policy establishing guidelines for responsible social behaviour in determining prices and incomes of groups, and machinery for administering these guidelines and ensuring compliance where necessary.

This last area was administered by the Anti-Inflation Board, implemented on October 14, 1975; it started phasing out its control operation on April 14, 1978, with all businesses and employee groups to be free of controls by December 31, 1978.

By early 1978 progress had been made on all four fronts of the anti-inflation program. Growth of the money supply (defined as currency plus demand deposits) in 1976 and again in 1977 was held to just over 8%, down from the rate of almost 14% in 1975. Government spending growth decelerated during the life of the anti-inflation program from more than 20% per year in 1974 and 1975 to just over 12% in 1976 and 11% in 1977.

There was improvement in the growth rates of the underlying costs and prices in Canada. For example, in 1976 consumer prices, excluding the more volatile food prices, grew by 9.4%, down from 10.1% in 1975. A further decline to a rate of 7.9% was achieved in 1977 and through the first few months of 1978 this rate dropped to 6.9%.

Although food costs are a significant cost item in the budget of the average consumer, food prices have not been amenable to, nor subject to, the influence of the monetary, fiscal and controls policies of the anti-inflation program. It is only by looking at general price developments aside from those in the area of food that a reasonable assessment of underlying cost movements can be obtained.

The progressive deceleration in underlying price changes underway since late 1975 was closely related to the deceleration in the underlying cost increases that also were taking place. Labour, one of the more significant underlying cost items, showed dramatic improvement during two and a half years between the implementation of the program in October 1975, and the beginning of the phase-out of controls in April 1978. For example, labour department base wage-rate settlements data indicate that the average first year increase obtained by groups of 500 or more in the organized sector, excluding construction, declined from 21% in 1975 to 12.3% in 1976 and 8% in 1977. As a result of this deceleration, and an equally significant change in compensation increases in the unorganized sector, increases in average weekly earnings in the industrial sector declined from an average rate of 14.2% in 1975 to 12.1% in 1976 and 9.6% in 1977.

Sources

21.1 Gross National Product Division, System of National Accounts (Current) Branch, Statistics Canada.

21.2 Industry Product Division, System of National Accounts (Current) Branch, Statistics Canada.

21.2.1 Input-Output Division, System of National Accounts (Structural) Branch, Statistics Canada.

21.3 Prices Division, General Statistics Branch, Statistics Canada.

21.4 - 21.5 Balance of Payments Division, System of National Accounts (Current) Branch, Statistics Canada; Department of Banking and Financial Analysis, Bank of Canada.

21.6 Financial Flows and Multinational Enterprise Division, System of National Accounts (Current) Branch, Statistics Canada.

21.7 Editing Services, Communications Division, Anti-Inflation Board.

Tables

21.1 Gross national product in current and constant (1971) dollars, and index of gross national expenditure in constant (1971) dollars, 1961-77

Year	Gross national product		Index of gross national expenditure in constant (1971) dollars[r] (1971 = 100)
	Millions of current dollars	Millions of constant (1971) dollars	
1961	39,646	54,741	58.2
1962	42,927	58,475	62.1
1963	45,978	61,487	65.3
1964	50,280	65,610	69.7
1965	55,364	69,981	74.4
1966	61,828	74,844	79.5
1967	66,409	77,344	82.2
1968	72,586	81,864	87.0
1969	79,815	86,225	91.6
1970	85,685	88,390	93.9
1971[r]	94,450	94,450	100.0
1972[r]	105,234	100,248	106.1
1973[r]	123,560	107,812	114.1
1974[r]	147,175	111,766	118.3
1975[r]	165,445	112,955	119.6
1976	190,027	118,484	125.4
1977	207,714	121,566	128.7

21.2 National income and gross national product, by component, 1974-77 (million dollars)

Item	1974[r]	1975[r]	1976	1977
Wages, salaries and supplementary labour income	80,086	93,562	107,612	118,324
Military pay and allowances	1,203	1,336	1,495	1,645
Corporation profits before taxes[1]	19,811	20,159	20,102	22,400
Deduct: dividends paid to non-residents[2]	-1,646	-1,835	-1,729	-1,823
Interest, and miscellaneous investment income[3]	7,733	8,334	10,626	12,360
Accrued net income of farm operators from farm production[4]	3,823	3,813	3,147	2,720
Net income of non-farm unincorporated business, incl. rent[5]	7,084	7,600	8,613	9,345
Inventory valuation adjustment	-4,244	-2,938	-2,028	-3,213
Net national income at factor cost	113,850	130,031	147,838	161,758
Indirect taxes less subsidies	18,257	17,665	21,143	23,410
Capital consumption allowances and miscellaneous valuation adjustments	15,769	17,921	20,177	22,700
Residual error of estimate	-701	-172	869	-154
Gross national product at market prices	147,175	165,445	190,027	207,714

[1]Excludes profits of government business enterprises.
[2]Includes withholding tax.
[3]Includes profits (net of losses) of government business enterprises and other government investment income.
[4]Includes value of physical change in farm inventories and accrued earnings of farm operators arising out of operations of the Canadian Wheat Board.
[5]Includes net income of independent professional practitioners and imputed net rent on owner-occupied dwellings.

21.3 Gross national expenditure, 1974-77 (million dollars)

Item	1974[r]	1975[r]	1976	1977
Personal expenditure on consumer goods and services	83,441	97,016	110,543	121,955
Government current expenditure on goods and services	27,797	33,553	38,641	42,516
Gross fixed capital formation	34,260	40,120	44,309	47,650
Government	5,462	6,323	6,482	7,226
Business	28,798	33,797	37,827	40,424
Residential construction	8,776	9,308	12,258	12,594
Non-residential construction	9,178	11,691	11,977	13,522
Machinery and equipment	10,844	12,798	13,592	14,308
Value of physical change in inventories	3,052	-247	2,112	48
Government	26	31	41	43
Business				
Non-farm	3,330	-511	1,597	155
Farm and grain in commercial channels	-304	233	474	-150
Exports of goods and services	38,930	40,364	45,385	52,482
Deduct: imports of goods and services	-41,006	-45,533	-50,094	-57,092
Residual error of estimate	701	172	-869	155
Gross national expenditure at market prices	147,175	165,445	190,027	207,714

21.4   Gross national expenditure in constant (1971) dollars, 1974-77 (million dollars)

Item	1974ʳ	1975ʳ	1976	1977
Personal expenditure on consumer goods and services	67,357	70,784	75,105	77,186
Government current expenditure on goods and services	20,656	21,571	21,757	22,225
Gross fixed capital formation	25,732	26,744	26,949	26,800
Government	3,957	4,131	3,899	4,023
Business	21,775	22,613	23,050	22,777
Residential construction	5,943	5,547	6,523	6,133
Non-residential construction	6,898	7,828	7,328	7,598
Machinery and equipment	8,934	9,238	9,199	9,046
Value of physical change in inventories	2,334	-526	1,257	-64
Government	18	16	22	22
Business				
Non-farm	2,421	-583	1,068	122
Farm and grain in commercial channels	-105	41	167	-208
Exports of goods and services	25,570	23,930	26,060	28,067
Deduct: imports of goods and services	-30,453	-29,707	-32,132	-32,760
Residual error of estimate	570	159	-512	112
Gross national expenditure in constant (1971) dollars	111,766	112,955	118,484	121,566

21.5   Year-to-year percentage change in gross national expenditure, 1974-77

Item	1974ʳ	1975ʳ	1976	1977
Personal expenditure on consumer goods and services				
Value	17.1	16.3	13.9	10.3
Volume	5.4	5.1	6.1	2.8
Price	11.0	10.7	7.4	7.3
Government current expenditure on goods and services				
Value	20.7	20.7	15.2	10.0
Volume	4.3	4.4	0.9	2.2
Price	15.6	15.5	14.2	7.7
Gross fixed capital formation				
Value	23.0	17.1	10.4	7.5
Volume	5.5	3 9	0.8	-0.6
Price	16.5	12.7	9.6	8.2
Government				
Value	26.9	15.8	2.5	11.5
Volume	5.5	4.4	-5.6	3.2
Price	20.2	10.9	8.6	8.1
Business				
Value	22.3	17.4	11.9	6.9
Volume	5.5	3.8	1.9	-1.2
Price	16.0	13.0	9.8	8.2
Residential construction				
Value	18.8	6.1	31.7	2.7
Volume	-0.4	-6.7	17.6	-6.0
Price	19.3	13.6	12.0	9.3
Non-residential construction				
Value	25.3	27.4	2.4	12.9
Volume	7.6	13.5	-6.4	3.7
Price	16.4	12.2	9.4	8.9
Machinery and equipment				
Value	22.8	18.0	6.2	5.3
Volume	8.2	3.4	-0.4	-1.7
Price	13.6	14.1	6.7	7.0
Exports of goods and services				
Value	26.7	3.7	12.4	15.6
Volume	-2.2	-6.4	8.9	7.7
Price	29.6	10.8	3.3	7.3
Imports of goods and services				
Value	32.5	11.0	10.0	14.0
Volume	9.4	-2.4	8.2	2.0
Price	21.1	13.8	1.7	11.8
Gross national expenditure at market prices				
Value	19.1	12.4	14.9	9.3
Volume	3.7	1.1	4.9	2.6
Price	14.9	11.2	9.5	6.5

21.6   Personal income, by source and by province, 1974-77 (million dollars)

Source and province or territory	1974[r]	1975[r]	1976	1977
Source				
Wages, salaries and supplementary labour income	80,086	93,562	107,612	118,324
Military pay and allowances	1,203	1,336	1,495	1,645
Net income received by farm operators from farm production	3,320	3,919	3,112	2,694
Net income of non-farm unincorporated business including rent	7,084	7,600	8,613	9,345
Interest, dividends and miscellaneous investment income	10,971	12,227	14,870	16,518
Current transfers				
From government				
Transfer payments to persons (excl. interest on public debt)	13,880	17,059	19,451	22,304
Capital assistance	71	147	119	95
From corporations (charitable and other contributions and bad debts)	215	239	265	279
From non-residents	225	256	258	281
Total, personal income	117,055	136,345	155,795	171,485
Province or territory				
Newfoundland	1,902	2,246	2,564	..
Prince Edward Island	398	486	558	..
Nova Scotia	3,379	3,876	4,420	..
New Brunswick	2,546	3,096	3,484	..
Quebec	29,010	33,795	39,037	..
Ontario	47,151	53,988	61,375	..
Manitoba	4,994	5,881	6,535	..
Saskatchewan	4,537	5,637	6,273	..
Alberta	9,052	11,043	12,781	..
British Columbia	13,687	15,823	18,230	..
Yukon Territory and Northwest Territories	292	347	400	..
Foreign countries[1]	107	127	138	..

[1]Income of Canadians temporarily abroad, including pay and allowances of Canadian Armed Forces abroad.

21.7   Disposition of personal income, 1974-77 (million dollars)

Item	1974[r]	1975[r]	1976	1977
Personal expenditure on consumer goods and services	83,441	97,016	110,543	121,955
Current transfers				
To government				
Income taxes	16,155	18,023	21,127	22,992
Succession duties and estate taxes	178	149	143	138
Employer and employee contributions to social insurance and government pension funds	4,864	5,886	7,033	7,499
Other	1,127	1,153	1,463	1,591
To corporations (transfer portion of interest on the consumer debt)	1,498	1,637	1,842	2,124
To non-residents	224	235	246	258
Personal saving	9,568	12,246	13,398	14,928
Total, personal income	117,055	136,345	155,795	171,485

21.8   Personal expenditure on consumer goods and services, 1974-77 (million dollars)

Item	1974[r]	1975[r]	1976	1977
Food and non-alcoholic beverages	13,057	15,230	16,787	18,297
Tobacco and alcoholic beverages	4,804	5,529	6,097	6,649
Clothing, footwear and accessories	5,944	6,782	7,716	8,198
Gross rent, fuel and power	14,209	16,399	19,072	21,701
Furniture, furnishing and household equipment and operation	8,868	10,251	11,410	12,251
Transportation and communication	12,167	14,196	16,440	18,149
Medical care and health services	2,404	2,790	3,200	3,517
Other	21,988	25,839	29,821	33,193
Total	83,441	97,016	110,543	121,955
Durables	13,650	16,026	17,804	19,167
Semi-durables	10,685	12,173	13,733	14,763
Non-durables	26,322	30,364	34,460	37,806
Services	32,784	38,453	44,546	50,219

21.9 Federal, provincial and local government revenue and expenditure[1], 1974-77 (million dollars)

Item	1974[r]	1975[r]	1976	1977
Revenue				
Direct taxes: persons and unincorporated business				
Income taxes	16,155	18,023	21,127	22,992
Succession duties and estate taxes	178	149	143	138
Employer and employee contributions to social insurance				
and government pension funds	4,864	5,886	7,033	7,499
Direct taxes: corporate and government business enterprises	6,943	7,891	7,907	7,930
Direct taxes: non-residents (withholding taxes)	430	465	504	533
Indirect taxes	20,876	21,518	24,465	26,714
Other current transfers from persons	1,127	1,153	1,463	1,591
Investment income				
Interest and royalties	5,257	6,378	7,386	8,411
Remitted profits of government business enterprises	676	726	730	1,033
Total, revenue	56,506	62,189	70,758	76,841
Current expenditure				
Purchases of goods and services	27,797	33,553	38,641	42,516
Transfer payments to persons	13,880	17,059	19,451	22,304
Current transfers to non-residents	407	591	540	629
Interest on the public debt	5,425	6,548	7,974	8,990
Capital assistance	326	482	501	549
Subsidies	2,619	3,853	3,322	3,304
Saving	6,052	103	329	-1,451
Total, current expenditure	56,506	62,189	70,758	76,841
Surplus or deficit (on a national accounts basis)				
Saving	6,052	103	329	-1,451
Add: capital consumption allowances	2,130	2,483	2,830	3,169
Deduct: gross capital formation	-5,488	-6,354	-6,523	-7,269
Equals: surplus or deficit	2,694	-3,768	-3,364	-5,551

[1]Excludes current transfers from other levels of government.

21.10 Annual growth rates of real domestic product, by industry, selected periods, 1961-77

Industry	1961-66	1967-77	1961-77	1971-77
Agriculture	5.5	1.2	1.1	0.5
Forestry	6.6	2.1	3.5	1.5
Fishing and trapping	2.6	-0.8	0.1	1.2
Mines (incl. milling), quarries and oil wells	5.9	3.7	5.0	1.4
Manufacturing	8.5	4.0	5.2	3.1
Non-durables	6.3	3.7	4.5	2.7
Durables	11.0	4.3	5.9	3.5
Construction	7.0	3.3	3.9	2.6
Electric power, gas and water utilities	7.3	7.3	7.6	6.0
Transportation, storage and communication	6.9	5.9	6.3	5.2
Transportation	7.4	5.1	6.0	3.6
Trade	6.8	5.4	5.6	5.0
Wholesale	8.3	4.7	5.5	3.8
Retail	5.8	5.9	5.6	5.7
Finance, insurance and real estate	4.8	5.1	5.0	5.0
Community, business and personal services	7.0	5.0	5.9	4.9
Public administration and defence	1.9	4.0	3.5	3.9
Real domestic product	6.7	4.6	5.2	4.0

21.11 Quantity indexes of real domestic product at factor cost, by industry of origin, 1973-77 (1971=100)

Industry	1973[r]	1974[r]	1975[r]	1976	1977
Agriculture	90.2	80.6	89.2	100.4	96.4
Forestry	122.1	119.6	97.5	108.5	119.7
Fishing and trapping	100.4	89.8	86.0	102.0	112.3
Mines (incl. milling), quarries and oil wells	118.9	118.3	109.3	110.4	114.1
Manufacturing	116.1	120.1	114.2	120.0	122.1
Non-durables	113.7	116.7	111.8	117.3	120.9
Durables	118.7	123.4	116.8	122.7	126.4
Construction	107.3	112.1	116.3	116.6	113.5
Electric power, gas and water utilities	119.6	127.4	126.7	137.8	149.3
Transportation, storage and communication	115.7	122.8	125.2	130.5	136.3
Trade	117.6	125.5	125.9	133.1	134.4
Wholesale	117.6	126.2	121.2	126.2	126.4
Retail	117.6	125.0	129.0	137.6	139.7
Finance, insurance and real estate	112.1	118.1	122.3	128.2	134.6
Community, business and personal services	110.5	115.9	122.0	127.7	132.4
Public administration and defence	109.8	114.1	118.8	122.8	125.0
Real domestic product	113.0	117.6	118.7	124.2	127.8

21.12 Percentage share of gross domestic product by goods- and service-producing industries, 1949, 1961 and 1971

Item	1949	1961	1971
Goods-producing industries	53.080	44.401	40.572
Service-producing industries	46.920	55.559	59.428
Gross domestic product	100.000	100.000	100.000

21.13 Census value added in goods-producing industries, by industry and province, 1975

Industry	Newfoundland		Prince Edward Island		Nova Scotia		New Brunswick	
	$'000	%	$'000	%	$'000	%	$'000	%
Agriculture	..		42,891	28.7	58,942	4.2	50,537	3.8
Forestry	36,628	3.5	—	—	20,364	1.4	63,961	4.9
Fisheries	45,728	4.4	12,410	8.3	91,010	6.4	25,515	1.9
Trapping	91	..	70	..	340	..	465	..
Mining	296,144	28.6	—	—	77,674	5.5	79,454	6.0
Electric power	123,830	11.9	7,254	4.8	78,771	5.5	71,993	5.5
Manufacturing	224,139	21.6	36,741	24.5	700,019	49.5	610,085	46.1
Construction	310,568	30.0	50,577	33.7	388,347	27.5	420,845	31.8
Total	1,037,128	100.0	149,943	100.0	1,415,467	100.0	1,322,855	100.0

Industry	Quebec		Ontario		Manitoba		Saskatchewan	
	$'000	%	$'000	%	$'000	%	$'000	%
Agriculture	798,240	4.9	1,534,804	5.5	638,622	25.1	1,884,767	50.7
Forestry	308,411	1.9	176,263	0.6	19,924	0.8	16,017	0.4
Fisheries	15,433	0.1	11,052	..	5,961	0.2	2,238	0.1
Trapping	3,793	..	6,142	..	2,555	0.1	2,007	0.1
Mining	630,831	3.8	1,367,275	4.8	202,357	7.9	752,089	20.2
Electric power	807,071	4.9	919,619	3.2	158,584	6.2	71,403	1.9
Manufacturing	10,458,512	63.6	20,122,934	70.6	1,080,398	42.5	455,185	12.2
Construction	3,414,187	20.8	4,366,672	15.3	438,466	17.2	533,349	14.4
Total	16,436,478	100.0	28,504,761	100.0	2,546,867	100.0	3,717,055	100.0

Industry	Alberta		British Columbia		Yukon Territory and Northwest Territories		Canada	
	$'000	%	$'000	%	$'000	%	$'000	%
Agriculture	1,371,885	12.9	249,186	3.6	..	..	6,629,874	9.1
Forestry	23,900	0.2	460,059	6.6	—	—	1,125,527	1.5
Fisheries	975	..	79,681	1.2	717	0.3	290,720	0.4
Trapping	2,829	..	1,350	..	1,889	0.9	21,531	..
Mining	5,362,642	50.4	790,740	11.2	190,826	90.3	9,750,032	13.4
Electric power	188,847	1.8	266,301	3.8	11,911	5.7	2,705,584	3.7
Manufacturing	1,638,347	15.4	3,383,285	48.6	5,956	2.8	38,715,600	53.1
Construction	2,056,492	19.3	1,738,216	25.0	1	1	13,717,719	18.8
Total	10,645,917	100.0	6,968,818	100.0	211,299	100.0	72,956,587	100.0

[1]Included with British Columbia.

21.14 Aggregate productivity measures, for selected years, 1961-77 (1971=100)

Year and industry		Output	Persons employed	Man-hours	Output per person	Output per man-hour	Labour compensation	Unit labour cost
Commercial industries								
1961		57.0	80.1	86.6	71.1	65.8	44.4	78.0
1966		79.8	93.6	97.7	85.3	81.7	66.3	83.0
1971		100.0	100.0	100.0	100.0	100.0	100.0	100.0
1972		105.9	103.2	102.3	102.6	103.5	111.1	104.9
1973		114.3	108.3	107.6	105.5	106.2	129.6	113.4
1974		119.3	113.7	112.2	104.9	106.3	154.6	129.6
1975		119.8	114.8	112.6	104.3	106.4	178.7	149.2
1976		126.1	116.8	114.0	108.0	110.6	203.6	161.4
1977		130.1	119.7	114.7	108.7	113.4	223.1	171.5
Annual rate of change[1]								
1961-77	%	5.3	2.5	1.7	2.8	3.5	10.8	5.2
1961-71	%	5.7	2.3	1.5	3.3	4.2	8.8	2.9
1971-77	%	4.3	3.1	2.4	1.2	1.8	15.1	10.3
Commercial non-agricultural industries								
1961		56.5	75.1	79.7	75.3	71.0	43.1	76.2
1966		79.0	92.3	96.2	85.6	82.1	65.6	83.0
1971		100.0	100.0	100.0	100.0	100.0	100.0	100.0
1972		106.7	104.0	103.5	102.6	103.1	111.5	104.5
1973		115.3	109.9	109.4	105.0	105.4	129.6	112.4
1974		121.0	115.7	114.4	104.6	105.7	155.1	128.2
1975		121.1	116.8	114.6	103.7	105.6	179.3	148.0
1976		127.3	119.1	116.6	106.8	109.1	204.1	160.4
1977		131.5	122.4	117.9	107.5	111.5	223.2	169.7
Annual rate of change[1]								
1961-77	%	5.5	3.0	2.4	2.4	3.0	11.0	5.2
1961-71	%	5.9	3.0	2.4	2.8	3.5	9.1	3.0
1971-77	%	4.5	3.4	2.8	1.0	1.6	15.1	10.2

21.14 Aggregate productivity measures, for selected years, 1961-77 (1971=100) (concluded)

Year and industry	Output	Persons employed	Man-hours	Output per person	Output per man-hour	Labour compen-sation	Unit labour cost
Commercial goods-producing industries							
1961	56.5	91.7	97.3	61.6	58.1	47.2	83.5
1966	82.1	101.7	105.4	80.7	77.9	70.5	85.9
1971	100.0	100.0	100.0	100.0	100.0	100.0	100.0
1972	104.6	100.9	100.1	103.7	104.7	109.2	104.4
1973	113.0	104.8	104.4	107.8	108.3	129.2	114.3
1974	115.7	107.5	106.8	107.6	108.3	153.2	132.5
1975	112.5	106.1	105.1	106.0	107.0	174.1	154.8
1976	117.8	107.1	106.1	110.0	111.1	196.4	166.6
1977	120.2	107.5	104.5	111.8	115.0	215.7	179.5
Annual rate of change[1]							
1961-77 %	4.8	0.9	0.3	3.9	4.4	10.0	5.0
1961-71 %	5.7	0.9	0.3	4.7	5.4	8.2	2.3
1971-77 %	2.8	1.3	0.9	1.6	1.9	14.4	11.3
Commercial service-producing industries							
1961	57.5	69.2	75.7	83.1	76.0	41.5	72.2
1966	77.3	85.8	90.0	90.1	85.9	61.8	79.9
1971	100.0	100.0	100.0	100.0	100.0	100.0	100.0
1972	107.2	105.4	104.7	101.8	102.4	113.1	105.5
1973	115.5	111.6	110.8	103.5	104.3	130.1	112.7
1974	123.0	119.5	117.7	102.9	104.5	156.1	126.9
1975	127.1	123.1	120.3	103.3	105.7	183.6	144.5
1976	134.4	125.9	122.1	106.8	110.0	211.3	157.2
1977	139.9	131.2	125.1	106.6	111.9	230.8	165.0
Annual rate of change[1]							
1961-77 %	5.8	4.1	3.2	1.7	2.5	11.6	5.5
1961-71 %	5.8	3.9	2.9	1.8	2.8	9.6	3.6
1971-77 %	5.7	4.6	3.9	1.0	1.8	15.8	9.5
Agriculture							
1961	66.8	133.6	142.2	50.0	47.0	78.8	117.9
1966	96.7	106.5	110.0	90.7	87.9	83.0	85.8
1971	100.0	100.0	100.0	100.0	100.0	100.0	100.0
1972	87.9	94.3	92.8	93.2	94.7	101.8	115.8
1973	90.2	91.5	92.9	98.6	97.1	129.5	143.5
1974	80.6	92.9	94.6	86.8	85.2	143.1	177.6
1975	89.2	94.1	96.5	94.8	92.4	164.0	183.9
1976	100.4	91.8	93.5	109.3	107.4	193.1	192.4
1977	96.4	91.1	89.0	105.8	108.3	221.9	230.2
Annual rate of change[1]							
1961-77 %	1.1	−2.5	−2.8	3.7	4.0	6.2	5.1
1961-71 %	2.1	−3.0	−3.7	5.3	6.0	3.0	0.9
1971-77 %	0.5	−1.1	−1.1	1.6	1.6	15.0	14.4
Manufacturing							
1961	54.5	83.6	84.4	65.1	64.5	46.9	86.0
1966	81.6	101.3	103.4	80.5	78.9	71.9	88.2
1971	100.0	100.0	100.0	100.0	100.0	100.0	100.0
1972	106.9	102.9	102.5	103.9	104.3	110.6	103.4
1973	116.1	107.4	107.0	108.1	108.6	126.7	109.1
1974	120.1	109.5	108.7	109.6	110.4	148.9	124.0
1975	114.2	106.8	105.2	106.9	108.6	165.0	144.4
1976	120.0	107.9	107.0	111.2	112.2	189.1	157.6
1977	123.7	108.1	106.0	114.5	116.7	206.3	166.8
Annual rate of change[1]							
1961-77 %	5.2	1.5	1.2	3.6	3.9	9.6	4.2
1961-71 %	6.3	2.1	1.9	4.2	4.3	8.3	1.9
1971-77 %	3.1	1.2	0.9	1.9	2.2	13.4	10.0

[1]Annual rates of change are calculated from the least squares of the logarithms of the index numbers for all years in the time span.

21.15 Consumer price indexes for specific groups, 1966-77 (1971=100)

Year	Food	Housing	Clothing	Transpor-tation	Health and personal care	Recreation, education and reading	Tobacco and alcohol	All-items index
Group weight as a percentage of total[1]	24	32	11	16	4	7	6	100
1966	88.7	79.5	87.0	82.6	81.8	80.1	83.7	83.5
1967	89.9	82.9	91.4	86.1	86.0	84.1	85.8	86.5
1968	92.8	86.7	94.1	88.3	89.5	88.3	93.6	90.0
1969	96.7	91.2	96.7	92.4	93.8	93.5	97.2	94.1
1970	98.9	95.7	98.5	96.1	98.0	96.8	98.4	97.2
1971	100.0	100.0	100.0	100.0	100.0	100.0	100.0	100.0
1972	107.6	104.7	102.6	102.6	104.8	102.8	102.7	104.8
1973	123.3	111.4	107.7	105.3	109.8	107.1	106.0	112.7
1974	143.4	121.1	118.0	115.8	119.4	116.4	111.8	125.0
1975	161.9	133.2	125.1	129.4	133.0	128.5	125.3	138.5
1976	166.2	148.0	132.0	143.3	144.3	136.2	134.3	148.9
1977	180.1	161.9	141.0	153.3	155.0	142.7	143.8	160.8

[1]These weights, indicating the components' relative importance, are based on 1967 expenditures and have been incorporated since May 1973; prior to May 1973, the weights reflected 1957 expenditures.

21.16 Consumer price indexes reclassified by goods and services[1], 1966-77 (1971=100)

Year	Goods					Total services	All-items
	Non-durable		Semi-durable	Durable	Total		
	Food	Other					
Group weight as a percentage of total[2]	25	17	12	14	67	33	100
1966	88.7	84.5	87.1	92.2	87.8	76.1	83.5
1967	89.9	86.9	91.6	94.7	90.0	80.2	86.5
1968	92.8	91.5	94.5	96.2	93.4	84.4	90.0
1969	96.7	94.7	97.1	97.2	96.3	90.0	94.1
1970	98.9	97.0	98.7	98.4	98.2	95.3	97.2
1971	100.0	100.0	100.0	100.0	100.0	100.0	100.0
1972	107.6	102.9	102.4	101.2	104.6	105.2	104.8
1973	123.3	108.3	107.0	102.6	113.7	111.7	112.7
1974	143.4	120.4	117.2	110.4	128.1	120.5	125.0
1975	161.9	136.0	124.0	118.9	142.0	133.4	138.5
1976	166.2	147.6	129.6	125.3	149.0	149.6	148.9
1977	180.1	159.0	138.3	131.7	160.0	163.2	160.8

[1]The previous supplementary classification (by type of commodity and service) has been revised. Historical series relating to the revised classification replace the previously published indexes.
[2]These weights, indicating the components' relative importance, are based on 1967 expenditures and have been incorporated since May 1973; prior to May 1973, the weights reflected 1957 expenditures.

21.17 Consumer price indexes for regional cities, 1966-77 (1971=100)

Year	City and province							
	St. John's, Nfld.	Hali-fax, NS	Saint John, NB	Que-bec, Que.	Mont-real, Que.	Ottawa, Ont.	Toronto, Ont.	Thunder Bay, Ont.
1966	87.4	85.0	86.1	–	86.8	84.5	86.2	–
1967	89.8	87.0	88.7	–	90.2	86.5	88.8	–
1968	93.8	90.4	91.9	–	93.3	90.6	92.2	–
1969	96.6	94.6	95.7	–	96.2	94.2	95.9	–
1970	98.5	98.5	98.6	98.3	98.2	97.5	98.4	98.6
1971	100.0	100.0	100.0	100.0	100.0	100.0	100.0	100.0
1972	105.0	103.7	104.5	102.8	103.8	104.1	104.1	103.7
1973	113.7	110.9	112.2	110.1	110.7	111.9	111.3	110.8
1974	128.3	121.6	123.7	122.5	123.0	123.9	123.0	122.6
1975	143.0	133.9	138.1	135.0	136.4	135.8	136.1	136.4
1976	154.2	145.1	148.0	144.1	145.6	146.1	146.0	148.4
1977	165.7	156.4	159.2	156.6	157.8	158.0	157.3	159.9

Year	Winni-peg, Man.	Saska-toon, Sask.	Regina, Sask.	Edmon-ton, Alta.	Calgary, Alta.	Van-couver, BC
1966	85.0		87.8	83.9		84.3
1967	88.1		90.2	87.2		87.4
1968	91.9		93.8	91.0		90.6
1969	95.7		97.0	94.8		93.7
1970	98.8		99.0	97.6		96.9
1971	100.0		100.0	100.0		100.0
1972	103.8		104.0	103.9		104.0
1973	110.4		110.0	110.6		110.0
1974	122.2	120.4	120.4	122.5	120.9	123.9
1975	137.4	133.9	133.1	135.8	134.6	137.7
1976	149.3	144.9	145.0	146.7	145.8	151.0
1977	161.4	157.3	159.2	159.9	158.0	161.8

21.18 Percentage change in consumer price indexes in Canada and other countries, 1975-76

Country	% change	Country	% change
North America		Africa	
Canada	7.5	Kenya (Nairobi)	10.9
Mexico (Mexico City)	15.8	South Africa (European population)	11.1
United States	5.8	Zaire (Kinshasa)	86.2
South America		Asia	
Argentina (Buenos Aires)	444.1	India	-7.8
Brazil (São Paulo)	35.3	Indonesia (Jakarta)	19.9
Chile (Santiago)	211.9	Korea, Republic of	15.3
		Pakistan (Karachi) — industrial workers	7.2
Europe		Sri Lanka (Colombo)	1.2
Belgium	9.2		
Denmark	9.0		
France	9.2	Australasia	
Germany, Federal Republic of	4.5	Australia	13.6
Greece	13.3	New Zealand	17.0
Ireland	18.0		
Netherlands	8.8	Middle East	
Sweden	10.3	Iran	11.3
Switzerland	1.7	Israel	31.3
		Turkey (Ankara)	15.3
United Kingdom	16.5		

21.19   Industry selling price index (manufacturing), 1958-77 (1971=100)

Year	Index	Year	Index	Year	Index	Year	Index
1958	81.4	1963	84.4	1968	92.3	1973	116.2
1959	82.1	1964	85.1	1969	95.8	1974	138.3
1960	82.2	1965	86.2	1970	98.1	1975	153.7
1961	82.4	1966	88.7	1971	100.0	1976	161.6
1962	83.3	1967	90.4	1972	104.5	1977	173.9P

21.20   General wholesale index annual averages, 1954-77 (1935-39=100)

Year	Average	Year	Average	Year	Average	Year	Average
1954	217.0	1960	230.9	1966	259.5	1972	310.3
1955	218.9	1961	233.3	1967	264.1	1973	376.9
1956	225.6	1962	240.0	1968	269.9	1974	461.3
1957	227.4	1963	244.6	1969	282.4	1975	491.6
1958	227.8	1964	245.4	1970	286.4	1976	512.4
1959	230.6	1965	250.4	1971	289.9	1977	559.3P

21.21   Chemical and mineral process plant price indexes, 1972-77 (1971=100)

Year	Machinery and equipment	Field erection	Buildings	Engineering and administration	Total
1972	103.1	110.2	106.9	106.6	105.3
1973	108.7	120.2	117.1	113.4	112.5
1974	133.5	133.6	134.4	128.5	132.9
1975	160.8	153.7	147.9	148.4	155.9
1976	170.5	172.8	161.6	163.3	168.5
1977	182.2	192.4	175.5	174.7	181.8

21.22   Price indexes of residential and non-residential building materials and wage rates, 1968-77 (1971=100)

Year	Residential input indexes			Non-residential input indexes		
	Building materials	Labour	Total	Building materials	Labour	Total
1968	91.5	71.1	79.9	90.0	71.7	79.7
1969	96.4	76.5	85.1	94.0	77.2	84.5
1970	95.3	87.9	91.2	96.6	88.5	92.0
1971	100.0	100.0	100.0	100.0	100.0	100.0
1972	109.8	110.6	110.1	104.9	111.0	107.8
1973	124.0	121.8	123.2	113.1	122.3	117.5
1974	135.2	133.9	134.7	137.3	134.7	136.1
1975	139.8	151.6	144.0	147.0	154.1	150.4
1976	153.7	172.7	160.5	156.6	175.9	165.7
1977	165.4	193.8	175.6	165.5	195.1	179.5

21.23   Price indexes of inputs into residential building construction, 1973-77 (1971=100)

Province and trade	1973	1974	1975	1976	1977
Atlantic provinces	127.3	143.8	151.3	169.0	185.4
Materials	125.1	140.8	142.9	156.1	168.3
Labour	131.1	149.1	166.3	191.3	214.8
Quebec	123.3	133.6	147.1	163.0	178.2
Materials	127.1	136.5	143.3	158.9	171.5
Labour	116.2	128.3	154.4	170.8	190.9
Ontario	123.5	134.5	141.6	156.4	170.2
Materials	123.5	134.9	138.8	151.0	161.3
Labour	123.5	133.8	146.2	165.8	185.3
Prairie provinces	122.5	134.8	145.4	167.6	184.6
Materials	121.4	132.8	138.9	154.4	167.4
Labour	124.7	138.7	158.1	193.3	217.8
British Columbia	121.3	132.4	142.0	157.9	174.8
Materials	124.0	134.1	136.4	152.0	165.7
Labour	116.4	129.2	152.1	168.4	191.0
Canada	123.2	134.7	144.0	160.5	175.6
Materials	124.0	135.2	139.7	153.6	165.4
Labour	121.8	133.9	151.6	172.7	193.8

21.23 Price indexes of inputs into residential building construction, 1973-77 (1971=100) (concluded)

Province and trade	1973	1974	1975	1976	1977
Trade					
Rough carpentry	143.0	141.7	144.1	163.1	179.6
Finished carpentry	134.3	143.3	146.7	164.5	180.3
Concrete	119.8	135.6	157.4	176.6	192.5
Masonry	116.9	130.3	145.9	160.8	178.9
Plumbing	113.0	139.8	148.1	159.5	172.2
Drywall	116.3	130.6	141.8	156.6	169.6
Windows	121.1	127.5	137.9	164.7	181.9
Cabinetry	111.3	122.8	116.4	123.3	131.3
Painting	118.0	131.6	148.6	168.7	186.6
Electrical	114.1	129.0	136.2	149.5	161.4
Forced warm air heating	109.1	119.2	130.7	142.6	152.0
Flooring, resilient	108.4	126.7	136.9	146.4	156.5
Roofing	117.6	138.7	152.1	176.1	194.9
Insulation and vapour barriers	113.6	127.9	148.1	158.6	178.0
Flooring, hardwood	136.9	150.8	149.0	162.8	176.5

21.24 New housing price indexes, for six metropolitan areas, 1972-77 (1971=100)

Year	Montreal	Toronto	Ottawa–Hull	Winnipeg	Calgary	Edmonton
1972	107.6	110.2	112.7	105.2	110.0	109.1
1973	125.8	137.6	138.2	128.4	126.4	132.6
1974	177.7	171.6	171.2	163.5	162.3	172.8
1975	190.3	171.0	178.3	177.5	195.0	205.2
1976	200.9	180.7	192.5	199.8	243.1	245.8
1977	211.7	180.2	198.1	211.8	259.9	262.8

21.25 Highway construction price indexes, 1956-76 (1971=100)

Province and item	Year[1]	0	1	2	3	4	5	6	7	8	9
Newfoundland											
Total	1950	...	...	...	...	...	...	77.3	65.0	74.0	67.5
	1960	70.8	56.8	61.9	57.4	61.5	67.8	73.6	65.4	68.6	66.1
	1970	82.6	100.0	101.1	109.6	129.0	128.0	111.3	...	...	...
Grading	1970	...	100.0	101.4	103.1	124.6	125.3	114.6	...	...	...
Granular base course	1970	...	100.0	100.9	117.6	130.5	123.3	94.3	...	...	...
Paving	1970	...	100.0	100.4	117.3	139.6	142.2	124.5	...	...	...
Total contract work	1970	...	100.0	101.1	109.3	127.9	126.5	108.5	...	...	...
Total supplies	1970	...	100.0	101.4	114.4	149.5	157.7	165.6	...	...	...
Nova Scotia											
Total	1950	...	...	...	...	...	...	77.6	70.5	69.9	74.2
	1960	79.8	67.4	66.2	64.7	65.0	78.8	78.3	82.7	80.8	83.0
	1970	90.4	100.0	107.2	118.8	164.1	185.0	176.2	...	...	...
Grading	1950	...	...	...	...	...	...	71.9	63.7	70.3	76.0
	1960	86.8	70.7	67.1	66.5	65.6	74.8	78.2	80.9	76.9	82.5
	1970	90.8	100.0	106.5	120.8	170.6	190.7	179.5	...	...	...
Granular base course	1950	...	...	...	...	...	...	68.4	64.8	60.2	66.9
	1960	67.7	61.3	59.7	55.5	58.9	82.3	78.9	85.2	83.3	83.3
	1970	88.7	100.0	107.5	117.6	162.8	175.6	159.6	...	...	...
Paving	1950	...	...	...	...	...	...	103.4	93.0	84.4	82.0
	1960	85.2	70.6	74.6	75.5	73.5	81.4	77.5	82.7	85.0	83.9
	1970	92.4	100.0	107.9	118.2	159.5	188.1	187.7	...	...	...
Total contract work	1970	...	100.0	108.3	119.4	163.3	181.6	168.2	...	...	...
Total supplies	1970	...	100.0	100.0	114.8	169.6	208.2	231.4	...	...	...
New Brunswick											
Total	1950	...	...	...	...	...	...	87.0	84.6	89.9	89.3
	1960	84.3	87.1	86.5	89.0	90.3	89.9	90.2	89.7	88.7	89.0
	1970	105.7	100.0	123.9	139.3	191.8	188.6	209.4	...	...	...
Grading	1950	...	...	...	...	...	...	85.7	84.5	95.2	93.8
	1960	84.0	91.2	93.4	94.2	91.3	95.2	86.1	88.2	93.3	90.5
	1970	105.9	100.0	142.9	163.1	220.6	211.7	225.1	...	...	...
Granular base course	1950	...	...	...	...	...	...	87.9	85.4	87.2	89.1
	1960	86.2	89.0	80.9	85.6	88.3	88.9	94.7	92.6	84.5	92.6
	1970	106.7	100.0	125.7	129.2	163.3	165.4	187.7	...	...	...
Paving	1950	...	...	...	...	...	...	88.7	83.1	81.7	78.2
	1960	81.4	73.0	80.2	82.7	91.9	78.5	92.1	87.9	85.1	77.9
	1970	103.2	100.0	103.4	118.6	173.7	174.3	202.2	...	...	...
Total contract work	1970	...	100.0	126.2	142.0	190.1	185.7	202.8	...	...	...
Total supplies	1970	...	100.0	104.1	115.2	206.5	214.4	267.4	...	...	...

21.25 Highway construction price indexes, 1956-76 (1971=100) (continued)

Province and item	Year[1]	0	1	2	3	4	5	6	7	8	9
Quebec											
Total	1960	...	...	...	...	79.6	76.9	82.6	80.8	80.5	85.4
	1970	87.2	100.0	106.6	120.5	154.0	180.7	190.5	...	...	...
Grading	1960	...	...	...	...	80.2	74.7	81.8	77.3	79.0	84.2
	1970	84.4	100.0	109.6	125.8	160.2	188.7	197.3	...	...	...
Granular base course	1960	...	...	...	...	78.6	76.7	82.5	83.3	79.3	85.8
	1970	89.2	100.0	101.9	116.4	147.5	173.3	193.3	...	...	...
Paving	1960	...	...	...	...	79.4	81.3	84.4	85.3	84.4	87.1
	1970	90.8	100.0	103.8	112.2	145.1	168.7	173.1	...	...	...
Total contract work	1970	...	100.0	106.6	120.4	149.8	177.7	186.9	...	...	...
Total supplies	1970	...	100.0	106.8	122.0	210.1	221.1	239.2	...	...	...
Ontario											
Total	1950	...	...	...	...	...	...	81.6	71.4	66.4	69.2
	1960	65.2	60.8	67.0	76.9	75.2	87.5	95.7	95.0	92.1	93.6
	1970	96.8	100.0	106.3	114.5	149.5	164.2	172.4	...	...	...
Grading	1950	...	...	...	...	...	...	76.3	63.7	58.8	58.7
	1960	58.5	53.4	60.4	71.7	71.1	81.5	93.6	93.2	90.4	90.6
	1970	95.5	100.0	108.9	117.0	155.8	171.4	180.0	...	...	...
Granular base course	1950	...	...	...	...	...	...	82.9	71.7	67.0	71.5
	1960	62.8	63.3	66.4	76.1	73.5	90.5	96.1	96.4	93.0	96.5
	1970	96.9	100.0	101.2	108.1	140.7	154.0	165.6	...	...	...
Paving	1950	...	...	...	...	...	...	89.4	86.6	80.2	85.4
	1960	83.9	70.4	82.0	89.2	87.3	93.2	98.8	95.5	93.8	93.5
	1970	99.2	100.0	108.0	117.5	150.1	165.2	169.5	...	...	...
Total contract work	1970	...	100.0	106.7	115.1	148.1	162.7	170.8	...	...	...
Total supplies	1970	...	100.0	100.2	105.3	172.1	188.7	198.1	...	...	...
Manitoba											
Total	1950	...	...	...		...	...	83.3	93.4	69.8	68.7
	1960	72.9	62.6	67.6	75.3	77.5	83.5	95.7	96.3	88.2	90.5
	1970	100.9	100.0	111.3	126.3	166.8	172.5	197.6	...	...	...
Grading	1950	...	...	...	...	...	...	80.9	79.3	61.2	58.4
	1960	63.0	51.4	57.6	80.7	75.6	82.6	95.3	98.4	85.6	86.2
	1970	102.1	100.0	117.8	115.6	158.9	161.1	169.7	...	...	...
Granular base course	1950	...	...	...	...	...	...	115.2	134.4	82.1	83.8
	1960	85.3	77.4	80.0	72.6	95.1	97.1	111.1	101.2	91.2	100.2
	1970	107.4	100.0	113.4	139.3	176.7	180.3	213.0	...	...	...
Paving	1950	...	...	...	...	...	...	71.3	93.2	76.0	75.9
	1960	80.6	71.2	75.7	68.9	71.4	78.1	88.9	91.0	90.3	92.0
	1970	95.9	100.0	99.9	124.7	165.2	178.2	215.9	...	...	...
Total contract work	1970	...	100.0	113.3	127.7	166.6	170.8	194.9	...	...	...
Total supplies	1970	...	100.0	99.5	118.0	167.8	182.2	213.6	...	...	...
Saskatchewan											
Total	1950	...	...	...	...	...	...	103.7	106.5	82.6	75.5
	1960	71.6	68.0	66.9	69.8	79.4	98.2	114.3	93.5	84.9	89.8
	1970	98.5	100.0	104.4	129.6	175.8	210.4	206.6	...	...	...
Grading	1950	...	...	...	...	...	...	94.9	87.7	68.2	68.7
	1960	73.1	64.2	63.5	73.1	78.5	94.7	110.5	96.3	81.3	88.5
	1970	96.0	100.0	101.8	126.6	167.3	212.8	213.4	...	...	...
Granular base course	1950	...	...	...	...	...	...	117.5	135.7	101.0	81.2
	1960	65.8	63.9	65.6	60.0	76.5	103.3	121.7	86.2	85.1	88.1
	1970	98.7	100.0	103.7	126.1	167.8	201.2	181.6	...	...	...
Paving	1950	...	...	...	...	...	...	98.9	97.9	83.9	81.6
	1960	78.8	85.5	77.9	80.5	87.4	96.9	109.3	100.6	93.8	96.3
	1970	104.4	100.0	107.9	135.6	191.3	214.8	218.8	...	...	...
Total contract work	1970	...	100.0	103.8	130.9	173.2	212.5	204.0	...	...	...
Total supplies	1970	...	100.0	108.3	121.5	192.2	196.9	223.3	...	...	...
Alberta											
Total	1970	...	100.0	99.5	128.6	184.7	209.7	200.5	...	...	...
Grading	1970	...	100.0	98.9	129.2	202.4	234.0	230.2	...	...	...
Granular base course	1970	...	100.0	99.2	136.6	178.2	217.0	200.2	...	...	...
Paving	1970	...	100.0	100.3	122.2	174.1	183.4	175.2	...	...	...
Total contract work	1970	...	100.0	98.6	128.5	184.5	210.7	199.0	...	...	...
Total supplies	1970	...	100.0	108.9	129.5	186.5	199.0	216.4	...	...	...
British Columbia											
Total	1950	...	...	...	...	...	...	102.4	95.4	80.3	82.1
	1960	81.8	71.9	68.5	69.7	76.3	91.9	93.3	85.9	91.1	103.0
	1970	96.7	100.0	95.7	101.6	170.2	183.2	213.2	...	...	...
Grading	1950	...	...	...	...	...	...	102.7	93.2	76.2	79.0
	1960	76.1	64.4	68.3	71.2	72.2	92.7	91.8	80.1	90.7	103.7
	1970	96.4	100.0	95.2	102.3	171.1	175.2	237.6	...	...	...
Granular base course	1950	...	...	...	...	...	...	107.1	98.8	82.4	89.4
	1960	91.0	89.1	70.0	71.0	90.5	95.7	99.6	96.3	94.6	106.4
	1970	96.8	100.0	96.1	99.2	179.4	197.8	187.7	...	...	...
Paving	1950	...	...	...	...	...	...	94.1	98.2	92.0	82.3
	1960	88.8	73.6	67.2	61.8	69.7	82.8	88.7	91.8	87.0	95.4
	1970	97.6	100.0	96.4	102.9	158.2	183.7	190.9	...	...	...
Total contract work	1970	...	100.0	95.3	100.6	169.5	181.4	212.5	...	...	...
Total supplies	1970	...	100.0	103.7	122.8	185.5	220.7	228.7	...	...	...

21.25 Highway construction price indexes, 1956-76 (1971=100) (concluded)

Province and item	Year[1]	0	1	2	3	4	5	6	7	8	9
Canada											
Total[2]	1950	...	...	...	...	...	...	87.1	80.9	73.0	73.2
	1960	72.1	65.0	67.6	72.2	76.2	83.0	89.4	86.0	84.8	88.7
	1970	92.7	100.0	105.1	118.3	158.7	177.5	185.1	...	...	...
Grading	1950	...	...	...	...	...	...	85.5	74.8	69.8	68.7
	1960	68.4	60.4	65.7	72.6	75.7	81.1	88.6	84.8	84.1	87.7
	1970	71.4	100.0	107.3	120.1	162.2	181.7	193.2	...	...	...
Contract work	1970	...	100.0	107.3	120.0	162.3	181.9	193.3	...	...	...
Supplies	1970	...	100.0	108.0	126.1	152.3	166.2	181.6	...	...	...
Granular base course	1950	...	...	...	...	...	...	88.6	84.9	72.7	74.8
	1960	71.8	68.2	66.0	68.9	76.5	85.0	91.5	87.7	84.5	90.1
	1970	93.6	100.0	102.5	118.4	154.9	173.0	177.2	...	...	...
Paving	1950	...	...	...	...	...	...	92.7	92.7	83.5	82.3
	1960	83.7	72.5	76.0	77.0	79.2	83.7	88.7	88.0	87.1	88.3
	1970	94.7	100.0	104.2	117.0	156.6	174.7	179.4	...	...	...
Contract work	1970	...	100.0	104.3	116.9	145.9	164.8	164.6	...	...	...
Supplies	1970	...	100.0	103.8	117.2	190.1	205.6	225.3	...	...	...
Total contract work	1970	...	100.0	105.2	118.3	156.4	175.4	182.1	...	...	...
Total supplies	1970	...	100.0	104.1	117.9	187.2	202.6	222.0	...	...	...

[1]Within decade.
[2]The Canada index includes seven provinces from 1956 to 1964, when Quebec was added. Alberta was included from 1971. Prince Edward Island was excluded throughout.

21.26 Price indexes of electrical utility distribution systems, transmission lines, transformer stations, hydroelectric generating stations and steam electric generating stations, 1972-77 (1971=100)

Year	Distribution systems			Transmis-sion lines	Transformer stations			Generating stations	
	Total direct costs	Equip-ment	Total		Support structures and fixtures	Equip-ment	Total	Hydro-electric	Steam electric
1972	103.3	99.7	104.2	106.1	108.1	101.2	104.1	106.3	106.1
1973	113.4	100.7	113.8	115.3	120.5	107.3	111.2	116.1	115.9
1974	138.6	125.7	137.5	137.7	148.8	134.9	136.3	137.9	139.6
1975	153.6	136.7	153.2	161.2	165.2	162.5	159.1	157.6	158.4
1976r	162.4	126.5	163.0	173.7	182.1	176.6	171.8	171.6	174.1
1977	172.9	130.1	173.9	188.4	192.5	179.7	176.9	183.4	187.7

21.27 Price indexes of machinery and equipment, 1968-77 (1968=100)

Year	Construction	Forestry (east of the Rockies)	Year	Construction	Forestry (east of the Rockies)
1968	100.0	100.0	1973	114.9	117.0
1969	104.4	102.9	1974	130.0	133.4
1970	106.4	106.4	1975	161.0	155.8
1971	108.4	109.3	1976	168.0	162.8
1972	110.8	112.3	1977P	191.3	176.4

21.28 Farm input price index total, 1968-77 (1961=100)

Year	East	West	Canada	Year	East	West	Canada
1968	124.3	125.6	124.9	1973	170.1	162.9	166.7
1969	128.1	130.2	129.1	1974	199.1	190.7	195.2
1970	131.4	131.0	131.2	1975r	214.2	224.6	219.0
1971	135.8	135.9	135.9	1976	226.4	238.4	231.9
1972	141.7	145.2	143.3	1977P	241.3	251.0	245.8

21.29 Summary of the Canadian balance of international payments between Canada and all non-residents, 1971-77 (million dollars)

Year	Current account balances			Capital account flows			Allocation of SDRs	Net official monetary movements
	Merchan-dise	Non-merchan-dise	Balance	Long-term	Short-term	Net		
1971	+2,563	−2,132	+431	+664	−318	+346	+119	+896
1972	+1,857	−2,243	−386	+1,588	−983	+605	+117	+336
1973r	+2,735	−2,627	+108	+385	−960	−575	−	−467
1974r	+1,689	−3,202	−1,513	+871	+666	+1,537	−	+24
1975r	−534	−4,245	−4,779	+3,848	+526	+4,374	−	−405
1976	+1,089	−5,276	−4,187	+7,874	−3,165	+4,709	−	+522
1977	+2,907	−7,145	−4,238	+4,346	−1,529	+2,817	−	−1,421

21.30 Canada's net official international monetary assets (IMF), 1975-77

Item	1975	1976	1977
	US $'000,000		
CANADIAN ASSETS			
Official holdings of foreign exchange:			
United States dollars	3,207	3,446	2,299
Other convertible currencies	16	16	16
Monetary gold[1]	899	879	935
Special Drawing Right (SDR)[1]	555	558	505
Reserve position in IMF[1]	648	944	852
Total, official international reserves[2]	5,325	5,843	4,607
CANADIAN LIABILITIES			
Use of IMF credit[3]	- -	- -	- -
Foreign exchange deposit liabilities	- -	- -	- -
Reported use of central bank reciprocal credit facilities	- -	- -	- -
Total, official monetary liabilities	- -	- -	
Net official monetary assets	5,325	5,843	4,607
Change in net official monetary assets	−7,500	+518	−1,236
	Canadian $'000,000		
Net official monetary assets	5,411	5,894	5,040
Change in net official monetary assets	−359	+483	−854
Change due to:			
Valuation adjustments[4]	+46	−39[5]	+567
Transactions	−405	+522[5]	−1,421
Net official monetary movements[6]			
Total, official international reserves	−405	+522	−1,421
Total, official monetary liabilities	—	—	—

[1]From July 1, 1974 the basis of valuation for Canada's gold-based assets was changed from US$42.22 per ounce of fine gold to a formal link with a basket of 16 currencies as calculated by the IMF and thus became SDR-based.
[2]As published by the finance minister.
[3]Represents transactions with the IMF when that institution held Canadian dollars in excess of 100% of Canada's quota.
[4]Valuation adjustments represent changes in the external value of the Canadian dollar in relation to gold, the SDR, the United States dollar and other convertible currencies.
[5]Includes revaluation from book value to market value of gold transferred from the reserves to the olympic coin program.
[6]Excludes valuation adjustments.

21.31 Canadian balance of international payments, by area, 1976 and 1977 (million dollars)

Item	United States		United Kingdom		Other non-residents		All non-residents	
	1976	1977	1976	1977	1976	1977	1976	1977
CURRENT ACCOUNT								
Current receipts								
Merchandise exports (adjusted)	25,616	30,983	2,017	2,215	10,342	11,367	37,975	44,565
Service receipts								
Travel	1,346	1,509	139	110	445	387	1,930	2,006
Interest and dividends	497	465	29	48	327	338	853	851
Freight and shipping	959	1,141	181	197	851	942	1,991	2,280
Other service receipts	1,082	1,128	437	455	1,117	1,197	2,636	2,780
Total, service receipts	3,884	4,243	786	810	2,740	2,864	7,410	7,917
Total, exports of goods and services	29,500	35,226	2,803	3,025	13,082	14,231	45,385	52,482
Transfer receipts								
Inheritances and immigrants' funds	105	98	106	91	516	446	727	635
Personal and institutional remittances	172	189	23	25	63	67	258	281
Withholding tax	. .	. .	. .	. .	. .	. .	504	533
Total, current receipts	29,777	35,513	2,932	3,141	13,661	14,744	46,874	53,931
Current payments								
Merchandise imports (adjusted)	25,223	29,277	1,301	1,588	10,362	10,793	36,886	41,658
Service payments								
Travel	1,956	2,278	288	373	877	1,010	3,121	3,661
Interest and dividends	2,580	3,181	145	156	619	927	3,344	4,264
Freight and shipping	1,198	1,354	210	225	756	827	2,164	2,406
Other service payments	2,749	3,076	306	341	1,020	1,153	4,075	4,570
Withholding tax	. .	. .	. .	. .	. .	. .	504	533
Total, service payments	8,483	9,889	949	1,095	3,272	3,917	13,208	15,434
Total, imports of goods and services	33,706	39,166	2,250	2,683	13,634	14,710	50,094	57,092
Transfer payments								
Inheritances and emigrants' funds	86	89	32	34	63	67	181	190
Personal and institutional remittances	109	116	38	40	184	194	331	350
Official contributions	—	—	—	—	455	537	455	537
Total, current payments	33,901	39,371	2,320	2,757	14,336	15,508	51,061	58,169

21.31 Canadian balance of international payments, by area, 1976 and 1977 (million dollars) (concluded)

Item	United States		United Kingdom		Other non-residents		All non-residents	
	1976	1977	1976	1977	1976	1977	1976	1977
CURRENT ACCOUNT (concluded)								
Current account balance								
Merchandise trade	+393	+1,706	+716	+627	−20	+574	+1,089	+2,907
Service transactions	−4,599	−5,646	−163	−285	−532	−1,053	−5,798	−7,517
Net transfers	+82	+82	+59	+42	−123	−285	+522	+372
Total, current account balance	−4,124	−3,858	+612	+384	−675	−764	−4,187	−4,238
CAPITAL ACCOUNT								
Direct investment								
In Canada	−490	+516	+72	−61	+123	−45	−295	+410
Abroad	−234	−497	−12	−24	−309	−269	−555	−790
Portfolio transactions								
Canadian securities								
Outstanding bonds	+20	+44	+274	+42	+275	+213	+569	+299
Outstanding stocks	−69	+101	−33	−74	+23	−118	−79	−91
New issues	+5,609	+2,951	+230	+111	+3,251	+2,716	+9,090	+5,778
Retirements	−706	−562	−45	−86	−177	−238	−928	−886
Foreign securities								
Outstanding issues	+55	+142	+1	−5	+6	+38	+62	+175
New issues	−12	−3	−	−	−25	−19	−37	−22
Retirements	+42	+8	−	−	+18	+72	+60	+80
Loans and subscriptions, Government of Canada								
Advances	−	−	−	−	−410	−556	−410	−556
Repayments	−	−	−	+24	+18	+10	+18	+34
Export credits directly or indirectly at risk of the Government of Canada	+11	+5	−3	+22	−271	−559	−263	−532
Other long-term capital transactions	+351	+80	+21	+109	+270	+258	+642	+447
Balance of capital movements in long-term forms	+4,577	+2,785	+505	+58	+2,792	+1,503	+7,874	+4,346
Resident holdings of foreign currencies:								
Chartered bank net foreign currency position with non-residents	−611	+881	−9	+1,187	−322	−684	−942	+1,384
Non-bank holdings of foreign currencies abroad	−79	−140	−46	−338	−105	−19	−230	−497
Non-resident holdings of Canadian:								
Dollar deposits	+3	+144	+109	−141	+40	+222	+152	+225
Government demand liabilities	−	−	−	−	+7	+172	+7	+172
Treasury bills	+27	+147	+193	−81	+478	+176	+698	+242
Commercial paper	+286	−121	−	−	−	+12	+286	−109
Finance company paper	+18	+33	+2	−1	−	+5	+20	+37
Other short-term paper	+146	+270	+21	−	+21	−18	+188	+252
Other finance company obligations	+40	−54	+8	−5	−3	−	+45	−59
Other short-term capital transactions	−3,392	−3,004	−144	−81	+147	−91	−3,389	−3,176
Balance of capital movements in short-term forms	−3,562	−1,844	+134	+540	+263	−225	−3,165	−1,529
Total, net capital balance	+1,015	+941	+639	+598	+3,055	+1,278	+4,709	+2,817
Balance settled by exchange transfers	+3,377	..	−1,251	..	−2,126	..	−	−
Allocation of Special Drawing Rights	...	...	...	...	−	..	−	−
Net official monetary movements								
Official international reserves	+268	..	−	..	+254	..	+522	−1,421
Official monetary liabilities	−	..	−	..	−	..	−	−
Net official monetary movements	+268	..	−	..	+254	..	+522	−1,421

21.32 Canadian balance of international indebtedness, selected years, 1939-75 (billion dollars)

Item	1939	1950	1960	1970	1973r	1974	1975
CANADIAN ASSETS							
Direct investment	0.7	1.0	2.5	6.2	7.8	9.3	10.7
Portfolio investment	0.7	0.6	1.3	2.7	3.7	3.9	4.3
Miscellaneous investment[1]	−	−	−	1.0	2.2	3.1	3.7
Government of Canada credits[2]	−	2.0	1.5	1.5	1.8	2.0	2.3
Government of Canada subscriptions to international financial agencies	−	0.1	0.1	0.3	0.5	0.6	0.6
Total, Canadian long-term investment abroad	1.4	3.7	5.3	11.7	16.0	18.9	21.6
Net official monetary assets	0.5	2.0	2.0	4.7	5.8	5.8	5.4
Other Canadian short-term holdings of foreign exchange	−	0.1	1.2	3.6	3.0	2.7	2.6
Gross assets[3]	1.9	5.7	8.5	20.0	24.8	27.4	29.6
Net official monetary assets	0.5	2.0	2.0	4.7	5.8	5.8	5.4
United States[3,4]	0.9	1.2	3.7	8.6	8.5	8.4	8.2
United Kingdom[3,4]	0.1	1.6	1.5	4.1	3.9	4.6	5.5
Other countries[3,4]	0.4	0.9	1.3	2.6	6.6	8.6	10.5
Short-term receivables (not included elsewhere)[5]	..	0.2	0.5	3.6	8.3	10.1	10.9
Gross assets	1.9[3]	5.9	8.9	23.6	33.1	37.5	40.5
CANADIAN LIABILITIES							
Direct investment	2.3	4.0	12.9	26.4	32.7	36.1	39.8
Government bonds	1.7	2.0	3.3	7.9	10.0	11.6	15.1
Other portfolio investment	2.6	2.4	4.6	6.9	8.7	9.2	10.1
Miscellaneous investment	0.3	0.3	1.4	2.9	3.3	3.4	3.6
Total, foreign long-term investment in Canada	6.9	8.7	22.2	44.0	54.7	60.3	68.6

21.32 Canadian balance of international indebtedness, selected years, 1939-75 (billion dollars) (concluded)

Item	1939	1950	1960	1970	1973r	1974	1975
CANADIAN LIABILITIES (concluded)							
Non-resident equity in Canadian assets abroad	0.2	0.3	1.1	2.7	3.1	3.3	3.9
Official Special Drawing Right liabilities	...	...	...	0.1	0.4	0.4	0.4
Total, long-term liabilities	7.1	9.0	23.3	46.9	58.2	64.0	72.9
Non-resident holdings of Canadian dollars	0.3	0.6	0.6	0.8	1.3	2.0	2.7
Gross liabilities[3]	7.4	9.6	24.0	47.7	59.5	66.0	75.6
United States[3]	4.5	7.1	18.0	37.4	45.0	49.9	56.4
United Kingdom[3]	2.6	2.0	3.5	4.3	5.5	5.7	6.3
Other countries[3,6]	0.3	0.5	2.4	6.0	9.0	10.4	12.9
Short-term payables (not included elsewhere)[5,7]							
Finance company obligations				1.3	1.2	1.4	1.6
	..	0.8	1.6				
Other				3.0	4.1	5.4	5.6
Gross liabilities	7.4[3]	10.4	25.6	52.1	64.8	72.8	82.8
CANADA'S INTERNATIONAL INDEBTEDNESS							
Canadian net international indebtedness	5.5[3]	4.5	16.6	28.5	31.7	35.3	42.3
Net official monetary assets	−0.5	−2.0	−2.0	−4.7	−5.8	−5.8	−5.4
United States	3.6	5.9	14.3	28.8	36.5	41.5	48.2
United Kingdom	2.5	0.4	2.0	0.2	1.6	1.1	0.8
Other countries	−0.1	−0.4	1.1	3.4	2.4	1.8	2.4
Short term (not included elsewhere)	..	0.6	1.1	0.8	−3.0	−3.3	−3.7

[1] Includes export credits by government and private sectors less reserve against government inactive assets.
[2] Includes medium-term non-marketable United States government securities held under the Columbia River Treaty arrangements since 1964.
[3] Excludes short-term receivables and payables.
[4] Excludes net official monetary assets.
[5] Country distribution not available.
[6] Includes international financial agencies.
[7] At the end of 1964 about $450 million previously classified as long-term investment was reclassified to short-term finance company obligations.

21.33 Canadian long-term investment abroad[1], by country and by type of investment, selected years, 1951-74 (million dollars)

Location and type of investment	1951	1960	1970r	1971r	1973r	1974
United States						
Direct investment	912	1,618	3,262	3,399	3,924	4,909
Portfolio investment						
Stocks	289	827	2,115	2,185	2,640	2,765
Bonds	87	120	224	217	203	202
Miscellaneous investment	9	18	234	277	365	500
Government of Canada credits[2]	—	—	26	—	—	—
Government of Canada subscriptions to international investment agencies	—	—	—	—	—	—
Total, United States	1,297	2,583	5,861	6,078	7,132	8,376
United Kingdom						
Direct investment	74	257	586	590	797	879
Portfolio investment						
Stocks	17	26	60	65	75	90
Bonds	17	16	20	20	35	34
Miscellaneous investment	13	18	74	78	158	180
Government of Canada credits[3]	1,394	1,092	1,017	997	954	932
Government of Canada subscriptions to international investment agencies	—	—	—	—	—	—
Total, United Kingdom	1,515	1,409	1,757	1,750	2,019	2,115
Other Commonwealth countries[4]						
Direct investment	88	299	791	843	1,160	1,254
Portfolio investment						
Stocks	6	10	15	20	25	30
Bonds	8	18	23	27	36	44
Miscellaneous investment	..	...	...	...	...	...
Government of Canada credits	—	35	243	354	473	605
Government of Canada subscriptions to international investment agencies	—	—	—	—	—	—
Total, other Commonwealth countries	102	362	1,072	1,244	1,694	1,933
Other countries						
Direct investment	92	293	1,549[5]	1,706	1,929	2,265
Portfolio investment						
Stocks	155	187	155	220	315	365
Bonds	30	111	207	256	365	394
Miscellaneous investment	−80	−54	669	1,013	1,668	2,271
Government of Canada credits	528	335	204	214	412	518
Government of Canada subscriptions to international investment agencies	66	85	268	319	485	581
Total, other countries	791	957	3,052	3,728	5,174	6,394

21.33 Canadian long-term investment abroad[1], by country and by type of investment, selected years, 1951-74 (million dollars) (concluded)

Location and type of investment	1951	1960	1970r	1971r	1973r	1974
All countries						
Direct investment	1,166	2,467	6,188	6,538	7,810	9,307
Portfolio investment						
Stocks	467	1,050	2,345	2,490	3,055	3,250
Bonds	142	265	474	520	639	674
Miscellaneous investment	−58	−18	977	1,368	2,191	2,951
Government of Canada credits[6]	1,922	1,462	1,490	1,565	1,839	2,055
Government of Canada subscriptions to international investment agencies	66	85	268	319	485	581
Total, all countries	3,705	5,311	11,742	12,800	16,019	18,818

[1]Figures include the equity of non-residents in assets abroad of Canadian companies but exclude investment of insurance companies and banks (held mainly, against liabilities to non-residents).
[2]Medium-term non-marketable United States government securities acquired under the Columbia River Treaty arrangements are shown from 1964.
[3]Includes deferred interest on the United Kingdom 1946 loan agreement starting from 1956 and amounting to $101 million in 1973.
[4]Includes investment in Newfoundland prior to 1949.
[5]New series not strictly comparable with previous years.
[6]Includes United Nations bonds from 1962, which amounted to $4 million in 1973.

21.34 Foreign long-term investment in Canada, by type of investment and by country, as at Dec. 31, 1974 and 1975 (million dollars)

Year and type of investment	United States[1]	United Kingdom[1]	Other countries	Total investments of non-residents
1974				
Government securities				
Government of Canada	303	53	194	550
Provincial	7,156	274	1,982	9,412
Municipal	1,189	44	355	1,588
Total, government securities	8,648	371	2,531	11,550
Manufacturing				
Vegetable products	1,496	181	198	1,875
Animal products	345	10	36	391
Textiles	304	75	23	402
Wood and paper products	2,733	276	484	3,493
Iron and products	4,582	221	155	4,958
Non-ferrous metals	1,834	120	115	2,069
Non-metallic minerals	423	119	247	789
Chemicals and allied products	1,969	274	121	2,364
Miscellaneous manufactures	310	3	10	323
Total, manufacturing	13,996	1,279	1,389	16,664
Petroleum and natural gas	7,986	956	1,146	10,088
Other mining and smelting	4,095	364	663	5,122
Utilities				
Railways	501	316	100	917
Other (excl. public enterprises)	1,765	89	101	1,955
Total, public utilities	2,266	405	201	2,872
Merchandising	1,995	382	268	2,645
Financial	4,320	1,255	911	6,486
Other enterprises	1,117	158	141	1,416
Miscellaneous investments	2,382	160	905	3,447
Total, investments, 1974	46,805	5,330	8,155	60,290
1975				
Government securities				
Government of Canada	346	74	226	646
Provincial	9,259	305	2,804	12,368
Municipal	1,473	62	560	2,095
Total, government securities	11,078	441	3,590	15,109
Manufacturing				
Vegetable products	1,715	197	157	2,069
Animal products	364	8	48	420
Textiles	336	82	17	435
Wood and paper products	2,877	308	491	3,676
Iron and products	5,069	250	183	5,502
Non-ferrous metals	2,054	122	139	2,315
Non-metallic minerals	495	119	310	924
Chemicals and allied products	2,255	288	133	2,676
Miscellaneous manufactures	368	3	9	380
Total, manufacturing	15,533	1,377	1,487	18,397

21.34   Foreign long-term investment in Canada, by type of investment and by country, as at Dec. 31, 1974 and 1975 (million dollars) (concluded)

Year and type of investment	United States[1]	United Kingdom[1]	Other countries	Total investments of non-residents
1975 (concluded)				
Petroleum and natural gas	8,592	981	1,234	10,807
Other mining and smelting	4,409	367	716	5,492
Utilities				
Railways	566	326	115	1,007
Other (excl. public enterprises)	1,878	99	113	2,090
Total, public utilities	2,444	425	228	3,097
Merchandising	2,261	411	324	2,996
Financial	4,879	1,332	1,318	7,529
Other enterprises	1,316	166	187	1,669
Miscellaneous investments	2,423	175	955	3,553
Total, investments, 1975	52,935	5,675	10,039	68,649

Common and preferred stocks are at book values as shown in the balance sheets of the issuing companies; bonds and debentures are valued at par; and liabilities in foreign currencies are converted into Canadian dollars at par of exchange.
[1]Includes some investments held for residents of other countries.

21.35   Price of the United States dollar in Canada, by month, 1970-77 (Canadian cents per US dollar)

Month	1970	1971	1972	1973	1974	1975	1976	1977
January	107.28	101.16	100.59	99.91	99.14	99.48	100.64	101.09
February	107.31	100.75	100.46	99.55	97.67	100.05	99.37	102.79
March	107.27	100.63	99.84	99.66	97.20	100.03	98.58	105.11
April	107.28	100.76	99.56	100.06	96.73	101.11	98.33	105.11
May	107.28	100.87	98.87	100.05	96.21	102.81	98.00	104.85
June	103.84	102.12	97.94	99.83	96.64	102.64	97.36	105.75
July	103.20	102.11	98.39	99.94	97.61	103.07	97.22	106.10
August	102.14	101.33	98.22	100.38	97.98	103.53	98.53	107.49
September	101.59	101.29	98.29	100.81	98.63	102.62	97.50	107.33
October	102.14	100.44	98.26	100.09	98.30	102.50	97.26	109.88
November	102.00	100.37	98.72	99.88	98.72	101.37	98.57	110.92
December	101.74	99.92	99.67	99.94	98.81	101.38	101.87	109.72
Annual average	104.40	100.98	99.05	100.01	97.80	101.73	98.61	106.35

Rates published by Bank of Canada. Noon average market rate for business days in period.

21.36   Foreign exchange rates, 1975-77

Year	Canadian dollar					
	Vis-à-vis United States dollar[1]				Premium (+) or discount (−) on forward US$[2]	
	Spot rates (US cents)					
	High	Low	Close	Noon average		
1975	100.95	96.15	98.43	98.30	+0.96	
1976	103.89	95.88	99.13	101.41	+3.95	
1977	99.85	89.63	91.41	94.03	+1.53	
	Vis-à-vis other currencies[3]					
	Spot rates (foreign currencies)					
	Pound sterling	French franc	Deutsche mark	Swiss franc	Japanese yen	SDR[3]
1975	0.44	4.21	2.41	2.54	291.55	0.81
1976	0.56	4.84	2.55	2.53	300.57	0.88
1977	0.54	4.62	2.18	2.25	251.26	0.81

[1]Calculated on the basis of rates prevailing on the interbank market in Canada.
[2]Rates per annum, computed on basis of average 90-day forward spread on the spot noon rates.
[3]Calculated on the basis of average spot rates based (except for SDRs) on nominal quotations in terms of United States dollars.

21.37 Intercity indexes of retail price differentials for selected commodities and services, 1976 and 1977 (combined cities average = 100)

Commodity grouping	St. John's, Nfld.	Charlotte-town, PEI	Halifax, NS	Saint John, NB	Montreal, Que.	Ottawa, Ont.	Toronto, Ont.	Winnipeg, Man.	Regina, Sask.	Edmonton, Alta.	Vancouver, BC
As at June 1976											
Food at home	114	104	103	108	97	98	99	103	102	100	109
Household operation[1]	106	101	104	101	101	99	98	97	97	99	108
Transportation	108	104	101	105	105	98	100	93	97	92	101
Health and personal care	102	90	95	97	95	104	104	99	90	99	106
Recreation, education and reading	98	92	99	98	104	98	102	94	96	91	99
Tobacco and alcohol	121	110	106	103	101	99	100	103	97	92	99
As at September 1977											
Food at home	114	107	101	106	99	98	98	102	101	102	107
Household operation[1]	105	101	106	103	103	100	96	97	97	100	109
Transportation	106	104	101	104	107	97	100	92	95	94	99
Health and personal care	99	89	94	98	96	104	103	97	93	102	108
Recreation, education and reading	98	91	98	98	105	96	100	95	97	92	98
Tobacco and alcohol	120	106	106	103	101	102	101	100	102	91	96

[1]Excludes fuel and utilities.

21.38 Summary of the financial market, 1976 and 1977 (million dollars)

Category	1976f Jan. 1-Mar. 31	Apr. 1-June 30	July 1-Sept. 30	Oct. 1-Dec. 31	Total	1977 Jan. 1-Mar. 31	Apr. 1-June 30	July 1-Sept. 30	Oct. 1-Dec. 31	Total
Funds raised in credit markets										
Persons and unincorporated business	3,863	4,964	4,363	5,102	18,292	4,878	5,715	5,522	6,074	22,189
Consumer credit	197	1,589	1,036	1,038	3,860	358	1,223	811	1,073	3,465
Bank loans	1,100	329	207	887	2,523	726	419	105	805	2,055
Other loans	296	39	84	-31	388	418	-281	207	266	610
Short-term paper	-	11	-7	2	6	7	11	4	-5	13
Mortgages	2,262	2,992	3,034	3,219	11,507	3,356	4,336	4,394	3,924	16,010
Bonds	8	4	9	-13	8	13	11	1	11	36
Non-financial private corporations	2,135	2,434	1,302	2,231	8,102	2,568	2,791	1,568	753	7,680
Bank loans	1,113	663	394	800	2,970	1,432	610	462	-323	2,181
Other loans	525	91	-239	343	720	304	267	-160	427	838
Short-term paper	-188	602	160	-221	353	220	-120	236	-430	-94
Mortgages	119	154	208	276	757	106	258	107	178	649
Bonds	490	748	529	365	2,132	393	614	487	382	1,876
Stocks	76	176	250	668	1,170	113	1,162	436	519	2,230
Non-financial government enterprises	2,841	420	1,975	1,101	6,337	986	718	1,352	487	3,543
Bank loans	668	4	392	-176	888	-11	58	-90	63	20
Other loans	66	-14	503	-79	476	-38	24	31	-128	-35
Short-term paper	171	-103	-98	-20	-50	-99	4	44	10	-41
Mortgages	15	42	3	22	82	24	9	34	54	121
Bonds	1,907	420	1,189	1,373	4,889	963	689	1,300	511	3,463
Stocks	14	71	-14	-19	52	71	-66	33	-23	15

21.38 Summary of the financial market, 1976 and 1977 (million dollars) (concluded)

Category	1976					1977				
	Jan. 1-Mar. 31	Apr. 1-June 30	July 1-Sept. 30	Oct. 1-Dec. 31	Total	Jan. 1-Mar. 31	Apr. 1-June 30	July 1-Sept. 30	Oct. 1-Dec. 31	Total
General government	*1,782*	*1,404*	*941*	*4,016*	*8,143*	*2,838*	*1,903*	*1,947*	*5,635*	*12,323*
Bank loans	376	-291	-229	315	171	202	-250	-166	16	-198
Other loans	52	20	1	101	174	14	443	-49	43	451
Treasury bills	295	440	440	470	1,645	410	595	615	850	2,470
Short-term paper	106	-116	-145	-20	-175	62	-143	-22	-46	-145
Mortgages	5	4	5	5	19	5	4	5	5	15
Bonds	948	1,347	869	3,145	6,309	2,145	1,254	1,564	4,767	9,730
Total borrowing by domestic non-financial sectors	10,621	9,222	8,581	12,450	40,874	11,270	11,127	10,389	12,949	45,735
Rest of the world	*332*	*25*	*530*	*506*	*1,393*	*542*	*134*	*214*	*499*	*1,389*
Bank loans	164	82	263	278	787	182	70	172	70	494
Other loans	151	94	247	178	670	309	207	69	509	1,094
Mortgages	—	—	—	—	—	—	—	—	—	—
Stocks	17	-151	20	50	-64	51	-143	-27	-80	-195
Total borrowing, excluding domestic financial institutions	10,953	9,247	9,111	12,956	42,267	11,812	11,261	10,603	13,448	47,124
Domestic financial institutions	*-90*	*1,513*	*369*	*1,524*	*3,316*	*403*	*1,428*	*550*	*1,265*	*3,646*
Bank loans	-388	507	-485	121	-245	54	321	72	-73	374
Other loans	-385	239	256	243	353	-389	220	15	57	-97
Short-term paper	110	249	-127	470	702	289	339	-8	401	1,021
Mortgages	—	14	7	15	36	22	11	-4	16	45
Bonds	475	432	470	426	1,803	241	502	366	411	1,520
Stocks	98	72	248	249	667	186	35	109	453	783
Total funds raised	10,863	10,760	9,480	14,480	45,583	12,215	12,689	11,153	14,713	50,770
Funds supplied directly to credit markets										
Persons and unincorporated business	387	-86	251	2,557	3,109	1,512	1,139	725	3,324	6,700
Non-financial private enterprises	-725	284	64	55	-322	196	175	-130	347	588
Public sector (general government and non-financial government enterprises)	806	284	299	277	1,666	890	754	289	912	2,845
Public financial institutions	468	742	448	747	2,405	557	474	619	664	2,314
Rest of the world	3,461	1,943	2,697	2,241	10,342	1,377	932	1,949	1,106	5,364
Bank of Canada	105	-329	382	330	488	68	134	672	722	1,596
Chartered banks	3,603	3,679	2,172	3,442	12,896	3,516	3,724	3,080	3,078	13,398
Private domestic financial institutions (excl. chartered banks)	2,758	4,243	3,167	4,831	14,999	4,099	5,357	3,949	4,560	17,965
Total funds supplied	10,863	10,760	9,480	14,480	45,583	12,215	12,689	11,153	14,713	50,770

The table above is an aggregation of financial flows data designed to provide a synopsis of Canadian credit market activity. The top portion of the table focuses on, but is not limited to, borrowing by the non-financial sectors of the economy. Only flows through what may be termed organized credit markets are included. Funds supplied by the immediate lending sector appear in the bottom portion of the table. This part of the table measures the asset increase of the immediate lending sectors in respect of the same conventional credit market instruments encountered in the top portion of the table.

21.39 Book value, ownership and control of capital employed[1] in non-financial industries, 1969-76

Item and year	Ownership Investment owned in			Percentage of capital employed owned in			Control Investment controlled in			Percentage of capital employed controlled in			Total capital employed
	Canada $ billion	United States $ billion	Other countries $ billion	Canada %	United States %	Other countries %	Canada $ billion	United States $ billion	Other countries $ billion	Canada %	United States %	Other countries %	$ billion
Manufacturing													
1969	10.9	10.6	1.9	47	45	8	9.3	11.1	3.0	40	47	13	23.5
1970	11.7	11.3	2.1	47	45	8	9.8	11.7	3.4	39	47	14	25.0
1971	12.7	11.7	2.3	47	44	9	11.2	11.7	3.8	42	44	14	26.7
1972	13.3	12.5	2.6	47	44	9	11.8	12.5	4.2	41	44	15	28.4
1973	14.7	13.7	2.8	47	44	9	13.1	13.6	4.6	42	43	15	31.3
1974	16.8	15.3	3.1	48	43	9	15.0	15.2	5.0	43	43	14	35.2
1975[2]	:	:	:	:	:	:	:	:	:	43	43	14	:
1976[2],e	:	:	:	:	:	:	:	:	:	:	:	:	:
Petroleum and natural gas													
1969	4.2	5.9	1.3	37	51	12	2.9	6.8	1.7	26	60	14	11.4
1970	4.8	6.2	1.4	39	50	11	3.0	7.5	1.9	24	61	15	12.4
1971	5.6	6.5	1.5	41	48	11	3.2	8.3	2.2	23	61	16	13.6
1972	6.4	6.8	1.7	43	46	11	3.9	8.6	2.4	26	58	16	15.0
1973	6.9	7.6	1.9	42	46	12	4.1	9.6	2.7	25	59	16	16.4
1974	7.7	8.0	2.1	43	45	12	4.4	10.5	2.8	25	59	16	17.7
1975[2]	:	:	:	:	:	:	:	:	:	26	58	16	:
1976[2],e	:	:	:	:	:	:	:	:	:	27	57	16	:
Other mining and smelting													
1969	2.5	3.1	0.6	40	50	10	1.8	3.7	0.7	30	59	11	6.2
1970	2.7	3.2	0.7	41	48	11	2.0	3.9	0.7	30	59	11	6.5
1971	3.1	3.4	0.7	43	47	10	2.1	4.3	0.8	29	59	12	7.2
1972	3.4	3.6	0.7	45	46	9	3.3	3.6	0.8	43	47	10	7.7
1973	3.8	3.8	0.8	45	45	10	3.7	3.8	0.9	44	45	11	8.4
1974	3.9	4.1	1.0	43	45	12	3.8	4.1	1.2	42	45	13	9.0
1975[2]	:	:	:	:	:	:	:	:	:	43	45	13	:
1976[2],e	:	:	:	:	:	:	:	:	:	:	44	:	:
Railways													
1969	4.8	0.5	0.5	83	8	9	5.6	0.1	—	98	2	—	5.7
1970	5.0	0.4	0.5	84	7	9	5.8	0.1	—	98	2	—	5.9
1971	5.1	0.4	0.5	85	7	8	5.9	0.1	—	98	2	—	6.0
1972	5.1	0.4	0.5	85	7	8	5.9	0.1	—	98	2	—	6.0
1973	5.2	0.4	0.5	85	7	8	6.0	0.1	—	98	2	—	6.1
1974	5.3	0.5	0.4	84	9	7	6.2	0.1	—	99	1	—	6.3
1975[2]	:	:	:	:	:	:	:	:	:	99	1	—	:
1976[2],e	:	:	:	:	:	:	:	:	:	:	:	—	:
Other utilities													
1969	16.1	3.4	0.3	81	17	2	18.7	0.8	0.4	94	4	2	19.8
1970	17.3	3.7	0.4	81	17	2	20.0	0.8	0.6	93	4	3	21.4
1971	19.2	3.9	0.5	81	17	2	21.9	1.0	0.7	93	4	3	23.6
1972	21.0	4.2	0.7	81	16	3	23.9	1.2	0.8	92	5	3	25.9
1973	23.0	4.6	0.8	81	16	3	26.4	1.2	0.9	93	3	3	28.5
1974	26.0	5.2	1.2	80	16	4	31.1	1.3	0.1	96	4	—	32.4
1975[2]	:	:	:	:	:	:	:	:	:	96	4	—	:
1976[2],e	:	:	:	:	:	:	:	:	:	96	4	—	:

21.39 Book value, ownership and control of capital employed¹ in non-financial industries, 1969-76 (concluded)

| Item and year | Ownership | | | | | | Control | | | | | | Total capital employed² $ billion |
| | Investment owned in | | | Percentage of capital employed owned in | | | Investment controlled in | | | Percentage of capital employed controlled in | | | |
	Canada $ billion	United States $ billion	Other countries $ billion	Canada %	United States %	Other countries %	Canada $ billion	United States $ billion	Other countries $ billion	Canada %	United States %	Other countries %	
Total³													
1969	55.3	24.7	5.2	65	29	6	55.0	23.7	6.5	64	28	8	85.2
1970	59.2	26.1	5.6	65	29	6	57.9	25.5	7.5	64	28	8	90.9
1971	64.6	27.3	6.2	66	28	6	62.6	26.9	8.5	64	27	9	98.0
1972	69.2	29.0	6.8	66	28	6	68.4	27.6	8.9	65	26	9	104.9
1973	76.5	32.0	7.5	66	27	7	75.9	30.1	9.8	66	25	8	115.9
1974	86.0	35.2	8.5	..	..	..	86.5	33.3	9.9	67	25	8	129.7
1975ª	..	..	..	..	..	..	..	..	..	67	..	8	..
1976ª,e	..	..	..	..	..	..	..	..	..	..	..	..	..

¹The book value of long-term debt and equity (including retained earnings) employed in enterprises in Canada.
ªRatios for 1975 and 1976 are projections based on the adjustment of 1974 data to reflect subsequent major identified changes.
³Includes data for merchandising and construction.

Sources

21.1 - 21.9 Gross National Product Division, System of National Accounts (Current) Branch, Statistics Canada.
21.10 - 21.13 Industry Product Division, System of National Accounts (Current) Branch, Statistics Canada.
21.14 Input-Output Division, System of National Accounts (Structural) Branch, Statistics Canada.
21.15 - 21.28 Prices Division, General Statistics Branch, Statistics Canada.
21.29 - 21.34 Balance of Payments Division, System of National Accounts (Current) Branch, Statistics Canada.
21.35 Bank of Canada.
21.36 Balance of Payments Division, System of National Accounts (Current) Branch, Statistics Canada.
21.37 Prices Division, General Statistics Branch, Statistics Canada.
21.38 Financial Flows and Multinational Enterprises Division, System of National Accounts (Current) Branch, Statistics Canada.
21.39 Balance of Payments Division, System of National Accounts (Current) Branch, Statistics Canada.

Appendices

Government organizations and related agencies

A summary organization chart of the federal government appears in Chapter 3. That chapter also explains how executive acts of the federal government are carried out, how legislature is introduced, amended, approved and proclaimed. Some titles of the federal identity program are included in brackets following the statute names of departments and agencies.

Advisory Council on the Status of Women. The council received official status by order-in-council PC 1976-781 on April 1, 1976. It advises the government and informs the public on matters pertaining to the status of women. It recommends changes in legislation and other actions to improve the position of women, and publishes research papers which are available on request.

The council consists of a president and two vice-presidents who are full-time members and 27 part-time members, appointed from each province and territory by the Governor-in-Council for three-year terms. It reports to Parliament through the minister responsible for the status of women.

Agricultural Products Board. This board was established under authority of the Emergency Powers Act by order-in-council PC 3415 in 1951 to administer contracts with other countries to buy or sell agricultural products, and to carry out other commodity operations considered necessary or desirable for Canada's needs and requirements. The board was re-established under the Agricultural Products Board Act in 1952 and operates now under RSC 1970, c.A-5. Under the act the minister may require any staff of the agriculture department to provide services for the board.

Agricultural Stabilization Board. Established in 1958 as a Crown corporation under the Agricultural Stabilization Act (RSC 1970, c.A-9), the board is empowered to stabilize prices of agricultural products both to assist the industry in realizing fair returns for labour and investment and to maintain a fair relationship between the prices received by farmers and the costs of goods and services that they buy. Programs under the act are administered by board staff with assistance from the agriculture department. The board reports to Parliament through the minister of agriculture.

Air Canada. Formerly Trans-Canada Air Lines, Air Canada was incorporated by an act of Parliament in 1937 (RSC 1970, c.A-11) to provide a publicly owned air transportation service, with powers to carry on its business throughout Canada and outside Canada. The corporation maintains passenger, mail and commodity traffic services over nationwide routes and to the United States, Britain, Ireland, France, Switzerland, the Federal Republic of Germany, Denmark, Bermuda, the Bahamas, Jamaica, Antigua, Barbados, the French Antilles, Cuba and Trinidad. Air Canada is responsible to Parliament through the minister of transport.

Anti-dumping Tribunal. Under the Anti-dumping Act (RSC 1970, c.A-15, as amended by SC 1970-71, c.3), the tribunal is a court of record and makes formal inquiry into the impact of dumping on production in Canada. Within 90 days of a preliminary determination of dumping by the deputy minister of national revenue for customs and excise, the tribunal must make an order or finding on the question of material injury, threat of material injury or retardation to production in Canada of like goods. The tribunal may at any time after the date of an order or a finding made by it review, rescind, change, alter or vary the order or finding or may rehear any matter before deciding it. The Governor-in-Council may ask the tribunal to investigate and report on any matter relative to importation of goods that may cause or threaten injury to production of goods in Canada.

The tribunal has a chairman, four other members, a secretary, and research and support staff, with offices in Ottawa. The tribunal conducts public and closed hearings, personal interviews, in-house research, statistical and financial analysis, interviews with Canadian manufacturers and associations, and inspection of facilities. It reports to Parliament through the minister of finance.

Anti-Inflation Appeal Tribunal. The tribunal was established by the Anti-Inflation Act, SC 1974-75-76, c.75; amended by SC 1974-75-76, c.98 to hear appeals resulting from orders issued by the administrator under the Anti-Inflation Act. The act provides for the chairman of the Tax Review Board to be appointed chairman of the Anti-Inflation Appeal Tribunal and for members of the Tax Review Board to hold office as members of the Anti-Inflation Appeal Tribunal. The Governor-in-Council may appoint more members as he considers necessary to deal effectively with appeals taken to the tribunal. The principal office is at Ottawa and the tribunal sits at such times and places as the chairman considers necessary. The tribunal has all powers, rights and privileges of a superior court of record. Its decisions or orders are subject to review and are to be set aside by the Federal Court of Appeal. The tribunal is under the jurisdiction of the minister of justice but is independent of the justice department.

Anti-Inflation Board. An interim board was created by order-in-council 1975-2429, October 14, 1975. It was replaced December 15, 1975 by the Anti-Inflation Board (AIB) which was established by the Anti-Inflation Act (SC 1974-75-76, c.75 amended by SC 1976, c.98). Under the act, the board administers guidelines enacted to restrict price and wage increases. It is required to monitor changes in prices, profits, compensation and dividends; to consult and negotiate with the parties involved to bring such changes within the guidelines; and to inform the public about inflation and its causes. These controls were to be phased out between April 14 and December 31, 1978. The board consists of a chairman, vice-chairman and five regional members appointed by the Governor-in-Council and reports to Parliament through the minister of finance.

Army Benevolent Fund Board. The board, established by the Army Benevolent Fund Act (SC 1947, c.49, as amended by SC 1974-75-76, c.3), administers the Army Benevolent Fund and other like funds, from special accounts set up in the Consolidated Revenue Fund. The board awards grants from the special account to veterans or their dependents for relief, if none is available from government sources, and for educational assistance, contingent on need and continued progress. The board has five members appointed by the Governor-in-Council, one of them nominated by the Royal Canadian Legion and one by the National Council of Veterans Associations in Canada. Head office is in Ottawa. The board reports to Parliament through the minister of veterans affairs.

Atlantic Development Council (Atlantic Development Council Canada). Created under the 1969 Government Organization Act (SC 1968-69, c.28), the council is composed of 11 members, including a chairman and vice-chairman, appointed by the Governor-in-Council to reflect the economic structure of New Brunswick, Nova Scotia, Prince Edward Island and Newfoundland. Its function is to advise the minister of regional economic expansion, in respect of the Atlantic region, on matters to which his duties, powers and functions extend, and particularly on plans, programs and proposals for fostering the economic expansion and social adjustment of the Atlantic region, and the feasibility and merits of particular programs and projects.

Atomic Energy Control Board. By act of Parliament (RSC 1970, c.A-19) proclaimed October 1946, the regulation and control of atomic energy in Canada was placed under the Atomic Energy Control Board. The board reports to Parliament through the minister of energy, mines and resources.

Atomic Energy of Canada Limited. This Crown company was incorporated in February 1952 under the Atomic Energy Control Act, 1946 (RSC 1970, c.A-19) to take over from the National Research Council in April 1952 the operation of the Chalk River project. Its main activities include scientific research and engineering development in the atomic energy field, the development, design and marketing of nuclear power systems, and the production of radioactive isotopes and associated equipment, such as cobalt-60 beam therapy units for the treatment of cancer. AECL is responsible for construction and operation of heavy water plants and research and development involving heavy water production methods. The company reports to Parliament through the minister of energy, mines and resources.

Bank of Canada. Legislation of 1934 (RSC 1970, c.B-2) provided for the establishment of a central bank in Canada to regulate credit and currency, to control and protect the external value of the Canadian dollar and to stabilize the level of production, trade, prices and employment as far as possible within the scope of monetary action. The Bank of Canada acts as the fiscal agent of the Government of Canada, manages the public debt and has the sole right to issue notes for circulation in Canada. It is managed by a board of directors appointed by the government, composed of a governor, a deputy governor and 12 directors; the deputy minister of finance is also a member of the board (ex officio). The bank reports to Parliament through the minister of finance.

Blue Water Bridge Authority. Created by the Blue Water Bridge Authority Act (SC 1964, c.6), this non-profit organization is responsible for the operation of the Canadian portion of the bridge spanning the St. Clair River from Point Edward, Ont., to Port Huron, Mich. Tolls set are subject to approval of the Canadian Transport Commission. All toll moneys must be used for the operation and maintenance of the present bridge or for building a new one. The authority is not an agent of the Crown but its members are appointed by the Governor-in-Council on the recommendation of the minister of transport for terms ranging from one to five years.

Board of Examiners for Dominion Land Surveyors. Established under the Canada Lands Survey Act (RSC 1970, c.L-5), the board examines candidates: for admission as articled pupils; for commissions as dominion land surveyors; or for certificates as dominion topographical surveyors. It is also responsible for the discipline of dominion land surveyors. The board has three members appointed by the Governor-in-Council, one of whom, the chairman, is the surveyor general of Canada lands; it is part of the energy, mines and resources department.

Bureau of Pensions Advocates (Bureau of Pensions Advocates Canada). The bureau was established in 1971 by amendments to the Pension Act (SC 1970-71, c.31). Composed of a chief pensions advocate appointed by the Governor-in-Council, and pensions advocates, officers and employees appointed under the Public Service Employment Act, it provides an independent professional legal aid service to applicants for awards under the Pension Act. The bureau's head office is in Ottawa; there are district offices in 18 major centres across Canada. It reports to Parliament through the minister of veterans affairs.

Canada Council. The council was established by order-in-council dated April 15, 1957, under the terms of the Canada Council Act (RSC 1970, c.C-2 assented to March 28, 1957) to encourage the arts, humanities and social sciences in Canada, mainly through a broad program of fellowships and grants. Its principal sources of income are an annual grant from the government ($59.7 million for the year ended March 31, 1977) and an endowment account (originally of $50 million) which yielded over $7 million in 1976-77. On April 1, 1978 a separate Social Sciences and Humanities Research Council began operation, under the terms of the Government Organization (Scientific Activities) Act, 1976. The Canada Council retains responsibility for encouragement of the arts. The proceedings of the council are reported each year to Parliament through the secretary of state.

Canada Deposit Insurance Corporation. The corporation was established by legislation (RSC 1970, c.C-3), which received royal assent on February 17, 1967. It is empowered to insure Canadian currency deposits other than those belonging to the Government of Canada, up to $20,000 a person, in banks, federally incorporated trust and loan companies that accept deposits from the public, and in similar provincially incorporated institutions authorized by their provincial governments to apply for such insurance. The corporation is also empowered to act as a lender of last resort for member institutions. Its board comprises a chairman, appointed by the Governor-in-Council, and four other directors who hold the positions of governor of the Bank of Canada, deputy minister of finance, superintendent of insurance and inspector general of banks. It reports to Parliament through the minister of finance.

Canada Development Corporation. The corporation (CDC) was established in 1971 by the Canada Development Corporation Act (SC 1970-71, c.49) to develop and maintain strong Canadian-controlled and managed corporations in the private sector of the economy, to give Canadians greater opportunities to invest and participate in the economic development of Canada, and to operate profitably and in the best interests of all its shareholders. Administration of CDC is vested in a board of 19 directors. CDC is neither an agent of the Crown nor subject to the Financial Administration Act.

CDC concentrates on control-position equity investments in leading corporations in selected industries. Industries characterized by large, longer-range development projects, an upgrading of Canadian resources, a high technological base, and a good potential for building a Canadian presence in international markets are considered. Investments have been made in petrochemicals, mining, oil and gas, health care and venture capital.

Polysar Ltd. is CDC's wholly owned operating company in petrochemicals. CDC and Polysar together own 60% of Petrosar Ltd. which operates the first world-scale crude oil topping and naphtha cracking unit in Canada. CDC's interest in the mining industry is represented by 30% ownership of Texasgulf, Inc., one of Canada's largest mineral producers.

CDC Oil & Gas Ltd., 100% owned, is CDC's operating company in oil and gas exploration and production. Through Connlab Holdings Ltd., also 100% owned, CDC is developing a Canadian presence in the health care and pharmaceutical field. CDC's associated venture capital companies, Venturetek International Ltd. of Toronto, Innocan Investments Ltd. of Montreal, and Ventures West Capital Ltd. of Vancouver, together represent the largest pool of venture capital in Canada and have themselves invested in 26 small and medium-sized businesses.

As at September 30, 1977, CDC's consolidated assets amounted to $1,882 million and shareholders equity was $722 million.

Canada Employment and Immigration Advisory Council. This council was established by the Employment and Immigration Reorganization Act — Part II, the Canada Employment and Immigration Advisory Council Act (SC 1976-77, c.54) proclaimed on August 15, 1977. The council replaces the Canada Manpower and Immigration Council and the Unemployment Insurance Advisory Committee. The act provides for a chairman and no fewer than 15 or more than 21 other members to be appointed by the Governor-in-Council, to advise the minister of employment and immigration on all matters related to labour market resources, employment services, unemployment insurance and immigration.

Canada Employment and Immigration Commission (Employment and Immigration Canada). The Employment and Immigration Reorganization Act — Part I, the Employment and Immigration Department and Commission Act (SC 1976-77, c.54) passed in August 1977 created the Canada Employment and Immigration Commission by integrating the former Unemployment Insurance Commission and the former Department of Manpower and Immigration. The legislation also created the Department of Employment and Immigration which provides services to the commission.

The employment and insurance objective of the commission is to further the attainment of national economic and social goals by realizing the full productive potential of Canada's human resources, while supporting the initiatives of individuals to pursue their economic needs and, more generally, their self-fulfilment through work.

The immigration objective of the commission is to administer the admission of immigrants in accordance with the economic, social and cultural interests of Canada.

Canada Labour Relations Board. Established under the authority of the Canada Labour Code Part V (RSC 1970, c.L-1), this board administers provisions of the code with respect to workers in industries under federal jurisdiction. It consists of a chairman, a vice-chairman, an additional vice-chairman where considered advisable by the Governor-in-Council and not less than four or more than eight other members.

Canadian Arsenals Limited (Arsenals Canada). The principal function of this Crown corporation is to operate government-owned facilities for the production of certain defence material and other complementary items. It was established under the Companies Act in September 1945, and is subject to the Government Companies Operation Act (RSC 1970, c.G-7) and certain provisions of the Financial Administration Act (RSC 1970, c.F-10). It reports to Parliament through the minister of supply and services.

Canadian Broadcasting Corporation. The CBC is a Crown corporation established by an act of Parliament in 1936, replacing an earlier public broadcasting agency, the Canadian Radio Broadcasting Commission, created in 1932. The Broadcasting Act of 1968 (RSC 1970, c.B-11) describes the CBC as "established by Parliament for the purpose of providing the national broadcasting service".

The corporation has a president and 14 other directors appointed by the Governor-in-Council. The president is the chief executive officer. The executive vice-president is appointed by the corporation on the recommendation of the president and with the approval of the Governor-in-Council. He is responsible to the president for the management of broadcasting operations in accordance with corporation policies.

CBC operations are financed by public funds voted annually by Parliament, with supplementary revenue obtained from commercial advertising. The CBC's accounts are audited annually by the auditor general of Canada and the corporation reports to Parliament through the secretary of state.

Canadian Commercial Corporation. This corporation, wholly owned by the Government of Canada, was established in 1946 by an act of Parliament (RSC 1970, c.C-6) to assist in the development of trade between Canada and other nations. The corporation may act either as the principal or agent in the import or export of goods and commodities to or from Canada.

Under this broad charter, it acts primarily as the contracting agency when other countries and international agencies wish to purchase from Canada on a government-to-government basis.

Management and staff are provided by the supply and services department, which is responsible for the central procurement of goods and related services for all Canadian government departments and agencies. The corporation reports to Parliament through the minister of supply and services.

Canadian Consumer Council (Consumer Council Canada). The council was established in 1968 (RSC 1970, c.C-27) to advise the minister of consumer and corporate affairs on all facets of consumerism. It meets with the minister three or four times a year and consists of members representing all segments of the population and all areas of Canada.

Canadian Council on Rural Development (Council on Rural Development Canada). This council was established in 1965 under authority of the Agricultural and Rural Development Act (RSC 1970, c.A-4) and now reports to the minister of regional economic expansion. In advising the minister on rural development problems and issues, the council provides a forum for the expression of views by organizations and persons sharing its concerns and, in general, facilitates public understanding of rural development needs and programs. Its membership, maximum 40, is made up of representatives from about 20 national organizations and individuals with rural and regional development expertise. The council meets three times a year in plenary session, but its four regional and other committees meet more frequently. Members are served by a secretariat drawn from the public service.

Canadian Dairy Commission. This commission, which reports to Parliament through the minister of agriculture, was established in December 1966 (RSC 1970, c.C-7) to provide efficient producers of milk and cream the opportunity of obtaining a fair return for labour and investment and thus ensure consumers of dairy products a continuous and adequate supply. The commission consists of three members appointed by the Governor-in-Council and operates with the advice of a nine-member advisory committee appointed by the minister. Since 1970, the commission has chaired a milk supply management committee, comprised of provincial milk marketing agencies and provincial government agencies, which manages the market share quota system under the terms of a federal-provincial milk marketing plan.

Canadian Film Development Corporation. This corporation, established by an act of Parliament in March 1967 (RSC 1970, c.C-8), fosters and promotes the development of a feature film industry in Canada through investment in productions, loans to producers, awards for outstanding accomplishments, and advice and assistance in distribution and administrative matters. It co-operates with other federal and provincial departments and agencies having like interests and as of April 1, 1977, is financed by a yearly appropriation. The corporation consists of the government film commissioner (ex officio) and six other members appointed by the Governor-in-Council for terms of five years. The corporation reports to Parliament through the secretary of state.

Canadian Government Specifications Board (Specifications Board Canada). Created in 1934 under the authority of the National Research Council Act (RSC 1970, c.N-14) as the Government Purchasing Standards Committee, this interdepartmental agency's name was changed in 1948 to Canadian Government Specifications Board (CGSB).

In 1965, responsibility for the CGSB's operation was transferred by order-in-council to the defence production department, now part of supply and services. Membership of the board was revised to include the secretary of the Treasury Board, the president of the National Research Council, and the deputy ministers of consumer and corporate affairs, national defence, public works, supply and services, transport, and industry, trade and commerce. The deputy minister of supply and services was designated chairman of the board.

The role of the CGSB is to provide standards for both public and private sectors for procurement, consumer requirements, legislation, technical practices, test procedures and to support international standardization in more than 100 fields. It has compiled more than 1,800 standards in both official languages. The technical process of developing and revising standards is performed by some 300 committees and about 3,000 members representing governments, producers, consumers, research and testing agencies, educational institutions, professional, technical and trade societies. The board works closely with the Standards Council of Canada and Metric Commission Canada in relation to national and international standardization and metric conversion. It is accredited by the council as a national standards writing organization.

Canadian Grain Commission. The Canada Grain Act (SC 1970-71, c.7) came into force in April 1971, repealing the Canada Grain Act, 1930 (RSC 1952, c.25) and replacing the former Board of Grain Commissioners for Canada. The commission reports to Parliament through the minister of agriculture. It provides general supervision over the physical handling of grain in Canada by licensing elevators and elevator operators, by inspecting, grading and weighing grain received at and shipped from terminal elevators, and by other services associated with regulating the grain industry. It manages and operates the six Canadian government elevators in Western Canada and administers the Grain Futures Act, which provides for grain futures trading.

The commission consists of a chief commissioner and two commissioners. Its objects are, in the interests of grain producers, to establish and maintain standards of quality for Canadian grain, to ensure a dependable commodity for domestic and export markets and to regulate grain handling in Canada. It has authority to conduct investigations and hold hearings, and to undertake, sponsor and promote research in relation to grain and grain products. The commission is part of the agriculture department, but submits a separate report to the minister.

Canadian Human Rights Commission. This commission was established on July 14, 1977 by the Canadian Human Rights Act (SC 1976-77, c.33) to deal with complaints regarding discriminatory practices and to develop and conduct information programs to foster public understanding of this act. The commission may designate an investigator to examine a complaint and may appoint a conciliator to bring about a settlement. At any stage after a complaint is filed, the commission may appoint a human rights tribunal to inquire into it.

The commission consists of two full-time members, the chief commissioner and the deputy chief commissioner, and from three to six other members who may be appointed by the Governor-in-Council as full- or part-time members for a term of seven years.

The minister of justice designates one member of the commission to be privacy commissioner to receive, investigate and report on complaints from individuals who allege that they are not being accorded the rights they are entitled to in relation to personal information recorded in a federal information bank.

The head office of the commission is in the National Capital Region. The commission reports to Parliament through the minister of justice.

Canadian Indian Rights Commission. This commission was established by order-in-council PC 1977-702 and works under the Joint Committee of Cabinet and the National Indian Brotherhood for the purpose of assisting the committee in resolving issues that come before it.

Canadian International Development Agency. The operation of Canada's international development programs is the responsibility of the Canadian International Development Agency. CIDA was originally established by order-in-council PC 1960-1476 and until 1968 was known as the External Aid Office. The agency is under the direction of a president and a governing body — the Canadian International Development Board — and reports to Parliament through the secretary of state for external affairs.

Canadian International Development Board. The board is the governing council responsible for directing the operations of the Canadian International Development Agency (CIDA). It is made up of the under-secretary of state for external affairs, the deputy ministers of the departments of finance and industry, trade and commerce, the governor of the Bank of Canada, the secretary of the treasury and the president of the International Development Research Centre. It meets under the chairmanship of CIDA's president.

Canadian Livestock Feed Board. This board is a Crown corporation reporting to Parliament through the minister of agriculture. Established under the Livestock Feed Assistance Act in 1967, its objectives are to ensure: the availability of feed grain in Eastern Canada and British Columbia, adequate storage space in Eastern Canada, and reasonable stability and fair equalization of feed grain prices in Eastern Canada and in British Columbia. The board administers a feed freight equalization program which pays part of the transportation costs of feed grains. The act stipulates that the board must make a continuing study of feed grain requirements and availability and study and make recommendations to the minister on requirements for additional feed grain storage facilities in Eastern Canada. The board must advise the government on all matters pertaining to stabilization and fair equalization of feed grain prices to livestock feeders and, to the greatest extent possible, to consult and co-operate with all federal departments, branches or other agencies or any province with similar duties, aims or objects.

The board has been assigned responsibilities under the national feed grain policy, effective since August 1974. It examines selling practices east of Thunder Bay and supervises the domestic market outside the designated region of the Canadian Wheat Board. The board designates the chairman of the committee supervising reserve stocks of feed grains presently held at Thunder Bay, Vancouver and various locations in Eastern Canada. If it finds bad pricing or supply practices, it can intervene directly as buyer or seller of feed grain. The Livestock Feed Assistance Act stipulates that the board may buy, transport, store and sell feed grains in Eastern Canada and British Columbia when authorized to do so by the Governor-in-Council.

The board is composed of three to five members with headquarters in Montreal and a branch office in Vancouver. A seven-member advisory committee, appointed by the Governor-in-Council, and representing livestock feeders in Eastern Canada and British Columbia, meets periodically with the board to review and discuss all aspects of feed grain supplies and prices, and related policies. This committee may make recommendations to the minister and the board.

Canadian National Railways. The Canadian National Railway Company was incorporated to administer an undertaking made up mainly of railway and other service facilities and activities. It includes the assets of the former Grand Trunk Railway Company of Canada and its subsidiaries, and of the Canadian Northern System, as well as certain Crown-owned properties which Canadian National manages and operates.

Primary statutes governing its organization and operation are the Canadian National Railways Act (RSC 1970, c.C-10) and the Railway Act (RSC 1970, c.R-2). Direction and control of the company and its undertaking are vested in a board of directors; its principal officers are the chairman of the board and the president, who is the chief executive officer.

Canadian Patents and Development Limited (CPDL) is a Crown corporation, wholly subsidiary to the National Research Council of Canada (NRC). CPDL was established in 1947 to handle patentable material of NRC and other government-financed research. The Public Servants Inventions Act in 1954 made CPDL the prime patenting and licensing agency for public servant inventions which by that act belong to the Canadian government.

CPDL may receive ideas and inventions from federal public servants and from the professional staff and employees of universities. The ideas and inventions are assessed for patentability and commercial use. Patent applications may be filed in various countries for those which are considered commercially exploitable and patentable. Some which are not patentable may be licensed independently, or together with patents. That portion of the licence fees and royalties paid under licence agreements and retained by CPDL is used to defray CPDL commercial operating expenses.

CPDL has also entered into agreements with many universities, provincial research organizations, and other publicly financed institutions to assess, patent and license their industrial and intellectual property.

The corporation's board of directors is composed of members from industry, universities and the federal government. The head office is in Ottawa. CPDL reports to Parliament through a designated minister.

Canadian Penitentiary Service. The penitentiary service operates under the Penitentiary Act (RSC 1970, c.P-6) and is under the jurisdiction of the solicitor general of Canada. It is responsible for all federal penitentiaries and for the care and training of persons committed to those institutions. The commissioner of penitentiaries, under the direction of the solicitor general, has control and management of the service and all matters connected therewith.

Canadian Pension Commission. This commission, established in 1933 by amendments to the Pension Act (RSC 1970, c.P-7), replaced the Board of Pension Commissioners, the first organization created to deal solely with war pensions for service in Canada's armed forces. The commission's main function is administration of the Pension Act under which it adjudicates on all claims for pensions in respect of disability or death arising out of service in Canada's armed forces; and parts of the Civilian War Pensions and Allowances Act, which provide for payment of pensions in respect of death or disability arising out of civilian service directly related to the prosecution of World War II. It also adjudicates on claims for pension under various other measures, authorizes and pays monetary grants accompanying certain gallantry awards bestowed on members of the armed forces and administers various trust funds established by private individuals for the benefit of veterans and their dependents. The commission consists of eight to 14 commissioners and up to 10 ad hoc commissioners appointed by the Governor-in-Council. Its chairman has the rank of a deputy minister and it reports to Parliament through the minister of veterans affairs.

Canadian Permanent Committee on Geographical Names. This committee deals with all questions of geographical nomenclature affecting Canada and advises on research and investigation into the origin and use of geographical names. Its membership includes representatives of federal mapping agencies and other federal offices concerned with nomenclature and a representative appointed by each province. The committee's functions were redefined in 1969 (order-in-council PC 1969-1458). The order-in-council recognizes that the provinces have exclusive jurisdiction to make decisions on names in lands under their jurisdiction. The committee is administered by the energy, mines and resources department.

Canadian Radio-television and Telecommunications Commission. This commission, established as the Canadian Radio-Television Commission under the provisions of the Broadcasting Act (RSC 1970, c.B-11), regulates and supervises all aspects of the Canadian broadcasting system. The Canadian Radio-television and Telecommunications Commission Act, promulgated April 1, 1976, amended the Broadcasting Act to assign regulatory responsibility to the Canadian Radio-television and Telecommunications Commission (CRTC) over federally-regulated telecommunications carriers.

CRTC regulates and supervises a single Canadian broadcasting system, mainly through the process of licensing broadcasting undertakings and administering a body of regulation and policy statements with a view to implementing the policies set out in the Broadcasting Act.

One of the commission's methods of satisfying the concerns as set out in the Railway Act is the holding of public hearings in connection with applications for general rate revisions submitted by the telecommunications carriers under its jurisdiction.

The CRTC consists of an executive committee of up to nine full-time members composed of a chairman, two vice-chairmen and six other full-time members. The full commission includes the executive committee and up to 10 part-time members chosen regionally. All are appointed by the Governor-in-Council. The commission reports to Parliament through the minister of communications.

Canadian Saltfish Corporation. Established under the Saltfish Act (SC 1969-70, c.32) and operative since May 1970, this corporation's main purpose is to improve the earnings of fishermen and other primary producers of salt-cured fish, through production or purchase, processing and marketing of salt cod from participating provinces.

The head office is at St. John's, Nfld. The board of directors is composed of a chairman whose office is in Ottawa, a president who is general manager, one director for each participating province and not more than five other directors, all appointed by the Governor-in-Council. It is assisted by an advisory committee of 15 members, at least half of them fishermen or representatives of fishermen. The corporation's financial obligations are limited to $15 million and it is required to operate without grant appropriation from Parliament. It reports to Parliament through the minister of fisheries and the environment.

Canadian Transport Commission. The commission, a court of record created in 1967 by the National Transportation Act (RSC 1970, c.N-17), took over powers formerly vested in the Board of Transport Commissioners, the Air Transport Board and the Canadian Maritime Commission. Its regulatory and judicial functions apply to almost all aspects of railway, commercial air, merchant marine, and commodity pipeline services. The act also provides for the regulation of extra-provincial motor vehicle transport but applicable sections of the act were not in effect as at September 1977, except with respect to the Roadcruiser autobus service operated by the Canadian National Railways in Newfoundland. On July 16, 1976, the CNR bus service was exempted by the Governor-in-Council from the provisions of the Motor Vehicle Transport Act (RSC 1970, c.M-14), and came under the jurisdiction of the Motor Vehicle Transport Committee of the Canadian Transport Commission, pursuant to Part III of the National Transportation Act. In all other cases, regulatory control over extra-provincial motor vehicle undertakings is exercised by provincial highway transport boards, acting as agents of the federal government, as provided for in the Motor Vehicle Transport Act. The commission is responsible for undertaking studies and research into the economic aspects of all modes of transport within, into or from Canada.

Five committees perform the regulatory duties: the railway transport committee, the air transport committee, the water transport committee, the motor vehicle transport committee, and the commodity pipeline transport committee.

The commission consists of not more than 17 members, including a president and two vice-presidents, appointed by the Governor-in-Council for a maximum of 10 years; it reports to Parliament through the minister of transport.

Canadian Wheat Board. The board was incorporated in 1935 under the Canadian Wheat Board Act (RSC 1970, c.C-12) to market, in the interprovincial and export trade, grain grown in Canada. Its powers include authority to buy, take delivery of, store, transfer, sell, ship or otherwise dispose of grain. Except as directed by the Governor-in-Council, the board was not originally authorized to buy grain other than wheat but since August 1949 it has also been authorized to buy barley and oats. Only grain produced in the designated area, Manitoba, Saskatchewan, Alberta and parts of British Columbia, is purchased by the board, which controls the delivery of grain into elevators and railway cars in that area as well as the interprovincial movement for export of wheat, oats and barley generally. The board reports to Parliament through a designated minister, at present the minister of transport.

Cape Breton Development Corporation. This proprietary Crown corporation was created by an act of Parliament in July 1967 (RSC 1970, c.C-13) and came into existence by proclamation on October 1, 1967. It was set up to rationalize the coal industry of Nova Scotia's Cape Breton Island and to broaden the base of the area's economy by assisting financing and development of industry to provide employment outside the coal mines.

The corporation acquired former interests of the major coal producer in the Sydney coalfield and is operating three mines, two of them new, a modern coal preparation plant and other ancillaries. It is active in development of tourism, primary products and secondary industries.

The act provides for a board of directors, comprising a chairman, a president and five other directors. Head office is located in Sydney. The corporation reports to Parliament through the minister of regional economic expansion. Its operations are financed by the federal government.

Central Mortgage and Housing Corporation. This Crown agency was incorporated by an act of Parliament (RSC 1952, c.46) in December 1945 to administer the National Housing Act. Under the National Housing 1954 Act (RSC 1970, c.C-16), the corporation insures mortgage loans made by approved lenders for new and existing housing and makes direct loans in resource communities and rural areas; guarantees home improvement loans made by banks; undertakes subsidized rental housing projects and land assembly developments under federal-provincial arrangements; offers loans and subsidies for public housing projects; makes loans for land assembly projects to be used for general residential development; makes loans to individuals or organizations for low-rental housing projects; makes loans to provinces and municipalities, with provincial concurrence, for sewage and water treatment projects designed to eliminate water and soil pollution; makes contributions and loans to provinces and municipalities for neighbourhood improvement; conducts housing research; encourages urban planning; and owns and manages rental housing units including those built for war workers and veterans. The corporation arranges for and supervises construction of housing projects on behalf of other government departments and agencies. It is responsible to Parliament through the minister of state for urban affairs. The national office of the corporation is in Ottawa with branch offices in major urban areas.

Columbia River Treaty Permanent Engineering Board. The permanent engineering board, consisting of two Canadians and two Americans, was established under the 1964 Columbia River Treaty between Canada and the United States. The board assembles records and inspects and reports at least annually on matters within the scope of the treaty. It reports to Parliament through the minister of energy, mines and resources.

Commissioner of Official Languages. Appointed by Parliament pursuant to the Official Languages Act (RSC 1970, c.O-2), the commissioner holds office for a term of seven years, renewable until age 65. He is responsible to Parliament for ensuring recognition of the equal status of French and English as Canada's official languages and for ensuring compliance with the spirit and intent of the act in all the institutions of the Parliament and Government of Canada. The commissioner is empowered to receive and investigate complaints from the public and, on his own initiative, to conduct investigations into possible violations of the act. The results of investigations must be communicated to the complainants and the institutions concerned and may, at the commissioner's discretion, be the subject of a special report to Parliament. The commissioner reports annually to Parliament on the conduct of his office and may make recommendations for changes in the act as he deems necessary or desirable.

Copyright Appeal Board (Copyright Appeal Board Canada). The board was established to provide an agency to which people using music protected by copyright could direct appeals against the fees proposed by performing rights societies for the use of the music. The Copyright Act (RSC 1970, c.C-30) empowers the board to deal only with the amount of the fees that the societies propose to collect for an ensuing calendar year. It has no authority to draft terms and conditions of the tariffs. Hearings before the board are conducted in a quasi-judicial manner. After considering an appeal the board makes such alterations to the proposed statements of fees as it thinks appropriate and transmits the statements thus altered, revised or unchanged to the minister of consumer and corporate affairs. The decision of the board is final and binding. The Copyright Appeal Board consists of three members appointed by the Governor-in-Council, one of whom, as chairman, must hold or have held high judicial office.

Correctional Investigator. Appointed by order-in-council PC 1973-1431 on June 5, 1973 the correctional investigator has the powers of a commissioner under the Inquiries Act. This officer investigates problems of inmates on subjects for which the solicitor general is responsible and reports to him. The office consists of the correctional investigator and three complaint officers. It is located in Ottawa and is independent of the solicitor general's department.

Court Martial Appeal Court. This court was established as a superior court of record under the National Defence Act (RSC 1970, c.N-4). Accused persons found guilty by a court martial have the right to direct an appeal to the Court Martial Appeal Court on the legality of any or all findings, or of the whole or any part of the sentence. The court is composed of not fewer than four judges of the Federal Court of Canada and additional judges of a superior court of criminal jurisdiction as designated by the Governor-in-Council, with one judge appointed as president. Appeals are heard by a minimum of three judges. The Court Martial Appeal Court may sit and hear appeals at any place under direction of the president. An appellant whose appeal has been wholly or partially dismissed by the court may, under certain circumstances, appeal to the Supreme Court of Canada; where the Court Martial Appeal Court has wholly or partially allowed an appeal, the minister of national defence may similarly enter an appeal to the Supreme Court of Canada.

Crown Assets Disposal Corporation (Surplus Canada). This agency corporation was established in 1944 as the War Assets Corporation under the Surplus Crown Assets Act (RSC 1970, c.S-20) and is subject to the Financial Administration Act (RSC 1970, c.F-10). Its name was changed to Crown Assets Disposal Corporation in 1949. The corporation is responsible for the sale of federal government surplus movable assets located in Canada and at Canadian government establishments throughout the world. It also acts as agent on behalf of foreign governments in selling their surplus property located in Canada and has an agreement with a European agency for marketing Canadian military surplus assets located abroad. While the corporation's normal method of sale is to invite written offers, on occasion it sells by public auction and through retail outlets. The act provides for a board of directors, comprising a chairman and a minimum of five other directors. Its head office is in Ottawa. Regional offices are in Halifax, Montreal, Toronto, Ottawa, Edmonton and Vancouver. The corporation is responsible to Parliament through the minister of supply and services.

Defence Construction (1951) Limited (Defence Construction Canada). This Crown corporation contracts for major construction and maintenance projects required by the defence department. It was incorporated in May 1951 under the authority of the Defence Production Act. In April 1965 its control and supervision were transferred from the minister of defence production to the minister of national defence.

Defence Construction (1951) Limited (DCL) obtains tenders, makes recommendations regarding awards, awards and administers major construction and maintenance contracts. This includes supervision of construction work and the certification of contractors progress claims for completed work.

The company provides technical and administrative assistance to government departments and agencies. Head office is in Ottawa and branch offices are in Halifax, Montreal, Toronto, Winnipeg, Vancouver and Lahr, Federal Republic of Germany.

Defence Research Board. The board, established in 1947 by an amendment to the National Defence Act (RSC 1970, c.N-4), advises the minister of national defence on scientific matters relating to defence and evaluates the contribution of science and technology to the achievement of defence objectives. The functions of the board were redefined in 1974 when its research and administrative activities and staff were absorbed within the framework of the defence department.

The board consists of a full-time chairman, a vice-chairman and 12 members appointed by the Governor-in-Council for three-year terms. The deputy minister of national defence, the president of the National Research Council and three senior officers of the Canadian forces are ex officio members. The board has its headquarters in Ottawa.

Department of Agriculture (Agriculture Canada). This department was established in 1867 (SC 1868, c.53) and now operates under authority of RSC 1970, c.A-10. It undertakes work on all phases of agriculture. Research and experimentation are carried out by the research, health of animals and economics branches, and a grain research laboratory. Maintenance of standards and protection of products are the responsibility of the food production and marketing and health of animals branches. The Canada Grain Act, as it pertains to the inspection, weighing, storage and transportation of grain, is administered by the Canadian Grain Commission, a part of the department. Programs concerning farm income security and price stability are provided under the Crop Insurance Act, the Canadian Dairy Commission Act, the Agricultural Stabilization Act and the Agricultural Products Board Act. The Agricultural Stabilization Board, the Agricultural Products Board, the Farm Credit Corporation, the Canadian Dairy Commission, the Canadian Livestock Feed Board and the National Farm Products Marketing Council report to Parliament through the minister of agriculture.

Department of Communications. The department was established under the 1969 Government Organization Act and operates under authority of the Department of Communications Act (RSC 1970, c.C-24). The minister of communications is responsible for fostering the orderly operation and development of communications for Canada. This includes recommending national policies and programs regarding communications services for Canada, promoting the efficiency and growth of Canadian communications systems and helping them adjust to changing conditions, and encouraging development and introduction of new communication facilities and resources. Responsibilities also include managing the radio frequency spectrum to permit orderly use of radio communications, protecting Canadian interests in international telecommunications matters, and co-ordinating telecommunications services for departments and agencies of the federal government.

Teleglobe Canada, the Canadian Radio-television and Telecommunications Commission and Telesat Canada report to Parliament through the minister of communications.

Department of Consumer and Corporate Affairs (Consumer and Corporate Affairs Canada). This department was established in 1967 (RSC 1970, c.C-27) replacing the Department of the Registrar General of Canada. The duties, powers and functions of the minister extend to and include all matters over which Parliament has jurisdiction, not by law assigned to any other department, branch or agency of the federal government, relating to: consumer affairs; corporations and corporate securities; combines, mergers, monopolies and restraint of trade; bankruptcies and insolvencies; and patents, copyrights, trade marks and industrial design.

The functions of the department are divided into five main areas. The consumer affairs bureau co-ordinates government activities in this field; the corporate affairs bureau administers the government's corporate activities; a bureau of intellectual property administers laws and regulations pertaining to patents, trade marks, industrial designs and copyrights; and the field operations service supervises the department's operations across Canada, staffing regional and district offices in five cities from coast to coast and district offices in 27 others. Competition policy is regulated by the competition policy bureau. As registrar general of Canada, the minister of consumer and corporate affairs is the custodian of the Great Seal of Canada, the Privy Seal of the Governor General, the seal of the administrator of Canada and the seal of the registrar general of Canada. The Restrictive Trade Practices Commission (Combines Investigation Act) is part of the department and reports directly to the minister.

Department of Employment and Immigration (Employment and Immigration Canada). This department was established in 1977 to provide services to the Canada Employment and Immigration Commission.

Department of Energy, Mines and Resources (Energy, Mines and Resources Canada). The department was created in 1966 by the Government Organization Act (RSC 1970, c.E-6). In addition to its administrative services and the office of energy conservation, it is organized into three sectors. The energy policy sector has responsibilities relating to the development of plans and policies for all forms of energy, the development of programs, legislation and agreements to implement those policies, the direction of studies relating to energy sources and requirements, and the co-ordination of policy advice. A major responsibility of the sector is research on and formulation of a national energy policy. The mineral development sector gathers economic data on non-renewable resources for use by government, industry and the public. It also develops policy proposals for the government and the mineral industry to help determine policies and decisions that will ensure an adequate, dependable and timely flow of minerals to meet the country's needs at reasonable cost. The science and technology sector includes the Geological Survey of Canada, the Canada Centre for Mineral and Energy Technology (CANMET), the Surveys and Mapping Branch, the Earth Physics Branch, the Canada Centre for Remote Sensing and the Polar Continental Shelf Project, all engaged in research and the provision of information, the Office of Energy Research and Development, which co-ordinates federal research and development related to energy policies, the Explosives Branch, which controls, under the provisions of the Explosives Act, the production and handling of explosives, and the Canada Centre for Geoscience Data.

Atomic Energy of Canada Limited, Eldorado Nuclear Limited, Eldorado Aviation Limited, the Atomic Energy Control Board, the National Energy Board, Uranium Canada Limited, Petro-Canada, and the interprovincial boundary commissions report to Parliament through the minister of energy, mines and resources. Operationally the international boundary commission reports through the minister of energy, mines and resources; in dealing with its counterpart in the United States it is responsible to the secretary of state for external affairs.

Department of External Affairs (External Affairs Canada). The main function of the department, established in 1909 (RSC 1970, c.E-20), is the protection and advancement of Canadian interests abroad. The responsible minister is the secretary of state for external affairs. The senior permanent officer (deputy minister) of the department, the under-secretary of state for external affairs, is assisted by four deputy under-secretaries and by four assistant under-secretaries and is advised by officers in charge of bureaus, offices and divisions. Directors-general or directors of these units are each responsible for part of the department's work and are assisted by foreign service officers, specialists in various occupational groups and an administrative staff. Officers serving abroad are formally designated as high commissioner, ambassador, minister, minister-counsellor, counsellor, first secretary, second secretary, third secretary and attaché at diplomatic posts and consul general, consul and vice-consul at consular posts. Canada maintains approximately 185 diplomatic, consular and other missions (200 including the permanent delegations to international organizations), 74 of which are non-resident.

In Ottawa the department's work is conducted by regional, functional and administrative bureaus and a number of operational units. The five regional bureaus administer 13 geographical divisions, each responsible for the countries of a region. Eight functional bureaus, including 22 divisions are concerned with commercial and commodity relations; consular services; co-ordination; defence and arms control; development, industry and science relations; legal affairs; public affairs; and United Nations affairs. Five administrative bureaus are responsible for personnel, finance and management, communications and general services, and security and intelligence liaison.

In addition, there are an inspection service, a policy analysis group, a protocol division, an operations centre, an interdepartmental committee on external relations, a chief air negotiator, an adviser on bilingualism, a passport office and a departmental press office.

The International Joint Commission reports to the secretary of state for external affairs of Canada as well as to the secretary of state of the United States. The secretary of state for external affairs reports to Parliament for the Canadian International Development Agency.

Department of Finance (Department of Finance Canada). Created by an act of Parliament in 1869, this department now operates under the Financial Administration Act (RSC 1970, c.F-10 as amended). It is primarily responsible for advising the government on the economic and financial affairs of Canada. The department's work is carried out in five branches. The tax policy and federal-provincial relations branch helps to form tax policy and maintain the tax structure. It deals with personal income and commodity taxes, taxes on corporations and Canada's international tax relations. Fiscal relations with the provinces are the responsibility of a federal-provincial relations division. The branch also administers grants to municipalities in lieu of taxes on government property and advises on the government's social development and manpower policies. The economic programs and government finance branch is concerned with resource development, various government programs of broad economic development and the financing of Crown corporations and government agencies. The international trade and finance branch is concerned with trade policy and development, the Canadian aid program and customs tariffs. The fiscal policy and economics analysis branch monitors the economy, analyzes the potential impact of various alternative courses for government fiscal policy and participates in a number of international organizations, including the International Monetary Fund. A capital markets division is responsible for monitoring developments in capital markets and advising on the government's debt operations. The inspector general of banks is an office of the department. An economic planning branch is responsible for co-ordinating, planning and developing medium- and long-term economic measures and policies. The following agencies report to Parliament through the minister of finance: the Anti-dumping Tribunal, the Bank of Canada, the Canada Deposit Insurance Corporation, the Department of Insurance, the Tariff Board and the Anti-Inflation Board. The minister of finance acts as spokesman in Parliament for the auditor general.

Department of Fisheries and the Environment (Fisheries and Environment Canada). Established by an act of Parliament in June 1971 (SC 1970-71, c.42), the environment department, now called the Department of Fisheries and the Environment, carries the main federal responsibility for attacking pollution and ensuring proper management and development of Canada's renewable resources. The department is organized into two principal components, the fisheries and marine service and environmental services, each headed by a senior assistant deputy minister.

The fisheries and marine service is composed of three major divisions — fisheries management, ocean and aquatic sciences and small craft harbours — which function in conjunction with an international directorate and other policy and liaison groups at headquarters in Ottawa.

Fisheries management is responsible for conservation and restoration of fish stocks, allocation and control of access to fishery resources, biological and technical research of fish and other aquatic flora and fauna, a national program of fish inspection, marketing and promotion, fishing vessel insurance and vessel construction subsidy administration.

Ocean and aquatic sciences conducts research programs on the physical, chemical and biological properties and processes of marine waters in coastal and offshore waters of concern to Canada; undertakes measurements and forecasts of tides, currents and water levels; provides advice and information on aspects of the marine environment; conducts research and development of oceanographic techniques and instruments; undertakes hydrographic surveys and produces nautical charts and publications for navigation, fisheries and resource exploitation.

Small craft harbours administers harbours and marine facilities used by fishing vessels and recreational craft at some 2,300 locations across Canada.

Environmental services comprises the atmospheric environment service, the environmental protection service and the environmental management service.

The atmospheric environment service acquires and processes weather and climate data and provides climatological and meteorological information, including weather forecasts. It carries out research on meteorological air quality and environmental matters.

The environmental protection service is responsible for taking action in preventing or combating environmental problems that fall within the department's terms of reference. These activities include problem surveillance, air and water pollution control, ocean dumping control, waste management, control and disposal of environmental contaminants, assistance in controlling activities having an ecological impact, noise control, response to environmental emergencies and management of the federal government's own clean-up program. Its responsibilities include the development and enforcement of environmental regulations, codes, protocols and other protection and control instruments used to implement federal environmental legislation. The service is a point of contact for the public and other government departments on matters relating to the implementation of environmental protection measures.

The environmental management service co-ordinates activities related to terrestrial renewable resources, their use, and the impact of their use on the environment. Its five staff directorates — forestry, inland waters, wildlife, lands, and policy and planning development — are all located in the Ottawa region. Line management operations are decentralized in five regional directorates covering all Canada.

The planning and finance service provides policy and planning direction and co-ordinates the government's relationships in environmental and resources matters with the provinces and with other countries.

Advice to the minister is provided by an environmental advisory council and separate fisheries and forestry advisory councils which include representatives from industry, the universities and the scientific community.

Department of Indian Affairs and Northern Development (Indian and Northern Affairs). This department was established in June 1966, superseding the Department of Northern Affairs and National Resources; it now operates under authority of RSC 1970, c.I-7. In 1968 the department was reorganized, creating, in addition to departmental support services and an engineering and architectural branch, three distinct program areas. The Indian and Inuit program is responsible for programs for Canada's 288,938 registered Indians and some 4,000 Inuit of Northern Quebec. These include education, economic development, local government and social assistance. The northern affairs program covers management of all natural resources north of the 60th parallel except game, the protection of the northern environment, government activities in economic development and support of the territorial governments in providing social and other local services. Parks Canada is responsible for national parks, national historic parks and sites, and joint federal-provincial agreements for recreation and conservation. In 1972 a corporate policy group was formed to advise the deputy minister on broad policy questions, in particular those involving co-ordination among the programs and co-operation with other departments and agencies.

The office of native claims, established in the department in 1974, represents the government in both comprehensive and specific claims negotiations with native groups.

The commissioner of the Northwest Territories and the commissioner of the Yukon Territory report to Parliament through the minister of Indian affairs and northern development. The minister is also responsible to Parliament for the Northern Canada Power Commission, the National Battlefields Commission and the Historic Sites and Monuments Board of Canada.

Department of Industry, Trade and Commerce. In 1969, the departments of industry and of trade and commerce were merged to form the Department of Industry, Trade and Commerce (ITC), which operates under authority of RSC 1970, c.I-11. ITC promotes establishment, growth and efficiency of manufacturing, processing and tourist industries in Canada and fosters development of trade. Programs assist manufacturing and processing industries in adapting to new technology and changing market conditions, in developing potential and in rationalizing productivity, greater use of research, modern equipment, improved industrial design, the application of advanced technology and modern management techniques, and development and application of sound industrial standards in Canada and in world trade.

The department's functions include: improving access of Canadian goods and services into external markets through trade negotiations; contributing to improvement of world trading conditions; providing support services for industrial and trade development, including information, import analysis and traffic services; analyzing implications for Canadian industry, trade and commerce and for tourism of government policies; contributing to the formulation and review of those policies; and compiling information on trends and developments in Canada and abroad related to manufacturing and processing and tourist industries.

The department is organized into eight major functional groups: policy planning, enterprise development, industry and commerce development, international trade relations, trade commissioner service and international marketing, tourism, finance and administration, and human resource planning. The department operates 12 regional offices across Canada and a trade commissioner service which has 89 offices in 65 countries.

The minister also reports to Parliament on behalf of Statistics Canada, the Federal Business Development Bank, and the Export Development Corporation. Boards and other organizations reporting to the minister are the Machinery and Equipment Advisory Board, Design Canada, the Standards Council of Canada, the Textile and Clothing Board, Metric Commission Canada, the Foreign Investment Review Agency, the minister's advisory council and the Canadian Footwear and Leather Institute.

Department of Insurance. The minister of finance is responsible for this department, which originated in 1875 as a branch of the finance department but was constituted a separate department in 1910. It is authorized and governed by the Department of Insurance Act (RSC 1970, c.I-17). Under the superintendent of insurance, who is the deputy head, the department administers statutes applicable to federally-incorporated insurance, trust, loan and investment companies; provincially-incorporated insurance companies registered with the department; British and foreign insurance companies operating in Canada; small loans companies and money-lenders; co-operative credit societies registered under the Co-operative Credit Associations Act; pension plans organized and administered for the benefit of persons employed in connection with certain federal works, undertakings and businesses; and life insurance issued to certain members of the public service prior to May 1954.

Under the relevant provincial statutes, the department examines trust and loan companies incorporated in Nova Scotia, trust companies incorporated in New Brunswick and insurance and trust companies incorporated in Manitoba. It reports to Parliament through the minister of finance.

Department of Justice (Department of Justice Canada). This department, established by SC 1868, c.39, now operates under authority of the Department of Justice Act (RSC 1970, c.J-2). The minister of justice is the official legal adviser of the Governor General and the Queen's Privy Council for Canada. It is his duty to see that administration of public affairs is in accordance with law, to superintend all matters connected with the administration of justice in Canada that are not within the jurisdiction of the provincial governments, to advise upon the legislation and proceedings of the provincial legislatures, and generally to advise the Crown on all matters of law referred to him by the Crown. The minister of justice is, ex officio, Her Majesty's attorney general of Canada. In this capacity it is his duty to advise the heads of the departments of the federal government on all matters of law connected with such departments, to settle and approve all instruments issued under the Great Seal of Canada, and to regulate and conduct all litigation for or against the Crown in the right of Canada. The minister also recommends to Cabinet the selection of judges for the Supreme Court and the Federal Court of Canada as well as judges of superior, county and district courts. When amendments to the Judges Act come into force, the justice department will no longer be responsible for the administration of the Supreme Court of Canada and the Federal Court of Canada, or for the administration of the salaries and pensions of other federally-appointed judges. However, the minister will submit the estimates for such courts and judges to Parliament. The minister of justice reports to Parliament for the Tax Review Board and the Law Reform Commission of Canada.

Department of Labour (Labour Canada). The department was established in 1900 by an act of Parliament (SC 1900, c.24) and now operates under the authority of the Department of Labour Act (RSC 1970, c.L-2). The department administers legislation dealing with: fair employment practices; hours of work, minimum wages, annual vacations, holidays with pay, equal wages, group and individual terminations of employment, severance pay and the regulation of fair wages and hours of labour in contracts made with the federal government for construction, remodelling, repair or demolition of any work; government employee compensation, merchant seamen compensation, and employment safety; and transitional assistance benefits for auto workers and adjustment assistance benefits for textile workers and for footwear and tanning workers. It promotes joint consultation with industries through labour management committees and operates a women's bureau. The department publishes the *Labour Gazette* and other publications as well as general information on labour management, employment and manpower.

The Merchant Seamen Compensation Board reports to the minister of labour. The department is the official liaison agency between the Canadian government and the International Labour Organization. The Canada Labour Relations Board reports to Parliament through the minister of labour.

Department of National Defence. The department and the Canadian forces operate under the authority of the National Defence Act (RSC 1970, c.N-4). The minister of national defence is responsible for the control and management of the Canadian forces, the Defence Research Board and all matters relating to national defence. He is also responsible for construction and maintenance of all defence establishments and facilities required to defend Canada.

The deputy minister is the senior public servant in the department and the principal civilian adviser to the minister on all departmental affairs. He is responsible for ensuring that all policy direction from the government is reflected in the administration of the department and in military plans and operations. The chief of the defence staff is the senior military adviser to the minister and is charged with the control and administration of the forces. He is responsible for the effective conduct of military operations and the readiness of the Canadian forces to meet the commitments assigned to them by the government.

A defence council, consisting of the minister of national defence as chairman, the deputy minister of national defence, the chief of the defence staff, the chairman of the Defence Research Board, the vice-chief of the defence staff, the assistant deputy minister (policy), and the deputy chief of defence staff, meets as required to consider and advise on major policy matters. The Crown corporation Defence Construction (1951) Limited reports to Parliament through the minister of national defence.

Department of National Health and Welfare (Health and Welfare Canada). This department was established in October 1944 under the Department of National Health and Welfare Act (RSC 1970, c.N-9). The deputy minister of national health and welfare administers 10 branches: health programs, health protection, medical services, long-range health planning, administration, social services programs, income security programs, policy research and long-range planning (welfare), and welfare information systems; fitness and amateur sport is a branch of the department reporting to Parliament through the minister of state for fitness and amateur sport.

Departmental programs on health include hospital insurance and diagnostic services, medical care insurance, health resources, food and drug supervision, narcotics control, federal emergency health services, environmental health, adverse drug reaction reporting, operation of a central clearing house for poison control centres, health, medical and hospital services to Indians and Inuit across Canada and all residents of the Yukon Territory and Northwest Territories, government employee health services and leprosy control as well as assistance and consultation services to the provinces on request.

Welfare programs include the Canada Pension Plan, old age security and guaranteed income supplements, family allowances, the Canada Assistance Plan and emergency welfare services. There are also developmental programs, including national welfare grants, family planning grants and information and grants to groups of retired persons.

The National Council of Welfare reports directly to the minister who also reports to Parliament for the Medical Research Council.

Department of National Revenue (Revenue Canada Customs and Excise, Revenue Canada Taxation). From Confederation until May 1918, customs and inland revenue acts were administered by separate departments; after that date they were amalgamated under one minister as the Department of Customs and Internal Revenue. In 1921 the name was changed to the Department of Customs and Excise. In April 1924 collection of income taxes was placed under the minister of customs and excise and, under the Department of National Revenue Act, 1927, the department became known as the Department of National Revenue. It operates now under the Income Tax Act, SC 1970-71-72, c.63, as amended.

The customs and excise component is responsible for assessment and collection of customs and excise duties as well as of sales and excise taxes. The taxation component is responsible for assessment and collection of income taxes, Part I of the Canada Pension Plan, and collection of premiums and administration of the coverage provisions of the Unemployment Insurance Act through its 28 district taxation offices, head office and taxation data centres in Ottawa and Winnipeg.

Department of Public Works (Public Works Canada). This department was constituted in 1867 and operates under the legislative authority of the Public Works Act (RSC 1970, c.P-38). The department is the primary agent of the federal government in the development and management of real property, providing office accommodation for some 90 federal departments and agencies, together with architectural, engineering, construction management and realty services for special purpose facilities. The department also has responsibilities in transportation (roads) and marine (dredging) works. It is decentralized, with regional headquarters at Halifax, Montreal, Ottawa, Toronto, Edmonton and Vancouver, and subsidiary offices in all but the Ottawa region. Main line functions are design and construction, realty planning and development and realty services plus departmental planning and co-ordination (including policy research) and technological research and development; in addition, the dominion fire commissioner operates under the authority of the minister of public works, with responsibility for protection of life of occupants of government property and for the minimization of property loss as a result of fire.

Department of Regional Economic Expansion. This department was established in 1969 (RSC 1970, c.R-4). Its objective is to facilitate economic growth and social adjustment in Canada through federal-provincial agreements, special programs and other activities designed to increase, and improve access to, development opportunities in various regions of the country. DREE's present activities are divided into three major areas: general development agreements, industrial incentives and other programs.

Development agreements signed between the federal government and a provincial government provide a 10-year formal framework for co-ordinated action aimed at exploiting the potential for socio-economic development. Specific development activities are carried out under subsidiary agreements. These have been signed with every province except Prince Edward Island, which has a comparable comprehensive development plan. The range of development activities includes forestry, agriculture, fisheries, transportation, tourism, industrial development, northlands, mineral development and planning.

The Regional Development Incentives Act (RDIA), passed in 1969 and now extended to 1981, provides grants to business and industry to establish, expand or modernize manufacturing and processing facilities in designated regions which cover all the Atlantic provinces, Manitoba, Saskatchewan and the Northwest Territories, and parts of Ontario, Quebec, Alberta and British Columbia.

Special rural development activities are carried out under the Agricultural and Rural Development Act (ARDA). Programs to improve the economic circumstances of people of Indian ancestry are provided under a special ARDA program in some provinces. The department is also responsible for the Prairie Farm Rehabilitation Act (PFRA), designed to combat drought and soil drifting in the Prairies.

The minister of regional economic expansion reports to Parliament for the Cape Breton Development Corporation. He is advised by the Atlantic Development Council on policies and programs for future economic development and social adjustment in the Atlantic region, and by the Canadian Council on Rural Development, on rural development policy and programming. The department has headquarters in Ottawa, regional offices at Moncton, Montreal, Toronto and Saskatoon, a provincial office in each provincial capital and various branch offices.

Department of the Secretary of State of Canada. The duties, powers and functions of the secretary of state (RSC 1970, c.S-15) extend to and include all matters over which Parliament has jurisdiction not by law assigned to any other department, branch or agency of the federal government relating to: citizenship; elections; state ceremonial, conduct of state correspondence and custody of state records and documents; encouragement of the literary, visual and performing arts, learning and cultural activities; and libraries, archives, historical resources, museums, galleries, theatres, films and broadcasting.

Responsibilities include those pertaining to the administration of the following branches: cultural affairs including education support, research and liaison, Canada student loans, language programs, state protocol and special events, movable cultural property export control, grants, film festivals, certification of Canadian films, translation bureau; citizenship programs including citizenship registration, multiculturalism, native citizens, women, citizenship participation, official language minority groups and human rights.

The secretary of state reports to Parliament for the Canadian Cultural Property Export Review Board, the Canadian Film Development Corporation, the National Arts Centre Corporation, the National Film Board, the National Library, the Public Archives, the National Museums of Canada, the Canada Council, the Social Sciences and Humanities Research Council, the Canadian Broadcasting Corporation and the Public Service Commission and acts as spokesman for the Office of the Representation Commissioner.

Department of the Solicitor General (Solicitor General Canada). Before 1936, the office of the solicitor general was either a Cabinet post or a ministerial post outside the Cabinet. From 1936 to 1945 the position did not exist, the duties of the office being wholly absorbed by the attorney general of Canada. The Solicitor General Act of 1945 re-established the solicitor general as a Cabinet officer. In 1966 a new Department of the Solicitor General was created (RSC 1970, c.S-12); the solicitor general became the Cabinet minister with primary responsibility in the fields of correction and law enforcement. He is responsible for the Royal Canadian Mounted Police and the Canadian Penitentiary Service and also reports to Parliament for the National Parole Board, an independent agency.

Department of Supply and Services (Supply and Services Canada). This department was established on April 1, 1969 (RSC 1970, c.S-18) to furnish certain services previously provided by other departments, in line with recommendations of a royal commission on government organization (Glassco commission). The minister of supply and services is also the receiver general for Canada and exercises all the duties, powers and functions assigned to that office by law.

The department is organized into two major administrations, each headed by a deputy minister. The supply administration acquires and provides goods and services required by federal government departments and agencies. It maintains federal government equipment and provides printing facilities. Since the 1973-74 fiscal year, the supply administration has been on a cost recovery basis for services rendered to its customers. The supply administration has 18 regional or district supply offices across Canada and an overseas supply office in London, England, and in Koblenz, Federal Republic of Germany. At various locations it provides purchasing and warehousing services and other services such as field contract administration, equipment maintenance, security, emergency supply planning, assets management and printing. With the disbanding of Information Canada in 1976, two functions, publishing and expositions, became the responsibility of the department. The supply administration is organized into commercial supply service, science and engineering procurement and corporate management service.

The services administration provides payment or cheque-issuing services for all federal departments, maintains the fiscal accounts of Canada and prepares the public accounts. It offers departments and agencies a broad range of services in management consulting, auditing and computer services. It also provides administrative services for pay, pensions and other employee benefit plans, together with financial management reports and statistical information. Service functions are carried out through regional and district offices throughout Canada and abroad.

The minister of supply and services reports to Parliament for the Canadian Commercial Corporation, Canadian Arsenals Limited, Crown Assets Disposal Corporation and the Royal Canadian Mint. The minister also has the responsibilities of the custodian of enemy property under the Trading with the Enemy (Transitional Powers) Act, which consists of receiving, managing, releasing and disposing of properties seized from enemy interests during wartime.

Department of Transport (Transport Canada). The department is a corporate structure of operating administrations and Crown corporations, having varying degrees of autonomy, plus separate agencies for development and economic regulation. A departmental headquarters staff supports the minister and deputy minister in planning, policy formulation and assessment of program achievements.

The marine transportation administration co-ordinates activities of the Canadian Coast Guard, National Harbours Board, the St. Lawrence Seaway Authority and four pilotage authorities. Its operations include management of the St. Lawrence Seaway through the St. Lawrence Seaway Authority and direct supervision of 12 major Canadian harbours and other facilities through the National Harbours Board, 300 public harbours and 11 others administered by commissions under the supervision of the department. The Canadian Coast Guard is responsible for navigational aids and waterways, ship safety and regulations, pilotage, telecommunications and electronics systems, and the fleet. The duties of the fleet include ice-breaking and ice escort, search and rescue, fishing zone patrol and maritime pollution control.

The air transportation administration provides and operates domestic airway facilities, a national air terminal system, and the regulatory services needed for aviation safety. It is responsible for the provision of air traffic control and international airway facilities and services. The administration owns 160 and operates 90 of the 875 licensed land airports in Canada.

The surface transportation administration is responsible for the federal government's involvement in railways, urban transportation, motor vehicle safety, highways and ferries.

An Arctic transportation directorate provides liaison with the two territorial governments, other federal departments and various interests within the territories. It co-ordinates transportation functions to respond to Arctic needs and to maintain general liaison with bodies interested in solving transportation problems in the North.

The transportation research and development centre conducts mission-oriented research and demonstrations in all modes of transport, working with government agencies and the academic and scientific community to provide the national focus for changing technology and development.

The department includes Air Canada, Canadian National Railways, and Northern Transportation Company Limited, three autonomous Crown corporations which maintain close consultation with the minister. The minister of transport also reports to Parliament for the Canadian Transport Commission, the National Harbours Board and the St. Lawrence Seaway Authority.

Department of Veterans Affairs (Veterans Affairs Canada). This department, established in 1944 (RSC 1970, c.V-1), is concerned exclusively with the well-being of veterans and with the dependents of veterans who died during active service or as a result of disability attributable to war service. The department provides treatment services (hospital, medical, dental and prosthetic), counselling services, education assistance, life insurance, and land settlement and home construction assistance. It has treatment institutions and facilities in six major urban centres and three veterans homes in Canada and maintains administrative offices in the larger cities and in London, England.

The Canadian Pension Commission, the War Veterans Allowance Board, the Bureau of Pensions Advocates, the Pension Review Board and the Army Benevolent Fund Board report to Parliament through the minister of veterans affairs.

Director of Soldier Settlement and Director of the Veterans' Land Act. The director of soldier settlement (SC 1919, c.71) is also director of the Veterans' Land Act (RSC 1970, c.V-4), and in each capacity is legally a corporation sole. For administrative purposes the programs carried on under both acts constitute integral parts of the services provided by the veterans affairs department.

Economic Council of Canada. This corporation, established under legislation passed on August 2, 1963 (RSC 1970, c.E-1), consists of a full-time chairman and two full-time directors appointed for a term not to exceed seven years, and not more than 25 additional members to serve part-time and without remuneration. The council is to be as representative as possible of the private sector, labour, agriculture, primary industry, secondary industry, commerce and the general public. Its functions are to study and recommend measures to achieve the highest possible levels of employment and efficient production. The council reports to Parliament through the prime minister and publishes various reports and studies.

Eldorado Aviation Limited. This company is a wholly owned subsidiary of Eldorado Nuclear Limited and was incorporated in 1953 to carry air traffic, both passenger and freight, for Eldorado Nuclear Limited. It reports to Parliament through the minister of energy, mines and resources.

Eldorado Nuclear Limited. Set up in 1944 (RSC 1952, c.53) under the name of Eldorado Mining and Refining (1944) Limited (the date was omitted in June 1952 and the name changed in 1968), the Crown company's business is the mining and refining of uranium and production of nuclear fuels in Canada. The company acts as a custodian of concentrates purchased under stockpiling contracts. It reports to Parliament through the minister of energy, mines and resources.

Export Development Corporation. EDC operates under authority of the Export Development Act (RSC 1970, c.E-18, as amended). A federally owned enterprise that is commercially self-sustaining, it provides financial facilities to assist Canada's export trade. EDC reports to Parliament through the minister of industry, trade and commerce. Its affairs are administered by a 12-member board of directors chaired by the corporation's president. To reflect the nature of a publicly owned corporation involved with the Canadian business and banking community, the board consists of senior representatives of government and the Canadian financial and private business sectors. The services of EDC are designed to help Canadian exporters remain as competitive as possible in world markets. The principal services are: export credits insurance, to insure Canadian exporters of goods and services against non-payment by foreign buyers due to credit or political events over which neither buyer nor seller has any control; a comprehensive insurance package for performance bonds and guarantees, available to Canadian exporters, banks, surety companies and other financial institutions; long-term export loans to foreign buyers in respect of the purchase of capital goods or major services from Canada when extended terms are necessary to meet international credit competition; and foreign investment guarantees, to guarantee Canadian investments abroad against non-commercial risks such as war or revolution, expropriation or confiscation, or the inability to repatriate capital or earnings. EDC may also guarantee financial institutions against loss when they are involved in an export transaction by financing either the Canadian supplier or the foreign buyer.

Farm Credit Corporation. This Crown corporation, established in 1959 (RSC 1970, c.F-2) is responsible to Parliament through the minister of agriculture. Under the Farm Credit Act it makes long-term mortgage loans to farmers. It also administers the Farm Syndicates Credit Act and acts as an agent of the agriculture department in administering the land transfer plan of the small farm development program.

Federal Business Development Bank. The bank was established by an act of Parliament in 1974 (SC 1974-75-76, c.14) as a federal Crown corporation to succeed the Industrial Development Bank. Under the act which came into force in October 1975, FBDB assists in establishing and developing business enterprises by providing financial and management services and by supplementing services available from other sources. The bank gives particular attention to the needs of small enterprises.

The board of directors consists of the president, four persons from the public service, and 10 persons from outside the public service. The bank's authorized capital is $200 million, but it may raise additional funds by the issue and sale of debt obligations, provided that the total of the bank's direct and contingent liabilities shall not exceed 10 times its capital.

Federal Environmental Assessment Review Office. Following a Cabinet decision in December 1973, this office was formed to administer the environmental assessment and review process on behalf of the minister of fisheries and the environment. The process requires federal departments and agencies to screen their activities for adverse environmental effects, and refer those which may have significant impact to the review office for formal review. The office establishes separate environmental assessment panels to undertake the formal review and recommend appropriate action to the minister of fisheries and the environment for each project submitted.

In developing its report, each panel must provide guidelines for, and review, an environmental impact statement prepared by the project initiator, obtain response to the statement through public hearings, and secure any other information considered necessary. Panel recommendations are implemented through ministerial direction.

The office is directed by an executive chairman who reports to the minister of fisheries and the environment. He (or his delegate) chairs the environmental assessment panels formed for each project. He is also responsible for providing guidelines, procedures and methodologies to agencies and individuals involved.

Federal-Provincial Relations Office. For administrative purposes, the office is regarded as a department of government under the prime minister. The office came into being in January 1975 under legislation passed by Parliament in December 1974. For some years prior to the creation of the new office, its functions had been the responsibility of a division of the Privy Council Office. The office is headed by the secretary to Cabinet for federal-provincial relations.

The office assists the prime minister in his overall responsibility for federal-provincial relations; assists the Cabinet in examining federal-provincial issues of concern, including co-ordination and support activities for the Cabinet committee on federal provincial relations; assists the minister of state for federal-provincial relations in the performance of his duties; assists ministers, departments, and agencies in the conduct of their relations with provincial governments; undertakes special studies as required; monitors provincial views on federal policies and programs and the evolution of provincial policies as they affect federal policies; and co-ordinates federal participation in conferences of first ministers.

Fisheries Prices Support Board. Under the Fisheries Prices Support Act (RSC 1970, c.F-23) the board is responsible for investigating and, where appropriate, recommending action to support prices of fishery products where declines have occurred. Subject to approval of the Cabinet, it is empowered to purchase fishery products at prescribed prices or to make deficiency payments to producers of fishery products equal to the difference between a prescribed price and the average price at which such products were sold. The board functions under the direction of the minister of fisheries and the environment.

Fisheries Research Board of Canada. The board is a research body operating under an act of Parliament (RSC 1970, c.F-24) to advise the minister of fisheries and the environment on national fisheries and marine research and development policies, plans and programs. The majority of the board's 18 members are senior scientists from universities and provincial agencies; the other members are senior executives from Canada's fisheries and marine industries.

Foreign Claims Commission. By order-in-council PC 1970-2077 of December 1970, the Canadian government established this commission to inquire into property claims made by Canadian citizens and the federal government against foreign countries which may, from time to time, be referred to the commission by the government. The reference is made after the government has negotiated a financial agreement with the foreign country. The commissioners submit reports and recommendations regarding each claim to the secretary of state for external affairs and the minister of finance, stating whether, in the opinion of the commissioners, each claimant is eligible to receive a payment under regulations promulgated from time to time by order-in-council. Up to December 31, 1976 claims against Hungary, Romania, Poland and Czechoslovakia had been referred to the commission. The claims against the two first-mentioned countries have been dealt with and in 1977 the commission was nearing completion of its work on claims against Poland and Czechoslovakia.

Foreign Investment Review Agency. The agency was established on April 9, 1974 by proclamation of the Foreign Investment Review Act (SC 1973-74, c.46). It assesses whether there is or will be significant benefit to Canada in proposals by non-Canadians regarding acquisition of control of Canadian business enterprises or establishment of new businesses in Canada. The agency is responsible to the minister of industry, trade and commerce.

Freshwater Fish Marketing Corporation. This corporation was established under the Freshwater Fish Marketing Act of 1969 (RSC 1970, c.F-13) and given the function of marketing and trading in fish, fish products and fish byproducts in and out of Canada with the objectives of ensuring more orderly marketing for the benefit of the whole fishery and achieving higher and more stable prices for the catch. The corporation received a grant for initial operating and establishment expenses but conducts its operations on a self-sustaining basis without parliamentary appropriations; it is financed by bank loans with government guarantee of repayment, or by direct loans. The corporation consists of a board of directors composed of a chairman, a president, one director for each participating province and four other directors appointed by the Governor-in-Council for a term not exceeding five years. The corporation reports to Parliament through the minister of fisheries and the environment.

Grains Group. In 1970 the minister responsible for the Canadian Wheat Board (the minister of transport at present) organized a special advisory group on grains (Grains Group) to co-ordinate, review and recommend federal policies for grain production, transportation and handling, and marketing. The minister responsible for the wheat board serves as the chairman of the group. A group co-ordinator and three advisers for the areas of production, transportation and handling, and marketing are drawn from the federal departments of agriculture, industry, trade and commerce, and transport. Offices of the Grains Group are in Ottawa.

Heritage Canada. Established under the Canada Corporations Act (RSC 1970, c.C-32), Heritage Canada is a national trust independent of government. It is concerned with the conservation of buildings, sites and natural areas of importance to the country's heritage. Its work is financed by memberships, contributions and the interest on an endowment fund to which the federal government granted $12 million. Heritage Canada seeks to enlist the support of the general public, foundations and corporations; membership is open to anyone.

Historic Sites and Monuments Board of Canada. This board, established in 1919, and now operated under the Historic Sites and Monuments Act (RSC 1970, c.H-6, as amended), is the advisory body to the minister of Indian affairs and northern development on the commemoration of persons, places and events of national historical and architectural significance.

The act provides for 17 members — two representatives each from Ontario and Quebec and one each for the eight other provinces, the Yukon Territory and the Northwest Territories — appointed by the Governor-in-Council, together with the dominion archivist, one representative from the National Museums of Canada and one from the Indian affairs and northern development department. The board is comprised for the most part of professional historians, archivists and architects.

Immigration Appeal Board. The board was established in 1967 by the Immigration Appeal Board Act (RSC 1970, c.I-3) as a court of record with broad discretionary powers. Changes were effected by the Immigration Act of 1976 in August 1977. The acts provide for the operation of the board and in particular for the legal and administrative processes involved in appeals by individuals against deportation or exclusion, detention and the refusal of admission of sponsored relatives ordered under the provisions of the Immigration Act or regulations. A decision of the board is appealable to the Federal Court of Canada by leave. The board is also empowered to consider applications for redetermination of claim of refugees.

International Boundary Commission. The commission functions by virtue of a 1925 treaty between Canada and the United States and the International Boundary Commission Act (RSC 1970, c.I-19). The commissioners, one for Canada and one for the United States, are empowered to inspect the boundary, to repair, relocate and rebuild monuments, to keep boundary vistas open, to regulate all work within 3.05 metres of the boundary including structures of any kind or earthwork, to maintain at all times an effective boundary line and to determine the location of any point of the boundary line which may become necessary to settle any question that may arise between the two governments. Each country pays the salaries of its commissioner and his assistants and the costs of maintaining the boundary are shared equally. The Canadian section comes under the energy, mines and resources department for administrative purposes but the Canadian commissioner reports functionally to the secretary of state for external affairs. The commissioners meet at least once annually, alternately in Ottawa and Washington.

International Development Research Centre. Established as a public corporation by act of Parliament (RSC 1970, c.21, 1st Supp.), the IDRC is an international organization supported financially by Canada. Its objectives are to initiate, encourage, support and conduct research into the problems of developing countries and into methods of applying and adapting scientific and technical knowledge to their socio-economic advancement. A chief purpose is to help them develop their own research skills and facilities.

The board of governors consists of a chairman, a president and not more than 19 other members, nine of whom must be Canadian citizens. The IDRC reports to Parliament through the secretary of state for external affairs.

International Fisheries Commissions. The minister of fisheries and the environment reports to Parliament on Canadian participation in the several international fisheries commissions of which Canada is a member.

International Joint Commission. This commission was established under a Britain–United States treaty signed in January 1909 and ratified by Canada in 1911 (RSC 1970, c.I-20). The commission, composed of six members (three appointed by the president of the United States and three by the Government of Canada), is governed by five specific articles of the Boundary Waters Treaty of 1909. The commission's approval is required for any use, obstruction or diversion of boundary waters affecting the natural level or flow of boundary waters in the other country; and for any works which, in waters flowing from boundary waters or below the boundary in rivers flowing across the boundary, raise the natural level of waters on the other side of the boundary.

Problems arising along the common frontier are also referred to the commission by either country for examination and report, such report to contain appropriate conclusions and recommendations. Provided both countries consent, questions or matters of difference between the two countries may be referred to the commission for decision.

The commission was given responsibilities under the Canada–United States Great Lakes Water Quality Agreement of 1972 to assist in the implementation of the agreement by co-ordinating the various programs referred to therein and monitoring their effectiveness. The commission established a Great Lakes regional office at Windsor, Ont., staffed by American and Canadian public servants; operating costs are shared equally by the two governments.

The commission reports to the secretary of state for external affairs of Canada and to the secretary of state of the United States.

Interprovincial and Territorial Boundary Commissions. The Manitoba–Saskatchewan Interprovincial Boundary Commission and the Alberta–British Columbia Boundary Commission, each consisting of a commissioner from the respective provinces and the surveyor general of Canada, are at present the only commissions concerned with boundaries between provinces. The latter was established in 1974 by federal and provincial Alberta–British Columbia boundary acts to deal with resurveys of the sinuous boundary, the settlement of problems or disputes, and the establishment, restoration and maintenance of survey monuments. However, there are also boundary commissions responsible for the borders between Manitoba and the Northwest Territories; Saskatchewan and the Northwest Territories; Alberta and the Northwest Territories; and British Columbia, the Yukon Territory and the Northwest Territories. All report to Parliament through the minister of energy, mines and resources.

Law Reform Commission of Canada. This commission was established (RSC 1970, c.23, 1st Supp.) as a permanent body to study and keep the laws of Canada under continuing and systematic review. The commission makes recommendations for the improvement, modernization and reform of federal laws including, without limiting the generality of the foregoing: the removal of anachronisms and anomalies in the law; the reflection in and by the law of the distinctive concepts and institutions of the common law and civil law legal systems in Canada, and the reconciliation of differences and discrepancies in the expression and application of the law arising out of differences in those concepts and institutions; the elimination of obsolete laws; and the development of new approaches to and new concepts of the law in keeping with and responsive to the changing needs of Canadian society and its individual members. The commission reports to Parliament through the minister of justice.

Library of Parliament. This library was established by an act in relation to the Library of Parliament (SC c.21) now the Library of Parliament Act (RSC 1970, c.L-7). The library had been formed initially by the amalgamation of the legislative libraries of Upper and Lower Canada following their unification as the Province of Canada in 1841. The library is designated as a department within the meaning and purpose of the Financial Administration Act, the parliamentary librarian holding the rank of deputy minister. The parliamentary and the associate parliamentary librarians are appointed by the Governor-in-Council. The parliamentary librarian under the speaker of the Senate and the speaker of the House of Commons, assisted by a joint committee appointed by the two houses, is responsible for the control and management of the library including the Confederation Branch Library, the parliamentary reading room and the Confederation Building reading room. Persons entitled to borrow books from the Library of Parliament are the Governor General, members of the Privy Council, members of the Senate and the House of Commons, officers of the two houses, judges of the Supreme Court of Canada and the Federal Court of Canada, and members of the Parliamentary Press Gallery. The library serves the Senate and the Commons in both a reference and a research capacity, and is responsible for all books, paintings, maps, and other library effects in the joint possession of the Senate and the Commons. In addition the library indexes Senate committee minutes of proceedings and reports, provides an extensive clipping service to Parliament and is also the public's information centre for parliamentary information. Its collection is accessible to other libraries through interlibrary loan.

Loto Canada Inc. (Loto Canada). Established June 29, 1976 by Appropriation Act No. 4, 1976 (SC 1974-75-76, c.103) Loto Canada is a Crown corporation which began operation in September 1976 on the termination of the Olympic lottery. It manages and conducts a lottery, primarily to assist, until the end of 1979, in financing the deficits of the 1976 Olympics at Montreal and the 1978 Commonwealth Games at Edmonton. A small portion of the net revenue is divided among the provinces (12.5%) and the federal government (5%). The corporation consists of a board of directors with up to seven members representing all regions of Canada. Its head office is at Ottawa. The corporation reports to Parliament through the minister of state for fitness and amateur sport.

Machinery and Equipment Advisory Board. This board, established in 1968, is responsible for considering applications for remission of duty on certain machinery and equipment and advising the minister of industry, trade and commerce as to the eligibility of such machinery for remissions. The board is composed of a chairman and the deputy ministers of industry, trade and commerce, finance and national revenue. It is assisted by branches of the industry department concerned with individual industries, including machinery manufacturing. The objective of the machinery program is to increase efficiency in Canadian industry by enabling machinery users to acquire advanced equipment at the lowest possible cost while affording tariff protection on machinery produced in Canada.

Maritime Pollution Claims Fund. Under the Canada Shipping Act (SC 1971, c.27), a strict liability is created on the part of a shipowner discharging oil from a ship in Canadian waters without need to prove fault or negligence; this liability covers the cost of remedial action if authorized by the Governor-in-Council, preventive action by the minister of transport and damages suffered by any person. Proceedings are taken against the shipowner and served on the administrator of the fund to make him a party to the litigation; upon failure to recover from the shipowner, the administrator is to the claimant in the position of a guarantor or unsatisfied judgment fund. If the ship cannot be identified, suit may be taken against the administrator. There is also a special claim made directly to the administrator by fishermen suffering a loss of revenue resulting from an oil discharge attributable to a ship and not otherwise recoverable at law. The administrator reports to Parliament through the minister of transport.

Medical Research Council. Established in 1969 and operating under authority of RSC 1970, c.M-9, the council is a departmental Crown corporation of the federal government. It is composed of a president, a vice-president and 20 members. The primary aim of the council is to support and develop research in the health sciences in Canadian universities and affiliated institutions. It reports to Parliament through the minister of national health and welfare.

Merchant Seamen Compensation Board (Merchant Seamen Compensation Board Canada). The board was established by authority of the Merchant Seamen Compensation Act (RSC 1970, c.M-11, as amended) and reports to the minister of labour. The three members are appointed by the Governor-in-Council. The board adjudicates claims for compensation made by injured seamen employed on ships registered in Canada when they are not entitled to worker compensation under any provincial worker compensation act or the Government Employees Compensation Act.

Metric Commission Canada. The commission was established by Metric Commission Order (PC 1971-1146) June 1971. It consists of a full-time chairman and up to 20 part-time commissioners, all appointed by the Governor-in-Council for a term of three years. An executive director acts for the commission in directing the full-time staff.

The commission advises the minister of industry, trade and commerce on conversion to the metric system and assists all sectors to prepare conversion plans and disseminate information. It includes over 100 committees covering all areas of the economy. The staff and 12 steering committees play a co-ordinating role for these sector committees, with the major impetus for conversion coming from the committee members who represent industry, labour, consumer, trade, standards and service associations, governments and other concerned bodies.

Each sector committee develops a conversion plan; after liaison with related sectors, the committee recommends the sector plan to a steering committee for concurrence, and the plan is reviewed and approved by the commission. Both sector plans and national guidelines follow a four-phase program of guideline dates (investigation, planning, scheduling and implementation) to ensure, as far as possible, that programs are phased in and co-ordinated to obtain the benefits of metric conversion with minimal costs.

The steering committees and sector committees monitor the progress of conversion and suggest any necessary modifications to meet changing conditions.

Ministry of State for Science and Technology (Ministry of State Science and Technology Canada). This ministry was established by order-in-council PC 1971-1695 on August 11, 1971, with the primary purpose of formulating and developing policies in relation to federal government activities that affect the development and application of science and technology. It is organized into a government branch, industry branch, university branch and an administrative division, corporate services branch. The minister is also responsible for the Science Council Act and is the Cabinet member to whom the Science Council of Canada reports.

Ministry of State for Urban Affairs (Ministry of State Urban Affairs Canada). The ministry was created in June 1971 in accordance with the Government Organization Act 1970 (SC 1971, c.42). The ministry's objective is to develop appropriate means by which the federal government may beneficially influence the evolution of urbanization in Canada, through the integration of urban policy and objectives with other federal policies, objectives and programs. The ministry fosters co-operative relationships in urban affairs with the provinces and, through them, their municipalities, and with private organizations and the public. Under the direction of the secretary and two assistant secretaries, the ministry is divided into two branches: urban analysis, responsible for initiating research on urbanization and developing federal urban policies and objectives, and ministry operations, responsible for liaison with other levels of government and the public, undertaking special federal urban projects, communications and internal administration.

National Advisory Council on Fitness and Amateur Sport. The council was established in 1961 by the Fitness and Amateur Sport Act (RSC 1970, c.F-25) to advise the minister of national health and welfare on matters relating to fitness and amateur sport. The council advising the minister is an autonomous organization, composed of 30 members appointed by the Governor-in-Council, who represent every Canadian province and territory. Its three committees — fitness, recreation and sport — meet periodically to discuss and examine matters related to their areas of concern. At least twice a year, a general council meeting is held and recommendations to the minister are formulated. The administrative arm for fitness and amateur sports is the fitness and amateur sport branch of the national health and welfare department. Through numerous programs and operations, it is involved in improving the participation of all Canadians in physical recreation and amateur sport as well as supporting Canadian athletes.

National Arts Centre Corporation. The act establishing the corporation (RSC 1970, c.N-2) received assent in July 1966. The corporation consists of a board of trustees composed of a chairman, a vice-chairman, the mayors of Ottawa and Hull, the director of the Canada Council, the president of the Canadian Broadcasting Corporation, the government film commissioner and nine other members appointed by the Governor-in-Council for terms not exceeding three years, except for the first appointees whose terms ranged from two to four years. The objects of the corporation are to operate and maintain the National Arts Centre, to develop the performing arts in the capital region and to assist the Canada Council in development of the performing arts elsewhere in Canada. The corporation reports to Parliament through the secretary of state.

National Battlefields Commission. This commission was established by an act of Parliament in 1908 (SC 1908, cc.57-58, as amended) to acquire, restore and maintain the historic battlefields at Quebec City to form a National Battlefields Park. Composed of nine members, seven appointed by the federal government and one each by Ontario and Quebec, the commission is supported by the federal government through annual appropriations and is responsible to Parliament through the minister of Indian affairs and northern development.

National Capital Commission. This commission, successor to the Federal District Commission, is a Crown agency created by the National Capital Act (RSC 1970, c.N-3), proclaimed February 6, 1959. Headed by a chairman, it is made up of 20 members, representing the 10 provinces and the capital region.

The commission is responsible for acquisition, development and maintenance of public land in the capital region; it co-operates with municipalities by providing planning aid or financial assistance in municipal projects of benefit to the region; and it advises the public works department on the siting and appearance of all federal government buildings in the 4 662 km² capital region. The commission reports to Parliament through the minister of state for urban affairs.

National Council of Welfare. The council is an advisory body of 21 private citizens, drawn from across Canada and appointed by the Governor-in-Council. Its members include past and present welfare recipients, public housing tenants and other low-income citizens, as well as lawyers, professors, social workers and others involved in voluntary service associations, private welfare agencies and social work education. The council advises the minister of national health and welfare on matters related to welfare. The office of the council carries out research and other support activities for the council.

National Design Council. The council was established by an act of Parliament in 1961 (RSC 1970, c.N-5) to promote and expedite improvement of design in the products of Canadian industry. The council makes recommendations on design policies and programs. Design Canada (industry, trade and commerce), serves as the administrative arm of the council. Policies and programs recommended by the council are implemented by departments and agencies of the federal government, regional governments and other private and institutional bodies. The council has 17 members appointed by the Governor-in-Council and reports through its chairman to the minister of industry, trade and commerce.

National Emergency Planning Establishment (Emergency Planning Canada). In April 1974, Canada Emergency Measures Organization (EMO), the federal co-ordinating agency for civil emergency planning, became the National Emergency Planning Establishment, commonly known since 1975 as Emergency Planning Canada (EPC). EMO was originally created to initiate and co-ordinate the civil aspects of defence policy delegated to federal departments and agencies to meet the threat of nuclear war.

Emergency Planning Canada has an extended role to co-ordinate and assist planning to ensure that the federal government is ready to meet the effects of natural or man-made disasters. Such planning is part of the normal responsibilities of federal government departments, Crown corporations and agencies. An EPC regional director in each provincial capital maintains contact with other federal departments and with provincial and municipal governments.

EPC promotes emergency preparedness of the federal government and encourages other levels of government to plan by providing grants for approved emergency planning projects; making arrangements for federal assistance to provinces to offset costs resulting from emergencies; sponsoring courses for representatives from the public and private sectors; and conducting an information and research program.

Civil emergency preparedness extends beyond the borders of Canada to nations abroad, including the US and NATO countries. The director general of emergency planning represents Canada on NATO committees for senior civil emergency planning and civil defence. Although attached for purposes of administration to the defence department, the agency receives functional direction from the Privy Council Office.

National Energy Board. This board was established under the National Energy Board Act, 1959 (RSC 1970, c.N-6) to assure the best use of energy resources in Canada. The board, composed of nine members, is responsible for regulating the construction and operation of oil and gas pipelines that are under the jurisdiction of Parliament, the tolls charged for transmission by oil and gas pipelines, the export and import of gas and oil, the export of electric power, and the construction of the lines over which power is exported or imported. Under the Petroleum Administration Act, 1975, the board administers the export charge on crude oil and certain refined petroleum products and administers, on behalf of the minister of energy, mines and resources, the pricing of natural gas entering interprovincial and international trade.

The board is required to study and keep under review all matters relating to energy under the jurisdiction of Parliament and to recommend measures it considers necessary and advisable. It reports to Parliament through the minister of energy, mines and resources.

National Farm Products Marketing Council. Established in 1972 under the Farm Products Marketing Agencies Act (SC 1972, c.65), the council consults with producers, commodity boards, and provincial and federal governments and co-ordinates their views on the establishment and operation of national marketing agencies. It assists and supervises the operations of agencies and promotes more effective marketing of farm products in interprovincial and export trade. The goal is to maintain and promote an efficient, competitive and expanding agricultural industry.

The council consists of a chairman, a vice-chairman, two full-time and four part-time members appointed by the Governor-in-Council and is directly responsible to the minister of agriculture. Council headquarters is in Ottawa.

National Film Board. The board, established in 1939, operates under the National Film Act (RSC 1970, c.N-7) which provides for a board of governors of nine members — a government film commissioner, appointed by the Governor-in-Council, who is chairman of the board, three members from the public service of Canada and five members from outside the public service. The board reports to Parliament through the secretary of state. It is responsible for advising the Governor-in-Council on film activities and is authorized to produce and distribute films in the national interest and, in particular, films designed to interpret Canada to Canadians and to other nations. The board is responsible for the production and processing of films for government departments. Its head office is in Ottawa and its operational headquarters is in Montreal.

National Harbours Board (Harbours Board Canada). The board was established by an act of Parliament in 1936 (RSC 1970, c.N-8). It is responsible for the administration of port facilities at the harbours of St. John's, Nfld.; Halifax, NS; Saint John and Belledune, NB; Sept-Îles, Chicoutimi, Baie-des-Ha! Ha!, Quebec City, Trois-Rivières and Montreal, Que.; Churchill, Man.; Vancouver and Prince Rupert, BC; the Jacques Cartier and Champlain bridges at Montreal; and the grain elevators at Prescott and Port Colborne, Ont. The board reports to Parliament through the minister of transport.

National Library of Canada. The library came into existence in January 1953 with the proclamation of the National Library Act (RSC 1970, c.N-11). It publishes *Canadiana,* a monthly catalogue of new publications relating to Canada, with an annual cumulation. The library also publishes other bibliographies. Its reference branch maintains the Canadian Union Catalogue which embodies the author catalogues of major libraries in the 10 provinces and is thus a key to the book collections of the whole country. The library's own bookstock totals more than 500,000 volumes. The national librarian reports to Parliament through the secretary of state.

National Museums of Canada. This is a departmental Crown corporation established in April 1968 by the National Museums Act (RSC 1970, c.N-12) to join under one administration the National Gallery of Canada; the National Museum of Man (including the Canadian War Museum); the National Museum of Natural Sciences; and the National Museum of Science and Technology (including the National Aeronautical Collection). The corporation reports to Parliament through the secretary of state.

The national museums corporation is governed by a board of trustees, consisting of a chairman, vice-chairman and 12 members, as well as two ex officio members — the director of the Canada Council and the president of the National Research Council. The secretary general is responsible for directing and managing the business of the corporation except for those matters which are the responsibility of the board or of the four museum directors. Museum directors are responsible to the board for the overall activities of their respective operations.

The purposes of the corporation are to demonstrate the products of nature and the works of man, with special but not exclusive reference to Canada, so as to promote interest therein through Canada and to disseminate knowledge thereof. The corporation is empowered to collect, classify, preserve and display objects; undertake or sponsor research; arrange for and sponsor travelling exhibitions of materials in, or related to, its collections and to arrange for publication or acquisition and sale to the public of books, pamphlets, replicas and other relevant materials; undertake or sponsor programs for training persons in the professions and skills involved in the operation of museums; and arrange for or provide professional and technical services to other organizations whose purposes are similar to any of those of the corporation, as approved by the minister.

National Parole Board (Parole Board Canada). The board was established in 1959 by the Parole Act (RSC 1970, c.P-2), which gives it absolute authority for parole of inmates under sentence of imprisonment under an act of Parliament or for criminal contempt of court. It also has authority for all unescorted temporary absences and for escorted temporary absences of certain inmates in penitentiaries. The board has 26 full-time members, appointed by the Governor-in-Council, who may also appoint temporary members to assist the board. Additionally, regional representatives of police forces, provincial and municipal governments, professional, trade, and community associations may participate in the consideration of releases for certain inmates. The board reports to Parliament through the solicitor general.

National Research Council of Canada (National Research Council Canada). This is an agency of the federal government established in 1916 to promote scientific and industrial research. The council operates science and engineering laboratories in Ottawa, Halifax and Saskatoon; gives direct financial support to research carried out in Canadian university and industrial laboratories; sponsors associate committees co-ordinating research on specific problems of national interest; and develops and maintains the nation's primary physical standards. The federal government has designated NRC as the co-ordinating body for development of a national scientific and technical information system under the general direction of the national librarian. Other activities include provision of free technical information to manufacturing concerns; publication of research journals; and representation of Canada in international scientific unions. Patentable inventions developed in the council's laboratories are made available for manufacture through a subsidiary company, Canadian Patents and Development Limited. The council consists of a president, three vice-presidents and 17 members representing Canadian universities, industry and labour. NRC is incorporated under the National Research Council Act (RSC 1970, c.N-14) and reports to Parliament through a designated minister.

Natural Sciences and Engineering Research Council. This council was established as a Crown corporation under the terms of the Government Organization (Scientific Activities) Act, 1976 (SC 1976-77 c.24) to promote and assist research in the natural sciences and engineering other than the health sciences.

Northern Canada Power Commission. The commission was established by an act of Parliament in 1948 (RSC 1970, c.N-21) to provide power to points in the Northwest Territories where a need developed and where power could be supplied on a self-sustaining basis; the act was amended in 1950 to give the commission authority to provide similar services in the Yukon Territory. The name of the commission (formerly the Northwest Territories Power Commission) was changed in 1956. It is composed of a chairman and four members appointed by the Governor-in-Council. Of the additional members, one each is appointed on the recommendation of the commissioners of the Northwest Territories and the Yukon Territory.

Northern Transportation Company Limited. The company was originally formed in 1934 under the Alberta Company's Act and the charter surrendered in 1947 to be replaced by a new entity under the name of Northern Transportation Company (1947) Limited, incorporated under the laws of Canada. In 1949 it was declared to be a proprietary company to which the Government Companies Act applied and placed under the Financial Administration Act. In 1952 the date was eliminated from the name.

The shares were acquired by Eldorado Nuclear Limited in 1947 and the company remained a wholly owned subsidiary until late 1975 when the equity was transferred to the minister of transport in trust for Her Majesty in right of Canada.

This Crown company conducts the business of a common carrier in the Mackenzie River watershed, the Western Arctic and Hudson Bay and operates a wholly owned subsidiary trucking company with operations in Alberta and the Northwest Territories. It is responsible to Parliament through the minister of transport.

Office of the Administrator under the Anti-Inflation Act. The office was established on December 15, 1975 by the Anti-Inflation Act (SC 1974-75-76, c.75, amended by SC 1974-75-76, c.98). The administrator enforces the Anti-Inflation Board guidelines with orders that are binding when the guidelines are disputed or contravened. In price and profit matters the administrator may order excess revenues to be returned to the buyers, the market or the Crown. In compensation matters, he may order that excess payments be recovered from the employer, the employee or both. The administrator, appointed by the Governor-in-Council, may appoint one or more deputy administrators. He reports to Parliament through the minister of national revenue.

Office of the Auditor General. This office originated in 1878 and currently functions under the Auditor General Act (SC 1976-77, c.34) which was proclaimed in August 1977. The auditor general is responsible for examining accounts of Canada including those related to the Consolidated Revenue Fund and to public property, and for reporting annually to the House of Commons the results of his examinations. In his report he calls attention to anything of significance that he considers should be brought to the attention of the Commons including cases in which he has observed that money has been expended without due regard to economy or efficiency, or satisfactory procedures have not been established to measure and report the effectiveness of programs, where such procedures could appropriately and reasonably be implemented. He also audits the accounts of various Crown corporations and other organizations. The minister of finance acts as spokesman in Parliament for the auditor general.

Office of the Chief Electoral Officer. This office was established in 1920 under the provisions of the Dominion Elections Act, now the Canada Elections Act (RSC 1970, c.14, 1st Supp.) as amended by the Election Expenses Act (SC 1973-74, c.51), the Statute Law (Status of Women) Amendment Act, 1974 (SC 1974-75-76, c.66) and the Judges Act (SC 1974-75-76, c.48), and is responsible for the conduct of all federal elections as well as the elections of members of the Northwest Territories Council and of the Yukon Territory Council. In addition, it conducts any vote taken under the Canada Temperance Act. The chief electoral officer is responsible directly to the House of Commons, the president of the Privy Council acting as spokesman for him in the Cabinet.

Office of the Co-ordinator Status of Women. The office received official status in April 1976 by order-in-council PC 1976-779. The co-ordinator reports to and assists the minister responsible for the status of women; monitors the activities of federal departments to ensure that they are in line with the policy of promoting equality between men and women; and co-ordinates new initiatives to improve the status of women within the federal government. The office, located in Ottawa, carries on work begun in 1970 in the Privy Council Office.

Office of the Representation Commissioner. The office was established in 1963 under the provisions of the Representation Commissioner Act (RSC 1970, c.R-6). After each decennial census, the representation commissioner is responsible for preparing maps showing the population distribution in each province and setting out alternative proposals respecting the boundaries of electoral districts. These maps are supplied to the 11 electoral boundaries commissions (one for each province and one for the Northwest Territories) established under the provisions of the Electoral Boundaries Readjustment Act (RSC 1970, c.E-2). The representation commissioner is a member of each of the commissions. The secretary of state acts as spokesman for the office in the Cabinet and the House of Commons.

Panarctic Oils Ltd. This corporation is a consortium of mining and oil and gas companies, individuals and Petro-Canada Exploration Inc., formed in 1967 to explore for oil and gas in the Arctic. Panarctic Oils Ltd. is not a Crown corporation and does not report to Parliament.

Patent Appeal Board (Patent Appeal Board Canada): This is an advisory body established in 1970 under the Patent Act (RSC 1970, c.P-4). Its function is to review final rejections of applications for patents of invention when applicants request review, to conduct hearings to consider arguments of applicants, and to make recommendations to the commissioner of patents for ultimate disposition of the applications. It acts in a similar capacity with delegated powers from the minister of consumer and corporate affairs under the Industrial Design Act (RSC 1970, c.I-8) to consider final rejections of industrial design applications made by the registrar of copyright and industrial design. The board consists of a chairman, a vice-chairman and one other member.

Pension Appeals Board. This board, established under the Canada Pension Plan Act (RSC 1970, c.C-5) hears appeals under the Canada Pension Plan and under certain provincial pension plans. It also hears appeals from certain decisions of the umpire under the Unemployment Insurance Act (SC 1971, c.48) as amended. The board consists of two judges of the Federal Court of Canada or of a superior court of a province appointed as chairman and vice-chairman, and not less than one and not more than eight other persons, each of whom must be a judge of the federal court or of a superior, district or county court of a province. For purposes of appeals under the Canada Pension Plan, the board reports to Parliament through the minister of national health and welfare.

Pension Review Board (Pension Review Board Canada). The board was created under the minister of veterans affairs by the amendments to the Pension Act 1971 (SC 1970-71, c.31). Further amendments were made May 12, 1977, by the Act to Amend the Pension Act. Composed of a chairman, deputy chairman and five other members, the board is an independent and autonomous body that hears appeals in the Ottawa region from pension applicants dissatisfied with decisions of an entitlement board or two members of the Canadian Pension Commission under Section 67. The board is also the responsible body when matters of interpretation of the acts are at issue.

Petro-Canada. On July 30, 1975 the Petro-Canada Act (SC 1974-75-76, c.61) established Petro-Canada as a Crown corporation to increase the supply of energy available to Canadians, to assist the government in formulating its national energy policy and to increase the Canadian presence in the petroleum industry. The corporation consists of a board of directors composed of a chairman, president and not more than 13 other persons appointed by the Governor-in-Council. Its head office is at Calgary, Alta. The corporation reports to Parliament through the minister of energy, mines and resources.

Pilotage Authorities. The Pilotage Act (SC 1971, c.52) established pilotage authorities for the Atlantic, Laurentian, Great Lakes and Pacific regions as proprietary corporations as specified in the Financial Administration Act. The objects of each authority are to establish, operate, maintain and administer in the interests of safety an efficient pilotage service within the region set out in respect of the authority. Each of the four authorities has a chairman and not more than six other members appointed by the Governor-in-Council for a term not exceeding 10 years. The pilotage authorities report to Parliament through the minister of transport.

Post Office Department (Canada Post). Administration and operation of the post office, by virtue of the Post Office Act (RSC 1970, c.P-14) and under the postmaster general, comprises all phases of postal activity, personnel, mail handling, transportation of mails by land, water, rail and air and the direction and control of financial services including the operation of the money-order service.

Department headquarters is in Ottawa, with regional headquarters in Halifax, Montreal, Toronto and Vancouver, and district offices in St. John's, Halifax, Saint John, Quebec City, Montreal, Ottawa, North Bay, Toronto, London, Winnipeg, Saskatoon, Edmonton and Vancouver.

Prairie Farm Rehabilitation Administration. The Prairie Farm Rehabilitation Administration (PFRA) was established in 1935 (RSC 1952, c.214) to help rehabilitate agricultural lands seriously affected by drought and soil drifting in Manitoba, Saskatchewan and Alberta. It has developed 101 community pastures on 1.0 million hectares (2.5 million acres) of marginal and submarginal land and continues to operate 96 of them. It has also been responsible for construction of many large irrigation and water storage projects. PFRA has assisted technically or financially in construction of 135,000 dugouts, dams, wells and irrigation projects for on-farm water supplies. PFRA operates a tree nursery which each year distributes several million trees free to farmers for development of farm and field shelterbelts.

Privacy Commissioner. A member of the Canadian Human Rights Commission appointed by the minister of justice on the recommendation of the chief commissioner, acts as privacy commissioner. The office was established by the Canadian Human Rights Act (SC 1976-77, c.33) to receive, investigate and report on complaints from individuals who allege that they have not been accorded the rights stipulated in the Human Rights Act to which they are entitled; namely, the right of access to, correction of, or comment upon personal information about them in federal information banks. Every investigation by the privacy commissioner is conducted in private. The commissioner reports to Parliament through the minister of justice.

Privy Council Office. For administrative purposes, the office is regarded as a department of government for which the prime minister has responsibility as set forth in PC 1962-240. The clerk of the privy council, under whose direction its functions are carried out, is considered as a deputy head and takes precedence among the chief officers of the public service. The authority of the office is in Sections 11 and 130 of the British North America Act, 1867, which constituted a council to aid and advise in the Government of Canada, to be styled the Queen's Privy Council for Canada. In 1940, with the wartime development of Cabinet committees and the consequent need for orderly secretarial procedures such as agenda, explanatory memoranda and minutes, the clerk of the privy council was designated secretary to the Cabinet, and the Cabinet secretariat was brought into being in the Privy Council Office. Since 1946, the office has been further reorganized, developed and enlarged and certain of its administrative functions and those of the prime minister's office have been closely integrated in the interests of efficiency and economy.

The organization consists primarily of the Cabinet secretariat with two divisions reporting to the clerk of the privy council and secretary to the Cabinet: deputy secretary to the Cabinet (operations); and deputy secretary to the Cabinet (plans). Each division contains a number of secretariats that support the Cabinet and its committees. The secretariats prepare and circulate agenda and necessary documents to ministers, and record and circulate decisions. They communicate with government departments and agencies and provide advisory support for the prime minister. Other sections of the office advise the prime minister on senior appointments, constitutional matters, emergency and long-range planning, and the exercise of his prerogative to allocate responsibilities between ministers. Submissions to the Governor-in-Council are received, draft orders and regulations are prepared, approved orders are circulated and the federal statutory regulations are edited, registered and published in the *Canada Gazette*.

Public Archives of Canada (Public Archives Canada). The public archives was founded in 1872 and is administered under the Public Archives Act (RSC 1970, c.P-27) by the dominion archivist who has the rank of a deputy minister and reports to Parliament through the secretary of state. Its purpose is to assemble and make available a comprehensive collection of source material relating to the history of Canada. It also has broad responsibilities to promote efficiency and economy in the management of federal government records. The archives branch in the National Library and Archives Building is a centre for research on the development of Canada. In addition to selected records of the federal government, it possesses an extensive collection of private papers of individuals and societies, a map collection which is the most important of its kind in the country, and extensive collections of paintings, drawings, prints, photographs, sound recordings and films relating to Canada. A specialized library is also at the disposal of searchers. The records management branch operates a large records centre in Ottawa and regional centres in Toronto, Montreal, Vancouver, Edmonton, Winnipeg and Halifax where non-current departmental records are centralized, stored and serviced and assists departments in their records management programs. The administration and technical services branch operates the central microfilm unit for federal departments.

Under the terms of the Laurier House Act (RSC 1952, c.163), the public archives is responsible for the administration of Laurier House in Ottawa as a museum.

Public Service Commission. Arrangements were made for civil service appointments under the first Civil Service Act of 1868 but the first civil service commission was not created until 1908. This established the beginnings of the merit system in the public service. The Civil Service Act of 1918 gave the commission authority to control recruitment, selection, appointment, classification and organization and to recommend rates of pay. The next Civil Service Act in 1961 strengthened the principles of the merit system, clarified the commission's role in other areas of personnel administration, and gave staff associations the right to be consulted on remuneration and conditions of employment.

The Public Service Employment Act (RSC 1970, c.P-32) which came into force in March 1967, redefined the commission's role as the central staffing agency and extended its authority to cover certain groups of employees exempt from the previous acts. The public service is specified in the Public Service Staff Relations Act. It does not include Crown corporations, such as the Canadian Broadcasting Corporation, Central Mortgage and Housing Corporation, Canadian National Railways and Air Canada. The new act reaffirmed the merit principle, and permitted delegation of the commission's authority, although not its responsibility to Parliament. Under the act, the commission was relieved of responsibility for recommending rates of pay and conditions of service to the government, for classification, and for consultation with staff associations on matters that are now the subject of collective bargaining.

In November 1972 the commission was assigned the duty, by order-in-council PC 1972-2569, of investigating cases of alleged discrimination on grounds of sex, race, national origin, colour or religion with respect to the application and operation of the Public Service Employment Act; the appeals and investigation branch is responsible for this function.

The Public Service Commission reports directly to Parliament. The secretary of state has traditionally been the minister who presents the commission's report to the House of Commons, and answers parliamentary questions on the commission's behalf.

Public Service Staff Relations Board. Established in 1967 by the Public Service Staff Relations Act (RSC 1970, c.P-35, as amended by SC 1972, c.18, SC 1973-74, c.15 and SC 1974-75-76, c.67), the board is an independent body responsible for determining bargaining units, certifying bargaining agents, dealing with complaints of unfair practices and generally overseeing the administration of legislation providing for collective bargaining in the public service. The board's full-time chairman, vice-chairman and not less than three deputy chairmen hold office for a period not exceeding 10 years; such other full-time members and part-time members as the Governor-in-Council considers necessary hold office for a period not exceeding seven years. Information on compensation and other conditions of employment is provided to employers and bargaining agents, primarily in the public service, by the pay research bureau which is under the administrative direction of the board. The Public Service Staff Relations Board reports to Parliament through a minister of the Crown designated by the Governor-in-Council. At present the responsible minister is the president of the privy council.

Queen Elizabeth II Canadian Research Fund. The Queen Elizabeth II Canadian Research Fund Act (SC 1959, c.33) established a fund of $1 million to be administered by a board of trustees to aid in research on children's diseases. The prime minister reports to Parliament on operations of this fund.

Regional Development Incentives Board. This board was established under the Regional Development Incentives Act 1968-69 (RSC 1970, c.R-3). It provides advice to the minister of regional economic expansion on matters respecting the administration of the act, particularly on applications for incentives relating to projects over a specified size or involving loan guarantees or sensitive industries. The board meets monthly and consists of representatives of various federal departments and agencies including environment, finance, the foreign investment review agency, employment and immigration, and industry, trade and commerce.

Restrictive Trade Practices Commission. The commission was established by the Combines Investigation Act (RSC 1970, c.C-23 as amended by SC 1974-75-76, c.76). In respect of trade practices contained in Part IV. 1 of the act, on application of the director of investigation and research and after holding a hearing at which evidence is submitted by the director and by the party against whom an order is sought, the commission may issue an order prohibiting the practice. In respect of restrictive trade practices contained in Part V of the act, the commission may hold hearings and appraise evidence submitted by the director and the parties under investigation, to report to the minister of consumer and corporate affairs.

Roosevelt Campobello International Park Commission. Established by the Roosevelt Campobello International Park Commission Act (SC 1964-65, c.19), the commission consists of six members, three appointed by the Government of Canada and three by the government of the United States, to administer the Roosevelt Campobello International Park at Campobello, NB. The Canadian section of the commission reports to Parliament through the secretary of state for external affairs.

Royal Canadian Mint. In operation since January 1908, the mint was first established as a branch of the Royal Mint under the United Kingdom Coinage Act of 1870. In December 1931, by an act of the Canadian Parliament, it became the Royal Canadian Mint and operated as a branch of the finance department. By the Government Organization Act of 1969, the mint became a Crown corporation, reporting to Parliament through the minister of supply and services. It operates under authority of RSC 1970, c.R-8.

The latter change was made to provide for a more industrial type of organization and for flexibility in producing coins of Canada and other countries; buying, selling, melting, assaying and refining gold and other precious metals; and producing medals, plaques and other devices. The mint has a seven-man board of directors appointed by the Governor-in-Council — the master of the mint is its chief executive officer; the chairman is appointed for a four-year period, subject to re-appointment; five other directors, two from inside and three from outside the public service, are appointed for terms of three years. The mint operates basically as a manufacturing enterprise. Financial requirements are provided through loans from the Consolidated Revenue Fund.

Royal Canadian Mounted Police. This civil force, organized and administered by the federal government, was established in 1873 as the North-West Mounted Police. It now operates under authority of the Royal Canadian Mounted Police Act (RSC 1970, c.R-9) and is responsible for enforcing federal laws throughout Canada. By agreement with the governments of eight provinces (all except Ontario and Quebec) it is also responsible for enforcing the Criminal Code of Canada and provincial laws within those provinces under the direction of the respective attorneys general. In these provinces the force provides police services to 192 municipalities, assuming enforcement responsibilities for criminal, provincial and municipal laws. The Yukon Territory and Northwest Territories are policed exclusively by the Royal Canadian Mounted Police. The commissioner, appointed by the Governor-in-Council, has control and management of the force and of all matters connected therewith, under the direction of the solicitor general of Canada.

St. Lawrence Seaway Authority. This authority was established by an act of Parliament in 1951 (RSC 1970, c.S-1) and came into force by proclamation on July 1, 1954. It was incorporated for the purposes of constructing, maintaining and operating all such works as may be necessary to provide and maintain, either wholly in Canada or in conjunction with works undertaken by an appropriate authority in the United States, a deep waterway between the Port of Montreal and Lake Erie. The Crown corporation, Seaway International Bridge Corporation Limited, is subsidiary to the St. Lawrence Seaway Authority. The authority is composed of a president, a vice-president and a member, and reports to Parliament through the minister of transport.

Science Council of Canada. This council was established in 1966 (RSC 1970, c.S-5) and became a Crown corporation on April 1, 1969. It consists of 30 members, each having a specialized interest in science or technology. Members normally hold office for three years. All are appointed by the Governor-in-Council. The duties of the council are to assess in a comprehensive manner Canada's scientific and technological resources, requirements and potential and to make recommendations, to increase public awareness of requirements and of interdependence of various groups in society in the development and use of science and technology. The council reports to Parliament through a designated minister, at present the minister of state for science and technology.

Seaway International Bridge Corporation Limited. The Seaway International Bridge Corporation Limited was established under the Companies Act, by Letters Patent, November 13, 1962. It operates the international toll bridge system between Cornwall, Ont. and Rooseveltown, NY on behalf of the owners, the St. Lawrence Seaway Authority and the Saint Lawrence Seaway Development Corporation. It reports to Parliament through the minister of transport.

Social Sciences and Humanities Research Council. The council was established by the Government Organization (Scientific Activities) Act, 1976 (SC 1976-77, c.24) as a Crown corporation to promote and assist research and scholarship in the social sciences and humanities and to carry out research.

Standards Council of Canada. The council was established by an act of Parliament (RSC 1970, c.41, 1st Supp.) which received royal assent on October 7, 1970. Its objectives are to foster and promote voluntary standardization in fields relating to the construction, manufacture, production, quality, performance and safety of buildings, structures, manufactured articles and products and other goods, including components thereof, not expressly provided for by law, as a means of advancing the national economy, benefiting the health, safety and welfare of the public, assisting and protecting consumers, facilitating domestic and international trade and furthering international co-operation in the field of standards. To this end, the council sponsors the national standards system, a federation of accredited independent Canadian standards-writing organizations and the Canadian national committees responsible for international standardization, to which will be joined accredited certification organizations and testing laboratories. The council holds membership in the International Organization for Standardization and sponsors the Canadian national committee of the International Electrotechnical Commission. The council has been responsible for co-ordinating the planning and execution of a program for the development of standards in the metric (SI) system. This activity is in support of the overall program being carried out by Metric Commission Canada.

The council consists of not more than 57 members; including six federal representatives, 10 representing the provinces and 41 other members. Membership is broadly representative of all levels of government, primary and secondary industries, distributive and service industries, trade associations, labour unions, consumer associations and the academic community. The council reports to Parliament through the minister of industry, trade and commerce.

Statistics Canada. The Dominion Bureau of Statistics was set up by statute in 1918 as the central statistical agency for Canada (SC 1918, c.43). In 1948 this statute, which had been consolidated as the Statistics Act (RSC 1927, c.190), was repealed and replaced by the Statistics Act (RSC 1952, c.257) which was amended by SC 1952-53, c.18, assented to March 31, 1953. The 1971 Statistics Act (SC 1971, c.15) replaced that statute.

The functions of Statistics Canada are to compile, analyze and publish statistical information relative to the commercial, industrial, financial, social and general condition of the people and to conduct regularly a census of population, housing and agriculture as required under the act.

Statistics Canada is a major publication agency of the federal government; its reports cover all aspects of the national economy and social conditions of the country. The administrative head of the bureau is the chief statistician of Canada who has the rank of a deputy head of a department and reports to Parliament through the minister of industry, trade and commerce.

Statistics Canada has offices in St. John's, Halifax, Montreal, Ottawa, Toronto, Winnipeg, Regina, Edmonton and Vancouver with facilities to provide information collected by the bureau and to explain how such data can be used.

Tariff Board. Constituted in 1931, the board derives its duties and powers from four statutes: the Tariff Board Act (RSC 1970, c.T-1); the Customs Act (RSC 1970, c.C-40); the Excise Tax Act (RSC 1970, c.E-13); and the Anti-dumping Act (RSC 1970, c.A-15).

Under the Tariff Board Act, the board looks into and reports on any matter in relation to goods that, if brought into Canada, are subject to or exempt from customs duties or excise taxes. Reports of the board are tabled in Parliament by the minister of finance. It is also the duty of the board to inquire into any other matter in relation to trade and commerce that may be referred to it by the Governor-in-Council.

Under the provisions of the Customs Act, the Excise Tax Act and the Anti-dumping Act, the Tariff Board acts as a court to hear appeals from decisions on customs and excise rulings by the national revenue department in respect of excise taxes, tariff classification, value for duty, drawback of customs duties and determination of normal value or export price in dumping matters. Declarations of the board on appeals are final and conclusive but the acts contain provisions for appeal on questions of law to the Federal Court and thence to the Supreme Court of Canada.

Tax Review Board. This board, formerly the Tax Appeal Board, was created and operates under the provisions of the Tax Review Board Act (SC 1970-71, c.11). The board has jurisdiction to hear appeals by taxpayers against their assessments, under the Income Tax Act and the Estate Tax Act as well as appeals under the Old Age Security Act, certain sections of the superannuation plan, the Unemployment Insurance Act, and in other acts of Parliament that specify the right to appeal to the board. It has, for the exercise of its jurisdiction, such powers, rights and privileges as are vested in a superior court of Canada. The board consists of no less than three nor more than seven members and at its full complement includes a chairman, an assistant chairman and five members. Its principal office is at Ottawa; the board sits at such times and places throughout Canada as it considers necessary. The board is under the jurisdiction of the minister of justice but is independent of the justice department.

Teleglobe Canada. Created in 1950 by an act of Parliament (RSC 1970, c.C-11), under the name of the Canadian Overseas Telecommunication Corporation, this Crown agency operates all overseas communications to and from Canada — whether by undersea cable or international satellite. By means of international switching-centres in Montreal, Toronto and Vancouver, Teleglobe Canada provides public telephone service to over 200 overseas territories. The corporation also provides public message telegraph service, Telex, private wire service, data and video transmissions to many points around the world. Teleglobe Canada is the designated operating entity for Canadian participation in Intelsat and represents Canada on the Commonwealth Telecommunications Council. It reports to Parliament through the minister of communications.

Telesat Canada. Telesat Canada was incorporated in 1969 by an act of Parliament (RSC 1970, c.T-4) to establish and operate a domestic satellite telecommunication system. It is a commercial venture whose ownership is shared by Canadian telecommunications carriers and the federal government, with possible public participation. It provides telecommunications services for the transmission of television, radio, telephone, teletype and data communications through a microwave link between earth stations and satellites in orbit. Its annual report is tabled in the House of Commons by the minister of communications.

Textile and Clothing Board. This board was established (SC 1971, c.39) to receive complaints and conduct inquiries about textile and clothing goods imported into Canada under such conditions as to cause or threaten serious injury to Canadian production. After its investigative procedures are completed, the board makes written recommendations to the minister of industry, trade and commerce. The board consists of three members appointed by the Governor-in-Council and maintains its head office in the Ottawa region.

Treasury Board. The board was established as a committee of the Queen's Privy Council for Canada by order-in-council PC 3 of July 2, 1867, and was made a statutory committee in 1869. The minister of finance was appointed chairman of the board, with four other privy councillors to be designated as members by the Governor-in-Council. The secretary of the board and the members of his staff were employed by the finance department.

By the Government Organization Act, 1966 (SC 1966, c.25) the board was established as a separate department of government with its own minister, the president of the board. The committee constituting the board includes, in addition to the president, the minister of finance and four other privy councillors.

The Financial Administration Act (RSC 1970, c.F-10) defines the board's responsibilities as the central management agency of government. These responsibilities include the organization of the public service, financial management, annual and longer-term expenditure planning, and expenditure control, including allocation of resources among departments and agencies of government; management of personnel functions in the public service; and improvement in the efficiency of management and administration in the public service.

The Treasury Board secretariat is divided into seven branches: administrative policy, efficiency evaluation, financial administration, official languages, personnel policy, planning and program.

Uranium Canada, Limited. This Crown company, incorporated in June 1971 under the Canada Corporations Act (RSC 1970, c.C-32) pursuant to the Appropriation Act No. 1, 1971, and the Atomic Energy Control Act, (RSC 1970, c.A-19) is an agency corporation under the Financial Administration Act (RSC 1970, c.F-10). For all purposes it is an agent of Her Majesty and its powers may be exercised only as such. The shares of the company, with the exception of the qualifying shares of the directors, are held by the minister of energy, mines and resources in trust for Her Majesty. Registered under the trade mark UCAN, the company acted as an agent on behalf of the federal government in the acquisition and sales of the joint stockpile of uranium concentrates established under an agreement with Denison Mines Ltd. dated January 1, 1971. UCAN also holds title to the general stockpile of uranium concentrates acquired by the federal government during the years 1963-70. The corporation's head office is in Ottawa.

VIA Rail Canada, Inc. Incorporated on January 12, 1977, VIA Rail Canada, Inc., is a subsidiary of Canadian National and is financed directly by the federal government. However, it operates at arm's length from CN, being neither comprised in CN nor consolidated in its accounts. Its functions are to market and manage all railway passenger services in Canada. It took over marketing June 1, 1977 and management April 1, 1978. The corporation consists of a board of directors composed of a minimum of three to a maximum of 15 members including a chairman and president. The head office is located in Montreal and the company reports to the minister of transport.

War Veterans Allowance Board. This board, established under the authority of the War Veterans Allowance Act, is a quasi-judicial body of eight members, including a chairman and a deputy chairman, appointed by the Governor-in-Council. The board acts as an appeal court for an applicant or recipient aggrieved by a decision of a district authority and may, on its own motion, review and alter or reverse any adjudication of a district authority. The board is responsible for advising the minister of veterans affairs with respect to regulations concerning the War Veterans Allowance Act and part of the Civilian War Pensions and Allowances Act.

Yukon Territory Water Board. The Northern Inland Waters Act, which came into effect in 1972, established the Yukon Territory Water Board whose objects are to provide for the conservation, development and utilization of the water resources of the Yukon Territory to provide the optimum benefit for all Canadians and for residents of the Yukon Territory in particular. The board achieves these objects by licensing water users. The licences contain terms and conditions which regulate the quantity of water to be used and the quality of waste water returned to the environment.

The board consists of nine members, six of them private citizens nominated by the commissioner-in-council of the Yukon Territory. Three are federal government members appointed by the minister of Indian affairs and northern development.

Synopsis of legislation

Appendix 2

Synopsis of legislation of the Second Session of the 30th Parliament, October 12, 1976 to October 12, 1977, passed in the 25th and 26th years of the reign of Her Majesty Queen Elizabeth II. In summarizing this material it is not possible to convey the full content of the legislation. For further details the reader should refer to the *Statutes of Canada, 1976-77.* The date of royal assent follows each chapter number; private members' bills are not assigned chapter numbers until the end of each session.

Chapter 1 (October 22, 1976) *Port of Halifax Operations Act* provides for the resumption and continuation of longshoring and related operations at the port of Halifax.

Chapter 2 (December 15, 1976) *Appropriation Act No. 5, 1976* grants certain sums of money for the public service for the financial year ending March 31, 1977.

Chapter 3 (December 22, 1976) *Government Expenditures Restraint Act* amends certain sections of the Adult Occupational Training Act, Family Allowances Act, 1973, Industrial Research and Development Incentives Act, Western Grain Stabilization Act, and repeals the legislation which provided for Information Canada and the Company of Young Canadians.

Chapter 4 (February 24, 1977) *An Act to amend the statute law relating to income tax.*

Chapter 5 (February 24, 1977) *An Act to amend the Customs Tariff* specifies changes in the rates of duty set for certain items imported into Canada, implementing a ways and means motion tabled by the finance minister on October 13, 1976.

Chapter 6 (February 24, 1977) *An Act to amend the Excise Tax Act* provides for excise taxes to be paid on certain items by licensed wholesalers, jobbers or other dealers, payable when the goods are delivered to a purchaser or retained by the wholesaler for rental or his own use. The act extends the list of items exempt from sales tax, and includes energy conservation equipment such as heat pumps, solar cells for charging batteries, solar collectors for heating, thermal insulation materials, storm doors and windows and timer-controlled thermostats.

Chapter 7 (March 29, 1977) *Appropriation Act No. 1, 1977* grants certain sums of money for the public service for the financial year ending March 31, 1977.

Chapter 8 (March 29, 1977) *Appropriation Act No. 2, 1977* grants certain sums of money for the public service for the financial year ending March 31, 1978.

Chapter 9 (March 29, 1977) *An Act to amend the Old Age Security Act* provides for payment of partial Old Age Security pensions to persons 65 or over who cannot qualify for full pensions but who have lived in Canada for at least 10 years since the age of 18; applicants must be Canadian citizens or legally in Canada under the Immigration Act. The act also provides for drawing up reciprocal arrangements between Canada and other countries to make social security benefits portable.

Chapter 10 (March 31, 1977) *Federal-Provincial Fiscal Arrangements and Established Programs Financing Act 1977* provides for certain fiscal payments to provinces for financing established programs in post-secondary education, hospital insurance, medical care and extended care health services; for income tax collection agreements; and for guarantee of payment of income tax revenue to provinces despite changes effected by any amendment to the Income Tax Act.

Chapter 11 (May 12, 1977) *An Act to provide for the consideration of certain unemployment insurance entitlements.*

Chapter 12 (May 12, 1977) *Advance Payments for Crops Act* facilitates making advance payments for crops up to a maximum of $15,000 for one producer in any crop year, for a crop that he has actually produced; the advance is to be repaid when the crop is marketed.

Chapter 13 (May 12, 1977) *An Act to amend the Pension Act* provides for changes in the constitution of the Pension Review Board.

Chapter 14 (June 16, 1977) *An Act to amend the Customs Tariff (No. 2)* implements a ways and means motion relating to the Customs Tariff that was tabled by the finance minister on March 31, 1977.

Chapter 15 (June 16, 1977) *An Act to amend the Excise Tax Act (No. 2)* among other provisions extends the air transportation tax to amounts paid outside of Canada for travel using air transport facilities in Canada and, after October 31, 1977, for travel with departure from an airport in Canada and destination outside of Canada.

Chapter 16 (June 16, 1977) *An Act to amend the Bank Act and the Quebec Savings Banks Act* extends the period during which banks in Canada may carry on the business of banking from July 1, 1977 to March 31, 1978.

Chapter 17 (June 16, 1977) *An Act to amend the Export Development Act* increases the liability of the Export Development Corporation under contracts of insurance and guarantees, issued and outstanding, from $750 million to $2.5 billion on behalf of the corporation and from $750 million to $1 billion on behalf of the government.

Chapter 18 (June 16, 1977) *An Act to amend the Financial Administration Act and to repeal the Satisfied Securities Act* adds to the Financial Administration Act a description of how securities to the Crown, which have been satisfied, may be released or discharged, and transfers the power to do so from the Governor-in-Council to the appropriate minister.

Chapter 19 (June 16, 1977) *An Act to amend the Motor Vehicle Safety Act* requires that records be kept to show compliance with safety standards as a condition for a manufacturer to use any national safety mark; requires that notice of defects be given to distributors and current owners of defective motor vehicles and to the minister of transport; and increases fines for offences under the act.

Chapter 20 (June 16, 1977) *An Act to amend the Historic Sites and Monuments Act* increases the number of members on the Historic Sites and Monuments Board from 15 to 17 by adding one representative for each of the Northwest Territories and the Yukon Territory.

Chapter 21 (June 16, 1977) *An Act to amend the Railway Act* changes regulations affecting a proposed deviation, change or alteration of a railway.

Chapter 22 (June 29, 1977) *Appropriation Act No. 3, 1977* grants certain sums of money for the public service for the financial year ending March 31, 1978.

Chapter 23 (June 29, 1977) *An Act to amend the Farm Improvement Loans Act, the Small Business Loans Act and the Fisheries Improvement Loans Act* increases the maximum principal amount of the loans from $25,000 to $75,000; redefines "small business enterprise" as a business with an estimated gross annual revenue not exceeding $1,500,000; and extends the limitations of liability.

Chapter 24 (June 29, 1977) *Government Organization (Scientific Activities) Act, 1976* provides for the organization of certain scientific activities of the federal government by establishing the Social Sciences and Humanities Research Council as a corporation to promote and assist research and scholarship in the social sciences and humanities, the Canada Council for the encouragement of the arts, the Natural Sciences and Engineering Research Council to promote and assist research in the natural sciences and engineering other than the health sciences, and an advisory council to be known as the Defence Research Board. The act amends the Canada Council Act, the Science Council of Canada Act, the National Research Council Act, the National Defence Act, the Medical Research Council Act and the National Library Act.

Chapter 25 (June 29, 1977) *An Act to amend the Judges Act and other acts in respect of judicial matters* includes time spent as a magistrate as a consideration of eligibility for appointment as a judge of the superior or county court, in addition to time spent as a barrister or advocate at the bar; provides for increases in salaries and allowances for federally-appointed judges; establishes an advisory committee to the Canadian Judicial Council and provides for the appointment of a commissioner for federal judicial affairs.

Chapter 26 (June 29, 1977) *An Act to amend the Aeronautics Act and the National Transportation Act* prohibits the transfer of shares of an international or interprovincial air carrier to a provincial government or one of its agents without approval of the Governor-in-Council.

Chapter 27 (June 29, 1977) *An Act to amend the Canada Deposit Insurance Corporation Act* authorizes the Canada Deposit Insurance Corporation to redeem the shares of its capital stock by paying the total par value to the Receiver General, to guarantee payment of fees and costs of a person appointed to act as liquidator or receiver of a member institution, and to make premium rebates to member institutions for any financial year when the amount credited to the Deposit Insurance Fund is more than adequate to maintain a satisfactory reserve.

Chapter 28 (June 29, 1977) *Miscellaneous Statute Law Amendment Act, 1977* corrects certain anomalies, inconsistencies, archaisms, errors and other matters of a non-controversial and uncomplicated nature in the Revised Statutes of Canada 1970 and other acts subsequent to 1970.

Chapter 29 (June 29, 1977) *An Act to implement conventions between Canada and Morocco, Canada and Pakistan, Canada and Singapore, Canada and the Philippines, Canada and the Dominican Republic and Canada and Switzerland for the avoidance of double taxation with respect to income tax.*

Chapter 30 (June 29, 1977) *An Act to amend the Canada Lands Surveys Act* replaces the designation Dominion land surveyor with Canada lands surveyor; substitutes Canada lands for public lands; permits application of the act to legal surveys in the offshore area; and empowers commissioners of the Yukon Territory and the Northwest Territories to have surveys made of Canada lands under their administration.

Chapter 31 (June 29, 1977) *Diplomatic and Consular Privileges and Immunities Act* defines the privileges and immunities for diplomatic and consular representatives in Canada, in effect for all countries including Commonwealth countries, and repeals the Diplomatic Immunities (Commonwealth Countries) Act.

Chapter 32 (July 14, 1977) *James Bay and Northern Quebec Native Claims Settlement Act* approves, gives effect to and declares valid certain agreements between the Grand Council of the Crees (of Quebec), the Northern Quebec Inuit Association, the government of Quebec, la Société d'énergie de la Baie James, la Société de développement de la Baie James, la Commission hydro-électrique de Québec and the Government of Canada; and certain other related agreements to which the federal government is a party.

Chapter 33 (July 14, 1977) *Canadian Human Rights Act* applies in areas coming under legislative authority of the federal government — the federal government itself, federal Crown corporations, and the federally regulated private sector. It prohibits discrimination in the provision of goods, services, facilities, or accommodation or in employment practices, based on race, national or ethnic origin, colour, religion, age, sex or marital status, or conviction for an offence for which a pardon has been granted and, in the case of employment practices, based on physical handicap; establishes a Canadian Human Rights Commission; provides for investigation, settlement and conciliation of complaints; provides for the appointment of human rights tribunals; provides for establishing public information programs and carrying out or sponsoring research; provides for protection of privacy of individuals and for their right of access to records containing personal information for any purpose including ensuring accuracy and completeness; and establishes the office of a privacy commissioner. The part on protection of personal information applies solely to information banks within federal government departments, agencies, boards, commissions, corporations and other bodies listed in the schedule to the act.

Chapter 34 (July 14, 1977) *Auditor General Act* clarifies the duties and responsibilities of the Auditor General of Canada, with respect to the independence of the office and the scope of the audit that he is required to perform.

Chapter 35 (July 14, 1977) *An Act to amend the Fisheries Act and to amend the Criminal Code in consequence thereof* redefines "fish" and "fishing" more precisely; gives fishery officers the powers of peace officers under the Criminal Code; clarifies regulations on the alteration, disruption or destruction of fish and aquatic habitat; and authorizes the Governor-in-Council to make regulations extending the coastal area beyond 12 nautical miles from shore on Canada's Atlantic Coast, for prohibited fishing operations.

Chapter 36 (July 14, 1977) *An Act to amend the Canada Pension Plan* provides, among other things, for the division of unadjusted pensionable earnings of former spouses, after divorce or annulment of marriage, for the period of cohabitation during the marriage; allows payment of full benefits to or on behalf of all dependent children of deceased or disabled contributors; allows applications to be made on behalf of a deceased person who, prior to his death, would have been entitled to benefits; and provides for the retroactive payment of retirement pensions up to 12 months.

Chapter 37 (July 14, 1977) *An Act to amend the Bretton Woods Agreement Act* increases Canada's subscription to the International Monetary Fund and to the International Bank for Reconstruction and Development reflecting the replacement of the US dollar by the Special Drawing Right (SDR); requires an annual adjustment of the valuation of the assets and liabilities of Canada in relation to the International Monetary Fund, to reflect the then current rate of exchange for Canadian dollars; implements changes for the International Monetary Fund dealing with the exchange arrangements that fund members may apply, providing a legal framework for present exchange rate practices, including floating, reducing the role of gold in the international monetary system, and improving the effectiveness of the fund's surveillance of international financial affairs.

Chapter 38 (July 14, 1977) *An Act to amend the Currency and Exchange Act and to amend other acts in consequence thereof* permits the use in Canadian contracts of units of account that are defined in terms of the currencies of two or more countries such as the special drawing rights issued by the International Monetary Fund; permits the Governor-in-Council to provide for the determination of the Canadian dollar equivalent of foreign currencies; provides that changes in valuation of the Exchange Fund's net assets in any year be credited or debited to the Consolidated Revenue Fund over the following three years; requires the auditor general to report to the finance minister instead of directly to Parliament in certifying the annual audit of the Exchange Fund Account.

Chapter 39 (July 14, 1977) *An Act to amend the Canadian and British Insurance Companies Act and the Foreign Insurance Companies Act* modifies and clarifies a number of provisions in these two acts and among other things extends the powers of an insurance company under the Canadian and British Insurance Companies Act to carry on business that is reasonably ancillary to the business of insurance.

Chapter 40 (July 14, 1977) *An Act respecting the Electoral Boundaries Readjustment Act (Beauharnois–Salaberry)* changes the name of the electoral district of Beauharnois in Quebec to Beauharnois–Salaberry.

Chapter 41 (July 14, 1977) *An Act respecting the Electoral Boundaries Readjustment Act (Blainville–Deux-Montagnes)* changes the name of the electoral district of Deux-Montagnes in Quebec to Blainville–Deux-Montagnes.

Chapter 42 (July 14, 1977) *An Act respecting the Electoral Boundaries Readjustment Act (Brampton–Georgetown)* changes the name of the electoral district of Brampton–Halton Hills in Ontario to Brampton–Georgetown.

Chapter 43 (July 14, 1977) *An Act respecting Electoral Boundaries Readjustment Act (Cochrane)* changes the name of the electoral district Cochrane North in Ontario to Cochrane.

Chapter 44 (July 14, 1977) *An Act respecting the Electoral Boundaries Readjustment Act (Huron–Bruce)* changes the name of the electoral district Huron in Ontario to Huron–Bruce.

Chapter 45 (July 14, 1977) *An Act respecting the Electoral Boundaries Readjustment Act (Kootenay East–Revelstoke)* changes the name of the electoral district of Kootenay East in British Columbia to Kootenay East–Revelstoke.

Chapter 46 (July 14, 1977) *An Act respecting the Electoral Boundaries Readjustment Act (Laval)* changes the name of the electoral district of Mille-Îles in Quebec to Laval.

Chapter 47 (July 14, 1977) *An Act respecting the Electoral Boundaries Readjustment Act (Lethbridge–Foothills)* changes the name of the electoral district of Lethbridge in Alberta to Lethbridge–Foothills.

Chapter 48 (July 14, 1977) *An Act respecting the Electoral Boundaries Readjustment Act (London–Middlesex)* changes the name of the electoral district of Middlesex East in Ontario to London–Middlesex.

Chapter 49 (July 14, 1977) *An Act respecting the Electoral Boundaries Readjustment Act (Saint-Jacques)* changes the name of the electoral district of Saint-Henri in Quebec to Saint-Jacques.

Chapter 50 (July 14, 1977) *An Act respecting the Electoral Boundaries Readjustment Act (Saint-Léonard–Anjou)* changes the name of the electoral district of Saint-Léonard in Quebec to Saint-Léonard–Anjou.

Chapter 51 (July 14, 1977) *An Act respecting the Electoral Boundaries Readjustment Act (Wellington–Dufferin–Simcoe)* changes the name of the electoral district of Dufferin–Wellington in Ontario to Wellington–Dufferin–Simcoe.

Chapter 52 (August 5, 1977) *Immigration Act, 1976* states the basic principles of immigration policy: non-discrimination, family reunion, humanitarian concern for refugees and the promotion of national social, demographic and cultural goals. It provides for a link between the immigration movement and Canada's population and labour market needs and for an annual forecast of the number of immigrants Canada can comfortably absorb, to be made in consultation with the provinces. It establishes a new family class, allowing Canadian citizens and permanent residents to sponsor a wide range of relatives. It confirms Canada's protective obligations to refugees under the United Nations Convention Relating to the Status of Refugees and establishes a new refugee class. It requires that immigrant and visitor visas be obtained abroad, and prohibits visitors from changing their status within Canada.

Chapter 53 (August 5, 1977) *Criminal Law Amendment Act, 1977* amends the Criminal Code, the Customs Tariff, the Parole Act, the Penitentiary Act and the Prisons and Reformatories Act. Amendments to the Criminal Code and Customs Tariff, related to gun control, regulate the sale and possession of firearms. The Parole Act is amended to provide for a larger National Parole Board, for provincial parole boards and for community participation in regional panels in parole review for murderers or those serving indeterminate sentences; the act introduces procedural safeguards for inmates being considered for parole.

Chapter 54 (August 5, 1977) *Employment and Immigration Reorganization Act* merges the former Unemployment Insurance Commission and the Department of Manpower and Immigration into one organization, the Canada Employment and Immigration Commission; establishes the Department of Employment and Immigration and the Canada Employment and Immigration Advisory Council; makes a number of amendments to the Unemployment Insurance Act, 1971 and restructures the general scheme of unemployment benefits.

Chapter 55 (August 5, 1977) *An Act to facilitate conversion to the metric system of measurement* amends the Canadian Wheat Board Act, the Consumer Packaging and Labelling Act, the Gas Inspection Act, the Oil and Gas Production and Conservation Act, the Prairie Grain Advance Payments Act, the Regional Development Incentives Act, the Two-Price Wheat Act, the Weights and Measures Act and the Western Grain Stabilization Act. Words denoting metric units of measurement are substituted in these acts for words denoting Canadian units of measurement.

Chapter 56 (August 5, 1977) *An Act to amend the Canadian Wheat Board Act respecting the establishment of marketing plans and to amend the Western Grain Stabilization Act in consequence thereof.*

Chapter 57 (August 9, 1977) *Air Traffic Control Services Continuation Act* provides for the immediate continuation or resumption of air traffic control services, invalidates a strike, and amends a collective agreement and extends its term to December 31, 1977.

Chapter 58 (July 14, 1977) *An Act to incorporate Continental Bank of Canada* provides for the organization of the bank as a wholly owned subsidiary corporation of the Industrial Acceptance Corporation Ltd., with an authorized capital stock of $100 million and head office in Toronto.

Canadian chronology 1977

Appendix 3

Events in the general chronology from 1497 to 1866 were published in the *Canada Year Book 1951*, pp 46-49; from 1867 to 1953 in the *Canada Year Book 1954*, pp 1259-1264; and annually from that year in successive editions. The following listing covers the year 1977; it should be noted that certain dates are approximate. Acknowledgment is given to the publication *Canadian News Facts*, Toronto, which has served as a reference.

January

Jan. 4, Agriculture Minister Eugene Whelan announced a $26 million program to stabilize calf prices by paying subsidies to beef cow-calf producers. *Jan. 7*, Preliminary figures indicated a national record of 270,000 housing starts in 1976, up from 231,456 in 1975, Central Mortgage and Housing Corp. announced. *Jan 10*, Canada expelled four Cubans (including two diplomats) following an RCMP investigation into charges that the Cuban consulate in Montreal was being used to train spies. *Jan. 11*, Average seasonally-adjusted unemployment rate for 1976 was 7.1% and the average estimated actual unemployed was 736,000, Statistics Canada reported; during 1975 the average seasonally-adjusted rate was 6.9% and the average estimated actual unemployed was 697,000. The Saskatchewan Court of Appeal overturned a lower court decision and upheld regulations enabling the provincial government to set quotas on potash production and floor prices on potash sales. *Jan. 12*, The Federal Court of Canada upheld a federal order restricting the use of French in Canadian airspace. Air Canada announced suspension of unprofitable regular flights to Moscow, Prague and Brussels, a revised domestic schedule, and increased domestic fares to offset an operating loss in 1976. Karen Kain and Frank Augustyn of the National Ballet, first Canadian dancers invited to perform with the Bolshoi Ballet, received a standing ovation in Moscow. *Jan. 13*, Average inflation rate for 1976 was 7.5% compared to 10.8% in 1975, Statistics Canada reported. *Jan. 14*, Metropolitan Toronto police announced formation of an ethnic squad to deal with Asian community problems in the wake of racially-motivated attacks. *Jan. 18*, McClelland and Stewart Ltd. in a joint venture with Bantam Books of Canada Ltd. said paperback books by Canadian authors would be published under the name Seal Books. *Jan. 21*, Prime Minister Pierre Trudeau proposed to provincial premiers a revised formula for amending the constitution. The Canadian government requested an exemption for Canada from new US legislation making it more difficult for Americans to obtain tax deductions for expenses in attending conventions outside their own country. *Jan. 25*, Quebec Premier René Lévesque told the Economic Club of New York that Quebec independence is inevitable, but assured bankers and investors that his government intended no large-scale nationalization of private industry. *Jan. 26*, External Affairs Minister Don Jamieson returned from Brazil, Peru and Colombia, after signing technical, scientific and financial co-operation agreements with the three countries. *Jan. 28*, US President Jimmy Carter thanked the Canadian government for selling extra emergency fuel supplies to the US to alleviate shortages during the unusually severe winter. *Jan. 31*, Joint winners of the Chalmers Canadian Play Award were W.O. Mitchell for *Back to Beulah* and Larry Fineberg for *Eve*.

February

Feb. 1, The three Maritime provinces and the federal government signed a memorandum of agreement which would give the three provinces 100% of royalties from future offshore mineral discoveries within 5 km (3 miles) of their coasts and 75% from resources outside that distance. *Feb. 3*, First allocations of a $200 million Canada Works program to combat unemployment were announced by the Department of Manpower and Immigration. Death of Gregory Clark, 84, veteran Toronto journalist and raconteur. *Feb. 5*, Top honours at the Canadian figure skating championships won by Lynn Nightingale, ladies' singles, and Ron Shaver, men's singles. *Feb. 8*, Estimated actual unemployment in January rose from December's 754,000 to 889,000, the highest level since Statistics Canada began collecting labour force statistics in 1953; the seasonally-adjusted jobless rate remained at 7.5%. *Feb. 14*, Nova Scotia homeowners were offered financial assistance to improve insulation in a $79.6 million energy conservation program announced jointly by the federal and provincial governments. A merger was announced for Unity Bank of Canada and Montreal-based Provincial Bank of Canada, reducing the number of chartered banks in Canada from 12 to 11. *Feb. 15*, Royal Bank of Canada announced the transfer of three head-office departments from Montreal to

Toronto, but said the move was not based on political or linguistic considerations. *Feb. 18,* The US government suspended work on the Garrison diversion project in North Dakota, which Canada saw as endangering water quality in Manitoba, pending a report of the International Joint Commission. *Feb. 22,* In the first speech by a Canadian prime minister to the United States Congress, Mr. Trudeau said Canada would remain united. Reversing a decision by the Appeal Division of the Federal Court of Canada, the Supreme Court of Canada ruled that Alberta was not obliged to seek Canadian Transport Commission approval for its purchase in 1974 of Pacific Western Airlines. *Feb. 24,* A record travel deficit of $1.18 billion was recorded in 1976 as Canadians travelling abroad spent $3.12 billion, almost 23% more than in 1975, while visitors from the US and other countries spent $l.94 billion in Canada. *Feb. 25,* Price increases for gasoline and home-heating fuel of between 3.1 and 4.6 cents a gallon effective March 2, were announced by the Anti-Inflation Board (AIB). *Feb. 26,* Death of Watson Kirkconnell, 81, author and former president of Acadia University. *Feb. 28,* Transport Minister Otto Lang announced creation of a new Crown corporation, VIA Rail Canada Inc., to operate the country's passenger rail service. The gross national product (GNP) for 1976 increased by 4.6% over its 1975 level to $190 billion, although it was in a decline at the close of the year, Statistics Canada reported; most of the increase was due to inflation.

March
Mar. 2, A treaty permitting the exchange of prisoners between Canada and the US was signed in Washington by Solicitor General Francis Fox. *Mar. 3,* Department of National Health and Welfare tests showed high blood levels of toxic mercury in two residents of Tuktoyaktuk, NWT, indicating mercury contamination. Keith Spicer submitted his resignation as Canada's Official Languages Commissioner. *Mar. 4,* In response to a federally-commissioned report critical of the legal abortion system, Health Minister Marc Lalonde announced plans to launch a publicity campaign for planned parenthood and to press the provinces to improve abortion services. Prime Minister Trudeau asked the Canadian Radio-television and Telecommunications Commission (CRTC) to investigate charges of pro-separatist bias in the CBC French-language service. *Mar. 7,* The Potash Corp. of Saskatchewan purchased its second potash mine, the Sylvite mine at Rocanville, bringing the province's ownership of the productive capacity of the Saskatchewan potash industry to 20%. *Mar. 8,* Premier Lévesque, opening the Parti Québécois government's first full legislature session, announced introduction of new language legislation to replace the Official Language Act passed by the Liberal government in 1974. *Mar. 9,* The federal health department announced saccharin would be

banned from use in foods, cosmetics and as a sweetening agent in drugs because of studies on rats indicating it may cause cancer. *Mar. 15,* Canada's balance of international payments showed a $4.329 billion deficit in 1976, about $600 million less than in 1975, Statistics Canada reported. *Mar. 16,* Finance Minister Jacques Parizeau of Quebec announced abolition of the province's Anti-Inflation Board. Top winners of the 1977 Juno awards for the Canadian recording industry were Gordon Lightfoot, folksinger and composer, and Burton Cummings, vocalist. *Mar. 17,* Prime Minister Trudeau said Quebec would be barred from delegations attending international conferences if provincial officials continued to renounce ties with the rest of Canada. *Mar. 21,* Greenpeace Foundation members were forced to abandon the protest against the annual seal hunt off Newfoundland because of bad weather, hazardous ice conditions and lack of funds. *Mar. 31,* In a budget he said was aimed at continuing the fight against inflation, Finance Minister Donald Macdonald announced tax incentives for business in an effort to create jobs and tax reductions for low-wage earners.

April
Apr. 4, The federal government announced a new dairy policy under which consumers would pay 10 cents a pound (0.454 kg) more for butter and the government would provide $266 million in subsidies to farmers. *Apr. 5,* Senate appointments were announced for Willy Adams (first Inuit senator), H.A. (Bud) Olsen, Royce Frith and Peter Bosa. *Apr. 7,* In the Toronto debut of American League baseball, Toronto Blue Jays defeated Chicago White Sox 9-5. *Apr. 13,* The estimated actual number of unemployed in March was 944,000, the highest figure recorded since Statistics Canada began collecting unemployment statistics in 1953; the seasonally-adjusted unemployment rate rose to 8.1% from 7.9% in February when estimated actual unemployment was 932,000. *Apr. 18,* The US government said work on the Garrison diversion project in North Dakota would be abandoned. *Apr. 20,* Air Canada announced a deficit of $10.4 million in 1976, down from about a $13 million loss in 1975. A $41 million program to upgrade Canada's East Coast fishery was announced by federal fisheries services. A ban by customs officials on the May issue of *Penthouse* magazine on grounds of indecency was upheld by the Federal Court. *Apr. 21,* Alberta MP Jack Horner was sworn in as minister without portfolio, severing a 19-year tie with the Progressive Conservative Party to join the Liberal government. *Apr. 27,* The Parti Québécois government introduced its Charter of the French Language as Bill One in the Quebec National Assembly; the bill was aimed at making French supreme in Quebec and limiting the use of English in education, business, the courts and public administration. Winners of the 1976 Governor General's literary awards: in fiction, Marian Engel

for *Bear* and André Major for *Les Rescapés;* in poetry, Joe Rosenblatt for *Top Soil* and Alphonse Piché for *Poèmes 1946-1968;* in non-fiction, Carl Berger for *The Writing of Canadian History* and Fernand Ouellet for *Le bas Canada 1791-1840, changements structuraux et crise. Apr. 28,* Jean-Luc Pepin said he would resign May 6 as chairman of the Anti-Inflation Board, to be succeeded by Harold Renouf. *Apr. 29,* Ontario Premier William Davis called an election for June 9, saying the opposition had made minority government unworkable.

May

May 8, Team Canada finished fourth in the world hockey championships, won by Czechoslovakia. *May 9,* The Berger Commission recommended a 10-year moratorium on construction of a Mackenzie River Valley pipeline to allow time to settle and implement native land claims and to solve technical and environmental problems, and also a permanent ban on any pipeline from Alaska across the environmentally-sensitive northern Yukon. The Bank of Canada announced reduction of its lending interest rate from 8.0% to 7.5%, the lowest level in three years, because of slow growth in the money supply; the rate had already been reduced from 8.5% in January. *May 10,* CBC president Al Johnson said some staff were reprimanded following allegations by federal politicians of a pro-separatist bias in the CBC's French-language service. *May 12,* In Canada's first fishing agreement with another country since extending its coastal zone to 200 nautical miles, Cuba received the right to fish within the zone and in return recognized Canada's interest in fish stock beyond that limit. *May 14,* Montreal Canadiens won their second consecutive Stanley Cup, defeating Boston Bruins in four straight games. *May 16,* The Hall Commission report on grain handling and transportation recommended retention of subsidized Crowsnest Pass transportation rates, establishment of a Prairie rail authority, gradual abandonment of more than 3 200 km (2,000 miles) of branch lines, and construction of a new Arctic railway. *May 17,* Ray Guy, St. John's, Nfld. columnist, won the 1977 Stephen Leacock award for humour for *That Far Greater Bay. May 21,* Dave Barrett, former BC premier, was returned as leader of the BC New Democratic Party at its annual convention. *May 24,* Liberals won five of six federal byelections, including the PEI riding of Malpeque held by the Conservatives for the last 25 years. *May 25,* Transport Minister Lang announced a $125 million program to aid highway, passenger and freight efforts in the Atlantic provinces in the next three years. The $50,000 annual Royal Bank of Canada award to outstanding Canadians for exceptional contribution to human welfare and the common good was split between Dr. Gordon W. Thomas and Dr. W.A.

Paddon, both of the International Grenfell Association. *May 26,* Acting on the Hall Commission report, Transport Minister Lang announced that more than 2 900 km (1,800 miles) of Prairie rail lines had been added to the basic network protected from abandonment to the year 2000. Quebec Nordiques won the World Hockey Association title, defeating Winnipeg Jets four games to three. *May 27,* Montreal actress Monique Mercure, first Canadian performer to win a top prize at the Cannes Film Festival, was co-winner of the best actress award in the National Film Board's *J.A. Martin, Photographe. May 31,* The Canadian Wheat Board announced the sale to China of 3 million tonnes (about 110 million bushels) of wheat with an estimated value of $330 million.

June

June 2, Quebec raised its minimum hourly wage from $3.00 to $3.15, the highest rate in Canada, in a decision to tie the minimum wage to the average industrial wage every six months. *June 7,* A special Commons subcommittee set up to examine the Canadian Penitentiary Service recommended restraints on power of employees over prisoners. *June 8,* Joey Smallwood, 76, premier of Newfoundland from 1949 to 1972, resigned as a Liberal opposition member of the Newfoundland legislature. *June 9,* In an Ontario provincial election Premier Davis' Progressive Conservative government retained power but again failed to win a majority; Liberals under Stuart Smith displaced the NDP as the official opposition. *June 11,* Electoral boundary changes came into effect to increase by 18 the number of seats in the House of Commons to 282 after the next general election. *June 14,* Agriculture Minister Whelan announced Cabinet approval of a national chicken broiler marketing agency; national marketing boards were already operating for eggs and turkeys. The Canadian Labour Relations Board ruled that unions representing bank employees could organize workers branch by branch rather than in a cross-country unit and certified the new Canadian Union of Bank Employees in three Ontario branches of the Bank of Nova Scotia. Jacques Lavoie, MP for the Montreal riding of Hochelaga, quit the Progressive Conservatives to join the Liberals. *June 15,* The final report of Ontario's Royal Commission on Violence in the Communications Industry, headed by Judy LaMarsh, rejected censorship, but recommended tougher controls over all broadcast and print media. *June 16,* The Quebec government ordered a commission of inquiry into illegal police activities, headed by labour lawyer Jean Keable, after three policemen, from the RCMP, the Quebec Provincial Police and the Montreal police, pleaded guilty to charges of failing to obtain a search warrant before a break-in at the office of Agence de Presse Libre du Québec. *June 20,* The federal government announced the end of a five-year freeze after August 1, making more than 0.4 billion hectares (one billion acres) of land available for exploration and development in northern and offshore areas. *June 24*, Death of André Fortin, 33, federal leader of the Social Credit party, in a car accident. *June 28*, Gilles Caouette, son of the late Réal Caouette, former federal Social Credit party leader, named interim national leader of the Social Credit party. *June 30*, All opposition members walked out of the Quebec National Assembly after the Parti Québécois government dismissed the Liberal opposition's proposal of Canada Day birthday wishes.

July

July 1, Canada Day celebrations were held across the country marking the 110th anniversary of Confederation; highlighting the day was a nationally televised spectacular from Parliament Hill; in 1976 the festivities had been cancelled as an economy measure. *July 4*, After 18 months of hearings, the National Energy Board recommended federal Cabinet approval of a proposal by Foothills Pipe Lines (Yukon) Ltd. for a pipeline along the Alaska Highway to move natural gas from Alaska to US markets. *July 5*, Prime Minister Trudeau announced the formation of a special group to study national unity headed by Jean-Luc Pepin, former Liberal Cabinet minister, and John Robarts, former Progressive Conservative premier of Ontario. *July 6*, Solicitor General Fox announced the establishment of a royal commission headed by Mr. Justice David C. McDonald to investigate alleged illegal RCMP activities across the country. *July 8*, Recommendations to expand the use of French in Quebec in air traffic control of small planes under visual flight rules were supported by all parties in the House of Commons, the Canadian Air Traffic Control Association, and the Canadian Air Line Pilots Association. At a federal-provincial consumer ministers meeting, a code of rust resistance standards for car manufacturers and importers, announced by Consumer and Corporate Affairs Minister Anthony Abbott, received full provincial backing. Federal and British Columbia governments signed an agreement providing $170 million in two five-year plans for industrial and agricultural development in the province. *July 12*, The Parti Québécois government introduced a new language bill, Bill 101, which eased some of the proposed provisions of the earlier Bill One, but stood firm on a proposal that only children whose mother or father had attended English elementary school in Quebec could be admitted to English schools. Agreement in principle reached for the $86.5 million purchase by the Potash Corp. of Saskatchewan of the Alwinsal Potash of Canada Ltd. mine, the third potash mine acquired by the province. *July 15*, Defence Minister Barney Danson said the armed forces would be increased by 4,700 persons in the next few years to 83,000, its strength prior to inflation-forced reductions. *July 16*, Death

of Marg Osburne, 49, nationally-known singer from Sussex, NB, of a heart attack following a concert. *July 18*, The House of Commons passed a bill to control the purchase and use of firearms and to expand police wiretapping powers. *July 19*, The House of Commons passed a bill increasing the qualifying period for unemployment benefits from eight weeks to a period ranging from 10 to 14 weeks depending on regional unemployment rates, and reducing the time a person can receive benefits to a period ranging from 10 to 50 weeks depending on a formula based on regional unemployment and number of weeks worked. The Department of Manpower and Immigration reported 149,429 immigrants came to Canada in 1976, a decline of 20.5% from 1975. *July 20*, A special inquiry by the CRTC found no specific separatist bias in the French-language service of the CBC but said that all media failed to cover regions of Canada adequately and to bridge the gap between English and French cultures. *July 21*, The Alberta government released a policy report under which up to 50% of the Rocky Mountain foothills would be kept in their natural state. *July 25*, The federal government announced creation of an information centre to counteract separatist propaganda. The House of Commons adjourned after passing an immigration bill, the first major changes to the Immigration Act since 1952, and a metric conversion bill, amended to allow grain producers to continue using the acre as their official unit of land measurement.

August

Aug. 2, Foothills Pipe Lines (Yukon) Ltd. proposal for an Alaska Highway natural gas pipeline was supported by the Lysyk inquiry, but a two-year delay was recommended to settle and implement native land claims, select a route through the southern Yukon, and allow Yukon communities to prepare for the development. The report of the Quebec Police Commission inquiry into organized crime recommended greater co-operation between police and the provincial revenue department and tougher bank regulations to control granting of credit to suspected criminals to help eradicate organized crime. *Aug. 5*, Max Yalden, deputy minister of the communications department, was named commissioner of official languages succeeding Keith Spicer. *Aug. 8*, Prime Minister Trudeau announced that the government was tentatively backing the Foothills Pipe Lines (Yukon) Ltd. proposal for an Alaska Highway pipeline through the southern Yukon, but that procedure with the project depended on negotiations with Washington. *Aug. 10*, Parliament passed emergency legislation to halt a three-day strike by the country's 2,200 air traffic controllers over a wage dispute. *Aug. 11*, Harry Boyle resigned as chairman of the CRTC effective September 16, to be succeeded by Pierre Camu, former president of the Canadian Association of Broadcasters. Gordon Fairweather, Progressive Conservative MP for

Fundy–Royal, resigned from the House of Commons to become head of the new Human Rights Commission. *Aug. 15,* Manpower and Immigration Minister Bud Cullen officially became the minister of employment and immigration with the merger of his department with the unemployment insurance commission; new legislation set up the Canada Employment and Immigration Commission, the Department of Employment and Immigration, and the Canada Employment and Immigration Advisory Council. *Aug. 19,* Nine provincial leaders rejected Quebec Premier Lévesque's call for reciprocal language agreements to guarantee English-language education for English-speaking newcomers to Quebec from other provinces in return for the same rights for French-speaking minorities in the other provinces; the nine premiers asserted that language rights were not to be subject to deals. *Aug. 22,* Rev. George Milledge Tuttle, Edmonton, elected Moderator of the United Church of Canada. *Aug. 24,* Northern Quebec Inuit communities demonstrated against Bill 101 in Fort Chimo, forcing closure of all provincial offices and removing all provincial flags. *Aug. 26,* The Quebec National Assembly passed Bill 101, the Charter of the French Language.

September

Sept. 6, The Canadian Wheat Board announced its first sale of wheat to Vietnam, approximately 120,000 tonnes (4.4 million bushels), to be shipped from Pacific Coast ports between October 1977 and March 1978. Federal Finance Minister Donald Macdonald resigned from the Cabinet for personal reasons and said he would not run in the next federal election. The Manitoba legislature was dissolved and a provincial election set for October 11. Death of Leslie McFarlane, 74, playwright, producer, and author of the first 20 books of The Hardy Boys series, in Whitby, Ont. Conversion of highway signs from miles to kilometres was begun in all provinces except Quebec and Nova Scotia. *Sept. 7,* Cindy Nicholas, Scarborough, Ont., became the first woman to complete a two-way, non-stop swim of the English Channel; her time of 19 hours, 55 minutes chopped 10 hours off the world record. General Motors of Canada declared it would not commit itself to new federal-provincial voluntary rust resistance standards on its 1978 models despite government plans to legislate the standards if the industry failed to meet them. *Sept. 9,* The federal government and the Public Service Alliance of Canada agreed on a bilingual bonus of $800 a year for bureaucrats required to use both official languages. The Quebec government launched a 10-year program to educate Quebecers on nutrition and counteract the growing number of malnutrition-related diseases. *Sept. 13,* Quebec appointed Yves Michaud as permanent delegate to international organizations to promote an exchange of views with the rest of the world. *Sept. 14,* The

federal government announced spending of $13.6 million in the next 18 months for films, exhibits and other projects designed to tell Canadians more about their country. *Sept. 16,* Prime Minister Trudeau announced 11 changes in the Cabinet; among ministers appointed were Jean Chrétien to Finance, Jack Horner to Industry, Trade and Commerce, and Marc Lalonde to the newly created Ministry of State for Federal-Provincial Relations. *Sept. 20,* Canada and the US formally signed an agreement in Ottawa to build a natural gas pipeline across the Yukon to move Alaskan fuel to US markets. Justice Minister Ron Basford announced the appointment of the first women named to the BC and Nova Scotia supreme courts: Constance Glube, Halifax city manager, to the Supreme Court of Nova Scotia and County Court Judge P.M. Proudfoot to the Supreme Court of BC. *Sept. 29,* Appointed to the Supreme Court of Canada were William Estey, head of the Ontario Court of Appeal and chief justice of Ontario, replacing retired Mr. Justice Wilfred Judson, and Yves Pratte, former chairman of Air Canada, replacing Mr. Justice Louis-Philippe de Grandpré. *Sept. 30,* As part of an overhaul of bilingualism policies in the public service, the federal government announced a phasing out of language training and bilingualism pay bonuses by 1983; those wishing to advance in the public service would still have to be bilingual, said Robert Andras, president of the Treasury Board. The Supreme Court of Canada upheld a New Brunswick lower court ruling that two or more breath analysis samples are necessary to convict a person of driving while under the influence of alcohol.

October

Oct. 3, Finance Minister Chrétien announced government plans to transfer 2,500 full-time and 1,500 part-time public service jobs out of Ottawa in the next five years. The federal government announced a formal inquiry into possible violations of the Combines Investigation Act by members of a government-approved cartel in the marketing of uranium. *Oct. 6,* Quebec Superior Court Judge Perry Meyer ruled unconstitutional a section of the province's Charter of the French Language requiring all court documents to be filed in French; the BNA Act assures that either English or French may be used in Quebec courts. *Oct. 7,* Stephen Juba, mayor of Winnipeg since 1956, announced he would not seek re-election in the October civic elections. *Oct. 11,* In Manitoba's provincial election Sterling Lyon's Progressive Conservative Party ended the eight-year NDP government of Ed Schreyer, winning 33 seats; the NDP won 23 and the Liberals one. *Oct. 14,* The Senate agriculture committee recommended stiff new restrictions on beef imports to protect Canadian beef producers from low-priced world competition. *Oct. 15,* William Rowe elected leader of the Newfoundland Liberal Party, defeating incumbent Ed Roberts.

Oct. 17, The first televised coverage of House of Commons proceedings began with the reopening of Parliament. *Oct. 18,* The third session of the 30th Canadian Parliament was opened with the throne speech read by the Queen for the first time since 1957; in the speech the government promised increased jobs, tax relief for business, continuation of wage and price controls through the first part of the year, and efforts to preserve national unity by working with the provinces on constitutional reform. *Oct. 19,* The Northern Ontario Heritage Party received official recognition from the provincial government as Ontario's newest political party. *Oct. 20,* Finance Minister Chrétien announced the removal of wage and price controls beginning April 14, 1978. INCO Ltd. announced a reduction of working staff in Canada by 3,450 by mid-1978, paring 2,800 jobs at Sudbury, Ont., and 650 jobs at Thompson, Man. The federal government announced a three-year program to protect the domestic textile and clothing industry from cheap imports. *Oct. 21,* Quebec Premier Lévesque announced a $470.6 million program of economic recovery and job creation for the next 18 months. *Oct. 24,* The Canadian dollar dropped to 89.88 US cents, the first time since 1939 it had closed below the 90-cent level; the federal government took steps to support the dollar by arranging a standby credit of $1.5 billion in US funds. *Oct. 31,* The James Bay land claims agreement, the first major modern treaty with Canadian native people, became law, providing Cree and Inuit in Northern Quebec $255 million over 20 years, community ownership of small areas, and exclusive hunting, fishing and trapping rights over large tracts of land, in return for surrendering aboriginal rights to about 60% of Quebec territory. Solicitor General Fox said in the House of Commons that the RCMP participated in two alleged illegal acts, a barn burning and the theft of dynamite near Montreal; Mr. Fox had said three days earlier that the RCMP had acted without a warrant in 1973 in taking computer tapes containing Parti Québécois membership lists and financial information from a Montreal office.

November

Nov. 2, Transport Minister Lang announced a government order from Bombardier-MLW Ltd. of Montreal for 22 locomotives and 50 coaches of the Canadian-designed LRC (light, rapid and comfortable) for VIA Rail's intercity passenger service. *Nov. 3,* French President Valery Giscard d'Estaing conferred on Quebec Premier Lévesque the honour of Grand Officer of the Legion of Honour, the highest decoration France could bestow on anyone below the rank of head of state. *Nov. 4,* Progressive Conservative leader Joe Clark was given a 93.1% vote of confidence by delegates at the party's national policy conference in Quebec City. *Nov. 5,* Death of bandleader Guy Lombardo, 75, a native of London, Ont., whose New Year's Eve

performances had been a tradition since 1929. *Nov. 9,* The government asked Parliament for an extra $1.91 billion to create jobs, aid provinces suffering from reduced tax income, and make up for the falling value of the dollar. *Nov. 10,* The Ontario Supreme Court ruled that the federal government could not prevent opposition MPs from discussing in the House of Commons or disclosing to journalists information about an international cartel formed to control the world price of uranium. *Nov. 16,* The return of Jean Lesage, former Liberal premier of Quebec, to provincial politics as chairman of a special Liberal party committee on Quebec's independence referendum was announced by interim party leader Gérard D. Lévesque. *Nov. 17,* Death of Nova Scotia industrialist Col. Sidney C. Oland, 91, in Halifax. *Nov. 22,* Canada and Mexico signed a prisoner exchange treaty allowing prisoners serving jail sentences to finish their terms in their own country. *Nov. 23,* The Supreme Court of Canada ruled unconstitutional the Saskatchewan government's mineral income tax and royalty surcharge on crude oil, but Premier Allan Blakeney's initial position was not to return any of the $580 million in taxes already paid by oil companies. *Nov. 26,* Death in Winnipeg of Tommy Prince, 64, Canada's most decorated Indian; in World War II and the Korean War he won 10 medals. *Nov. 27,* Montreal Alouettes won the Canadian Football League championship by defeating Edmonton Eskimos 41-6 before a record Grey Cup crowd of 68,205 in Montreal. *Nov. 29,* A bill to set up Société Nationale de l'Amiante (National Asbestos Corp.) was introduced in the Quebec National Assembly. *Nov. 30,* The Supreme Court of Canada upheld a Quebec Appeals Court ruling that Quebec legislation regulating cable television was illegal, and reaffirmed the federal government's exclusive authority over cable TV.

December

Dec. 1, The federal government announced a three-year program to limit footwear imports threatening Canadian production and employment. *Dec. 2,* The government of the Northwest Territories announced imposition from Jan. 1, 1978, of personal and corporation income taxes formerly levied by the federal government. The 1977 wheat harvest was estimated at 19.6 million tonnes (722 million bushels), down from 23.6 million tonnes (866.7 million bushels) in 1976. *Dec. 6,* The Canadian Transport Commission ruled that Air Canada, CP Air, and the five regional air carriers be permitted advance booking charters in Canada on a trial basis in 1978. The seasonally-adjusted unemployment rate in November reached a post-1940 high of 8.4%; estimated actual unemployment rose to 840,000 from 787,000 in October when the seasonally-adjusted rate was 8.3%. *Dec. 8,* Falconbridge Nickel Mines Ltd. announced a cut of 750 jobs at Sudbury, Ont., by March 1978 because of reduced nickel production. *Dec. 9,* Minister of

Supply and Services Jean-Pierre Goyer announced that Canadair Ltd. of Montreal would receive $100 million in contracts from Lockheed Corp. of California to construct components for Aurora and P-3C aircraft. *Dec. 12,* Death of Frank Boucher, 76, member of the National Hockey League's Hall of Fame and seven-time winner of the Lady Byng Trophy, in Kemptville, Ont. *Dec. 13,* The Consumer Price Index rose by 0.7% in November after a 1.0% rise in October, lifting the 12-month inflation rate to 9.1%, the highest since February 1976. *Dec. 16,* Iona Campagnola, federal minister of state for fitness and amateur sport, named Thunder Bay, Ont., as the site of the 1981 Canada Summer Games. *Dec. 18,* Two Nova Scotia companies and one from Quebec announced shared contracts up to $125 million for Canada's participation in an international railway-building

project in Venezuela. *Dec. 19,* External Affairs Minister Don Jamieson announced that the European Economic Community had agreed to all the aspects of the 1974 Canadian policy on nuclear safeguards and that uranium shipments to the EEC could be resumed. Mr. Jamieson also said Canada would end government support for commercial relations with South Africa to underline its opposition to South African racial policies. *Dec. 21,* The Quebec National Assembly approved the Quebec government-operated no-fault car insurance plan which would insure against bodily injuries while leaving insurance for property damage to the private sector. *Dec. 22,* Defence Minister Danson announced Cabinet approval of $63 million for design and cost details for the first of six new frigates in a $1.5 billion program to bolster Canada's navy and shipbuilding industry.

Canadian honours

An exclusively Canadian honours system was introduced in 1967 with the establishment of the Order of Canada. The honours system was enlarged in 1972 with the addition of the Order of Military Merit and three decorations to be awarded in recognition of acts of bravery.

The Order of Canada, instituted on July 1, 1967, the 100th anniversary of Confederation, is designed to honour Canadians for outstanding achievement and service to their country or to humanity at large. Originally, two levels of membership were provided: Companions of the Order and recipients of the Medal of Service. The order was revised in 1972 and now comprises three categories of membership: companions, officers — which includes all those who received the Medal of Service — and members. The last category is intended especially to recognize service in a locality or in a particular field of activity. Not more than 15 persons may be appointed in any one year as companions and the total number of companions is not to exceed 150. Officers of the order may be appointed to the number of 40 persons a year and up to 80 persons may be appointed yearly as members.

All members of the order are entitled to have letters placed after their names as follows: for the companion CC, for the officer OC and for the member CM.

The Queen is sovereign of the order and the Governor General holds office as chancellor and principal companion. Appointments to the order are made, with the Queen's approval, by the Governor General assisted by an advisory council which meets twice a year under the chairmanship of the chief justice of Canada. Members of the advisory council include the clerk of the Privy Council, the under-secretary of state, the chairman of the Canada Council, the president of the Royal Society of Canada, the president of the Association of Universities and Colleges of Canada and not more than two other members who may be appointed by the Governor General from among members of the order.

While Canadians are the primary recipients of the order, the constitution provides that persons who are not Canadian citizens and whom Canada desires to honour may be appointed as honorary members at any of the three levels of membership.

The Order of Military Merit has been established to provide a means of recognizing conspicuous merit and exceptional service by members of the Canadian Armed Forces, both regular and reserve. The order has three levels of membership: commander (CMM), officer (OMM) and member (MMM).

The Queen is the sovereign of the order and the Governor General is the chancellor as well as a commander of the order. The chief of the defence staff is the principal commander of the order. Appointments to the order are made by the Governor General on the recommendation of the minister of national defence; nominations are made by the chief of the defence staff assisted by an advisory committee for the order.

The number of appointments made annually will vary, depending on the number of nominations submitted and approved. The order's constitution stipulates, however, that the total number of appointments made annually will not exceed a tenth of one per cent of the forces' average strength. Members of foreign armed forces who render particularly meritorious service to Canada or the Canadian Armed Forces in the course of their military duties may be made honorary members of the order at any of the three levels.

Canadian bravery decorations. A Medal of Courage was included in the Order of Canada in 1967 but it was found that a single medal would not serve to recognize in an equitable manner acts of bravery which entail varying degrees of risk. Consequently, no awards were made and the medal has now been superseded by a series of three decorations: the Cross of Valour (CV), the Star of Courage (SC) and the Medal of Bravery (MB). Instances of extraordinary heroism in circumstances of extreme peril will be marked with the award of the Cross of Valour; other outstandingly courageous actions may qualify for the award of the Star of Courage or the Medal of Bravery. The bravery decorations are awarded with the approval of the sovereign by the Governor General on the advice of a decorations advisory committee. They may be awarded to civilians, members of the Canadian Armed Forces and of the protective services, and may be awarded posthumously.

Honours and decorations announced in 1977 and January 1978 and the dates of appointment of their recipients are as follows.

ORDER OF CANADA
Appointed July 11, 1977
Companions

The Honourable George Alexander Gale, CC, QC

Roland Giroux, CC "elevated"
The Honourable Jean-Luc Pepin, PC, CC

Officers

Roger Baulu, OC
James M.R. Beveridge, OC, MD
Sidney Thomson Fisher, OC
David A. Golden, OC
Thérèse Gouin Décarie, OC
G. Sydney Halter, OC, QC
Gordon Fripp Henderson, OC, QC
Robert Hamilton Hubbard, OC
Harold Elford Johns, OC
Sol Kanee, OC
Paul Lacoste, OC
Charles Philippe Leblond, OC, MD
Charles B. Lynch, OC
H. Ian Macdonald, OC
Fernand Nault, OC
P.K. Page, OC
M. Vera Peters, OC "elevated", MD
Joseph A. Rouleau, OC
Robert Bruce Salter, OC, MD
G. Hamilton Southam, OC
Sam Steinberg, OC
Harold M. Wright, OC

Members

C.J. Sylvanus Apps, CM
Elsje Armstrong, CM
Iphigenie Arsenault, CM
Pitseolak Ashoona, CM
Sister Marcelle Boucher, CM
Eleanor Boyce, CM
Lawrence Brown, CM
Robert Cauchon, CM
Hugh Clifford Chadderton, CM
Florence Cottee, CM
William H. Cranston, CM
Glenn W. Drinkwater, CM, CD
Ovila Duval, CM (deceased: October 15, 1977)
Colonel Frederick H. Ellis, CM, CD
Isidore-M. Gauthier, CM, MD
Alexandrina Goolden, CM
J. King Gordon, CM
Mildred Gottfriedson, CM
Marion Margaret Graham, CM
David Georges Greyeyes, CM
Murray B. Koffler, CM
Muriel Kovitz, CM
Gérard Lamarche, CM
Kresmir Louis Lukanovich, CM
Francis Alexander L. Mathewson, CM, MD
George Robert McLaughlin, CM
Alfred W.H. Needler, OBE, CM
Elizabeth Reid, CM
Judith Richard, CM
Gérard Roussel, CM
Claude St-Jean, CM
Anna Wilson Sharpe, CM, MD
David Macdonald Stewart, CM
Arthur H. Sweet, CM, MD
Edith Tufts, CM
Cairine R.M. Wilson, CM
Peter Bowah Wong, CM

Appointed January 11, 1978

Companions

General Jacques A. Dextraze, CC, CBE, CMM, DSO, CD
John Robert Evans, CC, MD

Officers

Paul-E. Auger, OC
Alfred Goldsworthy Bailey, OC
Lloyd Barber, OC
Henri Bergeron, OC
Harry J. Boyle, OC
S.D. Clark, OC
Robert Clark Dickson, OC, OBE, CD, MD
G. Campbell Eaton, OC, MC, CD
Nicholas Goldschmidt, OC
Jacques Hébert, OC
Bobby Hull, OC
Marsh Jeanneret, OC
Walter H. Johns, OC
Larkin Kerwin, OC
Roy K. Kiyooka, OC
Jean P.W. Ostiguy, OC, DSO
Sylvia Ostry, OC
Robert Prévost, OC "elevated"
Sol Simon Reisman, OC
Keith Spicer, OC
W. Garfield Weston, OC

Members

Louise Cimon Annett, CM
Anthony G. Anselmo, CM
David B. Archer, CM
Edmund Charles Bovey, CM
John W.D. Broughton, CM
June Callwood, CM
Alfredo F.M. Campo, CM
Fulgence Charpentier, MBE, CM
Samuel N. Cohen, CM
Maxwell Cummings, CM
Jeanne Cypihot, CM
Lieutenant-Colonel Ian Douglas, CM, CD, QC
James B. Driscoll, CM
Aida Flemming, CM
Edith Fulton Fowke, CM
Dorothy Maquabeak Francis, CM
Marion Fulton, CM
Captain Georges Edouard Gaudreau, CM
Robert Gauthier, CM
Reverend Father André-M. Guillemette, CM, CD
Guy Henson, CM
Michael John Kindrachuk, CM
Paul Lacoursière, CM
Albert Landry, CM
Leonard Miller, CM, MD
Edwin Mirvish, CM
Pipe Major Malcolm Nicholson, CM
Ernie Richardson, CM
Aldoria Robichaud, CM, MD
Jeanne-Marguerite Saint-Pierre, CM
Joseph Harvey Shoctor, CM, QC
Captain Charles Szathmary de Kovend, CM
Myrtle E. Tincombe, CM
Joseph W. Tomecko, CM

Guy Toupin, CM
Catharine Robb Whyte, CM
Milton Donovan Williams, CM, MD
Walter R. Wood, CM
Ben Wosk, CM

ORDER OF MILITARY MERIT
Appointed June 20, 1977
Commanders
Vice-Admiral Andrew Lawrence Collier, CMM, DSC, CD
Brigadier-General Joseph Lucien Roger Lacroix, CMM, CD

Officers
Major Nadine Enid Demitre, OMM, CD
Major Robert Harold Easby, OMM, CD
Lieutenant-Colonel James Burbeck Fay, OMM, CD
Major Georges Joseph Roméo Gauthier, OMM, CD
Major John Richard Hosang, OMM, CD
Major Joseph Eric Montambault, OMM, CD
Captain (N) Christopher Gratrix Pratt, OMM, CD
Colonel Charles Eugène Savard, OMM, CD
Lieutenant-Colonel Peter James Taggart, OMM, CD
Lieutenant-Colonel Ronald Thacker, OMM, CD

Members
Master Corporal Andrew Crouse Allen, MMM, CD
Sergeant Alexander David Armstrong, MMM, CD
Petty Officer First Class Donald Brown, MMM, CD
Master Warrant Officer Sam Brown, MMM, CD
Warrant Officer Frank Edward Clarke, MMM, CD
Sergeant Harvey Earl Copeland, MMM, CD
Master Corporal Ian Barry Culbertson, MMM
Chief Warrant Officer Frank Albert Darton, MMM, CD
Chief Warrant Officer Guy Joseph Arthur Denis, MMM, CD
Chief Warrant Officer Enrico Dilio, MMM, CD
Master Corporal Clifford Martin Doak, MMM, CD
Chief Warrant Officer Eric Gordon Edgar, MMM, CD
Sergeant Gerald Clement Ekstrom, MMM, CD
Sergeant Robert Allan Fralic, MMM, CD
Sergeant John Clifford Gallant, MMM, CD
Captain Nelson Borden Gesner, MMM, CD
Captain Robert James Goldie, MMM, CD
Master Corporal Joseph Edouard Lagacé, MMM, CD
Master Warrant Officer André Laurent Levesque, MMM, CD
Master Warrant Officer Shun Hunk Lowe, MMM, CD
Master Warrant Officer Gérard Mayer, MMM, CD
Sergeant Roderick Duncan McLeod, MMM, CD

Petty Officer First Class Richard Floyd Miller, MMM, CD
Chief Warrant Officer Alexander Garfield Morran, MMM, CD
Chief Warrant Officer Walter James Neve, MMM, CD
Master Warrant Officer Maurice Parent, MMM, CD
Sergeant William Henry Perkins, MMM, CD
Master Warrant Officer Robert John Pharoah, MMM, CD
Captain Duncan James Phillips, MMM, CD
Chief Warrant Officer John Edward George Reichel, MMM, CD
Master Warrant Officer Albert Jacques Russell, MMM, CD
Captain Joseph Allan Eugène Sauvé, MMM, CD
Chief Warrant Officer William Donald Schussler, MMM, CD
Chief Warrant Officer John Thomas Speirs, MMM, CD
Lieutenant-Commander Frank Frederick Stockwell, MMM
Sergeant Robert Vézina, MMM, CD
Chief Warrant Officer Robert Moore Wadden, MMM, CD

Appointed December 12, 1977
Commanders
Major-General Maurice Gaston Cloutier, CMM, CD
Major-General Bruce Jarvis Legge, CMM, ED, CD
Lieutenant-General Ramsay Muir Withers, CMM, CD

Officers
Colonel Willard George Ames, OMM, CD
Lieutenant-Colonel Robert Roy Cooper, OMM, CD
Lieutenant-Commander George Thomas Fraser, OMM, CD
Commander Rex Geoffrey Guy, OMM, CD
Major Jean-Louis André Langlais, OMM, CD
Major John Alexander McTavish, OMM, CD
Lieutenant-Colonel George Edward Miller, OMM, CD
Major Walter Murray, OMM, CD
Lieutenant-Commander Edward Palmer Tracy, OMM, CD
Lieutenant-Commander Lawrence Truelove, OMM, CD

Members
Chief Warrant Officer Arthur Milton Anderson, MMM, CD
Master Warrant Officer Charles Patrick Caldwell, MMM, CD
Sergeant Raymond Dean Calver, MMM, CD
Chief Warrant Officer Donald Clarke, MMM, CD
Captain Barry Emerson Crozier, MMM, CD
Chief Petty Officer First Class Ernst Ebner, MMM, CD
Corporal Oscar Gordon Estabrooks, MMM, CD
Sergeant Rudolph Faessler, MMM, CD

Chief Warrant Officer Louis Folliet, MMM, CD
Captain Jules Fortin, MMM, CD
Chief Warrant Officer Hugh Dewar Fraser,
MMM, CD
Lieutenant Silas Robert Gallant, MMM
Master Warrant Officer Gary William Handson,
MMM, CD
Warrant Officer Ernest Kenneth Homeniuk,
MMM, CD
Lieutenant (N) Duncan Ewing Leslie, MMM,
CD
Chief Warrant Officer John Ronald MacDonald,
MMM, CD
Master Warrant Officer Joseph Francis
MacIntyre, MMM, CD
Captain Olga Joyce McEvoy, MMM, CD
Captain Michael Gregory McKeown, MMM, CD
Warrant Officer John Francis Madison, MMM,
CD
Chief Warrant Officer Clermont Morin, MMM,
CD
Master Corporal Earl Octave Morrell, MMM,
CD
Chief Warrant Officer Roy Herbert Pepper,
MMM, CD
Chief Warrant Officer Valdin Roy Clare
Pomeroy, MMM, CD
Sergeant Dolly Ethel Pond, MMM, CD
Master Warrant Officer Marcel Proulx, MMM,
CD
Captain Henry Joseph Rice, MMM, CD
Warrant Officer Jean Claude Rhéaume, MMM,
CD
Warrant Officer Vernon Howard Robison,
MMM, CD
Warrant Officer Michael Joseph Simmons,
MMM, CD
Chief Warrant Officer William Frederick Switzer,
MMM, CD
Master Seaman David Stewart Tyre, MMM
Master Warrant Officer Gerald Richard Venn,
MMM, CD
Chief Warrant Officer James Edgar Walker,
MMM, CD

BRAVERY DECORATIONS
Appointed March 7, 1977
Star of Courage
Robert Lorne Bell, SC
Terrance Henry Creelman, SC (posthumous)
Kathryn Louise Panton, SC (posthumous)

Medal of Bravery
Ralph Edward Barrett, MB
Louise Boyce, MB
Luigi Demitri, MB
Howard Hatton, MB
Perry Wayne McKinnon, MB
David Wesley Meyers, MB
Wylie Simmonds, MB

Appointed May 2, 1977
Star of Courage
Palmer Ferguson, SC
Roger Joseph La Brie, SC (posthumous)

Medal of Bravery
Vincent Bastien, MB
Rhonda Dawn Kennedy, MB
Brian Robert Leslie, MB
Robert Irwin Longdo, MB
Corporal John Robert Meads, MB
Clément Adélard Moisan, MB
Judi Ann Skinner, MB
Kyle Mitchell Sutthery, MB
John Hodgson Tooley, MB
Jari Seppo Vainionpaa, MB

Appointed July 25, 1977
Star of Courage
Chris Ann Bishop, SC
Gail Flynn, SC
Patrick Harrington, SC (posthumous)

Medal of Bravery
James Wilmont Ashton, MB
Corporal Kenneth William Caswell, MB
Ronald Dundas, MB
Corporal Marvin Douglas Hewitt, MB
Sergeant Winston Churchill Hurry, MB, CD
William E. O'Neill, MB
Gary Rowe, MB

Appointed October 3, 1977
Star of Courage
Susumu Nomura, SC
Anna Gertrude Swayze, SC
Neil Albert Swayze, SC

Medal of Bravery
Constable Diane Mary Margaret Brock, MB
Ted Bruce, MB
Gordon Fuller, MB
William Gaunt, MB
Corporal Thomas Phillip Griffin, MB
David Allan MacFarlane, MB
Wayne Takasaki, MB
Ralph Weick, MB

Appointed December 5, 1977
Star of Courage
Michael Patrick Weaver Cox, SC
Captain Gary Wayne Fulton, SC (posthumous)
Kenneth W. Gilders, SC
Jacob Etzerd Greidanus, SC
Officer Cadet Robert Alan Hansen, SC
(posthumous)
Sharon Pearl Jackson, SC (posthumous)
Joseph Alphonse Gérard Maillé, SC

Medal of Bravery
Captain Lorne Odin Elwood Bakke, MB
Warrant Officer John Kurt Boehne, MB
Constable (RCMP) Richard Allan Burns, MB
Kelly Devenport Dunn, MB
Lionel Lapointe, MB
George William Maybury, MB
Glenda Mary Meshake, MB
Richard Allan Smith, MB

Diplomatic and consular representation   Appendix 5

Canada maintains diplomatic, consular or trade representation with the following countries and organizations. This list, giving the status of representatives and their mailing addresses, was updated to January 1978 by Information Services Division, Department of External Affairs.

Canadian representatives abroad

Afghanistan
Ambassador: c/o Canadian Embassy, GPO Box 1042, Islamabad, Pakistan.

Algeria
Ambassador: PO Box 225, Gare Alger.

Argentina
Ambassador: Casilla de Correo 1598, Buenos Aires.

Australia
High Commissioner: Commonwealth Ave., Canberra, ACT, 2600.

Austria
Ambassador: Luegerring 10, A-1010 Vienna.
Head of Delegation, Ambassador: The Canadian Delegation to the Mutual and Balanced Force Reduction Talks, Vienna.

Bahamas
High Commissioner: c/o Canadian High Commission, PO Box 1500, Kingston 10, Jamaica.

Bahrain
Ambassador: c/o Canadian Embassy, PO Box 1610, Tehran, Iran.

Bangladesh
High Commissioner: GPO Box 569, Dacca-2.

Barbados
High Commissioner: PO Box 404, Bridgetown.

Belgium
Ambassador: rue de Loxum 6, 1000 Brussels.

Belize
Commissioner: c/o Canadian High Commission, PO Box 1500, Kingston 10, Jamaica.

Benin, People's Republic of
Ambassador: c/o Canadian High Commission, PO Box 1639, Accra, Ghana.

Bermuda
Commissioner: c/o Canadian Consulate General, 1251 Ave. of the Americas, New York, NY, 10020, USA.

Bolivia
Ambassador: c/o Canadian Embassy, Casilla 1212, Lima, Peru.

Botswana
High Commissioner: c/o Canadian Embassy, PO Box 26006, Arcadia, Pretoria 0007, South Africa.

Brazil
Ambassador: Caixa Postal 07-0961, 70000, Brasilia DF.

Britain
High Commissioner: Canada House, Trafalgar Square, Cockspur St. SW, 1Y 5BJ London.

Bulgaria
Ambassador: c/o Canadian Embassy, Proleterskih Brigada 69, 11000 Belgrade, Yugoslavia.

Burma
Ambassador: c/o Canadian Embassy, PO Box 2090, Bangkok, Thailand.

Burundi
Ambassador: c/o Canadian Embassy, PO Box 8341, Kinshasa, Republic of Zaire.

Cameroon
Ambassador: PO Box 572, Yaoundé.

Cape Verde Islands
Ambassador: c/o Canadian Embassy, PO Box 3373, Dakar, Senegal.

Central African Empire
Ambassador: c/o Canadian Embassy, PO Box 572, Yaoundé, Cameroon.

Chad
Ambassador: c/o Canadian Embassy, PO Box 572, Yaoundé, Cameroon.

Chile
Ambassador: Casilla 427, Santiago.

China, People's Republic of
Ambassador: No. 10, San Li Tun Road, Chao Yang District, Peking.

Colombia
Ambassador: Apartado Aéreo 53531, Bogota 2.

Comores (Islands)
Ambassador (designate): c/o Canadian High Commission, PO Box 1022, Dar-es-Salaam, United Republic of Tanzania.

Congo, People's Republic of the
Ambassador: c/o Canadian Embassy, PO Box 8341, Kinshasa, Republic of Zaire.

Costa Rica
Ambassador: Apartado Postal 10303, San José.

Cuba
Ambassador: c/o PO Box 499 (HVA), Ottawa, K1N 8T7.

Cyprus
High Commissioner: c/o Canadian Embassy, PO
   Box 6410, Tel Aviv, Israel.

Czechoslovakia
Ambassador: Mickiewiczova 6, Prague 6.

Denmark
Ambassador: Prinsesse Maries Allé 2, 1908
   Copenhagen V.

Dominican Republic
Ambassador: c/o Canadian Embassy, Apartado del
   Este No. 62302, Caracas, Venezuela.

Ecuador
Ambassador: c/o Canadian Embassy, (airmail)
   Apartado Aéreo 53531, Bogota 2, Colombia;
   (surface) Apartado Nacional 696, Bogota 2,
   Colombia.

Egypt, Arab Republic of
Ambassador: Kasr el Doubara Post Office, Cairo.

El Salvador
Ambassador: c/o Canadian Embassy, Apartado
   Postal 10303, San José, Costa Rica.

Ethiopia
Ambassador: PO Box 1130, Addis Ababa.

European Communities
The European Economic Community
The European Atomic Energy Community
The European Coal and Steel Community
Head of Mission: The Mission of Canada to the
   European Communities, 5th floor, rue de
   Loxum 6, 1000 Brussels, Belgium.

Fiji
High Commissioner: c/o Canadian High
   Commission, PO Box 12-049, Wellington
   North, New Zealand.

Finland
Ambassador: P Esplanadi 25B, 00100 Helsinki 10.

France
Ambassador: 35, avenue Montaigne, 75008 Paris
   VIIIᵉ.

Gabon
Ambassador: c/o Canadian Embassy, PO Box 572,
   Yaoundé, Cameroon.

Gambia
High Commissioner: c/o Canadian Embassy, PO
   Box 3373, Dakar, Senegal.

Germany, Democratic Republic of
Ambassador: c/o Canadian Embassy, Ulica Matejki
   1/5, Warsaw 00-481, Poland.

Germany, Federal Republic of
Ambassador: Friedrich-Wilhelm-Strasse 18, 53
   Bonn.

Ghana
High Commissioner: PO Box 1639, Accra.

Greece
Ambassador: 4 Ioannou Ghennadiou St. and
   Ypsilantou, Athens 140.

Grenada
High Commissioner: c/o Canadian High
   Commission, PO Box 404, Bridgetown,
   Barbados.

Guatemala
Ambassador: PO Box 400, Guatemala, GA.

Guinea
Ambassador: c/o Canadian Embassy, PO Box
   3373, Dakar, Senegal.

Guinea – Bissau
Ambassador: c/o Canadian Embassy, PO Box
   3373, Dakar, Senegal.

Guyana
High Commissioner: PO Box 660, Georgetown.

Haiti
Ambassador: CP 826, Port-au-Prince.

Holy See
Ambassador: 7 Via della Conciliazione 4/D, 00193
   Rome, Italy.

Honduras
Ambassador: c/o Canadian Embassy, Apartado
   Postal 10303, San José, Costa Rica.

Hong Kong
Commissioner: PO Box 20264, Hennessy Road
   Post Office.

Hungary
Ambassador: Budakeszi, Ut 55/D P/8, H-1021
   Budapest.

Iceland
Ambassador: c/o Canadian Embassy, Postuttak,
   Oslo 1, Norway.

India
High Commissioner: PO Box 5207, New Delhi.

Indonesia
Ambassador: PO Box 52/JKT, Jakarta.

Iran
Ambassador: c/o Canadian Embassy, PO Box
   1610, Tehran.

Iraq
Ambassador: c/o Canadian Embassy, PO Box
   6112, Baghdad.

Ireland
Ambassador: 65 St. Stephens Green, Dublin 2.

Israel
Ambassador: PO Box 6410, Tel Aviv.

Italy
Ambassador: Via GB de Rossi 27, 00161 Rome.

Ivory Coast
Ambassador: CP 21194, Abidjan.

Jamaica
High Commissioner: PO Box 1500, Kingston 10.

Japan
Ambassador: 3-38 Akasaka 7-chome, Minato-ku,
   Tokyo 107.

Jordan
Ambassador (designate): c/o Canadian Embassy, CP 2300, Beirut, Lebanon.

Kenya
High Commissioner: PO Box 30481, Nairobi.

Korea
Ambassador: PO Box 6299, Seoul 100.

Kuwait
Ambassador: c/o Canadian Embassy, PO Box 1610, Tehran, Iran.

Laos
Ambassador: c/o Canadian Embassy, PO Box 2090, Bangkok, Thailand.

Lebanon
Ambassador (designate): c/o Canadian Embassy, CP 2300, Beirut.

Lesotho
High Commissioner: c/o Canadian Embassy, PO Box 26006, Arcadia, Pretoria 0007, South Africa.

Liberia
Ambassador: c/o Canadian High Commission, PO Box 1639, Accra, Ghana.

Libyan Arab Republic
Ambassador: c/o Canadian Embassy, Kasr el Doubara Post Office, Cairo, Arab Republic of Egypt.

Luxembourg
Ambassador: c/o Canadian Embassy, rue de Loxum 6, 1000 Brussels, Belgium.

Macao
Consul: c/o Commission for Canada, PO Box 20264, Hennessy Road Post Office, Hong Kong.

Madagascar, Democratic Republic of
Ambassador: c/o Canadian High Commission, PO Box 1022, Dar-es-Salaam, United Republic of Tanzania.

Malawi
High Commissioner: c/o Canadian High Commission, PO Box 1313, Lusaka, Zambia.

Malaysia
High Commissioner: PO Box 990, Kuala Lumpur.

Mali
Ambassador: c/o Canadian Embassy, PO Box 21194, Abidjan, Ivory Coast.

Malta
High Commissioner: c/o Canadian Embassy, Via GB de Rossi 27, 00161 Rome, Italy.

Mauritania
Ambassador: c/o Canadian Embassy, PO Box 3373, Dakar, Senegal.

Mauritius
High Commissioner: c/o Canadian High Commission, PO Box 1022, Dar-es-Salaam, United Republic of Tanzania.

Mexico
Ambassador: Melchor Ocampo 463-7, Mexico 5, DF.

Monaco
Consul General: c/o Canadian Consulate General, 24, avenue du Prado, 13006 Marseille, France.

Mongolia
Ambassador: c/o Canadian Embassy, 23 Starokonyushenny Pereulok, Moscow, USSR.

Morocco
Ambassador: CP 709, Rabat-Agdal, Maroc.

Mozambique
Ambassador: c/o Canadian High Commission, PO Box 1313, Lusaka, Zambia.

Nepal
Ambassador: c/o Canadian High Commission, PO Box 5207, New Delhi, India.

Netherlands
Ambassador: Sophialaan 7, The Hague.

New Zealand
High Commissioner: PO Box 12-049, Wellington North.

Nicaragua
Ambassador: c/o Canadian Embassy, Apartado Postal 10303, San José, Costa Rica.

Niger
Ambassador: c/o Canadian Embassy, PO Box 21194, Abidjan, Ivory Coast.

Nigeria
High Commissioner: PO Box 851, Lagos.

North Atlantic Council
Permanent Representative and Ambassador: Léopold III Blvd., 1110 Brussels, Belgium.

Norway
Ambassador: Postuttak, Oslo 1.

Oman
Ambassador: c/o Canadian Embassy, PO Box 1610, Tehran, Iran.

Organization of American States
Ambassador and Permanent Observer: 2450 Massachusetts Ave. NW, Washington, DC, 20008, USA.

Organization for Economic Co-operation and Development
Ambassador and Permanent Representative: 19, rue de Franqueville, Paris XVIe.

Pakistan
Ambassador: GPO Box 1042, Islamabad.

Panama
Ambassador: c/o Canadian Embassy, Apartado Postal 10303, San José, Costa Rica.

Papua New Guinea
High Commissioner: c/o Canadian High Commission, Commonwealth Ave., Canberra, ACT, 2600 Australia.

Paraguay
Ambassador: c/o Canadian Embassy, Casilla de
Correo 1598, Buenos Aires, Argentina.

Peru
Ambassador: Casilla 1212, Lima.

Philippines
Ambassador: PO Box 971, Commercial Centre,
Makati, Rizal, Manila.

Poland
Ambassador: Ulica Matejki 1/5, Warsaw 00-481.

Portugal
Ambassador: Rua Rosa Araujo 2, 6th floor,
Lisbon 2.

Qatar
Ambassador: c/o Canadian Embassy, PO Box
1610, Tehran, Iran.

Romania
Ambassador: PO Box 2966, Post Office No. 22,
Bucharest.

Rwanda
Ambassador: c/o Canadian Embassy, PO Box
8341, Kinshasa, Republic of Zaire.

San Marino
Consul: c/o Canadian Embassy, Via GB de Rossi
27, 00161 Rome, Italy.

Saudi Arabia
Ambassador: PO Box 5050, Jeddah.

Senegal
Ambassador: PO Box 3373, Dakar.

Seychelles
High Commissioner: c/o Canadian High
Commission, PO Box 1022, Dar-es-Salaam,
United Republic of Tanzania.

Sierra Leone
High Commissioner: c/o Canadian High
Commission, PO Box 851, Lagos, Nigeria.

Singapore
High Commissioner: PO Box 845, Singapore 1.

Somali Democratic Republic
Ambassador: c/o Canadian High Commission, PO
Box 1022, Dar-es-Salaam, United Republic of
Tanzania.

South Africa, Republic of
Ambassador: PO Box 26006, Arcadia, Pretoria
0007.

Spain
Ambassador: Apartado 587, Madrid.

Sri Lanka
High Commissioner: PO Box 1006, Colombo.

Sudan
Ambassador: c/o Canadian Embassy, Kasr el
Doubara Post Office, Cairo, Arab Republic of
Egypt.

Suriname, Republic of
Ambassador: c/o Canadian High Commission, PO
Box 660, Georgetown, Guyana.

Swaziland
High Commissioner: c/o Canadian Embassy, PO
Box 26006, Arcadia, Pretoria 0007, South
Africa.

Sweden
Ambassador: PO Box 16129, S-10323
Stockholm 16.

Switzerland
Ambassador: 88 Kirchenfeldstrasse, 3000 Berne.

Syrian Arab Republic
Ambassador (designate): c/o Canadian Embassy,
CP 2300, Beirut, Lebanon.

Tanzania, United Republic of
High Commissioner: PO Box 1022, Dar-es-Salaam.

Thailand
Ambassador: PO Box 2090, Bangkok.

Togo
Ambassador: c/o Canadian High Commission, PO
Box 1639, Accra, Ghana.

Tonga
High Commissioner: c/o Canadian High
Commission, PO Box 12-049, Wellington
North, New Zealand.

Trinidad and Tobago
High Commissioner: PO Box 1246, Port of Spain.

Tunisia
Ambassador: CP 31, Belvédère, Tunis.

Turkey
Ambassador: Nenehatun Caddesi No. 75,
Gaziosmanpasa, Ankara.

Uganda
High Commissioner: c/o Canadian High
Commission, PO Box 30481, Nairobi, Kenya.

Union of Soviet Socialist Republics
Ambassador: 23 Starokonyushenny Pereulok,
Moscow.

United Arab Emirates
Ambassador: c/o Canadian Embassy, PO Box
1610, Tehran, Iran.

United Nations
Ambassador and Permanent Representative: The
Permanent Mission of Canada to the United
Nations, 866 United Nations Plaza, Suite 250,
New York, NY, 10017.
Ambassador and Permanent Representative:
Permanent Mission of Canada to the Office of
the United Nations at Geneva, and to the
Conference of the Committee on
Disarmament, 10A, avenue de Budé, 1202
Geneva.
Ambassador and Permanent Representative:
Permanent Mission of Canada to the
Secretariat of the General Agreement on
Tariffs and Trade, 10A, avenue de Budé, 1202
Geneva.
Note: The Permanent Mission in Geneva is
accredited to the United Nations Specialized
Agencies having their headquarters in Geneva:

International Labour Organization (ILO);
International Telecommunications Union
(ITU); World Health Organization (WHO);
World Meteorological Organization (WMO);
World Intellectual Property Organization
(WIPO).

Permanent Representative: Permanent Mission of
Canada to the United Nations Environment
Program, Comcraft House, Hailé Sélassie
Avenue, PO Box 30481, Nairobi.

Ambassador and Permanent Delegate: Permanent
Delegation of Canada to the United Nations
Educational, Scientific and Cultural
Organization, 1, rue Miollis, Paris XVe, CP
3.07, Paris VIIe.

Permanent Representative: Permanent Mission of
Canada to the Food and Agriculture
Organization, Via GB de Rossi 27, 00161
Rome.

Permanent Representative: Permanent Mission of
Canada to the United Nations Industrial
Development Organization, Luegerring 10, A-
1010 Vienna.

Permanent Representative: Permanent Mission of
Canada to the International Atomic Energy
Agency, Luegerring 10, A-1010 Vienna.

Note: Canada is also a member of the following UN
Specialized Agencies to which there are no
accredited permanent representatives:
Universal Postal Union (UPU), Berne; Inter-
governmental Maritime Consultative
Organization (IMCO), London; International
Bank for Reconstruction and Development
(IBRD), Washington; International Finance
Corporation (IFC), Washington; International
Development Agency (IDA), Washington;
International Monetary Fund (IMF),
Washington.

United States of America
Ambassador: 1746 Massachusetts Ave. NW,
Washington, DC, 20036.

Upper Volta
Ambassador: c/o Canadian Embassy, CP 21194,
Abidjan, Ivory Coast.

Uruguay
Ambassador: c/o Canadian Embassy, Casilla de
Correo 1598, Buenos Aires, Argentina.

Venezuela
Ambassador: Apartado del Este No. 62302,
Caracas.

Vietnam, Socialist Republic of
Ambassador: c/o Canadian Embassy, San Li Tun
No. 16, Peking, People's Republic of China.

West Indies Associated States and Montserrat
Commissioner: c/o Canadian High Commission,
PO Box 404, Bridgetown, Barbados.

Western Samoa
High Commissioner: c/o Canadian High
Commission, PO Box 12-049, Wellington
North, New Zealand.

Yemen Arab Republic
Ambassador: c/o Canadian Embassy, PO Box
5050, Jeddah, Saudi Arabia.

Yemen, People's Democratic Republic of
Ambassador: c/o Canadian Embassy, PO Box
5050, Jeddah, Saudi Arabia.

Yugoslavia
Ambassador: Proleterskih Brigada 69, 11000
Belgrade.

Zaire, Republic of
Ambassador: PO Box 8341, Kinshasa.

Zambia
High Commissioner: PO Box 1313, Lusaka.

Representatives of foreign countries in Canada

Afghanistan
Ambassador: 2341 Wyoming Ave. NW,
Washington, DC, 20008, USA.

Algeria
Ambassador: 435 Daly Ave., Ottawa, K1N 6H3.

Argentina
Ambassador: 130 Slater St., 6th floor, Ottawa,
K1P 5H6.

Australia
High Commissioner: 130 Slater St., 13th floor,
Ottawa, K1P 5H6.

Austria
Ambassador: 445 Wilbrod St., Ottawa, K1N 6M7.

Bahamas
High Commissioner: c/o Embassy of the Bahamas,
600 New Hampshire Ave. NW, Suite 865,
Washington, DC, 20037, USA.

Bangladesh
High Commissioner: 85 Range Rd., Suite 402,
Ottawa, K1N 8J6.

Barbados
High Commissioner: 151 Slater St., Suite 700,
Ottawa, K1P 5H3.

Belgium
Ambassador: 85 Range Rd., Suites 601-604,
Ottawa, K1N 8J6.

Benin, People's Republic of
Ambassador: 58 Glebe Ave., Ottawa, K1S 2C3.

Bolivia
Ambassador: 350 Sparks St., Suite 308, Ottawa,
K1R 7S8.

Botswana
High Commissioner: c/o Embassy of the Republic
of Botswana, Van Ness Centre, 4301
Connecticut Ave. NW, Suite 404,
Washington, DC, 20008, USA.

Brazil
Ambassador: 255 Albert St., Suite 900, Ottawa,
K1P 6A9.

Britain
High Commissioner: 80 Elgin St., Ottawa, K1P 5K7.

Bulgaria
Ambassador: 325 Stewart St., Ottawa, K1N 6K5.

Burma
Ambassador: 116 Albert St., Royal Trust Bldg., 2nd floor, Ottawa, K1P 5G3.

Burundi
Ambassador: 2717 Connecticut Ave. NW, Washington, DC, 20008, USA.

Cameroon
Ambassador: 170 Clemow Ave., Ottawa, K1S 2B4.

Central African Empire
Ambassador: 381 Wilbrod Ave., Ottawa, K1N 6M6.

Chad
Ambassador: 2600 Virginia Ave., Suite 410, Washington, DC, 20037, USA.

Chile
Ambassador: 56 Sparks St., Suite 414, Ottawa, K1P 5A9.

China, People's Republic of
Ambassador: 411-415 St. Andrew St., Ottawa, K1N 5H3.

Colombia
Ambassador: 140 Wellington St., Suite 112, Ottawa, K1P 5A2.

Congo, People's Republic of the
Ambassador: c/o Permanent Mission of the Congo to the United Nations, 14 E 65th St., New York, NY, 10021, USA.

Costa Rica
Ambassador: 2112 S St. NW, Washington, DC, 20008, USA.

Cuba
Ambassador: 388 Main St., Ottawa, K1S 1E3.

Cyprus
High Commissioner: c/o Embassy of Cyprus, 2211 R St. NW, Washington, DC, 20008, USA.

Czechoslovakia
Ambassador: 171 Clemow Ave., Ottawa, K1S 2B3.

Denmark
Ambassador: 85 Range Rd., Suite 702, Ottawa, K1N 8J6.

Ecuador
Ambassador: 2535-15th St. NW, Washington, DC, 20009, USA.

Egypt, Arab Republic of
Ambassador: 454 Laurier Ave. E, Ottawa, K1N 6R3.

El Salvador
Ambassador: The Driveway Place, 350 Driveway, Suite 101, Ottawa, K1S 3N1.

Fiji
High Commissioner: c/o Fiji Mission to the United Nations, 1 United Nations Plaza, 26th floor, New York, NY, 10017, USA.

Finland
Ambassador: 222 Somerset St. W, Suite 401, Ottawa, K2P 2G3.

France
Ambassador: 42 Sussex Dr., Ottawa, K1M 2C9.

Gabon
Ambassador: 4 Range Rd., Ottawa, K1N 8J5.

Germany, Democratic Republic of
Ambassador: 1717 Massachusetts Ave. NW, Washington, DC, 20036, USA.

Germany, Federal Republic of
Ambassador: 1 Waverley St., Ottawa, K2P 0T8.

Ghana
High Commissioner: 85 Range Rd., Suite 810, Ottawa, K1N 8J6.

Greece
Ambassador: 76-80 MacLaren St., Ottawa, K1M 0G3.

Grenada
High Commissioner: The Driveway Place, 350 Driveway, Suite 605, Ottawa, K1S 3N1.

Guatemala
Ambassador: The Driveway Place, 350 Driveway, Suite 105, Ottawa, K1S 3N1.

Guinea
Ambassador: c/o Embassy of the Republic of Guinea, 2112 Leroy Place NW, Washington, DC, 20008, USA.

Guyana
High Commissioner: Burnside Bldg., 151 Slater St., Suite 309, Ottawa, K1P 5H3.

Haiti
Ambassador: 150 Driveway, Suite 111, Ottawa, K2P 1C7.

Holy See
Pro-Nuncio: Apostolic Nunciature, 724 Manor Ave., Rockcliffe Park, K1M 0E3.

Honduras
Ambassador: 350 Sparks St., Suite 403, Ottawa, K1R 7S8.

Hungary
Ambassador: 7 Delaware Ave., Ottawa, K2P 0Z2.

Iceland
Ambassador: c/o Embassy of Iceland, 2022 Connecticut Ave. NW, Washington, DC, 20008, USA.

India
High Commissioner: 200 MacLaren St., Ottawa, K2P 0L6.

Indonesia
Ambassador: 255 Albert St., Suite 1010, Kent Square Building "C", Ottawa, K1P 6A9.

Iran
Ambassador: 85 Range Rd., Suites 307-308,
Ottawa, K1N 8J6.

Iraq
Ambassador: 377 Stewart St., Ottawa, K1N 6K9.

Ireland
Ambassador: 170 Metcalfe St., Ottawa, K2P 1P3.

Israel
Ambassador: 410 Laurier Ave. W, Suite 601,
Ottawa, K1R 5C4.

Italy
Ambassador: 170 Laurier Ave. W, Ottawa,
K1P 5V5.

Ivory Coast
Ambassador: 9 Marlborough Ave., Ottawa,
K1N 8E6.

Jamaica
High Commissioner: 85 Range Rd., Suites 203-204,
Ottawa, K1N 8J6.

Japan
Ambassador: 75 Albert St., Suite 1005, Ottawa,
K1P 5E7.

Jordan
Ambassador: 100 Bronson Ave., Suite 701, Ottawa,
K1R 6G8.

Kenya
High Commissioner: c/o Permanent Mission of
Kenya to the United Nations, 866 United
Nations Plaza, Room 486, New York, NY,
10017, USA.

Korea
Ambassador: 151 Slater St., Suite 608, Ottawa,
K1P 5H3.

Kuwait
Ambassador: c/o Embassy of Kuwait, 2940 Tilden
St. NW, Washington, DC, 20008, USA.

Laos
Ambassador: c/o Embassy of People's Democratic
Republic of Laos, 2222 S St. NW, Washington,
DC, 20008, USA.

Lebanon
Ambassador: 640 Lyon St., Ottawa, K1S 3Z5.

Lesotho
High Commissioner: 350 Sparks St., Suite 503,
Ottawa, K1R 5A1.

Liberia
Ambassador: c/o Embassy of Liberia, 5201-16th St.
NW, Washington, DC, 20011, USA.

Libyan Arab Republic
Ambassador: c/o Permanent Mission of the Libyan
Arab Republic to the United Nations, 866
United Nations Plaza, New York, NY, 10017,
USA.

Luxembourg
Ambassador: c/o Embassy of Luxembourg, 2210
Massachusetts Ave. NW, Washington, DC,
20008, USA.

Madagascar, Democratic Republic of
Ambassador: c/o Permanent Mission of the
Malagasy Republic to the United Nations, 801
Second Ave., Suite 404, New York, NY,
10017, USA.

Malawi
High Commissioner: c/o Permanent Mission of the
Republic of Malawi to the United Nations, 777
Third Ave., New York, NY, 10017, USA.

Malaysia
High Commissioner: 60 Boteler St., Ottawa,
K1N 8Y7.

Mali
Ambassador: c/o Permanent Mission of the
Republic of Mali to the United Nations, 111 E
69th St., New York, NY, 10021, USA.

Malta
High Commissioner: c/o Embassy of Malta, rue
Jules Lejaune, 44-1060 Brussels, Belgium.

Mauritania
Ambassador: c/o Permanent Mission of the Islamic
Republic of Mauritania to the United Nations,
600 Third Ave., 37th floor, New York, NY,
10016, USA.

Mauritius
High Commissioner: c/o Embassy of Mauritius,
Van Ness Centre, 4301 Connecticut Ave. NW,
Suite 134, Washington, DC, 20008, USA.

Mexico
Ambassador: 130 Albert St., Suite 206, Ottawa,
K1P 5G4.

Mongolia
Ambassador: 7 Kensington Court, London, W8
5DL, England.

Morocco
Ambassador: 38 Range Rd., Ottawa, K1N 8J4.

Nepal
Ambassador: c/o Embassy of Nepal, 2131 Leroy
Place NW, Washington, DC, 20008, USA.

Netherlands
Ambassador: 275 Slater St., Ottawa, K1P 5H9.

New Zealand
High Commissioner: 77 Metcalfe St., Suite 804,
Ottawa, K1P 5L6.

Nicaragua
Ambassador: c/o Embassy of Nicaragua, 1627 New
Hampshire Ave. NW, Washington, DC,
20009, USA.

Niger
Ambassador: 190 Lisgar St., Ottawa, K2P 0C4.

Nigeria
High Commissioner: 320 Queen St., Suite 2000,
Place de Ville, Tower A, Ottawa, K1R 5A3.

Norway
Ambassador: 140 Wellington St., Suite 700,
Victoria Bldg., Ottawa, K1P 5A2.

Oman
Ambassador: c/o Embassy of Oman, 2342
    Massachusetts Ave. NW, Washington, DC,
    20008, USA.

Pakistan
Ambassador: 170 Metcalfe St., Ottawa, K2P 1P3.

Panama
Ambassador: c/o Embassy of Panama, 2862 McGill
    Terrace NW, Washington, DC, 20008, USA.

Papua New Guinea
High Commissioner: c/o Permanent Mission of
    Papua New Guinea to the United Nations, 801
    Second Ave., New York, NY, 10017, USA.

Paraguay
Ambassador: c/o Permanent Mission of Paraguay
    to the OAS, 2400 Massachusetts Ave. NW,
    Washington, DC, 20008, USA.

Peru
Ambassador: 539 Island Park Dr., Ottawa,
    K1Y 0B6.

Philippines
Ambassador: 130 Albert St., Suite 607, Ottawa,
    K1P 5G4.

Poland
Ambassador: 443 Daly Ave., Ottawa, K1N 6H3.

Portugal
Ambassador: 645 Island Park Dr., Ottawa,
    K1Y 0C2.

Qatar
Ambassador: c/o Permanent Mission of Qatar to
    the United Nations, 747 Third Ave., 22nd
    floor, New York, NY, 10017, USA.

Romania
Ambassador: 473-475 Wilbrod St., Ottawa,
    K1N 6N1.

Rwanda
Ambassador: 130 Albert St., Suite 1203, Ottawa,
    K1P 5G4.

Saudi Arabia
Ambassador: 99 Bank St., Suite 901, Ottawa,
    K1P 6B9.

Senegal
Ambassador: 57 Marlborough Ave., Ottawa,
    K1N 8E8.

Sierra Leone
High Commissioner: 1701-19th St. NW,
    Washington, DC, 20009, USA.

Singapore
High Commissioner: c/o Permanent Mission of
    Singapore to the United Nations, 1 United
    Nations Plaza, 26th floor, New York, NY,
    10017, USA.

Somali Democratic Republic
Ambassador: c/o Embassy of the Somali
    Democratic Republic, 600 New Hampshire
    Ave. NW, Suite 710, Washington, DC, 20037,
    USA.

South Africa
Ambassador: 15 Sussex Dr., Ottawa, K1M 1M8.

Spain
Ambassador: 350 Sparks St., Suite 802, Ottawa,
    K1R 5A1.

Sri Lanka
High Commissioner: 85 Range Rd., Suites 102-104,
    Ottawa, K1N 8J6.

Sudan
Ambassador: c/o Embassy of the Democratic
    Republic of the Sudan, 600 New Hampshire
    Ave. NW, Suite 400, Washington, DC, 20037,
    USA.

Swaziland
High Commissioner: c/o Embassy of the Kingdom
    of Swaziland, Van Ness Centre, 4301
    Connecticut Ave. NW, Suite 441,
    Washington, DC, 20008, USA.

Sweden
Ambassador: 140 Wellington St., Suite 604,
    Ottawa, K1P 5A2.

Switzerland
Ambassador: 5 Marlborough Ave., Ottawa,
    K1N 8E6.

Syria
Ambassador: c/o Permanent Mission of the Syrian
    Arab Republic to the United Nations, 150 E
    58th St., Suite 1500, New York, NY, 10022,
    USA.

Tanzania, United Republic of
High Commissioner: 50 Range Rd., Ottawa,
    K1N 8J4.

Thailand
Ambassador: 85 Range Rd., Suite 704, Ottawa,
    K1N 8J6.

Togo
Ambassador: 220 Laurier Ave. W, Ottawa,
    K1N 6P2.

Trinidad and Tobago
High Commissioner: 75 Albert St., Suite 508,
    Ottawa, K1P 5R5.

Tunisia
Ambassador: 515 O'Connor St., Ottawa, K1S 3P8.

Turkey
Ambassador: 197 Wurtemburg St., Ottawa,
    K1N 8L9.

Uganda
High Commissioner: 170 Laurier Ave. W,
    Suite 601, Ottawa, K1P 5V5.

Union of Soviet Socialist Republics
Ambassador: 285 Charlotte St., Ottawa, K1N 8L5.

United Arab Emirates
Ambassador: 747 Third Ave., New York, NY,
    10017, USA.

United States of America
Ambassador: 100 Wellington St., Ottawa, K1P 5T1.

Upper Volta
Ambassador: 48 Range Rd., Ottawa, K1N 8J4.

Uruguay
Ambassador: c/o Embassy of Uruguay, 1918 F St. NW, Washington, DC, 20006, USA.

Venezuela
Ambassador: 320 Queen St., Suite 2220, Place de Ville, Tower A, Ottawa, K1R 5A3.

Vietnam, Socialist Republic of
Ambassador: 290 Clemow Ave., Ottawa, K1S 2B8.

Western Samoa
High Commissioner: 300 E 44th St. and Second Ave., 3rd floor, New York, NY, 10017, USA.

Yemen Arab Republic
Ambassador: Watergate Six Hundred, 600 New Hampshire Ave. NW, Suite 860, Washington, DC, 20037, USA.

Yemen, People's Democratic Republic of
Ambassador: c/o Permanent Mission of the People's Democratic Republic of Yemen, 413 E 51st St., New York, NY, 10022, USA.

Yugoslavia
Ambassador: 17 Blackburn Ave., Ottawa, K1N 8A2.

Zaire, Republic of
Ambassador: 18 Range Rd., Ottawa, K1N 8J3.

Zambia
High Commissioner: 130 Albert St., Suite 1610, Ottawa, K1P 5G4.

Delegation of the Commission of the European Communities
Head of Delegation: 350 Sparks St., Suite 1110, Ottawa, K1R 7S8.

This list of books about Canada, compiled by the National Library of Canada, is a selective bibliography of works published during 1975 and 1976 in the social sciences and humanities. It is intended to assist all readers interested in Canadian materials. The publications are arranged alphabetically by author or title under the following subject headings: general reference works; biography; fine arts and performing arts; linguistics and literature; country and people; sports; general history; regional history; economics; government and politics, law; sociology; education; religion; environment; science and technology. Titles are listed in the language in which they are published. This listing does not replace the national bibliography. For additional works, the reader may consult the monthly and annual compilations of *Canadiana*.

General reference works

Alberta. Task Force on Urbanization and the Future. *Index of urban and regional studies, province of Alberta, 1950-1973.* Edmonton: Task Force on Urbanization and the Future, 1973-1975. (Collection Issues; 1-7)

Amtmann, Bernard. *Early Canadian children's books, 1763-1840: a bibliographical investigation into the nature and extent of early Canadian children's books and books for young people/Livres de l'enfance et livres de la jeunesse au Canada, 1763-1840: étude bibliographique.* Montréal: B. Amtmann, 1976. xv, 150 p.

Atlas des oiseaux de mer de l'est du Canada. Par R.G. Brown et al. Ottawa: Service canadien de la faune, Service de la gestion de l'environnement, Ministère de l'environnement, 1975. 220 p. (Comprend un résumé en anglais)

Atlas of eastern Canadian seabirds. By R.G. Brown, et al. Ottawa: Canadian Wildlife Service, Environmental Management Service, Dept. of the Environment, 1975. 220 p. (Contains a French summary)

Ball, John L. and Richard Plant. *A bibliography of Canadian theatre history, 1583-1975.* Toronto: Playwrights Co-op, 1976. 160 p.

Bibliographie des chroniques de langage publiées dans la presse au Canada. v.1- . Sous la direction d'André Clas. Montréal: Observatoire du français moderne et contemporain, Dép. de linguistique et philologie, Université de Montréal, 1975- . (Matériaux pour l'étude du français au Canada)

Bibliothèque nationale du Québec. *Bibliographie du Québec: index 1968-1973.* Montréal: Bibliothèque nationale du Québec, Ministère des affaires culturelles, 1975. 2 v.

The book trade in Canada. Directory/Industrie du livre au Canada. Annuaire. v.- . Toronto: Ampersand Pub. Services, 1976- .

Canadian book review annual 1975. Edited by Dean Tudor, Nancy Tudor and Linda Biesenthal. Toronto: P. Martin, 1976. viii, 304 p.

Canadian business & economics: a guide to sources of information/Sources d'informations économiques et commerciales canadiennes. Edited by Barbara E. Brown. Ottawa: Canadian Library Association, 1976. xviii, 636 p.

Canadian essays and collections index, 1971-72. Edited by Joyce Sowby, et al. Ottawa: Canadian Library Association, 1976. 219 p. (Continued by *Canadian essay and literature index*)

Canadian serials directory/Répertoire des publications sériées canadiennes, 1976. Edited by Martha Pluscauskas. Toronto: University of Toronto Press, 1976. xii, 534 p.

Cinémathèque québécoise. Musée du Québec. *Canadian film production index/Index de la production cinématographique canadienne, 1976.* Montréal: Cinémathèque québécoise, 1976. 97 p.

Colombo, John Robert. *Colombo's Canadian references.* Toronto: Oxford University Press, 1976. viii, 576 p.

Fee, Margery and Ruth Cawker. *Canadian fiction: an annotated bibliography.* Toronto: P. Martin, 1976. xiii, 170 p.

Flitton, Marilyn G., comp. *An index to the Canadian monthly and national review and to Rose-Belford's Canadian monthly and national review, 1872-1882.* Toronto: Bibliographical Society of Canada, 1976. xxiv, 151 p.

Hamel, Réginald, John Hare et Paul Wyczynski. *Dictionnaire pratique des auteurs québécois.* Montréal: Fides, 1976. xxv, 723 p.

Houle, Ghislaine. *Les sports au Québec, 1879-1975: catalogue d'exposition.* Montréal: Bibliothèque nationale du Québec, Ministère des affaires culturelles, 1976. xiii, 185 p. (Bibliographies québécoises; no 4)

Index de Parti pris, 1963-1968. Sous la direction de Joseph Bonenfant. Sherbrooke, Qué.: C.E.L.E.L., Université de Sherbrooke, 1975. 116 p.

Jeux olympiques, Montréal, Québec, 1976. Comité organisateur. Centre de documentation. *Bibliographie traitant des jeux olympiques et sujets connexes.* Montréal: Comité organisateur des Jeux olympiques de 1976, Direction générale de l'administration, Centre de documentation, 1975. xxi, 509 p.

Lande, Lawrence Montague. *Canadian imprints: a checklist.* Montreal: McGill University, 1975. 62 p. (Lawrence Lande Foundation for Canadian Historical Research. Publication; no. 13)

Lande, Lawrence Montague. *Canadiana miscellanies: a checklist.* Montreal: McGill University, 1975. 68 p. (Lawrence Lande Foundation for Canadian Historical Research. Publication; no. 12)

Lauzier, Suzanne et Normand Cormier. *Les ouvrages de référence du Québec: supplément 1967-1974.* Montréal: Bibliothèque nationale du Québec, Ministère des affaires culturelles, 1975. xv, 305 p.

McDonough, Irma, ed. *Canadian books for children/Livres canadiens pour enfants.* Toronto: University of Toronto Press, 1976. x, 112 p.

Monière, Denis et André Vachet. *Les idéologies au Québec.* Montréal: Bibliothèque nationale du Québec, Ministère des affaires culturelles, 1976. 156 p.

Moyles, Robert G. *English-Canadian literature to 1900: a guide to information sources.* Detroit, Mich.: Gale Research, 1976. xi, 346 p. (Gale information guide library) (American literature, English literature, and world literatures in English; v. 6)

National Capital Commission. *A bibliography of history and heritage of the National Capital Region/Une bibliographie de l'histoire et du patrimoine de la région de la capitale nationale.* Ottawa: National Capital Commission, 1976. xv, 310 p.

National Library of Canada. Task Group on Library Service to the Handicapped. *Task group on library service to the handicapped: report presented to the National Librarian, Dr. Guy Sylvestre.* Ottawa: National Library of Canada, 1976. xxx, 206, xxx, 225 p. (Text in English and French; French text on inverted pages under title: *Groupe de travail sur le service de bibliothèque aux handicapés*)

Néologismes, canadianismes. v.1- . Sous la direction d'André Clas. Montréal: Observatoire du français moderne et contemporain, Dép. de linguistique et philologie, Université de Montréal, 1976- . (Matériaux pour l'étude du français au Canada)

Notre Dame University of Nelson. Library. *Kootenaiana: a listing of books, government publications, monographs, journals, pamphlets, etc., relating to the Kootenay area of the province of British Columbia and located at the libraries of Notre Dame University of Nelson, BC and/or Selkirk College, Castlegar, BC, up to 31 March 1976.* Edited by R.J. Welwood. Nelson, BC: Notre Dame University Library; Castlegar, BC: Selkirk College Library, 1976. 167 p.

Olivier, Réjean. *Notre polygraphe québécois: Frédéric-Alexandre Baillairgé, prêtre.* L'Assomption, Qué.: Collège de L'Assomption, Bibliothèque, 1976. 43 f.

Québec. Assemblée nationale. Bibliothèque. Service de documentation politique. *Bibliographie politique du Québec pour l'année 1973-74.* Québec: Bibliothèque de la législature, Assemblée nationale, 1975. xi, 346 p.

Québec. Conseil du statut de la femme. *Les québécoises: guide bibliographique suivi d'une filmographie.* Québec: La Documentation québécoise, Ministère des communications, Éditeur officiel du Québec, 1976. 160 p. (Collection Études et dossiers)

Québec. Ministère des affaires culturelles. Direction générale du patrimoine. *Bibliographie thématique sur les Montagnais-Naskapi.* Par Richard Dominique. Québec: Centre de documentation, Service de l'inventaire des biens culturels, 1976. 102 p. (Collection Dossier; 21)

Québec. Ministère des affaires culturelles. Direction générale du patrimoine. *Inuit du Nouveau-Québec: bibliographie.* Par Pierrette Pageau. Québec: Centre de documentation, Service de l'inventaire des biens culturels, 1975. 175 p. (Collection Dossier; 13)

Sedgwick, Dorothy. *A bibliography of English-language theatre and drama in Canada, 1800-1914.* Edmonton: Nineteenth Century Theatre Research, 1976. 48 p. (19th century theatre research: Occasional publications; no. 1)

Theberge, C.B. *Canadiana on your bookshelf: collecting Canadian books.* Toronto: J.M. Dent (Canada), 1976. 134 p.

Biography

Allison, Susan. *A pioneer gentlewoman in British Columbia: the recollections of Susan Allison.* Edited by Margaret A. Ormsby. Vancouver: University of British Columbia Press, 1976. li, 210 p.

Armstrong, Clarence Alvin. *Flora MacDonald.* Don Mills, Ont.: J.M. Dent (Canada), 1976. vi, 217 p.

Batt, Elisabeth. *Monck: Governor General, 1861-1868.* Toronto: McClelland and Stewart, 1976. 191 p.

Beattie, Jessie L. *A walk through yesterday: memoirs of Jessie L. Beattie.* Dictated to Jean T. Thomson. Toronto: McClelland and Stewart, 1976. 320 p.

Bell, George Kenneth et Henriette Major. *Un homme et sa mission: le Cardinal Léger en Afrique.* Montréal: Éditions de l'Homme, 1976. 190 p.

Bell, George Kenneth and Henriette Major. *A man and his mission: Cardinal Léger in Africa.* Scarborough, Ont.: Prentice-Hall of Canada, 1976. 190 p.

Cercle des femmes journalistes. *Vingt-cinq à la une: biographies.* Montréal: La Presse, 1976. 189 p.

Collins, Maynard. *Norman McLaren.* Ottawa: Canadian Film Institute, 1976. 119 p. (Canadian film series; 1)

Desbarats, Peter. *René: a Canadian in search of a country.* Toronto: McClelland and Stewart, 1976. 223 p.

Desbarats, Peter. *René Lévesque ou le projet inachevé.* Montréal: Fides, 1977. 249 p.

Galbraith, John S. *The little emperor: Governor Simpson of the Hudson's Bay Company.* Toronto: Macmillan of Canada, 1976. 232 p.

Helmcken, John Sebastian. *The reminiscences of Doctor John Sebastian Helmcken.* Edited by Dorothy Blakey Smith. Vancouver: University of British Columbia Press; Victoria, BC: Provincial Archives of British Columbia, 1975. xlii, 396 p.

Hodgson, Maurice. *The squire of Kootenay West: a biography of Bert Herridge.* Saanichton, BC: Hancock House, 1976. 232 p.

Hutchison, Bruce. *The far side of the street.* Toronto: Macmillan of Canada, 1976. 420 p.

Kaye, Vladimir Julian. *Dictionary of Ukrainian Canadian biography pioneer settlers of Manitoba, 1891-1900.* Toronto: Ukrainian Canadian Research Foundation, 1975, xxv, 249 p.

Klinck, Carl F. *Robert Service: a biography.* Toronto: McGraw-Hill Ryerson, 1976. 199 p.

Leclerc, Jean. *Le Marquis De Denonville: gouverneur de la Nouvelle-France, 1685-1689.* Montréal: Fides, 1976. xxii, 297 p. (Collection Fleur de lys) (Études historiques canadiennes)

Lloyd, Woodrow S. *The measure of the man: selected speeches of Woodrow Stanley Lloyd.* Edited by C.B. Koester. Saskatoon, Sask.: Western Producer Prairie Books, 1976. 129 p.

Massey, Raymond. *When I was young.* Toronto: McClelland and Stewart, 1976. 269 p.

Melhuish, Martin. *Bachman Turner Overdrive: rock is my life this is my song: the authorized biography.* Toronto: Methuen; New York: Two Continents, 1976. 178 p.

Mercer, Ruby. *The tenor of his time: Edward Johnson of the Met.* Toronto: Clarke, Irwin, 1976. xv, 336 p.

Neatby, H. Blair. *William Lyon Mackenzie King: v. 3, The prism of unity, 1932-1939.* Toronto: University of Toronto Press, 1976. x, 366 p.

Parisé, Robert. *Georges-Henri Lévesque, père de la renaissance québécoise.* Montréal: Stanké, 1976. 172 p.

Pellerin, Jean-Marie. *Maurice Richard: l'idole d'un peuple.* Montréal: Éditions de l'Homme, 1976. 517 p.

Raddall, Thomas H. *In my time: a memoir.* Toronto: McClelland and Stewart, 1976. 365 p.

Remarkable women of Newfoundland and Labrador. Presented by St. John's Local Council of Women. St. John's, Nfld.: Valhalla Press Canada, 1976. 78 p.

Riel, Louis David. *The diaries of Louis Riel.* Edited by Thomas Flanagan. Edmonton: Hurtig, 1976. 187 p.

Rousseau, Normand et Jean-Guy Chaussé. *Réal Caouette: Canada!* Montréal: Éditions Héritage, 1976. 196 p.

Sweeny, Alastair. *George-Étienne Cartier: a biography.* Toronto: McClelland and Stewart, 1976. 352 p.

Taylor, Stephen and Harold Horwood. *Beyond the road: portraits & visions of Newfoundlanders.* Toronto: Van Nostrand Reinhold, 1976. 144 p.

Fine arts and performing arts

Agnes Etherington Art Centre. *From women's eyes: women painters in Canada.* Edited by Dorothy Farr and Nathalie Luckyj. Kingston, Ont.: Agnes Etherington Art Centre, 1975. xii, 81 p.

Barber, Philip and Brian Swarbrick. *The road home: sketches of rural Canada.* Scarborough, Ont.: Prentice-Hall of Canada, 1976. unpaged

Boulizon, Guy. *Les musées du Québec.* Montréal: Fides, 1976. 2 v. (Collection Loisirs et culture)

Cameron, Christina and Jean Trudel. *The drawings of James Cockburn: a visit through Quebec's past.* Agincourt, Ont.: Gage Pub., 1976. 176 p.

Cameron, Christina et Jean Trudel. *Québec au temps de James Patterson Cockburn.* Québec: Éditions Garneau, 1976. 176 p.

Canada on stage: Canadian theatre review yearbook 1975. By Don Rubin. Downsview, Ont.: CTR Publications, 1976. 379 p.

Canadian music: a selected checklist 1950-73/La musique canadienne: une liste sélective 1950-73. Edited by Lynne Jarman. Toronto: University of Toronto Press, 1976. xiv, 170 p.

Canadian Music Centre. *Catalogue of Canadian music for orchestra/Catalogue de musique canadienne pour orchestre.* Toronto: Canadian Music Centre, 1976. unpaged

Le comique et l'humour à la radio québécoise: aperçus historiques et textes choisis, 1930-1970. Par Pierre Pagé. Montréal: La Presse, 1976. 677 p. (L'ouvrage complet comprendra 2 v.)

Cotnam, Jacques. *Le théâtre québécois, instrument de contestation sociale et politique.* Montréal: Fides, 1976. 124 p. (Études littéraires) (Cet essai devait être publié dans le cinquième volume de la collection "Archives des lettres canadiennes", intitulé: *Le théâtre canadien-français: évolution, témoignages, bibliographie,* sous la direction de Paul Wyczynski)

De Visser, John and Harold Kalman. *Pioneer churches.* Toronto: McClelland and Stewart, 1976. 192 p.

Drabinsky, Garth. *Motion pictures and the arts in Canada: the business and the law.* Toronto: McGraw-Hill Ryerson, 1976. xix, 201 p.

Falardeau, Victor. *La musique du Royal 22ᵉ Régiment: 50 ans d'histoire, 1922-1972.* Québec: Éditions Garneau, 1976. 243 p.

Forsey, William C. *The Ontario community collects: a survey of Canadian painting from 1766 to the present.* Toronto: Art Gallery of Ontario, 1976. 247 p.

Gagnon, François-Marc. *Paul-Emile Borduas.* Ottawa: Galerie nationale du Canada, 1976. 96 p. (Collection Artistes canadiens; nᵒ 3)

Gauthier, Raymonde. *Les manoirs du Québec.* Montréal: Fides; Québec: Éditeur officiel du Québec, 1976. 245 p. (Collection Loisirs et culture)

Godsell, Patricia. *Enjoying Canadian painting.* Don Mills, Ont.: General Pub., 1976. viii, 275 p.

Handling, Piers. *Canadian feature films: 1913-1969. Part 3: 1964-1969.* Ottawa: Canadian Film Institute, 1976. vi, 64 p. (Canadian filmography series; no. 10)

Harper, Russell. *William Hind.* Ottawa: Galerie nationale du Canada, 1976. 92 p. (Collection Artistes canadiens; nᵒ 2)

Hatton, Warwick and Beth Hatton. *A feast of gingerbread from our Victorian past/Pâtisserie maison de notre charmant passé.* Montreal: Tundra Books/Les livres Toundra; Plattsburgh, NY: Tundra Books of Northern New York, 1976. 96 p.

Jouvancourt, Hugues de. *Henri Masson.* Montréal: Éditions La Frégate, 1976. 147 p.

Julien, Pierre-André, Pierre Lamonde et Daniel Latouche. *L'avenir des métiers d'art au Québec: Québec 2001, phase II.* Montréal: INRS-Urbanisation, 1976. x, 283 f.

Kurelek, William. *The last of the Arctic.* Toronto: McGraw-Hill Ryerson, 1976. 94 p.

Moissan, Stéphane. *A la découverte des antiquités québécoises.* Montréal: La Presse, 1976. 238 p.

Molinari, Guido. *Guido Molinari: écrits sur l'art, 1954-1975.* Compilés et présentés par Pierre Théberge. Ottawa: Galerie nationale du Canada, 1976. 112 p. (Documents d'histoire de l'art canadien; nᵒ 2) (Comprend du texte en anglais)

Musée national de l'homme. *"Bo'jou, Neejee!": regards sur l'art indien du Canada.* Par Ted J. Brasser. Ottawa: Musée national de l'homme, 1976. 204 p.

National Museum of Man. *"Bo'jou, Neejee!": profiles of Canadian Indian art.* By Ted J. Brasser. Ottawa: National Museum of Man, 1976. 204 p.

Priamo, Carol. *Mills of Canada.* Toronto: McGraw-Hill Ryerson, 1976. 192 p.

Québec. Ministère des affaires culturelles. Direction des communications. *Premiers peintres de la Nouvelle-France.* Par François Gagnon. Québec: Direction des communications, Ministère des affaires culturelles, 1976. 2 v. (Civilisation du Québec. Série Arts et métiers; 16, 17)

Reid, Dennis. *Edwin Holgate.* Ottawa: Galerie nationale du Canada, 1976. 88 p. (Collection Artistes canadiens; nᵒ 4)

Robert, Guy. *Marc-Aurèle Fortin, l'homme et l'œuvre.* Montréal: Stanké, 1976. 300 p.

Saint-Pierre, Angéline. *Emélie Chamard, tisserande.* Québec: Éditions Garneau, 1976. 194 p.

Saint-Pierre, Angéline. *L'œuvre de Médard Bourgault.* Québec: Éditions Garneau, 1976. 141 p.

Le théâtre canadien-français: évolution, témoignages, bibliographie. Sous la direction de Paul Wyczynski. Montréal: Fides, 1976. 1005 p. (Archives des lettres canadiennes; t. 5)

United States & Canada: an illustrated guide to textile collections in United States and Canadian museums. Edited by Cecil Lubell. New York: Van Nostrand Reinhold, 1976. 336 p. (Textile collections of the world; v. 1)

Urquhart, Olive. *Bottlers and bottles, Canadian.* S.l.: s.n., 1976. iii, 243 p.

Vézina, Raymond. *Napoléon Bourassa, 1827-1916: introduction à l'étude de son art.* Montréal: Éditions Elysée, 1976. 262 p.

Linguistics and literature

Dallard, Sylvie. *L'univers poétique d'Alain Grandbois.* Sherbrooke, Qué.: Éditions Cosmos, 1975. 134 p. (Profils; 9)

East of Canada: an Atlantic anthology. Edited by Raymond Fraser, Clyde Rose and Jim Stewart. Portugal Cove, Nfld.: Breakwater Books, 1976. xiii, 239 p.

Farley, Tom. *Exiles & pioneers: two visions of Canada's future, 1825-1975.* Ottawa: Borealis Press, 1976. 302 p.

Kattan, Naïm. *Écrivains des Amériques: t.2, Le Canada anglais.* Montréal: Hurtubise HMH, 1976. 207 p. (Collection Constantes; 33)

The Leacock medal treasury: 3 decades of the best of Canadian humour. Edited by Ralph L. Curry. Toronto: Lester and Orpen, 1976. xi, 302 p.

Leonard Cohen: the artist and his critics. Edited by Michael Gnarowski. Toronto: McGraw-Hill Ryerson, 1976. v, 169 p. (Critical views on Canadian writers)

McCullagh, Joan. *Alan Crawley and contemporary verse.* Vancouver: University of British Columbia Press, 1976. xxvi, 92 p.

Major, Jean-Louis. *Anne Hébert et le miracle de la parole.* Montréal: Presses de l'Université de Montréal, 1976. 114 p. (Collection Lignes québécoises textuelles) (Prix de la revue Études françaises)

Marcotte, Gilles. *Le roman à l'imparfait: essais sur le roman québécois d'aujourd'hui.* Montréal: La Presse, 1976. 194 p. (Collection Échanges)

Northey, Margot Elizabeth. *The haunted wilderness: the gothic and grotesque in Canadian fiction.* Toronto: University of Toronto Press, 1976. 131 p.

The poetry of the Canadian people, 1720-1920: two hundred years of hard work. Edited by Brian Davis. Toronto: NC Press, 1976. 288 p.

Rabotin, Maurice. *Le vocabulaire politique et socio-ethnique à Montréal de 1839 à 1842.* Montréal: Didier, 1975. 122 p.

Royer, Jean. *Pays intimes: entretiens 1966-1976.* Montréal: Leméac, 1976. 242 p. (Collection Documents)

The Toronto book: an anthology of writings past and present. Edited by William Kilbourn. Toronto: Macmillan of Canada, 1976. 290 p.

Twelve prairie poets. Edited by Laurence Ricou. Ottawa: Oberon Press, 1976. 198 p.

Young, Alan R. *Ernest Buckler.* Toronto: McClelland and Stewart, 1976. 64 p. (Canadian writers; no. 15) (New Canadian library)

Country and people

Adachi, Ken. *The enemy that never was: a history of the Japanese Canadians.* Toronto: McClelland and Stewart, 1976. vi, 456 p.

Anderson, Grace M. and David Higgs. *A future to inherit: Portuguese communities in Canada.* Toronto: McClelland and Stewart; Ottawa: Multiculturalism Programme, Department of the Secretary of State of Canada, 1976. 202 p. (Generations: a history of Canada's peoples)

Bovay, E.H. *Le Canada et les Suisses, 1604-1974.* Fribourg, Suisse: Éditions Universitaires, 1976. xii, 334 p. (Études et recherches d'histoire contemporaine)

Breckenridge, Muriel. *The old Ontario cookbook: over 420 delicious and authentic recipes from Ontario country kitchens.* Toronto: McGraw-Hill Ryerson, 1976. 249 p.

Bruce, Jean. *The last best west.* Toronto: Fitzhenry & Whiteside; Ottawa: Multiculturalism Programme, Department of the Secretary of State of Canada, 1976. x, 176 p.

Le Canada: images d'un grand pays. Par Pierre Berton et al. Paris: Éditions Vilo, 1976. 214 p.

Canada: pictures of a great land. By Pierre Berton, et al. Agincourt, Ont.: Gage Pub., 1976. 214 p.

Dunkelman, Ben. *Dual allegiance: an autobiography.* Toronto: Macmillan of Canada, 1976. xiii, 336 p.

Folklore of Canada. Compiled by Edith Fowke. Toronto: McClelland and Stewart, 1976. 349 p.

Johnston, Basil. *Ojibway heritage.* Toronto: McClelland and Stewart, 1976. 171 p.

Laforte, Conrad. *Poétiques de la chanson traditionnelle française: ou, Classification de la chanson folklorique française.* Québec: Presses de l'Université Laval, 1976. ix, 162 p. (Les Archives de folklore; 17)

Latzer, Beth Good. *Myrtleville: a Canadian farm and family, 1837-1967.* Carbondale, Ill.: Southern Illinois University Press, 1976. xiii, 312 p.

MacLaren, Roy. *Canadians in Russia, 1918-1919.* Toronto: Macmillan of Canada, 1976. viii, 301 p.

Malouin, Paul. *Le livre du trappeur québécois.* Montréal: Éditions de l'Aurore, 1976. 215 p. (Collection Connaissance des pays québécois: Le Jardin naturel)

Markotic, Vladimir, ed. *Ethnic directory of Canada.* Calgary: Western Publishers, 1976. 119 p.

Marler, George C. *The Edward VII issue of Canada: a detailed study.* Ottawa: National Postal Museum, 1975. viii, 211 p.

Marler, George C. *L'émission Edouard VII du Canada: une étude détaillée.* Ottawa: Musée national des postes, 1975. xii, 233 p.

National Film Board of Canada. Still Photography Division. *Between friends/Entre amis.* Toronto: McClelland and Stewart, 1976. 261 p.

Radecki, Henry and Benedykt Heydenkorn. *A member of a distinguished family: the Polish group in Canada.* Toronto: McClelland and Stewart; Ottawa: Multiculturalism Programme, Department of the Secretary of State of Canada, 1976. 240 p. (Generations: a history of Canada's peoples)

The Scottish tradition in Canada. Edited by W. Stanford Reid. Toronto: McClelland and Stewart; Ottawa: Multiculturalism Programme, Department of the Secretary of State of Canada, 1976. xi, 324 p. (Generations: a history of Canada's peoples)

Séguin, Robert-Lionel. *L'injure en Nouvelle-France.* Montréal: Leméac, 1976. 250 p. (Collection Connaissance)

Sévigny, P.-André. *Les Abénaquis: habitat et migrations, 17e et 18e siècles.* Montréal: Éditions Bellarmin, 1976. 247 p. (Cahiers d'histoire des Jésuites; no 3)

Trigger, Bruce G. *The children of Aataentsic: a history of the Huron people to 1660.* Montreal: McGill-Queen's University Press, 1976. 2 v.

Sports
Devaney, John and Burt Goldblatt. *The Stanley Cup: a complete pictorial history.* Chicago; New York: Rand McNally, 1975. 285 p.

Groote, Roger de. *Sports olympiques: album officiel, Montréal, 1976/Olympic sports: official album, Montreal, 1976.* Montréal: Éditions Martel, 1976. 303 p.

Lennox, Muriel. *E.P. Taylor: a horseman and his horses.* Toronto: Burns and MacEachern, 1976. 192 p.

Montréal '76: les jeux olympiques d'été/Montreal '76: the summer Olympic Games. Par Toby Rankin et al. Montréal: ProSport, 1976. ca. 200 p. (Textes en français, en anglais, en allemand et en italien)

General history
Berger, Carl. *The writing of Canadian history: aspects of English-Canadian historical writing, 1900-1970.* Toronto: Oxford University Press, 1976. x, 300 p.

Berton, Pierre. *My country: the remarkable past.* Toronto: McClelland and Stewart, 1976. 320 p.

Evans, Allan S. and Lawrence A. Diachum. *Canada: towards tomorrow.* Toronto: McGraw-Hill Ryerson, 1976. 405 p. (Canadian studies program)

Reader's Digest scenic wonders of Canada: an illustrated guide to our natural splendours. Montreal: Reader's Digest Association (Canada); Ottawa: Canadian Automobile Association, 1976. 384 p.

Visitons le Canada: guide illustré de Sélection du Reader's Digest et de la CAA. Montréal: Sélection du Reader's Digest (Canada); Ottawa: Association des automobilistes, 1976. 436, 23 p.

Regional history
The Arctic in question. Edited by Edgar J. Dosman. Toronto: Oxford University Press, 1976. 206 p.

Broadfoot, Barry and Rudy Kovach. *The city of Vancouver.* North Vancouver: J.J. Douglas, 1976. 182 p.

Brunet, Michel. *Notre passé, le présent et nous.* Montréal: Fides, 1976. 278 p. (Bibliothèque canadienne-française. Histoire et documents)

Buxton, Bonnie and Betty Guernsey. *Great Montreal walks.* Ottawa: Waxwing Productions, 1976. 176 p.

Byers, Mary, et al. *Rural roots: pre-Confederation buildings of the York region of Ontario.* Toronto: University of Toronto Press, 1976. xvi, 248 p.

Collard, Edgar Andrew. *Montreal: the days that are no more.* Toronto: Doubleday Canada; New York: Doubleday, 1976. xiii, 316 p.

Essays on western history: in honour of Lewis Gwynne Thomas. Edited by Lewis H. Thomas. Edmonton: University of Alberta Press, 1976. xi, 217 p.

Fournier, Rodolphe. *Lieux et monuments historiques de Québec et environs.* Québec: Éditions Garneau, 1976. 339 p.

Frégault, Guy. *Chronique des années perdues.* Montréal: Leméac, 1976. 250 p.

Frontier Calgary: town, city, and region, 1875-1914. Edited by Anthony W. Rasporich and Henry C. Klassen. Calgary: University of Calgary: McClelland and Stewart West. 1975. 306 p.

A harvest yet to reap: a history of prairie women. Edited by Linda Rasmussen, et al. Toronto: Women's Press, 1975. 240 p.

Histoire du Québec. Sous la direction de Jean Hamelin. St-Hyacinthe, Qué.: Edisem; Toulouse, France: E. Privat, 1976. 538 p. (Univers de la France et des pays francophones: Série Histoire des provinces)

Historical essays on British Columbia. Edited by J. Friesen and H.K. Ralston. Toronto: McClelland and Stewart; Ottawa: Institute of Canadian Studies, Carleton University, 1976. xxvi, 293 p. (The Carleton library; no. 96)

Hudson's Bay Company. *Hudson's Bay miscellany, 1670-1870.* Edited by Glyndwr Williams. Winnipeg: Hudson's Bay Record Society, 1975. xi, 245, xviii p. (Publications of Hudson's Bay Record Society; 30)

Jackson, John N. *St. Catharines, Ontario: its early years.* Belleville, Ont.: Mika Pub., 1976. 416 p.

Macleod, Roderick C. *The NWMP and law enforcement, 1873-1905.* Toronto: University of Toronto Press, 1976. xi, 218 p.

Malenfant, Robert. *Guide de Montréal en jeans.* Paris: Éditions de Cléry; Montréal: Éditions du Jour, 1976. 144 p.

Mead, Robert Douglas. *Ultimate North: canoeing Mackenzie's great river.* Garden City, NY: Doubleday, 1976. 312 p.

Mowat, Farley. *Canada North now: the great betrayal.* Toronto: McClelland and Stewart, 1976. 191 p.

The Nova Scotia Association of Architects. *Exploring Halifax and the South shore of Nova Scotia.* Toronto: Greey de Pencier, 1976. 127 p.

Paterson, Thomas W. *British Columbia shipwrecks.* Langley, BC: Stagecoach Pub., 1976. 195 p.

Sharpe, Errol. *A people's history of Prince Edward Island.* Toronto: Steel Rail Pub., 1976. 252 p.

Trudel, Marcel. *Montréal: la formation d'une société, 1642-1663.* Montréal: Fides, 1976. xxviii, 328 p. (Collection Fleur de lys) (Études historiques canadiennes)

Tulchinsky, Gerald, ed. *To preserve & defend: essays on Kingston in the nineteenth century.* Montreal: McGill-Queen's University Press, 1976. xiv, 402 p.

The West and the nation: essays in honour of W.L. Morton. Edited by Carl Berger and Ramsay Cook. Toronto: McClelland and Stewart, 1976. 335 p.

Wilson, Stuart R. *Vancouver Island reflections.* Nanaimo, BC: Vancouver Island Real Estate Board, 1976. 96 p.

Wright, Allen A. *Prelude to bonanza: the discovery and exploration of the Yukon.* Sidney, BC: Gray's Pub., 1976. x, 321 p.

Zinck, Jack. *Shipwrecks of Nova Scotia.* v.1- . Windsor, NS: Lancelot Press, 1975- .

Economics

Babe, Robert E. *Cable television and telecommunications in Canada: an economic analysis.* East Lansing, Mich.: Division of Research, Graduate School of Business Administration, Michigan State University, 1975. xxiv, 287 p. (MSU international business and economic studies)

Bennett, James E. and Pierre M. Loewe. *Women in business: a shocking waste of human resources.* Toronto: Maclean-Hunter, 1975. 150 p. (A Financial Post book)

Beyond industrial growth. Edited by Abraham Rotstein. Toronto: University of Toronto Press, 1976. xii, 131 p. (Massey College lectures; 1974-75)

The big tough expensive job: Imperial Oil and the Canadian economy. Edited by James Laxer and Anne Martin. Erin, Ont.: Press Porcépic, 1976. xi, 256 p.

British-Canadian Symposium on Historical Geography, Queen's University, Kingston, Ont., 1975. *The settlement of Canada: origins and transfer: proceedings of the 1975 British-Canadian Symposium on Historical Geography.* Edited by Brian S. Osborne. Kingston, Ont.: Queen's University, 1976. xi, 239 p.

British Columbia Forest Products Limited. *First growth: the story of the British Columbia Forest Products Limited.* Edited by Sue Baptie. Vancouver: British Columbia Forest Products: distributed by J.J. Douglas, 1975. xi, 286 p.

Chatillon, Colette. *L'histoire de l'agriculture au Québec.* Montréal: Éditions l'Étincelle, 1976. 125 p.

Courchene, Thomas J. *Money, inflation, and the Bank of Canada: an analysis of Canadian monetary policy from 1970 to early 1975.* Montreal: C.D. Howe Research Institute, 1976. xiv, 290 p. (A special study of the C.D. Howe Research Institute)

L'Économie québécoise: histoire, développement, politiques. Sous la direction de Rodrigue Tremblay. Montréal: Presses de l'Université du Québec, 1976. xv, 493 p.

Gould, Ed. *Oil: the history of Canada's oil & gas industry.* Saanichton, BC: Hancock House, 1976. 288 p.

Green, Alan G. *Immigration and the postwar Canadian economy.* Toronto: Macmillan of Canada, 1976. xvi, 312 p.

Hatch, James E. *The Canadian mortgage market.* Toronto: Ministry of Treasury, Economics and Intergovernmental Affairs, 1975. 273 p.

Joron, Guy. *Salaire minimum annuel $1 million!: Ou, la course à la folie: essai.* Montréal: Quinze, 1976. 159 p.

Laxer, Robert M. *Canada's unions.* Toronto: James Lorimer, 1976. xvi, 341 p.

Le Borgne, Louis. *La CSN et la question nationale depuis 1960.* Montréal: Éditions A. St-Martin, 1976. vii, 208 p. (Recherches & documents)

Legget, Robert F. *Canals of Canada.* Vancouver: Douglas, David and Charles; Newton Abbot, England: David and Charles, 1976. ix, 261 p. (Canals of the world)

Lessard, Diane. *L'agriculture et le capitalisme au Québec.* Montréal: Éditions l'Étincelle, 1976. 176 p.

MacEwan, Paul. *Miners and steelworkers: labour in Cape Breton.* Toronto: Samuel Stevens Hakkert, 1976. xiii, 400 p.

Marshall, Herbert. *Canadian-American industry: a study in international investment.* Toronto: McClelland and Stewart; Ottawa: Institute of Canadian Studies, Carleton University, 1976. xiii, 360 p. (The Carleton library; no. 93)

Milette, Marc. *Les marchés d'exportation et les coopératives du Québec.* Sherbrooke, Qué.: Université de Sherbrooke, 1976. 190 p. (Dossiers sur les coopératives; 3)

Natural resource revenues: a test of federalism. Edited by Anthony Scott. Vancouver: University of British Columbia Press for the British Columbia Institute for Economic Policy Analysis, 1976. xvii, 261 p. (British Columbia Institute for Economic Policy Analysis series)

Nish, Cameron. *François-Etienne Cugnet, 1719-1751: entrepreneur et entreprises en Nouvelle-France.* Montréal: Fides, 1975. xxxi, 185 p. (Histoire économique et sociale du Canada français)

Paterson, D.G. *British direct investment in Canada, 1890-1914.* Toronto: University of Toronto Press, 1976. xii, 147 p.

La publicité québécoise: ses succès, ses techniques, ses artisans. Par Raymond Beaulieu et al. Montréal: Éditions Héritage, 1976. xiii, 302 p.

Rannie, William F. *Canadian whisky, the product and the industry.* Lincoln, Ont.: W.F. Rannie, 1976. 169 p.

Regehr, T.D. *The Canadian Northern Railway: pioneer road of the northern prairies, 1895-1918.* Toronto: Macmillan of Canada, 1976. xv, 543 p.

Roby, Yves. *Les québécois et les investissements américains, 1918-1929.* Québec: Presses de l'Université Laval, 1976. 250 p. (Les Cahiers d'histoire de l'Université Laval; 20)

White, Clinton O. *Power for a province: a history of Saskatchewan power.* Regina: Canadian Plains Research Centre, University of Regina, 1976. xii, 370 p. (Canadian Plains studies; 5)

Working in Canada. Edited by Walter Johnson. Montreal: Black Rose Books, 1975. 162 p. (Black Rose Books; no. E25)

Government and politics, law

Bernard, André. *La politique au Canada et au Québec.* Montréal: Presses de l'Université du Québec, 1976. xxiii, 516 p.

Bernard, André. *Québec: élections 1976.* Montréal: Hurtubise HMH, 1976. 173 p. (Les Cahiers du Québec; 29: Science politique)

Britain and Canada: survey of a changing relationship. Edited by Peter Lyon. London: Cass, 1976. xxix, 191 p. (Studies in Commonwealth politics and history; no. 4)

Brossard, Jacques. *L'accession à la souveraineté et le cas du Québec: conditions et modalités politico-juridiques.* Montréal: Presses de l'Université de Montréal, 1976. 800 p.

Brown, Patrick, Robert Chodos and Rae Murphy. *Winners, losers: the 1976 Tory leadership convention.* Toronto: James Lorimer, 1976. 138 p. (A Last post book)

Canada and the United States: transnational and transgovernmental relations. Edited by Annette Baker Fox, Alfred O. Hero, Jr., and Joseph S. Nye, Jr. New York: Columbia University Press, 1976. xiii, 443 p.

Canada-United States relations. Edited by H. Edward English. New York: Academy of Political Science, 1976. xii, 180 p. (Proceedings of the Academy of Political Science; v. 32, no. 2)

Creighton, Donald. *The forked road: Canada, 1939-1957.* Toronto: McClelland and Stewart, 1976. 319 p. (The Canadian centenary series)

Doyle, Arthur T. *Front benches & back rooms: a story of corruption, muckraking, raw partisanship and intrigue in New Brunswick.* Toronto: Green Tree Pub., 1976. 306 p.

Dupont, Pierre. *15 novembre 76 . . . .* Montréal: Quinze, 1976. 205 p.

Hockin, Thomas A. *Government in Canada.* Canadian ed. Toronto: McGraw-Hill Ryerson, 1976. xiii, 252 p. (McGraw-Hill Ryerson series in Canadian politics)

Holmes, John W. *Canada: a middle-aged power.* Toronto: McClelland and Stewart; Ottawa: Institute of Canadian Studies, Carleton University, 1976. viii, 293 p. (The Carleton library; no. 98)

Journalism, communication and the law. Edited by G. Stuart Adam. Scarborough, Ont.: Prentice-Hall of Canada, 1976. x, 245 p.

Kesterton, Wilfred H. *The law and the press in Canada.* Toronto: McClelland and Stewart; Ottawa: Institute of Canadian Studies, Carleton University, 1976. 242 p. (The Carleton library; no. 100)

Kornberg, Allan and William Mishler. *Influence in Parliament: Canada.* Durham, NC: Duke University Press, 1976. xxi, 403 p. (Publications of the Consortium for comparative legislative studies)

Lacasse, Jean-Paul. *Le claim en droit québécois.* Ottawa: Éditions de l'Université d'Ottawa, 1976. 254 p. (Travaux de la faculté de droit de l'Université d'Ottawa: Monographies juridiques; nº 10)

Lagrave, Jean-Paul de. *Le combat des idées au Québec-Uni, 1840-1867.* Montréal: Éditions de Lagrave, 1976. 150 p. (Collection Liberté)

Langford, John W. *Transport in transition: the reorganization of the federal transport portfolio.* Toronto: Institute of Public Administration of Canada; Montreal: McGill-Queen's University Press, 1976. xviii, 267 p. (Canadian public administration series/Collection Administration publique canadienne)

MacKinnon, Frank. *The Crown in Canada.* Calgary: Glenbow-Alberta Institute: McClelland and Stewart West, 1976. 189 p.

McRoberts, Kenneth and Dale Postgate. *Quebec: social change and political crisis.* Toronto: McClelland and Stewart, 1976. viii, 216 p. (Canada in transition series)

Matthews, Keith. *Collection and commentary on the constitutional laws of seventeen century Newfoundland.* St. John's, Nfld.: Maritime History Group, Memorial University of Newfoundland, 1975. vi, 218 p.

Murray, Vera. *Le Parti québécois: de la fondation à la prise du pouvoir.* Montréal: Hurtubise HMH, 1976. 242 p. (Les Cahiers du Québec; 28: Science politique)

Nationalism, technology and the future of Canada. Edited by Wallace Gagné. Toronto: Macmillan of Canada, 1976. 166 p.

Ogelsby, J.C.M. *Gringos from the far North: essays in the history of Canadian–Latin American relations, 1866-1968.* Toronto: Macmillan of Canada, 1976. xiv, 346 p.

Partis politiques au Québec. Sous la direction de Réjean Pelletier. Montréal: Hurtubise HMH, 1976. 299 p. (Les Cahiers du Québec; 23: Science politique)

Political corruption in Canada: cases, causes and cures. Edited by Kenneth M. Gibbons and Donald C. Rowat. Toronto: McClelland and Stewart; Ottawa: Institute of Canadian Studies, Carleton University, 1976. x, 307 p. (The Carleton library; no. 95)

Le processus électoral au Québec: les élections provinciales de 1970 et 1973. Sous la direction de Daniel Latouche, Guy Lord et Jean-Guy Vaillancourt. Montréal: Hurtubise HMH, 1976. 288 p. (Les Cahiers du Québec; 22: Science politique)

The provincial political systems: comparative essays. Edited by David J. Bellamy, Jon H. Pammett and Donald C. Rowat. Toronto: Methuen, 1976. vi, 394 p.

Roche, Douglas J. *The human side of politics.* Toronto: Clarke, Irwin, 1976. xii, 209 p.

Roy, Jean-Louis. *La marche des québécois: le temps des ruptures, 1945-1960.* Montréal: Leméac, 1976. 383 p.

Winn, Conrad and John McMenemy. *Political parties in Canada.* Toronto: McGraw-Hill Ryerson, 1976. xi, 291 p. (McGraw-Hill Ryerson series in Canadian politics)

Woodward, Calvin A. *The history of New Brunswick provincial election campaigns and platforms, 1866-1974: with primary source documents on microfiche.* Toronto: Micromedia, 1976. vi, 89 p.

Sociology

André, Anne. *Je suis une maudite sauvagesse/Eukuan nin matshimanitu innu-iskueu.* Montréal: Leméac, 1976. 238 p. (Collection Dossiers) (Texte montagnais et traduction française en regard)

Bibeau, Gilles. *Les Bérets blancs: essai d'interprétation d'un mouvement québécois marginal.* Montréal: Éditions Parti pris, 1976. 187 p. (Collection "Aspects"; n⁰ 29)

Chandler, David. *Capital punishment in Canada: a sociological study of repressive law.* Toronto: McClelland and Stewart; Ottawa: Institute of Canadian Studies, Carleton University, 1976. xxiv, 224 p. (The Carleton library; no. 94)

Clark, Samuel D. *Canadian society in historical perspective.* Toronto: McGraw-Hill Ryerson, 1976. vii, 144 p. (McGraw-Hill Ryerson series in Canadian sociology)

Cohen, Anthony P. *The management of myths: the politics of legitimation in a Newfoundland community.* St. John's, Nfld.: Institute of Social and Economic Research, Memorial University of Newfoundland, 1975. viii, 146 p. (Newfoundland social and economic studies; no. 14)

Desmarais, Richard. *Le clan des Dubois.* Montréal: Stanké, 1976. 258 p.

L'emprisonnement au Québec. Montréal: Presses de l'Université de Montréal, 1976. 245 p. (Criminologie, v. 9, n⁰s 1-2)

Greeland, Cyril. *Vision Canada: les besoins non satisfaits des aveugles du Canada.* v.- . S.l.: s.n., 1976- . (L'ouvrage complet comprendra 5 v.)

Greeland, Cyril. *Vision Canada: the unmet needs of blind Canadians.* v.- . S.l.: s.n., 1976- . (The complete work will comprise 5 v.)

Griffiths, Naomi. *Penelope's web: some perceptions of women in European and Canadian society.* Toronto: Oxford University Press, 1976. 249 p.

Hiller, Harry H. *Canadian society: a sociological analysis.* Scarborough, Ont.: Prentice-Hall of Canada, 1976. xvi, 200 p.

Introduction to Canadian society: sociological analysis. Edited by G.N. Ramu and Stuart D. Johnson. Toronto: Macmillan of Canada, 1976. vi, 530 p.

Julien, Pierre-André, Pierre Lamonde et Daniel Latouche. *Québec 2001: une société refroidie.* Sillery, Qué.: Éditions du Boréal express, 1976. 213 p.

Kelly, William H. and Nora Kelly. *Policing in Canada.* Toronto: Macmillan of Canada, 1976. viii, 704 p.

Matthews, David Ralph. *"There's no better place than here": social change in three Newfoundland communities.* Toronto: P. Martin, 1976. xi, 164 p.

Ouellet, Fernand. *Le Bas-Canada, 1791-1840: changements structuraux et crise.* Ottawa: Éditions de l'Université d'Ottawa, 1976. 541 p. (Cahiers d'histoire de l'Université d'Ottawa; n⁰ 6)

The proper sphere: woman's place in Canadian society. Edited by Ramsay Cook and Wendy Mitchinson. Toronto: Oxford University Press, 1976. 334 p.

Pross, A. Paul. *Planning and development: a case of two Nova Scotia communities.* Halifax, NS: Institute of Public Affairs, Dalhousie University, 1975. viii, 109 p.

Québec. Commission d'enquête sur le crime organisé. *CECO: rapport officiel: la lutte au crime organisé.* Montréal: Stanké; Québec: Éditeur officiel du Québec, 1976. 376 p.

Québec. Ministère des affaires culturelles. *Pour l'évolution de la politique culturelle: document de travail.* Par Jean-Paul L'Allier. Québec: Ministère des affaires culturelles, 1976. 258 p.

Robin, Martin. *The bad and the lonely: seven stories of the best — and worst — Canadian outlaws.* Toronto: James Lorimer, 1976. 221 p.

Shortt, Samuel E.D. *The search for an ideal: six Canadian intellectuals and their convictions in an age of transition, 1890-1930.* Toronto: University of Toronto Press, 1976. viii, 216 p. (Canadian university paperbooks; 170)

Sutherland, Neil. *Children in English-Canadian society: framing the twentieth-century consensus.* Toronto: University of Toronto Press, 1976. viii, 336 p.

Symposium sur les Amérindiens, Montmorency, Québec, 1974. *Les facettes de l'identité amérindienne/The patterns of Amerindian identity.* Publié sous la direction de Marc-Adélard Tremblay. Québec: Presses de l'Université Laval, 1976. xi, 316 p.

Women in the Canadian mosaic. Edited by Gwen Matheson. Toronto: P. Martin, 1976. x, 353 p.

Education

Bedford, Allan G. *The University of Winnipeg: a history of the founding colleges.* Toronto: Published for the University of Winnipeg by University of Toronto Press, 1976. xii, 479 p.

From quantitative to qualitative change in Ontario education: a festschrift for R.W.B. Jackson. Edited by Garnet McDiarmid. Toronto: Ontario Institute for Studies in Education, 1976. xv, 190 p. (Ontario Institute for Studies in Education. Symposium series; 6)

Harris, Robin S. *A history of higher education in Canada, 1663-1960.* Toronto: University of Toronto Press, 1976. xxiv, 715 p. (Studies in the history of higher education in Canada; 7)

Johnston, Charles M. *McMaster University.* v.1- . Toronto: Published for McMaster University by University of Toronto Press, 1976- .

Rowe, Frederick W. *Education and culture in Newfoundland.* Toronto: McGraw-Hill Ryerson, 1976. xiv, 225 p.

Religion

Jaenen, Cornelius J. *The role of the Church in New France.* Toronto: McGraw-Hill Ryerson, 1976. x, 182 p. (The Frontenac library; 7)

Penton, M. James. *Jehovah's Witnesses in Canada: champions of freedom of speech and worship.* Toronto: Macmillan of Canada, 1976. xi, 388 p.

Religion in Canadian society. Edited by Stewart Crysdale and Les Wheatcroft. Toronto: Macmillan of Canada, 1976. xi, 498 p.

Rousseau, Louis. *La prédication à Montréal de 1800 à 1830: approche religiologique.* Montréal: Fides, 1976. 269 p. (Collection Héritage et projet; 16)

Environment

Carver, Humphrey. *Compassionate landscape.* Toronto: University of Toronto Press, 1975. viii, 251 p.

Dansereau, Pierre. *Le cadre d'une recherche écologique interdisciplinaire.* Montréal: Presses de l'Université de Montréal, 1976. xviii, 343 p.

Land use and development/Développement et aménagement du territoire. Edited by Georges Le Pape. Montreal: Federal Publications Service, 1976. xxxiii, 451 p. in various pagings

Nelson, James G. *Man's impact on the Western Canadian landscape.* Toronto: McClelland and Stewart; Ottawa: Institute of Canadian Studies, Carleton University, 1976. xiv, 205 p. (The Carleton library; no. 90)

Victoria: physical environment and development. Edited by Harold D. Foster. Victoria, BC: Dept. of Geography, University of Victoria, 1976. xvii, 334 p. (Western geographical series; v. 12)

Science and technology

Barrett, Harry B. *The 19th-century journals & paintings of William Pope.* Toronto: M.F. Feheley, 1976. 175 p.

Blood, Donald A. *Rocky mountain wildlife.* Saanichton, BC: Hancock House, 1976. 169, 129 p.

Churcher, C.S., ed. *Athlon: essays on palaeontology in honour of Loris Shano Russell.* Toronto: Royal Ontario Museum, 1976. 286 p. (Life sciences miscellaneous publication)

Clark, Lewis J. *Wild flowers of the Pacific Northwest from Alaska to Northern California.* Sidney, BC: Gray's Pub., 1976. xi, 604 p.

Evans, Anna Margaret and C.A.V. Barker. *Century one: a history of the Ontario Veterinary Association, 1874-1974.* Guelph, Ont.: A.M. Evans: distributed by C.A.V. Barker, 1976. xi, 516 p.

Ferguson, Mary and Richard M. Saunders. *Canadian wildflowers.* Toronto: Van Nostrand Reinhold, 1976. 192 p.

Frison-Roche, Roger. *Les seigneurs de la faune canadienne.* Saint-Laurent, Qué.: Flammarion, 1976. 302 p.

Froom, Barbara. *The turtles of Canada.* Toronto: McClelland and Stewart, 1976. 120 p.

Harrington, Lyn and Richard Harrington. *Covered bridges of Central and Eastern Canada.* Toronto: McGraw-Hill Ryerson, 1976. viii, 88 p.

Hewlett, Stefani and K. Gilbey Hewlett. *Sea life of the Pacific Northwest.* Toronto: McGraw-Hill Ryerson, 1976. 176 p.

Lansdowne, James Fenwick. *Birds of the West Coast.* v.1- . Toronto: M.F. Feheley, 1976- .

MacEwan, John W. Grant. *Memory meadows: horse stories from Canada's past.* Saskatoon, Sask.: Western Producer Prairie Books, 1976. 212 p.

Mackenzie, John P.S. *The complete outdoorsman's guide to birds of Canada and eastern North America.* Toronto: McGraw-Hill Ryerson, 1976. 240 p.

Mulligan, Gerald A. *Common weeds of Canada/Les mauvaises herbes communes du Canada.* Toronto: McClelland and Stewart; Ottawa: Information Canada: Department of Agriculture, 1976. 140 p.

Pouliot, Paul. *Arbres, haies et arbustes.* Montréal: Éditions de l'Homme, 1976. 347 p.

Provencher, Paul. *Mes observations sur les mammifères.* Montréal: Éditions de l'Homme, 1976. 158 p. (Collection Sport)

Provencher, Paul. *Mes observations sur les poissons.* Montréal: Éditions de l'Homme, 1976. 115 p. (Collection Sport)

Street, David. *Horses: a working tradition.* Toronto: McGraw-Hill Ryerson, 1976. 160 p.

Veilleux, Christian et Bernard Prévost. *Les papillons du Québec.* Montréal: Éditions de l'Homme, 1976. 142 p.

Young, Elrid Gordon. *The development of biochemistry in Canada.* Toronto: University of Toronto Press, 1976. vii, 129 p.

Canada Year Book special articles

Appendix 7

(Published in previous editions)

Agriculture

Historical background of Canadian agriculture, G.S.H. Barton. 1939. pp 187-90.

The major soil zones and regions of Canada, P.C. Stobbe. 1951. pp 352-6.

The Board of Grain Commissioners, W.J. MacLeod. 1960. pp 957-8.

The Canadian Wheat Board and its role in grain marketing, C.B. Davidson. 1960. pp 958-60.

Changes in Canadian agriculture as reflected by the Census of 1961. 1963-64. pp 409-15.

Contribution of the Canada Department of Agriculture to modern agricultural science. 1966. pp 457-61.

Federal assistance in livestock improvement. 1967. pp 453-7.

The role of government in the grains industry. 1972. pp 1021-8.

Banking and finance

The Bank of Canada and its relationship to the financial system. 1937. pp 881-5.

Historical sketch of currency and banking. 1938. pp 900-6.

Wartime control under the Foreign Exchange Control Board, R.H. Tarr. 1941. pp 833-5; 1942. pp 830-3.

Commercial banking in Canada, J. Douglas Gibson. 1961. pp 1115-20.

Citizenship

Early naturalization procedure and events leading up to the Canadian Citizenship Act. 1951. pp 153-5.

Climate and meteorology

Factors which control Canadian weather, Sir Frederick Stupart. 1925. pp 36-40.

Temperature and precipitation of northern Canada, A.J. Connor. 1930. pp 41-56.

Droughts in western Canada, A.J. Connor. 1933. pp 47-59.

The climate of Canada, C.C. Boughner and M.K. Thomas. 1959. pp 23-51; 1960. pp 31-77.

The climate of the Canadian Arctic, H.A. Thompson. 1967. pp 55-74.

Communications

The democratic functioning of the press, W.A. Buchanan. 1945. pp 744-8.

History and development of the Canadian Broadcasting Corporation, Augustin Frigon. 1947. pp 737-40.

A history of Canadian journalism, 1752-(circa) 1900, W.H. Kesterton. 1957-58. pp 920-34.

A history of Canadian journalism (circa) 1900-1958, W.H. Kesterton. 1959. pp 883-902.

The development of telecommunications in Canada, M.E. Callin. 1967. pp 862-9.

Constitution and government

Provincial and local government. 1922-23. pp 101-15.

The evolution of the Constitution of Canada down to Confederation, S.A. Cudmore and E.H. Coleman. 1942. pp 34-40.

The British North America Act, 1867. 1942. pp 40-59.

Canada's growth in external status, F.H. Soward. 1945. pp 74-9.

The Constitutional development of Newfoundland prior to Union with Canada, 1949. 1950. pp 85-92.

The terms of Union of Newfoundland with Canada, 1949. 1951. pp 56-7.

The Privy Council Office and Cabinet Secretariat in relation to the development of Cabinet government, W.E.D. Halliday. 1956. pp 62-70.

Amendment of the Canadian Constitution, J.R. Mallory. 1961. pp 51-7.

Yukon Territory and Northwest Territories (historical and current administration of). 1968. pp 110-6.

The Cabinet Committee System. 1970-71. pp 79-84.

Crime and delinquency

A historical sketch of criminal law and procedure, R.E. Watts. 1932. pp 897-9.

The influence of the Royal Canadian Mounted Police in the building of Canada, S.T. Wood. 1950. pp 317-31.

Education

Report of the Royal Commission on national development in the Arts, Letters and Sciences. 1952-53. pp 342-5.

Structural changes in tertiary education in Canada, Miles Wisenthal and Eve Kassirer. 1972. pp 370-7.

Fauna and flora

Faunas of Canada, P.A. Taverner. 1922-23. pp 32-6.

Faunas of Canada, R.M. Anderson. 1937. pp 29-52.

Migratory bird protection in Canada. 1951. pp 38-43.

The barren-ground caribou. 1954. pp 33-6.

Migratory bird legislation. 1955. pp 41-5.

The musk-ox. 1957-58. pp 28-30.

Provincial government wildlife conservation measures. 1963-64. pp 46-52.

The flora of Canada, H.J. Scoggan. 1966. pp 35-61.

Animal life in Canada today, Scientists of the Zoology Division, National Museum of Canada. 1968. pp 47-60.

Fisheries

Groundfish species in the Canadian fisheries, T.H. Turner. 1957-58. pp 591-5.

Canada's commercial fishery resources and their conservation. 1960. pp 625-30.

Forestry

Physiography, geology and climate as affecting the forests. 1934-35. pp 311-3.

The pulp and paper industry in Canada. 1952-53. pp 467-75.

The federal-provincial forestry agreements, H.W. Beall. 1956. pp 459-66.

The Pulp and Paper Research Institute of Canada, Rielle Thomson. 1957-58. pp 489-91.

Canadian forest products and changing world markets, J.T.B. Kingston. 1965. pp 511-7.

Fur trade

The development of marshlands in relation to fur production in the rehabilitation of fur-bearers, D.J. Allan. 1943-44. pp 267-9.

The fur industry, W.M. Ritchie. 1961. pp 618-22.

Geology

Geology in relation to agriculture, Wyatt Malcolm. 1921. pp 68-72.

The Geological Survey of Canada, J.M. Harrison. 1960. pp 13-9.

Geology and economic minerals of Canada, W.D. McCartney. 1967. pp 19-32.

History

Canadian chronology, 1497-1960. 1951-60.

History (Chapter 2). 1973. pp 48-68.

Labour

History of the labour movement in Canada, Eugene Forsey. 1967. pp 773-81.

Manufactures

Changes in Canadian manufacturing production from peace to war, 1939-44. 1945. pp 364-81.

Manufacturing production during the period 1945-59, A. Cohen. 1962. pp 600-9.

The petrochemical industry in Canada, G.E. McCormack. 1962. pp 609-15.

Secondary manufacturing in Canada, W.L. Posthumus. 1963-64. pp 637-43.

Manufacturing and the changing industrial structure of the Canadian economy, 1946-65. 1967. pp 665-78.

Technology, markets and costs in manufacturing. 1968. pp 689-94.

Origin and destination of Canadian manufacturers' shipments. 1970-71. pp 794-9.

Trends in the number of manufacturing establishments. 1972. pp 768-78.

Metric conversion

Canada converts to the metric system. 1976-77. pp vi-xiv.

Mining

Mines and minerals — a historical sketch. 1939. pp 309-10.

The coal deposits and coal resources of Canada, B.R. MacKay. 1946. pp 337-47.

History of pipeline construction in Canada, G.S. Hume. 1954. pp 861-9.

Canadian metallurgical development, John Convey. 1961. pp 513-22.

Geology and economic minerals of Canada, W.D. McCartney. 1967. pp 19-32.

Fuels in Canada, A. Ignatieff. 1969. pp 637-45.

Federal research advances Canadian mineral development, John Convey. 1970-71. pp 723-30.

Physiography and related sciences

Physical geography of the Canadian eastern Arctic, R.A. Gibson. 1945. pp 12-9.

Hydrographical features, F.C.G. Smith. 1947. pp 3-12.

Physical geography of the Canadian western Arctic, R.A. Gibson. 1948-49. pp 9-18.

The drainage basins of Canada. 1961. pp 16-8.

Economic regions of Canada, N.L. Nicholson. 1962. pp 17-23.

Main physical and economic features of the provinces and territories. 1963-64. pp 4-20.

Geophysics, G.S. Garland. 1963-64. pp 57-60.

Federal government surveying and mapping, Mary J. Giroux. 1965. pp 17-24.

Astronomy in Canada, Ian Halliday. 1965. pp 47-55.

Growth of geographical knowledge of Canada, Trevor Lloyd. 1967. pp 1-6.

Geology and economic minerals of Canada, W.D. McCartney. 1967. pp 19-32.

Archaeology in Canada, Scientists of the National Museum of Man. 1968. pp 20-9.

Resource and economic development north of the 60th parallel. 1970-71. pp 58-64.

Regional geography of Canada, J. Lewis Robinson. 1972. pp 1-26.

Population

Occupational trends in Canada, 1891-1931, A.H. LeNeveu. 1939. pp 774-8.

Developments in Canadian immigration. 1957-58. pp 154-76.

Integration of postwar immigrants. 1959. pp 176-8.

Use of the English and French languages in Canada, A.H. LeNeveu. 1965. pp 180-4.

Mobility of Canada's population, 1956-1961, Y. Kasahara. 1966. pp 179-87.

Recent trends in urbanization and metropolitan growth, Leroy O. Stone and Frances Aubry. 1969. pp 156-65.

Trends in population growth in Canada with special reference to the decline in fertility, M.V. George. 1970-71. pp 213-20.

Research

The International Geophysical Year, D.C. Rose. 1957-58. pp 35-8.

The Fisheries Research Board, J.L. Kask. 1959. pp 584-8.

Geophysics, G.S. Garland. 1963-64. pp 57-60.

The Fisheries Research Board of Canada. 1963-64. pp 612-4.

Astronomy in Canada, Ian Halliday. 1965. pp 47-55.

A selection of Canadian achievements in science and technology, 1800-1964, John R. Kohr. 1965. pp 398-401.

Trade

Canada's participation in the changing pattern of world trade, 1953-66, A.M. Coll. 1967. pp 953-66.

Canada's international trade after the Kennedy Round of trade negotiations, G.A. Richardson. 1968. pp 946-54.

Canada's trade with the European Economic Community, D. Paul Ojha. 1969. pp 977-85.

Canada's trade with the Pacific Rim Countries, M.P. Mathew. 1970-71. pp 1069-78.

The Canada–United States Automotive Products Agreement, W.M. MacLeod. 1972. pp 1058-69.

Transportation

The development of aviation in Canada, J.A. Wilson. 1938. pp 710-2.

Pre-war civil aviation and the defence program, J.A. Wilson. 1941. pp 608-12.

International Civil Aviation Organization and Canada's participation therein, C.S. Booth. 1952-53. pp 820-7.

Canals of the St. Lawrence waterway. 1954. pp 830-3.

History of the Canadian National Railways. 1955. pp 840-51.

The St. Lawrence Seaway. 1955. pp 885-8.

Traffic on the Great Lakes–St. Lawrence Seaway. 1956. pp 821-9.

The St. Lawrence Seaway in operation, S. Judek. 1960. pp 851-60.

Revolution in Canadian transportation, A.W. Currie. 1962. pp 753-8.

An outline of the development of Civil Air Transport in Canada, J.R.K. Main. 1967. pp 838-43.

The first decade of the Seaway, Pierre Camu. 1969. pp 841-5.

Political update

Appendix 8

To supplement Chapter 3 lists of governors general, federal and provincial cabinets and the table of members of Parliament elected in the general election of July 1974, this appendix records changes to January 1979. It gives results of federal byelections, a federal Cabinet shuffle of November 1978 and changes in provincial executive councils and territorial government representatives following 1978 elections.

Governor General. The Right Honourable Edward Schreyer, the 22nd Governor General of Canada, was appointed by Queen Elizabeth on December 7, 1978, to take office on January 22, 1979.

Federal byelections by electoral district, elected member and political affiliation:

October 14, 1975
Restigouche (NB), Maurice Harquail, Lib.
Hochelaga (Que.), Jacques Lavoie, Lib.

October 18, 1976
St. John's West (Nfld.), John C. Crosbie, PC
Ottawa–Carleton (Ont.), Jean E. Pigott, PC

May 24, 1977
Malpeque (PEI), Donald Wood, Lib.
Témiscamingue (Que.), Gilles Caouette, SC
Louis-Hébert (Que.), Dennis Dawson, Lib.
Langelier (Que.), J.-Gilles Lamontagne, Lib.
Terrebonne (Que.), J.-Roland Comtois, Lib.
Verdun (Que.), Raymond Savard, Lib.

October 16, 1978
Humber–St. George's–St. Barbe (Nfld.), Fonse Faour, NDP

Halifax–East Hants (NS), Howard Crosby, PC
Fundy–Royal (NB), Bob Corbett, PC
Lotbinière (Que.), Richard Janelle, SC
Saint-Hyacinthe (Que.), Marcel Ostiguy, Lib.
Westmount (Que.), Donald J. Johnston, Lib.
Hamilton–Wentworth (Ont.), Geoff Scott, PC
Ottawa Centre (Ont.), Robert de Cotret, PC
Broadview (Ont.), Bob Rae, NDP
Eglinton (Ont.), Rob Parker, PC
Parkdale (Ont.), Yuri Shymko, PC
Rosedale (Ont.), David Crombie, PC
York–Scarborough (Ont.), Paul McCrossan, PC
St. Boniface (Man.), Jack Hare, PC
Burnaby–Richmond–Delta (BC), Thomas Siddon, PC

Cabinet changes as of November 24, 1978:
President of the Board of Economic Development Ministers, Hon. Robert Knight Andras
Minister of Labour, Hon. Martin O'Connell (replacing Hon. John Carr Munro who resigned on September 8, 1978)
Minister of State for Federal-Provincial Relations, Hon. John M. Reid
Minister of Supply and Services, Hon. Pierre De Bané (replacing Hon. Jean-Pierre Goyer)
President of the Treasury Board, Hon. Judd Buchanan
Minister of Public Works and Minister of State for Urban Affairs (department to be eliminated March 31, 1979), Hon. André Ouellet

Minister of Justice and Attorney General of Canada, Hon. Marc Lalonde (replacing Hon. Otto Emil Lang who had been appointed following the resignation of Hon. Stanley Ronald Basford on August 2, 1978)
Minister of Energy, Mines and Resources and Minister of State for Science and Technology, Hon. Alastair William Gillespie
Minister of National Revenue (replacing Hon. Joseph-Philippe Guay) and Minister of State (Small Business), Hon. Anthony Chisholm Abbott

Prince Edward Island. As a result of the provincial election of April 24, 1978, party standings were: Liberals 17, Progressive Conservatives 15.

The Executive Council of Prince Edward Island in April 1978

Premier, President of the Executive Council, Minister of Justice, Attorney and Advocate General and Minister responsible for Cultural Affairs, Hon. Alexander B. Campbell
Minister of Health and Minister of Social Services, Hon. A.E. (Bud) Ings
Provincial Secretary, Minister of Municipal Affairs and Minister of Labour, Hon. George Proud
Minister of Tourism, Parks and Conservation and Minister of Environment, Hon. Arthur J. MacDonald

Minister of Public Works and Minister of Highways, Hon. George R. Henderson
Minister of Fisheries, Hon. Robert E. Campbell
Minister of Industry and Commerce and Minister of Development, Hon. John H. Maloney
Minister of Education and Minister of Finance, Hon. W. Bennett Campbell
Minister of Agriculture and Forestry, Hon. Edward W. Clark
Minister without portfolio (and responsible for the Housing Corporation), Hon. James B. Fay

Premier Alexander B. Campbell resigned September 11, 1978. Hon. W. Bennett Campbell was elected interim leader of the Liberal party September 16, 1978 and was sworn in as premier two days later.

Cabinet changes by November 1978 were as follows:

Premier, President of the Executive Council and Minister of Finance, Hon. W. Bennett Campbell

Minister of Education, Minister of Industry and Commerce and Minister of Development, Hon. John H. Maloney

Minister of Justice and Attorney General and Minister responsible for the Housing Corporation and for Cultural Affairs, Hon. James B. Fay

Minister without portfolio, Hon. Allison Ellis

Nova Scotia. Following the provincial election of September 19, 1978, party standings were: Progressive Conservatives 31, Liberals 17, New Democrats 4.

The Executive Council of Nova Scotia in October 1978

Premier, President of the Executive Council and Minister of Finance, Hon. John M. Buchanan

Minister of Development, Chairman of the Treasury Board, Minister responsible for Inter-Governmental Affairs and Minister in charge of Administration of the Research Foundation Corporation Act, Hon. Roland J. Thornhill

Minister of Lands and Forests and Minister in charge of Administration of the Liquor Control Act, Hon. George Henley

Minister of Agriculture and Marketing, Minister of Health, Minister in charge of Administration of the Drug Dependency Act and Registrar General, Hon. Gerald Sheehy

Attorney General, Provincial Secretary and Minister in charge of Administration of the Regulations Act, Hon. Henry W. How

Minister of Mines and Minister in charge of the Nova Scotia Energy Council, Hon. Ronald Barkhouse

Minister of Tourism, Minister of the Environment and Minister in charge of the Administration of the EMO (NS) Act and Regulations, Hon. Roger S. Bacon

Minister of Fisheries and Minister of Recreation, Hon. Donald Cameron

Minister of Consumer Affairs, Minister in charge of Administration of the Housing Development Act, Minister in charge of Administration of the Communications and Information Act and Minister in charge of the Residential Tenancies Act, Hon. Bruce Cochran

Minister of Social Services and Minister in charge of Administration of the Civil Service Act, Hon. John MacIsaac

Minister of Labour and Minister of Public Works, Hon. Kenneth Streatch

Minister of Municipal Affairs and Minister in charge of Administration of the Human Rights Act, Hon. Ronald Giffin

Minister of Education and Minister in charge of Administration of the Advisory Council on the Status of Women Act, Hon. Terence R.B. Donahoe

Minister of Highways and Minister in charge of the Office of Communications Policy, Hon. Thomas J. McInnis

New Brunswick. Following the provincial election of October 23, 1978, party standings were: Progressive Conservatives 30, Liberals 28.

The Executive Council of New Brunswick in November 1978

Premier and President of the Executive Council, Hon. Richard Hatfield

Minister of Justice, Hon. Rodman E. Logan

Minister of Finance, Hon. Fernand Dubé

Chairman of the Treasury Board, Hon. Jean-Maurice Simard

Minister of Supply and Services, Hon. Harold Fanjoy

Minister of Transportation, Hon. Wilfred G. Bishop

Minister of Natural Resources, Hon. J.W. (Bud) Bird

Minister of Agriculture and Rural Development, Hon. Malcolm MacLeod

Minister of Health, Hon. Brenda Robertson

Minister of Social Services, Hon. Leslie Hull

Minister of Labour and Manpower, Hon. Mabel M. DeWare

Minister of Education, Hon. Charles Gallagher

Minister of Municipal Affairs, Hon. Horace B. Smith

Minister of Commerce and Development, Hon. Gerald S. Merrithew

Minister of Fisheries, Hon. Jean Gauvin

Minister of the Environment and Minister responsible for housing and the New Brunswick Housing Corporation, Hon. Eric J. Kipping

Minister of Youth, Recreation and Cultural Resources, Hon. Jean-Pierre Ouellet

Minister of Tourism, Hon. Leland McGaw

Chairman of the New Brunswick Electric Power Commission, Hon. G.W.N. Cockburn

Saskatchewan. Following the provincial election of October 18, 1978, party standings were: New Democrats 44, Progressive Conservatives 17.

The Executive Council of Saskatchewan in January 1979

Premier and President of the Executive Council, Hon. A.E. Blakeney

Attorney General, Hon. R.J. Romanow

Minister of Mineral Resources, Hon. J.R. Messer

Minister of Finance, Hon. W.E. Smishek

Minister of Labour and Minister of Government Services, Hon. G.T. Snyder

Minister of the Environment, Hon. G.R. Bowerman

Minister of Northern Saskatchewan, Hon. N.E. Byers

Minister of Municipal Affairs, Hon. G. MacMurchy

Minister of Highways and Transportation, Hon. E. Kramer

Provincial Secretary, Hon. E. Cowley

Minister of Health, Hon. E. Tchorzewski

Minister of Revenue, Supply and Services and Minister of Co-operation and Co-operative Development, Hon. W.A. Robbins

Minister of Consumer Affairs, Hon. E.C. Whelan

Minister of Agriculture, Hon. E.E. Kaeding

Minister of Tourism and Renewable Resources, Hon. A. Matsalla

Minister of Social Services and Minister of Continuing Education, Hon. H.H. Rolfes

Minister of Culture and Youth and Minister of Education, Hon. E.B. Shillington

Minister of Industry and Commerce, Hon. N. Vickar

Minister of Telephones, Hon. Don Cody

Yukon Territory. Ione Christensen became commissioner on January 20, 1979, replacing Frank Fingland who had been appointed interim commissioner in November 1978 following the resignation of Dr. Arthur Pearson.

On November 20, 1978, in the first Yukon Territory election to involve political parties, party standings were: Progressive Conservatives 11, Liberals 2, Independents 2, New Democrats 1. Ridings, members elected and their political affiliation were:

Klondike, Meg McCall, PC

Old Crow, Grafton Njootli, PC

Mayo, Swede Hanson, PC

Kluane, Alice McGuire, Lib.

Tatchun, Howard Tracy, PC

Hootalinqua, Allen Falle, PC

Faro, Maurice Byblow, Ind.

Campbell, Robert Fleming, Ind.

Watson Lake, Don Taylor, PC

Riverdale North, Chris Pearson, PC

Riverdale South, Iain Mackay, Lib.

Porter Creek West, Doug Graham, PC

Porter Creek East, Dan Lang, PC

Whitehorse North Centre, Jeff Lattin, PC

Whitehorse South Centre, Jack Hibberd, PC

Whitehorse West, Tony Penikett, NDP

Index

Page

Page

Page

Page

Page

Page

Page

Page

Page

Page

Page